TRANSACTIONS

OF THE

INTERNATIONAL

ASTRONOMICAL UNION

VOLUME XXIIA

REPORTS

INTERNATIONAL COUNCIL OF SCIENTIFIC UNIONS

INTERNATIONAL ASTRONOMICAL UNION

UNION ASTRONOMIQUE INTERNATIONALE

TRANSACTIONS

OF THE

INTERNATIONAL ASTRONOMICAL UNION

VOLUME XXIIA

REPORTS ON ASTRONOMY

Edited by

JACQUELINE BERGERON

General Secretary of the Union

SPRINGER-SCIENCE+BUSINESS MEDIA, B.V.

Library of Congress Cataloging-in-Publication Data

Reports on astronomy / edited by Jacqueline Bergeron.
p. cm. -- (Transactions of the International Astronomical Union ; v. 22A)
ISBN 978-94-010-4481-3 ISBN 978-94-011-1100-3 (eBook)
DOI 10.1007/978-94-011-1100-3
1. Astronomy--Congresses. 2. International Astronomical Union--Congresses. I. Bergeron, J. (Jacqueline) II. Series.
QB1.I6 vol. 22A
520 s--dc20
[520] 93-50750

ISBN 978-94-010-4481-3

Printed on acid-free paper

Preface

Volume XXIIA of IAU Transactions contains the Reports on Astronomy by the Presidents of Commissions. These triennial Reports highlight the main progress in astronomical research during the period July 1990-June 1993. They concentrate on topics considered to be most important by each Commission and thus differ from exhaustive reviews or abstracts.

As the topics of IAU Commissions widely cover the various fields of astronomy, the Reports should be a useful tool for researchers wishing to gain an overview of a certain area, not necessarily their own, and should be helpful for subsequent, specific searches of the literature.

The Reports on Astronomy are compiled mostly by Presidents of Commissions or by Secretaries or Organizing Committee members of Commissions with material supplied by a team of Commission members, and thus varied in style. I am most thankful to all of those involved, for the efforts made into the preparation of these Reports.

I hope this volume will prove useful to the Astronomical Community and look forward for your comments as to improve them in the future.

Jacqueline Bergeron
General Secretary, IAU

CONTENTS

REPORTS OF COMMISSIONS

*** No report provided

COMMISSION 4: EPHEMERIDES (EPHÉMÉRIDES)

President: B. D. Yallop

Vice President: H. Kinoshita

Organising Committee: V. K. Abalakin, J. Chapront, L. E. Doggett, Y. Kubo, B. L. Morando, H. Schwan, P. K. Seidelmann, E. M. Standish, Fu Tong

I. Introduction

This report covers the period 1 July 1990 to 30 June 1993. The ephemerides that have been published during this period have in general made use of the IAU (1976) system of astronomical constants, apart from minor modifications introduced to allow a better fit to observations. The majority of the ephemerides are based on the fundamental ephemerides of the Moon and planets prepared by the Jet Propulsion Laboratory. At the 21st General Assembly of the IAU, the Working Group on Reference Systems produced nine resolutions, mostly aimed at ensuring that the definitions of the space time reference system will meet future demands as techniques improve.

II. International and National Ephemerides

1. The Fundamental System

The computation and publication of *Apparent Places of Fundamental Stars* (APFS) has continued at the Astronomisches Rechen-Institut, Heidelberg, Germany. The volume for the year 1994 was published in 1993. The number of copies was 750 of which 329 were distributed free to various astronomical institutions.

2. Reference Frame Studies

The Jet Propulsion Laboratory (JPL) have been working on the establishment of the Dynamical Reference Frame of the Lunar/Planetary Ephemerides, and on the determination of ties between the various reference systems.

Reference frame studies have included the establishment of the JPL Radio Frame and the Dynamical Reference Frame of the Lunar/Planetary Ephemerides, determination of frame ties, and development of the concept of the dynamical equinox as a reference point for the modern ephemerides and the connection of other coordinate systems to this reference point. Ties between VLBI and optical frames are being established via the observations of radio stars (in collaboration with French colleagues); a link has been determined between the VLBI and the ephemeris frame through a variety of methods, including differential spacecraft/quasar VLBI and ground ties.

J. Bergeron (ed.), Reports on Astronomy, Vol. XXIIA, 1–6.

3. Printed Ephemerides

The Bureau des Longitudes (BdL), Paris, France has published yearly the *Connaissance des Temps,* the *Ephémérides Astronomiques (Annuire du Bureau des Longitudes)* and the *Ephémérides Nautiques.* BdL also published three supplements to *Connaissance des Temps*: Ephemerides of the satellites of Jupiter, Saturn and Uranus, that give the differential coordinates of the satellites and tables for the computation of the phenomena of the Galilean satellites; Phenomena and configurations of the Galilean satellites of Jupiter; Configurations of the first eight satellites of Saturn. Two main improvements were introduced in the Supplements to Connaissance des Temps after 1990; Ephemerides of the Martian moons and predictions of mutual phenomena for the Galilean satellites. Floppy discs containing satellite ephemerides are provided with the printed ephemerides.

The Hydrographic Department of Japan (JHD), Tokyo, Japan has continued to publish the *Japanese Ephemeris,* the *Nautical Almanac* and the *Abridge Nautical Almanac* for the years 1992, 1993 and 1994, with no major changes since 1985. The *Polaris Almanac* for Azimuth determination, the *Altitude and Azimuth Observation Almanac* for Antarctic Observation and the *Abstract* from the *Japanese Ephemeris* have also been compiled annually by JHD.

The Institute of Theoretical Astronomy (ITA), St Petersburg, Russia published the *Astronomical Yearbook* for 1993 and 1994. The 1994 edition onwards is in two parts. Part 1: The ephemerides of the Sun, Moon and planets. Part 2: The mean and apparent places of the stars. Part 1 is based on DE200/LE200.

The *Navy Astronomical Yearbook* and the *Air Navigation Astronomical Yearbook* were published for 1991 to 1993 and the *Astronavigation Almanac for 1991-1995.*

HM Nautical Almanac Office, Royal Greenwich Observatory, (HMNAO) and Nautical Almanac Office, US Naval Observatory (USNO) have continued their cooperative preparation and annual publication of *The Astronomical Almanac, The Nautical Almanac, The Air Almanac* and *Astronomical Phenomena.*

A new edition of *Sight Reduction Tables for Air Navigation*, (epoch 1995), was produced in collaboration by HMNAO, USNO and Defense Mapping Agency.

The Astronomy & Atmospheric Research Unit, Malaysia is creating an almanac and data facility using an IBM RISC 6000 workstation.

4. Electronic Dissemination of Ephemerides

Starting from 1992 BdL have established an ephemerides service on the French "Mintel Service" carried by the public telephone network.

To plan observing runs, astronomers using STARLINK now have access to information on sky brightness, Daylight and Moonlight Diagrams and Observers Calendar produced by HMNAO.

Recent trials with exporting the JPL planetary and lunar ephemerides on CD-ROM'S have been successful. These will soon become widely available. A new, improved ephemeris for all of the planets and the Moon will soon be integrated, extending over a number of millennia.

All JPL ephemerides are available for export via the JPL Navigation and Ancillary Information Facility.

5. Computerised Ephemerides

The US Naval Observatory (USNO) has produced *MICA*, a high-precision almanac for microcomputers. This menu-driven program provides positions of the Sun, Moon, planets and catalogue objects for 1990 to 1999 inclusive. Versions are available for machines running MS-DOS and for Apple Macintoshes. *MICA* can be obtained from National Technical Information Service, Springfield, VA 22161, USA.

The USNO has also completed a new version of the Satellite Almanac. This MS-DOS program is available on a floppy disc and provides accurate ephemerides for all planetary satellites for the period 1990-1999.

At HMNAO a software package based on the data and algorithms published in *Compact Data for Navigation and Astronomy* called NAVPAC is being prepared. The package fits onto a single floppy disc. The program is interactive and the data used by the program is displayed on the monitor screen where it may easily be updated. The program has also been carefully error trapped. It is intended to market the product with the next edition of the book.

At present the first version of the *Astronomical Yearbook* Part 1, has been prepared by ITA on a floppy disc.

The program system EPHRAT has been elaborated for the ephemeris maintenance of the observations of various radio sources (stars, major planets, Sun, Moon, including surface details and the Galilean satellites of Jupiter) on RATAN-600. The system for prediction of occultations of stars by objects of the solar system has been also elaborated, in Proceedings of the 1st Spain-USSR Workshop on Positional Astronomy (1991).

The CERES software package, which can generate various ephemerides of minor planets and comets, was produced for IBM PC's and compatibles, see Proceedings of the 2nd Spain-USSR Workshop on Positional Astronomy (1992).

The Ephemeris Service of the Institute of Applied Astronomy, Russia, (IAA), has produced a programming system ERA, which is designed to solve a wide range of problems that require automation of computations involving ephemerides. The object is to create a PC-based "work station" for both theoretical and observational astronomers who have to deal with ephemerides and problems in dynamical astronomy. It may also be used as a teaching aid. The system, which is extremely versatile, would now benefit from independent testing by those astonomers in Commission 4 who are qualified to suggest standards for the dedicated computer language, and make proposals for further developments and improvements to the program package.

6. Ephemerides for Special Purposes.

The following ephemerides have been produced by BdL and published in the *Notes Scientifiques et Techniques du Bureau des Longitudes*: Ephémérides de petites planèts de 1989 à 1996 (S018, S022, S023, S031, S032, S039, S040, S041) – Détermination d'orbits de comètes de 1989 à 1994 (S019, S020, S027, S033, S034, S042) ELP 2000/85: Une solution au movement de la Lune couvrant la période historique (S021) – Construction d'une ephémérides de Phobos issue de la thèorie ESAPHO (S024) – Modélisation des phénomènes mutuels (S030), fonction perturbatrice astéroïdale et méthodes de perturbation en mécanique céleste (S035, S037).

Ephemerides of 48 Hipparcos minor planets for 1992 and for 1993 have been published by Bec-Borsenberger in Astron. & Astrophys. Suppl. Series, **93**, 1992 and **98**, 1993.

The following ephemerides and reference materials have been prepared by ITA and published in Russia: The astronomical data for all types of calendars, "The ephemerides for comparing and setting instruments" for astronomical observatories, ephemerides for observations of solar eclipses for observatories and expeditions, the ephemerides for occultations of stars by the Moon, "The instructions on definition of azimuth of reference point by the Moon's observations", the ephemerides of approaches of fifteen selected minor planets to FK5-stars were computed for the period 1991-2000.

7. Special Publications

The *Explanatory Supplement to the Astronomical Almanac*, edited by PK Seidelmann, was published by University Science Books.

US Naval Observatory Circular No.176, "Central Solar Eclipses of 1992", by JA Bangert, AD Fiala and WT Harris was published. This is the last of the eclipse circulars to be published by USNO.

The Canon of Lunar Eclipses 1500BC - AD3000 by Bao-Lin and Fiala AD, containing 10990 lunar eclipses, was published in 1992 by Willmann-Bell.

III. Observational Data for Improving the Ephemerides

The services of the International Lunar Occultation Centre have continued since 1981. The number of the timing data collected at the Centre was 34,646 from 35 countries during the years 1990 to 1992. Reports containing all the reduced data as well as the station coordinates are published annually. Nos. 9 to 11, listing observations in the years 1988 to 1990, were published in the period.

Observations of the 1991 July 11 total solar eclipse was made by JHD in Mexico for the purpose of examining the values for the position of the Sun listed in the current ephemeris and the reduction was completed. The result is published in Report of Hydrographic Researches, No. 29 (1993).

At HMNAO DB Taylor has continued to improve the orbits of the satellites of Saturn and has begun reducing and analysing La Palma CCD observations of the five major satellites of Uranus to improve their ephemerides.

JPL is actively involved in the development of models, ephemerides and the determination of the associated constants to be used as standards by the astronomical community.

The planetary ephemerides at JPL continue to improve as the result of new and improved observational data. Since the masses of the Jovian planets and Pluto are now well determined from the Voyager Mission, these no longer introduce significant uncertainties into the outer planet orbits. Accurate photoelectric observations of the outer planets from La Palma and Bordeaux have now been taken for the past decade, including since 1989, observations of Pluto. Recent radar ranges to Mercury, Venus and Mars have significantly extended the observational coverage for those planets, providing a greater measure on their mean motions. Venus topography, determined from the Magellan Mission at Venus, is being used to reduce the uncertainties normally introduced into the Venus radar ranges. Improved modelling of the perturbations of the asteroids upon the orbits of Mars and the Earth reduces the deterioration of the orbits of these two planets. It is expected that the information available from all of the navigational tracking files from outer planet encounters will soon be incorporated into the ephemeris data set.

JPL continues to maintain ephemerides for all of the natural planetary satellites including the small satellites discovered from the Voyager imaging observations. The ephemerides are periodically updated as new observations become available. The most recent work was the generation of new ephemerides for the major Saturnian satellites based on a fit of numerical integration to observations from 1971 to 1991. Currently under investigation are upgrades to the ephemerides for the Martian and Galilean satellites and the minor Jovian and Saturnian satellites.

Development of Galilean satellite ephemerides in support of NASA's Galileo Mission continues. Earth based observations of eclipse timings, photographic data, radar data and mutual event observations are being combined with Voyager optical observations to produce the new ephemerides.

The analysis of lunar laser ranging (LLR) data, now consisting of over 8,000 normal points, from 1969 to 1993, has influenced a broad spectrum of studies including astronomy, ephemeris development, lunar science, geodynamics, relativity and gravitational physics. As the major focus here, the lunar ephemeris benefits from recent accurate ranges, which show a post-fit scatter of ~3 cm. Solutions using these data improve the lunar ephemeris; determine the mutual orientations between the equator, lunar orbit, and ecliptic planes; determine the mass ratio Sun/(Earth+Moon); improve the precession and nutations of the Earth's pole; monitor Earth rotation; determine Earth station and lunar reflector locations; improve lunar physical librations; and test theories of relativity. The determination of the tidal secular acceleration of the Moon has improved, and the separate contributions of the Earth's diurnal and semi-diurnal tides can now be distinguished. The sensitivity to lunar physical librations has provided improved values for the lunar moment of inertia combinations $(C-A)/B$ and $(B-A)/C$, third-degree gravitational harmonics, Love number k, and rotational dissipation. The lunar and planetary orbits are integrated simultaneously with the lunar physical librations. The partial derivatives of the orbits and physical librations are also integrated numerically. Modelling improvements have been made for small tidal effects which displace the stations and change the rotation of the Earth.

A new foundation of ephemeris calculations has been elaborated by ITA. A new numerical model of the barycentric motion of the Sun, major planets, Moon and major axes of inertia of the Moon has been constructed, which reproduce very accurately the DE200/LE200 ephemeris. A highly accurate method, using truncated Chebyshev series, to integrate and approximate objects positions and velocities has been developed. The errors for the approximation to the radius-vector do not exceed 1mm over a time span of 10 years.

The theoretical considerations and the accuracy estimations for the method of determining planetary positions from photographic observations of planetary satellites have been completed.

The method of calculation of ephemerides for mutual events between close planetary binaries, including shadow effects has been developed.

ITA has coordinated the work on the astrometric observations of the Sun, major and minor planets and the satellites of major planets, by providing the observers with the necessary observational information and by processing the observations.

From the analysis of 30,000 meridian observations of the inner planets the correction to Newcomb's constant of precession and the value of non-precessional motion of the FK4 equinox were taken out by the comparison with the long-term numerical ephemerides of the major planets and the Moon (EPMBB). Also estimates of the constancy of *G* and an upper limit on the solar dynamical oblateness were obtained. This work appeared in a joint paper by Krasinsky and Pitjeva of IAA, and Sveshnikov and Chunayeva of ITA, 1993, Celest. Mech., **55**, 1-23.

IV Working Group Activities

The IAU Working Group on Astronomical Standards, lead by T Fukushima, has been organised into sub-groups on Constants, Procedures, Time and Electronic Distribution. A report is expected at the next General Assembly.

The International Earth Rotation Service has issued a new IERS Standards (1992). IERS Technical Note 13, edited by DD McCarthy.

The IAU/IAG/COSPAR Working Group on Cartographic Coordinates and Rotational Elements of the Planets and Satellites published its latest report in Celestial Mechanics and Dynamical Astronomy, Davies, ME et al, 1992, **53**, 377-397. The Working Group continues to gather relevant data and information and is working on the contents of the next report that will be presented at the 22nd General Assembly.

5. DOCUMENTATION AND ASTRONOMICAL DATA (DOCUMENTATION ET DONNEES ASTRONOMIQUES)

PRESIDENT: B. Hauck
VICE PRESIDENT: O. Dluzhnevskaya
ORGANIZING COMMITTEE: M. Crézé, P. Dubois, J. Grosbol, A. Heck, Li Qi-Bin J. Mead, L. Schmadel, K. Turner, W.H. Warren, G. Westerhout, G. Wilkins

1. INTRODUCTION

The scope of our commission may appear very broad, with ill-assorted topics. However, this is a wrong impression and I think that in the very near future the convergence of all our activities will appear clearer. With the development of computer resources and possibilities, all our activities lead to what we could call information science. Two books published lately (Intelligent Infomation Retrieval, A. Heck & F. Murtagh, Eds., Kluwer Acad. Publ., 1993 and Databases and On-line Data in Astronomy, M. Albrecht & D. Egret, Eds., Kluwer Acad. Publ., 1991) give a very good idea of this tendency.

Dr. A. Heck was the IAU representative to the General Assembly of CODATA, held in Beijing in October 1992. The next General Assembly will take place in Chambéry, France, 18-22 September 1994. The scientific theme is Data and Knowledge in a Changing World.

The past triennium has been marked by a lot of activities related to Commission 5, the major ones being listed below. I would just emphasise at this point those activities which I consider to be most important.

First is the archiving of current observational data. A JCM was foreseen during the last General Assembly in Buenos Aires, but due to the fire which broke out at the San Martin Cultural Center, it was not possible to hold it. However, the proceedings are published in Highlights of Astronomy, **9**, 1992. In some observatories [53.013.039, 53.013.040, 53.013.041] steps have been taken to organise observational archives. The proceedings of a workshop on this topic held in April 1993 at Trieste will be published soon by M.A. Albrecht and F. Pasian.

Next may be mentioned as an important change the apparition of two information systems ESIS [53.013.042] and ADS [53.013.043], the latter being more developed at the time of the preparation of this report than the first.

The last point I would mention is the astronomical literature that is accessible on-line. The main project in this field is STELLAR (Study of Electronic Literature for Astronomical Research), a project developed at GSFC which provides abstracts of articles in ApJ, ApJS, AJ, PASP, A&A, AAS, MN and JGR.

2. WORKSHOPS, COLLOQUIUMS, CONFERENCES

Various meetings relating to the scope of Commission 5 were held, of which the following may be mentioned:

J. Bergeron (ed.), Reports on Astronomy, Vol. XXIIA, 7–12.

Desktop Publishing in Astronomy and Space Sciences, Strasbourg, 1991 [55.012.053]
Astronomy from Large Databases II, Haguenau, 1992 [A. Heck & F. Murtagh (Eds.)]
Handling and Archiving Data from Ground-based Telescopes, Trieste, 1993
Astronomical Data Analysis Software and System I, Tucson, 1991 [D.M. Worrall et al. (Eds.)]
Astronomical Data Analysis Software and System II, Boston, 1992 [R.J. Hanisch et al. (Eds.)]
Astronomical Data Analysis Software and System III, Victoria, 1993
Databases for Galactic Structure, Swarthmore, 1993 [A.G.D. Philip et al. (Eds.)]

3. REPORTS FROM DATA CENTRES

3.1 NASA ADC (by N. Roman)

The NASA ADC produced and distributed more than 2000 copies of the first ADC CDROM which contained 114 catalogues in ASCII and FITS formats. The ADC prepared new printed versions of the Bonner Durchmusterung, the Cordoba Durchmusterung, the Cape Photographic Durchmusterung and the WDS/HD/DM/ADS cross-index.

As of April, 1993, the NASA ADC had 669 catalogues in its archive of which 82 were added since July 1991, besides those received from other data centres. The ADC is cooperating with the CDS in archiving smaller data sets by acquiring these from North American publications. In addition to the ADC CDROM, the NSSDC is distributing CDROMs of data from the Einstein Observatory, the IRAS Sky Survey Atlas, and Halley and planetary observations.

3.2 CDS, Strasbourg (by M. Crézé)

Over the last two years the CDS activity has been developing along its usual lines of action, namely:

The Simbad Database

Now widely used as a reference for cross-identification of astronomical objects, it provides cross-identifications, bibliographic references and a selection of basic data for over 1 million astronomical objects including galactic and extra galactic sources observed at various wavelengths.

- Improvements to the new software are going on, including assistance to Simbad users and updates of the user's guide.

- The bibliographical effort has been continued, references to astronomical objects are available nearly at publication time.

- There are 880 registered accounts actually used. The number of countries with Simbad users was 24 by the fall of 1990, 36 by the fall of 1992.

- Provisional free access has been open for countries with currency difficulties.

- A major update of the Simbad content has been the cross-identification of the IRAS Point Source Catalogue.

- New developments currently investigated : graphical user interface, Computer Assisted Quality Control.

The catalogue exchange and distribution facility

- Over 80 new catalogues have been incorporated in the CDS files.

- A new facility has been implemented to make most catalogues (and all the recent ones) freely available through an anonymous ftp procedure, being the currently most widely used process.

- From January 1993 onwards, an agreement has been passed with the editor of the journal "Astronomy and Astrophysics" to incorporate in the CDS files data related to papers published in "Astronomy & Astrophysics Supplements" and to make them available via the ftp procedure.

- Experience has been gained in re-digitising printed data through scanner and character recognition, which will greatly help to create new compilations .

Methodologies and techniques for information retrieval

The CDS has published "Star Guides" : A Directory of Astronomy, Space Sciences and Related Organisations of the World (1993, CDS Special Pubs n° 20), and "Star Briefs" Dictionary of Abbreviations, Acronyms and Symbols in use in Astronomy, Space Sciences and related fields (1993, CDS Special Pubs n° 21).

International Co-operation

The CDS has been permanently fostering the exchange of experience and data between the 7 Astronomical Data Centres inside the working group created by the IAU. Main efforts have been dedicated to bring computer-readable data availability up to equivalent levels in all parts of the world. This is far from being achieved mainly due to the uneven level of technical development. Current efforts aim at creating highly compatible tools for handling, analysing and formatting new data sets in a number of ADCs. Such tools would enable Data Centres to share the task of incorporating and distributing data sets into and from their own archives and eventually create mirror archives at different places.

The CDS also collaborates in the development of new facilities to access data such as the ESIS project of ESA and the ADS project of NASA.

3.3 Astronomical Data Analysis Center, National Astronomical Observatory of Japan (by S. Nishimura)

A data storage and distribution system for astronomical catalogues was developed on UNIX workstations based on magneto-optical disks. The older system on open-reel magnetic tapes is also running on the main-frame computer.

Validation, documentation and reorganisation of astronomical catalogues, especially those made in Japan, have been done at the Center. A CD-R writer was recently installed at the Center, and it is planned to distribute custom-made sets of catalogues, such as the Photometric Atlas of Galaxies or data sets of special fields of astronomy.

3.4 Centre of Astronomical Data, Institute of Astronomy of the Russian Academy of Sciences (INASAN) (by O. Dluzhnevskaya)

The Centre (CAD) was established as a branch centre of the Strasbourg Data Centre (CDS) and has been active since 1977 on the basis of the Agreement signed by both Centres.

The CAD is responsible for data activity and coordination in the former SU and provides the CDS with new catalogues and other astronomical data prepared at different astronomical institutes.

Since 1980 the World Data Center A of the Goddard Space Flight Center (ADC) and CAD have also begun to exchange machine-readable versions of astronomical catalogues.

In the frame of the agreements, the CAD is responsible for the preparation of machine-readable astronomical catalogues, software etc. in the FSU.

During the last 3 years 11 astronomical catalogues were sent to the CDS and 9 are ready.

The CAD participated in the preparation of the list of astrophysical catalogues on CD-ROM prepared by the ADC and after receiving the CD-ROMs (Test disk, Selected Astronomical Catalogs, vol. 1) they were disseminated throughout the astronomical observatories and institutes of the former SU.

The CAD provides the CDS and ADC with information about the preparation of new catalogues compiled in observatories of the FSU.

During the past year a compilation was made of different astronomical data sets in the form of archives, data bases (already existing and those being created), miscellaneous astronomical data sets and plate vaults stored at the various observatories. This compilation contains information about the instruments used for observations, areas of the sky and time intervals they cover, photometric systems, limiting stellar magnitudes and computer facilities available at the observatories. It has been sent to the CDS for publication.

The Center has a present staff of 12 people, using 4 work stations and 5 PCs connected via the LAN Ethernet.

Projects in progress include: completion of a machine-readable version of the catalogue of observational data of star-forming regions; the creation of an updated collection of observational data of stars in stellar associations; studies in the field of galactic stellar population evolution (initial mass function, star clusters, young stellar complexes); 26 papers on this subject were published this year.

3.5 Centro de Datos Astronomicos, Argentina (by E.F. Arias)

The Centro de Datos Astronomicos Argentina (CDA) is installed at the Facultad de Ciencias Astronomicas y Geofisicas de La Plata, Argentina. It works under the responsibilitiy of two local astronomers, Dr. Elisa Felicitas Arias and Lic. Horacio Luis Tignanelli. It was created by means of an agreement between the Universidad Nacional de La Plata and the Strasbourg Stellar Data Centre. The CDA distributes astronomical data to all Argentine observatories. It also provides on request information to some South American colleagues. Periodically the CDA sends by e-mail information bulletins to Argentine users.

4. REPORTS FROM WORKING GROUPS AND TASK GROUPS

Before reporting on the activity of the Commission's WGs and TGs I would mention the activity of two WGs of two other commissions that receive our support. The first is the *WG on "Radioastronomical Databases"* (Commission 40), chaired by Dr. H. Andernach, IAC. This WG held its first WG meeting (ESO-OAT Workshop in Trieste in April 1993). Recommendations emerged on electronic availability of published data, on the educational needs for the age of "Network Astronomy", the preparation of a master index of all machine-readable astronomical catalogues, and the nomenclature of radio sources. Electronic access to radio source catalogues has been vastly improved by the WG: 45 tables (~420,000 entries) are now searchable on-line in the "Einstein On-line Service" and 40 more tables (~80,000 entries) are in the waiting list. With Commissions 5 and 29 it is proposed to hold an "Interdisciplinary Discussion on the Status of Archiving of Astronomical Data" at the GA in 1994.

The second is the *WG on Spectroscopic Data Archives*, chaired by Dr. E. Griffin, which was established in mid-1992, its *raison d'être* having been enshrined in IAU Resolution C13 (1991). The 12 members come from all continents except Antarctica; most are research astronomers who are personally committed to the benefits of archiving, and who try to encourage, stimulate, rationalize and sometimes help organise suitable archiving policies in all observatories and institutions. The WG took an active part in various workshop meetings both in Europe and the USA and was especially in evidence at the workshop in Trieste (April 1993), where its own EC grant helped support the attendance of 19 people. The WG is currently tackling the cataloguing of spectroscopic plates and is considering the idea of putting all such plates in one place as a long-term safeguard.

The *Astronomy Thesaurus Task Group,* headed by Mrs. Robyn Shobbrook, has published an astronomical thesaurus. Its main aim is to standardize the terminology for the purposes of aiding unambiguous library cataloguing and more precise recall of data from computer databases. The thesaurus can be ordered directly from Mrs. R. Shobbrook, Anglo-Australian Observatory, Epping Laboratory, P.O. Box 296, 296 Epping N.S.W. 2121, Australia.

The *WG on Libraries*, co-chaired by Dr. W.H. Warren and Mrs. B. Corbin, is now planning a conference on Library and Information Services in Astronomy (LISA II)

The *WG on Designations,* chaired by Dr. P. Dubois, has been very active since the last GA, in two directions.

First, a new version of the "Dictionary of Nomenclature of Astronomical Objects", compiled by M.-C. Lortet and collaborators, is a very important tool in the field of designation. It is now possible to access it on-line on the SIMBAD computer (simbad.u-strasbg.fr or 130.79.128.4) at Strasbourg and without password under the login "info". Also to be found there is other information such as the IAU document on "Specifications...". The dictionary is also on-line in the SIMBAD database and is regularly updated. A printed version will also be available at the end of the year. Designation problems were also the subject of discussion during the Trieste meeting.

Second, the WG has been active in promoting the IAU Specifications (see e.g. A&A 267, A12, 1993). This document should serve as a reference for problems encountered in the designation of astronomical objects. A new version is now in preparation and the chairman of the WG welcomes all suggestions.

B. Hauck
President

6. ASTRONOMICAL TELEGRAMS (TELEGRAMMES ASTRONOMIQUES)

(Committee of the Executive Committee)

PRESIDENT: J. E. Grindlay

VICE-PRESIDENT: R. M. West

ORGANIZING COMMITTEE: S. Isobe, B. G. Marsden, E. Roemer

DIRECTOR OF THE BUREAU: B. G. Marsden, Smithsonian Astrophysical Observatory, 60 Garden Street, Cambridge, MA 02138, U.S.A. (TWX 710-320-6842 ASTROGRAM CAM, e-mail MARSDEN@CFA.HARVARD.EDU)

ASSOCIATE DIRECTOR OF THE BUREAU: D. W. E. Green

I. INTRODUCTION

The need for rapid communication of astronomical discoveries continues to increase, as the number of observers able to participate and report rapidly by electronic mail inexorably rises. Both the increased number and power of observatories in space (e.g., HST, ROSAT, CGRO, ASCA) and the increased access to powerful observing and reduction tools for ground-based observations have contributed in a significant way. The past three years (1990-93), or the triennium covered by the present report, has seen the launch of *all* of the above-named space observatories, so that it has been an especially active period. With long operational lifetimes planned, and the addition of new major satellite-borne observing capabilities (e.g., XTE, ISO, AXAF) planned for the next 3-6 years, this will increase even further. On the ground, the rapid dissemination of CCDs and computer data acquisition (and processing) systems among the amateur astronomical community that has taken place over the past three years is also expected both to increase the rate of discovery of novae and dwarf novae outbursts as well as to provide increased longitudinal and temporal coverage for the increasing number of multi-wavelength monitoring campaigns being carried out. Announcement and coordination of these campaigns, particularly with the satellite observatories, has become an increasingly important function of the *Circulars* issued by the Central Bureau.

The report of the Central Bureau, given below, charts the increase in *Circulars* issued and the understandable demise of the telegrams in this era of electronic communication. The Commission owes, once again, a special debt of gratitude to Dr. Brian Marsden for performing such capable service to the entire astronomical community in his role as Director of the Bureau. This year marks the 25th anniversary of the directorship of Dr. Marsden and thus should be noted with even more fanfare than the rapidly approaching issuance of the 6000th *Circular*. The Commission wishes to thank also Mr. Daniel Green, Associate Director of the Bureau, for his capable assistance to Dr. Marsden and for his proven judgement on publication of a majority of the recent *Circulars*.

The next triennium promises transient events of great interest that are both predicted (e.g., the periastron passage of the millisecond pulsar PSR 1259–63 into the wind of its Be star companion in January 1994; or the probable impact of comet 1993e with Jupiter in July 1994) and of course only desirable (e.g., the first identification of a counterpart of a γ-ray burst source, perhaps after the launch of the HETE satellite in 1994-95). It is certain that the Commission, and the Central Bureau, will have a significant role in the rapid dissemination of these and other discoveries. It may be desirable to augment the membership of the Commission in anticipation of this activity.

J. E. Grindlay

President of the Commission

J. Bergeron (ed.), Reports on Astronomy, Vol. XXIIA, 13–14.

II. REPORT OF THE CENTRAL BUREAU FOR ASTRONOMICAL TELEGRAMS

During the triennium the Central Bureau issued 'telegram books' and *Circulars* as follows:

	Telegrams	*Circulars*
1990 July-Dec.	16	Nos. 5046-5152
1991 Jan.-June	28	Nos. 5153-5295
1991 July-Dec.	18	Nos. 5296-5419
1992 Jan.-June	16	Nos. 5420-5551
1992 July-Dec.	16	Nos. 5552-5685
1993 Jan.-June	1	Nos. 5686-5827

Comparison with the report of the previous triennium shows that there has been almost a 25-percent increase in the number of *Circulars* issued—but a corresponding *decrease* in the number of telegrams. The fact that only one telegram book was issued after the end of 1992 exaggerates this difference, because the Central Bureau deliberately did not issue any telegrams after 1993 Jan. 5. Some three months later an enquiry was made as to whether the telegrams were actually needed. Most telegram subscribers did not respond, a few thought that to stop the telegrams was an excellent idea, and only one (a U.S. amateur who receives the telegrams by mailgram) felt it necessary that they continue! Clearly, most subscribers are now receiving the complete *Circulars* themselves by electronic mail, and the often-very-terse telegrams have become prohibitively expensive in comparison to their value. There are still a few important observatories that are outside the electronic-mail circuit, but their number continues to decrease. Accordingly, steps are being taken to discontinue use of the telegrams (except in the titles of the Commission and the Central Bureau!), and any credit remaining in the telegram accounts is being transferred to the corresponding *Circulars* accounts. The fact that the previous report mentioned that the telegrams "continue to be essential" and that "they are likely to continue to be so for the foreseeable future" merely illustrates the difficulty of making predictions and the speed with which international communication practice has been changing in the aftermath of the Cold War. Although the Central Bureau continues occasionally to *receive* information by telegram (as opposed to e-mail), there seems now no point in going forth with the promised reprinting of the telegraphic code from *IAU Inf. Bull.* No. 46 (1981).

Scientifically, most of the material handled by the Central Bureau continues to be in the four traditional areas of comets, earth-approaching minor planets, novae and supernovae. Records for the number of supernovae were broken successively in 1990, 1991 and 1992 (with 72 objects designated in that year), but the discovery rate was noticeably less in 1993. A dozen galactic novae appeared in less than two years, and 1991 had record numbers of comets and near-earth asteroids. Beginning with the discovery of (5145) Pholus in early 1992 there has been an extraordinary surge of interest in objects in the outer solar system, including two that appear to be in stable orbits entirely beyond Neptune.

Concern has been voiced about the decreasing number of actual subscribers to the Central Bureau's services and the indiscriminate redistribution of the electronic versions of the *Circulars* on computer bulletin boards. Paid subscriptions are necessary for the continuation of the service, and some astronomers—professional and amateur—have rallied to the cause. The number of subscribers to the printed *Circulars* seems to have stabilized at a little more than 700, and of these some 250 also subscribe electronically, an increase by some 20 percent during the past year.

B. G. Marsden
Director of the Bureau

COMMISSION 7: CELESTIAL MECHANICS (MÉCANIQUE CÉLESTE)

President: A.Deprit

Vice-President: S.Ferraz-Mello

Organizing Committee: K.B.Bhatnagar, J.Chapront, C.Froeschlé, J.Henrard, He Miaofu, K.V.Kholshevnikov, H.Kinoshita, J.H.Lieske, A.Milani, S.J.Peale, A.E.Roy, A.G.Sokolsky

The Vice-President assumed the duty of writing this report only one month before the deadline and this circumstance certainly affected its completeness. Because of the great diversity of research under way in the various fields of Celestial Mechanics, triennial reports have been usually divided into chapters critically describing in some detail the achievements in specific domains of Celestial Mechanics. The choice of the chapters reflects, this time, also the areas for which it has been possible to find a colleague willing to stop other duties and give priority to writing a report on the progress of the area. The Vice-President acknowledges that, without the kind cooperation of several colleagues and the members of the Organizing Committee, this task would not have been possible and we would have to regret the absence of a Celestial Mechanics report in this volume.

Since the last report, Commission 7 has sponsored IAU Symposium No. 152 "Chaos, Resonance and Collective Dynamical Phenomena in the Solar System" (Angra dos Reis, Brazil, July 1991) and co-sponsored IAU Symposium No. 160 "Asteroids, Comets, Meteors 1993" (Belgirate, Italy, June 1993). It has also co-sponsored two joint discussions during the XXIst General Assembly (Buenos Aires, 1991): "Origin of Stars and Planetary Systems" and "Reference Systems: What are they and what's the Problem".

Other international conferences of interest to Commission 7 were: "Hamiltonian Systems and Celestial Mechanics" (Guanajuato, Mexico, October 1991); "The dynamics and evolution of minor bodies with galactic and geological implications" (Kyoto, Japan, October 1991); 3rd Alexander von Humboldt Colloquium for Celestial Mechanics: "Qualitative and Quantitative Behaviour of Planetary Systems" (Ramsau, Austria, April 1992); "Interactions between Physics and Dynamics of Solar System Bodies" (Val André, France, June 1992); "Ergodic Concepts in Stellar Dynamics: Theory, Computer Experiments, Observations" (Geneva, Switzerland, March 1993); the International Conference on Hamiltonian Dynamics: "Integrability and chaotic behaviour", (Toruń, Poland, June 1993); the NATO Advanced Study Institute "From Newton to Chaos: Modern Techniques for Understanding and Coping with Chaos in N-Body Dynamical Systems" (Cortina d'Ampezzo, Italy, July 1993); "Dynamics and Astrometry of Natural and Artificial Celestial Bodies" (Poznań, Poland, September 1993); "Seventy-five years of Hirayama Collisional families: The role of collisions in the Solar System history", (Sagamihara, Japan, November 1993) and "Advances in Non-Linear Astrodynamics" (Minneapolis, USA, November 1993)

Several regional or national meetings were also held during the past three years: the 22nd, 23rd and 24th annual meetings of the Division of Dynamical Astronomy of the American Astronomical Society (Key Biscayne, May 1991; Chicago, June 1992 and Santa Barbara, May

J. Bergeron (ed.), Reports on Astronomy, Vol. XXIIA, 15–28.

1993); the 24th and 25th Japanese Symposiums on Celestial Mechanics (Tokyo, January 1991 and December 1992); the 5th and 6th Brazilian Colloquiums on Orbital Dynamics (Curitiba, November 1990 and Aguas de São Pedro, November 1992); the first Italian Conference on Celestial Mechanics (L'Aquila, May 1993); the 15th Spring School of Goutelas (France): "Interrelations between Physics and Dynamics of Minor Bodies in the Solar System" (April 1991); the Chamonix winter schools: "Dynamique relativiste appliquée à l'Astrometrie, à la Mécanique Céleste et aux modèles d'Univers" (February 1992) and "Integrabilité des Systèmes Differentiels, techniques symplectiques en physique et applications à la Mécanique Céleste" (February 1993) and the workshop "Artificial Satellite Theory" (Washington, November 1993).

Several books directly related to Celestial Mechanics were published since the last report:

Benest, C. & Froeschlé, C. (eds.): 1990, "Modern Methods in Celestial Mechanics", Ed.Frontières, Gif-sur-Yvette.

Benest, C. & Froeschlé, C. (eds.): 1992, "Interrelations between Physics and Dynamics for Minor Bodies in the Solar System", Ed.Frontières, Gif-sur-Yvette.

Bois, E., Oberti, P. & Henrard, J., (eds.): 1993, "Interactions between Physics and Dynamics of the Solar System Bodies", Kluwer A.P., Dordrecht.

Brumberg, V.A.: 1991, "Essential Relativistic Celestial Mechanics", Adam Hilger.

Capitaine, N., (ed.): 1991, "Journées 1991: Systèmes de reference spatio-temporels". Observatoire de Paris.

Capitaine, N., (ed.): 1992, "Journées 1992: Systèmes de reference spatio-temporels". Observatoire de Paris.

Clube, S.V.M., Yabushita, S. & Henrard, J., (eds.): 1992, "Dynamics and Evolution of Minor Bodies with Galactic and Geological Implications"

Dvorak, R. & Henrard. J., (eds.): 1992. "Qualitative and Quantitative Behaviour of Planetary Systems", Kluwer, Dordrecht.

Ehlers, J. & Schaefer, G. (eds.), "Relativistic Gravity Research with Emphasis on Experiments and Observations". Lecture Notes in Physics, 410, 1992, Springer-Verlag, Berlin.

Ferraz-Mello, S. (ed.): 1991, "Chaos, Resonance and Collective Dynamical Phenomena in the Solar System", Proceedings of IAU Symposium No. 152, Kluwer, Dordrecht.

Gerasimov, I.A.: 1990, "Weierstrass Functions and applications to Mechanics and Astronomy", in Russian, Moscow University.

Hughs, J.A., Smith, C.A. & Kaplan, G.H. (eds.): 1991: "Reference Systems", Proceedings of IAU Colloquium No. 127, U.S. Naval Observatory.

Kinoshita, H. & Yoshida, H. (eds.): 1991, Proceedings of the 24th Symposium on Celestial Mechanics, Tokyo.

Kinoshita, H. & Nakai, H. (eds.): 1992, Proceedings of the 25th Symposium on Celestial Mechanics, Tokyo.

Lacomba, E.A. & Llibre, J. (eds.): 1993, "Hamiltonian Systems and Celestial Mechanics", World Scientific, Singapore.

Lieske, J.H. & Abalakin, V.K., (eds.): 1990, "Inertial Coordinate Systems on the Sky", Proceedings of IAU Symposium 141", Kluwer, Dordrecht.

Liu, B.-L. & Fiala, A.D.: 1992, "Canon of Lunar Eclipses 1500 BC - AD 3000", Willmann-Bell, Richmond.

Neutsch, W. & Scherer, K.: 1992. "Celestial Mechanics. An Introduction to Classical and Contemporary Methods", BI, Mannheim.

Phillips, J.A., Thorsett, S.E. & Kulkarni, S.R. (eds.): 1992, "Planets around Pulsars", Astron. Soc. Pacific Conference Series, vol. 36.

Poincaré, H.: 1993, New Methods of Celestial Mechanics, American Institute of Physics, New York (History of Modern Physics and Astronomy, vols. 12–14).

Rastall, P.: 1991, "Postprincipia: Gravitation for Physicists and Astronomers", World Scientific, Singapore.

Roy, A.E., (ed.): 1991, " Predictability, Stability and Chaos in N-Body Dynamical Systems", Proceedings of the 1990 Cortina NATO ASI, Plenum, New York.

Schneider,M.: 1992, "Himmelsmechanik. BI, Mannheim.

Seidelmann, P.K., (ed.): 1992, "Explanatory Supplement to the Astronomical Almanac", University Science Books.

Wright, J.L.:1992, "Space Sailing" Gordon and Breach, Philadelphia.

PLANETARY COMPANIONS

Besides the specific progresses reported in the forthcoming sections, we have to report some new problems open by recent discoveries in Astrophysics concerning possible planetary companions of stars.

Two pulsars have shown residuals in the time of pulse arrivals leading to the suspect of the existence of planetary companions: PSR 1829-10 and PSR 1257+12 (Bailes *et al.*, *Nature* **352**, 311; Wolszcan & Frail, *Nature* **355**, 145). In the past, other similar cases (the Crab nebula pulsar PSR 0531+21 and PSR 0329+54) were announced but were not confirmed by subsequent observations and the residuals of PSR 1829-10 are now known to be of terrestrial origin. However, in the case of PSR 1257+12, the residuals fit a model consisting of two planets in almost circular orbits and no alternative hypothesis can so easily reproduce the observations. The minimum mass of these planets (corresponding to the case where the system would be edge on) is 3.4 and 2.8 Earth masses. The periods of 66.56 and 98.23 days are very close to a commensurability 2/3. The mutual perturbation of the orbits must reflect in additional variations in the pulse timings (Rasio *et al.*, *Nature* **355**, 325; Malhotra *et al.*, *Nature* **355**, 583). Peale (AJ **105**, 1562) and Malhotra (ApJ **407**, 266) indicate that 3 to 5 years of continuous observations will be enough to support or refute the planetary interpretation of the residuals. The current observations are already enough to exclude the possibility of planets larger than a dozen Earth masses.

The existence of planetary companions was also hypothesized by astrophysicists to explain several features of the gas cloud about the star β *Pic*: the inner disk gap at 20–30 AU, deduced from IR data, and the finite thickness of the cloud (30–50 AU at 200 AU from the star). The inner gap could be due to large bodies shepherding and clearing the fine dust and the finite thickness could be maintained by the gravitational perturbation of embedded bodies (see Norman & Paresce, in *The formation and evolution of Planetary Systems*, Cambridge, 1989). There are also some unusual redshift spectral absorption events attributed to the evaporation of infalling solid bodies (comets?). In this case, the presence of a planet has been invoked to perturb passing-by bodies and making some of the star-grazers detectable by spectroscopic means. A model showing planetary perturbations able to put numerous small bodies into star-grazing orbits has been established by Beust *et al.* (AA **247**, 505). The shepherding of the gas in the inner border has been studied by several authors (Scholl *et al.*, CM **56**, 381; Lazzaro *et al.*, CM **56**, 395 and *Icarus*, in press; Sicardy *et al.*, CM **57**, 373; Roques *et al.*, *Icarus*, in press and Beaugé & Ferraz-Mello, *Icarus*, submitted). The main result in these papers is the temporary capture of the particles in some exterior resonances with the planet. They are reviewed in section E of this report.

A. Relativistic Celestial Mechanics

(Victor A. Brumberg)

The starting point of many investigations on relativistic Celestial Mechanics during the period 1991-1993 was the adoption by the 21st General Assembly of IAU (Buenos-Aires, 1991) of the new recommendations on reference systems and time scales based, for the first time in the IAU history, on General Relativity. These recommendations aroused a lot of questions. Some authors tried to answer these questions and to discuss problems related with practical using of new recommendations in Celestial Mechanics, Astrometry and Geodesy (Brumberg, *Highlights of Astronomy*, **9**, 133; Seidelmann & Fukushima, AA **265**, 833; Soffel & Brumberg, CM **52**, 355). Main new results have been obtained in the following domains:

1. RELATIVISTIC THEORY OF REFERENCE SYSTEMS

In a series of papers, Damour, Soffel & Xu (*Phys.Rev* D, **43**, 3273; **45**, 1017; **47**, 3124) have elaborated a new theory of relativistic reference systems. Its main difference from the earlier theory by Brumberg & Kopejkin is the introduction of the physically more plausible Blanchet-Damour moments to describe the multipole structure of gravitational fields, instead of formally Newtonian multipole moments . This enables one to present the relativistic space-time transformations between different reference systems and the corresponding metric tensors in more compact form and gives an efficient tool to handle the problems of relativistic Celestial Mechanics on the post-Newtonian level of accuracy. Not too technical description of this (DSX) approach may be found in Damour *et al.* (IAU Coll 127, 50; Lect.Notes in Physics, 410, 46). Along with this, the earlier theory by Brumberg & Kopejkin has been re-formulated in the closed form (Brumberg, IAU Coll 127, 36; Klioner & Voinov, *Phys. Rev.* D **48**, 1451). Klioner (AA, in press) has investigated in detail the notion of relativistic kinematically non-rotating reference system (as prescribed by the IAU recommendations) and found that this notion involves ambiguities (being dependent on a place of a reference system in the whole hierarchy of reference systems) which may be removed only by specific conventions.

2. RELATIVISTIC THEORY OF TIME SCALES

This topic is in intimate relation with the previous one because the new time scales are considered as the coordinate times of the corresponding reference systems. Based on the planetary and lunar semi-analytical theories of motion of the Bureau des Longitudes, Brumberg, Bretagnon & Francou (Journées Obs. Paris 1991, p.141) deduced the analytical relationships between these time scales and the physically registered proper time of an observer. In order to actually implement such time scales, one needs to solve the problem of remote clock synchronization. The algorithms to realize several specific types of such relativistic synchronization have been analyzed by Klioner (CM **53**, 81) and Klioner & Fukushima (*Manuscr. Geodaetica*, in press). These algorithms have an accuracy of 10–100 ps, which should be sufficient for the decade to come.

3. RELATIVISTIC REDUCTION OF ASTRONOMICAL OBSERVATIONS

Increasing precision of modern astronomical observations demands the elaboration of highly accurate models of observations. Such models involve the second-order effects in light propagation. Such effects, including the coupling of rotational and translational motion of the gravitating masses and the contribution due to the quadrupole gravitational field, have been studied in detail by Klioner (*Sov.Astron.* *35*, 523; NOAA Tech.Rep. No. 137, 188), Klioner & Kopejkin (AJ **104**, 897) and Hartmann *et al.* (*Veroeff. Bayer. Akad Wiss.* A No.108). The accurate model for optical observations of the accuracy of 1 microarcsecond has been proposed by Klioner & Kopejkin (AJ **104**, 897). VLBI observations demands, even now, relativistic models at the level of 1 ps in time delay and 1 fs/s in delay rate. Several versions of such models have been constructed, for instance, by Klioner (NOAA Tech.Rep.No.137) and Soffel *et al.* (AJ **101**, 2306). A significant progress has been achieved in the relativistic model for analyzing LLR data (Mueller *et al.*, ApJ **382**, L101; Lect.Notes in Phys. 410, 87) and GPS data (Schwarze *et al.*, *Manuscr. Geodaetica* **18**, 306).

4. RELATIVISTIC EQUATIONS OF EARTH SATELLITE MOTION IN BARYCENTRIC AND GEOCENTRIC COORDINATE SYSTEMS

Post-Newtonian equations of Earth satellite motion in the closed form and taking into account all necessary perturbing factors (rotation and non-sphericity of the Earth, solar-lunar action, superposition of different effects) were obtained in different papers (Brumberg, AA **257**, 777; Klioner & Voinov, *Phys.Rev.* D **48**, 1451; Damour et al., *Phys.Rev.* D, in press).

5. RELATIVISTIC DESCRIPTION OF THE EARTH ROTATION

The results obtained in this domain are of preliminary nature. Some expressions for the angular velocity of the Earth rotation in different relativistic coordinate systems were obtained by Brumberg *et al.* (AA **275**, 651). An approach to construct the relativistic model of the Earth rotation has been developed by Bizouard *et al.* (Journées Obs. Paris 1992, p.76).

6. POST-POST-NEWTONIAN EFFECTS IN CELESTIAL MECHANICS PROBLEMS

Investigation of the post-post-Newtonian effects in the two-body problem is of importance with respect to the binary pulsar motion. Detailed analysis of these effects including their relation with gravitational radiation has been performed in a series of papers written or co-written by Schaefer (Schaefer, *Ann.d.Physik* **48**, 601; Damour & Schaefer, *Phys.Rev.Lett.* **66**, 2549, *J. Math. Phys.* **32**, 127; Junker & Schaefer, MNRAS **254**, 146; Schaefer & Wex, *Phys. Lett.* A **174**, 196; **177**, 461; Blanchet & Schaefer, *Class.Quant.Grav.*, in press). Some of these effects have no analogies in the Newtonian N-body problem and are of interest both from mathematical and physical points of view. Post-post-Newtonian treatment of the two-body problem was also done in a different manner by Kopeikin & Potapov (*Astron.Zh.*, in press).

B. Planetary Obliquities

(Sylvio Ferraz-Mello)

A major breakthrough was recorded in our knowledge of the evolution of the obliquities of the planets. The most striking results concern the Earth whose obliquity would have been frozen to its current value by the capture of the Moon. Indeed, the dynamics of the rotation of an oblate planet under the action of secular planetary perturbations is chaotic. In the case of the Earth, the integration of the equations of the precession over 18 million years shows a large chaotic zone. In its present state, the Earth avoids this chaotic zone and its obliquity is essentially stable, exhibiting only small variations of $\pm 1.3°$ around the mean value of 23.3°. If the Moon were not present, the chaotic zone would extend from nearly 0° to about 85°. Thus, even if the initial obliquity of the Earth was very small, this chaotic behaviour could have raised it to a large value in a few million years (Laskar *et al.*, *Nature* **361**, 615). Precise solutions for the obliquity and the precession of the Earth were published by Quinn *et al.* (AJ **101**, 2287) and by Laskar *et al.* (AA **270**, 522). These solutions provide the necessary data for the computation of the insolation at the surface of the Earth in the past. The insolation at the latitude 65°, whose long term variations is associated to major variations in the Earth climate, varied in the past million years between 380 and 500 W/m^2. Without the Moon, this variation range increases by a factor 2.5. Laskar *et al.* claim that the Moon is a climate regulator for the Earth and, if it were not present, the climate on Earth would suffer drastic changes. We may wonder if organized life would be possible in such circumstances.

The planetary obliquity of Mars has been studied by Ward & Rudy (*Icarus* **94**, 160), Touma & Wisdom (*Science* **259**, 294) and Laskar & Robutel (*Nature*, **361**, 608). The main results are as follows: There is a strong secular resonance due to near equality of the frequency of precession with s_2'. The obliquity of Mars is chaotic with possible variations from 0° to 60°.

In what concerns the other planets the results are the following (Laskar *et al., Nature* **361**, 608): The present spin of Mercury is very low and apparently trapped in a 2 : 3 spin-orbit resonance; however, provided its primordial period was smaller than 300 h, it must have suffered large-scale chaotic behaviour during its history. In the same way, if the primordial rotation of Venus was prograde, it has suffered large-scale chaotic behaviour in the past, being able to diffuse from 0° to nearly 90° in 1 Myr. When the obliquity is $\sim 90°$, it can evade the chaotic zone as the planet slows down and dissipative effects can then drive the spin axis to its present position (178°). The situation with the outer planets is different: The spectrum of the planetary forcing terms consists only of well isolated lines. In general, it is difficult to destroy the stability of the obliquities of these planets until their precession constant reaches 26 arcsec/yr. The obliquities of the outer planets are essentially stable and should be considered as primordial.

The primordial obliquities of the planets have been recently studied by Tremaine (*Icarus* **89**, 85), Lissauer & Safronov (*Icarus* **93**, 288), Tanikawa *et al.* (*Icarus* **94**, 112), Lissauer & Kary (*Icarus* **94**, 126) and Dones & Tremaine (*Science*, **259**, 350; *Icarus* **103**, 67). In general, the origin of nonzero primordial obliquities is unknown and even the sign of the rotation induced by accretion is difficult to determine. Planets which accrete from large planetesimals have, on the average, shorter spin periods and greater obliquities than those accreted exclusively from small bodies. Present values for the obliquity of the outer planets require implausible large impacting bodies; for instance, to produce Saturn's obliquity, that planet should have accreted at least one planetesimal several times as massive as the Earth.

The study of the obliquities of the planets required important methodological advances: (a) Touma & Wisdom extended the symplectic N-body integration algorithms of Wisdom & Holman (AJ, **102**, 1528; **104**, 2022) to include the rotational dynamics of extended bodies; (b) Laskar extended the frequency analysis technique used to detect layers of chaotic behaviour in the orbital motion of the planets to the study of global dynamics (Laskar, *Icarus* **88**, 266; Laskar *et al., Physica* D **56**, 253; see also Dumas & Laskar, *Phys.Rev.Letters* **70**, 2975).

C. Dynamics of Asteroids

(Andrea Milani & Sylvio Ferraz-Mello)

Very significant progress has occurred, since the last report, on many subjects related to the dynamics of the asteroid belt; we are now beginning to have a global understanding of the long-term dynamics of a very large portion of the phase space, which includes the orbits of most real asteroids. Although the different problems are strictly related, and the same techniques can be used for several tasks, we shall separate for ease of presentation the following subjects: 1. Proper elements and secular resonances; 2. Chaos and stability; 3. Mean motion resonances; 4. The model problem; 5. Binary asteroids; 6. Discovery and identification.

1. PROPER ELEMENTS AND SECULAR RESONANCES

The state of the art, as reported at the last IAU Assembly, was characterized by the development, then in progress, of new techniques of analytical and semianalytical investigation, with the numerical ("synthetic") techniques more and more used for confirmation and/or extrapolation of the analytical results. These methods are now fully developed and many results have been published. The location of the main secular resonances was derived with a purely analytical technique by Knežević *et al.* (*Icarus* **93**, 316). The extension of this calculation to much higher inclinations and eccentricities was possible with the use of a semianalytic method by Morbidelli & Henrard (CM **51**, 131; CM **51**, 169). The problem which were formerly intractable, because of the overlapping of a mean motion and a secular resonance, can now use the formalism of successive elimination of harmonics by Morbidelli (CM **55**, 101) and Morbidelli & Giorgilli (CM **55**, 131); in this way the occurrence of secular resonances inside and besides the Kirkwood gaps was investigated by Morbidelli & Moons (*Icarus* **102**, 316), Morbidelli *et al.* (AA, in press). Of course, there are always dynamical behaviours too complicated to be investigated only with analytical and semianalytical methods, but it is now possible to use the theories as a guide for selective numerical experiments on the most complicated orbits, such as the meteorite delivery route investigated by Farinella *et al.* (*Icarus* **101**, 174; CM **56**, 287).

A more detailed analysis of the dynamical structure of the phase space is possible for the "core" region of the asteroid belt, the one at low to moderate inclination, bounded by the secular $g - g_6$ resonance and by the 2 : 1 Kirkwood gap; this is where most real asteroids are. In this region we can apply a fully analytical theory, expanded to degree 4 (and more) in the eccentricities and inclinations and to order 2 in the planetary masses, developed by Milani & Knežević (*Icarus* **98**, 211; *Icarus*, in press). In this way, it has been possible to explore the web of higher degree secular resonances (with divisors formed by combinations of 4, 6 and even 8 fundamental secular frequencies), and to assess the reliability of proper elements computed with this theory. Proper elements in the core belt are now stable enough to identify even small asteroid families, as confirmed by numerical experiments spanning several Myr. A recent development has been the development of reliable and reasonably accurate proper elements even outside the core belt: with semianalytical methods, Lemaître & Morbidelli (CM, in press) have computed proper elements (also expanded to order 2 in the masses) for high-inclination asteroids and Morbidelli (*Icarus*, in press) has computed proper elements for secularly resonant asteroids. Proper elements for Hildas have been computed by Schubart (AA **241**, 297) and for Trojans by Milani (CM, in press), with synthetic methods based upon the post-processing of the output of a numerical integration; although these methods are still based on computational brute force, they do allow to understand the dynamical structure and to locate the main resonances, as in Milani (IAU Symp 160, in press). In conclusion, both the location of the main resonances and the computation of proper elements stable over millions of years are essentially solved problems for almost the entire asteroid belt. Reviews on these topics can be found in Lemaître (CM **56**, 103), Knežević & Milani (IAU Symp 160, in press), Froeschlé & Morbidelli (IAU Symp. 160, in press).

2. CHAOS AND STABILITY

At the time of the previous IAU assembly, the main focus of the investigations of chaos and long-term stability was on the orbits of the major planets; more recently, the same problems have been raised for the asteroids. There are so many asteroids that an investigation of the stability of a significant proportion of their population amounts almost to a global exploration of the phase space. Moreover, the motion of the asteroids can be described within a simplified model, although some caution is required (see also 4.). The analytical, and fully rigorous, results on the stability of some asteroids are still scarce; however, the first results applicable for a time span as long as the age of our Solar System have been obtained for Trojan type orbits: see Celletti &

Giorgilli (CM **50**, 31); this is the continuation of earlier work by Giorgilli *et al.* (*J.Diff.Eq.* **77**, 167) and Simó (*Mem.R.Acad.Cienc.Ar.Barcelona* **48**, 303). Unfortunately, these results based upon the Nekoroshev method still apply only to the restricted 3D problem and to very small libration amplitudes. On the contrary, numerical experiments on larger and larger sets of initial conditions and for longer and longer times are done by many. Lecar *et al.* (*Icarus* **96**, 251) report that there is a correlation between the Lyapunov time and the crossing time obtained in a number of restricted 3-body problem experiments; they compute a best fitting power law, e.g., for the outer asteroid belt the crossing time is most likely to be proportional to the Lyapunov time raised to power 1.7. Holman & Wisdom (AJ **105**, 1987) and Levison & Duncan (ApJ **406**, L35) find different best fitting exponents (viz. 1.9 and 1.4) for experiments in the outer Solar System and point out that the dispersion around the best fitting curve amounts to two orders of magnitude in the time span for instability. The stability of orbits in the Trojan region, and also for hypothetical Trojans of other planets, has been investigated by Mikkola & Innanen (AJ **104**, 1641), Holman & Wisdom (AJ **105**, 1987) and De la Barre (PhD. UCLA, 1993): they all report a very complicated dynamical structure with gaps and peculiar behaviours still to be explained. On the other hand, Milani & Nobili (*Nature* **357**, 569) report that the asteroid *522 Helga* is chaotic but macroscopically stable for a time span of 1000 Lyapunov times; examples of this puzzling "stable chaos" are also reported by Mikkola & Innanen (AJ **104**, 1641), Milani (CM, in press) and Milani & Knežević (*Icarus*, in press). In conclusion, the relationship between chaos and stability over time spans comparable to the age of the Solar System is far from clear and is currently the object of intense investigations. The long term instability for another class of orbits, more typical of comets, has been established by Bailey *et al.* (AA **257**, 315): a significant fraction of the very high inclination objects ends up in sungrazing orbits.

3. MEAN MOTION RESONANCES

The knowledge of the global dynamics of the mean motion resonances, from 3:1 to 4:3, recorded a rapid progress in the past triennium. The state of this knowledge, at the end of the past decade, was reviewed by Froeschlé & Greenberg (Asteroids II, 1989, 827). It consisted, mainly, of the explanation of the 3:1 and 5:2 gaps, by orbital intermitencies increasing the eccentricity to values high enough to allow the asteroid orbit to intersect the orbit of Mars (Wisdom, *Icarus* **72**, 241; Yoshikawa, *Icarus* **87**, 78) and the understanding of the low-eccentricity chaos of the resonances 2:1 and 3:1 as due to separatrix-crossing mechanisms associated to secondary resonances (Lemaître & Henrard, *Icarus* **83**, 391; Henrard & Caranicolas, CM **47**, 99).

The resonance 3:1 has been revisited by several authors. Ferraz-Mello & Klafke (Cortina 1990 NATO ASI, p.177) mapped the structure of the phase space up to $e = 0.9$ and showed that, besides the chaotic region found by Wisdom, there is another one at high eccentricities and these two chaotic regions are not always disjoint. Intermitencies in this region allow the asteroid to have close approaches to the Earth which, because its larger mass, may have played an important role in the scattering of the 3:1 resonant orbits. This structure was also found by Hadjidemetriou (CM **53**, 151) and orbits of this kind were computed by Saha (*Icarus* **100**, 434). The overlapping of the 3:1 resonance and the secular resonance ν_6 was studied by Ch.Froeschlé and Scholl (CM, **56**, 163).

The previous models were extended to the resonances 5:2 and 7:3 by Šidlichovský (CM **56**, 143; AA, in press), Klafke *et al.* (IAU Symp 152, 153), Yokoyama & Balthazar (IAU Symp 152, 159; *Icarus* **99**, 175) and Yoshikawa (*Icarus* **92**, 94). These studies show that the dynamics of the 5:2 resonance is very similar to that of the 3:1 resonance and the chaotic origin of the corresponding gap was confirmed by Ipatov (*Sov.Astron.Lett.* **15**, 324; *Icarus* **97**, 309). The 7:3 resonance presents confined low-eccentricity orbits (like the 2:1 resonance) and the action of Jupiter is not sufficient to explain the gap.

The existence of a gap at the 2:1 resonance and a group of asteroids at the 3:2 resonance was explained on the basis of different degrees of chaoticity under the perturbations of the outer planets (Ferraz Mello, IAU Symp 160, in press; Ferraz-Mello *et al.*, Cortina 1993 NATO ASI, in press). The Lyapunov times of the resonant orbits in the 2:1 and 3:2 resonances lie, respectively, in the ranges $10^{3.5} - 10^{5.5}$ and $10^6 - 10^7$. The averaged phase space of these two resonances, when the perturbations of Jupiter and Saturn are considered, has been studied by Morbidelli & Moons (*Icarus* **102**, 316). The results show sets of seemingly regular motions consistent with large Lyapunov times. Yoshikawa (*Icarus* **92**, 94) showed that the boundary of the 2:1 Kirkwood gap coincides with the region of resonant motions and Michtchenko (Dr. Thesis, USP) has shown that these limits coincide with the limits of the region of low-eccentricity chaotic orbits in the restricted three-dimensional problem. Simulations showing chaotic depletion in the 2:1 resonance were presented by Scholl in IAU Symposium 160. High Lyapunov times for the 3:2 resonant asteroids were also found by Franklin *et al.* (AJ **105**, 2336). Alfimova & Gerasimov (IAU Symp 152, 139) reconsidered the statistical hypothesis and associated the non-existence of low-eccentricity asteroids in the resonance 2:1 to their fast perihelion motion. Ferraz Mello (IAU Symp 160, in press) and Ferraz-Mello *et al.* (Cortina 1993 NATO ASI, in press) have also shown that the maximum value allowed for the eccentricity of an asteroid in the resonance 3:2 is ~ 0.4. Orbits with higher eccentricities are generally driven to close approaches to Jupiter and are scattered from this resonance.

Some partial results were obtained concerning other resonances. The structure of the phase space of the 4:1 resonance has been mapped by Klafke *et al.* (IAU Symp 152, 153). It overlaps with the secular resonances ν_6 and ν_{16} in a very complex way; Scholl & Ch.Froeschlé (AA **245**, 316) have shown that this overlapping is responsible for the huge depletion in the whole region 2.07 – 2.13 AU. Ries (BAAS, in press) studied the resonance 4:3 in the frame of the planar restricted three body problem and found results pointing towards the survival of low eccentricity objects ($e < 0.075$). An extended numerical survey of all resonances was done by Dvorak *et al.* (AA, **274**, 627; see also IAU Symp 152, 145)

The regular solutions of the averaged problem were the subject of many papers (for a review see *Rev.Mex.Astr.Astrof.* **21**, 569). Ferraz-Mello *et al.* (IAU Symp 152, 167; CM **55**, 25) studied the solutions of the restricted elliptic problem where secular and mean-motion resonances happen simultaneously (corotations). Tsuchida (*Rev.Mex.Astr.Astrof.* **21**, 585), studied the regular librations of *279 Thule*. Moons & Morbidelli (CM, **56**, 273; CM, in press) studied the phase portrait of several resonances in the restricted circular problem up to very high eccentricities. They also included the secular resonances ν_5 and ν_6 and determined their positions in the plane of initial conditions. These papers use local representations of the disturbing functions valid for high eccentricities, to deal with the convergence problem (see Ferraz-Mello & Sato, AA **225**, 541; Morbidelli & Giorgilli, CM **47**, 145). In this respect, Ferraz-Mello (CM, in press) revisited Sundman's convergence criterion and extended it for the expansions in the inclination. According to Sundman's criterion the convergence radius at the main asteroidal resonances: 3:1, 2/1 and 3/2, is, respectively, ~ 0.35, ~ 0.20 and ~ 0.09 (in eccentricity). These limits put strong constraints to the domain of validity of analytical theories of the mean motion resonances.

Schubart (AA **241**, 297) and Michtchenko & Ferraz-Mello (CM **56**, 121) used synthetic theories to study the regular librational motion of the Hildas. Hadjidemetriou (CM **56**, 201) studied the families of periodic orbits of the first and second kind in the 2:1, 3:1 and 4:1 resonances. The stability of the solutions of the circular problem in the 2:1 and 3:1 resonances was studied by Dvorak & Kribbel (IAU Symp 152, 171).

4. THE MODEL PROBLEM

As the accuracy of the methods available to observe the asteroid orbits increases (see Ostro, AJ **102**, 1491; Yeomans et al. AJ **103**; Bec-Borsenberger AA **257**, 844; AA **258**, 94), the need arises not only for more and more sophisticated computational methods, but also to take into account more and more subtle perturbing forces. This process is analogous to what happened for the orbits of artificial satellites between the 60's and the 80's (for a review see Milani et al. *Non gravitational perturbations and satellite geodesy*, Hilger 1987). A controversy has arisen on the orbit of asteroid *1566 Icarus*, with Yeomans (AJ **101**, 1920) claiming that non gravitational forces were needed to fit the observations and Sitarski (AJ **104**, 1226) claiming that general relativistic effects were enough, as later acknowledged by Yeomans (AJ **104**, 1266). The use of mutual perturbations of asteroids as a way to estimate the mass of the largest asteroids has been continued (see Schubart AA **264**, 719; Kuzmanoski & Knežević, *Icarus* **102**, 93). It has even been proposed by Plastino & Vucetich (AA **262**, 321) and Orellana & Vucetich (IAU Symp 152, 185; AA **273**, 313) to use the orbits of some Trojan asteroids as a way to search for deviations from the known theories of gravitation. Another problem is the choice of a simplified model to be used to explain the occurrence of a given dynamical behaviour in the orbits of real objects. More care should be used in extrapolating, to the complexity of real orbits, the conclusions obtained from tests and/or theories based upon very simple models, such as the restricted 3-body problem. As an example, Schubart (CM **56**, 153) points out that the results on some chaotic orbits in the outer belt strongly depend upon the model and cannot be described in terms of the restricted problem.

5. BINARY ASTEROIDS

The interest for this subject was of course raised by possible observational evidence for contact binary asteroids and also by the need to assess the safe fly-by distance from an asteroid for space probes (Hamilton & Burns, *Icarus* **96**, 43). Some recent investigations focused on the long term stability of possible binary asteroids against the perturbations occurring at close approaches (Chauvineau *et al., Icarus* **94**, 299; Farinella, *Icarus* **96**, 284). A method to identify former binary asteroids even after they split has been proposed for Trojans (Milani, IAU Symp. 160, in press). There are many open problems in this field.

6. DISCOVERY AND IDENTIFICATION

Although this subject concerns more Commission 20 and the WGNEO, it is important to point out that the renewal of interest on these topics has resulted in new investigations on the problems of accuracy and reliability of orbit determination (see e.g. Marsden, AJ **102**, 1539; Whipple & Hemenay, AJ **102**, 816; Kristensen, AA **262**, 606; Bowell & Muinonen, *Icarus*, in press). Some of the problems raised about the algorithms to be used to solve the problem of identification of newly observed asteroid orbits with some of the known ones, and to assess the reliability of these procedures, imply non trivial investigations in the fields of Celestial Mechanics, Probability Theory, Numerical Analysis and Image Processing; the specialists of Celestial Mechanics should follow more closely these discussions.

D. Dynamics of the Kuiper Belt

(Martin J. Duncan & Harold F. Levison)

In 1951, Kuiper pointed out that it seems unlikely that the disk of planetesimals that formed the planets would have abruptly ended at the current location of the planet Neptune. He suggested that originally there was a significant number of planetesimals in circular orbits about the Sun outside the planetary region. The objects that formed in this region presumably resemble present-day comets. This region is now called the *Kuiper belt.* Over the years, many researchers have attempted in vain to discover objects in the Kuiper belt (Levison & Duncan, AJ **100**, 1669). However, within the last year, Luu & Jewitt have discovered two large (radius on the order of 100 km) objects in this region of the solar system: 1992 QB1 (IAU Circular 5611) and 1993 FW (IAU Circular 5730).

Perhaps the most intriguing argument for the existence of a populous Kuiper belt is one put forth by Fernández (MNRAS **192**, 481), who suggested that it is the source of the known Jupiter-family comets (with periods, $P \lesssim 20yrs$, hereafter JFCs). Indeed, the most comprehensive numerical integrations done to date, performed by Quinn, Tremaine, & Duncan (ApJ **355**, 667), show that the inclination distribution of JFCs is inconsistent with them being captured long-period comets, but is consistent with JFCs originating in the Kuiper belt. If so, there should be hundreds of millions of comet-sized (1–10 km radius) objects in the Kuiper belt, although most would be too faint to be discovered using conventional ground based surveys (Levison & Duncan, AJ, **100**, 1669). We should note, however, that parts of Quinn, Tremaine, & Duncan's argument are still contested (Stagg & Bailey, MNRAS **241**, 507).

One link in the theory of the Kuiper belt as the origin of JFCs has been missing. Neither Fernández (MNRAS **192**, 481) nor Quinn, Tremaine, & Duncan (ApJ **355**, 667) could directly model the mechanism responsible for injecting Kuiper belt objects into Neptune-crossing orbits. In 1990, Torbett & Smoluchowski (*Nature* **345**, 49) showed that most particles with initial values of perihelion distance less than $\sim 45AU$ and eccentricity greater than ~ 0.01 are chaotic due to the gravitational perturbations of the giant planets and therefore can in principle leave the Kuiper belt. However, since the expected depletion timescales at the current epoch are on the order of the age of the solar system, they could not follow this evolution in their 10 Myr simulations.

The missing link was filled within the last year with the publication of two papers, by Levison & Duncan (ApJ **406**, L35) and Holman & Wisdom (AJ **105**, 1987), on the long-term dynamical evolution of objects in the Kuiper belt. Both sets of authors integrated the orbits of more than a thousand massless test particles on initially low-inclination, low-eccentricity orbits within the Kuiper belt. Holman & Wisdom's integrations lasted for 10^8 years and focussed on initially circular test particle orbits. Levison & Duncan's lasted for 10^9 years, using an approximation for the planetary perturbations but studying a range of initial eccentricities for the Kuiper belt objects. Integrations of this magnitude have only recently become possible because of advances of computer hardware as well as the development of powerful new integration techniques by Wisdom & Holman (AJ **102**, 1528).

Both groups found that particles which eventually crossed Neptune's orbit often showed long periods of relatively low-eccentricity oscillations punctuated by a very rapid jump to Neptune-crossing eccentricity. This flux may be the ultimate source of present-day short-period comets. For initially near-circular orbits an intricate structure in the region between 35 and 45 AU is found at the end of the simulations. However, an inner edge of the Kuiper belt can be said to be at approximately 36AU. Also, there is a low density region between 40 and 42 AU due to the presence of several resonances in this region of the solar system. For higher initial eccentricities the depletion is much more extensive interior to 42 AU.

Both groups also found that there is a rough correlation between the Lyapunov time of a Kuiper belt object and the time it takes to become a Neptune-crosser (although there is a wide dispersion about the mean crossing time for a given Lyapunov time). Levison & Duncan (ApJ **406**, L35) found that the correlation is in good agreement with the results of Soper, Franklin & Lecar (*Icarus*, **87**, 265). The currently active regions are found to be those in which the Lyapunov timescales are between about 0.2 and 0.8 Myr.

The minor bodies 1992 QB1 and 1993 FW may well be the first of a large number of objects to be discovered in the Kuiper belt. The object 1992 QB1, which is currently estimated (IAU Circular 5855) to have a semi-major axis of 44AU, an eccentricity of 0.09 and an inclination of 2 degrees is located in a relatively stable region according to the simulations of Levison & Duncan. Similar comments apply to 1993 FW, for which the best available estimates give orbital elements very similar to 1992 QB1. Very recently, 4 more objects have been found beyond Neptune; 1993 RO, 1993 RP (IAU Circular 5867), 1993 SB and 1993 SC (IAU Circular 5869), but their orbits are very poorly constrained at this time. It is possible that some or all of these objects may be resonantly trapped in regions near Neptune's Trojan points.

E. Capture into Resonance

(Jacques Henrard)

The effect of small non-conservative forces in shaping the solar system as we see it now has been brought to the attention of dynamicists by Goldreich (MNRAS **130**, 159). This effect is particularly dramatic when a resonance is encountered. It is now well established that tidal dissipation can lead to capture into orbit-orbit resonance and is responsible for at least some of the commensurabilities among the satellites of Jupiter and Saturn (see the review by Peale in *Natural Satellites*, Univ. Arizona Press, 1986). Recently, the scenario of capture into resonance of three of the Galilean satellites (Yoder, *Nature* **279**, 767) has been rewritten by Malhotra (*Icarus* **94**, 399) to include episodes of high eccentricity for Ganymede which could be responsible for its resurfacing.

Tidal dissipation is also responsible for the spin-orbit resonances in which most of the regular satellites and Mercury are locked (see also the review by Peale). Recent contributions concern the stability of the spin-orbit resonance (Celletti & Falconi, CM **53**, 113; Celletti, CM **57**, 329) and the consideration of gas drag rather than tides to provide the "small" dissipation mechanism (Winter & Sessin CM **57**, 329).

The evidence of melting of the small satellites of Uranus revealed by Voyager II, has spurred interest in the tidal evolution of these satellites (Peale, *Icarus* **74**, 153) although they are not, at present, locked in resonance. Tittemore & Wisdom (*Icarus* **74**, 174) investigated numerically the passage of Ariel and Umbriel through a 5/3 resonance and found evidence of possible *temporary* resonance, something which had not been encountered in the studies concerning the satellites of Jupiter and Saturn. Dermott *et al.* (*Icarus* **76**, 295) found also, in their numerical investigation of the 3/1 Miranda-Umbriel resonance, evidence of temporary capture and they attributed to it the anomalously large inclination of Miranda and its post-accretional resurfacing. Tittemore & Wisdom (*Icarus* **78**, 63) identified the mechanism responsible for the disruption of the resonance: capture into *secondary* resonances.

Indeed, the small oblateness of Uranus (compared with the oblateness of Jupiter and Saturn) makes possible the apparition of secondary resonances inside the primary resonance. When Miranda evolves inside the 3/1 primary resonance, it encounters these secondary resonances and upon capture by one of them is dragged into the chaotic layer surrounding the primary resonance and then escapes. Malhotra & Dermott (*Icarus* **85**, 444) confirmed and extended these results. The extent of the chaotic layer and the location of secondary resonances were also confirmed by Henrard & Sato (CM **47**, 391) with a semi-analytical perturbative analysis. Two building blocks of the basic scenario, the probability of capture into secondary resonance and the mechanism of escape from the chaotic layer were further investigated by Malhotra (*Icarus* **87**, 249; **94**, 524), Henrard & Moons (*Icarus* **95**, 244), Henrard & Henrard (*Physica* D **54**, 135) and Henrard & Morbidelli (*Physica* D **68**, 187).

Evolution through other possible resonances (2/1 Ariel-Umbriel and 4/1 Ariel-Titania) were numerically simulated by Tittemore & Wisdom (*Icarus* **85**, 394) and Tittemore (*Icarus* **87**, 110) as possible scenario leading to tidal heating of Ariel.

The effect of gas drag on a swarm of boulders or planetesimals orbiting in a protosolar nebula can lead them into capture in resonance with an embryo of planet. This is possible only for exterior resonances as shown by Weidenschilling & Davis (*Icarus* **62**, 12) and Patterson (*Icarus*

70, 319). Weidenschilling & Davis see in this mechanism a way to feed the embryo, Patterson see in it a way to form new embryos. The neoadiabatic theory (see Henrard, *Dynamics Reported*, Springer-Verlag 1993), the extension of the adiabatic theory to separatrix crossing used with success in the case of tidal dissipation, could be applied here also but only for particles 100 meters in radius or larger. Beaugé & Ferraz-Mello (IAU Symp. 152, 355; *Icarus* **103**, 301) consider the effect of Stokes drag dissipation upon smaller particles (10 meters in radius) and confirm the above mentioned results. They point out that the particles captured in resonance acquire eventually an equilibrium eccentricity which curiously enough depends only on the angular velocity of the gas and not on the drag coefficient or on the mass of the embryo. Nevertheless the probability of capture into resonance and the time scale of the capture do depend upon these other parameters.

A similar problem, connected with the evolution of dust grains around the star β-Pic under the influence of Poynting-Robertson drag, has been investigated by Scholl *et al.* (CM **56**, 381) and Sicardy *et al.* (CM **57**, 373). Very recently (*Icarus* submitted) Beaugé & Ferraz-Mello have applied the neoadiabatic theory to this situation and computed probability of capture in exterior resonances. As the equilibrium eccentricity is large in this case, the capture is only temporary; the dust grains collide with the protoplanet before reaching this equilibrium.

F. Symplectic Integrators

(Haruo Yoshida)

The exact time evolution (solution) of a Hamiltonian system is symplectic, i.e., the mapping from $(q(0),p(0))$ to $(q(t),p(t))$ conserves the symplectic 2-form $dp \wedge dq$. Traditional integration methods, such as the classical Runge-Kutta method, do not respect this fact and after a long-term integration, fictitious damping or excitation occurs which makes the result of integration unreliable. Recently there have been much interest in the integration methods which are designed to keep the symplectic property of the original Hamiltonian flow. These integration methods are in general called symplectic integrators (symplectic integration methods). With use of symplectic integrators, the error of energy does not grow monotonically. This comes from the existence of a conserved quantity (interpolating Hamiltonian) which is close to the original Hamiltonian. General reviews of this field of research are found in Sanz-Serna (*Acta Numerica* **1**, 243) and Yoshida (CM **56**, 27).

1. IMPLICIT SCHEME

For all Hamiltonian systems, implicit symplectic integration method is always possible. There are essentially two different kinds of implicit schemes. The first one is by the generating function (of Von Zeipel type). The generating function of mixed variable near the identity can be found to arbitrary order. These schemes are found by Feng & Qin (Lect.Notes in Math. 1297, 1), Channell & Scovel (*Nonlinearity* **3**, 231), etc. The second one is by the implicit Runge-Kutta method. Sanz-Serna (BIT **28**, 877) and Lasagni (ZAMP **39**, 952) found the condition of an implicit Runge-Kutta method to be symplectic when applied to a Hamiltonian system. The Gauss-Legendre type of Runge-Kutta methods fall in this category and the simplest 2nd order one is the implicit midpoint rule.

2. EXPLICIT SCHEME

For a Hamiltonian which is a sum of two trivially integrable parts, explicit scheme are possible to design. The typical example is a Hamiltonian of the form $H = T(p) + V(q)$. The second-order one has been known as the leap-frog method. The third-order one is found by Ruth in 1983 and the fourth order one by Forest in 1987. Yoshida (*Phys.Lett.* A **150**, 262) and Suzuki (*J.Math.Phys.* **32**, 400; *Phys.Lett.* A **165**, 387) found the way to obtain the explicit scheme of higher orders. Other contributions in this direction are found in Forest (*J.Comp.Phys.* **99**, 209) and McLachlan & Atela (*Nonlinearity*, **5**, 541)

3. APPLICATION TO DYNAMICAL ASTRONOMY

Kinoshita, Yoshida & Nakai (CM **50**, 59) applied the 4th and 6th order explicit integrator to the Kepler problem and confirmed several advantages over the non-symplectic methods. Gladman *et al.* (CM **52**, 221) also tested symplectic integrators to the Kepler problem and they found that the introduction of variable steps decrease the advantage of symplectic methods. Gladman & Duncan (AJ **100**, 1680) simulated the evolution of test particles in the outer solar system. Wisdom & Holman (AJ **102**, 1528) divide the Hamiltonian of the solar system into two parts (pure Keplerian part and the perturbation) and applied the idea of 2nd order explicit scheme to obtain a map which they call the N-body map. This N-body map has been discussed also in Wisdom & Holman (AJ **104**, 2022) and Holman & Wisdom (AJ **105**, 1987). See also Saha & Tremaine (AJ **104**, 1633).

4. APPLICATIONS IN OTHER FIELDS OF RESEARCH

Pullin & Saffman (*Proc.R.Soc.London* **432**, 481) applied the symplectic implicit Runge-Kutta method to the motion of point vortices successfully. Cary & Doxas (*J.Comp.Phys.* **107**, 98) used explicit scheme for plasma simulation. Some partial differential equations such as sine-Gordon and nonlinear Schrödinger equations are infinitely dimensional Hamiltonian systems. For the numerical integration of these partial differential equations, Herbst & Ablowitz (*J.Comp.Phys.* **105**, 122) tested symplectic schemes and found their advantages.

Sylvio Ferraz-Mello
Vice-President of the Commission

COMMISSION 8: POSITIONAL ASTRONOMY (ASTRONOMIE DE POSITION)

PRESIDENT: L. V. Morrison
VICE-PRESIDENT: T. E. Corbin
ORGANIZING COMMITTEE: P.Benevides-Soares, L. Helmer, Hu Ning-sheng, J Kovalevsky, L. Lindegren, J. A. Lopez, M. Miyamoto, F. Noël, D. Polojentsev, Y. Requieme, H. Schwan, M. Yoshizawa

1. Introduction

The scope of this report embraces activities in wide-angle optical astrometry, and specifically excludes photographic astrometry, which is the province of Commission 24, and radio-astrometry, which is the province of Commission 40.

This somewhat arbitrary sub-division of positional astronomy by discipline rather than subject is not particularly satisfactory scientifically, and this point is being addressed by the General Secretary who has expressed the wish to merge commissions.

The report is based on information supplied by members of the Commission. Where no information was forthcoming, I tried to compensate for the deficiency by adding material from other sources. Nevertheless, there are areas where the report fails to do justice. For this I apologise.

References to papers are quoted in square brackets where they are listed in *Astronomy & Astrophysics Abstracts*; otherwise simplified abbreviations for journals are used, as recommended by Commission 5.

2. Meetings, Symposia, Colloquia

Problemy astrometrii: Kosmicheskoj geodimamikj, Kiev, Ukraine, September 1990 [54.012.049].

IAU Colloquium 127, Reference Systems, Virginia Beach, USA, October 1990 [53.012.086].

Linkage of terrestrial, dynamical, stellar and extragalactic reference frames (III), Japanese Symposium on Earth Rotation, Astrometry and Geodesy, Naruko, Japan, January 1991 [53.012.079].

IAU Symposium 156, Developments in astrometry and their impact on Astrophysics and Geodynamics, Shanghai, China PR, Sept. 1992, (eds. Mueller & Kolaczek), Kluwer Acad. Press.

Galactic and Solar System Optical Astrometry: Observation & Application, Cambridge, UK, June 1993, (eds. Morrison & Gilmore), Cambridge Univ. Press.

J. Bergeron (ed.), Reports on Astronomy, Vol. XXIIA, 29–37.

3. Ground-based astrometry

ARGENTINA. The Repsold meridian circle of the Felix Aguilar Astronomical Observatory continued its programme of observation of fundamental stars, and the Fourth San Fernando Meridian Circle Catalogue was published [54.041.026]. Observations of 16 minor planets were compiled in order to study the behaviour of the FK5 around -30° [54.002.127].

BRAZIL. Astrolabe observations continued at Rio Grande and Valinhos. The Second Rio Grande Catalogue of 122 FK4 and FK4S stars by the Danjon OPL01 astrolabe was published [53.002.091]. Optical positions of the radio star α Scorpii by the astrolabe at Valinhos were published [53.041.036].
The reduction of meridian circle observations by an overlap method was tested [53.041.034], and the colour-index and mag. dependence of declinations from three astrolabe catalogues observed at Valinhos were investigated [53.041.037].

CHILE. The Repsold Meridian Circle at Santiago continued its programme of fundamental stars and published the Second Santiago Declination Catalogue of 493 FK5 stars [54.002.031]. The Santiago Fundamental Catalogue of 1105 FK5 stars (J2000.0) was published (A&AS,95,355). The astrolabe at Santiago continued its programme of radio stars, and positions of α Scorpii and 9 Sagitarrii were published (A&AS,96,567).

CHINA(PR). Photoelectric astrolabes and transit instruments continued in use at the Observatories of Beijing, Shanghai, Shaanxi, Yunnan and Purple Mountain. Their programmes centered on extending the fundamental optical reference frame to fainter magnitudes and linking the optical and radio frames. A new type III astrolabe (A&A,271,360) was installed at Xin Long station, Beijing Observatory. The low latitude meridian circle of Yunnan Observatory was being manufactured, and the horizontal meridian circle built co-operatively by Shaanxi and Copenhagen Observatory was adjusted optically and mechanically.

In December 1991 the type II photoelectric astrolabe of Beijing Observatory was transferred to San Juan, Argentina, to participate in the southern astrolabe programme. In November 1992 the type II photoelectric astrolabe of Shanghai Observatory was moved to Beijing Observatory.

The following photoelectric astrolabe catalogues were published: Third Beijing PACP3 [52.002.081], Second General Catalogue of 2606 stars GCPA2 [55.002.131], Third General Catalogue of 2577 stars GCPA3 [53.002.106], Second Yunnan [53.041.033], Fourth Preliminary Beijing PACP4 [55.002.132], Third PPCP3 and Fourth PPCP4 Beijing [54.002.060, 54.002.061], Yunnan 606 geodetic stars [54.041.008].

The Purple Mountain Catalogue of right ascensions of 508 stars obtained with the photoelectric transit instrument [55.041.014], and an analysis of the General Catalogue (CGS) of 4949 geodetic stars [55.041.017] were published. Precise positions and proper motions of 1980 General Catalogue stars were obtained [54.002.123].

DENMARK, GREAT BRITAIN & SPAIN

The Carlsberg Automatic Meridian Circle (CAMC) continued in regular use at the international observatory of the Roque de los Muchachos on the island of La Palma in the Canaries. It was operated jointly by Copenhagen University Observatory, the Royal Greenwich Observatory and the Real Observatorio y Instituto de la Armada en San Fernando.

The new and more accurate micrometer designed by Helmer which was fitted to the CAMC in March 1988 was descibed in [54.034.107]. This new micrometer was first used in the production of Carlsberg Meridian Catalogue Number 5 [53.002.129]. It contains positions and magnitudes of ~17 000 stars north of -45°, ~15 000 proper motions and 2847 positions of 66 solar system objects. The mean error of a catalogue position in the zenith is 0.″07 in right ascension and declination and 0.04 in magnitude. The same accuracy applies to Carlsberg Meridian Catalogue Number 6 [55.002.071] which contains positions and magnitudes of ~12 800 stars, ~11 600 proper motions, and 1368 positions and magnitudes of 69 solar system objects.

Carlsberg Meridian Catalogues 1-6 contain ~28 000 of the ~36 000 International Reference Stars north of -45°. Catalogues 5 & 6 contain the beginning of the observation (~6400) of the ~36 000 faint reference stars (V~12) selected at Copenhagen University Observatory and the Real Instituto y Observatorio de la Armada from the Astrographic Catalogue zones between -45° and $+90^{\circ}$ [53.041.053].

In October 1991 the divided circle was replaced by a new glass circle from Heidenhain with Diadur division lines. The six photoelectric cameras for reading the declination circle were replaced by CCD cameras. Using these new cameras in fast read-out mode enabled the telescope to be set more accurately, and stars were routinely scanned in the smallest slits which are 12″ long.

A joint discussion of the CAMC observations of FK5 stars with those made at Bordeaux was published [52.041.011]. This provided evidence of systematic errors in the FK5 reaching 0.″1 in some parts of the sky. Another joint discussion with Bordeaux – the optical positions of 23 radio stars – was published [52.041.004]. The first meridian circle observations of Pluto were published (A&A,262,347).

Positions of 53 radio stars obtained with the San Fernando Meridian Circle (SFMC) were published [53.041.054]. Following this programme the SFMC was taken out of service and automated, with a similar scanning slit and operational system as the Carlsberg AMC. Discussions began to find a suitable site in South America for this telescope.

FRANCE. Since 1990 the Bordeaux automatic meridian circle was mainly used for the determination of faint reference star positions from the Faint Reference Star list selected by the Carlsberg group in the range $11<V<12$. Measurements of 8000 stars in the zones +10 to +25 and +50 to +76 deg. have been completed. Positions of 221 radio stars were published [54.041.004]. Some of these were compared with VLA radio positions and the alignment of the FK5 and radio frames was deduced [54.041.024].

The global treatment of the Bordeaux AMC observations was investigated [52.041.028]. Taking as reference the preliminary Hipparcos positions of FK5 stars obtained by NDAC and FAST Consortia after processing of 18 months of data, a new reduction of eight years of Bordeaux differential observations was carried out. The results showed a remarkable improvement of the reference star residuals, with, however, small systematic waves (amplitude 0.″04) corresponding to unexpected systematic errors.

Development of a CCD micrometer was started and it was planned to mount it on the Bordeaux AMC in early 1994. Measurements of faint stars down to 15th magnitude could then be measured with an accuracy of 0.″02-0.″03, observing in narrow strips and taking as reference the final Hipparcos positions of stars with V>9 when they were released.

Improvements of the prototype photoelectric astrolabe were discussed and a new programme of observation with respect to FK5 was started [52.041.021]. The catalogue of stars observed with the CERGA photoelectric astrolabe March 1988 to July 1991 was published (A&AS,96,477). Systematic deformations of the apparent almucanter as derived from Danjon astrolabes in Paris and Santiago-de-Chile were investigated (A&AS,274,621). The result of eight campaigns of astrolabe observations of the radio star β Persei were published (A&AS,87,513).

GERMANY. The FK5 extension, comprising 3117 stars down to magnitude 9.5 taken from the FK4S and IRS, was compiled by the Astronomisches Rechen-Institut and US Naval Observatory [55.002.039].

Work continued on the data base ARIGFH at the Astronomisches Rechen-Institut. The main aim is to collect all the relevant astrometric data on stellar positions and proper motions in a comprehensive data bank. During the period of this report the main task was to bring older catalogues into a machine-readable form. There are now more than 1400 of such astrometric catalogues available, 1200 of which have been typed into the computer at Heidelberg. The cross-identifications of all these observations with the stars in the master catalogue and the formation of the data base was started.

ITALY. The programme of observation with the astrolabe of the Cagliari Astronomical Observatory continued, and the positions of radio stars were published [53.041.047].

JAPAN. The Tokyo Photoelectric Meridian Circle (Tokyo PMC) continued in operation. The third and fourth annual catalogues (PMC87, PMC88) of 5748 and 3800 stars, respectively, in the FK5 system were published [53.002.142, 55.002.152]. The global adjustment of star positions for the compilation of independent catalogues was investigated [53.041.031]. Systematic errors in declination arising from secular changes in the graduated circle were also investigated [53.041.045].

Work continued on developing a CCD micrometer for the PMC. Test observations were carried out on the Mitaka Gautier meridian circle. The first experimental model, used in drift-scanning mode (DISC=DIgital Strip scanning Ccd), enabled magnitude 14.5 to be reached. Work started on a second version (DISC-II), using a larger chip ($\sim$ 1200 $\times$ 1200 pixels) with reduced read-out noise which would permit 16 magnitude to be reached. It was planned to stop the first PMC programme in 1993 and replace the photoelectric micrometer by DISC-II. The second PMC programme of observation would begin by strip-scanning the declination band –25 to +35 (IAU Symp.156, 71).

A method of improving fundamental star catalogues using astrolabe observations was applied to the Mizusawa observations of FK4 stars [52.041.004].

LATVIA. The catalogues, Ri-80 and Ri-86, containing the right ascensions of stars observed for the time service of the University Observatory, were published [53.002.041].

POLAND. Systematic differences in zenith distance from two successive groups of stars in the OPL astrolabe at Borowice were investigated [54.041.016]. The influence of standard and anomalous refraction on astrolabe results was also investigated [54.041.019].

ROMANIA. An analysis was carried out of the separation in right ascension of wide doubles observed with the Bucharest Meridian Circle [54.041.017].

RUSSIA. At Pulkovo Observatory the Zverev Photographic Vertical Circle (PVC) after its reconstruction in 1987, carried out ~8000 absolute observations of 2050 FK5 star declinations from the North Pole to -15°. It was planned to complete the first version of the PVC catalogue in 1994.

The absolute declination PVC catalogue which was compiled from 12 762 observations made in Chile in 1964-66 was finished and made ready for publication. It contains ~700 FK4 stars and ~700 FKSZ stars (IAU Symp.141, 68).

In 1990 the observations of two star lists with the Pulkovo Horizontal Meridian Circle (HMC) were finished. The Catalogue of 170 reference stars around 63 radiosoures – Lazorenko list –

and the Catalogue of 911 stars (502 FK5 and 409 FKSZ stars) were compiled and published. Also the results of the differential observation of declinations with the HMC made in 1982-83 were finished and published.

At the Kislovodsk Mountain Station (h=2100 m) regular meridian observations of the Sun, Mercury, Venus, Mars and daytime stars were made using the Struve-Erthel Vertical Circle and Struve-Erthel Large Transit Instrument. In 1990-1992 more than 1500 observation of planets and Sun and about 4700 observations of stars were made with an accuracy more than twice as good as that obtained with the same instument in Pulkovo.

Work continued on new meridian instruments and modernising old ones. Experiments with CCD micrometers were carried out with the Struve-Erthel Great Transit Instrument on Kislovodsk Mountain Station (IAU Coll.136), with the Pulkovo Suharev HMS (IAU Symp.156, 119). Experimental CCD-observations with short-focus telescopes were started in 1990 as a part of the development of the REGATTA-ASTRO space astrometry project (Space Research Institute and Pulkovo Observatory). Astrometric and photometric results obtained with very small CCD-cameras (D=80, F=212 mm) shows that they are suitable for the purpose.

More precise simultaneous astrometric and photometric CCD-observations with a short-focus astrograph (D=100, F=712) were started in December 1992 at Pulkovo. Preliminary results of differential observations (exposure 2.5 min., field 0.8×1.1 deg.) were as follows: accuracy of position determination, $0''.1$ to $0''.3$ for stars from 8 to 14 mag. (series of 5 frames with 2.5 min. exposure); accuracy of photometry, 0.02 to 0.1 for a single observation (frame).

A new Pulkovo Meridian Reflecting System by A.A.Nemiro and Yu.S.Strletsky was in the process of being developed.

The construction of the Meridian Automatic Horizontal Instrument (MAHIS) was well advanced. In 1991-92 a new mirror for the MAHIS was made from carbide silicon. All optical and mechanical parts of the MAHIS were built in triplicate during 1990-1993. Some investigations of MAHIS were carried out (IAU Symp.156, 117).

Two papers concerning some problems of the determination of systematic errors in declination, especially of the FK5 catalogue were prepared for publication in the Izvestia GAO. Also considered was the problem of the influence of reference catalogue errors on the parameters of the fundamental system using minor planet observations. The analysis and generalization of the methods of compilation of secondary reference systems using meridian observations were carried out. The methods for unification of meridian catalogues, which were created without the strict principles of differential connection to the fundamental system, were investigated.

A detailed comparative analysis of two general catalogues of bright stars (BSC and KGZ) was completed. The work of amalgamating these two catalogues was preparatory to obtaining the united catalogue of positions and proper motions of 6474 stars down to 7.9 mag. in the FK5 system.

The absolute catalogue of declinations of 249 stars based on the Sanders-Raimond method and observed with the Sternberg Institute GOMZ's transit circle and zenith telescope was published [52.002.100]. The external accuracy was estimated to be $\pm0''.19$ [52.002.101].

UKRAINE. As part of the CONFOR programme, the positions of 2149 stars around 238 radio sources were observed with the Nikolaev Repsold meridian circle. Observations of Uranus and Neptune were continued. The Sun and planets were observed with two instruments: Freiberg-Kondratiev transit instrument in Nikolaev: Sun (145 observations), inner planets (136), external planets (29); Struwe-Erthel large transit instrument at Kislovodsk Mountain Station (see above).

Nikolaev Observatory in collaboration with Pulkovo finished observation of two differential catalogues made with the Pulkovo Horizontal Meridian Circle (IAU Symp.156, 119). The Pu(HMC)88 catalogue consists of 911 selected stars from FK5B, FK4S and FKFS, of which 502 belong to the FK5. It was shown that the systematic accuracy of the Pulkovo HMC was $\sim 0\farcs02$–$0\farcs03$. It was confirmed that the FK5 has systematic errors of about $0\farcs1$ in the declination zone +40 to +60. The Pu(HMC)89 catalogue, consisting of 170 AGK3 reference stars around 63 extragalactic radio sources, was published in VINITI, 1992.

The semi-automization of the Nikolaev Repsold meridian circle was finished. Four CCD matrices were used for the circle-reading, having an accuracy of $\pm0\farcs05$ for one microscope. Two computers were used for the registration and reduction of the observations. The Axial Meridian Circle (AMC) underwent investigation and automization. The previous value of the flexure was $0\farcs09$. Collimation stability with temperature was $0\farcs026$ per 1°C. The systematic errors of the AMC were not expected to be more than $0\farcs02$. (Pinigin *et al.* , 1993, in Kinematics and Physics of Celestial Bodies, in press).

The Nikolaev Observatory took part in the joint project MAHIS, involving Pulkovo and Kazan Observatories. Nikolaev Observatory finished the meteorological data collection and time service for MAHIS (IAU Symp.156, 117).

At Kiev the orientation of the FK5 and VLBI frames was investigated using the positions of 251 radio sources [54.043.002, 54.043.003]. The RRS2 list of 2575 meridian reference stars in the fields of 238 radio sources was prepared [55.002.065]. The observation of the position of planets continued at the Kiev State University Astronomical Observatory [55.041.015].

USA. The U.S. Naval Observatory continued with its program to produce an absolute catalog of observations of the FK5, IRS and radio stars from pole to pole. The observations made with the 6-inch transit circle in Washington, DC and the 7-inch transit circle on Black Birch, New Zealand were over 90% complete. Large numbers of observations of solar system objects were being made as part of that program. The 7-inch catalog of 23 000 stars observed in 1967-73 was compared with the FK4 [55.041.008].

The positions of 12 FK5 stars obtained with the Mark III optical interferometer were published [52.041.006]. The accuracies were 10 mas in RA and 6 mas in Dec. The Navy Prototype Optical Interferometer was in the process of being constructed at Anderson Mesa, Flagstaff, Arizona. First fringe detection was expected in late 1993, with observing commencing in the spring of 1994. The first program would be aimed at producing a catalog of 1000+ bright stars with accuracies of better than 4 mas.

The observations made with the 6-inch transit circle in Washington from 1977 to 1982 were being reduced to form the W1J00 catalog. The instrumental system was compared with the FK5, showing differences less than 150 ms of time in RA and 180 mas in Dec. Equator and equinox points were determined with respect to DE200.

The compilation of the Washington Fundamental Catalog (WFC) was in progress. Most of the effort was concentrated on the determination of zero points of the WFC system using observations of the Sun, Mercury and Venus. These data were taken from the absolute catalogs observed since 1900. Investigations into the possible use of observations of the minor planets,

Mars and Jupiter to strengthen the solution were also being carried out. Error estimates in astrometric catalogues were investigated [52.002.091].

The work on the W1J00 and WFC zero points was also used to evaluate W. Fricke's solution for the correction to the FK4 equinox. Using observations of the Sun, Mercury and Venus from eight Washington catalogs and six Cape catalogs (1907-1971), it was found that the FK5 proper motion system needed a correction of $-0\overset{''}{.}73$/cy. The observations of the Sun made for the W1J00 confirmed this result.

At the USNO Flagstaff Station, the modernization of the 8-inch transit circle continued, and was renamed the Flagstaff Astrometric Scanning Transit Telescope (FASTT). The Texas Instruments 800×800 thinned CCD was replaced by a CRAF/Cassini 1024×1024 12μ pixel, thick front-illuminated CCD. A laser interferometer was installed to monitor motion in azimuth in real-time. The internal accuracy obtained with FASTT was $\sim0\overset{''}{.}03$ for V<14, rising to $\sim0\overset{''}{.}25$ at the limiting magnitude around 17.5. Besides instrumental development, observations of FK5 and radio stars, and QSOs were made (IAU Symp.141, 369; IAU Symp.156, 65).

The CCD/Transit Instrument (CTI) of the Steward Observatory produced relative positions of stars brighter ~17 mag. with standard errors less than ~40 mas over a test region of 8×12 arcmin [53.041.001].

At the Space Telescope Science Institute a method of infinitely overlapping circles to obtain estimates of the systematic differences between fundamental astrometric catalogues was developed [55.041.013].

YUGOSLAVIA (former). A catalogue of 290 stars around 78 radio sources was produced with the large Meridian Circle of Belgrade Observatory [53.041.002]. Observation of solar system bodies continued [53.041.050; 55.041.006]. The positions of 223 PZT stars and 173 fundamental stars made with the MC in 1985-90 were analysed (A&AS,97,417).

4. Wide-angle space astrometry

ESA – HIPPARCOS. The ESA astrometry satellite Hipparcos completed more than three years of high quality data collection since its launch in August 1989, before operations were terminated on 15 August 1993. After functioning for some months with only two gyros, observations were suspended in March 1993 following the failure of the fourth gyro. Although attempts were made to continue observations without gyros, a series of other problems were encountered (attributable to the particle radiation damage resulting from its highly elliptical orbit), which made further satellite operations impossible. The operational lifetime foreseen pre-launch was 2.5 years.

The Hipparcos mission comfortably exceeded all of the original mission goals, and significantly exceeded many of them. Data reductions completed the treatment of the first 18 months of data, including iterations, in both of the main data analysis teams. A first treatment of all three years of the mission data was expected by the end of 1993. Final positions, parallaxes and annual proper motions were expected to be below the 2 mas target for the great majority of stars. Photometric results were also outstanding, with accurate photoelectric photometry for all 120 000 main mission objects, and all one million Tycho Catalogue objects, at 200 or so epochs throughout the observational period. Several thousand new double star systems were detected. Results from the two main reduction groups were in excellent agreement. Catalogue completion was foreseen for around 1996. Many of the results reported in the following paragraphs by the reduction consortia can be found in A&A,258,No.1(1992).

The FAST data-reduction consortium received and analysed all data collected by Hipparcos in the period 1989 November 27 to 1992 June 6. Several solutions using successively 6, 12 and

18 months of data were computed, showing a steady and rapid increase in the precision of the results. Significant parallaxes were obtained even from the 12-month solution, and proper motions became computable at the level of the 18-month reduction.

The next phase was to perform an iteration on the 18-month data span. About 100 000 stars were kept in that solution, the recognised and suspected double and multiple stars being temporarily left out of the solution, awaiting a determination of their elements. That iteration differed from the first treatment due to the fact that astrometric parameters obtained during the first treatment were used to recompute the attitude and the projections on reference great circles.

The results showed the following mean precisions for stars of magnitude 9:

1.4 mas in ecliptic longitude
1.2 mas in ecliptic latitude
3.3 mas/yr in proper motion in longitude
2.7 mas/yr in proper motion in latitude
1.6 mas in parallax.

Finally, the remaining 13 months of data were added to the iterated solution, showing a dramatic improvement in proper motions and a significant improvement in the three other parameters. Preliminary results for this global 31-month solution gave the following precisions for stars of magnitude 9:

1.3 mas in ecliptic longitude
1.1 mas in ecliptic latitude
1.7 mas/yr in proper motion in longitude
1.5 mas/yr in proper motion in latitude
1.5 mas in parallax.

In addition, the precison in determining the separation of the recognised (known or newly discovered) double stars was better than 12 mas. Most of the newly discovered double stars had separations between 0.″15 and 0.″4 and magnitude differences smaller than 3.

The data reductions in the NDAC consortium were shared between the Royal Greenwich Observatory in Cambridge (processing of raw data, attitude determination and photometry), Copenhagen University Observatory (determination of star coordinates on great circles) and Lund Observatory (sphere solution and treatment of non-single stars). A succession of preliminary catalogues for over 100 000 stars were produced, using the data collected during the first 6, 12, 18 and 30 months of the mission. These catalogues showed the expected progressive improvement in accuracy and clearly indicated that the mission goals of 2 mas average accuracy in the positions, annual proper motions and parallaxes would be reached, and in many cases exceeded, when all the data (approximately 37 months) were incorporated in the solutions.

Much effort was devoted to systematic comparisons with the results of the FAST consortium and with external data, which on the whole showed very satisfactory agreement. Calibrations and studies of the spacecraft dynamics performed by the NDAC group at the Royal Greenwich Observatory provided essential input for operating the satellite towards the end of its lifetime, when the on-board gyroscopes could not be used as intended.

The data reductions in the TDAC consortium were shared between the Observatory of Strasbourg, Astronomisches Rechen-Institut at Heidelberg, Astronomisches Institut at Tuebingen and Copenhagen University Observatory. The reductions made use of the satellite attitude and the Hipparcos reference catalogue received from NDAC. Working from a TYCHO Input Catalogue containing about the 3 million brightest stars on the sky, the star mapper signals were extracted and analysed in order to derive B and V magnitudes and astrometry for stars brighter than 12 mag.

The first year of observations from the TYCHO experiment were analysed to give a preliminary catalogue (TICR) of one million stars with an accuracy of 0."06. Astrometric processing of the first four months of observations showed that the complete 3-yr TYCHO mission would result in positions, annual proper motions and parallaxes for these stars with an accuracy of ~0."03 *rms* at 11 mag., improving to 0."006 for stars brighter than 9 mag. (See Høg *et al.* in *Galactic & Solar System Optical Astrometry*, eds. Morrison & Gilmore, Cambridge Univ. Press).

The INCA consortium completed its task of publishing the Hipparcos Input Catalogue in 1992 (ESA SP-1136) [55.002.054].

Volumes 1-5 of the main catalogue lists for the ~118 000 stars in the final Hipparcos observing programme their positions, proper motions, magnitudes and colours, and where available, spectral types, radial velocities, multiplicity and variable information. The catalogue is composed of two main parts. The first is the survey of ~52 000 bright stars which is complete to the magnitude limit in V of 7.3 to 7.9, depending on galactic latitude and spectral type. The second is the list of ~66 000 'faint' stars selected from the proposed observing programmes according to the priorities which were allocated by the Hipparcos Scientific Selection Committee.

Volume 6 (Annex 1) gives information and data for all known individual components of ~14 000 double and multiple systems for which at least one component is identified in the main Hipparcos Input Catalogue (Vols. 1-5).

Volume 7 (Annexes 2-4) is an atlas of ~10 900 identification charts for stars in the main catalogue.

5. Proposals for future wide-angle astrometric space missions

The ROEMER mission has been proposed to ESA by a group of European astronomers (see Høg & Lindegren, in *Galactic & Solar System Optical Astrometry*, eds. Morrison & Gilmore, Cambridge Univ. Press). The proposed mission aims at accurate astrometric and photometric measurements of 100 million stars down to 18th magnitude. Based on the well-proven concepts of the ESA Hipparcos mission, but employing a vastly superior CCD detection system, it would give the positions, parallaxes and annual proper motions to 0.1–1.5 mas, depending on magnitude, and multi-colour photometry to 0.001–0.02 mag. The programme would include a complete survey of the 50 million stars brighter than $V = 15.5$.

The AIST (Astrometric Integrating Space Telescope) project was proposed in 1989 at Pulkovo (Chubey *et al.* , IAU Symp.141, 77) as the 'second epoch of space astrometry' experiment. In essence, the design consists of two telescopes symmetrically placed around a central beam combiner unit. It will survey the positions, proper motions and parallaxes of 10-15 million stars brighter than 15th mag. In 1992-93 the computer simulation of the dynamic features of spacecraft and the numerical estimates of the level of errors of the output data were obtained. The arrangement and optical scheme and the focal plane ensemble were optimized (Chubey *et al.* , IAU Symp.156, 415). The plans for the mission are being developed in a collaboration between the Applied Mechanical Consortium in Krasnoyarsk, the Space Instrumentation Institute in Moscow and Space Flight Control Centre in Zelenograd, with contacts in the Hipparcos and Roemer scientific groups.

extracted and analysed in order to derive [illegible] and astrometry [illegible] 17 mag.

The first [illegible] [illegible]

[illegible]

[illegible]

INTERNATIONAL ASTRONOMICAL UNION

COMMISSION 9

INSTRUMENTS AND TECHNIQUES (INSTRUMENTS ET TECHNIQUES)

President: J.C.Bhattacharyya **Vice-President:** G.Lelievre

Organizing Committee: M. Cullum, J. Davis, C. Humphries, I.S. McLean, F. Merkle, F. Roddier, W.J. Tango, R.M. West.

1. Introduction

Considerable progress could be achieved in the realization of super large telescopes with apertures of 8 m plus during the period covered by this report. The 10 m Keck I telescope, the first of the twin facility utilizing active mosaic reflector has entered the operational phase, while the second unit progressed in its construction phase aiming towards its completion in 1995. Although the Columbus project of constructing a twin binocular telescope ran into some organization problems, the other projects e.g. JNLT and VLT progressed satisfactorily. Successful casting of large spun mirrors was achieved, and the possibilities of casting 8 m aperture light weight mirrors for future telescopes became very bright. The idea of diffraction limited image formation from large ground based telescopes became much nearer target with the employment of active mirrors in ESA's NTT. The utilization of larger fields in multi-object spectroscopy became possible with the help of multiple fibre-optic image transferring devices. Remote automatic operation of telescopes located in sites posing difficult logistics, came much nearer to realization.

Detector arrays continued becoming larger and finer with development of new chips; keeping pace with the development of computer hardware and image processing software. The spectral coverage extending deep into infra-red, created new classes of instruments in IR imaging and spectroscopy. Further developments in wide-field imaging were made by creating large area detectors by composite mosaics of buttable chips, as well as by improving scanning techniques.

In the area of high resolution imaging, several new long baseline interferometers were created, while some of the large telescope facilities nearing completion have been designed to use them as interferometers.

There have been three IAU Symposia (No. 154, 158 and 161) and one IAU Colloquium (No. 136) in which different aspects of astronomical instruments and techniques were discussed. An international conference dealing with particle streams in nature has been co-sponsored by us; it is scheduled to be held in Bangalore, India, during January 2-9, 1994. The new series of one day symposia to be held during the XXII General Assembly will also cover some of the areas of current interest in the topics covered by the Commission. Besides the above, a series of scientific meetings were held by different organizations which dealt extensively on the topics covered by our Commission. In preparation of the report, information of 21 such meetings has been taken into account. Lack of space prevents us from listing these here.

J. Bergeron (ed.), Reports on Astronomy, Vol. XXIIA, 39–52.

2. Telescopes

2.1 Telescopes commissioned, or under construction

Keck I telescope

Telescopes with apertures exceeding 8 m became a reality when the 10 m Keck telescope was completed in April 1992, at Mauna Kea in Hawaii, USA. The telescope has several novel features; the primary consists 36 hexagonal segments, each 1.8 m across which are positioned by a computer controlled active mirror support system. The basic optical design is a f/1.75 Ritchey-Chretien, whose hyperboloidal primary is made up of a mosaic of these 1.8 m segments, each figured separately by following a novel technique of stressed polishing. Additional corrections to the figure are provided by mechanical stresses on individual mirrors on the telescope itself. The segments are only 75 mm thick, thereby achieving the ease in warping the mirror surfaces for desired figures and at the same time making it much lighter than conventional thick mirrors. Precision sensors can detect minute shifts of edges between neighbouring segments, and correct them by electromechanical actuators. The design aims at 0.25 arc second FWHM, and preliminary tests have justified the expectation. The project is a joint effort of the University of California and the California Institute of Technology; they propose to have another 10 m telescope in an adjoining dome, so that the twin can be used as a powerful interferometer probing ultra-high resolution structures in the celestial sources. Funding for Keck II telescope has already been assured and the telescope is expected to be completed by 1996.

The Columbus project of constructing a 8.4 m twin telescope to be located on Mount Graham has run into organizational problems. The project which was a joint venture between the University of Arizona, Ohio State University and Arcetri Astrophysical Observatory in Italy, was to use spin-cast light weight mirrors from the University of Arizona's mirror laboratory. But because of the Ohio State University's withdrawal from the consortium, some uncertainty of funding has crept in.

J.N.L.T. (Adapted from a report by Masanori Iye)

The Japanese project of building a 8 m aperture Japanese National Large Telescope has progressed well. It has been finally decided that the mirror will have an aperture of 8.2m. The project has been christened "Subaru"; the telescope will be located on Mauna Kea next to the Keck telescopes.

The project work has started from 1991 and expected to be completed in 9 years. Contracts were awarded in 1991 to procure the primary mirror and to design the telescope structure. The process of polishing is expected to start in 1994 (some details of the mirror in sec. 2.4).

The ground breaking at the construction site was performed in July 1992. The telescope pier and the lower part of the building are under construction at site. The manufacturing of the mechanical structure of the telescope, including center section, mirror cell and 264 actuators to support the 8 m mirror is in progress.

V.L.T.

The VLT project is near the half way mark, both in time and in the resources contractually committed. In this process ESO has engaged many industries in a remarkably extensive set of R & D contracts, while continuing preliminary inquiries with many for immediate future requirements. Several path breaking types of developments in new technology, have been achieved in international collaborations; major examples can be seen in the development of adaptive optics with several French institutions, and of remote control systems in collaboration with Trieste.

The site for the VLT has been selected at Cerro Paranal, near the NTT. It is a plateau 2632 m above sea level in which specific excavations for the foundations of the four 8 m telescopes and the control building have been dug out.

2.2 Solar telescopes (Prepared by K.R.Sivaraman)

The medium sized national solar telescope facility operating on Tenerife and La Palma yielding near diffraction limited images of the sun are the 50 cm aperture Swedish vaccum telescope, and the two German telescopes; the 45 cm Gregory coudé telescope and the 70 cm Kiepenheuer vaccum tower telescope. Another telescope projected by the French at La Palma is a dedicated telescope of 90 cm aperture for polarimetric measurements (THEMIS). The Earth-based Solar Telescope also planned to be located at La Palma is presently financially insecure. Similarly the Indian National Solar Vaccum Telescope project although passed through the preliminary stages of acceptance tests by the national funding agencies, has now been shelved. The Uttar Pradesh State Observatory, Nainital, India has plans for a 50 cm vaccum telescope.

Telescopes of similar aperture (7 cm) for installation at the network of six stations of the Global oscillation Network Group (GONG) together with the Michelson interferometer units are getting ready at the National Solar Observatory (NSO), Tucson. A new project titled 'Precision Solar Photometric Telescope' of the NSO, Tucson using a 15 cm aperture refractor and a precision photometer to study the sun's variable radiative output in the Ca II K line has received initial funding for the development of the telescope and the photometer. Images of the Fe XIV solar corona have been obtained during trials with the 15 cm aperture reflecting coronagraph (Mirror Advanced Coronagraph) developed at the Sacramento Peak by joint efforts from several agencies within the U.S. and the Institut d' Astrophysique, Paris. This instrument will be tested and modified further to optimise its design.

2.21 New instrumentation facilities

The 512 channel magnetograph which has been in operation at the Kitt Peak vaccum telescope has been replaced by the spectromagnetograph which provides better measurements. The Mc Math-Pierce telescope has now the new facility of a dual grating spectrometer with IR and visible options. At the NSO at Sacramento Peak, the Advanced Stokes Polarimeter developed by the High Altitude Observatory has been installed at the vaccum tower telescope. Another vector magnetograph using the Zeeman sensitive line (Ca I 6122.2 Å) is installed at the existing 25 cm telescope.

A new prototype solar vector magnetograph was given preliminary trials at the Kodaikanal solar telescope. The arrangement uses a CCD at the spectrograph focal plane and measures all stokes vectors along the 6302 line; an attached P.C. computes magnetic field parameters, strengths, inclinations and azimuthal directions on a rectangular strip on the sun.

2.22 Birefringent filters in solar observations (Contributed by A.Bhatnagar)

With the invention of birefringent filters by Lyot in 1933 and independently by Ohman in 1935 and by

Sloc in early thirties, the two dimensional monochromatic solar observations have been revolutionized. Since then significant contributions have been made to improve their performance by Evans at Sac Peak Observatory and building a narrow pass band filter by Ramsey, Smartt and by Giovanelli at CSIO, Australia. In recent past Alan Title at the Lockheed Observatory has made important contributions using partial polarizer to develop tunable narrow pass band (0.1A) birefringent filters. During the last few years the chinese solar astronomer Ai Guoxiang has developed narrow band (1/8 Å) and multichannel tunable birefringent filters.

2.3 Telescope environment and operation

It has been observed that the "seeing" measured at a new site invariably deteriorates after the installation of the telescope. The reasons for such degradations have been traced to "dome seeing", by which the trapped air inside the telescope enclosure contributes to the deteriorating seeing quality. With a view to avoid such situations, a study group for the VLT project have designed "open" type of enclosure, in which the telescope would be largely exposed to the undisturbed windflow during observations.

In the very first conception, it was visualised that all four units of the VLT will be completely exposed during observations and protected during the day by movable roll-on/off shelters. A large wind screen, nevertheless, had the function of decreasing the wind force on the telescope. On closer examination, the idea was found not very practical and abandoned.

In the modified design, the lower part of the telescope is kept well protected in a cylindrical recess. The retractable enclosure consists of a fixed base and a rotating part. The fixed base is made of a metal space frame ring-shaped structure and supports the rotating part on a number of roller bearings. The upper rotating part is made of an approximately cylindrical panel-clad space frame, which constitutes a wind shielded recess in which the lower part of the telescope is contained. The top part, in this design, composes of a retractible hemispherical dome. This may be in the form of an inflatable dome made up of pressurized double fabric supported by rigid hoops, or a dome consisting of overlapping shell sections connected along a common axis to the rotating cylindrical part.

An alternative enclosure type which is being considered is a scaled up simplified version of the NTT building installed at La Silla. The additional advantage in this design is that this does not limit the observing elevation to 10° - 15° above the horizon, but reaches much further down. It features large upside down L-shaped doors to cover the observing slit as well as louvres around the periphery of the building which allow some direct ventilation of the telescope at any azimuth.

Other studies aiming towards improvement in seeing have been reported from the 2.4 m Hiltner Telescope of the Michigan-Dartmouth - MTT Observatory. The primary and secondary optics have been refigured so that ~80% of the light falls within a 0.4 arc second diameter circle, and to take full advantage of these the thermal properties of the telescope and dome have been remodelled. The inside dome temperature can rise 15° C above the night time ambient thereby heating the telescope and its optics. To offset this problem, active dome vents and an airconditioning system to cool the primary mirror have been installed. A computer collects data from thermocouples which measure the temperature of the primary mirror and other key points. The primary mirror is cooled to the predicted ambient night time air temperature using graphical data from previous nights and a computer controlled weather station. The telescope which is located on the south-west ridge of Kitt-Peak yields images of 0".7 FWHM on CCD frames exposed for 600^s.

2.4 Telescope mirror manufacture

For large aperture telescopes two lines of approach are being developed in two centres. ESO is experimenting on solid thin mirrors whose figure will be controlled by a closed loop servo system through electromechanical actuators. They have already employed this technique to 3.58 m NTT at La Silla.

The first of the four 8.2 m VLT mirrors has already been cast, annealed and rough ground at Schott Glaswerke, Mainz and delivered to REOSC (Recherches et études ďoptique et des sciences connexes) at Ballainvilliers, near Paris, France, who will produce the final mirror. Schott have set up three furnaces, the idea being that all three will be used simultaneously and the subsequent blanks will be delivered at one year intervals. REOSC have set up two machines, one for coarsely figuring the mirrors, after which the blanks will be transferred to the second machine with which the final, highly delicate polishing will be performed. After thorough testing the mirrors are scheduled to be transported to Chile at the new site at Cerro Paranal. The scientists are confident that very accurate figuring as they have achieved for the NTT mirror will also be possible.

The second line of approach is by the Steward Observatory's Mirror Laboratory (SOML) at the University of Arizona, where the mirror of borosilicate glass, a material with much lower melting point is moulded in a form with honey comb network at the back for stiffness as well as for reduced thermal inertia. Dr Angel also visualizes a system of cooling by forced draughts on the telescope. The largest blank moulded by this technique is the 6.5 m meant for replacing the six 1.8 m mirrors in the Multi-Mirror Telescope at Mount Hopkins. The two 8 m blanks for the Columbus Twin telescope project are expected to be prepared in this laboratory.

Still a third technique of mirror making has been adopted for the "Subaru" project – the Japanese Large National Telescope, scheduled to be put in operation by 1998 at the Mauna Kea Observatory. The 20 cm thick primary mirror will have its surface generated on a composite meniscus comprising of a top plano-plano surface backed by a bank of 44 boules of hexagonal cross section fused to this meniscus. The meniscus will be generated by sagging the top plate down in an oven; it is expected that f/1.8 will be achieved in one year.

Not much progress have been reported about other techniques of producing light weight mirrors using all metal or metal foam sandwiches. The success attained by the thin glass mirrors assisted by active corrections perhaps have reduced the urgency of finding alternative approaches.

One notable success in mirror fabrication techniques has been the experiment of stressed polishing adopted in Keck telescope mirrors. All laboratory tests on the individual segments showed that the idea has been totally sound, and giant aspheric suraces can be generated by this technique. Reports on sky observations are awaited.

3. Instruments and techniques

The availability and promise of newer and larger telescopes have promoted activities in design and fabrication of many new instruments. A comprehensive cover of all these activities is not feasible in this small report. Attempts have been made to cover some of them, which unfold glimpses of scientific ingenuities and foresight among the instrumentation scientists.

3.1 Instrumentation (Prepared by S.K.Jain)

The LOOM (Laboratoire d Optique de l Observatoire de Marseille) team reports the installation of a laser machine, LAMA, which will be used for producing 2D masks. This will allow greater accuracy at the slit edge and more reliability under operation. A new version of CORAVEL radial velocity meter-SUPERCORAVEL - has been designed and built jointly with Observatories de Geneve and de Haute-Provence. This instrument will operate in 1993 at OHP. The very high resolution available increased by use of optical fibers is expected to provide radial velocity accuracies upto 10 or 5 m/s. An active optics method has been developed to obtain plane-aspheric reflective gratings which now form part of a number of high-throughput spectrographs.

Pulkovo Observatory, Russia, has started work on the development of two star-measuring mechines, "FANTASIA" and "PARSEC". The former can accommodate a plate of 240 mm x 240 mm, and measure 800-1500 stars/hour, with a precision of 0.00033 mm; the latter can accommodate a plate of 160 mm x 160 mm, and measure 400 stars/hour with a precision of 0.001 mm.

Two infrared array camera systems-IRCAM and MIRAC (Middle Infrared Camera) - for ground based astronomy have been reported by Shivanandan. IRCAM is based on 10 x 50 elements Si: As BIBIB/ SWIFET HYBRID array operating at 10K. MIRAC uses a 20 x 64 elements Si:As with a capacitance trans-impedance amplifier readout and operates at 10K in a liquid helium cooled cryostat. It has an operating wavelength range of 2 to 26 microns.

Infrared Imaging Detector Lab. (UCLA, USA) has constructed a K-band survey camera for small telescopes which provides 0.5" pixels and a 128" x 128" field of view. This lab is also working on the development of a general-purpose, two-channel camera system designed for larger telescopes.

A high-resolution slitless spectrograph-Heterodyned Holographic spectrometer (HHS) - has been developed at Kapteyn Observatory (The Netherlands). It combines the high throughput of a Fourier Transform Spectrometer with the convenience of a CCD detector.

The instrument laboratory at Paris-Meudon Observatory, France, has developed a single object fiber spectrograph, ISIS-BIS which is the second generation version of ISIS. This spectrograph is able to achieve a resolution of 60000. Its infrared version, ISIS-IR, has been updated with a 2-D mode and can go upto 2.2 microns. A multi-object fiber positioner has been developed for the prime focus of the 3.60 m telescope at La Silla, Chile. This has 30 independent arms which sweep a field of one degree. This laboratory has also developed a second generation cassegrain double spectrograph for the CFHT. This instrument employing MOS/SIS, has a field of 10' x 10'.

In an interesting paper (1), a team at Space Science Department, ESTEC, proposes a novel approach to the physics of optical photon detection. They demonstrate the feasibility of optical photon counting using a superconducting substrate in combination with an array of widely spaced superconducting tunnel junctions of lower energy gap. Apart from high quantum efficiency over a very broad wavelength range, without readout noise, with high time resolution, the new proposed detector will have intrinsic higher wavelength resolution, claim the researchers.

VULCAN: A low resolution spectrophotometer for measuring the integrated colors of galaxies has been developed jointly by Astronomisches Institut, Austria and Department of Astronomy, University of Michigan (USA). The main purpose of this instrument is to obtain narrow-band optical colors for low surface brightness galaxies, which can then be applied to the study of stellar populations in these galaxies.

Indian Institute of Astrophysics (IIA), Bangalore, has developed a Rotational Shear Interferometer for high angular resolution imaging. The program has been done jointly with Raman Research Institute (RRI), Bangalore.

A dual polarization receiver operating in the frequency range 85-115 GHz has been installed on the 10.4 m millimeter wave telescope of RRI.

3.11 Developments in observatories

A few other progress reports received from observatory groups are summarized below:
The SOFIN Spectrograph has been constructed for the 2.5 Nordic Optical Telescope (La Palma, Canaries) to provide Spectroscopic observations with very high spectral resolution. The spectrograph has been designed and manufactured at the Crimean Astrophysical Observatory in collaboration with the observatory of Helsinki University. The Spectrograph has echelle grating, cross dispersion prism and three camera of 2000, 1000 and 350 mm focal lengths. The maximum spectral resolution with the first camera is about 1,50,000. It is installed at the Cassegrain focus at the telescope and operates under full remote control through the CAMAC interface from the Gateway-2000 personal computer. The Astromed CCD Cameras are used for spectra detection (Contributed by P.P.Petrov).

The TV complex using intensified isocon D Digital and 0.5 m telescope are in progress. TV photometry has been developed for objects upto 17^m with exposure time 1-6 minutes and accuracy about $0^m.02$ digital TV Spectrophotometry optical unit is created by VI Pronik and L M Sharipova where astigmatism and coma have been suppressed in the slitless spectrograph using transparent gratings. The original digital TV Spectrograph of moderate dispersion created by A N Abramenko permits seeing spectra and slit images on the TV screen simultaneously. The TV method and apparatus for the photometrical observations of near-earth objects have been developed (Report by VV Prokof'eva)

3.2 Optical fibres

Use of optical fibres in astronomical instruments continued to increase. Remarkable advantages in manouvering beams in instruments like multiple object spectrographs are being fully utilised. Fibres in the improved transmission characteristics have been developed and many uses of these have been described in the previous section. Combination of images obtained by separate telescocpes are also being investigated in some novel projects. A special meeting on these topic was held during the period (3).

3.3 Active and adaptive optics (adapted from a report by J.M.Beckers)

The two functions of collection of large photon flux and concentrating them into sharp images by optical telescopes have undergone changes in their cycle of domination in optical astronomy. Initially the angular resolution was the dominant requirement, until the atmospheric seeing limit was reached, and then the collecting area and detection efficiency became the most important factor. With the introduction of techniques which overcame the limits set by atmospheric seeing the emphasis is changing back to angular resolution.

The 240 cm diameter Hubble Space Telescope has overcome the seeing limit by locating it above the earth's atmosphere. It is radical way of solving the atmospheric seeing problems. It is expensive and is almost certainly not affordable at least for some time to come for modern 8-10 meter diameter telescopes. Except for wavelengths inaccessible from ground astronomers and military programs have therefore, been developing techniques which makes diffraction limited imaging possible with earth based telescopes. One of these, speckle image construction has been very successful in giving diffraction limited images. The most powerful and promising technique for all astronomical applications is, however, the one using adaptive optics. For an unresolved image it puts most of the collected photons in as small an image as possible, which benefits in discriminating against the sky background both in high spectral and spatial resolution spectroscopy and in doing interferometric imaging. For resolved objects adaptive optics allows imaging without the complications of image reconstruction techniques applied to short exposure, noisy speckled images, thereby reaching to much fainter and more complex objects.

Modern large telescopes have been designed to give image qualities by themselves of a 0.2 arc seconds FWHM. A number of telescope imperfections contribute to this: like i) tracking errors ii) mechanical flexure of telescope structure supporting the optical components resulting in misalignment and aberrations of the optics iii) deviations from the perfect optical figure resulting from polishing errors, mirror support and thermal inhomogeneities in the mirror blanks, and iv) seeing generated by the telescope itself, by creating a temperature gradient in its vicinity.

Active optics is a technique which removes some of these. A good example is the ESO New Technology Telescope (NTT) where the shape of the rather thin primary mirror is controlled. A number of actuators in the mirror support adjusts its shape until the mirror figure, as measured by a Hartmann Shack wavefront sensor using a stellar image in the telescope field of view, is optimum. This correction is done with a time constant of a few minutes, so that only slow tracking errors, mechanical flexure effects, mirror support and large scale (> actuator spacing) optical figure errors are removed. This has resulted in a very good telescope image quality in which only fast tracking errors, small scale optical figure errors and mirror seeing remain to limit the telescope image quality. Active mirror supports are now being incorporated in all 8 to 10m telescopes; it is an essential ingredient in the achievement of the 0.2 arc sec. FWHM image quality.

Adaptive optics is very similar to active optics in principle; the significant difference being that it works much faster, in the millisecond range. This removes not only the slowly changing telescope wavefront disturbances, but also the rapidly changing atmospheric variations.

Over the past few years a few astronomical systems using adaptive optics have become operational. These include the 3.6 m ESO telescope at La Silla, the Canada-France-Hawaii 3.6 m telescope at Mauna Kea, the 6.8 m MMT at Mount Hopkins, Arizona and a few other smaller systems. Full details of the activities are given in (4)

In adaptive optics, the necessity of using a resonably bright unresolved star becomes necessary. While some of them have utilized nearby stars in the field, a new idea of creating a small spot of light high up in the atmosphere has been successfully employed. The method is known as LGS (Laser Guide Star), where a strong laser beam is sent up after collimation by the telescope. The artificial stars are created by Rayleigh scattering of laser radiation off molecules and aerosols in the high (10–15 km) atmosphere.

A still better idea for measuring the wave front distortions is to utilize the Mesospheric layer of neutral sodium atoms at around 90 km height, which can create a brighter and more distant light source by resonance scattering. Preliminary experiments with Na-LGS have been performed; their use in adaptive optics operations are expected shortly.

Even with the use of adaptive optics angular resolution achievable will be limited by telescope diameters. The largest telescopes with diameters of 8 to 10 meters, presently in existence or being planned approach probably the limit of what can be achieved in single mirror telescopes. The way out of this limitation is probably interferometric imaging with arrays of telescopes; major efforts are presently under way to implement interferometric imaging in optical astronomy. Interferometric arrays are in operation in Australia (Culgoora) England (Cambridge), France (CERGA) and in the USA (Mt. Wilson, MMT). Others are under construction or in advanced planning stages in Europe (the VLT interferometer or VLTI) and in the USA (Mt Hopkins, Anderson Mesa, the CHARA array, the Keck Interferometric array or KIA). Both the VLTI and the KIA involve the use of large 8-10 m telescopes to achieve the highest possible sensitivity; to perform optimally these telescopes will be equipped with adaptive optics.

Even with interferometric arrays angular resolution will be probably limited to milli-arc seconds because of atmospheric and possibly physical limitations. Interferometers in space or even on the moon have been proposed to remove these limitations. Indirect ways to achieve sub-milli arc seconds resolution can be found through temporal photometric spectroscopic and astrometric variations. It has been proposed to discuss some aspects of such experiments in the symposium No. S166 during XXII General Assembly in August 1994.

4. Working groups

Commission 9 has three working groups, Wide Field Imaging, Detectors and High Angular Resolution Interferometry. The name of the working group on Wide Field Imaging has been changed from its earlier one of Astronomical Photography, because of new developments in techniques. Reports about the activities of these groups are given below:

4.1 Wide Field Imaging (Report by R.M.West, Chairman of the Working Group)

The IAU Working Group on Wide Field Imaging (WG WFI) was established at the 1991 General Assembly in Buenos Aires. It is the successor of the WG on Astronomical Photography, but covers more areas Sky surveys and Patrols, Wide Field Detector (e.g. Photographic Techniques and CCD Mosaics), Digitization Techniques, Archival and Retrieval of Wide Field Data. By the end of June 1993 it had about 200 members and consultants. The organizing committee (OC) consists of: J. Guivert (Paris, France), R.M. Humphreys (Minneapolis, USA), K. Ishida (Tokyo, Japan), R. Kron (Chicago, USA), B.M. Lasker (Baltimore, USA) H. Lorenz (Potsdam, Germany), H.T. McGillivray (Edinburgh, UK, Secretary), D. Malin (Epping, Australia), N. Reid (Pasadena, USA), M. Tsvetkov (Sofia, Bulgaria) and R.M. West (Garching, Germany, Chairman). The WG WFI issues a semi-annual Newsletter (Editor, H.T. McGillivray).

Wide Field Imaging is characterized by the rapid and efficient acquisition of very large data sets. This leads to considerable handling problems, concerning both the initial extraction of relevant information for scientific interpretation, as well as the time- robust data storage and easy retrieval at a latter date. It is a major goal of the WG to bring together all available expertise on the related repidly progressing technology (e.g. panoramic detectors and storage media) and those major research projects which are dependent on large-scale observations (e.g. cataloguing and structural studies). The WG is particularly concerned with 1) integration of techniques and science, 2) coordination (e.g. of the photography and CCD groups), 3) standardization (procedures, formats etc.) and dissemination and sharing of information to the larger community.

The WG activities have included the organization of IAU Symposium 161 on "Astronomy from Wide-Field Imaging" , (Potsdam, Germany, 23–27 August, 1993) with about 200 participants. A major WG Project (WFPA) is the registration in a computer readable form of the plate catalogues over the world (in-charge: M. Tsvetkov). A large fraction of plates in storage are now on file. The possibility of large scale digitisation of (a subset of) these plates is being studied.

A recent decision by Kodak to discontinue the production of several important astronomical emulsions, notably IIa and 103a, has led to severe problems for various current large scale projects. The WG now attempts to ensure the continued availability of at least the IIIa and IVN emulsions, and also to further development of a blue-sensitive Tech-Pan film similar to the 4415 red-sensitive emulsion.

4.2 Detectors (Report by M.Cullum, Chairman of the Working Group)

The possibility of CCD detectors with 2048 x 2048 pixels was first mentioned in a Detector Working Group report in 1985, although the available technology did not then permit the manufacture of useful devices of this size. In the intervening period CCD technology has steadily evolved and matured to the extent that today, chips of this format are becoming one of the mainstay optical detectors in all the major observatories.

Several trends can be observed in present developments, both on the part of manufacturers as well as users. These include the increased popularity of foundry runs, improvements in thinning techniques and back surface treatment, and the development of large mosaics.

Many of the large format CCDs currently in operation at the telescope have been produced by so- called silicon foundries. To date almost all of these have been produced by Loral, although other firms, such as Orbit, are now offering similar services. Instead of marketing individual tested CCD chips, foundries manufacture batches of wafers on a best effort basis according to standard or custom designs. Although it is likely that the majority of the chips produced in this way will not be suitable for astronomical applications, the significantly reduced unit cost and the generally acceptable production yields in recent years have made it worthwhile for many large institutes to go in this direction. A few have also designed their own CCDs to cater for special instrumental requirements. Even circular CCDs have been produced in this way.

The largest CCDs that are in widespread astronomical use have 2048 x 2048 pixels of 15 micron square. Larger single chips have been manufactured, and several institutes have tested chips of 4096 x 4096 pixels with 7.5 or 15 micron pixels. CCDs of the latter dimension are about the largest that can be accommodated within the 4 inch wafers used by many manufacturers. Efforts to achieve even larger formats have mainly concentrated on making mosaics of chips that are designed to be buttable with a gap of less than 0.5 mm between the butted sides. These devices may either be two-side buttable (either adjacent sides for 2 x 2 mosaics, or opposite sides for 1xn mosaics) or three sides (for 2xn mosaics). Several institutes are developing 2 x 2 mosaics of Loral chips giving 4096 x 4096 pixels (ESO, NOAO,

IFA/Hawaii, RGO, amongst others). Existing plans even include a 2 x 5 mosaic with a total of 8190 x 10240 pixels for direct imaging applications (Japanese Nat. Obs.). For some specialised applications 'sparse mosaics' have been assembled using conventional CCD designs.

Major efforts have been made by a number of manufacturers and individuals to improve the sensitivity of CCDs, especially at UV wavelengths. Nevertheless, the fact that the large format CCDs now in use still have widely varying sensitivity figures show that these techniques have not yet been fully mastered. Of the major chip manufacturers only Tektronix produces thinned large format CCDs on a commercial basis, although even here the supplies have been erratic. EEV, Reticon and Thomson all have on-going thinning development programmes for their large format devices. M. Lesser (Steward Obs.) has successfully thinned Loral CCDs and supplied these to a number of observatories. It is worth noting here that if the rear surface has not received a permanent implant during manufacture, thinned CCDs usually require a UV flood prior to use to achieve a good short-wave sensitivity, thus adding to operational headaches. Also important for improving the sensitivity of a back-illuminated CCD is the anti-reflection coating. Both Tektronix and Lesser have demonstrated considerable advances in this area. Peak RQE figures exceeding 90% have been recorded in some cases, and over 50% at 300 nm. Even 40% RQE at 400 nm has been measured with front illuminated CCDs using the thin gate technology developed at JPL. Sadly, the RQE of CCDs in operational use at observatories today is generally significantly less than these figures suggest.

The readout noise of CCDs measured at the telescope at observatories around the world show a fairly consistent picture. Values in the range of 3 - 4 e- are routinely observed with many recent types of CCD, with values up to 7 - 10 e- with some chip types. The use of mpp implants in many chip designs to reduce the dark current allows integrations of several minutes at room temperature.

There are several avenues of development that promise performance benefits in the near future. These include the improvement of skipper readout techniques to the point where single photo-electrons can be unambiguously detected (although at the expense of longer readout times), improved low signal CTE using super-notch designs (JPL), and advances in fast readout and/or low noise CCDs (EEV, JPL, Kodak, MIT Lincoln Labs, Sarnhoff). Fast parallel readout CCDs with smaller formats (32 x 32 to 256 x 256) are available from several manufacturers for applications such as fast guiding and wavefront sensing.

Developments in photon-counting detectors have generally been motivated by far UV applications or for those requiring high temporal resolution, for example adaptive optics and speckle interferometry. At the time of writing, the 1024 x 1024 MAMA detectors for the STIS have successfully passed the NASA technology demonstration test, and two engineering tubes are being assembled. The STIS will be equipped with two types of MAMA tube with solar blind CsI and Cs2Te cathodes. The DQE of the demonstration Cs2Te tube was 17% at 200 nm. The MAMA is also available with visible light cathodes.

LEP in Paris has developed an electron bombarded CCD tube (EBCCD) which is able to distinguish up to 10 electrons per pixel per CCD readout, thus increasing the dynamic range by the same amount compared to a normal ICCD detector with the same frame rate. The device is presently employed as a wavefront sensor in the Come- On-Plus adaptive optics system at ESO. The format is 604 x 288 pixels, although larger formats are planned. At the present time CCDs still cannot compete with photon-counting detectors for fast wavefront-sensing applications at very low signal levels, but the recent advances in fast readout CCDs indicate that this situation may change in the future.

In the field of infrared array detectors substantial progress has been made during the last three years. Detectors with up to 256 x 256 pixels are now in fairly widespread use at observatories. In the wavelength range between 1 and 5 microns, two types of material are dominant: HgCdTe and InSb. Rockwell's 256 x 256 NICMOS-3 arrays (HgCdTe) are employed for wavelengths up to 2.5 microns. They have a pixel size of 40 microns, an RQE of 40 - 60%, and a read noise of around 30 e- for a single readout, and down to 15 e- with multiple non-destructive readouts. The full-well capacity is about 2.10^5 e-. The dark current can be under 1 e-/sec/pixel, but existing versions of these 256 x 256 devices have shown a slight persistence effect in which the dark current is dependent on the previous light exposure. This can be problematic, for example, in instruments that switch between imaging and spectroscopy on the same detector.

The SBRC 62 x 58 InSb array, which has been a workhorse for infrared astronomy over a number of years, has now been displaced by a new 256 x 256 array. This detector, which has 30 micron pixels, has a very high RQE of typically 80% out to 5 microns. The read noise, dark current and full well capacity are broadly similar to those of the NICMOS-3 detectors. Both the Rockwell and SBRC arrays are in use at many observatories, in many cases in instruments that have been specifically designed for the new large arrays.

Another InSb detector that has recently appeared on the market is a 256 x 256 array from Cincinnati Electronics. Although this device has not yet reached the astronomer, an interesting feature is the promised high well capacity of 10^6 e-. Amber Engineering also produces an InSb array with 512 x 512 pixels, but this has a higher dark signal and is primarily intended for thermal imaging applications. Both Rockwell and SBRC have begun to develop 1024 x 1024 infrared arrays. One potential problem with arrays of this size is that the full-well capacity is likely to be similar to, or possibly even smaller than, present 256 x 256 arrays. With 16-times as many pixels to readout, this will require complex parallel readout electronics when used at longer wavebands.

In the mid-infrared region (10 - 30 microns), the situation has also changed rapidly. Doped silicon photoconductor arrays have now been displaced by more sensitive Impurity Band Conduction (or Blocked Impurity Band) detectors. Gallium, Arsenic or Antimony doped silicon devices are common, and the readout structures are similar to the near IR arrays. Aerojet Electrosystems has produced a 96 x 96 array, Hughes has made a 256 x 256 array, and Rockwell can provide 128 x 128 or 256 x 256 detectors. Available reports indicate that the performance of these detectors is excellent, although the larger devices have yet to be used on a telescope. Germanium IBC arrays are also being developed for the far infrared region.

Two major meetings on infrared array detectors took place in 1993. The first of these was a SPIE meeting in Orlando, Florida in April on 'Infrared Detectors and Instrumentation'. The proceedings of this meeting are published as SPIE Vol. 1946 and edited by A Fowler. The second conference took place in July 1993 at UCLA under the title 'Infrared astronomy with arrays; the next generation'. The proceedings of this meeting will be edited by I. McLean and published by Kluwer by mid-1994. The IAU Symposium 167 ('Advances in Array Technology and Applications') will take place between 15 - 19 August 1994, and will cover both optical and infrared array detectors. The Chairman of this symposium is D. Philip.

Much of the data on which this report is based was obtained from an e-mail survey carried out in September 1993. Thanks are due to the following who supplied information on the status of detectors at their respective institutes: J. Barton, A. Danks, G. Finger, J. Geary, F. Harris, O. Iwert, M. Iye, J. Janesick, P. Jordan, G. Lelievre, G. Luppino, I. McLean, R. Reed, P. Sinclaire, R. Stover and A. Walker

Infrared detectors (Contributed by I.McLean)

The astronoshing progress in the development of infrared array detectors and their impact on astronomy was demonstrated in July 1993 at a conference entitled "Infrared Astronomy with Arrays": The News Generation organized by Prof Ian McLean and held at the University of California, Los Angeles (UCLA). Almost 300 registered participants from all over the world gathered to present observational results with arrays, describe new instruments and hear from detector manufacturers about the future prospects. About 180 papers were presented including invited reviews, contributed talks and posters. The conference covered astronomical applications of IR arrays, from near to the far infrared, new instruments, future detectors, space missions, the impact of IR astronomy on very large telescopes and even the importance of IR arrays for astronomy from Antarctica.

Infrared array detectors for wavelengths less than 5 microns now have 256 x 256 pixels, high QE (> 60%) and low noise (< 40 electrons). Numerous cameras are in operation, such as the two color, twin detector system at UCLA. One of the highlights of the meeting was the extent to which high resolution (R>10,000) IR spectroscopy is benefitting from these sensitive devices. For longer wavelengths (10-20) microns) where the problems of thermal background are worse, arrays of 128 x 128 pixels have arrived.

Manufacturers of near IR arrays (made from indium antimonide on from mercury-cadmium telluride) announced at the meeting that they plan to develop 1024 x 1024 pixels arrays within the next 2-3 years. Noise performance is expected to reach 10 electrons or less.

In other words, infrared array detectors will provide a performance comparable with optical instruments, and many optical astronomy projects - especially spectroscopic ones - can now be carried out at infrared wavelengths. This is ideal for cooler sources, heavily reddened objects and galaxies at high redshifts.

The Proceedings of the conference at UCLA will be edited by McLean and published in 1994 by Kluwer.

4.3 Working Group on High Angular Resolution Interferometry (Chairman, W.J.Tango)

Major focus of activities in the WG has been organization of IAU symposium No. 158 on "Very High Angular Resolution Imaging" held on 11-15 January 1993 in Sydney, Australia. Principal topics discussed in the symposium cover imaging techniques, optical, radio and millimetre wave domains, speckle interferometry both in optical and infrared bands, and the results and future prospects of the experiments.

5. Extra-visible band observations

In the course of evolution of observational astronomy, many channels of information have been opened up, as new laboratory techniques of detecting and measuring these have been recognized and developed. This is not a new process; in the history of science we find opening up of IR channel (Herschel 1800), UV (Ritter 1801), Radio (Hertz 1888), X-rays (Roentzen 1896) and γ-rays (Rutherford 1901) in the electromagnetic spectra. As carriers of astronomical information even fundamental particle streams have come into the fold with cosmic rays (Hess 1912), solar wind (Bierman 1951) and neutrinoes (Davis 1966). The newly discovered channel of the gravitational waves (Weber 1962) has technically entered the area of astronomical techniques.

The recognition of the new channels by astronomers have been rather slow and hesitant. IR could not be effectively used until developments in the solid-state physics in the middle of the present century brought forth sensitive detectors; neither could the ultra-violet channel used before we could send the instruments and detectors beyond the absorbing layers of our atmosphere. Years after Jansky had stumbled upon the new sources through radio waves, the astronomers were slow to recognize the technique as a powerful one complementary to existing optical method.

Radio Astronomy was only a technique when started, but later astrophysics flooded the whole scene. The specialization which had distinguished the branch have largely disappeared. Many of the developments in the field e.g. image processing techniques have been adopted in other branches of astronomy; and several innovations in technique developed elsewhere have been assimilated in this branch. Logically the technique - part of Radio Astronomy should come under the fold of Instruments and Techniques, as it has happened in the case of IR astronomy. So also the innovations in High Energy Astrophysics techniques, which do not have a proper organized group in the IAU should logically be taken under the wings of Commission 9. A multinational collaborative project where development of a rocket-borne EUV imaging spectrograph for solar studies were needed now nearing completion (5); the first experiment is expected to be performed in late 1993.

We have been approached by a group of particle physicists, to co-sponsor a meeting where astronomical techniques of detecting and measuring particle streams may be discussed. Such experiments have been performed by several astronomical groups and important bits of information about several forms of physical processes happening in the universe have been obtained. With the idea of the need to develop any new techniques of observation, which complements our existing methods, we have extended our co-sponsorship for the meeting. The meeting is going to be held in January 1994, in Bangalore, India.

Acknowledgements

In preparation of this report I have received help from many scientists, who had sent reports, reprints etc, based on which bulk of the report has been written; I am grateful to all of them. Wherever possible I have acknowledged their help in the body of the report. I specially appreciate the useful advices sent by J. Davis, I. McLean, M. Cullum, R.M. West, C. Humphries, P. Mack, P. Felenbok, A. Bhatnagar and Shen Panan which have been of enormous help in my efforts. I am grateful to several of my colleagues for help and assistance in the preparation of this report; they include K.R. Sivaraman, S.K. Jain, A.K. Saxena, A.K. Pati, A. Vagiswari, K.T. Rajan, S.B. Ramesh and Pramila Mohan.

References

1. Perryman M.A.C., Foden C.L. and Peacock A: Nuclear Instruments and Methods in Physical Research, A 325 (1993)
2. Izv. Krim astrophys obs. 86, (1992)
3. Optical Fibres in Astronomy II (Sydney, Australia, Nov 10-13, 1991)
4. J.M.Beckers, "Adaptive Optics for Astronomy: Principles, Performance and Applications" Ann. Rev. Astron & Astrophys, 31, 1993.
5. S.K.Jain et al. High Resolution Extreme Ultraviolet (EUV) Spectrometer: XV meeting of Astronomical Society of India, Bombay (1993)

COMMISSION 10 : SOLAR ACTIVITY (ACTIVITÉ SOLAIRE)

PRESIDENT V Gaizauskas
VICE PRESIDENT O Engvold
ORGANISING COMMITTEE G Ai, E Antonucci, A Benz, TG Forbes, AA Galal, E Hildner, J Jakimiec, M Karlicky, ME Machado, T Sakurai, B Schmieder, M Schüssler, AV Stepanov, EA Tandberg-Hanssen

The Sun's activity may have declined in the three years since the last report, but the solar community carries on its research at a vigorous pace. During that interval there have been six IAU Symposia/Colloquia sponsored (or co-sponsored) by this commission on topics related to solar activity. Numerous activity-related conferences and workshops supported by other bodies attest to the vibrant interest in, and committement to, research on the wide range of topics which encompass this discipline. Flare research continues as the dominant topic, and is now fueled by a new generation of ground-based instruments and a highly successful solar satellite - Yohkoh (formerly Solar-A). Other recent space missions, although not devoted entirely to solar flares, are likely to have a profound bearing on our understanding of flare issues. They include, among others, the continuing series of sounding rocket flights of the Normal Incidence X-ray Telescope (NIXT), the Compton Gamma Ray Observatory (CGRO), designed to observe energetic phenomena in the Universe, and GRANAT, which is also dedicated to high-energy astronomy. The flare problem spurs research into many related subjects: active-region formation, physical proceses in prominences and filaments, coronal mass ejections, even periodicities in solar activity. This last topic is given separate attention here in order to draw attention to the still obscure link between the Sun's interior and the activity at its surface. It is a pleasure to thank the individual reviewers who took time from their busy schedules to write separate sections of this report.

As well as moments of triumph, there have been moments of anxiety, disappointment, and sadness for the solar community. Substantial numbers of solar astronomers and solar physicists in the states of the former Soviet Union live under stressful circumstances as struggles go on to form new political systems in their homelands. Elsewhere ambitious plans for new ground-based or space facilities have been delayed as a worldwide recession continues. And the community has been saddened by the loss of pioneering figures who had a major impact on shaping our discipline: Prof. Keizo Kai, former chief of the solar radio group at the Nobeyama Radio Observatory, Prof. Zenzaburo Suemoto, former director of the Tokyo Astronomical Observatory, Prof. Giuseppe Vaiana, best known in his role as principal investigator of the S054 experiment on Skylab, Dr. Mstislav N. Gnevyshev, founder and director of the Kislovodsk Solar Station of the Central Astronomical Observatory of the Russian Academy of Sciences. They will remain alive in our memory and in our work, which builds upon their considerable achievements.

1. GROUND-BASED OPTICAL INSTRUMENTS

(Guoxiang Ai)

1.1 HIGH SPATIAL RESOLUTION DEVICES

The following new devices or techniques have been introduced on the German Vacuum Tower Telescope (=VTT, Soltau 1991), on the Swedish Vacuum Solar Telescope (VST), and on the VTT of NSO/SPO (Dunn & Smartt 1991; Acton & Smithson 1992): real-time frame selection (Scharmer & Lofdahl 1991); a correlation tracker (von der Lühe 1991); adaptive optics (Dunn & Smartt 1991); and a speckle polarimeter-interferometer (Keller 1993). Dramatically improved resolution has been realized from these improvements, e.g., diffraction-limited performance of about 0.3" and successful reconstruction, using speckle polarimetry, of a narrow-band Stokes V image in which magnetic elements have a FWHM of about 200 km (Keller 1993).

New devices under construction include a continuous face-plate mirror with 218 mm clear aperture and 61 actuators (Dunn *et al.* 1991), and a coronagraph with adaptive optics using a membrane mirror with 91 actuators (Clampin *et al.* 1991).

J. Bergeron (ed.), Reports on Astronomy, Vol. XXIIA, 53–84.

1.2 MAGNETOGRAPHS AND POLARIMETERS

Since April 1992 the NASA/NSO spectromagnetograph (Jones *et al.* 1992) has been measuring magnetic flux, line-of-sight velocity, continuuum intensity, equivalent widths, and line depths (spatial pixels of 1.1", spectral pixels of 30-40 mÅ). It scans the full disk in about 1 h and local regions in <5 min. The HAO/NSO Advanced Stokes Polarimeter (Lites 1993) became available to users in mid-1993 (~0.4", 12mÅ/pixel, integration time per spectrum ~4s, time resolution (for a scan 85"x85") ~16 min).

Videomagnetographs are now as popular as Hα filters were in the 1950s and 1960s. A high-resolution videomagnetograph using cross-correlation techniques to achieve 0.1"/pixel is operating in the Swedish VST (Lundstedt *et al.* 1991). A solar flare telescope (four tubes: Hα and white light (aperture 15 cm); vector magnetic field and velocities (20 cm)) is operating at Mitaka since 1991 (Ichimoto *et al.* 1993). A similar telescope is under development in the Republic of Korea. A vector videomagnetograph with a universal filter (range 4600-6600 Å; bandpass 0.05-0.125 Å) is being developed in Taiwan. Two full-disk videomagnetographs with time-resolution ~1 min have been built in tandem in Beijing and Tokyo (Ai 1993a; Sakurai 1993). Two videomagnetographs with a lithium niobate solid Fabry-Perot etalon are operational in the U.S.A.: at the Mees Solar Obs. and by JHU/APL at NSO/SPO (Rust & O'Byrne 1991); a similar unit is under development at the Udaipur Solar Observatory in India. Comparison of magnetographs built around narrow-band filters with grating instruments based on Stokes polarimetry show a high degree of consistency in the morphology of both longitudinal and transverse fields (Wang *et al.* 1992). Measurements of longitudinal fields agree better than those of transverse fields which show scatter originating in different sources for different instruments (Wang *et al.* 1992; Ronan *et al.* 1992).

Polarimetry at IR wavelengths enjoys the advantage of enhanced Zeeman splitting. The Stokes I, Q, U, and V profiles have been obtained in the lines 1.56μ FeI and 12.32μ MgI with infrared detector arrays having 640x488 pixels (1-1.5 μ), 256x256 pixels (1-5.5 μ) and 58x62 pixels (5-24 μ; Deming *et al.* 1991). Two-dimensional observations of sunspot magnetic fields have been made with a 128x128 format HgCdTl infrared array working at wavelengths from 1 to 2.5μ (McPherson & Kuhn 1992). A 0.5 Å birefringent filter at 10830 Å is being built (Ai 1993a).

An electrograph for measuring the dependance of polarization on line width in Stark-broadened Paschen lines around 8500 Å has been tested by Moran & Foukal (1991).

1.3 TWO-DIMENSIONAL IMAGING SPECTROGRAPHS

Grating-based 2-D spectrographs are gaining popularity with the improvement in quality of CCD cameras. The imaging spectrograph of the Swedish Solar Telescope (0.1"x 0.22"/pixel) is reported to have 5 s time resolution over a 25"x 25" area (Scharmer *et al.* 1991). Similar imaging spectrographs have been completed in the Solar Tower Telescope of Nanjing Univ. (Fang *et al.* 1993) and in a new 50-cm solar telescope for the Yunnan Observatory (Zhang *et al.* 1993). A 2-D spectrometer consisting of a scanning Fabry-Pérot interferometer with a universal birefringent filter as a monochromator (FWHM 0.02 Å) has been tested and will be used for THEMIS; a similar device is under way for the German VTT (Soltau 1991).

A third generation Multichannel Subtractive Double Pass (=MSDP) imaging spectrograph is presently working at the German VTT. It provide images simultaneously in many wavelengths (typically 20), covering two line profiles with spectral resolution of 0.26, 0.12, or 0.06 Å and local bandwidths 0.13, 0.04, 0.03 Å respectively (Mein & Rayrole, 1991). A multi-slit spectrograph for prominences, flares, etc. has been running at the Udaipur Solar Obs. A 2-D real-time polarizing spectrograph with 64 channels (tunable range: 4200-7000 Å, FWHM: 18-5 mÅ, wavelength coverage: 1-3.2 Å) is under development in China (Ai, 1993a).

1.4 LARGE OPTICAL SOLAR TELESCOPES

Detailed engineering design with integrated adaptive optics for LEST has been underway since 1991 (LEST Ann. Rep. 1991). Two imaging Stokes polarimeters are being prepared for it: ZIMPOL I, to obtain simultaneously I, Q, U, V with three demodulating CCD sensors (Keller *et al.* 1992); ZIMPOL II, to measure the four Stokes parameters simultaneously with a single demodulating CCD and a microlens array (Stenflo *et al.* 1992). THEMIS is expected to be operational 1995-1996 (Rayrole & Mein 1993).

A coronagraph with a 15-cm aperture, superpolished Zerodur mirror objective, now under development, will serve as a prototype for a meter-class research coronagraph (Dunn & Smartt 1991). An Open Tower Telescope (both the tower and the telescope have an open framework construction) with a 45 cm aperture is nearing completion (Hammerschlag & Zwaan 1992). It is an innovative concept aimed at decreasing the cost and complexity of high angular resolution telescopes.

REFERENCES

I: "The Magnetic and Velocity Fields of Solar Active Regions", IAU Colloq. 141 (Ref. I), Zirin, H, Ai, G, and Wang, H (eds.), APS Conf. Series (San Francisco), **46,** 1993
Acton, D and Smithson, R: 1992, *Appl. Opt.* **31**, 3161
Ai G: 1993a, Ref.I, p. 149
Ai G: 1993b, in T Sakurai (ed), "Proc. Japan-China Workshop on Solar Phys."
Clampin M, Durrance S, and Golimowski D: 1991, SPIE Conf. "Active and Adaptive Opt. Syst.", 165
Deming D, Hawagama T, Jennings D, and Wiedemann G: 1991, NSO/SP Summer Workshop **11**, 341
Dunn R and Smartt R: 1991, *Adv. Space Res.* **11**, No. 5, 139
Dunn R, Streander G, Hull E, and Wilkins L: 1991, SPIE Conf. "Active and Adaptive Opt. Syst.", 88
Fang C: 1993, in T Sakurai (ed), "Proc. Japan-China Workshop on Solar Phys."
Hammerschlag R, and Zwaan C: 1992, Ann. Rep. of the Sterrekundig Inst. Utrecht
Ichimoto K, Sakurai T, Nishino Y, Shinoda K, Noguchi M Kumagai, K Imai H, Irie, M, Miyashita M, Tanaka N, Sano I, Suematsu Y, and Hiei E: 1993, Ref. I, p. 166
Jones H, Duvall T, Harvey J, Mahaffey C, Schwitters J, and Simmons J: 1992, *Solar Phys.* **139**, 211
Keller C, Aebersold F, Egger U, Povel H, Steiner P and Stenflo J: LEST Tech. Rep. No. 53
Keller C:1993, Ref. I, p. 3
Lites B: 1993, Ref. I, p. 173
Lundstedt H, Johannesson A, Scharmer G, Stenflo J, Kusoffsky U, and Larsson B: *Solar Phys.* **132,** 233
McPherson M, Lin H, and Kuhn J: 1992, *Solar Phys.* **139**, 255
Mein P and Rayrole J: 1991, *Astron. Astrophys.* **248**, 669
Moran T and Foukal P: 1991, *Solar Phys.* **135**, 179
Rayrole J and Mein P:1993, Ref. I, p. 170
Ronan S, Orrall F, Mickey D, West E, Hagyard M, and Balasubramaniam K: 1992, *Solar Phys.* **138,** 49
Rust D and O'Byrne J: 1991, NSO/SP Summer Workshop **11**, 74
Scharmer G and Lofdahl M: 1991, *Adv. Space Res.* **11**, No.5, 129
Soltau D: 1991, *Adv. Space Res.* **11,** No.5, 133
Stenflo J, Keller C, and Povel H: LEST Tech. Rep. No. 54
von der Lühe O: 1991, *Adv. Space Res.* **11**, No.5, 275
Wang H, Varsik J, Zirin H, Canfield R, Leka K, and Wang J: 1992, *Solar Phys.* **142,** 11
Zhang B, Deng Y, and Li Y: 1993, in T Sakurai (ed), "Proc. Japan-China Workshop on Solar Phys."

2. ADVANCES IN SOLAR MICROWAVE INSTRUMENTATION
(Shinzo Enome)

In preparation for the maximum of solar activity in 1991, one major new instrument was constructed and existing observing facilities were upgraded extensively. These new designs enhance the multi-dimensional character of microwave observations in the spatial, temporal, and frequency domains. A new 17-GHz T-shaped solar-dedicated array, consisting of 84 parabaloidal elements of 80-cm diameter each, was constructed at Nobeyama Radio Observatory. It began regular observations in late June, 1992. The major specifications of this unique instrument are: spatial resolution, 10"; temporal resolution 1 s with 50 ms option; 40' field of view; daily observation time of eight hours, from 2300UT-0700UT; and dynamic range of image, or quality of image, 25 dB or 300:1 for 1 s or snapshot mode (Enome 1992, Enome *et al.* 1993, Nakajima *et al.* 1993). Initial observational results have been presented for the the quiet Sun component, for active regions, and for flare sources in their impulsive and post-flare phases (Enome *et al.* 1993).

Ultra wide-band imaging systems were installed on the RATAN-600 antenna (Bogod *et al.* 1993) and on the Owens Valley Solar Array (Hurford & Gary, 1989). They take different approaches - multi-channel vs. frequency-agile systems - to cover the range 1-18 GHz in order to provide us with very rich spectral information for constraining the choice of emission mechanisms. Diagnostics and examples have been published for emission from active regions and flares (Gelfreikh *et al.* 1992, Alissandrakis *et al.* 1993, Stähli *et al.* 1990).

A five-element multi-beam system feeding a 48 GHz receiver was installed on the 13.7-m Itapetinga antenna. It produces five beams overlapping by one HPBW or 2' and covers a field of view ~2' x~4'. It is claimed to be able to track the centroid of flare sources with an angular accuracy of 5" to 19" with 1 ms time resolution (Costa *et al.* 1992; Herrmann *et al.* 1992)

Synthesis of solar images using the E-W fan-beam output of the Siberian Solar Radio Telescope has been performed for active regions (Alissandrakis *et al.* 1992). Two-dimensional maps computed from these one-dimensional observations clearly show the quiet-Sun background, sunspot- and plage-associated emission, as well as compact sources above the neutral line in some active regions.

REFERENCES

I: "The Magnetic and Velocity Fields of Solar Active Regions", IAU Colloq. 141 (Ref. I), Zirin H, Ai G, and Wang H (eds.), APS Conf. Series (San Francisco), **46**, 1993

Alissandrakis CE, Lubyshev BI, Smolkov GYa, Krissinel BB, Treshkov TA, Miller VG, and Kardapolova NN: 1992, *Solar Phys.* **142**, 341

Alissandrakis CE, Gelfreikh GB, Borovik VN, Korzhavin AN, Bogod VM, Nindos A, and Kundu MR: 1993, *Astron. Astrophys.* **270**, 509

Bogod VM, Vatrushin SM, Abramov-Maximov VE, Tsvetkov SV, and Dikij VN: 1993, Ref. I, p. 306

Bogod VM: 1992, *Astron. Nachr.* **313**, 97

Costa JER, Correia E, Kaufmann P, Herrmann R, and Magun A: 1992, Ref. I, p. 177

Enome S: 1992, in "Eruptive Solar Flares", Svestka Z, Jackson BV, and Machado ME (eds.), IAU Colloq. 133, Lecture Notes in Physics **399**, Springer-Verlag (Berlin), p. 314

Enome S, Nakajima, H, Shibasaki K, Nishio M, Takano T, Hanaoka Y, Torii C, Sekiguchi H, Kawashima S, Bushimata T, Shinohara N, Irimajiri Y, Koshiishi H, Kosugi T, Shiomi Y, Sawa M, and Kai K: 1993, Ref. I, p 310

Gelfreikh GB: 1991, *Adv. Space Res.* **11**, No. 1, 89

Gelfreikh GB. Bogod VM, Abramov-Maximov VE, and Tsvetkov SV: 1993, Ref. I, p. 271

Herrmann R, Magun A, Costa JER, Correia E, and Kaufmann P: 1992, *Solar Phys.* **142**, 157

Hurford GJ and Gary DE: 1989, Proc. Max '91 Workshop #1, p. 328

Nakajima H, Nishio M, Enome S, Shibasaki K, Takano T, Hanaoka Y, Torii C, Sekiguchi H, Bushimata T, Kawashima S, Shinohara N, Irimajiri Y, Koshiishi H, Kosugi T, Sawa M, and Kai K: 1993, Special Issue, *Proc. IEEE*: "Design and Instrumentation of Antennas for Deep Space Telecommunications and Radioastronomy"

Stähli M, Gary DE, and Hurford GJ: 1990, *Solar Phys.* **125**, 343

3. PERIODICITIES IN SOLAR ACTIVITY

(Jose Luis Ballester)

Throughout this century the existence has been claimed of periodicities in different manifestations of solar activity. Some of the reported periodicities are close to the rotation period, others lie close to that of the 11- or 22-yr cycles, while still others have been proposed with entirely different values. The interest in this topic was renewed by the discovery, in data obtained by the Solar Maximum Mission, of a periodicity between 152-158 days in the rate of flare occurrence during solar cycle 21. Since then most of the research in this field has assessed the existence of this periodicity in various forms of solar activity (flares, flare-related data, sunspots, etc). A variety of techniques have been applied to search for and characterize periodicities in time series: Rayleigh power spectrum, maximum likelihood method, Scargle periodogram, complex demodulation, etc. Also, some tentative models have been set up in order to explain their existence.

3.1 THE NEAR 155-DAY PERIODICITY

Studies by several authors using flare and flare-related data have shown that the above periodicity has been present for some time. Dröge *et al.* (1990) examined the occurrence of energetic (>10 MeV) solar flare electron events from 1978 to 1982, finding strong evidence for a periodicity of 153 ± 2 days. Bai & Cliver (1990) studied the occurrence rate of proton flares during solar cycles 19 to 21 and identified two epochs exhibiting a 154-day periodicity. Those epochs are: a 14-year interval from January 1958 till December 1971, and a 5.5-year interval from February 1978 till August 1983. The best determined period is 154.6 (± 0.6) days. They found evidence that, between these epochs, the periodicity suffered a phase shift of 180°. Gabriel *et al.* (1990) searched for the periodicity in time-series of solar proton events during solar cycles 19 to 21. They found a prominent periodicity around 152-156 days in the occurrence of solar proton events during cycles 19 and 21, but only inconclusive evidence for it in solar cycle 20. Kile & Cliver (1991), having analysed Ottawa 2.8 GHz burst data from cycles 19, 20 and 21, concluded that the periodicity is only statistically significant during the maximum of solar cycle 21; for the other periods the evidence is contradictory. A study of flare rates during cycle 22 made by Bai (1992) shows that a periodicity around 77 days, third subharmonic of the fundamental period of ~25.5 days (the 154-day period is the sixth subharmonic), was in operation during a 15-month interval (November 1988 - February 1990) and that it reappeared in 1991. He too found a 180° phase shift of the periodicity between those epochs. The presence of a 155-day periodicity in sunspot areas during cycles 12 to 21 has been examined by several authors. Lean (1990) found that this periodicity is only present during epochs of maximum activity and that it occurs in episodes of 1 to 3 years. Carbonell & Ballester (1990, 1992) showed that a periodicity between 150-160 days seems to be significant during all solar cycles from 16 to 21; during previous cycles it is only found in some time intervals, but with high significance. They also found evidence suggesting that, in cycles characterized by a strong north-south asymmetry of solar activity, the periodicity is only present in the hemisphere most favoured by solar activity. There is also slight evidence for the intermittent presence of a 155-day periodicity in auroral data going back till the sixteenth century (Silverman 1990), which could indicate that it is a persistent feature of solar activity.

3.2 MODELS

Some explanations for the existence of a periodicity of 155 days, most of them qualitative, have been advanced. Wolff (1992) argues that some of the reported periodicities can be explained as aliases of real periodicities sampled in the 11-year window of the solar cycle. He proposes that the 155-day periodicity can be understood in terms of the normal modes of oscillation of a nearly spherical, slowly rotating star, when two r-modes (inertial modes) couple with an interior g-mode. This suggestion agrees qualitatively with the fact that the periodicity is intermittent and stronger around the maximum of solar activity; detailed calculations are needed to confirm this hypothesis. From spectrum analysis of long records of sunspot numbers, areas, and flares, Bai & Sturrock (1991) and Sturrock & Bai (1992) propose that the Sun contains a "clock", which they model by means of an oblique rotator or oscillator, with a period of 25.8 days. They suggest that the periodicity of 154 days is just a subharmonic of that fundamental period. Later, Bai & Sturrock (1993) analyzed the longitudinal distribution of major flares and found that it exhibits the largest modulation for a period 25.50 days for an oblique rotator tilted at 40°. From helioseismic evidence, however, Goode & Thompson (1992) conclude that the Sun's convective zone and at least the outer part of its radiative interior rotate on the same axis. Another explanation links the increase in flare-rate occurrence to a periodic emergence of magnetic flux through the photosphere (Lean, 1990; Brueckner & Cook, 1990; Carbonell & Ballester, 1990, 1992). In order to develop theoretical models, further studies of the 155-day periodicity are needed to focus on key points: intermittency, phase shifts, simultaneity of its presence in different primary indicators of solar activity, and the relationship to the N-S asymmetry of solar activity.

3.3 OTHER PERIODICITIES

The existence of longer periodicities has also been studied. Oliver *et al.* (1992) analyzed sunspot numbers (cycles 6-21) and sunspot areas (cycles 12-21); they found no evidence, except in solar cycle 21, for a proposed periodicity of 323 days. Another one of 540 days, first detected in flare occurrence in the northern hemisphere during cycle 19, appears clearly. It shows up in northern hemisphere data during cycles 18-19 but less prominently in cycles 12, 14 and 17. This last detection needs to be confirmed from other indicators of solar activity. A 600-day periodicity in the areas of coronal holes during 1977-1989 has been reported by McIntosh *et al.* 1992; this periodicity is prominent for southern coronal holes but does not appear in the data for northern coronal holes. This again suggests a relationship between periodicity and asymmetry of solar activity. However, we should remember that flares and sunspots are, in general, associated with closed magnetic fields, while coronal holes are related to open magnetic fields, which might have some bearing on the origin of both periodicities.

REFERENCES

Bai T: 1992, *Astrophys. J*, **388**, L69
Bai T and Cliver EW: 1990, *Astrophys. J.* **363**, 299
Bai T and Sturrock P: 1991, *Nature* **350**, 141
Bai T and Sturrock P: 1993, *Astrophys. J.* **409**, 486
Brueckner GE and Cook JW: 1990, *Astrophys. J.* **350**, L21
Carbonell M and Ballester JL: 1990, *Astron. Astrophys.* **238**,377
Carbonell M and Ballester JL: 1992, *Astron. Astrophys.* **255**, 350
Dröge W, Gibbs K, Grunsfeld JM, Meyer P, Newport BJ, Evenson P, and Moses D: 1990, *Astrophys. J., Suppl. Ser.* **73**, 279
Gabriel S, Evans R, and Feynman J: 1990, *Solar Phys.* **128**, 415
Goode PR and Thompson MJ: 1992, *Astrophys. J.* **395**, 307
Kile JN and Cliver EW 1991, *Astrophys. J.* **370**, 442
Lean J.1990, *Astrophys. J.*, **363** 718
McIntosh PS, Thompson RJ, and Willock EC: 1992, *Nature* **360**, 322
Oliver R, Carbonell M, and Ballester JL: 1992, *Solar Phys.* **137**, 141
Silverman SM: 1990, *Nature* **347**, 365
Sturrock P and Bai T: 1992, *Astrophys. J.* **397**, 337
Wolff CL: 1992, *Solar Phys.* **142**, 187

4. ACTIVE REGION EVOLUTION

(Jingxiu Wang)

Studies of active-region evolution have proceeded vigorously as a result of intense interest in a wide range of topics during the last three years. This is evident in the special colloquium on active regions (Zirin *et al.* 1993, I), and in three workshops which cover most of the topics in this area (November 1991, II; Harvey, 1992, III; Thomas & Weiss, 1992, IV). Some advances in flare-associated changes in magnetic fields are found in flare meetings (Schmieder & Priest 1991, V; Svestka *et al.* 1992, VI). A colloquium on the Sun and cool stars reflects the mutual interest in solar active regions of solar and stellar astronomers (Tuominen *et al.* 1991, VII). A major impact on active-region studies from Yohkoh is evident in the initial results published in an issue of *PASJ* (1992, VIII). Relevant topics are also contained in recent monographs (Priest & Hood, 1991, IX; Cox *et al.*, 1991, X; Schmelz & Brown, 1991, XI).

4.1 FLUX EMERGENCE

The process of flux emergence has been mapped in detail by vector magnetographs, e.g. at Huairou Observatory (Zhang *et al.* VII, p. 271; Zhang & Song, 1992; Wang & Shi, 1993). Wang and Shi conclude that the emergence of new flux drives flux cancellation with pre-existing flux and is a wholly inseparable, elementary process in an active region; the observed flux cancellation is most likely a slow reconnection in the lower atmosphere. In one example this reconnection is inferred to take place below the photosphere, making an emerging flux region (EFR) show up with only one pole, but with a remarkable bundle of enhanced transverse field. High resolution Hα filtergrams have revealed EFR-surges (Kurokawa, I, p. 507). Newly discovered X-ray jets from Yohkoh's Soft X-ray Telescope (SXT) are partly associated with EFRs (Shibata *et al.*, VIII, p. L173). The first comparison between Hα and SXT images of EFRs (Kawai *et al.*, VIII, p. L193) shows bright SXT features spatially coincident with Hα arch filament systems (AFSs), indicative of fast reconnection between EFRs and overlying magnetic fields. Numerical simulations by Shibata *et al.* (1992) have reproduced the above observational features.

The unusual emergence of new flux in the sheared corridors of large, compact, δ sunspots (Wang *et al.* 1991; Tang & Wang, 1993; Wang & Tang, 1993) challenges the traditional picture of EFR

(Chou, I, p. 471). This kind of unusual emergence was first discovered in two superactive regions, March 1989 and June 1991, during this solar cycle. It is characterized by (1) penumbral motions in the direction of sheared penumbral fibrils, (2) newly emerged sunspots often with only one polarity in the midst of penumbra, (3) disordered or irregular AFSs (Kurokawa, 1991), and (4) major flares mostly with big surges. Wang *et al.* (1991) report that 80% of the flux in the March 1989 region is of following polarity. Shi & Wang (I, p. 71) find that for all δ sunspots associated with X class flares in this cycle, the average flux ratio between the two polarities is larger than 5. Kurokawa (1991) and Tanaka (1991) suggest a topology of tightly twisted or kotted flux ropes for compact δ sunspots. Zirin & Wang (1993) identify a transverse field as high as 0.4 T parallel to the magnetic neutral line within δ sunspots.

In a global sense, the injection of intense magnetic fields in bursts at a few tightly defined locations over a long period is shown to be a basic property of solar activity (Gaizauskas, I, p. 479; Zwaan, IV. p. 75). van Driel-Gesztelyi *et al.* (III, p. 89) show that more than one third of all sunspot groups between 1940-1976 appear in compact active nests. For a large sample of dipole active regions throughout cycle 21, Harvey (1993) finds that 40 - 55% of new dipole regions occur in nests, and new dipoles emerge within existing sunspot regions at a rate at least 22 times greater than they do outside. Wentzel & Seiden (1992) have reproduced the quasi-regular distributions of active regions in clusters by using percolation theory.

4.2 MAGNETIC TOPOLOGY

Solar flare research has greatly boosted the study of active regions by promoting an awareness of the coupling between flare sites and the evolving topology of global magnetic field (Gaizauskas, 1993). Basic topological structures are seen as cells of magnetic connectivity defined by separatrices and separators (Priest 1991; Hénoux *et al.*, I, p. 333). Practical 3-D representation of these structures by potential or force-free fields using distributions of charges or dipoles based on observations have been developed by Démoulin *et al.* (1992). Mandrini *et al.* (1991, 1993) find that the topological evolution corresponding to flux emergence and disappearance in NOAA 2372 is favourable to reconnection at the separator to which the Hα flare kernels are connected. In an activity complex, Schmieder *et al.* (1991), van Driel-Gesztelyi *et al.* (1993), and Démoulin *et al.* (1993) find that an old region with less magnetic shear is highly active, while a young region with strong shear is inactive. They establish that a separator is present in the old region, but not in the young one. Topologically, a complex and an active region may be equivalent. If the interaction between two regions in a complex does not create a separator, they show little similarity in activity. But if a separator is present between two regions, sympathetic activity is triggered between the two regions and is interpreted in terms of interaction between different flux loop systems. This is analogous to the interacting dipole representation for individual flares in a single region (Polletto *et al.* 1993). Loop interactions take place at various energy scales when a topological singularity appears in the evironment. They are shown to be the cause of 'active region transient brightenings' (Shimizu *et al.*, VIII, p. L147) and are, perhaps, the ultimate source of the continual expansion of the active region corona (Uchida *et al.*, VIII, p. L155).

The topology deduced by Démoulin *et al.* (1992) applies only in a global sense. Current-aligned magnetic loops may well be interwoven or twisted. Another topological concept, helicity, becomes important in the study of free energy build-up (Berger, IX, p. 241; Seehafer, I, p. 435).

4.3 MAGNETIC NON-POTENTIALITY

Greatly improved observations of vector magnetic fields during flare campaigns have given a fresh impetus to the study of magnetic non-potentiality, defined as the departure of the observed vector magnetic field from the corresponding potential configuration (Hofmann & Kalman, 1991; Gary *et al.*, II, p. 65; Canfield *et al.*, II, p. 296; Abramenko *et al.*, 1992; Wang, I, p. 323; Wang, I, p. 425). Intriguing increases of shear after major flares are reported by Wang (1992) and in other recently submitted studies. Shear increase preceding flares and relaxation following flares are shown by Hagyard *et al.* (1993) and by Wang (I, p. 425). These have revived Tanaka's old classification of shear evolutionary modes which are associated with flare activity (Tanaka, 1986). Tang & Wang (I, p. 323) find that the neutral line may not have strong magnetic shear when flare ribbons start quite far away from the neutral line. On the other hand, strong shear is not a sufficient condition for flares to occur

(Hagyard & Heyvaerts, V, p. 1). Possible non-potential features in the chromosphere are suggested by Wang and Shi (1992), Zhang (I, p. 263). These improved observations lead to a recognition of the shear evolution in both magnitude and height, in accordance with general field evolution and energy build-up.

With regard to how magnetic non-potentiality is created, three categories of evidence are reported. First, motions at the surface: shear motion of magnetic footpoints (Zhang, III, p. 124; Wang, 1992), collision of opposite polarity sunspots (Ai *et al.*, 1991; Gaizauskas and Harvey, V, p. 25), and sunspot rotation (Hofmann & Kalman, 1991; Abramenko *et al.* 1991). Second, highly non-potential features may break out from below during flux emergence (Tanaka, 1991; Kurokawa, 1991; Tang & Wang 1993; Zirin & Wang; 1993). And third, flux cancellation converts the line-of-sight magnetic field into transverse field parallel to the neutral line, favourable to filament formation (Martin & Livi, VI, p. 33). To understand these, Wang (1992) has demonstrated that the changes of the non-potential degree are caused by either the generation of local shear due to the interaction between magnetic fields and plasma motion, or the transport of shear.

4.4 SUBSURFACE DYNAMICS

Recent statistical work on the global behaviour of active regions represents an important trend in studying the subsurface dynamics of flux loops which rise to the surface to form active regions. Mount Wilson daily magnetogram data from 1967 to 1988 (in its coarse form) and digitized sunspot data from 1917 to 1985 have been used by Howard in extensive studies of global properties. Among many interesting results, the polarity separation is shown to be a property of subsurface flux loops (Howard 1992a). The relation between magnetic axial tilt and polarity separation signals the surface effect of Coriolis forces that act to twist rising flux loops (Howard 1993a). The difference between the magnetic fields of average sunspots and plages in many aspects are explored for the first time (Howard, 1991a-d; 1992b,c; 1993b). They are indicative of the different depth and nature of the connections to the subsurface flux system. Harvey selects a sample of 978 emerging bipole regions throughout cycle 21, from Kitt Peak full disc magnetograms, for her systematic analysis (I, p. 488). The shape of the characteristic distribution of active-region sizes is shown to be a fundamental invariant property of solar magnetism, in revealing the process that releases or triggers flux loops to emerge from the toroidal flux system in the solar interior (Harvey 1993). The patterns in both the bipole orientation and latitude distribution point to magnetic fields in the subsurface flux system that are strong enough to resist disordered convection and Coriolis forces.

Numerical simulation of a toroidal flux ring in a rotating spherical geometry with the inclusion of the Coriolis force has given interesting new results. D'Silva & Choudhuri (1993) are able to predict the tilt-separation relations of active regions (Howard 1993a), and to get an estimation of 6-15 kG for the initial field strength of the toroidal loops. Fan *et al.* (1993) have carried out a series of numerical simulations to study the dynamical evolution of emerging flux loops in the solar convection envelope with their innermost portions anchored beneath the base of the convection zone. They have shown nicely how the combined action of Coriolis force and the anchoring of the innermost portions to a rising loop creates the well-observed morphological asymmetry in bipolar active regions.

REFERENCES

I: "The Magnetic and Velocity Fields of Active Regions", IAU Colloq. 141, Zirin H, Ai G, and Wang H (eds.), ASP Conf. Series (San Francisco), **46**, 1993.
II: "Solar Polarimetry", Proc. NSO/SP Summer Workshop, November L (ed.) New Mexico, 1991.
III: "The Solar Cycle", Proc. NSO/SP Summer Workshop, Harvey KL (ed.), ASP Conf. Series (San Francisco), **27**, 1992.
IV: "Sunspots: Theory and Observations", NATO ASI Series C: **375**, Thomas JH and Weiss NO (eds.), Kluwer (Dordrecht), 1992.
V: "Dynamics of Solar Flares", Proc. Flares22 Chantilly Meeting , Schmieder B, and Priest ER (eds.), Meudon, 1991.
VI: "Eruptive Solar Flares", IAU Colloq. 133, Svestka Z, Jackson BV, and Machado ME (eds.), Lect. Notes Phys. **399**, Springer (Berlin), 1992.
VII: "The Sun and Cool Stars: Activity, Magnetism, Dynamos", IAU Colloq. 130, Tuominen I, Moss D, and Rüdiger G (eds.), Lect. Notes Phys.**380**, Springer (Berlin), 1991.

VIII: "Initial Results from Yohkoh", Special Feature, *Publ. Astron. Soc. Japan* **44**, 1992.
IX: "Advances in Solar System Magnetohydrodynamics", Priest ER and Hood AW (eds.), Cambridge U. Press, 1991.
X: "Solar Interior and Atmosphere", Cox AN, Livingston WC, and Matthews MS (eds.), The Univ. of Arizona Press (Tucson), 1991.
XI: "The Sun: A Laboratory for Astrophysics", Schmelz JT and Brown JT (eds.), Kluwer (Dordrecht), 1992.
Abramenko VI, Gopasyuk SI, Ogir MB: 1992, *Bull. Crimean Astrophys. Obs.* **81**, 1
Ai G, Zhang H, Li W, Li J , and Chen J: 1991, *Chinese Scien. Bull.* **36**, 1275
Chargeishvili B, Zhao J, and Sakai J -I: 1993, *Solar Phys.* **145**, 297
Démoulin P, Hénoux J-C, and Mandrini CH: 1992, *Solar Phys.* **139**, 105
Démoulin P, van Driel-Gesztelyi L, Schmieder B, Hénoux J-C, Csepura G, and Hagyard MJ: 1993, *Astron. Astrophys.* **271**, 292
D'Silva S and Choudhuri AR: 1993, *Astron. Astrophys.* **272**, 621
Fan Y, Fisher GH, and Deluca EE: 1993, *Astrophys. J.* **405**, 390
Gaizauskas V: 1993, *Adv. Space Res.* **13**, No. 9, 5
Hagyard MJ, West EA, and Smith JE: 1993, *Solar Phys.* **144**, 141
Harvey KL: 1993, "Magnetic Bipoles on the Sun", Thesis, Univ. Utrecht
Hofmann A and Kalman B: 1991, *Astron. Astrophys.* **241**, 203
Howard RF: 1991a-d, *Solar Phys.* **134**, 233; **135**, 43; **135**, 327; **136**, 251
Howard RF: 1992a-c, *Solar Phys.* **142**, 233; **137**, 205; **142**, 47
Howard RF: 1993a,b, *Solar Phys.* **145**, 105; **145**, 95
Kurokawa H: 1991, *Adv. Space Res.* **11**, No. 5, 233
Mandrini CH, Démoulin P, Hénoux J-C, and Machado ME: 1991, *Astron. Astrophys.* **250**, 541
Mandrini CH, Rovira MG, Démoulin P, Hénoux J-C, Machado ME, and Wilkinson LK: 1993, *Astron. Astrophys.* **272**, 609
Priest ER: 1991, *Phil. Trans. R. Soc. Lond.* **A366**, 363
Polletto G, Gary GA, and Machado ME: 1993, *Solar Phys.* **144**, 133
Sakai J-I and Koide S: 1992, *Solar Phys.* **134**, 329
Schmieder B, van Driel-Gesztelyi L, Hénoux J-C and Simnett GM : 1991, *Astron. Astrophys.* **244**, 533
Shibata K, Nozawa S, and Matsumoto R: 1992, *Publ. Astron. Soc. Japan.* **44**, 265
Tanaka K: 1986 in "Hydrodynamic and Magnetohydrodynamic Problems in the Sun and Stars", Osaki,Y (ed.) Tokyo, p. 1
Tanaka K: 1991, *Solar Phys.* **136**, 133
Tang F and Wang H: 1993, *Solar Phys.* **143**, 107
van Driel-Gesztelyi L, Rayrole J, and Démoulin P: 1993, *Solar Phys.* **145**, 77
Wang H, Tang F, Zirin H, and Ai G: 1991, Astrophys. J. **380**, 282
Wang H: 1992, *Solar Phys. Solar Phys.* **140**, 85
Wang H and Tang F: 1993, *Astrophys. J.* **407**, L89
Wang J: 1992, *Acta Astrophysica Sinica,* **12**, 75
Wang J and Shi Z: 1992, *Solar Phys.* **140**, 67
Wang J and Shi Z: 1993, *Solar Phys.* **143**, 119
Wentzel DG and Seiden PE: 1992, *Astrophys. J.* **390**, 280
Zhang H and Song M: 1992, *Solar Phys.* **138**, 69
Zhang Z and Smartt R: 1992, *Acta Astronomica Sinica,* **32**, 233
Zirin H and Wang H: 1993, *Solar Phys.* **144**, 37

5. FILAMENTS AND PROMINENCES

(Oddbjørn Engvold and Jack B. Zirker)

The past 3 years have brought a significant shift in interest, away from traditional subjects of plasma properties and magnetic fields of prominences and filaments and towards studies of their fine-structure, motions, and formation. A major IAU colloquium, "Dynamics of Quiescent Prominences" held in Hvar, Yugoslavia, Sept 25-29, 1989, attracted a large group of investigators and sparked a renewed interest in the subject.

5.1 DYNAMICS

Schmieder (1990) and Rompolt (1990) have reviewed current understanding of the motions in and around prominences. While much information on the details of these motions has been obtained, we still do not have a convincing model of the circulation of mass to and from the corona and the photosphere.

The search for wave heating in filaments has motivated a number of studies of oscillations. Oscillations in Hα filaments (periods of 250, 200 and 150 s) were reported by Thompson & Schmieder (1991). Yi *et al.* (1991), in contrast, saw periods of 5 to 15 min in He I 10830. These oscillations occurred in fibril-like, horizontal structures that appear to be aligned with the horizontal magnetic field. Suematsu *et al.* (1990) found 240 and 830 s oscillations in prominences. Wiehr *et al.* (1990) recorded periods spread between 3 and 60 min. Mashnich & Bashkirtev (1990) also saw periods as long as an hour.

Joardar & Roberts (1992a and b) and Oliver *et al.* (1993) have studied possible MHD oscillation modes in prominences based on static, uniform (slab) models. The authors show that both slow-, fast- and Alfvén-modes can give rise to the observed short-period oscillations. Longer period oscillations may be produced by hybrid slow modes. Bakhareva *et al.* (1992) have investigated how the build up time of oscillations and their period depend on factors such as the transverse magnetic component, again in the framework of magnetostatic 'slab' models. The spatial characteristics of the short period oscillations suggest that they are due to Alfvén waves (Jensen *et al.* 1993).

Persistent flows, particularly in the vertical direction, have been studied in an attempt to clarify the geometry of prominence structures. Zirker *et al.* (1993) analyzed Dunn's (1960) Hα prominence movies using objective local correlation techniques and found consistency between the vertical velocities in prominence threads and in filament downflows. Kim (1990) found two types of downward flows in filaments: those with constant acceleration (less than gravitational) and those with uniform velocity.

Filaments undergo large-scale shearing motions. Athay (1990) showed that active-region filaments lie on long-lived shear lines. Schmieder *et al.* (1990) compared Hα and C IV spectra and found that significant shear appears in a filament before it erupts.

5.2 PROMINENCE FINE-STRUCTURE

Very little new data has been presented recently on the thermodynamic properties of the fine-structure, but Hirayama (1990) has reviewed recent determinations of mean density and temperature. Wiehr & Stellmacher (1991) have obtained magnetic measurements in structures 1" in diameter. They found field strengths between 7.5 and 18 mT in an active-region prominence. Kim *et al.* (1990) found values between 0.3 and 2.4 mT in quiet prominences and discuss statistics of the field direction. Zirker & Koutchmy (1990, 1991) obtained Hα spectra of 1"-resolution and showed that they could be modelled by random clusters of sub-arcsecond threads. Seven to twenty threads in the line of sight are typically inferred. Mein *et al.* (1990) and Mein &Mein (1991) used similar modelling to account for the observed distributions of doppler velocity and line width in well-resolved quiescent prominences.

The assumed significance of the fine-structure in the physics of prominences has led to a series of theoretical studies of its formation. Analytical solutions have been attempted by Hood & Anzer (1990), Priest *et al.* (1991), and Hood *et al.* (1992). These authors state that the conditions for Rayleigh-Taylor instabilities are not fulfilled in prominence regions. Thread structure is likely to occur as a result of radiative instability during prominence formation. Steele & Priest (1992) have modelled the prominence fine-structure as many cool slabs aligned parallel to the prominence axis in the frame of a magneto-static model. van der Linden & Goossens (1991) suggested that non-adiabatic terms in the energy equation may be applied to model the growth of prominence fine-structure. The non-static, strongly dynamic behaviour of prominence threads are not yet understood theoretically.

5.3 FORMATION AND SUPPORT

A number of authors have investigated idealized, sheared magnetic arcades in search of magnetic topologies that will provide support of prominence plasma (Fiedler & Wood 1993). Amari *et al.* (1991) found that sheared motion at the footpoints of magnetic arcades cannot produce magnetic field lines concave upwards. Ridgway & Priest (1993), on the other hand, demonstrated that converging photospheric motions acting on an already sheared force-free magnetic arcades will lead to helical fields which possibly provide the the needed support. Choe & Lee (1992) showed that quiescent prominences can be formed as the result of shearing motion at the photospheric level. Cool prominence matter collects from a combination of radiative cooling, provoking a thermal instability, and siphon upflow from photosphere and chromosphere. van Hoven *et al.* (1992) showed that rapid radiative cooling at the apex of coronal loops could give rise to siphon inflow from the chromosphere and subsequently to prominence formation. Similar results were reached by Wu *et al.* (1992).

Martin (1990) summarized several conditions for formation of prominences, as inferred from optical observations. Important new criteria include long term (hours to days) convergence of small field patches toward the neutral line and the cancellation of opposite polarity patches. Martin has proposed a conceptual model of filament formation based on her results (1992).

5.4 FILAMENT FIELDS AND THE SOLAR CYCLE

Martin *et al.* (1992) have reported a number of extremely interesting regularities of the direction of the magnetic field in quiescent filaments. Filaments are either "dextral" or "sinistral". Dextral (sinistral) filaments have their horizontal magnetic fields pointing to the right (left) as seen from the adjacent region of positive photospheric field. All northern (southern) hemisphere filaments are dextral (sinistral). The absolute direction of filament fields (ie East-West) reverses in successive cycles. The physical bases of these rules remain unknown.

5.5 PROMINENCE-CORONA TRANSITION REGION (P-CTR)

Chiuderi & Chiuderi-Drago (1991) were able to reconcile observations in EUV and at radio wavelengths. Chiuderi-Drago *et al.* (1992) modelled the P-CTR associated with individual magnetic threads. They concluded that dissipation of Alfvén-wave energy would seem able to explain the observed EUV flux radiated from the temperature layers below 100 000 K. Wiik *et al.* (1992) have performed a detailed analysis of a quiescent polar crown prominence from observations in UV and visible wavelengths.

5.6 RADIATIVE TRANSFER

Thermodynamic modelling of the prominence plasma is critically dependent on non-LTE effects in radiative transfer, where also the small-scale nature of the prominence plasma becomes important. Paletou *et al.* (1993) have studied the formation of resonance lines of H I, Mg II and Ca II in a 2-D atmospheric model. Improved codes for treatment of PRD scattering give good agreement with observed emission line profiles.

REFERENCES

IAUC117 ="Dynamics of Quiescent Prominences", IAU Colloq. 117, Ruzdjak V, Tandberg-Hanssen E (eds.), Lect. Notes Phys. **363**, Springer-Verlag (Berlin), 1990.
Amari T, Démoulin P, Browning P, Hood A, and Priest ER:1991, *Astron. Astrophys.* **241**, 604
Athay, RG: 1990, *Solar Phys.* **126**, 135.
Bakhareva NM, Zaitsev VV, and Khodachenko ML: 1992, *Solar Phys.* **139**, 299
Chiuderi C and Chiuderi-Drago F: 1991, *Solar Phys.* **132**, 81
Chiuderi-Drago F, Engvold O, and Jensen E: 1992, *Solar Phys.* **139**, 47
Choe GS and Lee LC: 1992, *Solar Phys.* **138**, 291
Dunn RB: 1960, Doctoral Dissertation, Harvard University
Fiedler RAS and Hood AW: 1993, *Solar Phys.* **146**, 297
Hirayama T: 1990, IAUC117, p. 106
Hood AW and Anzer U: 1990 *Solar Phys.* **126**, 117

Hood AW, Priest ER, and Anzer U: 1992, *Solar Phys.* **138**, 331
Jensen E, Yi Z, and Engvold O: 1993 *Solar Phys.* (in press)
Joardar PS and Roberts B: 1992a, *Astron. Astrophys.* **256**, 264
Joardar PS and Roberts B: 1992b, *Astron. Astrophys.* **261**, 625
Kim IS: 1990, *J.Korean Ast. Soc.* **23**, 31
Kim IS, Yu V, Klepikov M, Koutchmy S, Stepanov AI, and Stellmacher G: 1990, *Sov. Astron. Lett.* **16**, 3.
Martin SF:1990, IAUC117, p.1
Martin SF: 1992, in "Eruptive Solar Flares", IAU Colloq. 133, Svestka Z, Jackson BV, and Machado ME, (eds.), Lect. Notes Phys., **399**, Springer (Berlin), p. 331
Martin SF, Marquette WH, and Billimoria, R: 1992 in "The Solar Cycle", Harvey KL (ed.), ASP Conf. Series (San Francisco), **27**, p. 53
Mashnich GP and Bashkirtev VS: 1990, *Astron. Astrophys.* **235**, 428
Mein P, Mein N, Schmieder B, and Noens J-C: 1990, IAUC117, 113
Mein, P and Mein N: 1991, *Solar Phys.* **136**, 317
Oliver R, Ballester JL, Hood AW, and Priest ER: 1993, *Astrophys. J.* **409**, 809
Paletou F, Vial, J-C, and Auer LH: 1993, *Astron. Astrophys.* **274**, 571
Priest ER, Hood AW, and Anzer U: 1991, *Solar Phys.* **132**, L199
Ridgway C and Priest ER: 1993, *Solar Phys.* **146**, 277
Rompolt B: 1990, *Hvar Obs. Bull.* **14**, 37
Schmieder, B: 1990, IAU117, p.85
Schmieder B, and Mein P: 1990 IAUC117, p.235
Schmieder B, Dere K, Raadu M, Démoulin P, Alissandrakis C: 1990, *Adv.Space Res.* **10**, No. 9, 195
Steele CDC and Priest ER: 1992, *Solar Phys.* **140**, 289
Suematsu Y, Yoshinaga R, Terao N, and Tsubaki T: 1990, *Publ. Astron. Soc. Japan* **42**, 187
Thompson WT and Schmieder B: 1991, *Astron. Astrophys.* **243**, 501
van Hoven G, Mok Y, and Drake JF: 1992, *Solar Phys.* **140**, 269
van der Linden RAM and Goossens M: 1991 Solar Phys. 134, 247
Wiehr E., Balthasar H., and Stellmacher G: 1990, *Hvar Obs. Bull.* **13**, 131
Wiehr E and Stellmacher G: 1991, *Astron. Astrophys.* **247**, 91
Wiik JE, Heinzel P, and Schmieder B: 1992, *Astron. Astrophys.* **260**, 419
Wu ST, Bao JJ, An CH, and Tandberg-Hanssen E: 1992, *Solar Phys.* **125**, 277
Yi Z, Engvold O, and Keil SL: 1991, *Sol.Phys.* **132**, 63
Zirker JB and Koutchmy S: 1990, *Solar Phys.* **127**, 109
Zirker JB and Koutchmy S: 1991, *Solar Phys.* **131**, 107
Zirker JB, Engvold O, and Yi Z, 1993, *Solar Phys.* (in press)

6. CORONAL MASS EJECTIONS

(David F. Webb)

6.1 INTRODUCTION

Coronal Mass Ejections (CMEs) are ejected away from the Sun as self-contained structures of plasma and magnetic fields. CMEs apparently involve a significant restructuring of the corona on a large scale. They can significantly perturb the solar wind and disrupt the Earth's environment. The primary observations of CMEs have been obtained from orbiting white-light coronagraphs on the Skylab (1973- 74), P78-1 (1979-85), and SMM (1980 and 1984-89) spacecraft. These data have been complemented by observations of the inner corona from the K-coronameter at Mauna Loa Solar Observatory (MLSO) in Hawaii and, occasionally, in Fe emission lines. From 1975-83 white-light observations of CMEs in the inner heliosphere were obtained by the zodiacal light photometers on the Helios spacecraft.

During this reporting period, a number of Proceedings were published from meetings where important CME work was presented or reviewed. These include the Flares 22 meeting at Chantilly (Schmeider & Priest 1991, I); COSPAR at The Hague (Antonucci & Somov 1991, II); IAU Colloquium 133 in Iguazu, Argentina (Svestka *et al.* 1992, III); COSPAR in Washington, D.C. (Pick & Machado 1993, IV); Solar- Terrestrial Predictions Workshop:IV in Ottawa (Hruska *et al.* 1993, V). Review papers emphasizing CME-related observational results include Harrison (1991a, b), Kahler

(1992), Hildner (1992), Webb (1992; 1993), Dryer (1993) and Hundhausen (1993b), while those on CME modelling include Low (1990; 1993), Steinolfson (1990; 1991) and Somov (1991).

The Japanese Yohkoh satellite was launched in August 1991 and preliminary results were published (Ref. VI) with several papers discussing the coronal signatures of erupting prominences and other evidence of mass ejecta. The scope of this review does not include the considerable work done recently on interplanetary aspects of CMEs, particularly magnetic clouds, driver gas signatures, energetic particles, and shocks; some summaries appear in References III, VII-X.

6.2 OBSERVATIONAL RESULTS

Significant new results on CMEs during this period have come from analyses of the SMM, MLSO and HELIOS data. The most easily measured properties of CMEs are their occurrence rates, central latitudes, angular widths and speeds. A fundamental empirical conclusion that must be addressed by models is the large range in these parameters exhibited by CMEs. Their speeds (and accelerations), masses and energies all range over 2-3 orders of magnitude. Although difficult to determine accurately, the masses and kinetic energies of SOLWIND CMEs extended from 10^{14} - 5×10^{16} g and 10^{29} - 6×10^{31} erg, respectively (Howard *et al.* 1985), and some SMM-observed CMEs exceeded these values (Hundhausen 1993b). For the first time Gopalswamy & Kundu (1992) were able to calculate the mass of a CME from radio observations assuming only thermal emission. Parameters which vary by an order of magnitude are: CME occurrence rates over the solar cycle (Webb & Howard 1993), and CME widths, which exceed by factors of 3-10 the sizes of flares and active regions. The annual occurrence rate of CMEs is well correlated with other solar activity and tracks the sunspot cycle in both amplitude and phase (Webb & Howard 1993).

The basic structure of many CMEs consists of a bright leading arc followed by a dark cavity and a bright core of denser material, suggesting the eruption of a pre-event prominence, its overlying coronal cavity, and the ambient corona. However, there is a disparity in the fraction of loop-like CMEs reported for Skylab, SOLWIND and SMM, ranging from only 1% of the SOLWIND CMEs to nearly half of the SMM-observed CMEs, 2/3 of which contained bright cores. The leading CME structures are likely the skyplane projections of three-dimensional structures such as arcades (Steinolfson 1992a; Gopalswamy & Kundu 1993) or shells (Hundhausen 1993b).

Preliminary studies indicate that CMEs arise from large-scale, closed structures, most (~75%) from pre-existing coronal streamers. This is not consistent with the suggestion of Hewish and coworkers that CMEs arise in open field regions, i.e., coronal holes (Harrison 1990; Hundhausen 1993a). The temporal and latitudinal distributions of CMEs are similar to those of streamers and prominences, being confined to low latitudes about the current sheet near cycle minimum and becoming very broad near maximum (Hundhausen 1993a). This evolution is very different from that of active regions, flares or sunspots. Many of the most energetic CMEs are actually the disruption of a pre-existing streamer, which increases in brightness and size for days before erupting as a CME. Afterwards the streamer and CME are gone, often replaced by a thin ray, probably a current sheet. These events appear on white light synoptic charts as "bugles", portions of the streamer belt which brighten and widen then disappear (Hundhausen 1993a).

The speeds of the leading edges of SMM-observed CMEs ranged from 10 to 2100 km/s. The average speeds of SOLWIND CMEs were much higher near solar maximum than minimum, but Hundhausen emphasizes that SMM CME speeds did not vary much over the cycle. Significant progress has been made in simulating the MHD response to CMEs propagating through the corona (reviewed by Steinolfson (1992b) and by Hundhausen (1993b)). Fast MHD shocks apparently driven by fast CMEs are the most common type in the solar wind. However, in the lower corona the speeds of typical CME outer loops are supersonic but sub-Alfvénic, averaging 445 km/s for SMM. 80% of the trajectories of these CMEs were consistent with constant speed. Numerical simulations of the disruption of model streamers indicates that slow, intermediate and fast shocks should form ahead of CMEs with speeds of 200-300, 300-900, and >900 km/s, respectively. Thus, slow and intermediate shocks might be associated with most CMEs, and some CMEs have been observed that exhibit the predicted flattened fronts. Although the observation of metric type II bursts and their strong association with faster (>400 km/s) CMEs indicates that fast shocks exist in the lower corona (Kahler, 1992), it remains unclear whether they compress sufficient material to be detected optically as bow waves in front of CMEs. Some CMEs do show significant accelerations over large distances, which argues against an impulsive thermal driving force. Often such CMEs are associated with large prominence eruptions.

Statistical studies continue to show a stronger association between CMEs and erupting prominences and X-ray long duration events (LDEs) than with optical flares (reviewed by Webb 1992). But a third to a half of all CMEs cannot be associated with any near-surface activity. Previous studies have shown that CMEs are well associated with soft X-ray flares of longer duration, but Harrison (1991c) disputes this, finding that CMEs can be associated with flares of any duration. St. Cyr & Webb (1991) found that the distribution of the various forms of activity related to CMEs does not change over the solar cycle, although near minimum a lower fraction of SMM CMEs had apparent associations.

Recent results from SMM continue to indicate that, even when associated, flares are likely a by-product of the CME process rather than directly related to its origin. The departure times of flare-associated CMEs typically precede the flare onsets, even for the fastest SMM/MLSO events (Harrison 1991c; Webb 1992; Hundhausen 1993b). Some of these CMEs began their motion within the MLSO field of view before flare onset and some energetic CME/erupting prominences occurred in the absence of any GOES X-raybursts. Recent studies indicate that CME/flares can lie anywhere under the CME span, that associated active regions are offset towards the CME legs, and that even large HeI double-ribbon events have smaller spans and poor spatial overlaps with accompanying CMEs (Harrison 1991b,c; Kahler 1991; Webb 1992).

In previous years much theoretical work on CMEs was focussed on the reconnection of the magnetic fields which close after the CME has erupted. The recent models of the LDE/2-ribbon flare process describe this late phase reasonably well, but results based on CME observations have lagged. Recently, Kahler & Hundhausen (1992) found that the bright structures following many SMM CMEs are newly-formed streamers. A recent Yohkoh/MLSO observation of the reformation of a giant helmet streamer also provides strong evidence of reconnection following CMEs (Hiei *et al.* 1993). McComas *et al.* (1992) reported evidence of moving concave-outward structures which might be the upper U-shaped loops resulting from this process. They interpret these structures as the disconnection and eruption of previously open fields in streamers.

6.3 THEORETICAL MODELLING

Of the many models intended to describe the origin and propulsion of CMEs, most are over-simplified and not sufficiently developed to allow reasonable comparison with observations. The class of models which require a thermal or pressure pulse (i.e., flare) as driver no longer seem viable. Recently there has been intensive work on the origin of CMEs based on the slow evolution of particular coronal structures through metastable states or sequences of stable equilibria until the stability or equilibrium breaks down, resulting in the mass ejection and opening of the field. Low (1990; 1993), Steinolfson (1991) and Dryer (1993) have recently reviewed such analytic models and numerical simulations. Causes of the evolution of these coronal structures, especially streamer configurations, include the motion of loop or arcade footpoints (Low *et al.* 1982), the emergence of magnetic flux (Steinolfson 1992a; Guo *et al.* 1992), the dynamical evolution of arcades (Steinolfson 1991; Inhester *et al.* 1992) or the shear of field lines across neutral lines (Steinolfson 1991; Wolfson & Low 1992).

However, Kilmchuk (1990) pointed out that arcades in general must be sheared and their evolution does not readily lead to instability or loss of equilibrium. In addition, to be tractable, most of these models involve force-free equilibria which, as pointed out by Hundhausen (1993b), cannot realistically describe the complex evolution of the pressure, magnetic and gravitational forces acting on a magnetically closed coronal structure.

Another and very basic problem with both potential and force-free fields was pointed out by Aly (1991) who showed that the energy in an open field configuration always exceeds that of any closed state. Thus, there is not sufficient free magnetic energy available to open up a closed, force-free configuration to produce a mass ejection. Recently, Low & Smith (1993; reviewed by Low, 1993) found that an evolving streamer configuration can contain sufficient free energy to open the field if it contains cross- field currents in a magnetic bubble contained within the streamer core. Such a bubble is detached from the surface and probably resides in the dark cavity surrounding prominences where strong fields may exist. The internal energy of the trapped plasma is also available for the eruption and once the system becomes unstable, pressure gradients may drive the system to eruption (e.g., Wolfson & Low, 1992; Low & Smith, 1993; Low, 1993).

REFERENCES

I: "Dynamics of Solar Flares", Proc. Flares 22 Chantilly Workshop, Schmeider B and Priest ER (eds.), Obs. de Paris, 1991.
II: "Solar Corona and Solar Wind", 28th COSPAR Symp. 9, Antonucci E and Somov BV (eds.), *Adv. Space Res.* **11**, No. 1,1991.
III: "Eruptive Solar Flares", IAU Colloq. 133, Svestka Z, Jackson BV, and Machado ME (eds.), Lect. Notes Phys. **399**, Springer (Berlin), 1992.
IV: "Fundamental Problems of Solar Activity", 29th COSPAR Symp. E3, Pick M and Machado ME (eds.), *Adv. Space Res.* **13**, No. 9, 1993.
V: "Solar-Terrestrial Predictions Workshop:IV", Hruska J, Shea MA, Smart DF, and Heckman G (eds.), NOAA (Boulder), 1993.
VI: "Initial Results from Yohkoh", Special Feature, *Publ. Astron. Soc. Japan* **44**, 1992.
VII: Proc. 22nd Int. Cosmic Ray Conf., Vol. 3, Dublin, 1991.
VIII: "Solar Wind Seven", Marsch E and Schwenn R (eds.), Pergamon (Oxford), 1992.
IX: Proc. SOLTIP No. 1, Fischer S and Vandas M, (eds.),Liblice, 1992
X: Proc. 23rd Int. Cosmic Ray Conf., Vol. 3, Calgary, 1993.
Aly JJ: 1991, *Astrophys. J.* **375**, L61.
Dryer M: 1993, *Space Sci. Rev.*, in press.
Gopalsawmy N and Kundu MR: 1992, *Astrophys. J.* **390**, L37
Gopalsawmy N and Kundu MR: 1993, *Adv. Space Res.* **13**, No. 9, 75
Guo WP, Wang JF, Liang BX, and Wu ST: 1992, in Ref. III, p. 381
Harrison RA: 1990, *Solar Phys.* **126**, 185
Harrison RA: 1991a, in Ref. I, p. 165
Harrison RA: 1991b, *Phil. Trans. R. Soc. Lond. (A)* **336**, 401
Harrison RA: 1991c, *Adv. Space Res.* **11**, No. 1, 25
Hiei E, Hundhausen AJ and Sime DG: 1993, *Geophys. Res. Lett.*, in press.
Hildner E: 1992, in Ref. III, p. 227
Howard RA, Sheeley NR Jr., Koomen MJ, and Michels DJ: 1985, *J. Geophys.Res.* **90**, 8173
Hundhausen AJ: 1993a, *J. Geophys. Res.* **98**, 13177
Hundhausen AJ: 1993b, in "The Many Faces of the Sun", Strong K, Saba J and Haisch B (eds.), in press
Inhester B, Birn J, and Hesse M: 1992, *Solar Phys.* **138**, 257
Kahler S: 1991, *Astrophys. J.* **378**, 398
Kahler SW: 1992, *Ann. Rev. Astron. Astrophys.* **30**, 113
Kahler SW and Hundhausen AJ: *1992, J. Geophys. Res.* **97**, 1619
Klimchuk JA: 1990, *Astrophys. J.* **354**, 745
Low BC: 1990, *Ann. Rev. Astron. Astrophys.* **28**, 491
Low BC: 1993, *Adv. Space Res.* **13**, No. 9, 63
Low BC and Smith DF: 1993, *Astrophys. J.* **410**, 412
Low BC, Munro RH, and Fisher, RR: 1982, *Astrophys. J.* **254**, 335
McComas DJ, Phillips JL, Hundhausen AJ, and Burkepile JT: 1992, in Ref. IX, p. 225
Somov BV: 1991, *J. Geomag. Geoelectr. Suppl.* **43**, 31
St. Cyr OC and Webb DF: 1991, *Solar Phys.* **136**, 379
Steinolfson R: 1990, Max '91 Workshop #3, Univ. of Colorado, Boulder, 123
Steinolfson R: 1991, in Ref. I, p. 171
Steinolfson, RS: 1992a, *J. Geophys. Res.* **97**, 10811
Steinolfson RS: 1992b, in Proc. of 26th ESLAB Symposium, ESA SP-346, 51
Webb DF: 1992, in Ref. III, p. 234
Webb DF: 1993, in Ref. V
Webb DF and Howard, RA: 1993, *J. Geophys. Res.*, in press
Wolfson R and Low BC: 1992, *Astrophys. J.* 391, 353

7. HIGHLIGHTS FROM SPACE - YOHKOH

(Yutaka Uchida)

The Yohkoh satellite was launched on 1991 August 30. Developed under the responsibility of the Institute of Space and Astronautical Science (ISAS), Japan, with the cooperation of other Japanese institutions, and with important contributions from UK and US institutions, Yohkoh aims at studying high-energy solar phenomena (Ogawara *et al.* 1991). As the only solar-dedicated mission launched during Cycle 22, Yohkoh is the successor to previous solar observatories in space: Skylab, P78, SMM and Hinotori. It carries four scientific packages (Tsuneta *et al* 1991; Kosugi *et al* 1991; Yoshimori *et al.* 1991; Culhane *et al.* 1991): a soft X-ray telescope (SXT); a hard X-ray telescope (HXT); a set of wideband spectrometers (WBS) whose energy range spans from soft X-rays (~1 keV) to γ-rays (<100MeV); and a set of Bragg crystal spectrometers (BCS).

The following sections provide a condensed overview of the main preliminary results from Yohkoh; extended analyses are in press or submitted at the time of this report.

7.1 ACTIVE REGION BEHAVIOR

The conventional concept of the active-region corona, as a region of a hot, dense gas tied down by the magnetic field of sunspots to the photosphere by some uniform heating process, is not supported by the Yohkoh observations. Instead, the active region corona is found to be supplied with mass *already heated from below* along magnetic loops (Shimizu *et al.* 1992). These have $5x10^6$ K-type high temperatures (Hara *et al.* 1992; Watanabe *et al.* 1992). Yohkoh was able to reveal these transient brightenings in active regions because of the high cadence of its observations and its low-scattering SXT optics. The discovery may affect basic ideas about coronal heating in general.

Furthermore, the active-region corona expands intermittently and, in many cases, almost continually with a velocity ~10 km s^{-1}, showing that active regions are not in magnetohydrostatic equilibrium (Uchida *et al.* 1992). Because there is prior evidence for interplanetary magnetic clouds being ejected without any corresponding flare or dark filament disappearances on the Sun (Nakagawa et al.1989), an examination of the interrelation of these events to active-region expansion is under way. If it can be proved that active-region expansion extends into interplanetary space, concepts about solar mass loss, and especially mass loss for stars showing magnetic activity, may be significantly affected.

7.2 DYNAMISM OF THE BACKGROUND CORONA

One of the most important findings by Yohkoh is the dynamism of the background corona, a behavior recognized in its full form for the first time by viewing the high-cadence Yohkoh-SXT images as video-movies. The background corona forms an impressive dynamic system with active regions. Many examples are seen of interesting interplay between them (Uchida 1993) and of flare-related phenomena (Tsuneta & Lemen 1993).

The coronal dynamism discovered by Yohkoh is most evident in: (i) the disappearance of dark filaments far from active regions; (ii) very small changes in active regions whose effects can cover a very wide region of the corona; (iii) ejection of heated mass from the brightenings in active regions into magnetic tubes comprising the overlying magnetic structure without changing the magnetic structure appreciably. These ejecta appear jet-like when they are injected into an open flux tube (Shibata *et al.* 1992). Fast disturbances propagating with velocities greater than 1000 km s^{-1} may exist but cannot be captured effectively with the present cadence of SXT images.

Disappearing Hα dark filaments outside active regions are accompanied by the well-known, low-energy, Hα double-ribbons. Skylab captured a few examples of arcade-formation related to these events as "still" pictures. Now Yohkoh has clarified the detailed dynamic behavior of X-ray arcade-formation. New findings, such as those by McAllister *et al.* (1992), seem to contradict models proposed earlier by Sturrock (1966), by Hirayama (1974), or by Kopp & Pneuman (1976), and will contribute to the derivation of more realistic ones.

A larger version of closed-loop arcade-formation at high latitudes far removed from the active-latitude belt was seen for the first time by Yohkoh (Tsuneta *et al.* 1992b). This immense structure grew progressively in height (3x10^5 km) and in length (10^6 km). It formed in a dark area, seemingly a coronal hole. But it was later confirmed that a field-polarity reversal line was there in the photosphere, so that the coronal hole was not a magnetically unipolar area. The inside of the arcade appears bright in X-rays when the structure is observed on the west limb. The global structural change of the coronal magnetic field appears to take place through a non-explosive, quasi-steady magnetic reconnection.

7.3 GLOBAL AND LONG-TERM BEHAVIOR OF THE CORONA, CORONAL HOLES, AND X-RAY BRIGHT POINTS

Synoptic analyses of Yohkoh data reveal that the high latitude part of coronal holes are not rotating rigidly; the rigid rotation of a coronal hole is restricted only to that portion which is in the active-latitude zones. It is suggested (Uchida 1993) that the rigid rotation of coronal holes, noted earlier in Skylab results, merely reflects the rigidly rotating active-longitude belts.

The behavior of X-ray bright points (XBP) was examined with Yohkoh (Strong *et al.* 1992; Nitta *et al.* 1992), but their number is considerably less than found by Skylab (Golub *et al.* 1974). The number does not thus far show a marked dependence on the phase of the cycle. There is, however, a possibility that the temperature range of the Yohkoh-SXT is not suited for finding XBPs.

7.4 FLARES

By the end of June 1993, Yohkoh observed 834 flares, of which 14 are powerful GOES X-class, and 230 are M-class flares. Yohkoh observations make the detailed structure and changes in preflare structures available both in soft and hard X-rays for the first time. Knowledge of preflare-to-flare changes of magnetic structures is essential to the understanding of flare mechanisms. Results are now available from coordinated observations with ground-based optical and radio observations (Ichimoto *et al.* 1992; Sakurai *et al.* 1992; Kurokawa *et al.* 1992).

New observations with the Yohkoh-HXT in four separate high-energy bands are free of the contamination from lower energy bands in which the thermal emission from the superhot plasma could be dominant. It was found that sources in the high energy bands are from the footpoints of loops and flicker rapidly, whereas sources in the lowest energy band tend to come from the higher parts of the loops (Kosugi *et al.* 1992; Sakao *et al.* 1992).

Yohkoh-SXT results reveal several very different types of flares. One clear category is the arcade-type flare which is large, long in duration and low in energy, although the total energy yield can be large. Another clear type, which occurs only in active regions, is the simple loop flare: small, short in duration, but sometimes high in energy. There may be other categories related to the interaction of loops and emerging flux regions.

7.4.1 EVOLUTION OF ARCADE-LIKE FLARES

Arcade flares can have very large flux as measured by GOES, sometimes go up to X-class, but the emission is not necessarily contained in a high-energy range. A very instructive example is a long-enduring limb flare described by Tsuneta *et al.* (1992a) which grew steadily in a helmet-streamer arch. The development of this type of event seems at first sight to be consistent with the classical model in which a dark filament rises, and cuts open the field. The reclosing of the severed field lines through magnetic reconnection is supposed to release energy, to evaporate chromospheric plasma which fills the structure, thus causing the Hα double-ribbon flare. The preflare magnetic structure revealed by Yohkoh in both of two well-observed limb flares, however, turn out to have configurations suggesting a quadrupolar field, not a dipolar one as expected in the conventional (classical) model. A quadrupole model was proposed for the magnetic configurations of dark filaments and applied to related flares by Uchida *et al.* (1980), but the idea was not received too well that time because it lacked experimental connfirmation. Investigations are now being pursued along this line.

7.4.2 ENERGETIC FLARE WITH HARD X-RAY EVENTS

The large explosive flare of 1991 November 15 is one of the most comprehensively studied events captured by Yohkoh. This single event enables some far-reaching conclusions to be drawn

(Acton *et al.* 1992; Canfield *et al.* 1992). The morphology of X-ray and Hα emissions from the flare provide supporting evidence that energy release takes place through driven rather than spontaneous reconnection. High-velocity upflows were observed (Culhane *et al.* 1993) as soon as brightenings began in SXT images. But strong hard X-ray bursts did not begin until more than 2 min later (Sakao *et al.* 1992). This creates a problem for supplying mass to the soft X-ray emitting loops by electron bombardment leading to chromospheric evaporation. It supports the view that in some flares a thermal conduction front is responsible for evaporation. Nevertheless, the simultaneous occurrence of a white-light flare with the hard X-ray sources during this event seems to support the electron bombardment model (Hudson *et al.* 1992). In a study of X-ray emission with height for nearly 100 flares, Matsushita K, *et al.* (1992) found no evidence to support thin-target emission from flaring coronal loops.

The evolution of the hard X-ray source from a single to a double source, with separation increasing with time (Sakao *et al.* 1992), is evidence for a multiple loop system in which particle acceleration proceeds on successively higher and higher field lines. The changing shapes of the HXT images in different energy bands indicate a preferential loss of higher energy electrons from the source region as the flare progresses.

REFERENCES

Acton LW *et al.*: 1992, *Science* **258**, 618

Canfield RC, *et al.*: 1992, *Publ. Astron. Soc. Japan* **44**, L111

Culhane JL *et al.*: 1991, *Solar Phys.* **136**, 89

Culhane JL *et al.*: 1993, *Adv. Space Res.*, **13**, No. 9, 303

Golub L, *et al.*: 1976, *Solar Phys.* **49**, 79

Hara H, *et al.*: 1992, *Publ. Astron. Soc. Japan* **44**, L135

Hirayama T: 1974, *Solar Phys.* **34**, 323

Hudson HS, *et al.*: 1992, *Publ. Astron. Soc. Japan* **44**, L77

Ichimoto K, *et al.*: 1992, *Publ. Astron. Soc. Japan* **44**, L117

Kopp RA and Pneuman GW: 1976, Solar Phys., **50**, 85

Kosugi T *et al.*: 1991, *Solar Phys.* **136**, 17

Kosugi T *et al.*: 1992, *Publ. Astron. Soc. Japan* **44**, L45.

Kosugi T: 1993, in "Physics of Solar and Stellar Coronae", G. S. Vaiana Memorial Symp., Linsky JL and Serio S (eds.), Kluwer (Dordrecht), in press

Kurokawa H *et al.*: 1992, *Publ. Astron. Soc. Japan* **44**, L129

Matsushita K, *et al.*: 1992, *Publ. Astron. Soc. Japan* **44**, L89

McAllister AH *et al.*: 1992, *Publ. Astron. Soc. Japan*, **44**, L205

Nakagawa T, *et al.*: 1989, *J. Geophys. Res.*, **94**, 11761

Nitta N, *et al.*: 1992, *Publ. Astron. Soc. Japan*, **44**, L167

Ogawara Y, Takano T, Kato T, Kosugi T, Tsuneta S, Watanabe T, Kondo I,and Uchida Y: 1991, *Solar Phys.* 136, 1

Sakao T, *et al.*: 1992, *Publ. Astron. Soc. Japan*, **44**, L83

Sakurai T, *et al.*: 1992, *Publ. Astron. Soc. Japan*, **44**, L123

Shibata K, *et al.*: 1992, *Publ. Astron. Soc. Japan*, **44**, L173

Shimizu T, *et al.*: 1992, *Publ. Astron. Soc. Japan*, **44**, L147

Strong KT, *et al.*: 1992, *Publ. Astron. Soc. Japan*, **44**, L161

Sturrock PA, 1966, *Nature*, 221, 695

Tanaka, K., *et al.*: 1983, *Solar Phys.*, **86**, 91

Tsuneta S *et al.*: 1991, *Solar Phys.* **136**, 37

Tsuneta S *et al.*: 1992a, *Publ. Astron. Soc. Japan*, **44**, L63

Tsuneta S *et al.*: 1992b, *Publ. Astron. Soc. Japan*, **44**, L211

Tsuneta S and Lemen J: 1993, in "Physics of Solar and Stellar Coronae", G. S. Vaiana Memorial Symp., Linsky JL and Serio S (eds.), Kluwer (Dordrecht), in press

Uchida Y and Sakurai T: 1980, in "Solar Flares", Skylab Workshop, Sturrock PA (ed.), U. Colorado Press, 110, p. 67

Uchida Y *et al.*: 1992, *Publ. Astron. Soc. Japan*, **44**, L155

Uchida Y: 1993, *Adv. Space Res.*, **13**, No. 9, 205

Watanabe T, *et al.*: 1992, *Publ. Astron. Soc. Japan*, **44**, L141

Yoshimori M, *et al.*: 1991, *Solar Phys.* **136**, 69

8. HIGHLIGHTS FROM SPACE - COMPTON GAMMA-RAY OBSERVATORY

(James Ryan and Richard A. Schwartz)

The instruments on-board the Compton Gamma Ray Observatory (CGRO) are far larger and more sensitive than any flown previously. The measurements of solar flare high energy fluxes have, as expected, yielded new insights (or questions) regarding X- and gamma-ray emissions from flares. The progress brought about by the CGRO measurements are in the following areas: (1) long-duration, high-energy flares, (2) neutron measurements, (3) microflares, (4) bremsstrahlung-dominated, gamma-ray flares, and (5) good statistics in smaller hard X-ray flares. The four instruments on the CGRO are BATSE, COMPTEL, OSSE, and EGRET. They are briefly described elsewhere (Gehrels *et al.* 1993). BATSE and OSSE are spectrometers operating in the 100 keV to 10 MeV range, while COMPTEL and EGRET are gamma-ray telescopes that can resolve the Sun in gamma rays as a point source in the 1 - 30 MeV and 30 - 30 000 MeV ranges, respectively.

8.1 LONG-DURATION, HIGH-ENERGY FLARES

The active region 6659 during the month of 1991 June produced a series of enormous flares during its transit across the disk, at least four of which one can classify as long duration high-energy flares. Each of these flares (1991 June 4, 9, 11 and 15) was measurable for at least two spacecraft orbits and each exhibited pion-related gamma-ray emission. The flares on 11 (Kanbach *et al.* 1993) and 15 (Leikov *et al.* 1993) June both produced gamma-ray emission with energies on the order of 1 GeV. The 11 June flare emitted gamma rays for a period of over 8 hours. The flare of 1991 June 9 (Ryan *et al.*, 1993) was of much lower intensity than the others but by virtue of the sensitivity of the instruments on CGRO, the measurements reveal that it too was a long-duration, high-energy event. This phenomenon of pion- related high-energy gamma-ray emission was observed earlier on SMM (Dunphy & Chupp, 1993), but the new measurements push theoretical models to the limit. There is a consensus that energetic protons must be stored for long periods of time (Mandzhavidze & Ramaty, 1992), but it is not clear whether there is continued acceleration in the traps high in the solar corona (Ramaty & Mandzhavidze, 1993).

8.2 NEUTRON MEASUREMENTS

The COMPTEL instrument brings a new aspect to solar studies - neutron spectroscopic capabilities. Being able to measure the energy of individual neutrons means that neutron emission can be studied in the same manner as gamma-ray emission, i.e., measuring the neutron spectrum at the time of emission. The neutron emission from the 1991 June 9 event (Ryan *et al.* 1993) primarily occurs after the impulsive phase. This clearly identifies the long-duration emission as having a proton rather than an electron origin.

8.3 MICROFLARES

The BATSE instrument's ability to detect weak cosmic gamma-ray bursts makes it an ideal instrument for measuring solar microflares. Flares three times smaller than those detected by HXRBS are being recorded by data inspection, while an automated search (Biesacker *et al.* 1993) has a sensitivity a factor of ten below the published HXRBS results. The sensitivity translates into a database of good statistics for microflare studies. Although the HXRBS results for flare-size distribution break down into one at solar maximum and another at solar minimum one, the BATSE data can be broken down to time scales as short as one solar rotation. Early results indicate no striking change in the flare-size distribution other than total frequency as a function of solar activity.

8.4 BREMSSTRAHLUNG-DOMINATED GAMMA-RAY FLARES

One of the many surprises from the Solar Maximum Mission was the discovery of gamma-ray flares which exhibited no nuclear line features. Typically, in the MeV range, proton- (ion-) produced nuclear lines from carbon, nitrogen, etc. dominate the photon energy spectrum. The spectra of these flares, however, are consistent with an emission process dominated by the bremsstrahlung of primary solar flare electrons up to energies of tens of MeV (Rieger and Marschhäuser, 1991). The CGRO has been able to observe several more of these events in an attempt to understand why the population of accelerated particles can change its composition so dramatically.

8.5 GOOD COUNT RATE STATISTICS FOR SMALL X-RAY FLARES

With the new sensitive measurements by CGRO come opportunities to correlate these data with those from other spacecraft and ground-based observatories. A graphic example is the correlation of the BATSE data with those of the BCS instrument on the YOHKOH spacecraft. One of the first observable quantities of soft X-ray observations of flares is the non-thermal broadening of the Ca XIX line. In the SMM data, the line width is large (doppler widths corresponding to ~100 km/s) as soon as the line is measurable. The BCS instrument on YOHKOH is much more sensitive than that on SMM, so it is of great interest to combine the sensitive hard X-ray data of BATSE and those of the YOHKHO/BCS instrument, possibly revealing the nature of the interplay in lower coronal mass motions with the energy deposited by fast electrons. Similarly, the cross comparison of metric and decimetric radiation with high time resolution X-ray data of BATSE can be more productive now that the sensitivity of the X-ray measurements is closer to that of the radio (Aschwanden *et al.*, 1993). The analysis of these data will improve our understanding of how the energetic electron population evolves, preferentially being transported upwards to produce Type III bursts or being transported downwards to produce the hard X-ray bursts.

REFERENCES

Aschwanden MA, Benz AO and Schwartz RA: 1993, *Astrophys. J.* **417**, in press

Biesecker D, Fishman GJ and Ryan JM: 1993," High Energy Solar Phenomena", AIP Conf. Proc. **294**, in press

Dunphy P and Chupp E: 1993, "High Energy Solar Phenomena", AIP Conf. Proc. **294**, in press

Gehrels N, Chipman E and Kniffen DA: 1993, "Compton Gamma-Ray Observatory", AIP Conf. Proc. **280**, 3

Kanbach G, Bertsch DL, Fichtel CE, Hartman RC, Hunter SD, Kniffen DA, Kwok PW, Lin YC, *et al.*: 1993, *Astron. Astrophys., Suppl. Ser.* **97**, No. 1, 349

Leikov N, Akimov VV, Volzhenskaya VA, Kalinkin LF, Nesterov VE, Galper AM, Zemskov VM, Oserov YV, *et al.*: 1993, *Astron. Astrophys., Suppl. Ser.* **97**, No. 1, 345

Mandzhavidze N and Ramaty R: 1992, *Astrophys. J.* **389**, 739

Ramaty R and Mandzhavidze N: 1993, "High Energy Solar Phenomena", AIP Conf. Proc. **294**

Rieger, E and Marschhäuser, H: 1991, Max '91 Workshop #3 Proc., p. 68

Ryan J, Forrest D, Lockwood J, Loomis M, McConnell M, Morris D, Webber W, Rank G, *et al.*: 1993, AIP Conf. Proc. **280**, 631

ADVANCES IN FLARE PHYSICS

The final four sections of this report bring together the significant results in solar flare research, from theory as well as from observations, achieved in the past 3 years. Because the subject of flares permeates many aspects of solar activity, it is unavoidable for some overlap to occur between the following and preceding sections. Overlaps have been eliminated except in those cases where different reviewers bring an entirely different perspective to the significance of any single publication.

9. PREFLARE ENERGY STORAGE

(Jean-Claude Hénoux and Boris V. Somov)

It is widely believed that flare energy is stored in electric current systems. Two possible origins of these currents have been advocated: 1) stressed magnetic fields, carrying currents, emerge through the photosphere; 2) currents are formed through motions of the photosphere that displace field line footpoints. In the latter case we can distinguish localised electric currents formed by twisting motions from extended current sheets due to shear motions. Vector magnetic field measurements show unneutralized electric current islands. As discussed by Melrose (1991), such observations support a subphotospheric origin of the currents. However the ambiguity in the sign of the transverse field may prevent a detection of the neutralizing currents.

9.1 THEORY OF ELECTRIC CURRENT GENERATION

Theoretical investigations of the formation of current sheets have been pursued. Low (1991) and Vekstein *et al.* (1991) showed analytically that shearing motions can produce current sheets along separatrices with or even without neutral points of magnetic field. Numerical computations of Karpen *et al.* (1991) confirmed the formation of currents in the line-tying hypothesis. However, applying a force to photospheric field lines not strictly line-tied, these authors found more distributed currents that cannot be considered as sheets. In these numerical computations time-dependent MHD equations were solved. They showed that currents are continuously evolving and are due to variations in plasma response time, with the line-tying condition corresponding to an infinite response time.

Other numerical studies of the dynamics of solar coronal magnetic fields resulting from footpoint motions at the photosphere have been made. Dahlburg *et al.* (1991) twisted the field of an arcade and followed the evolution of the field after stopping the photospheric motions. No instability and no current-sheet formation were observed; the stresses were minimized by an extension of the magnetic field. In an attempt to understand coronal mass ejection rather than to study current-sheet formation, Steinolfsen (1991) sheared an initial dipole configuration by azimuthal motions in opposite directions on each side of the equatorial plane. Resistive slippage was prevented, leading to an eruptive behavior; rapid expansion of the coronal magnetic field at velocities exceeding the Alfvén speed was obtained.

9.2 FREE MAGNETIC ENERGY STORAGE

Demonstrations that a force-free configuration with a completely open field has lower magnetic energy than a closed-field state were given by Aly (1991) and by Sturrock (1991). This property indicates either that the coronal mass ejection (CME) phenomenon cannot be strictly driven by force-free magnetic fields or that the final configuration associated with a CME is not in the open state; it gives the maximum value of the energy that can be released by current dissipation. Then using a Clebsch-variables representation of the magnetic field, Klimchuk & Sturrock (1992) and Porter *et al.* (1992) computed numerically the amount of free magnetic energy built up by photospheric shearing or twisting motions. The computations were done for magnetic configurations either of arcade type or cylindrically symmetric around a vertical axis. They suggest that a completly open-field configuration corresponds to the maximum of energy storage and indicate that the energy buildup can satisfy the energy requirements of solar flares. Wolfson & Low (1992) show that, by shearing, a closed force-free field can become partially open. LaRosa (1992) has found, in contrast to the previous studies, that a laminar reconnecting current sheet cannot be responsible for the activity associated with emerging flux. This result comes from a study of the resistivity in a neutral current sheet and seems to agree with a conclusion by Somov (1992) that there is no thermal equilibrium for such a sheet with classical heat conduction along and across the magnetic field.

9.3 THE OBSERVED ROLE OF PHOTOSPHERIC MOTIONS IN FLARE OCCURENCE

Observations of shear motions, and of head-on collision of bipoles, leading to flares have been reported by Wang *et al.* (1991), Wang (1992) and Tang & Wang (1993). Generally the shear derived from the shape of Hα fibrils is reduced after a flare. However some observational results contradict this simple behavior. Simaravan *et al.* (1992) relate flare occurence to the change in the shear, decrease or increase, a day prior to the flare. Using a vector magnetograph, Wang (1993) measured the change in shear angle (angle between the observed transverse field and the direction of an extrapolated transverse potential field) after a flare and observed an increase of this shear angle. Then Forbes (1993) pointed out that rotations of the transverse photospheric magnetic field are not sufficient by themselves to determine the change in the stored energy.

9.4 OBSERVATIONAL EVIDENCE OF THE ROLE OF ELECTRIC CURRENTS IN SOLAR FLARES

Calculations of the longitudinal component of electric currents in active regions rely on the measurement of transverse fields; one has then to solve the 180° ambiguity to determine the direction of the transverse fields (Wang 1993; Linxiang 1993). The development of vector magnetographs (Wang *et al.* 1992) leads to new observational results on the association between electric currents and solar flares. The morphological relationship between vertical electric currents in the photosphere and flare kernels has been confirmed. Abramenko et al. (1991) suggest that the direction of the current affects the Hα brightness. Using not exactly simultaneous observations of currents and flares, Lin *et al.* (1993) suggested that Hα kernels were located near the peaks of the vertical currents. On the other hand, Canfield *et al.* (1991) in four flares found that the non-thermal precipitation sites were at the edges of the major vertical currents, whereas the sites of high coronal pressure were within them. In a class X Flare, the hard X-ray emission (23-33 keV) was found to be located at the edges of sites of high vertical current density (Canfield *et al.* 1992), a result confirmed in more recent work by the same group (Leka et al, 1993; de la Beaujardière et al. 1993). All these observations suggest that flares begin at the edges of current systems and spread towards their interiors.

Comparison of the location of Hα kernels with the location of separatrices of magnetic flux in active regions has been pursued and developed (Mandrini *et al.* 1991, 1993; Démoulin *et al.* 1993; Hénoux *et al.* 1993). In the ten flares that have been studied, the Hα kernels were located on the separatrices. For all events for which longitudinal current densities were measured, two of the four Hα kernels partially overlapped the current-density kernels. Then two stages are possible. In the first, currents are produced by photospheric shear motions; magnetic energy is stored in field aligned currents on separatrices. In the second, flare-energy release takes place when a strong current system approaches the separator and is disrupted by field line reconnection in the region of the separator.

REFERENCES

IAUC141= "The Magnetic and Velocity Fields of Solar Active Regions", IAU Colloq. **141**, Zirin H, Ai G, and Wang H (eds.), Astron. Soc. Pacific
Abramenko VI, Gopasyuk SI and Ogir MB: 1991, *Solar Phys.* **134**, 287
Aly JJ, 1991: *Astrophys. J. Lett.* **375**, L61
Canfield RC, de la Beaujardière JF and Leka KD: 1991, *Phil. Trans. Royal Soc. London,* **336**, 381
Canfield *et al.*: 1992, *Publ. Astron. Soc. Japan,* **44**, L111
Dahlburg RB, Antiochos SK, and Zang TA: 1991, *Astrophys. J.* **383**, 420
De la Beaujardière JF, Leka KD, and Canfield RC: 1993, *Astrophys. J.* **411**, 378
Démoulin P, van Driel-Gesztelyi L, Schmieder B, Hénoux JC, Csepura G and Hagyard MJ: 1993, *Astron. Astrophys.* **271**, 292
Forbes T: 1993, IAUC 141, p. 415-424
Hénoux JC, Démoulin P, Mandrini C and Rovira M, 1993: IAUC141, p. 333-342
Karpen JT, Antiochos SP, and DeVore CR: 1991, *Astrophys. J.* **382**, 327
Klimchuk JA and Sturrock PA: 1992, *Astrophys. J.* **385**, 344
LaRosa TD: 1992, *Astrophys. J.* **396**, 289
Leka KD, Canfield RC, McClymont AN, de la Beaujardière JF, Fan Y: 1993, *Astrophys. J.* **411**, 370
Lin Y, Wei X and Zhang H: 1993, IAUC141, p. 446-449
Linxiang, W: 1993, IAUC141, p. 439-442
Low, B C, 1991: *Astrophys. J.* **381**, 295
Mandrini, CH, Démoulin, P, Hénoux, JC, and Machado ME: 1991, *Astron. Astrophys.* **250**, 541
Mandrini, CH, Rovira, MG, Démoulin, P, Hénoux, JC, Machado, ME and Wilkinson, LK: 1993, *Astron. Astrophys.* **272**, 609
Melrose DB: 1991, *Astrophys. J.* **381**, 306
Porter LJ, Klimchuk JA and Sturrock PA: 1992, *Astrophys. J.* **385**, 738
Rausaria RR, Aleem SM and Sundara Raman K: 1992. *Solar Phys.* **142**, 131
Sivaraman KR, Rausaria RR and Aleem M: 1992, *Solar Phys.* **138**, 353
Somov BV: 1992, "Physical Processes in Solar Flares", Kluwer (Dordrecht)
Steinolfson RS: 1991, *Astrophys. J.* **382**,677
Sturrock PA: 1991, *Astrophys. J.* **380**, 655
Tang F and Wang H: 1993. *Solar Phys.* **143**, 107
Wang H: 1992, *Solar Phys.* **140**, 85
Wang H: 1993, IAUC141, p. 323-332
Wang H, Tang F, Zirin H, and Ai G: 1991, *Astrophys. J.* **380**, 282
Wang H, Varsik J, Zirin H, Canfield RC, Leka KD, and Wang J: 1992, *Solar Phys.* **142**,11
Wang J: 1993, IAUC141, p. 425-434
Wolfson R and Low BC: 1992, *Astrophys. J.* **391**, 353

10. THE 'COOL' PLASMA IN FLARES

(Petr Heinzel)

The relationship between the cool flare plasma observed at optical and UV wavelengths and the hotter plasma 'seen' in EUV, X-ray and radio wavebands has become clearer in studies based on SMM and ground-based data. Now we are entering a new phase of flare research where more complex questions can be addressed with high-resolution YOHKOH observations. Substantial research effort has been reported in the proceedings of several workshops and colloquia, in monographs, and in extensive reviews (see Refs. I-XI).

10.1 CHROMOSPHERIC FLARE MODELS

Several important papers have appeared which deal with non-LTE aspects of flares. Their discussion in this section is organized around three categories: (i) semi-empirical models, (ii) synthetic models, and (iii) general line-formation problems related to flares.

New semi-empirical flare models have been constructed by Mauas *et al.* (1990) for an observed case of a white-light flare (WLF) in which two bright kernels had the same level of continuum emission, but different line spectra in which the line-formation depths spanned from the photosphere to the chromosphere. They found the origin of the continuum emission to be photospheric, not chromospheric, for both observed kernels. The observations and the models derived from them show that continuum emission can occur in the absence of Hα emission, thus leading the authors to suggest that some WLFs are inconsistent with the conventional picture for WLFs: energy released in the corona and transported downward by some mechanism. On the other hand, Sotirovski *et al.* (1992) found from another WLF spectrum that the Paschen continuum is responsible for WLF emission. A combination of both possibilities for WLF emission is suggested by Baranovsky (1990) on the basis of semi-empirical modelling. A drawback of semi-empirical models is that they do not take into account non-thermal excitation and ionization of atoms by beamed particles; this naturally leads to an overestimation of empirically derived temperatures. Also, soft X-ray irradiation from the overlying coronal loop is also neglected in these models. The models are static while, in fact, asymmetries in the observed line profiles may indicate departures from hydrostatic equilibrium. The effect of non-thermal excitation and ionization by electron resp. proton beams on Hα and CaII K lines was recently investigated by Fang *et al.* (1993) and by Hénoux *et al.* (1993) for two earlier semi-empirical models. Hydrogen lines are greatly strengthened and broadened, showing an obvious central reversal. The effects are weaker for the CaII K line. Similar calculations have been performed by Zharkova & Kobylinsky (1993).

Synthetic models (class (ii)) of physically self-consistent coronal loops which cover a wide range of physically relevant conditions, have been presented for the case of giant flares on M dwarf stars by Hawley & Fisher (1992). This work has recently been extended to the case of the Sun (Hawley & Fisher, submitted 1993). These models are also static, but the energy balance problem is solved consistently by taking into account electron-beam heating, coronal X-ray heating and thermal conduction. In contrast to previous non-LTE models of this kind, e.g. Falchi *et al.*(1990), the electron flux, conductive flux and the coronal pressure are mutually connected which clearly shows that certain combinations of these parameters used in previous work are not physically allowed. For example, high coronal pressure occurs in tandem with high conductive flux thus forming a deep transition region with reduced chromospheric heating. High coronal pressure is not consistent with very bright Hα, as previously assumed; typically observed Hα fluxes can only be produced under conditions of a low pressure corona with strong beam heating. Surprisingly, the computed ratio of total Hα flux to the flux of non-thermal electrons with energies > 20 Kev (F_{20}), differs strongly from that derived for well observed flares by Canfield et al. (1991). Therefore, the authors suggest that the pressure is a more relevant physical parameter than the beam flux F_{20}.

The third class of non-LTE computations is represented by the work of Gayley (1991) and Gayley & Canfield (1991). They studied semi-analytically the conditions under which hydrogen Lyman and Balmer lines are formed in chromospheric flares, taking into account the effects of partial redistribution. A rough proportionality was found between Stark wing-intensity and the total energy deposited in the partially ionized layers. Therefore, under favourable conditions (i.e. without additional masking opacity effects), one can infer the magnitude of the nonthermal heating in the chromosphere from enhancements in the Stark wings of Lyman and Balmer lines. The blue wing is suggested for such a diagnostic. However, intense electron beams produce impulsive Lα emission which is much brighter than that so far detected (Gayley & Canfield 1991). This will be a challenging application for the SUMER experiment on the SOHO satellite mission.

10.2 RADIATIVE-HYDRODYNAMICAL MODELS OF CHROMOSPHERIC HEATING

Time-dependent simulations of pulsed beam heating of the solar atmosphere have been done by Karlicky (1990) and by Karlicky & Hénoux (1991). A new particle-representation of an electron beam enabled these authors to take into account the finite travel time of non-thermal electrons and to simulate the general form of the beam distribution function. A chromospheric response is treated by solving the standard set of hydrodynamical equations. As shown by Heinzel & Karlicky (1992a), for a series of subsecond beam pulses, one gets a correlation between computed hard X-ray spikes and Hα variations. However, it was found that subsecond Hα pulses exhibit an intensity decrease rather than the expected increase. Relevant HXR observations for testing such models are now being made on GRO (BATSE instrument) and on YOHKOH. High temporal resolution Hα data come from new CCD cameras in Boulder (Kiplinger *et al.* 1990) and in Ondrejov (Kotrc *et al.* 1993). More consistent radiation-hydrodynamical simulations have been presented by Heinzel & Karlicky (1992b), who demonstrated the importance of hydrogen recombination relaxation for modelling subsecond heating pulses. Using time-dependent numerical simulations and SMM observations of a soft X-ray line of CaXIX, Mariska (1992) examined the ability of electron-beam heating to reproduce the rise phase of flares.

The correspondence between coronal upflows (CaXIX from YOHKOH) and chromospheric downflows (Hα) are under extensive discussion. For the hydrodynamical model of a flare loop Gan *et al.* (1991) and Gan *et al.* (1992) have evaluated synthetic CaXIX spectra as well as Hα and CaII K line profiles. The conclusion of these authors is that a thermal hydrodynamical model can describe the behaviour of these lines rather well. In another paper, Gan *et al.* (1990) computed hydrodynamical models of the impulsive phase. Efficiency of chromospheric heating by soft X-rays is stressed; the authors propose a method for evaluating X-ray irradiation of the chromosphere by a flaring loop.

10.3 FLARE DYNAMICS FROM SPECTRAL DIAGNOSTICS

All semiempirical models (based on spectral data) published so-far are static. The most sophisticated synthetic models (like those of Hawley & Fisher 1992) are also treated in hydrostatic equilibrium. On the other hand, the above-mentioned hydrodynamical simulations naturally predict the macroscopic velocity fields (evaporation, downward moving chromospheric condensation), but only rather limited effort has been devoted to a direct comparison with asymmetrical optical or UV line profiles. Using the bisector method for Hα, Wülser *et al.* (1992, 1993) have deduced the downflow velocities of the chromospheric condensation from typical red asymmetryies of this line. The velocities, of the order of 40-50 km s^{-1}, give the plasma momentum which seems to be consistent with that derived from soft X-ray data (SMM, YOHKOH). This would confirm the idea of explosive evaporation during thick-target heating.

Heinzel *et al.* (1993) analyzed a blue asymmetry in the Balmer and CaII lines which was clearly detected with the Ondrejov multichannel flare spectrograph. An upflow of cool material cannot be excluded (Canfield *et al.* 1990; Karlicky & Hénoux 1992). However, an attractive explanation for the blue asymmetry at flare-onset is that the downward moving plasma with a certain velocity gradient produces an excess of blue-wing emission, simply due to non-LTE transfer effects. In such a case the bisector method, indicating an upflow, fails. A velocity gradient was also deduced from SPO spectral observations by Falchi *et al.* (1992). These authors derived downward velocities ~30 km s^{-1} from Balmer lines, ~20 km s^{-1} from the CaII K line, and ~10 km s^{-1} from the Na D2 line. Simultaneous upflows were detected for this flare on SMM. The line asymmetries were also studied by Fang *et al.* (1991) using CaII line observations and by Fang *et al.* (1992) with an approximate non-LTE model which incorporates an ad hoc velocity field around the temperature minimum region. However, the velocities considered by these authors are too large for models in hydrostatic equilibrium. New dynamical non-LTE models are thus required.

10.4 Hα POLARIZATION

Linear polarization of the hydrogen Hα line has been suggested as an indicator of the precipitation of downward-directed electron and/or proton beams. This diagnostic can give us an idea about the ratio of the particle number densities in the proton and electron beams and about their energy distribution, both clues to understanding the acceleration mechanisms. Impact linear polarization was studied by Hénoux *et al.* (1990) as a diagnostic of 100 keV proton beams, using new Hα polarization observations made at Meudon. A polarization degree as high as 2.5 % was detected in some Hα kernels. Preference is made for non-local proton acceleration which leads to a bombardment of the chromosphere. It was shown that transport processes in the chromosphere strengthen the anisotropic velocity distribution of protons, which is the cause of the collisional polarization of the Hα line. When Hα polarization is observed, X-ray emission appears to be more thermal (soft) than impulsive (hard). For example, polarization above 2 % correlates well with the Hα mean intensity during 20-30 min. (Aboudarham *et al.* 1992a). It resembles the temporal variation of soft X-ray emission, while no hard X-ray bursts were detected during this period by the HXRBS instrument on SMM. Another plausible explanation for the Hα linear polarization is the existence of neutral beams with the same speed for both protons and electrons (Aboudarham *et al.* 1992a). Similar Hα polarization characteristics have also been deduced from measurements made with the Hawaii Stokes Polarimeter (Metcalf *et al.* 1991). A new CCD Hα polarimeter is currently being completed at Meudon (Aboudarham *et al.* 1992a).

Further investigation of the electron impact excitation of Hα by Aboudarham *et al.* (1992b) shows that, while the anisotropy of a proton beam can be caused or maintained by transport effects, in the case of electrons the anisotropy may result from large temperature gradients or from acceleration within electric fields. Although the electrons may be equally responsible for the observed Hα linear polarization, their anisotropy due to heat conduction has to be treated with caution. New detailed computations for this case are required. Another possibility is acceleration by an ambipolar electric field (Aboudarham *et al.* 1992b). A general quantum-mechanical scheme of electron impact polarization was given by Fineschi & Landi Degl'Innocenti (1992).

10.5 COOL FLARE LOOPS

The term 'cool' flare loops was introduced instead of 'post'-flare loops because these loops form a natural part of a developing flare complex and appear during the flare. Moreover, they are obviously cool (10^4 K) relative to hot flare loops. The morphology, physical and evolutionary characteristics of cool flare loops have been reviewed by Simnett & Forbes (1991), Svestka (1991) and by Schmieder (1992). Heinzel *et al.* (1992) have made rather complex non-LTE and hydrodynamical analysis of the loops observed in Hα by the MSDP instrument at Meudon and derived the values of several plasma parameters. The basic conclusion is that the downward motions along both legs have lower velocities than a free-fall motion. Inspired by this finding, Karlicky & Simnett (1992) performed radiation-hydrodynamical simulations of a cool plasma blob moving inside the hot loop and showed that the blob can indeed be decelerated by pressure gradients. The evolution of cool flare loops system was studied by Gu *et al.* (1992), who found the whole system rising with a velocity of about 5-10 km s^{-1}. Effects of heating cool loops by Alfvén waves were considered by Lin and Zhang (1991). Finally, we mention an interesting observation by Hiei *et al.* (1992) of a loop system on the limb in white light, indicating electron densities of 10^{12}-10^{13} cm^{-3}.

REFERENCES:

I: "Dynamics of Solar Flares", Flares 22 Workshop, Schmieder B & Priest ER (eds.), Obs. de Paris, DASOP, 1991

II: "Max '91/SMM Solar Flares: Observations and Theory", 3rd Max'91 Workshop, Winglee RM and Kiplinger AL (eds.), U. Colorado (Boulder), 1991

III: "Spectral Line Formation in Plasmas Under Extreme or Unusual Conditions", Peyraud N, Heinzel P & Hooper CF (eds.), *J.Quant.Spectrosc.Radiat.Transfer* **44**, No. 1, 1990

IV: "Eruptive Solar Flares", IAU Colloq. 133, Svestka Z, Jackson BV and Machado ME (eds), Lect. Notes Phys. **399**, Springer-Verlag (Berlin), 1992

V: "Flare Physics in Solar Activity Maximum 22", Uchida Y, Canfield RC, Watanabe T and Hiei E (eds.), Lect. Notes Phys. **387**, Springer-Verlag (Berlin), 1991

VI: "Coronal Streamers, Coronal Loops, and Coronal and Solar Wind Composition", Proc. First SOHO Workshop, Mariska JT, Domingo V, and Poland AI (eds.), ESA SP-348 (ESTEC), 1992

VII: "Initial Results from Yohkoh", Special Feature, *Publ. Astron. Soc. Japan* **44**, No. 5, 1992.

VIII: "Physical Processes in Solar Flares",Somov BV, Kluwer (Dordrecht), 1992

IX: "Plasma Loops in the Solar Corona", Bray RJ, Cram LE, Durrant CJ and Loughhead RE, Cambridge Univ. Press (Cambridge), 1991

X: Haisch, B and Strong, KT, *Ann. Rev. Astron. Astrophys.* **29**, 275, 1991

XI: Zirin H, MacKinnon A and McKenna-Lawlor SMP, in "Solar Interior and Atmosphere", Cox AN, Livingston WC & Matthews MS (eds.), U. Arizona Press (Tucson), 1991
Aboudarham J, Chambe G, Feautrier N & Hénoux, JC: 1992a, in Faurobert-Scholl M, Frisch H, and Mein N, (eds.), "Methodes de determination des champs magnetiques solaires et stellaires", Obs. de Paris, p. 163
Aboudarham J, Berrington K, Callaway J, Feautrier N, Hénoux JC *et al.*:1992b, *Astron. Astrophys.* **262**, 302
Baranovsky, EA: 1990, *Soln. Dannye Bull.* **8**, 79
Canfield RC, Kiplinger AL, Penn MJ and Wülser JP: 1990, *Astrophys. J.* **363**, 318
Canfield RC, Zarro DM, Wülser JP and Dennis BR: 1991, *Astrophys. J.* **367**, 671
Falchi A, Falciani R and Smaldone LA: 1990, *Astron. Astrophys., Suppl. Ser.* **84**, 601
Falchi A, Falciani R and Smaldone LA: 1992, *Astron. Astrophys.* **256**, 255
Fang C, Hénoux JC and Gan WQ: 1993, *Astron. Astrophys.* **274**, 917
Fang C, Hiei E and Okamoto T: 1991, *Solar Phys.* **135**, 89
Fang C, Hiei E, Yin SY and Gan WQ: 1992, *Publ. Astron. Soc. Japan* **44**, 63
Fineschi S and Landi Degl'Innocenti E: 1992, *Astrophys. J.* **392**, 337
Gan WQ and Fang C: 1990, *Astrophys. J.* **358**, 328
Gan WQ, Fang C and Zhang HQ: 1991, *Astron. Astrophys.* **241**, 618
Gan WQ, Rieger E, Fang C & Zhang HQ: 1992, *Astron. Astrophys.* **266**, 573
Gayley KG: 1991, *Astrophys. J.* **374**, 773
Gayley KG and Canfield RC: 1991, *Astrophys. J.* **380**, 660
Gu XM, Lin J, Luan T and Schmieder B: 1992, *Astron. Astrophys.* **259**, 649
Hawley SL and Fisher GH: 1992, *Astrophys. J., Suppl. Ser.* **78**, 565
Heinzel P, Kotrc P, and Svestka Z: 1993, *Solar Phys.*, in press
Heinzel P and Karlicky M: 1992a, in ref. IV, p. 359
Heinzel, P and Karlicky, M: 1992b, in ref. VI, p. 237
Heinzel P, Schmieder B and Mein, P: 1992, *Solar Phys.* **139**, 81
Hénoux JC, Chambe G, Smith DF, Tamres D, Feautrier N. *et al.*: 1990, *Astrophys. J. Suppl.* **73**, 303
Hénoux JC, Fang C and Gan WQ: 1993, *Astron. Astrophys.* **274**, 923
Hiei E, Nakagomi Y and Takuma H: 1992, *Publ. Astron. Soc. Japan* **44**, 55
Karlicky M and Hénoux JC: 1991, *Bull. Astron. Inst .Czech.* **42**, 22
Karlicky M and Hénoux JC: 1992, *Adv. Space. Res.* **13**, No. 9, 195
Karlicky M and Simnett GM: 1992, in ref. VI, p. 205
Karlicky M: 1990, *Solar Phys.* **130**, 347
Kiplinger AL, Labow G and Orwig LE: 1990, in ref. II, p. 210
Kotrc P, Heinzel P and Knizek M: 1993, *JOSO Annual Report 1992*
Lin J and Zhang Z: 1991, in ref. I, p. 159
Mariska JT: 1991, in ref. V, p. 153
Mauas PJD, Machado ME and Avrett EH: 1990, *Astrophys. J.* **360**, 715
Metcalf TR, Wülser JP, Canfield RC and Hudson HS: 1991, in *Proc. of 2nd GRO Workshop*
Schmieder B: 1992, in ref. IV, p. 124
Simnett GM and Forbes TG: in ref. I, p. 121
Sotirovski P, Boyer R, Hiei E and Vince I: 1992, *Astron. Astrophys.* **262**, 597
Svestka Z: 1991, in ref. V, p. 165
Wülser JP *et al.*: 1993, *IFASG Preprint 78*
Wülser JP, Zarro DM and Canfield RC: 1992, *Astrophys. J.* **384**, 341
Zharkova VV and Kobylinsky VA: 1993, *Solar Phys.* **143**, 259

11. THE HOT PLASMA COMPONENT IN FLARES

(Giannina Poletto)

11.1 GENERAL OVERVIEW

An overall description of the flare phenomenon can be found in chapters of recently published textbooks on solar physics (e.g. Refs. I, II). Theoretical aspects have been dealt with in some detail in more advanced texts (Refs. III, IV). Additional material apears in reviews on solar flares by Phillips (1991), Sakurai (1991), and Rust (1992). But the wide interest in solar flare research is best documented in the numerous meetings dedicated to flares held in the past three years (e.g. the proceedings in Refs. V-IX). All include papers which focus on the hot plasma component of flares.

When we speak of the hot plasma component of flares, we refer to plasma at temperatures $\sim 10^7$ K or higher and densities $\sim 10^{10}$ cm^{-3}. In spite of extensive work, we still have not arrived at an unambiguous solution to the problem of how such high densities and temperatures are attained. The primary energy release site has not yet been observed - although the sudden energy release is generally ascribed to some sort of magnetic reconnection process - and there is no consensus on how energy is transported from the energy release site to other places within the flaring flux tube (see, e.g., Zarro 1992). The controversy about the origin of high densities is still very much alive: see, e.g. Feldman 1990; Bornmann & Lemen, 1992; Wülser *et al.* 1992; Seely & Feldman 1992.

11.2 MORPHOLOGY OF THE HOT PLASMA COMPONENT

From an analysis of 10 large non-impulsive flares observed by the SXT instrument on Yohkoh, Acton *et al.* (1992) showed that these events usually consist of a single or, at most, a few loops where plasma at $\sim 2 \times 10^7$ K is concentrated in a small volume (2-3 " diameter) at loop tops. As the flare decays, this hot region stays bright and concentrated in a fairly small volume, while the legs of the loops do not appear to contain any significant amount of hot plasma ablated from the chromosphere. Analogously, Doschek *et al.* (1992), from an analysis of ~60 events (most of X-ray class M1) observed by the BCS experiment on Yohkoh, conclude that most flares have weak blue-shifted signatures which are not consistent with the standard evaporation scenario. These observations support the suggestion that plasma is heated *in situ* and compressed via the pinch effect of a large electric current or via a "sweeping pinch effect" (Uchida & Shibata 1988). However, large plasma motions indicative of chromospheric evaporation have been observed by BCS as well (Culhane *et al.* 1992) while other flares do not present conclusive evidence in favor of one mechanism or the other (Ichimoto *et al.* 1992). It certainly looks like the transport of energy in flares is more complicated than previously assumed; further BCS spectra, taken during the initial phase of a flare, may help discriminate between different scenarios.

11.3 MODELLING

Flare modelling progressed along two main lines: by numerical modelling as pursued by Gan *et al.* (1991), and by using the Palermo-Harvard hydrodynamic code (Jakimiec *et al.* 1992; Sylwester *et al.* 1993). A simplified analytical treatment has been developed by Serio *et al.* (1991). These models provide a good description of flares where standard evaporation occurs. Takakura (1992) has made numerical simulations of impulsive flares caused by transient heat conduction along a loop with an applied axial electric current. His model predicts X-ray emission from the two footpoints and a coronal source expanding along the loop. A different approach has been used by Zaitsev & Stepanov (1992) who, while further developing the circuit theory of solar flares, point out that a powerful energy release at chromospheric levels should accompany the energy release in the coronal section of the loop. It is difficult to establish how realistic these predictions are and what percentage of flares are represented by them. Further analysis of Yohkoh data will possibly allow us to define the most representative characteristics of flares, thus allowing modelers to focus on these aspects of flare events.

The model of Melrose (1992) takes into account a feature which is usually overlooked, i.e. the propagation of energy from the photosphere into the corona that is required prior to or during the flare. However, his model pertains to the class of *current models*; hence it favors energy dissipation through double layers rather than through reconnection. This may occur only in a minority of flares, since many Yohkoh observations support magnetic reconnection in a neutral sheet at loop tops as the source of the flare energy (Tsuneta *et al.* 1992a; Ichimoto *et al.* 1992). In addition, Yohkoh showed that magnetic reconnection plays a role also in non-flaring conditions: the global solar corona continually undergoes a process of global and local restructuring which is ascribed to quasi-steady magnetic reconnection (Tsuneta *et al.* 1992b). Larger spatial dimensions and a smaller energy release seem to be the only differences between flare events and this large-scale non-explosive restructuring of the corona. This new result from Yohkoh points towards a unified view of flares and low-level large-scale coronal activity; it was anticipated over the past few years by a number of works which tried to establish a link between flares and large scale events.

In fact, careful analysis of SMM and Skylab flares has shown that secondary excitation of large-scale structures - X-ray giant arches and flaring X-ray arches - may develop after the primary energy release (Svestka 1991). The latter features - with lengths extending up to 300 Mm - may give us the unique opportunity of observing the effect of coronal heating, as through a magnifying lens, by following the progressive X-ray brightening along the arch. A remote and delayed brightening with respect to the impulsive phase of the flare has been detected also by Martens *et al.* (1990) in NIXT flare data and ascribed to sudden topological changes occurring when one of the flare ribbons, in its motion away from the neutral line, crosses a magnetic separatrix. Analogously, an analysis of Yohkoh X-ray bright points revealed the presence of many bright point flares associated with the enhancement of larger scale structures (Strong *et al.* 1992). Reconnection of emerging flux with ambient fields is invoked to explain this unexpected phenomena. Topological models in which flares are ascribed to reconnection of large-scale magnetic structures have been developed by Mandrini *et al.* (1991. All these studies show that the flare community starts considering the hot flaring plasma no longer as a local event, but as the outcome of the energy released by the interaction of an ensemble of large-scale loops, in a scenario which couples flare activity to a global background.

11.4 MICROFLARES

A unified view of the flare phenomenon may result also from studies of coronal heating. The possibility that microflares are the high-energy tail of a hidden distribution of events (nanoflares) possibly responsible for the hot solar corona, has become popular lately (see Parker 1993 and references therein; Chouduri *et al.* 1993; Kopp & Poletto, 1993). Although individual nanoflares supposedly release too little energy to be detectable by present instrumentation, Yohkoh's observations of repeated, transient brightenings of compact loops (Shimizu *et al.* 1992) are consistent with the view of the solar corona as a swarm of nanoflares. Thus the corona, rather than being interpreted as a medium where impulsive energy release occurs at the time of flares, will itself be the outcome of a myriad of these events.

We may say that emphasis is shifting from individual powerful events to the overall effect of multitudes of tiny events. Traditionally, flares occur because of some instability causing the release of magnetic energy stored in a large-scale structure. Lately, however, the suggestion has been advanced that flares occur because of a chain reaction which involves many small-scale sites (see, for instance, Lu & Hamilton 1991; Zirker & Cleveland 1993). In the so-called "avalanche model" of flares, a small-scale event creates a disturbance that propagates close-by, triggering additional energy release events. The traditional and the avalanche scenarios are not necessarily in conflict, but stress different aspects of the same phenomenon.

REFERENCES

I: "Plasma Loops in the Solar Corona", Bray RJ, Cram LE, Durrant CJ, and Loughhead RE: Cambridge University Press 1991.
II: "Guide to the Sun", Phillips KJH: Cambridge University Press, 1992.
III: "Advances in Solar System Magnetohydrodynamics", Priest ER and Hood AW (eds.): Cambridge University Press, 1991.
IV "The Sun: A Laboratory forAstrophysics", Schmelz JT and Brown JC (eds.): NATO ASI Series, Kluwer (Dordrecht) 1992.
V: "Dynamics of Solar Flares", Schmieder B and Priest E.(eds.): Obs. de Paris, DASOP, 1991.
VI "Flare Physics in Solar Activity Maximum 22", Uchida Y, Canfield RC, Watanabe T and Hiei E (eds.): Lect. Notes Phys. **387**, Springer (Berlin), 1991.
VII "The Physics of Solar Flares", Culhane JL. and Jordan C. (eds.): 1991, Cambridge University Press.
VIII "Eruptive Solar Flares", IAU Colloq. 133, Svestka Z, Jackson BV, and Machado ME (eds.), Lect. Notes Phys. **399**, Springer (Berlin) 1992.
IX "Fundamental Problems in Solar Activity", Pick M and Machado ME (eds.), *Adv. Space. Res.* **13**, No. 9, 1993.
Acton LW *et al.*: 1992, *Publ. Astron. Soc. Japan* **44**, L71
Bornman PL and Lemen JR: 1992, *Astrophys. J.* **385**, 363
Chouduri AR, Auffret H., and Priest ER: 1993, *Solar Phys.* **143**, 49
Culhane JL *et al.*: 1992, *Publ. Astron. Soc. Japan* **44**, L101
Doschek GA *et al.*: 1992, *Publ. Astron. Soc. Japan* **44**, L95
Feldman, U.: 1990, *Astrophys. J.* **364**, 322
Gan WQ, Fang C, and Zhang HQ: 1991, *Astron. Astrophys.* **241**, 618

Ichimoto K *et al*: 1992, *Publ. Astron. Soc. Japan* **44**, L117
Jakimiec J, Sylwester B, Sylwester J, Serio S, Peres G, and Reale F: 1992, *Astron. Astrophys.* **253**, 269
Kopp RA and Poletto G: 1993, *Astrophys. J.* **418**
Lu ET and Hamilton JR: 1991, *Astrophys. J.* **380**, L89
Mandrini CH, Démoulin P, Hénoux J-C, and Machado ME: 1991, *Astron. Astrophys.* **250**, 541
Mandrini C, Rovira MG, Hénoux J-C, Machado ME and Wilkinson LK: 1993, *Astron. Astrophys.* **272**, 609
Martens PCH, Golub L, and Herant M: 1990, in "Plasma Phenomena in the Solar Atmosphere", Dubois, MA and Bely-Dubau F (eds.), 153
Melrose DB: 1992, *Astrophys. J.* **387**, 403
Ogawara Y, Takano T, Kato T, Kosugi T, Tsuneta S, Watanabe T, Kondo I, and Uchida Y: 1991, *Solar Phys.* **136**, 1
Parker EN: 1993, *Astrophys. J.* **407**, 342
Phillips KJH: 1991, *Vistas in Astronomy* **34**, 353
Rust DM: 1992, *Adv. Space Res.* **12**, No. 2-3, 289
Sakurai T: 1991, *Adv. Space Res.* **11**, No. 5, 17
Seely JF and Feldman U: 1992, *Astrophys. J.* **394**, 697
Serio S, Reale F, Jakimiec J, Sylwester B, and Sylwester J: 1991, *Astron. Astrophys.* **241**, 197
Shimizu T, Tsuneta S, Acton LW, Lemen JR, and Uchida Y: 1992, *Publ. Astron. Soc. Japan* **44**, L147
Strong KT, Harvey K, Hirayama T, Nitta N, Shimizu T, and Tsuneta S: 1992, *Publ. Astron. Soc. Japan* **44**, L95
Svestka Z: 1991, in Ref. VI, 165
Sylwester B, Sylwester J, Serio S, Bentley RD, and Fludra A: 1993, *Astron. Astrophys.* **267**, 586
Takakura T: 1992, *Solar Phys.* **142**, 327
Tsuneta S, Hara H, Shimizu T, Acton LW, Strong KT, Hudson HS and Ogawara Y:1992a, *Publ. Astron. Soc. Japan* **44**, L63
Tsuneta S, Takahashi T, Acton L, Bruner M, Harvey K, and Ogawara Y:1992b, *Publ. Astron. Soc. Japan* **44**, L211
Uchida Y and Shibata K: 1988, *Solar Phys.* **116**, 291
Wülser J-P, Zarro DM, and Canfield RC: 1992, *Astrophys. J.* **384**, 341
Zaitsev VV and Stepanov AV: 1992, *Solar Phys.* **139**, 343
Zarro DM: 1992, in Ref. VIII, 96
Zirker JB and Cleveland FM: 1993, *Solar Phys.* **145**, 119

12. PARTICLE ACCELERATION IN FLARES
(Arnold O. Benz)

The 3-year period covered by this report has seen major conferences devoted entirely to acceleration phenomena in astronomy. The Bartol conference (Zank & Gaisser 1992) and the IAU Colloquium 142 (Chupp & Benz 1993) on this general topic had substantial sessions on particle acceleration in solar flares. Solar flares have become a laboratory for acceleration processes in the Universe. New solar observing facilities put into operation during the period include the Yohkoh satellite (Kosugi *et al.* 1991) with an imaging hard X-ray telescope, the X-ray and γ-ray spectrometers on board the Compton Gamma Ray Observatory (Fishman *et al.* 1992), and Granat (Talon *et al.* 1993). New imaging solar radio interferometers in Nobeyama (Enome *et al.* 1993) and Siberia (Alissandrakis *et al.* 1992) have seen 'first light', and the BIMA (Kundu *et al.* 1990) and Itapetinga (Herrmann *et al.* 1992) millimeter telescopes have become operational. A broadband radio spectrometer, Phoenix (Benz *et al.* 1991), has been finished near Zürich.

A general overview on acceleration processes in coronal plasmas has appeared in Benz (1993); more detailed reviews on specific theoretical ideas have e.g. been published by Melrose (1990) on bulk energisation of electrons, by Reames (1990) on heavy ion acceleration, and in the proceedings of the above conferences.

12.1 ROLE OF ACCELERATION IN FLARES

The role of acceleration in the energy release during flares has been studied in emissions of energetic (high temperature or non-thermal) particles and thermal particles, respectively. The correlation observed in some solar flares between hard X-rays (bremsstrahlung of energetic electrons) and the time-integrated soft X-rays (thermal) has been called the Neupert effect (Hudson 1991). It supports a flare model in which the primary energy is initially released into energetic electrons at a high (or at least constant) efficiency. The electrons lose their energy in the lower corona and chromosphere emitting hard X-rays and heating the plasma to radiate enhanced thermal soft X-rays. Feldman (1990) has questioned a causal relationship in view of non-impulsive flares where the hard X-ray emission varies more gradually and the Neupert effect is often absent. Dennis & Zarro (1993) have used the large data base of X-ray flares observed on board the SMM (Dennis *et al.* 1991) and GOES (Speich *et al.* 1991) satellites to determine the fraction of flares that show the Neupert effect. They find that 80% of the large flares (peak count rate > 1000 counts s^{-1}) show good correlation.

12.2 FRAGMENTATION OF ACCELERATION

Hamilton *et al.* (1994) have shown that the peak intensity distribution of hard X-ray events associated with a type III radio burst is significantly different from the distribution of all hard X-ray events. This is inconsistent with the statistical independence of the two emissions and suggests that they are caused by electrons drawn from the same population. Aschwanden *et al.* (1990) presented observations of a flare that shows a linear correlation between the type III burst rate and the hard X-ray flux. The occurrence of up to ten type III bursts per second, each representing an independent electron beam, suggests a high degree of fragmentation in the acceleration region and an approximately constant rate of electrons per type III burst. Sawant *et al.* (1990) find a one-to-one correlation in some reverse drifting type III bursts (corresponding to downward moving beams) and hard X-ray peaks. Narrowband radio spikes of a few tens of millisecond duration occur during some flares and show an order of magnitude higher fragmentation than type III bursts. Spikes in the 0.3-3 GHz range have been found to be well associated with an enhanced hard X-ray flux and are usually correlated with a peak in hard X-rays (Güdel *et al.* 1991). Aschwanden & Güdel (1992) find an average delay of the radio emission of the order of 1s, decreasing with increasing flare importance. The spikes have been a target of intensive study during the period (Güdel, 1990, Bruggmann *et al.*, 1990, Güdel & Zlobec, 1991, Benz *et al.* 1992, Csillaghy & Benz 1993), and have inspired new interpretations of the emission process by Melrose (1991), Tajima *et al.* (1991) and Wentzel (1993). Yet, the nature of narrowband spikes remains controversial and is an interesting field for future research.

12.3 ACCELERATION SITE

Matsushita *et al.* (1992) observed a decrease of the average height of hard X-ray sources with increasing photon energy. At higher energies, the sources become more compact and patchy. These observations are compatible with an acceleration height near the top of the loop. Chupp *et al.* (1993) examined the evolution of electron acceleration (as observed by the Nançay Radioheliograph at meter waves) and ion acceleration (GRS on SMM in γ-ray lines). Gamma-rays have been detected before the main peak of the flare. As new radio sources appear at different locations, the spectra of hard X-rays and γ-ray lines change, reflecting changes in the electron and ion acceleration process.

12.4 THEORY AND MODELS

The ideas on particle acceleration in solar flares may be divided into three groups, each having many variants: (i) acceleration by constant or slowly varying electric fields parallel to the magnetic field, (ii) interaction of particles with waves may stochastically increase the particles' energy, (iii) acceleration by shock waves. The period of this report has seen the first attempts at modelling specific observations, although the controversy over the energisation of the hard X-ray emitting electrons continues.

Schindler *et al.* (1991) have related magnetic field-aligned electric potentials in non-ideal plasma flows to magnetic field properties. Holman & Benka (1992) have explored the consequences of electric field acceleration and heating in current channels. For electrons, Joule heating by the current and the acceleration of runaway particles are both crucially important. The radiative consequences are seen in microwave gyrosynchrotron radiation and the hard X-ray emissions from flares. In hard X-rays, the presence of a superhot thermal component at low photon energies, together with nonthermal emission at higher energies, is a natural consequence of the presence of currents. Tajima *et al.* (1991) have simulated acceleration by electric fields using a particle code. They find that the acceleration process is accompanied by strong, narrowband radio emission. Karlicky (1993) simulated the propagation of an electron beam in a loop and finds acceleration of back-scattered electrons by the electric field of the return current.

An alternative acceleration process for hard X-ray emitting electrons has been modelled by LaRosa & Moore (1993). They suggest the dissipation of MHD turbulence in solar flares in order to accelerate electrons by the second-order Fermi process. Each energy-release fragment (possibly visible as a radio spike) can be produced by an MHD turbulent cascade. Smith & Brecht (1993) have modelled proton acceleration by MHD waves in a loop geometry and, including the related proton transport, find it possible to reach the observed short duration of γ-ray bursts.

The observed fragmentation of the energy release has been taken up by Kliem (1990) who modelled particles acceleration by a multitude of current sheets. He has also found that a perpendicular electric field can accelerate electrons efficiently. Alternatively, Anastasiadis and Vlahos (1993) have studied the effect of a large number of randomly propagating shocks on the evolution of the energy distribution of electrons.

With only a small fraction of the observations of the passed maximum analyzed, it is clear that progress in the field of particle acceleration will continue. High resolution imaging and spectral hard X-ray telescopes and dedicated solar radio instruments will be needed to make further progress, but will have to wait for the next maximum.

REFERENCES

Alissandrakis CE, Lubyshev BI, Smolkov GYa, Krissinel BB, Treshkov TA, Miller VG and Kardapolova NN: 1992, *Solar Phys.* **142**, 341
Anastasiadis A and Vlahos L: 1991, *Astron. Astrophys.* **245**, 271
Aschwanden MJ, Benz AO, Schwartz RA, Lin RP, and Pointon L: 1990, *Solar Phys.* **130**, 39
Aschwanden MJ and Güdel MJ: 1992, *Astrophys. J.* **401**, 736
Benka S and Holman G:1992, *Astrophys. J.* **391**, 854
Benz AO: 1993, "Plasma Astrophysics, Kinetic Processes in Solar and Stellar Coronae", Kluwer (Dordrecht) .
Benz AO, Güdel M, Isliker H, Miszkowicz S, and Stehling W: 1991, *Solar Phys.* **133**, 385
Benz AO, Su H, Magun A, and Stehling W: 1992, *Astron. Astrophys. Suppl. Ser.* **93**, 539
Bruggmann G, Benz AO, Magun A, and Stehling W: 1990, *Astron. Astrophys.* **240**, 506
Chupp EL and Benz AO: 1993, *Astrophys. J. Suppl.* **89**
Chupp EL, Trottet G, Marschhäuser H, Pick M, Soru-Escaut I, Rieger E, and Dunphy PP: 1993, *Astron. Astrophys.* **275**, 602
Csillaghy A and Benz AO: 1993, *Astron. Astrophys.* **274**, 487
Dennis BR and Zarro DM: 1993, *Solar Phys.* **146**, 177
Dennis BF, Orwig LE, Kennard GS, Labow GJ, Schwartz RA, Shaver AR, and Tolbert AK: 1991, NASA TM-4332
Enome S, Nakajima H, Shibasaki K, Nishio M, Takano T, Hanaoka Y, Torii C, Sekiguchi H,
Kawashima S, Bushimata T, Shinohara N,Irimajiri Y, Koshiishi H, Kosugi T, Shiomi S, Sawa M, and Kai K: 1993, APS Conf.Series **46**, 310
Feldman U: 1990, *Astrophys. J.* **364**, 322
Fishman G, Meegan C, Wilson, R, Paciesas W, and Pendleton G: 1992, NASA CP-3137, 26
Güdel M: 1990, *Astron. Astophys.* **239**, L1
Güdel M and Zlobec P: 1991, *Astron. Astrophys.* **245**, 299
Güdel M, Aschwanden MJ, and Benz AO: 1991, *Astron. Astrophys.* **251**, 285
Hamilton RJ, Petrosian V, and Benz AO: 1990, *Astrophys. J.* **358**, 644
Herrmann R, Magun A, Costa JER, Correia E, Kaufmann P: 1992, *Solar Phys.* **142**, 157
Holman G and Benka S: 1992, *Astrophys. J.* **400**, L79
Hudson HS: 1991, *Bull. Am. Astron. Soc.* **23**, 1064
Karlicky M: 1993, *Solar Phys.* **145**, 137
Kliem B: 1990, *Astron. Nachr.* **311**, 399
Kosugi T, Makishima K, Murakami T, Sakaom T, Dotani,T, Inda M, Kai K, Msuda S, Nakajima H, Ogawara Y, Sawa M, and Shibasaki K: 1991, *Solar Phys.* **136**, 17
Kundu MR, White SM, Gopalswamy N, Bieging JH and Hurford GJ: 1990, *Astrophys. J.* **358**, L69
La Rosa TN and Moore RL: 1993, *Astrophys. J.* **411**, 617

Matsushita K, Masuda S, Kosugi T, Inda M, and Yaji K: 1992, *Publ. Astron. Soc. Japan* **44**, L89
Melrose DB: 1991, *Astrophys. J.* **380**, 256
Sawant HS, Lattari CJB, Benz AO and Dennis BR: 1990, *Solar Phys.* **130**, 57
Schindler K, Hesse M, and Birn J: 1991, *Astrophys. J.* **380**, 293
Smith DF and Brecht SH: 1993, *Astrophys. J.* **406**, 298
Speich DM, Nelson JJ, Licata JP, and Tolbert AK: 1991, NASA TM-4287
Talon R, Trottet G, Vilmer N, Barat C, Dezalay JP, Sunyaev R, Terekhov O, and Kuznetstov A: 1993, *Solar Phys.* **147**, 137
Tajima T, Benz AO, Thaler M, and Leboeuf JN: 1991, *Astrophys. J.* **353**, 666
Wentzel DG: 1993, *Astrophys. J.* **407**, 380
Zank GP and Gaisser TK: 1992, "Particle Acceleration in Cosmic Plasmas," AIP Conf.Proc. **264**, Am. Inst. of Physics (New York)

COMMISSION 12: SOLAR RADIATION AND STRUCTURE (RADIATION ET STRUCTURE SOLAIRES)

PRESIDENT: J.O. Stenflo
VICE PRESIDENT: F.L. Deubner
ORGANIZING COMMITTEE: T. Ayres, J.W. Harvey, V. Karpinsky, E. Landi Degl'Innocenti, Å. Nordlund, K.R. Sivaraman, J. Staude, Wang Jing-Xiu, P.R. Wilson, H. Yoshimura

1. INTRODUCTION (J.O. Stenflo)

At the XXIst IAU General Assembly in Buenos Aires in 1991 the name of Commission 12 was changed from "Radiation and Structure of the Solar Atmosphere" to "Solar Radiation and Structure". The new name better represents our interests, since we are dealing not only with the atmosphere but with the whole sun. An example is the growing field of helioseismology, which during the past decade has played a prominent role in the life of our commission.

As usual the topics for the present report have been selected in coordination with Commission 10 (Solar Activity), to make the two reports as complementary as possible. Due to space limitations we of course cannot attempt to cover all important areas of solar physics. Instead Commission 12 has limited itself to cover six topics of particular current interest, in the form of brief but critical reviews documenting progress made over the past three years (1 July 1990 – 30 June 1993).

This scientific report may in fact be the last one made by Commission 12 as a separate commission. The IAU Executive Committee is currently considering a general overhaul of the now largely outdated commission structure of the union. Through coordinated mergers of previous commissions a much smaller set of new commissions with redefined goals should result. After various consultations within and between the solar-oriented commissions a proposal has been made to merge (in the context of other concerted mergers) Commissions 10, 12, and 49 (The Interplanetary Plasma and the Heliosphere) into a single, new solar commission with the name "The Sun and the Heliosphere". This would allow us to deal with the solar issues in a more unified way.

I would like to thank the authors of this report for their excellent reviews in spite of the severe page limit constraints. I am also grateful to the members of the Organizing Committee and all commission members who have provided feedback and support in dealing with the various issues.

J. Bergeron (ed.), Reports on Astronomy, Vol. XXIIA, 85–104.

2. HELIOSEISMOLOGY (J. Toomre)

The oscillations of the sun are being used to study its internal structure and dynamics. A substantial component of the velocity and intensity fluctuations observed in the solar atmosphere result from the interference between about 10^7 resonant modes of oscillation of the interior. The frequency of any given mode is determined mainly by the stratification and dynamics within that portion of the spherical shell where the mode is resonant. Accurate measurement of very large sets of mode frequencies can be used as inputs to inverse theory to obtain detailed deductions about flows and inhomogeneities over a range of depths and latitudes in the solar interior.

From the outset a hallmark of helioseismology has been the close interplay between observations and theory, and this has contributed to rapid advances in the subject in the past three years. This is clearly evident in the recent major conference proceedings dealing with seismic investigations of the sun and stars edited by Berthomieu & Cribier (1990), Osaki & Shibahashi (1990), Gough & Toomre (1991a), Weiss & Baglin (1993) and Brown (1993), and in the large tome assessing advances in solar physics (Cox, Livingston & Matthews 1991). Recent accomplishments in helioseismology, and fundamental issues that can only be addressed by the several new international observational programs now being implemented, are discussed by Gough & Toomre (1991b).

2.1 Ground-Based Networks and Instruments in Space

The central observational issue in helioseismology is how to obtain nearly uninterrupted observations of long duration so that oscillation mode frequencies and their splittings can be determined to high accuracy. Intense efforts have been made in the past few years to addressing this issue, leading to the development of three ground-based networks of observing stations that span the globe, thus mitigating the effects of sidelobes in temporal power spectra (e.g. Hill, Deubner & Isaak 1991). The most ambitious of these is the Global Oscillations Network Group (GONG) project, which has been working to place identical Doppler imaging instruments using solid Michelson interferometers at six sites around the world to study intermediate- and high-degree modes with $\ell \leq 250$ (Harvey *et al.* 1993). Prototype testing is nearly completed, and the GONG stations should become operational in late 1994. Accurate frequencies of low-degree modes are being provided by whole-disk observations from two other ground-based networks called IRIS and BISON (Schmider *et al.* 1990, Elsworth *et al.* 1990).

The ground-based observations must be complemented by space-based instruments in order to remove effects of atmospheric seeing distortions that limit high spatial resolution, and thus major effort has also gone into building these instruments in the past three years. The study of detailed dynamics in the convection zone requires access to very high-degree modes (ℓ up to 4000) that will be provided by the Solar Oscillations Investigation (SOI) using a Michelson Doppler imaging instrument. It will fly on the *SOHO* spacecraft, to be placed at the L_1 Lagrangian point between the earth and the sun (Scherrer, Hoeksema & Bush 1991). The continuous sunlight at L_1 will also be exploited by low-degree helioseismic observations to be conducted with the full-disk atomic resonance scattering instrument (GOLF) (Gabriel *et al.* 1991) and the active cavity radiometer together with photometers (VIRGO) (Andersen 1991). The *SOHO* spacecraft is expected to arrive on station in 1995.

2.2 Theoretical Progress

There have also been substantial advances in many theoretical components that bear on helioseismology, with major refinements in the equation of state and opacity that influence directly the accuracy of solar reference models (e.g. Däppen, Keady & Rogers 1991, Ulrich & Cox 1991, Christensen-Dalsgaard & Däppen 1992, Rogers & Inglesias 1993, Däppen 1993, Seaton 1993), in assessing the sensitivity of mode eigenfunctions and frequencies to changes in solar models (e.g. Christensen-Dalsgaard & Berthomieu 1991), in methods of inversion to deduce mean structure, asphericity and differential rotation (e.g. Christensen-Dalsgaard, Schou & Thompson 1990, Gough & Thompson 1991, Kosovichev *et al.* 1992, Schou, Christensen-Dalsgaard & Thompson 1993, Thompson 1993, Sekii 1993), and in evaluating mechanisms for the excitation and decay of the acoustic modes (e.g. Balmforth & Gough 1990, Osaki 1990, Goldreich & Kumar 1991, Cox *et al.* 1991, Balmforth 1992, Murray 1993). The anticipated arrival of voluminous data sets from GONG and SOI has led to extensive data analysis preparation on diverse aspects such as implementing two-dimensional inversions, devising peak fitting procedures for power spectra with mode crosstalk, and studying the implications of gap filling in time series, with the work carried out by collaborative teams of scientists (e.g. Anderson, Duvall & Jefferies 1990, Brown & Christensen-Dalsgaard 1990, Gough & Toomre 1991c, Harvey *et al.* 1993, Bogart *et al.* 1993).

2.3 Selected Research Highlights from Observations and their Interpretation

The diagnostic potential of helioseismology has been affirmed recently in a number of novel ways, and there have been notable observational discoveries concerning properties of the oscillations. The depth of the solar convection zone has been determined from helioseismic measurement of the gradient in sound speed associated with the transition in the temperature gradient from being adiabatic within the zone to subadiabatic in the stable stratification below. That depth was deduced to be 0.287 solar radii (Christensen-Dalsgaard, Gough & Thompson 1991). The helium abundance within the solar convection zone has been assessed by helioseismic measurements that respond to the variation of the adiabatic compressibility of the material in the second helium ionization zone. The uncertainties in deducing the helium abundance by such methods are substantially less than by any other available observational method (Kosovichev *et al.* 1992). Using the most sophisticated equation of state available, based on the computations of Mihalas, Däppen & Hummer (1988), a value $Y \simeq 0.23$ was found; this is considerably less than the values obtained from stellar evolution calculations with the best estimates of Z/X, suggesting there has been substantial gravitational settling of helium.

Stratification within the deep solar interior has been evaluated by inversion methods applied to both low-degree p-mode frequency data of whole-disk measurements from the ground (Elsworth *et al.* 1991, Anguera Gubau *et al.* 1992) and from the IPHIR space experiment (Toutain & Fröhlich 1992), joined with the intermediate-ℓ data (Libbrecht, Woodard & Kaufman 1990). The structure inversions by Gough & Kosovichev (1993) indicate that the stratification in the solar radiative envelope is similar to that of a solar model which accounts for helium settling against microscopic diffusion in the absence of turbulent mixing (Christensen-Dalsgaard, Proffitt & Thompson 1993). Although there is still a paucity of data, the inversions suggest that within the energy-generating core the spherically-symmetric hydrostatic structure is largely in accord with that of standard models. This appears to rule out the WIMP models offered to reduce the flux of higher-energy neutrinos from the core. The recent GALLEX and SAGE experimental counts are nearly consistent with standard-model predictions of the lower-energy neutrinos produced by the main $p-p$ chain, leaving a continuing puzzle to explain the apparent low flux of higher-energy neutrinos (e.g. Gough 1993). Though the answer may lie in particle physics, such as neutrino matter-induced transitions, it may also be influenced by local material redistribution in the solar core through dynamical processes, the consequences of which on the stratification in the core should be testable with the low-degree helioseismic data from VIRGO and GOLF.

A clear consensus has emerged that there is a systematic shift in oscillation frequencies associated with changes in solar magnetic activity, increasing by as much as 0.4 μHz from solar activity minimum to maximum (e.g. Elsworth *et al.* 1990, Libbrecht & Woodard 1990, Anguera Gubau *et al.* 1992, Bachmann & Brown 1993). The oscillation frequencies can show shifts over even monthly intervals as the solar activity changes (Woodard *et al.* 1991). The signature of the frequency shifts suggests that the oscillations are sensing changes in conditions near their upper turning point just below the solar surface (e.g. Gough 1990, Kuhn 1993). Rotational frequency splitting also shows temporal variations of statistical significance, indicating that there are small changes (about 1%) in the sun's subsurface angular velocity with the solar cycle, the rotation rate at the higher latitudes having the greater variation (e.g. Woodard & Libbrecht 1993a,b, Gough *et al.* 1993a).

Another recent observational discovery concerns acoustic oscillations of intermediate and high degree that extend as discernible ridges in power spectra to frequencies as great as 10 mHz, well above the acoustic cutoff frequency of about 5.3 mHz for the photosphere (e.g. Duvall *et al.* 1991, Woodard & Libbrecht 1991, Harvey 1991, Fernandes *et al.* 1992, Ronan & LaBonte 1993, Milford *et al.* 1993). Further, the long-noted 3-min chromospheric oscillations, though elusive in helioseismic data, have now been observed to form a broad background feature that underlies the extended ridges at the higher frequencies (Harvey *et al.* 1993). It has been proposed that such high-frequency ridges may result from interference patterns of initially upward and downward-directed waves emitted from subphotospheric sources near the top of the convection zone (e.g. Kumar *et al.* 1990, Brown 1991, Kumar & Lu 1991, Kumar 1993), or from a chromospheric cavity which couples to the deeper acoustic cavity to achieve the resonances (e.g. Balmforth & Gough 1990). The intriguing possibility of directly measuring travel times and distances of individual acoustic waves, now demonstrated using temporal cross-correlations of intensity fluctuations on the solar surface (Duvall *et al.* 1993), may permit detailed studies of local subsurface structures, such as near sunspots. Localized analysis of wave fields has revealed that there is apparent absorption and scattering of p modes both by sunspots and their surrounding active regions (e.g. Braun, LaBonte & Duvall 1990, Braun *et al.* 1992, Braun *et al.* 1993), and this may provide tools for detecting magnetic structures before they emerge through the surface.

The solar differential rotation profile inferred seismologically (e.g. Brown *et al.* 1989, Dziembowski *et al.* 1989, Schou 1991, Sekii 1991, Gough *et al.* 1993b) is still strikingly at variance with predictions based on global simulations of solar convection (e.g. Glatzmaier 1987). Convection models generally favor an angular velocity Ω which is nearly constant on cylinders aligned with the rotation axis, whereas seismic inversions suggest that within the convection zone Ω is more nearly constant on radial lines. However, the resolution provided by current data is poor. Analogy with the differential rotation of the giant planets Jupiter and Saturn suggests that the sun may also possess far more structured zonal flows, and there is some hint of this from local-area ring diagram analyses by Hill (1990), Hill *et al.* (1991), and Patrón *et al.* (1993). The high-resolution data from GONG and SOI will no doubt reveal more subtle details of the dynamics of the convection zone.

REFERENCES

Andersen, B.N. 1991, *Adv. Space Res.*, **11**, 93.
Anderson, B.N., Duvall, D.L., Jr., Jefferies, S.M. 1990, *Ap. J.*, **364**, 699.
Anguera Gubau, M., Pallé, P.L., Pérez Hernández, F., Régulo, C., Roca Cortés, T. 1992, *Astr. Ap.*, **255**, 363.
Bachmann, K.T., Brown, T.M. 1993, *Ap. J. (Letters)*, **411**, L45.
Balmforth, N.J. 1992, *M.N.R.A.S.*, **255**, 639.
Balmforth, N.J., Gough, D.O. 1990, *Ap. J.*, **362**, 256.
Berthomieu, G., Cribier, M., eds. 1990, *Inside the Sun, IAU Colloq. No. 121*, Kluwer, Dordrecht, 518 pp., = VERSAILLES.
Bogart, R.S., Hill, F., Toussaint, R.,, Hathaway, D.H., Duvall, T.L., Jr. 1993, in BOULDER, pp. 429-432.
Braun, D.C., Duvall, T.L., Jr., LaBonte, B.J., Jefferies, S.M., Harvey, J.W., Pomerantz, M.A. 1992, *Ap. J. (Letters)*, **391**, L113.
Braun, D.C., LaBonte, B.J., Duvall, T.L., Jr. 1990, *Ap. J.*, **354**, 372.
Braun, D.C., LaBonte, B.J., Duvall, T.L., Jr., Jefferies, S.M., Pomerantz, M.A., Harvey, J.W. 1993, in BOULDER, pp. 77-80.
Brown, T.M. 1991, *Ap. J.*, **371**, 396.
Brown, T.M., ed. 1993, *GONG 1992: Seismic Investigation of the Sun and Stars, ASP Conf. Ser.*, vol. **42**, Astron. Soc. Pacific, San Francisco, 495 pp., = BOULDER.
Brown, T.M., Christensen-Dalsgaard, J. 1990, *Ap. J.*, **349**, 667.
Brown, T.M., Christensen-Dalsgaard, J., Dziembowski, W.A., Goode, P., Gough, D.O., Morrow, C.A. 1989, *Ap. J.*, **343**, 526.
Christensen-Dalsgaard, J., Berthomieu, G. 1991, in SIA, pp. 401-478.
Christensen-Dalsgaard, J., Däppen, W. 1992, *Astr. Ap. Rev.*, **4**, 267.
Christensen-Dalsgaard, J., Gough, D.O., Thompson, M.J. 1991, *Ap. J.*, **378**, 413.
Christensen-Dalsgaard, J., Proffitt, C.R., Thompson, M.J. 1993, *Ap. J. (Letters)*, **403**, L75.
Christensen-Dalsgaard, J., Schou, J., Thompson, M.J. 1990, *M.N.R.A.S.*, **242**, 353.
Cox, A.N., Chitre, S.M., Frandsen, S., Kumar, P. 1991, in SIA, pp. 618-660.
Cox, A.N., Livingston, W.C., Matthews, M.S., eds. 1991, *Solar Interior and Atmosphere*, Univ. Arizona Press, Tucson, 1416 pp., = SIA.
Däppen, W. 1993, in VIENNA, pp. 208-221.
Däppen, W., Keady, J., Rogers, F. 1991, in SIA, pp. 112-139.
Duvall, T.L., Jr., Harvey, J.W., Jefferies, S.M., Pomerantz, M.A. 1991, *Ap. J.*, **373**, 308.
Duvall, T.L., Jr., Harvey, J.W., Jefferies, S.M., Pomerantz, M.A. 1993, *Nature*, **362**, 430.
Dziembowski, W.A., Goode, P.R., Libbrecht, K.G. 1989, *Ap. J.*, **337**, L53.
Elsworth, Y., Howe, R., Isaak, G.R., McLeod, C.P., New, R. 1990, *Nature*, **345**, 322.
Elsworth, Y., Howe, R., Isaak, G.R., McLeod, C.P., New, R. 1991, *M.N.R.A.S.*, **251**, 7p.

Fernandes, D.N., Scherrer, P.H., Tarbell, T.D., Title, A.M. 1992, *Ap. J.*, **392**, 736.
Gabriel, A.H., and the GOLF team, 1991, *Adv. Space Res.*, **11**, 103.
Glatzmaier, G.A. 1987, in *The Internal Solar Angular Velocity*, eds. B.R. Durney, S. Sofia, Reidel, Dordrecht, pp. 263-274.
Goldreich, P., Kumar, P. 1991, *Ap. J.*, **374**, 366.
Gough, D.O. 1990, *Nature*, **345**, 768.
Gough, D.O. 1993, *Phil. Trans.*, in press.
Gough, D.O., Kosovichev, A.G. 1993, *M.N.R.A.S.*, in press.
Gough, D.O., Kosovichev, A.G., Sekii, T., Libbrecht, K.G., Woodard, M.F. 1993a, in VIENNA, pp. 93-96.
Gough, D.O., Kosovichev, A.G., Sekii, T., Libbrecht, K.G., Woodard, M.F. 1993b, in BOULDER, pp. 213-216.
Gough, D.O., Thompson, M.J. 1991, in SIA, pp. 519-561.
Gough, D.O., Toomre, J., eds. 1991a, *Challenges to Theories of the Structure of Moderate-Mass Stars*, *Lec. Notes Phys.*, vol. **399**, Springer, Heidelberg, 414 pp., = ITP.
Gough, D.O., Toomre, J., 1991b, *Ann. Rev. Astr. Ap.*, **29**, 627.
Gough, D.O., Toomre, J. 1991c, in ITP, pp. 1-9.
Harvey, J.W., 1991, in HAKONE, pp. 115-128.
Harvey, J., Hill, F., Kennedy, J., Leibacher, J. 1993, in BOULDER, pp. 397-409.
Hill, F. 1990, *Solar Phys.*, **128**, 321.
Hill, F., Rhodes, E.J., Jr., Korzennik, S.G., Cacciani, A., Brown, T.M. 1991, in ITP, pp. 271-275.
Hill, F., Deubner, F.-L., Isaak, G. 1991, in SIA, pp. 329-400.
Kosovichev, A.G., Christensen-Dalsgaard, J., Däppen, W., Dziembowski, W.A., Gough, D.O., Thompson, M.J., 1992, *M.N.R.A.S.*, **259**, 536.
Kuhn, J.R. 1993, in BOULDER, pp. 27-39.
Kumar, P. 1993, in BOULDER, pp. 15-26.
Kumar, P., Duvall, T.L., Jr., Harvey, J.W., Jefferies, S.M., Pomerantz, M.A., Thompson, M.J. 1990, in HAKONE, pp. 87-92.
Kumar, P., Lu, E. 1991, *Ap. J. (Letters)*, **375**, L35.
Libbrecht, K.G., Woodard, M.F. 1990, *Nature*, **345**, 779.
Libbrecht, K.G., Woodard, M.F., Kaufman, J.M. 1990, *Ap. J. Suppl.*, **74**, 1129.
Mihalas, D., Däppen, W., Hummer, D.G. 1988, *Ap. J.*, **331**, 815.
Milford, P.N., Scherrer, P.H., Frank, Z., Kosovichev, A.G., Gough, D.O. 1993, in BOULDER, pp. 97-100.
Murray, N. 1993, in BOULDER, pp. 3-14.
Osaki, Y. 1990, in HAKONE, pp. 75-86.
Osaki, Y., Shibahashi, H., eds. 1990, *Progress of Seismology of the Sun and Stars*, *Lec. Notes Phys.*, vol. **367**, Springer, Heidelberg, 467 pp., = HAKONE.
Patrón, J., Hill, F., Rhodes, E.J., Jr., Korzennik, S.G., Cacciani, A., Brown, T.M. 1993, in BOULDER, pp. 437-440.
Rogers, F.J., Inglesias, C.A. 1993, in BOULDER, pp. 155-167.
Ronan, R.S., LaBonte, B.J. 1993, in BOULDER, pp. 93-96.
Scherrer, P.H., Hoeksema, J.T., Bush, R.I. 1991, *Adv. Space Res.*, **11**, 113.
Schmider, F.-X., Fossat, E., Gelly, B., Grec, G. 1990, in HAKONE, pp. 241-251.
Schou, J. 1991, in ITP, pp. 81-86.
Schou, J., Christensen-Dalsgaard, J., Thompson, M.J., 1993, in VIENNA, pp. 72-74.
Seaton, M.J. 1993, in VIENNA, pp. 222-235.
Sekii, T. 1991, *Pub. A.S. Japan*, **43**, 381.
Sekii, T. 1993, in BOULDER, pp. 237-240.
Thompson, M.J. 1993, in BOULDER, pp. 141-154.
Toutain, T., Fröhlich, C. 1992, *Astr. Ap.*, **257**, 287.
Ulrich, R.K., Cox, A.N. 1991, in SIA, pp. 162-191.
Weiss, W.W., Baglin, A., eds. 1993, *Inside the Stars*, *IAU Colloq. 137*, *ASP Conf. Ser.*, vol. **40**, Astron. Soc. Pacific, San Francisco, 882 pp., = VIENNA.
Woodard, M.F., Kuhn, J.R., Murray, N., Libbrecht, K.G. 1991, *Ap. J. (Letters)*, **373**, L81.
Woodard, M.F., Libbrecht, K.G. 1991, *Ap. J. (Letters)*, **374**, L61.
Woodard, M.F., Libbrecht, K.G. 1993a, *Ap. J. (Letters)*, **402**, L77.
Woodard, M.F., Libbrecht, K.G. 1993b, *Science*, **260**, 1778.

3. MAGNETOCONVECTION (M. Schüssler)

The domain of magnetoconvection embraces all kinds of interaction between convective flows and magnetic fields in an electrically conducting fluid. The origin of the subject lies in the search for an explanation for the darkness of sunspots in the 1930s although the word 'magnetoconvection' was coined only much later by W.V.R. Malkus (1959). Apart from its astrophysical implications the subject has developed a life of its own as an example of structure formation in dissipative dynamical systems. The relevance of magnetoconvection to solar physics is obvious: for investigating processes like the solar dynamo mechanism in the convection zone, the distribution and physical properties of magnetic flux elements in the observable layers of the photosphere, and the structure and dynamics of sunspot umbrae and penumbrae, we need to understand the interaction between convection and magnetic fields. At present, three 'mainstream' lines of research in magnetoconvection can be identified:

a) analytical studies of linear problems and nonlinear dynamics of restricted or truncated systems,
b) numerical experiments for idealized problems,
c) comprehensive simulations.

In what follows these approaches will be discussed briefly. Emphasis will be laid on more recent developments and on work which has implications for the magnetic structure in the convection zone and photosphere of the Sun. More comprehensive reviews have been given by Proctor & Weiss (1982), Hughes & Proctor 1988, Weiss (1990, 1991), and by Proctor (1992).

3.1 Linear analysis and nonlinear dynamics

Historically, this line of research began with the linear analysis of convective instability of a Boussinesq fluid in the presence of a magnetic field (for a summary see the book by Chandrasekhar 1961). Linear theory and simple nonlinear models for compressible convection in oblique magnetic fields have been investigated recently by Matthews et al. (1992) who found that the excitation of travelling waves is a general feature in such a situation. Weakly nonlinear theory has been utilized (e.g. Jones et al. 1990; Matthews & Rucklidge 1993) and methods for the analysis of dynamical systems have been applied to determine nonlinear bifurcation structures (e.g. Knobloch et al. 1992; Rucklidge et al. 1993) and the onset of chaos in magnetoconvective flows (Rucklidge 1992, 1993). Kerswell & Childress (1992) have provided an analytical model for the equilibrium structure of a magnetic flux tube sustained by a convective cell in a compressible medium with possible application to solar photospheric magnetic structures.

3.2 Numerical experiments

This approach was initiated by the work of Weiss (1966) on the kinematic expulsion and concentration of magnetic fields by prescribed velocity patterns. Numerical experiments are not intended to directly simulate or represent real (astro-)physical systems; they are used as a tool to investigate basic physical processes which are beyond the limit of analytical tractability. Therefore, complications like radiative energy transport are ignored and simple boundary conditions are used. The results, however, may still have direct implications for astrophysical problems insofar as important processes can be identified.

Oscillations, pulsating and travelling waves in two-dimensional compressible convection with an imposed *horizontal* magnetic field have been found by a number of authors (Brownjohn et al. 1993; Lantz & Sudan 1993; see also Weiss 1991). These results may possibly apply to sunspot penumbrae and the bottom layers of the solar convection zone while the case of an imposed *vertical* field may be relevant for sunspot umbræ (Hurlburt et al. 1989; Weiss et al. 1990; Proctor et al. 1993) and for the formation of concentrated magnetic fields in the photosphere (Hurlburt & Toomre 1988; Fox et al. 1991). Calculations with both kinds of imposed field, vertical and horizontal, have been performed by Fox et al. (1991) and by Hanami & Tajima (1991).

Results of three-dimensional calculations of compressible magnetoconvection in an imposed vertical field have been presented by Matthews (1993). Brandenburg et al. (1990) have studied the case with rotation and determined values for the turbulent α-effect which is important for mean-field dynamo theory. Spontaneous local dynamo action has been found by Meneguzzi & Pouquet (1989) in a three-dimensional simulation of Boussinesq magnetoconvection and by Nordlund et al. (1992) for a compressible medium including a stably stratified overshoot layer. In the latter simulation most of the generated magnetic field appears in the form of isolated magnetic flux tubes associated with strong, rotating downdrafts (vortex tubes). The basic field generation process is akin to the stretch-twist-fold dynamo sequence proposed by Vainshtein & Zeldovich (1972). Saturation is reached when the growing magnetic curvature force inhibits further twisting of the flux tube by the flow. The downflows push the generated field towards the bottom of the computational box and into the overshoot layer where it resides in the form of a mainly horizontal field (Stein et al. 1992). The evolution of such a horizontal magnetic flux tube in two-dimensional penetrative convection has been studied by Jennings et al. (1992). Cattaneo et al. (1991) discuss possible pitfalls in numerical simulations of hydromagnetic dynamos and propose a criterion for dynamo action in such simulations.

3.3 Comprehensive simulations

Although the numerical experiments discussed in the preceding section become increasingly sophisticated and 'realistic' we may still distinguish them from simulations which directly aim at modelling structures in the real solar convection zone and atmosphere. Since comparison with observational data is an important aspect for this approach, diagnostic information like intensity or polarization maps and profiles of spectral lines has to be drawn from the numerical results. A problem for this kind of simulations is that due to limited computer speed and storage capacity the whole range of length scales of the real flow and field cannot be encompassed and 'turbulent' or 'sub-grid' diffusivities (e.g. Canuto 1992; Theobald et al. 1993) have to be employed (see also Cattaneo & Vainshtein 1991). The hydrodynamic and magnetic Reynolds numbers which can be effectively achieved by a simulation are orders of magnitude smaller than those of the real astrophysical system. This is probably one reason for the failure of early attempts to comprehensively simulate the solar convection zone and dynamo. Another complication concerns the boundaries of the computational box which in most cases have to be placed at locations where no physical boundaries are. Therefore, boundary conditions are often formulated to be 'transmitting' and 'open' in order to avoid artificial reflection of waves and stagnation of flows. However, since only information from the inside of the computational domain is available during the simulation, some degree of arbitrariness is necessarily introduced and careful tests concerning the influence of the boundary conditions on the results are necessary (cf. Chan & Serizawa 1991). In any case the validity of a code should be ascertained by extensive series of test calculations. A set of test problems well suited for MHD codes has been proposed by Stones et al. (1992).

Simulations of non-magnetic granular convection near the solar surface including non-grey radiative transfer and sophisticated diagnostics (e.g. Steffen 1989, Lites et al. 1989) have evolved quite successfully in recent years (Stein & Nordlund 1989; Steffen et al., 1989; Steffen & Freytag 1991; Rast et al. 1993). Nordlund & Stein (1989, 1990) have included a magnetic field in their three-dimensional code and performed simulations of a sunspot umbra and a very strong plage region. The umbra simulation shows suppression of convective energy transport interrupted by episodic eruptions of convection. In the plage simulation the magnetic flux is rapidly swept to the convective downflows in the intergranular regions. In accordance with observations (e.g. Title et al. 1992) the granular structure is disturbed by the magnetic field and the velocity fluctuations are decreased with respect to non-magnetic granular convection.

The influence of convective motions on the eruption of an initially horizontal flux sheet in the solar atmosphere due to the Parker instability has been investigated by Kaisig et al. (1990). Structure and dynamics of two-dimensional magnetic sheets as models of magnetic flux elements in the solar (sub)photosphere have been simulated and compared to observational data by Knölker et al. (1991) and by Grossmann-Doerth et al. (1993). A strong influence of the magnetic structures on the surrounding convective flow pattern is found: Radiative cooling of the flux sheet drives an external flow which advects heat towards it and forms a high-velocity downflow immediately adjacent to the magnetic structure.

REFERENCES

Brandenburg, A., Nordlund, Å., Pulkkinen, P., Stein, R.F., Tuominen, I. 1990, *Astr. Ap.* **232**, 277
Brownjohn, D.P., Hurlburt, N.E., Proctor, M.R.E., Weiss, N.O. 1993, *J. Fluid Mech.*, in press
Canuto, V.M. 1992, *Ap. J.* **392**, 218
Cattaneo, F., Hughes, D.W., Weiss, N.O. 1991, *Mon. Not. R. astr. Soc.* **253**, 479
Cattaneo, F., Vainshtein, S.I. 1991, *Ap. J.* **376**, L21

Chan, K.L., Serizawa, K. 1991, in *The Sun and Cool Stars: activity, magnetism, dynamos*, IAU-Kolloq. No. 130, eds. I. Tuominen, D. Moss, G. Rüdiger, Springer, Berlin, p. 15
Chandrasekhar, S. 1961, *Hydrodynamic and Hydromagnetic Stability*, Clarendon Press, Oxford
Fox, P.A., Theobald, M.L., Sofia, S. 1991, *Ap. J.* **383**, 860
Grossmann-Doerth, U., Knölker, M., Schüssler, M., Solanki, S.K. 1993, *Astr. Ap.*, in press
Hanami, H., Tajima, T. 1991, *Ap. J.* **377**, 694
Hughes, D.W., Proctor, M.R.E. 1988, *Ann. Rev. Fluid Mech.* **20**, 187
Hurlburt, N.E., Proctor, M.R.E., Weiss, N.O., Brownjohn, D.P. 1989, *J. Fluid Mech.* **209**, 587
Hurlburt, N.E., Toomre, J. 1988, *Ap. J.* **327**, 920
Jennings, R.L., Brandenburg, A., Nordlund, Å., Stein, R.F. 1992, *Mon. Not. R. astr. Soc.* **259**, 465
Jones, C.A., Roberts, P.H., Galloway, D.J. 1990, *Geophys. Astrophys. Fluid Dynam.* **53**, 145
Kaisig, M., Tajima, T., Shibata, K., Nozawa, S., Matsumoto, R. 1990, *Ap. J.* **358**, 698
Kerswell, R., Childress, S. 1992, *Ap. J.* **385**, 746
Knobloch, E., Proctor, M.R.E., Weiss, N.O. 1992, *J. Fluid Mech.* **239**, 273
Knölker, M., Grossmann-Doerth, U., Schüssler, M., Weisshaar, E. 1991, *Adv. Space Res.* **11**, (5)285
Lantz, S.R., Sudan, R.N. 1993, *Ap. J.*, in press
Lites, B.W., Nordlund, Å., Scharmer, G.B. 1989, in *Solar and Stellar Granulation*, eds. R. J. Rutten, G. Severino, Kluwer, Dordrecht, p. 349
Malkus, W.V.R. 1959, *Ap. J.* **130**, 259
Matthews, P.C. 1993, in *Theory of Solar and Planetary Dynamos*, eds. M.R.E. Proctor, P.C. Matthews, A.M. Rucklidge, Cambridge Univ. Press, Cambridge, p. 211
Matthews, P.C., Hurlburt, N.E., Proctor, M.R.E., Brownjohn, D.P. 1992, *J. Fluid Mech.* **240**, 559
Matthews, P.C., Rucklidge, A.M. 1993, *Proc. R. Soc. Lond.* A **441**, 649
Meneguzzi, M., Pouquet, A. 1989, *J. Fluid Mech.* **205**, 297
Nordlund, Å., Brandenburg, A., Jennings, L., Rieutord, M., Ruokolainen, J., Stein, R.F., Tuominen, I. 1992, *Ap. J.* **392**, 647
Nordlund, Å., Stein, R.F. 1989, in *Solar and Stellar Granulation*, eds. R. J. Rutten, G. Severino, Kluwer, Dordrecht, p. 453
Nordlund, Å., Stein, R.F. 1990, in *Solar Photosphere: Structure, Convection and Magnetic Fields*, IAU-Symp. No. 138, Kluwer, Dordrecht, p. 191
Proctor, M.R.E. 1992, in *The Physics of Sunspots*, eds. J.H. Thomas, N.O. Weiss, Kluwer, Dordrecht, p. 221
Proctor, M.R.E., Weiss, N.O. 1982, *Rep. Progr. Phys.* **45**, 1317
Proctor, M.R.E., Weiss, N.O., Brownjohn, D.P., Hurlburt, N.E. 1993, *J. Fluid Mech.*, in press
Rast, M.P., Nordlund, Å., Stein, R.F., Toomre, J. 1993, *Ap. J.* **408**, L53
Rucklidge, A.M. 1992, *J. Fluid Mech.* **237**, 209
Rucklidge, A.M. 1993, *Physica* **62D**, 323
Rucklidge, A.M., Weiss, N.O., Brownjohn, D.P., Proctor, M.R.E. 1993, *Geophys. Astrophys. Fluid Dynam.* **68**, 133
Steffen, M. 1989, in *Solar and Stellar Granulation*, eds. R. J. Rutten, G. Severino, Kluwer, Dordrecht, p. 425
Steffen, M., Freytag, B. 1991, in *Rev. in Mod. Astron.* **4**, ed. G. Klare, Springer, Berlin, p. 43
Steffen, M., Ludwig, H.-G., Krüss 1989, *Astr. Ap.* **213**, 371
Stein, R.F., Nordlund, Å. 1989, *Ap. J.* **342**, L95
Stein, R.F., Brandenburg, A., Nordlund, Å. 1992, in *Seventh Cambridge Workshop on Cool Stars, Stellar Systems, and the Sun*, eds. M. S. Giampapa, J. A. Bookbinder, ASP Conference Series, Vol. 26, p. 148
Stones, J.M., Hawley, J.F., Evans, C.R., Noeman, M.L. 1992, *Ap. J.* **388**, 415
Theobald, M.L., Fox, P.A., Sofia, S. 1993, *Phys. Fluids B*, in press
Title, A.M., Topka, K.P., Tarbell, R.D., Schmidt, W., Balke, C., Scharmer, G. 1992, *Ap. J.* **393**, 782
Vainshtein, S.I., Zeldovich, Ya. B. 1972, *Sov. Phys. Usp.* **15**, 159
Weiss, N.O. 1966, *Proc. R. Soc. Lond.* A **293**, 310
Weiss, N.O. 1990, *Comp. Phys. Rep.* **12**, 233
Weiss, N.O. 1991, *Geophys. Astrophys. Fluid Dynam.* **62**, 229
Weiss, N.O., Brownjohn, D.P., Hurlburt, N.E., Proctor, M.R.E. 1990, *Mon. Not. R. astr. Soc.* **245**, 434

4. THE SOLAR CYCLE (P.R. Wilson)

Some progress towards an understanding of the solar cycle has been made in several areas but many problems remain.

4.1 Synoptic Observations

4.1.1 The Extended Activity Cycle

K. Harvey (1992) has investigated the overlap of the wings of the butterfly diagram using the criteria of magnetic orientation and latitude of emergence. In an extensive study of magnetic bipoles of all sizes, she has determined that the period between first and last magnetic bipoles of Cycles 14-21 lay in the range 13.2-14.8 yr.

Other groups working in areas such as the coronal green line emission (Altrock 1992), the enhancement of geomagnetic disturbances (Leroy and Neons 1983) and the torsional shear (Snodgrass 1991) have independently suggested that their data were evidence for a high-latitude component of the new solar activity cycle, which begins some 4-5 yr before the appearance of the first sunspots of the new cycle, i.e., shortly after the maximum of the old one, and progresses equatorwards in synchrony with the sunspot butterfly diagram. Their conclusions are supported by synoptic contour maps which exhibit dominant ridges extending from high latitudes through the butterfly diagram to the equator.

Some of these charts also exhibit a polewards bifurcation and several authors, e.g. Sivaraman and Makarova (1992), have studied the poleward evolution of various high-latitude activity phenomena during the cycle. Stenflo (1992) has constructed a 'modally clean' butterfly diagram using the superposition of the first seven odd modes of the spherical harmonic expansion of the radial component of the axisymmetric magnetic field. This clearly shows two components, or branches, one which first appears at latitudes of 40°–50° and extends equatorwards over ~22 yr, overlapping with the butterfly pattern of the cycle associated with the latter part of this period and the other which extends polewards during the first cycle of this period. While this apparently confirms the extended overlap of the cycles, Stenflo (1992) has emphasised that each harmonic of his expansion contributes to *both* the polar and the equatorial branches, noting that such patterns are not uncommon in dynamo theory.

4.1.2 The Evolution of the Large-scale and Polar Fields

The observed evolution of the large-scale field patterns at different phases of the cycle have been compared with the simulated evolution based on solutions of the flux transport equation (Wilson and McIntosh 1991, Murray and Wilson 1993). These have shown several cases in which the large scale patterns cannot be explained in terms of the diffusion and meridional transport of decaying active region fields. Stenflo (1992) has also argued that the large-scale patterns must be maintained by the emergence of magnetic bipoles of all sizes rather than by the decay of surface flux. Legrand and Simon (1991), having reviewed both the geomagnetic and the solar activity data of recent cycles, conclude that the global dipole field is not a surface phenomenon, but one which originates deep within the solar interior. They propose a *two-component* cycle in which the dipole field is generated by a mechanism which is related to the *following* rather than the preceding sunspot cycle.

4.2 Stellar Activity

While the Sun permits the detailed two-dimensional study of cyclic activity phenomena, it exhibits only a single set of stellar parameters. Using the emission cores of the H and K lines of Ca II, and, more recently, of Mg II, as common indicators of stellar activity, many recent studies (e.g. Radick 1992), have revealed important correlations between stellar activity and stellar parameters such as rotation rate, age, and the depth of their convection zones.

Young, fully convective stars exhibit strong but non-cyclic activity which decreases in older stars concurrently with a decrease in their rotational velocity. The *Rossby number* N_R, (the ratio of the rotation period to the turnover time of the largest convective eddies) provides the connection, a low Rossby number (indicating a greater influence of rotation on convection) correlates remarkably well with strong surface activity. Cyclically varying activity is found in older stars with rotation periods of 20 days or longer. Luminosity variations are found to correlate with activity variations in older, cyclically active stars, but anti-correlate in younger, irregularly active, stars.

Baliunas and Jastrow (1990) have proposed that some solar-type stars which exhibit anomalously low levels of activity are, in fact, normal solar-type stars passing through grand minima similar to the Maunder minimum.

These comparisons between solar and stellar activity have implications for the mechanisms which are responsible for the solar activity cycle.

4.3 Dynamos and Chaos

While the above correlations strongly suggest that dynamo action, in one of its several modes, is responsible for stellar activity and activity cycles, the identification of the detailed nature of the solar dynamo remains elusive. Although mathematically attractive, kinematic $\alpha - \omega$ dynamos involve assumptions which do not apply to solar conditions, while the development of non-linear and 'fast' dynamos has far to go (Rosner and Weiss 1992).

Recent advances in the understanding of the *chaotic* properties of solutions to the equations governing non-linear systems, and the recognition that solar magnetoconvective activity is such a system, has initiated studies of the solar cycle as an example of a chaotic system (e.g. Ruzmaikin 1990; Schmalz and Stix 1991). The irregular structure of the sunspot number plot since 1650 and, in particular, the Maunder minimum and the earlier grand minima which have been inferred from proxy data are strongly suggestive of chaotic behaviour. However, attempts to identify the *strange attractor* and to determine its fractal dimension have been frustrated by the inadequate length of the available data string.

4.4 Forecasting and the Parameters of Cycle 22

The possibility that the solar cycle is a chaotic system and, therefore, inherently unpredictable, did not deter a great many forecasters from making predictions regarding the significant parameters of Cycle 22. With the benefit of hindsight one may assess the success or otherwise of the various methods employed.

Cycle 22 began in September 1986, from the highest minimum so far recorded, $R_Z(min) = 12.3$, and the early values of the Zurich sunspot number, R_Z were the highest on record. It also featured the shortest rise time from minimum to maximum (35 months), but the maximum of $R_Z(M) = 158.5$, which was achieved in July, 1989, was disappointing, being ranked third of the modern era, after Cycles 19 ($R_Z(M) = 201.3$), and 21 ($R_Z(M) = 164.5$).

The more successful predictions were those based on the magnitude of the peak in the indices of geomagnetic disturbances occurring during the declining phase of Cycle 21 (Thompson 1988), Thompson's estimate of 159 being the closest. Spectacularly unsuccessful were those forecasts based on the observed polar fields at minimum or on the sunspot numbers observed during the early years of Cycle 22. Indeed, after such a spectacular beginning, the early and disappointingly small maximum of the cycle is more suggestive of a chaotic system than of a regular oscillator. Nevertheless, if the rise of the geomagnetic disturbances during the decline of the old cycle is interpreted as arising from some fundamental changes occurring within the Sun during this phase, then the success of predictions based on these indices supports the arguments of Legrand and Simon regarding the polar field reversals and suggests an interesting avenue for future studies.

REFERENCES

Altrock, R.C., 1992, *Bull. A.A.S.*, **24**, 746.
Baliunas, S.L., Jastrow, R., 1990, *Nature*, **348**, 520.
Harvey, K.L., 1992, in *Proc. NSO/Sacramento Peak 12th Summer Workshop*, ed. K.L. Harvey, *PASP Conf. Ser.*, **27**, p. 335.
Legrand, J.P., Simon, P.A., 1991, *Solar Phys.*, **131**, 187.
Leroy, J.L., Neons, J.C., 1983, *Astr. Ap.*, **120**, L1.
Murray, N., Wilson, P.R., 1992, *Solar Phys.*, **142**, 221.

Radick, R.R., 1992, in *Proc. NSO/Sacramento Peak 12th Summer Workshop*, ed. K.L. Harvey, *PASP Conf. Ser.*, **27**, p. 450.
Rosner, R., Weiss, N.O., 1992, in *Proc. NSO/Sacramento Peak 12th Summer Workshop*, ed. K.L. Harvey, *PASP Conf. Ser.*, **27**, p. 511.
Ruzmaikin, A. A., 1990, in *Solar Photosphere: Structure, Convection, and Magnetic Fields*, ed. J.O. Stenflo, ed., *IAU Symp.*, **138**, p. 343.
Schmalz, S., Stix, M., 1991, *Astr. Ap.*, **245**, 654.
Sivaraman, K.R., Makarova, V.V, 1992, in *Proc. NSO/Sacramento Peak 12th Summer Workshop*, ed. K.L. Harvey, *PASP Conf. Ser.*, **27**, p. 415.
Snodgrass, H.B., 1991, *Astrophys. J.*, **383**, L85.
Stenflo, J.O., 1992, in *Proc. NSO/Sacramento Peak 12th Summer Workshop*, ed. K.L. Harvey, *PASP Conf. Ser.*, **27**, p. 421.
Stenflo, J.O., 1992, in *Proc. NSO/Sacramento Peak 12th Summer Workshop*, ed. K.L. Harvey, *PASP Conf. Ser.*, **27**, p. 83.
Thompson, R.J., 1988, *IPS Techn. Rep.* IPS-TR-88-01, IPS Radio and Space Services, Sydney, Australia.
Wilson, P.R., McIntosh, P.S., 1991, *Solar Phys.*, **136**, 221-237.

5. HEATING OF THE OUTER SOLAR ATMOSPHERE (U. Narain)

Since the discovery that the temperature of the outer solar atmosphere (corona) is about 10^6 K and is much higher than the regions lying above and below it, various mechanisms have been put forward to explain this phenomenon (see e.g. Narain & Ulmschneider 1990; Ulmschneider et al. 1991 and references contained therein). Because of complexities introduced by magnetic fields the proposed mechanisms could explain this problem with limited success. The attempts in this direction are continuing, theoretically as well as observationally.

It is now believed that acoustic waves can heat the inner solar atmosphere via shock formation (non-magnetic regions only), but magnetic fields play an important role in the heating of the outer solar atmosphere. Different MHD wave modes generated in the presence of magnetic fields are, e.g., slow, fast, and Alfvén modes. If there is a boundary in the medium, surface waves come into existence. These surface waves propagate along the boundary such as the surface of solar coronal loops. Body waves exist in a homogeneous medium. All these waves derive their energy from the subphotospheric convection zone. If the medium is non-uniform, the complexity of the generated wave modes increases (Califano et al. 1990, 1992).

X-ray pictures of the corona contain a variety of structures. The coronal loops are characterized by closed magnetic field lines, whereas the coronal holes have open magnetic field lines extending to interplanetary space. These are the basic building blocks of the corona. Once we know with certainty the basic mechanism(s) of their heating, the coronal heating problem would come to its logical conclusion. These basic structures may be heated through the dissipation of MHD waves as well as currents/magnetic fields.

5.1 Heating of Coronal Loops

Unlike other MHD modes, the Alfvén waves reach coronal heights even after passing through the photosphere, chromosphere and transition region. In coronal loops these waves may dissipate via resonant absorption, phase mixing, mode conversion, Kolmogoroff turbulent cascade, etc., to heat the coronal plasma contained in them.

For simplicity, loops are approximated by straight, cylindrical, axisymmetric plasma columns with equilibrium quantities varying only in the radial direction. There exist a continuum of oscillation frequencies of such a plasma. If this plasma is forced to oscillate at some frequency by photospheric mass motions (convection), resonance may occur in some layer where the local Alfvén frequency equals the frequency of the external (photospheric) source. Energy is thus transferred from the photospheric source to the coronal loop. Since the spectrum of photospheric frequencies is not known, Poedts et al. (1990) could estimate only the efficiency of this heating mechanism and found it to be a viable mechanism for heating coronal loops.

Alternatively, coronal loops may be heated via episodic dissipation of magnetic fields in current sheets (Parker 1991a). Granular motions shuffle the coronal fields' footpoints, and braiding of coronal loops takes place. As a consequence of this non-potential energy is stored in the system. Eventually, a critical threshold is reached and reconnection of coronal fields occurs in thin current sheets. The released energy heats the loop impulsively creating a tiny flare. Parker calls it a nanoflare, in which about 10^{23}–10^{25} ergs of energy is released. Thus the X-ray corona is heated by 'swarms' of nanoflares. Taking 10^{23} ergs per nanoflare as an average, the mean output of 10^7 ergs cm^{-2} s^{-1} requires about one new nanoflare per second per area 1 Mm × 1 Mm, where Mm stands for 10^3 km. Nanoflares may, alternatively, be produced by the twisting of loop footpoints. Both mechanisms, twisting and braiding, are equally effective in producing nanoflares (Zirker & Cleveland 1993). Sturrock et al. (1990) also propose episodic heating, where each short burst of heating is followed by a long period of radiative cooling but they do not specify the mechanism. The spectrum, computed by Raymond (1990) for this model, agrees with the observations qualitatively.

Due to microturbulence the coronal medium may be endowed with an effective viscosity and resistivity (different from the molecular one). When the boundary motions stress such a dissipative medium, turbulence is generated. After some time equilibrium is established. This approach is different from the usual frozen-in assumption. There is a net flow of energy from the photosphere to the coronal loop, where it is dissipated. In 2-D MHD Heyvaerts & Priest (1992) find a DC flux of 5×10^5 erg cm^{-2} s^{-1} for quiet region loops and 2×10^6 erg cm^{-2} s^{-1} for active region loops. The rms velocities are 20 and 30 km s^{-1}, respectively. There is a free parameter, C, in the expression for energy flux, which cannot be specified unambiguously.

5.2 Heating of Coronal Holes

It is obvious from the observational analysis of Withbroe (1988) that a coronal hole requires a heat input of about 5×10^5 erg cm^{-2} s^{-1}, the major portion of which is deposited within 1–2$R_\odot$. A recent review by Parker (1991b) infers an upper limit of 1×10^5 erg cm^{-2} s^{-1} for the Alfvén wave flux in a coronal hole. The agitation observed in the photosphere suggests characteristic wave periods in the range 50–300 s, which correspond to a granule size of about 500 km and characteristic velocities of 1 km s^{-1}. Phase mixing and other known mechanisms are not able to damp these waves within 1–2$R_\odot$ (Parker 1991c). Kolmogoroff turbulent cascade could deposit energy in the first 1–2 solar radii provided the turbulence could be produced by standing Alfvén waves. However, for open-ended coronal holes inward flux is not expected at periods of 100 s, which would be essential for the formation of standing waves. Photospheric motions generate an estimated Alfvén flux of the order of 5×10^5 erg cm^{-2} s^{-1}. Obviously, the entire generated flux cannot be transported to the corona. Thus a coronal hole is not heated principally by the dissipation of Alfvén waves produced in the subphotospheric convection. Only in the fibril photospheric field of 10^3 G an adequate amount of Alfvén wave flux could be generated.

For a simple class of spherical sources, adopted as a model for oscillating solar granules with a five minute period, Collins (1992) finds that the generated Alfvén fluxes are comparable to those required to heat active regions. This may not be true for coronal holes.

Under isothermal hydrostatic conditions Moore et al. (1991) find that for a base radius of 1.15 $R_\odot$, base field strength of 10 G and a base electron density of 3×10^7 cm^{-3} the waves with periods of about 5 minutes are trapped within the hole and heat it if its temperature is slightly less than one million Kelvin. They escape to the solar wind and accelerate it for coronal holes of higher temperatures. The potential magnetic field is assumed to decrease according to $B \sim R^{-m}$, where R is the radial distance from the Sun centre, and m is an even integer such that for $m = 2$ the field spreads radially.

A new mechanism called 'intermittent magnetic levitation' for the dissipation of reflected Alfvén waves has been proposed (Moore et al. 1992), for which the waves propagate from the sites of their generation to the coronal hole plasma in a flux tube. The waves with periods exceeding a critical period get reflected by the density gradients and push up against the coronal plasma. Consequently a quasi-hydrostatic vertical distribution of plasma could result such that there would be more supported plasma than in the absence of the reflecting waves. Because the events (granules, microflares and spicules) that generate the Alfvén waves are episodic and intermittent, it is expected that the coronal plasma in the flux tubes at any instant would be rising or falling. During rise the plasma gains gravitational potential energy which, during fall, gets converted into thermal energy. The viability of this mechanism is yet to be established.

The principal source for coronal holes could be the dissipation of quasi-static magnetic fields, which contain thin current sheets to provide enhanced dissipation. The statement that almost all field topologies produce internal current sheets as an intrinsic part of static equilibrium is not true for coronal holes, because the transverse components of the field are free to propagate away to infinity along the field. Since the energy requirement of a coronal hole is smaller (about 1/20) than that of the active X-ray corona it is quite reasonable to look for small-scale magnetic fields that appear at the supergranule boundaries. This idea has already been put to practice (Parker 1991b and references contained therein). The emerging picture is that the small-scale ($2–4 \times 10^3$ km) magnetic bipoles are jostled together by subphotospheric convection. They develop internal, tangential discontinuities, where magnetic dissipation heats the gas trapped in the bipole. The small bipoles reconnect where they are pushed against other bipoles or against unipolar fields. The magnetic energy is continually replenished by the emergence of new flux bundles and by the continual deformation of the bipolar fields by the subphotospheric mass motions. This magnetic energy of the small-scale fields is converted into thermal energy at thin current sheets in the usual way. Porter & Moore (1988) estimate an average overall magnetic dissipation rate of 5×10^5 erg cm^{-2} s^{-1} in coronal holes. Most of this energy is injected into the coronal hole, in the form of jets of gas, lashing flux bundles and superheated plasma from a bipole interior suddenly freed by reconnection into the ambient unipolar field.

5.3 Heating in General

The smallness of dissipative coefficients in many astrophysical situations requires that large gradients should develop in the dynamical evolution of the system (i.e., small spatial scales are formed), so that efficient dissipation may take place. Within the framework of normal mode analysis and incompressible MHD, Califano et al. (1990) establish the existence of a new class of resistive (nonresonant) solutions, which are characterized by the explicit appearance of resistivity in their asymptotic form and the formation of small scales over the entire inhomogeneous region. This feature of the solutions distinguishes them from the more familiar resonant solutions that obey ideal asymptotic boundary conditions and develop large gradients at particular spatial locations.

In the more realistic compressible case, adopting slab geometry (where relevant quantities depend on a single coordinate normal to the magnetic field direction) Califano et al. (1992) arrive at similar solutions. For shear Alfvén waves the smallest damping length turns out to be of the order of the scale of the non-uniformity. The existence of small spatial scales is supported by the numerical simulation of Malara et al. (1992) for shear Alfvén waves with phase-mixing as the damping mechanism, within the frame work of incompressible MHD. Further support will be available when numerical simulation of the more realistic compressible case is attempted.

5.4 Observational Aspect(s)

Cook & Ewing (1990), by examining Kitt Peak magnetograms and spectroheliograph observations at 1600 Å, find that the brightness temperature above 4400 K is linearly related to the magnetic field strength. Alfvén waves seem to be the most obvious candidate, as their flux $\rho v^2 V_A$ (where ρ is the matter density, v is the driving, photospheric velocity, and V_A is the Alfvén velocity) is proportional to B, the magnetic field strength. The structures so heated are found to be bright points.

As a result of an analysis of Solar Maximum Mission data on coronal X-ray line broadening of the Mg XI ion in a nonflaring active region Saba & Strong (1991) are tempted to conclude that the heating of the active region takes place either by Alfvén waves directly or as a side effect of magnetic reconnection and current dissipation.

Parker (1993) has suggested that resistive dissipation and rapid reconnection at the current sheets is the major heat source for the active X-ray corona of the Sun or any star, through flaring on all scales from nanoflares to microflares to major flares. In view of this, observations providing insight into small-scale magnetic activity seem quite crucial, although they are hampered by limited spatial and temporal resolution (Stenflo 1989).

Acknowledgement: The author is grateful to the Inter-University Centre for Astronomy & Astrophysics (IUCAA), Pune, India, for making its facilities available for the writing of the present review.

REFERENCES

Califano, F., Chiuderi, C., Einaudi, G. 1990, *Ap. J.*, **365**, 757.
Califano, F., Chiuderi, C., Einaudi, G. 1992, *Ap. J.*, **390**, 560.
Collins, W. 1992, *Ap. J.*, **384**, 319.
Cook, J.W., Ewing, J.A. 1990, *Ap. J.*, **355**, 719.
Heyvaerts, J., Priest, E.R. 1992, *Ap. J.*, **390**, 297.
Malara, F., Veltri, P., Chiuderi, P., Einaudi, G. 1992, *Ap. J.*, **396**, 297.
Moore, R.L., Musielak, Z.E., Suess, S.T., An, C.H. 1991, *Ap. J.*, **378**, 347.
Moore, R.L., Hammer, R., Musielak, Z.E., Suess, S.T., An, C.-H. 1992, *Ap. J.*, **397**, L55.
Narain, U., Ulmschneider, P. 1990, *Space Sci. Rev.*, **54**, 377.
Parker, E.N. 1991a, in *Mechanisms of Chromospheric and Coronal Heating*, eds. P. Ulmschneider, E.R. Priest, R. Rosner, Springer-Verlag, Heidelberg, pp. 615–629.
Parker, E.N. 1991b, *Ap. J.*, **372**, 719.
Parker, E.N. 1991c, *Ap. J.*, **376**, 355.
Parker, E.N. 1993, *Ap. J.*, **407**, 342.
Poedts, S., Goossens, M., Kerner, W. 1990, *Ap. J.*, **360**, 279.
Porter, J.G., Moore, R.L. 1988, in *Proc. NSO/Sacramento Peak 9th Summer Workshop*, ed. R.C. Altrock, Sunspot, NM, pp. 125–130.
Raymond, J.C. 1990, *Ap. J.*, **365**, 387.
Saba, J.L.R., Strong, K.T. 1991, *Ap. J.*, **375**, 789.
Stenflo, J.O. 1989, *Astr. Ap. Rev.*, **1**, 3.
Sturrock, P.A., Dixon, W.W., Klimchuk, J.A., Antiochos, S.K. 1990, *Ap. J. (Letters)*, **356**, L31.
Zirker, J.B., Cleveland, F.M. 1993, *Solar Phys.*, **144**, 341.

6. ADVANCES IN SOLAR POLARIMETRY (C.U. Keller)

Solar polarimetry is most often employed to detect and measure magnetic fields via the Zeeman or the Hanle effect. Polarimetry can also be used to study the resonance polarization in spectral lines near the solar limb, the electron impact polarization in flares (Hénoux *et al.* 1990), or electric fields via the Stark effect (Foukal and Hinata 1991). This report covers selected highlights and recent work done in the context of solar polarimetry published in the period from July 1, 1990 to June 30, 1993. I concentrate on instruments and observational techniques. All instruments that measure, in one or another way, the polarization of solar light will be called polarimeters. Due to the lack of space the report remains necessarily incomplete, and only some references are cited. Most of the recent reviews and papers on solar polarimetry can be found in the proceedings of the 11th Sacramento Peak Workshop (November 1991) and the recent IAU Colloquium at Beijing (Zirin *et al.* 1993). Detailed information on the fundamentals and applications of polarized light can be found in Collett (1993).

6.1 Instrumentation

Most solar telescopes have large effective focal lengths to achieve high spatial resolution. This results in rather complicated optical constructions with beam-folding mirrors inside the telescope or in front of static image-forming optics. Hence, we often face complicated instrumental polarization patterns in the final focus due to oblique reflections. In addition, solar post-focus instruments often have strongly polarizing properties, e.g. spectrograph gratings or birefringent filters. This requires that the polarization analysis is done, at least, in front of the post-focus instrument. Modern solar telescope designs even place the polarization analysis as early in the optical train as possible where the optics is still cylindrically symmetric. This spatial separation of the polarization analysis optics from the detector system in the final focus avoids instrument-induced cross-talk between the various Stokes parameters, which is essential for highly sensitive and accurate solar polarimetry.

Polarization measurements always involve the detection of intensity differences, i.e. at least two measurements are needed to determine a polarization signal. Reliable polarimetry requires that the two intensities are detected within a period that is short with respect to the fastest changes of seeing. This can be achieved either by spatial modulation (with our terminology including the case of static beam splitters), which guarantees absolutely simultaneous measurements, or temporal modulation. Spatial modulation has the drawback that the two (or more) beams do not pass the optics along the same optical path, and differential aberrations may become crucial. Furthermore, spatial modulation requires that the various beams are detected by different detectors or detector areas, which makes the measurements susceptible to differential gain effects. Temporal modulation should be performed at frequencies well above the fastest changes due to seeing (typically a few 100 Hz). Unfortunately this is well above the current read-out rate of large CCD array detectors, which are now employed in almost all solar polarimeters. In this context we may distinguish between slow and fast temporal modulation schemes, where the latter uses frequencies well above the seeing frequencies. Finally any purely temporal modulation scheme requires that about half of the light remains unused to eventually achieve a temporal intensity variation. In conclusion, technological limitations always impose some compromises in the design, which has resulted in a large variety of polarimeter schemes. In the following I discuss some of the most recently designed or constructed solar polarimeters.

Most modern, precise solar polarimeters employ a temporal modulation scheme in combination with sensitive CCD array detectors. Single exposures of CCD detectors have the disadvantage of a relatively low S/N ratio (typically several hundred). Therefore high frame read-out rates (often at video frequency) and on-line digital summation is combined to obtain the high S/N ratio required for solar polarimetry. Fast tip-tilt mirrors can significantly reduce image motion, which is important when the temporal modulation frequency is comparable to the seeing frequencies. Powerful real-time image processing hardware may perform sophisticated on-line analysis.

The High Altitude Observatory/National Solar Observatory Advanced Stokes Polarimeter (Elmore *et al.* 1992) is a spectrograph-based vector polarimeter. A mechanically rotating retarder and a polarizing beam-splitter system combine temporal and spatial modulation with CCD imagers that are read out at video frequency. The combination of the two modulation schemes leads to a nearly optimum use of the photons entering the polarimeter. Two spectral lines may be observed simultaneously and the full line profiles recorded by the instrument makes it particularly useful to determine physical quantities of solar magnetic features.

Another spectrograph-based polarimeter is the NASA/NSO Spectromagnetograph (Jones *et al.* 1992), which has recently replaced the old 512-channel Diode Array Magnetograph that produced the widely distributed daily Kitt Peak magnetograms. This instrument employs a Kerr cell to modulate the circular polarization. A video CCD array records alternate states of polarization in the spectrogram. A sophisticated real-time image processing system extracts the line-of-sight velocity and magnetogram signals, the continuum intensity, the equivalent width, and the line depth.

An interesting polarimetry set-up was used by Rust and Keil (1992). A Ronchi ruling provides a mask that covers every other pixel row of the CCD camera from direct illumination. The adjacent lithium niobate polarizing beam-splitter produces two orthogonally polarized beams, one of which is displaced by the width of one pixel row. The tiny displacement of the two beams minimizes differential aberrations. However, differential gain effects remain and the exact positioning of the ruling with respect to the CCD array structure is crucial.

The Zürich imaging Stokes Polarimeter ZIMPOL I (Keller *et al.* 1992) combines modulation of all Stokes parameters at 42 and 84 kHz with CCD arrays by synchronous charge shifting in specially masked CCD arrays. It will be capable of measuring all four Stokes parameters through the same optics. While this design allows extremely sensitive and accurate polarimetry down to the 10^{-4} level, the opaque mask covering every other CCD row and the beam-splitting system needed to feed three CCD arrays to measure all four Stokes parameters simultaneously makes the system rather inefficient as compared with other designs. Based on ZIMPOL I a new imaging vector polarimeter scheme (ZIMPOL II) has been presented by Stenflo *et al.* (1992). Only a single array is needed to detect all four Stokes parameters. Together with a micro-lens array in front of the CCD imager this design increases the sensitivity with respect to ZIMPOL I by at least a factor of six.

The National Solar Observatory's Near Infrared Magnetograph (NIM) is a spectrograph-based polarimeter (Rabin 1992b, private communication). It builds up a two-dimensional array of polarized line profiles by scanning the solar image across the slit of a high-dispersion grating spectrograph. Polarization analysis is accomplished by liquid-crystal variable retarders and an infrared linear polarizer. It is mostly used to record Stokes I and V, but it can also be used to record the full set of Stokes parameters. The detector, a 256×256 indium antimonide array, is fed by anamorphic transfer optics that match the spatial and spectral scales to the 38×38 μm pixels. At each slit position, 8 polarization pairs for each Stokes parameter are acquired at 7 Hz, averaged, and recorded on tape.

6.2 Observational methods

We see a continuous increase in spatial and temporal resolution as well as coverage in the visible part of the solar spectrum. Most observations focus on sunspot regions. The Lockheed group continuously provides high spatial resolution results from La Palma (e.g. Title *et al.* 1993). A significant improvement in the spatial resolution of magnetic field measurements has been achieved by eliminating the influence of the Earth's atmosphere with speckle-interferometric methods (Keller and von der Lühe 1992). Speckle polarimetry is based on a two-channel system where one channel measures the instantaneous point spread function while the other records the polarization in the wing of a Zeeman sensitive spectral line. It has led to the first resolution of solar magnetic fluxtubes (Keller 1992). Almost continuous time series of vector magnetic field measurements over several day periods are now collected by the combination of the Big Bear and Huairou Solar Observatories (Zhang *et al.* 1992). Finally it has been recognized that spectral lines without intrinsic linear polarization due to the Zeeman effect (e.g. FeII 6149 Å) might play a key role in measuring instrumental polarization (Lites 1993).

The near-infrared has the major advantage of being able to directly indicate the field strength in the deep solar photosphere. NIM has found weak fields, i.e. fields with strengths below 1 kG (Rabin 1992a). NIM exploits the high Zeeman sensitivity of the line Fe I 15648.5 Å to measure $\vec{B}$. For $|B| \gtrsim 800$ G, the magnitude of the field is derived, without adjustable parameters, from the strong splitting of the Zeeman components. The relative strengths of the Stokes components indicate the direction of the field. Livingston (1991) has brought back into operation the two decades old Baboquivari infrared detector at Kitt Peak. Although this instrument contains only a single InSb diode detector, its simplicity and ease of use has led to a variety of scientific results (e.g. Solanki *et al.* 1992).

A recently developed far infrared detector system has been used to sequentially record the full Stokes vector in the extremely Zeeman-sensitive MgI 12.3 μm line (Hewagama *et al.* 1993). Since the spatial resolution of the 1.5 m McMath-Pierce facility is limited by the diffraction at these wavelengths, and not by seeing, the sequential recording of Fourier Transform Spectrometer interferograms in six polarization states (12 minutes for the full Stokes vector) is not crucial when observing sunspots. Furthermore, instrumental polarization due to oblique reflections is very small as compared to the visible.

6.3 Developments

Solar polarimetry will mostly be ground-based in the rest of this decade. An instrument that will be particularly suited for polarimetry is the 90 cm THEMIS (Rayrole 1991). It belongs to a new class of large-diameter solar telescopes that are virtually free from instrumental polarization and instrument induced cross-talk between the Stokes parameters. Among these designs is also LEST, the Large Earth-based Solar Telescope (Engvold and Andersen 1990). It will become the largest solar telescope and it is designed with high priority on sensitive and accurate polarimetry. Instead of going into space it is cheaper to use balloons to leave most of the atmosphere behind. The GENESIS telescope (Rust *et al.* 1991) will circle the South Pole during 2 weeks and collect magnetograms with an 80 cm telescope. This instrument may deliver the highest spatial resolution magnetograms in the near future.

REFERENCES

Collett, E. 1993, Polarized Light, Fundamentals and Applications, Marcel Dekker, Inc., New York.
Elmore, D.F., Lites, B.W., Tomczyk, S., Skumanich, A.P., Dunn, R.B., Schuenke, J.A., Streander, K.V., Leach, T.W., Chambellan, C.W., Hull, H.K., Lacey, L.B. 1992, *SPIE* **1746**, 22.

Engvold, O., Andersen, T. 1990, *Status of the Design of the Large Earth-based Solar Telescope*. LEST Foundation.
Foukal, P., Hinata, S. 1991, *Solar Phys.*, **132**, 307.
Hénoux, J.C., Chambe, G., Smith, D., Tamres, D., Feautrier, N., Rovira, M., Sahal-Bréchot, S. 1990, *Ap. J. Suppl.*, **73**, 303.
Hewagama, T., Deming, D., Jennings, D.E., Osherovich, V., Wiedemann, G., Zipoy, D., Mickey, D.L., Garcia, H. 1993, *Ap. J. Suppl.*, **86**, 313.
Jones, H.P., Duvall, T.L. Jr., Harvey, J.W., Mahaffey, C.T., Schwitters, J.D., Simmons, J.E. 1992, *Solar Phys.*, **139**, 211.
Keller, C.U. 1992, *Nature* **359**, 307.
Keller, C.U., von der Lühe, O. 1992, *Astr. Ap.*, **261**, 321.
Keller, C.U., Aebersold, F., Egger, U., Povel, H.P., Steiner, P., Stenflo, J.O. 1992, *LEST Technical Report* **53**.
Lites, B.W. 1993, *Solar Phys.*, **143**, 229.
Livingston, W. 1991, in *Solar Polarimetry*, ed. L.J. November, New Mexico, USA, pp. 356.
November, L.J. 1991, Proc. of the Eleventh Sacramento Peak Summer Workshop, Solar Polarimetry. New Mexico, USA.
Rabin, D. 1992a, *Ap. J.*, **390**, L103.
Rabin, D. 1992b, *Ap. J.*, **391**, 832.
Rayrole, J. 1991 , in *Solar Polarimetry*, ed. L.J. November, New Mexico, USA, pp. 496.
Rust, D.M., Keil S.L. 1992, *Solar Phys.*, **140**, 55.
Rust, D.M., Murphy, G.A., Strohbehn, K., Hochheimer, B., Henshaw, R., Hayes, J.R., Lohr, D.A., Harris, T.J. 1991, *Design Study of a Balloon-borne Solar Vector Magnetograph* (preprint).
Solanki, S.K., Rüedi, I., Livingston W. 1992, *Astr. Ap.*, **263**, 312.
Stenflo, J.O., Keller, C.U., Povel, H.P. 1992, *LEST Technical Report* **54**.
Title, A.M., Frank, Z.A., Shine, R.A., Tarbell, T.D., Topka, K.P., Scharmer, G.B., Schmidt., W. 1993, *Ap. J.*, **403**, 780.
Zirin, H., Ai, G., Wang, H., 1993, Proc. of the IAU Coll. No. 141, *The Magnetic and Velocity Fields of Solar Active Regions*, ASP Conference Series, Vol. 46.
Zhang, H., Ai, G., Wang, H., Zirin, H., Patterson A. 1992, *Solar Phys.*, **140**, 307.

7. ADVANCES IN INFRARED SOLAR PHYSICS (D. Rabin)

Although infrared observations have long played a role in the study of the Sun, it is only during the last five years that infrared solar physics has emerged as a vigorous subdiscipline. The explosive development of infrared array technology has contributed to this development by greatly expanding the range of feasible measurements in the 1–30 μm spectral band. However, as with infrared astronomy generally, unifying physical themes are at least as important as techniques—for example: a simple continuum opacity dominated by H^- and H free-free absorption; an approximately linear relationship between temperature and continuum intensity; the importance of rotation and vibration-rotation transitions of simple molecules; the quadratic wavelength dependence of the Zeeman effect.

IAU Symposium 154, the first international meeting devoted to infrared solar physics, took place during 1992. The symposium proceedings (I) provide the most comprehensive picture to date of current research and possible initiatives in this young field. The review by Deming *et al.* (1991) is a good introduction to the physics of the infrared spectrum.

7.1 Total Solar Eclipse of 11 July 1991

There was unprecedented coverage of the 1-μm to 3-mm spectral region during this favorable eclipse. Three experiments searched for thermal emission from dust in the range 1–6 $R_\odot$ (Kuhn *et al.* 1993; MacQueen *et al.* 1993; Tollestrup *et al.* 1993); each group employed an infrared array camera and filters in the 1–2.5 μm range. Kuhn *et al.* and MacQueen *et al.* found no evidence for coronal dust emission; Tollestrup *et al.* reported a marginal detection. MacQueen *et al.* hypothesize that the much stronger emission observed during the 1966 eclipse arose from dust supplied by one or more Sun-grazing comets.

Jennings *et al.* (1993) studied the limb profile of the 12.32 μm Mg I emission line at the IRTF. They found the emission peaks within ~ 300 km of the 12-μm continuum limb, consistent with new radiative-transfer models of the formation of this line (section 7.2). However, line emission can be traced to heights as great as 2000 km, indicating spatial inhomogeneities or departures from hydrostatic equilibrium.

Limb observations at submillimeter and millimeter wavelengths directly probe the vertical structure of the temperature minimum region and chromosphere. There was excellent coverage of the 1991 eclipse by four observatories: JCMT (Lindsey *et al.* 1992), CSO (Ewell *et al.* 1993), OVRO (Belkora *et al.* 1992), and BIMA (White & Kundu 1993). The results confirm and extend previous indications that chromospheric emission extends 1.5–3 times higher (depending on wavelength) than predicted by standard one-dimensional models in hydrostatic equilibrium. It is not yet certain that spicules (as seen in off-band Hα images) cause the extended submillimeter emission; if so, the spicules are rather cool, 6000–7000 K.

7.2 Atomic Physics and Line Formation

A highlight of the last triennium has been the explanation of the 12-μm emission lines and related high-*l* atomic Rydberg transitions. Detailed radiative transfer models establish that the Mg I emission lines are formed in the upper photosphere (Chang *et al.* 1991; Carlsson *et al.* 1992). The NLTE processes that affect them are now understood, as are their susceptibilities to the Zeeman and Stark effects (Chang 1993). This new understanding enhances the value of the 12-μm lines for magnetic observations (section 7.3).

The ATMOS atlas of the solar spectrum between 2 μm and 16 μm has proven to be an invaluable source of high-resolution spectra that are free of atmospheric contamination (Farmer 1993). ATMOS profiles of lines unobservable from the ground were instrumental in confirming the interpretation of the 12-μm lines. ATMOS profiles of some H I lines differ significantly from the predictions of current radiative transfer models (Avrett *et al.* 1993; Carlsson & Rutten 1993). As the models are notably sensitive to atmospheric structure, the infrared H I lines will be a valuable chromospheric diagnostic when they are better understood. Biémont (1993) and Sauval & Grevesse (1993) have reviewed recent progress in solar infrared spectroscopy of atoms and molecules, respectively.

7.3 Magnetic Fields

In an important series of papers, Solanki and colleagues have investigated the diagnostic power of near-infrared spectral lines, especially Fe I 15648.5 Å and 15652.9 Å, and used these diagnostics to probe solar magnetic features (Muglach & Solanki 1992; Solanki *et al.* 1992a,b,c; Rüedi *et al.* 1992a,b). The results include: evidence for the existence of stable "weak-field" flux tubes ($B \lesssim 1200$ G); evidence against a significant return flux in either small-scale flux tubes or sunspots; detection of a siphon flow in a small magnetic arch; and confirmation of the nonlinear field-strength vs. temperature relationship in sunspots reported by Kopp & Rabin (1992), together with a linear relationship between magnetic inclination angle and temperature.

An investigation of the spatial and statistical distribution of the properties of plage flux tubes was initiated by Rabin (1992a,b) using an array-based Near Infrared Magnetograph (described in section 4). This instrument produced the first two-dimensional map of plage magnetic fields in the deep photosphere. Field strengths in the range 1000–1600 G accounted for more than 90% of the magnetic flux, as expected from the work of earlier investigators. Within this range, the observed field strength varied coherently on several spatial scales within the field of view. A statistical relationship between magnetic field strength and magnetic flux was apparent, in the sense that strong magnetic fields are found at all magnetic flux densities, but weak fields ($B \lesssim 1200$ G) are found only in weak-flux areas; Rüedi *et al.* (1992a) report a similar result.

Polarimetric observations of the Mg I emission line at 12.32 μm have been used to study the structure of sunspot fields down to a field strength of 300 Gauss, well outside the radius of the penumbra-photosphere boundary, R_p (Hewagama 1991; Deming *et al.* 1993). The precise (±50 G) measurements of field strength clearly show departures from azimuthal symmetry. The average field strength at R_p is weaker than predicted by the relation of Beckers & Schröter (1969), but the field in the high photosphere persists to $R \sim 1.5R_p$, consistent with inferences from visible-light polarimetry. Measurements of the inclination and azimuth of sunspot fields from 12-μm data have been carried out but are still preliminary in character. An array-based imaging magnetometer operating at 12 μm is under construction (Deming *et al.* 1993). Now that the formation of these lines is understood (section 7.2), simultaneous observations at 12 μm and 1.56 μm should allow accurate measurements of flux-tube expansion from the bottom to the top of the photosphere.

Infrared observations have provided important constraints on theoretical models of magnetic flux tubes, as reviewed by Steiner (1993). Future measurements of thermal inhomogeneities in the temperature-minimum region and chromosphere should prove particularly useful; imaging observations in the millimeter and submillimeter continuum (Lindsey 1993) and in the carbon monoxide lines near 4.7 μm are now possible.

7.4 The Opacity Minimum

The solar atmosphere is most transparent at $\sim$ 1.6 μm. Although continuum radiation at 1.6 μm is formed only 30–40 km below the $\tau_{0.5\mu m} = 1$ level, this region is interesting because the balance between convective transport and radiative transport changes rapidly with height. Observations of granules and other convective structures at the opacity minimum have so far revealed small but significant differences from their visible-light properties (Koutchmy 1993; Keil *et al.* 1993); Stein & Nordlund (1993) review theoretical simulations. Moran *et al.* (1992) have investigated the 1.6-μm contrast of faculae and pores observed near disk center and find that the contrast depends on magnetic flux: $< 0.2\%$ for values below 2×10^{18} Mx but increasingly dark for stronger flux concentrations. These observations include both unresolved flux tubes and the atmosphere immediately around them, so a comparison with flux-tube models is not straightforward. Continuum observations in the 1–4 μm range are also valuable diagnostics for differentiating among models of sunspot atmospheres (Maltby 1993).

7.5 Solar Activity

Beyond extensive and useful observations in He I 10830 Å (Jones 1993), observations of solar activity and flares in the infrared and submillimeter spectrum are scarce. Falchi *et al.* (1993) review the potential value of various diagnostics, emphasizing infrared continuum windows. Petrosian (1993) discusses flare emission mechanisms in the range 10^{11}–10^{14} Hz (3 μm–3 mm), including non-thermal synchrotron emission in the impulsive phase and, in the gradual phase, free-free and thermal gyrosynchrotron emissions. According to recent observations with the Itapetinga antenna (Correia *et al.* 1993), $\sim$ 25% of solar bursts show a rising spectrum (flux increasing with frequency) in the range 35–90 GHz. The BIMA interferometer detected both impulsive and gradual emission at 86 GHz from many flares during a Max '91 campaign (White & Kundu 1992). Thus, there appears to be a promising future for millimeter and submillimeter observations of flares.

REFERENCES

I. *Infrared Solar Physics*, eds. D. M. Rabin, J. T. Jefferies, C. Lindsey. 1993, IAU Symp. 154, Kluwer, Dordrecht, in press.

Avrett, E. H., Chang, E. S., and Loeser, R. 1993, I, 323–341.
Belkora, L., Hurford, G. J., Gary, D. E., Woody, D. P. 1992, *Ap. J.*, **400**, 692.
Beckers, J., Schröter, E. H. 1969, *Solar Phys.*, **10**, 384.
Biémont, E. 1993, I, 503–512.
Carlsson, M., Rutten, R.J., Shchukina, N.G. 1992, *Astr. Ap.*, **253**, 567.
Carlsson, M., Rutten, R.J. 1993, I, 343–348.
Chang, E. S. 1993, I, 297–307.
Chang, E. S., Avrett, E. H., Mauas, P. J. Noyes, R. W., Loeser, R. 1991, *Ap. J. (Letters)*, **379**, L79.
Correia, E., Kaufmann, P., Magun, A. 1993, I, 93–101.
Deming, D., Jennings, D. E., Jefferies, J., Lindsey, C. 1991, in *Solar Interior and Atmosphere*, eds. A. N. Cox, W. C. Livingston, M. S. Matthews, University of Arizona Press, Tucson, pp. 933–963.
Deming, D., Hewagama, T., Jennings, D. E., McCabe, G., Wiedemann, G. 1993, I, 381–394.

Ewell, M. W., Jr., Zirin, H., Jensen, J. B., Bastian, T. S. 1993, I, 161–166.
Falchi, A., Falciani, R., Mauas, P. 1993, I, 113–123.
Farmer, C. B. 1993, I, 513–523.
Hewagama, T. 1992, Ph.D. Thesis, University of Maryland.
Jennings, D. E., Deming, D., McCabe, G., Noyes, R., Wiedemann, G., Espenak, F. 1993, I, 151–160.
Jones, H. P. 1993, I, 49–58.
Keil, S., Kuhn, J., Lin, H., Reardon, K. 1993, I, 251–257.
Kopp, G., Rabin, D. M. 1992, *Solar Phys.*, **141**, 253.
Koutchmy, S. 1993, I, 239–250.
Kuhn, J. R., Lin, H., Lamy, P., Koutchmy, S., Smartt, R. N. 1993, I, 185–197.
Lindsey, C. A., Jefferies, J. T., Clark, T. A., Harrison, R. A., Carter, M., Watt, G., Becklin, E. E., Roellig, T. L., Braun, D. C., Naylor, D. A., Tompkins, G. J. 1992, *Nature*, **358**, 308.
Lindsey, C. 1993, I, 85–92.
MacQueen, R. M., Hodapp, K.-W., Hall, D. N. B. 1993, I, 199–203.
Maltby, P. 1993, I, 425–437.
Moran, T. Foukal, P., Rabin, D. 1992, *Solar Phys.*, **142**, 35.
Muglach, K., Solanki, S. K. 1992, *Astr. Ap.*, **263**, 301.
Petrosian, V. 1993, I, 103–112.
Rabin, D. M. 1992a, *Ap. J. (Letters)*, **390**, L103.
Rabin, D. M. 1992b, *Ap. J.*, **391**, 832.
Rabin, D. M. 1993, I, 451–459.
Rüedi, I., Solanki, S. K., Livingston, W., Stenflo, J.O. 1992a, *Astr. Ap.*, **263**, 323.
Rüedi, I., Solanki, S. K., Rabin, D. 1992b, *Astr. Ap.*, **261**, L21.
Sauval, A. J., Grevesse, N. 1993, I, 551–558.
Solanki, S. K., Rüedi, I., Livingston, W. 1992a, *Astr. Ap.*, **263**, 312.
Solanki, S. K., Rüedi, I., Livingston, W. 1992b, *Astr. Ap.*, **263**, 339.
Solanki, S. K., Walther, U., Livingston, W. 1992c, *Astr. Ap.*, in press.
Stein, R. F., Nordlund, Å. 1993, I, 225–237.
Steiner, O. 1993, I, 409–424.
Tollestrup, E. V., Fazio, G. G., Woolaway, J., Blackwell, J., Brecher, K. 1993, I, 179–183.
White, S. M., Kundu, M. R. 1992, *Solar Phys.*, **141**, 347.
White, S. M., Kundu, M. R. 1993, I, 167–172.

COMMISSION 14: ATOMIC AND MOLECULAR DATA (DONNEES ATOMIQUES ET MOLECULAIRES)

PRESIDENT: W. L. Wiese
VICE PRESIDENT: W. H. Parkinson
ORGANIZING COMMITTEE: S. J. Adelman, J. Dubau, A. H. Gabriel, T. Kato, R. W. Nicholls, H. Nussbaumer, Z. Rudzikas, S. Sahal-Brechot, P. L. Smith

In recent years the interaction between astrophysics and atomic physics has entered a new stage. On one hand, increasingly powerful, high resolution astronomical spectrometers and other observational instruments, especially those for recent space and astronomy missions, have created vastly increased demands for more detailed and more accurate data on atomic and molecular processes. On the other hand, the advent of large-scale computing capabilities has been an important development for the atomic physics community, giving it opportunities to embark on comprehensive data generation projects that were unthinkable until recently. Indeed, several huge data projects have now come to successful conclusion. These projects have not only generated vast amounts of new data, but have also in many cases far surpassed the earlier data in their accuracy.

A principal task of this commission is to inform the astrophysics community of these new developments and provide assessments and recommendations. By its active participation in several recent and planned Joint Discussions with other IAU commissions, and by its sponsorship of relevant IAU workshops, this commission is providing the astrophysical community with information on the vast new atomic physics data capabilities that have been generated. Also, the commission provides the working group reports appearing below which highlight these new developments and list all important recent literature references in a convenient, annotated manner. The commission has expanded its scale of data gathering activities by adding a new working group on molecular reactions on solid surfaces and by adding in the collision part a section on rate coefficients for chemical reactions. The commission is fortunate to have engaged some energetic working group chairpersons for this task.

WORKING GROUP 1: ATOMIC SPECTRA AND WAVELENGTH STANDARDS

W. C. Martin, Chairperson

An extensive review of atomic spectroscopic data for astrophysics as available in 1991 has been published (Martin 1992). The references cited here are limited mainly to publications for the period 1991-1993 that include data for the elements having $Z \leqq 30$. Table 1 shows most of these references sorted according to spectrum. The references are designated by the first letter of the first author's last name and a serial number; these designations precede the appropriate full citations in the reference list at the end of this report.

A. Energy-Level Analyses, Line Classifications.

A few of the references comprise extensive new observations and analyses of particular spectra included in Table 1. Persson *et al.* (1991) give some 750 classified lines of Ne III in the range 430–12000 Å; their energy levels can be used to obtain new Ritz-principle extreme UV wavelengths down to almost 200 Å, with uncertainties smaller than 0.001 Å. Ekberg's (1993) investigation of Fe III has almost doubled the number of classified lines to about 3200, mainly in the range from 600 to 2100 Å. The complete line list resulting from an extension of the Ni III analysis by Garcia-Riquelme and Rico (1992) covers the range 1290–8680 Å. Forsberg's (1991) revision and extension of the Ti I analysis yielded more than 3300 lines in the range from 2000 Å to 5.5 μm. Similar research on Ni I by Litzén *et al.* (1993) has given about 2000 classified lines from 1700 Å to 5.35 μm. Most of the resulting level values for both Ti I and Ni I were determined from high-accuracy Fourier-Transform spectroscopy (FTS); the typical uncertainties in the range 0.002 to 0.005 cm^{-1} also apply to transition wavenumbers calculated from these levels.

J. Bergeron (ed.), Reports on Astronomy, Vol. XXIIA, 105–134.

TABLE 1

Selected references on energy levels, wavelengths, and line classifications for spectra of elements Z ≤ 30

C I	G3	Ar I	B1	Co I	P1
		Ar VI	R1	Co VIII-XXVII	S2
N III	B4	Ar VII	B6	Co XIII	S8
N IV	B5				
		K VII	L1	Ni I	L2,P1
O I	Z2			Ni III	G1
O II	M1,M2,W1	Ca I	G3,U1	Ni X-XVIII,	S9
O IV	B4	Ca VIII	R2	XX-XXVI	
O V	B5			Ni XIV	S8
O VII	I1	Ti I	F1,P1	Ni XVI	P3
		Ti X	P3	Ni XVI,XVII,	H1
F V	B4,E1	Ti XII	W2	XXI-XXIII	
F VI	B5				
		V I	P1	Cu X-XXIX	S3
Ne III	P2	V VI-XXIII	S5	Cu XIV	K4
		V XIX-XXII	B3	Cu XV	S8
Mg I-XII	K1			Cu XVI	S7
Mg I	A1,G3,S1	Cr I	B2		
Mg II	B7	Cr II	C3	Zn III	G2
Mg VIII	R2	Cr V-XXIV	S6	Zn XV	K4
		Cr XVI-XXII	S10	Zn XVI	S8
Al I-XIII	K2			Zn XVII	S7
Al I	B8	Mn VII-XXV	S4		
Al II	C2				
		Fe I	G3,J2,N1,		
Si I	G3		N2,N3,O1,		
Si VI	R2,T1,T2		P1,Z1		
Si VII	R2				
		Fe II	N1,N2,R3		
S I-XVI	K3	Fe III	E2		
S III	J1	Fe XII	S8		
S VI	W3	Fe XIV	P3		
S VIII	R2	Fe XXV	D1		
Cl I	C1				
Cl VII	W3				

Recent investigations of Fe I and Fe II are based on, or include, FTS measurements of a hollow-cathode spectrum. Johansson and Learner's (1990) analysis of Fe I in the infrared yielded about 360 classified lines, and Nave and Johansson's (1993) further extension of this analysis includes some 2200 classified lines between 1700 Å and 5 μm. Improved wavenumbers for many Fe I lines can also be obtained from the new values given for 442 levels by O'Brian *et al.* (1991). Rosberg and Johansson (1992) give 220 newly classified Fe II lines in the 9350-10670 cm^{-1} range.

A considerable amount of ongoing work on the first and/or second spectra of iron-group elements is centered at the University of Lund (Johansson 1993) and at Imperial College, London (Learner 1993); reports from these laboratories mention spectra of Ti, V, Cr, Mn, Fe, and Co. Most of the new observations include both FTS and high-dispersion vacuum-UV measurements, with several of the investigations involving collaborations with other laboratories.

A reference to Geller's (1992) list of solar lines in the 5000−600 cm^{-1} infrared region is included in Table 1 for C I, Mg I, Si I, Ca I, and Fe I. These solar observations made during the 1985 ATMOS experiment yielded spectra that may include more extensive and/or more accurately measured wavelengths for some atomic species than any available laboratory data. Similarly, the wavelengths of several infrared coronal lines as determined by Reconditi and Oliva (1993) from spectra of a planetary nebula are the most accurate available values.

Recent high-resolution observations of the spectra of chemically peculiar stars with the Hubble Space Telescope have stimulated laboratory research on the spectra of a number of elements having Z > 30. The current program at Lund, for example, includes new measurements and analysis of Zr II, Nb II, Mo II, Ru II, Pd II, Ag II, Ir II, and Au II (Johansson 1993), all of which are known or expected to be present in the spectrum of the B_p star χ Lupi (Leckrone *et al.* 1993). Recently published papers on Ru II (Joueizadeh and Johansson 1993), Pt II (Sansonetti *et al.* 1992, Blaise and Wyart 1992), Pt III (Ryabtsev *et al.* 1993), Au II (Wyart *et al.* 1992), and Hg III (Wyart *et al.* 1992, Sansonetti and Reader 1993) are all of interest for analysis of the spectrum of χ Lupi. An extensive analysis of Y II has been published (Nilsson *et al.* 1991).

Several references in this report are to papers presented at an international colloquium held in September 1992; additional reviews and presentations of atomic data of interest for astronomy will be found in the published proceedings (Leckrone and Sugar 1993).

B. Wavelength Standards.

The group at Imperial College has measured FTS spectra of an iron-neon hollow-cathode lamp and selected about 1100 Fe I and Fe II lines as recommended reference wavelengths in the range from 1830 Å to 4.25 μm. The most recent papers are by Nave *et al.* (1991, 1992); the 1991 paper gives wavelengths for 534 lines in the region 1830−3841 Å, with an uncertainty of 0.002 cm^{-1} (0.00008 to 0.00015 Å), and the 1992 paper has 290 lines from 4235 Å to 4.25 μm, with uncertainties of 0.001 to 0.0005 cm^{-1}.

Brault (1993) has measured the wavenumbers of selected Ar I lines excited in a low-pressure emission source with precision better than 0.00001 cm^{-1}. The absolute transition wavenumbers and energy-level values from this work will give new standards for FTS emission spectrometry in the 4170-14350 cm^{-1} region.

Users of ultraviolet reference lines from platinum-neon hollow-cathode lamps now have available an atlas of the spectrum from 1130−4330 Å (Sansonetti *et al.* 1992). The wavelengths of some 3000 of the 5600 lines included in an accompanying table are accurate to 0.0020 Å, with some of the Pt II Ritz-type wavelengths having uncertainties down to 0.0004 Å.

C. Compilations of Laboratory Data, Databases and Bibliographies.

Morton's (1991) compilation of data for 2031 resonance lines covers all spectra of hydrogen through germanium having allowed absorption transitions originating on ground-term levels with wavelengths longer than 911.75 Å. This compilation includes transition probabilities and other data, but is cited here because Morton tabulated the most accurate available wavelengths.

New compilations of wavelengths with energy-level classifications have been published for all spectra of magnesium, aluminum, and sulfur (Kaufman and Martin, 1991a, 1991b, 1993), and similar compilations for sodium and silicon are underway. The data are mainly based on earlier NIST compilations of the energy levels, but these are being updated and extended for many of the spectra to produce more accurate and complete wavelengths as calculated from the level values. A recent compilation of energy levels and wavelengths for O II has data based on newly optimized level values (Martin *et al.*, 1993). C. E. Moore's tables of energy levels and multiplets for hydrogen, carbon, nitrogen, and oxygen spectra, originally published during the period 1970−1985, have been collected into a single volume that includes data for O II based on the new compilation (Moore, 1993).

The most recent compilations by Shirai *et al.* of wavelengths and energy-level classifications for high ionization spectra of several elements ranging from vanadium to copper are also included in Table 1.

A fairly recent description of several of the larger and more inclusive collections of computer-readable atomic data (Martin 1992) need not be repeated here. As an addition, however, it may be noted that the library of computer-accessible data maintained by the Collaborative Computational Project No. 7 includes an extensive collection of atomic data (Jeffery 1993).

A biannual series of bulletins on atomic and molecular data for fusion includes references for recent papers with data on atomic energy levels and wavelengths (Botero 1992).

References

A1 Andersen, T., *et al.*, 1991. J. Phys. B **24**, 905-920.
B1 Baig, M. A., *et al.*, 1992. Phys. Rev. A **45**, 7963-7968.
B2 Baig, M. A., *et al.*, 1990. J. Phys. B **23**, 3489-3509.
B3 Beiersdorfer, P., *et al.*, 1991. Phys. Rev. A **44**, 396-409.
Blaise, J., Wyart, J.-F., 1992. J. Res. Natl. Inst. Stand. Tech. **97**, 217-223.
B4 Blanke, J. H., *et al.*, 1992. Phys. Scr. **45**, 430-435.
B5 Blanke, J. H., *et al.*, 1991. Phys. Scr. **44**, 436-441.
B6 Boduch, P., *et al.*, 1992. Phys. Scr. **45**, 203-211.
Botero, J., (Ed.), 1992. International Bulletin on Atomic and Molecular Data for Fusion, Nos. 42-45. Vienna: International Atomic Energy Agency.
B7 Brage, T., Gaardsted, J. O., 1991. Phys. Scr. **44**, 336-342.
Brault, J. W., 1992. in High Resolution Fourier Transform Spectrroscopy, Technical Digest **21**, 11-13. Washington, DC: Optical Society of America. See also W. Whaling and J. W. Brault, 1992. *Ibid.*, p.10.
B8 Buurman, E. P., *et al.*, 1990. J. Phys. B **23**, 2687-2695.
C1 Cantu, A. M., Mazzoni, M., 1991. Phys. Scr. **43**, 472-474.
C2 Costello, J. T., *et al.*, 1992. J. Phys. B **25**, 5055-5068.
C3 Costello, J. T., *et al.*, 1991. Phys. Rev. A **43**, 1441-1450.
D1 Decaux, V., *et al.*, 1991. Phys. Rev. A **43**, 228-234.
E1 Engström, L., Reistad, N., Westerlind, M., 1991. Phys. Scr. **44**, 548-554.
E2 Ekberg, J. O., 1993. Astron. Astrophys. Suppl. (in press).
F1 Forsberg, P., 1991. Phys. Scr. **44**, 446-476.
G1 Garcia-Riquelme, O., Rico, F. R., 1992. Phys. Scr. **45**, 212-230.
G2 Gayasov, R. R., Ryabtsev, A. N., 1992. Phys. Scr. **45**, 322-327.
G3 Geller, M., 1992. A high-resolution atlas of the infrared spectrum of the sun and the earth atmosphere from space. Washington: Natl. Aeron. Space Admin., NASA Ref. Publ. 1224, Vol. III.
H1 Hinnov, E., *et al.*, 1990. J. Opt. Soc. Am. B **7**, 2002-2005.
I1 Isler, R. C., Jupén, C., Martinson, I., 1993. Phys. Scr. **47**, 32-37.
Jeffery, C. S., (Ed.), 1993. Daresbury Laboratory Newsletter on Analysis of Astronomical Spectra, No. 18. Fife, Scotland: Dept. of Physics and Astron., Univ. of St. Andrews.
J1 Johansson, L., *et al.*, 1992. Phys. Scr. **46**, 221-224.
Johansson, S., 1993. private communication. Lund, Sweden: Physics Dept., University of Lund.
J2 Johansson, S., Learner, R. C. M., 1990. Astrophys. J. **354**, 755-762.
Joueizadeh, A., Johansson, S., 1993. NIST Spec. Publ. **850**, 131-133 (Sugar, J., Leckrone, D., Eds.) Washington, D.C.: U.S. Gov't. Printing Office.
K1 Kaufman, V., Martin, W. C., 1991a. J. Phys. Chem. Ref. Data **20**, 83-152.
K2 Kaufman, V., Martin, W. C., 1991b. J. Phys. Chem. Ref. Data **20**, 775-858.
K3 Kaufman, V., Martin, W. C., 1993. J. Phys. Chem. Ref. Data **22**, 279-375.
K4 Kaufman, V., Sugar, J., Rowan, W. L., 1990. J. Opt. Soc. Am. B **7**, 1169-1175.
Learner, R. C. M., 1993. private communication. London SW7 2BZ, U.K.: Blackett Laboratory, Imperial College. See Murray, J. E., *et al.* 1993. NIST Spec. Publ. **850**, 154-156 (Sugar, J., Leckrone, D., Eds.) Washington, D.C.: U.S. Gov't. Printing Office.
Leckrone, D. S., and Sugar, J., Eds., 1993. The 4th International Colloquium on Atomic Spectra and Oscillator Strengths for Astrophysical and Laboratory Plasmas, Phys. Scr. **T47**. Poster

papers: Sugar, J., and Leckrone, D., Eds., 1993. NIST Spec. Publ. **850**. Washington, DC: U.S. Gov't. Printing Office.
Leckrone, D. S., *et al.*, 1993. Phys. Scr. **T47**, 149-156.
L1 Levashov, V. E., Ryabtsev, A. N., Churilov, S. S., 1990. Opt. Spectrosc. (USSR) **69**, 20-22.
L2 Litzén, U., Brault, J. W., Thorne, A. P., 1993. Phys. Scr. **47**, 628-673.
Martin, W. C., 1992. Lecture Notes in Physics **407**, 121-147. Chapter 8 in Smith, P. L., Wiese, W. L., (Eds.) Atomic and molecular data for space astronomy. Berlin: Springer-Verlag.
M1 Martin, W. C., Kaufman, V., Musgrove, A., 1993. J. Phys. Chem. Ref. Data **22**, 1179.
M2 Moore, C. E., 1993. Tables of spectra of hydrogen, carbon, nitrogen, and oxygen atoms and ions. Boca Raton, FL: CRC Press.
N1 Nave, G., *et al.*, 1991. J. Opt. Soc. Am. B **8**, 2028-2041.
N2 Nave, G., *et al.*, 1992. J. Phys. II (France) **2**, 913-929.
N3 Nave, G., Johansson, S., 1993. Astron. Astrophys. **274**, 961-967; also, Astron. Astrophys., Suppl. Series (in press).
Nilsson, A. E., Johansson, S., Kurucz, R. L., 1991. Phys. Scr. **44**, 226-257.
Morton, D. C., 1991. Astrophys. J., Suppl. Ser. **77**, 119-202.
O1 O'Brian, T. R., *et al.*, 1991. J. Opt. Soc. Am. B **8**, 1185-1201.
P1 Page, R. H., Gudeman, C. S., 1990. J. Opt. Soc. Am. B **7**, 1761-1771.
P2 Persson, W., *et al.*, 1991. Phys. Rev. A **43**, 4791-4823.
P3 Pinnington, E. H., *et al.*, 1990. Z. Phys. D **17**, 5-9.
R1 Raineri, M., *et al.*, 1992. Phys. Scr. **45**, 584-589.
R2 Reconditi, M., Oliva, E., 1993. Astron. Astrophys. **274**, 662-666.
R3 Rosberg, M., Johansson, S., 1992. Phys. Scr. **45**, 590-597.
Ryabtsev, A. N., *et al.*, 1993. Phys. Scr. **47**, 45-58.
Sansonetti, C. J., Reader, J., 1993. Phys. Rev. A **47**, 3080-3084.
Sansonetti, J. E., *et al.*, 1992. J. Res. Natl. Inst. Stand. Tech. **97**, 1-211.
S1 Schinn, G. W., Dai, C. J., Gallagher, T. F., 1991. Phys. Rev. A **43**, 2316-2327.
S2 Shirai, T., *et al.*, 1992a. J. Phys. Chem. Ref. Data **21**, 23-122.
S3 Shirai, T., *et al.*, 1991. J. Phys. Chem. Ref. Data **20**, 1-81.
S4 Shirai, T., *et al.*, 1993b. J. Phys. Chem. Ref. Data, to be published.
S5 Shirai, T., *et al.*, 1992b. J. Phys. Chem. Ref. Data **21**, 273-390.
S6 Shirai, T., *et al.*, 1993a. J. Phys. Chem. Ref. Data **22**, Issue No. 5.
S7 Sugar, J., Kaufman, V., Rowan, W. L., 1990. J. Opt. Soc. Am. B **7**, 152-158.
S8 Sugar, J., Kaufman, V., Rowan, W. L., 1991. J. Opt. Soc. Am. B **8**, 22-26.
S9 Sugar, J., Kaufman, V., Rowan, W. L., 1992. J. Opt. Soc. Am. B **9**, 344-346.
S10 Sugar, J., Kaufman, V., Rowan, W. L., 1993., J. Opt. Soc. Am. B **10**, 13-15.
T1 Trigueiros, A. G., Jupén, C., Redfors, A., 1991. Phys. Scr. **44**, 191-194.
T2 Trigueiros, A. G., Jupén, C., Redfors, A., 1992. J. Phys. B **25**, 2765-2770.
U1 Ueda, K., *et al.*, 1990. Phys. Scr. **41**, 75-77.
W1 Wenåker, I., 1990. Phys. Scr. **42**, 667-684.
W2 Westerlind, M., 1990. Phys. Rev. A **42**, 3996-4000.
W3 Westerlind, M., *et al.*, 1991. Phys. Scr. **44**, 582-586.
Wyart, J.-F., *et al.*, 1992. J. Phys. II (France) **2**, 895-912.
Z1 Zhu, Y., Knight, R. D., 1992. J. Opt. Soc. Am. B **9**, 27-31.
Z2 Zink, L. R., *et al.*, 1991. Astrophys. J. **371**, L85-L86.

WORKING GROUP 2: ATOMIC TRANSITION PROBABILITIES

W. L. Wiese, Chairperson

The Data Center on Atomic Transition Probabilities at the National Institute of Standards and Technology (NIST), formerly the National Bureau of Standards, in Gaithersburg, Maryland 20899, USA, is continuing its critical data compilation work and is also engaged in developing a comprehensive numerical database. In addition, the existing bibliographical database is continuously updated. The current compilation work concentrates on tabulating transition probabilities for allowed, as well as forbidden lines of light elements.

Extensive data tables for all ions of carbon, nitrogen, and oxygen are near completion. Work is also progressing on other light elements, specifically hydrogen, helium, lithium, beryllium, boron, fluorine, and neon.

TABLE 1
Recent literature sources for atomic
transition probability data of astrophysical interest

This table covers the 3 year period since the publication of our last IAU report (Reports on Astronomy, Volume XXIA, 111 (1991); preparation date: July, 1990) to the present (August 1993). The table is arranged in alphabetical order of element symbols, with further subdivisions according to stage of ionization (I, II, etc.). The numbers are the running numbers of the reference list following this table.

Al I: 40,53,106,121
Al II: 52,85,114
Al III: 65

Ar I: 81
Ar II: 35

B II: 52,85
B IV: 43

Ba II: 55,102,123

Be I: 85
Be II: 14,55
Be III: 43,85

Bi I: 44

C I: 69,72,117
C III: 7,8,17,85
C IV: 17,37,50,92
C V: 43,85,129

Ca I: 85
Ca II: 11,20,55,70,102,135

Cd I: 106,127
Cd II: 20,88,126

Cl II: 134

Co I: 106

Cr I: 106,115
Cr II: 1,112,115,118
Cs I: 13,74,102,106,123

Cu I: 2,27,106,125

F I: 38

Fe I: 12,57,99,111,115
Fe II: 4,10,29,57,71,
73,101,104,111,115
Fe III: 111
Fe IV: 111
Fe VII: 107,108,109
Fe VIII: 109,135
Fe X: 64

Ga I: 40
Ga II: 52

He I: 3,43,45,47,62,
113,120,122,128

Hg I: 95,106
Hg II: 41,132
Hg III: 41

In II: 52

K I: 74,88,97,135

Kr I: 81,106
Kr II: 35

Li I: 13,85,92,97
Li II: 43,85

Mg I: 9,18,28,48,63,
79,84,85,106
Mg II: 55,88,93

N I: 21,25,26,67,68,
69,75,106,117
N II: 15,21,24,42,60,
61,134
N III: 33,80,82,83,119,130
N IV: 7,8,34,49,85,
87,124
N V: 14,50,85,86
N VI: 43

Na I: 13,44,74,85,88,
93,97

Ne I: 2,46,66,81,91,100
Ne II: 32,35,38
Ne III: 134

O I: 22,30,31,39,76,
94,116,117
O II: 23,134
O III: 5,6,16
O IV: 33
O V: 7,34,85,124
O VI: 50,92
O VII: 43

P I: 96

Rb I: 13,74,123

S I: 54,56,96
S II: 78,134
S III: 134
S V: 85

Si I: 96,98,133
Si II: 19,58,59,77
Si III: 19,85
Si IV: 19

Sr I: 131
Sr II: 55,70,102,123

Ti I: 89,90,103,105
Ti II: 110
Ti IV: 135
Ti VI: 64

Tl II: 52

Xe I: 81

Zn I: 36,51,95
Zn II: 20

Some outstanding developments in this field during the last three years have been: (a) the completion of the Opacity Project including the publication of several more papers summarizing the very extensive and sophisticated calculations of transition probabilities for various light element sequences [7,8,21,39,109,124]; (b) calculations of accurate transition probabilities with the configuration-interaction computer code CIV 3 by Hibbert and co-workers for neutral carbon, nitrogen and oxygen and some of their lower ions [5,23,24,75,119]; (c) very precise calculations for ions of the helium sequence [43], and (d) various very accurate measurements of atomic lifetimes [15-17,60,83,113] including those of levels that combine to the ground state only via intersystem lines [42].

The main part of this report is a bibliography containing selected recent literature references which contain new transition probability data of astrophysical interest produced during the last three-year period. Thus, this new selected bibliography continues where the last working group report left off. As in the previous reports, the bibliographical material is ordered to element and state of ionization. Table 1 provides an overview of the bibliographical data by spectrum. The references are identified by a running number which refers to the general reference list at the end of this report. In the general reference list the literature is ordered alphabetically according to first author and each reference contains one or more code letters indicating the method applied by the authors. These code letters are defined as follows:

THEORETICAL METHODS:
- Q - quantum mechanical (including self-consistent field) calculations.
- CA- Coulomb approximation
- I - interpolation within isoelectronic sequences, spectral series, or homologous atoms; also, data that are presented in graphical, rather than tabular form.

EXPERIMENTAL METHODS:
- E - measurements in emission (arc, furnace, discharge tube, shock tube, etc.).
- A - measurements in absorption (King furnace, absorption tube, etc.).
- L - lifetime measurements (including Hanle-effect).
- H - anomalous dispersion (hook) measurements.
- M - miscellaneous experimental methods (for example, Stark effect, astrophysical measurements, etc.).

OTHER:
- CM- additions or suggested revisions to data in previous articles, comments on particular theoretical or experimental methods, etc.
- CP- data compilation.
- R - relative (non-absolute) oscillator strengths have been determined.
- F - data on forbidden (i.e. other than electric dipole) transitions have been determined.

References

1. Aashamar, K., Luke, T. M., Talman, J. D.: Phys. Scr. **42**, 145 (1990). Q
2. Abdel Rahman, A. N., Abdallah, M. H., El Ghandour, M. A.: Egypt. J. Phys. **11**, 79 (1980). L
3. Abrashkevich, A. G., Abrashkevich, D. G., Gaysak, M. I., Lendyel, V. I., Puzynin, I. V., Vinitsky, S.I.: Phys. Lett. A **152**, 467 (1991). Q
4. Adam, J., Baschek, B., Johansson, S., Nilsson, A. E., Brage, T.: Astrophys. J. **312**, 337 (1987). Q
5. Aggarwal, K. M., Hibbert, A.: J. Phys. B **24**, 3445 (1991). Q
6. Aggarwal, K. M., Hibbert, A.: J. Phys. B **24**, 4685 (1991). Q
7. Allard, N., Artru, M.-C., Lanz, T., Le Dourneuf, M.: Astron. Astrophys., Suppl. Ser. **84**, 563 (1990). CP
8. Allard, N., Artru, M.-C., Lanz, T., Le Dourneuf, M.: Astron. Astrophys., Suppl. Ser. **91**, 399 (1991). CP
9. Andersen, T., Brage, T., Froese-Fischer, C., Sorensen, L. Eg.: J. Phys. B **24**, 905 (1991). L
10. Ansbacher, W., Li, Y., Pinnington, E. H. 1990. Atomic Spectra and Oscillator Strengths for Astrophysics and Fusion Research. J. E. Hansen (Ed.), Amsterdam: North-Holland, 172-173.

11. Arbes, F., Gudjons, T., Kurth, F., Werth, G., Marin, F., Inguscio, M.: Z. Phys. D **25**, 295 (1993). LF
12. Bard, A., Kock, A., Kock, M.: Astron. Astrophys. **248**, 315 (1991). E
13. Barrientos, C., Martin, I.: Can. J. Phys. **67**, 996 (1989). Q
14. Barrientos, C., Martin, I.: J. Quant. Spectrosc. Radiat. Transfer **48**, 287 (1992). Q
15. Bastin, T., Baudinet-Robinet, Y., Garnir, H. P., Dumont, P. D.: Z. Phys. D **24**, 343 (1992). L
16. Baudinet-Robinet, Y., Dumont, P. D., Garnir, H. P.: Phys. Rev. A **43**, 4022 (1991). L
17. Baudinet-Robinet, Y., Dumont, P.-D., Garnir, H.-P., El Himdy, A.: Phys. Rev. A **40**, 6321 (1989). L
18. Beck, D. R.: Phys. Rev. A **40**, 2887 (1989). Q
19. Becker, S. R., Butler, K.: Astron. Astrophys. **235**, 326 (1990). Q
20. Belinskii, S. V., Parashchuk, A. V., Smol'skii, V. N., Yudin, G. L.: Institute of Physics and Power Engineering Report No. 2162 (Obninsk, 1991). Q
21. Bell, K. L., Berrington, K. A.: J. Phys. B **24**, 933 (1991). Q
22. Bell, K. L., Hibbert, A.: J. Phys. B **23**, 2673 (1990). Q
23. Bell, K. L., Hibbert, A., McLaughlin, B. M., Higgins, K.: J. Phys. B **24**, 2665 (1991). Q
24. Bell, K. L., Ramsbottom, C. A., Hibbert, A.: J. Phys. B **25**, 1735 (1992). Q
25. Bengtsson, G. J., Hansen, K., Larsson, J., Schade, W., Svanberg, S.: Z. Phys. D **22**, 397 (1991). L
26. Bengtsson, G. J., Larsson, J., Svanberg, S., Wang, D. D.: Phys. Rev. A **45**, 2712 (1992). L
27. Bengtsson, J., Larsson, J., Svanberg, S., Wahlstroem, C.-G.: Phys. Rev. A **41**, 233 (1990). L
28. Beverini, N., Maccioni, E., Pereira, D., Strumia, F., Vissani, G., Wang, Y.-Z.: Opt. Commun. **77**, 299 (1990). L
29. Biemont, E., Baudoux, M., Kurucz, R. L., Ansbacher, W., Pinnington, E. H.: Astron. Astrophys. **249**, 539 (1991). L
30. Biemont, E., Hibbert, A., Godefroid, M., Vaeck, N., Fawcett, B. C.: Astrophys. J. **375**, 818 (1991). Q
31. Biemont, E., Zeippen, C. J.: Astron. Astrophys. **265**, 850 (1992). Q
32. Blagov, K. B., Borisov, E. N., Burshtein, M. L.: Opt. Spectrosc. (USSR) **67**, 601 (1989). L
33. Blanke, J. H., Fricke, B., Heckmann, P. H., Traebert, E.: Phys. Scr. **45**, 430 (1992). L
34. Blanke, J. H., Fricke, B., Heinemann, D., Heckmann, P. H., Moeller, G., Traebert, E.: Phys. Scr. **44**, 436 (1991). L, Q
35. Bogdanovich, P. O., Kanyauskas, Yu. M., Karaziya, R. I., Merkelis, G. M., Rudzikas, Z. B., Tutlis, V. I.: Sov.Phys. J. **33**, 631 (1990). Q
36. Brage, T., Froese Fischer, C.: Phys. Scr. **45**, 43 (1992). Q
37. Burke, V. M.: J. Phys. B **25**, 4917 (1992). Q
38. Burshtein, M. L., Vujnovic, V.: Astron. Astrophys. **247**, 252 (1991). CM, CP
39. Butler, K., Zeippen, C. J.: J. Phys. IV, Colloq. C1 1, C1-14 (1991). Q
40. Buurman, E. P., Doenszelmann, A.: Astron. Astrophys. **227**, 289 (1990). L
41. Calamai, A. G., Johnson, C. E.: Phys. Rev. A **42**, 5425 (1990). LF
42. Calamai, A. G., Johnson, C. E.: Phys. Rev. A **44**, 218 (1991). L
43. Cann, N. M., Thakkar, A. J.: Phys. Rev. A **46**, 5397 (1992). Q
44. Carlsson, J.: Z. Phys. D **9**, 147 (1988). L
45. Chan, W. F., Cooper, G., Brion, C. E.: Phys. Rev. A **44**, 186 (1991). M
46. Chan, W. F., Cooper, G., Guo, X., Brion, C. E.: Phys. Rev. A **45**, 1420 (1992). M
47. Chan, W.-F., Cooper, G., Sze, K.-H., Brion, C. E.: J. Phys. B **23**, L523 (1990). M
48. Chang, T. N.: Phys. Rev. A **41**, 4922 (1990). Q
49. Chang, T. N., Mu, Y.: J. Quant. Spectrosc. Radiat. Transfer **44**, 413 (1990). Q
50. Charalambidis, D., Brenn, R., Koulen, K. J.: Phys. Rev. A **40**, 2359 (1989). L, LF
51. Czajkowski, M., Bobkowski, R., Krause, L.: Spectrochim. Acta, Part B **46**, 1161 (1991). L
52. Das, B. P., Idrees, M.: Phys. Rev. A **42**, 6900 (1990). Q
53. Davidson, M. D., Volten, H., Doenszelmann, A.: Astron. Astrophys. **238**, 452 (1990). E, L
54. Delalic, Z., Erman, P., Kaellne, E.: Phys. Scr. **42**, 540 (1990). L
55. Djerad, M. T.: J. Phys. II **1**, 1 (1991). Q
56. Doering, J. P.: J. Geophys. Res., Sect. A **95**, 21313 (1990). M
57. Drake, J. J., Smith, G.: Mon. Not. R. Astron. Soc. **250**, 89 (1991). M

58. Dufton, P. L., Keenan, F. P., Hibbert, A., Ojha, P. C., Stafford, R. P.: Astrophys. J. **387**, 414 (1992). Q
59. Dufton, P. L., Keenan, F. P., Hibbert, A., Stafford, R. P., Byrne, P. B., Agnew, D.: Mon. Not. R. Astron. Soc. **253**, 474 (1991). Q, QF
60. Dumont, P.-D., Garnir, H.-P., Baudinet-Robinet, Y.: Z. Phys. D **21**, 209 (1991). L
61. Ellis, D. G.: Phys. Rev. A **47**, 161 (1993). Q
62. Erman, P., Sundstroem, G.: Phys. Rev. A **43**, 5790 (1991). L
63. Farooq, W. A., Nawaz, M., Connerade, J. P., Marangos, J. P.: J. Phys. B **25**, 4141 (1992). M
64. Fawcett, B. C.: At. Data Nucl. Data Tables **47**, 319 (1991). Q
65. Gaardsted, J. O., Brage, T., Froese Fischer, C., Sonnek, D.: Phys. Scr. **42**, 543 (1990). L, Q
66. Garcia, G., Ortiz, M., Blanco, F., Sanchez, J. A., Campos, J.: Can. J. Phys. **67**, 977 (1989). L
67. Godone, A., Novero, C.: Phys. Rev. A **45**, 1717 (1992). L
68. Goldbach, C., Ludtke, T., Martin, M., Nollez, G.: Astron. Astrophys. **266**, 605 (1992). E
69. Goldbach, C., Nollez, G.: J. Phys. IV, Colloq. C1 **1**, C1-83 (1991). CM
70. Guet, C., Johnson, W. R.: Phys. Rev. A **44**, 1531 (1991). Q
71. Guo, B., Ansbacher, W., Pinnington, E. H., Ji, Q., Berends, R. W.: Phys. Rev. A **46**, 641 (1992). L
72. Haar, R. R., Curtis, L. J., Kvale, T. J., Beideck, D. J., Martinson, I., Hellborg, R.: Astron. Astrophys. **241**, 321 (1991). L
73. Hannaford, P., Lowe, R. N., Grevesse, N., Noels, A.: Astron. Astrophys. **259**, 301 (1992). L
74. He, X., Li, B., Chen, A., Zhang, C.: J. Phys. B **23**, 661 (1990). Q
75. Hibbert, A., Biemont, E., Godefroid, M., Vaeck, N.: Astron. Astrophys., Suppl. Ser. **88**, 505 (1991). Q
76. Hibbert, A., Biemont, E., Godefroid, M., Vaeck, N.: J. Phys. B **24**, 3943 (1991). Q
77. Hibbert, A., Ojha, P. C., Stafford, R. P.: J. Phys. B **25**, 4153 (1992). Q
78. Ho, Y. K., Henry, R. J. W.: Astrophys. J. **351**, 701 (1990). Q
79. Hoang-Binh, D.: Astron. Astrophys., Suppl. Ser. **97**, 769 (1993). CA
80. Kastner, S. O., Bhatia, A. K.: Astrophys. J. **381**, L59 (1991). Q
81. Korotkov, A. I., Khromov, N. A.: Opt. Spectrosc. (USSR) **69**, 721 (1990). QF
82. Kotze, T. C., van der Westhuizen, P., Visser, K.: Phys. Rev. A **44**, 4180 (1991). L
83. Kotze, T. C., van der Westhuizen, P., Visser, K.: J. Quant. Spectrosc. Radiat. Transfer **49**, 213 (1993). L
84. Larsson, J., Svanberg, S.: Z. Phys. D **25**, 127 (1993). L
85. Laughlin, C. 1990. Atomic Spectra and Oscillator Strengths for Astrophysics and Fusion Research. J. E. Hansen (Ed.), Amsterdam: North-Holland, 108-113.
86. Laughlin, C.: Z. Phys. D **16**, 9 (1990). Q
87. Laughlin, C.: Phys. Scr. **42**, 551 (1990). Q
88. Laughlin, C.: Phys. Scr. **45**, 238 (1992). Q
89. Lawler, J. E.: Astron. Astrophys. **252**, 853 (1991). CM, L
90. Lowe, R. M., Hannaford, P.: Z. Phys. D **21**, 205 (1991). L
91. Luke, T. M.: J. Phys. B **19**, 843 (1986). Q
92. Martin, I., Karwowski, J.: J. Phys. B **24**, 1539 (1991). Q
93. Martin, I., Karwowski, J., Lavin, C., Diercksen, G. H. F.: Phys. Scr. **44**, 567 (1991). I, Q
94. Mason, N. J.: Meas. Sci. Technol. **1**, 596 (1990). L
95. Migdalek, J., Stanek, M.: Phys. Rev. A **41**, 2869 (1990). Q
96. Mukherjee, P. K., Ohno, K.: Phys. Rev. A **40**, 1753 (1989). Q
97. Nawaz, M., Farooq, W. A., Connerade, J.-P.: J. Phys. B **25**, 5327 (1992). M
98. O'Brian, T. R., Lawler, J. E.: Phys. Rev. A **44**, 7134 (1991). CP, L
99. O'Brian, T. R., Wickliffe, M. E., Lawler, J. E., Whaling, W., Brault, J. W.: J. Opt. Soc. Am. B **8**, 1185 (1991). E, L
100. Ortiz, M., Garcia, G., Blanco, F., Campos, J.: Phys. Scr. **42**, 689 (1990). L
101. Pauls, U., Grevesse, N., Huber, M. C. E.: Astron. Astrophys. **231**, 536 (1990). E
102. Poirier, M.: Z. Phys. D **25**, 117 (1993). I, Q, QF
103. Ramaswami, S., Reeves, R. R., Rutten, M., Halstead, J. A.: J. Quant. Spectrosc. Radiat. Transfer **49**, 303 (1993). L
104. Rosberg, M., Johansson, S.: Phys. Scr. **45**, 590 (1992). Q

105. Salih, S., Lawler, J. E.: Astron. Astrophys. **239**, 407 (1990). L
106. Saloman, E. B.: Spectrochim. Acta, Part B **46**, 319 (1991). CP, Q
107. Saraph, H. E.: J. Phys. IV, Colloq. C1 1, C1-15 (1991). Q
108. Saraph, H. E., Storey, P. J. 1990. Atomic Spectra and Oscillator Strengths for Astrophysics and Fusion Research. J. E. Hansen (Ed.), Amsterdam: North-Holland, 128-129.
109. Saraph, H. E., Storey, P. J., Taylor, K. T.: J. Phys. B **25**, 4409 (1992). Q
110. Savanov, I. S., Huovelin, J., Tuominen, I.: Astron. Astrophys., Suppl. Ser. **86**, 531 (1990). CP
111. Sawey, P. M. J., Berrington, K. A.: J. Phys. B **25**, 1451 (1992). Q
112. Schade, W., Mundt, B., Helbig, V.: Phys. Rev. A **42**, 1454 (1990). L
113. Schmoranzer, H., Roth, H., Volz, U., Marger, D.: J. Phys. B **24**, 595 (1991). L
114. Serrao, J. M. P.: J. Quant. Spectrosc. Radiat. Transfer **45**, 121 (1991). Q
115. Sigut, T. A. A., Landstreet, J. D.: Mon. Not. R. Astron. Soc. **247**, 611 (1990). CM
116. Smith, A. V., Raymond, T. D., Riley, M. E.: Phys. Rev. A **45**, 4688 (1992). M
117. Sohns, E., Kock, M.: J. Quant. Spectrosc. Radiat. Transfer **47**, 335 (1992). E
118. Sperger, R., Neger, T.: Opt. Commun. **80**, 401 (1991). MR
119. Stafford, R. P., Hibbert, A., Bell, K. L.: Mon. Not. R. Astron. Soc. **260**, L11 (1993). Q
120. Tang, J. Z., Watanabe, S., Matsuzawa, M.: Phys. Rev. A **46**, 3758 (1992). Q
121. Theodosiou, C. E.: Phys. Rev. A **45**, 7756 (1992). Q
122. Tiwary, S. N., Kumar, P.: Acta Phys. Pol. A **80**, 23 (1991). Q
123. Tiwary, S. N., Kumar, P., Singh, D. D., Sharma, R. J.: Fizika (Zagreb) **22**, 577 (1990). Q
124. Tully, J. A., Seaton, M. J., Berrington, K. A.: J. Phys. IV, Colloq. C1 **1**, C1-16 (1991). CM
125. v.d. Veer, W. E., Doenszelmann, A.: Z. Phys. D **17**, 159 (1990). E, L
126. Vaeck, N., Godefroid, M., Froese Fischer, C.: Phys. Rev. A **46**, 3704 (1992). Q, QF
127. Vaeck, N., Godefroid, M., Hansen, J. E.: J. Phys. B **24**, 361 (1991). Q, QF
128. van der Hart, H. W., Hansen, J. E.: J. Phys. B **25**, 41 (1992). Q
129. van der Hart, H. W., Hansen, J. E.: J. Opt. Soc. Am. B **26**, 641 (1993). Q
130. van der Westhuizen, P., Coetzer, F. J., Kotze, T. C., 1990. Atomic Spectra and Oscillator Strengths for Astrophysics and Fusion Research. J. E. Hansen (Ed.), Amsterdam: North-Holland, 182-183.
131. Werji, H. G. C., Greene, C. H., Theodosiou, C. E., Gallagher, A.: Phys. Rev. A **46**, 1248 (1992). EF, ER, Q, QF
132. Wilson, M. 1990. Atomic Spectra and Oscillator Strengths for Astrophysics and Fusion Research. J. E. Hansen (Ed.), Amsterdam: North-Holland, 144-145.
133. Wilson, M.: Z. Phys. D **21**, 7 (1991). QF
134. Yang, F., Cunningham, A. J.: J. Quant. Spectrosc. Radiat. Transfer **49**, 53 (1993). ER
135. Zilitis, V. A.: Opt. Spectrosc. (USSR) **67**, 595 (1989). Q

WORKING GROUP 3. ATOMIC AND MOLECULAR COLLISIONS

J. W. Gallagher, Chairperson

A. ELECTRON COLLISIONS WITH ATOMS AND ATOMIC IONS

New general resources include Volume 32 of "Advances in Atomic, Molecular, and Optical Physics" edited by M. Inokuti [56], scheduled for publication in late 1993, which assesses the current status of collisional cross section data. It includes surveys of collisional cross-section data sources, but not extensive listings of data. The proceedings of the Third Atomic Data Workshop, held at Meudon Observatoire de Paris, September, 1989, provides a variety of useful reports [108].

An annotated bibliography by Itikawa [57] lists papers on experimental and theoretical studies of electron-impact excitation and ionization of atomic ions. A bibliography by Pradhan and Gallagher [90] reviews the quality of theoretical electron impact excitation papers.

The Iron Project, an outgrowth of the international Opacity Project, will provide high-accuracy calculated data for electron collisions with astrophysically prominent atoms and atomic ions. Specifically, the current focus is the calculation of cross sections and rate coefficients for the fine structure transitions for infra-red spectral diagnostics, collisional data for all iron ions, and collisional and radiative data for the low ionization stages of iron (Fe I-VI), nickel and other iron-group elements. The status of this work is described in three new publications [55,65,109].

Important reviews by Cartwright et. al. [33] and by Trajmar et. al. [102] summarize and review data for electron impact excitation of helium. Another important article by Pradhan [88] reviews electron impact excitation for Fe I - Fe VIII. Many individual studies of electron-impact excitation of neutral atoms and atomic ions have been reported in the last three years and these are listed in Table 1. New studies of electron-impact ionization are listed in Table 2.

Table 1. References on electron impact excitation of neutral atoms and atomic ions.

H seq	38	O II	75	Si X	109
H I	68,69,3,34	O III	9,22,2	S I	110
H I	4,48,98	O IV	66,25	S II	54,32
He I	33,102,70,47	O V	60	S VII	81
He I	95,97,84,50	O VI	17	Cl IX	77
He I*	71	F IV	40	Ar I	28,29,24
He II	8	Ne I	28	Ar III	59
Li seq	111	Ne V	64	Ar VII	52
Li I	26	Ne VI	109	Ar VIII	17
Li II	19	Na I	82,23,27	Ar IX	80
Be seq	110	Na VI	40	Ar XIB	109
Be I	49	Mg I	39	Ca II	105,106
Be II	63	Mg II	63	Ca XV	5,6
Be ions	18	Mg IV	76	Ca XVI	109
B I	83	Mg VII	30	Ca XX	11
B ions	18	Mg VIII	109	Fe II	89,91
C II	66,25	Al III	82	Fe III	20
C IV	17,31	Al IX	109	Fe VII	61
C VI	10	Si II	45	Fe IX	46
N I	44	Si III	52	Fe XVII	21
N III	66,25,100	Si IV	17,104	Fe XXI	1
O I	62,58,72,103	Si VI	79,78	Fe XXVI	12

Table 2. Studies of electron-impact ionization.

He I	94	Ar I	74
C IV, N V, O VI	93,94	Ar III	67
O I	37	Ar VII	92,113
Ne VIII	41	Ar IX	125
Mg I	73	Ar XVI, Fe XXIV	112
Mg II	85	Ca II	15
Cl II	42	Fe ions	16
Si II, Si III	43	Kr ions	67,35,16,36
Si VII, Si VIII	107		

Progress has been made in dielectronic and radiative recombination, both experimentally and theoretically. Pequignot et al. present parametric fits for all H, He, C, N, and O ions [86].

Data references contain data in the form of cross sections, collision strengths, and/or rates. Each reference is annotated with a (T) theoretical; (E) experimental; (E/T) both.

References

1. Aggarwal, K. M., Astrophys. J., Suppl. Ser. **77**, 677 (1991). (T)
2. Aggarwal, K. M., Astrophys. J., Suppl. Ser. **85**, 197 (1993). (T)
3. Aggarwal, K. M. et. al., J. Phys. B **24**, 1385 (1991). (T)
4. Aggarwal, K. M. et. al., J. Phys. B **24**, 1411 (1991). (T)
5. Aggarwal, K. M. et. al., J. Phys. B **24**, 2201 (1991). (T)
6. Aggarwal, K. M. et. al., Astrophys. J., Suppl. Ser. **72**, 815 (1990) (T); Erratum: Astrophys. J., Suppl. Ser. **77**, 441 (1991).
7. Aggarwal, K. M. et. al., J. Phys. B **24**, 1757 (1991). (T)
8. Aggarwal, K. M. et. al., Astrophys. J., Suppl. Ser. **80**, 473 (1992). (T)
9. Aggarwal, K. M. and A. Hibbert, J. Phys. B **24**, 3445 (1991). (T)
10. Aggarwal, K. M. and A. E. Kingston, J. Phys. B **24**, 4583 (1991). (T)
11. Aggarwal, K. M. and A. E. Kingston, J. Phys. B **25**, 751 (1992). (T)
12. Aggarwal, K. M. and A. E. Kingston, Astrophys. J., Suppl. Ser. **85**, 187 (1993).
13. Andersen, L. H. and J. Bolko, J. Phys. B **23**, 3167 (1990). (E)
14. Andersen, L. H. et. al., J. Phys. B **25**, 277 (1992). (E)
15. Badnell, N. R. et. al., J.Phys. B **24**, L275 (1991). (T)
16. Badnell, N. R. and M. S. Pindzola, Phys. Rev. A **47**, 2937 (1993). (T)
17. Badnell, N. R. et. al., Phys. Rev. A **43**, 2250 (1991). (T)
18. Berrington, K. A. and R. E. H. Clark, Nucl. Fusion **34**, 87 (1993). (T)
19. Berrington, K. A. and S. Nakazaki, J. Phys. B **24**, 1411 (1991). (T)
20. Berrington, K. A. et. al., J. Phys. B **24**, 3467 (1991). (T)
21. Bhatia, A. K., G. A. Doschek, Atomic Data Nuc. Data Tables **52**, 1 (1992).
22. Bhatia, A. K., S. O. Kastner, Atomic Data Nuc. Data Tables **54**, 133 (1993).
23. Bielschowsky, C. E. et. al., PRA **43**, 5975 (1991). (T)
24. Blanco, F. et. al., J. Phys. B **25**, 3531 (1992). (T)
25. Blum, R. D. and A. K. Pradhan, Astrophys. J., Suppl. Ser. **80**, 425 (1992). (T)
26. Bray, I. et. al., Phys. Rev. A **47**, 1101 (1993). (T)
27. Bray, I. et. al., Phys. Rev. A **44**, 7179 (1991). (T)
28. Bubelev, V. E. and A. N. Grum-Grzhimailo, Opt. Spectrosc. (USSR) **69**, 178 (1990). (T)
29. Bubelev, V. E. and A. N. Grum-Grzhimailo, J. Phys. B **24**, 2183 (1991). (T)
30. Burgess, A.et. al., Astrophys. J. **376**, 803 (1992). (T)
31. Burke, V. M., J. Phys. B **25**, 4917 (1992). (T)
32. Cai, W. and A. K. Pradhan, Astrophys. J., Suppl. Ser. (1993) (in press). (T)
33. Cartwright, D. C. et. al., Phys. Rev. A **45**, 1602 (1992). (E/T)
34. Chang, E. S. et. al, Astron. Astrophys. **247**, 580 (1991).
35. Chen M. H. and K. J. Reed, Phys. Rev. A **45**, 4525 (1992). (T)
36. Chen, M. H. and K. J. Reed, Phys. Rev. A **47**, 1874 (1993). (T)
37. Chung, S. et. al., Phys. Rev. A **47**, 3867 (1993). (T)
38. Clark, R. E. H., Astrophys. J. **354**, 382 (1990). (T)
39. Clark, R. E. H. and J. Abdallah, Jr., Phys. Rev. A, **44**, 2874 (1991). (T)
40. Conlon, E. S. et. al., PS **45**, 309 (1992). (T)
41. DeFrance, P. et. al., J. Phys. B **23**, 2333 (1990). (E)
42. Djuric, N. et. al., Phys. Rev. A **46**, 270 (1992). (E/T)
43. Djuric, N. et. al., Phys. Rev. A **47**, 4786 (1993). (E/T)
44. Doering, J. P., J. G. Res **96**, 16021 (1991). (E)
45. Dufton, P. L. and A. E. Kingston, Mon. Not. R. Astron. Soc. **248**, 827 (1990). (T)
46. Fawcett, B. C. and H. E. Mason, Atomic Data Nuc. Data Tables **47**, 17 (1991). (T)
47. Fon, W. C. et. al., J. Phys. B **24**, 2161 (1992). (T)
48. Fon, W. C. et. al., J. Phys. B **25**, 2625 (1992). (T)
49. Fon, W. C. et. al., J. Phys. B **25**, 507 (1992). (T)
50. Fon, W. C. et. al., J. Phys. B **26**, 305 (1993). (T)
51. Ganas, P. S., J. Chem. Phys. **92**, 2374 (1990). (T)
52. Griffin, D. C. et. al., Phys. Rev. A **47**, 2871 (1993). (T)

53. Hayes, M. A., J. Phys. B **25**, 2649 (1992). (T)
54. Ho, Y. K. and R. J. W. Henry, Astrophys. J. **351**, 701 (1990). (T)
55. Hummer, D. G. et. al., Astron. Astrophys. (1993) (in press). (T)
56. Inokuti, M., Editor, *Advances in Atomic and Molecular Physics*, Academic Press (1993) (in press).
57. Itikawa, Y., Atomic Data Nuc. Data Tables **49**, 209 (1991). (E/T)
58. Itikawa, Y. and A. Ichimura, JPCRD **19**, 637 (1990). (E/T)
59. Johnson, C. T. and A. E. Kingston, J. Phys. B **23**, 3393 (1990). (T)
60. Kato, T. et. al., Atomic Data Nuc. Data Tables 44, 133 (1990). (T)
61. Keenan, F. P. and R. H. Norrington, Astrophys. J. **368**, 486 (1990). (T)
62. Laher, R. R. and F. R. Gilmore, JPCRD **19**, 277 (1990). (E/T)
63. Lengyel, V. I. et. al., J. Phys. B **23**, 2847 (1990). (T)
64. Lennon, D. J. and V. M. Burke, Mon. Not. R. Astron. Soc. **251**, 628 (1991).
65. Lennon, D. J. and V. M. Burke, Astron. Astrophys. (1993) (in press). (T)
66. Luo, D. and A. K. Pradhan, Phys. Rev. A **41**, 165 (1990). (T)
67. Man, K. F. et. al., J. Phys. B **26**, 1365 (1993). (E)
68. Mansky, E. J. and M. R. Flannery, J. Phys. B **23**, L501 (1990). (T)
69. Mansky, E. J. and M. R. Flannery, J. Phys. B **23**, 4549 (1990). (T)
70. Mansky, E. J. and M. R. Flannery, J. Phys. B **23**, 4573 (1990). (T)
71. Mansky, E. J. and M. R. Flannery, J. Phys. B **25**, 1591 (1992). (T)
72. Mantas, G. P. and H. C. Carlson, Geophys. Res. Lett. **18**, 159 (1991). (T)
73. McCallion, P. et. al., J. Phys. B **25**, 1051 (1992). (E)
74. McCallion, P. et. al., J. Phys. B **25**, 1061 (1992). (E)
75. McLaughlin, B. M. and K. L. Bell, J. Phys. B **26**, 1797 (1993). (T)
76. Mohan, M. and K. L. Baluja, Zeit. Phys. D **14**, 135 (1990). (T)
77. Mohan, M. et. al., J. Phys. B **24**, 3889 (1991). (T)
78. Mohan, M. and M. LeDourneuf, Astron. Astrophys. **227**, 285 (1990). (T)
79. Mohan, M. and M. LeDourneuf, Phys. Rev. A **41**, 2862 (1990). (T)
80. Mohan, M. et. al., J. Phys. B **24**, 299 (1991). (T)
81. Mohan, M. et. al., J. Phys. B **23**, 989 (1990). (T)
82. Msezane, A. Z. and P. Awuah, J. Phys. B **23**, 4615 (1990). (T)
83. Nakazaki S. and K. A. Berrington, J. Phys. B **24**, 4263 (1991). (T)
84. Nakazaki, S. et. al., J. Phys. B **24**, L27 (1991). (T)
85. Peart, B. et. al., J. Phys. B **24**, 4453 (1991). (E)
86. Pequingnot, D. et. al., Astron. Astrophys. **251**, 680 (1991).
87. Pindzola, M. S. et. al., Phys. Rev. A **44**, 5628 (1991). (T)
88. Pradhan, A. K., Atomic Data and Nucl. Data Tables (1993) (in press). (T)
89. Pradhan, A. K. and K. A. Berrington, J. Phys. B **26**, 157 (1993). (T)
90. Pradhan, A. K. and J. W. Gallagher, Atomic Data Nuc. Data Tables **52**, 227, (1992). (T)
91. Pradhan, A. K. and H. L Zhang, Astrophys. J. **409**, L77 (1993). (T)
92. Rachafi, S. et. al., J. Phys. B **24**, 1037 (1991). (E)
93. Reed, K. M. and M. H. Chen, Phys. Rev. A **45**, 4519 (1992). (T)
94. Rudd, M. E., Phys. Rev. A **44**, 1644 (1991). (E)
95. Sawey, P. M. J. et. al., J. Phys. B **23**, 4321 (1990). (T)
96. Schmidt, H. T. et. al., J. Phys. B **25**, 3165 (1992). (E)
97. Scholz, T. T. et. al., Mon. Not. R. Astron. Soc. **242**, 692 (1991). (T)
98. Scott, M. P. and P. G. Burke, J. Phys. B **26**, L191 (1993). (T)
99. Smith, S. J. et. al., Phys. Rev. Lett. **67**, 30 (1991). (E)
100. Stafford, R. P. et. al., J. Phys. B **25**, 5449 (1992). (E)
101. Tayal, S. S., J. Phys. B **25**, 2639 (1992). (T)
102. Trajmar, S. et. al., J. Phys. B **25**, 4889 (1992). (E/T)
103. Wang, S. and J. W. McConkey, J. Phys. B **25**, 5461 (1992). (E)
104. Whalin, E. K. et. al., Phys. Rev. Lett. **66**, 157 (1991). (E)
105. Zatsarinnyi, O. I. et. al., Opt. Spectrosc. (USSR) **67**, 10 (1989). (T)
106. Zatsarinnyi, O. I. et. al., Phys. Rev. A **44**, 7343 (1991).
107. Zeijlmans van Emmichoven, P. A., et. al., Phys. Rev. A **47**, 2888 (1993).

108. Zeippen, C. J. and M. LeDourneuf, Eds. J. de Physique IV, Coll. C1 (1991). (T)
109. Zhang, H. L. et. al., Astron. Astrophys.(1993) (in press). (T)
110. Zhang, H. L. and D. H. Sampson, Atomic Data Nuc. Data Tables **52**, 143 (1992). (T)
111. Zhang, H. L. et. al., Atomic Data Nuc. Data Tables **44**, 31 (1990). (T)
112. Zhang, Y. et. al., Phys. Rev. A **44**, 4366 (1991). (E)
113. Zhang, Y. et. al., Phys. Rev. A **45**, 2929 (1992). (E)

Rate Coefficients for Chemical Reactions
(W. F. Huebner)

Chemical reactions, in particular their rate coefficients, are a form of atomic and molecular data that have not been reviewed by Commission 14 in the past. Chemical reactions can be categorized according to the phases of the reactants: (1) gas phase (neutral - neutral reactions), (2) plasma (ion - molecule reactions), (3) solutions, in particular non-ideal solutions (liquids and solids), and (4) heterogeneous or catalytic reactions (usually gas - solid). Catalytic reactions in which the third body is a gas are usually listed under gas-phase reactions. In astrochemistry, gas-phase reactions are of prime interest in modeling cold dense systems such as planetary atmospheres (Levine, 1985), protostars, the solar nebula, and preplanetary subnebulae (Howe et al., 1992). Plasma or ion - molecule reactions are of interest in circumstellar envelopes (Glassgold et al., 1992), the interstellar medium including diffuse (van Dishoeck, 1990) and dense clouds (Hasegawa et al., 1992), the solar nebula, ionospheres of planets and satellites, and comet comae (Schmidt et al., 1988).

This first review of rate coefficients for chemical reactions centers on basic definitions to guide potential users and introduce them to existing databases.

Chemical kinetics deals with the time rate of chemical change and with the use of measured and calculated rates to explain and clarify the mechanisms by which chemical systems approach thermodynamic equilibrium or a steady state. A complication arises when a reaction proceeds through a *kinetically stable* (i.e., thermodynamically metastable) state which is separated from the *thermodynamically stable* state by a potential energy barrier and the reaction energy is insufficient to overcome the barrier. In this case, true chemical (thermodynamic) equilibrium may be reached only after an infinite time.

Chemical reactions depend on the concentration of the reactants and the speed of the reactions, i.e., the rate coefficients. The species concentrations depend also on the rate of the gas flowing through the reaction volume. The rate of a chemical reaction can be described by

$$\partial n_i/\partial t = P_i - L_i - \nabla \cdot v_i n_i \,, \qquad (1)$$

where n_i is the concentration (number density) of a product species i, P_i and L_i are the chemical production and loss terms for species i, and the last term in Eq. (1) is the divergence of the flux of particles having velocity v_i. It is this last term which may prevent reactions from reaching equilibrium. The rates of most chemical reactions increase rapidly with increasing temperature. The rate coefficient data measured or calculated at temperature T are normally fitted to the standard Arrhenius form for collisional processes

$$k = A\,(T/300)^B \exp(-C/T) \,. \qquad (2)$$

Here T is either the gas, ion, or electron temperature depending on the physical process. The dimensions of the A coefficient contain concentrations for all except the first-order reactions (e.g., photo reactions). The coefficient C contains the activation energy. The rate coefficient k is independent of concentration of the reactants, but may depend strongly on temperature.

Rate coefficients for a forward (exothermic) reaction and its reverse (endothermic) reaction are related by the equilibrium coefficient

$$K = k_f / k_r \,. \tag{3}$$

Equilibrium coefficients as a function of temperature and relative to reference states are given in the JANAF Thermochemical Tables (Chase et al., 1985). Additional information about the relationship of rate coefficients, equilibrium coefficients, and thermodynamic variables can be found in Cohen and Westberg (1983, 1991).

Databases for chemical reaction networks are available in several formats. The NIST (National Institute of Standards and Technology) database (Mallard et al., 1993) contains measured data, means of entering additional data, and computer codes for analyzing and fitting data to an Arrhenius equation. The data in the current version (5.0) include reactions for 6400 distinct reactant pairs in thermal gas-phase kinetics, particularly for combustion, i.e., hydrocarbon chemistry. The least squares fit Arrhenius parameters (A, B, and C in Eq. 3) can be saved to a file together with the reaction species (reactants and products) and the temperature limits for the data fit. The NIST Chemical Kinetics Database provides a powerful and flexible tool, but requires manipulative effort on a PC.

Ion - molecule gas phase reactions have been compiled by Anicich (1993). The compilation includes rate coefficients for reactions of the first 30 elements in the periodic table (except Li and Be) plus bromine, krypton, zirconium, silver, xenon, and barium for quoted temperatures, usually around 300 K. Certain classes of ion - molecule reactions are not sensitive to temperature and can thus be presented by a single parameter. Another compilation of rate coefficients for astrophysical and planetary applications is the UMIST rate file (Millar et al., 1991). The latest version (UMIST Database for Astrochemistry Rate 92) contains about 390 chemical species and over 3700 reactions. Rate coefficients for astrochemistry in general have been published by Millar and Williams (1988).

Reaction rates for aeronomy were discussed by Fox (1991). Solar photo rate coefficients for dissociation, ionization, and dissociative ionization have been calculated and evaluated on a uniform basis for about 100 atomic and molecular species (Huebner et al., 1992).

Finally, for calculating rate coefficients for reverse reactions, tables of equilibrium coefficients have been compiled on a uniform basis at temperatures up to 6000 K in the JANAF Thermochemical Tables by Chase et al. (1985). Equilibrium coefficients have been incorporated in the NIST Chemical Kinetics Database and are also available on magnetic tape from NIST. In addition, the equilibrium coefficients are useful to determine astrophysical chemical equilibrium compositions including condensation sequences, as for example, in the solar nebula (Sharp and Huebner, 1990).

References

1. Anicich, V. G., 1993: Astrophys. J., Suppl. Ser., **84**, 215.
2. Chase Jr., M. W., Davies, C. A., Downey Jr., J. R., Frurip, D. J., McDonald, R. A., and Syverud, A. N., 1985: "JANAF Thermochemical Tables," third edition, J. Phys. Chem. Ref. Data **14**, Suppl. 1.
3. Cohen, N., and Westberg, K. R., 1983 and 1991: J. Phys. Chem. Ref. Data **12**, 531 and **20**, 1211.
4. Fox, J. L., 1991: Rev. Geophys. Suppl. 1110.
5. Glassgold, A. E., Omont, A., and Guélin, M., 1992: Astrophys. J. **396**, 115.
6. Hasegawa, T. I., Herbst, E., and Leung, C. M., 1992: Astrophys. J., Suppl. Ser. **82**, 167.
7. Howe, D. A., Millar, T. J., and Williams, D. A., 1992: Mon. Not. R. Astron. Soc. **255**, 217.
8. Huebner, W. F., Keady, J. J., and Lyon, S. P., 1992: Kluwer Academic Publishers, Dordrecht.
9. Levine, J. S., 1985: <u>The Photochemistry of Atmospheres: Earth, the Other Planets, and Comets,</u> Academic Press, Orlando.
10. Mallard, W. G., Westley, F., Herron, J. T., Hampson, R. F., and Frizzell, D. H., 1993: <u>NIST Chemical Kinetics Database: Version 5.0,</u> National Institute of Standards and Technology, Gaithersburg, MD.
11. Millar, T. J., Rawlings, J. M. C., Bennett, A., Brown, P. D., and Charnley, S. B., 1991: Astron. Astrophys. **87**, 585.

12. Millar, T. J., and Williams, D. A., 1988: Rate Coefficients in Astrochemistry. Kluwer Academic Publishers.
13. Schmidt, H. U., Wegmann, R., Huebner, W. F., and Boice, D. C., 1988: Comp. Phys. Comm. **49**, 17.
14. Sharp, C. M., and Huebner, W. F., 1990: Astrophys. J., Suppl. Ser. **72**, 417.
15. van Dishoeck, E. F., 1990: "The chemistry of the diffuse interstellar gas." In The Evolution of the Interstellar Medium, L. Blitz (ed.), Astron. Soc. Pac. Conf. Ser.

WORKING GROUP 4 : LINE BROADENING

N. Feautrier, Chairperson

The field of line broadening has continued to be very active during the last three years, mostly stimulated by the interpretation of recent observations or by the preparation of space investigations. It is not the aim of this report to be exhaustive, so we will simply give a number of useful results for astrophysical purposes.

We will mention an important and very useful Bibliography on Atomic Line Shapes and Shifts (1), jointly established by the Département d'Astrophysique Stellaire et Galactique of the Observatoire de Paris-Meudon (A. Lesage) and by the NIST Data Center (J. Fuhr). This important piece of work is concerned with Stark broadening and shifts as well as Van der Waals broadening and covers the literature from July 1978 to March 1992. Concerning these topics, the present report is mainly confined to a selection of references on the basis of their astrophysical interest.

1. Stark Broadening

1.1 STARK BROADENING FOR PLASMAS OF LOW DENSITIES: THE IMPACT LIMIT

First considering low to moderately dense plasmas, the Opacity Project constitutes a major effort to produce extensive calculations of atomic data required for opacity determination. This work includes Stark broadening of atomic and ionic lines. For atoms (except H) and non hydrogenic ions, the dominant contribution to the broadening is due to electronic collisions. Use of the R-Matrix method leads to very accurate results for transitions involving states of low excitation, and approximate formulae have been developed for transitions between higher excited states.

Approximate methods (semiclassical perturbation theory) have been used extensively by several groups to provide broadening parameters for a large number of lines of astrophysical interest, with a mean accuracy of about 20 to 30 percent. Furthermore, much work has been done experimentally.

Table 1 presents references containing results for many transitions of astrophysical interest. The table is arranged in alphabetical order of element symbols with subdivisions according to stage of ionization. The numbers refer to the running number of the reference list.

Another interesting and complementary direction concerns investigations of systematic trends of Stark broadening parameters. Regularities of these parameters within a multiplet or within spectral series and homologous atoms have been investigated (6, 21, 32-34).

1.2 BEYOND THE QUASISTATIC LIMIT FOR IONIC PERTURBERS: HOT AND DENSE PLASMAS AND/OR HYDROGENIC EMITTERS

In the case of dense and hot plasmas, interactions with ionic perturbers become a major source of broadening. Due to the linear Stark effect, it is the same situation when the emitter is either the H atom or an hydrogenic ion. The quasistatic approximation used for ionic broadening at low densities is no longer valid in these cases. Simultaneous strong interactions between the radiator and the surrounding ions may occur. The relatively high velocities of the perturbers preclude the consideration of these interactions as static and much effort, both experimental (36-39) and theoretical (10, 40-44), has been devoted these last

Table 1.

Ar I	2	Cd II	12	O I	25
Ar III	3			O II	26
Ar IV	3,4	F I	6		
		F II	13	S II	3,27
B I	5	F III	13	S III	27
B II	5				
B III	5	He I	14,15,16,17		
				Si II	28
Br I	6	Hg I	18,19	Si III	22,29
Br II	7	Hg II	18	Si IV	29,30
Br III	7				
Br IV	7	Li I	20	Xe I	31
				Xe II	31,32
C I	2,8	N II	21		
C II	8	N IV	22		
C IV	9,10	N V	22,23		
Ca II	11	Na I	5,24		

years to this many body dynamical problem. The physical limit between bound and free states becomes a function of the density and the temperature, leading to the disappearance of higher spectral series numbers. Such studies lead to interesting improvements of the Inglis Teller formula. Consequences for the determination of opacities have been analyzed (45). In the infrared, a correct modeling of the broadening of the solar hydrogen lines (46) gives profiles in a generally good agreement with the observational data of ATMOS.

2. Line Broadening by foreign gases and molecular line broadening

2.1 BROADENING OF ATOMIC LINES

Pressure broadening of neon (47), argon (48) and caesium (49) lines by various rare gases has been investigated experimentally. Many calculations involving various degrees of sophistication in the determination of interatomic interactions have appeared in the literature: self broadening of helium lines (50), Ca lines perturbed by H (51), alkali resonance lines perturbed by H and rare gases (52). One has to mention interesting results on the broadening of Rb Rydberg lines by rare gases (53).

2.2 FAR WINGS

Far wing profiles have received less attention. New data are available on Ba (54, 55) and Ar, Kr, Xe perturbed by Ne (56, 57). The temperature and density dependence of the Lyman a line wing in hydrogen-rich white dwarf atmospheres have been investigated (58).

2.3 MOLECULAR BROADENING

Concerning pressure broadening of molecular lines, many important results have been obtained; for astrophysical purposes we will quote particularly the calculation of the broadening parameters of CO (59), HD perturbed by He and H_2 (60), and line width measurements in H_2O (61, 62) . Many theoretical or experimental investigations in this domain have been greatly stimulated by new spatial observations of planetary atmospheres: particular attention has been given to collisional broadening of C_2H_4 (63), C_2H2 (64), CD_4 (65), NH_3 (66, 67), CH_4 (68,69). Very important results were reported on CO_2(70-73).

A special attention has been paid to the difficult problem of line mixing in CO_2 (74), CO and CO_2 (75), HCN and C_2H2 (76). This considerable effort in molecular line broadening has contributed to the improvement of molecular databases (HITRAN (77), 1991GEISA (78)) which provide the necessary fundamental parameters to represent molecular properties in atmospheric spectroscopic analysis. Although focussed on terrestrial atmospheric studies, many of the listed data are very important for outer planets.

2.4 COLLISION INDUCED SPECTRA

Collision induced absorption is known to be the main source of opacity of the outer planets, which are composed of non polar molecules like hydrogen and helium. Recently a new interest for collision induced spectra has appeared with the need of molecular opacities for the modeling of cool and dense stars: white dwarfs, brown dwarfs, protostars...

Accurate calculations of collision-induced infrared absorption spectra of H_2-H2 and H2-He pairs (79-82) have been carried out. The agreement between the theoretical model and the experimental data is usually good to within a few percent, except for the H_2-He spectra calculated at 296 K which does not agree with the latest measurements (83). The reason for this discrepancy is still unsolved. New measurements of the collision induced emission in H_2-H_2, Ne, Ar, Xe are available (84). Calculations of the far infrared spectra of N_2-CH_4 pairs are presented in (85). This work is of interest for the modeling of Titan's atmosphere.

Collisional interference between electric dipole allowed and collision induced transitions has been investigated (86).

References

1. Fuhr J. R., Lesage A., NIST Special Publication 366, Suppl. 4, Bibliography on Atomic Line Shapes and Shifts (1993).
2. Hahn T. D., Woltz L. A., Phys. Rev. A **42** 1450 (1990).
3. Kobilarov R., Konjevic N., Phys. Rev. A **41** 6023 (1990).
4. Hey J. D., Gawon A., Xu X. J., Breger P., Kunze H. -J., J. Phys. B **23** 241 (1990).
5. Djenize S., Sreckovic A., Labat J., Platisa M., Phys. Scr. **45** 320 (1992).
6. Djurovic S., Konjevic N., Dimitrijevic M. S., Z. Phys. D **16** 255 (1990).
7. Labat O., Djenize S., Puric J., Labat J. M., Sreckovic A., J. Phys. B **24** 1251 (1991)
8. Pérez C., de la Rosa I., de Frutos A. M., Mar S., Phys. Rev. A **44** 6948 (1991)
9. Dimitrijevic M. S., Sahal-Bréchot S., Bommier V., Astron. Astrophys., Suppl. Ser. **89** 581 (1991)
10. Schöning T., Astrophys. J. **267** 300 (1993)
11. Dimitrijevic M. S., Sahal-Bréchot S., J. Quant. Spectrosc. Radiat. Transfer **49** 157 (1993).
12. Djenize S., Sreckovic A., Labat J., Konjevic R., Popovic L., Phys. Rev. A **44** 410 (1991).
13. Djenize S., Labat J., Sreckovic A., Labat O., Platisa M., Puric J., Phys. Scr. **44** 148 (1991).
14. Suemitsu H., Ikawi K., Takemoto Y., Yoshida E., J. Phys. B **23** 4119 (1990).
15. Perez C., de la Rosa I., de Frutos A. M., Mar S., Phys. Rev. A **44** 6785 (1991).
16. Uzelac N. I., Stefanovic I., Konjevic N. J., J. Quant. Spectrosc. Radiat. Transfer **46** 447 (1991).
17. Adler H. G., Piel A., J. Quant. Spectrosc. Radiat. Transfer **45** 11 (1991).
18. Djenize S., Sreckovic A., Platisa M., Labat J., Konjevic R., Puric J., J. Quant. Spectrosc. Radiat. Transfer **44** 405 (1990).
19. Dimitrijevic M. S., J. Quant. Spectrosc. Radiat. Transfer **47** 315 (1992).
20. Dimitrijevic M. S., Sahal-Bréchot S., J. Quant. Spectrosc. Radiat. Transfer **46** 41 (1991).
21. Djenize S., Sreckonic A., Labat J., Astron. Astrophys. **253** 632 (1992).
22. Djenize S., Labat J., Platisa M., Sreckovic A., Puric J., Astron. Astrophys. **227** 291 (1990).
23. Huang Y. W., Böttcher F., Wang J. S., Griem H. R., Phys. Rev. A **42** 2322 (1990).
24. Dimitrijevic M. S., Sahal-Bréchot S., J. Quant. Spectrosc. Radiat. Transfer **44** 421 (1990).
25. Alexiou S., Baird J. C., Phys. Rev. A **42** 7413 (1990).
26. Djenize S., Sreckovic A., Labat J., Platisa M., Z. Phys. D **21** 295 (1991)
27. Djenize S., Sreckovic A., Platisa M., Konjevic R., Labat J. Puric J., Phys. Rev. A **42** 2379 (1990).

28. Pérez C., de la Rosa M. I., de Frutos A. M., Gonzalez V. R., Mar S., Ann. Phys. Paris Colloq. No 3 **15** 115 (1990).
29. Djenize S., Sreckovic A., Labat J., Puric J., Platisa M., J. Phys. B **25** 785 (1992).
30. Dimitrijevic M. S., Sahal-Bréchot S., Bommier V., Astron. Astrophys. Suppl. Ser. **89** 591 (1991).
31. Konjevic N., Uzelac N. I., J. Quant. Spectrosc. Radiat. Transfer **44** 61 (1990).
32. Bertucelli D., Bertucelli G., Di Rocco H. O., Phys. Scr. **43** 469 (1991).
33. Puric O., Cuk M., Dimitrjevic M. S., Lesage A., Astrophys. J. **382** 353 (1991).
34. Djenize S., Sreckovic A., Labat J., J. Quant. Spectrosc. Radiat. Transfer **46** 433 (1991).
35. Wiese W. L., Konjevic N., J. Quant. Spectrosc. Radiat. Transfer **47** 185 (1992).
36. Benredjen D. E., Nguyen Hoe, Coulaud G., J. Quant. Spectrosc. Radiat. Transfer **43** 415 (1990).
37. Uhlenbusch J., Viöl W., J. Quant. Spectrosc. Radiat. Transfer **44** 47 (1991).
38. Helbig V., Contrib. Plasma Phys. **31** 183 (1991).
39. Mijatovic Z., Pavlov M., Djurovic S., Phys. Rev. A **43** 6095 (1991).
40. Halenka J., Z. Phys. D **16** 1 (1990).
41. Seaton M. J., J. Phys. B **23** 3255 (1990).
42. Stehlé C., J. Quant. Spectrosc. Radiat. Transfer **44** 135 (1990).
43. Günter S., Hitzschke C., Röpke G., Phys. Rev. A **44** 6834 (1991).
44. Stehlé C., J. Phys. IV Colloq. (Suppl. II) 121 (1991).
45. Stehlé C., Jacquemot S., Astron. Astrophys. **271** 348 (1993).
46. Carlsson H., Rutten R. J., Astron. Astrophys. **259** L53 (1992)
47. Bielski A., Brym S., Szudy J., Trawinski R. S., Wolnikowski J., J. Phys. B **24** 4909 (1991).
48. Bielski A., Trawinski R. S., Wolnikowski J., Szudy J., Phys. Scr. **44** 141 (1991).
49. Inoue Y., Uchida K., Hori H., Sakwrai T., J. Phys. Soc. Jpn. **59** 516 (1990).
50. Mullanphy D. F. T., Peach G., Whittingham J. P., J. Phys. B **24** 3709 (1991).
51. Spielfiedel A., Feautrier N., Chambaud G., Lévy B., J. Phys. B **24** 4711 (1991).
52. Andretta V., Gomez M. T., Severino G., Solar Phys. **131** 1 (1991).
53. Sun J.-Q., West P. J., J. Phys. B **23** 4119 (1990).
54. Ueda K., Sotome H., Sato Y., J. Chem. Phys. **94** 1903 (1991).
55. Ueda K., Sotome H., Sato Y., J. Chem. Phys. **94** 1907 (1991).
56. Zagrebin A. L., Pavlevskaya N. A., Opt. Spectrosc. (USSR) **68** 19 (1990).
57. Zagrebin A. L., Pavlevskaya N. A., Opt. Spectrosc. (USSR) **68** 299 (1990).
58. Allard N. F., Kielkopf J., Astron. Astrophys. **242** 133 (1991).
59. Belov S. P., Tretyakov M. Yu., Svenram R. D., Astrophys. J. **383** 848 (1992).
60. Schaefer J., Monchick L., Astron. Astrophys. **265** 859 (1992).
61. Bauer A., Godon M., J. Quant. Spectrosc. Radiat. Transfer **46** 211 (1991).
62. Godon M., Carlier J., Bauer A., J. Quant. Spectrosc. Radiat. Transfer **47** 275 (1992).
63. Brannon Jr J. F., Varanasi P., J. Quant. Spectrosc. Radiat. Transfer **47** 237 (1992).
64. Varanasi P., J. Quant. Spectrosc. Radiat. Transfer **47** 263 (1992).
65. Millot G., Lavorel B., Steinfeld J. I., J. Quant. Spectrosc. Radiat. Transfer **47** 81 (1992).
66. Varanasi P., Gopalan A, J. Quant. Spectrosc. Radiat. Transfer **49** 383 (1993).
67. Markov V. N., Pine A. S., Buffa G., Tarrini O., J. Quant. Spectrosc. Radiat. Transfer **50** 167 (1993).
68. Benner D. C., Malathi Devi V., Smith M. A. H., Rinsland C. P., J. Quant. Spectrosc. Radiat. Transfer **50** 65 (1993).
69. Margolis J. S., J. Quant. Spectrosc. Radiat. Transfer **49** 71 (1993).
70. Phillips W. J. J. Quant. Spectrosc. Radiat. Transfer **48** 91 (1992).
71. Rothman L. S., Hawkins R. L., Wattson R. B., Gamache R. R., J. Quant. Spectrosc. Radiat. Transfer **48** 557 (1992) and **48** 567 (1992).
72. Malathy Devi V., Benner D. C., Rinsland C. P., Smith M. A. H., J. Quant. Spectrosc. Radiat. Transfer **48** 581 (1992).
73. Sirota J. M., Reuter D. C., Mumma M. J., **50** 193 (1993).
74. Bonamy L., Robert D., Boissoles J., Boulet C., J. Quant. Spectrosc. Radiat. Transfer **45** 305 (1991).
75. Filippov N. N. , Tonkov M. V., J. Quant. Spectrosc. Radiat. Transfer **50** 111 (1993).
76. Pine A. S., J. Quant. Spectrosc. Radiat. Transfer **50** 149 (1993).
77. Rothman L. S., *et al.*, J. Quant. Spectrosc. Radiat. Transfer **48** 469 (1992).

78. Husson N., Bonnet B., Scott N. A., Chedin A., J. Quant. Spectrosc. Radiat. Transfer **48** 509 (1992).
79. Borysow A., Frommhold L., Astrophys. J. **348** L41 (1993).
80. Borysow A., Icarus **92** 273 (1991).
81. Birnbaum G., Borysow A., Mol. Phys. **73** 57 (1991).
82. Borysow A., Icarus **96** 169 (1992).
83. Bouanich J. P., Brodbeck C., Nguyen-Van-Thanh, Drossart P., J. Quant. Spectrosc. Radiat. Transfer **44** 393 (1990).
84. Caledonia G. E., Krech R. H., Wilkerson T. D., Taylor R. L., Birnbaum G., Phys. Rev. A **43** 6010 (1991).
85. Borysow A., Tang C., Icarus *in press* (1993).
86. Gao B., Tabisz G. C., Trippenbach M., Cooper J., Phys. Rev. A **44** 7379 (1991).

WORKING GROUP 5: MOLECULAR STRUCTURE AND TRANSITION DATA

W. H. Parkinson, Chairperson

This report has been compiled in large part from edited contributions sent to me in the summer of 1993.

Laboratory and theoretical studies in molecular spectroscopy and the resulting molecular data and parameters have continued to grow with the needs of chemists, physicists and astronomers. The number and holdings of atomic and molecular databases,[1,2,3] have increased. The expansion and inclusion of atomic and molecular data (AMD) on the interactive, rational database, Astrophysical Data System (ADS) will soon begin at the Harvard-Smithsonian Center for Astrophysics (CfA). Information about AMD in ADS is available from Peter L. Smith. The useful molecular bibliography, Berkeley Newsletter, continues to be available from Davis and colleagues.

The period 1990-1993 could be remembered for the basic spectroscopic studies of H_3^+ and for the use of its infrared spectrum as an astronomical probe.[4] The infrared H_3^+ emission of the v_2 fundamental band has been established as a powerful ground based observational tool to study plasma activities in planetary ionospheres. H_3^+ spectra have been observed in the Jovian atmosphere[5,6,7,8,9] and in the atmospheres of Uranus[10] and Saturn.[11] Recent improvements in spectrometer sensitivity are expected to lead to detection of H_3^+ in Neptune.

Takeshi Oka has reviewed the indispensable, close collaboration of astronomers, laboratory spectroscopists, and theorists for the study of H_3^+. The H_3^+ spectrum of hot bands $2v_2(2) \leftarrow v_2$, $2v_2(0) \leftarrow v_2$, and $v_1 + v_2 \leftarrow v_2$ have been reported[12] along with laboratory measurements of the first overtone band[13] $2v_2(2) \leftarrow 0$ and the second overtone band[14] $3v_2(1) \leftarrow 0$. Rotational levels up to J=K=15, 5092 cm^{-1} above the J=0 state, have been observed and the data have been summarized.[15] Forbidden transitions $v_1 \leftarrow 0$, $v_1 + v_2 \leftarrow v_2$ have been observed.[16] The v_2 fundamental band of H_3^+ has now been observed in absorption by using a Fourier transform spectrometer (FTS).[17]

Basic theoretical calculation[18,19,20] of the vibration-rotation levels of H_3^+ played a crucial role in the identification of the laboratory spectrum and the Jupiter emission spectrum. The theory has been extended to forbidden[21] and overtone[22] transitions. Recently, Watson[23] has modified the *ab initio* potential of Meyer, Botschwina, and Burton by a least squares procedure to fit the observed frequencies. A new *ab initio* calculation of the potential has been given by Frye and others[24,25], and the potential recently modified[26] to fit the observed frequencies and predicted higher vibration rotation states.

The very large discrepancy in values of the H_3^+ electron recombination rate seems to have been settled by laboratory measurements. Amano[27] published details of his earlier spectroscopic experiment and concluded that the recombination rate is 1.8 x 10^{-7} cm^3 s^{-1} at 273 K. A flowing after glow/Langmuir probe experiment[28] gives 1.1 x 10^{-7} cm^3 s^{-1} at 650 K. The rate below 1 ev of vibrationally cooled H_3^+ is 6 x 10^{-8} cm^3 s^{-1} from measurements[29] made at the ion storage ring, Manne Siegbahn Institute of Physics, Sweden and is in satisfactory agreement with the merged-beam experimental value[30] of 2 x 10^{-8} cm^3 s^{-1}.

A. R. McKellar has reported results from the Herzberg Institute of Astrophysics. Among diatomic molecules, electronic spectra have been studied for N_2,[31,32] O_2,[33] OH,[34] and NO^+.[35] In the infrared region, work on the spectra of weakly-bound dimers include experiments on the H_2 dimer[36,37] and the NO dimer[38]. A new technique involving high resolution infrared detection has been applied to measuring dissociative recombination rates for the ions H_3^+, HN_2^+, and HCO^+.[26] In addition to these molecular ions, a number of other unstable polyatomics have been studied in various wavelength regions. These include infrared[39] and vibration-rotation[40] spectra of NH_2, a near-infrared electronic spectrum of triplet C_3,[41] vibration-rotation spectra of C_2D [42] and $^{13}C_5$,[43] and pure rotational spectra of NCS[44] and HCCCO.[45] Work has continued on the precise measurement of infrared intensities of CO_2.[46,47] A large number of stable polyatomic molecules have been studied at high resolution in the infrared region, including H_2O_2,[48] ketene,[49] acetaldehyde,[50] HNCO,[51] OCCCS,[52] C_2H_2,[53] C_2H_6,[54] C_4H_2,[55] C_6H_6,[56] CH_3OH,[57] C_3O_2,[58] and DNO_3.[59]

Investigators at the CfA have reported studies of fundamental molecular parameters and processes. Basic calculations have been carried out by K. Kirby and colleagues[60] of all the potential curves and transition moments relevant to the photodissociation of HCl^+ and the charge transfer of H^+ + Cl. Direct photodissociation cross sections, transition moments, and radiative lifetimes of NH have been computed by Kirby and Goldfield.[61]

The millimeter-wave and far infrared spectra of HOCO and DOCO have been detected in the gas phase and studied in the CfA laboratory.[62,63] In combination, the investigations yield a complete set of precise spectroscopic constants for the molecules. Quantitative studies at high wavelength resolution of molecules of atmospheric importance have been continued by Yoshino and colleagues. Predissociation linewidths of the O_2 Schumann-Runge (S-R) bands have been obtained from the high resolution cross section measurements of the isotopes, $^{16}O_2$,[64] $^{16}O^{18}O$,[65] and $^{18}O_2$.[65] Theoretical studies of the predissociation of the S-R bands and their rotational dependency have been carried out.[66,67] Cross sections in regions of the transmission windows of the S-R bands have been obtained along with the weak Herzberg continuum cross sections under the S-R bands.[68] The oscillator strengths of the hot bands (from $v' = 1$) of S-R system have been obtained.[69] Polynomial coefficients for calculating the S-R cross sections have been determined[70] and the results agree with the high resolution measurements over the temperature range 78 K to 570 K. The absolute cross sections of ozone have been obtained[71] in the wavelength region 185-254 nm at temperature, 295 K, 228 K, and 195 K. No temperature dependency is observed. The solution to the need for absorption measurements at higher wavelength resolution has been pursued by Yoshino and colleagues with the vacuum UV FTS at Imperial College, London, England. Measurements have been completed at a resolution of 0.06 cm^{-1} of absorption cross sections and band oscillator strengths of the $\delta(0,0)$, $\delta(1,0)$, and $\beta(7,0)$ bands of NO. Work is also underway to measure the absorption cross sections of the O_2 Herzberg I band system (A $^3\Sigma_u^+$ - X $^3\Sigma_g^-$) at a resolution of 0.06 cm^{-1} with the FTS in combination with a White cell. A number of other high resolution measurements of VUV molecular cross sections have been made: C_2H_2, 147-201 nm,[72] CO, E(0,1) - X(0),[73] N_2 95.5-100 nm,[74] and CO between 96.7 and 98.8 nm at 20 K.[75] The work includes the first supersonic jet-cooled absorption cross section measurement on molecules cooled to about 30 K. Additional absorption cross section data (CO 92.5-96 nm @ 30 K and HCl C-X bands) will soon be available. Adjunct studies of CO at CfA and independently by Ubachs[76] disclosed a small systematic error in published wavelengths/wavenumbers for CO;[77,78] Eidelsberg *et al.*[79] discuss the recalibration.

C. E. Brion and colleagues have recently applied the electron-impact (EI), dipole (e,e) excitation method[80] to the measurement of absolute oscillator strengths for transitions in H_2,[81] N_2,[82] O_2,[83] NO[84] and CO[85]. The method appears to have some advantages over direct optical measurements and over other EI experiments. When results are compared, there are both satisfactory agreements[86,74] and puzzling discrepancies.[73,87]

Theoretical calculations of optical and generalized oscillator strengths for relevant transitions in CO have been carried out by R. E Buenker and colleagues.[88,89]

P. K. Carroll at University College, Dublin has reported the measurement and analysis of the $^1\Sigma_u^+$ states of N_2^{15} from high resolution absorption spectra in the 89.5-77.0 nm region.[90]

F. Rostas has outlined the activities in molecular research of the Meudon group. These include a new determination of the ground state potential of N_2 [91] and the vibrational analysis[92] of low pressure emission spectra of N_2 between 107.7 and 124.7 nm. An atlas of the EUV absorption spectrum of CO[77,79] has been published as well as band-integrated optical absorption cross sections for the A $^1\Pi$ (v-12) - X $^1\Sigma^+(v)$ transitions.[93] Studies of the photodissociation process involving the CO B $^1\Sigma^+$ - D $^1\Sigma^+$ interaction,[94] a two photon study of the E $^1\Pi$ state[95] and an analysis of the perturbation of the D $^1\Delta$ state,[96,97] have been published. The constants of the X $^1\Sigma^+$ state of CO have been improved.[98] Line positions and emission probabilities of the Lyman and Werner band systems have been calculated taking into account the interactions of the B, B$'$, C and D states.[99,100] Nine rotationally resolved bands of CO_2 have been analyzed[101] and *ab initio* calculations support the analysis.[102] Low lying linear Rydberg states have also been calculated, providing a model for the CO_2 "Rathenau bands" in the 11.4-12.6 eV range[103] and Rydberg-valence interactions for the lowest lying Rydberg states around 11 eV.[104] The vibration-rotation spectrum of Si_2C [105] and CO_2^+(X $^2\Pi_g$)[106] have been calculated. The electronic spectrum of Fullerenes has been recorded between 190 and 750 nm and analyzed in terms of Herzberg-Teller couplings;[107,108] triplet-triplet transitions have been observed in pulsed experiments.[109]

R. W. Nicholls (Center for Research in Earth and Space Science, York University) reports on new NbO bands[110] from shock excitation of powered Nb_2O_5. A theoretical basis for an extended Birge-Mecke rule of molecular spectroscopy,[111] augmented sum rules for progressions of Franck-Condon factors for bound-free and free-free transitions,[112] and new line parameters for the O_2 Herzberg I system,[113] have been obtained. A new identification atlas of the atmospheric and infrared atmospheric band system of O_2 is available from Nicholls.

Franck-Condon factors and the r-centroids for the bands A$^2\Pi$- $^2\Sigma^+$ of CP, C$^3\Pi$-X$^3\Pi$ of SiC, and B $^2\Sigma^+$-X $^2\Sigma^+$ of CO^+, and potential energy curves and dissociation energies of NbO, SiC, CP, PH^+, SiF^+, and NH^+ have been published by Reddy and colleagues.[114,115]

Sumner Davis and John Phillips have reported on a variety of important molecular spectra. They have observed and measured highly excited rotational states of the OH Meinel band system,[116] as produced in a N_2-afterglow. These results were incorporated into a theoretical study.[117] Davis and colleagues[118] have extended measurements on the OD(1,0) band and measured the (2,0) band for the first time. A CN spectrum[119] from 9000 to 31,000 cm^{-1} was recorded in a N_2-afterglow containing a trace of CH_4; the data contain C^{12} and C^{13}. Work on the iron hydride spectrum has continued with the band in the 700 nm region considered to be HFeH.[120] The quartet and sextet states of FeH have been studied theoretically by Langhoff and Bauschlicher.[121] The calculations are entirely consistent with the earlier observations of the near-infrared system by Phillips and Davis, but the spectroscopic designation should be F $^4\Delta$-X $^4\Delta$. Theoretical studies of FeH^+ have also been presented by Langhoff and colleagues.[122] Earlier work by Littleton and Davis[123] on transition strengths in the singlet and triplet systems of ZrO as measured from emission spectra was inconsistent with stellar data and did not agree with the calculations of Langhoff and Bauschlicher.[124] New absorption spectra have been analyzed.[125] A synthetic spectrum, calculated with the new parameters, matches the observed spectrum and removes inconsistencies. Langhoff and colleagues[126] have studied the Phillips band system of C_2 and have calculated the radiative lifetime of the A $^1\Pi_u$ state ($v' = 0$, $J' = 1$); these calculations suggest a large systematic error is possible in the measurements of Bauer *et al.*[127] The line strengths for the ro-vibrational bands of X$^1\Sigma^+$ state of SiO have been calculated by Langhoff & Bauschlicher.[128]

U. G. Jorgensen has computed the line strengths for 4 x 10^6 lines of the A $^2\Pi$-X $^2\Sigma^+$ Red system of CN.[129] Line lists for TiO, CH, H_2O, and HCN are under construction and a review of existing astrophysical molecular data has been published.[130] A complete and very accurate computation of the 8 x 10^4 existing vibrational transitions in the H_2O ground electronic state is in press.[131]

Michael E. Mickelson, Denison University, Ohio, has described measurements of the absorption coefficient for the 619 nm visible CH_4 band between 300 K -150 K.[132,133] A new cold absorption cell and a tunable dye laser (linewidth 0.06 cm^{-1}) are used. Molecular parameters for H_2, including line positions, strengths, pressure-broadening coefficients and shifts, have also been determined[134,135] from measurements with a long path absorption cell (2-6 Km) and a spectrometer (resolution .015 cm^{-1}) or tunable dye laser (resolution 3×10^{-5} cm^{-1}).

Frank J. Lovas has published[136] a new revision to the 1985 edition of the recommended rest frequencies for observed, interstellar molecular microwave transitions. He and colleagues have measured low-J rotational transitions[137] in SO, C_2S, C_3S, c-HC_3, CH_2CC and c-C_3H_2. The frequencies measured are sufficient to allow molecular dynamics of cold interstellar clouds to be determined with improved accuracy. Molecular electric dipole moments have also been determined for SO, C_3S, c-HC_3 and c-C_3H_2. New measurements and analysis of the v_5 and $2v_9$ infrared absorption bands of HNO_3 have been made;[138] in the range 850 cm^{-1} - 920 cm^{-1}, the bands are important in the spectrum of the Earth's upper atmosphere.

References

1 Parkinson, W. H., 1992, in Atomic and Molecular Data for Space Astronomy, ed. Peter L. Smith & Wolfgang L. Wiese (Springer-Verlag, Berlin, Heidelberg) 149.
2 Rothman, L. S. *et al.* 1992, Special Issue, J. Quant. Spectrosc. Rad. Transfer **48**, 469.
3 Kurucz, R. L. 1992, Rev. Mex. Astron. Astrofis. **23**, 45.
4 Oka, T. 1992, Rev. Mod. Phys. **64**, 1141
5 Oka, T. & Geballe, T. R. 1990, Astrophys. J. **351**, L53.
6 Miller, S., Joseph, R. D. & Tennyson, J. 1990 Astrophys. J. **360**, L55.
7 Maillard, J.-P., Drossard, P., Watson, J. K. G., Kim, S. J. & Caldwell, J. 1990, Astrophys. J. **363**, L37.
8 Kim, S. J., Drossard, P., Caldwell, J., Maillard, J.-P., Herbst, T. & Shure, M. 1991, Nature **353**, 536.
9 Baron, R., Joseph, R. D., Owen, T., Tennyson, J., Miller, S. & Ballester, G. E. 1991, Nature **353**, 539.
10 Trafton, L. M., Geballe, T. R., Miller, S., Tennyson, J. & Ballester, G. E. 1993, Astrophys. J. **405**, 961.
11 Geballe, T. R., Jagod, M.-F. & Oka, T. 1993, Astrophys. J. **408**, L109.
12 Bawendi, M. G., Rehfuss, R. D. & Oka, T. 1990, J. Chem. Phys. **93**, 6200.
13 Xu, L.-W., Gabrys, C. M. & Oka, T. 1990, J. Chem. Phys. **93**, 6210.
14 Lee, S. S., Ventrudo, B. F., Cassidy, D. T., Oka, T., Miller, S. & Tennyson, J. 1991, J. Mol. Spectrosc. **145**, 222.
15 Kao, L., Oka, T., Miller, S. & Tennyson, J. 1991, Astrophys. J., Suppl. Ser. **77**, 317.
16 Xu, L.-W., Rösslein, M., Gabrys, C. M. & Oka, T. 1992, J. Mol. Spectrosc. **153**, 726.
17 Nakanaga, T., Itoh, F., Sugawara, K., Takeo, H. & Matsumura, C. 1990, Chem. Phys. Lett. **169**, 269.
18 Tennyson, J. & Miller, S. 1992, Chem. Soc. Rev. **21**, 91.
19 Miller, S. & Tennyson, J. 1992, Chem. Soc. Rev. **21**, 281.
20 Tennyson, J. 1992, J. Chem. Soc., Faraday Trans. **88**, 3271.
21 Miller, S., Tennyson, J. & Sutcliffe, B. T. 1990, J. Mol. Spectrosc. **141**, 104.
22 Dinelli, B. M., Miller, S. & Tennyson, J. 1992, J. Mol. Spectrosc. **153**, 718.
23 Watson, J. K. G., Can. J. Chem., in press.
24 Frye, D., Preiskorn, A., Lie, G. C. & Clementi, E. 1990, J. Chem. Phys. **92**, 4948.
25 Lie, G. C. & Frye, D. 1992, J. Chem. Phys. **96**, 6784.
26 Dinelli, B. M., Miller, S. & Tennyson, J., J. Mol. Spectrosc., in press.
27 Amano, T. 1990, J. Chem. Phys. **92**, 6492.
28 Canosa, A., Gomet, J. C., Howe, B. R., Mitchell, J. B. A. & Queffelec, J. L. 1992, J. Chem. Phys. **97**, 1028.
29 Larsson, M., Danared, H., Mowat, J. R., Sigray, P., Sundström, G., Broström, L., Filevich, A., Källberg, A., Mannervik, S., Rensflet, K. G. & Datz, S. 1993, Phys. Rev. Lett. **70**, 430.
30 Yousif, F. B., van der Donk, P. J. T., Orakzai, M. & Mitchell, J. B. A. 1991, Phys. Rev. A **44**, 5653.
31 Huber, K. P. & Jungen, Ch. 1990, J. Chem. Phys. **92**, 850.
32 Huber, K. P., Stark, G. & Ito, K. 1993, J. Chem. Phys. **94**, 4471.
33 Fink, E. H., Setzer, K. D., Wildt, J., Ramsay, D. A. & Vervloet, M. 1991, Int. J. Quantum Chem. **39**, 287.
34 Huber, K. P., Holland, F. & Coxon, J. A. 1992, J. Chem. Phys. **96**, 1005.
35 Huber, K. P. & Vervloet, M. 1991, J. Mol. Spectrosc. **146**, 188.
36 Schaefer, J. & McKellar, A. R. W. 1990, Z. Phys. D. **15**, 51.
37 McKellar, A. R. W. & Schaefer, J. 1991, J. Chem. Phys. **95**, 3081.

38 Howard, B, J. & McKellar, A. R. W. 1993, Mol. Phys. **78**, 55.
39 McKellar, A. R. W., Vervloet, M., Burkholder, J. B. & Howard, C. J. 1990, J. Mol. Spectrosc. **142**, 319.
40 Dixon, R. N., Irving, S. J., Nightingale, J. R. & Vervloet, M. 1991, J. Chem. Soc. Faraday Trans. **87**, 2121.
41 Sasada, H., Amano, T., Jarman, C. & Bernath, P. F. 1991, J. Chem. Phys. **94**, 2401.
42 Yan, W.-B., Warner, H. E. & Amano, T. 1991, J. Chem. Phys. **94**, 1712.
43 Moazzen-Ahmadi, N., Flatt, S. D. & McKellar, A. R. W. 1991, Chem. Phys. Lett. **186**, 291.
44 Amano, Takayoshi & Amano, Takako 1991, J. Chem. Phys. **95**, 2401.
45 Cooksey, A. L., Watson, J. K. G., Gottlieb, C. A. & Thaddeus, P. 1992, J. Mol. Spectrosc. **153**, 610.
46 Johns, J. W. C. & Noël, M. 1992, J. Mol. Spectrosc. **156**, 403.
47 Watson, J. K. G. 1992, J. Mol. Spectrosc. **153**, 211.
48 Camy-Peyret, C., Flaud, J.-M., Johns, J. W. C. & Nöel, M. 1992, J. Mol. Spectrosc. **155**, 84.
49 Johns, J. W. C., Nemes, L., Yamada, K. M. T., Wang, T. Y, Domenech, J. L., Santos, J., Cancio, P., Bermejo, D., Ortigoso, J. & Escribano, R. 1992, J. Mol. Spectrosc. **156**, 501.
50 Kleiner, I., Godefroid, M., Herman, M. & McKellar, A. R. W. 1990, J. Mol. Spectrosc. **142**, 238.
51 Yamada, K. M. T., Winnewisser, M. & Johns, J. W. C. 1990, J. Mol. Spectrosc. **140**, 353.
52 Holland, F., Winnewisser, M. & Johns, J. W. C. 1990, Can. J. Phys. **68**, 435.
53 Kabradi, Y., Herman, M., DiLonardo, G., Fusina, L. & Johns, J. W. C. 1991, J. Mol. Spectrosc. **150**.
54 Moazzen-Ahmadi, N., McKellar, A. R. W., Johns, J. W. C. & Ozier, I. 1992, J. Chem. Phys. **97**, 3981.
55 Arié, E. & Johns, J. W. C. 1992, J. Mol. Spectrosc. **155**, 195.
56 Pliva, J., Johns, J. W. C. & Goodman, L. 1991, J. Mol. Spectrosc. **148**, 427.
57 Xu, L.-H., Lees, R. M., Mukhopadhyay, I., Johns, J. W. C. & Moruzzi, G. 1993, J. Mol. Spectrosc. **157**, 447.
58 Vander Auwara, J., Johns, J. W. C. & Polyansky, O. L. 1991, J. Chem. Phys. **95**, 2299.
59 Maki, A. G., Tan, T. L., Looi, E. C., Lua, K. T., Johns, J. W. C. & Nöel, M. 1993, J. Mol. Spectrosc. **157**, 248.
60 Pradhan, N., Kirby, K. & Dalgarno, A. 1991, J. Chem. Phys. **95**, 9009.
61 Kirby, K. & Goldfield, E. M. 1991, J. Chem. Phys. **94**, 1271.
62 Radford, H. E., Wei, Wang & Sears, Trevor J. 1992, J. Chem. Phys. **97**, 3989.
63 Sears, Trevor J., Radford, H. E. & Moore, Mary Anne 1993, J. Chem. Phys. **98**, 6624.
64 Cheung, A. S.-C., Yoshino, K., Esmond, J. R., Chiu, S. S.-L., Freeman, D. E. & Parkinson, W. H. 1990, J. Chem. Phys. **92**, 842.
65 Chiu, S. S.-L., Cheung, A. S.-C., Yoshino, K., Esmond, J. R., Freeman, D. E. & Parkinson, W. H. 1990, J. Chem. Phys. **93**, 5539.
66 Chiu, S. S.-L., Cheung, A. S.-C., Finch, M., Jamieson, M. J., Yoshino, K., Dalgarno A. & Parkinson, W. H. 1992, J. Chem. Phys. **97**, 1787.
67 Cheung, A. S.-C., Mok, D. K.-W., Jamieson, M. J., Finch, M., Yoshino, K., Dalgarno, A. & Parkinson, W,. H. 1993, J. Chem. Phys. **99**, 1086.
68 Yoshino, K., Esmond, J. R., Cheung, A. S.-C., Freeman, D. E. & Parkinson, W. H. 1992, Planet. Space Sci. **40**, 185.
69 Yoshino, K., Esmond, J. R., Cheung, A. S.-C., Freeman, D. E. & Parkinson, W. H. 1990, J. Geophys. Res. **95**, 11,743.
70 Minschwaner, K., Anderson, G. P., Hall, L. H. & Yoshino, K. 1992, J. Geophys. Res. **97**, 10,103.
71 Yoshino, K., Esmond, J. R., Freeman, D. E. & Parkinson, W. H. 1993, J. Geophys. Res. **98**, 5205.
72 Smith, Peter L., Yoshino, K., Parkinson, W. H., Ito, Kenji & Stark, Glenn 1991, J. Geophys. Res. **96**, 17,529.
73 Stark, G., Yoshino, K., Smith, P. L., Ito, K. & Parkinson, W. H. 1991, Astrophys. J. **369**, 574.
74 Stark, G., Smith, P. L., Huber, K. P., Yoshino, K., Stevens, M. H. & Ito, K. 1992, J. Chem. Phys. **97**, 4809.
75 Stark, G., Yoshino, K., Smith, P. L., Esmond, J. R., Ito, K. & Stevens, M. H. 1993, Astrophys. J. **410**, 837.
76 Levelt, P., Ubachs, Wm. & Hogervorst, Wm. 1992, J. Phys. II, France **2**, 801.
77 Eidelsberg, M., Benayoun, J. J., Viala, Y. & Rostas, F. 1991, Astron. Astrophys., Suppl. Ser. **90**, 231.
78 Eidelsberg, M. & Rostas, F. 1990, Astron. Astrophys. **235**, 472.

79 Eidelsburg, M., Benayoun, J. J., Viala, Y., Rostas, F., Smith, P. L., Yoshino, K., Stark, G., & Shettle, C. A. 1992, Astron. Astrophys. **265**, 839.
80 Chan, W. F., Cooper, G. & Brion, C. E. 1991, Phys. Rev. A **44**, 186.
81 Chan, W. F., Cooper, G. & Brion, C. E. 1992, Chem. Phys. **168**, 375.
82 Chan, W. F., Cooper, G., Sodhi, N. R. S. & Brion, C. E. 1993, Chem. Phys. **170**, 81.
83 Chan, W. F., Cooper, G. & Brion, C. E. 1993, Chem. Phys. **170**, 99.
84 Chan, W. F., Cooper, G. & Brion, C. E. 1993, Chem. Phys. **170**, 111.
85 Chan, W. F., Cooper, G. & Brion, C. E. 1993, Chem. Phys. **170**, 123.
86 James, G. K, Ajello, J. M., Franklin, B. O. & Shemansky, D. E. 1990, J. Phys. B. **23**, 2055.
87 Letzelter, C., Eidelsberg, M. & Rostas, F. 1987, Chem. Phys. **114**, 273.
88 Chantranupong, L., Hirsch, G., Bhanuprakash, K. & Buenker, Robert J. 1992, Chem. Phys. **164**, 183.
89 Chantranupong, L., Bhanuprakash, K., Honigmann, M. & Buenker, Robert J. 1992, Chem. Phys. **164**, 351.
90 Mose, A. A. 1993, Thesis, National University of Ireland.
91 Edwards, S. A., Roncin, J.-Y., Launay, F. & Rostas, F. 1993, J. Mol. Spectrosc., in press.
92 Roncin, J.-Y., Launay, F, Subtil, J.-L. & Yoshino, K. 1991, Planet. Space Sci. **39**, 1301.
93 Eidelserg, M., Rostas, F., Breton, J. & Thieblemont, B. 1992, J. Chem. Phys. **96**, 5585.
94 Tchang-Brillet, W.-Ül, Julienne, P. S., Roble, J. M., Lettzelter, C. & Rostas, F. 1992, J. Chem. Phys. **96**, 6735.
95 Baker, J., Lemaire, J. L., Couris, S., Vient, A., Malmasson, D. & Rostas, F. 1993, Chem. Phys., in press.
96 Garetz, B. A., Kittrel, C. & Le Floch, A. 1991, J. Chem. Phys. **94**, 843.
97 Kittrell, C., Le Floch, A., Garetz, B. A. 1993, J. Chem. Phys. **97**, 2221.
98 Le Floch, A. 1991, Mol. Phys. **72**, 133.
99 Abgrall, H., Roueff, E., Launay, F., Roncin, J. Y. & Subtil, J. L. 1993, J. Mol. Spectrosc. **157**, 512.
100 Abgrall, H., Roueff, E., Launay, F., Roncin, J. Y. & Subtil, J. L. 1993, Astron. Astrophys., Suppl. Ser., in press.
101 Cossart-Magos, C., Launay, F., & Parkin, J. E. 1992, Mol. Phys. **75**, 835.
102 Spielfiedel, A., Feautrier, N., Chambaud, G., Rosmus, P. & Werner, H. J. 1992, J. Chem. Phys. **97**, 8382.
103 Werner, H. J., Spielfiedel, A., Feautrier, N., Chambaud, G., & Rosmus, P. 1990, Chem. Phys. Lett. **175**, 203.
104 Spielfiedel, A., Feautrier, N., Chambaud, G., Rosmus, P. & Werner, H. J. 1991, Chem. Phys. Lett. **183**, 16.
105 Gabriel, W., Chambaud, G., Rosmus, P., Spielfiedel, A. & Feautrier, N. 1992, Astrophys. J. **398**, 706.
106 Brommer, M., Chambaud, G., Reinsch, E.-A., Rosmus, P., Spielfiedel, A. & Feautrier, N. 1991, J. Chem. Phys. **94**, 8070.
107 Bensasson, R., Hill, T., Lambert, C., Land, E. J., Leach, S. & Truscott, T. G. 1993, Chem. Phys. Lett., in press.
108 Leach, S., Vervloet, M., Despres, A., Breheret, E., Hare, J. P., Dennis, T. J., Kroto, H. W., Taylor. R. & Walton, D. R. M. 1992, Chem. Phys. **160**, 451.
109 Bensasson, R., Hill, T., Lambert, C., Land, E. J., Leach, S. & Truscott, T. G. 1993, Chem. Phys. Lett. **201**, 326.
110 Gertner, M. R., Shin, J. B. & Nicholls, R. W. 1993, Spectrosc. Lett **26**(4), 677.
111 Nicholls, R. W. 1992, J. Quant. Spectrosc. Radiat. Transfer **48**, 315.
112 Nicholls, R. W. 1991, J. Quant. Spectrosc. Radiat. Transfer **45**, 262.
113 Cann, M. W. P. & Nicholls, R. W. 1991, Can. J. Phys. **69**, 1163.
114 Reddy, R. R., Vishwanath, R. & Rao, T. V. R. 1990, Astrophys. Space Sci. **164**, 113.
115 Reddy, R. R., Rao, T. V. R. & Vishwanath, R. 1992, Astrophys. Space Sci. **189**, 29.
116 Abrams, Mark C., Davis, Sumner P., Rao, M. L. P. & Engleman Jr., Rolf 1990, Astrophys. J. **363**, 326.
117 Chackerian, C., Goorvitch, D., Benidar, A., Farreno, R., Guelachvili, G., Martin, P. M., Abrams, M. C. & Davis, S. P. 1992, J. Quant. Spectrosc. Radiat. Transfer **48**, 667.
118 Engleman Jr., R., Davis, S. P., & Rao, M. L. P. 1992, Pramana **39**, 163.
119 Davis, S. P., Abrams, M. C., Rao, M. L. P. & Brault, J. W. 1991, J. Opt. Soc. Am. **138**, 198.
120 Phillips, John G. & Davis, Sumner P. 1993, Astrophys. J. **409**, 860.

121 Langhoff, Stephen R. Bauschlicher Jr., Charles W. 1990, J. Mol. Spectrosc. **141**, 243.
122 Langhoff, S. R. & Bauchlicher, C. W. 1991, Astrophys. J. **375**, 843.
123 Hammer, P. D. & Davis, S. P. 1979, J. Mol. Spectrosc. **78**, 337 & Littleton, J. E. & Davis, S. P. 1985, Astrophys. J. **296**, 152.
124 Langhoff, Stephen R. & Bauschlicher Jr., Charles W. 1990, Astrophys. J. **349**, 369.
125 Littleton, J. E., Davis, Sumner P. & Song, Mei 1993, Astrophys. J. **404**, 412.
126 Langhoff, Stephen R., Bauschlicher Jr., Charles W. & Rendell, Allistair, P. 1990, J. Chem. Phys. **92**, 6599.
127 Bauer, W., Becker, K. H., Bielefeld, M. & Meuser, R. 1986, Chem. Phys. Lett. **123**, 33.
128 Langhoff, S. R. & Bauschlicher, C. W. 1993, Chem. Phys. Lett. **211**, 305.
129 Jorgensen, U. G. & Larsson, M. 1990, Astron. Astrophys. **238**, 424.
130 Jorgensen, U. G. 1992, Rev. Mexicana Astron. Astrophys. **23**, 49.
131 Jorgensen, U. G. & Jensen, P. 1993, J. Mol. Spectrosc. **161**, 219.
132 Mickelson, M. E., Larson, L. E. & Schubert, A. 1991, J. Geophys. Res. **96**(E2), 17,507.
133 Mickelson, M. E. & Larson, L. E. 1992, Bull. Am. Astron. Soc. **24**, 990.
134 Mickelson, M. E. & Trauger, J. T. 1990, J. Mol. Structure **217**, 85.
135 Ferguson, David W., Rao, K. Narahari, Mickelson, Michael E. & Larson, Lee L. 1993, J. Mol. Spectrosc. **60**, 315.
136 Lovas, Frank J. 1992, J. Phys. Chem. Ref. Data **21**, 181.
137 Lovas, F. J., Suenram, R. D., Ogata, T. & Yamamoto, S. 1992, Astrophys. J. **399**, 325.
138 Maki, A. G. & Wells, J. S. 1992, J. Mol. Spectrosc. **152**, 69.

WORKING GROUP 6: MOLECULAR REACTIONS ON SOLID SURFACES

Sydney Leach, Chairperson

Molecular reactions on solid surfaces is a new subject for a report to Commission 14. From the astrophysical viewpoint it concerns the physical chemistry of dust and grains. These entities are chemically active, the major illustration of which is the association of hydrogen atoms at interstellar dust surfaces to form molecular hydrogen. Observations of H_2 in diffuse clouds imply that this is an efficient process. Several other recent observations of interstellar molecular abundances appear to be incompatible with purely gas phase reactions, so that the intervention of grain chemistry is surmised. Development of this subject is hampered by the relatively small amount of knowledge we have as to the nature and composition of interstellar and solar system grains, and the structure and properties of the gas-grain interface. This makes it hard to evaluate key parameters such as sticking coefficients and adsorption, condensation, desorption and evaporation properties, as well as the thermal, photochemical and ionizing-radiation induced chemical reactions occurring on the surfaces and which can all contribute to gas-grain exchange. The present non-exhaustive report is intended to give some key indications to the physico-chemical background, model experimental and theoretical studies and to the astrophysical aspects of the subject.

1. Reviews

1.1. Reviews on the nature and properties of dust and on material processing in space.

Laboratory, observational and theoretical studies of interstellar ices (1); Models of interstellar grains (2); Ion-solid interactions (3); Meteoritic, interplanetary and cometary dust (4); Irradiation of solids (5); Mass-spectrometric in situ analysis of solid-state extraterrestrial samples (6); Spectroscopy of solid materials simulating cosmic dust (7); Molecule formation by fast particles in astronomical objects (8); Nucleation and growth of solid particles from high-temperature gas mixtures (9); Spectroscopy of ion-irradiated carbonaceous materials (10); Theory of Cosmic Grains (11); Observations and Theories of Interstellar Dust (12); Dust and its Chemistry in Astronomy (13-15).

1.2. Reviews on Surface Science.

Surveys of concepts and experimental techniques in surface chemical physics (16-19). Discussion on the Structure of Surfaces and Interfaces as studied using Synchrotron Radiation (20). Photochemistry and Radiation chemistry of Adsorbed Molecules (21)

2. Studies of Dust Grain Properties and Models.

Dust grain properties in Reflection Nebulae (22); Dust grain processing in the Orion nebula (23); Electric charge on grains in diffuse clouds (24); Dust coagulation (25); Dust grain composition, size and processing models (22-31). Cosmic -ray induced desorption (32); Carbon dust formation on interstellar grains (33). Hydrogenated Amorphous Carbon Grains and Mantles in the ISM (34-37).

3. Astrophysical observations and models suggesting molecule formation or gas-grain exchange.

NH in diffuse clouds (38, 39); H_2CO in exterior parts of dense clouds (40); D_2CO observations (41); Astrophysical models involving adsorption and chemistry on grains (42-48); Return or freezing out of gas material in the ISM(49-51); Chemical activity of dust (52-54).

4. Models of the Formation of H_2 on Dust Grains and its Excitation

Formation processes and possibility of observing infrared emission from H_2 formed on surfaces (55); ortho:para H_2 ratio (56); H_2 rotational level distribution (57); possibility of detecting H_2 on interstellar dust by infrared absorption (58).

5. Laboratory studies of hydrogen recombination and hydrogenation reactions on carbonaceous and on analogous materials.

Surface diffusion of hydrogen atoms on a silicon surface (59); Hydrogen-surface interactions (60); Chemical erosion of graphite by hydrogen atoms (61); Hydrogen atom chemisorption on a graphite surface (62); energetics of hydrogenated surfaces of diamond (63); Light-enhanced hydrogen motion in hydrogenated amorphous silicon (64); Hydrogen recombinative desorption on silicon and other surfaces (65-69).

6. Surface Properties and Dynamics.

Surface melting and role of surfaces in melting (70-72); Dynamics at Surfaces (73); Vibrations and Sticking of a particle on a Surface (74); Direct Observation of Chemical Bond dynamics on Surfaces (75); Hydrogen atom site-to site hopping rates (76); Heat capacities of physisorbed films on graphite (77); Experimental and theoretical studies of NH_3 adsorbed on graphite (78-81); Mechanisms and energetics of photofragmentation of molecules condensed on surfaces (82,83); Oxidation on carbon surfaces (84); Photon-induced Desorption of Molecules from Surfaces (85-87).

7. Laboratory Studies of Astrophysical-type Ices

CO_2 in astrophysical ice analogs (88); Clathrate hydrate formation in amorphous cometary ice analogs (89); H_2CO and organic molecule production in astrophysical ices (90); Condensation and structure of amorphous ice (91); Amorphization of cubic ice by UV irradiation (92).

References

1. Tielens, A.G.G.M., Allamandola, L.J. and Sandford, S.A.: 1991, in "Solid-State Astrophysics," Proc. Int. Sch. Phys. Enrico Fermi, Vol. 111, Ed. E. Bussolletti and G. Strazzulla, North Holland, Amsterdam, p. 29.

2. Duley, W.W.: 1991, "Solid-State Astrophysics," in Proc. Int. Sch. Phys. Enrico Fermi, Vol. 111, Ed. E. Bussolletti and G. Strazzulla, North Holland, Amsterdam, p. 59.
3. Foti, G.: 1991, "Solid-State Astrophysics," in Proc. Int. Sch. Phys. Enrico Fermi, Vol. 111, Ed. E. Bussolletti and G. Strazzulla, North Holland, Amsterdam, p. 73.
4. Jessberger, E.K.: 1991, "Solid-State Astrophysics," in Proc. Int. Sch. Phys. Enrico Fermi, Vol. 111, Ed. E. Bussolletti and G. Strazzulla, North Holland, Amsterdam, p. 107.
5. Johnson, R.E.: 1991, "Solid-State Astrophysics," in Proc. Int. Sch. Phys. Enrico Fermi, Vol. 111, Ed. E. Bussolletti and G. Strazzulla, North Holland, Amsterdam, p. 129.
6. Kissel, J.:1991, "Solid-State Astrophysics," in Proc. Int. Sch. Phys. Enrico Fermi, Vol. 111, Ed. E. Bussolletti and G. Strazzulla, North Holland, Amsterdam, p. 169.
7. Wdowiak, T.J.:1991, "Solid-State Astrophysics," in Proc. Int. Sch. Phys. Enrico Fermi, Vol. 111, Ed. E. Bussolletti and G. Strazzulla, North Holland, Amsterdam, p. 279.
8. Pirronello, V.:1991, "Solid-State Astrophysics," in Proc. Int. Sch. Phys. Enrico Fermi, Vol. 111, Ed. E. Bussolletti and G. Strazzulla, North Holland, Amsterdam, p. 381.
9. Stephens, J.R.: 1991, "Solid-State Astrophysics," in Proc. Int. Sch. Phys. Enrico Fermi, Vol. 111, Ed. E. Bussolletti and G. Strazzulla, North Holland, Amsterdam, p. 391.
10. Strazzulla, G.A., Baratta, G. and Magazzu.: 1991, "Solid-State Astrophysics," in Proc. Int. Sch. Phys. Enrico Fermi, Vol. 111, Ed. E. Bussolletti and G. Strazzulla, North Holland, Amsterdam, 403.
11. Hoyle, F. and Wickramasinghe, N.C.: 1991,"The Theory of Cosmic Grains," Kluwer, Dordrecht.
12. Mathis, J.S.: 1993, Rep. Prog. Phys., **56**, 605.
13. Williams, D.A.: 1991, in "Molecular Clouds," Ed. R.A. James and T.J. Millar, Cambridge Univ. Press, Cambridge, 295.
14. Whittet, D.C.B.: 1992, "Dust in the Galactic Environment," IOP Publishing, Bristol.
15. Millar, T.G. and Williams, D.A., (Ed.): 1993, "Dust and Chemistry in Astronomy," IOP Publishing, Bristol.
16. Block, J.H., Bradshaw, A.M., Gravelle, C., Haber, J., Hansen, R.S., Roberts, M.W., Sheppard, N. and Tamaru K.: 1990, Pure Appl. Chem., **62**, 2297.
17. Sheppard, N.: 1991, Pure Appl. Chem., **63**, 887.
18. Somorjai, G.A.: 1990, J. Phys. Chem., **94**, 1013.
19. Joyner, R.W.: 1990, J. Chem. Soc. Faraday Trans., **86**, 2675.
20. Pershan, P.S.: 1990, Faraday Discuss. Chem. Soc., **89**, 231.
21. Thomas, J.K.: 1993, Chemical Reviews, **93**, 301.
22. Pendleton, Y.J., Tielens, A.G.G.M. and Werner, M.W.: 1990, Astrophys. J., **349**, 107.
23. Sorrell, W.H.: 1992, Mon. Not. R. Astron. Soc., **255**, 594.
24. Taylor, S.D.: 1991, Mon. Not. R. Astron. Soc., **248**, 148.
25. Chokshi, A., Tielens, A.G.G.M. and Hollenbach, D: 1993, Astrophys. J., **407**, 806.
26. Chlewicki, G. and Greenberg, J.M.: 1990, Astrophys. J., **365**, 230.
27. Rowan-Robinson, M.: 1992, Mon. Not. R. Astron. Soc., **258**, 787.
28. Désert, F.-X., Boulanger, F. and Puget, J.L.: 1990, Astron. Astrophys., **237**, 215.
29. Siebenmorgen, R. and Krügel, E.: 1992, Astron. Astrophys. **259**, 614.
30. Voit, G.M.: 1991, Astrophys. J. **379**, 122.
31. Pei, Y.C.: 1992, Astrophys. J. **395**, 130.
32. Hartquist, T.W. and Williams, D.A.: Mon. Not. R. Astron. Soc., **247**, 343.
33. Jenniskens, Baratta, G.A., Kouchi, A., de Groot, M.S., Greenberg, J.M. and Strazzulla, G.: 1993, Astron. Astrophys., **273**, 583.
34. Jones, A., Duley, W.W. and Williams, D.A.: 1990, Quart. J. Roy. Astron. Soc., **31**, 567.
35. Duley, W.W., Jones, A., Taylor, S.D. and Williams, D.A.: 1993, Mon. Not. R. Astron. Soc., **260**, 415.
36. Duley, W.W.: 1992, Mon. Not. R. Astron. Soc., **258** , 773.
37. Witt, A. and Boroson, T.A.: 1990, Astrophys. J., **355**, 182.
38. Meyer,D.M. and Roth, K.C.: 1991, Astrophys. J. **376**, L49.
39. Wagenblast, R., Williams, D.A., Millar, T.J. and Nejad, L.A.M.: 1993, Mon. Not. R. Astron. Soc., **258**, 773.
40. Federman, S.R. and Allen, M.: 1991, Astrophys. J., **375**, 157.
41. Turner, B.E.: 1990, Astrophys. J. **362**, L29.
42. Brown, D.: 1990, Mon. Not. R. Astron. Soc., **243**, 65.

43. Brown, D. and Charnley, S.B.: 1990, Mon. Not. R. Astron .Soc., **244**, 69.
44. Brown and Charnley, : 1991, Mon. Not. R. Astron. Soc., **249**, 432.
45. Pineau des Forêts, G., Flower, D.R. and Chièze, J.: 1992, Mon. Not. R. Astron. Soc., **256**, 247.
46. Hasegawa, T.I., Herbst, E. and Leung, C.-M.: 1992, Astrophys. J., Suppl. Ser., **82**, 167.
47. Hasegawa, T.I. and Herbst, E.: 1993, Mon. Not. R. Astron. Soc., **261**, 83.
48. Hasegawa, T.I. and Herbst, E.: 1993, Mon. Not. R. Astron. Soc., **263**, 589.
49. Charnley, S.B., Tielens, A.G.G.M. and Millar, T.J.: 1992, Astrophys. J. **399**, L74.
50. Mauersberger, R., Wilson, T.L., Mezger, G., Gaume, R. and Johnston, K.J.: 1992, Astron. Astrophys., **256**, 640.
51. Willacy, K. and Williams, D.A.: 1993, Mon. Not. R. Astron. Soc., **260**, 635.
52. Williams, D.A.: 1993, in "Dust and Chemistry in Astronomy," Ed. T.J. Millar and D.A. Williams, IOP Publishing, Bristol, 143.
53. Duley, W.W.: 1993, in "Dust and Chemistry in Astronomy," Ed. T.J. Millar and D.A. Williams, IOP Publishing, Bristol, 71.
54. Tielens, A.G.G.M.: 1993, in "Dust and Chemistry in Astronomy," Ed. T.J. Millar and D.A. Williams, IOP Publishing, Bristol, 103.
55. Duley, W.W. and Williams, D.A.: 1993, Mon. Not. R. Astron. Soc., **260**, 37.
56. Flower, D.: 1991, Comments At. Mol. Phys., **26**, 33.
57. Wagenblast, R.: 1992, Mon. Not. R. Astron. Soc., **269**, 155.
58. Sandford, S.A. and Allamandola, L.J.: 1993, Astrophys. J., **409**, L65.
59. Reider, G.A., Höfer, U and Heinz, T.F.: 1991, Phys. Rev. Lett., **66**, 1994.
60. Hopman, H.J.: 1990, in "Nonequilibrium Processes in Partially Ionized Gases," Ed. M. Capitelli and J.N.Bardsley, Plenum, New York, 241.
61. Biener, J., Schubert, U.A., Schenk, A., Winter, B., Lutterloh, C. and Küppers, J.: 1993, J. Chem. Phys., **99**, 3125.
62. Klose, S.: 1992, Astron. Astrophys., **260**, 321.
63. Yang, Y.L. and D'Evelyn, M.:1992, J. Am. Chem. Soc., **114**, 2796.
64. Santos, V., Johnson, N.M. and Street, R.A.: 1991, Phys. Rev. Lett., **67**, 2686.
65. Sinniah, K., Sherman, M.G., Lewis, L.B., Weinberg, W.H., Yates Jr., J.T. and Janda, K.C.: 1990, J. Chem. Phys., **92**, 5700.
66. Reider, G.A., Höfer, U. and Heinz, T.F.: 1991, J. Chem. Phys., **94**, 4080.
67. Schröter, L., David, R. and Zacharias, H.: 1990, J. Electron. Spectrosc. Relat. Ph. **54/55**, 143.
69. Kasai, H. and Okiji.: 1990, J. Electron, Spectrosc. Relat. Ph. **54/55**, 153.
70. Yang, K. and Rahman, T.S.: 1990, J. Chem. Phys., **93**, 6834.
71. Frenken, J.W.M.: 1990, Endeavour, New Series, **14**, 2.
72. van der Veen, J.F.: 1991, in "Phase Transitions in Surface Films," Volume 2, Ed. H. Taub et al., Plenum Press, New York, 289.
73. van der Veen, J.F. and Frenken, J.W.M.: 1991, Surf. Sci. **251/252**, 1.
74. Tully, J.C.: 1990, J. Electron Spectrosc. Relat. Ph. **54/55**, 1.
75. Harris, J.: 1990, J. Electron Spectrosc. Relat. Ph. **54/55**, 115.
76. Yates, Jr., J.T., Alvey, M.D., Dresser, M.J., Henderson, M.A., Kiskinova, M., Ramsier, R.D. and Szabo, A.: 1992, Science, **255**, 1397.
77. Haug, K., Wahnströhm, G. and Metiu, H.: 1990, J. Chem. Phys., **92**, 2083.
78. Inaba, A., Shirakami, T. and Chihara, H.: 1991, J. Chem. Thermodyn. **23**, 461.
79. Cheng, A. and Steele, W.A.: 1990, J. Chem. Phys., **92**, 3858.
80. Cheng, A. and Steele, W.A.: 1990, J. Chem. Phys., **92**, 3867.
81. Rowntree, P., Scoles, G and Xu, J.: 1990, J. Chem. Phys.,**92**, 3853.
82. Yang, Q. and Osgood Jr., R.M.: J. Chem. Phys., **97**, 8855.
83. Hatch, S.R., Zhu, X.-Y., White, J.M. and Campion, A.: 1990, J. Chem. Phys. **92**, 2861.
84. Meldrum, B.J. and Rochester, C.H.: 1990, J. Chem. Soc. Faraday Trans., **86**, 3647.
85. Vertes, A.: in "Methods and Mechanisms for Producing Ions from Large Molecules," Ed. K.G. Standing and W. Ens, Plenum, New York.
86. Cavanagh, R.R., Buntin, S.A., Richter, L.J. and King, D.S.: 1990, Comments At. Mol. Phys., **24**, 365.
87. Budde, F., Heinz, T.F., Loy, M.M.T., Misewich, J.A., de Rougemont, F. and Zacharias, H.: 1991, Phys. Rev. Lett., **66**, 3024.

88. Sandford, S.A. and Allamandola, L.J.: 1990, Astrophys. J., **355**, 357.
89. Blake, D., Allamandola, L., Sandford, S., Hudgins, D. and Freund, F.: 1991, Science, **254**, 548.
90. Schutte, W.A., Allamandola, L.J. and Sandford, S.A.: 1993, Science, **259**, 1143.
91. Zhang, Q. and Buch, V.: 1990, J. Chem. Phys., **92**, 1512.
92. Kouchi, A. and Kuroda, T.: 1990, Nature, **344**, 134.

COMMISSION 15: PHYSICAL STUDY OF COMETS, MINOR PLANETS AND METEORITES (L'ETUDE PHYSIQUE DES COMETES, DES PETITES PLANETES ET DES METEORITES)

PRESIDENT: Alan W. Harris
VICE-PRESIDENT: Michael F. A'Hearn
PAST PRESIDENT: Jurgen Rahe
SECRETARY: Duncan I. Steel

ORGANIZING COMMITTEE: M. A. Barucci, M. J. S. Belton, J. C. Brandt, E. Grün, W. F. Huebner, D. Hughes, H. U. Keller, N. N. Kiselev, D. F. Lupishko, H. R. Rickman, E. F. Tedesco, Z. Knezevic

I. COMETS

The level of activity during the period covered, July 1990 to June 1993, was comparable, if not higher than in the previous period. Work continued on interpretation and modeling of the data from the 1986 apparition of comet P/Halley. Many of the research results reported in Bamberg (IAU Colloquium No. 116) were transferred to text book level (Mason 1990, Huebner 1990). Space mission research continued with the Giotto extended mission and preparation of CRAF and Rosetta. CRAF was eventually canceled, putting more importance on international collaboration for Rosetta. New lines of research were opened with the discovery of activity of 2060 Chiron, improved ground based CCD observations of distant comets, the spectacular breakup of the comet Shoemaker-Levy 1993e, and the detection of trans-Neptunian planetary bodies - potential Kuiper belt members. An important review of cometary science has been written by Festou *et al.* (1993). It contains a very complete list of references of literature up to 1992, including proceedings and monographs. Many international meetings as well as sections of larger Symposia dealt with the subjects of Commission 15. Most important for the community the *Asteroids, Comets, Meteors* series continued with its meetings in Flagstaff, USA, June 24-28, 1991 and in Belgirate, Italy, June 14-18, 1993 (IAU Symposium 160). Information on meetings relevant to Commission 15, can be also found in the IAU Information Bulletins and/or in the Astronomy and Astrophysics Abstracts.

1. Nucleus: properties, origin, and evolution - H. U. Keller and W. J. Markiewicz

Possibly the most important outcome of the period covered is the formulation of what is still not understood about the nature of cometary nuclei. Continued analysis of data from the last apparition of P/Halley and ground based observations of distant short period comets show that nuclei are highly irregular in overall shape, and may be inhomogeneous in physical structure or composition, are highly porous, and have very low albedos. Most of their surface is not active. Whipple's icy conglomerate model of the nucleus was confirmed in its basic idea. Significant conceptual differences, however, are emerging in various models (Houpis 1991). In the model of Greenberg and Hage (1990) the individual sub-micron particles are described in detail by relating elemental abundances to those of interstellar dust. These authors also predict porosity to be at least 95 %. Keller (1990) advocates a model of a solid matrix (microstructure) with volatile ices filling the internal voids (icy dirt ball) - in a sense reversing Whipple's original model. Still lacking is quantification of the mass and size distribution of agglomerates making up the solid matrix. The question of the size distribution is crucial for understanding the activity of comets, the relations between comets and meteors, as well as the scenario of nucleus formation itself.

Giotto and Vega data were used to create ever more complex models of P/Halley's nucleus. Detailed cartographic models of P/Halley's topography were constructed by Stooke and Abergel (1991). Improvements beyond their model are limited by the lack of understanding P/Halley's rotation (see below). The study of P/Halley's rotational properties continued (Belton *et al.* 1991, Samarashinha and A'Hearn 1991). This complex problem important for the coma morphology as well as the nongravitational effects on the orbit is still not completely resolved. The irregular shape of the nucleus and the asymmetric outgassing are too complex to fully constrain models that try to describe the rotational, precessional and nutational motion. Effects of random jets on the spin state of P/Halley were considered by Julian (1990). Existing photometric and spectral data suggest a remarkable similarity between Trojan asteroids and cometary nuclei (Jewitt and Luu 1990). Modern CCD instruments have made it possible to observe with ground based telescopes the spectroscopic properties of nuclei of distant (inactive) comets (Luu 1993). Jewitt and Luu (1992a) developed a novel observational method in the submillimeter continuum. Whipple (1992) introduced a volatility index for comets. His study of many long and short period comets resulted in the same average values of volatility, and showed no correlation with usual orbital

J. Bergeron (ed.), Reports on Astronomy, Vol. XXIIA, 135–156.

and physical parameters. This result suggests that active comets are basically alike regardless of their orbits or ages.

Laboratory studies of ices and ice/dust mixtures continued. The results of the KOSI (cometary simulations) project were reported in a special issue of Geophysical Research Letters (Grün *et al.* 1991). A review of sublimation experiments in USSR was written by Ibadinov *et al.* (1991). The cosmic-ray irradiation experiments have been reviewed by Strazzulla and Johnson (1991). Experiments with amorphous ices or ice mixtures have been reviewed by Klinger (1991).

A number of additional physical processes have been included in a new generation of thermal models of the nucleus. Porosity, heat transport by water vapor, chemical differentiation, amorphous - crystalline transition of the H_2O are some of the points discussed in this context (Fanale and Salvail 1990, Espinasse *et al.* 1991, Prialnik and Mekler 1991, Prialnik 1992). Similar models were calculated by Spohn and Benkhoff (1990) and applied to KOSI experiments resulting in estimates of various thermal diffusivities of a ice/dust porous mixture.

The restricted surface activity of comets continues to challenge our understanding of the surface composition, presence and formation of dust mantles, and the ways in which the dust component is agglomerated (glued) within the nucleus. The formation of dust mantles on nuclei was inferred from statistical analysis of nongravitational effects of short period comets (Rickman *et al.* 1991). Baratta *et al.* (1991) derive from photometry of P/Brorsen-Metcalf the existence of tiny grains of blue color with diameters < 0.1 μ. Their results supposedly support the existence of organic cometary crust. Sekanina (1993) analyzed the evolution of active sites as inferred from nongravitational perturbations of cometary orbits. The rôle of POM and CHON particles as cometary gluing agents was discussed by Boehnhardt *et al.* (1990). H_2CO rotational lines in mm range were predicted by (Bockelée-Morvan and Crovisier 1992) and detected by microwave observations in three different comets (Colom *et al.* 1992). Rickman *et al.* (1990) show by numerical simulations that the formation of stable dust mantles is possible for specific circumstances on parts of a cometary surface. Carusi *et al.* (1991) studied the last 47 orbits of P/d'Arrest (since 1678). They showed that the strength of nongravitational forces was stable and that the absolute brightness and the shape of the light curve have not changed during this period.

Cometary activity at large heliocentric distances received much attention. West *et al.* (1991) reported the observation of P/Halley's outburst at 14 AU, the farthest cometary outburst to date. An analysis and models are given by Sekanina *et al.* (1992), requiring CO as the volatile agent. Hughes (1991) summarized possible outburst mechanisms. Polymerization of HCN could also be a possible cause of cometary outbursts (Rettig *et al.* 1992). Wallis and Wickramasinghe (1991) draw an analogy between Triton's geysers and their "frozen sea" model of cometary outbursts. 2060 Chiron was shown to be a giant comet by photometric observations (Hartmann *et al.* 1990, West 1991). From submillimeter photometry the diameter of Chiron was constrained to be ≤300 km (Jewitt and Luu 1992b). Chiron's activity was reported to vary on time scales of hours (Luu and Jewitt 1990). Hahn and Bailey (1990) calculated past and future orbits of 2060 Chiron. They conclude that Chiron was likely a short period comet in the past and will be one in the future. CN emission from Chiron was detected by Bus *et al.* (1991). Interest in P/Halley's outburst and the cometary activity of 2060 Chiron gave rise to the *Workshop on the Activity of Distant Comets* (Huebner *et al.* 1993).

IAU Symposium No. 152, *Chaos, Resonance and Collective Dynamical Phenomena in the Solar System*, in Angra dos Reis, Brazil (15-19 July 1991) covered the new developments in the study of evolution of cometary orbits, the Oort cloud and related matters. Bailey (1992) concluded from a numerical study, that long-period comets with initially high-inclination (i~90°) orbits and moderately small perihelion (q < 2 AU) frequently become sun grazers. The mass involved in their destruction is an important source of interplanetary matter. The observed asymmetry in aphelia of new (long-period) comets is attributed to the motion of the solar system through its local galactic neighborhood (Brunini 1993). Revising data on orbits to include statistical effects of outbursts and nucleus splitting Kresák (1992) confirmed that there is no evidence that any of the recorded comets has a hyperbolic orbit (*i.e.* comes from interstellar space). Contrary to Whipple's nongravitational model Sekanina (1993) showed that the change in orbital period is not directly correlated with the sense of rotation of the nucleus. Asymmetry in the production rate relative to perihelion and the location of single jets influence the orbital period.

A review of the origin and evolution of the Oort cloud was written by Weismann (1990). Numerical studies of the formation of the Kuiper belt by scattering of comets by planets was reported by Ip and Fernández (1991). Abundance ratios of key gaseous molecular components of P/Halley were used to infer under what conditions this comet was formed (Engel *et al.* 1990). Carbon species could have originated from within the solar nebula, but ammonia must have come from outside. A surrounding molecular cloud is proposed as a possible source. Sublimation and reformation of icy grains in the primitive solar nebula are discussed by Lunine *et al.* (1991). It is shown that comets are likely to be made up of unprocessed grains. Growth of fluffy cometary nuclei by random accumulation of grains was studied by Donn (1991). The general aspects of cometary origin and relation to solar system cosmogony were reviewed in several chapters of *Protostars and Planets III* (Levy et al., eds., The University of Arizona Press, 1993). The first detection of a potential member of the Kuiper belt, object 1992QB$_1$, was reported by Jewitt and Luu (1993) and was followed by their discovery of a second, 1993 FW (IAU Circular No. 5730).

The possibility of observing line emission in OH, CN, and C_2 from cometesimals around millisecond pulsars as a means of detection of another planetary system was proposed by Chakrabarti and Krishna Swami (1992). Mitrofanov and Sagdeev (1990) have shown that gamma ray bursts from neutron stars could be accounted for by considering encounters with comets. Stern *et al.* (1990) explain the observation of water masers and complex molecules in envelopes of giant and super-giant stars in terms of an evaporating circumstellar comet disk. This model provides additional evidence for planetary systems about other stars.

2. Dust coma and tail - S. M. Larson

This period was punctuated by the Giotto flyby of P/Grigg-Skjellerup, the first Hubble Space Telescope comet observations, an outburst of P/Halley at 14 AU, the development of the coma and tail of 2060 Chiron, and the appearance of P/Swift-Tuttle.

The 1992 Giotto Extended mission to within 200 km of P/Grigg-Skjellerup on July 10 detected 3 meteoroid impacts indicating a dust mass distribution dominated by large particles (McDonnell *et al.* 1993). Modeling of nearly simultaneous ground based observations also indicated large particles as well as to provide additional dust parameters including a hemispherical emission pattern (Fulle *et al.* 1993).

Cometary observations with the Hubble Space Telescope started with Comet Levy (1990c) whose coma was characterized by expanding shells with the very low apparent velocity of 160 m/s (Weaver *et al.* 1992). Since that time, the dust comae of several other comets have been observed.

Observations at ESO showed that P/Halley underwent an outburst when at 14 AU from the sun. The expanding crescent-shaped dust coma was observed for several months, and could be modeled as a short-term emission from the sunlit side (Sekanina *et al.* 1992). The derived mass loss in this event was ~10^{12} g, and the particles could have been driven by carbon monoxide to their 45 m/s terminal velocity, but the mass loading would have been enormous. Although there are several possibilities (*e.g.* Schmidt *et al.* 1993) the triggering mechanism may never be known.

The coma of 2060 Chiron has become more pronounced as it approaches perihelion (*e.g.* Luu and Jewitt 1993), and a tail was reported in 1992 (Larson and Marcialis 1992).

Dust trails for 8 periodic comets near perihelion, and others with no known parents, were identified on IRAF images (Sykes and Walker 1992). These features likely consist of large refractory particles.

The characteristic polarization-phase angle dependence of cometary dust, and a wavelength dependence of polarization at larger phase angles, observed in 13 comets by Chernova *et al.* (1993), was found to be consistent with large fluffy particles. The deviation from a 1/r radial profile of dust in 10 of 14 comets observed by Baum *et al.* (1992) was interpreted as evidence for grains fading with time. The particle size distribution has been modeled in several comets using a numerical inversion technique to fit observed isophotes and neck-line structures (Fulle 1992a, 1992b; Fulle *et al.* 1992, 1993). They show that the dust size distribution of dynamically newer comets tends to have a higher power index than for periodic comets.

P/Swift-Tuttle, the parent of the Perseid meteor stream, went through perihelion in 1992 and displayed spectacular dust jets (Yoshida *et al.* 1993) whose evolution was consistent with the one observed in 1862 (Sekanina 1981).

3. Gas coma and photochemistry - M. C. Festou

The data collected during the spacecraft encounters with comet P/Halley were investigated in more detail. Geiss *et al.* (1991) using IMS-HIS data found that all ions observed in the 25-35 amu range could be explained as the result of the protonation of methanol, formaldehyde, hydrogen cyanide and hydrogen sulfide molecules. The IMS data were used by Boice *et al.* (1990) to infer the presence of an extended source of coma species and they derived an upper limit of the CH_4 production relative to water of 2%. The PICCA data revealed the presence of many sulfur-bearing ions (Marconi *et al.* 1991). Huntress *et al.* (1991) discussed the possibility that carbon suboxide could exist in comet P/Halley. Using TKS spectra Krasnopolsky and Tkachuk (1991) derived the NH and NH_2 productions and Krasnopolsky *et al.* (1991) studied the spatial distributions of CN and C_3. Krasnopolsky (1991) discussed the possible nature of the parents of C_3 and CN. Chaizy *et al.* (1991) have presented for the first time evidence of the presence of negative ions in the inner coma of P/Halley. The water production rate in P/Grigg-Skjellerup has been determined by Johnstone *et al.* (1993) from the JPA experiment data.

Jets of various coma species, including ions, have been found in comet P/Halley as well as in comets observed more recently. H_2O^+ and C_2 jets were studied by Formisano *et al.* (1991). Images of the H_2O^+ emission, an indicator of gas phase reactions with water molecules, were obtained by Disanti *et al.* (1990) at the time of the spacecraft encounters with P/Halley. Fabry-Perot observations by Scherb *et al.* (1990) can provide the velocities of these ions. Velocities of H_2O^+ ions were measured in comet Levy (1990v) by Rauer and Jockers (1993).

The structure of the CN jets was studied by Schulz (1991, 1992) and Schulz *et al.* (1993), who found that CN jets were evolving into CN shells. Column densities were derived by Schulz (1993). The jets and shells structures were also modeled by Kömle (1990). Klavetter and A'Hearn (1992) deduced from their images, which revealed the existence of CN jets in P/Halley, that the velocities of these jets exhibited a wide range of values and little evidence of acceleration. The connection between the C_2, CN, C_3 jets and active regions was studied by Jockers *et al.* (1993). The effect of solar activity on CN and C2 coma scales was investigated by Meredith *et al.* (1992).

Images of the L-α emission obtained of P/Halley from the Pioneer-Venus probe were analyzed by Smyth *et al.* (1991) using a comprehensive model of the kinematics and dynamics of the H coma. Images of the L-α coma at a time close to the spacecraft encounters were analyzed by McCoy *et al.* (1992). The morphology and characteristics of the H coma were reviewed by Shimizu (1991) and Hirao (1990). Combi and Feldman (1992) showed that the L-α observations of comet P/Giacobini-Zinner obtained with IUE in 1985 were fully compatible with an origin of the H atoms from the photodestruction of water and OH molecules.

Rickman *et al.* (1991) did a statistical study of nongravitational forces in short period comets in relation to their absolute brightness and orbital history. Spinrad (1991) measured center of light displacements in the comae of comet 1983 VII and P/T2, and Medvedev (1991) investigated the relationship between non gravitational effects and such shifts. Chernetenko (1991) found that the light shift in comet P/Encke did not depend on heliocentric distance.

The brightness of comets, closely related to the CN, C_2 and dust contents of their comae has been the object of many investigations. Kamél (1991) found that the absolute magnitude of P/Encke did not significantly vary since the discovery of the comet whereas Kresák and Kresáková (1990) found the opposite result in their more extensive study of the evolution of the absolute magnitude of comets. Svoren (1991) found that the decrease of the brightness of short period comets was linked to the comet orbital elements and indicated rather short orbital lifetimes. Roettger *et al.* (1990) found a clear correlation between gas production rates and visual magnitudes, although departures from this perfect behavior are frequent. Donnison (1990) investigated the properties of the distribution of comet magnitudes. Morris and Hanner (1992) have attempted for the first time to correlate the IR and visual light curves of comets. The brightness of the coma of P/Halley in 1910 was investigated by Schleicher and Bus (1991) who found the same 7.4 day period as was derived from photometric data collected in 1986. Howell *et al.* (1991) analyzing 1910 spectra of P/Halley found that gas and dust productions were slightly larger then than during the 1986 apparition.

Spectrophotometric observations of comets were interpreted in terms of gas production rates of radicals observed in the optical by De Almeida (1991), Sanwal and Rautela (1991), Schleicher *et al.* (1991)(comet Levy, 51 nights), Fink *et al.* (1991), Wyckoff *et al.* (1991a)(NH_2), Wyckoff *et al.* (1991b)(N depleted in P/Halley by about a factor of 6 relative to the solar composition), Landaberry *et al.* (1991), De Almeida (1992), Tegler *et al.*

(1992)(NH_2 in comet P/Brorsen-Metcalf), Ellis and Neff (1992), Beaver *et al.* (1990)(P/Giacobini-Zinner depleted in NH_2), Schleicher *et al.* (1990)(P/Halley, from Sept. 85 till June 86), Fink and DiSanti (1990)([O I] lines), Boehnhardt *et al.* (1990)(P/Temple 2), Jockers *et al.* (1993) and Vid'machenko *et al.* (1990)(P/Halley pre-perihelion). Spectrophotometric and in situ measurements that led to the determination of the composition of comets were discussed in Krankowsky (1991), isotopic ratios are presented by Vanysek (1991). The various water production rate evaluations were compared by Festou (1990a) who found them to be sometimes slightly discrepant. The size and homogeneity of the IUE data base allowed the intercomparison of UV spectra of comets and established that all comets look similar, except for their CO/water and dust/water production ratios (Festou 1990b).

Upper limits for the production of SO and SO_2 molecules (Kim and A'Hearn 1991) and SH (Kim and A'Hearn 1992) were derived from the interpretation of IUE spectra. The fluorescence of S_2 molecules was re-evaluated by Kim *et al.* (1990), which led to the determination of very low upper limits to the production of that species in many comets (Budzien and Feldman 1992). The origin of S_2 remains unknown. The IUE observations of comet Levy (1990c) revealed a short term variability of all emission features; the data were used to infer the velocity of the dust particles escaping from the nucleus (Feldman *et al.* 1992).

The existing reviews on the UV spectrum of comets have been updated by Feldman (1990,1991a) and the composition of volatiles was derived (Feldman 1991b). Budzien and Feldman (1991) present IUE observations of comet 1983 VII that indicate the presence of OH prompt emission longward of 3000 Å. Krishna Swamy and Tarafdar (1993) studied the rotational distribution of CS. First observations of the 900-1200 Å region was accomplished for comet Austin (1989c1) by Green *et al.* (1991) and for comet Levy (1990c) by Feldman *et al.* (1991). The optically thin H I L-β line (blended with an oxygen line) was observed for the first time, and upper limits for the production of some rare gases were obtained. Upper limits of He and Ar productions were derived by Stern *et al.* (1992) from comet Austin data. O'Dell *et al.* (1991) recorded the UV spectrum of comet P/Brorsen-Metcalf (1989o) in which they found numerous new lines or bands, most of which are CO_2^+ emissions. Valk *et al.* (1992) performed similar work with observations of comet Austin (1989c1). Measurements of the H_2O^+/CO^+-ratio have been obtained for comets Halley, Bradfield and Wilson (Lutz *et al.* 1993) and Prasad *et al.* 1991). The values were consistent with photoionization of H_2O and CO. However ratios of other species were found to disagree with theoretical predictions. Mamadov (1990) claimed the discovery of water lines in spectra of comet Okazaki-Levy-Rudenko (1989 XIX). Esipov *et al.* (1991) claim they have detected new bands of CN and C_2 in the near-IR region. The investigation of 13 CN spectra of comet P/Halley taken at high resolution led Jaworski and Tatum (1991) to a new determination of the ratio $^{12}C/^{13}C = 89\pm17$. While the observation of CO^+ emission in comet P/Schwassmann-Wachmann 1 by Cochran *et al.* (1991) was not surprising, that of CN by Cochran and Cochran (1991) was quite unexpected and is still unexplained. A possible detection of a CO^+ tail of that comet on CCD images was reported by Jockers *et al.* (1992).

The 3.4 micron feature has been given much attention since its discovery in P/Halley. It was imaged by Klavetter and Hoban (1992). The spectrum near 3.28 micron was studied by Davies *et al.* (1991). Hoban *et al.* (1991) tentatively attributed the 3.52 micron band to methanol emission, an identification that is supported by calculations by Reuter (1992) of the fluorescence of formaldehyde in that region of the spectrum. The nature of the organic compounds in comets is discussed by Delsemme (1991). Brooke *et al.* (1991) suggest that organics are present in all comets at the same level relative to water. Observations of comet Austin (1989c1) by DiSanti *et al.* (1992) placed a significantly lower upper limit for the production of OCS in comets than had been previously obtained from P/Halley . These molecules cannot be a significant source of sulfur atoms in comets.

Hu *et al.* (1991) determined the theoretical velocity profiles of parent molecules. Larson *et al.* (1991) measured the velocity profiles of water molecules in comet Wilson (1987 VII). Discussions of the IR and radio observations of comets are found in Weaver *et al.* (1991) and Crovisier and Schloerb (1991), respectively. Encrenaz and Knacke (1991) discussed the presence of carbonaceous materials in comets. Radio interferometric observations were covered by de Pater *et al.* (1991). Snyder (1992) presented the parent molecules that have been safely identified in comets.

The stable H_2S (hydrogen sulfide) and CH_3OH (methanol) molecules were discovered by Bockelée-Morvan *et al.* (1990a, 1991) in comets Austin (1989c1) and Levy (1990c). Observations of H_2S and searches for other parent molecules are also presented in Crovisier *et al.* (1991) for comets Austin (1989c1) and Levy (1990c).

Additional millimeter observations of various comets were presented by Colom *et al.* (1992), Crovisier *et al.* (1993) and Snyder *et al.* (1990). These data allow the investigation of the kinematics of parent molecules in comet comae and in general suggest a strong sunward ejection.

The observation of an occultation of a point like radio source by comet Okazaki-Levy-Rudenko (1989 XIX) allowed Crovisier *et al.* (1992) to study in detail the excitation mechanism that is used for observations of the OH radio lines. OH data collected at Nançay have been analyzed by Bockelée-Morvan *et al.* (1990b). From observations of the 1667 MHz line, Tacconi-Garman *et al.* (1990) found that the OH coma of various comets was highly asymmetric.

Formation of water dimers in the inner coma was investigated theoretically by Crifo and Slanina (1991). Crifo (1992) described a new coma model which includes homogeneous and ion-induced water recondensation. A complete investigation of the chemistry in comet P/Temple 2 was performed by Huebner *et al.* (1992). The structure of the coma of comets was reviewed by A'Hearn and Festou (1990). Ionization mechanisms that may be at work in comets were investigated by Ibadov (1991). The ion enhancement observed at the ionopause of P/Halley was modeled by Keller and Cravens (1990).

4. Plasma - H. Rauer

A number of books containing reviews on cometary plasma has been published in the past years. Reviews published in *Comets in the Post-Halley Era* (Newburn *et al.* eds. 1991) address the interaction of the solar wind with the comet, particularly the magnetic field structure, waves and instabilities and the ion population. These points are also addressed in *Comet Halley, Investigations, Results, Interpretations* (Mason ed. 1990) with detailed overviews on the results obtained by the various spacecraft experiments at comet Halley encounter. Ip and Axford (1990) discussed in detail the processes associated with ions in the vicinity of comets, like large-scale processes, plasma waves and ion acceleration. Many interesting aspects are covered in the proceedings of the 1989 AGU Chapman conference, *Cometary Plasma Processes* (Johnstone ed. 1991). Recent results of cometary observations of ions have been presented at a workshop, *Observations of Recent Comets*, held in 1990 (Huebner ed. 1990) and at *Asteroids, Comets, Meteors*, in 1991 (Harris and Bowell eds. 1992), and in 1993.

Observations of large scale structures in plasma tails have also been obtained. An asymmetric ion ray enhancement in comet Halley in December 1985 was described by Watanbe *et al.* (1990). Tail rays in comet Bennet 1970 were described by Miller (1992) who suggested magnetically channeled outflow for their creation mechanism. Moore (1991) studied the closure rates of tail rays, compared them to predictions of existing models of ray formation and concluded that non of the models considered can explain all aspects of the observed phenomena. Delva *et al.* (1991) correlated sector boundaries in the solar wind measured by the Vega spacecraft with disconnection events seen in the tail of comet Halley. A correlation of only 50 % was found; in half of the cases the sector boundary had no effect on the plasma tail.

Further measurements of ion velocities have been made. Herbig (1990) measured gas and ion velocities in the inner coma region. Rauer and Jockers (1993) combined velocity measurements using a Fabry-Perot etalon with imaging of comet Levy 1990c1. The 2-D distribution of the velocity field was obtained for the first time. It was possible to show that structures seen in the tail of comet Levy move due to particle motion. Determinations of ion production rates from ground generally gave values less then expected from theory.

Data obtained at the encounters of the ICE spacecraft with comet Giacobini-Zinner and of Giotto and the Vega spacecraft with comet Halley continued to be evaluated and updated. First results on the encounter of Giotto with the weak comet Grigg-Skjellerup have also been reported. The plasma parameters at Halley's bow shock were re-analyzed and data of the outbound path have been presented for the first time (Coates *et al.* (1990) and Amata *et al.* (1991)). The magnetic field observations obtained by Giotto at the outbound bow shock of comet Halley showed a new type of shock transition called a "draping shock" (Neubauer *et al.* 1990). A comparison of JPA data with data obtained by the IMS experiment was made by Neugebauer *et al.* (1992).

Results on the energization and pitch-angle scattering at comets Halley and Giacobini-Zinner have also been published and compared with models (Tan *et al.* 1990, Neugebauer *et al.* 1990, McKenna-Lawlor 1990, Staines *et al.* 1991a, Kecskeméty and Cravens 1992 and Tan *et al.* 1993a). Pick-up water group ions were also detected at the most recent comet encounter, Giotto's visit to comet Grigg-Skjellerup (Johnstone *et al.* 1993 and Coates *et al.* 1993). The magnetometer data (Neubauer *et al.* 1993) showed the crossing of a bow shock on the outbound and of a bow wave on the inbound path at this weakly outgassing comet. Waves produced by ion pick-up showed much simpler waveforms than found at comet Halley. Data from the EPONA experiment (McKenna-Lawlor (1993)) showed periodic variations in the ion intensities, indicating a strong coupling to the magnetic field close to the nucleus. Observations of plasma waves at comets Halley and Giacobini-Zinner have been presented by Mazelle *et al.* (1991), Staines *et al.* (1991b), Moses *et al.* (1992) and Tan *et al.* (1993b).

A reevaluation of the data obtained by the HERS experiment at comet Halley has been presented by Neugebauer *et al.* (1991) showing disagreement with predictions of models. The data were discussed together with the results of the HIS experiment by Goldstein *et al.* (1992). Heavy ions at large distances ($8 \cdot 10^6$ kilometers) from the nucleus were found by analysis of the EPONA data (Erdös *et al.* 1991). The effect of solar wind charge exchange in the cometary coma has been discussed by Fuselier *et al.* (1991) and Wallis (1991).

The spacecraft data initiated a large number of theoretical investigations. A gasdynamic model, describing the interaction of the solar wind with a comet was given by Baranov and Lebedev (1993). The authors found good agreement of their model with the spacecraft data of comet Halley in the outer shock layer and in the description of the shear layer, identified with the heavy ion mantle found by the experiments. Schmidt and Wegmann (1991) gave an overview on the results of a 3-D MHD approach of the solar wind interaction and the comparison with spacecraft data. A parametric study of the interaction has been performed by Russell *et al.* (1991a).

Pitch-angle scattering of cometary ions and the effect on the cometary environment was discussed in many papers. Flammer *et al.* (1991 and 1992) presented a self-consistent 1-D model to study particles and fields kinetically. The model is applicable to the unshocked region upstream of the comet. Both the ring and shell like ion velocity distribution and the role of pitch-angle scattering were considered. A stability analysis of the ion shell distribution was given by Yoon (1990). Yoon and Ziebell (1991) studied the dependence of pitch-angle diffusion on the injection angle α. For the quasi parallel case pitch angle diffusion remained small, for the quasi perpendicular case the diffusion rate increased. This was in general also found by Gary *et al.* (1991) in agreement with observations at comet Halley. The scattering was also found to be almost independent of the solar wind/cometary ion speed and to increase as the square root of the injection rate. Pitch-angle scattering into a mono- and bi-spherical distribution was discussed by Miller *et al.* (1991b) and the combined effect of pitch-angle and energy diffusion was studied by Ye and Cravens (1991) and Ye *et al.* (1993). As expected they found pitch-angle diffusion to occur much faster. The thermalisation of an isotropic spherical shell distribution was discussed by Yoon (1992). Huddleston *et al.* (1991) had studied the pick-up process at comet Halley to explain the measured development of the implanted ion distribution. They also compared the ion energy distribution obtained by their model with results obtained by Giotto (Huddleston *et al.* 1992).

A number of publications showed various aspects of waves and instabilities generated in the cometary environment (Miller *et al.* 1991a, Lee and Gary 1991, Kotelnikov *et al.* 1991, and in Johnstone ed. 1991). Khabibrakhmanov *et al.* (1991), Ip and McKenzie (1991) and Zank (1990) investigated cometary bow shocks. A review on the formation of cometary shocks was given by Omidi and Winske (1991).

The stability of the ionopause was also investigated. McKenzie *et al.* (1990) extended their analysis by including finite plasma pressure in addition to the effect of ion-neutral friction. Another stability analysis taking into account the effect of ion-neutral drag was given by Daohan and Linzhong (1992). The flow around the diamagnetic cavity was discussed by Isrealevich *et al.* (1992) by adding centrifugal force to the momentum balance to explain the magnetic field structure observed. The existence of energetic ions in the inner coma has been addressed by Gurgiolo and Winningham (1990) and Ip (1992), by suggesting influences of sudden movements of the contact surface and of a magnetic neutral line. The formation of an ion density enhancement at the diamagnetic cavity found by the Giotto measurements was studied by Keller and Cravens (1990).

Wang (1991) investigated mechanisms to produce the large-scale structures found in cometary plasma tails in ground-based observations. He suggested the formation of some of these structures due to streaming sausage, kink and tearing instabilities. Russell *et al.* (1991b) studied the existence of mirror waves at comet Halley which might become visible as rays under favorable viewing conditions.

REFERENCES - COMETS

The following books contain numerous papers on physical studies of comets. Papers appearing in these books are referenced by the abbreviations indicated.

Comet Halley, Investigations, Results, Interpretations. Volume 1: Organization, Plasma, Gas. (Mason, J.W. Ed.), 295 pp.; *Volume 2: Dust, Nucleus, Evolution.* (Mason, J. and P. Moore Eds.), 275 pp. 1990. Ellis Horwood Library, Chichester. (*CHIRI*)

Cometary Plasma Processes. (Johnstone, A.D., ed.) 1991. American Geophysical Union, Washington D.C., USA (*CPP*)

Comets in the Post-Halley Era (Newburn, R. and J. Rahe Eds.) 1991. Kluwer Academic Publishers, Dordrecht, The Netherlands. 2 volumes, 1360 pp. (*CPHE*)

Physics and Chemistry of Comets (Huebner, W.F. Ed.) 1990. Springer-Verlag, Berlin, New York, Tokyo, 376 pp. (*PCC*)

Workshop on the Activity of Distant Comets (Huebner, W.F., H.U. Keller, D. Jewitt, J. Klinger and R. West eds.) 1993. Southwest Research Institute, San Antonio, Texas, USA, 185 pp. (*ADC*)

1. Nucleus: properties, origin and evolution

Bailey, M.E. 1992. *Astron. Astrophys.* **257**, 315-322.
Baratta, G.A., F.A. Catalano, F. Leone, and G. Strazzulla 1991. *Astron. Astrophys.* **251**, 349-355.
Belton, M.J.S., W.H. Julian, A.J. Anderson, and B.E.A. Mueller 1991. *Icarus* **93**, 183-193.
Bockelée-Morvan, D. and J. Crovisier 1992. *Astron. Astrophys.* **264**, 282-291.
Boehnhardt, H., H. Fechtig, and V. Vanysek 1990. *Astron. Astrophys.* **231**, 543-547.
Brunini, A. 1993. *Astron. Astrophys.* **273**, 684-694.
Bus, S.J., M.F. A'Hearn, D.G. Schleicher, and E. Bowell 1991. *Science* **251**, 774-777.
Carusi, A., L. Kresák, M. Kresákova, and G.B. Valsecchi 1991. *Astron. Astrophys.* **252**, 377-384.
Chakrabarti, S.K. and K.S. Krishna Swamy 1992. *Astron. Astrophys.* **263**, L1-L2.
Colom, P., J. Crovisier, D. Bockelée-Morvan, D. Despois, and G. Paubert 1992. *Astron. Astrophys.* **264**, 270-281.
Donn, B. 1991. In *CPHE*, pp. 335-359.
Engel, S., J.I. Lunine, and J.S. Lewis 1990. *Icarus* **85**, 380-393.
Espinasse, S., J. Klinger, C. Ritz, and B. Schmitt 1991. *Icarus* **92**, 350- 365.
Fanale, F.P. and J.R. Salvail 1990. *Icarus* **84**, 403-413.
Festou, M.C., H. Rickman and R.M. West 1993. *Astron. Astrophys. Rev.* **4**, 363-447.
Greenberg, J.M. and J.I. Hage 1990. *Astrophys. J.* **361**, 260-274.
Grün, E., H. Kochan, and K.J. Seidensticker 1991. *Geophys. Res. Lett.* **18**, 245-248.
Hahn, G. and M.E. Bailey 1990. *Nature* **348**, 132-136.
Hartmann, W.K., D.J. Tholen, K.J. Meech, and D.P. Cruikshank 1990. *Icarus* **83**, 1-15.
Houpis, H.L.F. 1991. In *CHIRI 2*, pp. 173-188.
Hughes, D.W. 1991. In *CPHE*, pp. 825-851.
Ibadinov, Kh.I, A.A. Rahmonov and A. Sh. Bjasso 1991. In *CPHE*, pp. 299-311.
Ip, W.-H. and J.A. Fernández 1991. *Icarus* **92**, 185-193.
Jewitt, D.C. and J.X. Luu 1990. *Astron. J.* **100**, 933-944.
Jewitt, D. and J.X. Luu 1992a. *Icarus* **100**, 187-196.
Jewitt, D. and J.X. Luu 1992b. *Astron. Astrophys.* **104**, 398- 491.
Jewitt, D. and J.X. Luu 1993. *Nature* **362**, 730-732.
Julian, W.H. 1990. *Icarus* **88**, 355-371.
Keller, H.U. 1990. In *PCC*, pp. 13-68.
Klinger J. 1991. In *CPHE*, pp. 227-241.
Kresák, L. 1992. *Astron. Astrophys.* **259**, 682-691.
Lunine, J.I., S. Engel, B. Rizk, and M. Horanyi 1991. *Icarus* **94**, 333-344.
Luu, J.X. 1993. *Icarus* **104**, 138-148.

Luu, J.X. and D.C. Jewitt 1990. *Astron. J.* **100**, 913-932.
Mitrofanov, I.G. and R.Z. Sagdeev 1990. *Nature* **344**, 313-315.
Prialnik, D. 1992. *Astrophys. J.* **388**, 196-202.
Prialnik, D. and Y. Mekler 1991. *Astrophys. J.* **366**, 318-323.
Rettig, T.W., S.C. Tegler, D.J. Pasto, and M.J. Mumma 1992. *Astrophys. J.* **398**, 293-298.
Rickman, H., J.A. Fernández, and B.A.S. Gustafson 1990. *Astron. Astrophys.* **237**, 524-535.
Rickman, H., L. Kamél, C. Froeschlé, and M.C. Festou 1991. *Astron. J.* **102**, 1446-1463.
Samarasinha, N.H. and M.F. A'Hearn 1991. *Icarus* **93**, 194-225.
Sekanina, Z. 1993. *Astron. J.* **105**, 702-735.
Sekanina, Z., S.M. Larson, O. Hainaut, A. Smette, and R.M. West 1992. *Astron. Astrophys.* **263**, 367-386.
Spohn, T. and J. Benkhoff 1990. *Icarus* **87**, 358-371.
Stern, S.A., J.M. Shull, and J.C. Brandt 1990. *Nature* **345**, 305-398.
Stooke, P.J. and A. Abergel 1991. *Astron. Astrophys.* **248**, 656-668.
Strazulla, G. and R.E. Johnson 1991. In *CPHE*, pp. 243-275.
Wallis, M.K. and N.C. Wickramasinghe 1992. *Adv. Space Res.* **12**, 133-138.
Weissman, P.R. 1990. *Nature* **344**, 825-830.
West, R.M. 1991. *Astron. Astrophys.* **241**, 635-645.
West, R.M., O. Hainaut and A. Smette 1991. *Astron. Astrophys.* **247**, L77.
Whipple, F.L. 1992. *Icarus* **98**, 108-114.

2. Dust coma and tail

Baum, W., Kreidl, T., and Schleicher, D. 1992. *Astron. J.* **104**, 1216-1225.
Chernova, G., Kiselev, N., and Jockers, K. 1993. *Icarus* **103**, 144-158.
Fulle, M., 1992a. *Nature* **359**, 42-44.
Fulle, M., 1992b. *Astron. Astrophys.* **265**, 817-824.
Fulle, M., Cremonese, G., Jockers, K., and Rauer, H., 1992. *Astron. Astrophys.* **253**, 615-624.
Fulle, M., Bosio, S., Cremonese, G., Cristaldi, S.,Liller, W., and Pansecchi, L. 1993. *Astron. Astrophys.* **272**, 634-650.
Larson, S., and Marcialis, R., 1992. *IAUC* 5669.
Luu, J and Jewitt, D. 1993. In *ADC*, pp. 44-53.
McDonnell, J.A.M., N. McBride, R. Beard, E. Bussoletti, L. Colangeli, P. Eberhardt, J.G. Firth, R. Gérard, S.F. Green, J.M. Greenberg, E. Grün, D.W. Hughes, H.U. Keller, J. Kissel, B.A. Lindblad, J.-C. Mandeville, C.H. Perry, K. Rembor, H. Rickman, G.H. Schwehm, R.F. Turner, M.K. Wallis, and J.C. Zarnecki 1993. *Nature* **362**, 732-734.
Sekanina, Z. 1981. *Astron. J.* **86**, 1741-1773.
Sekanina, Z., Larson, S., Hainaut, O., Smette, A., and West, R.M. 1992. *Astron. Astrophys.* **263**, 367-386.
Schmidt, B., Espinasse, S., and Klinger, J. 1993. In *ADC*, 124-132.
Sykes, M., and Walker, R. 1992. *Icarus* **95**, 180-210.
Weaver, H., A'Hearn, M., Feldman, P., Arpigny, C., Baum, W., Brandt, J., Light, r., and Westphal, J. 1992. *Icarus* **97**, 85-95
Yoshida, S., Aoki, T., Soyano, T., Tarusawa, K., Van Driel, W., Hamabe, M., Ichikawa, T., Watanabe, J., and Wakamatsu, K. 1993. *Astron. Soc. Japan* **45**, L33-L37.

3. Gas coma and photochemistry

A'Hearn, M.F. and M.C. Festou 1990. In *PCC*, pp. 69-110.
Beaver, J.E., R.M. Wagner, .D.G. Schleicher, and B.L. Lutz 1990. *Astrophys. J.* **360**, 696-701.
Bockelée-Morvan, D., J. Crovisier, P. Colom, D. Despois, and G. Paubert 1990a. In *Proc. of the 24th ESLAB Symposium on the Formation of Stars and Planets, and the Evolution of the Solar System*. ESA- SP-315, pp. 143-148. ESA Publications Division, Noordwijk (Holland).
Bockelée-Morvan, D., J. Crovisier, and E. Gérard 1990b. *Astron. Astrophys.* **238**, 382-400.
Bockelée-Morvan, D., P. Colom, J. Crovisier, D. Despois, and G. Paubert 1991. *Nature* **350**, 318-320.
Boehnhardt, H., K. Beisser, V. Vanysek, B.E.A. Mueller, M. Weiss, M. Jäger, K. Reinsch, and E. Grün 1990. *Icarus* **86**, 58-68.

Boice, D.C., W.F. Huebner, M.J. Sablik, and I. Konno 1990. *Geophys. Res. Lett.* **17**, 1813-1816.
Brooke, T.Y., A.T. Tokunaga, and R.F. Knacke 1991. *Astron. J.* **101**, 268-278.
Budzien, S.A. and P.D. Feldman 1991. *Icarus* **90**, 308-318.
Budzien, S.A. and P.D. Feldman 1992. *Icarus* **99**, 143-152.
Chernetenko, Yu.A. 1991. In *Spain-USSR Workshop on Positional Astronomy and Celestial Mechanics*. (López Garzia, A., R.F. López Machi, A.G. Sokolsky Eds.) pp. 115-123. Universitat de Valéncia (Spain), Observatorio Astronomico.
Cochran, A.L. and W.D. Cochran 1991. *Icarus* **90**, 172-175.
Cochran, A.L., W.D. Cochran, E.S. Barker, and A.D. Storrs 1991. *Icarus* **92**, 179-183.
Colom, P., J. Crovisier, D. Bockelée-Morvan, D. Despois, and G. Paubert 1992. *Astron. Astrophys.* **264**, 270-281.
Combi, M.R. and P.D. Feldman 1992. *Icarus* **97**, 260-268.
Crifo, J.F. 1992. *Astrophys. J.* **391**, 336-352.
Crifo, J.F. and Z. Slanina 1991. *Astrophys. J.* **383**, 351-355.
Crovisier, J. and F.P. Schloerb 1991. In *CPHE,* pp. 149-173.
Crovisier, J., D. Despois, D. Bockelée-Morvan, P. Colom, and G. Paubert 1991. *Icarus* **93**, 246-258.
Crovisier, J., D. Bockelée-Morvan, G. Bourgois, and E. Gérard 1992. *Astron. Astrophys.* **253**, 286-291.
Crovisier, J., D. Bockelée-Morvan, P. Colom, D. Despois, and G. Paubert 1993. *Astron. Astrophys.* **269**, 527-540.
Davies, J.K., S.F. Green, and T.R. Geballe 1991. *Mon. Not. R. Astr. Soc.* **251**, 148-151.
de Pater, I., P. Palmer, and L.E. Snyder 1991. In *CPHE,* pp. 175-207.
De Almeida, A.A. 1991. *Earth, Moon, Planets* **53**, 205-216.
De Almeida, A.A. 1992. *Earth, Moon, Planets* **56**, 61-74.
Delsemme, A.H. 1991. In *CPHE,* pp. 439-462.
DiSanti, M.A., U. Fink, and A.B. Schultz 1990. *Icarus* **86**, 152-171.
DiSanti, M.A., M.J. Mumma, and J.H. Lacy 1992. *Icarus* **97**, 155-158.
Donnison, J.R. 1990. *Mon. Not. R. Astr. Soc.* **245**, 658-664.
Ellis, T.A. and J.S. Neff 1992. *Icarus* **97**, 99-110.
Encrenaz, T. and R. Knacke 1991. In *CPHE,* pp. 107-137.
Esipov, V.E., P.P. Korsun, O. Mamadov, and V.G. Parusinov 1991. *Solar System Research* **25**, No.5.
Feldman, P.D. 1990. In *CHIRI 1.*
Feldman, P.D. 1991a. In *CPHE,* pp. 139-148.
Feldman, P.D. 1991b. In *International School of Space Chemistry. 1. Course: Chemistry in space - a NATO ASI.* (Greenberg, J.M. and V. Pirronello Eds.) pp. 339-361. Kluwer Academic Publ., Dodrecht (Netherlands).
Feldman, P.D., A.F. Davidsen, W.P. Blair, C.W. Bowers, W.V. Dixon, S.T. Durrance, H.C. Ferguson, R.C. Henry, R.A. Kimble, G.A. Kriss, J.W. Kruk, K.S. Long, H.W. Moos, O. Vancura, and T.R. Gull 1991. *Astrophys. J. Lett.* **379**, L37-L40.
Feldman, P.D., S.A. Budzien, M.C. Festou, M.F. A'Hearn, and G.P. Tozzi 1992. *Icarus* **95**, 65-72.
Festou, M.C. 1990a. In *CHIRI 1.*
Festou, M.C. 1990b. In *Evolution in astrophysics. IUE astronomy in the era of new space missions*. ESA-SP-310, pp. 3-10. European Space Agency, 75-Paris (France).
Fink, U. and M.A. DiSanti 1990. *Astrophys. J.* **364**, 687-698.
Fink, U., M.R. Combi, and M.A. DiSanti 1991. *Astrophys. J.* **383**, 356- 371.
Formisano, V., G. Bellucci, F. Tripodi, and E. Amata 1991. *Nuovo Cimento C* **14**, 319-334.
Geiss, J., K. Altwegg, E. Anders, H. Balsiger, W.-H. Ip, A. Meier, M. Neugebauer, H. Rosenbauer, and E.G. Shelley 1991. *Astron. Astrophys.* **247**, 226-234.
Green, J.C., W. Cash, T.A. Cook, and S.A. Stern 1991. *Science* **251**, 408- 410.
Hirao, K. 1990. In *CHIRI 1.*
Hoban, S., M. Mumma, D.C. Reuter, M. DiSanti, R.R. Joyce, and A. Storrs 1991. *Icarus* **93**, 122-134.
Howell, E.S., B.L. Lutz, and V.M. Slipher 1991. *Astrophys. J.* **371**, 776- 784.
Hu, H-Y., H.P. Larson, and K.C. Hsieh 1991. *Icarus* **91**, 238-250.
Huebner, W.F., D.C. Boice, I. Konno, and P.D. Singh 1992. In *IAU Symposium No. 150: Astrochemistry of cosmic phenomena.* (Singh P.D. Ed.) pp. 449-450. Kluwer Academic Publishers, Dordrecht, The Netherlands.

Huntress, W.T., M. Allen, and M. Delitsky 1991. *Nature* **352**, 316-318.

Ibadov, S. 1991. *Astron. Tsirk.* **1549**, 23-24.

Jaworski, W.A. and J.B. Tatum 1991. *Astrophys. J.* **377**, 306-317.

Jockers, K., T. Bonev, V. Ivanova, and H. Rauer 1992. *Astron. Astrophys.* **260**, 455-464.

Jockers, K., N.N. Kiselev, H. Boehnhardt, and N. Thomas 1993. *Astron. Astrophys.* **268**, L9-L12.

Johnstone, A.D., A.J. Coates, D.E. Huddleston, K. Jockers, B. Wilken, H. Borg, C. Gurgiolo, J.D. Winningham, and E. Amata 1993. *Astron. Astrophys.* **273**, L1-L4.

Kamél, L. 1991. *Icarus* **93**, 226-245.

Keller, C.N. and T.E. Cravens 1990. *J. Geophys. Res.* **95**, 18755-18768.

Kim, S.J. and M.F. A'Hearn 1991. *Icarus* **90**, 70-95.

Kim, S.J. and M.F. A'Hearn 1992. *Icarus* **97**, 303-306.

Kim, S.J., F. A'Hearn, and S.M. Larson 1990. *Icarus* **87**, 440-451.

Klavetter, J.J. and M.F. A'Hearn 1992. *Icarus* **95**, 73-85.

Klavetter, J.J. and S. Hoban 1992. *Icarus* **95**, 60-64.

Kömle, N.I. 1990. In *CHIRI 1*.

Krankowsky, D. 1991. In *CPHE*, pp. 885-877.

Krasnopolsky, V.A. 1991. *Astron. Astrophys.* **245**, 310-315.

Krasnopolsky, V.A. and A.Y. Tkachuk 1991. *Astron. J.* **101**, 1915- 1919.

Krasnopolsky, V.A., A.Yu. Tkachuk, and O.I. Korablev 1991. *Astron. Astrophys.* **245**, 662-668.

Kresák, L. and M. Kresákova 1990. *Icarus* **86**, 82-92.

Krishna Swamy, K.S. and S.P. Tarafdar 1993. *Astron. Astrophys.* **271**, 326-334.

Landaberry, S.J.C., P.D. Singh, and J.A. de Freitas Pacheco 1991. *Astron. Astrophys.* **246**, 597-602.

Larson, H.P., Hu Hong-yao, K.C. Hsieh, H.A. Weaver, and M.J. Mumma 1991. *Icarus* **91**, 251-269.

Lutz, B.L., M. Womack, and R.M. Wagner 1993. *Astron. J.* **407**, 402- 411.

Mamadov, O.M. 1990. *Astron. Tsirk.* **1545**, 43-44.

McCoy, R.P., R.R. Meier, H.U. Keller, C.B. Opal, and G.R. Carruthers 1992. *Astron. Astrophys.* **258**, 555-565.

Medvedev, Yu.D. 1991. In *Spain-USSR Workshop on Positional Astronomy and Celestial Mechanics*. (López Garzia, A., R.F. López Machi, A.G. Sokolsky Eds.) pp. 101-113. Universitat de Valéncia (Spain), Observatorio Astronomico.

Meredith, N.P., M.K. Wallis, and D. Rees 1992. *Mon. Not. R. Astr. Soc.* **254**, 693-704.

Morris, C.S. and M.S. Hanner 1992. *Astron. J.* **105**, 1537-1546.

O'Dell, C.R., C.O. Miller, A.L. Cochran, W.D. Cochran, C.B. Opal, and E.S. Barker 1991. *Astrophys. J.* **368**, 616-621.

Prasad, C.D., K. Jockers, and E.H. Geyer 1992. *Icarus* **95**, 211-221.

Rauer, H. and K. Jockers 1993. *Icarus* **102**, 117-133.

Reuter, D.C. 1992. *Astrophys. J.* **386**, 330-335.

Rickman, H., L. Kamél, C. Froeschlé, and M.C. Festou 1991. *Astron. J.* **102**, 1446-1463.

Roettger, E.E., P.D. Feldman, M.F. A'Hearn, M.C. Festou, L.A. McFadden, and R. Gilmozzi 1990. *Icarus* **86**, 100-114.

Sanwal, B.B. and B.S. Rautela 1991. *Earth, Moon, Planets* **54**, 125-128.

Scherb, F., K. Magee-Sauer, F.L. Roesler, and J. Harlander 1990. *Icarus* **86**, 172-188.

Schleicher, D.G. and S.J. Bus 1991. *Astron. J.* **101**, 706-712.

Schleicher, D.G., R.L. Millis, D.T. Thompson, P.V. Birch, R. Martin, D.J. Tholen, J.R. Piscitelli, N.L. Lark, and H.B. Hammel 1990. *Astron. J.* **100**, 896-912.

Schleicher, D.G., R.L. Millis, D.J. Osip, and P.V. Birch 1991. *Icarus* **94**, 511-523.

Schulz, R. 1991. In *3. ESO/ST-ECF Data Analysis Workshop*. (Grosbel, P.J. and R,H. warmels Eds.) ESO, Conf. Worshop Proc. No.38, pp. 73-78. ESO, Garching (Germany).

Schulz, R. 1992. *Icarus* **96**, 198-203.

Schulz, R. 1993. *Astron. Astrophys.* **268**, 319-325.

Schulz, R., M.F. A'Hearn, and H. Samarashina 1993. *Bull. Amer. Astro. Soc.* **24**, 1004.

Shimizu, M. 1991. In *CPHE*, pp. 897-905.

Smyth, W.H., M.R. Combi, and A.I.F. Stewart 1991. *Science* **253**, 941- 1064.

Snyder, L.E. 1992. In *IAU Symposium No. 150: Astrochemistry of cosmic phenomena*. (Singh, P.D. Ed.) pp. 427-434. Kluwer Academic Publishers, Dordrecht, The Netherlands.

Snyder, L.E., P. Palmer, and I. De Pater 1990. *Icarus* **86**, 289-298.

Spinrad, H. 1991. *Astron. J.* **102**, 1207-1212.

Stern, S.A., J.C. Green, W. Cash, and T.A. Cook 1992. *Icarus* **95**, 157-161.

Svoren, J. 1991. *Adv. Space Res.* **11**, 141-148.

Tacconi-Garman, L.E., F.P. Schloerb, and M.J. Claussen 1990. *Astrophys. J.* **364**, 672-686.

Tegler, S., L.F. Burke, S. Wyckoff, M. Womack, U. Fink, and M. Disanti 1992. *Astrophys. J.* **384**, 292-297.

Valk, J.H., C.R. O'Dell, A.L. Cochran, W.D. Cochran, C.B. Opal, and E.S. Barker 1992. *Astrophys. J.* **388**, 621-632.

Vanysek, V. 1991. In *CPHE*, pp. 879-895.

Vid'machenko, A.P., V.K. Rozenbush, and K.I. Churyumov 1990. *Soviet Astron.* **34**, 316-319.

Weaver, H.A., M.J. Mumma, and H.P. Larson 1991. In *CPHE*, pp. 93-106.

Wyckoff, S., S. Tegler, and L. Engel 1991a. *Astrophys. J.* **368**, 279-286.

Wyckoff, S., S. Tegler, and L. Engel 1991b. *Astrophys. J.* **367**, 641-648.

4. Plasma

Amata, E., V. Formisano, P. Torrente, and R. Giovi 1991. In *CPP*, pp. 131-138.

Baranov, V.B. and M.G. Lebedev 1993. *Astron. Astrophys.* **273**, 695-706.

Chaizy, P., H. Réme, J.A. Sauvaud, C. d'Uston, R.P. Lin, D.E. Larson, D.L. Mitchell, K.A. Anderson, C.W. Carlson, A. Korth, and D.A. Mendis 1991. *Nature* **349**, 393-396.

Coates, A.J., A.D. Johnstone, and D.E. Huddleston 1993. *Geophys. Res. Lett.* **20**, 483-486.

Coates, A.J., A.D. Johnstone, R.L. Kessel, D.E. Huddleston, B. Wilken, K. Jockers, and F.M. Neubauer 1990. *J. Geophys. Res.* **95**, 20701-20716.

Daohan, C. and L. Linzhong 1992. *Astrophys. Space Sci.* **189**, 45-55.

Delva, M., K. Schwingenschuh, M.B. Niedner, and K.I. Gringauz 1991. *Planet. Space Sci.* **39**, 697-708.

Erdös, G., P.W. Daly, E. Kirsch, and S. McKenna-Lawlor 1991. *Ann. Geophys.* **9**, 291-297.

Flammer, K.R., D.A. Mendis, T.G. Northrop, and E.C. Whipple 1991. *J. Geophys. Res.* **96**, 15907-15916.

Flammer, K.R., D.A. Mendis, T.G. Northrop, and E.C. Whipple 1992. *J. Geophys. Res.* **97**, 8173-8181.

Fuselier, S.A., E.G. Shelley, B.E. Goldstein, R. Goldstein, M. Neugebauer, W.-H. Ip, H. Balsiger, and H. Réme 1991. *Astron. J.* **379**, 734-740.

Gary, S.P., R.H. Miller, and D. Winske 1991. *Geophys. Res. Lett.* **18**, 1067-1070.

Goldstein, B.E., R. Goldstein, M. Neugebauer, S.A. Fuselier, E.G. Shelley, H. Balsiger, G. Kettmann, W.-H. Ip, H. Rosenbauer, and R. Schwenn 1992. *J. Geophys. Res.* **97**, 4121-4132.

Gurgiolo, C. and J.D. Winningham 1990. *J. Geophys. Res.* **95**, 17051-17073.

Herbig, G.H. 1990. *Astron. J.* **99**, 1262-1267.

Huddleston, D.E., A.J. Coates, and A.D. Johnstone 1992. *J. Geophys. Res.* **97**, 19163-19174.

Huddleston, D.E., A.D. Johnstone, A.J. Coates, and F.M. Neubauer 1991. *J. Geophys. Res.* **96**, 21329-21341.

Ip, W.-H. 1992. *Geophys. Res. Lett.* **19**, 1327-1330.

Ip, W.-H. and I. Axford 1990. In *PCC*, pp. 177-232.

Ip, W.-H. and J.F. McKenzie 1991. *Planet. Space Sci.* **39**, 1045- 1048.

Israelevich, P.L., A.I. Ershkovich, and Z.M. Ioffe 1992. *J. Geophys. Res.* **97**, 17045-17050.

Johnstone, A.D., A.J. Coates, D.E. Huddleston, K. Jockers, B. Wilken, H. Borg, C. Gurgiolo, J.D. Winningham, and E. Amata 1993. *Astron. Astrophys.* **273**, L1-L4.

Kecskeméty, K. and T.E. Cravens 1992. *J. Geophys. Res.* **97**, 2891-2906.

Keller, C.N. and T.E. Cravens 1990. *J. Geophys. Res.* **95**, 18755-18768.

Khabibrakhmanov, I.Kh., A.J. Coates, and V.L. Galinsky 1991. *Geophys. Res. Lett.* **18**, 1509-1512.

Kotelnikov, A.D., A.N. Polyudov, M.A. Malkov, R.Z. Sagdeev, and V.D. Shapiro 1991. *Astron. Astrophys.* **243**, 546-552.

Lee, M.A. and S.P. Gary 1991. *J. Geophys. Res.* **96**, 21319- 21327.

Marconi, M.L., D.A. Mendis, D.L. Mitchell, R.P. Lin, A. Korth, and H. Reme 1991. *Astrophys. J.* **378**, 756-762.

Mazelle, C., G. Belmont, K.-H. Glassmeier, D. LeQuau, and H. Rme 1991. *Adv. Space Res.* **11**, 73-77.

McKenna-Lawlor, S.M.P., P. Daly, E. Kirsch, F.M. Neubauer, D. O'Sullivan, A. Thompson, and P.K. Wenzel 1990. *Icarus* **87**, 430- 439.
McKenna-Lawlor, S.M.P., P.W. Daly, E. Kirsch, D. O'Sullivan, A. Thompson, P.K. Wenzel, and V. Afonin 1993. *Nature* **363**, 326-329.
McKenzie, J.F., A.I. Ershkovich, and W.I. Axford 1990. *Astrophys. J.* **360**, 275-278.
Miller, F.D. 1992. *Astron. J.* **104**, 1207-1215.
Miller, R.H., T.I. Gombosi, S.P. Gary, and D. Winske 1991a. *J. Geophys. Res.* **96**, 9479-9490.
Miller, R.H., S.P. Gary, D. Winske, and T. Gombosi 1991b. *Geophys. Res. Lett.* **18**, 1063-1066.
Moore, E.P. 1991. *Astron. Astrophys.* **247**, 247-251.
Moses, S.L., F.V. Coroniti, E.W. Greenstadt, and B.T. Tsurutani 1992. *J. Geophys. Res.* **97**, 19157-19162.
Neubauer, F.M., K.-H. Glassmeier, M.H. Acuña, F. Mariani, G. Musmann, N.F. Ness, and A.J. Coates 1990. *Ann. Geophys.* **8**, 463-471.
Neubauer, F.M., H. Marshall, M. Pohl, K.-H. Glassmeier, G. Musmann, F. Mariani, M.H. Acuña, L.F. Burlaga, N.F. Ness, M.K. Wallis, H.U. Schmidt, and E. Ungstrup 1993. *Astron. Astrophys.* **268**, L5-L8.
Neugebauer, M., A.J. Coates, and F.M. Neubauer 1990. *J. Geophys. Res.* **95**, 18745-18753.
Neugebauer, M., B.E. Goldstein, R. Goldstein, S.A. Fuselier, F.M. Neubauer, H. Balsiger, and W.-H. Ip 1992. *Astron. Astrophys.* **258**, 549-554.
Neugebauer, M., R. Goldstein, B.E. Goldstein, S.A. Fuselier, H. Balsiger, and W.-H. Ip 1991. *Astrophys. J.* **372**, 291-300.
Omidi, N. and D. Winske 1991. In *CPP,* pp. 37-48.
Rauer, H. and K. Jockers 1993. *Icarus* **102**, 117-133.
Russell, C.T., G. Le, J.G. Luhmann, and J.A. Fedder 1991a. In *CPP,* pp. 65-72.
Russell, C.T., G. Le, K. Schwingenschuh, W. Riedler, and E.G. Eroshenko 1991b. In *CPP,* pp. 161-170.
Schmidt, H.U. and R. Wegmann 1991. In *CPP,* pp. 49-64.
Staines, K., A. Balogh, S.W.H. Cowley, R.J. Hynds, T.S. Yates, I.G. Richardson, T.R. Sanderson, K.-P. Wenzel, D.J. McComas, and B.T. Tsurutani 1991a. *Planet. Space Sci.* **39**, 479-506.
Staines, K., A. Balogh, S.W.H. Cowley, P.M. deForster, R.J. Hynds, T.S. Yates, T.R. Sanderson, K.-P. Wenzel, and B.T. Tsurutani 1991b. *Adv. Space Res.* **11**, 83-86.
Tan, L.C., G.M. Mason, G. Gloeckler, F.M. Ipavich, and A.B. Glavin 1990. *J. Geophys. Res.* **95**, 20685-20692.
Tan, L.C., G.M. Mason, I.G. Richardson, and F.M. Ipavich 1993a. *J. Geophys. Res.* **98**, 3613-3622.
Tan, L.C., G.M. Mason, and B.T. Tsurutani 1993b. *Geophys. Res. Lett.* **20**, 169-172.
Wallis, M.K. 1991. In *CPP,* pp. 125-129.
Wang, S. 1991. *Astron. Astrophys.* **243**, 521-530.
Watanabe, J.-I., T. Aoki, and T. Saito 1990. *Icarus* **87**, 421-429.
Ye, G. and T.E. Cravens 1991. *J. Geophys. Res.* **96**, 5479-5488.
Ye, G., T.E. Cravens, and T.I. Gombosi 1993. *J. Geophys. Res.* **98**, 1311-1324.
Yoon, P.H. 1990. *Geophys. Res. Lett.* **17**, 1033-1036.
Yoon, P.H. 1992. *J. Geophys. Res.* **97**, 6467-6478.
Yoon, P.H. and L.F. Ziebell 1991. *Phys. Fluids* **3**, 2124-2132.
Zank, G.P. 1990. *Planet. Space Sci.* **38**, 1355-1360.

II. MINOR PLANETS AND METEORITES - V. Zappalà and P. Farinella

In the following section, the main papers related to asteroid physical studies and the connections with meteorites will be briefly reviewed. The reader is also referred to the Comm. 20 report, in which studies of asteroid orbital dynamics are reviewed. As for meteorites, we will mainly refer to the aspects that are directly applicable to an astronomical perspective of these bodies. Large meetings devoted to asteroids (as well as to comets and meteors) were held in the U.S.A. (in 1991) and in Italy (in 1993). The proceedings of the 1991 meeting have been published (A. Harris and E. Bowell eds.), consisting of many short papers. Some are mentioned below, and the reader may find additional information in the many others not mentioned.

1. Asteroid Sizes and Masses

The mass of 1 Ceres was redetermined By Williams (1992) through careful analysis of the Ceres-induced perturbations undergone by the smaller asteroid 348 May. A similar result has been obtained by Goffin (1991) from the motion of 203 Pompeja, while Sitarski and Todorovic-Juchniewicz (1992) combined the data from both asteroids for the same purpose. The mass of 704 Interamnia has been derived by Landgraf (1992) from the perturbations of 993 Moultona. In order to extend the scope of this method, applied so far only to the few largest asteroids, Kuzmanoski and Knezevic (1993) compiled an extensive list of close encounters between large asteroids and all the other known ones that will take place in the next 50 years.

Two stellar occultations by large asteroids (2 Pallas and 4 Vesta) were well observed through organized *ad hoc* campaigns (Millis *et al.* 1990, Dunham *et al.* 1990). They led to new, more accurate estimates for the sizes, shapes and densities of these bodies, leading to important reassessments and calibrations of the other methods which are more frequently available for the same purposes. From the occultation of the star SAO 190531, Kissling *et al.* (1991) derived a diameter of 9 Metis in a good agreement with radiometric and polarimetric results.

Using albedos derived from the IRAS spacecraft survey of infrared sources, Cellino *et al.* (1991) analyzed the distribution of asteroid sizes taking into account the main observational selection effects. They found clear deviations from a power-law size distribution, both for the overall asteroid population and for several subsamples, sorted according to orbital semimajor axis and family membership.

2. Photometry, Shapes, Rotations

Lightcurve photometry is still a very active area of asteroid astronomy. Lightcurve observations are continuously published in the Minor Planet Bulletin as well as in other journals. A partial list of papers presenting lightcurves of individual or groups of asteroids includes Hoffmann and Geyer (1990), Velichko *et al.* (1990), Magnusson and Lagerkvist (1991), Dahlgren *et al.* (1991), Erikson *et al.* (1991), Belskaya and Dovgopol (1992), Dovgopol and Lisina (1992), Dovgopol *et al.* (1992), Lagerkvist *et al.* (1992), Dotto *et al.* (1992) and Wisniewski (1992). Other papers, such as Michalowski and Velichko (1990), Michalowski *et al.* (1991), Kristensen (1991), Michalowski (1992) and Kwiatkowski and Michalowski (1992), have presented lightcurves together with an analysis of the inferred shapes and/or poles of the corresponding objects. Ostro and Wisniewski (1992) compared lightcurve and radar data relevant for the shape of 1917 Cuyo.

In the last three years a particular effort has been focused on the targets of current or planned space missions, such as 951 Gaspra (Di Martino *et al.* 1990, Wisniewski *et al.* 1993) and 243 Ida (Gonano-Beurer *et al.* 1992), in order to compare lightcurve-derived physical properties with high-resolution imaging data. Photometric studies of individual asteroids can also indicate heterogeneities of composition and texture of the surface, such as indicated by phase relation and color variations for 51 Nemausa (Gammelgaard and Kristensen 1991, Kristensen and Gammelgaard 1993). Analyses of the data set contained in the *Asteroid Photometric Catalogue* have been presented by Magnusson and Lagerkvist (1990) and Magnusson (1991), who aimed at detecting the presence of albedo variegation on asteroid surfaces by analyzing the Fourier expansions of composite lightcurves. CCD photometry of selected asteroids centered on wavelengths of common emission features in comet spectra was applied by Rajamohan and Bhargavi (1992) to check for possible cometary activity features. Shkuratov and Muinonen (1992) developed a model to interpret asteroid photometry and polarimetry data in terms of shadowing and coherent backscattering effects.

Since currently the available lightcurve data base includes several hundreds of bodies, the emphasis of current research is shifting from studies of individual asteroids to general surveys of large samples of objects. In some cases, these surveys concerned the general asteroid population (Harris *et al.* 1992), but increasingly they are being designed to provide information on some subset of objects sharing some common property. For example, the "photometric geodesy" program of Weidenschilling *et al.* (1990) has focused on a number of large main-belt asteroids, with the aim of deriving accurate shape and polar direction estimates in addition to rotation periods, in order to assess how closely these objects match the axisymmetric or triaxial ellipsoidal equilibrium figures produced by the interplay of rotation and self-gravity (Drummond *et al.* 1991).

Another category of interesting bodies, which were under-represented in the observed sample until recently, are the outer-belt and Trojan asteroids. For these objects, an important research issue is whether they are genetically or compositionally related to comets (Hartmann and Tholen 1990). Following an earlier suggestion that distant asteroids may be more elongated than main belt objects, Di Martino *et al.* (1992) and Binzel and Sauter (1992) have obtained more data on distant asteroids. The newer data alone fail to show significantly greater elongations, although better statistical confirmation is needed.

The extension of the available data set is also important for small asteroids. Recent surveys aimed at these bodies (Wisniewski 1991, Binzel *et al.* 1992, Barucci *et al.* 1992a) have confirmed that these bodies have in general irregular, fragment-like shapes, and that there is a clear trend toward increasing spin rates for decreasing sizes, in agreement with theoretical predictions based on collisional evolution models (Farinella *et al.* 1992c). Hopefully, observations of km-sized bodies in the main belt will soon allow a meaningful comparison between the rotational properties of main-belt and near-Earth asteroids.

Theoretical studies of lightcurve inversion methods to derive poles and shapes have continued (Kaasalainen *et al.* 1992a,b; Kristensen 1992a,b). An important task now is that of obtaining a large enough sample of asteroid polar directions to assess whether their distribution is isotropic (*i.e.* randomized by precessional and collisional processes), or if there is some preferential clustering (*e.g.* about the pole of the ecliptic), that may be reminiscent of the asteroid accretion phase (Erikson and Magnusson 1993).

3. Radar, Radiometry, Polarimetry, Interferometry

In the last few years several observational techniques aimed at asteroids have been renewed and refined, providing a wealth of new information on the physical properties of these bodies and the composition and textures of their surfaces. The most outstanding example is probably the radar technique. Thanks to the availability of the large Goldstone and Arecibo antennas, the group led by S.J. Ostro has shown that radar data are not only extremely valuable for astrometry and orbit determination (Ostro *et al.* 1991a, Yeomans *et al.* 1992), but can also provide fairly accurate global shape estimates for the target asteroids. Some near-Earth objects (1672 Ivar and in particular 1989 PB, later named 4769 Castalia) clearly displayed bifurcated shapes hinting at a contact or quasi-contact binary structure (Ostro *et al.* 1990a, 1990b). For another body, 1986 DA, the very high radar reflectivity suggests a metal-rich composition (Ostro *et al.* 1991b).

Other techniques successfully applied to study asteroid physical properties include millimeter and submillimeter radiometry, used by Redman *et al.* (1992) to investigate the thermal behavior and the shape of 4 Vesta; speckle interferometry, applied again to 4 Vesta by Tsvetkova *et al.* (1991) to gather new shape and size data; and polarimetry, used both by the latter authors to estimate the contribution of albedo features to Vesta's lightcurve and by Chernova *et al.* (1993) to compare asteroid surfaces with dust particles emitted by comets. A search for rotational variations in the optical polarization of light coming from asteroids has been performed by Broglia and Manara (1990, 1992) first for 3 Juno and 7 Iris and then for 16 Psyche. An extensive campaign of polarimetric observations has been carried out on CMEU-type asteroids by Belskaya *et al.* (1991), who showed the possible different content of metal of different M-type objects.

4. Binaries and Families

While still not proven the case for the existence of binary asteroids was strengthened by radar observations of 1989 PB (Ostro *et al.* 1990b). Stern and Barker (1992) have conducted a further search for binaries among main belt asteroids. Their negative results may be due mainly to the limitations of the observational techniques available for searching satellites there.

At the same time a number of studies have addressed in detail the dynamical stability of these systems and the most important evolution mechanisms affecting them, including solar and Jovian perturbations, close encounters and collisions with passing asteroids and with planets (Chauvineau and Mignard 1990a, 1990b, 1990c; Chauvineau *et al.* 1991; Farinella 1992). The formation of doublet craters on Earth by collision of binary asteroids has been modeled by Melosh and Stansberry (1991). The possible existence of debris and dust particles around main-belt asteroids, which could pose a hazard for encountering spacecraft, has been investigated by Burns and Hamilton (1992) and Hamilton and Burns (1992). Orbital dynamics about a triaxial primary is complex, and numerical simulations on this problem have been carried out by German and Friedlander (1991).

Significant progress has been achieved in the last three years in studies concerning asteroid families. Williams (1992a) has published the results of his "initial search" for asteroid families, carried out on a sample of about 3000 objects (including both numbered and PLS asteroids) and showing a wide variety of family morphologies, including some peculiar, strongly non-isotropic structures (Williams 1991). The relationship to families of the Galileo target asteroids Gaspra and Ida was discussed by Williams (1992b). While Williams' search was based on human-eye pattern recognition in the proper element space, in recent years a number of automated clustering techniques have become available, leading to more objective and repeatable family identification procedures.

Zappalà *et al.* (1990) applied their "hierarchical clustering" method to a sample of 4200 proper element triplets derived by Milani and Knezevic (1990), and submitted the detected grouping to rigorous tests for statistical significance and robustness vs. (small) errors in the proper elements. The same proper elements sample was analyzed independently by Bendjoya *et al.* (1991), who used the "wavelet analysis" method. Although based on very different principles, this method yielded family memberships in fair agreement with those found by Zappalà *et al.* (1990). Another automated search was performed by Lindblad (1992) through the "D-criterion", and again the resulting list of most reliable families was similar to those found in the papers quoted above. The work of the Italian and French groups was later reviewed and reassessed by Milani *et al.* (1992), Bendjoya and Cellino (1992), Zappalà and Cellino (1992) and Bendjoya *et al.* (1993). In the latter paper, the two clustering methods were successfully applied to recover "artificial families" generated numerically by exploding asteroids overlapping different backgrounds of field objects. A comparison with Williams' results was carried out by Zappalà *et al.* (1992) who concluded that the main remaining discrepancies mostly involve the many families identified by Williams with a small number (<10) of members - hence least significant from the statistical point of view. To provide a simple tool to distinguish among families with differing degrees of significance and reliability, Farinella *et al.* (1992a) proposed the adoption of a new nomenclature. Williams (1992c) has discussed in some detail the different effects bearing on the reliability of families found with his and other techniques.

5. Spectra and Compositions

As with lightcurves, a particular effort has been made to investigate by spectroscopic observations the physical and chemical peculiarities of the outer-belt asteroids. In particular their possible relationship to comet nuclei has been studied with a variety of different techniques. CCD spectra of 32 Trojans allowed Jewitt and Luu (1990) to infer a remarkable similarity between Trojans and the comet nuclei. Vilas and McFadden (1992) performed a search for weak features in the reflectance spectra on a set of 35 asteroids. One of their main results is that the Trojan D-class asteroids show no clear absorption features, implying that aqueous alteration did not operate at that heliocentric distance. Smith *et al.* (1992) performed JHK photometry of selected Trojan and Hilda asteroids. They found that leading Trojans and Hildas show similar and homogeneous colors, while trailing Trojans appear more heterogeneous.

Several papers have addressed the problem of the composition of the (presumably) "primitive" asteroids, namely those belonging to the C, D, P, and G taxonomic classes. In particular the presence or absence of water was investigated in detail. Jones *et al.* (1990) presented an extensive laboratory and telescopic research on water distribution among the low-albedo outer-belt asteroids. Their main conclusion is that the C class is not a primitive group in the mineralogical sense, although its anhydrous members appear little altered. Coupled with the apparently anhydrous P and D surfaces, this fact points to an original outer-belt asteroid composition of anhydrous silicates, water ice, and complex organic material. Vilas *et al.* (1993) have identified a 0.43 micron absorption feature in 12 high-resolution UV/visible spectra of asteroids of the C, P and G classes, indicating an aqueous alteration. The largest asteroid, 1 Ceres, has been the target of searches for water. King *et al.* (1992) obtained spectra of this asteroid to be compared with theoretical calculations and laboratory measurements. They suggested that an ammoniated phyllosilicate is present on the surface of Ceres, rather than water frost as it had been reported previously. On the other hand, A'Hearn and Feldman (1992) performed long-exposure IUE spectra of the region around Ceres, searching for escaping OH as a result of the photodissociation of atmospheric water vapor. The amount of OH found close to the northern limb is consistent with a polar cap, which is replenished during winter and dissipates in summer.

4 Vesta has been studied by Festou *et al.* (1991) through visual and ultraviolet observations from the IUE satellite, with the aim of developing a self-consistent picture of that large asteroid.

Fitzsimmons *et al.* (1990) discovered three broad absorption features (at 416, 441 and 515 nm) in the reflectance spectrum of the outer-belt asteroids 1269 Rollandia and 279 Thule. Cruikshank *et al.* (1991b) identified cyano-group containing molecules in the very dark solids of the surfaces of a few D-class asteroids, the dust of some comets, the low-albedo hemisphere of Iapetus, and the rings of Uranus. The occurrence of this band may establish a link between outer solar system and interstellar materials. Grundy and Fink (1992) obtained high quality CCD spectra of Deimos, showing that its very red slope matches those of D- rather than C-type asteroids. On other hand, McFadden *et al.* (1993) showed that the dynamically chaotic Earth-crossing object 2201 Oljato has a fairly high albedo and a spectrum more typical for an inner-belt asteroid than for an extinct or dormant comet.

Stephens and Gustafson (1991) presented laboratory measurements of bi-directional reflectances from particle/ice mixtures representing "dirty" ice surfaces on atmosphereless solar system bodies. The differences found between their samples and solar system bodies may reflect different physical processes during mantle formation as well as differences in chemistry of silicate and carbonaceous materials.

A newly discovered distant asteroid, 5145 Pholus (1992 AD), has shown some very peculiar features. In particular it displays a reflection spectrum having a red slope much steeper than that of any other atmosphereless object. No absorption or emission feature has been detected. The interpretations given by several authors (Fink *et al.* 1992, Mueller *et al.* 1992, Binzel 1992, Hoffmann *et al.* 1993) are generally in good agreement, and indicate an association of exposed organics that are purer or more pristine than that found on the surfaces of C, P, D asteroids and comets. Davies *et al.* (1993) also suggested the presence of surface ices mixed with complex organic compounds.

A new method to detect mass loss from near-Earth asteroids has been presented by Luu and Jewitt (1992), using high-resolution surface photometry. The negative result obtained so far is probably due to the low expected rates of activity and to the small sizes of these bodies.

An extensive search for silicate minerals has been performed by Shestopalov *et al.* (1990). They concluded that the silicate minerals of some asteroids contain iron-rich and calcium pyroxenes. Hiroi and Takeda (1990) applied a model for investigating mineral assemblages in asteroidal surfaces, based on theoretical studies of spectral reflectance of olivine, pyroxene and their mixture, to show that the properties of 29 Amphitrite can be reproduced by a mineral assemblage of primitive achondrites.

6. Origin and Collisional Evolution

The problem of the origin of asteroids presents a complex mixture of dynamical, cosmochemical and cosmogonical puzzles. It has been attacked from a number of different starting points. Lecar *et al.* (1992) and Soper *et al.* (1992) tried to explain the current distribution of the asteroids in the outer portion of the belt and the lack of long-lived bodies between Jupiter and Saturn by investigating the dynamical mechanisms that may have depleted the solar system from the original population of minor bodies. Hartmann (1990) reviewed the evidence for an early intense bombardment from dark, carbon-rich planetesimals in the inner solar system. Grimm and McSween (1993) modeled the heating of primordial asteroids by the short-lived Al-26 isotope, showing that this provides a plausible alternative to the other often mentioned explanation for the partial differentiation of primordial asteroids via electromagnetic induction. The formation of metallic cores inside differentiating asteroids was modeled by Taylor (1992) and Haack and Scott (1992). Wetherill (1992) explored a new model for the formation of asteroids, in which the current belt results from the collisional and dynamical interaction of a number of primordial lunar-sized bodies, later lost from the system due to mutual perturbations into resonances. Safronov and Ziglina (1991) investigated the reasons for the absence of an ordinary planet in the asteroid zone, concluding that the combined action of three possible mechanisms increasing relative velocities and removing material should be taken into account. From the current abundance of metal-rich bodies, Hughes (1991) estimated that the largest bodies that ever existed in the asteroid belt had diameters of around 8600 km. The origin of Trojan asteroids has been investigated by Ruskol (1990).

Collisions - in particular those which catastrophically shatter the target objects - represent the most important process currently causing the asteroid population to evolve. Farinella and Davis (1992) have performed a detailed quantitative study on asteroid collision probabilities and impact velocities, and how these quantities depend on orbital parameters and family membership. The study of asteroid collisional evolution requires as an essential prerequisite a reliable model of how asteroid-sized bodies respond to energetic collisions, derived by scaling from laboratory experiments (Housen and Holsapple 1990, Housen *et al.* 1991) and from numerical modeling work (Cellino *et al.* 1990). Melosh *et al.* (1992) and Asphaug and Melosh (1993) have been developing complex hydrodynamical algorithms, both in two and three dimensions, for this work. Based on simple scaling relationships, a quantitative study then becomes possible of how the size and spin rate distributions of a primordial population of asteroidal bodies may have evolved into the current one (Farinella *et al.* 1992c).

However, extending the experimental data base on the outcomes of hypervelocity impacts is still a high-priority task. After the pioneering work carried out in the '70s and '80s, a refinement of the experimental techniques is now leading to exploring complex (albeit crucial for asteroid applications) issues such as the velocity distribution of fragments (Nakamura and Fujiwara 1991, Nakamura *et al.* 1992, Takagi *et al.* 1992), the effects of pre-fractured targets (Ryan *et al.* 1991), the rotation of fragments (Nakamura *et al.* 1992), the melting, vaporization and energy partitioning effects (Smither and Ahrens 1992), the angular momentum transfer in oblique impacts (Yanagisawa *et al.* 1991), the jetting of fragments about preferential planes and directions (Martelli *et al.* 1993).

7. Gaspra

(951) Gaspra has become the first asteroid subjected to close-up examination, by the space probe Galileo on 29 October 1991. These data and a first analysis have been presented by Belton *et al.* (1992). Pre-Galileo photometric observations of Gaspra had been performed by Di Martino *et al.* (1990) and Goldader *et al.* (1991), the latter authors deriving also an infrared reflectance spectrum. An extensive observational campaign to obtain Gaspra lightcurves was carried out prior to the encounter, as reported by Wisniewski *et al.* (1993). The shape, rotation rate and polar axis derived from these observations were presented by Magnusson *et al.* (1992). Barucci *et al.* (1992b) showed the pre-encounter shape models, obtained with different lightcurve inversion methods, are in agreement with the first Galileo image. Cartography techniques for asteroids and other small bodies imaged at low resolution were developed by Stooke (1992).

Estimates of the collisional lifetime and the surface evolution of Gaspra were obtained, before the encounter, by Namiki and Binzel (1991) and Farinella *et al.* (1992b). Both papers predicted a likely age of Gaspra on the order of several times 10^8 yr, a result consistent with the spacecraft observations. The possibility (unconfirmed by Galileo images) of finding small satellites around Gaspra was discussed by Van Flandern (1992). Williams (1992) analyzed the relationship of Gaspra to the populous Flora family.

8. Asteroids vs. Meteorites

Significant progress is being made in one of the most important issues in planetary science: the relationships between asteroid and meteorites and the identification of meteorite sources in the solar system. The dynamical side of the problem has been addressed by Froeschlé and Scholl (1991), Knezevic *et al.* (1991), Scholl and Froeschlé (1991) and Dahlgren *et al.* (1992), who studied the locations and effects of mean motion and secular resonances. Farinella *et al.* (1993a, 1993b) showed that both types of resonances provide efficient dynamical routes to deliver fragments of main-belt asteroids to Earth-crossing orbits. A clear example of this has been identified by Binzel and Xu (1992), whose spectroscopic observations of a a number of small Vesta family members, extending to the vicinity of the 3:1 mean motion resonance, showed the striking similarity of these objects to the basaltic achondrite meteorites. This finding is remarkable in that it provides strong evidence that collisional processes are capable of transporting debris from Vesta, which is far from any of the orbital resonance "escape hatches", to Earth-crossing orbits.

The comparison of asteroid and meteorite spectra has been carried out in a more extensive way than had been done previously (McFadden and Chamberlin 1992), including principal analysis classification of both data samples (Britt *et al* 1992). The genetic implications of these comparisons have been carefully assessed (Fanale *et al.* 1992), taking into account the effects of a number of possible thermal and collisional processes leading to alterations of the spectral features of asteroid surfaces (Clark *et al.* 1992). In particular, these processes are important to assess whether some main-belt asteroids for which visible and near-infrared spectra are available are plausible candidate parent bodies for the ordinary chondrites, the most common meteorite class (Keil *et al.* 1992, McSween *et al.* 1992, McSween 1992). Shestopalov and Golubeva (1991, 1992) have found a correlation between the u-x color index and a spectral absorption band at 505 nm. They concluded that the differences between stony meteorites and S-type asteroids are caused by a systematic difference in their surface material composition.

Specific comparisons between individual asteroid and meteorite classes, based on spectral data, showed that a number of intriguing similarities indeed exist, involving both main-belt (Hiroi and Takeda 1990, Burbine *et al.* 1992) and near-Earth asteroids (Cruikshank *et al.* 1991a, Gaffey *et al.* 1992). Laboratory work on meteorite spectra has included exploring such specific problems as the effects of mixing different mineralogical types (Hiroi *et al.* 1993), and using powdered meteorite samples (Salisbury *et al.* 1991). Geologic processes on asteroids required to produce individual meteorite classes have been modeled in some detail (Kargel 1991, Warren and Kallemeyn 1992, Rubin and Mittlefehldt 1993, Benoit and Sears 1993).

REFERENCES - MINOR PLANETS

The following proceedings volumes contain many additional papers relating to physical studies of minor planets and comets not mentioned in this summary. Papers which are mentioned that appear in these books are referenced by the abbreviations indicated.

Asteroids, Comets, Meteors 1991 (Harris, A.W., and E. Bowell, eds.) 1992. Lunar and Planetary Institute, Houston, 694 pp. (*ACM*)

Observations and Physical Properties of Small Solar System Bodies (A. Brahic, J.-C. Gérard and J. Surdej, eds.). University of Liège, Belgium. 385 pp. (*OPS*)

Interrelations between Physics and Dynamics for Minor Bodies in the Solar System (D. Benest and C. Froeschlé, eds.), Editions Frontieres, Gif-sur-Yvette, France. (*IPD*)

Papers:

A'Hearn, M.F., Feldman, P.D. 1992. *Icarus* **98**, 54-60.

Asphaug, E., Melosh, H.J. 1993. *Icarus* **101**, 144-167.

Barucci, M.A., Di Martino, M., Fulchignoni, M. 1992a. *Astron. J.* **103**, 1679-1686.

Barucci, M.A., Cellino, A., De Sanctis, C., Fulchignoni, M., Lumme, K., Zappalà, V., Magnusson, P. 1992b. *Astron. Astrophys.* **266**, 385-394.

Belskaya, I.N., Dovgopol, A.N. 1992. *ACM 91*, pp. 41-44.

Belskaya, I.N., Kiselev, N.N., Lupishko, D.F., Chernova, G.P. 1991. Kinem. Fiz. Nebesn. Tel. 7, 11-14.

Belton, M.J.S., Veverka, J., Thomas, P., Helfenstein, P., Simonelli, D., Chapman, C., Davies, M.E., Greeley, R., Greenberg, R., Head, J., Murchie, S., Klaasen, K., Johnson, T.V., McEwen, A., Morrison, D., Neukum, G., Fanale, F., Anger, C., Carr, M., Pilcher C. 1992. *Science* **257**, 1647-1652.

Bendjoya, Ph., Slezak, E., Froeschlé, C. 1991. *Astron. Astrophys.* **251**, 312-330.

Bendjoya, Ph., and Cellino, A. 1992. In *IPD*, pp. 19-43.

Bendjoya, Ph., Cellino, A., Froeschlé, C., Zappalà, V. 1993. *Astron. Astrophys.* **272**, 651-670.

Benoit, P.H., Sears, D.W.G. 1993. *Icarus* **101**, 188-200.

Binzel, R.P. 1992. *Icarus* **99**, 238-240.

Binzel, R.P., Sauter, L.M. 1992. *Icarus* **95**, 222-238.

Binzel, R.P., Xu, S., Bus, S.J., Bowell, E. 1992. *Icarus* **99**, 225-337 (1992).

Binzel, R.P., Xu, S. 1993. *Science* **260**, 186-190.

Britt, D.T., Tholen, D.J., Bell, J.F., Pieters, C.M. 1992. *Icarus* **99**, 153-166.

Broglia, P., Manara, A. 1990. *Astron. Astrophys.* **237**, 256-258.

Broglia, P., Manara, A. 1992. *Astron. Astrophys.* **257**, 770-772.

Burbine, T.H., Gaffey, M.J., Bell, J.F. 1992. *Meteoritics* **27**, 424-434.

Burns, J.A., Hamilton, D.P. 1992. *ACM 91*, pp.101-108.

Cellino, A., Zappalà, V., Davis, D.R., Farinella, P., Paolicchi, P. 1990. *Icarus* **87**, 391-402.

Cellino, A., Zappalà, V., Farinella, P. 1991. *Mon. Not. R. Astron. Soc.* **253**, 561-574.

Chauvineau, B., Mignard, F. 1990a. *Icarus* **83**, 360-381.

Chauvineau, B., Mignard, F. 1990b. *Icarus* **87**, 377-390.

Chauvineau, B., Mignard, F. 1990c. *Cel. Mech.* **47**, 123-144.

Chauvineau, B., Mignard, F., Farinella, P. 1991. *Icarus* **94**, 299-310.

Chernova, G.P., Kiselev, N.N., Jockers, K. 1993. *Icarus* **103**, 144-158.

Clark, B.E., Fanale, F.P., Salisbury, J.W. 1992. *Icarus* **97**, 288-297.

Cruikshank, D.P., D.J. Tholen, W.K. Hartmann, J.F. Bell, R.H. Brown 1991a. *Icarus* **89**, 1-13.

Cruikshank, D.P., Allamandola, L.J., Hartmann, W.K., Tholen, D.J., Brown, R.H., Matthews, C.N., Bell, J.F. 1991b. *Icarus* **94**, 345-353.

Dahlgren, M., Fitzsimmons, A., Lagerkvist, C.-I., Williams, I.P. 1991. *Mon. Not. R. Astron. Soc.* **250**, 115-118.

Dahlgren, M., Hahn, G., Lagerkvist, C.-I., Lundstrom, M. 1992. *ACM 91*, pp. 141-144.
Davies, J.K. Sykes, M.V., Cruikshank, D.P. 1993. *Icarus* **102**, 166-169.
Di Martino, M., Ferreri, W., Fulchignoni, M., De Angelis, G., Barucci, M.A., Lecacheux, J., Burchi, R., Di Paolantonio, A. 1990. *Icarus* **87**, 372-376.
Di Martino, M., Gonano-Beurer, M., Mottola, S., Neukum, G. 1992. *ACM 91*, pp. 157-160.
Dotto, E., Barucci, M.A., Fulchignoni, M., Di Martino, M., Rotundi, A., Burchi, R., Di Paolantonio, A. 1992. *Astron. Astrophys. Suppl. Ser.* **95**, 195-211.
Dovgopol, A.N., Lisina, L.R. 1992. *Kinem. Fiz. Nebesn. Tel.* **8**, 36-39.
Dovgopol, A.N., Krugly, Yu.,N., Shevchenko, V.G. 1992. *Acta Astron.* **42**, 67-72.
Drummond. J.D., Weidenschilling, S.J., Chapman, C.R., Davis, D.R. 1991. *Icarus* **89**, 44-64.
Duhnam, D.W. and 45 Coauthors, 1990. *Astron. J.* **99**, 1636-1662.
Erikson, A., Magnusson, P. 1993. *Icarus* **103**, 62-66.
Erikson, A., Cutispoto, G., Debehogne, H., Hahn, G., Lagerkvist, C.-I., Lindgren, M., Magnusson., P. 1991. *Astron. Astrophys. Suppl. Ser.* **91**, 259-264.
Fanale, F.P., Clark, B.E., Bell, J.F. 1992. *J. Geophys. Res.* **97**, 20863-20874.
Farinella, P. 1992. *Icarus* **96**, 284-285.
Farinella, P., and D.R. Davis 1992. *Icarus* **97**, 111-123.
Farinella, P., Davis, D.R., Cellino, A., Zappalà, V. 1992a. In *ACM 91*, pp. 165-166.
Farinella, P., Davis, D.R., Cellino, A., Zappalà, V. 1992b. *Astron. Astrophys.* **275**, 329-330.
Farinella, P., D.R. Davis, P. Paolicchi, A. Cellino, V. Zappalà 1992c. *Astron. Astrophys.* **253**, 604-614.
Farinella, P., Gonczi, R., Froeschlé, Ch., Froeschlé, C. 1993a. *Icarus* **101**, 174-187.
Farinella, P., Ch. Froeschlé, R. Gonczi 1993b. *Cel. Mech.* **56**, 287-305.
Festou, M.C., Stern, S.A., Tozzi, G.P. 1991. *Icarus* **94**, 218-231.
Fink, U., Hoffmann, M., Grundy, W., Hicks, M., Sears, W. 1992. *Icarus* **97**, 145-149.
Fitzsimmons, A., Williams, I.P., Lagerkvist, C.-I., Dahlgren. 1990. *Mon. Not. R. Astron. Soc.* **247**, 26-29.
Froeschlé, Ch., and H. Scholl 1991. In *ACM 91*, pp. 205-209.
Gaffey, M.J., Reed, K.L., Kelley, M.S. 1992. *Icarus* **100**, 95-109.
Gammelgaard, P., Kristensen, L.K. 1991. *Astron. Astrophys.* **244**, 544-552.
German, D., and A. Friedlander 1991. Proc. AAS/AIAA Spaceflight Mechanics Meeting, Houston, Texas (February 11-13, 1991).
Goffin, E. 1991. *Astron. Astrophys.* **249**, 563-568.
Goldader, J.D., Tholen, D.J., Cruikshank, D.P., Hartmann, W.K. 1991. *Astron. J.* **102**, 1503-1507.
Gonano-Beurer, M., Di Martino, M. Mottola, S., Neukum, G. 1992. *Astron. Astrophys.* **254**, 393-396.
Grimm, R.E., McSween, H.Y. Jr. 1993. *Science* **259**, 653-655.
Grundy, W.M., Fink, U. 1992. *ACM 91*, pp.215-218.
Haack, H., Scott, E.R.D. 1992. *J. Geophys. Res.* **97**, 14727-14734.
Hamilton, D.P., and J.A. Burns 1992. *Icarus* **96**, 43-64.
Harris, A.W., Young, J.W., Dockweiler, T., Gibson, J., Poutanen, M., Bowell, E. 1992. *Icarus* **95**, 115-147.
Hartmann, W.K. 1990. *Icarus* **87**, 236-240.
Hartmann, W.K., Tholen, D.J. 1990. *Icarus* **86**, 448-454.
Hiroi, T., Takeda, H. 1990. *Icarus* **88**, 205-227.
Hiroi, T., Bell, J.F., Takeda, H., Pieters, C.M. 1993. *Icarus* **102**, 107-116.
Hoffmann, M., Geyer, E.H. 1990. *Acta Astron.* **40**, 389-396.
Hoffmann, M., Fink, U., Grundy, W.M., Hicks, M. 1993. *J. Geophys. Res.* **98**, 7403-7407.
Housen, K.R., and K.A. Holsapple 1990. *Icarus* **84**, 226-253.
Housen, K.R., R.M. Schmidt, K.A. Holsapple 1991. *Icarus* **94**, 180-190.
Hughes, D.W. 1991. Q. J. R. Astron. Soc. **32**, 133-145
Jewitt, D.C., and Luu, J.X. 1990. *Astron. J.* **100**, 933-944.
Jones, T.D., Lebofsky, L.A., Lewis, J.S., Marley, M.S. 1990. *Icarus* **88**, 172-192.
Kaasalainen, M., Lamberg, L., Lumme, K., Bowell, E. 1992. *Astron. Astrophys.* **259**, 381-332.
Kaasalainen, M., Lamberg, L., Lumme, K. 1992. *Astron. Astrophys.* **259**, 333-340.
Kargel, J.S. 1991. *Icarus* **94**, 368-390.
Keil, K., Bell, J.F., Britt, D.T. 1992. *Icarus* **98**, 43-53.

King, T.V.V., Clark, R.N., Calvin, W.M., Sherman, D.M., Brown, R.H. 1992. *Science* **255**, 1551-1553.
Kissling, W.M., Blow, G.L., Allen, W.H., Priestley, J., Riley, P., Daalder, P., George, M. 1991. *Proc. Astron. Soc. Aust.* **9**, 150-152
Knezevic, Z., A. Milani, P. Farinella, Ch. Froeschlé, C. Froeschlé 1991. *Icarus* **93**, 316-330.
Kristensen, K.L. 1991. *Astron. Nachr.* **312**, 209-220.
Kristensen, K.L. 1992a. In *OPP*, pp. 189-194.
Kristensen, K.L. 1992b. *Astron. Nachr.* **313**, 1-10.
Kristensen, K.L., Gammelgaard, P. 1993. *Astron. Astrophys.* **272**, 345-354.
Kuzmanoski, M., Knezevic, Z. 1993. *Icarus* **103**, 93-103.
Kwiatkowski, T., Michalowski, T. 1992. *ACM 91*, pp. 333-336.
Lagerkvist, C.-I., Magnusson, P., Debehogne, H., Hoffmann, M., Erikson, A., De Campos, A., Cutispoto, G. 1992. *Astron. Astrophys. Suppl. Ser.* **95**, 461-470.
Landgraf, W. 1992. In IAU Symposium No. 152, pp. 179-182.
Lecar, M., Franklin, F., Soper, P. 1992. *Icarus* **96**, 234-250.
Lindblad, B.A. 1992. In *ACM 91*, pp. 363-366.
Luu, J.X., Jewitt, D.C. 1992. *Icarus* **97**, 276-287.
Magnusson, P. 1991. *Astron. Astrophys.* **243**, 512-520.
Magnusson, P., Lagerkvist, C.-I. 1990. *Astron. Astrophys. Suppl. Ser.* **86**, 45-51.
Magnusson, P., Lagerkvist, C.-I. 1991. *Astron. Astrophys. Suppl. Ser.* **87**, 269-275.
Magnusson, P. Barucci, M.A., Binzel, R.P., Blanco, C., Di Martino, M., Goldader, J.D., Gonano-Beurer, M., Harris, A.W., Michalowski, T., Mottola, S., Tholen, D.J., Wisniewski, W.Z. 1992. *Icarus* **97**, 124-129.
Martelli, G., P. Rothwell, I. Giblin, M. Di Martino, P. Farinella 1993. *Astron. Astrophys.* **271**, 315-318.
McFadden, L.A., Chamberlin, A.B. 1992. *ACM 91*, pp. 413-416.
McFadden, L., Cochran, A.L., Barker, E.S., Cruikshank, D.P., Hartmann, W.K. 1993. *J. Geophys. Res.* **98**, 3031-3041.
McSween, H.Y.Jr. 1992. *Icarus* **95**, 239-243.
McSween, H.Y. Jr., Bennett, M.E. Iii, Jarosewich, E. 1991. *Icarus* **90**, 107-116.
Melosh, H.J., and J.A. Stansberry 1991. *Icarus* **94**, 171-179.
Melosh, H.J., Ryan, E.V., Asphaug, E. 1992. *J. Geophys. Res.* **97**, 14,735-14,759.
Michalowski, T. 1992. *ACM 91*, pp. 417-420.
Michalowski, T., Velichko, F.P. 1990. *Acta Astron.* **40**, 321-332.
Michalowski, T., Velichko, F.P., Lindgren, M., Oja, T., Lagerkvist, C.-I., Magnusson, P. 1991. *Astron. Astrophys. Suppl. Ser.* **91**, 53-59.
Milani, A., and Z. Knezevic 1990. *Cel. Mech.* **49**, 347-411.
Milani, A., P. Farinella, Z. Knezevic 1992. In *IPD*, pp. 85-132.
Millis, R.L., Franz, O.G., Wasserman, Lebofsky, L.A., Aspahug, E., Hubbard, W.B., Hunten, D.M., A'Hearn, M., Schnurr, R., Klemola, A.R., Osborn, W., Vilas, F., Potter, A.E. Maley, P.D. Manly, P.L. 1990. *Bul. Amer. Astron. Soc.* **21**, 1247.
Mueller, B.E.A., Tholen, D.J., Hartmann, W.K., Cruikshank, D.P. 1992. *Icarus* **97**, 150-154.
Nakamura, A., and A. Fujiwara 1991. *Icarus* **92**, 132-146.
Nakamura, A., K. Suguiyama, A. Fujiwara 1992. *Icarus* **100**, 127-135.
Namiki, N., Binzel, R.P. 1991. *Geophys. Res. Lett.* **18**, 1155-1158.
Ostro, S.J., Wisniewski, W.Z. 1992. *ACM 91*, pp. 447-450.
Ostro, S.J., Campbell, D.B., Hine, A.A., Shapiro, I.I., Chandler, J.F., Weerner, C.L., Rosema, K.D. 1990a. *Astron. J.* **99**, 2012-2018.
Ostro, S.J., Chandler, J.F., Hine, A.A., Rosema, K.D., Shapiro, I.I., Yeomans, D.K. 1990b. *Science* **248**, 1523-1528.
Ostro, S.J., Campbell, D.B., Chandler, J.F., Shapiro, I.I., Hine, A.A., Velez, R., Jurgens, R.F., Rosema, K.D., Winkler, R., Yeomans, D.K. 1991a. *Astron. J.* **102**, 1490-1502.
Ostro, S.J., Campbell, D.B., Chandler, J.F., Hine, A.A., Hudson, R.S., Rosema, K.D., Shapiro, I.I. 1991b. *Science* **252**, 1399-1404.
Rajamohan, R., Bhargavi, S.G. 1992. *ACM 91*, 487-491.
Redman, R.O., Feldman, P.A., Matthews, H.E., Halliday, I., Creutzberg, F. 1992. *Astron. J.* **104**, 405-411.

Rubin, A.E., Mittlefehldt, D.W. 1993. *Icarus* **101**, 201-212.
Ruskol, E.L. 1990. *Astron. Vestn.* **24**, 232-243.
Ryan, E.V., W.K. Hartmann, D.R. Davis 1991. *Icarus* **94**, 283-298.
Safronov, V.S., Ziglina, I.N. 1991. *Astron. Vestn.* **25**, 190-199.
Salisbury, J.W., D'Aria, D.M., Jarosewich, E. 1991. *Icarus* **92**, 280-297.
Scholl, H., and Ch. Froeschlé 1991. *Astron. Astrophys.* **245**, 316-321.
Shestopalov, D.I., Golubeva, L.F. 1991. *Pis'ma Astron. Zh.* **17**, 453-461.
Shestopalov, D.I., Golubeva, L.F. 1992. *Astron. Vestn.* **26**, 77-88.
Shestopalov, D.I., Golubeva, L.F., Pogosbekov, D.G. 1990., Atai, A.A., Gorbanev, V.E. 1990. *Astron. Vestn.* **24**, 232-243.
Shkuratov, Yu.G., Muinonen, K. 1992. *ACM 91*, 549-552.
Sitarski, G., Todorovic-Juchniewicz, B. 1992. *Acta Astron.* **42**, 139-144.
Smith, D.W., Johnson, P.E., Buckingham, W.L., Shorthill, R.W. 1992. *Icarus* **99**, 485-488.
Smither, C.L., Ahrens, T.J. 1992. *ACM 91*, pp. 561-564.
Soper, P., Franklin, F., Lecar, M. 1990. *Icarus* **87**, 265-284.
Stephens, J.R., Gustafson, B.A.S. 1991. *Icarus* **94**, 209-217.
Stern, S.A., Barker, E.S. 1992. *ACM 91*, pp. 577-581.
Stooke, P.J. 1992. *ACM 91*, pp. 583-586.
Takagi, Y., Kato, M., Mizutani, H. 1992. *ACM 91*, pp. 597-600.
Taylor, G.J. 1992. *J. Geophys. Res.* 97, 14,717-14,726.
Tsvetkova, V.S., Dudinov, V.N., Novikov, S.B., Pluzhnik, Ye.A., Shkuratov, Yu.G., Vakulik, V.G., Zhelezniak, A.P. 1991. *Icarus* 92, 342-349.
Van Flandern, T. 1992. *ACM 91*, pp. 609-612.
Velichko,F.P., Kruglyj, Yu.N., Lupishko, D.F., Mohamed, R.A. 1990. *Astron. Tsirk.* **1546**, 39-40.
Vilas, F., McFadden, L.A. 1992. *Icarus* **100**, 85-94.
Vilas, F., Hatch, E.C., Larson, S.M., Sawyer, S.R., Gaffey, M.J. 1993. *Icarus* **102**, 225-231.
Warren, P.H., Kallemeyn, G.W. 1992. *Icarus* **100**, 110-126.
Weidenschilling, S.J., Chapman, C.R., Davis, D.R., Greenberg, R., Levy, D.H., Binzel, R.P., Vail, S.M., Magee, M., Spaute, D. 1990. *Icarus* **86**, 402-447.
Wetherill, G.W. 1992. *Icarus* **100**, 307-325.
Williams, G.V. 1992. *ACM 91*, pp. 641-643.
Williams, J.G. 1991. *Lunar Planet. Sci. XIX,* 1277-1288.
Williams, J.G. 1992a. *Icarus* **96**, 251-280.
Williams, J.G. 1992b. *ACM 91*, pp. 645-649.
Williams, J.G. 1992c. *ACM 91*, pp. 649-652.
Wisniewski, W.Z. 1991. *Icarus* **90**, 117-122.
Wisniewski, 1992. *ACM 91*, pp. 653-656.
Wisniewski, W.Z., Barucci, M.A., Fulchignoni, M., De Sanctis, C., Dotto, E., Rotundi, A., Binzel, R.P., Madras, C.D., Green, S.F., Kelly, M.L., Newman, P.J., Harris, A.W., Young, W., Blanco, C., Di Martino, M., Ferreri, W., Gonano-Beurer, M., Mottola, S., Tholen, D.J., Goldader, J.D., Coradini, M., Magnusson, P. 1993. *Icarus* 101, 213-222.
Yanagisawa, M., Eluszkiewicz, J., Ahrens, T.J. 1991. *Icarus* **94**, 272-282.
Yeomans, D.K., Chodas, P.W., Keesey, M.S., Ostro, S.J., Chandler, J.F., Shapiro, I.I. 1992. *Astron. J.* **103**, 303-317.
Zappalà, V., A. Cellino, P. Farinella, Z. Knezevic 1990. *Astron. J.*, **100**, 2030-2046.
Zappalà, V., Cellino, A. 1992. Celest. Mech. **54**, 207-227.
Zappalà, V., Cellino, A., Farinella, P. 1992. In *ACM 91*, pp. 675-678.

COMMISSION 16: PHYSICAL STUDY OF PLANETS & SATELLITES
ETUDE PHYSIQUE DES PLANETS & SATELLITES

PRESIDENT: David Morrison
VICE-PRESIDENT: Mikhail Ya. Marov
VICE-PRESIDENT: Catherine De Bergh

ORGANIZING COMMITTEE: Alexandr Basilevsky, Michael Belton, Andre Brahic, Dale Cruikshank, Merton Davies, Therese Encrenaz, Daniel Gautier, Vasily Moroz, Tobias Owen, Jurgen Rahe, Bradford Smith, David Tholen

Report prepared by David Morrison

1. INTRODUCTION

The study of the physical properties of planets and satellites has flourished since the beginning of direct space exploration in the 1960s. Although the pace of planetary exploration has declined since its peak (roughly from 1970 to 1986), this field remains strong and productive. The level of research activity was approximately the same as in the previous report period, with the publication of more than a thousand research papers. This activity is broadly international, and growth in planetary research has been especially notable in Europe. The two most important annual meetings of planetary scientists are the Division for Planetary Sciences (DPS) of the American Astronomical Society, held each autumn, and the Lunar and Planetary Science Conference, held each spring at Houston. Both meetings attract between 300 and 600 registrants, with a comparable number of papers presented. The primary journals for publication of planetary science are *Icarus* and the *Journal of Geophysical Research*, although many other astronomical and geoscience journals play an important role, as well as the two general-science journals *Science* and *Nature*.

Highlights of planetary exploration and research during this reporting period (1990-1992) included: (1) radar mapping of the surface of Venus at 100 m resolution by Magellan, generating a data set from which to determine the geology and evolutionary history of the planet; (2) continuing strong interest in Mars, in part focused toward eventual human presence on that planet; and (3) analysis of results from the 1989 Voyager flyby of Neptune and Triton. Other events of note include the end of the 13-year Pioneer Venus studies as the spacecraft entered the atmosphere of the planet, two productive flybys of the Earth-Moon system by the Galileo spacecraft, and a growing interest in the Pluto-Charon system. This field is increasingly interdisciplinary, involving astronomers, geoscientists, physicists, and biologists, and motivated by interest in understanding the origin and evolution of the planets.

Although the present review attempts to mention the most important new results in the physical study of planets and satellites, it can cite individually fewer than 25% of the papers actually published in this field during 1990-92. The selection of papers to cite is partly subjective. In general, there has been an effort to emphasize astronomical work over that which is primarily of a geological or geochemical nature, or that which can be classified as space physics. Major research areas dealing with the analysis of lunar samples and martian samples (the SNC meteorites) are not included due to limitation in space. For complementary reviews of current research the reader is referred to reports from Commissions 15, 22, and 51.

J. Bergeron (ed.), Reports on Astronomy, Vol. XXIIA, 157–168.

2. MERCURY

Mercury is a difficult planet to observe, and it has been visited by only one spacecraft, Mariner 10, which made 3 flybys in 1974/75. Limited progress has been made since Mariner 10 in understanding the geology and crustal evolution of the planet, but continuing interest in the dynamics of the inner solar system has connected the anomalous high density of Mercury with the era of catastrophic collisions. A giant impact could have removed much of the mantle of Mercury not long after the planet differentiated, leaving the present silicate-depleted object.

Earth-based spectroscopy of Mercury has revealed additional details of its tenuous atmosphere of sodium and potassium, which appears to be the result of diffusion upward through the crust (Sprague *et al., Icarus* **84**:93; Killian *et al., Icarus* **85**:145). Enhancement of K has also been reported above the Caloris basin (Sprague *et al., Science* **249**:1140). Sprague (*J. Geophys. Res.* **97**:18257) proposed a model for Na and K variability based on ion implantation into the regolith during the long mercurian night with subsequent diffusion to the atmosphere as temperatures rise. Ludlow *et al.* (*Astrophys. J.* **384**:640) have used the VLA to obtain high-resolution maps of thermal emission from the planet.

Perhaps the most surprising new result on Mercury in this triennium was the discovery of greatly enhanced radar reflectivity from high latitudes, which is interpreted as near-surface polar ice (Slade *et al., Science* **258**:635; Harmon *et al., Science* **258**:640; Paige *et al., Science* **258**:643). The observations, which were made with both the Arecibo and Goldstone radars, confirm the presence of such ice caps at both poles (Butler *et al., J. Geophys. Res.* **98**:15005), but the ability of the planet to retain ice over the history of the solar system is unexplained.

3. VENUS

During this reporting period, Venus received more attention than any other member of the planetary system. The Pioneer Venus orbiter continued its long series of observations of the upper atmosphere and space environment of the planet until its orbit decayed in October 1992. During the final months of the mission the Pioneer spacecraft sampled regions of the atmosphere at lower elevations than had been studied directly previously. Excellent atmospheric results were also obtained during the February 1990 flyby of Venus by the Galileo spacecraft (Johnson *et al., Science* **253**:1516). In the early part of this triennium, analysis continued of the radar maps of the planet at resolutions of 1-2 km obtained by the Venera 15 and 16 orbiters and the Arecibo and Goldstone ground-based radars. However, these radar results were quickly superseded when the Magellan spacecraft began its mapping mission in September 1990. The Magellan radar images have revolutionized our understanding of the geology and geological history of Venus, as described further below.

Results on atmospheric dynamics as derived from differential VLBI measurements of the VEGA balloons were reported by Sagdeyev *et al.* (*Astron. Astrophys.* **254**: 387), and Schinder *et al.* (*J. Atmos. Sci.* **47**:2037) discussed the formation of waves, advection, and cloud patterns. However, issues of chemistry, origin, and evolution continued to be the focus of most observational and theoretical work on the atmosphere of Venus. Infrared and millimeter observations yielded detection of HDO and determination of the abundance of deuterium. De Bergh *et al.* (*Science* **251**:547) obtained D/H = 120 ± 40 at altitudes near 40 km. Encrenaz *et al.* (*Astron. Astrophys.* **246**:L63) used IRAM data to derive a mixing ratio of water below 95 km of 3.5 ppm, while Crisp *et al.* (*Science* **253**:1263) used IR observations to derive a mixing ratio of 40 ± 20 ppm. In a summary of all the deuterium and water abundance data, Donahue & Hodges (*J. Geophys. Res.* **97**:6083) argued that the mixing ratio for atmospheric water increases by a factor of 4 between the surface and 10 km altitude to a value of 150 ppm and drops to 10 ppm above 50 km. These results require large-scale temporal and spatial variations; these authors also conclude that the early Venus must have been endowed with at least 2 orders of magnitude more water above the surface than is presently there. Fegley (*Astron. Vest.* **26**:3) reviewed the chemistry of the surface and lower atmosphere, using atmospheric data to infer information on surface chemistry. Kaula (*Science* **247**:1191) suggested that the current dry atmosphere is a consequence of low levels of crustal turnover, which in turn are related to the absence of a large satellite of Venus. Many authors considered the structure of the ionosphere and thermosphere, and the interaction of Venus with the solar wind, based on both Pioneer Venus and Galileo data (*e.g.*, Bougher *et al.*, *J. Geophys. Res.* **95**:6271; Kim *et al.*, *J. Geophys. Res.* **95**:6569, Kivelson *et al.*, *Science* **253**:1518; Hord *et al.*, *Science* **253**:1548). A detailed exposition of this work through about 1990 appears in the book *Venus Aeronomy*, edited by Russell and published in 1991 by Kluwer, while a more recent review is by Luhmann *et al.* (*J. Geophys. Res.* **96**:11019).

The Magellan spacecraft was launched on 4 May 1989 and began its radar mapping mission on 15 September 1990. The spacecraft utilized a single radar sensor to collect synthetic aperture images, radar altimetry, and radiometric data at 2.4 Ghz. Nominal imaging resolution was about 100 m. The first 8-month cycle imaged 84% of the planet with left-looking radar. The second cycle used right-looking radar and increased the coverage to 97%. Cycle 3 (ending in September 1992) was again left-looking and yielded considerable stereo coverage. In 1993 aerobraking was used to lower and circularize the orbit for enhanced mapping of the planet's gravity field. A series of papers summarizing the state of knowledge just before Magellan appeared in *Earth Moon Planets* **50/51**:3-558, and an account of much of the earlier Russian work on Venus is found in the book *Venus Geology, Geochemistry, and Geophysics: Research Results from the USSR*, edited by Barsukov *et al.* and published by the University of Arizona Press. An early summary of Magellan results was published by Saunders *et al.* (*Science* **252**:249); however, the bulk of the initial scientific results from Magellan (primarily interpretation of data from cycle 1) were published in 50 papers printed in two special editions of *J. Geophys. Res.* dated 25 August 1992 (**97**:13067-13689) and 15 October 1992 (**97**:15921-16380).

Magellan provided a fundamental data set for assessment of global volcanism, tectonism, impact processes, and surface processes on Venus. The mean radius is 6051.84 km, and more than 80% of the surface lies within 1 km of this elevation. The rotation period is 243.0185 days, and the north pole position is RA 272.76, dec 67.16 (Saunders *et al.*, *J. Geophys. Res.* **97**:13067). Most of the surface is either volcanic plains (85%) or complex ridged terrain or tessera, including highland plateaus and mountains. Volcanism is widespread, with thousands of individual volcanic constructs including shield volcanoes and cinder cones, down to the 100-m resolution of the images. The most extensive units consist of flood lavas. A number of distinctive volcanic features have been identified, including "pancake" domes (circular, flat-toped constructs of highly viscous lava) and lava channels or rivers of remarkable size. Hildr, the longest lava channel, is more than 6000 km long, yet of nearly uniform width. The planet has been subject to extensive tectonic deformation, and the resulting ubiquitous folds and cracks are readily seen in radar images at all scales. There are a few large offset faults, but most of the evidence suggests local shear and relatively small-scale tectonic features. While there is no indication of global plate tectonics such as that experienced by the Earth, there are regions of both extension (rifting) and convergence (producing folded mountains). The steep mountains of

Ishtar Terra suggest the presence of currently active tectonic deformation, presumably supported by a large mantle plume. Mantle upwelling is further suggested by correlation of large shield volcanoes with geoid anomalies. Among the most distinctive features of the geology of Venus are coronae, which are large circular tectonic features characterized by both concentric and radial cracks, often associated with volcanic activity. The coronae appear to be supported by mantle plumes, and many are probably an intermediate stage leading toward volcanism and surface rifting.

Nearly 1000 impact craters have been discovered on the surface of Venus. They range from large multiring basins (up to nearly 300 km diameter) to simple craters some tens of km in diameter, often with flooded floors and extensive surrounding flows due to fluidized ejecta or induced volcanic activity. Below a diameter of 35 km, there are fewer craters than would be expected from lunar comparisons, as a consequence of shielding by the atmosphere. At smaller sizes, craters are replaced by radar-dark (smooth) regions apparently resulting from atmospheric disintegration of incoming projectiles; the atmospheric breakup of meteoroids has been modeled by Zahnle (*J. Geophys. Res.* **97**:10243). Also distinctive are large dark paraboloidal surface features associated with some younger craters. Surprisingly, abundant wind streaks have also been identified on the surface. The estimated surface age from crater density is between 200 and 700 million yr with no clear indication of substantial local variations in age. Since only a small percentage of the craters have been modified by flooding or other geological activity since they were formed, one interpretation of the crater statistics is that there was a global resurfacing of Venus several hundred million years ago, in which most of the crust was renewed at about the same time. Intensive analysis of the Magellan data is continuing; it is notable that this mission has returned more data than all previous planetary missions combined.

4. MOON

Lunar science is a healthy and active part of solar system studies, with most research still focused on the analysis and interpretation of samples returned by the Apollo and Luna missions of the early 1970s. Dozens of papers in lunar science are presented at the Lunar and Planetary Science Conference held each spring in Houston. We will not attempt to review this body of information on lunar geology and geochemistry in the current brief overview; the interested reader is directed to the various publications and bibliographies available from the Lunar Science Institute in Houston.

The thin, transient atmosphere of the Moon has been studied by several researchers, stimulated in part by discovery of a similar atmosphere on the planet Mercury (Morgan & Shemansky, *J. Geophys. Res.* **96**:1351). A group of related papers deal with the presence of sodium and its possible time variability (Potter & Morgan, *Geophys. Res. Lett.* **18**:2089; Ip, *Geophys. Res. Lett.* **18**:2093; Mendillo *et al., Geophys. Res. Let.* **18**:2097; Hunten *et al., Geophys. Res. Lett.* **18**:2101).

Preliminary results from the Galileo spacecraft encounter with the Earth-Moon system of December 1990 were presented by Belton *et al.* (*Science* **255**:570). The geometry of the flyby was such that the western limb of the Moon was illuminated, in contrast to the Apollo missions, providing unique data on the Orientale basin and parts of the lunar farside highlands that had not been well observed previously. The primary data set discussed by Belton *et al.* consists of multispectral images obtained with the Galileo CCD camera, which provide sensitive discrimination of various mineralogic units. Observations of the South Pole-Aitken basin (which is more than 2000 km in diameter) suggest that this impact, one of the largest preserved in the lunar record, penetrated into the mantle, unlike the Orientale basin, which appears to sample only crustal material. A second Earth-Moon flyby took place in December 1992.

5. MARS

The study of Mars is central to the discipline of planetary science. This is probably the best observed of all the planets, and the Viking missions of 1976 provided a level of *in situ* data that is exceeded only for the Moon, where humans have landed. Mars is, in many respects, the planet most like the Earth, and during this triennium it took on added significance from efforts within the United States to promote efforts leading toward human exploration early in the next century. Although these efforts (the "Space Exploration Initiative") were largely abandoned in 1993, they served to focus both scientific and general interest on this planet.

A major book entitled *Mars* (edited by Kieffer, Jakosky, Snyder, and Matthews) was published in 1992 by the University of Arizona Press, consisting of 38 chapters by 114 authors. This volume serves as a concise scientific summary of our knowledge of Mars at the time of its publication. It is organized to present 4 introductory chapters, 6 chapters on solid body geophysics, 6 chapters on bedrock geology and geologic units, 7 chapters on surface properties and processes, 10 chapters on the atmosphere, 2 chapters on biology, and 3 chapters on the martian satellites Phobos and Deimos. It is impossible in this brief overview of progress in planetary studies to summarize the contents of this book, or of the dozens of subsequent papers dealing in even greater detail with many of these same topics. In this review we will briefly discuss a few of the additional publications on Mars that have appeared since the University of Arizona volume. A growing body of information is derived from analysis of the SNC meteorites, which are samples of martian material ejected and transported to Earth, but we will not review that material here, since meteorites are discussed under Commission 15.

Publication has continued in this triennium of data and interpretation from the Phobos-2 mission. Surkov *et al.* (*Pis'ma Astron. Zh.* **16**:355) reported on remote determination of the fundamental composition of the martian crust, concluding that much of the surface consists of subalkaline basalt, while interpretation of the VSK and KRFM investigations was presented by Murchie *et al.* (*J. Geophys. Res.* **96**:5925). A series of papers on the Phobos-2 plasma results appeared in *Geophys. Res. Lett.* **17**:869-900; e.g., a discussion by Luhmann *et al.* (*Geophys. Res. Lett.* **17**:869) of the similarities between the solar wind interactions at Mars and Venus. The general issue of the martian ionosphere and magnetosphere was reviewed by Luhmann (*Rev. Geophys.* **29**:121).

One of the surprises of the Viking mission was the discovery that only the southern polar cap of Mars is able to retain frozen carbon dioxide throughout the year. This difference and the general issue of the stability of the polar caps were addressed by Jakosky & Haberle (*J. Geophys. Res.* **95**:1359), who concluded that the fate of the caps depends critically on the rate of heat conduction into the underlying soil. Carr (*Icarus* **87**:210) discussed the interpretation of the recent measurements of the D/H ratio on Mars, examining the effects of surface processes such as flood volcanism, impacts, and the formation of polar deposits. He concluded that the observed value of D/H in Mars (5 times greater than on Earth) depends on the rate of exchange of water between the atmosphere and polar deposits. Jakosky (*Icarus* **94**:14) also considered volatile evolution of Mars, analyzing how the observed fractionation of oxygen, carbon, and hydrogen is produced by exchanges among the atmosphere, regolith, and caps.

An important focus of current Mars studies is the evolution of the surface and atmosphere through time. Haberle has edited a special report on this topic published in *J. Geophys. Res.* **98**:3091-3482 and **98**:10897-11121. Here more than 100 authors, in 38 papers, review such topics as surface mineralogy and geology, atmospheric origin and evolution, current climate and atmospheric dynamics, fluvial processes, ground ice distribution, and the SNC meteorites. Of particular value are three papers that discuss the NASA Ames general circulation model (GCM) for the martian atmosphere, a sophisticated computer code that is able to simulate the dynamics of the martian atmosphere using appropriate boundary conditions such as, for example, the correct distribution of surface topography (Haberle *et al.*, *J. Geophys. Res.* **98**:3093; Barnes *et al.*, *J. Geophys. Res.* **98**:3125; Pollack *et al.*, *J. Geophys. Res.* **98**:3149).

One reason for interest in the evolution of the atmosphere and climate of Mars is to assess the probability of the origin of life and to identify possible sites for martian fossil organisms. Many of these are exobiological issues, discussed in the report of Commission 51 on Bioastronomy. Rothshild *et al.* (*Icarus* **88**:246) discussed the possibility of preserving martian microbes in evaporites, which might be found in paleolakes on Mars. The more general issue of ancient aqueous sedimentation on Mars was explored from a geological perspective by Goldspiel & Squyres (*Icarus* **89**:392), who identified 36 sites, concentrated in ancient cratered terrain between the equator and latitude 30 S, where ponding of water probably occurred. The question of the possible duration of such liquid water habitats was addressed by McKay and Davis (*Icarus* **90**:214). The more remote possibility of currently extant life forms on Mars was discussed by Boston *et al.* (*Icarus* **95**:300) in an analysis of chemosynthetic ecosystems in subsurface habitats.

6. JUPITER

Extensive Earth-based observations of Jupiter continue to be made, in part to provide continuity between the Pioneer and Voyager flybys of the late 1970s and the arrival of the Galileo orbiter and probe in 1995. An International Jupiter Watch (IJW) is coordinating many of these observations. The Galileo mission was successfully launched and completed its tour of the inner planets preparatory to the flight to Jupiter. Failure of the spacecraft high-gain antenna to deploy will substantially restrict the quantity of data that can be transmitted from Jupiter; this will not be a problem for the probe mission, but it will certainly have an impact on the orbital operations. Russell has edited a comprehensive book called *The Galileo Mission* (Kluwer); the same papers are also available in *Space Sci. Rev.* **60**.

Jupiter remains the prototype giant planet, and the one for which we have the best data on atmospheric dynamics. Conrath *et al.* (*Icarus* **83**:255) discussed the temperatures and circulation in the stratospheres of the jovian planets, while Ingersoll (*Science* **248**:308) reviewed models comparing their atmospheric dynamics. All four jovian planets have longitudinal banding and high-speed jet streams. Jupiter is the most active, while Saturn and Neptune have the highest wind velocities. Two general classes of fluid dynamical models have been developed to reproduce many of the observed features of winds, temperatures, and cloud patterns: those that consider only the atmosphere itself, and those that include circulation in the deep liquid mantles. A comparative analysis of the interior structures of the jovian planets has been published by Zharkov (*Astron. Vest.* **25**:627)

There is considerable interest in the possibility of detecting global oscillations of Jupiter in order to apply the techniques of helioseismology to probing conditions in the planet's interior. Tentative detection of oscillations with periods of tens of minutes was reported by Schmider *et al.* (*Astron. Astrophys.* **248**:281) and discussed by Mosser *et al.* (*Astron. Astrophys.* **251**:356 and *Icarus* **96**:15).

Infrared observations of auroral emissions in the upper atmosphere were reported by Kim *et al.* (*Nature* **353**:536) and Baron *et al.* (*Nature* **353**:539), while Livengood *et al.* (*Icarus* **97**:26) provided a summary of jovian ultraviolet auroral activity as observed by IUE from 1981 to 1991. Carlson *et al.* (*Astrophys. J.* **388**:648) found from a reanalysis of Voyager IR spectra that it is possible in jovian hot spots to probe atmospheric composition down to a pressure level of 5 bars; they find a relatively high abundance of water vapor, probably about twice the solar mixing ratio. De Pater has published two extensive reviews of microwave spectra and imaging of the giant planets (*Phys. Rep.* **200**, 1; *Ann. Rev. Astron. Astrophys.*), discussing both their atmospheres and magnetospheres.

7. SATURN

Recent interest in Saturn has concentrated on its atmosphere, particularly on the appearance of a major equatorial disturbance called the "great white spot" (Westphal *et al.*, *Astrophys. J.* **369**:L51; Sanchez-Lavega *et al.*, *Nature* **353**:397; Beebe *et al.*, *Icarus* **95**:163; Westphal *et al.*, *Icarus* **100**:485). Other results on Saturn's atmosphere and magnetosphere are included in the papers comparing the jovian planets cited above in Section 6.

8. URANUS

A comprehensive book on this planet and its satellite and ring system was published in 1991: *Uranus,* edited by Bergstralh, Miner, and Matthews, University of Arizona Press. A total of 21 papers by 86 authors summarized the results from the 1986 Voyager-2 flyby and subsequent interpretation. A more popular account by Miner is found in *Uranus: The Planet, Rings, and Satellites* (Ellis Norwood Library). After the 1989 Voyager flyby of Neptune, models for the interiors and structure of Uranus and Neptune were compared by Hubbard *et al.* (*Science* **253**:648) and Podolak *et al.* (*Geophys. Res. Lett.* **17**:1737).

9. NEPTUNE

Voyager 2 completed its grand tour of the solar system with a flyby of Neptune and Triton in August 1989; thus most analyses of these new data fall within the current triennium. Many of the initial results were published in 32 papers in a special issue of *Geophys. Res. Lett.* **17**:1661-1776 (September 1990), and a later more extensive series of papers appeared in a special issue of *J. Geophys. Res.* **96**:18907-19268 (30 October 1991), and in *Icarus* **99**:241-435 (October 1992). A popular-level account of the Neptune encounter is to be found in *Far Encounter: The Neptune System* by Burgess (Columbia University Press), while a multi-author technical volume called *Neptune* is in preparation in the Space Science Series of the University of Arizona Press.

Unlike Uranus, Neptune has a substantial internal heat source. Voyager IR observations (Pearl *et al.*, *J. Geophys. Res.* **96**:1829) yield a bolometric bond albedo of 0.29, a calculated equilibrium temperature of 47 K, but an observed effective temperature of 59 K, implying an energy balance given by E = 2.6 ± 0.3. Voyager IR data also yielded abundances of acetylene and ethane (Bezard *et al., J. Geophys. Res.* **96**:18961), while Rosenqvist *et al.* (*Astrophys. J.* **392**:L99) used millimeter spectra obtained with the IRAM telescope to measure the mixing ratios of CO and HCN on Neptune. De Bergh *et al.* (*Astrophys. J.* **355**:661) discovered monodeuterated methane in the spectrum of Neptune and derived a D/H mixing ratio of 0.00012, indicating that deuterium is strongly enhanced (as it is also for Uranus). Voyager imaging data allow determination of atmospheric zonal winds from observed displacement of cloud features (Limaye *et al., J. Geophys. Res.* **96**:18941).

The orientation and strength of the magnetic field are similar to those of Uranus, displaying a dipole tilt of 47 degrees as well as a major offset from the center of the planet (Voigt & Ness, *Geophys. Res. Lett.*, **17**:1705; Connerney *et al., J. Geophys. Res.* **96**:19023). A great many authors report on aspects of the magnetosphere and plasma environment of Neptune in these two special journal issues. A comprehensive comparative study of the magnetospheres of all four jovian planets was published in 1991 by Baganel (*Ann. Rev. Earth Planet. Sci.* **20**:289)

10. PLUTO-CHARON

The outermost planet and its satellite continue to yield their secrets to studies with large telescopes, assisted by the fact that the planet passed through perihelion in 1989. Although Pluto is now the only planet not to have been visited by a spacecraft, there is currently considerable interest in the United States in flying a fast flyby that would reach the planet before increasing distance from the Sun causes its atmosphere to freeze out on the surface. A biography by Levy titled *Clyde Tombaugh: Discoverer of Planet Pluto* was published in 1991 by the University of Arizona Press, and an international meeting on Pluto was held in 1993.

A number of papers have dealt with results from the mutual occultation events of the late 1980s (e.g., Elliot & Young, *Icarus* **89**:244; Blanco *et al., Astron. J.* **101**:2262; Buie, *Icarus* **97**:211), which have yielded both fundamental constants of the system and albedo maps of Pluto. Observations with the Hubble Space Telescope (Albrecht *et al., Astrophys. J.* **374**:L65) improved the value for the separation of Pluto and Charon and hence the mass of the system, and they also hint at a difference in density between the two objects. The diameter of Pluto is now known to be 2300 km and its density is 2.1, suggesting a composition roughly 75% silicate and 25% ice. The diameter of Charon is 1200 km. The most spectrally active component of the surface of Pluto is methane, but this is only a trace constituent, with nitrogen emerging as the likely candidate for the most abundant molecule in the crust.

Although exceedingly tenuous, the atmosphere of Pluto continues to represent an area of research interest. Non-thermal atmospheric models were published by Hubbard *et al.* (*Icarus* **84**:1), while Clark *et al.* (*Icarus* **95**:173) carried out an analysis of the extended atmosphere predicting that dissociation of methane would lead to an extensive hydrogen corona; they searched for Lyman-alpha emission with the IUE but were able only to set an upper limit. A comprehensive review of the Pluto-Charon system was published by Stern in 1992 (*Ann. Rev. Astron. Astrophys.* **30**:185)

11. GALILEAN SATELLITES

Improved monitoring of the volcanic activity on Io has become possible with increases in spatial resolution achieved by infrared telescopes, which now permit the thermal emission from different "hot spots" to be measured separately. Spencer *et al.* (*Nature* **348**:618; *Science* **257**:1507) carried out a series of disk-resolved observations at better than 0.2 arcsec resolution with the IRTF that distinguished several eruptive centers during eclipses and allowed the brighter eruptions to be followed in full sunlight. McLeod *et al.* (*Astron. J.* **102**:1485) used the MMT to achieve a similar resolution; they were able to detect activity from Pele for the first time since the Voyager flybys. Such observations should provide a good baseline on Io volcanism as we approach the 1995 arrival of Galileo.

New infrared observations and laboratory spectroscopy have revealed minor constituents of the surface of Io, including both H_2O and H_2S, which are the first hydrogen-bearing molecules identified (Salama *et al.*, *Icarus* **83**:66). Ices containing 3% H_2S and 0.1% H_2O provide a good fit to observed spectra. The IUE (Ballwester *et al.*, *Icarus* **88**:1) and IRAM (Lellouch *et al.*, *Nature* **346**:639) have been used to measure the SO_2 atmosphere of Io; Lellouch *et al.* conclude that the neutral SO_2 is in equilibrium with surface frost, buffered by the colder (higher-albedo) surface traps. Structure of the extended corona of gas around Io was discussed by Schneider *et al.* (*Astrophys. J.* **368**:298), and ultraviolet observations of the Io torus from the 1990 ASTRO-1 mission were reported by Moos *et al.* (*Astrophys. J.* **382**:L105).

The extraordinary radar echoes (high reflectivities and a large amount of polarization in the opposite sense from that expected in a single reflection) from Europa, Ganymede, and Callisto are a continuing source of information on the crusts of icy satellites. Ostro & Shoemaker (*Icarus* **85**:335) discussed electromagnetic scattering for a variety of geological models of the regolith of icy objects. An alternative model including coherent backscatter was published by Hapke (*Icarus* **88**:407). New observational data from both Arecibo and Goldstone presented by Ostro *et al.* (*J. Geophys. Res.* **97**:18277) suggest that radar reflectivity may provide a crude indicator of relative silicate abundance, and also indicate first detection of several albedo features, including Galileo Regio (Ganymede) and Valhalla (Callisto).

After more than a decade of work, the Voyager images of the Galilean satellites continue to be used by planetary geologists. For example, Thomas and Squyres (*J. Geophys. Res.* **95**:19161) discussed crater palimpsests on Ganymede and offered a model that includes volcanism triggered by impact, and Shenk (*J. Geophys. Res.* **96**:15635) analyzed the cratered units of Callisto and Ganymede to find systematic differences in crater morphology.

12. TITAN

The atmosphere of Titan remains a subject of substantial interest, stimulated by plans for the Cassini mission with its Huygens entry probe. Sicardy *et al.* (*Nature* **343**:350) and Hubbard *et al.* (*Nature* **343**:353) discussed results on upper atmospheric structure revealed by the occultation of the star 26 Sag on 3 July 1989, which effectively probed an altitude range from 250 to 500 km and yielded a mesospheric temperature of 180 K. Toon *et al.* (*Icarus* **95**:24) and Cabane *et al.* (*Icarus* **96**:176) discussed physical and chemical models for Titan aerosols, and dynamical models for the ionosphere were presented by Ip (*Astrophys. J.* **362**:354).

Radar was used to distinguish among surface models of Titan on the basis of the observed reflectivity. Muhleman *et al.* (*Science* **248**:975) concluded that Titan is not covered with a deep global ocean of ethane, but that at least some areas of "dry land" are present. Surface temperatures were discussed by McKay *et al.* (*Science* **253**:1118), who considered the roles of both the greenhouse effect and an "antigreenhouse effect" in establishing thermal equilibrium at the surface.

Based on an analysis of the impact delivery of volatiles, Zahnle *et al.* (*Icarus* **95**:1) suggested that the retention of an atmosphere by Titan and the loss of atmospheres from Ganymede and Callisto (which have nearly the same mass and surface gravity) can be understood as the result of higher impact velocities in the jovian system: on the jovian satellites impact erosion dominated, while on Titan the impacting volatiles could largely be retained.

13. TRITON

Data analysis and interpretation from the 1989 Voyager-2 encounter have put Triton in lead place among the satellites for this triennial report. Much of the new work on Triton was published in special issues of *Geophys. Res. Lett.* **17**:1661-1776 (September 1990), *J. Geophys. Res.* **96**:18907-19268 (30 October 1991), and *Icarus* **99**:241-435 (October 1992). Voyager cameras yielded images with better than 1 km resolution; the global color and albedo were analyzed by McEwen (*Geophys. Res. Lett.* **17**:1765), Buratti *et al.* (*J. Geophys. Res.* **96**:19197), Hillier *et al.* (*Science* 250:419 and *J. Geophys. Res.* **96**:19203), and Thompson & Sagan (*Science* **250**:415). The diameter of Triton is 2705 km (Davies *et al. J. Geophys. Res.* **96**:1575), and its relatively high density of about 2.1 is suggestive of a composition about 75% silicate and 25% ice. The cratering history of the satellite was discussed by Strom *et al.* (*Science* **250**:437).

Perhaps the most unexpected discovery of the Triton flyby was the presence of dark plumes or geysers, rising to 8 km above the surface and then carried horizontally for more than 100 km; this discovery was initially described by Soderblom *et al.* (*Science* **250**:4979). The greenhouse effects of surface ices and their possible contribution to the energy source for the plumes were calculated by Brown *et al.* (*Science* **250**:431 and **251**:1465) and Kirk *et al.* (*Science* **250**:424), while an alternative "dust devil" model was suggested by Ingersoll (*Science* **250**:435). The surface energy balance and plume dynamics were further considered by Yelle *et al.* (*Icarus* **89**:34) and Stansberry *et al.* (*Icarus* **99**:242), while evidence for dark streaks and other associated surface features was presented by Hansen *et al.* (*Science* **250**:421). Imaging of clouds and haze in the atmosphere of Triton was described by Rages and Pollack (*Icarus* **99**:289), who concluded that clouds are present at high southern latitudes below 8 km altitude, in addition to a ubiquitous high-altitude haze of submicrometer particles.

Triton's atmospheric photochemistry was discussed by Strobel *et al.* (*Geophys. Res. Lett.* **17**:1729). Ingersoll (*Nature* **344**:315) modeled the atmosphere (primarily nitrogen) in equilibrium with surface frosts and concluded that surface temperature contrasts lead to westward winds at high altitudes. A massive early atmosphere for Triton was considered by Lunine *et al.* (*Icarus* **100**:221), created from the present inventory of volatiles at a time when the satellite was subject to substantial tidal heating. The ionosphere of Triton was the subject of reports by Ip (*Geophys. Res. Lett.* **17**:1713), Yung & Lyons (*Geophys. Res. Lett.* **17**:1717), and Majeed *et al.* (*Geophys. Res. Lett.* **17**:1721). The Triton torus and its interactions with the rest of the neptunian magnetosphere were discussed by Cheng (*Geophys. Res. Lett.* **17**:1669) and Richardson *et al.* (*Geophys. Res. Lett.* **17**:1673)

14. PLANETARY RINGS

Based in part on the extensive Voyager data set for the rings of all four jovian planets, studies of rings and ring dynamics were pursued by many researchers during this triennium. Further analysis of Voyager occultation data by Nicholson *et al.* (*Astron. J.* **100**:1339) has yielded a new absolute radius scale for the Saturn rings, while Shan & Gortz (*Astrophys. J.* **367**:350) have presented an analysis of the radial structure of the B Ring. Rosen *et al.* (*Icarus* **93**:3 and **93**:25) derived resonance structure in the Saturn rings from the Voyager radio occultation data and examined the role of a number of electromagnetic effects on ring structure. Dynamical analysis of Voyager ring data also led Showalter (*Nature* **351**:709) to the discovery of a new satellite of Saturn, named Pan (formerly 1981S13), orbiting within the Encke gap at a distance from Saturn of 133,584 km. Porco (*Science* **253**:995) proposed an explanation for the azimuthal structure in the Neptune rings (the "ring arcs" Adams, Leverrier, and Galle), attributing most of the observed structure to perturbations by the satellite Galatea.

A frequent theme of recent ring research has been the production and evolution of small ring particles that are presumably eroded from larger (often invisible) particles or moonlets. Cuzzi & Durison (*Icarus* **84**:467) studied the effects of bombardment of rings by meteoroids and concluded from calculated dynamical lifetimes that the ages of the C Ring of Saturn and the Alpha and Beta Rings of Uranus are each of order 100 million years. Dones (*Icarus* **92**:194) followed up on the implications of such short timescales and suggested a possible recent origin for the rings of Saturn through tidal disruption of a large comet. Kolvoord *et al.* (*Nature* **345**:695) analyzed periodic features in the F Ring and concluded on dynamical grounds that there must be undiscovered embedded satellites in this ring.

There is increasing interest in the dust rings of low optical depth discovered around each of the outer planets. Grün *et al.* (*Icarus* **99**:191) have reanalyzed all of the Voyager observations of the spokes caused by elevated dust in the B Ring of Saturn and, in contrast with earlier studies, find limited evidence for electromagnetic effects. For the most part the spokes rotate at Keplerian velocities, with an average lifetime of about 100 minutes. Colwell & Esposito (*Icarus* **86**:530; *J. Geophys. Res.* **97**:10227; and *J. Geophys. Res.* **98**:7387) studied the tenuous uranian dust rings discovered by Voyager and concluded that they are formed by micrometeoroid bombardment of unseen moonlet belts made of objects with diameters from about 10 cm to 1 km. They concluded that the small satellites of Uranus and Neptune are fragments or rubble pile agglomerations left over from some older, larger population of satellites. Showalter *et al.* (*Icarus* **94**:451) studied the faint E Ring of Saturn and concluded that the particles in the ring do not follow a power-law size distribution but cluster tightly about a radius of 1 micrometer. Grün (*Rev. Modern Astron.* **4**:157) has written a general review of dust rings around planets, and Showalter addressed similar topics in his paper on dust in planetary ring systems, presented at IAU Colloquium 126.

15. PLANETARY ORIGIN AND EVOLUTION

One of the primary purposes of studying individual planets and satellites is to gain insight into the processes that influence the long-term evolution of the solar system. Many aspects of solar system origin and early evolution are best approached through studies of its most primitive members: comets, asteroids, and meteorites -- all discussed in the report from Commission 15. Here we briefly review the results of cosmogonic research based primarily on the properties of the planets and satellites and upon more general issues of solar system dynamics.

Rudin & Pollack (*Astrophys. J.* **275**:740) discussed the dynamical evolution of the solar nebula, while Safranov (*Icarus* **94**:260) presented a general review of the formation of the planets in his 1991 Kuiper Prize Lecture. Korycansky *et al.* (*Icarus* **92**:234) discussed current theories for the formation of the giant planets, while Wetherill (*Science* **253**:535) argued for the critical role of impacts in the formation of the terrestrial planets. The survival of protoplanets in the inner solar system is intimately related to the general dynamical evolution of the system, and in particular to the role of Jupiter in modifying the orbits of objects in the inner solar system and in influencing the scattering of high-velocity projectiles from the outer solar system into the region of the terrestrial planets. Slattery *et al.* (*Icarus* **99**:167) modelled giant impacts on the early Uranus in order to explore conditions under which such impacts could produce the present obliquity and rotation period of the planet.

The detailed information now available on the elemental and isotopic compositions of planetary atmospheres offers a rich field for interpretation. Lutz *et al.* (*Icarus* **86**:329) discussed the expected enrichment of deuterium in the primitive ices of the protosolar nebula, in order to understand the observed D/H ratios in planetary atmospheres that may have been derived, at least in part, from these primitive ices. They concluded that D/H of 0.0001 to 0.001 should have been expected. Owen presented a more general review of the significance of deuterium in the solar system in his paper at IAU Colloquium 150. Cruikshank *et al.* (*Icarus* **94**:345) reviewed the evidence for solid C-N bearing material in a variety of primitive solar system bodies, including the D asteroids, comets, the dark side of Iapetus, and the rings of Uranus.

Chyba (*Nature* **343**:129) discussed the formation of planetary atmospheres through the balance between impact delivery of volatiles to the planets and the impact erosion of their atmospheres, showing how the relative importance of these two competing effects depends on planetary mass. Pepin (*Icarus* **97**:2) reviewed the origin and evolution of the atmospheres of the terrestrial planets through the processes of volatile mass fractionation and hydrodynamic escape, in an effort to understand the observed isotopic ratios in the atmospheres of Venus, Earth, and Mars; he also published a comprehensive review of the origin of the noble gases in the terrestrial planets in *Ann. Rev. Earth Planet. Sci.* **20**:389. Hunten (*Icarus* **85**:1) published a general review (in the form of his Kuiper Prize Lecture) of the escape and evolution of planetary and satellite atmospheres, while the Kuiper Prize Lecture by Pollack (*Icarus* **91**:173) discussed the evolution of climates in the terrestrial planets over the age of the solar system. Binzel's Urey Prize Lecture (*Icarus* **100**:274) also concerned physical evolution in the solar system, with emphasis on the smaller bodies. Finally, McKay (*Icarus* **91**:93), in his Urey Prize Lecture, considered issues of planetary evolution and the origin of life.

COMMISSION 19: EARTH ROTATION (ROTATION DE LE TERRE)

President: B.Kolaczek
Vice-president: J.Vondrak

Organizing Committee:

N.Capitaine	Jin Wen-Jing	L.Morrison
J.O.Dickey	D.D.McCarthy	D.Robertson
S.Dickman	W.G.Melbourne	T.Sasao
M.Feissel-past president	N.T.Mironov	P.Wilson

INTRODUCTION

In the term 1990.5-1993.5 the work of the IAU Commission 19 encompasses a broad range of activities with:
- the International Earth Rotation Service (IERS) (director of the Central Bureau: M.Feissel)
- the Commission 19 Working Group on Earth Rotation in Hipparcos Reference Frame (chairman: J. Vondrak) two intercommission working groups :
- on Astronomical Standards (chairman: T. Fukushima)
- on Reference Frames (chairman: Ch.de Vegt)
- the IAG/IAU Special Study Group on Rapid Earth Orientation Variations (chairman: J.O.Dickey)

and independent institutions as well as individual members of the IAU Commission 19. Their reports are shortly reviewed.

Investigations of Earth rotation, their correlations with variations of atmospheric angular momentum (AAM) and of terrestrial and celestial reference frames have been developed strongly. In addition to the development of Satellite and Lunar Laser Ranging (SLR and LLR), Very Long Base Interferometry (VLBI), the new observational technique, Global Positioning System (GPS) was introduced to routine determinations of the Earth rotation parameters(ERP) with a high accuracy of the milliarcsecond level and a high resolution of hours.Two international GPS observational campaigns GIG'91 and IGS'92 in conjunction with the SEARCH'92 were organized in 1991 and 1992 respectively. Diurnal and semidiurnal oscillations of the ERP were determined from GPS and VLBI observations, and also from SLR data and well explained by tidal effects. The influence of atmosphere, ocean and undergroundwater on high frequency variations of Earth rotation has been widely investigated and better understood. The appropriate theoretical studies were carried out in order to improve formula needed for determinations of the EOP (Earth Orientation Parameters) in a case of such highly precise observations as well as for geophysical interpretations of determined variations of Earth rotation. The nutation theory for the rigid Earth was improved and nutation observational models were worked out. In this period there were many international scientific conferences at which the new achievements in this field were presented:

1. Annual IERS Workshops held in Paris in 1991, in Mizusawa in 1992 and in Paris in 1993.
2. IAU Colloquium No 127th on Reference Systems held in Virginia Beach, VA, USA in October 1990.
3. XX IUGG General Assembly held in Vienna in August 1991 and its Symposia.
4. Chapman Conference on Geodetic VLBI: Monitoring Global Change held in Washington, DC, in April 1991.
5. The third Satellite Positioning Symposium held in Columbus, Ohio, USA, in March 1992.
6. IAU Symposium No 156 on Developments in Astrometry and Their Impact on Astrophysics and Geodynamics, held in Shanghai, China, in September 1992.
7. IAG Symposium No 112, 7th International Symposium Geodesy and Physics of the Earth, Potsdam, Germany,held in October 1992.
8. Annual Meetings of the European Geophysical Society in 1991, 1992, 1993 held in Wiesbaden, Germany and Edinburgh, Scotland.
9. Third Orlov's Conference on Investigations of the Earth as a Planet by Astronomical, Geodetical and Geophysical Methods, held in Odessa, Ukraine, in September 1992.
10. International Workshop for Reference Frame Establishment and Technical Development in Space Geodesy (IRIS'93) held in Tokyo, in January 1993.
11. The 1993 GPS Workshop, held in Bern, Switzerland, in March, 1993.
12. Spring and Autumn Meetings of the American Geophysical Union in 1990-1993, USA.
13. World Space Congress 92, The Orientation of the Planet Earth as Observed by Modern Space Techniques, held in Washington, DC,USA,1992.

About 200 papers concerning Earth rotation as well as terrestrial and celestial reference frames are indicated by the Astronomy and Astrophysics Abstracts, Vol. 51-55, from the considered period of time.

J. Bergeron (ed.), Reports on Astronomy, Vol. XXIIA, 169–183.

REPORT OF THE INTERNATIONAL EARTH ROTATION SERVICE (IERS)

INTRODUCTION

The IERS, a member of the Federation of Astronomical and Geophysical Services (FAGS) was created in 1988 by the International Union of Geodesy and Geophysics (IUGG) and the International Astronomical Union (IAU). It maintains: a terrestrial reference frame, a celestial reference frame, and it monitors the Earth Orientation Parameters (EOP), which relate the two frames as a function of time. The IERS plays a key role in the coordination of the observations and of their analysis, setting up and making available the indispensable basic references to the scientific community. Nowadays, the global observing activities involves VLBI, LLR, GPS and SLR with worldwide networks totaling 150 sites in 1992. The analyses performed by over 30 groups in 10 countries provide the daily monitoring of the Earth's rotation irregularities, the determinations of terrestrial reference frames, crustal motions, as well as the unification of celestial reference frames used in space geodesy and in space navigation with the directions of the extragalactic quasars. Most of the results are accurate and precise at the level of 0.0003", corresponding to 1 cm on the surface of the Earth. While these high accuracy results are used every day in space research programs such as orbit maintenance, space oceanography, space navigation, they also provide unique material for the study of the Earth's interior, crustal deformations, or global dynamics of the oceans and atmosphere, with time characteristics ranging from hours to centuries.

The IERS consists of Directing Board (DB), Central Bureau (CB) with three sections of Terrestrial Frame, Celestial Frame and Earth Orientation, two sub-bureaus(SB) for Rapid Service and Predictions (NEOS) and Atmospheric Angular Momentum (SBAAM), Coordinating Centers (CC) for each of the present observing techniques: VLBI, LLR, SLR and GPS. The IERS is supported by many organizations that contribute to the tasks of observations and data processing. There are about 90 corresponding members of the IERS. The IERS structure and responsibilities were described in details in the previous volume of the IAU Transaction Vol. XXIA,169-178pp and in the IERS Annual Reports. In 1992 the chairman of the DB was changed on the proposition of Dr. K.Yokoyama, the chairman of the DB in the term 1988-1992. Dr.Y. Yatskiv was elected for the next triennial by the DB. The DB met in the last term five times. The IERS Workshops were held each year: at the Observatoire de Paris in 1991 and 1993 and at the National Astronomical Observatory in Mizusawa in 1992. At the IERS Workshop the reports of the IERS activities and the recent achievements of determinations and investigations of Earth rotation and nutation variations were presented and future programs were discussed.

The IERS published regularly weekly Bulletin A, Monthly Bulletin B, Annual Report, Special Bulletin C,D and Technical Notes. The following IERS Technical Notes were published in the last triennial period:

No 8 - Earth orientation and reference frame determinations, atmospheric excitation functions, up to 1990, (Annex to the IERS Annual Report for 1990)",IERS Analysis Centres for VLBI,LLR,SLR,AAM
No 9 - ITRF 90 and other realizations of the IERS Terrestrial Reference System for 1990, C.Boucher, Z.Altamimi
No10 - The IERS GPS Terrestrial Reference Frame, C.Boucher, Z.Altamimi
No11 - Earth orientation, reference frames and atmospheric excitation functions submitted for the 1991 IERS Annual Report, VLBI, LLR, GPS, SLR and AAM IERS Analysis Centres, P.Charlot (ed.)
No12 - ITRF 91 and its associated velocity field, C.Boucher, Z.Altamimi, L.Duhem
No13 - IERS Standards (1992), D.D.McCarthy (ed.)

The mailing list of the IERS publications includes 600 addresses in 65 countries. The main results are also available through electronic means.

THE IERS CENTRAL BUREAU ACTIVITY

The IERS Standards (D.D.McCarthy -NEOS, Washington, D.C. USA)

The IERS reference System is composed of the IERS Standards, the IERS terrestrial and celestial reference frames and the corresponding series of the Earth Orientation Parameters (EOP) which are consistent with one another at the milliarcsecond level.

The IERS Standards consist of a set of constants and models used by the IERS. The new revised edition of the IERS Standards (1992) was completed by D.D.McCarthy and published in the IERS Technical Note No13. It replaced the IERS Standards published in 1989 and contains 14 chapters covering the following topics:
1. Numerical Standards, 2. Celestial Reference System, 3. Conventional Terrestrial Reference Frame, 4. Lunar and Planetary Ephemerides, 5. Transformation between the Celestial and Terrestrial Systems, 6. Geopotential, 7. Solid Earth Tides, 8. Ocean Tide Model, 9. Local Site Displacement, 10. Tidal Variations in the Earth's Rotation, 11. Tropospheric

Model, 12. Radiation Pressure Reflectance Model, 13. General Relativistic Models for Time, Coordinates and Equations of Motion, 14. General Relativistic Models for Propagation.

The IERS Terrestrial Reference Frame
(C.Boucher - CB Terrestrial Frame Section, Institut Geographique National, Paris, France)

The IERS terrestrial system is made geocentric through the use of LLR, GPS and SLR observation techniques. Its realization also includes VLBI results, which enables the tie of the terrestrial reference frame to the extragalactic reference frame. A key aspect of this combination is the collocation of observing stations of the various techniques operated in common sites, either in a permanent mode or by repeated occupations. A large proportion of the involved observing sites participate in long range Earth orientation monitoring programs, which strengthens the control of the long term deformations of the Earth's crust. The IERS Terrestrial Reference Frame (ITRF) includes of about 150 site positions derived from the four observing techniques, and site velocities derived from SLR and VLBI data. The velocity field has no rotation relative to the geological plate motion model NNR1-NUVEL. After the ITRF0, the initial IERS-ITRF, new improved systems were worked out each year: the ITRF90, 91 and , 92. They are described in detail in the IERS Technical Notes.The ITRF92 contains in total 281 stations located in 152 sites spread over 8 tectonic plates. Their position and velocity accuracies are of the order of 1-3 cm and 1-5 mm/year respectively.

The IERS Celestial Reference Frame (J.F.Lestrade - CB Celestial Frame Section, Observatoire de Paris, France)

The IERS Celestial Reference Frame (ICRF) is based on the directions of compact extragalactic radio sources observed with VLBI by several national and international programs. It is extended and improved as new information becomes available, while maintaining the stability of its reference directions within 0.1 milliarcsecond. The ICRF90,91,92 were worked out. The origin of the reference system is at the barycentre of the Solar System. Its polar axis and origin of right ascensions are consistent with the FK5 System at J2000.0, and the accuracy of their realization is better than 0.01", a two orders of magnitude improvement relative to the ground-based stellar realizations. Its tie with the dynamical frame used in space navigation is known within a few milliarcseconds, thanks to the participation of Lunar Laser Ranging together with VLBI in the Earth rotation program. Its tie with the galactic frame established on the Hipparcos mission as well as the time evolution of the tie will be known within the uncertainty of the Hipparcos frame itself. The core of the most regular observed objects in the ICRF, included in VLBI Earth rotation monitoring programs,insures an excellent tie of the satellite geodesy systems, through the unification of terrestrial frames and the Earth rotation determinations. The characteristics of the ICRF are summarized in Table 1.

Table 1. The IERS Celestial Reference Frame, ICRF91: 422 objects.
1a - The extragalactic frame, observing techniques: VLBI. Stability of the reference axes: ± 0.0001"

Type of object	%	Total flux (Jansky)	%	Redshift	%	Precision (0.001")	%
Quasars	68	0-1	29	0-1	47	<0.5	61
Galaxies	6	1-2	45	1-2	36	0.5< <1.0	21
BlLac	12	2-4	18	2-3	13	1.0< <2.0	9
Other	14	>4	8	3-4	4	>2.0	9

1b - Tie with other celestial frames

Scale	Observing techniques	Objects	Precision of the tie
Solar System Galaxy	Lunar Laser Ranging Hipparcos	Earth, Moon 11 radio stars	± 0.005" ± 0.001"; ± 0.0005"/y

The IERS Earth Orientation Parameters (M.Feissel - CB, Earth Orientation Section, Observatoire de Paris, France)

The spectrum of the Earth's rotation irregularities is extremely broad, from hour to centuries, reflecting the free and forced oscillations of the various layers of the planet. Most of the irregularities are aperiodic and poorly predictable, hence the need of monitoring their variations. At present they are represented by five Earth Orientation Parameters (EOP), four to describe the motion of the rotation axis relative to the Earth crust (x,y) and the inertial frame described by the present theories of the precession and nutation ($d\psi$, $d\epsilon$), and one (UT1) to describe the angular rotation of the Earth. The EOP series are determined by the IERS Analyzing Centers yearly with taking into account new observations. Detail information about these EOP series are published in the IERS Annual Reports. The main features of the IERS series of EOP are summarized in Table 2.
In 1993 the following number of EOP series were determined by the IERS Analyzing Centers:

VLBI: **14 series** (GIUB, SFC, JPL, NOAA, OPA, SHA, USNO)
LLR : **3 series** (JPL, SHA, UTXMO)
SLR : **13 series** (CLG, CSR, DUT, GAOUA, GFZ, GSFC)
GPS : **6 series** (CODE, CSR, EMR, ESOC, JPL, SIO)
COMBINE : 3 series (JPL, NEOS, IERS)

Glossary:

CODE Centre for Orbit Determination in Europe, Switzerland
CSR Center for Space Research, University of Texas, USA
DUT Delft Univ. of Techn., The Netherlands
EMR Energy, Mines and Resources, Canada
ESOC European Space Operation Centre
GAOUA Main Astron.Obs.of Ukraine Acad.of Sciences, Ukraine
GFZ GeoForschung Zenter, Potsdam, Germany
GIUB Geodetic Institute of the University of Bonn,Germany
GSFC Goddard Space Flight Center, MA, USA
IERS International Earth Rotation Service, Paris, France
JPL Jet Propulsion Laboratory, CA, USA
NEOS National Earth Orientation Service, Washington DC, USA
NOAA National Oceanic and Atmospheric Administration, MA ,USA
OPA Observatoire de Paris, France
SHA Shanghai Observatory, China
SIO Scrips Institution of Oceanography, CA, USA
USNO U.S. Naval Observatory, Washington,USA
UTXMO Dept. of Astronomy, Univ. of Texas, USA

Table 2. The IERS Earth Orientation Parameters.

Period (years)	PM, UT,CP	Observing technique contribution						Times resolution	Precision
		Opr.Ast.	LLR	Doppler	SLR	VLBI	GPS		
1900-1961	PM	*						month	0.05"
1962-1972	PM	**						weeks	0.02"
	UT	**							0.002 s
1972-1979	PM	**		**				week	0.01"
	UT	**	**						0.001 s
1980-1983	PM				***			week	0.002"
	UT		*		**	*			0.0004 s
1984-1991	PM				**	**		days	0.0005"
	CP					****			0.0005"
	UT				*	***			0.00005 s
1992	PM				**	**	**	day	0.0002"
	CP					****			0.0005"
	UT				*	***	*		0.00005 s

Notes. The number of * roughly reflect the level of contribution(precision and density of measurements).
PM: polar motion (x,y); UT: universal time; CP: celestial pole coordinates($d\psi \sin\epsilon$,$d\epsilon$)

THE IERS SUBBUREAUS ACTIVITY

The IERS Sub-bureau for Rapid Service and Predictions (D.D.McCarthy - NEOS, Washington, D.C. USA)

The U.S. Naval Observatory (USNO), as a partner in the National Earth Orientation Service (NEOS) with the National Oceanic and Atmospheric Administration (NOAA), continued to function as the Sub-bureau for Rapid Service and Prediction of the International Earth Rotation Service (IERS) publishing IERS Bulletin A weekly, NEOS Annual Report and distributing information by mail, electronic mail and computer bulletin board. The prediction of the correction to the current nutation theory was modified to account for the most recent observations. Predictions of polar motion, UT1-UTC, nutation were provided routinely. Weekly analysis of Earth orientation data derived from GPS orbits was begun by D.D. McCarthy. The information is provided to the International GPS Service (IGS) for distribution to its users. The GPS data were found to make a valuable contribution to the combination with other contributed data, and to improve the rapid service estimates and the predicted Earth orientation parameters. B.J. Luzum designed software to combine the different GPS Earth orientation series to provide a single GPS combination series. A special prediction file to be used by all participating in the IGS is now being produced routinely. The study of the application of atmospheric angular momentum data to the estimation of Earth orientation was continued. Corrections to the main zonal tide components in the variation of the length of day were derived in the process and used to derived corrections to the zonal tidal coefficients for UT1-UTC.

The IERS Sub-bureau for Atmospheric Angular Momentum of the IERS
(Reported by D.A.Salstein, Atmospheric and Environmental Research Inc.(AER),Cambridge, Mass.USA; D. Kann, Research and Data Systems Corp. Greenbelt,M.D.USA)

Since its start date of 1 October 1989, the operations of the Sub-bureau for Atmospheric Angular Momentum (SBAAM) have steadily come to maturity with regular inclusion of atmospheric excitation function from four meteorological data centers. Such functions relate directly to excitations of variations in length of day and polar motion of the Earth. The SBAAM, housed at the Climate Analysis Center of the U.S.National Meteorological Center (NMC) in Washington,D.C., calculates parameters based on wind and surface pressure analyses and forecasts from the NMC global model. Additionally, similar values are received via rapid transmission from the United Kingdom Meteorological Office, the European Centre for Medium-Range Weather Forecasts, and the Japan Meteorological Agency. The parameters are currently produced as frequently as four time per day. Data back to 1976 are available as well, at daily or twice-daily resolution. Studies based on geodetic and SBAAM atmospheric data have addressed agreement between variations in the momentum of solid Earth and atmosphere occurring on a range of time scale from interannual to interseasonal, with evidence for significant coherence between the two as short as 8 days. The Sub-bureau has studied the differences amongst the function produced by the various meteorological centers, which tend to agree down to even shorter time scales. Forecasts of angular momentum quantities out to ten days are also produced by the SBAAM and their skill has been estimated to be positive during these lead time.Such atmospheric data have already been used successfully in forecasts of length of day. Additional quantities produced by the SBAAM include values of zonal mean winds, which are a regional basis for the global axial angular momentum parameter, as well as zonal mean temperature fields. Also the impact of the global ocean is included through computations of pressure-based excitation values modified by the so-called inverted barometer (IB) approximation. Using the IB modification, the distribution of ocean mass responds to atmospheric pressure forcing, with the pure IB response implying a drop in local sea level of 1 cm per millibar rise in overlying pressure. Other parameters produced by the SBAAM include global mean surface pressure, and a number of low-order harmonics of surface pressure, useful to both geophysical studies and those of satellite orbit determination.
Angular momentum data are distributed to an international group of scientists for atmospheric and geodetic studies both directly from the Sub-bureau as well as from archives maintained at NMC and AER. Further information can be found in Salstein *et al.*, 1993, Bulletin of the American Meteorological Society, **74**, 67-80.

THE IERS COORDINATING CENTERS ACTIVITY

Satellite Laser Ranging - SLR (B.E.Schutz, Center for Space Research (CSR), Austin, Texas, USA)

In October 1992, Lageos-II was launched, an important addition to the existing set of SLR satellites used for Earth rotation and reference frame studies, namely, Lageos-I, Etalon-I and Etalon-II. The performance of the global network of SLR stations has been good, with about the same number of stations (32) contributing about 2 100 passes on both Lageos-I and Lageos-II in the same period. Lageos-II has an inclination of 53 degrees compared to 110 degrees for Lageos-I and both satellite have an altitude of about 5900 km. Tracking of the Etalon satellites has, unfortunately, not yet reached a level that would make these satellites regular contributors to Earth rotation. On the average, the Etalon tracking was slightly more than one pass per day. During 1992, scheduling modifications were recommended to allow partitioning of an Etalon pass into periods when other satellites can be tracked. The full implementation of these considerations has not been made at all stations. SLR data from the past decade have been a significant contributor to new gravity fields, for example, JGM-1 and JGM-2 prepared as a joint activity of the NASA Goddard Space Flight Center and the University of Texas at Austin. Data from Lageos-II is expected to further "tune" the gravity field and produce a new field for Earth rotation and reference frame applications.

Very Long Base Interferometry - VLBI (W.E.Carter - National Geodetic Survey -NGS; Gravity, Astronomy and Satellite Branch, USA Dept.of Commerce, Rockville MD,USA)

Since mid-1990 the number of VLBI observations has increased significantly, including in the southern hemisphere where VLBI facilities had been very sparse. In chronological order: the Italian Space Agency (ASI) built a new geodetic VLBI facility at Matera, Italy, which began operations in 1990; a Japan facility at Syowa, Antarctica, made its first observations in 1990; a format NASA tracking antenna at Santiago, Chile, was converted to geodetic VLBI use in 1991, through the joint efforts of Institute of Applied Geodesy (IFAG), Germany, the National Aeronautics and Space Administration (NASA) and the National Oceanic and Atmospheric Administration (NOAA), United States, and Center for Space Studies, University of Chile; IFAG established a new observatory at the Chilean Antarctic station O'Higgins during 1992; the Canada Center for Surveying upgraded the Algonquin Park station and began participating regularly in international Mark III observing sessions in 1992; and NOAA, the University of Sao Paulo, and the Brazilian Space Agency (INPE) made the first successful observations from a jointly operated new facility near Fortaleza, Brazil, in April, 1993.Work is continuing on the Russian (former USSR) Quasar network and joint Institute for Applied Astronomy, Russia and NASA experiments are planned for mid 1993. A new observatory is also under construction at Ny Alesund, Spitzbergen, by Statens Kartverk, Norwegian Mapping Authority. In the United States, the Mojave station was closed in July 1992, and the Richmond station was destroyed by Hurricane Andrew in August 1992. However, the U.S. Naval Observatory has nearly completed a new 20m telescope at Kokee Park, Hawaii, and similar telescopes will be built at Green Bank, West Virginia, and Richmond, Florida, within the next year. The 10-station Very Long Baseline Array (VLBA) network will become fully operational during 1993 with observatories spanning North America, from the Virgin Islands to Hawaii. When fully operational these new facilities will provide a global network of reasonably well distributed VLBI stations to define and maintain both the terrestrial and celestial reference frames. The new observations in the southern hemisphere have already been especially significant in expanding the celestial reference frame. Between 1990 and 1992, the IERS Celestial Reference Frame grew from 228 to 422 sources, much more uniformly distributed about the sky. The coordinates of 81 sources observed in the IERS programs, ranging in declination from 78 degrees north to 80 degrees south, have formal errors of a few tenths of a millisecond of arc (Robertson et al., 1993; A.J.,105,1,3353-3358).
Haystack Observatory is developing a next generation Mark IV VLBI data acquisition system, with joint support from NASA, NOAA, USNO and IFAG. The Mark IV system will have double the sensitivity of the Mark IIIA system by increasing the maximum data rate to 1 024 Mbits/sec. In addition, the Japan Communications Research Laboratory is developing an upgraded K4 and the Canadian Institute for Space and Terrestrial Science is developing the highly transportable S2 system, both of which use rotary-head cassette recorder systems.

Two large meetings with major VLBI emphasis have been held since mid-1990. The "Chapman Conference on Geodetic VLBI: Monitoring Global Change" was sponsored by NOAA and held in Washington, DC on 22-26 April 1991. The Japanese Science and Technology Agency held the "International Workshop for Reference Frame Establishment and Technical Development in Space Geodesy" (IRIS'93) in Tokyo on 18-21 January 1993. Proceedings volumes have been published for both meetings.

Dong and Herring (1990; EOS,71,482) reported unambiguous detection of EOP variations in the diurnal and semidiurnal bands due to oceanic tidal effects. The variations had been predicted a decade earlier by Yoder et al.(1981; JGR,86,881-891). Subsequent VLBI, GPS and SLR analyses have supported the Dong and Herring results and routine EOP analyses have begun to take these effects into account.

Global Positioning System - GPS (W.G.Melbourne - Jet Propulsion Laboratory, Pasadena, CA, USA)

During the past decade over two orders of magnitude in accuracy has been achieved in the application of GPS to Space Geodesy. The period 1990-1993 has been a time of expanded utility of GPS geodetic systems to a broader spectrum of scientific applications, not only because of their potential for high accuracy, but also for their technical and economic capability to sharpen spatial and time resolutions and to complement established space geodetic techniques. The figure shows baseline repeat abilities reported over the last decade by a number of investigators for a wide range of baseline lengths.

In 1990 the IERS Directing Board set up an interim Technique Coordinator for GPS to evaluate its potential for EOP applications and to encourage broader participation in the use of the technique.The GIG'91 campaign was a first attempt of coordinating,on a international basis,the efforts of a number of participants in developing a globally distributed network of GPS tracking stations. The campaign was executed over a three-week period in Jan-Feb 1991. Its primary purpose was to collect tracking data with sufficient strength to evaluate the accuracy of the GPS ephemeris products, the tracking station locations and the EOP information. A subsequent workshop sponsored by "Institute fur Angewandte Geodasie" in Ahrweiler, Germany, in Aug. 1991, verified the expected performance of GPS. There are now a total of eight GPS Analysis Centers contributing on a regular basis their EOP&TRF results to the IERS Central Bureau. As results of these IERS activities and partially out of recognition of the enormous potential of GPS, if properly coordinated, a small working group was formed in early 1990 with the goal of establishing a permanent GPS service under the IAG auspices for scientific applications (see the paragraph on the International GPS Service for further information). The IERS participated in the organization of the highly successful summer-long campaign IGS'92 to further evaluate GPS performance but also to evaluate the ability of the participating organizations to support on a permanent basis the requirements of the Service. An additional activity was the SEARCH'92 campaign, which coordinates these GPS activities with those from SLR and VLBI to support short period Earth rotation studies. The 1993 GPS Workshop reviewing results of the GPS'92 Campaign was held at the Technical University in Berne, in March 1993. The Berne meeting also provided the opportunity to establish agreements with the IERS regarding IERS/IGS interface questions and respective areas of responsibility. In effect, the IERS will depend on the IGS for operation of the GPS network and delivery of GPS data products. The IGS will use IERS Standards and depend on the IERS for EOP and Terrestrial Reference Frame information.

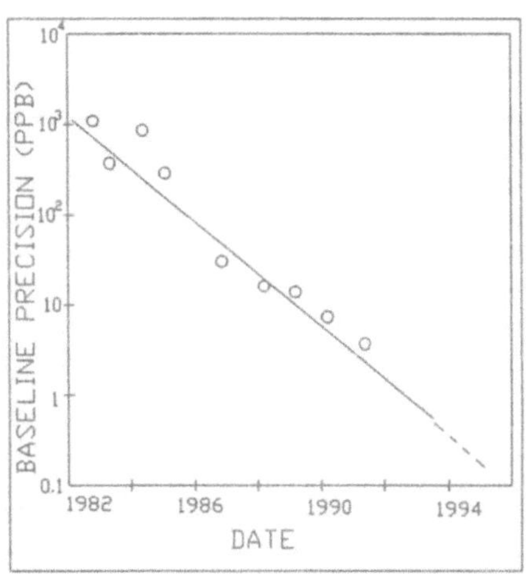

REPORTS OF THE WORKING GROUPS

IAG/IAU Special study group on rapid Earth orientation variations
(chairman: J.O.Dickey, JPL, Pasadena, USA)

Special Study Group, Rapid Earth Orientation Variations 5.143, a joint International Association of Geodesy and International Astronomical Union (IAG/IAU) SSG, has been formed to (1) interface with the IERS in the determination of rapid variations in Earth rotation by the space geodetic techniques; (2) advocate for the best possible auxiliary data from geophysical, oceanographic and atmospheric sources; (3) encourage improvements in measurement techniques (including geodetic, atmospheric , oceanographic and geophysical); and (4) encourage cooperative multidisciplinary studies.

A special measurement campaign, SEARCH'92 (Study of Earth Atmosphere Rapid Changes) was held during the Summer of 1992 (June 21 - September 22, 1992) in conjunction with the GPS experiment sponsored by the International GPS Service; the coordination was effected through the IERS. The campaign involved all space techniques and the best available complementary geophysical, oceanographic and atmospheric data were obtained. Spacial efforts were successful in obtaining 6-hourly AAM determinations as well as the routine calculation of atmospheric torques. Plans are now being planned for a complementary experiment to be held in Northern Hemisphere in the winter of 1995. The scientific benefits to be obtained from these campaigns include increased understanding of the properties and origin of short-period fluctuations in the Earth's orientation, improvements to the tidal model at sub-monthly periods, and improved ability to predict changes in the Earth's rotation up to a month in advance. Several sessions on this topic and of the SSG 5.142 have been held in conjunction with scientific conferences. Minutes of the SSG5.142 meetings were printed and distributed.

IAU Commission 19 Working Group on Earth rotation in Hipparcos Reference Frame
(Chairman: J.Vondrak, Astronomical Institute, Prague, Czech Republic)

The Working Group has continued in preparing the new global solution of Earth Orientation Parameters from optical astronomy since the beginning of the century, referred to the Hipparcos catalog now under preparation (Capitaine,1991; Vondrak,Ron, 1993). Individual star observations will be used, instead of group means used in the past, and a number of additional parameters will be estimated from a single adjustment.The main effort was on two directions:

i) Collection of the observations from the observatories and their preliminary analysis; ii) Preparation of algorithms and softwares to be used and their tests.

The final list of observatories selected to participate comprises 55 different instruments working at 37 observatories in 19 states. At present, about 50% of data is collected at the Astronomical Institute in Prague and the data flow proceeds well. The data files are analyzed for possible systematic effects and reformatted into standard formats, according to the types of instruments. The search for systematic effects allow to detect the systematic deformations of the almucantar of the order of 0.1" in Paris and Chile and significant color and magnitude effects of the same order for astrolabe anf transit instruments which have to be adjusted as a new parameters in the global solution (Hefty, 1991; Bougeard, 1992, 1993; Peek et al., 1993). The problems of refraction have been studied in Poland (Kurzynska, 1991) in order to improve the models of refraction originally used at the observatories to reduce their data in the past and to recalculate the observations in a unique manner. A relatively simple refraction formula was derived on the basis of using both the Laplace formula and a numerical integration. To use it, meteorological data changes during each night are needed (temperature, air pressure and also air humidity).

Statistical studies have been made in France (Bougard et al.,1991) in order improve modeling, detection and separation of perturbing effects, using the most recent findings of mathematical statistics. The main effort was concentrated on three topics:
a) multivariate statistical procedures to detect and separate sources of perturbing effects in optical observations,
b) a posteriori (linear or nonlinear) modeling of magnitude color and sidereal time effects in the reduction model,
c) elaboration of numerical algorithms for robust fitting on large data sets.

The problem of simultaneous determination of polar motion and UT1-UTC with celestial pole offset was studied. It was found that the use of celestial pole offset has certain advantage over classic z and w terms used in the past (Vondrak et al., 1993). The time evolution of the accuracy expected from the prepared solution was estimated from the known precision and distribution of the observations in space and time. Polar motion of a 5-day solution can be known to ±0.030" in the first half of the century and ± 0.009" after 1960, universal time can be determined to ± 0.0008 s after 1960. The expected accuracy of the celestial pole offset is of the order ± 0.02" - ± 0.08" at the beginning of the century and ± 0.01" during the last two decades.

The prepared software has been tested first with simulated data and then with the ILS observations between 1900 and 1978, using the orginal star catalog. The results proved that the proposed algorithms are capable of providing new interesting information on EOP (Vondrak, 1993).

The WG met three times at the period covered by this report; at the "Journees 1991 Systemes de Reference Spatio-Temporels - Geodynamique Globale and Systemes de Reference" (Paris, June 1991), at the IAU Symposium N.156 "Developments in astrometry and their impacts on astrophysics and geodynamics"(Shanghai, September 1992) and during the VIII Lohrmann Colloquium "Geodetic Astrometry"(Dresden, March 1993).

IAU Intercommission Working Group on Astronomical Standards
(Chairman: T. Fukushima, Hydrographic Department of Japan, Tokyo, Japan)

Since its establishment in 1991, the IAU WG on Astronomical Standards has worked actively through e-mail based correspondences and by issuing 3 questionnaires on the given themes: the way of electronic distribution, the collection of standard softwares and constants, and answering some questions on time. The responses seem to support the idea of establishing an international center plus a board of reviewing for the maintenance of "IAU Standards", which will be liked to the astronomical version of the IERS Standards. The differences between the IAU and IERS Standards are not only in their scopes. The IAU Standards will be distributed in a machine-readable form. The test collection of standard softwares was held in the first half of 1993 with a focus on the routines computing the precession in FORTRAN. The diversity of interfaces of submitted routines has raised a serious question on the policy of creating IAU Standards; whether we adopt a standard interface or not.

IAU Intercommission Working Group on Reference Frames
(Chairman: Chr. de Vegt, Hamburg Observatory, Germany)

Working Group on Reference Frames focus its work on preparation of a revised list of candidate extragalactic sources as well as a list of radio stars and stars for use with the new optical astrometric interferometer (USNO+NRL). They will be discussed in March 1994 and a preliminary report will be available at the IAU General Assembly 1994.

THE INTERNATIONAL GPS GEODYNAMICS SERVICE (IGS)

In March 1990, at the IAG Executive Committee meeting in Paris the IAG Planning Committee for the IGS was formally established. The call for participation was issued 1 February 1991. More than 100 proposals made it possible to structure the observation network (30 core stations equipped with precise P-Code receivers and efficient data links), and to organize the data flow (data centers on different levels) and processing. At the XX-th General Assembly of the IUGG in Vienna, in August 1991, the Planning Committee was reorganized and named the IGS Campaign Oversight Committee; The IUGG resolution No 5, recommending that the concept of the IGS be explored over the next four years, was adopted. More information about the early phase of IGS may be found in the papers: (Mueller,1992), (Mueller and Beutler, 1992), (Beutler, 1992).

The 1992 IGS Test Campaign (21 June - 23 September) and the intensive observation campaign called Epoch'92 were organized by the IGS Campaign Oversight Committee. The declared goal of the three-month 1992 IGS Test Campaign was the routine production of accurate GPS orbits and determinations of the rapid Earth rotation. IGS Epoch Campaigns will be organized about every second year. The primary goal is a first densification of the sparse IGS Core Network. At the 3rd IGS Oversight Committee meeting, in October 1992 at Goddard Space Flight Center, it was decided to establish the IGS, PILOT SERVICE to bridge the gap between the very successful 1992 IGS Campaign and the start of the routine IGS service. The routine service should start on 1 January 1994. The primary goal of the International GPS Geodynamics Service (IGS) is to give the scientific community high quality GPS orbits on the regular basis and related information like Earth rotation parameters, and to perform regional or local GPS analyses without further orbit improvement.

The 1993 GPS Workshop reviewing results of the GPS'92 Campaign was held at the Berne Technical University in March 1993 (Beutler, Brockman, 1993). At this meeting the agreement with the IERS regarding IERS/IGS interface questions and responsibilities was established.

REPORTS OF INSTITUTE *

This section summaries the reports received from institutes and IAU Commission members. The Jet Propulsion Laboratory, the U.S.Naval Observatory, the National Astronomical Observatory in Mizusawa, the Shanghai Astronomical Observatory have carried on wide investigations of Earth rotation and are reviewed first.

Jet Propulsion Laboratory, Pasadena, CA, USA (reported by J.O.Dickey). JPL has been actively involved in many areas which briefly can be outlined as follows:

o the determination of constants and the development of models and ephemerides for use by the community and the IERS;
o reference frame studies: 1) establishment of the JPL Radio Frame; 2) the establishment of the Dynamical Reference Frame of the Lunar/Planetary Ephemerides; 3) determination of ties between the various reference systems; 4) development of the concept of the dynamic equinox as a reference point; 5) development of the fiducial-free approach using GPS data and the realization of an independent GPS terrestrial reference frame tied to the IERS.
o acquisition, reduction and analysis of VLBI data, including the TEMPO program, for Earth rotation, precession, and nutation studies;
o reduction and analysis of lunar laser ranging (LLR) data for Earth rotation, precession and nutation studies;
o acquisition, reduction and analysis of GPS data in Earth rotation and reference frame studies;
o intercomparisons of Earth rotation data of the various techniques and combination of them with a Kalman filter;
o analysis of the scientific implications of these measurements.

Earth rotation and polar motion studies have clearly demonstrated that an unprecedented accuracies (0.3 mas for polar motion and 0.03 ms for UT1) are now being achieved by the modern space geodetic techniques using high-quality estimates of the atmospheric excitation of Earth rotation and polar motion. It was shown that changes in the (LOD) at seasonal and higher frequencies are dominated by the exchange of angular momentum between the atmosphere and the solid Earth except at tidal periods and are coherent down to 8 days, with lack of coherence at shorter periods caused by the deckling signal-to-measurement noise ratio in both the AAM and LOD data types (Dickey, Marcus, Steppe and Hide, 1992). Using the high resolution GPS measurements along with the non-axial components of the (AAM) chi-functions computed from general circulation models of the atmosphere, it has been shown that atmospheric wind and pressure fluctuations were largely responsible for exciting the rapid polar motions observed during the three-week period of the GIG'91 measurement campaign (Gross, Lindqwister, 1992).

Since 1992, JPL has been routinely producing daily estimates of polar motion with 0.3 mas accuracy using data from a 40-station global network of GPS receivers (Zumberge, Jefferson, Blewitt, Heflin, Webb, 1993).

---

* List of References indicated in these reports can be obtained on request from the IAU Commission 19. Some of them can be found in the Astronomy and Astrophysics Abstracts.

An expression was first derived which relates the location of the Celestial Ephemeris Pole (CEP) to that of the rotation pole which is reported by observing programs. The Liouville equation was rewritten in terms of the location of the CEP (Gross, 1992). This newly derived equation can then be used to directly interpret polar motion observations in terms of its geophysical excitation mechanisms (see also A.Brzezinski, at the paragraph of the Space Research Center, Warsaw, Poland).

Subdaily Earth orientation changes have been detected by both the GPS and VLBI observing programs at JPL and have been theoretical interpreted in terms of ocean tidal effects on the Earth's rotation (Lichten, Marcus, Dickey, 1992; Freedman, Ibanez-Meier, Dickey, Lichten, Herring, 1993; Lichten, Marcus, Dickey, 1992; Dickey, Marcus, Lichten, 1992; Sovers, Jacobs, Gross, 1993). The angular momentum carried by the ocean tidal current and sea level height variations can be generated with the angular momentum of the solid Earth, thereby affecting the solid Earth's rotation. The theoretical ocean tidal angular momentum results of Seiler were used to compute the theoretical effect of ocean tides on the Earth's rotation, accounting for the yielding of the solid Earth and the presence of the Free Core Nutation in the diurnal tidal band. These theoretical Earth orientation changes at semi-diurnal and diurnal tidal frequencies agree better with the polar motion observations than they do with the UT1 observations (which can differ with the theoretical predictions by up to a factor of two in amplitude), including the need for further refinements of the ocean tidal models.

JPL has explored the source of the intraseasonal "40-50 day" oscillation by analyzing LOD data in concert with AAM results from both operational analyses and the UCLA Atmospheric General Circulation Model (Dickey, Ghil and Marcus, 1991). Findings indicate that two intraseasonal oscillations exist in the Earth-atmosphere system: a tropical 50-day oscillation, associated with the convectively driven Madden-Julian wave and a mid-latitude 40-day oscillation associated with the interaction of non-zonal flow with topography. On the seasonal time periods, the contributions of the global oceans and in particular that of the Antarctic Circumpolar Current to Earth rotation were investigated (Dickey, Marcus, Hide, Thompson, Johns, 1993).

JPL has linked the El Nino /Southern Oscillation (ENSO) phenomenon to LOD variation. Significant variability of the LOD and Southern Oscillation Index (SOI) is centered at $\sim$ 4.2 years (traditional ENSO cycle and a distinct quasi-biennial component (at $\sim$2.3 years) is observed. During the 1982-1983 ENSO event 90% of the LOD variance can be explain by the atmosphere. Examining the latitudal redistribution of angular momentum within the atmosphere, we discovered a slow, global-scale poleward AAM propagation that originates in a quatorial regions, where westerly anomales lead the main ENSO anomalies by nearly two years. Coherence sygnals penetrate to latitudes higher than 60° in both hemispheres, where they lag behind the ENSO cycle by four years (Dickey, Marcus, Hide, 1992). High correlation between LOD and SOI is obtained from about 1930 to the present, indicating that LOD can be used as a proxy-index for interannual variations in global wind system after 1930 (Dickey, Marcus, Hide and Eubanks, 1992, 1993).

High-quality observations of the Earth's rotation are being combined using a Kalman filter in order to produce a suite of Earth rotation series of varying duration and temporal resolution (Gross, 1992). Singular Spectrum Analysis (SSA) is being applied to these series in order to separate the meaningful signal from the noise prior to interpreting the signal in terms of geophysical excitation mechanisms such as atmospheric fluctuations or core-mantle interactions. The theoretical effect on the Earth's rotation of the 30 greatest earthquakes to have occurred since 1950 has been computed and found to be large enough to contribute to the observed polar motion variations at detectable levels (Gross, 1992).

The analyses of lunar laser ranging (LLR) data, now over 8,000 normal points from 1969-1993 permit long term studies of variations in the Earth's rotation, as well as determination of many parameters of the Earth-Moon System. The LLR-determined value of Earth gravity constant is GM = 398600.443 $\pm$ 0.004 km^3/sec^2 and of the tidally driven secular acceleration of the Moon is -25.9 $\pm$ 0.5 $arcsec/cy^2$ corresponding to a linear increase of 3.8 $\pm$ 0.1 cm/yr in the mean distance of the Moon. The correction to the IAU-adopted precession constant is -3.3 $\pm$ 0.4 milliarcseconds/yr, giving the luni-solar precession constant as 50.3845 arcsec/yr at J2000. The amplitude of the 18.6-year nutation of the Earth's pole in space needs to be increased by 3 milliarcseconds (Williams, Newhall, Dickey, 1993). A joint analysis of LLR and VLBI data for precession and nutation corrections utilizes the complementary strength of both techniques (Charlot, Sovers, Williams and Newhall, 1991).

Mizusawa Astrogeodynamics Observatory/NAO, Japan *(reported by T.Sasao). Developments of instruments.* K-4 VLBI system has been successfully used in many geodetic applications including Earth rotation monitoring (Yoshino, 1991; Kiuchi et al., 1991). A variety of high-performance VLBI terminals and correlators are under development on the basis of the K-4 system (Kawaguchi et al., 1991; Kawaguchi, 1993; Kiuchi et al., 1993; Matsumoto and Kawaguchi, 1993; Tamura et al., 1993). Compatibility of VLBI systems is an increasingly vital problem for the international scientific community (Hama et al., 1993). A new 10 m VLBI antenna with the phase-stabilized optical-fiber signal transmission system (Sato et al., 1992, 1993) and an advanced digital servo-control system is now in use at Mizusawa Astrogeodynamics Observatory /NAO (Hara et al., 1993; Kameya et al., 1992a, b). The new 10 m antenna will succeed the role of 6 m antenna at Nobeyama (Fujishita, 1993) in the international Earth rotation observations. A concept of a new VLBI system with clusters of 4 antennas at each end of a VLBI baseline has been developed for Project VERA (VLBI for the Earth Rotation study and Astrometry). Simultaneous VLBI observations of 4 different radio sources and effective cancellation of irregular atmospheric refraction effect as well as other sources of systematic errors shall make the new system a promising tool for higher precision and high time-resolution geodesy and 10 microarcsecond level differential astrometry (Kawano et al., 1992; Sasao and Morimoto, 1991; Sasao et al., 1992; Sasao et al., 1993). Experiments for new types of VLBI observations such as Earth rotation monitoring with orthogonal VLBI baselines (Yoshino et al., 1990), burst mode observations (Yokoyama et al., 1990) and higher frequency observations (Takahashi et al., 1993) have been successfully performed. Effects of atmosphere on VLBI observations were studied in several aspects (Abe et al., 1991; Goto et al., 1991; Manabe et al., 1991a; Sasao et al., 1993). A new SLR station has started observation at Communications Research Laboratory, Koganei, Tokyo since 1990 and successfully obtained many passes for major geodetic satellites (Kunimori, et al., 1993). The station will serve as a collocation point with VLBI. Radio transmitters which are planned to be installed on the surface of the Moon for differential VLBI observations of lunar librations and tidal deformations have been developed and proven to survive after strong impacts with 10000 G decelerations (Hanada et al., 1992; Kawano et al., 1993). Measurable tidal deformations of the lunar surface were estimated in relation to the internal structure of the Moon (Ooe and Hanada, 1992).IERS VLBI Technical Development Center was established in 1990 in Communications Research Laboratory, promoting and coordinating the efforts for developing new technologies in space geodesy (Yoshino, 1993).

Participation in International Earth Rotation Observations. National Astronomical Observatory and Communications Research Laboratory have jointly conducted IRIS-Pacific VLBI observations with Kashima 34 m and Nobeyama 6 m antennas on the monthly basis with occasional burst mode observations (Yokoyama et al., 1990). The Hydrographic Department of Japan has carried out a large number of SLR observations at Simosato, which is active as an IERS site since 1988, and at several sites in Japan with a transportable SLR system HTLRS (Sasaki and Sengoku, 1993). Algorithm and physical models for the VLBI Earth Rotation analysis have been improved taking into account achievements in the relevant fields of science. VLBI data obtained at nearly 60 observation sites over the world since 1984 were reanalyzed with the improved software and the results were reported to the Central IERS Bureau. A software was developed to determine the Earth orientation parameters with arbitrary time interval and applied to an analysis of the short time scale variations of the polar motion and UT (Manabe et al., 1991b; Manabe, 1992; Manabe et al., 1993). SLR ETALON data collected during 5-month period including the IERS ETALON Autumn Intensive Campaign were analyzed to evaluate the scientific potential of ETALON in determining the Earth orientation and other geodynamic parameters (Murata and Tsujii, 1992). A new method was developed for improving the fundamental star catalog on the basis of the astrolabe data and applied to the past Danjon astrolabe data collected at Mizusawa. Precise positions were determined for all the observed stars (Manabe and Sakai, 1990). Recent activities of the IERS and VLBI contributions to IERS reference frames and Earth orientation parameters have been summarized (Yokoyama, 1991; Yokoyama, 1992; Yokoyama 1993). A number of Japanese institutions actively took part in observations and analyses for IGS/Epoch'92 and other global GPS campaigns in order to confirm the feasibility of GPS observations in determination of the Earth rotation parameters and establishment of the global terrestrial reference frame (Hatanaka et al., 1993; Kato and Nakano, 1993; Murata et al., 1993a, b). Results of precise time comparison with LORAN-C and GPS time signals have been carefully examined and some improvements were achieved in the GPS time transfer with application of an advanced analysis software (Horiai, 1992; Horiai and Tamura, 1993).

Physics of the Earth's Variable Rotation. The seasonal budget of the Earth's axial angular momentum has been studied using IRIS VLBI data and numerical models for global atmospheric motion provided by Japan Meteorological Agency. A new quasi-seven-month oscillation ('QSO') was found in both the LOD (length of day) and the tropospheric zonal wind.It was found that the quasi-biennial oscillation, which has been observed in the LOD and the stratospheric zonal wind, exists also in the tropospheric zonal wind. It is confirmed that the inter-annual variations of the LOD with time scales up to five years are mostly explained by the AAM (atmospheric angular momentum) changes due to the oscillations in the tropospheric and stratospheric zonal wind. LOD residual variations with time scales longer than five years, after removal of the inter-annual AAM contributions, are probably related with the core-mantle coupling process (Naito and Kikuchi, 1990, 1991, 1992a,b). Possible non-linear behavior of the oscillations in the AAM and the Earth's rotation was pointed out (Naito,1992). UT1 variations due to the long period tides were analyzed and the possible range of the mantle Q was discussed (Tamura, 1990, 1993). Free Core Nutation parameters were estimated on the basis of the superconducting gravimeter data obtained at three sites in Japan. The results show closer agreement with the VLBI results compared with other reported SCG-based parameters (Sato et al., 1992). Tidal evolution of the LOD, the lunar orbit and the obliquity of the Earth over the geological past was numerically calculated on the basis of a variable ocean-continent distribution model (Abe et al., 1992a,b). Variations of the plumb line direction are derived from the results of the astronomical time-and-latitude observations and their possible correlation with El Nino events was pointed out (Abe, 1992).

U.S.Naval Observatory, USA (reported by D.D.McCarthy). *PZT Observations.* Operation of the last PZT 7 in Washington ceased with the observation of 31 December 1991. The total statistics of all USNO PZT data are as follows: Observing sites: Washington, D.C. (PZT's 1, 3, 7) and Miami (Richmond),FL (PZT's 2,6); Observing period: October

18, 1915 to December 26, 1991; total number of stars observed: 472; total number of nights observing: 26 971; total number of star observations: 675 329. B. Archinal completed gathering USNO PZT data for distribution to the IAU Working Group on "Earth Rotation in the HIPPARCOS Reference Frame".

VLBI. The NEOS VLBI network was created from the merger of the NOAA IRIS and the USNO NAVNET networks. This became effective on 1 May 1990. In the three-year reporting period a total of 157 NAVNET experiments of 24-hour duration have been fully processed creating a data base of 71 668 delay observations. The accuracy appears to be 0.6 mas in x, 0.5 mas in y, and 0.05 ms in UT1-UTC.

The treatment of NAVNET observations was modified by M.Eubanks to include solving for the UT1 rate. Under the guidance of M. Eubanks, the new Hawaii telescope for use in the NAVNET observations was completed in 1993. In a study of NAVNET residuals, D. Matsakis found that the residual scatter is a strong function of the wet path length and it is site dependent as well. The average residual was a function of elevation for all baselines involving Kauai. Software was written to plot the average closure phase violation as a function of baseline and source.

In cooperation with personnel from NASA and Haystack Observatory,K. Kingham continues work on the MARK-IV correlator development. The NEOS Directing Board agreed that investigations should take place into using data from NRAO's new VLBA network for Earth orientation information.

Approximately 800 000 frequency-baselines were processed by the Washington Mark IIIA VLBI correlator between January 1991 and May 1993. In this time period (20 857 hours), the correlator operated for 21 347 hours at a load factor of 102% (greater than 100% load factor is made possible by splitting the Mark IIIA correlator to process two sessions simultaneously). The Washington correlator facility shipped and received approximately 9 500 tapes, equivalent to 16 000 miles or 70 tons of tape.

Prediction of Earth rotation. The prediction of the correction to the current nutation theory was modified by D.D.McCarthy taking into account the most recent observations. Predictions of polar motion, UT1-UTC, and nutation were provided to the IERS (McCarthy, and Luzum, 1991, 1992; McCarthy, 1993).

Reference systems. Work by D.D. McCarthy, M. S. Carter, B. Archinal and D. Florkowski on the construction of an extragalactic reference frame was continued in cooperation with the Naval Research Laboratory (NRL), Hamburg Observatory, and the National Aeronautics and Space Administration (NASA). D.D.McCarthy completed and distributed the second edition of the IERS Standards and B.Archinal wrote the chapter on Terrestrial Coordinates and the Rotation of the Earth for the Explanatory Supplement to the Astronomical Almanac.

Analyses of Earth orientation observations. D.D. McCarthy and B.J. Luzum (1991, 1992) derived corrections to the main zonal tide components in the variation of the LOD and used them to derive corrections to the zonal tidal coefficients for UT1-UTC. D.D. McCarthy investigated various ocean tide models by comparing the predicted variations in LOD with observations. Improved model coefficients nutation were determined by D.D. McCarthy and B.J. Luzum (1992) from observations of nutation. D.D. McCarthy and B.J. Luzum (1991) also completed work on extending the NEOS combination EOP time series back to 1979. B. Archinal continued his investigations on the use of the GEODYN/SOLVE software, with which it could be possible to do direct combination SLR/VLBI solutions. Recent software advances show great promise for allowing combination of SLR, VLBI, and GPS solutions, which in turn should provide much higher accuracy of Earth rotation and reference frames.

Shanghai Astronomical Observatory, Chinese Academy of Sciences, China (reported by Jin Wenjing, Wei Xue, Xu Tongqi). The following investigations were carried out:
- construction of new instruments: 25 m VLBI antenna with MKIII receiver in Urumqi station, third generation laser with precision of 2-3 cm in Shanghai and Beijing SLR stations, Urumqi SLR station is under construction;
- Shanghai Observatory is one of the IERS Analysis Centers for VLBI, SLR, LLR data;
- participation in the IGS'92 and SEARCH'92 campaigns with 10 GPS receivers and 3 laser ranging systems;
- collection and reduction of 75 293 optical astrometric observations made in 1988.0-1991.0 by 64-42 instruments located all over the world. (Quarterly and Annual Reports on Optical Astronomical Observations, Shanghai 1989-1992). Since 1991 no ERP have been computed from these observations;
- nine Chinese optical astrometric instruments are included in the list of 55 instruments of which ERP results will be rereduced in the Hipparcos reference frame;
- the effects of the systematic difference between FK5 and FK4, as well as the change of astronomical constant systems, on the EOP were studied extensively (Jin et al., 1991; Xu et al., 1991; Li, 1990a; Xu et al., 1990). After the elimination of systematic difference of the catalogs, the accuracy of comparisons between Chinese Joint Universal Time (CJS) and BIH or IERS one are better than old ones (Li, 1990);
- effective multistage-multiarc method was introduced into the determination of ERP series and geocentric coordinates by

processing SLR data;
-new methods like a differential filter, a deconvolution method of Fourier transformation, Marple algorythm of AR spectrum, adoptive AR model have been used for analyses of the pole motion and pole motion predictions;
-UTO data determined by LLR were studied and effects of lunar libration models, relativistic effects, uncertainties ephemeris equinox and the elliptic obliquity were discessed;
-the period and Q value (60-70) of Chandler wobble were derived using Zschau's mantle rheology model, and the effect of the equilibrium pole tide of the inelastic Earth on Chandler wobble was discussed;
- the excitation of atmosphere and a water storage distribution to the polar motion were analyzed showing the correlation of an annual variation of a water storage and of the polar motion;
- short periodical oscillations of the Earth rotation and their relationship with variations of the atmosphere, ocean and solar activities have been studied deeply. Oscillations with periods of 50, 90, 120 days in LOD and AAM variations were found. Big influence of zonal winds of atmospheric circulation on 50 days LOD oscillation was indicated. The study shows that high frequency oscillations of LOD are caused mainly by medium and low latitude regions of northern and southern hemisphere;
-study of the correlations between solar constant, near-Earth space environment, solar neutrino flux and Earth rotation indicated that the interannual changes of the Earth rotation and El Nino phenomena might be induced by solar activities;
-the results of sea level changes (SLC) with Earth rotation were studied on the basis of 30 years of observations of 50 tide gauges located along Pacific and the ERP data from the IERS. Positive correlation was obtained in the eastern Pacific and a negative one in the western Pacific. Dynamical analysis of the angular momentum of the tropical Pacific shows that the tropical Pacific influence account for 30% of the interannual change of the LOD and more during El Nino events, especially this in 1982-1983. The raise of a sea level in the Pacific with the rate of 0.2 cm/year in the last 30 years was found;
- the influence of changes of the zonal angular momentum of the Earth's atmosphere in the stratosphere and lower mesosphere on Earth rotation have been estimated with the method of balance of angular momentum. Results show that the solar activity causes the average deviations of the Earth's angular velocity of $1.35x10^{13}$/s in the time of the high and low solar activity;
- correlation between variations of the Earth rotation and earthquake in different time scales decades, interannual and seasonal ones were analysed statistically. Correlation of oscillations with four years period was found.
- anomalies of astronomically determined time and latitude residuals were found before and after the earthquakes around Shanghai, Beijing, Yunan as well as in other stations in the world located close to epicentrums. Four earthquakes were predicted by the use of such anomalies in China. A service of such earthquake prediction was proposed;
-the mean acceleration of the Earth rotation rate w = - 62.02/sec/cy/cy was obtained on the basis of old Chinese observations of lunar occultations value 960 AD.

Besides mentioned references about 60 papers were published in Chinese journals (with english abstracts) mostly in Annales of Shanghai Observatory and Acta Astronomica Sinica (partly translated into English in Chinese Astronomy and Astrophysics).

University of Texas at Austin, Department of Geological Sciences, Center for Space Research, Institute of Geophysics,Austin, Texas,USA (reported by C.Wilson). Analysis of atmospheric excitation of nonseasonal polar motion have been done by J.Kuehne, S.Johnson and Clark R. Wilson (1993). Analysis of non-seasonal polar motion excitation and atmospheric mass equatorial angular momentum (EAM) over land for the period 1980-89 reveals a clear pattern of high power and correlation during the northern hemisphere (NH) winter followed by low power and correlation during the NH summer. A special case of this pattern occurs for longer than 14 months (from January 1987 to March 1988) when the correlation throughout the NH summer remains statistically significant. During this epoch an average of 72 percent of the non-seasonal polar motion excitation power at frequencies between -30 and + 12 cycles per year is linearly related to atmospheric EAM over land. During the SH winter there is significant correlation between the atmospheric EAM over mid-latitude southern ocean and polar motion excitation indicating the existence of a non-equilibrium ocean excitation. The atmospheric excitation power is too small to explain the large correlation during the NH winter. The effects of winds probably account for the deficit in power.

University of Beograd, Mathematical Faculty, Yugoslavia (reported by N.Pejovic) N.Pejovic got a good agreement of the model of the Chandler wobble with modulated frequency dependent amplitude with observations in a nine year period. Agreement is better when influence of groundwater variations is taken into account (Pejovic, 1990, 1992).

Sternwarte der Universitat Bonn, Germany (reported by P.Brosche). P.Brosche in collaboration with oceanographers at Hamburg and geodesists at Bonn, were dealing with theoretical and observational aspects of oceanic influence on the Earth's rotation. Short-period tidal influences have been obtained from hydrodynamical models. The effects predicted in UT have been found in very precise VLBI-data (Brosche, Wünsch, Campbell and Schuh, 1991). The seasonal interaction between the atmosphere and the oceans, characterized by varying wind stress and pressure fields, is currently treated by a very complex model. First results and comparisons with LOD and AAM data show that oceanic contribution should not be neglected for a complete balance (Brosche, Wünsch, Frische, Sündermann, Maier-Reimer and Mikolajewicz, 1990).

P.Brosche and J.Wunsch found the periodic variations in the location of the center of mass of the solid Earth due to ocean tides from the ocean tide model of Sailer (1991). The sum of nine partial tides leads to changes in its coordinates X,Y,Z within a range of 1 cm (Brosche P., and Sündermann J., 1990).

P.Brosche served as a president of the Special Study Group on Long-term Variations in Earth Rotation, of the IAG and was the co-editor of a book on long-term aspects of Earth rotation (Brosche and Wünsch, 1990).

Slovak Technical University, Bratislava, Slovakia (reported by J.Hefty). The comparison of LOD data determined by VLBI, SLR and LLR observations with AAM series from four meteorological data centers carried out by J.Hefty (Hefty 1991a) shows differences in atmospheric series for annual, semiannual, monthly and fortnightly periods. The discrepancies are at 10 μs level and are most significant for a semiannual term.

The fortnightly and monthly zonal UT1 variations were analyzed in order to derive scaling factor k/C using optical astrometry, VLBI and SLR data. The frequency dependent k/C with "out of phase" constituent, obtained from VLBI and SLR, are consistent with the most recent theories of the Earth's mantle inelasticity and the non-equilibrium ocean tide response (Hefty and Capitaine, 1990).

In the VLBI "intensive" observations on the Westford-Wettzell baseline the UT1 diurnal and semidiurnal variations caused by non-equilibrium oceans were detected by Hefty (1990). Their interference with nutation errors is discussed. The long term stability of Westford - Wettzell UT1 series was analyzed by Hefty (1993a) showing the systematic effects of quarterly alteration of observing schedule reaching 50 μs biases. The analysis of short term variations in GPS polar motion and UT1 series proves the ability of GPS technique to monitor ERP high frequency oscillations for daily and weekly periods. Significant correlation coefficient (0.4-0.6) were obtained from comparison of polar motion and UT1 determined by GPS with those induced by the AAM (Hefty,1993b).

Observatoire Royal de Belgique, Bruxelles, Belgium (reported by V.Dehant and P.Paquet). New theoretical nutations have been computed and compared to the adopted nutation series, to observations and some other theoretical nutation series by V.Dehant (Dehant, 1992). This model accounts for mantle inelasticity, new core flattening, resonance effect at the periods of the Free Core Nutation (FCN) and the Free Inner Core Nutation (FICN), new rigid nutations and ocean effects (computed from literature). There are remaining discrepancies for which an explanation could be found in atmospheric variations; the atmospheric effects on the nutations have been computed for an oceanless Earth and have been shown to be very efficient (Dehant, Hinderer, Legros and Lefftz, 1993).

Defraigne, Dehant, Hinderer (1993) have used nutation observations jointly with tidal observations in order to determine the FCN period and damping (quality factor). A stacking method applied on both kinds of observations has given a FCN period around 431 days and a quality factor around three thousand.

The amplitudes of the fluctuations observed at 50 and 120 days in the Earth rotation as well as the atmospheric angular momentum, the solar wind, the interplanetary magnetic field were estimated; their mutual correlations and coherencies have been also established by Djurovic and Paquet (Djurovic and Paquet, 1991). The quasi biennial oscillation (QBO), well known in the Earth rotation fluctuations, has been also identified in the solar activity and the atmospheric angular momentum. In the shorter periods (50 and 120 days) a high level of coherency is observed between these signals (Djurovic and Paquet, 1991).

Institute of Astronomy, Academia Romana,Bucuresti, Romania (reported by M.Stavinschi). The long series of astrometric determinations of UT0 by the passage instrument have been analysed by M.Stavinschi and collaborators and the star coordinate corrections determined (Ciobanu, Stavinschi, Dinescu, 1992; Stavinschi, Dinescu, Vass, 1992). Theoretical studies of the movement of the Earth instantaneous axis of rotation within the Earth and of the inverse mechanical problem were carried out (Stavinschi, Dinescu, 1992a; Stavinschi 1990; Stavinschi, Mioc, 1993).

Institute of Physical Geodesy,Darmstadt, Germany (reported by E.Groten) E.Groten and H.Yan (Shanghai Observatory) elaborated a "non-rotating" reference frame and its implementation in terms of time. A Departure Point different from Guinot's concept on the moving true equator was studied in detail. It is not assumed to eliminate completely equinox systems. Details of the Departure Point systems with related formulas were published by H.Yan and E.Groten in 1992a, b, c; Groten and Yan, 1993, Yan and Groten, 1993).

Paris Observatory, Paris, France (reported by N.Capitaine). The developments of the celestial pole coordinates have been given as functions of time with an agreement of 0.5 mas with the classical developments of precession and nutation (Capitaine, 1990) and has been applied to the procedure for deriving the Earth Rotation parameters from VLBI (Capitaine, Gontier, 1991). An accurate procedure for deriving UT1 at ± 0.01 mas accuracy has been developed (Capitaine,

Gotnier,1993), based upon the concept of "non-rotating origin" (Guinot, 1979); it has been proposed in the new IERS Standards instead of the usual one and the consequences of its use have been estimated on the practical derivation of UT1. The accurate relationship between the spatial and terrestrial polar motion of the rotation axis and the observed changes of the Celestial Ephemeris Pole have been given (Brzezinski, Capitaine, 1993).

The study of the Earth rotation in the framework of General Relativity has provided the amplitudes of all the post-newtonian nutations in the kinematically non-rotating reference frame (Bizouard et al. 1992). A comparison between recent theories of nutation for a rigid-Earth model (Souchay,1993) shows the large improvement of these new nutation series with respect to those of Kinoshita (1977).

In 1991 the Paris Observatory organized the conference: "Journees 1991 Systemes de Reference Spatio-Temporels - Geodynamique Globale and Systemes de Reference".

GeoForschungs Zentrum, Potsdam, Germany (reported by H.Jochmann and M.Meinig). Investigations on Earth rotation of H.Jochmann, H.Greiner-Mai and their collaborators are related to the influence of geophysical processes on polar motion and the LOD. Especially, excitation functions caused by atmospheric, hydrospheric and core dynamics are discussed.

The relations between the LOD were investigated for different models of fluid motions within the outer core. The rigid rotation of the fluid core, following from the frozen field theory, was extended to the consideration of a non-axial rigid rotation of the Earth's core but it does not improve agreement with variations of the LOD. A better agreement was obtained applying zonal fluid motion of the fluid outer core (Jochmann, 1993).

A further improvement in investigating the relation between time series of geomagnetic field representations and variations of the LOD was obtained by lengthening both time series. Correlations between time series of the geomagnetic field representations were explained assuming that an inner core motion is indicated by the motion of the axis of the geomagnetic dipole. This hypothesis was confirmed by investigating a theoretical model of the relation between inner core rotation and the magnetic dipole axis (Greiner-Mai, 1992a,b; Greiner-Mai, 1993; Greiner-Mai, Felsmann, 1993).

Different periodic variations of the sea level of the Atlantic and Pacific oceans were found, which possibly influence polar motion. The periods of these variations are between 4 and 6 years (Ballani, Felsmann and Jochmann, 1992).

A possible influence of a climate change on Earth rotation was discussed using theoretical models of a climate change. It was found that the variation of atmospheric dynamics caused by climate change has no observable influence on the Earth rotation. A small observable influence on Earth rotation is caused by mass redistributions between the ocean and Antarctic ice sheet (Jochmann, 1992).

Long series of latitude observations together with PZT2 carried out in the Geodetic Institute in Potsdam in the last 100 years have been rereduced by M.Meinig. Linear draft of latitude in this century of +0.0050" ± 0.0004" per year was obtained against the plate tectonic model AMO-2 values + 0.0005" (Meining, 1991, Meining, 1992, Meining, 1993).

Royal Greenwich Observatory,University of Durham, UK *(*reported by L.Morrison and R.Stephenson). L.Morrison and R.Stephenson (1992) have completed an analysis of historical observations of solar and lunar eclipses in the period 700 BC to the present with the objective of determining long-term changes in UT and in the length of the day. The data set comprises timed and untimed data from Babylonian, Chinese and European Sources. They have added new material mainly in the period AD 500-1300 since their last study in 1984. These data have helped to define long-term changes in UT on a time scale of millennia (Morrison ans Stephenson, 1991):

1. The average change in the length of the day over the past 2500 years is + 1.7ms/century. This is about 25 percent less than the change expected on the basis of tidal friction alone (+ 2.3ms/century).
2. The length of the day fluctuates between -0.6 and + 4.0 ms/century about the average trend of +1.7 ms/century on a time scale of 700 years.

Institute of Astronomy, Moscow, Russia (reported by L.Rykhlova). Approximation of the observational data by quasipolynomial model and sero-crossing statistics for signal detection was used for analyses of the ERP and their predictions (Rykhlova et al.1990).

From analysis of 143-year series of the pole coordinates from 1846.0 till 1988.0 (Rykhlova et al. 1991) and the series from 1962.0 till 1982.0 in BIH system (Rykhlova et al. 1992) the following conclusions were obtained: there is a smooth change of the Chandler period within 411-438 days; oscillations with periods of 6.0, 8.85, 18.6, 26.0, 32.0, 43.3 and 65.0 years were detected in smoothly changing amplitude.

Analysis of 10-year series of inclination measurements in Finland showed the presence of long-period oscillations with periods similar to those in the pole motion, 0.5 years, 365 days and Chandler ones (Rykhlova et al. 1991).Correlations and interactions of various natural forces with the Earth rotation variations are discussed (Rykhlova 1991) as well as the common properties of the geodynamical and geophysical phenomena (Levitski,Rykhlova 1992, Bobova, Kurbasova, Rykhlova 1993, Rykhlova 1992).

The observations of latitude and longitude variations at Zvenigorod (PZT) and Simeiz (astrolabe) have been carried out.

Astronomical Institute in Prague, Czech Republic (reported by J. Vondrak). Atmospheric excitation of polar motion was studied by J.Vondrak (1992) in the interval 1976.5-1988.0. It was found that better agreement between observed and integrated polar motion is obtained if the frequency of the free Chandler wobble is assumed to be variable and amplitude-dependent. The method in which the integrated polar path is compared with the observed polar motion is shown to be superior to the one in which the differentiation of the observed polar motion is used to calculate the expected excitations and then compared with the observed ones. The reason for this is the transfer of the noise in the observed data sets and is significant especially for periods shorter than 100 days.

Institute of Geodesy and Geophysics, Wuhang, China (reported by Zhu Yaozhong). Considering the elliptical rotating Earth with inelastic mantle, a fluid core and equilibrium pole tide Zhu Yaozhong determined theoretically the Chandler wobble period and Q factor equal 434.8 days and 71 years respectively. The effects of the inelastic perturbation of the equilibrium pole tide on the Chandler wobble is much larger than the effects of the non-equilibrium part of the pole tide, and mantle inelasticity is likely to be the most important dissipative source of the Chandler wobble. He found that high frequency oscillations of LOD and AAM variations with periods ranging from 40 to 90 days are unstable.

Space Research Centre PAS, Warsaw, Poland (reported by B.Kolaczek). The relations between the EOP reported by the IERS and the dynamical quantities which are coordinates of the Earth's angular velocity vector were studied by A.Brzezinski. The most important result is the derivation of the more rigorous equation of polar motion in which the observed terrestrial coordinates of the Celestial Ephemeris Pole are used as variables instead of the rotation pole coordinates (Brzeziński A., 1992, Brzeziński and Capitaine, 1993). In view of the sub-diurnal resolution of the EOP Brzezinski discussed the problem of distinction between high-frequency polar motion and nutation and finally extended the equation of polar motion for high-frequency variations by including effects of the Free Core Nutation resonance at nearly diurnal retrograde frequency (Brzeziński, 1993).

The short periodical oscillations of polar motion, their predictions and their correlations with such oscillations of the AAM have been analyzed. Variations of the short periodical oscillations of polar motion with periods of 120, 100, 80, 60, 50 days were analyzed on the basis of SLR and VLBI data. It was found that they are elliptical oscillations with variable amplitudes, ranging 2-5 mas and modulated with the period of about 600 days (Kolaczek 1992a,b; Kosek, Kaczkowski 1992; Kolaczek, Kosek 1992). Correlations of the short periodical oscillations with periods of 80-120 days in polar motion with such oscillations of the AAM variations and the solar activity have been found (Kolaczek and Kosek, 1992).

In early 1988 strong short periodic oscillations with periods of about 35-95 days were detected in both, the χ_y component of polar motion excitation function and LOD variations (Nastula et al. 1990, Nastula 1992). These perturbations are correlated with AAM components variations (Nastula 1992). The χx components of atmospheric and geodetic excitation functions of polar motion are not significantly correlated (Nastula 1992). The common periods of about 28 and 40-50 days were detected in *y* component of both the geodetic and the atmospheric excitation functions of polar motion. These oscillations are influenced mainly by the atmosphere in the northern hemisphere, while the southern influences them only negligibly (Nastula, 1993). The distinct variation of LOD and χ_y components of polar motion excitation function, which occurred in early 1988 seems to be caused by the 40-50 day oscillation in the northern hemisphere atmosphere (Nastula, 1993). The prediction of polar motion coordinates, which now has the accuracy of several milliarcsecond for two weeks, can be improved by taking into account a prediction of short periodic oscillations. The new autocovariance prediction method (Kosek 1992a) or the autoregressive prediction method (Kosek 1992b) were applied to the short periodic oscillations of the polar motion showing that the prediction error depends mostly on the starting prediction moments.

Analyses of amplitude variations of the Chandler and annual wobbles of polar motion in 1846-1988 were carried out on the basis of the longest set of pole coordinates worked out by the Main Astronomical Observatory in Kiev (Nastula et al. 1993). The results show the strong dependence of amplitude modulations of the Chandler wobble on widths of FT filter pass bands. The most energetic oscillations of amplitude variations of the Chandler and annual wobbles of pole coordinates have common periods of about 40, 20-25 and 12-13 years what allows to predict the next deep minimum of the Chandler amplitude for 2 010 year(Nastula et al., 1993).

Investigations of the optimal terrestrial reference frame consisting of the minimum number of stations and the highest accuracy were, carried out. Results show that the Primary Terrestrial Reference Frame (PTRF) consisting of 64 SSC collocation points carefully chosen, with a good distribution and accuracy of their coordinates, defines the Terrestrial Reference Frame with almost the same accuracy as the ITRF89 consisting of 119 SSC collocation points (Kosek and Kolaczek, 1991a,b).

Besides these reported above investigations of Earth's rotation more investigations have been carried out especially by the IERS Analyzing Centers. Their results were presented at conferences, mentioned at the begining and can be found also in the papers indicated in the Astronomy and Astrophysics Abstracts.

IAU COMMISSION 20

Positions and Motions of Minor Planets, Comets and Satellites

REPORT ON ACTIVITY 1991–1994

President: A. Carusi
Vice-President: D.K. Yeomans
Secretary: H. Rickman

I. Introduction

During the past triennium, many events related to the activity of members of Commission 20 have taken place. On the side of discovery rate and observations of asteroids, we have had an outstanding increase, as is shown in Section II of this report. The augmented rate is best noted by the record 5802 pages of the *Minor Planet Circulars* and the 20% increase of pages of the *Efemeridy Malykh Planet*.

The existing surveys of Earth-crossing asteroids, and more generally of Near-Earth Objects (*NEOs*) have resulted in an increasing discovery rate. The activity in this field is illustrated in the Report by the IAU *Working Group on Near-Earth Objects*, contained in this volume. The growing interest in these researches, in which Commission 20 is deeply involved, has led to numerous meetings, reports, and workshops. It is easy to forecast an even larger increase in the NEO search activities for the future years.

In addition, the discovery and observation of comets has seen a sharp increase. Apart from the exceptional case of comet P/Shoemaker-Levy 9 – the first observed case of an object temporarily in orbit about Jupiter, with which the fragmented comet will collide just before the next General Assembly of the IAU – 20 short-period and 23 long-period comets have been observed in the last three years. Some new objects are being discovered with unusual orbital characteristics and their classification as asteroids or comets is rather difficult.

Within the naming Committees of the Commission (*Minor Planet Names Committee* and *Comet Nomenclature Committee*) the feeling is growing that a reform of the naming procedures is necessary, in order to cope with the increasing discovery rate of minor bodies in general and the mentioned classification problems. A closer cooperation with the *Working Group on Planetary System Nomenclature*, chaired by K. Aksnes, is advisable, and is already under discussion.

The *Asteroids, Comets and Meteorites '93* meeting, held in Belgirate (Italy), June 1993, had over 400 scientists in attendance, and this testifies to a growing interest on minor bodies studies within the planetary science community. Many review papers on all aspects of minor body studies will be found in the Proceedings of the meeting, when they are published in the IAU Symposium Series.

It is sad to announce the deaths of M.A. Dirikis, R.S. Harrington, E.I. Kazimirchak-Polonskaya. With them, Commission 20 has lost some of its most active and esteemed members.

II. Minor planets (B.G. MARSDEN)

II.1 General

Commission 20 continues to be responsible for obtaining and collecting astrometric observations and for computing and disseminating orbital and ephemeris data on minor planets, principally through the media of the *Minor Planet Circulars* (*MPCs*) and the *Efemeridy Malykh Planet* (*EMP*). The *MPCs* are issued monthly by the Minor Planet Center (director B. G. Marsden, associate director G. V. Williams, contributions also from C. M. Bardwell and D. W. E. Green), Smithsonian Astrophysical Observatory, Cambridge, Mass. The *EMP* is issued annually by the Minor Planets and Comets section (chief Yu. V. Batrakov, deputy V. A. Shor, contributions also from Yu. A. Chernetenko, L. L. Filenko, G. R. Kastel', O. M. Kochetova, N. K. Sumzina, O. O. Vasil'kova and T. A. Vinogradova), Institute for Theoretical Astronomy, St. Petersburg. S. Nakano serves as the liaison between the Minor Planet Center and the numerous amateur astronomers who carry out observational and orbital work in Japan. Following the recommendation of

J. Bergeron (ed.), Reports on Astronomy, Vol. XXIIA, 185–204.

Commission 20, the FK_5/J2000.0 system was to be used in this work, and all contributing orbit computers and most contributing observers cooperated in making the change around the stipulated date of 1992 Jan. 1.

During the triennium July 1990-June 1993 a total of 5802 pages of *MPC*s were issued, the record batch in Feb. 1993 having 324 pages that had to be issued in two parts. The number of pages in the *EMP* increased from 447 in the 1991 edition to 552 in that for 1994.

The *MPC*s and the *EMP* are now available in both printed and computer form. The collected observations and the actual *MPC* pages and files containing orbital elements have been issued on diskettes since 1986, but the diskettes are being phased out in favor of an extension of the dial-in Computer Service. *STAMP*, the diskette version of the *EMP*, was initiated in 1991 and includes the possibility of extracting orbital elements for the numbered minor planets according to particular criteria (56.098.223). The complete tape of observations issued in July 1992 contained 412 043 observations of numbered minor planets and 243 630 of unnumbered minor planets, the total up 28 percent over its 1989 predecessor.

At the end of June 1993 there were 5610 numbered minor planets, a 24% increase during the triennium. The 1992 edition of the *Catalogue of Orbits of Unnumbered Minor Planets* contained 14 367 entries, up by 23% in three years. Some 3023 of these entries referred to objects observed at multiple oppositions, and the number of unnumbered objects in this category increased by as much as 19% in only one year. In order to satisfy requests for more uniform data on such objects, a *Catalogue of High-Precision Orbits of Unnumbered Minor Planets* was issued for the standard epoch 1993 Aug. 1, and there are thoughts of making this (and the accompanying diskette edition) an annual publication.

A *Dictionary of Minor Planet Names* (55.002.001) was published by L. D. Schmadel in his capacity as chairman of the Commission 20 Study Group on Minor Planet Names. Explanations are provided for all but 132 of the names assigned to 3957 of the first 5012 numbered minor planets.

II.2 Observations and Orbits

A total of 139 156 astrometric observations made of minor planets during the triennium had been published in the *Minor Planet Circulars* or included in the Minor Planet Center's accessible files by Sept. 1993. Some 35% of these observations referred to the minor planets that had by then received permanent numbers. About 25% of the numbered objects were not observed at all during the triennium, while 34% were observed ten or more times. Twelve numbered minor planets were observed at least 100 times and three more than 200 times, the latter being the Galileo targets (243) Ida and (951) Gaspra and the Apollo object (4179) Toutatis (which at a geocentric distance of 0.024 AU in Dec. 1992 was making the closest predicted approach of any natural object other than the moon), with 312, 219 and 264 observations, respectively. Provisional designations were given to as many as 13 029 unidentified minor planets observed (still often, regrettably, on only one night) during the triennium, but 20% of these objects have since been definitely identified with other minor planets, and possible identifications have been suggested for a further 2%. Three of the newly-designated minor planets, the Earth-approaching objects 1992 AC, 1992 HE and 1992 LR, were observed at least 100 times.

The 139 156 observations were obtained at 129 different observatories, of which 31 have been utilizing CCDs to perform at least some of the astrometry, while the remainder still conduct programs that are exclusively photographic. Ninety-eight observatories were responsible for 10 or more observations, 56 for 100 or more, and the following 15 observatories made more than 1000 high-quality observations:

1. European Southern Observatory (ESO), with 35 143 observations, mainly by E. W. Elst from exposures with the 1-m Schmidt and by H. Debehogne with the 0.40-m GPO astrograph; both astronomers are on the Uccle staff and make a practice of obtaining three observations of each object on a given night.

2. Palomar, with 29 864 observations, mainly by E. Bowell and others in Flagstaff on films taken by C. S. and E. M. Shoemaker and others with the 0.46-m Schmidt; this project leads in observations of numbered minor planets, although the Shoemaker program, as well as the one conducted by E. Helin with the same instrument, is designed to discover near-Earth objects, two observations generally being obtained of each object on a given night.

3. Kitt Peak, 13 440 observations by T. Gehrels, D. L. Rabinowitz and J. V. Scotti with the scanning-CCD Spacewatch camera on the Steward Observatory's 0.9-m reflector, also for the discovery of near-Earth objects, with three observations per object per night; *in addition*, there were no fewer than 70 568 single-night (triple) detections that have not been designated and inserted in the official files but that are available for checking for possible identifications.

4. Oak Ridge, 10 298 CCD observations by R. E. McCrosky and C.-Y. Shao with the 1.5-m Wyeth reflector, two observations per object per night; this is the principal program designed for follow-up rather than discovery, about one-third of the minor planets currently being numbered on the basis of observations made exclusively at Oak Ridge at their latest oppositions.

5. Tautenburg, 6552 observations by F. Börngen (sometimes in collaboration by L. D. Schmadel, Heidelberg) as a "bad weather program" with the 1.3-m Schmidt, some concentration on follow-up.

6. Kushiro, 4275 observations by S. Ueda and H. Kaneda in the leading program by amateur astronomers, using 0.25-m and 0.16-m instruments.

7. Siding Spring, 4122 observations, measurements by R. H. McNaught from plates taken with the 1.2-m U.K. Schmidt and 0.5-m Uppsala Southern Schmidt, also CCD observations with a 1.0-m reflector, concentration on near-Earth objects.

8. Crimean Astrophysical Observatory (CrAO), 3307 observations in a long-standing program conducted by N. S. Chernykh and others with the 0.4-m double astrograph.

9. Geisei, 3114 observations by amateur astronomer T. Seki, using a 0.6-m reflector.

10. Kitami, 3014 observations by K. Endate, K. Watanabe and other amateurs using 0.25-m instruments.

11. La Palma, 2528 observations, mainly of numbered minor planets with the Carlsberg Meridian Circle, under the direction of L. V. Morrison.

12. Kleť, 2434 observations with two 0.6-m instruments, under the direction through 1991 of A. Mrkos and subsequently of J. Tichá.

13. Observatoire de la Côte d'Azur, 2347 observations with the 0.9-m Schmidt at Caussols, mainly by Elst and C. Pollas photographically, development work by A. Maury to convert to a CCD search for near-Earth objects.

14. Kitakoma, 2084 observations by amateur S. Otomo with a 0.25-m reflector.

15. Haute Provence, 1098 observations, 0.6-m Schmidt, another program by Elst.

Orbital elements were computed during the triennium for a total of 9666 minor planets. The number of orbits actually computed and published was actually well over 10 000, for the count refers only to the most recent results for each object. In addition, it should be noted that "Väisälä" (perihelion) orbits are automatically computed for all objects observed at least twice. Of the 9666 orbits, some 56% involved the linkage of observations at multiple oppositions or long-arc single oppositions and the need to consider planetary perturbations. In order to assist small-field observers, a special effort was made, beginning in Apr. 1993, routinely to recompute the orbits of multiple-opposition unnumbered minor planets whenever new observations are reported and to publish these results very concisely – rather than to accept that the available orbits might yield ephemerides in error by up to 5 arcmin yet delay the publication of a new orbit until enough observations are secured that the object can be numbered. The percentages of orbits computed by the nine most prolific orbit computers were: Williams 37, Bowell 19, Marsden 15, Nakano 13, Green 5 (mainly the principal share of the Palomar-Leiden T-1 survey of 1971, the astrometry for which has now been completed by I. van Houten-Groeneveld), Bardwell 3, Kaneda 2, Sumzina 1 (exclusively improvements of orbits of numbered minor planets), Filenko 1 (likewise); the remaining 4% were computed by 14 other astronomers.

Most of the multiple-opposition orbits involve the establishment of identifications by the same orbit computers, one significant success being Williams' identification of (878) Mildred, lost since its discovery on 1916, with a triplet of observations at ESO on 1991 Apr. 10 and a single CrAO observation on 1985 Nov. 11. Further images found on a 1984 Siding Spring plate and on plates taken at ESO in 1977 confirmed the identification; McNaught had in fact tentatively marked the 1984 image as a possible candidate for (878) Mildred. McNaught, and also Bowell, have made deliberate searches on isolated archival Siding Spring and Palomar survey plates for early observations of near-Earth objects; by recognizing an image of 1982 DB five months before discovery McNaught was then successfully able to recover this object, now numbered (4660) and named "Nereus", in 1990. Bowell traced (4015) 1979 VA to cometary images on the Palomar Sky Survey in 1949 that Marsden immediately noted as being P/Wilson-Harrington (1949

III), which had in fact been described as of asteroidal appearance on three other nights during its 6-day apparition; post-1979 observations show the object persistently to have asteroidal appearance, so the cometary appellation has been abandoned and the minor planet named "Wilson-Harrington".

The cometary nature of (2060) Chiron (perihelion and aphelion distances q = 8 AU, Q = 19 AU), approaching perihelion in 1996, became apparent in 1988-1989. When 1991 DA (q = 2 AU, Q = 22 AU) and 1992 AD (q = 9 AU, Q = 32 AU) were discovered – near perihelion – these objects were carefully scrutinized for cometary activity, but none was found, and the objects became (5335) Damocles and (5145) Pholus. Another object of the same general class is 1993 HA_2 (q = 12 AU, Q = 38 AU), observations of it now covering a 52-day arc. The discovery of 1992 QB_1 at a distance of more than 40 AU by D. Jewitt and J. Luu in the course of a careful search with the University of Hawaii's 2.2-m reflector for potential members of the supposed 'Kuiper Belt', in essentially stable orbits beyond Neptune, opened the possibility that this was an object in an eccentric orbit near aphelion. Nevertheless, subsequent observations indicated that 1992 QB_1 has q = 40 AU, Q = 48 AU. Its reality as a member of what must presumably be a quite populous Kuiper Belt was apparently confirmed when Jewitt and Luu discovered another object that was observed only sporadically for three months but that probably has a rather similar orbit. This second object, 1993 FW, is almost antipodal to 1992 QB_1 and was found when the search had yet to cover one square degree of sky! R. M. West, O. Hainaut and others at the European Southern Observatory contributed strongly with follow-up observations of both objects.

Bec-Borsenberger and Calaf (55.098.062, 55.098.065) have continued to provide ephemerides for the Hipparcos program and made new orbit improvements (53.098.056), as have Whipple *et al.* (54.098.054) in connection with HST. Batrakov *et al.* (54.098.038) have brought the Orel'skaya program up to date.

II.3 Theoretical Investigations

Dirikis (53.098.040) discussed the use of Väisälä orbits for searches and identifications after an interruption of observations, and Marsden (54.042.050, 56.098.226) showed how an inversion of Gauss' method generalizes the Väisälä procedure, helps in the identification process and readily produces meaningful orbits in indeterminate and uncertain cases. Shunning the use of orbital elements, Kristensen (56.098.015) described a statistical procedure to identify further single observations. Duma and Fedij (54.098.046) have considered the effect on orbit determinations of the neglect of mutual perturbations, an error in the mass of Jupiter and the difference between DE-200 and DE-202.

Goffin (54.098.035) and Williams (56.098.242) determined the mass of Ceres separately from observations of (203) Pompeja and (348) May, while Sitarski and Todorović-Juchniewicz (55.098.140) obtained a combined solution of $4.8(\pm 0.1) \times 10^{-10}$ solar mass. Schubart (56.098.002) discussed the astrometric observations used in the first mass determinations of Ceres, Pallas and Vesta; and Geffert (56.098.131) considered the mass-determination problem more generally. Landgraf (55.098.070) attempted a mass determination of (704) Interamnia from observations of (993) Moultona. Sitarski (56.098.028) used the Painlevé form to confirm the effect of relativity on the motion of (1566) Icarus, negating the need to include cometary-type nongravitational forces (53.098.013, 56.098.029).

Milani and Knežević (52.098.127, 53.098.001) extended their secular-perturbation theory to e^4 and m^2, determining proper elements and discussing secular resonances (56.098.005, 56.098.222), while Zappalà *et al.* (52.098.088, 56.098.141, 56.098.247) assessed the reliability of Hirayama families; J. Williams (55.098.145, 56.098.244) and Gordillo *et al.* (53.098.071) used clustering techniques to search for families, and Bendjoya *et al.* (54.098.047, 56.098.207) applied wavelet transform techniques. Klačka (54.098.051, 56.098.221) fully explained the Flora A jet stream as a selection effect. Froeschlé and Scholl (52.098.084), together with Morbidelli (54.098.034) and Henrard (54.042.012), have studied the secular resonances ν_5 and ν_{16}, partly in connection with the perihelion libration of (2335) James. Kozai (51.098.050) reviewed the stability of motion and secular perturbations using an averaged Hamiltonian.

Yoshikawa (51.098.049, 54.098.018) studied orbital behavior near the 5:2, 7:3 and 2:1 resonances in terms of the elliptical restricted three-body problem, showing the importance of the effect of Saturn, something Schubart (53.098.002, *Cel. Mech.* **56**, 153) has applied in the 3:2 and 2:1 cases, and Ferraz-Mello (*Cel. Mech.* in press) has considered in connection with the 3:2 resonance and depletion of the outer belt, a topic also discussed by Lecar *et al.* (55.098.144), as well as by Milani (55.098.072) in terms of "stable chaos". Tsuchida (53.098.068) determined the libration period of the 4:3 librator (279) Thule to be 190 years, with the apsides circulating in 450 years. Ipatov (51.098.044, 55.098.044, 56.098.106) and Hahn *et al.* (53.098.044) made numerical studies of motion near the 5:2 resonance, while the latter group also examined the 4:1 case (56.098.211). Hadjidemetriou (52.098.085) demonstrated a simple mapping model for the 3:1 resonance, and Ferraz-Mello (53.098.067) studied analytical results on regular solutions near resonances with Jupiter. Gerasimov and Mushailov (52.098.006) used the circular restricted problem to evolve orbital elements in terms of Weierstrass functions and with Vinnikov (53.098.020) made an application to the $(k+3){:}k$ resonance; they also discussed (54.098.089) Simonin's theory of the 2:1 case (108) Hecuba and (56.098.016) the effect of commensurabilities with Mars. Zagretdinov *et al.* (56.098.246) investigated the jovian 1:1 resonance, and Mikkola and Innanen (56.098.031) continued their numerical exploration of 1:1 resonances with various planets.

Zausaev and Pushkarev (53.098.016, 55.098.067, 56.098.019) integrated the orbits of Earth-approaching objects for some 11 500 years, and Zhang (53.098.110) and Hahn (53.098.061, 56.098.133) examined motions of Earth-approachers over a shorter term. Hahn and Bailey (56.098.218), as well as Steel *et al.* (56.098.238), found the eccentric orbit of (5335) Damocles to be very stable. Kuzmanoski (56.098.006, 56.098.007, 56.098.008) considered mutual encounters of minor planets. Zhang and He (52.098.126, 54.098.087) made an automatic determination of Hansen general perturbations for minor planets in the Flora group.

III. Comets (H. Rickman)

III.1 Discoveries, Recoveries and Astrometric Observations

The past triennium saw the discoveries of 23 long-period (LP) and 20 short-period (SP) comets, and 40 previously known SP comets were recovered. One of the new SP comets proved identical to the long-lost comet P/Metcalf, leading to the new double name P/Metcalf-Brewington. A drop in LP comet discoveries from the preceding triennium (from 38 to 23) is mainly due to the recent lack of space-based discoveries of Sun-grazers. All recent discoveries were Earth-based. The record number of provisional designations for a year (31 in 1989) was equalled in 1991. One of the new SP comets first received an asteroidal provisional designation (1990 UL3 = 1990 XVI P/Shoemaker-Levy 2).

The distinction between amateur and professional contributions to comet discoveries and recoveries is becoming less clearcut than before. Counting only work done at observatories affiliated to universities, academies or other national scientific institutions as "professional", the discoveries of LP comets continued to have a 1/3 contribution by amateurs, mostly visually at magnitudes 8–11. The 2/3 share of professional discoveries were photographic at magnitudes 15–18 with one exception – a CCD-aided discovery by the Spacewatch team. This and a SP comet discovery at magnitude 21 by the same team represent the introduction of CCD technique in comet discoveries.

The 19 remaining SP comets fall into two categories: three visual discoveries at magnitudes 8–10 by amateurs (P/Metcalf-Brewington, P/Levy and P/Brewington) and 16 photographic discoveries, mostly at Palomar (14 cases), where the magnitudes range from 13 to 18. Most remarkably, at Palomar, the Shoemaker-Levy team discovered nine SP comets (P/Shoemaker-Levy 1 – 9), which by far exceeds the number of SP comets ascribed to any previous discoverers. C.S. Shoemaker has now discovered 30 comets, thus rivalling J.-L. Pons as the all-time top comet discoverer. With large contributions also from the PCAS (Planet-Crossing Asteroid Survey) led by E. Helin (4 LP comets and 1 SP comet), and the Palomar Sky Survey II with comets searched for by J. Mueller (3 LP comets and 3 SP comets), this places nearly 60% of all recent comet discoveries at Palomar Observatory. The second most important discovery site was Siding Spring with 4 cases (9%). The latter constitutes the majority (2/3) of all Southern hemisphere discoveries.

Recoveries of SP comets continued to be done mainly with CCD detectors, the outstanding contribution coming from the Spacewatch team at Kitt Peak (J. Scotti: 17 recoveries + 2 shared with other observers). However, while the total number of CCD recoveries during the past triennium was nearly the same as previously, the number of photographic recoveries had a dramatic increase mainly due to T. Seki (11 recoveries). In total there were 22 CCD recoveries (at magnitudes 18–22) and 15 photographic ones (at magnitudes 14–20) plus one shared between the two techniques. In addition, two remarkable SP comets were recovered visually at 11th magnitude: 1991t P/Hartley 2 and 1992t P/Swift-Tuttle. In 28 cases the recovery circumstances can be compared with earlier ephemeris-aided recoveries, and nearly half of these occurred at larger pre-perihelion distances than ever before, e.g. P/Ashbrook-Jackson (4.7 AU), P/Tuttle (6.3 AU) and P/Tempel 1 (3.9 AU). The median heliocentric distance for the 35 pre-perihelion recoveries was [illegible]6 AU with quartiles at 2.0 and 3.8 AU. There is a clear trend toward more frequent near-aphelion observations of SP comets, of which there were four during the past triennium. The current total number of annual comets is thus 11 (P/Encke, P/Grigg-Skjellerup, P/Machholz, P/Tempel 2, P/Schwassmann-Wachmann 2, P/Arend-Rigaux, P/Gunn, P/Helin-Roman-Crockett, P/Smirnova-Chernykh, P/Helin-Roman-Alu 1 and P/Schwassmann-Wachmann 1), indicating that the identification of recovery observations is becoming less meaningful.

At the General Assembly in Buenos Aires in 1991, B.G. Marsden reminded Comm. 20 attendees of his 1973 prediction that P/Swift-Tuttle, the parent of the Perseid meteors, might return in late 1992. Two weeks later, an impressive display of Perseids within 0.1 day of the Earth's passage through the plane of the comet's current orbit suggested that the comet was nearby. Despite this, little in the way of professional searches were made for the comet, and it was finally found already at magnitude 11 in late September 1992 by the Japanese amateur T. Kiuchi (IAUC 5620). The predicted perihelion time proved to be only 17 days in error after 130 years (the longest period of any comet the return of which has been successfully predicted -- the record hitherto having been held by P/Halley). Subsequently, L. Kohoutek found an image of the comet on a single plate taken at Calar Alto in January 1992. This observation confirmed other pre-recovery observations then by Italian amateurs. The accuracy of the final predicted perihelion time is remarkable, when one considers that the Cape observations in October 1862 show large systematic residuals, smaller systematic effects also being noticeable in July 1862 and the January 1992 pre-recovery. The accuracy becomes astounding when one appreciates that an entirely gravitational orbit solution satisfies the general character of the comet's motion via 1737 back to the years +188 and -68 (Waddington, IAUC 5671). In a suggestion that was widely misunderstood, and before the +188 and -68 identifications were established, Marsden hinted that there was a tiny chance that P/Swift-Tuttle could collide with the Earth in 2126 and suggested that the problem posed by the October 1862 discordance could be resolved by making a series of observations of the comet with large telescopes in the southern hemisphere between mid-1993 and the end of the century.

Eleven comets were observed on their critical, second apparitions in addition to 1991a P/Metcalf-Brewington and 1992t P/Swift-Tuttle. Nine of these have orbital periods from 6 to 15 yrs and were discovered on their previous apparitions; for the two remaining ones (P/Schuster and P/Kowal 2), one intervening apparition had been missed. The latter was recovered by M. Ishikawa at a position yielding a large perihelion correction: $\Delta T = -54^{d}$. SP comets missed at expected perihelia were: P/Tritton in 1990; P/Russell 1, P/Haneda-Campos and P/Kowal-Mrkos in 1991; P/Shoemaker 2 and P/du Toit-Hartley in 1992; and P/Wiseman-Skiff and P/Lovas 2 in 1993. Most of them were badly placed for observation, and the last two had their perihelion passages only in June 1993; in addition, P/Kowal-Mrkos has undergone strong jovian perturbations and now has a too uncertain orbit to hope for recovery.

Observations of remote comets continued to be done mainly by K. Meech (summary in ADC, p. 12) including comets 1983 XII Černis at more than 19 AU and 1985 XII Shoemaker at 17 AU. Hainaut *et al.* (ADC, p. 54) reported on the last observation, at the detection limit, of 1986 III P/Halley at 16.2 AU and on negative results from searching for other comets at 17–32 AU. The possibility of detection of several very distant comets was also suggested by Bailey *et al.* (55.103.015). Observations of the P/Halley outburst in Dec. 1990 at 14 AU were described and interpreted by West *et al.* (53.103.106), Sekanina *et al.* (56.103.166) and Meech (55.103.120). Remote observations of P/Halley were reviewed by Boehnhardt (ADC, p. 3).

Astrometric observations were routinely published in the *MPC*'s, and the number during the past triennium was 7831, whereof 3279 referred to LP comets and 4552 to SP comets. The largest sets of observations were acquired for P/Swift-Tuttle (718), P/Schaumasse (479) and 1991 XXIV Shoemaker-Levy (419). Of the 110 observatories involved, more than 30 are using CCDs at least part of the time. The most prolific ones were Oizumi (1088 observations; however, generally involving dozens of observations of the same comet during each night), Oak Ridge (834), Spacewatch (530), Geisei (505), Chiyoda (424), Linz (385), Palomar (374) and Kleť (316). The 1st, 4th, 5th and 6th in this list are amateur observatories with small-field instruments. The Spacewatch programme was described by Gehrels (54.013.094), and accounts of observations made at Skalnaté Pleso were published by Svoreň (53.103.016; 56.103.014).

III.2 Orbits and Ephemerides; Orbital Evolution

New editions, the seventh and the eighth, of B.G. Marsden's Catalogue of Cometary Orbits were issued in 1992 and 1993, the seventh partially typeset, the eighth completely so. G.V. Williams co-authored these editions. The eighth edition contained orbits for 1397 apparitions of 855 comets, 174 of short period (of which 107 have been observed at more than one perihelion passages). 'Original' and 'future' orbits were given for 289 comets with high-quality orbit determinations. The *Comet Handbook*, edited by S. Nakano and D.W.E. Green, was published yearly by the *ICQ*, containing orbits and ephemerides for upcoming cometary apparitions. Yeomans and Wimberly (53.103.023) presented data for all SP comet apparitions expected in 1990-2010, and Kosai and Nakamura (53.103.024) made a statistical analysis of discovery/recovery conditions.

On the basis of Kresák and Kresáková's (54.103.011) analysis of the prospects for identification of old apparitions of known SP comets, Carusi *et al.* (54.103.067) were able to identify La Hire's comet of 1678 with P/d'Arrest. This accomplishment is remarkable for a comet with a relatively strong nongravitational effect and repeated encounters with Jupiter, but equally striking is the fact that the best fit requires the nongravitational perihelion delay of P/d'Arrest to have remained practically constant over three centuries.

Comet 1993e (P/Shoemaker-Levy 9), discovered some 4° from Jupiter by C.S. and E.M. Shoemaker and D.H. Levy, as well as independently by others, was remarkable in that it was seen to consist of from 11 to 20 separate nuclei over a 50" line. Evidently the comet had brightened because of the splitting, and the splitting had been brought about by a passage within Jupiter's Roche limit. From its present distance from Jupiter and the failure of observers to detect the comet at its opposition in early 1992, Marsden immediately suggested a Jupiter encounter in July 1992 (IAUC 5726), the nominal date being found in error by only 20 days when there were sufficient observations to refine the orbit computation. He then suggested that the comet was at least temporarily a satellite of Jupiter (IAUC 5744). By the end of April, computations by Marsden and independently by Nakano were refining the circumstances of the 1992 encounter and indicating that a 1994 encounter might even be closer. On May 22 (IAUC 5800) Marsden announced the likelihood that the center of the comet train would actually collide with Jupiter in July 1992, and A. Carusi (IAUC 5807) noted that, if there were a reasonably central collision, all of the nuclear train would collide with Jupiter, even if it were 30 times its observed length.

A new look at the problem of interstellar (hyperbolic) comets was taken by Kresák (55.103.021), who concluded that no positive evidence for such orbits exists, devoting special analysis to comet 1976 I Sato (55.103.044). Yabushita (53.103.014) had estimated the maximum nongravitational shift of $(1/a)_{\rm orig}$ and found comet Sato to be possibly interstellar. By contrast, Matese *et al.* (54.102.027; 56.102.089) found that negative $(1/a)_{\rm orig}$ correlates with small perihelion distance and retrograde orbit and offerred a possible explanation in terms of abnormally large nongravitational effects.

Todorović-Juchniewicz (56.103.005) determined original and future orbits of ten LP comets. Nongravitational orbit determinations were reviewed by Yeomans (53.102.045), and Sitarski (53.102.043) reviewed the problem of long-term linkages. Many attempts were made to model nongravitational effects in terms of physical parameters of the nuclei rather than using the standard (A_1, A_2, A_3) parameters. Sitarski (52.102.068) introduced a scheme involving the spin of the nucleus, applying it to P/Kopff, and then went on to study P/Grigg-Skjellerup (54.103.075; 55.103.102), P/Metcalf-Brewington (55.103.150), P/Swift-Gehrels (Bielicki and Sitarski 54.103.155), P/Kearns-Kwee and P/Gunn (Todorović-Juchniewicz and Sitarski 56.103.078). Szutowicz (56.103.247) similarly studied comet P/Wolf-Harrington. Very complex models also involving the light shift (displacement of the photocenter from the center of mass) as a physical parameter were discussed by Medvedev (54.102.013) and applied to P/Halley by Batrakov and Medvedev (52.103.161). Chernetenko (52.103.120; 54.103.068) applied such models to orbital linkages of comet P/Encke.

Investigations of the motion of P/Halley over its observed millennia were reviewed by Yeomans (53.103.155) and Ziołkowski (53.103.156), and the identification of ancient observations was discussed by Stephenson (53.103.154). The secular perturbations suffered by this comet over 10^5 yrs were investigated by Obrubov (56.104.070; MPB, p. 69). Numerical integrations of the orbital evolution of "giant comet" Chiron were performed by Hahn and Bailey (52.102.028), demonstrating that past visits into Earth-crossing orbits of the Jupiter family type are statistically likely. The temporary jovian satellite capture of P/Helin-Roman-Crockett shortly preceding its discovery was investigated by Tancredi *et al.* (52.103.187). Chaos in the long-term motions of comets P/Slaughter-Burnham (56.103.226) and P/Boethin (MPB, p. 97) was studied by Benest and Gonczi by estimating Lyapunov Characteristic Exponents.

III.3 Cometary Origin, Dynamics and Spatial Distribution

General reviews are found in the proceedings of IAU Coll. 116 and IAU Symp. 152 as well as in "Physics and Chemistry of Comets" (ed. W.F. Huebner; Springer-Verlag 1991). Among specific topics the most intense scrutiny was devoted to the dynamical evolution of the Oort cloud and the modelling of cometary capture.

The picture of Marochnik *et al.*, revisited in (55.102.053), of a massive Oort cloud whose angular momentum is too large for formation by means of gravitational scattering from the outer planets was criticized by Weissman (53.102.012) on the ground that the angular momentum of the cloud should be mostly acquired by the action of external perturbers. Later on, Lopatnikov *et al.* (54.102.020) modelled the "thermalization" of initially flat swarms of comets, concluding that the angular momentum of a swarm formed at the periphery of the solar nebula is little affected by such perturbations. Simulations of the long-term Oort cloud evolution predicting the influx rate into the inner Solar System were performed by Heisler (52.102.033), Matese and Whitman (56.102.033) and Chakrabarti (56.102.023). Brunini (*A&A*, **273**, 684,

1993) investigated the dynamical friction effects on the Oort cloud due to the solar motion by analytical means and concluded, in agreement with an earlier suggestion by Brunini and Canosa (55.103.019), that otherwise unexplained features of the aphelion distribution of new comets can be thus accounted for. The effects of deeply penetrating stellar encounters on the Oort cloud population and on the injection of comet showers were calculated by Weissman (56.102.049); see also Fernández (55.102.017). Possible associations of LP comets, earlier suggested by Biermann and coworkers to result from individual passing stars, were discussed by Whipple (53.102.081).

The flux of near-parabolic comets passing perihelia in the outer planetary zone was modelled by Yabushita and Tsujii (53.102.005; 54.102.026), thus evaluating the rate of captures into short-period orbits. This capture rate and its failure to balance the estimated loss rate of the Jupiter family was discussed by Rickman (52.102.040; 56.103.041). Further discussion of the origin of SP comets was given by Guliev (56.102.001). The choice between alternative sources of SP comets, supplementing the classical Oort cloud, was discussed by Levison (56.102.086) and Bailey (56.102.035). The transfer from the Kuiper belt was investigated using Markov chain modelling by Levison (54.091.024; 55.102.020), and using new efficient integrators to trace the orbits over 10^9 yrs by Levison and Duncan (*ApJ*, **406**, L35, 1993) and 10^8 yrs by Holman and Wisdom (*AJ*, **105**, 1987, 1993). Captures from the Oort cloud into Halley-type orbits were simulated by Gallardo and Fernández (PC, p. 35). Mapping techniques to model cometary chaos were reviewed by Froeschlé (55.042.053; 55.042.099). Manara and Valsecchi (54.102.011; 56.042.069) and Valsecchi (PC, p. 81) studied orbital transformations due to close encounters with the outer planets for comets in orbits of low or moderate eccentricities by accurate integration and found the speed-up technique of increasing the planetary masses to distort the perturbation distributions. Valtonen *et al.* (56.102.034) used numerical experiments with close encounters to find analytic formulae for the perturbation distributions. Emel'yanenko (52.102.035; 54.102.024; 55.102.046) developed an analytical theory for weak perturbations of cometary orbits in the circular restricted three-body problem and applied this (56.042.037; 56.102.013; 56.102.018) to study the orbital diffusion of LP comets.

Long-term integration samples based on real SP comets by Nakamura and Yoshikawa (53.102.084; 53.102.088; 56.102.037) and Tancredi and Rickman (55.102.019) gave indications of the statistics of jovian perturbations and the expected perihelion distance distribution of the Jupiter family. Lindgren (56.102.088) used a random sample of starting orbits for such integrations to find the typical dynamical residence time in the observable Jupiter family. Rickman *et al.* (SP, p. 55) performed Monte Carlo simulations of the physico-dynamical evolution of such comets, and Fernández *et al.* (SP, p. 143) investigated observational indications of their perihelion distance distribution. Evidence relevant to the physical evolution was found from statistics of nongravitational effects by Rickman *et al.* (54.102.018), indicating temporary, short-lived activation of SP comets after major reductions of the perihelion distance. Attention was focussed on Sun-grazing, or even sun-plunging, orbits as a relatively common cometary end state by Bailey *et al.* (55.102.010), mainly effective for LP and Halley-type comets.

The origin of comets by collisional aggregation in the transplutonian region was studied by Safronov and Gussejnov (52.102.015), and the dynamical criteria for the capture of the Oort cloud from a common cometary population of the stellar association where the Sun was formed were evaluated by Zheng *et al.* (52.102.070). Negative results from searches for the Kuiper belt were published by Levison and Duncan (52.102.014) and Cochran *et al.* (55.102.027), while the first positive detections may have been made by Jewitt and Luu (*Nature*, **362**, 730, 1993; see Sect. II.2). A search for comets in the vicinity of Jupiter was described by Tancredi and Lindgren (56.102.077; 56.103.081).

IV. Satellites (J.-E. Arlot)

IV.1 Observations for astrometric purposes – Photometry

At the USNO in Washington D.C., photographic observations of the Galilean satellites of Jupiter, satellites I-VIII of Saturn and the Martian satellites were continued by D. Pascu with the 63-cm refractor. This brings to 20 the number of apparitions observed for the Galilean satellites, 19 for Saturn's system and 12 for the satellites of Mars. Positions for Saturn's satellites are available only from 1974 to 1980. Positions for the Galilean satellites are available for 1967/68 as a result of the McCormick refractor and 1973-1993 as a result of the USNO 63-cm. Positions for the Martian satellites are available for 1967, some for 1969, 1971 and all from 1982 - 1988. Some of the Martian satellite observations were made with the 1.5-m Astrometric Reflector. The Galilean satellite observations are sent to JPL in support of the GALILEO project, and to the Bureau des Longitudes in Paris in support of ephemeris development. In addition to the photographic program for the brighter satellites, CCD observations are made of seven faint inner satellites: JV, JXIV, SXII, SXIII, SXIV, UV and NII (Rohde *et al.*, *BAAS*, **25**, 1993 ; Pascu *et al.*, *BAAS*, **24**, 1059, 1992). These observations have been continued for 12 years with the 1.5-m Astrometric reflector at Flagstaff by Pascu, P.K. Seidelman and J.R. Rohde. These observations provide both astrometric and photometric data (Pascu *et al.*, *BAAS*, **22**, 1117, 1990 ; Pascu *et al.*, *Icarus*, **98**, 38, 1992).

Astrometric positions of satellites Jupiter VI and VII were obtained on plates taken at Table Mountain Observatory in 1968 and are published by J.D. Mulholland (*PASP*, **102**, 1328, 1990). Shelus *et al.* obtained positions of Jupiter VI through Jupiter XIII at McDonald Observatory (53.099.006) as did Whipple *et al.* (55.099.015,105).

At JPL, R.A. Jacobson analyzed observations acquired from the Voyager missions to Uranus and Neptune. The results of these analyses were revised sets of dynamical constants, e.g., planet and satellites masses, for both the Uranian and Neptunian systems and numerically integrated ephemerides for the satellites. In addition, he converted the Voyager satellite observations to an equivalent astrometric form for general use (Jacobson *et al.*, 54.101.007 ; Jacobson, 54.101.003 ; Jacobson, *A&ASS*, **96**, 549,1992). From Voyager images, Showalter detected 1981 S 13 (Saturn XVIII) and studied its role in Encke gap (53.100.015).

T. Nakamura reported observations of close approaches between the Galilean satellites (Nakamura *et al.*, Proceedings of IAU Colloquium 127, "Reference Systems", p319, 1991) and published astrometric observations of the central satellites of Jupiter during 1986-1990 (Nakamura *et al.*, 53.99.001).

D.B. Taylor from the RGO reported CCD observations of the major satellites of Saturn, Uranus and Neptune that were made using the 1-m Jacobus Kapteyn Telescope on La Palma, Canary Islands during the 1990 and 1991 oppositions. Williams *et al.* (*MNRAS*, **250**, 1, 1991) used the 1990 CCD data for Nereid together with comparisons stars to estimate the period of rotation of Nereid. Beurle *et al.* (1993) have made a preliminary reduction of the Saturn satellite data for the 1991 opposition. They have shown these observations to be comparable in quality to the best photographic observations.

R. Viera-Martins reported that, using the Brazilian 1.6-m reflector, 668 CCD images of the Uranus system and 622 CCD images of the Neptune system were made and are now being reduced. A new photometric technique for the reduction of natural satellite images near a bright planet was also developed at the Observatorio Nacional (Rio de Janeiro, Brazil). These methods were applied to 173 photographic plates of the Uranium system. The results are presented in the Ph. D. dissertation of C.H. Veiga.

At Bordeaux Observatory (France) a CCD camera was built and used for the observation of the mutual events of the Galilean Satellites as well as for the astrometric measurements of positions of the satellites of Saturn.

At the Bureau des Longitudes (France), a campaign of CCD astrometric observations was continued and some data were published concerning the satellites of Mars (Colas and Arlot, 54.097.037 ; Colas, *A&ASS*, **96**, 485, 1992). and the satellites of Neptune (Colas and Buil, *A&A*, **262**, L13,1992).

At the Pulkovo Observatory, St. Petersburg, regular observations of the Galilean satellites with the 65-cm astrograph and the normal 33-cm astrograph were continued. 53 plates with satellites were taken altogether, but not yet processed. Moreover, the Saturn's satellites S2-S5 and S8, the satellites of Uranus U1-U4 and Neptune 1 were observed at the Byurakan Observatory by the staff members of Pulkovo Observatory using the 2.6-m reflector. 153 positions of Saturn's satellites with respect to Titan, 44 positions of Uranus' satellites with respect to the planet and 9 positions of Neptune 1 with respect to Neptune have been obtained. The mean error on the positions varies from 0.34" to 0.56". The results have been submitted for publication. 8 positions of Deimos in November 1990 were recorded and processed. A number of positions of Phobos and Deimos, observed in 1988, were also published (6.51.154, 6.51.155, 6.51.156).

S.V. Tolbin has published the observations of Saturn's satellites S1-S6 and S8 (220 plates), taken in 1977-1982 with the 65-cm refractor (11.51.102). The coordinates of the satellites were compared with modern theories. The mean residuals, equal to 0.20", is a sign of the high quality of these observations. Toblin published also the results of photographic observations of the Saturn's system (159 plates) taken at Pulkovo with the normal astrograph (11.51.103). The mean residuals for these observations (S2-S6, S8) are of 0.30".

At Goloseevo Observatory, Ukraine, the photographic observations made in preceding years were processed. The results of observations of Jupiter 6 at Goloseevo and at Maidanak observatories in 1987-1988 (12 positions) have been published (53.099.017, 54.099.039). The results of observations of the outer satellites of Jupiter in 1987-1989 at Majdanak (22 positions of Jupiter 6, 4 positions of Jupiter 7, and 2 positions of Jupiter 8) have been submitted for publication.

The observations of Saturn's satellites made at Goloseevo in 1980 with two instruments (the 40-cm double astrograph, and the 40-cm double large field astrograph) have been processed and published

(4.51.93) (altogether 305 positions of S1-S8), as well as the results of observations of S2-S9 taken in 1990 at Majdanak with the 60-cm reflector (altogether 232 positions). Also the results of observations of the satellites of Uranus and Neptune, including 13 positions of Triton made in 1990 at Majdanak, were deposited.

At Sternberg Astronomical Institute (SAI), Moscow, a new reduction of plates with Phobos and Deimos taken in 1988 at Majdanak by staff members of SAI (660 positions of Phobos and 639 positions of Deimos, referred to FK5/J2000.0) has been undertaken. The new positions have been compared with the theory of motion of the satellites, developed at the Space Flight Control Center (*Astronomicheskij Zhurnal*, **18**, 815-818, 1992).

At Nikolajev Observatory, Ukraine, 42 plates of the Galilean satellites and 42 plates with the bright Saturn's satellites (S3-S5, S8) were taken using the zone astrograph (D=210, F=74 cm). The observed positions with respect to the FOKAT catalogue have been processed and the results were deposited. The results of observations of these satellites in 1987-1987 were also deposited (11.51.99).

At Abastumani Observatory ,Georgia, observations of the satellites of Jupiter, Saturn, Uranus and Neptune with the large field astrograph (D=40cm, F=300cm) were continued. More than 100 plates with the Galilean satellites, more than 200 plates with Saturn's satellites (S3-S6, S8), more than 200 plates with Uranus 3 and Uranus 4, and more than 200 plates with Neptune 1 have been taken. A preliminary analysis of the results has been made. Mean errors of the positions, with respect to stars of the FOKAT-S catalogue, vary between 0.2" and 0.3".

The group of scientists involved in the Phobos-2 mission has published the list of on-board astrometric observations of Phobos and Deimos and other relevant data (53.097.140).

Nicholson *et al.* (52.100.031) and Hamilton *et al.* (54.100.037) reported observations of the co-orbital satellites of Saturn (52.100.031).

In 1991, an important campaign of observation of the mutual events of the Galilean satellites took place. Predictions were made by Aksnes and Franklin (*Icarus*, **84**, 542, 1990), by the British Astronomical Association in London (52.099.042) and by Arlot (*A&A*, **237**, 259, 1990). Some observations were published (Souchay *et al.*, *A&A*, **264**, 314, 1992; Le Campion *et al.*, *A&A*, **266**, 568, 1992; Mallama, *Icarus*, **97**, 298, 1992 and *Icarus*, **95**, 309, 1992; Oprescu *et al.*, *Rom. Astr. J.*, **2**, 61, 1992; Allen *et al.*, *Earth Moon and Planets*, **58**, 105, 1992; Froeschlé *et al.*, *A&A*, 1992). Descamps *et al.*, (*Icarus*, **100**, 235, 1992) put also into evidence the positions of hot spots on Io, the first satellite of Jupiter. At the same time, some results of the former campaign made in 1985 were also published (Arlot *et al.*, *A&ASS*, **80**, 1, 1990; Arlot *et al.*, 55.099.027; Franklin *et al.*, *AJ*, **102**, 806, 1991; Vasundhara, *J. Astrophys. Astr.*, *12*, **69**, 1991).

Blanco *et al.* (53.101.015) and Buie *et al.* (55.101.090) reported the observation of mutual events in the Pluto/Charon system.

IV.2 Comparison of observations with theories – Determination of elements

At JPL, the masses of Uranus and its satellites were determined using the recent Voyager data (Jacobson *et al.*, Meeting of the AAS, Division on Dynamical Astronomy, Key Biscayne, Fla, 1991; Jacobson *et al.*, paper G 32A-3, Spring Meeting of the AGU, Montreal, Canada, 1992 ; Jacobson *et al.*, 55.101.016). At RGO, in cooperation with Bordeau Observatory, Taylor *et al.* (*A&A*, **249**, 569, 1991) have analysed meridian circle observations of Titan and Iapetus made with the Carlsberg Automatic Meridian Circle on La Palma and the Meridian Circle at Bordeaux. From their analysis the corrections to the position of Saturn from the DE200 ephemeris agree well with the latest ephemeris DE202. The only significant correction to the elements of the satellites is to the mean longitudes of Iapetus. Harper and Taylor (*A&A*, **268**, 326, 1993) have determinded new improved orbital elements for the major satellites of Saturn from an analysis of observations spanning the period 1874 to 1988. New masses for the satellites Mimas, Enceladus, Tethys and Dione were derived. At USNO, an analysis of photographic astrometric observations of the Galilean Satellites was presented (55.101.091). At Bordeaux Observatory (France) Dourneau (*A&A*, **267**, 292, 1993) presented a theoretical work and fit the orbital elements of the eight major satellites of Saturn to 100 years of observations. At the Bureau des Longitudes, M. Chapront-Touze calculated the orbits of the Martian satellites from her theories (*A&A*, **240**, 159, 1990). From Voyager 2 data, Owen *et al.* (53.101.004) prsented the orbits of the six new satellites of Neptune. An analysis of the Pluto/Charon mutual events was made by Marcialis (52.101.043) and by Tholen and Buie (52.101.111).

N.V. Emel'yanov and S.N. Vashkov'yak (Sternberg Astronomical Institute, Moscow) have finished working on the determination of the dynamical and kinematic characteristics of the Martian satellite system from all available ground- and space-based observations. Among the determined parameters, are included the elements of Phobos and Deimos, the velocities of secular motions of angular elements, the secular accelerations of the satellites, the mass of the planet and the J2, J3, and J4 harmonics of its gravitational field, the position (coordinates) of the north pole of Mars' rotation axis. Using the absolute positions (w.r.t. reference stars) of Phobos and Deimos, the systematic differences between the dynamical system adopted in the DE118 ephemeris and the FK4 system were determined. The results are published in *A&A*, **267**, 634-642 (1993).

V.A. Shor (ITA, St.Petersburg) has published an improvement of the orbital parameters of Phobos and Deimos taking into account Phobos-2 on-board observations as well as all other available ground- and space-based observations of the satellites. The theoretical model used in that work was previously developed according to the analytical theory by Sinclair. Secular accelerations of both satellites were derived, with good precision, in agreement with the results of other investigators (54.097.011).

N.O. Kirsanov (ITA, St.Petersburg) has compared 473 positions of Titan and 93 positions of Hyperion, covering the time interval from 1972 to 1984, with the results of numerical integrations. The differential correction of initial elements of the two satellites has been fulfilled. The new orbits agree with the used observations with a mean error of 0.36". The work is aimed at the construction of a semi-analytical theory of the satellites, which is now under development.

I.A. Gerasimov (Sternberg Astronomical Institute) and B.R. Mushailov have studied the evolution of Hyperion's orbit (53.100.004, 53.100.020).

S.A. Gasanov (Sternberg Astronomical Institute) has used 314 observations of Iapetus, covering the time span from 1974 to 1984, to improve the parameters of the analytical theory of the satellite constructed with the Laplace method (54.100.010, 54.100.017). The observations are represented with a mean error of 0.73".

N.N. Gor'kavyj (Astronomical Institute, Moscow) and co-authors were involved in the elaboration of the theory of Neptunian arcs (54.101.126, 54.101.127, 55.101.007, 55.101.008, 55.101.009).

IV.3 Theoretical studies

Concerning the Galilean Satellites, the orbital resonance among JI, JII and JIII was studied by Changyin and Lin (54.099.056). At Lille University (France), Duriez and Vienne built a new theory of the eight major satellites of Saturn where all the dynamical parameters are taken into account at the same time (Duriez and Vienne, 53.100.006; Vienne and Duriez 53.100.014; Vienne and Duriez 55.100.009). A study of the secular acceleration of Mimas was presented (Vienne *et al.*, 55.042.043) and a synthetic theory of the motion for Titan-Hyperion was studied (Duriez, 55.042.041). Gerasimov and Mushailov studied the evolution of Hyperion's orbit (53.100.004,020) and an analytical theory of the motion of Iapetus was presented by Gasanov (54.100.017). Sessin made theoretical studies of the Mimas-Tethys system (55.042.044) and the influence of Enceladus in the motion of Helene was studied by Bevilacqua and Sessin (55.042.045). M. Chapront-Touze completed a theory of the motion of Phobos and Deimos (*A&A*, **235**, 447, 1990), Ivanov *et al.* (52.097.031) published a new theory of the motion of Phobos and Vashkov'yak and Emel'yamov (52.097.027) improved Phobos' orbit. Wnuk and Breiter also published a study on the motion of satellites in Mars' gravity field (53.097.200). Krasil Nikov (52.042.027) studied the spatial rotations of a satellite about the mass center in the circular 3-body problem in the case of a main resonance. Secular perturbations of fictitious satellites of Uranus were studied by Kinoshita and Nakai (54.042.092) and a semi-analytical theory for the motion of Uranus' satellites was presented by Lazzaro (54.101.012). At OCA (France) Oberti published an analytical model for an accurate solution for Nereid's motion (52.101.013).

IV.4 Ephemerides – Numerical Integrations – Predictions of events

At JPL, R.A. Jacobson began the development of numerically integrated Galilean and Saturnian satellite ephemerides in support of JPL's Galileo, and Cassini missions, respectively (Jacobson *et al.*, meeting of the AAS, Division on Dynamical Astronomy, Santa Barbara, Ca, 1993). Other activities included maintaining and updating the satellite observations database and providing ephemerides for a number of JPL and non-JPL users. Among the former were the JPL Mars Observer project (Phobos and Deimos) and the Pluto Fast Fly-by pre-project (Charon). Among the latter were the Space Telescope Science Institute (all natural satellites). At RGO, using the numerical integration of Hyperion by Sinclair and Taylor (1985), Taylor (*A&A*, **265**, 825, 1992) has developed a synthetic theory for the motion of this satellite and fitted it to the modern photographic observations. The coefficients of the synthetic long-period theory were used by Message (1993) as starting conditions in his differential correction procedure (based on a new theory of the long-period motion) to determine a set of dynamically consistent coefficients. A study of Saturnian satellite ephemeris calculations was presented by Rongchuan and Kaixian (54.100.035).

Data on the mutual events and eclipses by Saturn will help refine the motions of the Saturnian satellites. Preliminary predictions were made by Sema (*A&A*, **265**, L21, 1992) for eclipses of Iapetus and by Thuillot *et al.* (*BAAS*, **24**, 938, 1992) for the eclipses and mutual events.

IV.5 Dynamics of the Ring Systems

Goldreich (55.042.024) outlined some of the main processes that shape the planetary rings, and three-dimensional perturbations in a narrow planetary ring were studied by Kolvoord and Burns (55.091.005). A theory of the Neptunian ring arcs was made by Gor'kavyj *et al.* (54.101.126,127) and an explanation of those arcs was also given by Porco (54.101.005). Concerning the rings of Uranus, papers "Shepherding Satellites and Dynamical Structure of the rings of Uranus" were published by Kozai (*PASJ*, **44**, 135, 1992 and *PASJ*, **45**, 263, 1993). Colwell and Esposito have published a numerical model of Uranian rings (52.101.003). A paper on the orbits of shepherd satellites deduced from the structure of the rings of Uranus was also published by Murray and Thomson (52.101.025). French *et al.* presented a work on the dynamics and structure of uranian rings (54.101.104). Concerning the rings of Saturn, Kolvoord and Burns published a collisional modelling of Saturn's F ring (52.100.030) and the dynamics of Saturn's E ring particles was also published (55.100.069). Concerning observations, several efforts were completed: Turtle *et al.* (52.100.026) deduced the kinematics of Saturn's major narrow rings from Voyager data and ground-based observations and Nicholson *et al.* (52.101.011) reported on the observation of Neptune's ring arcs by stellar occultation as well as Sicardy *et al.* for the 1983-1989 period (53.101.006).

V. Radar Astrometry for Asteroids, Comets, and Natural Satellites (D.K. Yeomans)

Radar observations of the solar system's small bodies began with observations of asteroid (1566) Icarus in June 1968 and since that time there have been successful radar experiments involving over 60 different main belt and near-Earth asteroids. Although the focus of these radar experiments has been to infer the asteroids' physical characteristics from the strengths and properties of the returned signals, corrections to the predicted Doppler and time delay ephemerides are also possible. Once these corrections are made, Yeomans *et al.* (*AJ*, 94:189) showed that the resulting differences between received and transmitted frequencies (Doppler shifts) and the round trip time delays can provide extremely powerful data types for the orbit determination of asteroids and comets.

Astrometric radar data effectively measures the object's distance and velocity along the observer's line-of-sight and hence these data are complementary to optical, plane-of-sky measurements. Radar data taken during an object's close approach to the Earth are most powerful, and the orbit refinement most dramatic, if the object has only a short optical astrometric history.

In orbital solutions, radar observation residuals for near-Earth objects are often less than 1 Hz in Doppler and less than a microsecond in round-trip delay time. At the Arecibo transmitter S-band frequency (2380 MHz), these residuals correspond to range and velocity errors of less than 150 m and less than 6.5 cm/sec. For the Goldstone X-band frequency (8510 MHz), the corresponding velocity error is less than 2 cm/s. The power of the radar data becomes evident when one realizes that radar measurement errors are orders of magnitude smaller than the position and velocity uncertainties inherent in orbits based only upon optical data over short time intervals.

Although astrometric radar data have been published since the first asteroid radar experiment in 1968, few of these measurements have been used to improve asteroid orbits. Lieske and Null (*AJ*, **74**, 297) included the 1968 Goldstone radar data in their early analysis of the orbit of (1566) Icarus but it was more than two decades before additional "radar" orbits were published by S.J. Ostro and colleagues for close Earth approaching asteroids 1986 JK (*Icarus*, **78**, 382), 1989 PB (*Science*, **248**, 1523), (1627) Ivar (*AJ*, **99**, 2012), and 1986 DA (*Science*, **252**, 1399).

Since the population of short-period comets is roughly an order of magnitude smaller than the family of near-Earth asteroids, the opportunities for cometary radar observations are relatively scarce. Beginning with comet P/Encke in Nov. 1980 (Kamoun *et al.*, *Science*, **216**, 293), radar signals have been successfully returned from the nucleus regions of comets P/Grigg-Skjellerup in May-June 1982 (*BAAS*, **14**, 753), 1983 VII IRAS-Araki-Alcock in May 1983 (*AJ*, **89**, 1745; *ApJ*, **338**, 1071), 1983 V Sugano-Saigusa-Fujikawa in June 1983 (*BAAS*, **15**, 180), and from comet P/Halley in Nov. 1985 (*ApJ*, **338**, 1094). Doppler measurements are available for the first three comets but because of an apparent debris cloud surrounding comet P/Halley, there are no published astrometric radar measurements for this comet's nucleus.

With the relatively recent realization that a large population of near-Earth asteroids are on Earth approaching orbits, there is a critical need to monitor accurate their future motions. For the majority of these objects that lack a long history of optical astrometric data, accurate extrapolations of their motions will require the use of radar data in future orbital solutions.

A complete analysis of the existing asteroid and comet radar astrometric data was presented by Ostro *et al.* (*AJ*, **102**, 1490) for 30 asteroids and 4 comets. Using these radar data and the available optical observations, Yeomans *et al.* (*AJ*, **103**, 303) explained the process by which radar astrometric data are included in orbital solutions and presented orbits for those objects for which radar data existed at that time. Since these publications, Ostro *et al.* (*BAAS*, **23**, 1144) published radar astrometric data for near-Earth asteroids 1990 MF, 1991 AQ, and 1991 JX. In addition, astrometric data have been taken (but not yet published) by S.J. Ostro and colleagues for asteroids 1982 BB (July-Aug. 1991), (7) Iris, (324) Bamberga, and 1991 EE (Sept. 1991), (1981) Midas (Feb.-Mar. 1992), (4) Vesta (Mar.-Apr. 1992), 1990 UQ (May 1992) and (4179) Toutatis (Nov.-Dec. 1992).

Over the period between Nov. 27 and Dec. 18, 1992, the radar observation campaign undertaken by S.J. Ostro and colleagues provided impressive radar "images" of asteroid (4179) Toutatis. In the process, 34 astrometric delay measurements and 21 Doppler measurements were made from JPL's Goldstone radar facility and from the Arecibo radar facility in Puerto Rico. Ranging resolution of 1/8th microsecond (19 m) and Doppler resolution to 0.008 Hz (0.15 mm/s) was achieved, and at least through most of the observation period, the ephemeris prediction errors were smaller than the object itself (less than about 4 km) and primarily limited by the uncertainty in the time-dependent difference between the radar bounce point and the object's center of mass.

Note should also be made of the successful Toutatis radar experiment by Zaitsev *et al.* (ACM'93 abstract, p. 325) whereby the uplink transmitter frequency at the Yevpatoria antenna in the Crimea was modified in accordance with the Toutatis ephemeris so that the downlink signal was received at a constant frequency at the Effelsberg radio antenna in Bonn, Germany. Ten Doppler observations were obtained on Dec. 8–9. The existing optical and radar data, when fully reduced, will make the orbit of this object among the best known in the solar system. Additional radar opportunities will become available for this object during future Earth close approaches on Nov. 29, 1996 (to within 0.035 AU), on Oct. 31, 2000 (0.074 AU) and especially in 2004 when Toutatis will approach the Earth to within 0.010 AU) on Sept. 29.

De Pater *et al.* (in preparation) observed reflected radar transmissions at the VLA from main belt asteroids (7) Iris and (324) Bamberga as well as the near-Earth asteroid 1991 EE in Sept. 1991. On Dec. 17, 1992 a similar observation was made of asteroid (4179) Toutatis using the Goldstone – VLA configuration. Interferometric observations of this type allow two dimensional images to be developed and although the frequency resolution is orders of magnitude coarser than when Goldstone transmits and receives, the Goldstone – VLA configuration has the advantage of creating two dimensional images with 0.25 arcsec resolution at several different frequencies. By comparing the two-dimensional images thus created with the predicted ephemeris positions, interferometric radar observations also allow plane-of-sky angle astrometry to a fraction of an arc second.

In addition to the existing radar astrometric data for several asteroids and comets, Ostro *et al.* (*JGR*, **97**, no. E11, 18 227) presented Goldstone and Arecibo Doppler observations of Europa, Ganymede, and Callisto over the interval from June 1987 through Feb. 1991. Doppler observations were published for Europa (18 observations), Ganymede (23), and Callisto (35). The largest discrepancies between the predicted and observed Doppler observations implied that Callisto may have been lagging its ephemeris by an average of about 200 km.

Arecibo observations from Feb.–March 1992 have yielded the first successful radar ranging measurements to the Galilean satellites of Jupiter. In 1992, Harmon *et al.* (*AJ*, submitted) made range measurements to Ganymede on March 9 and Feb. 20 and to Callisto on 1992 March 3 and March 7. These observations suggest that the required along track position corrections to the ephemeris predictions in that time frame were approximately +122 km for Ganymede, −307 km for Callisto and about +11 km for Jupiter itself. J.H. Lieske is currently making the necessary program modifications to allow the processing of radar data in his Galilean satellite ephemeris development software at JPL.

VI. Occultations (L.H. WASSERMAN)

VI.I Identification and prediction of upcoming occultations

Computerized searches for future occultations are being performed independently as before by three investigators. Wasserman, Bowell and Millis published predictions for 1992 and 1993 (55.096.009). Dunham publishes predictions in the *Occultation Newsletter* (*ON*) and an annual summary appears in each January issue of *Sky and Telescope*. Goffin does an independent search. His results appear in *ON* and in the *European Asteroidal Occultation Network* (*EAON*). The chairman of the *EAON* Boninsegna, coordinates European observational activities.

VI.2 Prediction refinement

Astrometric measures for upcoming occultations were provided at various times by Klemola at Lick, by Thirionet at Uccle, by Van Vleck Observatory, and by Lowell Observatory.

VI.3 Observations

There have only been five occultations of minor planets observed during this reporting period which had a sufficient number of chords that a diameter could be determined. Kissling *et al.* observed an occultation by 9 Metis on 1989 Aug. 6 (54.098.088). Buchner reports on an occultation by 521 Brixia on 1989 Oct. 23 (52.098.068). Sato, Soma and Hirose (*AJ*, **105**,1553) report on an occultation by 381 Myrrha on 1991 Jan. 13. This event was especially unusual since the occultation was of the brightest star ever observed in occultation by a minor planet – 1.9 mag γ Geminorum. Observations were also made of an occultation by 4 Vesta on 1991 Jan. 4 and by 216 Kleopatra on 1991 Jan. 19, although the results are still unpublished. In addition, a larger number of other events were observed, but with insufficient chords to determine a diameter. Such events are periodically reported in *ON* and in *EAON*. Oddly, there have been no well-observed occultations in the two and a half years since the three events in January, 1991. A review paper by Millis and Dunham (51.098.149) was published in the book resulting from the Asteroid II conference held in Tucson in 1988.

VII. Asteroid and comet names (A. Carusi)

During the triennium there has been considerable discussion regarding the current procedure of giving names to minor planets and comets. Concerns about the suitability of the procedure were put forward, in view of the already increased number of discoveries and the prospect that even larger numbers of bodies will be automatically detected in the future.

Although the problem is probably not a very urgent one, some of the members of the Commission 20 naming committees feel that a start should be made on revising the naming rules. Other members are uncertain about the necessity of changing a system that apparently works. Nevertheless, the focus should not merely be on its inefficiency, but on its possible complete inadequacy to meet the requirements of reliability and fast response when a hundred thousand objects may be discovered each month.

In particular, it has been proposed that the naming rules for minor planets and comets should be unified, thereby providing a single system for all minor bodies. This unification could solve some of the problems currently presented by objects that are difficult to classify, being observed and classified sometimes as minor planets and sometimes as comets, e.g., the comet P/Wilson-Harrington (1949g = 1949 III), now known as minor planet (4015) Wilson-Harrington.

It is generally agreed that the need for thousands of new names may pose serious problems in the future. A possible solution is the creation of a name bank – including categories proposed by discoverers and scrutinized and accepted by the *MPNC* – from which new names should be extracted.

The name bank could be put on-line, so that discoverers may quickly and efficiently choose names they like. Categorization of special types of names, such as Trojans, AAAs, Centaurs, etc., should be clearly stated and maintained. This method may help resolve the conflicts that take place from time to time: the current naming rules could be applied to entire categories, instead of individual names, alleviating the bi-monthly discussions inside the *MPNC*. On the other hand, discoverers would maintain the privilege of proposing names, possibly assigning to them "tags" for specific objects.

IAU WORKING GROUP ON NEAR-EARTH OBJECTS

REPORT ON ACTIVITY 1991–1994

Convenor: A. Carusi
Members: A. Basilevsky, B. Gustafson, G. Hahn, A. Harris, S. Isobe, G. Lelièvre, A. Levasseur-Regourd, B. Marsden, D. Morrison, A. Milani, K. Seidelmann, E. Shoemaker, A. Sokolsky, D. Steel, Tong Fu, I. Williams.

I. Introduction

The Working Group on Near-Earth Objects (*WGNEO*) was created by the IAU at its XXIst General Assembly in Buenos Aires, Argentina, August 1991, to allow the interaction of astronomers working in a variety of fields. Members of the *WGNEO*, from Commissions 4, 7, 9, 15, 16, 20, 21, and 22, were convened by Andrea Carusi, President of Commission 20.

The main scope of the *WGNEO*, as outlined by the Resolution adopted at the XXIst General Assembly, was to "assess and quantify" the potential threat posed to our planet by impacts of asteroids and comets, to "pool resources" for their discovery and follow-up, and to "act as an international focal point", seeking cooperation with scientists in different disciplines.

The *WGNEO* has worked for three years, with active participation of its members. This Report contains an outline of its scientific evaluation of the problem, together with some suggestions for future activities.

The *WGNEO* concurs with the judgement that cosmic impacts represent a significant natural phenomenon that has played an important (perhaps critical) role in the past evolution of life on Earth, and that such impacts represent a continuing threat that should be considered and dealt with on an international basis. Both the accelerating rate of discovery of near-Earth asteroids (and comets) and studies of the potential environmental consequences of large impacts have enhanced scientific and public interest in cosmic impacts since the 1991 General Assembly. We anticipate that this interest and concern will continue as we learn more about NEOs.

II. Scientific overview (D. MORRISON)

During the past three years the known number of near-Earth asteroids (NEA) has nearly doubled, and one was discovered passing less than 150,000 km from Earth. The break-up of Comet P/Shoemaker-Levy 9 (1993e) and the anticipated collision of several of its fragments with Jupiter in July 1994 provide a visible example of planetary impacts by NEOs. Radar investigations of Toutatis during its close flyby of Earth in 1992 led to the first detailed images of a NEA, and the first spacecraft mission (Clementine, a US Navy spacecraft to fly past 1620 Geographos in 1994) is nearing readiness for launch. In addition, major international scientific meetings dealing with the NEOs and their impact threat have been held in Flagstaff (USA), San Juan Capistrano (USA), St. Petersburg (Russia), Liège (Belgium), Tucson (USA), Erice (Italy), Belgirate (Italy) and Sagamihara (Japan). Several more are planned for 1994. Detailed studies of the impact problem and possible ways of dealing with it have been completed by two working groups of NASA (USA) and presented to the U.S. Congress, and a comprehensive multi-author book on the subject is being published by the University of Arizona Press. All of these efforts have focused attention on the impact hazard and led to an improved understanding of the phenomena involved.

In any given year the probability of an impact large enough to do significant harm is small, but not negligible. In the long term, such impacts can be expected to result in catastrophes of a magnitude that equals or exceeds that of any other natural phenomenon, such as severe storms, volcanic eruptions, or earthquakes.

The collision of small asteroids or comets with the Earth (diameters less than a few tens of meters, impact energies less than about 10 megatons) can result in spectacular fireballs or bolides but generally do comparatively little damage on the ground. However, the atmospheric disintegration of such a body has the potential of being mistaken for a nuclear explosion at a time of international tension. Education concerning the nature of these events and their frequency is important if we are to avoid their misinterpretation as nuclear attacks.

Surface impacts or lower-atmosphere explosions of objects with energy greater than about 10 megatons (and perhaps 50 m in diameter) are capable of inflicting severe local or regional damage; such events are expected to take place approximately once in a few centuries. Ocean impacts of objetcs having energies greater than 1000 megatons and diameters of a few hundred meters have an especially large potential for damage to coastal regions resulting from tsunamis. Larger impacts, corresponding roughly to impacting objects having diameters greater than 1-2 km, generate global consequences, especially through injection of dust into the stratosphere. Such impacts, which are believed to occur on time-scales of hundreds of millennia, could produce short-term climate fluctuations, lead to widespread loss of crops and mass starvation, and destabilize modern civilization. Still larger impacts, occurring on time-scales of millions of years, could lead to mass extictions of species and alter the course of biological evolution.

Both the probabilities of impacts of various sizes and energies and their anticipated consequences have large uncertainties. It is estimated that fewer than 10% of NEAs with diameters greater than 1 km have been discovered, and our knowledge of the cometary population at these sizes is even more uncertain. Evaluation of the threat of cosmic impacts requires a better understanding of the population of comets and asteroids and also of their physical and chemical nature. Such information can be obtained from a combination of astronomical searches for NEOs, follow-up studies of their physical properties using optical and radar astronomy, and visits to a few representative objects by spacecraft. Such studies are of great scientific importance for understanding NEOs and their role in solar system history as well as for evaluating the impact threat. They are also a prerequisite for any efforts to develop strategies for deflecting an incoming object or mitigating the consequences of its impact.

III. Present and future search programs (E.L.G. Bowell, G. Hahn, A.W. Harris, D.I. Steel)

III.1 Present search programs

There are currently four dedicated near-Earth object search programs in operation (three in the U.S.A., one in Australia). From time to time, NEOs are discovered by other researchers, especially those using Schmidt telescopes. Indeed, wherever wide-field plates are taken, observers are encouraged to search them for tell-tale trails, since these represent a data source which is largely unexploited; the extensive plate libraries that exist also contain some thousands of unrecognized trails from NEOs yet to be discovered.

The longest-operating program is the *Planet-Crossing Asteroid Survey (PCAS)*, at Palomar, which was begun in 1972 under E.F. Helin (JPL) and E.M. Shoemaker (USGS). Ten years later, Shoemaker initiated the *Palomar Asteroid and Comet Survey (PACS)*, which has similar but distinct aims. The two programs use the same instrument for their data collection (the 0.46 m Schmidt at Palomar Mountain Observatory in California) and similar analytical techniques. Pairs of short-exposure photographs are inspected using stereomicroscopes to search for moving objects down to V = 17.5. Both teams cover about 40 000-50 000 square degrees of sky a year, and each discovers about a dozen near-Earth asteroids and several comets per annum as a result.

The third U.S. program is undoubtedly the most sophisticated world-wide. The *Spacewatch Telescope* (University of Arizona), with a long focus and a 0.91-m aperture, began operations in the late 1980s. This instrument is used in scan mode, with the RA drive switched off, so the sky is swept at sidereal rate. A large-format CCD is read out at the same rate to produce a scan about 40 arcmin wide and as long as the observation is produced (about 30 minutes of time). The target region is then twice re-scanned. A comparison in real time of the three scans then allows moving objects to be identified. Since the *Spacewatch* instrument is able to detect not only large NEOs at substantial geocentric distances but also nearby (cis-lunar) objects as small as 5-10 meters in diameter, it has been possible to fill the gap in the size-frequency plot between bright meteors and small asteroids. With good conditions the *Spacewatch* team may discover several NEOs in a month.

The only program located in the southern hemisphere is the *Anglo-Australian Near-Earth Asteroid Survey (AANEAS)*, begun in 1990. All plates and films taken with the 1.2-m U.K. Schmidt Telescope are searched for asteroid trails, those of unusual length or orientation being followed up. About 40 000 square degrees per year are covered to a limiting asteroidal magnitude of about V = 19. The discovery rate of NEOs is currently about 10 a year.

III.2 Future projects

For the future, there is a widespread belief among NEO researchers that an observational system like the proposed Spaceguard Survey is necessary. As described in *The Report of the Near-Earth-Object Detection Workshop* (NASA, 1992), an NEO survey lasting at least two decades would be carried out using a worldwide network of perhaps six 3-m-class telescopes. By scanning several thousand square degrees of sky up to ten times per month, and to a limiting magnitude V = 22, it should be possible to discover about 90% of NEAs larger than 1 km in diameter. Earth-crossing comets are harder to discover because many of them have long orbital periods. Overall, it is estimated that about three quarters of all NEOs larger than 1 km diameter could be discovered during the course of a 25-year Spaceguard Survey, though it is anticipated that improvements in technology will lead to even greater NEO discovery rates.

Meanwhile, a number of groups are undertaking the precursor research for a major effort like Spaceguard. In the U.S., the *Spacewatch* team is concentrating on installing a 1.8-m telescope; a group at JPL is investigating the use in NEO searches of one or more of the 1-m wide-field telescopes currently used by the U.S. Air Force for tracking Earth satellites; and at Lowell Observatory, a 0.41-m Schmidt telescope is being refurbished with a mosaic-CCD camera. In Europe, the *EUNEASO* initiative (*EUropean Near-Earth Asteroids Search Observatories*), consisting of groups at the OCA (France), DLR (Germany), Uppsala Observatory (Sweden), and Helsinki Observatory (Finland), is working to equip the 90-cm Schmidt telescope at Caussols (OCA) with a CCD-array camera, and run it as a search station. The possibility of a second search station at ESO (1-m Schmidt), La Silla, Chile, is currently being investigated. Several other stations are planning to participate as follow-up and/or physical observation sites; e.g., Ondřejov (Czech Republic). In addition, national efforts are being planned in Italy and Russia, as precursor projects. In Australia, the *AANEAS* program is being expanded to make use of the UKST during bright-of-Moon time when that instrument is normally unused, employing the stereo pair technique pioneered by the *PCAS* and *PACS* teams. There is broad recognition that ramping up toward a Spaceguard-like survey entails research in a number of areas, including the development of more sophisticated image-detection software than currently exists, the processing of large numbers of asteroid and comet detections, the efficient design of follow-up strategies, and the planning of physical observations. Consequently, healthy collaborations are developing among researchers to share their expertise and workload.

It should be noted that such programs, while aimed principally at the detection and follow-up of NEOs, will yield as a spinoff a huge increase in the inventory of solar system objects (main belt asteroids, Trojans, Centaurs, distant bodies). Thus, there are benefits to scientists other than planetary astronomers in designing and conducting extended NEO surveys.

III.3 Radar observations

Radar astrometry represents a very powerful mean for NEO studies. As a matter of fact, radar observations are capable not only to provide information on the physical characteristics of the bodies under study, but also to help in the refinement of their ephemerides.

A comprehensive review, by D.K.Yeomans, of the researches done in this field in recent years may be found in the report of IAU Commission 20 (this volume). Here we want only to stress the usefulness of this technique in NEO searches. The results obtained in 1992 with the radar observations of (4179) Toutatis, at its passage close to the Earth, have clearly shown that these studies may become of primary importance in the future.

IV. Follow-up programs (B.G. Marsden)

If and when a full-fledged NEO detection program like Spaceguard is undertaken, most of the necessary astrometric follow-up will occur automatically, as part of the scheduling of the observations. It is to be expected that the field in which a discovery is made will be reobserved a night or two later. Furthermore, there is merit to arranging for the field to be scanned on one of these nights both well east and well west of the meridian; not only will the shorter time difference simplify the linkage, but the resulting diurnal parallax will help the initial orbit determination. The same could be accomplished with observations from a different site, although unless there is a considerable difference in latitude the observations should be made at substantially different local times.

Two-night detections can automatically be inspected for identifications with earlier objects for which there are orbits in the database, even if these are of quite poor quality. New discoveries of particular interest can be flagged anyway. In automatic mode, the same general fields should also then be covered in the same way on two more nights later in the dark run. The database will at that point include some kind of orbital representations for the objects detected earlier in the month, so any rediscovery of the same object will immediately become apparent, after which a better orbit, compromising a 10-15-day arc, can obviously go into the database.

As noted in Sec. II.2 of the report of Commission 20, the current international effort on minor planets involves well over 100 observatories (15 or more very actively), the discovery of more than 4 000 unidentified objects each year, and an annual total that is well in excess of 40 000 published astrometric observations. Estimates made in connection with the discussion of Spaceguard suggest that this could became the rate of detection of actual or suspected NEOs alone, while the total activity might be two orders of magnitude higher.

If this should come to pass, there should be no real difficulty handling the data, even if the available computers were only one order of magnitude faster than those used in work on minor planets at present. Most of the time spent at present is caused by poor or unbusinesslike communications and the slowness of reducing photographic data and their frequent unreliability – notably the errors that often arise in the quoted times of observation. When CCD astrometry is carried out by an organized team – even a team of amateur astronomers – and the observations are reported in an efficient manner that minimizes any necessary reworking of the data, identifications can be made and orbits can be computed very rapidly. Another source of wasted time at the present is the manner in which the astrometric and orbital data are collected together specifically for publication in printed material according to a fixed schedule – like the monthly *Minor Planet Circulars*. Although this material is currently also available electronically, the orientation toward the printing schedule and the need to publish material that is internally consistent makes it more difficult to ensure that the very latest orbital elements and ephemerides are always available to those making further observations, particularly physical observations using telescopes that may have fields of view on the order of only 1 arcmin.

The reality of the moment is that much of the astrometric follow-up of NEOs is currently being done by the four principal discovery programs. Specific follow-up astrometry is carried out at the Oak Ridge Observatory in Massachusetts, at the University of Victoria and Dominion Astrophysical Observatory, at Mt. John University Observatory on New Zealand's south island, as well as by several amateur astronomers, notably in Japan and Italy. The growing amount of amateur involvement is an encouraging sign, but almost without exception, these are small-field instruments. Although extensive follow-up clearly detracts from the search activities themselves, it is therefore essential that accurate positions be quickly and reliably made available from the discovery data and that the wide-field instruments of the discoverers themselves be used for the second-night confimation – and further observations too if the initial recognition of a NEO is significantly delayed.

In the absence of other identifications, astrometric follow-up should be continued (on two nights each dark run) for as long as possible at the initial apparition. This minimizes the search area that will need to be covered to find the object in later years. Many NEOs are discovered on particularly favorable passages, and it is therefore these same apparitions that should be utilized for obtaining physical follow-up, notably three-colour (or more) photometry in order to establish quantities such as radius, albedo and rotation period. If an object comes close enough to the Earth, radar observations should also be made for both astrometric and physical purposes.

V. International context (A. CARUSI, B.G. MARSDEN, P.K. SEIDELMANN)

V.1 General considerations

The threat posed by NEOs makes no distinction by political boundaries: the entire planet is at risk. This international character of the hazard requires an equally international response.

There are at least three reasons why searches for NEOs should be overseen, in the long run, by an international agency:

- Scientific investigations devoted to the discovery and follow-up of NEOs, to the understanding of their physical nature and of the dynamical mechanisms that deliver them in the inner regions of the solar system, as well as to the possible phenomena associated with an impact, are best conducted in an international context, where a free exchange of knowledge and skills may allow a synergistic effort and therefore an optimal result. A close connection with international bodies other than the IAU should be established to broaden the interest in NEO searches. As an intermediate step, we recommend that an inter-union working group be formed as soon as possible.

- Research and experiments on defense systems are a very delicate issue, which involves people and assets formerly involved with the Cold War. It would be at least prudent that any decision and directive on these matters be discussed and agreed upon in an international forum.

- Political and social initiatives regarding NEO searches have a higher probability of success if the enterprise is international in nature. In particular, sharing resources and costs is especially applicable to efforts such as the proposed Spaceguard Survey, which require the establishment of observational and communication networks.

Since NEOs discovered in the northern hemisphere can quickly move to the southern (and vice versa), follow-up of NEOs is necessarily an international activity. The number of professional astronomers involved in NEO follow-up at present is extremely small. Further participation is actively encouraged, particularly by those with access to moderate-to-large-aperture telescopes that can follow discoveries to magnitude 20 and fainter. It is perhaps unfortunate that the principal discovery programmes also have to conduct a large part of the follow-up, but unnecessary duplication could even now be eliminated if there were greater cooperation among these programs.

A few words should be devoted to the prospects for radar observations. The Toutatis radar experiment by Zaitsev *et al.* (ACM'93 abstract), using the Yevpatoria antenna (Crimea) as the uplink transmitter and the Effelsberg radio antenna (Germany) as the downlink receiver, has opened a new field for international cooperation. No plans have yet been made, however, for coordinated efforts using radars.

Apart from the *WGNEO*, there are already other organizations at work which are international in character. This is the case, for example, of the scientific network *Role of Impact Processes in the Geological and Biological Evolution of Planet Earth*, of the European Science Foundation, chaired by Dr. D. Stöffler (Germany). It should also be stressed that the *Minor Planet Center* and the *Central Bureau of Astronomical Telegrams*, operated by the Smithsonian Astrophysical Observatory in Cambridge (USA), are in a way an international agency, operating under the auspices of the IAU.

V.2 WGNEO Recommendations

The IAU *Joint Working Group on Near Earth Objects* recognizes that:

1. our current knowledge of the quantity, distribution, and actual orbits of Near Earth Objects is very limited,

2. an inventory of NEOs as complete as possible with present techniques, is only possible through a cooperative, coordinated program of observation and data collection,

3. cosmic impact is an environmentally significant phenomenon which has played a major role in the evolution of life on Earth,

4. in any given year there is a very low probability that a large cosmic impact may occur which would destroy human civilization or even a significant fraction of life on Earth. However, the threat is real and requires further internationally coordinated, public educational efforts,

5. a significant near term cosmic impact threat is a naturally produced, atmospheric explosion of a small Near-Earth Object being mistaken for a nuclear explosion, at a time and place of international tension. These events have been observed. Such an event could be misinterpreted as a nuclear attack and trigger an unfortunate reaction,

6. the gathering of additional physical knowledge on Near Earth Objects and their effect on Earth is a scientifically and socially important endeavor. These multi-disciplinary efforts should be conducted in an open, coordinated, international manner. Dedicated international astronomical facilities similar to the proposed *Spaceguard System* should be developed. The assets and technologies of many countries can contribute to the gathering of these data, through ground and robotic space observations. The skills and technologies, necessary for any large, complex investigation, should be utilized,

7. unless a specific and imminent threat becomes obvious, actual construction and testing of systems that may have the potential to deflect or mitigate a threat may be deferred because technology and systems will improve,

and, therefore, recommends that:

1. a coordinated, international search program be initiated to detect all Near-Earth Objects, brighter than 20th magnitude,

2. the *Minor Planet Center* perform the functions of the collection of observational data and the determination of accurate ephemerides,

3. capabilities for obtaining accurate astrometric observations, as necessary to determine accurate ephemerides be established,

4. coordination between scientific and humanitarian organizations and governments be encouraged concerning NEOs,

5. an international source for accurate and responsible scientific information on NEOs be established and available to the media,

6. the *Working Group on NEOs* be continued to fulfill the above recommendations.

21. LIGHT OF THE NIGHT SKY / LUMIERE DU CIEL NOCTURNE

President: M. S. Hanner

Vice President: M. G. Hauser

Organizing Committee: S. Bowyer, R. Dumont, Yu. Galperin, S. S. Hong, Ph. Lamy, Ch. Leinert, A. C. Levasseur-Regourd, T. Mukai

I. INTRODUCTION

The diffuse components of the light of the night sky encompass a variety of physical phenomena over the full range of cosmic distance scales. From the Earth outward they include airglow in the Earth's atmosphere, scattering and thermal emission from the interplanetary dust cloud (zodiacal light), integrated starlight, diffuse galactic light, and extragalactic background radiation. These components are necessarily observed together along the line of sight. Thus, the successful measurement of any one component requires knowledge of all the diffuse components. Moreover, the brightness of the diffuse sky background limits the detection of very faint objects, both extended and point-like. Thus, the spectral and spatial characteristics of the diffuse components have to be known in order to maximize detectability of very faint objects. Commission 21 is the only organization in which all of the diffuse components are studied and compared.

The purview of Commission 21 covers the ultraviolet, visible, and infrared spectral regions. The main sources are the same and the observing techniques are generally similar in these spectral regions. Knowledge gained in one spectral region applies to the others. For example, the spatial distribution of interplanetary dust derived from visible light observations is applicable to analyzing the thermal emission from the dust.

Commission 21 periodically sponsors meetings related to the diffuse astronomical radiations. IAU Symposium 139 "The Galactic and Extragalactic Background Radiation" was held in Heidelberg in 1989 (GEBR) and IAU Colloquium 126 "Origin and Evolution of Interplanetary Dust" took place in Kyoto in 1990 (OED). A Symposium on "Examining the Big Bang and the Diffuse Background Radiations" will be held in conjunction with the IAU General Assembly in 1994. IAU Colloquium 154 "Physics, Chemistry and Dynamics of the Interplanetary Dust" is planned for 1995.

The following abbreviations are used in citing references.

A&A	Astronomy & Astrophysics	GRL	Geophys. Research Letters
Adv Sp Res	Advances in Space Research	JGR	Jour. Geophys. Research
AJ	Astronomical Journal	JGRP	Jour. Geophys. Research, Planets
ApJ	Astrophysical Journal		
GEBR	IAU Symposium 139, S. Bowyer and C. Leinert eds., Kluwer, Dordrecht, 1990		
OED	IAU Colloq. 126, A. C. Levasseur-Regourd ed. Kluwer, Tokyo, 1991		
MPB	Meteoroids and their Parent Bodies, J. Stohl and I. P. Williams eds., Slovak Acad. Sci., Bratislava, 1993		

J. Bergeron (ed.), Reports on Astronomy, Vol. XXIIA, 205–209.

II. AIRGLOW

Research on aeronomical topics related to atmospheric airglow are primarily in the domain of geophysics. However, the brightness and fluctuations in the airglow are important for low-light level astronomical observations, both from the Earth's surface and from low Earth orbit. The Berkeley group have shown that airglow forms a very significant background for diffuse imaging at 1400 to 1800 Å from the shuttle altitude (Lampton et al., GRL 20, 539, 1993).

Ultraviolet spectra of the tropical oxygen nightglow in the range of 830 to 1850 Å at 3 Å resolution were obtained with the Hopkins Ultraviolet Telescope during the Astro-1 shuttle mission in December 1990 (Feldman et al., GRL 19, 453, 1992). The dominant features of the spectrum, other than geocoronal hydrogen, are O I at 1304 and 1356 Å and the radiative recombination continuum near 911 Å. The continuum is resolved and found to be consistent with an electron temperature of 1000 to 1250 K. Weak O I emission at 989 Å is also detected, but there is no evidence for any similarly produced atomic nitrogen emissions. Upper limits on shuttle-induced ultraviolet emissions were published by Morrison, Feldman, and Henry (JGR 97, 1633, 1992).

III. ZODIACAL LIGHT

The zodiacal light is due to sunlight scattered by interplanetary dust particles. Thermal emission from the dust particles gives rise to the infrared zodiacal emission. The IAU Colloquium 126 "Origin and Evolution of Interplanetary Dust", August 1990, was covered in the previous triennial report.

Search for dust rings around the Sun during the 1991 solar eclipse

The past 25 years have witnessed many attempts to detect thermal emission from a possible dust ring around the Sun with contradictory results. Until the 1991 eclipse, all prior infrared experiments used single-element detectors to scan the corona along the ecliptic to look for excess surface brightness peaks. The availability of relatively large format infrared array detectors now provides the clear advantage of mapping the brightness and polarization of the corona. The 11 July 1991 eclipse path included the high altitude Mauna Kea Observatory, a further advantage to detect the corona at large distances. Two experiments were performed with HgCdTe array detectors through the broadband J, H and K filters (1.2 to 2.2 μm) (Lamy et al. Science 257, 1377, 1992; Hodapp, MacQueen, and Hall, Nature 355, 707, 1992). Although the sky conditions were not ideal, the H and K surface brightnesses clearly show the inhomogeneous structure in the K corona and the flattening of the F-corona toward the ecliptic but no evidence of a circumsolar, local dust component out to 15 solar radii. Thus, previous observations of thermal emission peaks may have been due to transitory phenomena, such as the breakup of sun-grazing comets. The eclipse data indicate a temperature of near-solar dust slightly below the blackbody temperature and a wide extended zone of sublimation from about 10 solar radii inward even to 3 solar radii (Mann & MacQueen A & A, in press, 1993).

Doppler shifts in the zodiacal light

Mukai and Mann (A&A 271, 530, 1993) have answered a puzzle of many years standing regarding the observed Doppler shifts in zodiacal light spectra: why dust particles within 0.7 AU of the Sun scatter sunlight as if they have orbital velocities greater than those for particles on Keplerian orbits. Mukai & Mann showed that, by taking into account the Doppler shift from the changing Sun-particle distance as well as the changing Earth-particle distance for particles on elliptical orbits and by allowing the force of radiation pressure to gravity (β) to vary with heliocentric distance, the predicted Doppler shifts are in agreement with the observations.

Dust bands and trails

The maps of zodiacal emission obtained by the IRAS satellite in 1983 continue to be analyzed for structure in the zodiacal dust cloud. Sykes and Walker (Icarus 95, 180, 1992) have published a survey of cometary dust trails found in the IRAS data. The dust trails consist of large refractory particles ejected from comets at low velocities. Examination of the entire sky seen by IRAS has resulted in the detection of eight trails associated with short-period comets, as well as many faint trails having no known parent comet. Trails tended to be associated with objects having low perihelion distances that were observed near perihelion.

Detection of a solar ring of asteroidal particles in resonant lock with the Earth has been reported by Dermott et al. (presented at Div Plan Sci meeting, Oct. 1993, Bull AAS in press). This is manifested in the IRAS maps as an asymmetry in the ecliptic brightness of the zodiacal emission when viewed at a constant elongation angle of 90 degrees, with the trailing direction about 3% brighter than the leading direction.

Love and Brownlee (AJ 104, 2236, 1992) have examined the zodiacal dust bands in the 60 μm and 100 μm IRAS bandpasses, where the contrast with the main zodiacal emission is larger than in the 12 and 25 μm bandpasses. From the 60 μm data, they estimate that the brightest dust bands contain approximately 0.2%–1.0% of the zodiacal dust. Analysis of the dynamical evolution of dust particles producing the bands is covered in Commission 22.

Zodiacal light modeling

The observed zodiacal light (or zodiacal emission) is an integral of the scattering (or emission) by all of the dust particles along the line of sight. From the observed properties of the zodiacal light, one wishes to derive the local values of fundamental quantities about the dust particles, such as their albedo, scattering function, degree of polarization, and temperature. Work on this topic has continued, using the "nodes of lesser uncertainty" method (Levasseur-Regourd, Renard, and Dumont, OED pp 131-138, 1991) to derive gradients of these quantities with heliocentric distance and with inclination from the dust symmetry plane. The local polarization degree decreases slightly with decreasing distance to the Sun between 1.5 and 0.3 AU in the symmetry plane, with a steeper decrease close to the Sun; the polarization decreases with inclination from the symmetry plane (Renard et al. OED pp 199-202, 1991). There is clear evidence for more than one population of dust, each population having distinctive optical properties and orbital characteristics. This is not surprising, since both comets and asteroids are sources of interplanetary dust. Divine has modeled the interplanetary dust cloud as a complex of five populations with differing mass distributions and orbital characteristics in order to interpret space probe observations of interplanetary dust (Divine JGRP, in press, 1993).

The phase functions and polarization of interplanetary dust, cometary dust, and asteroids of different compositional classes, are being intercompared and also compared with circumstellar dust, in order to give clues to the origin of interplanetary dust. (Levasseur-Regourd, Renard, and

Dumont, Adv Sp Res 11 No. 12, 175, 1991; Light scattering by irregular grains in comets and the interplanetary medium: Workshop on Polarization III, Levasseur-Regourd, Bastien, Hanner, eds, Aeronomie & IAP Paris 1992)

Modeling of the F corona by I. Mann (Lindau) shows that white light observations yield information about optical properties of the particles if connected with accurate polarization measurements. The near-infrared spectral range gives a better description of the spatial distribution. More than half of the contribution to the brightness of the F corona arises from dust near the Sun ($r < 0.2$ AU) (Mann A&A 261, 329, 1992; Mann in MPB, pp 387-390, 1992).

Optical Properties of Irregular Dust Particles

To interpret the observations of dust particles in the solar system and the Galaxy, the optical properties of non-spherical particles need to be understood. The radiation pressure force on irregularly shaped interplanetary dust particles was examined by T. Mukai et al. (A&A 262 315 1992) and their optical properties were studied in detail by Kozasa, Blum, and Mukai (A&A 263 423 1992) and Kozasa et al. (A&A in press 1993). Orbital evolution of porous particles from comets and asteroids was investigated by Ishimoto et al. (in MPB, pp 65-68, 1993). Angular scattering measurements on aggregate particles, with and without absorbing mantles, using the Bochum microwave scattering laboratory, have been published by Zerull et al. (Applied Optics 32, 4088, 1993).

IV. GALACTIC AND EXTRAGALACTIC COMPONENTS

Ultraviolet background

A wide field of view (8 degrees) far UV imager flown on the space shuttle in March 1992 obtained 22 images covering the bandpass 1400-1800 Å. These images have been analyzed to investigate Galactic and extragalactic diffuse radiation, as well as airglow and shuttle environment emission. (Lampton, Sasseen, Bowyer, Wu, GRL, 20, 539, 1993; Bowyer, Sasseen, Lampton, Wu, ApJ 415, in press, 1993). The Galactic FUV emissions show considerably more spatial structure than was previously thought, making the detection of the extragalactic emissions by their diffuse signals exceedingly difficult. Work in progress includes FUV properties of galaxies, FUV/IR correlations for diffuse emission from the ISM, Galactic stellar population studies for FUV stars, and FUV diffuse emission from dust.

The Far Ultraviolet Coronal Rocket established important new limits to diffuse spectral line emission in the 980 - 1080 Å bandpass. The first useful limit to the emission measure of O VI in the interstellar medium has been determined. The first directly measured limits to the filling factor of diffuse hot gas in the Galaxy have also been obtained (Edelstein and Bowyer, presented at COSPAR 29th Plenary Meeting, Adv Space Res, in press, 1993). The instrument was reflown in July 1992 as a secondary payload on the Delta launch vehicle as the Diffuse Ultraviolet Experiment (DUVE). The data obtained are now being analyzed and are expected to provide improved limits to FUV background emission lines, and to provide more stringent limits to diffuse FUV continuum radiation.

Optical Background

K. Mattila (Helsinki) and Ch. Leinert (Heidelberg) have utilized the 2.2 and 1.23 m telescopes of the Calar Alto Observatory in parallel to continue measurements of the extragalactic background radiation with the dark cloud method (dark clouds Lynds 134 and 183, and the Draco cloud). Previous data of Mattila and Schnur (Bochum) from the ESO 1m and 50 cm telescopes on Lynds 1642 have been further analyzed (Mattila, Leinert, Schnur, in Early Observable Universe from Diffuse Backgrounds. Proc. XIth Moriond Astrophysics Meeting, B. Rocca-Volmerange and J.-M. Deharveng, eds., Editions Frontieres, pp. 133-148, 1991).

In addition, the observations provide new information on the scattered light in dark nebulae and cirrus clouds for eight intermediate bands between 3500 and 8200 Å. Another byproduct is the determination of the sky brightness at these observatories and its variations during individual nights and from year to year.

Infrared Background

The COBE Science Working Group and their colleagues have been pursuing infrared background studies using data from the COBE Diffuse Infrared Background Experiment (DIRBE) (1.2 to 240 μm wavelength) and the COBE Far Infrared Absolute Spectrophotometer (FIRAS) (0.01 to 1 cm wavelength). These instruments surveyed the full sky from November 1989 to September 1990, at which time the liquid helium coolant was exhausted. Since that time, the DIRBE has continued to collect data at 1.2, 2.2, 3.5, and 4.9 μm with a planned completion of the mission in December 1993. Initial COBE data sets were released to the scientific community in July 1993. The FIRAS data show that the spectrum of the cosmic microwave background (CMB) radiation does not deviate from a Planckian by more than 0.03% of the peak CMB brightness over the interval 0.05 to 0.5 cm wavelength (Mather et al., to be published in ApJLett 10 Jan 1994). This also implies that any contributions from galaxy formation to the cosmic infrared background (CIB) in this spectral range can not significantly exceed this level. The FIRAS also provided a survey of Galactic diffuse continuum and line emission. These data included the first detection of the bright 205 μm [NII] line, as well as the most luminous cooling line of the interstellar medium, [CII] at 158 μm, an additional [NII] line, two [CI] lines, and 5 lines of CO (Wright et al., ApJ 381, 200, 1991).

DIRBE data have provided full sky absolute brightness maps in ten photometric bands from 1.2 to 240 μm, and linear polarization maps at 1.2, 2.2, and 3.5 μm. These measurements have high redundancy, providing detailed monitoring of the variation in sky brightness in every direction as the Earth moves within the interplanetary dust cloud. A preliminary measurement of the sky brightness in a relatively dark direction, the South Ecliptic Pole, which is a conservative upper limit on the CIB, was reported (Hauser et al., "After the First Three Minutes", AIP Conference Proc. 222, S. S. Holt, C. L. Bennett, & V. Trimble (eds.), New York, p. 161, 1991). Reports on studies of diffuse galactic emission, including the shape of the stellar bulge, the shape of the stellar and interstellar components of the disk, stellar colors and extinction, physical conditions in the interstellar medium, and early polarization results, have also been presented ("Back to the Galaxy", AIP Conference Proc. 278, S. S. Holt & F. Verter (eds.), New York, 1993). Present efforts focus on distinguishing foreground contributions from the solar system and Galaxy from the CIB.

COMMISSION 22: METEORS AND INTERPLANETARY DUST
(Météores et poussière interplanétaire)

ACTING PRESIDENT: I.P.Williams.
ORGANISING COMMITTEE: P.B.Babadzhanov, Z.Ceplecha, I.Hasegawa, J.Jones, C.S.L.Keay, C.Koeberl, R.K.Soberman, D.I.Steel, E.F.Tedesco, V.Porubčan (co-opted).

I. INTRODUCTION

The organization of Commission 22 was disrupted by the untimely and much-lamented death of our President, Jan Štohl, in 1993 March. An appreciation of Jan appeared in the newsletter of the commission issued in 1993 April, and obituaries have been published in various other places. The Vice-President, Iwan Williams, assumed the duties of President thereafter, and with help from the OC has taken responsibility for the production of this report, and providing the formal sanction of the commission for various meetings associated with the 1994 General Assembly, and other conferences.

As is the usual practice, members of this commission with expertise in different areas have prepared contributions reviewing their own fields, and these have been consolidated by the Acting President and Secretary to form the report as it appears below.

Over the past triennium there have been a number of conferences concerned either solely with meteors and/or interplanetary dust, or including these subjects amongst the topics covered; in fact, too many to list here. In the papers referenced below, many are taken from the proceedings of those conferences, and such proceedings may contain other specific papers of interest to the reader. Specific conferences in which members of Commission 22 have played a major part were the following:

Asteroids, Comets, Meteors 1991: Flagstaff, Arizona, USA, 1991 June (proceedings now published: Harris & Bowell 1992);

Meteoroids and their parent bodies: Smolenice, Slovak Republic, 1992 July (proceedings now published: Štohl & Williams 1993);

IAU Symposium 160, Asteroids, Comets, Meteors 1993: Belgirate, Italy, 1993 June (Reviews to be published in one volume by Kluwer, Dordrecht; contributed papers to be published in a special issue of *Planetary & Space Science*).

Commission 22 currently has membership of two inter-commission Working Groups: Interplanetary Pollution (which is chaired by C.S.L.Keay of Comm. 22) and Near-Earth Objects (WGNEO, chaired by A.Carusi of Comm. 20, with I.P.Williams and D.I.Steel being Comm. 22 representatives, Williams having replaced J.Štohl). The WGNEO has had several meetings, at approximately six-monthly intervals, and has prepared a report to the XXII*nd* GA; input of Comm. 22 to that WG is of importance since clearly much may be learnt about the large bodies which strike the Earth infrequently by studying the smaller bodies (meteoroids and dust) which strike it more frequently. The Commission also has one Working Group of its own (Amateur-Professional Co-operation in Meteors, WGAPCM, chaired by V.Porubčan who is a co-opted member of the OC). The WGAPCM has held several meetings, notably at Smolenice in 1992 when the professional conference mentioned above was held in conjunction with the (mainly amateur) International Meteor Conference of the International Meteor Organization (IMO). Good progress has been made in bringing people interested in meteors together, from many backgrounds and many nations, as the body of this report shows. Porubčan has issued a regular newsletter of the WGAPCM.

II. PHOTOGRAPHIC METEORS, FIREBALLS AND SPECTROSCOPY Z.CEPLECHA

A survey of theoretical and observational aspects of meteor phenomena, with applications to photographic meteors and fireballs, was published by Ceplecha and Borovička (1992).

The European Photographic Fireball Network (EN) has continued systematic operations from 54 stations in the Czech Republic (Spurný, in press), Slovakia, Germany, Austria and Switzerland, and in the

J. Bergeron (ed.), Reports on Astronomy, Vol. XXIIA, 211–224.

Netherlands. Precise data on 117 fireballs were obtained between 1991 January and 1993 May. [Note: all papers listed in the body of this section as being 'in press' were presented at the *Asteroids, Comets, Meteors '93* conference, and have been submitted for publication in *Planetary and Space Science*]. Preliminary computations of the most significant fireballs photographed within the EN have been regularly published in the *GVN Bulletin* (Smithsonian Institution, Washington, D.C.). Photographic recording of the deepest-ever penetration by a meteoroid was obtained for the 'Benešov' fireball (Spurný *et al.* 1991): its very steep trajectory was photographed down to an altitude of 16 km, with spectral records of resolution from 67 to 22 Å/mm also being obtained. It was predicted from the observations that several meteorites with masses of some kilograms were produced by this event, but searches in the unfavourable countryside have so far brought no recoveries. Spurný (1993) published data on photographic Geminids. The German section of the EN is now closely linked to the amateur organization, Vereinigung der Sternfreunde (VdS), but the Max-Planck-Institut für Kernphysik in Heidelberg has provided the necessary funds to keep that part of the EN running. An independent part of the photographic fireball network in Europe is operated by J.Rendtel in Germany and the results published in *Mitteilungen des Arbeitkreises Meteore* together with visual sightings of fireballs (Rendtel & Heinlein 1991). Three new photographic fireball stations are being built in Sardinia, Italy (Quesada & Uras, 1993).

The analysis of the atmospheric trajectory and orbit of the Peekskill (New York) meteorite fall, partly based on video records, is being carried out (see section VI). An extremely bright fireball exploded over northern Italy (K.Korlevic, in press) and was also recorded at the Bologna radar station (G.Cevolani, in press). One of the potentially meteorite-dropping events photographed by the EN, the 'Valeč' fireball, was thoroughly studied by Bronshten (1991a) and by Stulov & Chernenko (1991).

Data on an Earth-grazing fireball photographed in 1990 October were published by Borovička and Ceplecha (1992). The U.S. daylight Earth-grazing fireball from 1972 is continuously referred to with much greater mass than the observations indicate (Z.Ceplecha 1993, submitted to *Astron. Astrophys.*). M.Connors (private communication to Z.Ceplecha) carried out computations regarding the possible return of the 1972 Earth-grazing fireball body to the Earth's vicinity in 1997 taking into account all planetary perturbations and came to the conclusion that the most probable time of the closest approach to Earth will be 1997 July 31, at a distance of 0.08 AU. It would be of about 24th magnitude at that time, so that in principal it could be recovered as a very faint asteroid. Computer simulations of Earth-grazing fireballs were published by Olson et al. (1991).

The amateur organizations in Europe and in Japan are very active in photographing meteors. Meteor and fireball trajectories and orbits are regularly published in *WGN* (the journal of the International Meteor Organization, IMO: see Gyssens *et al.* 1991). Multi-station photographic trajectories and orbits of fireballs and meteors have been determined by members of the Dutch Meteor Society, in particular many Perseids and Geminids (Betlem & de Lignie, 1990; Betlem *et al.* 1993). Several prolific groups of Japanese amateurs continued photographic studies of meteors and fireballs (Ohtsuka *et al.* 1991; Miyashita *et al.* 1993). The capacity of many amateur groups for observing and measuring meteor trails, and then computing trajectories and orbits of meteoroids, is notable.

The Global Network for collecting all fireball sightings (previously the *SEAN Bulletin*, now the *GVN Bulletin*) has ceased publishing these observations. The service was taken over by *FIDAC* (the Fireball Data Centre of the IMO, directed by A.Knöfel). Statistics of brightnesses of 16,000 visually-observed meteors and fireballs were published by Martynenko & Levina (1991). Miles (1991) published visual data from a bright fireball observed over Ireland in 1991 February.

A model of meteoroid ablation taking into account gross fragmentation was presented and applied to the most precise photographic data on the Prairie Network (PN) fireballs (Ceplecha 1993). Meteoroid strength, density, ablation coefficient and amount of fragmented mass was determined, and a two-dimensional classification of fireballs based on the dynamic pressure at the fragmentation point and on the ablation coefficient was proposed. The structure and composition of meteoroids from photographic observations were studied by Babadzhanov (in press) and densities typical of stony bodies found for most of them. Adolfsson and Gustafson (in press) pointed to the importance of rotation of meteoroids for the onset of their ablation.

ReVelle & Ceplecha (in press) proposed methods for recognizing iron meteoroids among the PN fireballs: they found 7 such objects and investigated their possible relation to the M-type Earth-crossing asteroids.

In the area of very large meteoroid impacts into the atmosphere, Chyba *et al.* (1993) and Hills & Goda (1993) published new theoretical models for the atmospheric ablation of small cometary or asteroidal bodies, and applied them to the 1908 Tunguska explosion, arguing for a stony composition of the Tunguska body; the latter authors also studied a whole range of possible interactions between small asteroids and the atmosphere. Z.Sekanina (1993, submitted to *Science*) has applied the Chyba *et al.* model to the expected 1994 July collision of P/Shoemaker-Levy 9 with Jupiter. Korobejnikov *et al.* (1991) developed an ablation-explosion model and applied it to Tunguska fireball. Steel and Snow (1992) suggested that the extensive forest devastation of the South Island of New Zealand was due to a large body detonation in the atmosphere some 800 years ago, similar to the Tunguska event.

Padevět (1991) studied meteors and fireballs produced in Mars' atmosphere, and comparing these with the better-known terrestrial phenomena. Partly based on fireball observations, Padevět & Jakeš (1993) proposed that chondrites come also from inside cometary nuclei, though they originated in larger bodies than comets are today.

Pecina (1991) published a new method for computing fireball orbits and applied it to several well observed EN fireballs deriving orbits quite different to those obtained by using the classical method, in particular for low velocities. A method for determining the pre-atmospheric velocities of meteoroids from photographic records was proposed by Zausaev & Pushkarev (1992). Kazantsev & Sherbaum (1990) came to the conclusion that hyperbolic orbits amongst photographic meteors are mostly due to observational errors.

B.A.Lindblad continues in charge of the IAU Meteor Data Center at the Lund Observatory, Sweden. A catalogue of photographic and radar meteor orbits is available from him upon request (see section IV). Data from many photographic meteors and fireballs observed in different observational projects are also available from the Ondřejov Observatory, Czech Republic (contact: P.Spurný). A new search for streams amongst fireball orbits resulted in a lower number than previously determined (Porubčan & Gavajova, in press).

Partially based on photographic observations of meteors, and also on Spacewatch Telescope discoveries of small asteroids down to sizes of $\sim$10 m (Rabinowitz 1992) , Ceplecha (1992) computed influx to the Earth of meteoroids within the mass range from 10^{-21} to 10^{15} kg. He found a total annual mass influx of 1.7×10^{8} kg. Rates of impacts by different types of the large bodies recorded as photographic fireballs (up to 7 m) were determined by Ceplecha (in press); there exist only 14 bodies on precise records with pre-atmospheric sizes over 1 m. In the mass range of fireball bodies most of the influx belongs to carbonaceous bodies both of asteroidal and cometary origin; the relative importance of stony material decreases for bodies with sizes larger than 0.3 m, while the softest cometary material becomes the most important for bodies larger than 2 m in size. Based solely on their orbits, Bronshten (1991b) came to the conclusion that the majority of fireball-producing bodies photographed within PN and EN were asteroidal.

TV observations of faint meteor spectra were reported by Mukhamednazarov & Mal'tseva (1990). In Japan, near-infrared spectra of Perseids and Geminids were obtained (Murayama 1990). One Perseid spectrum was described by Evans & Ridley (1993). Kokhirova (1993) analyzed Perseid spectra quantitatively. Theoretical considerations on the intensities of spectral lines in meteors were done by Smirnov (1992). Three photographic spectra of meteor trains are compared in Rajchl et al. (1993). Padevět began a TV meteor spectrum observation program at the Ondřejov Observatory.

The most extensive work on meteor spectra is that of Borovička (1993a,b). The spectrum of a bright EN fireball was studied from the physical point of view along the trajectory between heights of 57 and 35 km. A synthetic spectrum was computed and compared with the observed spectrum. Thermal equilibrium with temperatures varying in the range 3 500–4 700 K was found to be a good first approximation to the physical conditions. Moreover the high excitation spectrum (T $\simeq$ 10 000 K) of the shock wave overlaid the main spectrum, showing that meteor spectra consist of two different components: low and high temperature. The refractory elements Ca, Al, Ti were underabundant in the radiating gas relative to chondritic values, pointing to incomplete evaporation. Melting was the favored ablation process. About 95% of the hot gas around the meteoroid was composed of atmospheric species.

III. METEOR SHOWERS/METEOROID STREAMS V.PORUBČAN

The formation and evolution of meteoroid streams in general was studied by several authors (Lebedinets *et al.* 1990; McIntosh 1991; Emel'yanenko 1992a,b; Williams 1992). According to Babadzhanov & Obrubov (1991a, 1992b), their investigation of the dynamics of the major meteoroid streams shows that up to eight related meteor showers may be produced by one stream. They have found five new complexes of small bodies. In a related study of 22 short-period meteoroid streams Babadzhanov & Obrubov (1991b) report that on the basis of theory these may produce 102 meteor showers, and of those 72 have been confirmed by photographic and radar observations. The mass distribution of comets and meteoroid streams and the shower/sporadic ratio in the incident visual meteoroid flux was studied by Hughes (1990).

Terenteva & Bayuk (1991) have studied the formation of short-period meteoroid streams and conclude that a transformation of cometary orbits into Geminid type orbits is possible. The association of 3200 Phaethon and the Geminids was discussed Kramer & Shestaka (1992) and Gorbanev *et al.* (1992). Since the discovery of Phaethon and its Geminid association the problem of meteoroid streams of asteroidal origin has become more intriguing. Several other associations with varying degrees of security have been identified (Hasegawa *et al.* 1992; Štohl & Porubčan 1993). Williams & Wu (1993) propose a new theoretical model for the Geminid stream which is based on the assumption that the parent of the Geminids is Phaethon, or at least a progenitor of that body.

The origin, dispersion mechanisms, evolution and structure of dense meteor storms are discussed by Kresák (1993), and predictions for future events of this kind are made. The formation of the Leonid meteoroid stream and its storms were studied by Wu & Williams (1992a) and by Brown & Jones (1993). Lindblad *et al.* (1993) determined the mean orbit and radiant motion of this stream.

Hasegawa (1992, 1993) has compiled a catalogue of historical records of meteor showers containing 331 good entries, and more than 100 uncertain records without the exact date of the shower being known. Hasegawa (1990) has also introduced a better method for the prediction of the meteor radiant associated with a parent body than that used previously.

The cross-section of the Quadrantid meteoroid stream based on radar observations over a period of 30 years was studied by Šimek & McIntosh (1991). Babadzhanov & Obrubov (1992a,b, 1993) have simulated the evolution of the P/Machholz meteoroid stream that may produce eight meteor showers: among them the Quadrantids, Ursids, N and S δ Aquarids, Daytime Arietids and α Cetids. Wu & Williams (1992b) integrated 118 photographic orbits of Quadrantids back in time to 5000 B.C., five distinct subsets of orbits being distinguishable. Jones & Jones (1993) have shown by a computer simulation that if P/Machholz was captured by Jupiter about 2,200 yrs ago, there has been sufficient time to produce most of the presently observed features of the meteor shower complex mentioned above.

Emel'yanenko (1991) analysed the recent bursts in Lyrid activity and concluded that these had appeared near the centres of resonances with Jupiter. He (and Kresák 1993) suggest that the next outburst will occur in 1994. The radiant and orbit of the Lyrid meteoroid stream has been determined from precisely-reduced photographic meteors observed in 1941–1985 (Lindblad & Porubčan 1991).

The approach of the parent comet of the Perseids (P/Swift-Tuttle) was indicated by the meteors themselves, a secondary maximum being visually observed first in 1988, with enhanced Perseid activity since then. Spectacular outburts have been observed in the past three years.

The evolution of the orbit of P/Halley and its associated meteoroid streams was studied by several researchers (Babadzhanov *et al.* 1991; Wu & Williams 1993). Obrubov (1993) indicated the possibility that P/Halley may produce two other showers in addition to the well-known η Aquarids and Orionids, and showed that the other two are indeed evidenced in radar meteor orbit data.

The structure and evolution of the Taurid complex has also been a topic of investigation (Steel *et al.* 1991; Štohl & Porubčan 1992; Asher *et al.* 1993). Steel *et al.* modelled the complex on the basis of the parent being a giant comet which entered the inner solar system some time 10,000–20,000 years ago.

IV. METEOROID ORBITS D.I.STEEL

Since much of the necessary material on orbits has been covered in other sections of this report, particularly sections II and III, the present section will be relatively short.

As noted elsewhere, a catalogue of about 68 000 individual meteoroid orbits in machine-readable form is available from B.A.Lindblad at the IAU Meteor Data Center, Lund Observatory, Box 43, 22100 Lund, Sweden (Lindblad 1991a). This includes over 60 000 radar meteor orbits from various surveys in the U.S.S.R., the U.S.A. and Australia, photographic meteor orbits from the U.S.A. and the U.S.S.R., faint TV meteor orbits from Canada, and all the fireball orbits from the Prairie Network (U.S.A.) and Meteorite Orbits and Recovery Program (Canada). Some fireball orbits are also available from the European Network, although others are available from the Ondřejov Observatory in the Czech Republic (see section II). Examples of the sorts of data available have been given by Steel (1991).

Almost all of the orbits archived at the IAU MDC were collected pre-1976. In the years since then a number of groups of amateurs, in particular, have determined smaller (but significant) numbers of orbits using 35 mm cameras and similar, and such efforts have been concentrated on specific showers such as the Geminids and the Perseids (*e.g.* Koseki *et al.* 1990). Some mention of these has also been made in sections II and III; see also Lindblad (1991b) for an assessment of the orbits measured from Japan.

However, in terms of the pure bulk of data throughput, the new meteor orbit radar (AMOR) recently built in New Zealand by Baggaley and co-workers is the most significant development for over two decades. To the time of writing over 300 000 individual meteoroid orbits have been determined, to a limiting magnitude of +12–13 (limiting size about 100 μm). Full details of the system should be published soon (W.J.Baggaley *et al.*, QJRAS, submitted), with preliminary information having been given in various shorter papers (Baggaley *et al.* 1992a,b; Baggaley *et al.* 1993a). A preliminary analysis of the data indicates that about 1% of the measured orbits may be hyperbolic (Baggaley *et al.* 1993b).

An example of the usefulness of this new radar is as follows. During the 1980's the very great interest in the apparition of P/Halley led to several investigations of the orbital evolution of the meteoroid stream associated with that comet, which gives rise to the η Aquarid and Orionid meteor showers. The best observed of these two is the η Aquarid shower, but Lindblad (1990) showed that the orbital evolution studies had made use of a 'mean' orbit based upon just one meteor! Lindblad extended that list to a total of 16 orbits; Baggaley *et al.* (1992b) have now presented a mean orbit based upon 361 η Aquarids, with over a thousand orbits from that shower now in hand from observation campaigns in 1990–93. In the context of the Halleyid stream, it should also be noted here that Obrubov (1993) has studied the orbital evolution of the comet, and finds two other theoretical showers (at quite different times of year, and different radiants), which apparently are evidenced in meteor surveys.

Other specific campaigns using the New Zealand radar have targetted the δ Aquarid showers and the times at which Apollo asteroid-related showers have been predicted; the data from these campaigns are now under analysis.

V. RADAR METEORS J.JONES

One of the most exciting advances in meteor radars has been the development of the AMOR radar meteor orbit system in New Zealand (Baggaley *et al.* 1992a,b, 1993a). The geometry of the system enables meteor trajectories to be determined without measurements of the Fresnel oscillations making reduction of the data fast enough to measure the orbits of ~1 500 meteors brighter than +13 magnitude per day. Some mention of the results obtained with the AMOR system has been made in section IV.

There has been a significant increase in the use of radiant imaging radars which use measurements of the echo direction to determine radiant distributions. Watanabe *et al.* (1991) have used data from an MST radar to study the structure of the Geminid radiant, while Jones & Webster (1991) have adapted the technique for use with a short-hop forward-scatter system. Andreev *et al.* (1991) have extended the method so that it can be used to recover sporadic meteor radiant distributions.

The revival of interest in meteor-burst communication systems has stimulated Jones & Jones (1990a,b,

1991) to develop a detailed theoretical treatment of the oblique-scatter from underdense trains. Jones (1992) has shown how the WKB approximation may be applied usefully to the case of very overdense trains. Jones & Webster (1992), Cevolani & Hajduk (1993) and Cevolani *et al.* (1993) have developed short-hop forward-scatter systems with the aim of improving meteor-burst communication and to find ways of making forward-scatter systems into useful astronomical tools.

The persistent problem of the attenuation associated with both the finite velocity effect and the initial radius of the train has received renewed attention. Multi-frequency studies by Steel & Elford (1991) have indicated that the attenuation may be much more severe than previously recognised so that the rate of mass influx from small high altitude meteoroids may have been greatly underestimated. The formation of the initial radius has been modelled by Grusha (1991) who has considered the effects of fragmentation, while the diffusion of meteor trains has been studied by Shodiev (1992) and ambipolar diffusion of a multi-ion plasma has been investigated by Jones & Jones (1990c). Jones (1992) has had another look at the problem of the ambipolar diffusion of a meteor train in the presence of a magnetic field and shown that the cylindrical geometry allows a closed-form solution which indicates that the effect of the magnetic field is felt when the train is very closely aligned with the magnetic field lines. Novikov & Pecina (1990) have investigated the effect of fragmentation on the Fresnel oscillations.

Šimek & McIntosh (1991) have investigated the cross-section of the Quadrantid stream using radar data. Kostylev (1992) has deduced many physical characteristic of the Draconid meteor stream from radio-meteor observations made in 1985. Bibarsov *et al.* (1991) have addressed the problem of determination of meteoroid mass and density from single station radar observations. Babadzhanov & Bibarsov (1992) have determined the mass distribution of sporadic meteors from the observations of overdense radio-meteors. The head-echo phenomenon has been explored again by Ol'khovatov (1991) and also by Jones & Webster (1991), who have been able to determine from an analysis of simultaneous radar and visual observations of meteors how the probability of getting a head echo depends on range and visual magnitude for several showers.

VI. TELEVISION METEORS R.L.HAWKES

Many researchers are now employing low light level television (LLLTV) methods for the detection of meteors, with active work underway in (at least) Canada, the Czech Republic, Japan, the Netherlands, Tajikistan, and the United States. The cost of such systems (and the microcomputer-based frame-grabbing cards needed for analysis) has fallen to the point where a number of amateur observers now use LLLTV equipment. For example, in Japan alone there are at least 13 different detection systems in operation for studies of the meteoroid mass distribution and flux, radiant points, trajectories and orbits, spectra and light curves (Fujiwara 1993). Indeed the distinction between 'amateur' and 'professional' now has little meaning in this field, with the the best amateur groups performing careful analyses of results and publishing in the scientific literature. Examples would be the two station work recently published by groups in the Netherlands (de Lignie *et al.* 1993; de Lignie 1993) and Japan (Ueda & Fujiwara 1993a; Suzuki *et al.* 1993). Specific advice regarding the selection and operation of LLLTV detection systems has been published (Hawkes 1990a,b).

While the number of active observing groups has grown significantly, and the number of partially processed and archived LLLTV meteors is probably in the tens of thousands, the number of precisely reduced LLLTV meteor atmospheric trajectories and orbits remain under one thousand, with little increase since the mid-1980's. As pointed out in a recent review (Hawkes 1993), the time required by current video-based analysis procedures is the main limiting factor. Hopefully, this roadblock will be overcome in the next few years through more automated and powerful processing software, and we will see an increasing number of TV meteor orbits archived in the IAU Meteor Data Center in Lund, Sweden. At the moment, it is important for those using LLLTV data to realize that only a small fraction of the year is represented in currently-published summaries, and some of the major showers are not represented at all. The Geminid shower has now been studied by two-station LLLTV (de Lignie *et al.* 1993) and additional observations of the Perseids have been published (Rozhilo *et al.* 1990). Television has not been extensively used for mass/magnitude distributions, and the recent study on the Geminid shower is significant (Ueda & Fujiwara 1993b). The 1993 Perseid shower was recorded by a variety of LLLTV equipment in Europe, North America and Asia, and the profile of this shower for faint magnitudes should soon be clarified. A D-criterion stream search has now been conducted (Jopek 1993) on the published television orbital data with 6 clear and 20 tentative associations, although the strong bias in the observational sample must be kept in mind. Techniques for the determination of meteor parameters from single station LLLTV measurements were independently developed in Canada and Tajikistan in the mid-1980's. Single station techniques have recently been applied to an analysis of 21 bright (to +2.4) video meteors (Malyshev 1992). Single station techniques offer a less time consuming way to obtain basic data on shower meteors with reasonable precision, and should be more widely used.

Recent published contributions in the area of television meteor spectra include a detailed study of three faint meteors (Mukhamednazarov & Mal'tseva 1990), the results suggesting a very similar chemical composition for all three. In addition, a sophisticated LLLTV spectrum system is now in routine operation in the Czech Republic. A search for wake in faint television meteors (Robertson & Hawkes 1992) has yielded the surprising result that very few LLLTV meteors exhibit wake within the limits of the equipment. This finding (which is supported by more recent unpublished work) has important implications for models of the structure and fragmentation of dustball meteors. Techniques for video photometry have been developed and applied to an analysis of light curves of faint, double station, sporadic meteors (Fleming *et al.* 1993). The light curves were, on average, symmetric about the point of maximum luminosity. There has been recent LLLTV detection of a meteor cluster with at least five almost-simultaneous meteors, and evidence of finer scale fragmentation (Piers & Hawkes 1993). It is probable that CCD astronomical cameras at slower than conventional video frame rates will increasingly be employed for meteor detection. Indeed, it was initially believed that a meteor outburst had been detected by the CCD camera on the Canada-France-Hawaii Telescope (Brown *et al.* 1992), although subsequently a non-meteoric explanation has been proposed (Jenniskens *et al.* 1993).

While there have, over the past several decades, been occasional simultaneous television-radar and television-visual meteor observations, LLLTV has not been as effectively employed as is possible for the calibration of different meteor observing methods. Recent excellent work has used LLLTV to estimate the errors in telescopic meteor observations (Pravec 1992; Pravec & Bocek 1992). A television zenithal hourly rate (TVZHR) standard has been defined (Hawkes 1990b).

The most spectacular video meteor records are, without a doubt, the 14 videotapes obtained from various locations in the eastern United States of the fireball associated with the Peekskill meteorite fall (1992 October 9). These are the first ever motion picture recordings of a fireball with an associated meteorite fall. There is a wealth of dynamic detail with a complex and rapidly changing fragmentation, the video records now being analysed. With the increasing availability of camcorders among the general public, serendipitous recording of bright, long lasting fireballs will become increasingly common, and will represent a new and valuable source of data on fireball atmospheric dynamics.

VII. DUST CHARACTERISTICS E.GRÜN

In the time period 1990 to 1993 new spacecraft data on interplanetary dust (IPD) have become available, both from interplanetary space and near-Earth. NASA's *Galileo* and *Ulysses* spacecraft made measurements of IPD from Venus to Jupiter. ESA's *Giotto* spaceprobe flew by its second comet: P/Grigg-Skjellerup. The Japanese *Hiten* spaceprobe explored the near-Earth and lunar environment. Analyses of the surfaces of NASA's Long Duration Exposure Facility (LDEF), which was retrieved in 1990 after six years in space, are still revealing new information on IPD (natural particles) and man-made space debris. In 1991 the proceedings of a conference on IPD were published (Levasseur-Regourd & Hasegawa 1991): this will be the reference book on IPD for some years to come.

IPD data from the *Galileo* and *Ulysses* spacecraft covering the heliocentric distance range 0.7–5.4 AU are described by Grün *et al.* (1992a): they show that zodiacal dust inside about 3 AU orbits the Sun in low eccentricity, low inclination orbits.

Results from the Jupiter flyby of *Ulysses* are reported by Grün *et al.* (1992b); the discovery of periodic dust streams originating within the jovian system and of interstellar dust are reported by Grün *et al.* (1993). Interstellar dust grains are identified by their retrograde trajectories and hyperbolic speeds at the position of *Ulysses*. First ideas about mechanisms for the generation of the *Ulysses* dust streams are discussed by Horanyi *et al.* (1993) and by Hamilton & Burns (1993).

The *Giotto* flyby of P/Grigg-Skjellerup in 1992 (McDonnell *et al.* 1993a; Levasseur-Regourd *et al.* 1993) showed that this comet is much less dusty than P/Halley.

The *Hiten* satellite was launched in 1990 and reached a maximum geocentric distance of 1.5 million km. After three years in orbit it crashed onto the Moon in 1993 April. The dust instrument on board monitored the dust environment near Earth and in interplanetary space (Iglseder *et al.* 1993).

An important question of the analyses of samples retrieved from LDEF is the separation of natural meteoroids from space debris. This is done both by studying the impact flux directions and by chemical analysis of residues in impact craters (*e.g.* McDonnell *et al.* 1993b). Both methods show that debris dominates the micron-sized particulates in low-Earth-orbit.

Progress has been made in the analysis and theory of IPD particles collected in the stratosphere. Significant enrichments in volatile elements compared to CI chondrites has been found (*e.g.* Gibson 1992). This observation may be interpreted as implying that these particles consist of very primitive solar system material that had never seen temperatures above 500°C. However, routine analyses of trace elements in nanogram particles (Jessberger *et al.*, 1992; Flynn & Sutton 1992) indicate that atmospheric effects may alter the volatile content of such particles.

Searches for the infrared signature of circumsolar dust during the 1991 solar eclipse failed to produce a positive result (Hodapp *et al.* 1992; Lamy *et al.* 1992). Such signatures were observed during the 1966 eclipse; these signatures may therefore be a time-dependent phenomenon.

Considerable progress has been made in determining the optical properties of fluffy porous dust aggregates (Kozasa *et al.* 1992; Mukai *et al.* 1992). Earlier approaches used simple spherical grains to model zodiacal light. However, studies of IPD particles and comet dust show that a significant fraction of dust grains is of more a complex nature and, therefore, the new results are more realistic.

VIII. TEKTITES AND IMPACT GLASSES — C.KOEBERL

Important progress has been made in the study of tektites and other glasses formed by meteorite impact. The study of tektites is closely linked with the investigation of impact craters and impact processes in general which are of great importance in planetary science. Impact craters are known for two of the four well-defined tektite strewn fields but no unequivocal association with a source crater has yet been possible for the Australasian and North American strewn fields. Hartung (1990) proposed Lake Tonle Sap in central Cambodia as a likely candidate for the Australasian tektite source crater. No direct evidence is yet available regarding the impact origin of Tonle Sap but this proposal is in agreement with a location derived from geographic variations in the abundance of microtektites (Glass 1993). Geochemical (Koeberl 1992a) and isotopic (Blum *et al.* 1992) analyses of Australasian tektites are also consistent with such a source location and favour a single large impact. Wasson (1991) proposed that the Australasian tektites originated in a multitude of small impacts; however, this proposal is difficult to reconcile with chemical and isotopic data. Schnetzler (1992) provided a detailed analysis of the distribution of some Australasian tektites and suggested that the source crater is situated in Laos.

Glass & Wu (1993) discovered shocked impact melts, quartz, and coesite and stishovite (high-pressure modifications of quartz) to be associated with microtektite-bearing layers at numerous deep-sea sites in the Australasian and North American tektite strewn fields, providing another direct link to impact processes. Schneider *et al.* (1992) and Glass *et al.* (1991) made a detailed study of the stratigraphy of microtektite-bearing layers in the Australasian and Ivory Coast strewn fields and concluded that the associated impact events were not related to the near-by geomagnetic polarity reversals. Regarding a source crater for the North American tektite strewn field, Poag *et al.* (1992) found a small structure on the continental shelf off the eastern coast of the U.S.A., but no definitive conclusions as to an impact origin of the structure are available so far. Using the distinctive characteristics of the rhenium-osmium isotope system, Koeberl & Shirey (1993) were, for the first time, able to provide unequivocal evidence for the presence of a small meteoritic component in tektites which otherwise have a composition indistinguishable from that of upper terrestrial crust. Glass & Wu (1992) and Fudali (1993) were able to show convincingly that there is no evidence for the occurrence of 'young' tektites (*i.e.* that are supposed to have fallen only recently, in contrast to their older radiometric age). These two important papers once again reiterate that the so-called 'age paradox' which is still occasionally advanced, is in fact a red herring.

Considerable progress was made regarding clarification of the nature of the famous Cretaceous-Tertiary (K-T) boundary event, 65 million years ago. It is now firmly established that a gigantic impact occurred at that time, and the 200–300 km diameter Chicxulub impact structure in Yucatan, Mexico, was identified as the source crater (Sharpton *et al.* 1992), making it the largest known impact crater on Earth. Glasses were discovered at a K-T boundary site in Haiti (Sigurdsson *et al.* 1991) and were confirmed to be of impact origin (Koeberl & Sigurdsson 1992; Koeberl 1992b). Isotope geochemistry confirmed that these impact glasses originated from the Chicxulub crater (Blum *et al.* 1993). As new impact craters and impact glasses are discovered (*e.g.* Claeys *et al.* 1992) it becomes clear how important impact processes have been in the evolution of the Earth.

IX. METEORS AND AERONOMY — W.J.BAGGALEY

The following is a summary of the work into the aeronomy of meteoric products and the fate of meteoric material subsequent to ablation. It is useful to consider the various topics in order of increasing distance from the immediate meteor trail invironment.

Meteor coma:

Work has continued on the phenomenon of the head echo: by collating visual and radar data Jones & Webster (1991) have shown how the plasma coma size depends on meteor parameters, while Ol'khovatov (1991) has developed a model for radar scattering as due to plasma turbulence resulting from the rapid injection of metals into the surrounding atmosphere. The probability of occurrence of visual trains as a function of shower type has been tabulated by Rubio (1992).

Initial expansion and diffusion:

Grusha (1991) considered fragmentation effects and micro-atmospheric density inhomogeneities to model the dependence of plasma initial expansion on meteor speed and height, and used two-wavelength data to test the model. The expansion of a meteoric plasma into the atmosphere depends on the ambient electric field, and the geomagnetic field (Jones 1992). Jones & Jones (1990c) gave solutions for the case of diffusion of a mixture of ions having different mobilities. Shodiev (1992) gave data on the dependence of the effective diffusion coefficient on both height and local time and the role of turbulent diffusion.

Aeronomy of the meteor region:

Since the ionic concentration of a meteoric plasma column is affected by reaction chemistry, it is possible, by studying enduring echo characteristics (and providing a correct understanding of the radar reflection process is available: Jones 1991), to realise a diagnostic for ozone profile determination (Jones *et al.* 1990).

Deposited products:

Kamijo (1991) modelled micrometeoroid ablation and subsequent re-condensation of vapour and coagulation. The general problem of the fate of ablation products and micrometeoritic material and of subsequent re-condensation and coagulation to form layers (Begkhanov *et al.* 1992) and aerosols (Lebedinets & Kurbanmuradov 1992) has received attention, as has the global balance of sulphate aerosols under the action of both cometary and volcanic sources (Lebedinets 1992). Thomas (1991) has provided a comprehensive review of mesospheric cloud aeronomy.

Meteoric ion layers:

Both rocket–borne mass spectrometers (Steinweg *et al.* 1992) and lidars (Kane *et al.* 1992; Kane & Gardner 1993), measuring positive ions and neutrals respectively, have yielded composition, abundances and vertical structure of metal layers while incoherent-scatter observations of have given high resolution profiles of ion layers (Bristow & Watkins 1993). With such detailed layer data and with improved chemical reaction schemes (*e.g.* Swider 1992) it is possible to model the formation and decay of the meteoric metal layers. Rocket-borne mass spectrometers have also permited detailed study of heavy negative ion clusters at the mesopause which have their source in meteoric smoke particles (Schulte & Arnold 1992). At lower altitudes, Krieger & Arnold (1992) have found aerosols at 40 km using a balloon-borne mass spectrometer and discussed the paths taken meteoric metallic species to reach this level.

Ionospheric effects:

Though it is known that there is little direct relation between meteoric influx and ionospheric E_s in the short term (hours), Baggaley (1990) presents evidence for close seasonal association. Some correlation was found between E_s and fireball events (Rajaram & Chandra 1991).

Atmospheric dust:

The effect of dust created by the 1908 Tunguska event in producing large scale sky luminescence increases (Bronshten 1992a) and noctilucent cloud formation (Romejko 1991) has been considered. On a global scale, a source of water and oxygen from the terrestrial influx of small comets and dust was suggested by Lebedinets (1991); the climatic effects of cometary/asteroid impact dust was discussed by Covey & Thompson (1991).

Electromagnetic effects:

Investigation of electrophonic sounds from large meteor fireballs has continued (Keay 1992a, 1993) and Astapovich's empirical conclusion that only fireballs brighter than mag −9 produce sustained electrophonic sounds has been substantiated on the basis of Keay's model (1992b). Keay and Ostwald (1991) have conducted laboratory experiments revealing that mundane objects can transduce audio-frequency electromagnetic emissions into sound. Keay (1992c) has shown that the reporting rate of electrophonic fireballs has greatly increased since the effect was explained physically, and he has examined the substantial contributions of Chant concerning electrophonic sounds produced by aurorae (Keay 1990). Bronshten (1992b) reviewed the models of the electrical and electromagnetic processes in meteoric plasma, including a discussion of train currents and electrophonic fireballs.

X. REFERENCES

Andreev V.V., Bel'kovich O.I., Filimonova T.K., Siderov V.V., 1991, in Harris & Bowell, 17–22.
Asher D.J., Clube S.V.M., Steel D.I., 1993, in Štohl & Williams, 93–96.
Babadzhanov P.B., Bibarsov R.Sh., 1992, Solar Sys. Res., 26, 103–108.
Babadzhanov P.B., Hajduk A., Obrubov Yu.V., Pushkarev A.N., 1991, Sol. Syst. Res., 25, 153–160.
Babadzhanov P.B., Obrubov Yu.V., 1991a, Astron. Vestn., 25, 387–407.
Babadzhanov P.B., Obrubov Yu.V., 1991b, Astron. Zh., 68, 1074–1085.
Babadzhanov P.B., Obrubov Yu.V., 1992a, in Harris & Bowell, 27–32.
Babadzhanov P.B., Obrubov Yu.V., 1992b, Celest. Mech. Dyn. Astron., 54, 111–127.
Babadzhanov P.B., Obrubov Yu.V., 1993, in in Štohl & Williams, 49–52.
Baggaley W.J., 1990, Planet. Space Sci., 38, 145–147.
Baggaley W.J., Steel D.I., Taylor A.D., 1992a, in Harris & Bowell, 37–40.
Baggaley W.J., Taylor A.D., Steel D.I., 1992b, *Periodic Comets*, eds. J.Fernández & H.Rickman, Universidad de la República, Montevideo, Uruguay, 115–124.
Baggaley W.J., Taylor A.D., Steel D.I., 1993a, in Štohl & Williams, 245–248.
Baggaley W.J., Taylor A.D., Steel D.I., 1993b, in Štohl & Williams, 53–56.
Begkhanov, M., Kurbanmuradov O., Lebedinets V.N., 1992, Sol. Syst. Res., 26, 212–220.
Betlem H., de Lignie H., 1990, in Lagerkvist *et al.*, 509–512.
Betlem H., ter Kuile C., de Lignie H., 1993, in Štohl & Williams, 161–164.
Bibarsov R.Sh., Narziev M., Chebotarev R.P, 1991, Solar Sys. Res., 24, 212–216.

Blum J.D., Chamberlain C.P., Hingston M.P., Koeberl C., Marin L.E., Schuraytz B.C., Sharpton, V.L., 1993, Nature, 363, 325–327.
Blum J.D., Papanastassiou D.A., Koeberl C., Wasserburg G., 1992, Geochim. Cosmochim. Acta, 56, 483–492.
Borovička J., 1993a, in Štohl & Williams, 303–306.
Borovička J., 1993b, Astron. Astrophys., in press.
Borovička J., Ceplecha Z., 1992, Astron. Astrophys., 257, 323–328.
Bristow W.A., Watkins, B.J., 1993, J. Atmos. Terr. Phys., 55, 873–894.
Bronshten V.A., 1991a, Sol. Syst. Res., 24, 102–110.
Bronshten V.A., 1991b, Sol. Syst. Res., 24, 206–211.
Bronshten V.A., 1992a, Sol. Syst. Res., 25, 93–104.
Bronshten V.A., 1992b, Sol. Syst. Res., 25, 369–381.
Brown P., Asher D., Steel D., 1992, WGN, 20, 28–31.
Brown P., Jones J., 1993, in Štohl & Williams, 57–60.
Ceplecha Z., 1992, Astron. Astrophys., 263, 361–366.
Ceplecha Z., 1993, Astron. Astrophys., in press.
Ceplecha Z., Borovička J., 1992, in *Interrelations between physics and dynamics for minor bodies in the solar system*, eds. D.Benest & C.Froeschlé, Editions Frontières, Gif-sur-Yvette, France, 309–368.
Cevolani G., Hajduk A., 1993, in Štohl & Williams, 253–256.
Cevolani G., Hajduk A., Kingsley S.P., 1993, in Štohl & Williams, 257–260.
Chyba C.F., Thomas P.J., Zahnle K.J., 1993, Nature, 361, 40–44.
Claeys P., Casier J.G., Margolis S.V., 1992, Science, 257, 1102–1104.
Covey C., Thompson S.L., 1991, Bull. Am. Astron. Soc., 23, 1150–1151.
de Lignie M., 1993, Radiant, 15, 44–46.
de Lignie M., Jobse K., Schievink R., 1993, in Proc. Int. Meteor Conf. 1992, eds. D.Očenáš & P.Zimnikoval, Observatory Banská Bystrica, Slovak Republic, 59–62.
Emel'yanenko V.V., 1991. Sol. Syst. Res., 24, 199–202.
Emel'yanenko V.V., 1992a, Astron. Vestn., 26, 58–66.
Emel'yanenko V.V., 1992b, Celest. Mech. Dyn. Astron., 54, 91–110.
Evans S.J., Ridley H.B., 1993, J. Brit. Astron. Assoc., 103, 27–29.
Fleming D.E.B., Hawkes R.L., Jones J., 1993, in Štohl & Williams, 261–265.
Flynn G.J., Sutton S.R., 1992, Proc. Lunar Planet. Sci. Conf., 22, 171–184.
Fudali R.F., 1993, Meteoritics, 28, 114–119.
Fujiwara Y., 1993, in Štohl & Williams, 265–268.
Gibson E.K., 1992, J. Geophys. Res., 97, 3865–3875.
Glass B.P., 1993, Meteoritics, 28, 354.
Glass B.P., Kent D.V., Schneider D.A., Tauxe L., 1991, Earth & Planet. Sci. Lett., 107, 182–196.
Glass B.P., Wu J., 1992, Meteoritics, 27, 605–608.
Glass B.P., Wu J., 1993, Geology, 21, 435–438.
Gorbanev Yu.M., Kramer E.N., Skoblikova L.Ya., Shestaka I.S., 1992, Astron. Vestn., 26, 113–120.
Grün E., *et al.*, 1992, Geophys. Res. Lett., 19, 1311–1314.
Grün E., *et al.*, 1992, Science, 257, 1550–1552.
Grün E., *et al.*, 1993, Nature, 362, 428–430.
Grusha G.V., 1991, Sol. Syst. Res., 25, 58–62.
Gyssens M., Knöfel A., Rendtel J., Roggemans P., 1991, in Levasseur-Regourd & Hasegawa, 335–338.
Hamilton D.P., Burns J.A., 1993, Nature, 364, 695–699.
Harris A.W., Bowell E.L.G., eds., 1992, *Asteroids, Comets, Meteors 1991*, Lunar and Planetary Institute, Houston, Texas, USA.
Hartung J.B., 1990, Meteoritics, 25, 369–370.
Hasegawa I., 1990, Publ. Astron. Soc. Japan, 42, 175–186; 42, 601.
Hasegawa I., 1992, Celest. Mech. Dyn. Astron., 54, 129–142.
Hasegawa I., 1993, in Štohl & Williams, 209–223.

Hasegawa I., Ueyama Y., Ohtsuka K., 1992, Publ. Astron. Soc. Japan, 44, 45–54.
Hawkes R.L., 1990a, WGN, 18, 145–151.
Hawkes R.L., 1990b, WGN, 18, 152–158.
Hawkes R.L., 1993, in Štohl & Williams, 227–234.
Hills J.G., Goda, M.P., 1993, Astron. J., 105, 1114–1144.
Hodapp K.W., MacQueen R.M., Hall D.N.B., 1992, Nature, 355, 707–710.
Horanyi M., Morfill G., Grün E., 1993, Nature, 363, 144–146.
Hughes D.W., 1990, Mon. Not. Roy. Astron. Soc., 245, 198–203.
Iglseder H., Grün, E., Münzenmayer R., Svedhem, H., 1993, in Štohl & Williams, 377–380.
Jenniskens P., Cailloux M., de Lignie M., Kuiper J., 1993, WGN, 21, 32–36.
Jessberger, E., *et al.*, 1992, Earth & Planet. Sci. Lett., 112, 91–99.
Jones J., Jones W., 1990a, Planet. Space Sci., 38, 55–66.
Jones J., Jones W., 1990b, Planet. Space Sci., 38, 925–932.
Jones J., Jones W., 1991, Planet. Space Sci., 39, 1289–1296.
Jones J., Jones W., 1993, Mon. Not. Roy. Astron. Soc., 261, 605–611.
Jones J., McIntosh B.A., Šimek, M., 1990, J. Atmos. Terr. Phys., 52, 253–258.
Jones J., Webster A.R., 1991, Planet. Space Sci., 39, 873–878.
Jones J., Webster A.R., 1992, in Harris & Bowell, 273–276.
Jones W., 1991, Planet. Space Sci., 39, 1283–1288.
Jones W., 1992, Planet. Space Sci., 40, 1487–1497.
Jones W., Jones, J., 1990c, J. Atmos. Terr. Phys., 52, 185–191.
Jopek T.J., 1993, in Štohl & Williams, 269–272.
Kamijo F., 1991, in Levasseur-Regourd & Hasegawa, 303-306.
Kane T.J., Gardner C.S., 1993, Science, 259, 1297–1300.
Kane T.J., Mui P.H., Gardner C.S., 1992, Geophys. Res. Lett., 19, 405–408.
Kazantsev A.M., Sherbaum L.M., 1990, Sol. Syst. Res., 24, 47–51.
Keay C.S.L., 1990, J. Roy. Astron. Soc. Canada, 84, 373–382.
Keay C.S.L., 1992a, in Harris & Bowell, 297–300.
Keay C.S.L., 1992b, Meteoritics, 27, 144–148.
Keay C.S.L., 1992c, Scientometrics, 23, 355–359.
Keay C.S.L., 1993, in Štohl & Williams, 315–318.
Keay C.S.L., Ostwald P., 1991, J. Acoust. Soc. Am., 89, 1823–1824.
Koeberl C., 1992a, Geochim. Cosmochim. Acta, 56, 1033–1064.
Koeberl C., 1992b, Geochim. Cosmochim. Acta, 56, 4329–4332.
Koeberl C., Shirey S.B., 1993, Science, 261, 595–598.
Koeberl C., Sigurdsson H., 1992, Geochim. Cosmochim. Acta, 56, 2113–2129.
Kokhirova G.I., 1993, Astron. Vestn., 27, 100–112.
Korobejnikov V.P., Chushkin P.I., Shurshalov L.V., 1991, Sol. Sys. Res., 25, 242–255.
Koseki M., Sekiguchi T., Ohtsuka K., 1990, in Lagerkvist *et al.*, 547–550.
Kostylev K.K., 1992, Astron. Vestn., 26, 112–115..
Kozasa T., Blum J., Mukai T., 1992, Astron. Astrophys., 263, 423–432.
Kramer E.N., Shestaka I.S., 1992, Astron. Vestn., 26, 85–90.
Kresák L., 1993, in Štohl & Williams, 147–156.
Krieger A., Arnold F., 1992, Geophys. Res. Lett., 19, 2301–2304.
Lagerkvist C.-I., Rickman H., Lindblad B.A., Lindgren M., eds., 1990, *Asteroids, Comets, Meteors III*, University of Uppsala, Sweden.
Lamy, P.L. *et al.*, 1992, Science, 257, 1377–1380.
Lebedinets V.N., 1991, Sol. Syst. Res., 25, 261–272.
Lebedinets V.N., 1992, Sol. Syst. Res., 26, 280–286.
Lebedinets V.N., Kulikova N.V., Pivnenko E.A., 1990, Sol Syst. Res., 24, 1–11.
Lebedinets V.N., Kurbanmuradov O., 1992, Sol. Syst. Res., 26, 61–69.

Levasseur-Regourd A.C., Hasegawa H., eds., 1991, *IAU Colloq. 126, Origin and Evolution of Interplanetary Dust*, Kluwer, Dordrecht, Holland.
Levasseur-Regourd A.C., *et al.*, 1993, Planet. Space Sci., 41, 167–169.
Lindblad B.A., 1990, in Lagerkvist *et al.*, 551–553.
Lindblad B.A., 1991a, in Levasseur-Regourd & Hasegawa, 311–314.
Lindblad B.A., 1991b, in Levasseur-Regourd & Hasegawa, 299–302.
Lindblad B.A., Porubčan V., 1991, Bull. Astron. Inst. Czechosl., 42, 354–359.
Lindblad B.A., Porubčan V., Štohl, J., 1993, in Štohl & Williams, 177–180.
Malyshev I.F., 1992, Astron. Vestn., 26, 112–114.
Martynenko V.V., Levina A.S., 1991, WGN, 19, 132–136.
McDonnell J.A.M., *et al.*, 1993a, Nature, 362, 732–734.
McDonnell J.A.M., *et al.*, 1993b, Adv. Space Res., 13(8), 87–101.
McIntosh B.A., 1991, in *Comets in the Post-Halley Era*, eds. R.L. Newburn, Jr., M. Neugebauer & J. Rahe, Kluwer, Dordrecht, 557–591.
Miles H., 1991, J. Brit. Astron. Assoc., 101, 291–292.
Miyashita A., Tanaka K., Watanabe H., Watanabe J., 1993, in Štohl & Williams, 319–322.
Mukai T. *et al.*, 1992, Astron. Astrophys., 262, 315–320.
Mukhamednazarov S., Mal'tseva N.V., 1990, Solar Syst. Res., 23, 179–183.
Murayama H., 1990, Tokyo Meteor Network Report No. 9, 53–57.
Novikov G.G., Pecina P., 1990, Bull. Astron. Inst. Czechoslov., 41, 380–387.
Obrubov Yu.V., 1993, in Štohl & Williams, 69–72.
Ohtsuka K., Murayama H., Hidaka E., Shigeno Y., 1991, Tokyo Meteor Network Rep. No. 10, 59–63.
Ol'khovatov A.Yu., 1991, Geomag. Aeron., 31, 606–607.
Olson D.W., Doescher R.L., Watson K.M., 1991, WGN, 19, 130–131.
Padevět V., 1991, Bull. Astron. Inst. Czechosl., 42, 206–215.
Padevět V., Jakeš P., 1993, Astron. Astrophys., 274, 944–954.
Pecina P., 1991, Bull. Astron. Inst. Czechosl., 42, 307–316.
Pecina P., 1993, in Štohl & Williams, 331–334.
Piers P.A., Hawkes R.L., 1993, WGN, 21, 168–174.
Poag C.W., Powars D.S., Poppe L.J., Mixon R.B., Edwards L.E., Folger D.W., Bruce S., 1992, Geology, 20, 771–774.
Pravec P., 1992, WGN, 20, 46–49.
Pravec P., Bocek J., 1992, WGN, 20, 70–83.
Quesada V., Uras S., 1993, in Štohl & Williams, 335–338.
Rabinowitz D.L., 1992, in Harris & Bowell, 481–485.
Rajaram G., Chandra H., 1991, Proc. Indian Acad. Sci. Earth Planet. Sci., 100, 255–65.
Rajchl J., Očenáš D., Zimnikoval P., Peresty R., 1993, in Štohl & Williams, 339–342.
Rendtel J., Heinlein D., 1991, Sterne, 67, 289–294.
Robertson M.C., Hawkes, R.L., 1992, in Harris & E. Bowell, 517–520.
Romejko V.A., 1991, Sol. Syst. Res., 25, 362–368.
Rozhilo A.A., Eremin S.S., Taranukha Yu.G., Kazantsev A.M., 1990, Komet. Tsirk., 417, 12.
Rubio L.R.B., 1992, WGN, 20, 140–144.
Schneider D.A., Kent D.V., Mello G.A., 1992, Earth & Planet. Sci. Lett., 111, 395–405.
Schnetzler C.C., 1992, Meteoritics, 27, 154–165.
Schulte P., Arnold F., 1992, Geophys. Res. Lett., 19, 2297–2300.
Sharpton V.L., Dalrymple G.B., Marin L.E., Ryder G., Schuraytz B.C., Urrutia-Fucagauchi J., 1992, Nature, 359, 819-821.
Shodiev U., 1992, Sol. Syst. Res., 26, 228-232.
Sigurdsson H., D'Hondt S., Arthur M.A., Bralower T.J., Zachos J.C., van Fossen M., Channell E.T., 1991, Nature, 349, 482–487.
Šimek M., McIntosh B.A., 1991, Bull. Astron. Inst. Czechosl., 42, 124–132.
Smirnov V.A., 1992, Astron. Vestn., 26, 116–120.

Spurný P., 1993, in Štohl & Williams, 193–196.
Steel D., 1991, in Levasseur-Regourd & Hasegawa, 291–298.
Steel D.I., Asher D.J., Clube, S.V.M., 1991, Mon. Not. Roy. Astron. Soc., 251, 632–648.
Steel D.I, Elford W.G., 1991, J. Atmos. Terr. Phys., 53, 409–417.
Steel D., Snow P., 1992, in Harris & Bowell, 569–572.
Steinweg A., Krankowsky D., Lammerzahl P., Anweiler B., 1992, J. Atmos. Terr. Phys., 54, 703–714.
Štohl J., Porubčan V., 1992, in *IAU Symp. 152: Chaos, resonance and collective dynamical phenomena in the solar system*, ed. S. Ferraz-Mello, Kluwer, Dordrecht, 315–324.
Štohl J., Porubčan V., 1993, in Štohl & Williams, 41–47.
Štohl, J., Williams, I.P., eds., 1993, *Meteoroids and their parent bodies*, Slovak Academy of Sciences, Bratislava, Slovak Republic.
Stulov V.P., Chernenko T.V., 1991, Sol. Syst. Res. 24, 203–205.
Suzuki K., Akebo T., Suzuki S., Yoshida, T., Ohtsuka K., 1993, WGN, 21, 214–215.
Swider W., 1992, Planet. Space Sci., 40, 247–253.
Terenteva A.K., Bayuk O.A., 1991, Bull. Astron. Inst. Czechosl., 42, 377–378.
Thomas G.E., 1991, Rev. Geophys., 29, 553–575.
Ueda M., Fujiwara Y., 1993a, in Štohl & Williams, 281–284.
Ueda M., Fujiwara Y., 1993b, WGN, 21, 215–217.
Wasson J.T., 1991, Earth & Planet. Sci. Lett., 102, 95–109.
Watanabe J., Nakamura T., Tsuda T., Tsutsumi M., Miyashita A., Yoshikawa M., 1991, in Harris & Bowell, 625–627.
Williams I.P., 1992, in *IAU Symp. 152: Chaos, resonance and collective dynamical phenomena in the solar system*, ed. S. Ferraz-Mello, Kluwer, Dordrecht, 299–313.
Williams I.P., Wu Z., 1993, Mon. Not. Roy. Astron. Soc., 262, 231–248.
Wu Z., Williams I.P., 1992a, in Harris & Bowell, 661–665.
Wu Z., Williams I.P., 1992b, Mon. Not. Roy. Astron. Soc., 259, 617–628.
Wu Z., Williams I.P., 1993, in Štohl & Williams, 77–80.
Zausaev A.F., Pushkarev A.N., 1992, Sol. Syst. Res., 25, 456-459.

I.P.WILLIAMS
Acting President, Commission 22

D.I.STEEL
Secretary, Commission 22

24. PHOTOGRAPHIC ASTROMETRY (ASTROMETRIE PHOTOGRAPHIQUE)

PRESIDENT: Christian de Vegt
VICE PRESIDENT: Catherine Turon
ORGANIZING COMMITTEE: A.N.Argue, P.D.Hemenway, H.Jahreiss, D.D.Polojentsev, J.Stock, W.F.van Altena, J.J Wang

1.0 Highlights

The ESA Astrometry Satellite Hipparcos collected regularly data in its operational mode from Nov. 27/1989 to Aug. 10/1992 with only minor problems. After this date problems with gyroscopes became serious but successful data collection until March 18/1993 was still possible. Due to additional serious hardware failures the observational phase of the Hipparcos mission will be terminated soon.
Data reduction has been started in May 1991 by the consortia FAST, NDAC and TDAC after major efforts in modifying and writing new software because of the non-geostationary satellite orbit.
Using data from 18 months observing, positions, preliminary proper motions and parallaxes of about 95,000 single stars and 6,000 double stars were obtained.In addition, some thousands of new double stars, (mainly Rho=0.15-0.4 arcsec) were found.
From these preliminary results it is already certain that the nominal goals of Hipparcos (2 mas in global positions, parallaxes and yearly proper motions for magnitude 9 stars) will be met.
In total, 36 month of data are available now for final treatment.
In addition about 1,000,000 stars up to B=11 have been measured by the Tycho mission.
The final Hipparcos and Tycho catalogs are expected to be released by ESA in the 1996/1997 time frame.

With view to the broad scientific research areas of our commission, the impact of the Hipparcos results can be assessed only marginally at present and will turn this decade into one of the most exiting periods in the long and successful scientific history of commission 24.

Besides the Hipparcos mission encouraging further progress has been achieved also in the fields of HST-FGS astrometry, ground-based optical interferometers and other instrumental developments as for example CCD detectors and new fast plate measuring machines.(see also the previous report).

Another large impact on our work is due to the mature astrometric capabilities of present VLBI which will provide the basis for the new inertial extragalactic radio-optical reference frame with the final goal of an all-frequency high precision reference frame which finally may be capable to align structural features of celestial objects from X-ray to IR and Radio bands unambiguously with sub-mas resolution and positional accuracy everywhere in the sky.

Commission 24 is part of the new Joint Working Group on Reference

J. Bergeron (ed.), Reports on Astronomy, Vol. XXIIA, 225–228.

Frames, established at the last GA (IAU Bull.67,33). In particular our commission will provide a substantial contribution to the problem of linking Hipparcos to the extragalactic frame (see 3.2)

Based on the excellent astrometric capabilities of the new generation of fast plate measuring machines and the huge data storage capacities of present computer systems there will be for the first time a realistic chance to digitize and preserve all old epoch astrometric plate material without any loss of intrinsic accuracy and information content.
In addition precise astrometry of very faint stars will be feasible when the digitization of several epochs of the large sky surveys will be finished soon.

In this context a negative highlight are the recent IAU attempts to force our commission to merge and to establish a new scheme of GA which scientifically is not very helpful to our work.
As it is obvious again from this report the broad and lively research fields of commission 24 clearly justifies its further existence (not excluding the possibility of minor organizational changes).
Because of very restricted space for this report, long name lists in particular for the wellknown large catalog projects had to be omitted.

2.0 Trigonometric Parallaxes and Nearby Stars

From Van Vleck Observatory A.Upgren reports continuation of the parallax program on nearby stars.In particular work concentrates on metal-poor subdwarfs and stars of intermediate metallicy. Upgren,Weis in collaboration with Gliese and Jahreiss are preparing a list of the Vyssotsky stars that have accurate positions, p.m. and photometric parallaxes. D.J.MacConnell (STI/CSC) and W.Osborn (Mich.U) report first observations with the HST-FGS to determine parallaxes for high velocity stars. Preliminary reductions indicate an accuracy <3mas for stars > 14th mag. Hanson(Lick),Lutz(Wash.State U.) and van Altena(Yale) have nearly completed their project to use the new Yale Parallax Catalog data to improve the luminosity calibration of field subdwarfs by color and metal abundance.

3.0 Catalogs of Proper Motions and Positions

At Hamburg/RGO the conventional plate adjustment solution and photographic photometry of the CPC2 was finished. The catalog contains 276,131 star positions and V-magnitudes for average epoch 1968 with average m.e. of 60mas. The block adjustment solution of the CPC2 is in progress. At Pulkovo Observatory work for the compilation of p.m. of the FOCAT-S catalog using AC and CPC2 data is finished (in coll. with Sternberg Institute, Astronomical Obs. St.Petersburg, ARI and Hamburg). Also at Pulkovo work for the creation of a catalog of positions and p.m. of about 5 million stars globally on the basis of photographic re-observations of the

AC zones using two similar astrographs on each hemisphere were begun. In the frame of this work at Bolivia already more than 1700 plates were taken in the declination zone +10 to -20 deg. and provisional plate measurements have started. Also at Pulkovo work on the KSZ-plan (absolute p.m. relative to galaxies) was continued. A working catalog data base of absolute p.pm. in the declination range -5 to +90 deg. was established on magnetic disks. At ARI, Heidelberg, S.Röser and U.Bastian have completed the final southern part of the PPM catalog . It contains 197,179 stars south of -2.5 deg. The average pos. error at present epoch is 110 mas, the p.m. accuracy is 3 mas/yr. The total PPM now contains 378,910 stars.
At Lick, Klemola, Hanson and Jones have completed work on Part I of the Lick NPM program (900 fields outside the Milky Way), the catalog gives absolute p.m., positions and photometry for 150,000 stars. Plans have begun for NPM Part II (350 Milky Way fields) to complete the sky down to -23 deg. Hanson is using the NPM to study the solar motion and galactic rotation.
At Straßbourg Astronomical Observatory A.Fresneau derived relative p.m. between northern AC zones(+1 to +31deg.) and the HST-GSC for 1 million stars in the magnitude range B=9.5-12.5. These data are being used for galactic p.m. studies.

3.1 Special Catalogs

At USNO D.Pascu has continued photographic and CCD observations of satellites of Mars,Jupiter and Saturn.(see also Comm.20 rep.) Klemola and Own(JPL) constructed a catalog of positions (4800 stars) in the field pf minor planet Ida for Galileo spacecraft encounter in August 1993.
D.Hoffleit (Yale) contributed to the updating of various Yale catalogs especially the Bright Star Catalog and the Catalog of Trigonometric Parallaxes she also wrote a historical account on "Yale and USNO Cooperation Especially in the Brouwer and Clemence Era".
J.Dommaget(ROB) in collaboration with O.Nys reports extensive work on double and multiple star systems especially in context with the Hipparcos Input Catalogue. The Catalog of Double and Multiple Stars(CCDM) has been completed.
At Lohrmann-Observatory (T.U.Dresden) A.Steinert reports cooperative work on minor planets with ITA St.Petersburg, about 50 plates/year were taken with the Zeiss-Sonnefeld astrograph.

3.2 The Extragalactic Reference Frame and Linking Procedures

The new WG on Reference Frames ,chaired presently by C.de Vegt, has established a first list of about 500 compact radio sources mainly with optical counterparts, optical and radio work on these objects is underway by many contributing observatories.
The Potsdam group(headed by E.Schilbach) reports work on optical positions of 50 radio sources, based on Tautenburg Schmidt plates. The plate material will be also used to derive p.m. relative to galaxies (joint prog. with Hamburg/MAMA/APM).

At Hamburg/USNO/NRL/CSIRO the extensive joint program on optical and VLBI radio positions of extragalactic sources has been continued, the program contains about 400 sources presently and will contribute in particular to the construction of a groundbased Hipparcos extragalactic link. At Bonn P.Brosche reports continuation of work on a Hipparcos p.m. link using plates of a small number of sources with up to 80 years epoch difference. At Shanghai Observatory positions of 85 radio stars and 3 extragalactic sources have been determined with the Zo-se refractor. At Pulkovo a program for determining positions of fainter stars of 13-16th mag. in 73 fields (dec. zone +30 to +90) around radio sources has been started.

4.0 Proper Motions and Star Clusters

At Potsdam absolute p.m. and B,V magnitudes for 10,000 stars up to 20th mag. were determined in two regions near the NGP. In addition extensive p.m. work in the Pleiades and the globular clusters M3 and M92 is reported.(coll. with Bonn,APM). In Taskent A.Latypov reports p.m. work on Cepheid EV Sct. At Bonn relative p.m. of selected open and globular clusters were determined.
At NAO Tokyo M.Miyamoto and M.Soma have finished an extensive study on the galactic velocity field of about 30,000 M-K giants.
At Shanghai Observatory J.J.Wang and collaborators report work on positions and p.m. of Presepe stars from plate material obtained with the Zo-se 40 cm refractor and other p.m. studies on NGC2286 and M67. Using a new kinematic method with the NPM data to correct rel. p.m. to absolute Hanson and Cudworth(Yerkes) have determined improved space velocities of 14 globular clusters. Klemola, Stauffer,Prosser and Probst(KPNO)
completed the first phase of an photometric and astrometric survey in the extended Pleiades field. Klemola and collaborators completed a study of the space motion for globular clusters NGC 362 and 6218. Jones,Klemola and Lin have completed the first phase in measurement and analysis of absolute p.m. of LMC red-giant stars in its remote halo around the globular cluster NGC 2257.

5.0 Instrumentation, Techniques and Reduction Methods

The Bordeaux Observatory is developing a CCD micrometer to be installed at its automatic meridian circle in 1994. Using Hipparcos star positions in narrow strips the expected accuracy of single stars to V=15 should be better than 50mas.
At Hamburg/USNO N.Zacharias has continued work on block adjustment simulations for complete sphere plate overlap configurations. Stable solutions with only a small number of reference stars <100 seem to be feasible. At Shanghai Observatory detailed studies of PDS astrometric capabilities have been made.
At Hamburg/USNO a new 5-lens astrograph lens system, optimized for the red spectral range (5500A-7000A), has been developed and is close to delivery.
Also at Hamburg work on the new CCD-based granite plate measuring machine is nearly finished, extensive test measurements are being prepared.

Commission 25: STELLAR PHOTOMETRY AND POLARIMETRY
Photométrie et Polarimétrie stellaires

PRESIDENT: Andrew T. Young
VICE-PRESIDENT: John D. Landstreet
ORGANIZING COMMITTEE: S. J. Adelman, M. Breger, R. M. Genet, J. K. Knude, A. U. Landolt, J. D. Landstreet, J. Lub, J. W. Menzies, I. S. McLean, T. Moffett, C. L. Sterken, V. Straizys, F. J. Vrba

1. INTRODUCTION

This is a summary of the essential work that was published since the previous report (McLean, 53.113.028), not a complete review. Most papers are cited by their volume, section, and reference number in *Astronomy and Astrophysics Abstracts*. Papers presented at the very successful IAU Colloquium 136 in Dublin (see report in IAU *Information Bulletin* 69, p. 26) are simply cited as (Dublin); they have just been published by Cambridge University Press.

The drafts of the sections below were edited by A. T. Young, who thanks the individual contributors, and is responsible for the published versions.

CCD techniques have almost completely replaced photography in photometry and polarimetry (52.012.014; 52.012.015). However, a surprising number of CCD instruments turn out to have nonlinearity and other problems of the sort long familiar to photographic photometrists. CCDs also pose transformation problems that can trap the unwary. The high detective quantum efficiency of these detectors has enticed users to press to the faintest possible limits, where systematic errors can lie hidden in the sky noise. There remain disturbingly large systematic problems in flat-fielding, despite strenuous efforts to overcome these difficulties. There are also technical problems in applying the modulation techniques of high-precision polarimetry to CCDs. Much remains to be done in all these areas.

The development of automatic photoelectric telescopes is also having a revolutionary effect on photometry. While the first generation of APTs involved only small telescopes and low-precison photometers, current ones are in the 1-meter class and have high-precision instruments that can produce excellent results. The high speed of these instruments offers the possibility — not yet fully exploited — of determining time-dependent extinction with great accuracy. Increasing numbers of these instruments are now competing with human observers.

Several meetings have dealt with these new developments. Besides the Joint Commission Meeting at Buenos Aires, Russ Genet's annual meetings on APTs have reported their progress (54.012.002), and a workshop on APTs was held at Kilkenny in conjunction with the Dublin Colloquium. Dave Philip's meeting (54.012.027) covered the scientific results of stellar photometry. The Dublin Colloquium reviewed the whole field. It seems safe to say that we now have powerful technologies available, but have not yet learned how to use them fully.

Finally, we must congratulate Dr. Alan Cousins on his 90th birthday! He has continued to establish standard stars for the benefit of Southern-hemisphere observers. In our enthusiasm for the newer technologies, we should not lose sight of what he has done by carefully applying the older ones.

J. Bergeron (ed.), Reports on Astronomy, Vol. XXIIA, 229–238.

2. PHOTOMETRY

(prepared by C. Sterken)

2.1 Instrumentation

Though CCDs are used more and more for photometry, new photoelectric photometers are being built at various places. Some of these designs are rather bulky, but there seems to be an increasing interest in portability, and lighter and versatile instruments are in high demand.

Several new photometer designs were reported, such as the high-speed two-star photometer by Rao et al. (52.034.100 and 54.034.168) and Venkatesha and Seetha (54.034.169), and multichannel photometers (Walker, 54.034.185, and Sullivan, 54.034.189).

A four-channel *WBVR* stellar photometer for observing bright stars described by Konilov and Krylov (51.034.018) was used for measuring 15 000 stars. A multicolor photometer described by Schoembs (54.113.021), and the four-star photometer by Le Contel (55.013.058) are novel designs. Caton and Hawkins (52.034.168) also describe a new two-star photometer. Higbie et al. (52.034.186) trace the design of a new imaging photometer instrument. Bell and Yoss (54.36.122) describe a multiband procedure that permits simultaneous star and sky exposures, and rapid acquisition of data over the entire sky. The method is faster, and more accurate, than the single-channel approach.

2.1.1 Photometry with scalar detectors. Bessell (54.36.364) reviews reduction techniques and effects of atmospheric extinction in stellar photometry. Tobin and Wadsworth (54.36.363) draw attention to the method and necessary equipment for the determination of photometer deadtime correction from measurements in daytime.

2.1.2 Photometry with CCDs. Gilliland (54.36.124) reviews recent efforts to set empirical limits on the precision of time-differential CCD photometry of rich stellar fields. Stetson (52.036.033) describes a computer program that generates the magnitude corrections required to relate profile-fitted or small-aperture magnitudes to total magnitudes, by means of the aperture-growth technique.

Jønch-Sørensen and Knude (52.113.017) present preliminary results of CCD-*uvby* photometry of field stars in four cardinal directions of the galaxy. Richtler (52.113.016) presents Strömgren colors of CCD photometry zero-point stars in the Magellanic Clouds and M67. Anthony and Twarog (54.113.012) illustrate the feasibility and value of CCD photometry in the *uvby* system.

Labhardt et al. (54.036.183) address the problems encountered in deep CCD photometry in the *UBV* system, particularly reproduction of the system and transformation to the standard system, in addition to the determination of instrumental magnitudes from model image profile fitting or from aperture photometry. Zickgraf et al. (52.113.036, 52.113.037) present *UBVR* sequences for CCD photometry in M83, NGC 5128, NGC 2403 and M81. Standardizing CCD images in the M67 "dipper asterism" is possible by using Cousins *VRI* data for 19 standard stars provided by Jones and Taylor (52.113.038). Jiang et al. (55.036.172, 55.036.173) apply a Schmidt telescope equipped with CCD to *BVRI* photometry.

Boyle et al. (52.113.001) present CCD photometry in the Vilnius system for 321 field stars in Lyra ($V \leq 17.5$).

2.1.3 Automatic Telescopes. Sterken and Manfroid (54.036.127) discuss some effects that have negative influences on the precision of automatically performed measurements.

Young (1992, in *Automated Telescopes for Photometry and Imaging*, p. 73) demonstrates that, since it is technically impossible to make identical instruments, one cannot avoid the general transformation problem. Thus, instead of using undersampled photometric systems, one must measure complete information on stellar spectra, using an appropriate well-sampled system (of 5 or 6 bands in the blue and visual region). Such will enable accurate transformations that are valid for all objects, no matter how pathological their spectra are. This is only possible when turning to the use of time-efficient automatic telescopes.

2.1.4 Atmospheric extinction. Methods for extinction determination are described by Schwarzenberg-Czerny (A&A 252, 425) and Forbes (1992, Southern Stars 35, 1). Poretti and Zerbi (1993, A&A in press) propose an analytical method for extinction determination that handles variable extinction. Reimann et al. (1993, A&A in press) analyse the variability of the extinction coefficient on the basis of data covering two decades. Sterken and Manfroid (1992, A&A 266, 619 with a sequel by Manfroid 1993) describe the evolution of atmospheric extinction at La Silla from *uvby* data covering almost 15 years. They find three episodes of enhanced extinction caused by volcanic aerosols (respectively due to the El Chichón, Nevado del Ruiz and Pinatubo eruptions), and — besides the well-known small-amplitude seasonal variations — also a short-timescale pseudo-periodic variation. Additional extinction effects caused by the Pinatubo eruption have been observed and described by Kilkenny and Westerhuys (Dublin).

Dravins et al. (Dublin) report on atmospheric scintillation variations on time scales of 100 ms to 100 ns.

2.1.5 Transformation. Young (1992, A&A) re-examines photometric transformation theory and demonstrates the importance of high-order terms that have previously been neglected, which have an appreciable magnitude because they depend on high-order moments of passband response profiles that are not well represented by a Taylor series. These transformation errors due to neglected high-order terms are likely to be larger for narrow-band photometry and spectrophotometry than for wideband systems. Transformation equations should at least be quartic, and this requirement is independent of spectral resolution.

Manfroid and Sterken (54.036.123 and 1992, A&A 258, 600) analyze systematic errors due to color transformations, and illustrate the reliability of transformations, emphasizing reduction methods, the validity of homogenisation procedures, and the choice of standard stars that define the system. Manfroid (1992, A&A 260, 517) discusses physical constraints on color transformations and transformation schemes, and gives directions to minimise those errors. Manfroid et al. (1992, A&A 264, 345) evaluate the impact of extrapolation errors in different color transformation schemes.

Alfaro and Delgado (54.113.031) analyse standard *uvby* indices of B stars in 17 young open clusters in order to determine systematic differences caused by interstellar reddening. Their results show that such systematic errors in c_1 increase with increasing amount of reddening, and they suggest that transformation procedures should incorporate reddening terms.

2.2 Standard stars

Voroshilov (52.113.025) discusses variability of some *WBVR* photometric standards. Menzies et al. (53.113.009) present $UBV(RI)_c$ photometry of equatorial standard stars and find systematic differences for $(B-V)$ and $(U-B)$ between the *UBV* system defined by Cousins's E region standards and that used by Landolt, who has just published equatorial *UBVRI* standards between mags. 11.5 and 16 (AJ 104, 340, 1992). Watanabe et al. (53.113.037) give values of $U-B$, $B-V$ and V for 223 stars selected from Nicolet's catalogue.

Osborn et al. (52.113.028) established a set of 63 standard stars in the bandpasses of the International Halley Watch system for cometary photometry. New Washington-system standards, suitable for observation with a CCD, were presented by Geisler (51.113.040). These stars are in four fields, each field including stars with a wide color range sufficiently faint to be observed with large telescopes.

2.3 Photometric Systems

Crawford (54.113.011) reviews the concept of Strömgren's *uvby* system, and elucidates its strong and weak points.

Pel (154.113.029) compares the systematic accuracy in four photometric systems (at the Schenectady conference).

Kiselman et al. (52.036.040) discuss the possibility of combining four filters of the HST Wide Field/Planetary Camera into a *uvby*-like photometric system. They calibrated the system (using ground-based observations) and obtained good estimates of effective temperatures and metallicities. Anthony-Twarog et al. (53.113.008) extended the standard *uvby* system with a FWHM 90 Å filter centered on CaII H and K. The extension is designed for application to stars where the m_1 index has reduced sensitivity, and leads to an alternative index for m_1.

Mendoza (54.113.009) discusses the $\alpha(16)$, $\Lambda(9)$ photometric system for normal stars, and also for WR, Be and other peculiar objects.

Straizys (52.113.043) reviews different medium-band systems from the viewpoint of determination of physical parameters, and demonstrates the abilities of the Vilnius system in the determination of spectral types, absolute magnitudes, metallicities, and different types of peculiarities in interstellar reddening.

2.4 Synthetic Photometry

Bessell (52.113.039) derived a good representation for the Johnson-Cousins *UBVRI* passbands by comparing synthetic photometry with observations. A major source of $(U-B)$ transformation problems can be traced to mismatched B bands. He also discusses suggested CCD filter combinations. Gocherman (54.036.294) attempted to reconstruct several *UBVRI* systems on the basis of flux-calibrated spectra.

Stift (52.036.043) presented a theoretical investigation of the influence of photometric passband effects on the parameters of close visual binaries obtained by area-scanning. He demonstrated that the use of broadband filters can lead to considerable errors in the observed distances and/or position angles.

Tripicco and Bell (54.113.034) explore the properties of the DDO system. Paltoglou and Bell (54.113.043) present a grid of synthetic surface brightness magnitudes for the 14 bandpasses of the Hubble Space Telescope Faint Object Camera. The synthetic colors have been used to examine transformations between ground-based and Faint Object Camera *UBV* and *uvby* systems.

2.5 Spectrophotometry

Papaj et al. (52.114.051) present mean energy distributions of samples of stars hotter than B5, and derive a new set of $(B-V)_0$ colours. The derived spectrophotometric standards are free of reddening effects in the visual and far-UV regions.

Taylor and Jones (52.113.012) present Cousins *VRI* photometry from Gunn and Stryker scans, and find systematic deviations in $R-I$, which are attributed to a declination effect in the Gunn-Stryker scans. They also find the Gunn-Stryker synthetic colours unexpectedly noisy. Glushneva (54.113.013) reviews spectrophotometric catalogues and stresses the need for increasing the number of standard stars and for elaborating the differences between catalogues in order to determine systematic differences.

2.6 Calibration

Reglero et al. (52.113.029) extend the *uvby* photometric calibration to spectral types later than G0 by using both the *uvby* and β systems to G5-K7 (V to III) stars. North and Kobi (54.113.019) compare the (T_{eff}, log g) calibrations of the *uvby*β system published by two different authors, and point out the existence of systematic differences. Gray (54.113.032) derives an empirical calibration of the *uvby*β system for the A, F and early G supergiants.

Arellano Ferro and Parrao (52.113.023) carried out an extensive compilation of non-Cepheid F0-G8 supergiants in open clusters and associations. On the basis of these data, the authors provide a reddening formula for F0-G3 supergiants that provides $E(B-V)$ within ±0.003, and an absolute magnitude calibration that predicts M_v with an estimated accuracy of ±0.5 for the most luminous F-G supergiants. Blackwell and Petford (54.113.013) derive the relationship between stellar integrated fluxes and the photometric indices B, V, I and K. Boyarchuk et al. (51.113.046) improve the absolute calibration of *UBV* photometry by comparing spectrophotometric energy distributions and catalogued *UBV* data.

Buser and Fenkart (52.113.024) present a new extended calibration of the 2-color and color-magnitude diagram and the metal-abundance parameter for the dwarf stars of the Basel-Potsdam 3-color photographic *RGU* survey of the galaxy.

Korotina et al. (52.113.013) propose a method for determining fundamental properties of late-type giants using Geneva photometry, and provide a catalog of T_{eff}, log$g$ and $[Fe/H]$ for 982 G5-K5 giants. New calibration of blanketing parameters Δm_2 (Geneva) and δm_1 (Strömgren) in terms of $[Fe/H]$ is given by Berthet (52.113.009). Kobi and North (52.113.010) give a new calibration of the Geneva d, m_2, and B_2-V_1 parameters, which allow estimation of T_e, log$g$, and $[Fe/H]$ of A4 V-III to G5 V stars.

Geisler et al. (54.113.036) present an improved metal-abundance calibration for the Washington photometric system.

2.7 Catalogs and Data Bases

An extension of the Geneva Catalogue of about 4200 additional stars, together with a statistical discussion and a bibliography has been presented by Cramer (54.113.016).

Oblak (51.113.022) gives *uvby* photometry for 367 southern stars of the Hipparcos Input Catalogue. Hauck and Mermillod (54.113.018) have collected from the literature all available *uvby*β data in a single file, and publish a homogeneous partition of this database. Arellano Ferro et al. (51.113.010) report *uvby*β photometry for 110 bright F0-K0 supergiants.

Kilkenny and Laing (52.113.048) give photoelectric UBVRI observations for 336 stars ($V \leq 11.5$) from the Hubble Space Telescope Guide Star Photometric Catalogue. Menzies et al. (52.113.052) report on $UBV(RI)_C$ photometry of TD1 satellite stars, many of which would be suitable for use as very blue standards. Abuladze (54.113.014) compiled a catalogue of 366 photoelectric *UBV* standard stars.

2.8 Books

Straizys's *Multicolor Stellar Photometry* (55.003.084, Pachart Publishing House) gives a detailed description of almost all photometric systems with their advantages and disadvantages, and a detailed description of the Vilnius system. Sterken and Manfroid's textbook *Astronomical Photometry, A Guide* (Kluwer 1992) covers the principles of astronomical photometry, and explains reduction and transformation methods. The Proceedings of IAU Colloquium 136, *Stellar Photometry — Current Techniques and Future Developments*, edited by Butler & Elliott, and a book on the history of photometry by Budding, have just been published by Cambridge University Press. We also call attention to the Schenectady *Precision Photometry* proceedings (54.012.027).

3. REDUCTION SOFTWARE

A general photometric reduction program has been distributed by ESO as the PEPSYS context of the MIDAS data-reduction system. It is intended for reductions in any system; information about the *UBVRI* and *uvby*β systems is built-in. Besides handling extinction and transformations, it includes a program for planning observations, an extensive User's Guide, and other documentation.

MIDAS is available free of charge to astronomical institutions, and runs on a wide variety of VAX/VMS and UNIX systems, including PC architectures running LINUX. For further information, send e-mail to resy@eso.org.

At the Dublin Colloquium, Peter Stetson announced that he has a reduction program that runs under IRAF. Potential users can obtain further information from him.

Both of these packages include multi-night reductions, time-dependent extinction, physically-based color terms, and other details that are often omitted or treated inadequately in "home-grown" reduction programs. They should be far superior to previous-generation programs like SNOPY. It is hoped that the widespread availability of standard reduction programs of high quality will help improve the consistency of photometry.

4. POLARIMETRY

(prepared by J. Landstreet)

4.1 Introduction

Polarimetry continues to grow as a technique that can provide useful information and constraints on the geometrical organization of complex objects, on the nature, location and sizes of scattering particles, and on magnetic field structures. There are now well over a dozen polarimeters in active service, some at major national observatories; polarimetry is gradually ceasing to be a specialized technique understood only by a few, and is becoming a generally available tool. More than 50 papers now appear each year that report some substantial optical, infrared or ultraviolet polarimetric result. In what follows, I offer a rather haphazard overview of some of the major polarimetric observations and modelling programmes, and a few specific projects of particular interest.

4.2 Current active research areas

4.2.1 Planetary atmospheres. Tinbergen has reported a study of 22-degree halos in the terrestrial atmosphere. Such halos reveal the presence of birefringent diffraction due to crystals in our atmosphere, and should provide a means of detecting such crystals in planetary atmospheres (54.082.002). This may indeed be the case for Venus (see Können et al., Icarus 102, 62, 1993). Polarimetry has long been used to study Mars (52.097.013), where crystal clouds in the atmosphere are detected.

4.2.2 Other Solar system objects. Eaton, Scarrott and collaborators, using the Durham imaging polarimeter, have found that imaging polarimetry of comet tails clearly reveals presence of a dust component that is barely visible on direct images. Their polarimetric images permit the study of the spatial distribution and nature of scattering particles (53.103.037). In a related study, the properties of the dust of the tail of comet P/Halley have been studied using a polarimeter on the Vainu Bappu Observatory 1-m telescope (53.103.094).

4.2.3 Clouds in star-forming regions and pre-main sequence stars. A large group of collaborators, using the UK IR Telescope and an IR polarimeter, have obtained IR polarization images of the Orion Molecular Cloud to study the distribution and nature of scattering matter around the new stars (53.131.080, 53.131.053). This work has been extended to other bipolar nebulae such as R Mon (53.131.113), GL 2591 (54.131.305), and several others (54.131.306).

Gledhill, Rolph, Scarrott and Wolstencroft and collaborators have been studying polarization of compact nebulae and circumstellar disks with the Durham imaging polarimeter (51.121.008, 51.121.012, 51.131.066, 53.131.056, 53.131.103, 53.131.128, 54.131.314, 54.112.098, 54.131.328, and MNRAS, 257, 485). These studies reveal locations of (often hidden) very young stars as well as giving information about the structure and particle sizes in the reflecting clouds and disks. A general review of this work may be found in 51.131.228.

Bastien and collaborators, using the optical polarimeter of the Observatoire du Mont Megantic, continue to study polarization of T Tau stars induced by the presence of scattering in circumstellar disks. They have carried out extensive modelling of such disks (51.121.028, 52.121.055, 54.063.036, 54.131.168, 55.121.010).

Polarization images in the IR of the nebula around T Tau have been obtained by Weintraub, Kastner and their collaborators, using the IR imager and polarimeter at Kitt Peak National Observatory (55.121.048). The data have been used to model the geometry of the disk around the star. Several other T Tau's have been observed (see also 55.131.083).

4.2.4 The Sun. Landi Degl'Innocenti, Bommier and Sachal-Brechot have carried out considerable theoretical analysis of the of the polarization of the Hanle effect in solar resonance lines (52.063.037, 53.063.050, 53.063.051, and 53.063.049). Faurobert-Scholl has also studied this problem in the case of partial frequency redistribution, especially with application to the Ca I 4227 A line (53.063.063, 55.063.090).

4.2.5 Envelopes around hot stars. A particularly exciting development has been the successful Wisconsin Ultraviolet Photo-Polarimeter Experiment (WUPPE), which has produced the first ultraviolet linear polarization spectra of a number of different types of hot stars and made possible a first exploration of the new information provided by such data. Observations are now available for hot supergiants (54.116.084), Be stars (54.116.092), and Wolf-Rayet stars (55.116.025). This experimental team has also been active in using optical linear polarization data to study the circumstellar environment of WR stars (52.116.048, 54.116.085, 55.116.011, ApJ 398, 286, and ApJ 401, L105).

The Montreal group of Moffatt, Drissen and Robert and their collaborators have been systematically exploiting linear polarimetry as a tool for modelling the winds and circumstellar environments of a number of WR stars, both single and in binary systems (51.117.065, 52.116.014, 55.116.009, and ApJ 397, 277).

Marlborough and his collaborators have been active in modelling the effects of scattering matter around hot stars. They have recently examined the polarized radiation produced by electron scattering in the winds of B[e] supergiants (53.064.042), and discussed the constraints which polarization data can provide for Be stars (55.112.039).

4.2.6 Magnetic fields of main sequence stars. The Zeeman analyser on the cassegrain echelle spectrograph of the 3.6-m ESO telescope has been used extensively by Mathys to obtain a wealth of data on the magnetic fields of Ap and Bp stars. Careful modelling of the spectropolarimetric data yield considerably stronger constraints on the magnetic geometries of the stars studied than have previously been obtained (54.116.002).

Donati, Semel and Rees and their collaborators have developed an extremely sensitive and precise method for obtaining high spectral resolution circular and linear polarimetry of magnetic stars. Such measurements have made possible the detection of very weak Ap star fields (52.116.001), and have particularly been used to detect and map fields on stars in active RS CVn binary systems (51.117.230, A&A 265, 669, and A&A 265, 685). This is the first time fields in such systems have been been detected, and the work provides by far the most convincing modelling of field structure in any cool star.

Bohlender, Landstreet and collaborators have continued their series of magnetic field strength measurements of magnetic Ap and Bp stars using the Western Ontario Balmer-line Zeeman analyser. Longitudinal field measurements and modelling have been presented for two mass-losing He-wk stars (51.116.010) as well as for a number of individual He-str (52.116.004, 53.116.017), Si (51.116.026, 55.116.019, and A&A 269, 355), Cr (52.116.003), and λ Boo (52.116.045) stars. This subject has been reviewed in detail by Landstreet (55.116.026).

In a very interesting new development, Leroy has begun publishing results from an extensive series of measurements of the transverse magnetic fields of a number of cool, large-field magnetic Ap stars. These are derived from broad-band linear polarization measurements made using the Sterenn polarimeter at the Observatoire du Pic-du-Midi (Leroy et al., 1993 A&A in press, Leroy 1993 A&A in press). These data provide extremely useful constraints on the gross geometrical structure, particularly the inclination and obliquity angles, of the stellar field.

4.2.7 Evolved stars. The intrinsic linear polarization of cool giants, which is produced either by asymmetric scattering in the stellar atmosphere or by circumstellar material, has been studied by LeBertre and Schwartz (A&A 229, 138) using the Pisco polarimeter at ESO, and by Raveendran (53.064.037, 53.112.052) with the 1.0 m telescope of the Vainu Bappu Observatory.

4.2.8 Magnetic fields in white dwarfs. Bailey, Ferrario, Hough, Wickramasinghe and collaborators, using the Hatfield Polytechnic polarimeter, have obtained broad-band filter polarimetry of a number of AM Her systems (close binary systems containing a white dwarf with a megagauss field; see 51.117.222, 53.117.073, 54.117.191, 55.117.228, and MNRAS 259, 583). Such data are very useful for modelling magnetic field structure.

Piirola, Reiz and collaborators have used a Finnish 5-channel polarimeter at ESO to study the time and wavelength dependence of circular and linear polarization in the very-large-field single magnetic white dwarf PG 1031 +234 (55.126.098) as well as in the AM Her system VV Pup (52.117.020).

The observed wavelength dependence of circular and linear broad-band polarization in strongly magnetic (B > 100 MG) white dwarfs certainly contains much useful information about the field structure in these stars, but has so far proven very difficult to interpret. A very interesting theoretical work by Whitney (53.063.053, 53.063.060) studies the simplest dichroism, that of free electrons, and succeeds reasonably well in fitting observations of one of the largest-field stars.

The polarimeters at Steward Observatory continue to provide much useful data on magnetic white dwarfs. New rotation ephemerides have been obtained for 4 magnetic white dwarfs showing broad-band continuum polarization (53.126.011). A weak field of some 2.3 MG is found in the most massive known single white dwarf (Schmidt et al. 1992, ApJ 394, 603).

4.2.9 Supernovae and supernova remnants. The optical polarization of the Crab SNR long ago provided direct evidence of synchrotron radiation from this intriguing object. Recently, an optical polarization study has been reported for a Crab-like SNR in the LMC (51.125.029). Further studies of the Crab itself continue to reveal details of the magnetic field structure and electron distribution in that remnant (52.125.055, 53.125.019). Polarimetry of SN 1987A has been analysed by Jeffreys (54.125.172); it appears that most of the polarization is due to asymmetric electron scattering.

4.2.10 The interstellar medium. The polarizing effect of aligned grains in the interstellar medium provides important information about the nature of these solid grains. New developments in the study of this polarization include determination of the wavelength dependence of polarization in the IR (51.131.004, 52.131.007, 55.131.039, 55.131.263), polarimetric studies in the visible of circular polarization and of the diffuse interstellar bands (52.131.194, ApJ 398, L69), and for the first time spectropolarimetry of the ultraviolet wavelength dependence of interstellar polarization by the WUPPE experiment team (55.131.020). The new data have provided considerable insight into variations in the ISM of grain composition and size distribution.

Polarimetry, particularly in the IR, has also been used to study the structure of the galactic magnetic field, both on a large scale (55.131.089) and near the galactic centre (54.155.096).

Leroy has recently completed measurements at Pic-du-Midi of linear polarization of a large number of nearby, intrinsically unpolarized stars to show that the local ISM out to a distance of at least 50 pc in all directions from the sun is remarkably free of dust (1993 A&A, in press).

4.2.11 External galaxies. A recent review of magnetic fields in galaxies by Wielebinski and Krause (51.157.149) naturally gives considerable prominence to the results of optical linear polarization studies of external galaxies.

Systematic studies, using the Hatfield Polytechnic polarimeter, of the nature of the polarization of Seyfert galaxies has recently been reported (51.158.256, 51.158.257). Polarimetry of starburst and interacting galaxies is reported in 54.157.274. The data suggest that the observed polarization is often produced by transmission through aligned grains, but in some cases scattering from the nucleus is also important.

Polarization images of a number of highly active galaxies have been studied by the group associated with the Durham imaging polarimeter. The polarization properties of a large sample of BL Lac objects ("blazars") are discussed in 51.158.227. More detailed studies of individual objects have yielded much information about polarizing mechanisms, such as reflection of central jets and aligned grains, in such objects as 3C368 (51.158.045), Cyg A (52.158.127), NGC 1068 (53.158.073), M82 (54.157.272) and NGC 4151 (MNRAS 257, 309). Optical polarization of intermediate and high redshift galaxies has revealed strong polarization of light which in the rest frame is in the ultraviolet; it appears that scattering from a beamed central source dominates the luminosity at these wavelengths (55.158.175).

Numerous groups in the U.S. have been using polarimetry to probe the geometrical structure and the nature of the central powerhouse and its immediate environment for a variety of active galaxies. There has been so much activity in this area that it is difficult to give even an overview. Studies of Seyfert galaxies include 51.158.085, 54.158.325, ApJ 397, 452, and ApJ 399, 50. Blazars are discussed in 51.158.085, 54.158.325, ApJ 400, 115, and ApJ 400, L17. Observations of quasars are reported in 51.159.086, 52.159.083, 53.159.054, ApJ 396, L19, ApJ 398, 454, and ApJ 400, 96. Optical polarization of radio galaxies is discussed in 53.158.013 and 54.158.013.

5. Report of the Working Group on Infrared Extinction and Standardisation

(prepared by E. F. Milone)

The Working Group, set up by President Ian McLean after a joint meeting of commissions 25 and 9 at the Baltimore IAU, and funded by University of Calgary and NSERC Research Grants, has followed recommendations put forward at that meeting as described in (51.012.122). We have examined the profiles of past and present versions of the passbands of the Johnson JHKLMNQ broadband photometric system along with atmospheric window transmissions calculated by MODTRAN, and have used a series of stellar flux models from Kurucz to probe the atmospheric extinction under different water vapor, height, and airmass conditions. The curvature of the resulting extinction line describes the sensitivity of each response function to variations in water-vapor extinction. On the basis of our simulations, and an extensive set of experiments to optimise S/N without seriously degrading the reproducibility and transformability, subject to field trials we recommend a set of passbands which should improve both extinction and standardization from all sites, and enable transformable infrared photometry from lower altitude sites than is currently the case, at least for the shorter passbands. The progress of a subcommittee of the working group has been reported at Dublin, and a detailed account of the work has been submitted to AA.

We have changed the specifications for the new N filter from the list which was circulated earlier at IAU Colloquium 136. The current one is broader (2 microns) and should meet the demands of those who want greater throughput, without greatly compromising the improved atmospheric transmission. The list of the new filters, optimized for minimum distortion due to water vapour, is shown in the table below. It should be noted that the designations 'z', 'J', etc. represent the windows, and are not intended to be confused with the previous and now completely unstandard Johnson designations. The filter nomenclature should be decided by the commission. It should also be noted that prior to formal adoption, while even though the list is labeled 'final', slight alterations in the specifications cannot be ruled out prior to the field trial tests any or further experimentation which may be required.

The triangular shapes of these filters aid transformability in most circumstances, but some flattening between 50% and 80% appears not to be badly detrimental. In any case, they need good blocking to the blue and red — 10^{-5} is preferred to 10^{-4} — especially for the thermal filters. The improvement in signal to noise that we expect is dependent on this, because we assumed *no* leakage in our simulations. Some of the working group prefers even stronger blocking but if the throughput is diminished by more than a few tenths of a magnitude, we will have more trouble convincing people working at longer WLs to use the new filters. Blocking is particularly important in K for new imaging work. The filter specifications are for cold filters: LN_2 for the short and LHe for the long. We would like to stay on the low side of the increasing slope of theta, which measures the distortion of stellar flux through the filters.

Since passbands in the longer wavelengths are more costly to manufacture, letters expressing interest in these filters to OCLI (attn. Mike Larro) and Barr (attn. Roger Heatley) from potential purchasers can only help to make them available.

Meanwhile, pending further funding, we plan to investigate further the effects of sky emission and aerosol scattering on the signal/noise ratio.

Final List of Recommended Passbands.

All wavelengths (μm) have nominal tolerances of 1%

Band	50% points		80% points		5% points		100%
	lower	upper	lower	upper	lower	upper	Peak
iz	0.996	1.069	1.016	1.047	0.970	1.099	1.032
iJ	1.201	1.280	1.225	1.254	1.170	1.315	1.240
iH	1.555	1.707	1.585	1.672	1.514	1.754	1.628
iK	2.100	2.288	2.141	2.240	2.047	2.353	2.196
iL	3.483	3.757	3.554	3.686	3.374	3.882	3.620
iL′	3.763	4.037	3.834	3.966	3.654	4.162	3.900
iM	4.618	4.732	4.652	4.698	4.564	4.784	4.675
in	8.873	9.196	8.968	9.101	8.731	9.339	9.030
iN	10.100	12.100	10.369	11.832	9.756	12.444	11.100
iQ	17.106	18.712	17.439	18.329	16.656	19.231	17.900

6. Report of the Working Group on Standardization of Filter Systems

(prepared by A. T. Young)

Investigation of the tolerances available from filter manufacturers has shown that passbands cannot be reproduced with enough accuracy to preserve the precision of good modern observations. Furthermore, existing undersampled systems do not measure enough information about stellar spectra to allow conventional transformations to preserve the inherent accuracy of a good instrumental system. To deal with this problem, this WG was asked to specify filters that would permit accurate transformations.

Subsequent investigation of transformation problems indicates that cosine-squared passband profiles, with bands placed so that each band has neighbors centered at the midpoints of its sides, are nearly ideal. However, the sharp-cut glasses in fact make edges that are much steeper than anything that is available for the long-wavelength sides of passbands; so *narrow*, symmetrical bands cannot be made with colored glasses. Broad, symmetrical passbands are possible. The cosine-squared profiles can be achieved better as a function of frequency than wavelength, if colored glasses are used.

Trial filter syntheses show that colored glasses can approximate such passbands between about 340 nm and 600 nm, provided that the FWHM is near 4000 cm^{-1}. This corresponds to about 100 nm at 500 nm, comparable to the width of the existing B and V bands. In the near UV, the FWHM is about 50 nm, similar to U. The whole UBV region can be completely sampled with about 7 filters. Typical total thicknesses are less than 1 cm, and peak transmittances of 50% or more are possible — similar to the existing U and B. There may be some difficulties with very thin elements, and/or with unstable glasses. It appears that satisfactory glass filters can be made, but the cost may be fairly high.

The alternative is interference filters. A new type of filter, using ion-assisted deposition of oxide coatings, is now commercially available. These are said to be completely immune to the ageing effects that have plagued conventional interference filters. However, because of the smaller range in refractive indices of the oxides used, compared to conventional coatings, many more layers are required. Consequently, the new filters are about twice as expensive as the older type. But they have the additional advantage of a much smaller temperature coefficient than glass filters.

Discussions are continuing with filter manufacturers to determine exactly what is possible in the two types of filters. Realistic tolerances must be balanced against cost. By the time of the General Assembly, it should be possible to make definite recommendations.

Until improved filters are available, transformations to existing systems will be optimal if the mean-square difference between standard and instrumental filters is minimized. The main difficulty is the very loose tolerance on sharp-cutoff glasses, which is typically ± 6 nm but as large as ± 10 nm for some glasses. The typical tolerance corresponds to a range of a factor of 4 in glass thickness, which is unacceptable.

That means that filter specifications should be stated in terms of spectral transmittance, not glass type and thickness. Individual pieces of such glasses must be polished, measured spectrophotometrically, and ground to the required thickness to meet a much tighter spectral transmittance specification than the catalog tolerance. Even with the bandpass colored glasses, it is advisable to use spectral-transmittance rather than thickness specifications.

The sharp-cutoff components of the *UBVRI* filters should be closely matched to the standard cutoffs. There is confusion in the literature, in addition to uncertainty about the original response functions. In particular, Johnson's recommended Schott glass for the V filter is incorrect; it should be GG 515, in accordance with Hardie's Table 3 in *Astronomical Techniques* and Bessell (52.113.039), not GG 495 as is often recommended. There is no Schott glass that closely matches the Corning/Kopp 5030 used to define the B passband; this has contributed to the difficulties in transforming B.

For the *uvby* system, there is no alternative to interference filters. Filter specifications should include the detailed transmittance curves tabulated in Crawford & Barnes (4.113.050), not just central wavelength and halfwidth values. The transformation problems described in several papers by Sterken and Manfroid indicate that much tighter filter specifications are needed than have generally been adopted in the past. Discussions with filter manufacturers indicate that nominal commercial tolerances can be improved by a factor of 2, with only a modest increase in price. Such tightening of specifications seems well justified.

COMMISSION 26: DOUBLE AND MULTIPLE STARS

PRESIDENT: H. A. Abt

VICE PRESIDENT: C. E. Worley

ORGANIZING COMMITTEE: Y. Y. Balega, P. L. Bernacca, F. C. Fekel, H. A. McAlister, C. D. Scarfe, E. L. Van Dessel, H. Zinnecker

Introduction

The nature of research on double and multiple stars has changed drastically during the past several decades. Prior to then the research was mostly concerned with visual measures for the purposes of obtaining orbital elements and masses. Then came spectroscopic studies that (1) when combined with the visual orbital elements allowed the determination of orbits in three dimensions and (2) started an emphasis on the astrophysical and evolutionary natures of the components in such systems. Unfortunately except for a few nearby systems, the systems with large enough separations for spectroscopic measures of individual components were just those with orbital periods too long for visual determinations, while those systems with short enough periods for spectroscopic determinations were too closely-spaced for visual measurements.

However even as far back as the work of Finsen, the importance of interferometric measurements for accurate measurements of double stars was realized down to the optical resolution of telescopes. With the development of the speckle technique of interferometry the efficiency of such measures led to the accumulation of exploratory measures (to search for close doubles) and repeated measurements (at rates of 200-300 stars per nights as practiced by McAlister) of discovered pairs.

More recently the limitation of the optical resolution of large telescopes is being overcome with the construction of long-baseline equipment being developed and used by groups under Townes at Mt. Wilson, Davis in Australia, Pan at Mt. Wilson, Mumma and others from Goddard and the U.S. Naval Research Laboratory.

J. Bergeron (ed.), Reports on Astronomy, Vol. XXIIA, 239–242.

The beginning of the windfall of results from the Hipparcos mission are starting to appear. We hope for the extension of that mission and another such mission in the future.

The recent launching of spacecraft for measurements in nearly all parts of the electromagnetic spectrum will undoubtedly have effects on double stars, such as the discovery of hot faint companions with IUE or cool companions with IRAS. The high resolution of the 2.3-meter HST is beginning to produce results on doubles under Franz and his colleagues.

The diversity of techniques now available to study double stars was well illustrated in IAU Colloquium No. 135: "Complimentary Approaches to Double and Multiple Star Research", held in April 1992 and published later that year in the Astronomical Society of the Pacific Conference Series as Volume 32. The conference was organized and the proceedings were edited by our commission Past President, Harold A. McAlister, and his colleagues, and it was dedicated to Alan H. Batten.

A sample of current programs by commission members is given below.

Astrometric Measures, Orbital Elements, Photometry, Spectroscopy, and Interferometry

We remain gratified that some astronomers continue to have the patience, critical eyesight, and observing time to continue with the discovery of new pairs and the continuation of observations of those systems known for decades or centuries; the latter increase the accuracy of the preliminary orbital elements and provide the basis of the stellar mass-luminosity relation.

In particular, we are grateful to Paul Couteau for his hundreds of new discoveries and elements of known systems, and his production and distribution of the commission's Circulaire d'Information. Those were started in 1954, with the first 90 issues produced by Paul Muller and the recent 30 issues by Couteau. Recently J. A. Docobo, assisted by J. F. Ling, has agreed to accept the responsibility for the circulars and he encourages all multiple-star astronomers to communicate to him for publication information on new discoveries, calculated orbital elements, planned observing programs, extensive publications, plans for meetings, and other general notes.

In the past three years P. Muller has made 2060 measures with the 50 cm telescope at the Observatoire de Nice. M. Scardia has made 350 micrometer measures of 108 southern pairs at La Silla. He also continues to make photographic measurements at Merate in connection with the Hipparcos Catalogue of the Components of Double and Multiple Stars. In this connection, R. Pannunzio reports that the astrometric group of the Astronomical Observatory of Torino has been setting up a chain of computer programs for the analysis of Hipparcos double and multiple systems and provided the database for their evaluation. The group tested the programs on simulated systems and are now working with actual Hipparcos data, obtaining encouraging results.

P. Brosche, D. Sinachopoulos and their colleagues at Hoher List in Bonn have been evaluating the existing data on wide binaries in connection with current Hipparcos measurements. They are obtaining photometry, area scanners, and relative proper motions to judge whether those wide pairs are physical systems. Their numerous papers are all in A&A and A&AS.

E. Oblak at the Observatoire de Besancon announces the establishment of a European Network on Visual Double Stars. It consists of coordinated photometric and astrometric measurement of several thousand visual doubles selected from Annex 1 of the Hipparcos program with conventional and CCD techniques. One aim is to obtain V and I (or R) photometry and another is to obtain relative astrometry (to several hundredths of an arcsec) for comparison with Hipparcos measures.

The accuracy of CCDs for the photometric and astrometric measurements of double stars has been explored by R. Müller in Berlin and E. H. Geyer in Bonn. Empirically, H. A. Abt and C. F. Corbally in Tucson have been studying the reality of 268 possible Trapezium systems, using CCDs for photometry and astrometry as well as photographic spectral classification. They obtain accuracies of ±0.020 mag, ±0.06^{o} in position angles, and ±0.07" in separations per measurement with a CCD.

At the Bosscha Observation in Indonesia, Jasinta, Sugiartini, and Raharto are analyzing photographic observations of 150 visual binaries along the lines of the work of Panjaitan and Herdiwidjaya (A&AS 88,63, 1991). From the latter list, 218 pairs showing no relative motions are being taken off the Bosscha observing lists while 68 pairs are showing physical motion. Southern bright close doubles are being observed for spectral classification, mostly to predict masses for stars above the main sequence.

The exciting double stars being discovered as central stars of planetary nebulae are being studied by Malasan at Bosscha under the guidance of K. Ishida in Japan. For V664 Cas the mass function yields a primary mass of 0.55-1.4 solar masses for the hot subdwarf and a secondary of 0.18-0.20 solar masses. For UU Sge the primary mass is similar while the secondary is between 0.27-0.48 solar masses.

Between 1975 and 1992 McAlister and his colleagues at CHARA (Center for High Angular Resolution Astronomy) obtained 12,000 speckle interferometric measures of over 3,000 stars with the 4-meter telescopes at KPNO and CTIO. Of those, 300 were newly-discovered interferometric pairs. That work is being transferred in 1993 to the newly renovated 100-inch (5 m) telescope on Mt. Wilson.

Catalogs

J. Dommanget and O. Nys of the Royal Observatory, Belgium are working on a two-part catalog of which the first (Definite) part consists of 31,305 systems with positional accuracies of 1" or better and contains all the multiple systems in the Hipparcos Input Catalogue. The second part will contain 31,178 systems with less accurate information.

J. Dommanget and O. Nys have completed their part in the preparation of Vol. 6 of the Hipparcos Input Catalogue, which lists double and multiple stars. They checked cross identifications of components, which is a difficult and important need.

27. VARIABLE STARS (ETOILES VARIABLES)

1. Introduction (J.R. Percy)

Commission 27 sponsors three special projects which are of great importance to the astronomical community: the Information Bulletin on Variable Stars (IBVS), the Archives of Unpublished Observations of Variable Stars, and the General Catalogue of Variable Stars (GCVS). The reports from the centers which organize these projects are found in section 2.

Several individuals and groups deserve the warm thanks of the IAU, of Commission 27, and of variable star astronomers everywhere: Katalin Oláh and Laszlo Szabados for editing the IBVS so effectively, and Christiaan Sterken and the Editorial Board of the IBVS for giving their time, effort and thought to support this project in difficult times; Nikolai N. Samus and his colleagues for continuing the progress with the GCVS in even more difficult times; Michel Breger for maintaining the Archives since their inception, and Ed Schmidt for willingly taking over this project. We thank all those named who have written the sections of this report with such promptness, diligence and good judgement. Don Fernie provided me with helpful comments. I am especially grateful to Elizabeth Kobluk who typed this lengthy report quickly and effectively, from very inhomogeneous input, at a busy time of year.

The sections of the report cover the time period mid-1990 to mid-1993. They are not intended to be comprehensive; space and time limitations make that impossible. Readers are referred to Astronomy and Astrophysics Abstracts for a complete bibliography. The journal abbreviations include: AJ = Astronomical Journal; Ap = Astrophysics; ApJ(S) = Astrophysical Journal (Supplement); AA(S) = Astronomy and Astrophysics (Supplement); ApSS = Astrophysics and Space Science; MN = Monthly Notices of the Royal Astronomical Society; PASJ = Publications of the Astronomical Society of Japan etc. Major conferences sponsored or supported by Commission 27 included: "Confrontation between Stellar Pulsation and Evolution, Bologna, Italy, May 28-31, 1990; proceedings: ASP Conference Series 11, 1991, ed. C. Cacciari and G. Clementini (hereinafter Bologna), IAU Colloquium #134, "Nonlinear Phenomena in Stellar Variability", Mito, Japan, Jan. 7-10, 1992, proceedings in press (hereinafter Mito); IAU Colloquium #139, "New Perspectives on Stellar Pulsations and Pulsating Variable Stars", Victoria, Canada, July 15-18, 1992, proceedings in press (hereinafter Victoria). The first European meeting of the AAVSO was held in Brussels, Belgium, July 24-28, 1990; the proceedings, "Variable Star Research: An International Perspective", were published by Cambridge Universities Press, ed. J.R. Percy et al., 1992. A symposium on "Variable Stars and Galaxies" was held in Cape Town, South Africa in honour of Michael W. Feast; proceedings: ASP Conference Series 30, 1992, ed. B. Warner. Other relevant conference proceedings are mentioned in the sections below.

2. Commission Activities

a) Information Bulletin on Variable Stars (L. Szabados and K. Oláh)

During the period covered by this report, the number of issues of the IBVS progressed from #3488 (July 1990) to #3902 (June 1993).

According to the decision of the Commission 27 meeting held during the GA in Buenos Aires, an Editorial Board (EB) of the IBVS has been established. The role of the board is to help maintain the scientific level of the Bulletin and assist in solving the financial problems of the distribution. The members of the EB are: L.A. Balona, M. Breger, M. de Groot, D.S. Hall, R. Koch, J.-M. LeContel, J.R. Percy, M. Rodono, J. Smak, N.N. Samus, C. Sterken (Chair) and the editors. This constitution also reflects the fact that IBVS is now co-sponsored by the IAU Commission 42 (Close binary stars).

In view of the budgetary situation at Konkoly Observatory, circulation of the IBVS had to be reduced in 1991. Following the first meeting of the EB (Vienna, November 1991) a new circulation policy was introduced in 1992. Instead of free circulation, IBVS can be obtained on a subscription basis for a nominal fee which covers only the mailing cost. This unavoidable measure did not cause

J. Bergeron (ed.), Reports on Astronomy, Vol. XXIIA, 243–276.

any interruption in the publication of the Bulletin: the number of subscribers was 328 in 1992, and for the time being there are 366 subscribers for 1993. The IAU kindly provided a grant of 1000 US$, a sum that allows to keep 25 East-European and Asian astronomy libraries on the mailing list for two years. The financial matters of the IBVS are handled by the international institute "Astronomy Research and Education Support" at the University of Brussels. The editors are grateful to Dr. Christiaan Sterken for his enormous help in the financial and administrative matters related to the IBVS. Thanks are also due to the other members of the EB for their help and co-operation.

In order to minimize the duration of the publication process, the majority of the manuscripts submitted for publication in the IBVS are refereed by the editors. In a number of cases, however, members of the EB, complemented by external referees, assist the editors with the refereeing process. It turns out that one out of two manuscripts can be accepted without revision, while publication of the other manuscripts is delayed for reasons of content, method or style. The editors strive for prompt editorial decisions concerning the acceptance of a paper, and — in order to avoid any delay — manuscripts can be sent via fax to referees. The percentage of the rejected manuscripts is not too high (about 4 per cent in 1992).

Our aim is to give the individual issues homogeneous appearance using LaTeX typesetting. Electronic submission of the text-file (even if it is in a plain ASCII or TeX is much appreciated. We are also planning for the electronic distribution of the Bulletin.

The address for editorial correspondence is: IBVS, Konkoly Observatory, P.O. Box 67, H-1525 Budapest XII, Hungary. e-mail addresses: H1036IBV@ella.hu *or* IBVS@ogyalla.konkoly.hu *or* IBVS@RMKI.KFKI.HU.

b) Archives of Unpublished Observations (Edward Schmidt)

The Archives of Unpublished Observations of Variable Stars is a depository for photometric observations which will not be published elsewhere. By placing data in the archives, they are available to other investigators without the need for lengthy and expensive tables in published journals. The archives also provide a means to preserve and make available observations which are never used in publication but may, nonetheless, be of use to other investigators.

There are now 247 completed files in the archives. Descriptions of older files have been published from time to time (see IBVS #3733 for references) and a description of more recent files, including those listed below, is in preparation for publication.

Astronomers wishing to obtain hard copies of files may request them from one of the addresses listed below. Some files are also available in electronic form. These can be obtained by electronic mail from the Centre de Données Stellaires. The addresses of the archives are:

Dr. P. Dubois
Centre de Données Stellaires
11, Rue de L'Université
F-67000 Strasbourg
France
e-mail:dubois@simbad.u-strasbg.fr

P.D. Hingley, Librarian
Royal Astronomical Society
Burlington House
London, W1V 0NL
United Kingdom

Dr. Yu. S. Romanov
Astronomical Observatory
Shevchenko Park
Odessa 270014
Ukraine

Astronomers who wish to submit new files should obtain a file number by contacting the coordinator at the following address:

Dr. Edward G. Schmidt
Department of Physics and Astronomy
University of Nebraska
Lincoln, NE 68588
USA
e-mail: eschmidt@unlinfo.unl.edu

Electronic transmission of data should be followed by a paper copy of both a descriptive cover sheet and the data. If electronic submission is not used, three paper copies of the data and the cover sheet should be submitted.

Since the previous report to IAU Commission 27, the following files have been assigned and completed:

File No.	Star	Contributor
155.	HD 155638	Hall
158.	GO Cyg	Hall
159.	BV Dra, BW Dra	Hall
160E.	SAO 4710	Poretti
201.	P Cyg	Percy
213.	Seven Algols	Caton
219.	50 suspected variables	Hooten and Hall
224E.	8 peculiar stars	Adelman
226E.	HD 101158	Poretti
227E.	V833 Tau	Oláh and Peterson
229.	UZ Psc	Hall
231.	τ Per	Hall
232E.	X Cae	Poretti
233E.	τ Per	Breger
234E.	VY Ari	Strassmeier and Bopp
235.	AB Dor	Anders
236E.	44 Tau	Poretti
238E.	9 Aur	Krisciunas et al.
239.	V365 Lac	Morris

File No.	Star	Contributor
242E.	LQ Hya	Strassmeier
243.	UL Leonis	Hegedüs
245E.	25 variables	Jerzykiewicz
246E.	UY Cam	Broglia
247E.	eclipsing binary	Broglia
248.	35 binaries	Caton and Burns
250E.	44 Boo	Jones
251E.	17 stars	Schmidt
252E.	18 stars	Schmidt
253E.	22 stars	Schmidt
254E.	23 stars	Schmidt
255E.	13 stars	Schmidt
258E.	28 And	Rodriguez
259E.	BF Phe, RX Sex, HY Vir	Rodriguez
260.	DH Leo	Hall
262E.	EN Lac, V360 Lac	Jerzykiewicz
264E.	44 Tau	Akan
265.	BF Dra	Diethelm
272E.	EP Cnc	Breger

c) General Catalogue of Variable Stars (N.N. Samus)

Our hopes expressed in the previous 3-year report to finish Volume V of the GCVS in 1991 were too optimistic. We have met serious problems, mainly connected again with co-ordinates of variables and cases of identical stars considered different by their discoverers. By now (May 1993) we have overcome these problems and expect to be able to send the manuscript to the printers later this year. It will contain a unified catalogue of variable stars in extragalactic systems. Simultaneously a magnetic tape for the Strasbourg centre will be prepared.

During the 3-year cycle we have published two name lists of variable stars, #70 and #71. We compiled and sent to Strasbourg a computer version of the name lists #67-70. Recently we have added the name list #71 to this computer version.

The magnetic tape versions of the GCVS Volumes I-III contain only the main table, without remarks and lists of references. We have prepared the earlier missing computer files. They still need some editing, and after it they will be sent to the Strasbourg centre.

We continue work on the GCVS data base in computer readable form. Unfortunately this work has been slow due to insufficient funding. There already operates the special GCVS data management and retrieval system developed by Dr. Yu. Fadeyev, but the fraction of our card catalogue available in computer readable form is not sufficient yet.

Our plans for the nearest future include an astrometric supplement to the GCVS, giving for the majority of GCVS stars co-ordinates good enough for automatic pointing of ground-based and space-borne telescopes. To prepare this supplement, we have to undertake effort-consuming checks of identifications of GCVS stars with positional catalogues (including the HST Guide Star Catalog) using published finding charts.

3. Early-Type Variable Stars (Luis A. Balona)

The workshop proceedings "Rapid Variability of OB Stars: Nature and Diagnostic Value", ed.

D. Baade, ESO, 1991, [hereinafter Munich] contains many relevant reviews and contributions.

Excitation of Pulsations. The most outstanding advance in the study of early-type stars in the last three years has been the realization that the same mechanism which causes classical Cepheids to pulsate also operates in certain B-type stars. The difference is that whereas the ionization zones of hydrogen and helium are responsible for driving the Cepheids, it is the sudden appearance of a tremendous number of iron lines at a temperature of about 150 000 K which is responsible for driving pulsations in β Cep stars and probably other pulsating B stars. This change in our understanding is due to the revision of metal opacities which was first suggested by Simon (ApJ 260, L87) on purely observational grounds and confirmed by Iglesias and Rogers (ApJ 371, L73) by opacity calculations. Calculations of pulsational driving in B-type stars by Cox et al. (ApJ 393, 272), Kiriakidis et al. (MN 255, 1p), Moskalik and Dziembowski (AA 256, L5), Dziembowski and Pamyatnykh (MN 262, 204) and Gautschy and Saio (MN 262, 213) all show that the mechanism which drives β Cep stars is now understood, but many problems remain unresolved. In particular, there is some discrepancy between the observed and calculated instability strip. More seriously, the mechanism does not explain long-period low spherical harmonic order pulsations found in hot B stars, though it predicts such pulsations for the cooler stars on the main sequence (see for example Degryse et al. AA 263, 137). Much of the predicting power of the models is lost because in all cases there is a fine balance between driving and damping and small changes to the opacities (or in the way they are treated in the models) can lead to large changes in the observed properties.

β Cep Stars. The dependence of β Cep pulsational driving on metal abundance predicted by the opacity mechanism has promoted observational studies to confirm this effect. Waelkens et al. (AA 251, 69) found a trend in the mean colour of these stars with galactocentric distance which may be explained in this way. Delgado and Alfaro (ApSS 169, 117) found a similar result. Perhaps the most convincing result of the metallicity effect is the total absence of β Cep stars in three Magellanic Cloud clusters: NGC 330 (Balona MN 256, 425) and NGC 2004, NGC 2100 (Balona & Jerzykiewicz MN 260, 782 and Balona MN 260, 795).

A connection between β Cep stars and Be stars was established when β Cep itself was found to have developed strong Hα emission (Mathias et al. Munich 193). While observing the well-known Be star 27 CMa, Balona and Rozowsky (MN 251, 66P) discovered that it had developed β Cep pulsations sometime during 1988 or 1989.

A major photometric study of β Cep stars which includes new period determinations for 57 stars has been completed by Heynderickx (AAS 96, 207). Other recent observational work on these stars can be found in Jerzykiewicz and Sterken (MN 257, 303), Jerzykiewicz (AAS 97, 421) and Sterken et al. (AAS 98, 383). Delgado et al. (AJ 103, 891) list suspected variables in two open clusters.

Mode identification in β Cep variables is necessary to test models of these stars and for asteroseismological studies. Waelkens and Aerts ('Workshop on the atmospheres of early-type stars', ed. Heber & Jeffery, Springer-Verlag, 159) and Aerts (same volume, 163) use the information contained in the moments of the line profile variations to determine the modes of selected β Cep stars. Aerts et al. (AA 266,294) use the same method to show that δ Cet is a radial pulsator. A different method based on visual and UV light amplitudes, radial velocities and phase shifts was used by Cugier and Boratyn (Acta Ast 42, 191).

53 Per And ς Oph Stars. This classification of intrinsic B-type variables has not been formalized and to some extent is a matter of personal taste. In this report we follow the classification scheme of Balona (Munich 249). The 53 Per group of stars may be identified with the 'mid B-type variables' or the 'slowly pulsating B stars' of Waelkens (AA 246, 453). This later work has considerably advanced the photometric study of 53 Per stars by showing the multiperiodic nature of seven long-period B-type stars. Other stars which may belong to this group are HD 80859 (van Genderen et al. AA 259, 574) and λ Col (Jerzykiewicz & Sterken MN 260, 826). An intensive photometric study of a member of this group, HR 2680, was made by Balona and Cuypers (MN 261, 1). The primary component of this eclipsing binary has two closely-spaced periods which do not appear to be stable. The lack of line profile observations of these photometrically detected stars is a serious obstacle to further progress. It is also abundantly clear that intensive photometry is required to resolve the periodicities, but it now appears that in some stars the stability of these periods must be questioned. This makes a definitive determination of their periods a very difficult process.

Many rapidly rotating B-type stars show 'moving bumps' in their line profiles, which defines them as belonging to the ζ Oph group. ζ Oph itself has been studied recently by Kambe et al. (PASJ 42, 687) who employ a novel method of detecting periodicity in the line profiles. Two NRP modes were found to be excited at the time of their observations. Yang et al. (ApJS 74, 595) performed a similar study of ζ Tau, while moving bumps in κ Dra were seen by Hill et al. (AA 246, 146). In all these stars the modes have very long periods in the co-rotating frame. One of the problems with line profile observations is that it is often not possible to obtain definitive periods for the oscillations owing to limited or ambiguous data particularly if more than one pulsation mode is involved. Furthermore, a theory which enables physical parameters to be derived (including the effect of temperature variations) is not available for rapidly rotating stars at present.

λ Eri Stars. The periodic light variations in Be stars are a characteristic feature of these objects and are in some unknown way connected with the enhanced mass loss in these stars. A statistical study by Balona (MN 245, 92) shows a strong correlation between the projected rotational velocity and the photometric period. He interprets these variations as due to some kind of rotational modulation, though the consensus of opinion is that non-radial pulsation (NRP) is involved. The idea of rotational modulation is strengthened by the discovery of many periodic Be stars in the SMC cluster NGC 330 (Balona MN 256, 425). No pulsational driving is expected in stars with low metal abundances.

Intensive photometric observations of these stars is presented in Balona et al. (AAS 92, 533). Balona et al. (MN 252,93) present detailed photometry of four λ Eri stars in NGC 3766 showing that the periods are strictly constant over five years.

Results of a multiwavelength campaign on Be stars aimed at determining the cause of short-term optical variability, are reported by Peters et al. ('Evolution in astrophysics: IUE astronomy in the era of new space missions', ed. E.J. Rolfe, ESA, p.257, 1990). The observations confirm that it is the modulation in the star's photospheric temperature that causes the flux variability. The authors find that NRP in an $\ell = 2$ sectorial mode can explain the results.

High resolution line profile studies of λ Eri itself in combination with near-simultaneous UV continuum observations have been analyzed by Smith and Polidan (ApJ 408, 323). They find several examples of a weakening in the C IV, N V resonance lines that coincides with the appearance of dimples in the λ6678 line profile. They interpret these results in terms of a condensation of high density structures above the photosphere. They speculate that small scale magnetic fields of several hundred gauss may exist in the atmospheres of mild Be stars. Evidence for violent magnetic activity in λ Eri comes from the detection of a giant X-ray flare on λ Eri (Smith et al. ApJ 409, L49). The results cannot be explained in terms of a cool secondary or degenerate companion.

O Stars, Supergiant Variables and Wolf-Rayet Stars. Balona (MN 254,404) conducted a photometric survey of several O-type stars to search for periodic variations which could be interpreted in terms of NRP. Although microvariability with a time scale of days was found in all supergiants and some dwarfs, evidence for NRP was not found. However, Baade ('Workshop on the atmospheres of early-type stars', ed. Heber & Jeffery, Springer-Verlag, 145) comes to the opposite conclusion from a spectroscopic point of view. Both viewpoints can be reconciled when it is realized that photometric observations are not sensitive to the high-order spherical harmonic modes seen in the line profiles of the stars. There is no doubt, however, that irregular or semi-regular variations due to variability of the winds of these stars play a role in both the photometry and line profiles.

Photometric observations of four supergiant variables (α Cyg stars) are presented by van Genderen et al. (AA 264, 88). Quasi-periods between 24 and 58 days are found. These variations are thought to be due to NRP, but the excitation of these modes is still a puzzle. Gautschy (MN 259, 82) has studied NRP in models of these stars and finds a number of excited low-order modes.

One of the most exciting developments in the study of Wolf-Rayet stars was the detection of a 627-second optical pulsation in the WN8 star HD 96548 (Blecha et al. Nature 360, 320). Whereas the reality of pulsations in this star seems to be well established, it is of great importance that this detection be independently confirmed. Unfortunately, this may not be a simple matter as the pulsations might be obscured most of the time by the optically thick expanding atmosphere. The importance of this work lies in the prediction that WR stars with masses exceeding 12-16 $M_\odot$ should be pulsationally unstable (Maeder QJRAS 32, 217). Good agreement is found between the period in HD 96548 and a model with a mass of 14-15 $M_\odot$.

An attempt to understand the periodic light variation in the peculiar WR star EZ CMa was made using photometry, polarimetry, spectroscopy and spectropolarimetry by Robert et al. (ApJ 397, 277). The linear polarization data are well fitted by an eccentric binary model, but the spectroscopic results are ambiguous and the question of the presence of a compact companion remains unresolved. Intensive photometry of HD 191765 (Antokhin et al. ApJS 82, 395) show irregular variations on different time scales, but there is some evidence for a 2-day quasi-periodicity. The cause of the periodic variations in this star is probably of the same nature as in EZ CMa.

4. Delta Scuti and roAp Stars (D.W. Kurtz)

In this report, I will recount some developments that have caught my attention and give a few opinions about needs in this research subfield; for a comprehensive review, consult Astronomy and Astrophysics Abstracts (AAA).

δ Scuti Star Newsletter = DSN. The "δ Scuti Star Newsletter" continues to be published by the Austrian Academy of Sciences with Michel Breger as editor. This newsletter is most interesting and useful; I encourage all who are interested in the field to read it and to submit notes, discoveries, abstracts, papers and opinions to it. The Editor's address is Institut für Astronomie, Türkenschanzstraße 17, A-1180, Wien, Austria and his email address is breger@avia.una.ac.at. There is no subscription charge and papers are indexed in AAA.

Catalogues. We all need access to up-to-date complete lists of δ Scuti stars. Various individuals and groups of astronomers have taken on the task of providing such lists. López de Coca et al. (AAS 83, 51) provided a list of 192 δ Scuti stars which has been used by others — probably because of its accessibility in the library. A more complete list of 254 stars is provided by Garcia et al. (CDS Bull. 34), but the latest, most up-to-date is a PC DOS data base compiled by Garcia et al. (see DSN 6, 14) which is available from the authors for a nominal charge. Contact them by email at jgarcia@icoper.edu.ar. This data base collates name, spectral type, pulsational data, photometric indices, v sin i, references to the literature and more. We all must thank the authors for their immense effort and I am sure we will all have many uses for their data base. On this point: It would be useful if workers could inform Garcia directly of publications of the discovery of new δ Scuti stars and other information relevant to keeping the data base up-to-date.

Reviews and other work. No comprehensive review of δ Scuti stars has appeared since Breger's 1979 (PASP 91, 5) or Wolff's 1983 monograph on the A stars. There is a need for this from both the theory and observational sides. Are there two collaborating volunteers (or one person who knows everything)?

At the Oji Seminar on Seismology of the Sun and Stars, Dziembowski (Lec. Not. Phys. 367, 359) gave a theoretical discussion of δ Scuti stars with synopses of much of his previous work with Królikowska and their co-workers (*e.g.* Dziembowski & Królikowska Acta Ast 40, 19). They repeatedly call for more mode identifications and frequency determinations from observers. They find mode trapping for non-radial modes primarily for $\ell = 1$, but generally not for $\ell = 2$ and 3; higher degree modes exhibit more and more trapping — an interesting result given line profile variability for $-m = \ell = 10$ to 20 found for 4 δ Scuti stars by Kennelly et al. (Bologna 320; see also Kennelly et al. PASP 103, 1250). Dziembowski and Królikowska suggest that detection of modes with $\ell \geq 2$ would imply that something more than mode trapping is acting in δ Scuti stars, but they are referring to the *photometric* detection of such modes which would imply large amplitudes; the detection of large ℓ modes spectroscopically does not imply these problematic large amplitudes. The $\ell = 1$ trapping is also interesting since that is what is unequivocally found in several roAp stars: HD 6532, HR 3831, and α Cir. Dziembowski also discussed the thorny problems of mode selection and amplitude modulation. Dziembowski and Pamjatnikh (AA 248, L11) calculate that one g-mode may be trapped by the boundary of the convective layer in the core for slightly evolved δ Scuti stars, and that observations of the frequency of this mode could constrain theories of convective overshoot. There have been suggestions from observers that g-modes may explain some observed frequencies. This is an unresolved issue which is strongly coupled to the issue of the degree to which frequencies claimed by observers can be believed — an issue which is discussed more below. Observers should keep this g-mode suggestion in mind; theoreticians should examine observers' claims of g-mode detections carefully.

Breger (Bologna 263) gives a review of observational results, concentrating on amplitude and frequency stability and their implications for stellar evolution. He lists a number of determinations of dP/dt in δ Scuti stars. These are often orders of magnitude larger than expected from theoretical predictions and both positive and negative values are found. The latter, if evolutionary, imply stars evolving bluewards. However, see Fernie's (AJ 103, 1647) suggestion that all of the δ Scuti stars are core burning and lie *below* the evolutionary blue loop (more about that paper below). Many of the period changes in δ Scuti stars are better interpreted as episodic, discrete changes, rather than continuous changes (*e.g.* CY Aqr — Coates et al. DSN 4,10; MN, in press). This is not necessarily in contradiction with stellar evolution. Discovery of significant dP/dt for main sequence stars where evolution is slow could refute the evolutionary origin of all dP/dt, as could the discovery of multiperiodic stars with very different dP/dt for different periods, although this must be modelled. Thompson and Coates (Proc. Ast. Soc. Aus. 9, 281) find a factor of 6 ratio in the rate of period change for P_0 and P_1 in SX Phe and both are an order of magnitude larger than expected from main sequence models. Both are also negative as are the dP/dt in three other Pop II δ Scuti stars. Breger points out that in the Blue Straggler Hypothesis for the SX Phe stars the decreasing periods need to be explained; in the post-HB hypothesis it is expected from the direction of evolution. Walraven et al. (MN 254, 59) find $dP_0/dt = 0$ and $dP_1/dt = 4.74 \times 10^{-9}$ from 38 years of observations of AI Vel.

Those working on frequency variability should look at interpretations other than stellar evolution for the apparent dP/dt that are being measured. Note that some of the roAp stars (see the last section of this report) show cyclic frequency variability which is intrinsic to the star and cannot be evolutionary. The sun is known to show pulsation frequency variability over the solar magnetic cycle. And the ZZ Ceti white dwarf G29-38 shows cyclic frequency variability in one pulsation frequency and not variability in another frequency (Winget et al. ApJ 357, 630).

Some years ago I claimed that it had not been shown that any periods or amplitudes were not constant in δ Scuti stars (how is that for a negative statement?). Breger in his review argues persuasively that both periods and amplitudes are not constant on a long time scale. These results come from many years of work and many multi-site campaigns (see Breger AA 240, 308).

The SX Phe stars are thoroughly reviewed by Nemec and Mateo (Bologna 64). They discuss this problem of the changing periods in SX Phe stars with a suggestion that the period of XX Cyg is probably increasing. They point out that the changing ratio of P_0/P_1 in SX Phe itself implies mass loss, if the change is evolutionary in origin. Most interestingly, they discuss their discovery of many SX Phe (or δ Scuti) stars in globular clusters and dozens of them in the Carina dwarf galaxy. (See their paper in ASP Conf. Ser. 10, 134 for a look at convincing δ Scuti light curves of 23rd magnitude stars.) The Carina galaxy contains stars of intermediate and old populations; it is not clear yet to which group the δ Scuti stars belong. Nemec and Mateo support the hypothesis that the SX Phe stars are blue stragglers which can be accounted for by binary coalescence. McNamara and Powell (Bologna 316) find a correlation of log P with [Fe/H] for SX Phe and δ Scuti stars which they suggest implies that SX Phe stars evolve from low mass, low metal stars; this explains why they are found only in the lower part of the instability strip. Gilliland and Brown (AJ 103, 1945) serendipitously discovered two blue straggler δ Scuti or SX Phe stars in M67 while searching for solar-type oscillations in cooler stars. A third blue straggler did not vary. Both the variables are multi-mode, but it was not possible to distinguish from the pulsation frequencies whether the stars had the high He content expected in the coalescence model for blue stragglers.

Frequency Analysis. This is an issue on which I have strong opinions with which some of you will disagree. I suggest that we argue this out in the pages of the DSN. The community of theoreticians for years has been calling for more frequency determinations and for mode identifications for δ Scuti stars. The great success of helioseismology has caused the large community of solar astronomers and helioseismologists to cast their eyes on the δ Scuti stars for more observations on which to apply their well-developed theory. The observational community continues to produce a large amount of material on the δ Scuti stars, and much of this is of a higher-than-previous quality, with much more effort going into multi-site campaigns (discussed further below). But we are not there yet, and many of our published frequencies are erroneous. This is worse than no frequencies at all, since it leads unsuspecting theoreticians astray. I urge conservatism in frequency analysis.

There are many methods of frequency analysis and a huge literature on the subject which I have no space here to review. Most techiques used for δ Scuti stars are equivalent to Discrete Fourier Transforms (DFTs) for unequally spaced data, and those are equivalent to least-squares fitting of sinusoids for well-sampled data sets. Some use the Maximum Entropy Method. I do not know, and do not claim, that these are optimum methods of frequency analysis. There may be better ways to extract information from time series yet to be discovered, so we must examine each new technique as it is developed.

The variations of the DFT that worry me are the many methods that treat the alias problems arising from the gaps in the time series in an automatic way; some of these fill in the gaps in the data with predictive methods, then revert to Fourier analysis. The most widely used and misused method is CLEAN, an algorithm to remove, or reduce, the alias patterns to leave a frequency spectrum with only the "true" frequencies.

Michel et al. (Bologna 332; AA 255, 139) observing GX Peg and Belmonte et al. (AA 246, 71) and Mangeney et al. (AA 244, 351) observing 63 Her report the results of multi-site campaigns of the STEPHI network and give theoretical interpretations. They have used CLEAN for the frequency analysis; while some of their identified frequencies are convincing, not all are believable. The amplitudes of some peaks differ by orders of magnitude more than the noise between the DFT and CLEANed spectra, and all of the amplitudes are reduced in the CLEANed spectra. Michel (DSN 6, 19) discusses this latter problem and its solution. Ostermann (DSN 4, 5) finds only 3 frequencies for 63 Her, instead of the 7 found and interpreted by the STEPHI networkers.

Another example of this can be found in the CLEAN analysis of κ Boo (Jones et al. DSN 5, 15) which found 11 frequencies, rotational splitting and changing frequencies between two high duty cycle data sets. Two independent frequency analyses by Balona and by me (both unpublished) using DFTs found two certain frequencies which were the same in both data sets, good evidence of a third frequency with an alias problem, and no evidence of any further frequencies. These are radically different results which need to be resolved amongst the observers before the theoreticians can realistically use the results to constrain their models.

My opinion is that CLEAN is being applied as a "black box" frequency analysis technique in some cases. It does help remove the confusing aliases from the frequency spectrum, but it must be remembered that those aliases tell us something about what we do not know, and where we might be confused; they must be examined, even where CLEAN is used. The helioseismology community had an extensive "hare and hound" exercise which was reported in the GONG newsletter, where "hares" made up realistic artificial solar time series with noise and time gaps. With no knowledge of the input the "hounds" then attempted to reconstruct the frequency spectra. They had major problems with data sets with duty cycles under 80%. We almost always have data sets with lower duty cycles, even for our multi-site campaigns. I suggest that regular workers in the field may wish to participate in similar "hare and hound" exercises to obtain experience with our frequency analysis techniques and confidence in the reliability and reproducibility of our results. Let the theoreticians be "hares" and the observers and frequency analysers be "hounds".

We need identified frequencies in δ Scuti stars which can be trusted with confidence by those who cannot easily judge frequency analyses independently. Suggestions are solicited for how we might provide that confidence to the astronomical community. As a start, I would urge more widespread availability of the data. Much data is available through Strasbourg, but much is not. As a start, I would urge Garcia et al. to add to their δ Scuti data base information about availability of observations. I would prefer that the observations actually be part of the data base, but this may not be acceptable for reasons of administration and expense. Other ideas are solicited. The widespread availability of all published observations would allow any interested researcher to reanalyse data to check published results, with old or new techniques. As an example of this: Walraven et al. (MN 254, 59) have reanalysed old data using the newer DFT technique and discovered two new frequencies in AI Vel, one of which they identify as 5th radial overtone (interesting because Stellingwerf's 1979 models showed highest growth rates for 5th overtone). Ostermann et al. (AA 245, 543) say that "a complete frequency solution should leave only white noise in the power spectrum of the residuals". I add to that *"and the solution must fit the data"*.

Multi-site Networks. There has been admirable success in instituting and running multi-site networks. The results of the observing campaigns of these networks have dramatically increased our knowledge of the frequencies in δ Scuti stars. Some of them are: The STEPHI (Stellar Photometry International) network (Michel et al. Bologna 332; AA 255, 139; Belmonte et al. AA 246, 71; Mangeney et al. AA 244, 351; Belmonte & Michel DSN 3, 16). The Delta Scuti Network (Breger AA 240, 308; Breger et al. AA 231, 56; Breger et al. AA 243, 160; DSN 5, 3). The newly constituted GLOBUS Network (DSN 6, 5) comprising the Delta Scuti Network and the Whole Earth Telescope (WET — see *e.g.* Clemens et al. ApJ 391, 773) has observed FG Vir with both standard differential photometry and high-speed photometry. Also newly constituted is A Small Telescope Array with CCD Cameras = STACC (Frandsen DSN 5, 12) which had its first campaign on δ Scuti stars in the cluster NGC 6134 in May 1993. I also must mention the prolific work of Poretti, Mantagazza and their coworkers at the Merate observatory, both in single-site work and as part of multi-site campaigns (AA 228, 350; 230, 91; 245, 136; 255, 153; 256, 113).

Eclipsing binary. Kollath and Nuspl (IAU Symp. 151, 289) discuss tidal excitation and damping of p-modes — an old topic much discussed for 1 Mon and 14 Aur, amongst other stars. Volkov (IBVS #3493) has announced δ Scuti variability in the primary of the eclipsing binary V 577 Oph. This is an equatorial star observable from both hemispheres. I here suggest a multi-site campaign on this star for June 1994 with both photometric and radial velocity observations. This would solve the binary orbit, solve for masses and radii, and study the pulsation frequency/ies. It might also be possible to watch the phase of the pulsation mode/s during ingress and egress from eclipse, leading to constraints on, and possible identification of, the degree of the mode. The orbital period is 6.07 days, so a two-week multi-site campaign would be ideal. At V = 11, small telescopes will be adequate.

X-ray Emission. Ayres et al. (ApJ 376, L45) announced the serendipitous discovery of soft X-ray emission from CN Boo — additional evidence of some chromospheric activity in A stars (*e.g.* see Simon & Drake ApJ 346, 303) and in a δ Scuti stars in particular.

PLC Relation. In an intriguing paper Fernie (AJ 103, 1647) suggested that the δ Scuti stars and Cepheids form a single class of pulsating stars with a single PLC relation. He says "If this [κ] mechanism is sufficient to connect the 2 and 100 day pulsators to the same Cepheid family, why should it not connect the <0.2 day pulsators as well?". My first thought is: because δ Scuti stars and Cepheids are not homologous. But Fernie's PL diagram is impressive. The problem is that it compares δ Scuti stars in the galaxy with Cepheids in the LMC. If there is no zero point shift between these two groups, then the linear relation he shows is correct and the zero point can be refined by this technique; but if there is an undetected zero point shift between the two groups, then the relation may not be linear. The test of this is to observe some δ Scuti stars in the LMC. With a distance modulus of 18.5, the δ Scuti stars should have V about 20 to 21 — well within the capabilities of the kinds of observations reported by Nemec and Mateo (ASP Conf. Ser. 10, 134) for 23rd magnitude δ Scuti stars in the Carina dwarf galaxy.

Rapidly Oscillating Ap (roAp) Stars. The roAp stars were last reviewed by Matthews (PASP 103, 5). HR 3831 has been discovered to pulsate in a distorted oblique dipole mode which gives rise to a frequency septuplet in the amplitude spectrum (Kurtz et al. MN 260, 343). Shibahashi and Takata (PASJ, in press) have shown that such a septuplet is expected for a dipole pulsation mode from the effects of magnetic and rotational second-order perturbations. It is not yet clear whether the observations and theory are in complete agreement.

HR 3831 also has an unstable pulsation frequency which I have recently shown (in preparation) varies cyclically with a period of 2 years. This is not a Doppler shift in a binary system, and its explanation is problematic. The amplitude of the frequency change is similar to that seen on the sun over the 11-year solar cycle, but A stars are not expected to have magnetic cycles because of their rigid rotation and lack of any significant surface convection zone. So far, all roAp stars that have been studied long enough show frequency variability (see Kriedl et al. MN 250, 477; Kurtz et al. MN 260, 343; Kurtz et al. MN 251, 152).

An important development is a multi-colour study of the pulsation of HR 3831 by Matthews et al. (ASP Conf. Ser. 44, 617) which requires a T-τ structure with a substantially steeper gradient than in a normal A-star atmosphere. This supports the same requirement independently suggested by Shibahashi and Saio (PASJ 37, 245) based on arguments concerning the critical frequency. If these steep temperature gradients are borne out, and if they apply to other Ap stars, then previous analyses will have overestimated the abundance peculiarities in these stars. That, in turn, has strong implications for diffusion theory.

Martinez finished a Ph.D. thesis with me in mid-1993 in which he discovered 10 new roAp stars (compare this to the 15 found during the 12 years prior to his thesis). A list of 23 of the 25 known roAp stars can be found in Kurtz and Martinez (ASP Conf. Ser. 44, 561). First results from, and a description of Martinez's survey are in Martinez et al. (MN 250, 666).

5. Cepheids (Nancy Remage Evans)

Fundamental Data. The first step in studying Cepheids is to identify them, particularly in directions useful for studying galactic structure. Such a search was performed by Caldwell et al. (AJ 101, 1763), followed up by photometry and velocity studies (listed below). Many papers containing photometry of Cepheids have appeared: Berdnikov (Pis'ma Astr Zhurnal 18, 325: UBVR), Laney and Stobie (AAS 93, 93: JHKL), Shobbrook (MN 255, 486: UBVRI), and Schechter et al. (AJ 104, 1930). Progress reports on photometric programs were given by Schmidt (AJ 102, 1766), DuPuy et al. (Victoria), Henden (Victoria) and Heiser (Victoria). Velocity data and discussions were published by Gorynya et al. (Pis'ma Astr Zhurnal 18, 777), Metzger et al. (ApJS 76, 803), Metzger et al. (AJ 103, 529).

Among the discussions of photometry, Fernie (Victoria) found no evidence of changes in light curves. He also found that the period change of Y Oph is well represented by a cubic polynomial (PASP 102, 905).

Other discussions of period changes and the techniques involved in determining them are provided by Berdnikov (Pis'ma Astr Zhurnal 18, 519), Szabados (Mitt. Sternw. Ugaris. Akad. Wissen. 96, 125), and Vinko (ApSS 183, 17). A particularly exciting development has been the application of high accuracy velocity techniques to Cepheids (Marcy & Butler PASP 104, 270). They reported a variable with an amplitude of 200 m/sec (Butler ApJ 394 L25), providing a new technique to investigate low-amplitude variability in the instability strip. Another techique to search for Cepheid candidates was presented by Sowell (AJ 100, 834) in his discussion of Hα profiles of supergiants. Polyakova (Pis'ma Astr Zhurnal 16, 916) reported polarimetry of 3 Cepheids.

Gray (AA 252, 237) derived colour excesses from Strongren photometry. Antonello et al. (Victoria) measured temperatures from spectrophotometry for 6 short-period Cepheids.

Individual Stars. New data for the following individual stars has been published: EW Sct (Figer et al. MN 249, 563 and Berdnikov et al. IBVS #3710), NSV 07453 (Shugarov IBVS #3778), V1334 Cyg (Usenko IBVS #3784), NO Cas and CN Tau (Schmidt & Gross PASP 102, 978), CE Cas (Berdnikov Soviet AJ 34, 400), EE Aql (Usenko IBVS #3519), V367 Sct (Berdnikov IBVS #3712), and AS Cas (Berdnikov et al. IBVS #3711).

Atmospheres. A number of studies have appeared discussing results pertaining to the details of Cepheid atmospheres. Sasselov and Lester (ApJ 360, 227) have investigated line ratios in the 1.1 micron region which provide very accurate relative temperatures, and reddenings. They (ApJ 362, 333) also provide continued studies of line profiles in the same wavelength region, showing asymmetries and occasional extra absorption components. Wallerstein et al. (MN 259, 474) presented velocity differences between Hα and metallic lines, and also Hα asymmetry for 7 Cepheids throughout their pulsation cycles, including the surprising result that the Hα asymmetry is in the same sense at all phases. Interpreting these results in atmospheric computations provides a significant challenge. Rodrigues and Böhm-Vitense (ApJ 401, 695) searched for mass loss signatures in ultraviolet lines in S Mus, but only determined an upper limit. Abundance analyses of Cepheids were performed by Klochova (Pis'ma Astr. Zhurnal 17, 732), Fenian et al. (IBVS #3595), Giridhar et al. (JApA, 12, 27) and Klochkova and Panchuk (Pis'ma Astr Zhurnal 17, 536).

Clusters and Associations. Four open clusters suggested to contain Cepheids were studied. Vazquez and Feinstein (AAS 86, 209) conclude that GH Car is a member of Tr 18. Turner et al. (AJ 104, 1132) reanalyzed NGC 129 containing the Cepheid DL Cas using additional photometry. Turner (AJ 104, 1865) concluded that SZ Tau is a member of the cluster on the basis of similar radial velocity and age, even though it is 2° from the cluster. This requires that it is pulsating in the first overtone mode. Claria et al. (MN 249, 193) presented new photometry of NGC 5662, from which they concluded that the Cepheid V Cen is a likely member. Turner et al. (ApJS 85, 119) discussed a search for clusters around southern Cepheids. Turner (AAS 97, 755) discussed unusual reddening in late B stars in the cluster Ros 3, which may also be found in clusters containing Cepheids.

Binaries. Evans (ApJ 384, 220) has used a survey of Cepheids brighter than 8th magnitude with the IUE satellite to determine that 21 per cent have blue companions. If corrected to the detection rate for radial velocity surveys, this is larger than the fraction of B stars of comparable mass which are binaries and have long period orbits. Szabados (Victoria) discussed other ways in which blue companions can be detected.

Orbital motion has been found or suspected for the following stars: VY Per (Szabados Observatory 112, 57); RX Cam (Szabados IBVS #3812); and MW Cyg (Gorynya et al. IBVS #3776). Gieren and Brieva (AA 253, 126), however, find that the suspected binary SV Per probably shows no orbital motion. Vinko (MN 260, 273) provides a now orbital solution for AW Per. Evans et al. (PASP 102, 981) observed suspected binaries FM Aql, FN Aql, RX Aur, Y Lac, and RS Ori with IUE and found that only Y Lac and possibly RS Ori may have blue companions.

The Period-Luminosity-Colour Relation and the Instability Strip. Wilson et al. (ApJ 378, 708) have calibrated the Cepheid period-luminosity (PL) relation using statistical parallax analysis. Fernie (AJ 103, 1647) derived a PL relation by treating Cepheids and δ Sct stars as one class. Hindsley and Bell (Victoria) reported a correction to Cepheid distances derived using a colour-surface brightness relation from model atmospheres. In a second paper, they discussed the absolute magnitudes derived from the very accurate Butler velocities. In a series of papers, Evans (ApJ 372, 597; ApJ 389, 657; AJ 104, 216) has developed the use of IUE observations of blue companions of Cepheids to determine Cepheid luminosities. These luminosities were combined with cluster luminosities to confirm the "wedge shape" of the instability strip. Evans (AJ 105, 1956) found minimal overlap in the HR diagram between Cepheids and nonvariables, but cautions about the interpretation of the blue edge. Schmidt (Observatory 111, 1103) reviewed recent determinations of the period-luminosity relation, and found a range in Mv of ± 0.1 magnitude at $\log P = 0.8$. Gieren and Fouque (Victoria) also found a dispersion in PL zero-points similar to that of Schmidt. Simon (Bologna), however, cautioned that intercomparing the results from Baade-Wesselink studies from different groups shows a very large scatter for individual stars.

The calibration of the period-luminosity-colour relation (PLC) is, of course, closely connected with extragalactic distance determination. A comprehensive review of the Cepheid distance scale is provided by Madore and Freedman (PASP 103, 933), which has extensive discussion of the PLC in many wavelength bands.

Fourier Analysis: Overtone Pulsation. Any use of Cepheids in distance determination requires the determination/assumption of their pulsation mode. Fourier analysis of light curves is a promising method to identify pulsation mode, and a number of studies have produced Fourier coefficients. Laney and Stobie (MN 260, 408) have analyzed infrared light curves. The use of Fourier coefficients to determine pulsation modes (especially for 's' Cepheids, suggested to be first overtone pulsators) has been discussed by Antonello et al. (AA 236, 138), Mantegazza and Poretti (AA 261, 137), and Antonello (Victoria). Berdnikov (Pis'ma Astr Zhurnal 18, 654) provides light curves for double mode Cepheids at many phases of the modulation cycle. Studies of a number of individual double mode or 's' Cepheids have been made: V950 Sco (Poretti & Mantegazza IBVS #3774), TU Cas (Matthews et al. AJ 104, 748, Andrievsky et al. Victoria) and also Szabados (Victoria), and V 473 Lyr (Fabregat et al. MN 245, 542). The colour-colour diagram of 's' Cepheids was discussed by Usenko (Kinematics and Physics of Celestial Bodies 6, 95).

So far, however, independent confirmation of the pulsation modes inferred from Fourier coefficients has been scarce. IUE studies (Evans ApJ 385, 680) of binary companions have shown that an 's' Cepheid (SU Cas) and a double-mode Cepheid (Y Car), and also V367 Sct lie on the blue edge of the instability strip as predicted by pulsation calculations. Furthermore, this study showed that Y Car is a second double-mode Cepheid to have a luminosity appropriate to its fundamental period, confirming that double-mode Cepheids are normal Population I Cepheids.

Pulsation Theory. See also section 11. New opacity calculations (Iglesias et al. ApJ 360, 221; Iglesias & Rogers ApJ 371, L73; Seaton, Victoria; Rogers, Victoria; Mihalas, Bologna) have significantly enhanced opacity near 250 000 K. These opacities provide a crucial alteration in pulsation calculations which removes or decreases the Cepheid mass discrepancy between beat and evolutionary masses.

An important aspect of research during this period has been the assessment of the effects of the revised opacities. Calculations by Moskalik et al. (ApJ 385, 685) found agreement between beat and evolutionary masses, and improved agreement between bump and evolutionary masses. Ultimately, however, some additional convection is required for complete agreement between bump and evolutionary masses according to Cox (Victoria) and Moskalik and Buchler (Victoria). Petersen (AA 265, 555; Victoria) discussed the sensitivity of period ratios to opacity, and the sensitivity of period-ratio to mass. Further work using the amplitude formalism has shown that 'resonances' produce additional detail in pulsating models (Buchler et al. ApJ 380, 195) including RV Tau characteristics in a Populaton I Cepheid (Moskalik & Buchler ApJ 366, 300). Rotation was found by Kovacs and Buchler (Bologna) to have no significant effect on Cepheid pulsation. Extensive calculations by the Japanese group have included investigation of the effects of the calculation grid mesh (Ishida & Takeuti PASJ 43, 795), a gray atmosphere (Zalewski PASJ 43, 345), modal interaction (Ishida ApSS 168, 243; Ishida & Saitou ApSS 182, 69; Takeuti et al. PASJ 44, 101; Ishida & Takeuti ApSS 178, 311; Tanaka et al. PASJ 44, 331) and period changes (Saitou & Takeuti PASJ 42, 341). Calculations from another group were discussed by Fadeyev and Muthsam (AA 260,

195), stressing different causes for light curve bumps in different period ranges. Li (Victoria) discussed the red edge of the instability strip; Morgan (PASP 105, 123) also discussed extensive calculations.

A number of computational improvements have been investigated. Gautschy and Glatzel (MN 245, 154) and also Dorfi and Feuchtinger (AA 249, 417) presented new methods of solving pulsation equations. The sensitivity of the models to the equation of state is discussed by Kanbur (AA 250, 395; 259, 175). Costa et al. (ApJ 373, 237) investigated the effect of radiation pressure on a one-zone pulsation model. Finally, future improvements in pulsation calculations were discussed by Kovacs (Victoria), Petersen (Bologna) and Davis (Bologna).

Evolution. Chiosi et al. (ARAA 30, 235) reviewed stellar evolution in detail, including considerable discussion of intermediate mass stars in the instability strip. The Padua group also discussed in detail the Cepheid instability strip (Chiosi et al. ApJ 387, 320) and the evolutionary/pulsation mass discrepancy (Chiosi et al. ApJ 385, 205). They found that they could duplicate the wedge shape of the instability strip and the morphology of LMC cluster NGC 2157 with models including core convective overshoot near the main sequence. A major summary of the Padua computations is presented by Chiosi et al. (ApJS 86, 541), who discussed in detail the instability strip and PL relations from this extensive grid of computations. However, these models were computed using old opacities; although evolutionary tracks are little affected by the change in opacity, it is not clear, for instance, how the edges of the instability strip are changed. The Padua group plan to look further into this in the future. Also relevant to the evolutionary state of Cepheids are the new evolutionary tracks by Schaller et al. (AAS 96, 269), using the new opacities, although they do not discuss Cepheids explicitly.

Observational results include the study of the Cepheid BP Cir (Evans et al. AJ 103, 1638) and Cepheids with evolved companions (Evans, Victoria). Cepheids with evolved main sequence companions are best matched with isochrones with at most a small amount of overshoot, at least for short period stars. Furthermore, some evolved companions are cooler than the ZAMS, possibly because of rapid rotation. Tsvetkov (ApSS 173, 31) discussed the relation between light curve morphology and instability strip crossing.

Extragalactic Cepheids. Studies of variable stars in the clusters in the Large and Small Magellanic Clouds (LMC and SMC) have progressed enormously through the use of CCD images and multi-aperture spectrographs. Detection of the variables is labourious, but automated techniques are being developed (Welch & Stetson AJ 105, 1813). The high relative accuracy of photometry of stars in the same cluster make cluster observations an excellent way to investigate long-standing questions such as the width of the PLC (very small: NGC 1866, Welch et al. AJ 101, 490), the identification of overtone pulsators (Welch et al. AJ 105, 146), the relation between variable and nonvariable stars (Mateo et al., Bologna; Balona MN 256, 425). Wesselink radii have also been determined for several Cepheids (Cote et al. AJ 101, 1681; Barnes et al. ApJ 405, L51). Photographic photometry of field Cepheids (short period) in the SMC (Smith et al. AJ 104, 1430) showed well-defined sequences of fundamental mode and overtone pulsators.

Investigations into non-uniform spatial distribution of Cepheids in the LMC have been made by Karimova (Pis'ma Astr Zhurnal 16, 48 and 16, 261). Vardanyan and Pogosyan (Astrofizika 34, 95) have determined a PLC relation.

Luck and Lambert (ApJS 79, 303) have extended abundance analyses to include Cepheids and nonvariable supergiants in both Magellanic Clouds, and conclude that there may be abundance differences between LMC stars, but differences between SMC stars are not measureable.

A number of studies have been done on extragalactic Cepheids more distant than the Magellanic Clouds using CCD photometry, particularly in the near IR, as summarized by Freedman and Madore (Victoria). These include M33 (Freedman et al. ApJ 372, 455), NGC 300 (Freedman et al. ApJ 396, 80), DDO 216 (Hoessel et al. AJ 100, 1151), NGC 3109 (Capaccioli et al. AJ 103, 1151) and NGC 6822 and WLM (Lee et al. Victoria).

In addition, Metcalfe and Shanks (MN 250, 438) have used CCD photometric sequences to reinvestigate variables in several galaxies. The distances from these studies have been used by Freedman (ApJ 355, L35) to calibrate an infrared Tully-Fisher relation. Freedman and Madore (ApJ 365, 186) studied Cepheids in three fields in M31 at 3, 10, and 20 kpc from the center to see whether differences in the PLC due to differences in metallicity could be detected. They concluded that differences in the distance moduli could be explained by differences in extinction, and were consistent with a small PL relation dependence on metallicity.

Special techniques have been used to observe Cepheids at larger distances. High resolution images from the ground of NGC 4571 in the Virgo cluster make it possible to search for Cepheids (Pierce et al. ApJ 393, 523), and several candidates have been identified (Pierce et al., Victoria). Using the Hubble Space Telescope, Cepheids have been detected in IC 4182 (Sandage et al. ApJ 401, L7), which in turn have been used to calibrate the distance to SN Ia 1937c. Freedman and collaborators (1993, AAS Abstract) have used the Hubble Space Telescope to increase the number of Cepheids measured in M81 from 2 to 31 and decrease the uncertainty in the distance (3.4 Mpc) to 10 per cent.

Finally, an excellent review of the use of Cepheids and other extragalactic distance indicators was provided by Jacoby et al. (PASP 104, 599). For example, they concluded that the uncertainty in the distance to the LMC as determined from Cepheids is dominated by the uncertainty in the calibration of the PLC, which they estimate to be ± 0.3 magnitude. They also discussed the distance to the LMC from the circumstellar-ring light echo of SN 1987A which has an uncertainty of 6 per cent and can be used to calibrate Cepheid luminosities. Pierce and Tully (ApJ 387, 47) also found that "the primary source of uncertainity in the extragalactic distance scale now lies with the Galactic calibration of the Cepheid and RR Lyrae variables". The application of the Cepheid PLC to other galaxies requires an understanding of its sensitivity to properties such as metallicity. Saha (1993, private communication), however, suggests that the accuracy of secondary calibrators must also be carefully scrutinized in assessing extragalactic distance accuracy.

Population II Cepheids. A good recent review of Population II variable stars was provided by Wallerstein (Bologna), starting with the difficulty of defining the group. This was also discussed by Nemec (Victoria) and Lutz and Nemec (Victoria). Observational and theoretical aspects of classifying short period variables are discussed by Diethelm (AA 239, 186) and Petersen (AA 272, 217) respectively.

Observational work on the group has included Stromgren photometry (Meakes et al. AJ 101, 1795), and identification of a possible new BL Her star (Sanwal & Sarma JApA 12, 119). High resolution spectral studies showing the complexity of the line profile changes during the cycle (Lebre & Gillet AA 255, 221; Lebre, Bologna) were particularly exciting. An introduction to the complexity of the phenomena needed to explain them was given by Wallerstein and Elgar (Science 256, 1531). The Cepheid in the AU Peg binary system had rapid changes in the pulsation period (Vinko, Victoria).

Theoretical models of the pulsation behaviour (Buchler & Moskalik ApJ 391, 736; Fadeyev & Muthsam AA 234, 188; and Moskalik & Buchler ApJ 406, 190) and the hydrogen lines (Fokin MN 250, 258; Huguet et al. AA 255, 233) have been produced.

Extragalactic Population II Cepheids (anomalous Cepheids) have been found in CCD observations of the Sextans dwarf spheroidal galaxy (Mateo et al. AJ 101, 892).

Acknowledgements: It is a pleasure to thank Drs. J. Matthews and J. Nemec for forwarding papers from the Victoria conference proceedings, and Dr. W. Freedman for up-to-the-minute information about M81 Cepheids. Drs. J.R. Buchler, H.C. Harris, and A. Saha provided helpful comments on the theoretical, Population II and extragalactic portions of this review respectively. Ms M. Cummins provided guidance and assistance in database searching.

6. RR Lyrae Stars in the Field (T.G. Barnes III)

This survey of field RR Lyrae variables covers the period from the last report through July 1993. Because of page limitations not all research can be cited. Additional reports on RR Lyrae research are contained in the sections on pulsation theory and variable stars in globular clusters and related systems.

Conferences and Reviews. The following conferences, which included significant discussion of RR Lyrae research, took place or appeared in print during the report interval: "Observation and interpretation of pulsating stars," a meeting of Subproject 3 of the Multinational Cooperation of the Academies of Sciences of Socialist Countries, Sep 4 - 8, 1989, Veroeff. Sternw. Sonneborg, Vol 13, 1990, ed. W. Wentzel; "New Results on Standard Candles," Proceedings of a workshop, Trani, Italy, Aug 26 - 30, 1991, Mem. Soc. Astron. Italy 63, 223, 1992, ed. F. Caputo (hereinafter Trani); "New Aspects of Magellanic Cloud Research, " Heidelberg, Germany, Jun 15 - 17, 1992; and "Stellar Photometry," IAU Coll. 136, Dublin, Ireland, Aug 4 - 7, 1992, as well as the Bologna, Mito and

Victoria conferences (see section 1).

Absolute Magnitudes. The Baade-Wesselink technique, modified in various ways, has continued to be used to determine RR Lyrae absolute magnitudes and their correlation with [Fe/H]. These analyses appear to have converged on a common solution for the absolute magnitudes and metallicity dependence. Within a few hundredths in each parameter, the B-W results published in the last three years agree with that given by Jones et al. (ApJ 386, 646): Mv = 1.02 + 0.16[Fe/H]. As noted below, this distance scale is about 0.3 mag fainter than the distance scale which is required to make the RR Lyrae and Cepheid variables in the LMC agree in distance.

Absolute magnitudes have been determined for UU Cet, W Tuc, and RV Phe using both the IR Flux and Surface Brightness versions of the B-W method by Clementini et al. (ApJ 396, 219; Trani 397). The two methods gave concordant results. When B-W results for 24 field stars were combined, they determined Mv = 1.01 + 0.19[Fe/H]. BVRI and CORAVEL data for these analyses were published in Cacciari et al. (AAS 85 865).

Fernley and collaborators continued to investigate RR Lyrae absolute magnitudes using their IR version of the B-W method. Supplementing previous work, they have applied it to the RR Lyrae variables V445 Oph, SS Leo, and VY Ser (Fernley et al. MN 247, 287).

Liu and Janes reviewed recent B-W methods for determining RR Lyrae absolutes (Trani 445; "Relativity and Gravitation: classical and quantum," Proc. 7th Latin Amer. Symp. on Rel. and Grav., Cocoyoc, Mexico, Dec 2 - 8, 1990, eds. J. D'Olivo et al. 278). They argued that the results from various groups are consistent with Mv = 1.06 + 0.20[Fe/H].

Jones et al. (ApJ 386, 646) used their IR version of the B-W method to determine the distances to SW And and DX Del. Combining these results with previous work using B-W methods, they find a best absolute magnitude calibration of Mv = 1.02 + 0.16[Fe/H]. Using a similar technique, Storm (Trani 387) determined the absolute magnitudes for three RR Lyraes in M5 and M92. His results for these stars agree with the field RR Lyrae Mv - [Fe/H] relation.

In work comparing B-W and other methods of Mv determination, evidence has appeared that the B-W method may be giving too faint a result. Carney reviewed the B-W method, main sequence fitting, and period-shift methods for calibrating the Mv - [Fe/H] relation (Trani 409; and "The Stellar Populations of Galaxies," IAU Symp. 149, Angra dos Ries, Brazil, Aug 5 - 9, 1991, eds. B. Barbuy & A. Renzini, 15). He suggested a best relation Mv = 1.01 + 0.15[Fe/H]. Fernley has also reviewed the determination of RR Lyrae absolute magnitudes by the usual methods (AA 268, 591; Trani 453). He calls attention to fainter absolute magnitudes found through B-W solutions than through the period-shift method and models of horizontal branch stars. Sandage (AJ 106, 703) makes the latter point as well. Clement et al. (ApJ 395, 192) and Simon and Clement (ApJ 410, 526) have use ϕ_{31} to establish a luminosity scale for RRc variables which is also brighter than the usual B-W scale. If indeed the B-W scale is too faint by 0.2 - 0.3 mag, as these latter works claim, and the true absolute magnitudes of RR Lyrae stars are that much brighter, the discrepancy between Cepheid and RR Lyrae distances to the LMC would be eliminated.

Use of the main sequence fitting method to determine absolute magnitudes of RR Lyraes has been explored by Fusi Pecci et al. using theoretical HB models (Bologna 31).

Carney (1991, in "Astrophysics with Infrared Arrays," Proceedings of a conference held in Tucson, Arizona, Feb 19 - 21, 1990, ASP Conf. Ser. 14, 115, ed. R. Elston) has reviewed the use of K-band photometry for distance determination of RR Lyrae stars. He discussed distance estimates to the Galactic Center, the LMC, the SMC and M31. Longmore et al. also discussed the role of IR photometry in RR Lyrae pulsation and distance studies, including the Mv - logP relation (Bologna 36). The period-luminosity relation in the K magnitude for globular cluster variables has been discussed by Buckley et al. (Trani 433) and by Janes and Liu (Trani 445).

Period-Shift Effect. The Sandage Period Shift measures the change in the mean period of pulsation with metallicity among RR Lyrae stars; it is closely related to the Oosterhoff dichotomy for globular clusters. An understanding of this phenomenon has eluded researchers for decades, but in the last three years progress has been made. Reviews on the topic have been given by Rood (Bologna 11), Lee (ApJ 363, 159); Bencivenni et al. (ApJ 380, 484); Carney et al. (ApJ 386, 663); Catelan and de Frietas Pacheco (AA 261, 457; AJ 106, 175); Fernley (AA 268, 591); and Sandage (AJ 106, 687).

The key to recent progress derived from realization that the period-shift effect arises, at least in part, from evolutionary effects along the horizontal branch, as discussed by Lee et al. (ApJ 350, 155), Lee (ApJ 363, 159), Bencivenni et al. (ApJ 380, 484) and Sandage (AJ 106, 687). The latter paper by Sandage gives a particularly good review of the history. Additional research on this topic includes that by Caputo (Bologna 22) who discussed the Sandage period-shift effect as a test of stellar evolution and pulsation theories. Lee (ApJ 373, L43) examined the absence of correlation between Mv and [Fe/H] for RR Lyrae stars in ω Cen in the context of his horizontal branch models. He argued that this is a consequence of the mean luminosity of the RR Lyrae stars depending sensitively on HB morphology as well as metallicity. Caputo and De Santis used reddening free parameters to determine the period-shift effect for field (AJ 104, 253) and globular cluster (ApJ 401, 260) variables. Sandage (AJ 106, 687 and 106, 703) proposed an interpretation of the period-shift effect and the Oosterhoff dichotomy which rests upon an increase in pulsation period (due to increased luminosity) and a change in the morphology of the horizontal branch which both come about as the metallicity decreases.

Catelan and de Freitas Pacheco (AA 261, 457; AJ 106, 175) have used new synthetic models of horizontal branch stars to study phenomena in RR Lyrae stars. In particular they have examined the effect of choosing Sweigart or Lee and Demarque opacities upon the Sandage period-shift. They concluded that these opacities lead to different predictions for the period-shift.

Multimode Behaviour, Period Ratios. Double-mode pulsators have been used for a long time to infer masses for RR Lyrae variables, but those masses have been somewhat less than masses inferred by other techniques as discussed by Simon and Cox (ApJ 376, 717). Recently this discrepancy has been investigated using new opacities for the models. Discussions of the effects of revised opacities on double-mode masses have been given by Cox (ApJ 381, L71), Kovacs et al. (AA 252, L27). Petersen (AA 243, 426; AA 265, 555), and Simon (ApJ 387, 162). Petersen has reviewed this issue and concluded that the pulsational models and the observations are now in reasonable agreement (Trani 345). The masses inferred from double-mode pulsation are increased to about 0.7 $M_\odot$ from the previous value near 0.6 $M_\odot$.

In addition, Kovacs et al. have explored the effect of changing the relative abundances of oxygen and other heavy elements in models of RR Lyrae star pulsation (AA 259, L46). They concluded that the masses of double-mode pulsators will be uncertain by about 0.1 $M_\odot$ until accurate abundances for the individual dominant elements are available.

Masses for RRc variables in five Oo I and five Oo II clusters have been determined by Bruzzi and Cacciari using Simon's ϕ_{31} method ("Star Clusters and Stellar Evolution," Mem. Soc. Astron. Italy, 63, 79). Their results suggested a mass difference of 0.1 $M_\odot$ between the Oosterhoff types and somewhat larger mean masses than determined by the double-mode analysis. Clement et al. also used this method to determine masses for six stars in M68 and concluded that the masses must lie in the range 0.75 - 0.90 $M_\odot$, depending on the metallicity and luminosity assumptions (ApJ 412, 183).

The number of known double-mode, field RR Lyrae stars has been increased from one to three by Clement et al. (ApJ 372, 273). They published periods, mean magnitudes, colours, metallicities and inferred masses for two stars.

Abundances. New ΔS values for 35 field RR Lyrae stars have been reported by Kinman and Carretta (PASP 104, 111). They also proposed that the anomalous colours and spectra of BB Her, and possibly of AR Her as well, may be a result of the star being binary with a non-variable, blue horizontal branch star. Suntzeff (ApJ 367, 528) has investigated the metallicities of RR Lyrae variables in the galactocentric distance range 4 - 30 kpc using ΔS. He found the metallicity to be a constant [Fe/H] = −1.65 outside the solar circle and to have a gradient −0.06 dex/kpc inside the solar circle. A calibration of equivalent width of the CaII K line with [Fe/H] has been made by Clementini et al. (AJ 101, 2168). They claim a tighter and more reliable relation than that based on ΔS.

Caputo (AJ 106, 301) has used light-to-mass ratios for RR Lyrae stars in the field over the galactocentric range 0.5 - 19 kpc to infer the field horizontal-branch morphology. He concluded from comparison of this morphology to models that the intial helium abundance had to be a constant Y = 0.23 for all field RR Lyraes with [Fe/H] < -0.7.

Radial Velocities. According to pulsation models constructed by Simon (MN 246, 70), the phase lag between light and velocity curves may be interpreted to give the distances of RRc stars from their respective blue edges. Liu (PASP 103, 205) has established a correlation between velocity curve and light curve amplitudes from a large number of RR Lyraes having high-precison data. From this correlation a synthetic radial velocity curve for RRab stars is constructed for use in determining systemic velocities. CORAVEL data for several stars were published by Cacciari et al. (AAS 85, 865). Fernley et al. (AAS 97, 815) published new CORAVEL radial velocities for 13 RR Lyrae stars.

Radii. Using a new CORS method, Cucurachi et al. (Bologna 116) determined radii for SW Dra (5.6 $R_{\odot}$) and X Ari (7.3 $R_{\odot}$). They argue from this result that X Ari is actually an anomalous Cepheid.

RR Lyrae Variables in Special Fields. The RR Lyrae variables in Baade's window have been studied using ΔS by Walker and Terndrup (ApJ 378, 119). They found the metallicity to be strongly peaked at [Fe/H] = −1.00 with very small dispersion. In a different window into the Galactic bulge, Blanco (AJ 103, 1872) found 112 new RR Lyrae stars, for which she gives light curves and finding charts.

Walker summarized work on photometry of RR Lyrae stars in the Magellanic Clouds ("The Magellanic Clouds," Proc. IAU Symp. 148, Sydney, Australia, Jul 9 - 13, 1990, eds. R. Haynes and D. Milne, 307). A review of RR Lyrae variables as standard candles for extragalactic work was given by Lee (Trani 331). From preliminary results of a new survey for RR Lyrae variables in the LMC, Reid and Freedman reported 550 likely candidates (Bologna 120). Kinman et al. discussed five new LMC field RR Lyrae variables (PASP 103, 1279). Walker used RR Lyrae variables in many LMC clusters to determine properties of the clusters and a mean absolute magnitude of the variables (Trani 479; ApJ 390, L81; AJ 105, 527). He found Mv = +0.44 mag at [Fe/H] = -1.9, which is nearly 0.3 mag brighter than suggested by statistical parallax and B-W studies of Galactic field RR Lyrae stars.

Saha and collaborators have found and measured photometrically numerous RR Lyrae variables in nearby galaxies: 32 in NGC 147 (AJ 100, 108), 30 in NGC 205 (AJ 103, 84) and 15 in IC 1613 (AJ 104, 1072). Finzi (AA 255, 115) has proposed, and discussed, the possibility that RR Lyrae stars in dwarf spheroidal galaxies, including NGC 147, have high luminosities.

Photometric Data, Light Curves. BVRI data for several stars were published by Cacciari et al. (AAS 85, 865). Observations of T Sex in the uvbyβ system are reported by Alaniya (Abastumani Ap. Obs. Bull 68, 4). Pena et al. (Rev. Mex. Astr. Ap 20, 139) reported uvbyβ photometry of several RR Lyrae variables in Serpens. Hobart et al. (Rev. Mex. Astr. Ap 22, 275) suggested that the light variation of T Sex is best represented by three periodicities. Barnes et al. (PASP 104, 514) published contemporaneous BVRIJHK data on four RR Lyrae stars. They also use these data to examine Saha's proposal that TU UMa may be binary. BV photometric data for six field stars has been published by Burki et al. (AAS 97, 827). Fernley et al. (AAS 97, 815) published JHK data on 110 variables in the first installment of a large program to obtain IR data on field RR Lyrae variables.

Simon (MN 246, 70) has used hydrodynamic models of RRc variables to study the Fourier decompostion parameters for these stars. He found the models to mimic the observations very well for ϕ_{31} but to fail for the parameter ϕ_{21}. Grieco and Antonello have used one and two-zone models to study the light curves of RR Lyrae stars (Bologna 101). They discuss the dependence of Te and radius estimates on the colour indices in the two-zone model.

In the first results from his continuing survey of poorly studied variable stars, Schmidt found six out of seventy-nine stars classified as RR Lyrae actually to be eclipsing or ellipsoidal variables (AJ 102, 1766). From this he concluded that there may be as many as 400 - 500 stars listed as RR Lyrae variables in the GCVS which may be short-period binaries instead.

Large numbers of RR Lyrae variables have had elements determined or revised in the last three years. The results have been published in BAV Rundbrief, IBVS, MVS, GEOS Circ, Astr. Tsirk., JAAVSO, Per. Zv., and JBAA. These sources also include references to period variations, to newly discovered RR Lyrae variables and to stars with newly discovered Blazhko effect. Ephemerides for RR Lyrae variables were published by Firmaniuk and Zakrzweski for 1990, 1991 and 1992 (Rocznik Astr. Obs. Krakow, Intern. Suppl. #61, 115, #62, 119, and #63, 121, respectively), and also by the AAVSO.

7. Variable Stars in Globular Clusters and Related Systems (Amelia Wehlau)

As far as possible this review covers material on this subject published in the three years ending in mid-summer 1993. As usual space limitations make it impossible to cite every paper published during this time. The reader is referred to the Comm. 38 report on globular cluster research and to the following conference proceedings: ASPCS (ASP Conference Series) 13, The Formation and Evolution of Star Clusters, ed. Janes, 1991; IAU Symp 149, The Stellar Populations of Galaxies, eds. Barbuy and Renzini, 1992; Star Clusters and Stellar Evolution, eds. Brocato, Ferraro and Piotto, MemSAIt 63, No. 1, 1992; ASPCS 48, The Globular Cluster-Galaxy Connection, eds. Brodie and

Smith, 1993 as well as the Bologna, Trani, and Victoria meetings mentioned in section 6.

Studies of Pulsating Variables In Galactic Globular Clusters. Listed in order of IAU cluster designation. CJS referes to Clement, Jankulak and Simon (ApJ 395, 192).

C0021-723 (NGC 104, 47 Tuc): Corwin and Carney (IAU Coll #139, 410) report on BV photometry of V9.

C0512-400 (NGC 1851): CCD observations for 22 of the 26 previously known variables are included in the data for a new CMD presented by Walker (PASP 104, 1063). Thirteen candidate RR Lyrae stars were also found.

C0522-245 (NGC 1904, M79): CCD photometry for a CMD by Ferraro et al. (MN 256, 391) includes observations of 10 known or possible variables. Among these, V5 apears to be a blend of two variables, both possible RR Lyrae stars.

C0734+390 (NGC 2419): Clement and Nemec (JRASC 84, 434) report V39 is a double-mode RR Lyrae star.

C1236-264 (NGC 4590, M68): Nine double-mode variables identified by Clement et al. (ApJ 412, 183) have been found to have masses and Fourier parameters similar to those found for the variables in M15.

C1310+184 (NGC 5024, M53): Kravtsov (Astr Tsirk 1544, 21) presents a finding chart and co-ordinates for 11 new variable stars in the central region. CJS give Fourier parameters for 15 RRc stars.

C1323-472 (NGC 5139, ω Cen): Simon (ApJ 360, 119) uses pulsation models to derive masses, temperatures and absolute luminosities for a large sample of RRc stars in the cluster and finds results which are not consistent with HB models. The lack of correlation between mass and metallicity of the RR Lyraes in this cluster (also found by Simon) is attributed by Lee (ApJ 373, L43) to effects of HB morphology. Frolov (IBVS #3566) determines masses for three pulsating blue stragglers to be comparable to RR Lyrae masses. Mukherjee et al. (PASP 104, 561) identify two new possible variables. Observations of the eclipsing blue straggler NJL5 are used by Helt et al. (AA 270, 297) to obtain tight constraints on the component masses and to determine $M_{V,HB} = 0.7$.

C1339+286 (NGC 5272, M3): CJS give Fourier parameters for six RRc stars. Frolov (IBVS #3566) finds a mass of 0.8 $M_\odot$ for the pulsating blue straggler and (Astr Tsirk #1553) shows that the most evolved RR Lyraes are well separated in the IR period-colour and HR diagrams. Photometry of variables 61 through 87 and V133 is presented by Meinunger (MittVerSt 12, 64, 67 & 70 (1990); 78, 95, 96 & 115 (1991)).

C1403+287 (NGC 5466): Photometry of three eclipsing blue stragglers is presented by Mateo et al. (AJ 100, 469). For 4 out of 5 blue stragglers Frolov (IBVS #3566) finds masses in accordance with those expected from coalesced binaries.

C1452-820 (IC 4499): A CCD CMD study by Sarajedini (AJ 105, 2172) includes photometry of 71 RR Lyrae stars, three newly identified. Reasons for the high frequency of these stars which comprise 68% of the HB stars are discussed. The study also turned up three probable variable blue stragglers.

C1516+022 (NGC 5904, M5): Cohen and Matthews (PASP 104, 1205) present visual CCD photometry for 8 RR Lyrae variables and IR photometry for four of these. The Baade-Wesselink analyses based on this photometry are given by Cohen (ApJ 400, 528). Fourier parameters for 13 RRc variables are given by CJS. Storm, Carney and colleagues (PASP 103, 1264) present BV CCD photometry for eleven RR Lyrae variables and for two of these, V8 and V28, radial velocity curves and K-band light curves are also given (PASP 104, 159 & 168). Photometry of two new RRc stars near the cluster center is given by Kravtsov (Soviet AJ Lett 17, 465).

C1620-264 (NGC 6121, M4): Liu and Janes (ApJ 360, 561) report on Baade-Wesselink analyses of four RR Lyrae stars. Yao presents CCD photometry of two low amplitude variables lying outside the instability strip (IBVS #3622 and IBVS #3846).

C1624-387 (NGC 6139): Photometry for 16 new variables, 10 within the tidal radius, is presented by Hazen (AJ 101, 170). Periods are given for 5 RR Lyraes.

C1629-129 (NGC 6171 M107): CCD photometry presented by Ferraro et al. (MN 252, 357) includes observations of 14 RR Lyrae stars. A study by Cudworth et al. (AJ 103, 1252) includes an investigation of cluster membership for 17 of the known RR Lyrae variables. CJS present Fourier parameters for six RRc stars.

C1645+476 (NGC 6229): Three possible new variables are identified by Carney et al. (AJ 101,

1699) as a result of their photometric study of the cluster.

C1715+432 (NGC 6341, M92): Cohen and Matthews (PASP 104, 1205) present visual CCD photometry for 5 RR Lyrae variables and IR photometry for two of these. Radial velocities and Baade-Wesselink analyses for V1 and V6 are given by Cohen (ApJ 400, 528). New proper motions given by Rees (AJ 103, 1573) confirm that almost all previously known or suspected variables are highly probable members of the cluster. His search for new variables revealed some possible candidates. Carney et al. (PASP 104, 44) discuss BV CCD photometry for seven RR Lyrae variables. For two of these, V1 and V3, Storm et al. (PASP 104, 159 & 168) present radial velocity curves and K-band light curves.

C1735-032 (NGC 6402, M14): Margon et al. (ApJ 369, L71) report HST U and B images of the field of the 1938 nova reveal no object with a marked UV excess.

C1742+031 (NGC 6426): Clement and Nemec (JRASC 84, 434) report V3 is a double-mode RR Lyrae star.

C1800-300 (NGC 6522): Abundances and radial velocities are given by Walker and Terndrup (ApJ 378, 119) for five RRc stars in this cluster, situated in Baade's window. Metallicities of the four variables which appear to be cluster members are close to that of the surrounding RR Lyraes in the nuclear bulge.

C1804-250 (NGC 6544): Hazen (AJ 106, Sept.1993) identifies a possible RR Lyrae star.

C1821-249 (NGC 6626, M28): Wehlau and Butterworth (AJ 100, 686) present photometry of the variables. A proper motion study by Rees and Cudworth (AJ 102, 152) includes four previously known variables and identifies four possible variables.

C1828-235 (NGC 6642): Hazen (AJ 105, 557) presents photometry for 18 variables within the tidal radius and 11 variables beyond it. Light curves are shown for 13 RR Lyrae stars, 10 of which fall within the tidal radius.

C1951+186 (NGC 6838, M71): Light curves based on deep CCD photometry by Hodder et al. (AJ 103, 460) are presented for four new variables, including one variable blue straggler and two candidate eclipsing binary systems.

C2059+160 (NGC 7006): Period change rates for 46 RR Lyrae variables given by Wehlau et al. (AJ 103, 1583) indicate a radial gradient in the rates.

C2127+119 (NGC 7078, M15): CJS give Fourier parameters for 24 RRc stars. Comparing theoretical models with observed data for 62 RR Lyrae stars, Simon (ApJ 387, 162) finds no single set of models can reproduce all the periods and luminosities and the ratio of blue to variable stars. Improved values for periods and masses of eight RRd stars are given by Clement and Walker (AJ 101, 1352). A CM study by Durrell and Harris (AJ 105, 1420) includes a comparison of the slope and zero point of the $M_{V,RR}$-metallicity relation derived by several methods. Meinunger (MittVerSt 12, 34, 58, 67 & 69) reports on observations of variables. HST observations of the inner region have discovered 19 new variables, including 15 probable RR Lyrae stars (Ferraro & Paresce AJ 106, 154).

C2137-234 (NGC 7099, M30): Observations presented by Machin et al. (MN 250, 602) indicate that the cataclysmic variable V4 is a not a cluster member.

Population II Variables in Magellanic Cloud Clusters and Local Group Dwarf Galaxies. Hazen and Nemec (AJ 104, 111; 105, 359) present photometry and light curves for 11 RR Lyrae stars within the tidal radius of the LMC cluster NGC 2210 and 52 field RR Lyraes. Walker, continuing his studies of variables in LMC clusters, presents light curves for RR Lyrae stars in NGC 1841 (AJ 100, 1532), Reticulum (AJ 103, 1166), NGC 1466 (AJ 104, 1395), and NGC 1835 (AJ 105, 527). Using his CCD photometry of 182 RR Lyraes in seven old LMC clusters he compares the magnitude-metallicity relation for these stars to that of galactic field RR Lyraes and finds the zero point for the LMC variables to be almost $0^{m}.3$ brighter (ApJ 390, L81). From the LMC relation he determines ages for galactic clusters and suggests an increase in the distance to the galactic center. Suntzeff et al. (AJ 104, 1743) review the properties of variables in old LMC clusters and find them to be very similar to those in outer galactic clusters, consistent with theories of the formation of the Galaxy from LMC-like clouds.

Observation of RR Lyrae stars in increasingly distant Local Group galaxies has become possible through the use of CCDs. Reid and Mould (AJ 101, 1299) report on R and I-band CCD photometry of stars in Leo I including observations of two variables, one of which seems to be a red variable instead of an anomalous Cepheid on the basis of its colour. Observations of variables in Leo I are also reported by Demers and Irwin (MN, in press) and M.G. Lee et al. (ASP CS 48, 1993; AJ, in press). For Leo II, 79 candidate HB variables and one possible anomalous Cepheid have been identified by

Demers and Irwin (MN 261, 657). Reports on studies in progress on the variables in Sculptor are given by Goldsmith (Victoria, 347) and in Sextans by Weller et al. (BAAS 877 (1991)) who report the discovery of approximately 40 RR Lyrae variables as well as possible anomalous Cepheids. Saha, Hoessel and colleagues present photometry and light curves for 42 variables in NGC 205 (AJ 103, 84) and 15 RR Lyraes in IC 1613 (AJ 104, 1072).

Red Variables, RRd and RRc Variables in Clusters. Perl and Tuchman (ApJ 360, 554) compare their theoretical mapping of the luminosity-period relation with the observed period-luminosity relation for red variables in 47 Tuc and suggest the dominant mode of pulsation is the second overtone for red semiregulars and the first overtone for Miras.

Clement and Walker (AJ 101, 1352) report that analyses of observations in three clusters found no RRd stars. Combining this result with an earlier search of 22 other clusters by Clement and Nemec (JRASC 84, 434) they suggest the occurrence of RRd stars is related to metallicity. Simon and Clement (ApJ 410, 526) match pulsation models with observations of cluster RRc stars to derive masses, luminosities and temperatures as functions of period and Fourier phase parameter and obtain a provisional RR Lyrae distance scale.

The Sandage Period-Shift Effect and Horizontal Branch Models. See also section 6. There is continuing controversy over the cause of the Oosterhoff dichotomy and the related Sandage period-shift effect, which appears to require an anticorrelation between helium abundance and metallicity. Still another aspect of this debate deals with the RR Lyrae luminosity-metallicity relation which directly impacts on use of these stars to yield information on distances, globular cluster ages and formation of the Galaxy. There are several papers dealing with these and related problems in the conference proceedings listed at the beginning of this report. A good review of the status of this controversy three years ago is given by Rood in ASP CS 11,11. Additional observational data has been published since then. A study of field RR Lyraes by Suntzeff et al. (ApJ 367, 528) finds an "Oosterhoff" gap and period-shift dependence on metallicity similar to that in the globular clusters. IR photometry of RR Lyrae stars in eight clusters is used by Longmore et al. (MN 247, 684) to establish a period-IR luminosity relationship and investigate the dependence of M_V upon metallicity. Recently published IR photometry of stars in M3, M5 and M15 is used by Fernley (AA 268, 591) to derive $M_{V,RR}$-metallicity and log P-metallicity relations. He compares methods of determining $M_{V,RR}$ and also finds the slope of the mass-metallicity relation from pulsation theory is in serious disagreement with that from HB theory. Nemec and Lutz (Victoria, 29) present new P-L-metallicity relations for four types of Pop II variables from observations of over 1200 stars in $\sim$40 stellar systems. Van den Bergh (MN 262, 588, AJ 105, 971) finds Oosterhoff type I clusters are significantly more likely to lie on retrograde orbits than type II clusters and suggests some second parameter effects are due to the way in which the protoGalaxy collapsed. Clement et al. (ApJ 395, 192) find, for RRc stars in 5 clusters, that the Fourier phase parameter ϕ_{31} is related to metallicity and period.

In an effort to understand the cause of these relationships, synthetic horizontal branches based on HB evolutionary tracks have been constructed by several groups and compared with observed correlations. Y.-W. Lee uses the synthetic HBs of Lee, Demarque and Zinn (LDZ) and his new HB morphology type in several investigations. The HB type, (B-R)/(B+V+R), (based on the numbers of blue HB, RR Lyrae and red HB stars) are listed for a number of clusters by Lee (ApJ 363, 159) who argues that the period shift arises from the fact that most of the RR Lyraes in Oosterhoff type II clusters are highly evolved stars from the blue side of the instability strip. Lee (ApJ 367, 524) finds agreement within the errors between observed secular period change rates of RR Lyrae stars with those predicted by the LDZ models. He suggests the mean luminosity of RR Lyraes in a cluster is a function of HB morphology and uses this to explain the lack of correlation between luminosity and metallicity of ω Cen RR Lyrae stars (ApJ 373, L43). See Simon (ApJ 360, 119) for a further comparison between observed parameters and predictions from HB models for RR Lyrae variables in this cluster. In several review papers presented at conferences Lee extends his discussion of cluster HB stars to derive information about cluster ages, galactic evolution and distances to other galaxies (ASPCS 13; Victoria). However a detailed study of the RR Lyrae stars in M15 by Simon (ApJ 387, 162) finds that the Lee and Demarque tracks cannot account for the observed properties of the variables in this cluster. The RRab/RRc transition in M15 is compared to that in ω Cen by Caputo (AA 239, 137).

Bencivenni et al. have also constructed synthethic HBs using enhanced abundances of O, Ne, Mg and Ca and have used them, along with the Lee HB type, to investigate the relations between metallicity and RR Lyrae mean luminosity and light-to-mass ratio as well as the age spread among clusters (ApJ 380, 484). Their results generally agree with those of LDZ. Caputo and De Santis (AJ 104, 253; ApJ 401, 260) investigate the correlations of light-to-mass ratio with metallicity and HB morphology for both RR Lyrae stars in the field and in clusters. They find agreement with the synthetic HB predictions, using a value of Y = 0.23 for all clusters. Carney et al. (ApJ 386, 663) compare the slope they obtain from Baade-Wesselink analyses and find it agrees within the errors with that obtained from other methods without requiring an anticorrelation between helium and heavy-element abundances. They extend their results to find there is a range in ages among intermediate-metallicity clusters and suggest the Oosterhoff dichotomy results from a sudden change in HB morphology at [Fe/H] = −1.7. Catelan (AA 261, 443 & 457; AJ 106, 175) discusses the use of several synthetic HB models in an attempt to reconcile the various values found for the slope of the HB luminosity-metallicity relation and to explain the cause of the period-shift effect.

8. Mira Variables, RCB Stars and Related Objects (M.W. Feast)

The application of interferometric techniques to obtain high resolution two-dimensional images of Miras opens a new and important phase in the study of these objects. *o* Ceti shows departures from spherical symmetry which may indicate non-radial pulsation modes co-existing with the radial modes which produce the photometric variability (cf. Karovska et al. ApJ 347, L54; Wilson et al. MN 257, 369; Haniff et al. AJ 103, 1662; Quirrenbach et al. AA 259, L19). Infrared (2 micron) diameters of *o* Ceti and R Leo have been published (from interferometry, Ridgway et al. AJ 104, 2224: and an occultaton, Di Giacomo et al. AAS 249, 397). A high-quality astrometric distance for R Leo (Gatewood PASP 104, 23) together with an interferometric diameter show that the star is pulsating in the first or higher overtone (Tuthill et al. preprint). Since R Leo fits the Mira PL relation closely (Whitelock, in IAU Symposium 153 in press) this would seem to establish that the Miras are not fundamental mode pulsators — a matter that has been under discussion for some time.

The galactic distribution and kinematics of Miras has much to tell us about galactic structure as well as about late stages of stellar evolution. A general overview of Miras in relation to stellar evolution can be gained from Highlights in Astronomy 9, 603-645. Whitelock (IAU Symposium 153 in press) reviews Miras in the Galactic Bulge and Whitelock and Catchpole (in The Center, Bulge and Disk of the Milky Way ed. Blitz 1992 (Kluwer)) show from Miras that the Bulge is bar-shaped. Jura et al. (ApJS 79, 105) discuss the space densities, lifetimes and mass loss rates of nearby Miras. Whitelock et al. (preprint) discuss the distribution and kinematics of high mass loss AGB stars (mainly Miras) in the South Galactic Cap. They conclude (contrary to one of the results of Jura et al.) that the period of a Mira is unlikely to undergo a large change during most of its life. The distribution and kinematics of OH/IR stars close to the Galactic Centre are correlated with envelope expansion velocity (Lundquist et al. AA 259, 118). Two high-velocity (about -350km/s) OH/IR stars have been found near the Centre (van Langevelde et al. AA 261, L17). The few OH/IR stars in the outer Galaxy are likely to be high mass objects (Blommaert et al. AA 267, 39). OH/IR stars in the Galactic Centre have been used to study the scattering properties of the interstellar medium (van Langevelde et al. ApJ 396, 686).

Interferometry at 11 microns shows that the dust in *o* Ceti reaches to within 3 stellar radii of the star (Bester et al. ApJ 367, L27). The acceleration of the circumstellar gas may begin at these small radii (Bowers ApJ 390, L27). Dust formation in Mira winds is discussed by Dorfi and Hofner (AA 248, 105; AA 265, 207) and Fleischer et al. (AA 266, 321). New candidates were found for C stars with silicate shells (Chan & Kwok ApJ 383, 837). Infrared emission and mass loss from circumstellar shells was modelled by Justtanont and Tielens (ApJ 389, 400).

The discovery of OH/IR variables in the Magellanic Clouds (Wood et al. ApJ 397, 552) is of great importance for the study of these objects. A number of large surveys of IRAS sources for OH and water masing have been published (Le Squeren et al. AA 254, 133; te Lintel Hekkert AA 248, 209; AAS 90, 327; Dickinson & Turner ApJS 75, 1323; Lewis & Engels MN 251, 391). The efficiency of OH masing increases with redness of the stellar envelope and probably also with decreasing distance from the Galactic Centre (Dickinson ApJ 379, L29). Le Bertre (AA 250, 351) suggests that all OH/IR stars with periods greater than 800 days are supergiants. A detailed study of the OH envelope of U Ori was made by Chapman et al. (MN 249, 227) and an ellipsoidal shell model proposed (Bowers ApJS 76, 1099); see also Collison and Fix (ApJ 390, 191). However Spaans and van Langeveldte (MN 258, 159) conclude that the average shape of OH lines in OH/IR stars can be explained by a spherically symmetric, homogeneous, constantly expanding shell and that the 'platelet' model of Alcock and Ross is not required. Studies of the circular polarizaton of OH emission suggest complex magnetic field structures with intensities of 1-100 microgauss (Zell & Fix ApJ 369, 506).

New, detailed maps of water masers around Miras and variable supergiants have been made by Bowers et al. (AJ 105, 284). In some cases there has been marked changes in structure over several years. The importance of the study of mm and submm water masers as probes of the physical state of the emitting gas was emphasised by Neufeld and Melnick (ApJ 368, 215); see also Menten and Melnick (ApJ 377, 647). Rapid time variations in astrophysical masers are considered by Scappaticci and Watson (ApJ 400, 351). A great deal of work has been done recently on maser and non-maser SiO emission in Miras. Sahai and Bieging (AJ 105, 595) made medium resolution images of non-maser SiO in Miras and found them to be largely spherically symmetric on angular scales of 3-9 arcsecs. Alcolea and Bujurrabal (AA 253, 475) made a systematic study of isotopic maser lines (see also Cernicharo et al. AA 249, L27; ApJ 401, L109) and new high excitation SiO maser lines have been detected in VY CMa (Cernicharo et al. ApJ 407, L33). McIntosh and Predmore (ApJ 404,L71) show that in *o* Ceti three different maser rotational transitions of SiO originate in the same volume. Detailed modelling by Lockett and Elitzur (ApJ 399, 704) shows that the primary pumping mechanism for SiO masers is collisional.

The carbon Mira IRS+10216 (CW Leo) continues to be intensively studied both for an understanding of the complex circumstellar chemistry it exhibits and for the detailed asymmetric shape of the molecular shell (cf. ApJ 388, L31; 388, L35; 396, 643; 407, L37; 382, 321; AA 259, L23; 266, 365; 249, 435; MN 254, 7). The extent of the molecular envelope in CO and SiO around several Miras is discussed by Bujarrabal and Alcoleo (AA 251, 536) and a fast bipolar outflow in CO found from the S-type variable π Gruis (Sahai AA 253, L33). The envelope of the interesting Mira OH231.8+4.2 contains both O-rich and C-rich molecules (including formaldehyde (Lindquist et al. AA 263, 183)). The time variations in the bipolar nebula (at 2 microns) around this star were studied by Kastner et al. (ApJ 398, 552). Brightness variations are out of phase in different parts of the nebula due to light travel time effects thus reducing the amplitude of the integrated light by about 2 magnitudes.

Pijpers (AA 267, 471) suggests that mass loss can significantly affect the pulsation periods of Miras and related stars whilst Icke et al. (AA 258, 341) find that some of the characteristics of these stars can be interpreted in terms of weak chaos. Two time scales in the variations of small amplitude red variables were found by Percy et al. (PASP 105, 287) and a number of periods (including 1190 days) identified in the SRc variable Y Lyn (Szatmary & Vinko MN 256, 321). Kerschbaum and Hron (AA 263, 97) discuss the relationship between Miras and SR variables. A detailed study (Goebel et al. ApJ 402, 680) shows that the results of the CNO tricycle dominate the photospheric abundances in the carbon star RX Psc. Wallerstein (PASP 104, 511) obtained the Rb/Zr ratio in the S-type Mira R And and deduced the neutron density at the time of s-processing. The effects of velocity stratification of a Mira photosphere on line strengths and curves of growth were evaluated by Scholz (AA 253, 203). Jonsson et al. (MN 258, 49p) identified the molecule TiS in S-type Miras whilst Hron (AA 252, 583) deduced a possible metallicity spread amongst short period Miras.

The relation between Miras and PNe was discussed by Whitelock and Feast (IAU Symposium 155 in press). Most old disc stars seem to go through both the Mira and PNe phases. However the most metal rich stars in the Galactic Bulge may go through neither phase. Schaefer (ApJ 366, L39) assembled evidence of 14 cases of flares (typically 0.5 magnitude, each lasting a few minutes) in Miras. In a sample of carbon Miras the optical radial velocities are consistently redshifted with respect to CO, possibly due to shock structure (Barnbaum ApJ 385, 694). VLA continuum observations show the existence of a hot plasma around carbon SR variables (Luttermoser & Brown ApJ 384, 634).

A large amount of work has been published on symbiotic stars, some of which contain Miras and it is impossible to review it all. Subarcsec structure in the R Aqr nebula has been studied from the ground and from space (Hege et al. ApJ 381, 543; Burgarella & Paresce ApJ 389, L29; Hollis et al. ApJ 386, 293) and a detailed binary model proposed (Burgarella et al. AA 262, 83). V1016 Cyg (already known to show bipolar structure) is surrounded by a 24 arcsec nebulosity (Bang et al. (MN 256, 59p). BI Cru is also found to have a bipolar nebula (Schwarz & Corradi AA 265, L37). A set of new near simultaneous UBVRI and JHKL observations of a large sample of symbiotics is given by Munari et al. (AAS 93, 383). Mm observations of some symbiotics suggest that free-free emission becomes optically thin at these wavelengths (Ivison et al. MN 257, 47) though this is apparently not the case for RX Pup (Seaquist & Taylor ApJ 387, 624). Analysis of emission line intensities in five symbiotics shows that helium abundances are similar to those in PN and correlated with N/O as expected from dredge-up theory. A discussion of the outburst of the symbiotic nova AS296 (Munari et al. AJ 104, 262; see also AA 257, 163; Gutierrez-Moreno et al. ApJ 395, 295) leads to the idea that such objects may be the progenitors of SNIa (see Munari & Renzini

ApJ 397, L87; but cf. Kenyon et al. ApJ 407, L81). The mechanism of jet formation discussed by Soker (ApJ 389, 628) may have relevance for symbiotics. Karovska et al. (ApJ 402, 311) provide an accurate stellar separation in the *o* Ceti system using speckle interferometry. The M giant components of symbiotics have similar IR colours to giants in the Galactic Bulge (Whitelock & Munari AA 255, 171). CH Cyg is in a triple system (Hinkle et al. AJ 105, 1074) and its pulsation period is increasing (Mikolajewski et al. AA 254, 127); see also Taranova and Yudin (AA 257, 615) for a possible obscuration phase of this star. H92α has been detected in the Mira symbiotic H1-36 (Bastian ApJ 387, L77). Murset et al. (AA 248, 458) find the hot component in symbiotic stars to be post-AGB whilst Schmid (AA 254, 224) presents models of Raman scattering in symbiotics.

Rao and Lambert (AJ 105, 1915) carried out detailed high resolution spectroscopy of the RCB star V854 Cen in deep minimum. Their results suggest a model in which the object is a bipolar nebula with a thick dust torus. It remains to be seen whether other RCB observations can be fitted into this picture. In minimum this variable shows polarization of the continuum but not the emission lines (Whitney et al. AJ 103, 1652) hence the latter are unobscured by (thin) dust . The UV spectrum seems unique (Clayton et al. ApJ 384, L19). Lawson (MN 258, 33p) suggests that the emission spectrum becomes obscured by optically-thick dust during decline. Evidence that the decline is linked to pulsation phase leads him (MN 256, 339) to the view that dust is formed close to the stellar surface (a view supported by Clayton et al. ApJ 397, 6527; see also Lawson in ASP Conf Series 30, 357). RCB light curve modelling has been carried out by Pugach (Sov AJ 67, 1280; 68, 122; 69, 1195) and a rather detailed physical model of dust formation and decline given by Goeres and Sedlmayer (AA 265, 216). Dick and Walker (AA 252, 701) claim that the declines of R CrB, RY Sgr and S Aps are random whereas that of SU Tau is not. Fernie (PASP 103, 1091) found R CrB at maximum with a period of 44 days in 1990. Lawson (MN 253, 625) discusses the best period at earlier times and concludes that declines are not related to periodicity in this star. The unusual hot RCB star DY Cen has H/He=0.1 (by numbers) and C/He=0.01 (Jeffery & Heber AA 270, 167).

9. RV Tauri Variables and Related Objects (J.R. Percy)

Whereas the pulsation of Cepheids can normally be defined by one or two periods, many yellow supergiant variables are less regular. These include RV Tau variables, and an assortment of stars which are usually classified as SRd variables. One subset of these are the UU Her stars - high-latitude yellow supergiants which continue to be puzzling and controversial as to their nature. While often superficially resembling normal population I stars, their high latitudes, unstable variability and spectral peculiarities at high resolution suggest that most are post-AGB stars. An extensive workshop encompassing the subject was held at the Center for Astrophysics, Cambridge USA in May 1992; the proceedings (Luminous High-Latitude Stars, ed. D.D. Sasselov, ASP Conference Series 45, 1993), hereinafter LHLS, contains useful reviews of UU Her and RV Tau variables, and related objects. There are also many interesting papers in "From Miras to Planetary Nebulae: Which Path for Stellar Evolution?" (ed. M. Mennessier and A. Omont; Editions Frontieres, 1990).

Light Curves. Long-term visual observations of R Sct have been published by Mattei et al. (AAVSO Monograph #3, Suppl. #1). Computerization of AAVSO visual observations (including RV Tauri stars) is now virtually complete. The French Association of Observers of Variable Stars (AFOEV) has also created a data bank of observations (Gunther Bull. AFOEV #60, 17). The long-term light curves of RVb stars (RV Tauri stars with long-term light variations) have been studied by Percy (LHLS 205). Long-term photoelectric photometry of bright RV Tauri stars has been obtained by Zsoldos (see references below).

Period Changes. Period changes in RV Tauri stars have been studied by Zsoldos (SS Gem: ApSS 181, 203; AC Her, R Sge, V Vul: AA 268, 149), Zsoldos and Kollath (TW Cam: ApSS 181, 251), Percy et al. (U Mon, R Sct: ApJ 375, 391; SS Gem, V Vul: JAAVSO, in press). There is a wide spectrum of behaviour, ranging from cyclic (AC Her) to abrupt (U Mon); it may be that the period changes are due to random cycle-to-cycle fluctuations as in the Mira stars (Percy et al. JAAVSO, in press). If these stars are truly post-AGB stars, some evolutionary period change should be detectable.

Spectroscopy. Gillet, Lebre and their collaborators (AA 251, 549; AA 246, 490; AA 237, 159) have used high-resolution spectroscopy to study the pulsational motions in the atmospheres of R Sct and AC Her. They were able to explain their results in terms of the propagation of a running wave, with two shock waves per (double) period. The main shock appears just after primary luminosity minimum, and produces strong Hα emission. A second weaker shock emerges from the photosphere at secondary luminosity minimum, when the main shock is disappearing in the high atmosphere. There is also some evidence in R Sct for a blue-shifted component in the sodium line, perhaps connected with the very high atmosphere, or with a circumstellar shell.

Abundances. Studies of the chemical abundances in RV Tauri stars (Wahlgren AJ 104, 1174; Baird BAAS 23, 944; Luck, in LHLS 87; Wahlgren, in LHLS 270; Bond, in Proc. IAU Symp. #145, 341) indicate that they are a rather heterogeneous group, with [Fe/H] ranging from near-solar to -1.7 or less, some with TiO absorption and/or Hα emission, at least at some phases.

Infrared Photometry. Infrared observations (Taranova & Torgovkina Astr Tsirk #1547, 11; Shenton et al. AA 262, 138) have revealed infrared excesses which can be explained in terms of post-AGB evolution. Shenton et al. postulate two dust shells — one, the AGB "fossil", the other, more recently formed. Both silicon carbide and silicates appear to be present, and the C12/C13 ratio appears to be consistent with post-AGB evolution.

Polarimetry. Several groups have obtained polarimetry of RV Tauri stars, including Polyakova (Sov AJ 36, 52) and especially Nook et al. (AJ 100, 2004), who conclude that, despite their extensive data, "we cannot determine the polarigenic mechanism responsible for the observed polarizations of these objects. The data do appear to rule out pure dust scattering and pure photospheric scattering as the polarigenic mechanisms. We suggest that the scattering of a time-variable anisotropic radiation field off an extended circumstellar dust layer could produce the magnitude of the variations that are observed".

Theory. Tuchman et al. (AA 271, 501) carried out a linear pulsation analysis of models of 0.5 to 0.8 $M_\odot$. The results were consistent with most observed features of RV Tauri stars. They suggested that the alternating deep and shallow minima in these stars are due to a resonance ($P_0 \sim 2 P_1$) between the fundamental (P_0) and first overtone (P_1) periods. Gillet (AA 259, 215), on the other hand, explained the alternating minima in AC Her and R Sct as being due to a radiative shock associated with the development of an extended running wave in the envelopes of these stars. Buchler, Kovacs, Moskalik and their collaborators (e.g. ApJ 366, 300; ApJ 385, 685; see also sections 5 and 11) have found evidence for period doubling and incipient chaos in models appropriate for RV Tauri stars. Attempts to find evidence for chaotic behaviour in these stars (Kollath MN 247, 377; Veldhuizen and Percy JAAVSO 18, 97; Saitou & Takeuti PASJ 41, 297) have concluded that the (photometric) data are not yet accurate or extensive enough for this purpose.

The above-mentioned mechanisms are not independent, and it is possible that all of them contribute to the alternating minima phenomenon — especially since the phenomenon is not always regular in any case.

10. Asteroseismology of Compact Pulsators (D.E. Winget)

We will consider pulsating white dwarf and hot pre-white dwarf stars collectively as the compact pulsators. This covers essentially the entire region below the main sequence from the pulsating planetary nebula nuclei (PNNV stars) down to the comparatively cool pulsating hydrogen surface layer white dwarf (DAV) stars; these stars span an incredible six orders of magnitude in luminosity, and more than a factor of ten in effective temperature.

Many articles of interest can be found in the proceedings of relevant meetings held during the period covered by this report. These include the NATO Advanced Research Workshop, "Seventh European Workshop on White Dwarfs", Toulouse, France, September 1990 (eds.: G. Vauclair, E. Sion, Kluwer, 1991; hereinafter referred to as WD); the IAU Coll #139, "New Perspectives on Stellar Pulsation & Pulsating Variable Stars", Victoria, Canada July 1992 (Ed. J. Nemec & J.M. Matthews, Cambridge University Press, in press, hereinafter Victoria; the NATO Advanced Research Workshop, "Eighth European Workshop on White Dwarfs", Leicester, England, July 1992 (ed.: M. Barstow, Kluwer, 1993; hereinafter WDAOT); Gong 1992: "Seismic Investigations of the Sun and Stars", Boulder, Colorado, August 1992 (ed.: T. Brown, ASP Conf. Ser., 42, 1993, hereinafter Gong) and the Second Whole Earth Telescope Conference, Vilnius, Lithuania, August 1993 (ed.: E. Meistas, Baltic Journal of Astronomy, in press; hereinafter WETCon); Recent progress has been reviewed by Bond and Ciardullo (WDAOT 491, hereinafter BC), Cox (Victoria), Wesemael et al.

(WD 159), Winget (WD 129; WDAOT 449; Gong 331); Brickhill (WDAOT 467); Bradley (Ph.D. Thesis, Univ. of Texas, 1993; hereinafter B93); Nather (WDAOT 499; WETCon), and Clemens (WETCon).

The primary goal of white dwarf asteroseismology is to determine basic structural and evolutionary parameters of the white dwarf stars. This will allow us to develop a chronology of star formation in our galaxy, and study the poorly understood phases of pre-white dwarf evolution. Considerable progress in this direction has been made since the last review in these pages (Winget, Trans. IAU, vol. XXIA, 267). Most significantly a number of controversial issues have been sorted out, some surprising discoveries made, and new puzzles uncovered. Given the space limitations at hand, we will focus our attentions on a sampling of these.

The count of PNNV stars has grown to 7. The PNNV stars have effective temperatures in excess of 100,000 K and surface gravities near log g = 6 (BC). All of these stars have complex, unresolved, and perhaps inherently unstable, dominant pulsation periods between 1,000 - 2,000 s. These stars are likely the progenitors of a significant fraction of the white dwarf stars and asteroseismological analysis of their interiors will be crucial to understanding both the post-main-sequence-planetary nebula connection and the planetary nebula-white dwarf connection (see discussion in BC). Of particular interest is their relationship to the DOV stars. In this regard the discovery of pulsations in the ROSAT soft x-ray source RX J2117.1+3412 (Motch et al. AA 268,561) reported by Watson (IAUC #5603) and Vauclair et al. (AA 267, L35) is extremely exciting. This object has pulsation periods intermediate between the PNNV and DOV stars and is embedded in a faint, extended planetary nebula. Asteroseismological analysis of its pulsation properties are currently underway (Vauclair, WETCon).

The DOV stars have effective temperatures between 100,000 K and 150,000 K and have luminosities between 100 and 1,000 times solar. The prototype of this class of stars, McGraw's star (PG1159-035, GW Vir), was the target of a Whole Earth Telescope (WET) run; the data and a preliminary asteroseismological analysis were presented by Winget et al. (ApJ 378, 326). The recent asteroseismological analysis of Bradley (B93) and Kawaler and Bradley (WDAOT, 459; preprint) give T_e = 141,000 K, surface gravity of log g = 7.4, mass = 0.59 $M_\odot$ and log L = 2.39 in solar units, in excellent agreement with the spectroscopically determined values reported by Warner and Heber (WDAOT, 303, and references therein). Matching the asteroseismologically and spectroscopically determined surface abundance ratios remains problematic. In a related way we note that the question of why some spectroscopically identical DO stars pulsate and others do not is still with us. Additionally, asteroseismologically determined parameters include the mass of the surface helium-rich layer which is between 0.003 and 0.006 $M_\odot$. The luminosity gives a distance of 500±40 pc. This value is consistent with that reported by Werner et al. (AA 244, 437) but with errors ten times smaller, pointing out the value of asteroseismologically determined distances.

Preliminary asteroseismological analyses of other DOV stars have also been reported. Fontaine et al. (ApJ 378, L49) report a mass of approximately 0.7 $M_\odot$ for PG1707+427 (Fontaine et al. ApJ 378, L49), although B93 reports a mass between 0.47 and 0.58 $M_\odot$ using a preliminary analysis of WET data. B93 also gives mass estimates for PG0122+200 near 0.8 $M_\odot$, and for PG2131+066 near 0.6 $M_\odot$. These values must all be regarded with some caution until more data is available; only the mass of McGraw's star is firm at this time.

The question of the location of the DBV instability region seems to be resolved in favour of the lower temperatures. The seven known DBV stars are found with luminosities of less than 0.1 solar and in the temperature range from 21,500 K to 24,000 K (Thejl et al. ApJ 370, 355).

We have achieved a real breakthrough in our understanding of these stars through the analysis of WET observations of the prototype GD358 (Nather, WETCon; Winget et al. IAU Coll. #137, 789: Winget et al. preprint). This work shows the dominant power is made up of ℓ = 1 multiplets of consecutive radial overtone number between 8 and 18. Based on comparison with the theoretical work of Bradley et al. (ApJ 406, 661) we derive a mean mass for GD358 of 0.61±0.03 $M_\odot$, and show that the mass of the surface He layer is log M < -5 in $M_\odot$ units. This value is much thinner than might naively be expected from stellar evolution calculations, and is even one to two orders of magnitude thinner than the values from the solution of the C pollution problem (Pelletier et al. ApJ 307, 242).

The surprises from this star don't end here. First, it is differentially rotating with a surface rotation period of 0.89 d for the helium envelope, and a core which rotates at least 1.8 times faster! Second, the fine structure frequency splitting is consistent with a magnetic field which is about B = 1300±300 G; this magnetic field is about two orders of magnitude smaller than the smallest magnetic field detectable from polarization work. A field of this magnitude may well be generated in the surface convection zone as suggested by Winget (WDAOT). The models of Van Horn et al.

(preprint) indeed indicate that a convection zone dynamo can generate magnetic fields of the observed magnitude, and even produce a dynamo cycle time on time scales of order months to years — in agreement with the observed time scale for amplitude changes in GD358. The results of Van Horn et al. suggest that magnetic fields may play a much larger role in the pulsation properties of the white dwarf stars than we had realized. Finally, the distance to GD358 derived from the asteroseismologically determined luminosity ($0.05 \pm 0.01\ L_{\odot}$) gives a distance of 42 ± 3 pc, in agreement with the observed parallax distance. This value confirms that the observed modes are $\ell = 1$ ($\ell = 2$ moves the distance almost twice as far and is completely inconsistent with the measured parallax), providing a fundamental calibration of asteroseismology of white dwarf stars.

This brings us to the coolest of the pulsating white dwarfs, the hydrogen atmosphere DAV stars. The twenty-two known DAV stars have temperatures between 11,000 K and 13,000 K (Wesemael et al. WD 159; Bergeron et al. ApJ 387, 288). In contrast to previous notions, the results of Dolez et al. (WD 361) and Kepler and Nelan (AJ 105, 608) suggest that there may be a small but significant number of nonvariable DA white dwarfs with temperatures inside the observed instability strip.

Kepler (WETCon) has measured the rate of period change in the dominant pulsation mode of the DAV G117-B15A. He finds $dP/dt = (3.2 \pm 2.8)e^{-15}$, some 4 times smaller than the previously reported value of Kepler et al. (ApJ 378, L45). This new value is completely consistent with the value expected for secular evolution in models with C/O cores (Bradley et al. ApJ 391, L33).

Asteroseismological determination of the surface hydrogen layer masses based on the distribution of pulsation frequencies has been carried out for several individual stars (Bergeron et al. AJ in press; Fontaine et al. WDAOT 479; Fontaine et al. ApJ 399, L91; B93). These authors find evidence that five DAV stars have hydrogen layer masses with $-4 < \log M > -7$. Vauclair et al. AA 264, 547) and B93 find three DAV stars with hydrogen layer masses $\log M << -7$.

These results, based on adiabataic theoretical calculations of eigenfrequencies, suggest that the pulsations occur in DA white dwarf stars over a range of hydrogen layer masses including relatively thick hydrogen layers, consistent with expectations based on the non-adiabatic calculations of Cox and collaborators (cf. Cox Victoria and references therein). This is in contrast to the results of Winget and collaborators (cf. Winget, IAU Symp. #123, 305 and references therein) who found that only thin hydrogen layers were pulsationally unstable. Recently, B93 and Bradley and Winget (preprint), have repeated the investigation of the dependence of the temperature of the blue edge on the surface hydrogen layer mass. They incorporated improved equilibrium models (see Wood ApJ 386, 539; and Wood PhD thesis Univ. Texas 1990) and better numerical techniques (Lee and Bradley preprint, and Brassard et al. in preparation) for solving the pulsation equations. They find that the hydrogen layer mass dramatically affects the growth rates of the modes, consistent with the previous results of Winget and collaborators, but not the overall stability of the mode, consistent with the results of Cox and collaborators. This latter result is also consistent with the conclusions of Brickhill (WDAOT 467 and references therein) and Brassard et al. (in preparation), as described in Fontaine et al. (WDAOT 479).

Thus the issue of the surface hydrogen layer masses and "thick" versus "thin" seems finally to be settled. In settling this mystery, however, we reopen the problem of how to understand significant numbers of nonvariables within the instability strip.

Some interesting progress has been made on the theory of the origin of the linear combination frequencies observed in many of the DAV stars. Brickhill (WDAOT 467) and Brassard et al. (WDAOT 485) have shown how the amplitude of the linear combination frequencies can be used to test models of how the pulsations originate (Brickhill), and independently determine the effective temperature (Brassard et al.).

Clemens (WETCon) has re-examined essentially all of the information we currently have on the DAV stars near the blue edge. He shows that all of these stars have surface hydrogen layer masses in the range from $-5 > \log M > -6$, a remarkably narrow range. His work also suggests that the longest period mode excited to large amplitude in each of these stars is a measure of the thermal time scale at the base of the partial ionization zone. This taken with the steady increase in observed amplitudes with decreasing effective temperature indicates that the amplitude limiting mechanism is saturation of the driving regions (the H partial ionization zone). If correct this solves one of the outstanding puzzles of the pulsating white dwarf stars.

11. Theory of Stellar Pulsation (Siobahn M. Morgan)

There were several publications of conference proceedings where papers on stellar pulsation theory were presented. Among the more prominent are the following: "Nonlinear Astrophysical Fluid

Dynamics," Gainesville, Florida, Oct. 1989, (eds. Buchler & Gottesmann, New York Academy of Sciences, 1990, [hereinafter referred to as Gainesville]), "The Numerical Modelling of Nonlinear Stellar Pulsations: Problems and Prospects," Les Arcs, France, Mar. 1989, (ed. Buchler, Kluwer Acad. Publ., 1990), "Physics of Luminous Blue Variables," Val Morin, Canada, Aug. 1988, (eds. Davidson, Moffat, & Lamers, Kluwer Acad. Publ., 1989), "Progress of Seismology of the Sun and Stars," Hakone, Japan, Dec. 1989, (eds. Osaki & Shibahashi, Springer-Verlag, 1990), in addition to the Bologna and Victoria meetings mentioned in section 1.

Other conferences which had several papers presented that dealt with stellar pulsation theory include the following: "White Dwarfs," Toulouse, France, Sept. 1990, (eds. Vauclair & Sion, Kluwer Acad. Publ., 1991, [hereinafter Toulouse]), "Challenges to Theories of the Structure of Moderate-Mass Stars," Santa Barbara, California, June 1990, eds. Gough & Toomre, Springer-Verlag, 1991, [hereinafter Santa Barbara]), "ESO Workshop on Rapid Variability of OB-Stars: Nature and Diagnostic Value," Garching bei München, Germany Oct. 1990, (ed. Baade, European Southern Observatory, 1991).

Chaos and Multimode Pulsations. Goupil et al. ("Applying Fractals in Astronomy," ed. Heck & Perdang, 1991) provide a thorough review of the theoretical and observational aspects of chaos in pulsating stars. Buchler (Gainesville 17) also reviews various aspects of chaos, including the causes of chaotic behaviour in stars and the occurrence of chaos in models that even include time-dependent convection.

Seya et al. (PASJ 42, 405) studied two-mode coupling using coupled van der Pol oscillators. They suggest that features such as double mode pulsation might be due to subharmonic phase-locking, rather than the superposition of two non-synchronized oscillations. In a similar study, Ishida and Takeuti (ApSS 178, 311) examine the criterion for double mode pulsations in oscillator model equations with two mode nonresonant coupling. Double mode pulsations in RR Lyrae models were examined by Kovács and Buchler (ApJ 404, 765). They find that double mode pulsations can be sustained in purely radiative models, mainly by reducing the artificial viscosity influence. Ishida (ApSS 168, 243) investigates the connection between three mode coupling and chaos. He finds that chaotic behaviour may result by modal coupling, even in classical Cepheids. In a followup paper, Ishida and Saitou (ApSS 182, 69) find other results of nonresonant three mode coupling, including second overtone pulsations and three mode behaviour.

Input Physics. The new opacity results from the OPAL calculations have been published in several papers including Iglesias et al. (ApJ 360, 221), Iglesias and Rogers (ApJ 371, L73), and Rogers and Iglesias (ApJS 79, 507). A significant number of pulsation studies have already made use of these updated opacity tables. The influence of diffusion on pulsation and vice-versa is reviewed by Guzik (Victoria 234).

The energies of adiabatic nonradial oscillations of nonrotating and uniformly rotating stars are investigated by Lee and Saio (ApJ 360, 590). They find that the oscillation energies of p- and g-modes of nonrotating stars are positive, as well as the g-modes of rotating stars. The effects of rotation on the pulsational characteristics of Cepheids were studied in hydrodynamic models by Kovács and Buchler (Bologna 226). They find that rotation causes the fundamental mode growth rate to increase. The effect of radiative transfer in a spherical, pulsating atmosphere is examined by Zalewski (PASJ 43, 345). Various models are produced, and the only major difference between the radiative transfer models and those with the diffusion approximation is in the amplitude of the flux variations. Further work by Zalewski (PASJ 44, 27) tested how the properties of the so-called strange modes are affected by the boundary conditions and radiative transfer. The effects of mass loss on the radial pulsations of stars is examined by Pijpers (AA 267, 471) who finds that the leakage of pulsation energy out from the star may help to stabilize the pulsation modes.

Numerical Methods. The various papers in the book "The Numerical Modelling of Nonlinear Stellar Pulsations" cover virtually every aspect of numerical modelling of stellar pulsations and is highly recommended. Dorfi and Feuchtinger (AA 249, 417) introduce a new adaptive grid method to compute nonlinear stellar pulsations. The main advantage of this method is the greater resolution obtained for regions with steep gradients. A similar method for the calculation of nonlinear radial oscillations in convective variable stars is presented by Gehmeyr (ApJ 399, 265 & 272). The method is applied to RR Lyrae models and the energy flow, light and velocity curve amplitudes, as well as the appearance of bumps in the light curves is examined. Li (AA 257, 133 & 145) introduces a linear oscillation theory for stars which includes the nonequilibrium effects of gas and radiation. Li uses this method to study the red edge of the Cepheid instability strip (Victoria 295). A modelling technique for calculating linear, nonadiabatic pulsations was introduced by Gautschy and Glatzel (MN 245, 154). They use the Riccati method to transform a linear first-order ordinary differential system describing a boundary eigenvalue problem into a numerically stable, nonlinear initial value problem. This technique is used to examine the origin of pulsations and strange modes in hydrogen deficient stars, and they find that the pulsations, as well as the strange modes, are due not to the κ mechanism, but rather to mode coupling. The Riccati method is also applied to RCrB stars (Glatzel, Santa Barbara 387), and is used in the calculation of nonradial, nonadiabatic oscillations in a helium star (Glatzel & Gautschy, MN 256, 209).

Massive Pulsating Stars. The new OPAL opacities have produced some rather significant results in the area of high mass pulsating stars. Several groups have modeled the pulsations of B stars using the new opacity results, including Cox et al. (ApJ 393, 272), Kiriakidis et al. (MN 255, 1P) and Moskalik and Dziembowski (AA 256, L5). They all find that the κ mechanism produces both radial and nonradial pulsations. One of the more thorough surveys of B star pulsations was done by Dziembowski and Pamyatnykh (MN 262, 204), who find the fundamental radial mode unstable in nearly all models, as well as a large number of nonradial modes. The pulsational characteristics of supergiant stars are studied by Schaller (Bologna 300 & 304). The effects of mass loss on the pulsations of Wolf-Rayet stars is noted. The radial pulsations of massive stars are examined by Glatzel et al. (MN 262, L7) and Glatzel and Kiriakidis (MN 262, 85). In both studies, they find that strange modes are the dominant instabilities, while the κ and ε mechanism instabilities are weaker and less prevalent. Stothers (ApJ 392, 706) uses the OPAL opacities to determine the maximum main sequence mass of a pulsationally stable star. For various compositions, the maximum mass is found to be significantly higher with the new opacities, up to nearly 150 $M_\odot$ for some compositions.

Long Period Variables (Miras, R CrB, AGB Stars). The pulsational characteristics of Mira stars is reviewed by Wood (Bologna 355), who discusses the problems of mode determination and the modelling of the atmosphere. Bowen (Gainesville 104) discusses models of Mira variables, which include the formation of shocks in the photosphere and dust grain formation in the cool outer layers of the atmosphere. Perl and Tuchman (ApJ 360, 554) constructed linear nonadiabatic models of red variables as observed in globular clusters. Initially the models are pulsationally unstable with the second overtone dominating, and would be observed as a semiregular red variable. As the luminosity is increased, the first overtone mode starts to dominate, and the models resemble Mira variables. The pulsations of hydrogen deficient stars were discussed by Saio (Bologna 557) who reviews the possible instability mechanisms.

Classical Pulsators (Cepheids, RR Lyrae Stars). Moskalik et al. (Gainesville 37) surveyed classical Cepheid models to investigate the relation between the fundamental and second overtone periods, as well as the shape of the velocity and light curves. The various Cepheid mass problems were discussed by Moskalik et al. (ApJ 385, 685) who used the new OPAL opacity tables. They find that the beat Cepheid mass discrepancy is no longer a problem, but the bump Cepheid mass differences are still present. The pulsational characteristics of BL Her stars are modeled by Buchler and Moskalik (ApJ 391, 736) and Moskalik and Buchler (ApJ 406, 190). They find the radial velocity Fourier parameters to be similar to the classical Cepheid models. The light Fourier coefficients do not follow the trends seen in the velocity coefficients, nor do they follow the coefficients found in classical Cepheid models.

The application of the new OPAL opacities to RR Lyrae stars was a very popular topic, particularly in trying to remove the mass discrepancy between the evolutionary and pulsational masses of RRd stars. Cox (ApJ 381, L71) finds better agreement between the pulsation and evolutionary masses for both Oosterhoff groups using a modified Stellingwerf opacity fit. Kovács et al. (AA 252, L27) used cubic spline interpolated values from the opacity tables, and their masses are slightly higher than those found by Cox for both Oosterhoff groups. In a followup, Kovács et al. (AA 259, L46) found that the double mode masses are influenced by variations in the heavy metal abundance as well as variations in the metal mixture.

White Dwarfs and Central Stars of Planetary Nebulae. A review of the theoretical work concerning the location of the ZZ Ceti instability strip was presented by Wesemael et al. (Toulouse 159). In a series of papers by Brassard et al. (ApJ 367, 601; ApJS 80, 369 & 725; ApJS 81, 747), the properties of DA white dwarfs are examined. They introduce a new method for solving the nonradial pulsation equations and perform a very thorough survey of the pulsational characteristics of the stars, including the effects of hydrogen and helium layer mass, convection, composition variations, mass and effective temperature on the pulsations. Stanghellini et al. (ApJ 383, 766) use post-asymptotic giant branch stellar models, with four different surface compositions to determine the location of the instability strips. The instability strips obtained correspond to the observed locations of such objects. Cox (Victoria 103) discusses possible pulsation mechanisms in DAV and DBV pulsators. It appears that with the effects of time-dependent convection included, only a thin CO convective layer can produce pulsations and stars with thick surface layers of hydrogen and helium will not pulsate. Bradley et al. (ApJ 391, L33) model the maximum rates of period change expected for ZZ Ceti stars with carbon and oxygen core compositions. The largest values for the rates of period change are found near the blue edge of the DAV instability strip.

Neutron Stars and Strange Stars. Models of neutron stars with a solid crust or core are examined by Finn (MN 245, 82). The crust or core leads to the appearance of nonradial elastic modes that are coupled to the star's g-modes. The pulsational characteristics of strange stars, which may be very similar to neutron stars, are examined by Benvenuto and Horvath (MN 250, 679). The damping mechanisms that can exist in these stars makes the detection of any pulsations very doubtful, though they could develop due to transient events. The radial oscillations of both neutron stars and strange stars are studied by Väth and Chanmugam (AA 260, 250). They find that the periods of all modes of strange stars go to zero as the central density approaches its smallest possible value.

12. Flare Stars (C.J. Butler)

Meetings. The field of active star research has continued to prosper in recent years with an increasing number of colloquia on the topic. The following list gives details of venues, dates and publications: Sixth Cambridge Workshop on Cool Stars, Stellar Systems and the Sun, Seattle, USA, September 1989 - hearafter 6CW; 'The Dynamic Sun' Debrecen, Hungary, May 1990, (ed. Dezsö, Pub. Debrecen Heliophysical Observatory Vol 7); 'The Sun and Cool Stars: Activity, Magnetism and Dynamos', IAU Coll. 130, Helsinki, Finland, July 1990, (ed. Tuominen et al., Springer-Verlag) - hereinafter AMD; 'Surface Inhomogeneities on Late-type Stars' Armagh, N. Ireland, July 1990, (ed. Byrne and Mullan, Springer-Verlag) - hereinafter SILTS; Flare 22 Workshop - 'Dynamics of Solar Flares' Chantilly, France October 1990. 'Eruptive Solar Flares', IAU Coll. 133, Buenos Aires, Argentina, August 1991, (ed. Jackson et al. Springer-Verlag) - hereafter ESF; Seventh Cambridge Workshop on Cool Stars, Stellar Systems and the Sun, Tucson, USA, October 1991, (Ed. Giampapa and Bookbinder, ASP Conf. Ser.); 'Physics of Solar and Stellar Coronae', Palermo, Italy, June 1992, (ed. Linsky and Serio, Kluwer); IAU Colloquium No 157, 'The Cosmic Dynamo', Potsdam, Germany, September 1992, (ed. Krause et al., Kluwer); Flare 22 Workshop - 'Recent Observations and Theories of Solar Flares', Ottawa, Canada, May 1993.

Reviews. Several excellent reviews have appeared including: a compendium edited by Petersen (Mem. del. Soc. Astron. Ital. 62, 211 - hereinafter MSAI) which contains articles on both observational and theoretical studies of stellar flares; a review by Haisch et al. (ARAA 29, 275); two by Pallavicini (NATO Adv. Study Inst. 'Sun - a Laboratory for Astrophysics', 509 and IAU Symp. 142, 'Basic Plasma Processes in the Sun', 77); and lastly one by Chugainov (Ap 34, 138) which reviews magnetic fields on stars and their effects.

Flare Stars in Stellar Aggregates. Our knowledge of the evolutionary status of FS rests largely on the study of those stellar aggregates that contain them. Mirzoyan (Ap Invest. 6, 71 and 'Phys. and Evol. of Stars - Star Clust. and Assoc.' hereinafter PESCA, 51) and Mirzoyan et al. (Ap 31, 567), from a comparison of flare frequency in late type dwarfs in the Orion Association, the Pleiades and the solar neighbourhood, have shown that the proportions of FS amongst main sequence dwarfs increases with decreasing luminosity and that the magnitude of the brightest FS in a group is indicative of age. Solar neighbourhood FS are found to be relatively old and are believed to have formed in stellar systems that have long since disintegrated.

Several studies of cluster FS have appeared including: a catalogue of Orion FS by Natsvlishvili (Astrofizika 34, 243), a reexamination of old Schmidt plate material of Orion by Chavira and Parsamyan (Rev. Mex. A. Ap. 22, 15) in which 38 new FS were discovered, and spectroscopic studies which examine the TiO bands and Hα respectively of Pleiades FS by Mirzoyan et al. (Ap 33, 291), Hamilton and Stauffer (AJ 105, 1855) and Szécsényi-Nagy (PESCA 55).

The relative energy involved in slow and fast flares (greater or less than 30m duration) has been investigated from U-band observations of Pleiades FS by Kelemen (Ap Invest. 6, 103) and shown to be approximately equal. Parsamyan and Oganyan (Ap 33, 438) find that flare amplitudes depend on the reciprocal of the rate of flare build-up and suggest that the more powerful flares lie deeper in the atmosphere.

Evolutionary studies of FS require surveys of a broader selection of young and not-so-young clusters which contain FS than are currently available. Searches for new FS in the southern open clusters: IC 2062, IC 2391, NGC 2516 and M7 have been undertaken by Jones and Page (Proc. A. Soc. Austr. 9, 277 and Southern Stars 34, 39) and a survey of the η Tau region was made by Szécsényi-Nagy (ApSS 170, 63).

Flare Stars in the Solar Neighbourhood. A comprehensive review of the properties of solar neighbourhood FS, by Pettersen (MSAI 62, 217), discusses the importance of duplicity, rotation and age to their activity levels. This review also incorporates an updated list of nearby FS. A substantial catalogue, by Panagi and Mathioudakis (AAS 100, 343), lists basic data for approximately 600 K and M dwarfs, including FS. In addition to V, (B-V), (R-I), radius etc., Hα, CaII, MgII and quiescent X-ray fluxes are given. A subset of these data has been analysed by Mathioudakis and Doyle (AA 262, 523) in a discussion of the relative importance of acoustic and magnetic heating in the quiescent atmospheres of dK-M stars. Weis (AJ 105, 1962) has compiled UBVRI photometry for 688 K-M dwarfs.

Observations by Rumsey (Southern Stars 34, 9) and Allen (Southern Stars 34, 15) suggest that all late M type stars are variable. Though the observational evidence at present is sparse, the appearance of flares on stars later than M5 which are believed to be fully convective, has significant repercussions for our understanding of how magnetic fields are generated in these stars.

X-Ray Observations. A review by Linsky (MSAI 62, 307) lists soft X-ray observations of stellar flares by Einstein, EXOSAT and some earlier satellites. It also provides references to earlier reviews.

EXOSAT and Einstein data on the double flare star Gliese 867 have been analysed by Pollock et al. (SILTS 331 and AA 241, 451) who propose that a previously suspected rotational modulation of the X-ray flux was due to flares.

Einstein observations of late type stars have been re-analysed by Schmitt et al. (ApJ 365, 704) who find that M dwarfs show evidence for the presence of material at both $T > 10^7$ K and $T \sim 3.10^6$ K. Schmitt and Snowden (ApJ 361, 207) estimate that late type stars contribute approximately 40% of the X-ray background at low galactic latitudes.

A correlation established earlier, between the integrated soft X-ray and Balmer emission from flares, has been extended by Butler (AA 272, 507) to include solar data (see also Haisch, AA 219, 317) and a flare on II Peg. A common relationship is found for flares on the Sun, dMe and RS CVn stars. A similar correlation between X-ray and Hγ flux for *quiescent* dM stars has been found by Mathioudakis and Doyle (AA 240, 357) which tends to confirm the suggestion that the quiescent X-ray emission of dMe stars arises from low level flaring.

Ultraviolet Observations. In spite of the fact that IUE observations have continued almost uninterrupted, over the past decade and a half, relatively few flares have been well observed in the optical and ultraviolet simultaneously. An exception is the giant flare seen on AD Leo in April 1985 by Hawley and Pettersen (ApJ 378, 725). Much of the difficulty arises from the poor time resolution possible with IUE and the problems associated with capturing a spasmodic event with satellite and ground-based telescopes. Though HST will help to overcome the first restriction, its relatively poor duty cycle and the difficutly of coordination with ground-based facilities means that we are unlikely to see a sudden improvement in the situation. The provision of optical detectors on board UV/X-ray satellites seems to be the best long-term solution.

From IUE observations, Phillips et al. (ApJ 385, 731) have concluded that the UV continuum of stellar flares arises from recombination of SiII which has been produced by ionization by strong UV lines. The same mechanism is believed to occur on the Sun and RS CVn stars. Elgaroy et al. (AA 234, 308) find the widths of the MgII h and k lines in M dwarfs to be narrower than expected from the Wilson-Bappu relationship for more luminous stars. They suggest that the optical thickness of the lines may depend on activity.

Mathioudakis et al. (AA 244, 155) reported a large UV flare on Gliese 182 which was simultaneously observed by the VLA. They estimate total radiative losses from $4 \leq \log T_e \leq 8$ of 6.10^{34} ergs. Byrne and Doyle (AA 238, 221), from IUE spectra of two dM and one dM(e) star, conclude that the transition region fluxes in these stars are more than an order of magnitude weaker than dMe stars and are more typical of solar values. Quin et al. (AA 272, 477) have confirmed the lack of any distinctive rotational modulation in ultraviolet (IUE) lines for the single dMe star AU Mic. It is becoming increasingly evident that large filling factors and frequent flaring obscure any inherent rotational modulation in chromospheric/transition region lines on the very active M type dwarfs.

Barstow et al. (Nature 353, 635) have detected a flare on BY Dra, with ROSAT, which has a substantial fraction of its total radiative energy in the EUV range. Doyle (AA 214, 258) and Butler (AA 272, 507) have estimated the EUV flux in quiescent dM stars and solar flares, as respectively 3.5 and 5 times the soft X-ray flux and have pointed to its possible role in exciting Balmer emission. For stellar studies, much of the EUV emission remains beyond the limits of current detectors.

From ROSAT observations reported by Jeffries and Bromage (MN 260, 132), Gliese 841A has been identified as a new short period, active, binary with a visual binary white dwarf companion.

Optical Observations. Optical photometry and spectroscopy of flares have been reviewed by Pettersen (MSAI 62, 217) and Butler (MSAI 62, 243).

Photometric monitoring, previously a tedious occupation, has received a new lease on life with the advent of automatic photoelectric telescopes (APTs). Several monitoring datasets have appeared, notably by Boyd et al. (IAPPP Comm. 42, 44), Ichimura (Pub Nat Obs Japan 1, 317), Cutispoto (AAS 84, 397), Mahmoud (ApSS 186, 109 & 113), Byrne et al. (AA 236, 455), Houdebine (IBVS #3631), Honeycutt et al. (PASP 104, 1039) and Liu Yanying et al. (Acta A Sinica 12, 42).

Two rather unusual papers describe how the complex structure of flare light curves may give constraints on the topology of magnetic fields (Houdebine IBVS #3643) and on the presence of nearby stellar or planetary bodies (Bromley PASP 104, 1049).

Several authors have pointed to the importance of time-resolved optical spectroscopy for our understanding of the detailed behaviour of stellar flares. Two recent studies have advanced our knowledge in this field. Firstly, Hawley (MSAI 62, 271) and Hawley and Pettersen (ApJ 378, 725) have described simultaneous IUE and optical spectroscopy of an exceptionally large flare ($\sim 10^{34}$ ergs) on AD Leo for which they have modelled the chromospheric and photospheric emission with considerable success. Secondly, Houdebine (Irish AJ 20, 213) has given a comprehensive description of spectroscopic signatures of stellar flares in the optical in which he has modelled a number of flares with a 'dense kernel' model atmosphere. He reports that the model can reproduce all the observed spectral features including broad band optical colours, Balmer and CaII line profiles etc. Cooling curves for several stellar flares were derived by Houdebine et al. (AAS 87, 33) which provided evidence that Balmer and CaII line fluxes were largely influenced by electron temperature. A gas-dynamic model was applied by Katsova et al. (MN 250, 402) to optical spectroscopic and soft X-ray observations of a flare on YZ CMi. Katsova (Sov AJ 34, 614) has discussed the use of Balmer decrements in the study of the physical conditions in the lower chromosphere and photosphere of active stars. Evidence for electron densities in the optically emitting regions $\geq 10^{14}$ cm^{-3} have been found in several of the above studies.

A 5^m (U) flare on UV Ceti was observed spectroscopically by Eason et al. (AJ 104, 1161) for which the total optical flare energy was $\sim 3\%$ of the quiescent bolometric luminosity. Line profiles showed evidence of moving chromospheric material with a turbulent velocity ~ 150 km/s and a systematic velocity of 70 km/s. The authors propose that Hα flare emission may arise from a relatively large volume rather than a compact high pressure region. Much higher bulk velocities (~ 5800 km/s) were observed by Houdebine et al. (AA 238, 249) in the Balmer lines from a 2.1^m flare on AD Leo. They point to the significance of such large mass ejections for stellar evolution. In a further paper, Houdebine et al. (AA 274, 245) report weak P Cygni profiles in the CaII lines, prior to the flare onset, which they explain in terms of a dark solar-like filament. During the decline of the flare a 2.68-minute oscillation developed in the position of the line centroids.

High resolution ($\Delta\lambda/\lambda \sim 8.10^4$) spectra of the CaII H line in late type dwarfs, by García López et al. (AA 262, 195), have shown evidence for chromospheric vertical velocity fields which persist for several years. Other time-resolved optical spectroscopic studies of FS were made by Falchi et al. (Ap Let Com 28, 15), McMillan and Herbst (AJ 101, 1788), Doyle et al. (AAS 86, 403) and Newmark et al. (AJ 100, 560). The latter detected some correlation between photometric and spectroscopic modulation which they interpreted as rotational modulation of plages associated with cool starspots.

Radio Observations. Three reviews of radio emission from FS have appeared: Stewart (ESF 301), Bookbinder (MSAI 62, 321) and Bastian (Sol Phys 130, 265). Bookbinder has stressed the relative paucity of radio observations of stellar flares and the importance of radio coverage in multiwavelength campaigns.

VLBI observations of YZ CMi by Benz and Alef (AA 252, L19) gave a radio diameter 1 ± 0.5 mas which strongly suggests emission by non-thermal particles. Lim (ApJ 405, L33) observed a steady, intense, radio emission from the rapidly rotating dM4e star Rositter 137B. Two flares were also observed.

Doyle and Mathioudakis (AA 241, L41) found weak quiescent emission, at the 3σ level, in millimetre radiation from two dM stars, in agreement with their finding of excess infrared emission believed to be associated with circumstellar matter (Mathioudakis & Doyle, AA 244, 433).

A combined photometric, spectroscopic and radio campaign on EV Lac was reported by Gershberg et al. (Sov AJ 35, 269). No microwave emission at 8 mm was detected from the 50 flares seen optically.

Peridicities, Oscillations and Microflares. Since the discovery of the 150^d periodicity in the occurrence of solar flares several authors have analysed stellar flare data gathered over recent decades for periodicities. No significant periodicities, in the range 1-14 days, were found in 228 flares on YZ CMi, AD Leo and EV Lac by Ishida et al. (ApSS 182, 227). On the other hand Cardona et al. (Rev Mex A Ap 22, 213) found two periods of 3060 days and 23.8 days in data for HII 2411. Kiang (Irish AJ 19, 161) discussed the probability that four flares observed on YY Gem occurred regularly spaced in time. For 22 years of data on UV Ceti, Pettersen et al. (6CW, 177) found 'definite' variations in flare frequency with time. Reviews of stellar activity cycles were published by Dravins (IUE Astron. in the Era of New Space Missions, 61) and Hall (NATO Adv. Stud. Inst. - Active Close Binaries, 95). Progress in this field is important for our knowledge of whether solar-type activity cycles are common in other active stars.

Andrews (AA 245, 219; AA 239, 235 and AA 235, 264) and Mullan et al. (ApJ 391, 265) have applied Fourier and autocorrelation techniques to search for low amplitude oscillations in dM-dMe stars. In some cases they report statistically significant variability with periods of tens of seconds to a few minutes. It is suggested that the oscillations originate in resonating coronal loops.

The existence of stellar microflares in optical and X-ray data is still a subject of discussion. Reviews by Haisch et al. Butler and by Bookbinder, previously referenced, discuss the evidence for optical, X-ray and radio microflares and their possible importance for coronal heating. Pustil'nik et al. (Int. Workshop on Reconnection in Space Plasma 179), Zhilyaev (Kinematic Phys. Celestial Bodies 7, 61) and Pustil'nik (AMD 504) have subjected rapid photometric observations to exhaustive statistical tests and conclude that, in general, stellar flares are not detected on time scales of less than a few tenths of a second. This is consistent with a thermal origin for the optical continuum emission.

Models of Stellar Flares. The successful development of computer codes for the interpretation of spectroscopic, photometric and X-ray data on flares has been one of the most productive areas of research over the past few years. A model by Hawley and Fisher (ApJS 78, 565; ApJS 81, 885; ApJ 357, 243) which included the effects of chromospheric evaporation and heating by coronal X-rays gave a good fit to observed transition region fluxes. However it predicted a continuum flux redder than that observed. Hawley and Fisher propose that the observed continuum is formed by photospheric reprocessing of the EUV and UV emission originating in the chromosphere.

The gas-dynamic model for a flare kernel, developed by Katsova and Livshits (Sov AJ 35, 65), has been used to describe the high time resolution observations of the CIV (1550Å) line by the ASTRON satellite and the rapid 1^s microflares seen by the 6m telescope. Houdebine (Irish AJ 20, 213) finds that a 'flare kernel' model is more successful in predicting the observed time profiles in optical emission lines etc. than soft X-ray backwarming or radiative pumping.

The X-ray emission from stellar flares is, on the solar model, believed to originate in giant magnetically confined loops. Cheng and Pallavicini (ApJ 381, 234) have modelled the process for dMe stars with good agreement of X-ray intensities, time profiles and coronal temperature with observations. The application of scaling laws to such models has been discussed by Reale et al. (AA 272, 486).

The primary source of the energy for solar and stellar flares is widely believed to lie in the disruption and reconnection of magnetic fields. However, this view is not universally held. Two papers on alternative schemes appeared: firstly a convective model by Grandpierre (MSAI 62, 401) and secondly a linear Z-pinch mechanism by Airapetyan and Nikogosyan (Ap 30, 329) and Airapetyan et al. (Ap 32, 230) in which rapid heating of the plasma follows a sudden compression by the magnetic field similar to that which occurs in 'Tokamak' machines.

The transport of the released energy down to the chromosphere and upper photosphere, where it is believed to be responsible for the optical emission, is also a controversial topic. Fast electrons are commonly considered to be responsible, however Mullan (Irish AJ 361, 215) and Grinin (MSAI 62, 389) have pointed to problems with this interpretation. Grinin and Sobolev (Ap 31, 729) have shown how the dependence of flare continuum colours on flare amplitude can be predicted by a proton heating theory. Direct evidence for proton beams during the impulsive phase of a flare on AT Mic, observed by the HRS on the HST, has been given by Woodgate et al. (ApJ 397, L95).

Flares on Non-Red-Dwarf Stars. Flare activity has, until recently, been largely the preserve of late type dwarfs and their close associates in the HR diagram. Evidence is now accumulating that magnetic activity is much more widespread than once thought; it has been observed (in one spectral range or another) in main sequence stars from B to M type and over a luminosity range from supergiants to dwarfs. Thus neither low temperature nor low luminosity are pre-requisites for flare activity. Flares have been recently reported on the following non-red-dwarf types: Miras (Shaefer, ApJ 366, L39), Low Mass X-ray Binaries (Kuijpers, NATO Adv. St. Inst. Active Close Binaries, 761), Algol systems (Stern et al. 6CW, 224) an FK Comae star (Cutispoto et al. AA 263, L3), an A type star (Wang IBVS #3863), a Be star λ Eri (Smith et al. ApJ 409, L49) and RS CVn stars (Stern et al. 6CW 227; Doyle et al. AA 262, 533; AAS 96, 351; MN 248, 503; Mathioudakis et al. MN 255, 48; Tagliaferri AA 251, 161; Henry and Hall, ApJ 373, L9; Huenemoerder et al. 6CW, 236). Though binarity is common in this list, it too is not a pre-requisite condition (witness FK Comae types and λ Eri). Other than dMe stars, the largest group of stars with well established flare activity are the RS CVn binaries on which reviews have recently been written by Neff (MSAI 62, 291), Kuijpers (ref above) and Ramsey (6CW, 195).

The enormous energies in some non-red-dwarf stellar flares ($\sim 10^{39}$ ergs on HD 32918, by Cutispoto) continues to stretch the solar analogy to its limits. The volume of plasma required to radiate the total energy observed in some RS CVn flares is comparable with the volumes of the systems themselves, suggestive of enormous magnetic loops linking the binary components. In this context the VLBI observations of a flare on HR 1099 by Triglio et al. (MN 260, 903) are important in that they confirm the involvement of a region with a linear size equal to the separation of the binary components.

Coordination of Multiwavelength Observations. The value of stellar flare observations is considerably increased if they are made over a broad spectral range and in particular if they include X-ray, UV, optical spectroscopy and radio coverage. Whilst the coordination of the various space and ground-based facilities poses many problems, it can and has been done, with dramatically successful results. A network of interested ground-based observers can be contacted via an email newsletter operated by the author. Those who would like further information are invited to contact CJB@STAR.ARM.AC.UK by Internet or the author at Armagh Observatory, BT61 9DG, N. Ireland by ordinary mail.

13. Variability of T Tauri Stars (L. Hartmann)

In the last year, great strides have been made in understanding the variability of T Tauri stars. Many pre-main sequence stars have large starspots on their surfaces, and these spots cause the apparent flux of the star to vary as the spot rotates into and out of view (Rydgren & Vrba ApJ 267, 191; Rydgren et al. AJ 89, 1015; Vrba et al. AJ 97, 483). Progress in understanding the light curves has been slowed by the difficulty of obtaining photometry extending over the typical one-week rotation periods. However, extensive campaigns of photometric monitoring have been mounted (Attridge & Herbst ApJ 398, L61; Edwards et al. AJ, in press), using telescopes at different sites (Bouvier et al. AA 272, 176). This work has led to a much better understanding of T Tauri variability. Most importantly, the determination of accurate rotational perods for T Tauri stars has led to a surprising result, with important implications for understanding the early angular momentum evolution of solar-type stars.

The extensive study of pre-main sequence stars in the Taurus-Auriga molecular cloud by Bouvier et al. (AA 272, 176), building on earlier work (Rydgren & Vrba; Rydgren et al.; Vrba et al.; ref. cit.), has led to the following picture of spot properties. First, many T Tauri stars exhibit spots. The clearest, most stable light curves are exhibited by the so-called "weak-emission" T Tauri stars or WTTS. Simple models suggest that the spots on WTTS are generally cooler than that of the surrounding photosphere. The "classical" or strong-emission T Tauri stars (CTTS), on the other hand, show much less clear photometric periods in general, and the optical colours suggest either hot or cool spots. In either case, the spot areas must be quite extensive, covering tens of percent of the photospheres of these stars.

These properties can be explained nicely with the standard model of T Tauri stars, in which the strong emission of the CTTS is powered by accretion from a circumstellar disk (Liseau et al. AA 183, 274; Bertout et al. AA 330, 350) while the WTTS show only chromospheric emission from solar-type activity, since they have no accretion disk (at least close to the star) (Walter et al. AJ 96, 297; Cabrit et al. ApJ 354, 687; Edwards et al. AJ, in press). In this model the WTTS show large spots due to enhanced solar-type magnetic activity. The CTTS also have spots, but the magnetic fields channel the accretion to the stellar surface, where it can create a hot spot for sufficiently rapid mass flow. Hence, depending upon the accretion rate, the spot may appear hot or cool. Moreover, it is likely that accretion is non-steady (Bouvier et al. AA 272, 176), which provides a natural explanation of the aperiodicity seen in some CTTS light curves.

The most interesting results concern the rotational periods. Attridge and Herbst (ApJ 398, L61) showed that the stars in the Orion Nebula Cluster (Trapezium cluster) exhibited a bimodal distribution in their photometric periods. Attridge and Herbst speculated that some sort of magnetic braking might produce this bimodal distribution. Edwards et al. (AJ, in press) followed up this work, identifying the spectral types of the stars observed by Attridge and Herbst. Combining these with previous results for stars in Taurus, the results showed that low-mass T Tauri stars (~ 0.8-$0.5 M_{\odot}$) rotate more slowly if they have infrared excesses thought to signify the presence of an inner accretion disk, while diskless stars were faster rotators. This was true even though the stars with and without disks had similar ages and masses. Simultaneously, Bouvier et al. (AA 272, 176) found similar results from their extensive monitoring of Taurus pre-main sequence stars, using a distinct but not completely independent sample. Bouvier et al. found that K7-M1 CTTS (stars with accretion disks) have rotational periods of 7.6 ± 2.1 days while the WTTS (weak-emission, stars without inner disks) have rotational periods of 4.1 ± 1.7 days. Thus, it appears that the presence of a circumstellar disk is strongly correlated with slower rotation in T Tauri stars.

This is an exciting result because it addresses a long-standing problem in the early evolution of solar-type stars. Current thinking suggests that T Tauri stars accrete substantial amounts of mass from circumstellar disks (Lynden-Bell & Pringle MN 168, 603; Bertout et al. AA 330, 350). This material contains substantial angular momentum, and would very likely spin up the central star to velocities close to breakup in the absence of any other angular momentum transfer (Hartmann et al. ApJ 309, 275). Yet most T Tauri stars rotate at $\sim 10\%$ of breakup or less (Bouvier et al. AA 165, 110; Hartmann et al. ApJ 309, 275; Bouvier AJ 99, 946). One standard explanation of this result is that a solar-type, magnetically-coupled stellar wind could carry away the needed angular momentum. However, this is difficult to explain in terms of the observed rotational velocities of young cluster stars, which suggest that they spun up as they contracted to the main sequence with very little, if any, angular momentum loss. In other words, the wind mechanism has to be very efficient, spinning down the T Tauri star as it rapidly acquires angular momentum on a typical time scale of 10^6 yr or less; but the subsequent stellar wind must be very inefficient in carrying away angular momentum during the $\sim 5 \times 10^7$ yr contraction to the main sequence.

Königl (ApJ 370, L39) (see also Camenzind (Rev. Mod. Ast. 3, 234) and Cameron & Campbell (AA, in press) proposed that T Tauri stars have extensive magnetic fields which couple to their circumstellar disks. In this situation it is possible for the star to transfer angular momentum to the disk, in a manner similar to that originally suggested for pulsars. This suggestion can explain the early spin evolution of solar-type stars. In the earliest phases, the disk-magnetosphere coupling prevents the star from spinning up to breakup as it accretes. Later on, when the disk dissipates or is all accreted, there is nothing to prevent the star from spinning up as it contracts to the main sequence, except for the stellar wind, which slows down the star on much longer time scales. The new rotational period determinations seem to provide some important support for a model in which the existence of a circumstellar disk provides a sink for stellar angular momentum via magnetospheric coupling.

Photometry and spectroscopy of the T Tauri star RY Lup by Gahm, Liseau, and collaborators (AA 183, 274; AA 211, 115; AA, in press) has suggested light variability due to variable extinction with a period of 3.75 days, comparable to the expected rotation period for this G8 star given the measured v sin i. It is conceivable that an asymmetric stellar magnetic field, rotating with the star, could disrupt the disk asymmetrically and produce variable disk extinction; more work should be done to test this possibility. Alternatively, if magnetic fields channel accretion as suggested by Bertout et al. (AA 330, 350) for DR Tau, then the light curves could be quite complicated, with a mixture of underlying dark, cool stellar photospheric spots along with bright, hot spots at the base of the accretion column. The resulting effects could be quite complex, and have not been thoroughly explored.

If spindown is accomplished by coupling the disk to the star by the stellar magnetic field, then the disk must be disrupted at several stellar radii. In turn, this implies that the accretion onto the star is through a magnetosphere, and not through the disk. The observations of ultraviolet excess emission must then be interpreted in terms of accretion columns rather than hot spots. Moreover, if the accreting gas is sufficiently hot, it might produce observable emission lines. In fact, Calvet and Hartmann (ApJ 386, 239) have proposed that the permitted emission lines of many T Tauri stars arise in infalling matter channeled by the stellar magnetic field, because infall models are able to explain many observed line profiles much better than wind models. In this picture most of the emission arises in infalling material in the magnetosphere, while the wind components arise in a much more tenuous flow further out. Further tests of this picture, with improved observational surveys and radiative transfer solutions, are forthcoming. For example, periodic behaviour in the wind of SU Aur has been found by Giampapa et al. (ApJ, in press). Observations of T Tauri line profiles, especially in conjuction with simultaneous photometry, may provide strong evidence for magnetospheric infall, and hence support the magnetosphere-disk interaction model for the spindown of young solar-type stars.

Commission 28: Galaxies

PRESIDENT: E. Ye. Khachikian
VICE-PRESIDENT: V. Trimble
ORGANIZING COMMITTEE: F. Bertola, G. Bruzual, J.-S. Chen, R. S. Ellis, K. C. Freeman, J. S. Gallagher, F. Israel, S. Okamura, G. A. Tammann, and M.-H. Ulrich

This report covers the period 1 July 1990 to 30 June 1993. It has, as usual, been prepared by the commission officers and committee and chairs of the Working Groups. The usual problems of covering a broad subject accurately and fairly have been compounded by the present very difficult situation in the President's home country of Armenia. In particular, the section on Active Galaxies, which he had hoped to prepare, is missing; and we apologize to colleagues whose work has gone unrecognized in the circumstances. References are given in abbreviated form, with no initials or years of publication, and with a symbol + replacing the usual "et al."

1. Highlights since Buenos Aires
(V. Trimble)

Many discoveries during the triennium were exciting and do not require extended discussion for their appreciation. This is a subset.

The Local Group has a new member, a dwarf spheroidal in the direction of Tucana (Lavery & Mighell AJ 103, 81) but has lost And IV which is really just an open cluster or association within M31 (Joseph AJ 105, 932). Van den Bergh (MNRAS 255, 29p) had never actually admitted it to the canonical list. Meanwhile, the LC dwarf irregular IC L613 has revealed its population of old stars in the form RR Lyrae variables (Saha + AJ 104, 1068). And the nuclear radio source in M31, corresponding to our own Sgr A*, has finally turned up, though the radio remnant of SN 1885A remains among the missing (Crane + ApJ 390, L9).

Gas in galaxies can take many forms. We note here only the existence of a complete bibliography of CO (Verter PASP 102, 1281), a fine review of the issue of gas in ellipticals (Volkov Astrofiz 32, 80), and the fact that it is possible to remove HI from a spiral while retaining the molecular gas, resulting, for instance, in the "anemic spirals" of Virgo (Valluri & Jog ApJ 357, 367; Cayatte + AJ 100, 604).

The gassiest galaxy is arguable IRAS 10214+4724 ($z = 2.286$) with $10^{11}\ h^{-1}$ solar masses of H_2 (Downes + ApJ 398, L25). It may also have the highest star $M_\odot$ formation rate (Tsuboi & Nakai PASJ 44, L241), but could also be a superposition of two separate objects along the line of sight, one of them a dusty qso (Soifer + ApJ 399, L55).

Valentijn's 1990 proposal (Nature 346, 153) that spiral disks are so opaque that we miss enough light to remove the need for dark matter, at least in individual galaxies, continued to provoke puzzlement. The disk of the Milky Way near us is quite transparent — the model used for the HIPPARCOS input catalogue has $A_V = 0.1$ mag above $b = 60°$ (Arenou + A&A 258, 104) and A_B is only 0.16 magnitude to Andromeda (Jablonka + A&A 260, 97). But many spirals seem to have very opaque central bulges (Huizinga & van Albada MNRAS 254, 677; Cunow MNRAS 258, 251). And while the space-averaged disks may be fairly transparent (e.g. $A_B = 0.46$ mag full thickness, Peletier & Willner, AJ 103, 1761), the fact that both luminosity and dust are concentrated in spiral arms may mean that effective extinction may be more like a whole magnitude (White & Keel Nature 359, 129). But the real puzzle is that the absorbed visible light must come back out as infrared emission (Phillips + MNRAS 253, 496) and most spiral galaxies are not more luminous

J. Bergeron (ed.), Reports on Astronomy, Vol. XXIIA, 277–303.

as IR sources than as optical ones.

Triggered star formation seems to be directly responsible for the high brightness of interacting galaxies. For instance, 60% of the brightest 10^{12} $L_{\odot}$) IRAS galaxies, but only 10% of the faintest ($L < 10^{10}$ $L_{\odot}$) appear to be interacting on ESO/SERC and Palomar sky survey images (Zou + MNRAS 252, 593) and pr. comm. from the same group. A sample of non-Seyfert Markarian galaxies also includes a large excess of close pairs (Keel & van Soest A&AS 94, 553). Radio emission from interacting galaxies is another symptom of vigorous star formation (Batuski + AJ 103, 1077). For IRAS interacting pairs, both galaxies typically show the high L_{FIR}/L_B ratio and infrared colors that are indicative of rapid star formation (Lutz A&A 259, 462). The same applies to other interacting pairs (Sekiguchi & Wolstencroft MNRAS 255, 581). The implication is that the detailed geometry of the encounter is not critical to triggering, though the morphology of the pair NGC 7714/7715 suggests that off-center encounters cause only core, not ring, star bursts. Other interesting specific cases include (a) the SMC whose youngest stars are mostly on the side toward the LMC (Gardiner & Hatzidimitriou MNRAS, 257, 195), (b) an interacting pair in the Bootes void with one starburst and one Seyfert 1 galaxy (Weistrop + ApJ 396, L23), (c) star formation caused by cannibalism of Cen A (Thomson MNRAS, 257, 689), and (d) two companions of M81, with about the same amount of gas, but very different star-formation rates (Thronson + MNRAS 252, 547; Keel & van Soest, A&A 94, 553).

Galaxy formation is an unsolved problem elegantly reviewed by Peebles and Silk (Nature 346, 233). It may be a gradual process (Larson PASP 102, 709) or an explosive one (Yoshioka & Ikeuchi ApJ 360, 352). One is perhaps a bit surprised at just how little difference the measured anisotropy of the 3K radiation has made to the discussion. The data base already included large scale structure (superclusters and voids) with internal velocity dispersions of a few hundred km/sec, coherent streaming over regions of 100 h^{-1} Mpc or so with somewhat large speeds, and the existence of qso's at redshifts of nearly 5. Even before COBE, none of the simple models did a very good job of dealing with the full range (Efstathiou + MNRAS 258, p. 1). At most one can now say that a few additional combinations are ruled out (Blanchard + A&A 267, 11; Dodelson & Jubas PRL 70, 2224; Kashlinsky ApJ 399, L1; Cen & Ostriker ApJ 399, L113). Strings are in (Bennett + ApJ 399, L5), and textures are out (Holman + PRL 69, 1489). Strangely, and owing to marginal contradictions between the COBE fluctuations and stringiest upper limits to ones on smaller angular scales as probed by ground-based data, a cosmic background of gravitational radiation is also "in" (Davis + PRL 69, 1856; Lucchin + ApJ 401, L49; Crittenden + PRL 71, 324).

Primordial galaxies, in the sense of large, bright high-redshift (pre-qso) objects undergoing their first episode of star formation continue to elude detection in both redshifted Lyman α and redshifted 21 cm (Lowenthal + ApJ 357, 3; Subramanyan & Swarup A&A 11, 221; Cowie + ApJ 360, L1). Suggestions of new search strategies persist (Carr + ApJ 367, 420). It is also true that for a significant subset of the galaxy formation mechanisms referenced in the previous paragraph, no such entities ever arise.

The youngest galaxies are perhaps those dwarf irregulars just now forming from gas torn loose in close encounters of larger spirals (Amram + A&A 266, 106; Henkel + A&A 273, L15; Elmegreen + ApJ 412, 90).

The lowest metallicity galaxy has long been the dwarf irregular I Zw 18. Its metallicity of [Fe/H] = -1.6 (Campbell ApJ 362, 100) is at least rivaled by that of SBS 0335-052, a couple of Byurakan galaxies, and UGC 4483 (Terlevich). Talk at 21st IAU GA; Izutov + A&A 247, 303; Skillman PASP 103, 919). A regression of He/H on O/H or O/N in these low-metallicity galaxies leads to a new, lower primordial helium abundance, $Y_p = 0.217 - 0.233$, which is awkwardly smaller than expected from big bang nucleosynthesis in a homogenous universe (Matthews + ApJ 403, 65).

Dynamos of the self-excited α–ω type have been the standard model for the origin of galactic magnetic fields for some time (Beck + IAU Symp. 140). Doubt has been cast on this conventional wisdom on both observational and theoretical groups. Wolfe + (ApJ 388, 17) find that galaxies at redshifts as large as $z = 2$ have fields about as strong as the Milky Way, meaning either that the field is largely primordial, or that a dynamo operated very rapidly. On the theoretical side, recent discussions suggest (a) that the most probable dynamo field geometry is not that seen most often in real galaxies (Elstner + ApJS 94, 587), (b) that most of the power quickly cascades down to small spatial scales, leaving no significant large-scale field (Kulsrud + ApJ 396, 606), while observations indicate roughly equal power on large and small scales (Jones + ApJ 389, 602), and (c) that galactic dynamos saturate at fields much smaller than the observed ones (Vainshtein & Cattaneo ApJ 393, 165). The alternative is a primordial field of sufficient strength that the observed ones can be produced form it by moderate amounts of winding and amplification. Possible sources for the necessary seed fields are vorticity in the early universe (Rebhan ApJ 392, 385) and inflation (Ratra ApJ 391, L1). I would probably find these more persuasive if I understood them better. Effects of star formation (Ko ApJ 360, 151) or an active galaxy phase (Daly & Loeb ApJ 364, 451) may be relevant.

Faint blue galaxies make up most of the fuzzy objects in the sky at $B_J = 24$, but not by $B_J = 22$ (Colless + MNRAS 253, 686). They typically have redshifts around unity (Cowie + Nature 354, 460; Tyson + ApJ 349, L1), and the shear numbers are rather mystifying (Ellis & Frenk Nature 346, 790). Only three sorts of answers are possible to the question, where have they gone? First, they have ceased to be blue and are now other kinds of faint galaxies. Second, they have ceased to be faint (by merging into the larger galaxies now seen). Or, third, they have faded completely out of local samples. All three have been proposed. Mergers of FBGs to the present population are favored by Guideroni and Rocca-Volerange (A&A 252, 449) on the basis of comparisons of blue and infrared counts and also by Broadhurst + (Nature 355, 54) who predict that the mergers should be accompanied by high star-formation rates at moderate redshift. In contrast, Lacey and Silk (ApJ 381, 14) propose that the modern forms will be stripped, low-surface-brightness galaxies in dense clusters and gas-rich galaxies with "retarded" star formation in the field. Babul and Rees (MNRAS 255, 346) propose nearly the opposite. Following a first star burst near $z = 1$, the FBGs in low-density environments expel their gas and so fade away, while those in clusters with high ambient gas pressure are unable to get rid of their gas. Thus they continue star formation somewhat longer, appearing now as the very extensive population of dwarf ellipticals characteristic of rich clusters (Binggeli & Cameron, A&A 252, 27). If Giraud (A&A 257, 501) is right that there are three physically distinct kinds of objects among the FBGs, then three different end points is not implausible. For one example, we apparently know the answer. An FBG on the line of sight to qso Pks 0454+0356 shows evidence of recent star formation, but no longer contains enough gas to impose an absorption line on the qso spectrum. It will, therefore, rapidly fade away (Steidel + ApJ 413, L77).

The largest-scale structures and deviations from smooth Hubble flow in the universe continue to be a topic of debate, but it is clear that homogeneity has not yet been reached at 100 h^{-1} Mpc (Broadhurst + Nature 343, 726; Picard AJ 102, 445; Bahcall ApJ 376, 43; Huchra + ApJ 365, 43). A candidate for the largest-scale streaming is the structure connected with the so-called Great Attractor. On beyond the local, Virgo, supercluster, recognized structures include major visible concentrations of galaxies in the Perseus-Pisces and Hydra-Centaurus regions, in roughly opposite directions in the sky, at distances of 3-5000 km/s. The surprise was that the Hydra-Centaurus galaxies were moving in the same direction as we are, implying a mass concentration (the GA) on the far side of them, but not as distant as the Shapley concentration on beyond at 14,000 km/s. It was expected that enlarged samples in several surveys would eventually all show the same structures with roughly the same degree of prominance. This has not happened. Some surveys have continued to show a concentration of galaxies in the vicinity of the GA (Scharf + MNRAS 256, 209; Strauss + ApJ 385, 421) or velocity streaming toward it from our side (Han ApJ 395, 75; Rauzy + A&A 256, 1). Others have found indeterminate results (Mould + ApJ 383, 467; Bothun + ApJ 388, 253). And some have found the GA unnecessary (Landy & Szalay ApJ 391, 494; Tully + ApJS 80, 479) even for the Seven Samurai's own data (Weigelt & Kates A&A 252, 1). Yet other investigations have concluded that the GA is actually forbidden, in the sense that galaxies in that part of the sky are not clumped at the right redshifts (Visvanathan & van den Bergh AJ 103, 1057) and that galaxies on the far side of the GA are not falling back into it toward us (Mathewson + ApJ 389, L5). The alternative is a bulk flow at about 600 km/s extending over a region 130 h^{-1} Mpc long, reaching from the GA across us and on to the Perseus-Pisces region, and perhaps beyond (Han & Mould ApJ 396, 453).

ROSAT has probably revealed a new class of X-ray clusters. Most of the well-known X-ray emitting clusters of galaxies are large and rich. The first of the new ones is a small group centered around NGC 2300 (Mulchaey + ApJ 404, 19). Surprises included the very large total mass, the low metallicity of the gas (about 6% of solar), and the very extended gas distribution. A Hickson compact group (HCG62) has turned up as the second such cluster (Ponman & Betrand Nature 363, 50). Two implications are that the cluster will merge (presumably to a giant elliptical galaxy with an enormous X-ray halo) in about 3 x 10^9 yr and that we would be in trouble with the nucleosynthesis limit to cosmic baryon abundance if all galaxies and clusters had as high a ratio of baryonic to total mass as this one does. Additional groups of this sort are in the unpublished data base (Hasinger IAU Symp. 161).

The triennium has seen detection of unexpectedly large numbers of active galaxies in two wavelength regions. First, owing to patchiness of the local interstellar medium, a few AGNs are part of the data base in the extreme ultraviolet from ROSAT (Pound + MNRAS 260, 77) and nine from EUVE, starting with Pks 2155-304 (Marshall + ApJ 414, L53). Of greater physical significance is the near-double-dozen of AGNs seen by EGRET on the Compton GRO. All show some sort of Blazar behavior (rapid variability, strong variable polarization, weak emission lines, Dermer & Schlickeiser Science 257, 1642). Virtually all models (e.g. Friedlander + AIP Conf Proc 280, Compton Gamma-Ray Observatory and the expected corresponding volume in AIP Conf series from the 1993 symposium) involve relativistic jets and strong beaming of the gamma rays, closely related to the strong beaming that makes the sources appear as Blazars in the first place. The closest source, Mkn 421 at z = 0.0308 has also been seen at TeV energies (Weekes + IAUC 5522). It is also the only one whose photons of such high energy can be expected to get through the sea of intergalactic infrared photons without colliding and being waylaid en route.

Alignments of the (presumably non-thermal) radio emission and (presumably stellar, thermal) optical emission of active and other galaxies at large redshifts was initially a considerable surprise. It is quite wide spread (McCarthy + AJ 100, 1014), and can include infrared emission (Chambers + ApJ 363, 21) and both line and continuous light (Eales ApJ 409, 578). The phenomenon is largely restricted to sources with the highest radio powers (Dunlop & Peacock MNRAS 263, 936). No less than five mechanisms have been proposed to explain such alignments (Daly ApJ 399, 426), of which the most straightforward are radio jet triggering of star formation and Thompson scattering of AGN core light by electrons in the radio jets. Optical synchrotron or inverse Compton emission in the jets themselves could also contribute, and I am pretty sure I don't understand the fifth one. Phenomena in nearby objects that seem to be analogous indicate that at least three of these do indeed occur. McNamara and O'Connell (AJ 105, 417) have found two cooling flow clusters of galaxies whose central galaxies have blue-lobe-shaped optical emission associated with the radio blobs. These argue for triggered star formation. The nearby Seyfert 1, Mkn 509, on the other hand, has its radio continuum, polarized H alpha, narrow line region, and optical and IR polarized continuum all more or less aligned. The combination strongly suggest that both scattering and synchrotron emission contribute to polarization at the same position angle (Singh & Westergaard A&A 264, 489).

QSO absorption lines have probably now been studied in enough different was that the territory has been fully surveyed (though many details remain to be painted in). Broad absorption lines with absorption redshift close to the emission one are produced in material closely associated with the qso itself, and the main dispute is whether the covering factor is small (Korista + ApJ 401, 529) or large (ApJ 413, 95). There are some data indicating that the BAL population is physically distinct (Boroson & Meyers ApJ 397, 442); Francis + AJ 106, 417. Metallic line and damped Lyman alpha lines are both produced in the gas in normal galaxies, the former in turbulent halo gas, the latter (along with 21 cm lines) in quiescent disk gas (Lu + ApJS 84, 1; Wolfe + ApJ 404, 480). The number of cases where the absorbing galaxy is seen independently continue to rise (Bowen & Blades ApJ 403, L55; Bahcall + ApJ 398, 495; Spinrad + AJ 106, 1; Moller & Warrent A&A 270, 43). And HST observations of absorption produced in our own Milky Way show that, on average, seen from far away, it would look very much like the qso systems (Savage + ApJ 404, 124 and 413, 116). The Lyman alpha forest comes from much smaller, less dense, lower-metallicity clouds. A major surprise was how many HST revealed at small redshift (Brandt + AJ 105, 831; Bahcall + ApJ 397, 68; Bahcall ApJ 405, 491; Bruhweiler + ApJ 409, 129), given how rapidly they "evolve" at higher redshifts (Frye + MNRAS 263, 575; Press + ApJ 414, 41). A continuing dispute concerns whether the clouds are confined by gravitational forces of "mini-halos" or by the pressure of a hotter surrounding intergalactic medium (Petitjean + MNRAS 262, 499; Miralda-Escude & Rees MNRAS 260, 617; Murakami & Ikeuchi ApJ 409, 42). In principle, both could happen, and the low redshift systems be different again (Charlton + ApJ 402, 493). Finally, continuous Gunn-Peterson absorption blueward of Lyman alpha in the qso rest frame has still not been seen (Giallongo + ApJ 398, L9). For further details on this topic, see Trimble and Leonard (PASP 106, 1).

The diffuse X-ray background was long the subject of debate on whether the primary emitter was diffuse intergalactic gas or a very large number of sources. Absence of distortion of the 3K spectrum as recorded by COBE ruled out hot gas (Rogers & Field ApJ 366, 22), focussing discussion on whether the sources were active galaxies of the sort seen around us or something else. When only a few X-ray spectra were known, it was difficult to tell whether they were typical or not (and if they were, the sum would not add up to the background). Ginga data (Morisawa + A&A 236, 299) now indicate that this is not a problem, because spectra harden above 10 keV, and statistical fluctuations across the sky at these energies (Martin-Mirones + ApJ 379, 507) also suggest resolution into sources with luminosities typical of AGNs. At 1-2 keV, the issue has been fully settled by resolving essentially all the background into individual sources in those fields where ROSAT has surveyed deepest (Hasinger + A&A 275, 1; Hasinger IAU Symp. 161). At least one model (Zdziarski + ApJ 414, L77) makes it seem reasonable that AGNs should add up to the background at all energies from 2 to 100 keV, and the same claim has been made for both gamma rays (Stechker + ApJ 410, L71) and the extreme UV (Denda & Ikeuchi PASJ 45, L1). Many other workers have claimed the opposite.

With Enrico Fermi, we find it difficult to make predictions, especially about the future. Nevertheless, it seems reasonable to hope that the next triennium will see (a) clarification of the issues connected with cooling flows in rich clusters (how much mass; where does it go) owing to the rich data base coming from ROSAT and ASCA, (b) a cleaner case of microlensing in a quasar than the classic objects 0957+561 (Falco + MNRAS 251, 698) and 2237+030 (Wambsganss & Paczynski AJ 102,

864), permitting at long last the promised independent measurement of H_o. (c) A good description of the correlations between properties of populations of globular clusters and their parent galaxies, (d) a statistically significant set of normal galaxies with and without central black holes, and (e) a clearer impression of the extent to which the luminosity (or mass) function of galaxies is universal.

2. Astronomy in the Former USSR
(A. Zasov)

In spite of the highly unfavorable conditions for astronomical researches in the former USSR caused by severe economical crisis and disintegration of the country, extragalactic investigations still have been developed in their basic directions. Here we consider four fields of research: 1. Stellar populations and interstellar medium. 2. Structure and dynamics of galaxies. 3. Active nuclei and QSO's. 4. Systems of galaxies.

I. STELLAR POPULATION AND INTERSTELLAR MEDIUM

Sharov, Lyutyi and Ikonnikova have published their new results of UBV-photometry of 22 globular clusters of Andromeda nebula (Let in Russ AJ 18, 99). Star formation efficiency (SFE) in this galaxy was the subject of the research work of Kolotovkina (Let in Russ AJ 18, 760). She compared SFE in the arms near the major axis of the galaxy and in the interarm space using HII regions count. She concluded that there is no increasing of SFE in the arms with the exception of the arm S4, distinguished by its large pitch angle. The distribution of blue stars in M31 was investigated in relation to spiral arms and HII regions of the galaxy by Efremov, Ivanov, Nedialkov (AAp Trans 3, 237). They found that only in the S4 arm there is a stellar age gradient across the arm. The most probable value of the radius of corotation in M31 is about 15 kpc and certainly larger than 10 kpc. Sharov presented new estimates of Nova rates in the disk of M31, its companions M32 and NGC 205 and in M33 on the basis of the date obtained in Crimea and Latvia in 1967-1990 years. The new data confirm a strong difference between the specific Nova rates in the disk and in the bulge of M31. A low specific Nova rate was found for M33 (< 1.0 per yr per 10^{10} of solar lum.) (Let in Russ AJ 19, 287). The results of new spectral observations of inner parts of 12 galaxies of E-Sb types were presented by Sil'chenko, Afanasiev and Vlasyuk (Russ AJ 69, 1121). For 7 out of the 12 galaxies a metallicity jump was found between nucleus and bulge. The nuclei are found to be more metal-rich by 0.6-1.3 dex. Jumps appear to be related to nuclei, which are dynamically decoupled, being distinguished by their high angular velocity. A catalog of 147 stellar associations and aggregates in S-galaxy NGC 628 was presented by Invanov, Popravko, Efremov, Tikhonov, and Karachentsev (AAp Sup 96, 645). Catalog gives positions of associations, their size, integrated magnitudes and HII identifications. Authors point out that stellar associations form groups with a length scale about 600 pc (star complexes).

Several papers were devoted to dwarf galaxies. Distances to three nearby dwarf galaxies (Ho I and II, UGCA 105) were determined from photometry of their brightest red and blue stars by Tichonov, Karachentsev, Bilkina and Sharina (AAph Trans 1, 269). An essential discrepancy was revealed between the distance obtained for Ho I ($m - M = 29.11$) and previous estimates. The distance they found exclude belonging of this galaxy to M 81 + NGC 2403 + IC 342 group. The record metal-poor BCD galaxy SBS 0335-052 was found by Izotov, Lipovetsky, Guseva and Stepanian (Let in Russ AJ 16, 609). The oxygen and neon abundance in this galaxy are shown to be 1.3 and 1.9 times lower than in the well-known metal-poor dwarf galaxy I Zw 18. Piluigin (AApC 260, 58) show that the basic observed peculiarities of N-O abundances of dwarf Irr-galaxies can be well reproduced within the framework of closed box model of chemical evolution if to admit that the giant HII-regions in these galaxies are self-enriched. The results of spectrophotometric studies of 9 giant extragalactic HII regions in dwarf galaxies of M 81 group were presented by

Izotov, Guseva, I. D. Karachentsev, V. E. Karachentseva (AAph Trans 1, 283). The heavy elements abundance they found are 3 - 10 times lower than solar value and are quite typical for blue compact dwarf galaxies. Kritsuk (AAp 261, 78) presented model of evolution of the hot X-ray gas in halo of E-galaxies considering it as an open system with mass and energy sources determined by the stellar mass losses, SN and radiative gas cooling. Numerical simulations showed that in a reasonable domain of the parameters the stable periodic solutions exist, which describe cyclic process of gas accumulation and heating due to the stellar sources with subsequent gas cooling and dropping out of the halo. The evolution scheme of disk galaxies within the framework of a gas turbulence energy balance between the SN input energy and the turbulence collisional dissipation was proposed by Firmani and Tutukov (AAp 264, 7). The IMF, angular momentum and the gas accretion rate determine the evolution of the models. As a result, authors predict the present gas content and star formation history of a galaxy. Sokoloff, Shukurov, and Krause (AAph 264, 396) used two-frequency polarization observations of S-galaxies to get the general pattern of the magnetic disk field in IC 342 and M 81. Regular field (several microgauss) appears to be axisymmetric for IC 342 and bisymmetric for M81. The flow of interstellar gas in the gravitational field of galactic density waves has been investigated by Mishurov (AAp Trans 3, 317) in the framework of the diffuse approximation. The galactic density waves are shown to stimulate the formation of molecules of hydrogen. The proposed theory explains the existence of two displaced spiral systems, discovered in M 51 by comparing CO, dust, HI and other spiral tracers.

II. STRUCTURE AND DYNAMICS OF GALAXIES

The line of sight velocity of ionized gas at different position angles and the rotation curves of the inner parts of spiral galaxies were presented for NGC 497, 895, 972, 3646, 4100, 4536, 5351, 6181, 7171 and 7721 by Afanasiev, Burenkov, Sil'chenko and Zasov (Rus. AJ 68, 1134 and 69, 19). It was shown that the central regions (R < 1 kpc) of the most of these galaxies as for galaxies presented in previous articles by this group, have peculiar shape of line-of-sight velocity curves being distinguished from their surroundings by abnormally low or high velocity gradients or by the presence of the inner maximum on the rotation curve. These features give evidence of the existence of the central mass concentration or -in some cases -of small central bar. In addition, local systemic deviations from circular rotation curves with amplitudes about 30 km/s were found. For NGC 972, 4100, 4536 and 6181 noncircular gas flows were discovered far beyond their nuclei which envelop regions of more than one kpc scale. Rotation curves parallel with UBVR-photometry were obtained for several galaxies of type Sc - Irr (NGC 959, 1156, 1160, 6643 and 7292) by Esipov, Kyasumov and Jafarov (Rus AJ 68, 909). In the frame of simplest model it was shown that the central depression of thin axisymmetrical disk is needed to account for the shape of their rotation curves. Sil'chenko and Vlasiuk (Let in Rus AJ 18, 643) presented the results of axes orientation and ellipticity investigations for the central parts of 5 S-galaxies with known gas kinematics. In the inner regions of NGC 615, 7013, 7331 whose nuclei are dynamically decoupled by their fast circular rotation there are no turning of major axes of nuclear isophotes relative to the line-of-nodes of global galactic disks. On the other hand NGC 2655 and 7217 which are suspected to have strong non-circular gas motions in their nuclei, demonstrate such a turn. Therefore they apparently possess bar-like non-axisymmetrical bulges which provoke the non-circular velocity motions. The small mini-bar-like feature parallel with the inner ring structure was found in the center of nearby galaxy M 81 by Georgiev and Getov (Let in Rus AJ 17, 393) from their high resolution isophotes maps. Dimensions of these central structures are 0.64 and 1.50 kpc correspondingly.

A series of articles devoted to edge-on disk galaxies had been published for the last years. Karachentsev and Xu obtained rotation curves for 41 thin galaxies having axes ratio a/b > 7 (Let in Rus AJ 17, 321 & 485). More than 90% of the rotation curves have a non-decreasing shape and about a quarter of them may be classified as "wave-like" curves. The optical rotation curve amplitude demonstrates a tight correlation with the 21 cm line width. This allows to use the optical rotation curves for flat galaxies as a tool to determine their distances by analogy with the Tully-Fisher relationship. According to Karachentsev (Let in Rus AJ 17, 671) the main parameters of thin edge-on galaxies are typical for Sc-Sd galaxies. Outside the Local Supercluster the linear diameter of these galaxies produce a standard deviation about 20% as a distance estimator. Isophotal magnitudes of 120 northern thin edge-on galaxies parallel with their angular diameters, luminosity profiles and some basic photometrical parameters obtained from CCD- observations at R'-band have been presented by Karachentsev, Georgiev, Kajsin, Kopylov, Ryadchenko and Shergin (AAp Trans 4, 265).

The stellar disk stability analysis and N-body numerical experiments were carried out by Zasov, Makarov and Mihailova (Let in Rus AJ 17, 884) to find relationship between mass of the spherical component and minimal thickness of collisionless gravitating disk. They conclude that the observed disk thickness of edge-on galaxies may be used to estimate the halo-disk mass ratio. For the thinnest observed edge-on galaxies their halo should be at least twice as massive as the disk. Simakov (Let in Rus AJ 16, 679) considered dynamical evolution of the differentially rotating marginally stable protogalactic disk. Density distribution profile is shown to tend to quasi-exponential law in accordance with observations provided the typical gas viscosity time scale is comparable to the star formation time scale. In the framework of the Lin - Pringle model of the evolution of protogalactic gaseous disks he showed that angular momentum redistribution and star formation lead to certain value of the central surface density of the disk which should be a function of halo parameters (Simakov AAp Trans 3, 327). Chuvenkov, Glukhov and Vainer (Ap Sp Sci 190, 243) have computed the model of evolution of a disk subsystem of a spiral galaxy paying the main attention to the radial abundance gradient. It was shown that accretion of intergalactic matter onto the disk results in smoothing the radial gradient, especially at the late stage of the evolution. The final gradient depends weakly on the age of the galaxy. The best agreement between calculated and observed abundance of He is revealed if the primordial yield Yo = 0.25. A stellar-dynamical explanation of the observed morphology of barred galaxies have been proposed by Pasha and Polyachenko (Let in Rus AJ 19, 3). It emphasizes the role of the resonance disk response of stars on nearly circular orbits to a slowly growing potential of the Lynden-Bell bar. A theory developed allows one to derive a correct ratio for the observed diameters of outer and inner rings which should be close to 2.2. Inner rings surrounding bars in SB(r) systems are related to the ILR with a dominant m=2 bar-mode, and the outer rings - to the m=4 resonance. Morphological difference between the SB(r) and SB(s) galaxies may reflect the different bar growth rate which is higher in the latter case. Analysis of the instability of rotating stellar systems with highly elongated orbits was presented by Polyachenko (Let in Rus AJ 18, 1066). He derived analytically the dispersion relationship for the small perturbations in the simplest model of rotating spherical cluster with elongated orbits of stars. This approach may account for the observational evidence of similarity of bulges to oblate spheroids and the absence of clear correlation between the rate of rotation and the shape of elliptical galaxies. The theoretical and observational data give evidence that the radial orbit instability is the very cause of ellipticity of E-galaxies. In this scheme the maximal flattening of E-galaxies (c/a = 0.3) may be naturally explained by the fire-hose instability (Polyachenko Rus AJ 69, 934).

III. ACTIVE NUCLEI AND QSO's

Spectral variations of the AGN NGC 7469 (Sy 1) were observed and analyzed by Chuvaev, Lyutyi and Doroshenko (Let in Rus AJ 16, 867). Data taken from the 1983-1990 yrs observations revealed correlation between H-beta/[OIII] - ratio and U-magnitude which agrees with model of photoionization. During several months AGN changed its type from Sy 1 to Sy 2 and back. The results of spectral and continuum variability observations for the other Seyfert galaxy - NGC 4151 - were given by Oknyansky, Lyutyi and Chuvaev in Let in Rus AJ 17, 238. For NGC 4151 significant variations in continuum intensity and broad H-alpha line profile have been found by Sergeev (Ap Sp Sci 197, 77). An attempt to find optical brightness periodicity of 3C 273 have been fulfilled by Belokon and Babadzanyants (Rus AJ 68, 1 & 70, 243). They claimed the existence of the 307-day period for optical flares about 0.3 magnitude amplitudes and the possible 13.4 year period during the whole observational time (1887-1991 yrs). Correlations are found between these periodicities and the ejections of radiojets observed at VLBI. Afanasiev, Balega, Orlov and Vasyuk (AAp 266, 15) used high resolution speckle spectrograph to find structure of the 1"-core of NGC 1068 in the [OIII] emission line. Accurate positions and cloud velocities were obtained. Their positions appear to coincide with radiocomponents on the VLA radiomap. Artyukh and Ogannisyan are presented their results of search of compact nuclear radiosources of nearby galaxies by their interplanetary scintillation at 102 MHz. Radiosources have been discovered in 44 out of 265 galaxies, 13 of them appeared to be compact (scintillating). None of compact sources was found in E-galaxies which gives evidence that there are different physical conditions in intrinsically weak active nuclei of E and S-galaxies. Bochkarev and Nazarova (Rus AJ 68, 918) estimated typical physical parameters of AGN's and QSO's by comparing observed relative intensities of hydrogen and He II - lines with calculated ones from photoionization models for the region of formation of high-excitation lines which are located above the accretion disk surface. Electron number density, temperature and velocity gradient for these extremely inner regions were found. Theoretical model of radioemission of AGN's which emerges from the regions of formation of their narrow emission lines (about 100 pc) was developed by Zentsova and Fedorenko. Gaseous clouds producing this emission, are moving in the surrounding hot gas and induce shock waves. The latter accelerate electrons which produce radio emission via synchrotron mechanism. According to Zentsova (Rus AJ 68, 652), emission clouds, radiating the permission lines, appear in the outer regions of accretion disk as a result of it's fragmentation due to the gravitational instability and move outward by the action of strong radiation force. Vil'koviskij (Rus AJ 68, 1150) proposed self-consisted model for the formation of absorption lines of QSO in a plasma flow basing on a joint solution of plasma dynamics and line radiation transfer equations. Model spectra are well consistent with observed QSO's spectra with broad absorption lines. The model of X-ray emission of QSO's was presented by Suleymanov (AASp Trans 2, 197). He considered high luminosity accretion disks around supermassive black hole in the zone where radiation pressure is dominant. The temperature of accretion disk is found to be a function of accretion rate.

IV. SYSTEMS OF GALAXIES

Ninkovich, Chernin, and Shakenov (Let in Rus AJ 16, 1059) have proposed a set of dynamical models for the Local Group to analyze the amount of and possible spatial distribution of dark matter (DM). Comparison with observational data shows that the mass of DM is as much as 5 - 10 times of the total stellar masses of the galaxies. It gave also arguments in favor of individual DM corona of the system as a whole. A model of chemical and photometrical evolution of clusters of galaxies explaining the observations of deuterium and CNO-elements in galaxies of different types and ages as well as evolution of galactic luminosities and colour indexes was proposed by Vajner, Glukhov and Chuvenkov (Rus AJ 68, 225). A possible presence of DM in the galaxy triplets was analyzed by Anosova, Kiseleva, Orlov and Chernin (Rus AJ 69, 461). A method of stimulated catalogues was suggested in order to estimate DM. The method was applied to the 46 probably physically connected galaxy triplets from the list by Karachentsev e.a. The most probable average estimate of hidden mass is $M_o = (4.6 \pm 1.1) M_t$, where M_t is the total mass of the galaxies in triplet.

3. Catalogues and Atlases

(S. Okamura)

This report covers the material which appeared in ApJS, ApJ, AJ, a&AS, and MNRAS during the period of mid-90 to mid-93 with several additions. I apologize for incompleteness.

1990

Lauberts+	The surface photometry catalogue of the ESO-Uppsala galaxies	ESO
Kodaira+	Photometric atlas of northern bright galaxies	Univ. Tokyo Press
Zabludoff+	The kinematics of Abell clusters	ApJS 74, 1
Ohta+	Surface photometry of barred spiral galaxies	ApJ 357, 71
Lu+	Identifying galaxies in the zone of avoidance	ApJ 357, 388
van den Bergh+	Classification of galaxies on CCD frames	ApJ 359, 4
Bushouse+	Near infrared imaging of interacting galaxies	ApJ 359, 72
Huchra+	A deep Abell cluster redshift survey	ApJ 365, 66
Cayatte+	VLA obs of HI in Virgo galaxies. I. The atlas	AJ 100, 604
Peletier+	CCD surface photometry of galaxies...	AJ 100, 1091
Seal+	...optical positions for 110 Byurakan objects...	AJ 100, 1028
Wegner+	Redshifts for fainter galaxies in CfA-I slice	AJ 100, 1405
Parker+	Galaxy redshifts with FLAIR	A&AS 84, 455
Stirpe	An atlas of Hα and Hβ profiles of AGN	A&AS 85, 1049
Gavazzi+	Near IR obs. of galaxies in Coma and Cancer	
Johansson+	...a complete sample of interacting galaxies...	A&AS 86, 167
Fouqué+	An HI survey of late-type galaxies...	A&AS 86, 473
Boyle+	A catalogue of faint, UV-excess objects	MNRAS 243, 1

1991

de Vaucouleurs+	Third reference catalogue of bright galaxies	Springer
Véron-Cetty+	A catalogue of quasars and active nuclei (5th ed.)	ESO
Dressler	The supergalactic plane redshift survey	ApJS 75, 241
Hewitt+	...optical catalog of emission-line objects	ApJS 75, 297
Kinney+	An ultraviolet atlas of quasar & blazer spectra	ApJS 75, 645
Roberts+	Interstellar matter in early-type galaxies. I...	ApJS 75, 751
da Costa+	Southern sky redshift survey: the catalog	ApJS 75, 935
Warren+	... high-redshift quasars, z>2.2. II. The catalog	ApJS 76, 23
Sparks+	Multiwavelength isophotal data for southern...	ApJS 76, 471
Stocke+	Einstein EMSS. II. The optical identifications	ApJS 76, 813
Pesche+	The Case low-dispersion northern sky survey XII..	ApJS 76, 1043
de Carvalho+	Surface photometry of..elliptical and SO galaxies	ApJS 76, 1067
Junkkarinen+	A catalog of absorption in the spectra of QSOs	ApJS 77, 203
Haynes+	HI observations of galaxies in superclusters	ApJS 77, 331
Struble+	...redshifts and velocity dispersions for clusters	ApJS 77 363
Lilly+	A deep imaging and..survey of faint galaxies	ApJ 369, 79
Iye+	A catalog of spin orientation of galaxies	ApJ 374, 112
Oegerle+	Fundamental parameters of..cluster galaxies	ApJ 375, 15
Fukugita+	The distance to the Coma cluster using...	ApJ 376, 8
Bothun+	Extremely low surface brightness galaxies in..	ApJ 376, 404
Rakos+	Narrow-band photometry..of galaxies. III...	ApJ 377, 382
Zepf+	...colors of early-type galaxies in compact groups	ApJ 383, 524
Hewett+	The large, bright QSO survey. III...	AJ 101, 1121
Gavazzi+	Multifrequency windows on spiral galaxies. I...	AJ 101, 1207
Porter+	CCD observations of Abell clusters. V...	AJ 101, 1561
Batuski+	...redshift obs. of in the Sextans-Leo region	AJ 101, 1983
Muriel+	Redshifts of southern clusters. II...	AJ 101, 1997
Weistrop+	Emission-line galaxies in the Case survey	AJ 102, 1680
Chaffee+	The large, bright QSO survey. IV...	AJ 102, 461
Morris+	The large, bright QSO survey. V...	AJ 102, 1627
Hummel+	An Effelsberg/VLA..survey of..edge-on galaxies	A&AS 87, 309

Richter+	HI obs.. of galaxies in nearby Zwicky clusters	A&AS 87, 425
Harnett+	Observations of HI galaxies at 843MHz	A&AS 88, 73
Wozniak+	Surface photometry of early-type barred galaxies	A&AS 88, 325
Cappi+	Redshifts of southern rich clusters	A&AS 88, 349
Nieto+	Isophotal shapes of early-type galaxies. I...	A&AS 88, 559
Maza+	Calar-Tololo survey. III. HII galaxies	A&AS 89, 389
Corradi+	Kinematical obs. of spiral galaxies: bibliography	A&AS 90, 121
Capelato+	...radial velocities in clusters of galaxies	A&AS 90, 355
Terlevich+	A spectrophotometric catalogue of HII galaxies	A&AS 91, 285
Paturel+	An extragalactic data base: errors & misprints	A&AS 91, 371
Wang+	Optical identification of IRAS point sources...	MNRAS 248, 112
Sutherland+	Finding charts for southern IRAS galaxies	MNRAS 248, 483
Owen+	Surface photometry of radio galaxies. II...	MNRAS 249, 164
Couch+	...catalogue of distant galaxy clusters	MNRAS 249, 606
James	An IR study of dwarf galaxies in the Virgo cluster	MNRAS 250, 544
Takase+	Kiso survey for UV-excess galaxies. XII-XIV	Publ Nat A Obs Japan 2, 7, 37, 239
Saito+	Catalogue of galaxies behind the Milky Way. I-II	Kyoto Univ.

1992

Dressler+	Spectroscopy of galaxies in distant clusters. IV.	ApJS 78, 1
Henning	A study of a 21cm-selected sample of galaxies. I.	ApJS 78, 365
Freudling+	HI obs. of galaxies in the Hercules supercluster	ApJS 79, 157
Tully+	Nearby galaxy flows modelled by the light distr.	ApJS 80, 479
Owen+	A 20cm VLA survey of Abell clusters of galaxies..	ApJS 80, 501
Fabbiano+	An X-ray catalog and atlas of galaxies	ApJS 80, 531
Schneider+	Northern dwarf & low surface brightness galaxies	ApJS 81, 5
Han+	I-band CCD surface photom. of...16 nearby clusters	ApJS 81, 35
Mathewson+	...peculiar velocities of 1355 spiral galaxies	ApJS 81, 413
Bozyan	Optical identification..of 3196 radio sources...	ApJS 82, 1
Stephenson+	The Case low-dispersion northern sky survey. XIII.	ApJS 82, 471
Edelson+	The Colorado IUE active galaxy survey. I. Blazers	ApJS 83, 1
Strauss+	A redshift survey of IRAS galaxies. VII...	ApJS 83, 29
Odewahn+	Mean galaxy luminosity classifications	ApJS 83, 65
Postman+	The distr. of nearby rich clusters of galaxies	ApJ 384, 404
Pierce+	Luminosity-line width relations and...	ApJ 387, 47
Lehnert+	Multicolor images of..around high redshift quasars	ApJ 393, 68
Fairall+	Redshift obs. in the Hydra-Centaurus region	AJ 103, 11
Visvanathan+	Redshifts of spiral galaxies...direction of the GA	AJ 103, 1057
Schombert+	A catalog of low surface brightness galaxies...	AJ 103, 1107
Visvanathan	Photom. of spiral galaxies...direction of the GA	AJ 103, 1501
Hutchings+	Optical imaging of QSOs with 0.5 arcsec resolution	AJ 104, 1
Thuan+	The far-IR properties of CfA galaxy sample. I...	A&AS 92, 749
Fouqué+	Groups of galaxies within 80 Mpc. II...	A&AS 93, 211
Prieto+	Multiband..surface brightness distr. of spirals...	A&AS 93, 557
Amram+	Hα velocity fields and rot. curves of cluster gal.	A&AS 94, 175
Scodeggio+	Multifrequency windows on spiral galaxies. II...	A&AS 94, 299
Trevése+	Properties of nearby clusters of galaxies. I...	A&AS 94, 327
Poulain+	Isophotal shapes of early-type galaxies. II...	A&AS 95, 129
Jörgensen+	CCD surface photometry of...galaxies in Coma	A&AS 95, 489
Garcia+	New HI obs. for some edge-on spiral galaxies	A&AS 96, 435
Green+	...active and normal gal. observed in IR & X-ray	MNRAS 254, 30
Lumsden+	Edinburgh-Durham...catalog IV. Cluster catalog	MNRAS 258, 1
Takase+	Kiso survey for UV-excess galaxies. XV-XVI	Publ Nat A Obs Japan 2 399, 573

1993

Nolthenius	A revised catalog of CfA1 galaxy groups in...	ApJS 85, 1
Ciliegi+	A catalog of X-ray spectra of BL Lac objects	ApJS 85, 111

Hamabe	Surface photometry of Hydra I cluster of galaxies	ApJS 85, 249
Kinney+	An atlas of UV spectra of star-forming galaxies	ApJS 86, 5
Price+	VLA observations of 91 quasars at 0.35<z<1	ApJS 86, 365
Owen+	A 20cm VLA survey of Abell clusters of galaxies...	ApJS 87, 135
Hewitt+	A revised and updated catalog of QSOs	ApJS 87, 451
Zucca+	All-sky catalogs of superclusters of Abell-ACO...	ApJ 407, 470
Scodeggio+	21 cm study of spiral galaxies in clusters III...	ApJ 409, 110
Lilly	A deep I-band selected galaxy sample	ApJ 411, 501
Wegnert+	A survey of the Pisces-Perseus supercluster V...	AJ 105, 1251
Giovanelli+	A survey of the Pisces-Perseus supercluster VI...	AJ 105, 1271
Huchra+	The Morphological Cat. of Gal. equatorial survey	AJ 105, 1637
Rice	An atlas of high-resolution IRAS maps of..galaxies	AJ 105, 67
Vogel+	Emission-line galaxies in Hamburg quasar survey	A&AS 98, 193
Binggeli+	Dwarf galaxies in the Virgo cluster. II...	A&AS 98, 297
Klaas+	Identification..of..extragalactic IRAS sources	A&AS 99, 71
Reshetnikov+	A photometric study of interacting galaxies. I..	A&AS 99, 257
Garilli+	Galaxy velocities in eight southern clusters	A&AS 100, 33
Colless+	Photoelectric & CCD photom. of E & SO galaxies	MNRAS 262, 475

4. Galaxy Clustering
(J. Huchra)

As our ability to observe the universe grows, so does both our knowledge of the properties of individual groups, clusters and superclusters of galaxies and our knowledge of the general pattern of the distribution of galaxies in space. Often this increase of knowledge does not bring with it a concomitant increase in understanding. Sometimes we astronomers get lucky. I will divide this report into three sections, one on groups and clusters, one on large-scale structure and one on large-scale velocity fields.

I. GROUPS AND CLUSTERS

The launch of ROSAT, the construction of large galaxy redshift databases, and the development of multiobject spectrographs all have had very large impacts on the study of groups and clusters over the last three years. ASCA is also just beginning to provide high quality x-ray data on clusters. Significant advances have been made in the determination of the cluster temperature function from x-ray data (e.g. Henry & Arnaud; Edge & Stewart) and the cluster velocity dispersion function from optical redshift surveys in clusters (e.g. Zabludoff +). These are just being both rectified with each other (Bahcall & Cen) and are beginning to be used in tests of cosmological simulations as proposed by Frenk + in 1990, and examined in more detail by other workers (e.g. Evrard +).

It is becoming increasingly clear that groups are low luminosity analogues of clusters — virialized systems but will smaller total masses and binding energies and thus velocity dispersions and temperatures. Ramella + have described the connection between groups and large-scale structures by examining the properties of groups in the ''Great Wall.'' Several analyses have been done of the properties of galaxies in groups including attempts to detect intergalactic extinction in groups. There is a puzzling observation that fainter galaxies that have been assigned to groups generally have larger than average redshifts. This is almost certainly a selection/interloper effect. Determinations of the mass-to-light ratios of groups still arrive at relatively small values (a few hundred in solar units with $H_0 = 100$ km/s/Mpc), which correspond to values of Ω of 0.1-0.2. The most exciting observations of groups over the last few years has been the detection of significant cool (≈ 1 kev) x-ray gas in many loose and compact groups (Mulchaey +; Ponman & Bertram; Mushotzky +; Ebeling +; Bohringer +). While the exact source of the gas is uncertain in any individual group (general potential versus individual galaxies), it is clear that many, if not all such objects are dynamically relaxed and belong at the low mass end of the continuum of galaxy clustering. Improved group catalogs have been constructed from homogeneous galaxy redshift surveys by Ramella +, Nolthenius, and others.

Work has also continued on two smaller but important fields in cluster research. Fabian +, Canizares + and others continue observations both in the x-ray and optical of cooling flows. There are indications from optical and IR photometry that low mass stars are forming in cooling flows, and HST observations of NGC1275 (Holtzman +) have revealed the existence of what appear to be young globular clusters. The study of gravitational lensing byclusters, pioneered by T. Tyson after the initial discoveries of arcs by Lynds and Petrosian, has become a major industry for the determination of cluster potentials and masses. Noteworthy work in this field has come particularly from the French group of LeFevre, Mellier, Soucail, Fort and collaborators.

Lastly, although severely hampered by the presence of spherical aberration, the HST observations of distant clusters by Dressler, Oemler, Gunn and Butcher are noteworthy for providing measures of the evolution of cluster galaxies and the probable resolution of the Butcher-Oemler effect.

ROSAT and redshift surveys have also improved our understanding of substructure in galaxy clusters (e.g. Forman & Jones; Fabricant +; Beers +; Zabludoff & Franx). Most clusters, even rich clusters like Coma, exhibit some substructure both in the optical galaxy distribution and in x-ray maps. Current estimates agree with the earlier estimates of Dressler and of Geller and Beers that 30-50% of all "rich" clusters are not completely dynamically relaxed. The actual fraction may be even higher.

II. LARGE-SCALE STRUCTURE

Observations of large-scale structure continue to be driven by the several major galaxy and cluster redshift surveys described in the section on Galaxy Redshifts. This three year period has seen the completion of several large surveys such as the IRAS QDOT survey (Saunders +) and 1.2 Jy Survey (Strauss +), the continuation of the CfA survey (Huchra & Geller) and the "Brick" survey of Kirshner +, and multiple analyses of same. There is now general agreement that the galaxy clustering distribution exhibits significant power on large scales (e.g. Vogeley +; Efstathiou +; Peacock, Fisher +). Discrepancies between the results from studies of infrared bright galaxies (IRAS selected) and optically selected galaxies may be the result of biases in the surveys or real differences in the clustering properties of early versus late type galaxies. This author favors the latter interpretation. In any case, the differences between the data sets are small compared to the difficulties of fitting proposed models of structure formation to the observations. When simulations are normalized at small scales, they fail to predict the observed large scale clustering by many model "sigma." Fits can only be obtained by introducing extra parameters, such as a scale dependent biasing parameter for galaxy formation. The importance of the initial results on periodicity in the pencil beam survey of Broadhurst + continues to be debated.

Observations of the clustering of clusters also contribute to the mismatch of theory and observation. Although there still is a debate about the exact correlation length of the cluster-cluster correlation function (e.g. Postman +; Efstathiou +, Plionis +, Scaramella +), which continues even when x-ray selected samples of clusters are studied (e.g. Nichol & Collins + versus Henry & Briel), independent of whether the scale length is 14 or 20 Mpc, there is still difficulty in fitting Cold Dark Matter simulations simultaneously to the small-scale galaxy and clustering and the clustering of clusters. Perhaps the best solutions to the above observational problem are offered by the ROSAT All Sky Survey of Hasinger,

Böhringer and Trumper, who, in collaboration with groups of optical observers in the northern and southern hemisphere, are systematically identifying clusters in the x-ray. These samples will provide deeper and more consistent cluster catalogs than have previously been available.

III. LARGE-SCALE VELOCITY FIELDS

Since the discovery (or, more correctly, the rediscovery and measurement) of the infall of the Local Group into the Virgo cluster and the rebirth of interest in the Rubin-Ford effect caused by the discovery of a possible Great Attractor in Hydra-Centuarus, less than a decade ago, there has been enormous interest in the measurement of large-scale flows in the nearby universe. The two major cosmological problems addresses by this work are (a) the attempt to discover the cause (gravitational) of the microwave background dipole anisotropy, which is usually attributed to a motion of the Local Group w.r.t. the cosmic reference frame, and (b) the use of the flow field with the local galaxy density field to determine Ω.

Several major efforts to measure the flow field have been completed in the last few years. For individual galaxies, Mathewson, Ford and Buchhorn have completed a survey of spirals over a large area in the direction of the GA, while Dressler and Faber and coworkers have surveyed a smaller area. Both groups were looking for signatures of the backside infall into the GA mass concentration. They disagree on their results, with Matthewson + failing to find convincing evidence for a spherical flow field, while Dressler and Faber, looking deeper over a much smaller field claim to see the infall. Willick and Courteau have completed detailed studies of the flow field as doctoral thesis projects; Willick examining the Perseus-Pisces region and Courteau working on a whole sky sample. Bertschinger and Dekel and collaborators have attempted to match the potential field as measured by the gradients in the galaxy flow field (from data assembled by Burstein and others) with the galaxy density field from the IRAS 1.936 Jy survey. Kaiser + have done similar analyses for the QDOT survey. Generally, the IRAS galaxy versus flow field results yield high values for Ω, near unity. Attempts to do the same for the optical surveys (e.g. Lahav & Scharf; Hudson) arrive at much lower values of Ω (0.3-0.5).

On larger scales, over the last few years there has been considerable debate as to the cause of the microwave background dipole, with a large group of astronomers favoring the Great Attractor hypothesis (that most of the 630 km/sec measured by COBE is developed locally, within 8000 km/s) and another large group (e.g. Tully & Scaramella & coworkers; Raychaudhury +) arguing that the Shapley Supercluster concentration, which lies behind Hydra-Cen by about a factor of 3 in distance, but appears to be much more massive) is also a major contributor to the acceleration of the Local group. Several groups have attempted to measure the Local group velocity w.r.t. the reference frame of galaxy clusters (e.g. Mould +; Lauer & Postman). Mould + used the Tully-Fisher technique to observe nearby clusters with Parkes and Arecibo and were not able to cleanly distinguish between a GA flow or a bulk flow on larger scales. Lauer and Postman used the luminosities of first ranked cluster galaxies to measure the motion of the Local Group w.r.t. a sample of Abell clusters inside 15,000 km/s and found, that to within 2σ, the LG appears to be at rest w.r.t that frame. This would imply, if the CMB dipole is due to a motion of the LG, that their entire Abell cluster inertial frame must be participating in a bulk flow w.r.t. the CMB frame with an amplitude of nearly 700 km/s. These results on large scales are, again, extremely difficult to reconcile with the current best CDM simulations of structure formation in the universe, especially when the simulations are normalized by either the now observed very low amplitude fluctuations in the CMB on degree scales or by the clustering amplitude of galaxies on small scales.

5. Population Synthesis and Spectral Evolution
(G. Bruzual)

During the last three years there has been excellent progress in our understanding of stellar populations in nearby early type galaxies. A long standing problem, the source of the UV light in elliptical galaxies and in the central bulge of early type spirals, and in particular, the very steep UV upturn observed in the spectral energy distribution (SED) of the most luminous and metal richest ellipticals, is now closer than ever to being understood. Despite its spherical aberration HST is finally resolving stars in the central bulge of M31 (King + ApJ 397, L35) and in M32. The Hopkins Ultraviolet Telescope (HUT) also acquired a spectrum of the central bulge of M31 (Fergusson & Davidsen ApJ, in press) and of the elliptical galaxy NGC 1399 (Fergusson + ApJ 382, L69) during the Astro-1 space shuttle mission. High quality IR photometry by Freedman (AJ 104, 1349) has revealed intermediate age AGB stars in M32. HST FOS-spectra of 6 early-type weak radio galaxies with $0.08 < z < 0.53$ have been obtained by Windhorst + (ST-ECF/STScI Workshop on Science with the HST, eds. Benvenuti & Schreier). All of these galaxies show UV upturns. These results should be taken with caution, however, since a good fraction, if not all, of the detected UV light may be FOS red grating scattered light (Windhorst +).

These data have permitted that the several possible sources of the UV light comprehensively reviewed by Greggio and Renzini (ApJ 364, 35) be examined by means of highly sophisticated and complete population synthesis models (Worthey, PhD Thesis; Bruzual & Charlot ApJ 405, 538; Magris & Bruzual ApJ, in press; Bressen + ApJ, in press). This last work summarizes the results of a large theoretical effort to compute the spectro-chemical evolution of elliptical galaxies. A clearer picture seems to be emerging in which PAGB stars can be responsible of the UV flux observed in the UV-coolest galaxies, whereas higher than solar metallicity AGB-Manqué stars are the most likely source of the UV photons in the steep UV galaxies. The lack of flux below 1100 Å in the SEDs of NGC 1399 and the central bulge of M31 has been used by Fergusson + and Bressan + to argue against hot PAGB stars as the only or most important contributor to the UV in these galaxies. The cooler and longer lived high metallicity AGB-Manqué stars produce enough light in Bressan + models to account for the measured fluxes. The presence of an intermediate age population in M32 seems to be an inescapable conclusion from the detection of stars at the tip of the AGB which belong to this population (Freedman). This is corroborated by spectral synthesis results.

I. EVOLUTION OF FIELD GALAXIES

Traditionally, the steep slope of the faint field galaxy counts and their very blue colors have been explained as the result of mild evolution in the SEDs of galaxies. The excess of counts above non-evolutionary models has been claimed to be a factor of two by $B \simeq 20$ (Maddox + MNRAS 247, 1P; Loveday + ApJ 390, 338) or by $B \approx 22.5$ (Colless + MNRAS 205, 1287). The large number of blue counts has been claimed to be incompatible with a flat ($\Omega = 1$) Freidmann universe (Koo ASP Conf No. 10, 268; Guiderdoni & Rocca-Volmerange A&A 227, 362), or with the observed near-infrared K band counts (Cowie + preprint), even if galaxy evolution is included in the models. In contradiction, recent redshift surveys show that galaxies fainter than 20th mag exhibit redshift distributions close to that predicted by no evolution models (Colless +; Lilly + ApJ 369, 79). These apparently contradictory facts have led several workers to propose that dramatic revisions to the conventional view are needed. The cosmology has been revised by Fukugita + (ApJ 361, L1) by adopting a non-zero cosmological constant, Λ. The non-conservation of galaxy number due to mergers has been proposed by Cowie + (Nature 354, 460) and Broadhurst + (MNRAS 355, 55). Cowie + and Babul and Rees (MNRAS 255, 346) have assumed a disappearing population of dwarf galaxies.

The review by Koo and Kron (ARAA 30, 612) summarizes current status of observation and interpretation on the subject of faint galaxy number counts, and color and redshift distributions. These authors show that the existing data on B, R, and K galaxy counts, (B–R) colors, and redshift distributions, is consistent with a simple picture in which the cosmological constant is zero, the number of galaxies is conserved over time, and the shape of the luminosity function for each galaxy class is constant. They argue that the combined uncertainties in the models and in the data so far preclude the necessity of more exotic assumptions. A no-evolution model accounts for most of the observed trends in the data, with the exception of the excess counts at faint visible-band magnitudes, which may indicate that mild luminosity evolution occurs in these galaxies. This controversial no-evolution model has been examined in more detail by Koo + (ApJ Letters, in press), who confirm Koo and Kron contentions. The essential difference in this interpretation with respect to previous models is the relaxation of the assumption that galaxies of all color classes share the same luminosity function (determined locally) and the use of an objective technique to derive from the same faint galaxy data that is being fitted the most plausible luminosity function for each color class. The derived total luminosity function is compatible with recent determinations by Lonsdale and Chokshi (AJ 105, 1333), Loveday + and Eales (ApJ 404, 51).

II. EVOLUTION OF BRIGHTEST CLUSTER GALAXIES

The simplest experiment to detect spectral evolution in galaxies is the comparison of SEDs of distant galaxies with those of nearby galaxies whose stellar population differ only in age. Up to $z \approx 0.5$ a red envelope seems to be well established, indicating that at any $z \leq 0.5$ the reddest galaxies (mostly BCG's) are as red at rest as nearby elliptical, allowing for essentially no spectral evolution during a large fraction of the age of the universe. The fraction of blue galaxies in clusters seems to increase with z (Dressler & Gunn ASP Conf No. 10, 200; Couch + MNRAS 249, 606). The selection of distant galaxies via optical luminosities and colors may exaggerate the importance of the Butcher–Oemler effect and discriminates against the early-type cluster galaxies (Aragón-Salamanca + MNRAS 262, 764). Comparison of the SEDs of the blue cluster galaxies, some of them undergoing short-term star formation, with nearby ellipticals does not provide conclusive detection of evolution of the red envelope galaxies, since the reddest galaxies in the clusters may have not been visible in the selecting optical band. Aragón-Salamanca + circumvent this problem by selecting the cluster galaxies at near-infrared wavelengths, which sample equally well both early and late type galaxies. By studying clusters up to $z \leq 0.9$ they conclude that they have detected the spectral evolution in the galaxies defining the red envelope: by $z \approx 0.9$ there are no cluster galaxies as red at rest as present day ellipticals.

6. Working Groups on Photometry, Kinematics, and Dynamics of Galaxies
(R. Buta)

Research the past three years on the topics related to this working group has been very extensive, and I cannot hope to summarize it all here. I only draw on a few topics that indicate the flavor of what has been done. This working group is a combination of what were formerly known as the "Working Group on Galaxy Photometry and Spectrophotometry" and the "Working Group on Internal Motions in Galaxies". The merger took place in 1991.

One of the most exciting results of the last three years has been the discovery of genuine counter-rotating systems in disk galaxies. Rubin and collaborators (ApJ 394, L9) discovered co-spatial counter-rotating stellar disk in NGC 4550. Baun and collaborators (Nature 360, 442) discovered counter-rotating and probably co-planar gaseous disks in the "Black Eye" galaxy NGC 4826. Up until these findings, kinematically-distinct, counter-rotating systems had only been found in ellipticals. These results show that even normal-looking disk galaxies can possess distinct kinematic subsystems indicative of the effects of recent or even old mergers.

Evidence for supermassive black holes in galactic nuclei increased with a detailed photometric and kinematic study of the edge-on SO galaxy NGC 3115 by Kormendy and Richstone (ApJ 393, 559). They regard this object as the "best BH candidate after M31".

The discovery of declining rotation curves in some nearby galaxies (Carnigan & Puche NGC 7793, AJ 100, 394; Casertano & van Gorkom NGC 2683 & 3521, AJ 101, 1231) has been important for the study of dark haloes. These observations may signal the end of the so-called "conspiracy" between disk and halo that has been hypothesized to explain why we do not normally see, in the rotation curve, the transition region between where the disk and halo each dominate the potential.

Polar ring galaxies continued to shed important light on the shapes of dark haloes around SO galaxies. An excellent atlas of polar ring galaxies by Whitmore + (AJ 100, 1489) has established the class well. Now, individual objects are being followed-up. In the case of NGC 4650A, a dynamical model which accounts for the mass of the polar ring (previously ignored) favors a halo as flattened as E6-7 (Sackett & Sparks ApJ 361, 408). In contrast, a study of the complex dust lane pattern in NGC 4753 has been interpreted as an inclined disk strongly twisted by differential precession in a total mass distribution no more flattened than E1.6 (Steiman-Cameron + AJ 104, 1339).

Numerical simulations by Athanassoula (MNRAS 259, 345) have clarified the factors that determine the shapes of dust lanes in bars. This led to the discovery of "centered shocks", that is, shocks that would give rise to dust lanes centered along the length of a bar, rather than greatly offset as is normally seen. Some late-type barred spirals show such centered dust lanes.

CO studies have led to some interesting findings. Molecular bars in the central regions of galaxies are now being routinely found. Since the beginning of the current triennium, molecular bars have been found in NGC 5457 (Kenney + ApJ 366, 432) and NGC 2251 (Kenney + AJ 103, 784). This is in addition to features found previously in IC 342, especially interesting because the molecular bar is a small feature aligned nearly perpendicular to the primary bar. In NGC 5457, the offset is only about 25 degrees between a molecular gas bar and a weak oval. These kinds of misalignments between a stellar and a gaseous bar could signify definitive detection of the inner Lindblad resonance (ILR) in these galaxies. The last triennium also saw the highest resolution CO observations yet obtained from M51 (Garcia-Burillo + AA 274, 123), which appears to have a molecular ring located near an ILR.

The new class of giant low surface brightness (LSB) galaxies, defined by Malin 1 in 1987, received an impetus with the discovery of a second example by Bothun + (ApJ 360, 427). Since these discoveries, much effort has gone into identifying more examples and establishing their properties over a variety of wavelengths (Bothun + AJ 106, 531 + references). These objects are important as a class because they appear to be disk galaxies whose rate of evolution of gas into stars has been slow compared to more normal Hubble types. It is argued that "the distinguishing characteristic between (high surface brightness) and LSB galaxies is the net number of tidal interactions that each has experienced over a Hubble time."

Infrared detectors have advanced over the past three years to the point where it is possible to image a moderate-sized galaxy in a single frame. Block and Wainscoat (Nature 353, 48) presented blue and 2.1 micron images of the prominent grand design spiral galaxy NGC 309 and highlighted the enormous difference in appearance. They argue that a near infrared classification of galaxies is now feasible and should be a major improvement over the existing Hubble system based on blue light images.

The last triennium also provided the first definitive identification of a leading spiral arm in a galaxy, the SA(r) ab spiral NGC 4622. G. Byrd recognized in 1988 a spiral feature in the inner regions of NGC 4622 that appeared to wind opposite to the winding of two outer spiral arms. A follow-up photometric BVI study (Buta + AJ 103, 1526) confirmed the reality of the feature as a stellar density enhancement. N-body simulations (Byrd + AJ 105, 477) have successfully reproduced the structure of NGC 4622 in terms of a small, plunging retrograde companion.

A stunningly large contribution to the database of surface brightness profiles and rotation curves of spiral galaxies has been the massive work of Mathewson and collaborators (ApJS 81, 413). The data were obtained to study peculiar velocities towards the so-called Great Attractor. Completely apart from this part of the study, there are now more homogeneous rotation curves and light profiles available than ever before, and surely they will be used for other studies on the structure and dynamics of galaxies.

The use of N-body simulations for the study of the evolution of galaxies has progressed immensely. The past triennium saw the first Astrophysical Journal videotape ever produced as a regular part of an issue. The papers associated with the video are in the July 10, 1992 issue and include simulations of galaxy clustering, mergers, off-center nuclei, and narrow-angle-tail radio sources. Other noteworthy studies not on video are 3D simulations of bars (Pfenniger & Friedli AA 252, 75), gaseous rings around triaxial systems (Pearce & Thomas MNRAS 248, 688), and mergers of disk galaxies producing counter-rotating gas disks (Hernquist & Barnes Nature 354, 210).

A flurry of activity concerning the optical thickness of galaxy disks followed a study by Valentijn (Nature 346, 153). He used data for 9,400 galaxies from the ESO-LV database to deduce that spiral disks are optically thick and suggested that this invalidates the mass-to-light ratios commonly used to infer the existence of dark matter. Burstein + (Nature 353, 515) dismissed Valentijn's conclusion as unreliable based on his analysis, owing to selection effects, and argued that one needs reliable distances to circumvent the confusion caused by these effects. They nevertheless also conclude that disks are optically thick. However, direct tests of optical thickness can be made for a small number of overlapping (spiral on elliptical) pairs of galaxies. White and Keel (Nature 359, 129) and Andredakis and van der Kruit (AA 265, 396) have used two such pairs to show that, except in the regions of bright spiral arms, disks are optically thin. The controversy continues. In a related near IR study (AJ 103, 1761), Peletier and Willner suggested that galaxy disks are semi-transparent in the H-band. They also suggest that "Freeman's Law" (constancy of central surface brightness) is not an intrinsic property of true disk mass distributions.

The triaxiality of elliptical galaxies has been the subject of numerous photometric and kinematic studies. The main evidence comes from isophotal twisting and minor axis rotation. A detailed review is given by Franx (Morphological and Physical Classification of Galaxies, Longo + eds, p. 23). The same volume provides work concerning stellar and gaseous disks in ellipticals. Concerning gaseous disks, Bertola + (ApJ 373, 369) found that a triaxial model could explain the velocities within a gaseous disk in the elliptical NGC 5077 whose rotation axis is misaligned with the apparent major axis by 23 degrees. Bertola + (ApJ 374, L13) also discuss triaxiality in disk galaxies by measuring misalignments between bulge and disk major axes.

Related to early-type galaxies, Schweizer and coworkers (ApJ 364, L33; AJ 104, 1039) discuss interesting correlations found between the line strengths and colors of E and S0 galaxies and the presence of fine structure in the morphology, such as ripples, plumes, boxy isophotes, shells, etc. The H-beta absorption line strength increases for ellipticals, and both ellipticals and S0's are systematically bluer, with increasing fine structure content. It is argued that these correlations are caused by mergers over at least 1/3 to 2/3 the present age of the universe.

The study of spiral structures in galaxies continues to be very active. A workshop was held at STScI that covered resonances in grand design and barred spiral galaxies (PASP 105, 638-674), and is a useful source of the most recent references. Perhaps the best identified resonance is the outer Lindblad resonance, which is obviously important in many early-type spiral galaxies (Buta & Crocker AJ 102, 1715). Other resonances are important; in grand design spirals, theoretical and observational techniques for identifying resonance locations were discussed by Elmegreen + (ApJS 79,37) and Patsis + (AA 243, 373). The origin of spiral structure is still an important problem. Byrd and Howard (AJ 103, 1089) argue that at least one third of a sample of Holmberg galaxies have tidal arms, based on a self-gravitating tidal perturbation survey.

The last triennium also saw some of the first extragalactic results from the Hubble Space Telescope. A possible black hole accretion disk was detected in the elliptical galaxy NGC 4261 (S&T 2/93), surface photometry and spectroscopy of an elliptical galaxy at $z = 2.39$ was reported (S&T 1/93), and an x-shaped dust pattern was discovered at the center of M51 (S&T 9/92). Most interesting was Dressler and Oemler's presentation and description of the first detailed images of cluster galaxies at redshifts of 0.4 (S&T 4/93, p. 22). The galaxies showed the familiar Hubble types of nearby galaxies, so these types were already in existence up to 4 billion years ago. Interestingly, the problem with the HST optics led to unprecedented advances in image processing and reconstruction (e.g., see King + AJ 102, 1553). Using such techniques, Lauer + (ApJ 369, L41) and Benedict + (AJ 105, 1369) presented some of the first surface photometry obtained with HST, of the S0 galaxy NGC 7457 and of the nuclear ring in NGC 4314.

Finally, the proliferation of data and papers in extragalactic research, particularly in the topics of this Working Group, and the development of sophisticated automatic plate scanners, has led to some important databases and other results since 1990. Firstly, the Third Reference Catalogue of Bright Galaxies (RC3), produced by de Vancouleurs and five other collaborators over the years since the 1976 publication of the Second Reference Catalogue (RC2), was finally published in 1991. Users will notice immediately how different RC3 looks from RC2: the former is in three volumes and includes more than 23,000 galaxies while the latter is one volume and includes only 4,400 galaxies. One volume of RC3 is devoted almost entirely to references. Secondly, the scanning of the Palomar Sky Survey for a massive star/galaxy catalogue has led to new techniques for automatic discrimination of star and galaxy images (see Odewahn + AJ 103, 318). Thirdly, G. Helou and B. Madore created the very useful NASA/IPAC Extragalactic Database (NED) which has changed "the way that extragalactic research is being done around the United States and around the world" (see Helou 1992, Proc. 27th Rencontre de Moriond, p. 25).

7. Galaxy Redshifts Working Group

Just as the field of mathematics and computing has its "Grand Challenges," so does astronomy. Some of ours have been set out in the 1991 Bahcall Committee Report on the next decade of astronomy. For extragalactic astronomy and cosmology, arguably the two largest projects for this decade are the mapping of the local galaxy density field (to a redshift of ≈ 0.1) and the local galaxy velocity field. Work on these projects has already begun through large programs to obtain uniform, whole sky galaxy samples, redshift surveys of large complete samples, and redshift independent distance measurements. Several of these will be described below.

In addition to large-scale nearby studies, the development of multiobject spectrographs combined with a new wealth of x-ray data, most notably from ROSAT, and deep optical surveys has continued the explosion of redshift data for galaxy clusters and for deep probes of galaxy evolution and cosmology.

Since this reporter's last report three years ago, the number of galaxies with measured redshifts has gone from about 38,000 to nearly 50,000. This reporter maintains the largest collection of published and private redshifts at the Harvard-Smithsonian Center for Astrophysics (the CfA Redshift Catalogue). This catalogue is periodically distributed to and obtainable from the National Space Science Data Center (NSSDC) in the U.S. and the Strasbourg Astronomical Data Center (CDS) in Europe. A somewhat more detailed catalog of galaxy properties, The Third Reference Catalogue, has been completed by G. and A. deVaucouleurs, H. Corwin and R. Buta. These catalogs contain several "complete" sub-catalogs, including the Revised Shapley-Ames catalog (RSA), the first CfA survey (CfA1), the Southern Sky Redshift Survey (SSRS), the Nearby Galaxies Catalog, the first CfA slice (CfA2), and the IRAS 1.936 Jy Redshift Survey (Strauss + 1992).

Similarly, the number of galaxy clusters from the lists of Abell, Zwicky, Corwin, Olowin and Abell and deeper surveys (e.g., Gunn & Oke, & Sandage +) with measured redshifts is over 1500. Cluster redshift catalogs can be obtained from H. Rood (USA), H. Andernach (Brazil), T. Festisova (Russia), M. Kalinkov (Bulgaria) and J. Huchra (USA).

In the last three years, several large groups continued to work on major large-area redshift surveys. The Arecibo Survey was continued by R. Giovanelli and M. Haynes and G. Wegner, who have recently completed a 21-cm survey of the Pisces-Perseus supercluster region (AJ 105, 1251 & 1271). The Southern Sky Redshift Survey was published by L. da Costa + (ApJS 75, 935), and he and M. Geller, A. Fairall, P. Pellegrini, D. Latham, J. Menzies and others have nearly completed work on the extension of the SSRS (SSRS2) to a limiting magnitude of $B_T \approx 15.0$. The QDOT redshift survey of 1-in-6 of the IRAS galaxies brighter than 0.6 Jy at 60μ has been completed by A. Lawrence, M. Rowan-Robinson, W. Saunders, G. Efstathiou, N. Kaiser and others. Although this survey contains only ≈ 2500 galaxies and is extremely sparsely sampled, it is the deepest of the whole-sky probes available and is currently being extended to include redshifts for all galaxies brighter than 0.6 Jy. The IRAS 1.2 Jy survey has been completed by M. Strauss, J. Huchra, M. Davis, A. Yahil, K. Fisher and J. Tonry. This sample contains over 5000 galaxies.

Both the QDOT sample and 1.2 Jy sample have been analyzed in long series of papers in MN and the ApJ. Primary results include the comparison of the IRAS galaxy density field with the density field determined by inverting galaxy flow field observations (most notably using the POTENT technique of Bertschinger, Dekel & Faber), attempts to estimate Ω from both this comparison and the microwave background dipole velocity — which generally give high values, 0.8–1.0, with estimated errors that are inconsistent with determinations of Ω from optical samples (Lahav +), and the derivation of the general clustering properties of late type galaxies $\xi(r)$.

The APM-Stromlo redshift survey of J. Loveday, G. Efstathiou, B. Peterson, S. Maddox + now contains nearly 1800 galaxies to a magnitude limit of ≈ 17. This sparse survey has been used to determine the clustering properties of galaxies on very large scales.

The CfA2 sruvey was continued by J. Huchra and M. Geller +. The final CfA2 survey will contain redshifts for all galaxies brighter than 15.5 in the Zwicky catalogue above the equator, a total of approximately 18,000 galaxies including those at galactic latitudes lower than originally covered in the CfA1 survey. They currently have redshifts for over 16,000 galaxies, including 2300 galaxies in the low latitude regions covered by Zwicky and Nilson. Completed CfA2 regions include 14 complete POSS strips, 6 in the north galactic cap between +8.5° and +44.5° declination, and 8 in the south galactic cap between -2.5° and +48°. Analyses of these samples by M. Vogeley + have strongly confirmed the results that standard Cold Dark Matter cosmological simulations cannot match the observed clustering properties of galaxies, in particular the clustering power spectrum, on large and small scales simultaneously. This group, with L. daCosta and others completed a redshift survey of 860 NGC galaxies between 20^h and 5^h and -17.5° and -2.5° to connect the CfA1 survey with the SSRS survey in the south galactic cap. With J. Thorstensen and G. Wegner of Dartmouth, they are also near completion of the "Century Survey" of 2400 galaxies in a 1° wide by 100° long strip through the north galactic pole.

The Las Campanas redshift survey of S. Shectman, R. Kirshner, G. Oemler and P. Schecter now also contains over 10,000 galaxies in a long strip in the south galactic cap. This survey indicates that astronomers might have finally found the scale on which homogeneity of the galaxy distribution is setting in. In previous shallow samples, the largest structures discovered were about as large as could have been discovered given the sample dimensions. In the LC survey, despite probing to more than 30,000 km/s, the largest void sizes appear to be only in the range of 5-6000 km s^{-1}.

Three major deep redshift probes have given preliminary results on the redshift distribution and spectroscopic properties of faint galaxies. These include the pencil beam surveys of T. Broadhurst, D. Koo, A. Szalay, T. Shanks +, who have also discovered an apparent periodicity in the spacing of structures along the line of sight, the deep surveys of optically selected galaxies by R. Ellis, C. Colless and collaborators, and the optical and IR based surveys of S. Lilly, L. Cowie and J. Gardner and collaborators. There yet remains considerable confusion over the nature of the number excess of faint galaxies, but it is now at least clear that the median redshift of galaxy samples as faint as 23nd is still much less than 0.5. Further progress on this front really awaits the 8 to 10-meter telescope now under construction.

Considerable progress has been made on the measurement of velocities for galaxies in clusters. Some, such as the surveys of A. Zabludoff, M. Geller and J. Huchra, D. Batuski +, and T. Beers +, have been done the old-fashioned way of one galaxy at a time. Others, such as those of W. Oegerle, J. Hill, P. Hintzen, M. Gitchett, J. Hoessel and collaborators, and P. Gray, R. Sharples, D. Carter, P. Teague and J. Lucey have used modern multifiber spectrographs on telescopes ranging from the U. of A. 90-inch to the AAT Velocity distributions in clusters are now considered not only important in the determination of the cluster masses and dynamical temperatures for comparison to X-ray observations, but are also useful in testing the small scale and deep potential properties of galaxy clustering simulations (e.g. Frenk + ApJ 351, 10).

In the near term, two major new projects that are about to get underway deserve mention. The first is the Sloan Digital Sky Survey, a project to produce a deep multicolor map of approximately 1 steradian followed by a redshift survey of 1 million galaxies in that region. The major goal of this program is to nail down, once and for all, the statistics of galaxy clustering such as the n-point correlation function, the topology of clustering and the power spectrum. The second is the Two Micron All Sky Survey, or 2MASS, which, in addition to providing fundamental information about the galaxy and stellar populations in the galaxy, will also provide a nearly whole sky map of the galaxy distribution selected at wavelengths minimally affected by extinction (i.e. the B and R bands) and dominated by normal as opposed to starburst galaxies (i.e. the IRAS surveys).

In closing, I would like to once again stress to all astronomers and Journal Editors the need to publish heliocentric velocities (cz) or redshifts (as $z = \Delta\lambda/\lambda$ in the optical convention), and also to publish accurate positions for the galaxies they observe. This policy was adopted by the members of the Working Group on Galaxy Radial Velocities present at our meeting during the 1988 IAU General Assembly. The publication of finding charts for confused fields is also good form. Authors wishing their data to be included rapidly in electronic databases for galaxy redshifts should send electronic versions of their tables and the correct reference to use to huchra@cfa.harvard.edu.

8. Magellanic Cloud Working Group

(N. Walborn)

Good overviews of recent Magellanic Cloud work may be found in the proceedings of a workshop at Heidelberg, "New Aspects of MC Research" (ed. Baschek + Springer 1993 - B93) and in IAU Symp. 148. Also, a review is presented by Westerlund in A&AR 2, 29, 1990. The best current Cepheid distance modulus of the LMC is 18.55 +/- 0.10 (Feast in B93). The distance modulus from SN 1987A is 18.50 with an uncertainty probably in excess of 0.13 (Panagia + ApJ 380, L23; Dwek & Felten ApJ 387, 551). Combining these distances with Walker's apparent magnitudes of MC RR Lyraes (ApJ 390, L81) significantly increases distances based on RR Lyraes and decreases the ages of galactic globular clusters. Reviews of specific areas of MC work are given by Frogel (Neugebauer Symposium, ASP in press) and Feast (Obs., in press).

Recent work on the chemical abundances of young objects in the MC confirms earlier work (a deficiency of about -0.2 dex (LMC) and -0.6 dex (SMC)), though the element ratios are non-solar (Russell & Dopita ApJ 384, 508). Adopting a higher reddening than previously for the young SMC cluster NGC 330 removes the discrepant abundances previously found (Bessell IAU Symp. 148 & Caloi + A&A 271, 109). Plez + (ApJ, in press) find a metallicity of -0.5 dex for bright SMC AGB stars.

There is much recent work on the structure and dynamics of the MC. Following earlier work which showed a large depth for old objects in the outer parts of the SMC, Hatzidimitriou + (MN 261, 873) have shown that there is a good correlation between velocity and line-of-sight distance for these stars. An interesting development is the discovery of carbon stars between the two Clouds (Demers + MN 260, 103). Westerlund + (A&AS 91, 425 & 97, 603; A&A 260, L4) find carbon stars in the SMC as faint as $M_{bol} = -1.7$. It is still not certain whether the LMC has a kinematic (i.e. non-rotating) halo of old objects. The very old clusters are too few to give an unambiguous answer (cf. Schommer + AJ 103, 447). The CH stars, previously thought to be very old and to rule out a halo, are in fact probably the most massive carbon stars in the LMC and are probably merged binaries of intermediate age or else young objects (Feast & Whitelock MN 259, 6; see also Suntzeff + PASP 105, 350). Other studies are of young stars between the Clouds (Irwin + AJ 99, 191; Grondin + AJ 103, 1234); kinematics of PN in outer LMC fields (Vassiliadis + ApJ 394, 489); and a major study of LMC structure and kinematics in H I (Luks & Rohlfs A&A 263, 41). Contrasting models of the Magellanic orbit and stream have been proposed (Shuter ApJ 386, 101; Liu A&A 257, 505; Lin, BAAS 25, 783). Fujimoto and Noguchi (PASJ 42, 505) have developed an interesting theory to explain the formation of massive young clusters in the LMC by collision between high-velocity clouds produced by LMC/SMC interactions.

Amongst other important MC results obtained recently are the discovery of OH/IR stars (Wood + ApJ 397, 552) and a methanol maser (Sinclair + MN 256, 33P); PN images from HST (Blades + ApJ 398, L41); the absence of LMC Beta Cepheid stars as evidence of a metallicity-related pulsation mechanism (Balona MN 260, 795); the detection of gamma-rays from the LMC (Sreekumar + ApJ 400, L67); and the MC as a possible source of gamma-ray bursters (Fabian & Podsiadlowski MN 263, 49).

Massive, hot stars in the Magellanic Clouds may be subdivided into the object-type categories of OB stars, Luminous Blue Variables and related objects, and Wolf-Rayet stars. Interest in the 30 Doradus region warrants a category of its own. Space does not permit listing the many discoveries and studies of individual interesting objects, so only major investigations of the categories or very detailed analyses of special objects are included. For the same reason, it will have to be assumed that massive X-ray binaries and SN 1987A are covered elsewhere.

Some of the most significant results concerning the OB stars were the detection of new structures in their composite HR diagram by Fitzpatrick and Garmany (ApJ 363, 119), widespread CNO anomalies in B-supergiant spectra by Fitzpatrick and Bohannan (PASP 103, 1123; ApJ 404, 734), and evolutionary interpretations of these effects by Langer (A&A 252, 669) or alternatively in terms of binary evolution by Ray and Rathnasree (MN 250, 453; ApJ 410, L99 & 411, 848). Advances in the study of O-type stellar winds and physical parameters were reported by Kudritzki + at a Workshop on HST results in Sardinia during 1992 (p. 279) and by Patriarchi and Perinotto from IUE data (A&A 258, 285). New abundance analyses of B supergiants were presented by Lennon + (A&A 234, 109) and Dufton + (ApJ 362, L59). Extensive studies of OB associations other than 30 Dor were done by Greve (A&AS 85, 895), Lortet (A&A 94, 359) and Parker + (AJ 103, 1205; ApJ 399, L87). Luminous O "stars" were resolved into multiple components by Heydari-Malayeri and Hutsemekers (A&A 243, 401 & 244, 64). Grebel (A&A 254, L5) found a large number of Be stars in NGC 330/SMC while Balona investigated variability of cluster B stars (MN 256, 425; 260, 782 & 795). Useful studies of early-type eclipsing binaries were presented by Jensen + (MN 250, 119; 254, 419 & 258, 527; 260, 777).

Only a few of the numerous studies of LBVs and related objects can be highlighted here. Wolf and Stahl (A&A 235, 340) detected inverse P Cyg profiles in the prototype S Dor. N82/LMC was found to be a very unusual Ofp object with strong carbon as well as nitrogen lines by Heydari-Malayeri + and Moffat (A&A 236, L21; 244, L9; 258, L13). A detailed spectral analysis of the Ofp/WN9 object R84 was presented by Schmutz + (ApJ 372, 664). Further observations or R127, currently in outburst, were made by Schulte-Ladbeck + and Clampin + (ApJ 407, 723 & 410, L35.)

Several systematic investigations of WR stars in the LMC were completed L. F. Smith + (ApJ 348, 471 & A&A 241, 77) studied the systematics of WC spectra and evolution, while Moffat (ApJ 348, 232) and Morris + (ApJ 412, 324) presented comprehensive results on WR colors, luminosities, and energy distributions. Koesterke + (A&A 248, 166) analyzed a number of WN spectra. Seggewiss + (A&AS 89, 105) and Walborn + (ApJ 393, L14) discussed HST spectroscopy of Melnick 42 and R136 in 30 Dor, while Hutchings + (ApJ 410, 803) did the same for R31. New WR objects were discovered by Morgan and Good (MN 243, 459 & 251, 51P) and by Testor and Schild (A&A 240, 299).

Significant advances in knowledge of the 30 Doradus complex were derived from new UV, optical and IR instrumental capabilities. HST images of R136 were presented by Weigelt + (ApJ 378, L21) and Campbell + (AJ 104, 1721), while improved ground-based interferometric observations of the object were made by Pehlemann + (A&A 256, 701). Extensive UV photometry of the region was performed with UIT by Cheng + and Hill + (ApJ 395, L29 & 413, 604). IR imaging by Hyland + (MN 257, 391) and Rubio + (A&A 261, L29) revealed a new stellar generation in 30 Dor and strongly suggests sequential star formation there. More traditional techniques also provided interesting results concerning star formation (Lortet & Testor, A&AS 89, 185), the stellar content of the surrounding associations (Schild & Testor A&AS 92, 729), and the IMF of the central cluster (Parker AJ 106, 560).

New discovery lists for planetary nebulae are given by Morgan and Good (A&AS 92, 571) and spectrophotometry of many PN is given by Meatheringham and Dopita (ApJ 75, 407 & 76, 1085) and Vassiliadis + (ApJS 83, 87). Dopita and Meatheringham (ApJ 367, 115 & 377, 480) have modelled the emission from PN in order to derive abundances and place the central stars in the HR diagram.

Abundance analyses of normal stars have been carried out by Spite and Spite (A&A 234, 67), Luck & Lambert (ApJS 79, 303), and Thevenin and Jasniewicz (A&A 266, 85), yielding [Fe/H] ~ -0.6 and -0.3 for young stars in the SMC and LMC respectively. Olszewski + (AJ 101, 515) give abundances for a large number of clusters and Kontizas + (A&A 269, 107) show that the abundance is spatially constant for radial distances from the LMC center less than 8 kpc but that the abundance decreases at larger radii. Smith and Lambert, (ApJ 361, L69) and Brett (MN 249, 538) have investigated Li and CNO abundances in upper AGB stars and suggested the abundance patterns observed in these stars result from hot-bottom burning.

The star formation history of the LMC has been analyzed in detail by Bertelli + (ApJ. 388, 400) who conclude that there was little star formation until the onset of a prolonged burst ~ 4 x 10^9 years ago, perhaps continuing until the present time. Star cluster formation in the LMC bar is discussed by Bica + (AJ 103, 1859). In the outer SMC, Gardiner and Hatzidimitriou (MN 257, 195) find 7% of the stars belong to a 1.5 x 10^{10} yr old population, while most stars have ages of ~ 10^{10} yr. Stars younger than 2 x 10^9 yr lie preferentially on the eastern (LMC) side of the SMC. Further searches for C stars in the Clouds have been carried out by Blanco and McCarthy (AJ 100, 674) and Rebeirot + (A&AS 97, 603

Kinematics of various populations have been studied. The oldest populations (age > 10^{10} yr) are the old LPVs (Hughes + AJ 101, 1304) and the old clusters (Schommer + AJ 103, 447) which seem to lie in a thick disk/flattened spheroid with velocity dispersion ~ 35 km/s. The intermediate age populations (planetary nebulae, clusters, LPVs, and CH stars all belong to a single rotating disk with a velocity dispersion of ~ 25 km/s. Kinematic models yield a total LMC mass in the range 0.5-2.5 x 10^{10} $M_{\odot}$.

Studies of AGB stars have been made by Frogel and Blanco (ApJ 365, 168,; LMC bar-west) and Reid and Mould (ApJ 360, 490; SMC). Reid (ApJ 382, 143) has obtained JHK photometry of a sample of dust-enshrouded IRAS sources while Wood + (ApJ 397, 552) monitored a sample of IRAS sources in JHKL to derive pulsation periods, and showed that some of them have OH maser emission characteristic of OH/IR stars.

Kinman + (PASP 103, 1279) have found new field RR Lyraes in the outer LMC and have estimated that the LMC has a stellar halo containing ~ 2% of its mass. Smith + (AJ 104, 1430) have discovered 133 new RR Lyraes and Cepheids in the SMC and find that the former are not concentrated to the SMC bar.

Searches for and monitoring of variable stars in clusters has been carried out by a number of groups. RR Lyraes have been found in old clusters by Hazen and Nemec (AJ 104, 11) and Walker (AJ 104, 1395 & 105, 527). Welch + (AJ 101, 490 & 105, 146), have obtained CCD photometry of Cepheids in the clusters NHC 1866 and 2164. The latter study also showed that objects previously described as "superluminous giants" are actually composite images. Chiosi + (ApJ 385, 205) and Bertelli + (ApJ 412, 160) have used the cluster Cepheids to discuss the LMC distance and the Cepheid mass discrepancy.

Two major absorption-line studies of the interstellar medium were completed. Wayte (ApJ 355, 473) presented surveys of both the LMC and SMC and correlated line strengths with polarization; he arrived at a depth distribution for gas and dust. Vladilo + (A&A 274, 37) completed a depth study of the gas in the field of 30 Dor and SN 1987A; they find various gas components and indicate that the lower velocity gas is rather behind most of the stars of the study.

Measurements of the CO radio emission are being continued with the SEST in both the LMC and the SMC (see contributions by Booth & Rubio, respectively, to the 1992 Heidelberg conference, pp. 26 and 36; A&A 271, 1 & 9). Radio-continuum studies of the LMC were completed by Haynes + (A&A 252, 475) and Xu + (A&A 257, 47), which include a correlation study of the radio and the far-IR fluxes. These and other data from many wavelength domains are available in digital form from Klein (MPIfR, Bonn). A CCD-based H-alpha survey of the SMC is in progress (le Coarer, 1992 Heidelberg conference, p. 77).

X-ray data on the LMC and SMC (Einstein) have been reinvestigated and overall studies were presented (Wang +, ApJ 374, 475 & ApJS 78, 391); it is clear that higher sensitivity data and better spatial resolution are needed in X-ray observations. Some H II regions show soft X-rays explained as due to asymmetrically placed supernova explosions (Chu & Mac Low ApJ 365, 510). The ROSAT produced its first measurements of the diffuse emission of the LMC and in particular the area of the supershell LMC4 appears to show an enhancement of soft X-ray emission, giving proof of hot gas inside the cavity (Bomans + preprint).

9. Supernova Working Group
(V. Trimble)

I. DISCOVERIES AND RATES

The triennium saw 170 SNe (1990R to 1993T) announced in the IAU Circulars. Of these 73 were Ia, 13 Ib/Ic, 5 I, 56 II, and 23 received no reported spectral type. Leading discoverers were Meuller (POSS II) and Antezana (CTIO Curtis Schmidt) with 35 each, followed by Pollas (CERGA) with 29, McNaught (UK Schmidt) with 27, Wischnjewski (CTIO) 11, the Berkeley APT projects with 9, Evans (visual) with 5, Gonzalez (CTIO) 2, Hartley (UK Schmidt) 1, and single discoveries made by 16 other individuals, projects, or observatories. An additional 29 archival events turned up, mostly on POSS I plates. And we lost two: 1950E turns out to have been a minor planet (PASJ 42, 597) and 1986J a luminous blue variable (AJ 101, 1275).

The corresponding SN rate, per unit galaxy luminosity and as a function of SN type and galaxy type, remains under vigorous discussion. Since the last major review of the subject (ARA&A 29, 363), the Asiago and Sternberg searches (A&A 273, 383) have reported 0.20 events/century/10^{10} L^B_0 in E galaxies (all type Ia), and for late spirals, 0.69 Ia, 0.48 Ib/c, and 2.63 II, leading to a prediction of 1.7 ± 0.9 events per century in the Milky Way. The Berkeley automated search (ApJ 384, L9) has found SNe only in late spirals, with rates 0.4 for Ia, 1.6 for Ic, and 1.1 for type II, and predicts at least 3.3 events/century in the Milky Way. Seyfert galaxies are the same as others of the same Hubble type (A&A 239, 63), and the expected high SN rate in starburst galaxies remains elusive (AJ 102, 875 & Nat. 350, 130). Disagreement about the importance for correcting for sin i of the parent galaxy contributes to discordance in the gradient of rates from early to late spirals (A&A 234, 84; ApJ 359, 277; PASP 102, 1318; AJ 101, 845; ApJ 384, L9). Within the Milky Way, the pulsar birthrate is near 4/century (MNRAS 245, 514), and the SNR birthrate in M31 is 1.25/century (A&AS 98, 327). Galaxies with several supernovae continue to proliferate (A&A 234, 84), records now belonging to M83 with six (23A, 45B, 50B, 57D, 68L, 83N), and a galaxy with two simultaneous Ia's (1992R, 1992ac, IAUC 5543).

The SN 1987A industry wound gradually down, with only about 150 papers in the major archival journals and a timely major review (ARA&A 31, 175). The radio emission turned back on (IAUC 5086 and 5088) and the gamma-ray driven X-ray emission turned off (IAUC 5112). Visible light continues! Early plates have been recalibrated (AJ 105, 1886 & 1892; A&A 256, 447) and the Ni^{56} model generally accepted (Sov Astron Lett 18, 1). But the leveling remains a mystery. Once good photometry from ground became impossible (MN 262, 313), FOC data continued to show fading with about the time scale expected for Co^{57} (ApJ 403, 736), but the absolute amount needed (AJ 102, 1118; ApJ 384, L33 & 389, L21) is larger than that expected from models (ApJ 368, L31) or revealed by IR and gamma ray spectroscopy (ApJ 399, L137; MN 261, 522; ApJ 407, L25; Sov Astr Lett 16, 171). Since the Co^{56} contribution turned off on schedule (ApJ 357, 638; ApJ 408, 277), one is left with Ti^{44}, a pulsar, or neutron star accretion as excess luminosity sources (A&A 243, L13, 273, 155 & 253, 459). Two putative pulsars have come and gone (Nature 349, 747 & 347, 534) and the absence of evidence for a neutron star remains puzzling (ApJ 376, 234; A&A 237, L9). Evolution of the progenitor has converged on a star of 15–20 $M_{\odot}$ that shed first a slow, dense wind, then a fast, tenuous one, enriched in N and He, as was the residual envelope, after making a loop back to the blue, as many LMC stars seem to have done (ApJ 383, 295; A&A 243, 155; ApJ 380, 575; ApJ 363, 255; A&A 245, 182 & 593). There is some evidence that it was rotating (ApJ 358, L9) and may once have been part of a binary system (PASP 104, 717; A&A 260, 273). Interaction between the two winds and other circumstellar material account for the complex morphology within about 1pc of the SN (MN 261, 391; ApJ 405, 337; A&A 262, L9; Nature 350, 683). When the ejecta hits the wind material, fireworks at radio IR, X-ray and other wavelengths are expected (ASS 190, 1; Nature 355, 147 & 617; ApJ 379, 659) at some time between early 1993 (probably wrong) and 2002. Some three dozen papers address various aspects of the efficient mixing and clumpiness of the ejecta at early and late phases, but no perfect fit to the spectra is possible even making use of this freedom (PASP 103, 787). The formation of dust that began after about a year (MN 261, 522) ceased after another year (A&A 249, 474). Iron whiskers have been claimed (A&SS 200, 97). Two geometrical methods lead to distances of 51.2 and 52.3 kpc (ApJ 380, L23; A&A 249, 36). And efforts continued to produce a second neutrino burst (Soc Astr 35, 42; Sov Astr Lett 18, 79; A&A 271, 187). Derived limits on neutrino physics were slightly relaxed from early claims (PRL 67, 2605; A&A 254, 121).

SN 1993J has become the event to watch over the next triennium. An authentic amateur discovery (IAUC 5731), it was quickly confirmed, typed (II), and seen by IUE (IAUC 5731-38). The progenitor has probably been identified (IAUC 5736, 5741), it has become the third X-ray SN (IAUC 5748), and a rapidly evoluving radio source (IAUC 5751 ff). The optical light curve is double peaked, with significant resemblances to that of SN 1987A, including evidence for a comparable amount of Ni^{56}, but also interesting differences. Models are in the preprint stage, but close binaries with mass transfer seem to be the winning strategy. That two such interesting SNe should have gone off just a couple of months before IAU commission reports have to be written is probably not causal, but watch out for spring 1996 and 1999.

Type Ia supernovae present essentially the same three problems they did last time around: what are the progenitors, what is the explosion like, and are they standard candles useful for cosmology? Cases have been made for symbiotic stars (ApJ 407, L81) or symbiotic novae (ApJ 397, L87) as progenitors. The difficulty is that H blown off the RG donor will pollute the SN spectrum (ApJ 388, 665). WD accretors that burn the H steadily present the same problem (ApJ 370, 324). Double degenerates remain the most attractive model (ApJ 366, 198). Current catalogues include some wide pairs (AJ 361, 196), but none both massive enough to explode and of small enough separation to merge in a Hubble time (ApJ 374, 281). Doubts have even been cast on the double degenerate nature of the AM CVn (low mass) pairs (ApJ 384, 220). Explosion of white dwarfs accreting helium remains in accord with the observations, but explosion may not actually occur (ApJ 371, 317; A Zh 36, 401). The possible explosion mechanisms are supersonic detonation, which does not leave necessary intermediate mass elements (A&A 254, 177), subsonic deflagration, which does not make the right line profiles and such (ApJ 371, L23), double detonation, with helium igniting first, which may fit 1991T (ApJ 384, L15) and perhaps others (ApJ 370, 272), and deflagration turning into detonation, called delayed detonation. The triennium has been marked by increased enthusiasm for this last (A&A 245, 114, 246, 383, 248, L7, 253, L9; ApJ 393, L55 & 397, 304, 386, L13). The enemy of all these processes is accretion-induced collapse (ApJ 396, 649; A&A 272, 446). The precise contributions of Ia's to galactic chemical evolution are still under discussion (APA 257, 534 & 270, 288; PASP 105, 444). On the issue of whether SN Ia can be used as standard candles for cosmology, there remain some resolute supporters of an unqualified "yes" answer (A&A 236, 9, 259, 63, 269, 104; AJ 101, 1281 & 104, 696), and also some equally firm "no's" (A&AS 89, 537; AJ 103, 6632 & 104, 1156). But the eruption of 1991 bg, which was unquestionably fainter than other Virgo Ia's and also peculiar in light curve and spectral development (AJ 104, 1542 & 105, 301), has led to some enthusiasm for the idea that they are not all the same brightness, but due regard to reddening (ApJ 401, 49 & 405, L5) nature of the parent galaxy (AJ 100, 530 & 102, 236; ApJ 390, 34), and the details of light curves and spectra (ApJ 387, L33, 384, L15, 362, 235, 400, 127; AJ 104, 1541, 105, 223 & 301) allow you to separate out subclasses that can function as standard candles.

Type Ib and Ic supernovae have become well established as classes distinct from the classical Ia's. They are fainter (AJ 100, 530), presumably because they make less Fe^{56} and occur mostly or wholly among Pop I stars (AJ 103, 1788), though not necessarily in spiral arms the way SN IIs do (PASP 102, 898). They are radio sources, though different from SN IIs (ApJ 409, 162). Many modelers have converged on progenitors initially of 12-18 $M_\odot$, stripped down to helium cores of 3-6 $M_\odot$, often in close binary systems (A Zh Lett 18, 242, MN 255, 267; ApJ 361, L23; 368, L27; 366, 198; AJ 100, 1575), though the Ic could be either the high or low mass end of the sequence (AJ 100, 1575; MN 255, 267). Reasonably successful models of light curves and spectra require both early mixing (like 1987A) and late retention of the Ni^{56} decay products (ApJ 368, L27 & 308; 379, L13), though no single choice of ejected mass and explosion energy seems to fit all phases (ApJ 409, 417).

Type II supernovae sit very much where they have for the last several reports. The progenitors are massive stars, though there remains some disagreement about the exact range (AA&A 261, 433; MN 250, 786) and whether the L-type light curves go with the largest (ApJ 389, L17) or smallest (ApJ 374, 266) masses. Some surely come from binaries (ApJ 391, 246), and most have shed significant wind material before the explosion, as revealed by radio and X-ray emission from collisions by the ejecta and reprocessing of the initial uv break-out flux (ApJ 398, 248 & L 107; A Zh 36, 63 & 164). AG Car, a potential progenitor, has looped back to the blue, based on its wind properties (A&A 269, 330). The Type II's are definitely not standard candles (A&AS 98, 443; AJ 100, 530), though the majority eject about 0.07 $M_\odot$ of Ni^{56} and roughly track the light curve of 1987A after several months (AJ 100, 771 & 105, 2237). It is generally agreed that mixing of the ejecta is important generically as well as in 1987A (ApJ 390, 230). The only trouble is that the models do not actually eject anything. Shocks stall, and this is a real, physical phenomenon, not the result of numerical viscosity, a slightly wrong equation of state, etc. (ApJ 364, L22 & 376, 678; A&AS 90, 283; A&A 268, 360). Improvements in the standard model may prove to be relevant, but so far only the infall phase has been followed (ApJ 410, 740). Some more desperate measures include rapid core rotation or strong magnetic field (A Zh Lett 18, 194; A Zh 34, 163), modified neutrino physics (ApJ 398, 190 & 389, 513; Ap Lett & Comm 28, 271), and dynamical instabilities or other two-dimensional processes (Sci 258, 430; ApJ 395, 642). The situation, like the total absence of observed progenitors for SN Ia events, remains profoundly unsatisfactory.

29. STELLAR SPECTRA (SPECTRES STELLAIRES)

President:	**D.L. Lambert**
Vice-President:	**M.S. Bessell**
Organizing Committee:	**D. Baade, B. Barbuy, A. Cassatella, P.S. Conti, M. Gerbaldi, B. Gustafsson, T. Nugis, C. Pilachowski, K. Sadakane**

In recent years, the report of Commission 29 has been assembled from short reports submitted by selected authorities at the request of the President. These reports have surveyed a variety of topics but no or little pretense has been made to provide thorough coverage of 'stellar spectra' over the preceding three years. This year, the coverage is even less thorough as a result of the President's firm conviction that the literature is surveyed supremely well by Astronomy and Astrophysics Abstracts, and by review papers that abound in proceedings of conferences, colloquia, symposia and workshops, as well as books and journals devoted to authoritative reviews. In addition, the growth of 'databases', including the valuable SIMBAD, would seem to lessen the usefulness of the traditional reports of an IAU commission. Finally, the sheer volume of material published on 'stellar spectra' and a page limit imposed on the report force a selection of topics and influence the style of presentation. (Presidential procrastination is another factor!)

A sociologist examining the field of stellar spectroscopy might conclude we are experiencing, in common with most fields of astrophysics, a period of rapid growth. The measure invoked might be a hard statistical one such as the number of papers published. A closer look would show a growth in the depth and scope of problems addressed; consider, for example, the following selection: tests of primordial nucleosynthesis, scrutiny of theories of stellar evolution, observational examination of stellar photospheres including inferences about granulation and spots, studies of layers exterior to the photospheres–chromospheres, coronae, hot and cool winds. Growth could also be measured by access to the electromagnetic spectrum; a noticeable fraction of papers today draw on spectra acquired from more than one region of the electromagnetic spectrum: for example, studies of red giants may integrate optical and infrared spectroscopy of the photospheres, IUE or HST spectra of the chromospheric emission lines, infrared spectra of the circumstellar dust and millimeter observations of circumstellar molecules. All would be a theoretical exercise without telescopes to collect the spectra; the renaissance of telescope building is welcomed and the new era has just begun for individuals fortunate to have access to the Keck 10m telescope! Tools to analyse spectra continue to develop at a pleasing rate. Model atmospheres built with the traditional set of core assumptions are generally available for all parts of the HR diagrams and even for more exotic objects such as neutron stars (non-traditional assumptions in this case!) Recently, there has been a movement to replace one or more traditional assumptions with a superior idea: for example, spherical for plane-parallel geometry or non-LTE for LTE. One notes too the increasing use of non-LTE instead of LTE in the analysis of the line spectra in the extraction of abundances and atmospheric parameters. In the following pages, I illustrate this summary of the growth of our subject with examples drawn from the last three years. I thank M. Briley, D. Kirkpatrick, and A. Hatzes for helpful commentaries on selected topics. I am indebted to R.E.M. Griffin for providing the report on the activities of the Working Group on Spectroscopic Data Archives.

J. Bergeron (ed.), Reports on Astronomy, Vol. XXIIA, 305–318.

Model atmospheres. Since a separate commission has responsibility for studies of stellar atmospheres, I restrict my remarks to three illustrative studies that are largely grounded in observations and aim to explore the validity of the theoretical atmospheres. A classical atmosphere supposes the plane-parallel layers to be homogeneous. In a valuable series of explorations of the photospheric line profiles of main and near-main sequence F and G stars, Dravins, Nordlund, and colleagues (see review by Dravins, 1992, in *High Resolution Spectroscopy with the VLT, ESO Workshop*) have analysed the line asymmetries found from high resolution ($\lambda/\Delta\lambda$ = 200,000) spectra. A recent study of the subgiant β Hyi (spectral type G2IV) compared observed profiles with those predicted from *ab initio* hydrodynamic models (Dravins et al. 1993, *ApJ*, 403, 385): observed and predicted line cores fit well but the predicted line wings are weaker than observed suggesting the presence of transonic or supersonic motions–a good fit to the entire profile is found in the case of the Sun. Extension of this work is to be welcomed.

Thanks to heroic efforts by a few individuals (notably Kurucz, Bell, Gustafsson), standard models are readily available for use by the spectroscopists. The path of least resistance is to adopt the models rather uncritically in the pursuit of the defining parameters (T_{eff} and log g) and the chemical abundances. The need for construction of empirical atmospheres should not be forgotten. A paper by Fuhrmann et al. (1993, *Astr. Ap.*, 271, 451) analysing the Balmer lines H α to H δ in the Sun, Procyon, and the metal-poor stars HD140283 and G41-41 reports that the line profiles can be fitted with theoretical models consistently with a mixing length much smaller than commonly supposed: $l= 0.5H_p$ instead of $l= (1\text{-}2)H_p$. This change affects the temperature profile in the deep layers and results in effective temperatures that differ quite significantly from other values: e.g., T_{eff} = 5810 K for HD140283 which the authors note is "definitely hotter than most values cited in the literature". Of interest here is the recent proposed temperature scale for metal-poor dwarfs (King 1993, *Astr. J.*, 106, 1206) which finds T_{eff} to be about 150-200 K hotter than propsed in the literature. This proposal based on empirical and model atmosphere based T_{eff}-color relations is in line with Fuhrmann et al.'s proposed changes. There is now a need for a full empirical analysis of the spectrum—lines and continuum fluxes—to establish the veracity of standard model atmospheres. Since conclusions about the primordial fireball and the early history of the Galaxy depend on abundances analyses of halo stars, the models must be checked!

Substitution of spherical for plane-parallel geometry and near-completeness of line blanketing are not the only advances seen recently in model atmosphere construction. The replacement of the simplifying assumption of local thermodynamic equilibrium by that of non-local thermodynamic equilibrium has been seen with increasing frequency in the recent literature. An example must suffice. Kilian et al. (1991, *Astr. Ap.*, 244, 419) derive effective temperatures and gravities for 19 unevolved B stars in the field and local associations. Non-LTE line formation for hydrogen (Balmer beta, gamma, and delta), Helium (He I and II), and Silicon (Si II, III, and IV) was considered in the context of LTE line-blanketed model atmospheres. It was found that the effective temperatures obtained are systematically hotter than those given by then available photometric calibrations. The sign of the systematic difference is confirmed by a thorough reassessment of the photometric calibrations by Napiwotski et al. (1992). A plausible speculation is that the origin of the difference lies in the use by Kilian et al. of rather lightly line blanketed model atmospheres. A reinvestigation with more completely line blanketed and non-LTE atmospheres is awaited with interest.

Substitution of non-LTE for LTE necessarily places more stringent demands on the stellar spectroscopist. In particular, there is the obvious need for more atomic and molecular data, as well as numerical techniques and associated computing power to handle the coupled equations of statistical equilibrium and radiative transfer. With the completion of the Opacity Project (Mendoza 1992, in *Atomic and Molecular Data for Space Astronomy, Lecture Notes in Physics*, Springer Verlag: Berlin, Vol. 405, p. 85) and the steady reporting of other theoretical calculations of *gf*-values, as well as experimental measurements of *gf*-values and radiative lifetimes, the stellar spectroscopist is now better supplied than ever before—of course, gaps remain in the identification and the classification of lines as well as the supply of accurate *gf*-values. The reader may wish to scan the proceedings of *The 4th International Colloquium on Atomic Spectra and Oscillator Strengths for Astrophysical and Laboratory Plasmas* edited by D. S. Leckrone and J. Sugar and published as volume T47 of *Physica Scripta*. But non-LTE demands more than *gf*-values and photoionisation cross-sections, rate constants for processes involving the atom and ions of concern with other particles in the atmosphere are needed. Customarily, the free electrons are exclusively taken to be those other particles. It is left as an exercise to the reader to recall why the electrons are awarded such an exclusive role when they may be outnumbered by factors of 10^4 or more by neutral hydrogen atoms in the atmospheres of cool stars. In a pioneering set of calculations of non-LTE effects on lines in spectra of cool stars, Holweger and his colleagues (see Holweger, 1988, *IAU Symposium No. 132*, p.411) included the effect of collisions with neutral H atoms in calculations of the non-LTE effects on the excitation of Li I, Ca I, Ca II, Fe I lines. The H-atom rates of excitation were estimated from a cross-section based upon Thomson's classical estimate for ionisation of atoms. Unfortunately, there is an understandable dearth of experimental data on collisions using H atoms at the low energies prevailing in stellar atmospheres. Less understandable is the almost complete lack of quantal calculations. In an assessment of the rate constants employed by the Kiel school, Lambert (1993, *Phys. Scripta*, T47, 186) found from an *ab initio* calculation on Li I and an experiment on H excitation of neutral Na atoms that the rate based upon Thomson's classical formula greatly overestimates the true rates. If these two selected examples are typical of the rate constants for collisions involving atoms and the H atoms, it appears that the electrons despite their low abundance in the atmospheres of cool stars will determine the departures from LTE.

Non-LTE effects on molecules in cool stars are reviewed by Johnson (1993, in *Proc. IAU Coll. No. 146*, in press). In light of the numerical techniques now available, it seems surprising that, relative to the extensive literature on non-LTE in atoms and ions, so little has been done on molecules despite their known importance as opacity sources, their potential as probes of the structure of the upper photospheres and low chromospheres of cool stars, and their exploitation as indicators of key elemental and isotopic abundances. From the few recent references cited by Johnson, I select two.

Hinkle et al. (1993, *ApJ*, in press) describe observations of the H_2 infrared quadrupole vibration-rotation lines in spectra of M, S, and C stars. As Johnson notes, the reappearance of the H_2 line in a Mira following passage of the pulsationally driven shock shows that recombination of H atoms into H_2 molecules occurs about as rapidly as the recombination of TiO and CO. Since it has been thought that H_2 molecules but not TiO and CO from by three-body recombination ($3H \rightarrow H_2 + H$) , this observation might seem to "demand higher densities of H I or consideration of processes other that three-body recombination." (It is possible that the shock will force consideration of non-steady state statistical equilibrium!)

Cores of the strongest CO lines in the fundamental and first-overtone vibration-rotation lines are formed high in the photosphere where one expects the temperature minimum between the photosphere and the chromosphere. As shown long ago by Carbon et al. (1976, *ApJ*, 207, 253), non-LTE effects in the CO molecule's ground electronic state are small and do not mask the flux reversal expected in the CO line cores from the chromospheric temperature rise. More extensive NLTE calculations (Wiedemann & Ayres 1991, *ApJ*, 366, 277) involving a larger model molecule and applied to F-K stars confirm that depratures from LTE are small for dwarfs and giants. Then, the non-appearance of the flux reversal may be due to the inhomogeneous nature of the atmosphere at heights around the temperature minimum. Around this topic has raged the debate of 'bifurcation'. As with non-LTE studies of atoms and ions, the stellar spectroscopist tackling non-LTE investigations of molecules finds available data on collisional processes to be sparse and incomplete.

Large-scale Inhomogeneities–Starspots: Doppler tomography is a technique applicable to rapidly rotating stars that exploits the fact that there is a one-to-one mapping between location on the stellar surface and a wavelength in a stellar line profile. Surface inhomogeneities whether due to temperature, composition, or velocity create distortions in a line profile that are a one dimensional projection of the stellar surface. If the surface inhomogeneities are long-lived with respect to the stellar rotation period, a time series of line profiles can be interpreted to yield a two dimensional map of the stellar surface. As introduced about a decade ago, Doppler tomography relied on trial and error to extract the surface maps. A sounder mathematical foundation has been supplied with recent advances including a reconstruction algorithm based on a CLEAN-like approach (Kürster 1993, *Astr. Ap.*, 274, 851) and a sequential linear least-squares technique (Allen &Vogt, 1993, *BAAS*, 25, 872).

Doppler tomography has been applied mostly to RS CVn stars (evolved late-type stars in close binaries) and a few FK Com stars (single stars generally identified as coalesced binaries. Recent references include Vogt & Hatzes (1991, in *IAU Coll. No. 130*, p297), Hatzes & Vogt (1992, *MNRAS*, 258, 387), and Strassmeier, *Astr. Ap.* in press) for RS CVn's and Kürster et al. (1992, 7th. *Cambridge Conf. on Cool Stars, Stellar Systems, and the Sun, ASP. Conf. Series* Vol 26, p249) on FK Com stars. Remarkably, the images show almost without exception the dominant inhomogeneity to be a polar spot with several equatorial features. (One exception is Sigma Gem (a RS CVn star)—see Hatzes (1993 *ApJ*, 410, 777). The polar spot can be long-lived (time scale = several months). There is no doubt that the polar spots are real. Surface maps constructed from the same data set by different techniques results are quite similar (Strassmeier et al. 1992, *Astr. Ap.*, 247, 130; Kürster et al. 1992). In the case of HR1099, magnetic maps derived from spectropolarimetry (Zeeman Doppler Imaging) show the magnetic field to be concentrated in an annulus around a rotation pole.

Time sequence of Doppler maps reveal the evolution and migration of spots. Although not firmly established, it seems that spots emerge at the equator and migrate poleward in sharp contrast to the Sun where migration is toward the equator and spots are never seen at the poles. Observations of spot migration provide measurements of differential rotation, a key ingredient in dynamo theory. RS CVn stars exhibit a smaller differential rotaton than the Sun: Vogt & Hatzes (1991) found UX Ari to a rate of differential rotation a factor of 10 less than the solar case and in the opposite sense, i.e., the stellar pole rotates faster than the equator; Hatzes & Vogt (1992) report EI Ari to have a rate at least a factor of 100 less than the Sun but in the solar sense.

Extension of Doppler tomography to other types of stars is awaited with interest. Kürster (1993, *Astr. Ap.*, 274, 851) obtained images of a pre-main sequence star AB Dor: spots were concentrated at a latitude of +25° and no polar spot. A development with a little history is the mapping of chromospheric/coronal structure. Donati & Catala (1993, *Astr. Ap.*, 277, 123) applied Doppler imaging to Hα, Ca II H and K emission to map chromospheric features. Rotational modulation of ultraviolet emission lines has been reported from series of IUE spectra of chromspherically active stars: Ayres (1991 *ApJ,* 375, 704) discusses the 'many faces' of Capella, Procyon and β Cas where rotational modulation is **not** seen. Doppler tomography is used in a slightly different sense by Bagnuolo & Gies (1991 *ApJ,* 376, 266) who use sequences of archival (IUE) spectra of double-lined spectroscopic binaries to separate cleanly the spectra of the two individual stars. Bagnuolo et al. (1992, *ApJ,* 385, 708) apply the technique to Plaskett's star which, since its discovery in 1922, has been widely considered to be the most massive binary known.

Brown Dwarfs The search for brown dwarfs has continued apace—without proven success. Two reviews may be usefully consulted: Burrows & Liebert ('The Science of Brown Dwarfs', 1993, *Rev. Mod. Phys.*, 65, 301) and Bessell & Stringfellow ('The Faint End of the Stellar Luminosity Function', 1993, *Ann. Rev. Astr. Ap.*, 31, 433). The search will continue because as Burrows & Liebert remark "... not a single, unambiguous case of a brown dwarf or planetary companion to a star other than the Sun has been established, with the exception of the cases that likely result from close binary evolution. It should also be clear, however, that many brown dwarf candidates have been found by a variety of methods. The lack of proven examples is due to the great difficulty of establishing accurate masses, temperatures, or ages of the objects." Spectroscopic examination of the candidates may lead to the identification of genuine brown dwarfs.

One candidate is the low mass very red companion to the white dwarf GD165. The companion is about 4 arc sec distant and provides the white dwarf with an apparent infrared excess. GD 165 A and B form a common proper motion pair (Zuckerman & Becklin 1992, *ApJ,* 386, 260). Kirkpatrick, Henry & Liebert (1993, *ApJ,* 406, 701) obtained a spectrum of GD165B from about 7500 to 9100Å finding this brown dwarf candidate to have 'a spectrum unlike that of any known M dwarf'. With this spectrum, and the fact that the object has redder colors and a fainter absolute magnitude than any known main sequence star maintain its status as a brown dwarf candidate. Unfortunately, the mass is poorly known: a definitive orbit for this wide pair will not be immediately forthcoming.

A spectroscopic distinction between a low mass star and a brown dwarf was proposed by Rebolo, Martín & Magazzù (1992, *ApJ,* 389, L83) and applied by Magazzù, Martín & Rebolo (1993, *ApJ,* 404, L17) to a sample of brown dwarf candidates. Lithium is completely destoyed in low mass stars because they are extensively convective and lithium is exposed to temperatures in excess of 2.5×10^6 K and so destroyed by protons. Brown dwarfs have lower central temperatures and below a certain mass lithium is expected to be retained at the surface. Magazzù et al. obtained spectra around the Li I 6707 Å doublet for three brown dwarf candidates and three low mass dwarfs. Lithium was not detected: depletion by a factor of at least $10^{4.5}$ was estimated. Predicted depletions as a function of a brown dwarf's mass and age are given in Bessell and Stringfellow (see also Magazzù et al.). The reported depletions imply a mass $>0.065\ M_\odot$ for ages of about 10^{10} yr. Brown dwarfs with masses of less than about $0.06\ M_\odot$ are predicted to deplete their surface Li by less than a factor of 10. This initial application has not changed the status of a brown dwarf candidate. Detection of Li in, for example, low mass candidates of young nearby clusters such as Alpha Persei and the Pleiades may require a very large telescope.

Dwarf Carbon Stars. The dwarf carbon star G77-61 was dicovered in 1977 and for nearly 15 years was a lone and peculiar curiosity. When it was found to be a spectroscopic binary with a long period (245 days), G77-61 was taken to be a product of mass transfer with the carbon dwarf receiving carbon-rich material from an AGB star whose core is now the unseen companion. The mass transfer scenario certainly creates the classical Ba giants, the CH giants, and various relatives. In the case of G77-61, one might have wondered why it was apparently unique. One need wonder no longer because about 9 are now known—see Green, Margon, & MacConnell (1991, *ApJ,* 380, L31); Green et al. (1992, *ApJ,* 400, 659); Warren et al. (1993, *MNRAS,* 261, 185), Heber et al. (1993, *Astr. Ap,.* 267, L31), and Liebert et al. (1993, *ApJ,* in press). Indeed, Green & Margon (1994, *ApJ,* March 10 issue) assert that "dwarf C stars may be the numerically dominant type of carbon star in the Galaxy"! Green & Margon find barium, a presumed tracer of the *s*-process in the AGB donor star, to be enhanced in each of the 6 dwarf carbon stars observed by them. The $^{12}C/^{13}C$ ratio spans a wide range including examples with a ratio similar to that of the giant J-type stars. Available limited evidence on composition and radial velocity variations is consistent with the mass-transfer scenario. Green & Margon suggest that the carbon dwarfs are the main sequence progenitors of CH giants and/or that CH giants are the high luminosity tail to a distribution dominated by dwarf carbon stars.

The Magellanic Clouds. Stars in our nearest neighbors in the extragalactic world offer one obvious advantage over similar and brighter stars in our Galaxy: the Magellanic Clouds are at a well determined distance such that the absolute magnitudes of stars in the Clouds may be well determined in sharp contrast to the uncertainty associated with most stars in the Galaxy. The advantage largely offsets the disadvantage of the much fainter apparent magnitudes of the Cloud stars. There is also the distinctive difference in evolution of the Clouds and the Galaxy: the Clouds are metal-poor relative to the solar neighborhood and the age-metallicity realtions appear to differ too. Recent developments in the spectrometers (and detectors) available at the southern hemisphere's have led to improvements in the wavelength coverage and limiting magnitude of spectroscopic observations of Cloud members. Nonetheless observations at high spectral resolution essential for a detailed abundance analysis remain limited to the most luminous stars in the H-R diagram: OB main sequence stars, and supergiants of all spectral classes from O to M. The evolved stars have extended and often unstable atmospheres. Non-LTE effects may often be important. It is most likely, as Bessell has argued (1992, in *High Resolution Spectroscopy with the VLT*) that the present generation of model atmospheres for supergiants are inadequate; he, and Jüttner & Wolf (1992, in *The Atmospheres of Early-type Stars,* ed. Heber and Jeffery, Springer-Verlag: Berlin, p.63) stress the need for differential analyses of Cloud and similar galactic stars.

Considerable effort has been devoted to establishing the overall composition of the Clouds and interpreting differences between the Small and Large Cloud, and the Galaxy in terms of chemical evolution of these three systems. Jüttner & Wolf review analyses of B stars in the Cloud—see also a paper by Howarth et al. in the same volume. Since the OB supergiants, as in the Galaxy, are apparently enriched with CNO-cycled material from their interiors, the more easily observed supergiants are not reliable indicators of the interstellar abundances of CNO but may be used to get 'metal' abundances. Fortunately, extant telescopes may provide high resolution spectra of B main sequence stars. (Another fortunate circumstance is that non-LTE calculations are available for the C, N, O and Si ions.) Jüttner & Wolf describe initial results from an ESO Key Programme for which the following are 'safe conclusions':

- The mean metallicity of B field stars in the SMC is -0.65 dex relative to the Sun.
- Stars in young blue globular clusters are underabundant by about 0.3 dex relative to Cloud field stars.

Howarth et al. are undertaking a similar programme on main sequence B stars and their preliminary conclusion is that the B stars have the composition previously adduced for late-type supergiants and H II regions and supernova remnants. A detailed analysis has just appeared (Rolleston et al. 1993, *Astr. Ap.*, 277, 10) (Freitas Pacheco et al.'s [1993, *Astr. Ap.*, 271, 429] analysis of the emission lines of Type I planetary nebulae in the LMC show that O, S and Ar—elements probably not affected by internal nucleosynthesis in PN progenitors—have the same abundances as in H II regions and stars.)

Russell & Dopita (1992, *ApJ*, 384, 508) discuss the collection of abundances for stars and gas in the Clouds. Abundance analyses of the stars in the clouds began several years ago with the late-type supergiants. New results have appeared. Spite, Richtler & Spite (1991, *Astr. Ap.*, 252, 557—see also Barbuy et al. 1991, *Astr. Ap.*, 247, 15) have rexamined supergiants in the young globular cluster NGC330 in the SMC. The four stars examined have effective temperatures from 3900 to 7500 K. The Fe abundance, the same for the 4 stars, is [Fe/H]= -0.9 to -1.0, a value similar to that for a B main sequence star in the cluster observed by Reitermann et al. (1990, *Astr. Ap.*, 234, 109). This metallicity, as noted by Jüttner & Wolf for this and other young globular clusters, is lower than for stars in the SMC field. (Earlier spectroscopy of NGC330 stars had suggested [Fe/H]=-1.3) Luck & Lambert (1992, *ApJ, Suppl.*, 79, 303) analysed 14 Cepheids and non-variable supergiants with a sample of galactic comparison stars of somewhat lower luminosity on average. The mean metallicities [Fe/H]= -0.37 for the LMC and -0.53 for the SMC are similar to previous estimates. It is interesting that Thévenin & Jasniewicz (1992, *Astr. Ap,*. 266, 85) from 5Å resolution spectra of about 100 field stars and an analysis using synthetic spectra obtain mean abundances of [Fe/H]= -0.19 (LMC) and -0.46 (SMC). The former abundance agrees well with several photometric determinations whereas high-resolution spectroscopy appears to favor a lower value. Luck & Lambert suggest that there may be real spread in the metallicities of the LMC field stars—one Cepheid was metal-rich or [Fe/H]=0.19—but the smaller spread for the SMC is probably no larger than the errors of measurement. The compositions of the Clouds appear not to be exact replicas of galactic field stars of the same metallicity. A clear example of this difference is shown by the heavy elements, those products of the neutron-capture *r* and *s* processes. As stressed by Russell & Bessell (1989, *ApJ Suppl.*, 70, 865) and discussed by Russell & Dopita, the lighter heavy elements (e.g. Sr and Zr) have galactic ratios with respect to Fe but the heavier elements (e.g. Ba) are enhanced in the Clouds relative to Fe. Another striking difference is shown by carbon from a comparison of the Cloud stars and H regions: the C abundance reported for the H II regions is significantly less than that of the stars: is this a real effect or does it reflect a systematic error in the analyses?

With continuing and modest advances in the instrumentation, it will be possible to extend the stellar analyses to less luminous late-type stars and lower mass main sequence stars. A quantum jump will come with ESO's VLT. One expects the lower luminosity stars to yield to abundance analyses less afflicted by systematic errors. But the lessons learnt in analysing the most luminous stars of the Clouds will be of especial value because, as Bessell stresses, the new generation of large telescopes will bring the very brightest stars in the next closest galaxies (NGC 6822, IC1613, M33 and M31) within the spectroscopists' grasp.

A wide variety of stars in the Clouds have been scrutinized by spectroscopists. A selection of results follows:

- Humphreys, Kudritzki & Groth (1991, *Astr. Ap.*, 245, 593) examined a group of 'anomalous A-type supergiants' in the LMC and SMC with Balmer lines and a Balmer jump too strong for their luminosities. NLTE model atmospheres are used to show that these strong features due to hydrogen are indicative of a high He abundance: N(He)/N(H)= 0.3 to 0.5. These stars with M_{bol} = -6 to -8 are assigned an initial mass of 10 to 20 $M_{\odot}$ and suggested to be He-burning post-red supergiants. Humphreys et al. claim that such He-rich A-type supergiants will die as exploding blue supergiants, as did SN1987A.

- Studies of Wolf-Rayet stars in the Clouds (and other nearby galaxies) has continued. One striking aspect of the SMC is the small number of WR stars in the SMC. In the previous report of the Commission, van der Hucht noted that WR's numbered 'still 8 in the SMC'. Morgan, Vassiliadis, and Dopita (1991, *MNRAS* 251, 51p) discovered a ninth WR star in the SMC. This like its predecessors is a WN-type with weak emission lines and is a binary like a majority of the other eight. These features of the SMC's WR stars are generally attributed to the low metallicity of the SMC. Koesterke et al. (1991, *Astr. Ap.*, 246, 166) report a model atmosphere analysis of 19 WR stars of the WN sequence in the LMC, and the results are compared with a sample of galactic WN stars analysed by the same techniques. One result is that the terminal wind velocities and mass-loss rates of LMC and galactic stars are similar for the same WN subtypes but, according to the theory of radiation pressure driven winds, the lower metallicity of the LMC should lead to differences in the winds.

- The most luminous hot stars in the Clouds are accessible to ultraviolet spectroscopy with HST. Initial results are appearing. Walborn et al. (1992, *ApJ*, 393, L13) discuss HST ultraviolet and ground-based optical spectra of the Of/WN object Melnick 42 and the multiple system Radcliffe 136a in the 30 Dor H II region of the LMC.

- Symbiotic stars are rare in the Clouds: 3 in the SMC and 3 in the LMC identified prior to 1991. Although the sample is small, half (3 of 6) have a carbon-rich red giant but such systems are rare in the Galaxy. The commonest systems in the galaxy are those with a M-type red giant but the 3 O-rich symbiotics in the Clouds have K-type giants. Morgan (1992, *MNRAS*, 258, 639) discovered on Schmidt plates 4 new symbiotics including the first with a M-type red giant but 2 of the newly discovered 4 symbiotics have C-rich giants so that the contrast with the Galaxy is maintained.

- Spectroscopy of the red giants in the Clouds has continued. Westerlund et al. (1991, *Astr. Ap Suppl.*, 91, 425, and 1992, *Astr. Ap.*, 260, L4) discuss the evolution of carbon stars using JHK photometry and spectra. Not all the carbon stars are products of thermal pulses on the AGB. Westerlund et al. confirm the finding of carbon stars with luminosities less than that of the red giant tip; the galactic R stars have similar luminosities. Westerlund et al. speculate that there are two evolutionary scenarios leading to formation of carbon stars: M-S-C(C2-poor)-C and M-J-C. Wood et al. (1992, *ApJ*, 397, 552) examined IRAS point source in the Clouds for OH maser emission and for infrared emission. Velocities of the stellar winds, as measured by the OH line profiles, are smaller than seen in comparable galactic stars. This velocity difference is attributed to the lower metallicity of the LMC resulting in less dust in the winds. Wood et al. find no evidence in either Cloud for AGB stars with a luminosity exceeding the AGB limit of M_{bol}= -7.1 attained for a classical AGB star with a C-O core at the Chandrasekhar mass; recent theoretical arguments suggest that luminous AGB stars burning H at the base of the deep convective envelope will have luminosities exceeding this classical limit. Wood et al. suggest the discrepancy with theory may be due to either an overestimate of the convective efficiency or to very severe mass loss by those stars that exceed the classical limit. There is no doubt that stars close to but below the classical limit burn H at the bottom of their envelopes—these stars are the super Li-rich stars.

Globular Clusters. Recent spectroscopic work on stars in globular clusters is covered in "The Globular Cluster-Galaxy Connection" (*ASP Conf. Ser. Vol. 48*) One highlight of recent years has been the considerable progress made in determining the abundances of the light elements—C, N ,O, Na and Al—in cluster giants. Despite this progress, the origins of these elements and, in particular, of the variation of their abundances within one cluster remain subject to considerable debate. Explanations proposed may be grouped into two broad categories: (1) the star to star abundance differences are due to primordial variations, possibly inhomogeneous pollution by an earlier generation of massive or intermediate mass stars, and (2) the star to star variations are the result of varying degrees of deep mixing that brings to the surface the fruits of nucleosynthesis achieved inside the star.

High-resolution surveys of bright giants in several globular clusters have revealed an interesting trend (Sneden et al. 1991, *Astr. J.*, 102, 2001; Kraft et al. 1992, *Astr. J.*, 104, 645; Sneden et al. 1992, *Astr. J.*, 104, 2121; and Kraft et al. 1993, *Astr. J.*, 106, 1490): Na enrichment is anti-correlated with the O abundance among the red giants in a number of clusters. Langer, Hoffman, & Sneden (1993, *PASP*, 105, 301) suggest that Na is synthesised from Ne in those layers where the ON-cycle burns hydrogen and reduces the O abundance. Should this prove correct, the observed correlations between the CN molecular band strengths and the Na and Al abundances in several clusters (e.g., Smith & Wirth 1991, *PASP*, 103, 1158; Brown & Wallerstein 1992, *Astr. J.*, 104, 1818; Drake, Smith & Suntzeff 1992, *ApJ*, 395, L95) may result from deep mixing. Observations of the NH 3360Å band in red giants of M13 and CN-weak giants of M5 (Briley & Smith 1993, *PASP*, in press) are consistent with this idea but Briley & Smith also demonstrate that the CN-strong stars of M5 may not be explainable by deep mixing alone. Large Na abundances, which do not follow the Na-O anticorrelation, have been observed in 47 Tuc (Brown & Wallerstein 1992, *Astr. J.*, 104, 1818) and ω Cen (Brown & Wallerstein 1993, *Astr. J.*, in press) but ω Cen with a large metallicity spread is unlike any other globular cluster and its chemical history may differ from that of other globular clusters. But deep mixing is not yet demanded by the observations which really only define the nature of the nucleosynthetic alterations but do not locate the site unambigously. Indeed CN variations from star-to-star have been seen in stars near the main sequence turn-off. Suntzeff (1993, *ASP Conf. Ser. Vol .48*) provides a recent review.

Low resolution spectroscopy of stars in NGC6752 and 47 Tuc have shown differences among the CN bands in stars belonging to the upper main sequence and the main sequence turn-off; recent work on these clusters includes Suntzeff & Smith (1991 *ApJ*, 381 160), and Briley, Hesser & Bell (1991, *ApJ*, 373, 482). It is most unlikely that deep mixing occurs in these low mass stars ($M = 0.6\ M_{\odot}$) and quite improbable that the ON-cycle can run effectively. Briley et al. (1991) suggest that the spread in the abundance ratios [N/Fe] and [C/Fe] among the main sequence stars is similar to that of the red giants, implying that the bi-modal distribution of CN strengths in 47 Tuc is not solely the result of a mixing process in the individual stars.

In the transition from a subgiant to a giant the star's convective envelope is predicted to dredge CN-cycled material to the surface. Since it is unlikely that this process operates simply as predicted and may operate differently in different stars due, say, to different rotational angular momentum, the spread of CN in giants cannot be exactly that of main sequence stars. Reasonably conclusive evidence of mixing in cluster red giants was presented prior to the current reporting period. More recently, a trend with luminosity along the red giant branch was reported in NGC6752 and M4 by Suntzeff & Smith (1991). The low $^{12}C/^{13}C$ ratio indicative of the CN-cycle in equilibrium have been reported for giants by Brown, Wallerstein & Oke (1991, *Astr. J.*, 101, 1693).

Despite recent progress in defining the abundances of the light elements, the origins of the star-to-star variations of these abundances within a cluster displaying no measureable variations in abundances of many other elements (ω Cen aside) remain unclear. It is possible that primordial enhancements and the alterations due to deep mixing both occur. Definition of the former should shed light on the origins of the globular clusters. Isolation of effects due to deep mixing provides a test for theories of the evolution of low mass metal-poor giants. Introduction of fiber-fed spectrographs and the next generation of large telescopes will permit quite detailed exploration of globular cluster main sequences. Perhaps, the next report of the commission will discuss spectroscopy that unambiguously identifies the roles of primordial enhancements and deep mixing.

Post-AGB Stars. The stellar population at high galactic latitudes is not purely comprised of low mass metal-poor main sequence and giant stars and white dwarfs. Population I OB stars (i.e., massive stars) are known at great distances from the plane; Keenan (1992, *Q. J. R. Astr. Soc.*, 33, 325) provides a concise review of these (and other high latitude) stars under the title 'Star formation in the Galactic Halo'. Normal Pop. I stars as verified by abundance analyses have been found up to about 20 kpc from the galactic disk. Conlon, Theissen, & Moehler (1993, *Astr. Ap.*, 269, L1) discuss PG 1708+142 which is at 10 kpc from the disk. Although some of these stars may have been ejected from the disk, several have locations and space motions that are incompatible with a birth in the disk. These stars formed either in clouds far from the disk or were accreted from outside. Keenan notes that the origins of these stars may become clearer when photometric and spectroscopic surveys of existing surveys of galactic stars and the halo of M31 are completed.

Not all the high latitude OB stars have a composition typical of Pop. I stars. Metal-poor B stars have now been found and identified as post-AGB stars, i.e., stars that have evolved from the red giants at the tip of the AGB and are nearing the top of the white dwarf cooling track. Discoveries are included in papers by Conlon et al. (1991, *MNRAS* 248, 820), McCausland et al. (1992, *ApJ*, 394, 298), and Conlon, Theissen, & Moehler (1993). On the basis of available data, it would seem that these hot post-AGB stars do not have the compositions of Population II dwarfs and red giants. Of course, C, N, and O have to be set aside in making this comparison. One difference (see McCausland et al. 1992) is that Ca appears more underabundant than Mg or Si: e.g., [Ca/H] = -2.9 but [Mg/H] = -1.8 for LS IV -4.01 for which [Fe/H] < -1.6. At this metallicity in the halo, [Ca/H] = [Mg/H]. One should note that in this cited example, the Ca and Mg abundances each depend on a single line. Further pursuit of these stars is warranted as abundance differences with halo dwarfs and giants (the progenitors of the B stars) should reveal aspects of the history of the post-AGB stars.

According to models of post-AGB stars, the B-type stars are evolving very rapidly. So rapidly that one in just 4 decades passed from the care of this Commission: Parthasarathy et al. (1993, *Astr. Ap.*, 267, L19) discuss CPD -59° 6926 as 'a post-AGB star which has turned into a planetary nebula within the last 40 years'—the star was classified as a Be star as late as about 1980 but the optical spectrum in 1990 shows an emission line spectrum of a planetary nebula. Cold dust in a detached shell is revealed by the IRAS fluxes. IUE spectra show a high velocity stellar wind with a central star of type O8V. An apparently more advanced example of the post-AGB to planetary nebula conversion may be IRAS 0652-0337 ('final mass-loss episodes before the formation of a planetary nebula?'—see García-Lario et al. 1993, *Astr. Ap.*, 267, L11) and a less advanced example may be LS II +34° 26 (Parthasarathy 1993, *ApJ*, 414, L109).

The hottest post-AGB stars are not the only ones to have attracted considerable attention this triennium. Abundance analyses of several high galactic latitude A, F and G supergiants show extreme deficiencies of iron. A few are apparently more metal-deficient ([Fe/H]= -4.5 or so) than the most metal-poor dwarfs and giants known. This is a remarkable finding for stars identified with a very rapid phase of evolution; Bond (199a, IAU Symposium No. 145, p. 341) suggested that these post-AGB stars are 'chemically peculiar' in the sense that the iron abundances do not reflect the stars' initial abundances. Adapting a published suggestion about Lambda Bootis stars that alluded to similarities with post-AGB stars, Bond suggested that the iron deficiency was created by a separation of dust and gas with the present atmospheres of the post-AGB stars being composed largely of gas from which grains have been removed; the abundance pattern of the post-AGBs resembles that of interstellar gas. Stars analysed recently include HD52961 with [Fe/H] possibly as low as -4.8 (Waelkens et al. 1991 *Astr. Ap.* 251, 495) and HD44179, the central star of the Red Rectangle, with [Fe/H]= -3.2 to -3.5 (Waelkens et al. 1992, *Astr. Ap.* 256, L15). These post-AGB stars as well as others with less extreme metal-deficiencies generally have near-solar abundances of C, N, O, and S. (The C, N, and O abundances show alterations due to H and He burning.) Van Winckel, Mathis & Waelkens (1992 *Nature* 356, 500) detect Zn in HD52961 with an abundance [Zn/H] = -1.3 or [Zn/Fe] $\simeq$ 3.5. Zn like S is only slightly depleted in interstellar and circumstellar gas. The high Zn/Fe ratio confirms that these stars are chemically peculiar. Mechanisms for achieving the gas-dust separation are discussed by Mathis and Lamers (1992, 259 L39) and Waters, Trams, & Waelkens (1992, *Astr. Ap.* 262, L37). Parthasarathy, García-Lario and Pottasch (1992, *Astr. Ap.* 264, 159) discuss HD52961 with [Fe/H] $\lesssim$ -1.0 but near-solar C, N, O, and S abundances. Waters et al. (1993, *Astr. Ap.* 269, 242) present a binary model for the well known F supergiant 89 Her including a highly asymmetric geomtery for its circumstellar envelope. A review of post-AGB stars, somethimes called proto-planetary nebulae is provided by Kwok (1993, *Ann. Rev. Astr. Ap.* 31, 63).

Testing Primordial Nucleosynthesis. Stellar spectroscopy is not a pursuit that is insulated from other major astrophysical activities. This claim is proven by the activity over the last three years in the field of the abundances of the light elements—Li, Be, and B—in halo dwarf stars and the interest taken in the results by cosmologists! Since the discovery in 1982 by M. and F. Spite of the Li 'plateau' observations of Li in halo (and other) stars have assumed a frenetic pace. As originally noted and initially confirmed, the plateau denotes that the warmer halo dwarfs have a uniform Li abundance near log (Li) = 2.1 on the usual scale. Recent observations have suggested that uniformity of Li abundances is not a universal attribute of halo dwarfs.

A few halo dwarfs are now known to be very deficient in Li, even stars in the effective temperature range associated with the plateau. Examples of Li-poor stars were reported by Hobbs, Welty, & Thorburn 1991, *ApJ*, 373, L47; Spite et al. 1993, *Astr. Ap.*,

271, L1; Thorburn 1992, *ApJ*, 399, L83; Thorburn & Beers 1993, *ApJ*, 404, L13). The Li deficiency of these stars is typically an order of magnitude with respect to the plateau. There is as yet no evidence of other peculiarities to be associated with the Li deficiency: for example, it seems that these stars are not exclusively binaries.

Early studies of the plateau suggested that the scatter in Li abundances was small with observational scatter the dominant contributor. But now Deliyannis, Pinsonneault, and Duncan (1993 *ApJ*, 414, 740) suggest that a real dispersion of 10 per cent in the Li abundances may be present. Thorburn (1994, *ApJ* in press) has obtained Li abundances for almost 100 halo dwarfs and finds 'a larger lithium abundance dispersion than can be explained by observational errors alone'. Thorburn also finds the Li abundance to decline with decreasing metallicity and attributes this slope of the plateau to lithium production by Galactic sources: excess scatter about the Li abundance—metallicity relation is identified with Li production and a 2 Gyr dispersion in the halo's age—metallicity relation. The primordial Li abundance from the hottest and most-metal poor stars is put at log(Li) = 2.22 ± 0.20 dex by Thorburn. Li production may be due to cosmic rays and the alpha on alpha reactions they induce in the halo gas. Since these fusion reaction may make ^{6}Li more than ^{7}Li, the 'extra' Li -rich stars of the plateau might be expected to show some ^{6}Li accompanying the primarily primordial ^{7}Li. Smith, Lambert, & Nissen (1993 *ApJ* 408, 262) reported a detection of ^{6}Li in HD84937 but not in a cooler dwarf HD19445. The detected isotopic ratio of ^{6}Li/^{7}Li= 0.05 ± 0.02 for HD84937 corresponds to an initial ratio of about 0.10 when destruction of ^{6}LI (relative to ^{7}Li) is considered according to standard (non-rotating) models. (The same models predict essentially the complete destruction of ^{6}Li in HD19445, a prediction consistent with the absence of ^{6}Li in this star.) The initial ^{6}Li content is consistent with the Be abundances of halo stars and the scenario in which cosmic rays produce Be from spallation reactions, primarily proton on oxygen collisions.

Knowledge of the Be abundances of halo stars has increased dramatically recently—see Gilmore, Edvardsson, & Nissen (1991, *ApJ*, 378, 17), Gilmore et al. (1992, *Nature* 357, 379), Ryan et al. (1992 *ApJ*, 388, 184), Rebolo et al. (1993, in *Origin and Evolution of the Elements*, p.149) , Molaro, Castelli & Pasquini (1993, in *Origin and Evolution of the Elements*, p. 153 and Boesgaard & King (1993, *ApJ* in press). Two results deserve deserve brief mention: (1) There is no evidence for a Be plateau to the metallicity limit presently reached ([Fe/H] ≃ -2.6); (2) the Be abundances scale linearly with metallicity or the oxygen abundance. The failure to locate the Be plateau is fully consistent with predictions of nucleosynthesis by the standard big bang that is identified with ^{7}Li production and the concomitant production of ^{2}H, ^{3}He, and ^{4}He: the predicted Be abundance is substantially below the lowest Be abundance detected or even detectable. Simple models of the production of Be by cosmic ray spallation reactions predict the Be abundance to scale approximately with the square of the oxygen (or iron) abundances, not linearly as observed—see Prantzos, Casse & Vangioni-Flam (1993 *ApJ* 403, 630 and in *Origin and Evolution of the Elements*, p. 156).

Boron was detected in three metal-poor stars by Duncan, Lambert & Lemke (1992 *ApJ* 401, 584) from spectra acquired with the Hubble Space Telescope. One B I resonance line was detected as a blended feature at 2496.8Å. A LTE analysis gave a B/Be ratio of about 10 and B follows Be in increasing approximately linearly with the oxygen or metal abundances. Since the ratio B/Be ≃ 10 is predicted directly for production by spallation reactions, it was suggested that Be and B in the early Galaxy are both fruits of interactions between cosmic rays and nuclei in the halo gas. This simple picture has been challenged by non-LTE calculations of the BI line (Kiselman 1993, *Astr.*

Ap. in press) showing that the LTE abundances underestimate the true abundance by a factor of 8 in the most metal-poor stars. The implied B/Be ratio of about 80 is in excess of what can be attributed to spallation. There is a possibility that the absorption attributed to the B I line is due instead to a weak transition of a more abundant species such as Fe I. But Lemke, Lambert & Edvardsson (1993, *PASP,* 105, 468) argue that this attribution in unlikely; this conclusion is based, however, on an LTE analysis of the behavior of alternative identifications. Alternatively, the B in the early Galaxy may come from another source, i.e., Type II supernovae where B (and Li) is a product of neutrino-induced nucleosynthesis. Calculations suggest that, Be is much less easily synthesised in Type II SN. Further observations of boron with HST and non-LTE analyses will presumably resolve the issue of the B/Be ratio in stars and, hence, of the origins of these light elements.

Working Group on Spectroscopic Data Archives

R.E.M. Griffin writes: The formation of a Working Group on Spectroscopic Data Archives was recommended in 1991 by IAU Resolution C13, in which "... the importance of safeguarding [spectroscopic] data, and the need to create an accessible archive of the observations" was formally recognized. In an admirable display of democracy, the issues were re-aired at a European venue later that year before a core membership of a Working Group was nominated. At a business meeting of its core members, plus others by invitation, the WG was eventually proposed in March 1992, and established by its nominal parent, Commission 29, in June of that year. The members of the WG now number 12, and represent Australia, Japan, Ukraine, Canada, USA and Argentina as well as Western Europe. The history and rationale of the project, and the need for a WG to oversee it, has been described in detail in 'Comments on Astrophysics' 16, 167, 1992.

The aims of the Working Group are to propose, stimulate and encourage the formation of archives of spectroscopic data, and also to devise a scheme for guaranteeing a future for spectroscopic plates. As individuals we have attended relevant meetings, visited observatories or other places where astronomical data are handled, and written numerous short articles, memoranda, reports and letters in places where they were intended to be noticed by project designers as well as data users. As a Group we held an open meeting in Haguenau in 1992 September (during "Astronomy from Large Databases II"), and another during the Trieste Workshop in April 1993. Our final business meeting for this triennium is planned for Spring 1994 and, together with the WG on Radioastronomy data bases, we are holding a Joint Discussion meeting at the 1994 GA.

The concept of generating a public route into a data bank through which spectra can be re-used, for whatever purpose, has not yet won universal acceptance, usually because the necessary resources are not made available. The concept of compiling archives of **reduced** data, and towards which the WG is firmly committed as a general aim, is even less popular because it will require far more time and effort. However, compelling arguments suggest long-term advantages that will be worth the initial cost if uniform and reliable calibration procedures can be set up in collusion with at least some of the team(s) who created the spectrograph(s). The enormous numbers of person-hours which are presently spent in re-reducing spectroscopic data will be saved, and errors made by those unfamiliar with routines will be avoided, not to mention the scientific advantages of panchromatic information for almost all research. The WG suggests as one solution the development of software to cope with many of the current instrumental setups in a robotic manner. It also proposes that the raw-data banks of those reduced spectra should still be available, if not on-line, for the purist.

In its first phase, the WG is concentrating on the fate of relatively high-dispersion spectra of stars. Several ground-based observatories adopt the philosophy that the data belong to the observer, so permanent copies of the raw data are not retained beyond an interim security version. Arguments relating to the scientific advantages for research and to the enhanced accountability of the instruments are beginning to bear fruit, and some new stores of raw data are now being started. Space missions, whose data are expected to arrive regularly for a relatively brief period and whose calibration procedures may have to be established with hindsight, are already geared towards servicing data requests from a data bank, though do not (yet) have facilities for creating uniform sets of reduced data.

The WG has also been concerned about the long-term safeguard for photographic spectra. There are about 35 stores of spectroscopic plates around the world; most are now closed, and very few have a service or loan scheme. For the would-be user of such data, an initial requirement is an on-line index of the contents. Most observatories can produce detailed logbooks of the observations or notebook equivalents from which index tables can be generated. After deliberating at Trieste on the details which that index should contain in order to keep it commensurate with the FITS header information that is creamed off automatically from digital observations, the WG issued a set of Guidelines for circulation. A subsequent test-run involving the historical spectra at the Cambridge Observatories, UK, produced a partial Index Table that can now be supplied as a sample upon request.

30. RADIAL VELOCITIES (VITESSES RADIALES)

PRESIDENT : G. Burki
VICE-PRESIDENT : C.D. Scarfe
ORGANIZING COMMITTEE : J. Andersen, L.N. da Costa, A.P. Fairall, K.C. Freeman, J.B. Hearnshaw, D.W. Latham

1. RADIAL VELOCITIES OF GALAXIES (A.P. Fairall, Cape Town University)

Cosmological implications, arising from the existence of large-scale structures and large-scale flows, have been major motivations for redshift surveys of galaxies and clusters of galaxies. This has been coupled with technical advances in the sensitivity of CCD detectors and multiple-object observations using fibres. Many fibre systems are now producing a hundred or more radial velocities of faint galaxies per night. Amongst current developments is the MEFOS system at the European Southern Observatory (where fibres are positioned via computer control, and which will supersede the older "plug in" OPTOPUS system). Also, at the Anglo-Australian Observatory, a system is under development that will enable 400 fibres (set in position by computer) to cover a two degree field on the 3.9m telescope, while a new Flair II (152 fibres) system has been commissioned on the wide-angle Schmidt telescope.

Although many investigators are publishing results and analyses of their redshift surveys (and showing plots of redshift "slices" at conferences), relatively few of the radial velocities are actually appearing in the literature. In part, this may be that the radial-velocity databases are intentionally being kept confidential for the time being. However, it may also indicate problems in mass reductions - since there are some fibre-redshift papers that show a greater proportion of discrepant redshifts than was the previous norm. A rough guess is that there are now "public" redshifts for approximately 30000 galaxies, with a similar number as yet unpublished.

Wide-angle surveys are not normally able to make use of fibres. The extension of the Centre for Astrophysics redshift survey (CfA2) to m=15.5 is complete over twelve declination slices. In the South, the complementary extension of the Southern Sky Redshift Survey (SSRS2) by L. da Costa & collaborators is virtually complete. M. Strauss & collaborators have published radial velocities for an all-sky survey of selected IRAS galaxies. The QDOT survey of Ellis, Rowan-Robinson & collaborators covers a similar sample.

Using fibres, the Las Campanas Deep Redshift Survey is coverering both a "Checkerboard" pattern of areas and declination strips in the southern skies, and has over 10000 redshifts. A "key-project" is under way at ESO by Vettolani & collaborators, using multi-fibres, that should obtain some 5000 redshifts of galaxies with $B(J) < 19.5$ in a strip of Declination 1.5 degrees wide. Koo & collaborators are continuing their deep pencil-beam surveys that revealed unexpected periodicity in distribution of redshifts in the directions of the galactic poles. A number of groups are looking at Abell and Abell-Corwin-Olowin clusters, such that the overlaps should in time provide a good comparison of radial velocity measurements. Kraan-Korteweg & collaborators have a large number of redshifts for galaxies partly obscured by the southern Milky Way.

The recent Principal Galaxy Catalogue (Paturel et al., 50.002.016) and the 3rd Reference Catalogue (de Vaucouleurs el al., 53.002.034) serve as useful sources of galaxy radial velocities. An ongoing database of galaxy redshifts (ZCAT) is maintained at the Center for Astrophysics by J. Huchra & collaborators, and an independent one for southern galaxies (SRC) is kept by A. Fairall in Cape Town. Redshifts can also be obtained through IPAC (NASA extragalactic database) and the Astronomical Data Centres at Goddard and Strasbourg. In respect of cataloguing, an appeal is made to all observers to publish accurate positions, and not just designations of galaxies, and to publish heliocentric velocities. Otherwise much of their efforts may be lost to the astronomical community.

2. STARS IN THE LOCAL GROUP OF GALAXIES (K.C. Freeman, Mount Stromlo Observatory)

2.1. Magellanic Clouds. IAU Symposium 148 (53.012.005) includes papers on velocities of Magellanic Cloud objects. Cowley & Hartwick (53.156.148) measured velocities for 40 CH stars in the LMC; these stars have an asymmetric velocity distribution suggesting two discrete kinematical groups. Hughes et al. (53.156.079) studied the kinematics of the long period variables in the LMC; the oldest

J. Bergeron (ed.), Reports on Astronomy, Vol. XXIIA, 319–324.

stars indicate the presence of a spheroidal population in the LMC. Vassiliadis et al. (1992) measured velocities of planetary nebulae in the outer LMC.

Hatzidimitriou et al. (1993) showed that the radial velocities of clump giants in the outer SMC are well correlated with the distance along the line of sight, over a line of sight distance of at least 10 kpc; this may be the result of tidal interaction of the SMC and LMC.

The kinematics of star clusters in the LMC were actively studied. Storm et al. (53.156.137) measured velocities of 3 distant LMC clusters. Schommer et al. (55.156.003) observed giants in 83 clusters at the Ca triplet and confirmed that the oldest clusters appear to lie in a rotating disk. Suntzeff et al. (1992) measured velocities for NGC 1841 and the Reticulum cluster, confirming their membership in the LMC and estimating their M/L ratios from the internal velocity dispersion. Fischer et al. (1993) observed 52 supergiants in the young LMC binary cluster NGC 1850 to estimate its M/L ratio.

2.2. Dwarf Spheroidal Galaxies. Pryor & Kormendy (52.157.024) reanalysed the kinematics of the Draco and UMi galaxies, and showed that their central dark matter densities are at least 0.05 $M_{\odot}pc^{-3}$ but could be as high as $1M_{\odot}pc^{-3}$. Da Costa et al. (53.157.225) measured radial velocities of 14 giants in the Sextans galaxy, and derived a systemic velocity of 230 kms^{-1}. Mateo et al. (54.157.198) made a major study of the Fornax galaxy, with new velocities for 44 giants and 4 globular clusters; they derived a central M/L ratio of 12. Mateo et al. (1993) measured velocities for 23 giants in the Carina galaxy; the velocity dispersion is 6.8 kms^{-1} and the M/L ratio is 39, so Carina's dark matter content is intermediate between Draco and UMi on one hand and the larger dwarf spheroidal galaxies on the other.

Munari (54.157.147) studied a symbiotic carbon star in the Draco galaxy. This star shows significant radial velocity variations and appears to be a binary (carbon giant with a hot compact companion).

2.3. M31. Federici et al. (52.157.044, 54.154.093, 55.157.305) studied the velocities of globular clusters in M31 to estimate the mass of M31 and investigate the orbital properties of the cluster system. The data suggest that the clusters have circular or isotropic orbit distributions. Huchra et al. (53.154.028) measured velocities and metallicities for a large sample of clusters in M31. They find that the inner metal rich clusters form a central rotating disk, but in the outer regions there is no distinction between the rotation of the metal rich and metal poor clusters.

2.4. M33. Schommer et al. (53.157.024) measured the kinematics of star clusters in M33. The younger clusters rotate with the HI disk, while the older clusters form a slowly rotating halo.

3. STRUCTURE OF THE MILKY WAY

(D.W. Latham, Center for Astrophysics, and
B.W. Carney, University of North Carolina)

Radial velocities have been used to study several aspects of the Galactic center, which has a diversity of structures that may even include a black hole. The gas near the center has been studied by Mehringer et al. (1992) in the vicinity of Sgr B1 and B2 . Herbst et al. (1993) have used infrared imaging spectroscopy to obtain high spatial (1") and low velocity (100 km/sec) resolution near the center. Binney et al. (54.155.109) have discussed the issues involved in the studies of gas dynamics of the center. Stars are immune to hydrodynamical effects, and new radial velocity work on OH/IR stars has been undertaken by Lindqvist et al. (55.155.096) and by Langevelde (Leiden) & his colleagues. Work is continuing on high-resolution IR spectroscopy of M giants by Sellgren (Ohio State) & the Riekes (Arizona). Blitz & Spergel (54.155.089) conclude that the H I kinematics in the Galaxy's central regions are best understood if there is a bar, following the earlier suggestions of Binney et al. (54.155.109).

The dynamics of the Galaxy's central bulge have also received considerable attention. Minniti et al. (1992) have studied the kinematics of K giants in two bulge fields, finding strong evidence for rotation and a velocity dispersion that decreases with increasing galactocentric distance . The transition between the bulge and the galactic halo is being studied by Morrison (1993), using objective prism techniques to identify K giants and radial velocities to help assign the stars to their respective stellar populations. The bulge or the thick disk appear to dominate, even at distances that used to be considered the province of the halo.

The study of stellar populations in the Galaxy has benefitted tremendously from extensive radial velocity programs. A survey of proper motion stars was completed by Norris & Ryan (54.155.095), who suggest that the thick disk may not be a discrete component but is instead an older, hotter extension of the thin disk. A survey of the kinematics and metallicity of an expanded survey of proper-motion stars by B. Carney (North Carolina), D. Latham (CfA), and J. Laird (Bowling Green) is nearing completion, as are surveys of southern hemisphere proper motion stars by Ardeberg et al. (53.115.046) and M. Grenon (Geneva). Kinematically unbiased samples are receiving attention, too, especially the south galactic pole K giants studied by Flynn & Freeman (1993), faint proper motion stars toward the north galactic pole studied by Majewski (55.155.008), and the faint stars in the galactic meridional plane being studied by R. Wyse (Johns Hopkins) & G. Gilmore (Cambridge). Radial velocities from these programs should help us understand the relative numbers and evolutionary histories of the major stellar populations.

The fundamental parameters describing galactic structure are becoming better defined from kinematical studies. Jaschek & Valbousquet (53.155.035) have observed 5800 nearby bright stars to study the Sun's motion and stellar velocity dispersions as a function of spectral type, while Kuijken & Gilmore (53.155.022) have discussed the Galactic force law and the disk's column mass density.

The Galaxy's total mass, based on radial velocities of distant stars and satellite systems, was discussed by Norris & Hawkins (54.155.093), and by Kulessa & Lynden-Bell (55.155.090), with the conclusion that the Galaxy's total mass exceeds $1.2\ 10^{12} M_{\odot}$.

4. STAR CLUSTERS (J.C. Mermilliod, Université de Lausanne)

4.1. Open clusters. The interest for the measurement of radial velocities has received new motivation due to the new instruments and detectors and the generalisation of the correlation technique. The results obtained are encouraging: it seems possible to obtain radial velocities with a good precision, 2 to 3 km/s, for early type stars. This has been demonstrated with observations of stars in the Pleiades by two different groups (Morse et al., 53.036.025; Liu et al., 54.111.004) and in other young open clusters (Liu et al., 54.153.056).

Old favorites have been revisited: Alpha Persei and IC 4665 (Morrell & Abt, 54.153.037; 1992). The situation in IC 4665 is even more confused, because further binaries have been rejected from binarity: only four of the 15 stars observed are spectroscopic binaries. The detection of binaries on the upper main sequence has been extended to the Orion cluster (Abt el al., 53.153.15) and to two clusters located in the Eta Carinae region, Tr 14 and Tr 16 (Levato et al., 54.153.016; 53.153.026).

Photoelectric scanners have been used to measure radial velocities of red dwarfs in nearby open clusters to search for binaries in Praesepe (Bolte, 54.153.017) and in the Pleiades (Mermilliod et al., 1992), to identify new member stars in the halo of the Pleiades (Rosvick et al., 55.153.002) and Praesepe clusters (Mermilliod et al., 52.153.006). Radial velocities have been obtained to confirm the membership of low-mass stars in M67 (Glushkova & Rastorguev, 53.153.022), Pleiades (Soderblom et al., 1993), Alpha Persei cluster and IC 4665 (Prosser, 55.153.010; 1993). The most remarkable result is the publication of the orbital elements for 22 spectroscopic binaries among red giants and upper main sequence stars in M67 (Mathieu et al., 52.153.020).

Red giants in numerous clusters have also been observed with photoelectric scanners (Mermilliod & Mayor, 52.153.009; Valitova et al., 52.153.43; Glushkova & Rastorguev, 53.153.022, 53.153.035, 53.153.036; Claria & Mermilliod, 1992). The aims of most of these studies are to determine the membership of the red giants, to discover new binaries or to define the global cluster velocity.

The cross-correlation technique was also used to obtain radial velocities and metallicity determinations for red giants in several old open clusters (Friel & Janes, 1993).

4.2. Globular clusters. Photoelectric scanner radial velocities have been obtained for 159 stars in M4, M5, M10, M12 and M71 (Ortolani et al., 54.154.127) to study the velocity dispersion. Meylan & Mayor (54.154.029) have used radial velocities for 127 stars in NGC 6397 in order to calculate the velocity dispersion and to compute a multi-mass, anisotropic dynamical (King-Michie) model from which the cluster characteristics have been determined. On the basis of radial velocities for 72 stars in the clusters NGC 5053 and 5466, Pryor et al. (54.154.105) have shown that five of these stars are probably spectroscopic binaries.

4.3. Magellanic Clouds clusters. Radial velocities with a precision of about 1 km/s have been obtained in NGC 1866 (Fischer et al., 55.154.015). The observed flattening of the cluster is consistent with the rotation deduced from the radial velocities.

5. SPECTROSCOPIC BINARIES (A. Duquennoy and M. Mayor, Observatoire de Genève)

Of course the domain of spectroscopic binaries is still a very rich, active and fruitful domain of astronomy. This can be seen from an inspection of the section 120 of the *A & A Abstracts* which lists each year more than 100 entries for spectroscopic binaries. Evidently one can only select here a few references, which we hope will well illustrate the rich and exciting new results and developments in the field of spectroscopic binaries.

The various leading groups, in terms of high precision velocities, are still the same as in the 1988 IAU report. Reaching a long-term precision better than 50 $m\ s^{-1}$ will be soon a common performance. At this level of precision one has difficulties to distinguish between velocity variations due to a companion and those due to intrinsic instabilities inside the star. Such high-precision radial-velocity surveys are in progress at McDonald, CFHT and Lick Observatories in order to detect planetary sized companions.

An important novelty in the domain of the SBs is the incursion of high angular resolution techniques, now capable to resolve SBs with periods of only a few tens of days and semi-major axis of the order of 0.01", thanks to long base interferometry: MARK III at Mt Wilson, USA, and next generation SUSI, in Australia (see IAU Coll. No 135 on *Complementary Approaches to Double and Multiple Star Research*, 1992, San Francisco), and thus opening the way to "massive" production of stellar masses determinations.

Spectroscopic binaries continue to be accumulated among various stellar populations: Pre-Main-Sequence stars (Andersen, 52.120.007; Mathieu et al., 53.120.011; Pasquini et al., 54.117.041), Wolf-Rayet stars (Williams & van der Hucht, 52.120.023; Annuk, 53.117.011), IAU velocity standards (Scarfe, 55.120.022), Post-AGB stars (Griffin, 53.120.016), White Dwarfs (Brabaglia et al., 52.126.074), F str λ4077 stars (North & Duquennoy, 53.113.010) and other peculiar stars (Howarth & Heber, 52.120.012; Morrell & Niemela, 52.120.022).

R. F. Griffin continues to produce spectroscopic orbits for a number of stars including composite spectra (Griffin, 52.120.019, 55.120.023), eclipsing binaries (Griffin et al., 55.119.001), long-P SBs, $P > 10yrs$ (Griffin & Herbig, 54.120.007), some of them having low-K amplitudes, $K_1 < 2km\ s^{-1}$ (Griffin, 53.120.016). At this point, it seems strange that with the increasing precision of the radial velocity determinations, no very low amplitude ($K_1 < 0.1\ km\ s^{-1}$) binaries have yet been found. This probably means that brown dwarf companions are rare if any.

However predictions of the rate of brown dwarf companions can be made from systematic surveys for duplicity in unbiased samples. In the solar neighbourhood, this proportion is estimated to $8 \pm 6\%$ of the G-dwarf primaries (Duquennoy & Mayor, 54.118.010).

Other recent systematic and fundamental spectroscopic surveys for duplicity include: nearby stars (Tokovinin, 54.120.026; Fischer & Marcy, 1992), halo stars (Latham et al., 1992), S stars (Jorissen & Mayor, 1992), open clusters (Mermilliod et al., 52.153.006; Mermilliod & Mayor, 52.153.009; Mathieu et al., 52.153.020; Mermilliod et al., 1992), globular clusters (Prior & al., 54.154.105). They mainly tend to obtain unbiased distributions of orbital elements, especially of Period, Eccentricity and Mass-Ratios, which are believed to be tracers of the binary star formation processes.

The importance of these systematic surveys is considerable, as they allow unprecedented insights into the (sometimes combined) effects of physical processes among SBs, such as tidal effects, AMLOSC, wind accretion, dynamical evolution, overshooting in stellar evolution model, and of course star formation mechanisms. This was the subject of a workshop held in Switzerland to honour R.F. Griffin's 100th paper in *The Observatory.* Twenty-three contributions were included in the proceedings (1992, in *Binaries as Tracers of Stellar Formation*, Cambridge).

6. PULSATING STARS (G. Burki, Observatoire de Genève)

The Baade-Wesselink method, requiring quasi-simultaneous photometric and radial velocity measurements of pulsating stars, allows the determination of mean radius and distance (if the effective temperature is also known) of the studied objects. Because of the increasing performances of the radial velocity spectrometers, it is now possible to derive high quality velocity curves for several tens of field RR Lyrae stars and for some of these stars in globular clusters.

The absolute magnitude of field RR Lyrae has been determined for SW And and DX Del by Jones et al. (55.122.027), for UU Cet, RV Phe and W Tuc by Cacciari et al. (1992) and for V445 Oph, SS Leo and VY Ser by Fernley et al. (55.122.077). Moreover, respectively 4, 2 and 2 RR Lyrae have been analysed in the globular clusters M4 (Liu & Janes, 52.122.022), M5 and M92 (Storm et al., 55.122.015). The impact of these determinations on the luminosity-metallicity relation of RR Lyrae stars has been discussed by Carney et al. (55.122.028).

The Baade-Wesselink method has also been applied to the cepheid stars of the remarkable young globular cluster NGC 1866 in the LMC (Welch et al., 53.122.001; Côté et al., 53.122.012). This is a unique case of a stellar cluster containing a lot of cepheids, allowing a very powerful test of the Baade-Wesselink method.

The distances to 101 cepheids were determined by Barnes et al. (53.122.163), by using the surface brightness version of the Baade-Wesselink method. An independant solution for the Period-Luminosity relation of the cepheids was then established.

The mean radial velocity (γ velocity) of 58 cepheids has been derived by Metzger el al. (53.111.010, 55.111.005); these new values have been used to improve the definition of the rotation curve of the Galaxy. Similarly, Fernley et al. (1993) got new radial velocity data for a sample of RR Lyrae stars in order to derive the mean absolute magnitude of these stars by means of the statistical parallaxes method.

The atmospheric motions in various types of pulsating stars have been analysed by using simultaneous radial velocity and high-resolution spectroscopic observations. A series of papers have been published by D. Gillet & collaborators on the RV Tauri stars AC Her and R Scu (Gillet et al., 52.122.012), on β Cephei stars β Cep, σ Sco and 12 Lac (Mathias et al., 54.122.063, 55.122.055) and on W Vir (Lèbre & Gillet, 55.122.009).

7. IAU STANDARD STARS

(R.P. Stefanik, Center for Astrophysics, and C.D. Scarfe, University of Victoria)

A meeting of the Working Group on Radial Velocity Standard Stars took place during the IAU General Assembly XXI. The results of that meeting, and velocity information on standard stars and candidates for standard stars, can be found in the Report of Commission 30: Radial Velocities, in Reports on Astronomy, IAU Transactions XXIB, 1992. The goal of the Working Group is to provide a list of standard stars whose velocities are known with an accuracy of 100 m/s. Problems requiring further attention include a color-dependence in the comparison of velocities from several observatories, velocity variables among the standard stars, and extending the list of standards to cover a wider range of spectral and luminosity classes.

7.1. Early-type velocity candidates.

F. Fekel continues to observe about 35 early-type stars as possible radial velocity standards (see the list in IAU Transactions XXIB, 1992). Most stars remain good candidates, but some show possible velocity variations. Two excellent candidates are HD 27962 (Cole et al., 55.118.008) and HD 128167 (Hartkopf et al., 55.118.013), while HD 147394 should be dropped because it shows asymmetries in its line profiles. Fekel suggests adding HD 48843 to the list.

J. Morse & R. Mathieu report that the Fekel list ties in well with the Pleiades and Alpha Persei velocities derived from late type stars (Morse et al., 53.036.025).

As part of an effort to use HIPPARCOS proper motions to set space motions and to study the dynamics of young stellar groups, W. Verschueren continues work on obtaining accurate radial velocities of early-type stars, including standards, from echelle spectra. Some results of this effort can be found in his PhD Thesis (1991, Free University of Brussels).

7.2. Late-type velocity candidates.

7.2.1 DAO work. At DAO, observations of IAU standards have been continued photoelectrically by R. McClure & C. Scarfe, and photographically by Scarfe & A. Batten. To search for color-dependencies, Scarfe has observed most of the northern standards photoelectrically with each of two masks, based on the spectra of Arcturus and Procyon. No color-dependent difference has been found between the masks, or between them and the photographic data in Scarfe et al. (51.111.008).

7.2.2. CfA Standard Program. The CfA team (Latham, 1992) continues to monitor IAU radial velocity standard stars, with nearly ten years of coverage and over 5000 observations with a precision of 0.3 to 0.4 km/sec. In addition, CfA monitors a number of stars which are candidates for radial velocity standards: a) 18 solar-type stars found to be constant by the CORAVEL team, with over 400 CfA observations and more than 500 days' coverage; and b) eight stars found to be constant by the Canadian HF-cell precise-velocity team, with over 130 CfA velocities and more than 500 days' coverage. Also being monitored are several CfA cluster standards and four Griffin (1969) standards. Lists of the candidate stars can be found in IAU Transactions XXIB, 1992. CfA also continues to observe minor planets, with more than 700 observations, in order to establish an absolute zero-point for the CfA system. In order to optimize the match between the observed star spectra and the cross-correlation templates, CfA has started using calculated, synthetic spectra for the templates. The synthetic spectra are based on Kurucz model atmospheres and cover a dense grid of effective temperature, surface gravity, metallicity and rotational velocity.

7.2.3. DAO-CfA comparison. The zero-point difference between Scarfe's combined DAO photoelectric data and the CfA system is independent of color, and has been used to shift the former to the latter's zero point. There is no difference in scale between the systems. The combined mean data do not confirm the existence of a difference in zero-point between the bright and faint groups of standards. However, the differences between them and the mean velocities from CORAVEL are correlated with the stars' color indices, becoming increasingly negative for redder stars.

Several apparently variable stars have been identified, and excluded from these comparisons. HD 156014 and HD 115521 are confirmed as low-amplitude variables; the latter is possibly triple. Among the Heard-Fehrenbach objects, in addition to HD 42397 (Scarfe, 55.121.022), HD 140913 and HD 171232 appear to be variable, the latter on a long time scale. A preliminary orbital solution for HD 140913 suggests some similarity to HD 114762 (Latham et al., 49.118.004).

7.2.4. The CORAVEL Standard Program. The CORAVEL team (Duquennoy & Mayor, 54.118.010) has been monitoring more than 100 standard stars for more than 12 years and have about 8000 velocity measurements with a precision of 0.2 to 0.3 km/sec. Several velocity variables have been found; those having orbital solutions are HD 3346, HD 35410, HD 44131, HD 115521 and HD 123782. In addition, they have been monitoring solar-type stars for more than a decade. Potential candidates drawn from this sample can be found listed in IAU Transactions XXIB, 1992.

7.2.5. Mt. John work. J. Hearnshaw & his students have been actively observing IAU standards from Mt. John University Observatory, New Zealand. Using a photoelectric velocity scanner and oscillating mask on a fiber-fed echelle spectrograph, LUCIFER, they have made about 1500 observations of 30 IAU standards. A precision of about 100 m/s has been achieved.

K. Murdoch & colleagues, using cross-correlation techniques and the fiber-fed echelle, have a program for obtaining radial velocities to a precision of 50 m/s. Among the stars being monitored are 12 IAU standards. After 2.5 years of observations five standards show evidence of low-amplitude variations: HD 18884, HD 81797, HD 108903, HD 146051, HD 204867 (Murdoch & Hearnshaw, 54.111.011; Murdoch et al., 55.111.007, 1993).

8. REFERENCES

Cacciari, C., Clementini, G., Fernley, J.A., 1992, Ap.J. **396**, 219.
Claria, J.J., Mermilliod, J.-C., 1992, A.A. Sup. **95**, 429.
Fernley, J.A., Skillen, I., Burki, G., 1993, A.A. Sup. **97**, 815.
Fischer, D.A., Marcy, G.W., 1992, Ap.J. **396**, 178.
Fischer, P., Welch, D.L., Mateo, M., 1993, A.J. **105**,938.
Flynn, C., Freeman, K.C., 1993, A.A. Sup. **97**, 835.
Friel, E.D., Janes, K.A., 1993 A.A. **267**, 75
Griffin, R.F., M.N.R.A.S. **145**, 163.
Hatzidimitriou, D., Cannon, R.D., Hawkins, M.R.S., 1993, M.N.R.A.S. **261**, 873.
Herbst, T.M., Beckwith, S.V.W., Forrest, W.J., Pipher, J.L., 1993, A.J. **105**, 956.
Jorissen, A., Mayor, M., 1992, A.A. **260, 115**.
Latham, D.W., 1992, IAU Coll. 135, 110.
Latham, D.W., Mazeh, T., Stefanik, R.P., Davis, R.J., Carney, B.W., Krymolowski, Y., Laird, J.B., Torres, G., Morse, J.A., 1992, A.J. **104**, 774.
Mateo, M., Olzewski, E.W., Pryor, C., Welch, D.L., Fischer, P., 1993, A.J. **105**, 510.
May, J., Bronfman, L., Alvarez, H., Murphy, D.C., Thaddeus, P., 1993, A.A. Sup. **99**, 105.
Mehringer, D.M., Yusef-Zadeh, F., Palmer, P., Goss, W.M., 1992, Ap.J. **401**, 168.
Mermilliod, J.C., Rosvick, J.M., Duquennoy, A., Mayor, M., 1992, A.A. **265**, 526.
Minniti, D., White, S.D.M., Olzewski, E.W., Hill, J.M., 1992, Ap.J. **393**, L47.
Morrell, N., Abt, H.A., 1992, Ap.J. **393**, 666.
Morrison, H.L., 1993, A.J. **105**, 539.
Murdoch, K., Hearnshaw, J.B., 1993, Ap.J., in press.
Prosser, C.F., 1993, A.J. **105**, 1441.
Soderblom, D.R., Stauffer, J.R., Hudon, J.D., Jones, B.J., 1993, Ap.J.Sup. **85**, 315.
Suntzeff, N.B., Schommer, R.A., Olzewski, E.W., Walker, A.R., 1992, A.J. **104**, 1743.
Vassiliadis, E., Meatheringham, S.J., Dopita, M.A., 1992, Ap.J. **394**, 489.

Commission 31. TIME: (heure)

President: E. Proverbio Vice-President: H. F. Fliegel

Organizing Committee: D. Allan V.A. Brumberg, M. Fujimoto, M. Granveaud, B. Guinot, W. Klepczynski, J. Kovalevsky, J. Luck, Y. Miao, I. Mueller, P. Paquet, Ye Shu Hua.

INTRODUCTION

The work of TAU Commission 31 is contained in the following sections contributed by the Members.

The two Working Groups installed during the period 1984-1987:

"The Use of Millisecond Pulsars and Timing of Pulsars"
and
"Time Transfer with Modern Techniques",

chaired by D. Allan and J. Luck, respectively, maintained their activities and will report at the GA Business Sections.

During the period 1991-1993 the following Meetings and Workshops of interest of Commission 31 took place:

6th European Frequency and Time Forum (17-19 march, 1992), Noordwijk, The Netherlands

Journees Systems de Reference Spatio-Temporels Geodinamique Globale et System de Reference (1-2 june, 1992), Paris, France

4th European Congress of Chronometry (29-30 october, 1992), Lausanne, Suisse
Meeting of the Time Laboratories Representatives (22-23 march, 1993), Paris France

7th European Frequency and Time Forum (16-18 march, 1993), Neuchatel, Suisse

Moreover it is right to mention the tutorial Workshops on Precise Time and Time Interval (PTTI) planned by the NASA Goddard Space Flight Center : 23th (1991), 24th (1992) and 25th (1993).

BUREAU INTERNATIONAL DES POIDS ET MESURES (BIPM) - Time Section

This report summarizes the activities of the Time Section of the Bureau International des Poids et Mesures. These activities include: current work on the production of TAI and UTC; some studies concerning time comparisons and time scale algorithms, necessary to maintain the high quality of international time scales; and more general studies, in particular on millisecond pulsars and general relativity.

1. Establishment of International Atomic Time (TAI) and Coordinated Universal Time (UTC)

The time scales TAI and UTC have been maintained and made available by the monthly Circular T, now distributed also through the BITNET and INTERNET

J. Bergeron (ed.), Reports on Astronomy, Vol. XXIIA, 325–336.

electronic networks. The Annual Report of the BIPM Time Section has been published each year since 1989.

On average, 190 clocks from 60 laboratories maintained by 45 national timing centres contribute to TAI. Presently, nearly all 45 centres are equipped with GPS time receivers and send regulary their GPS observations to the BIPM. This ensures the accuracy of nearly all time linhs involved in TAI computation. One main new feature in the generation of TAI is the use of data from hydrogen masers, ie. naturally drifting clocks. They presently account for 12% of the total weight.

As in the previous years, the conformity of the TAI scale interval with he SI second still rests almost entirely on the data of a single laboratory, the PTB. A tendency of the TAI frequency to decrease with respect to the PTB standards has been compensated by frequency "steering" corrections of 5 x 10-15 on ten occasions and 7,5 x 10-15 on two occasions since the beginning of 1989. New primary frequency standards are now developed at CRL, NIST, OP and PTB. Their evaluation is under way and, for some of them, first data are being reported to the BIPM.

2. Algorithms for time scales

Since 1989 no change has been made to the BIPM algorithm ALGOS which is considered to be properly optimized for long-term stability. Numerous studies have been carried out, all of which confirm this conclusion. These studies include a comparison of time scale algorithms[19], a report on the correlations among the frequency changes of the clocks[20], a study of the correlation between an individual clock and the ensemble and a study of the entries and exits of clocks. The use of the Kalman filter in time scale algorithms has been investigated.

3. Time links

GPS is the main technique of time comparison for TAI. Using strict common view measurements, it appears that the precision of one single measurement [UTC(k1) - UTC(k2)] is about 2ns for short distances and about 8ns for long distances [12]. The major error sources affecting the accuracy of GPS time links are the errors in antenna coordinates, in receiver internal delays, in determining ionospheric delays and in using broadcast satellite ephemerides. An important part of the research activity of the Time Section deals with these topics. The coordinates of all GPS receivers that provide data to the BIPM have been expressed in the IERS terrestrial frame. Ionospheric measurements are rou tinely used for the computation of the long time links, and the use of precise ephemerides is under study. It has been tested by the closure condition of 3 time links that these corrections greatly improve the precision and accuracy of GPS time transfer [14]. It has also been shown that Kalman prediction can overcome the voluntary degradation of the acces to GPS time (SA) [21].

In addition, a considerable amount of activity is directed towards better standardization of GPS time receivers, a necessary step to reach the subnanosecond level of accuracy. This work is done through the CCDS Group on GPS Time Transfer Standards (CGGTTS), the secretariat of which is in the hands of the Time section.

The russian system GLONASS is also routinely observed and its data reported in the BIPM publications. In 1991, the BIPM initiated an experiment whick allowed, for the first time, a direct comparison of GPS and GLONASS common-view time transfers, which demonstrates the capacity of GLONASS [14].

Finally the Time section participates in studies of two-way time transfer via geostationnary satellites and of the LASSO technique, in particular to compare them with the GPS technique.

4. Other activities

The time section participates in the work of the IAU Working Group on Astronomical Standards to update and improve the system of astronomical models and constants. It has also an active participation in the CCDS Working Group on the Applications of General Relativity to Metrology, created in 1993.

Work has been carried out to understand how millisecond pulsar data could be used to realize a pulsar time scale and what implications it would have for atomic time [17]. Collaboration has been established with the radioastronomy groups observing pulsars.

U.S. NAVAL OBSERVATORY, USA

1. Master Clock

Three reference systems are used to realize the coordinated cloch timescale of the Observatory. All time interval measurements are made against these reference systems which are designated MC1, MC2 and MC3. Each system is driven by a hydrogen maser directy. The frequency synthesizers of these masers are set once per day, if necessary, to keep the reference systems close to the computed mean timescale UTC(USNO), which in turn is close to the predicted UTC(BIPM). The unsteered internal reference is designated as A.1, while the reference of the actual Master Clock(s) is UTC(USNO). UTC(USNO) is kept within 100 ns of UTC(BIPM). An estimate of the slowly changing difference UTC(BIPM) - UTC(USNO,MC) is computed daily and published on the Automatic Data Service (ADS).

2. Timescale

The USNO timescale is generated as described by L. A. Breakiron, 1991, Proceedings of the 23rd Annual Time and Time Interval (PTTI) Applications and Planning Meeting, pp. 297-305, except for revisions in the weighting and steering. The ensemble consists of about 9 masers and 30+ cesium clocks. The clocks are included in the actual ensemble or rejected on the basis of both long-term and short-term performance. Most of the cesium clocks are of the new HP5071 type.

The clocks are distributed in 11 vaults, all of which are temperature-controlled. Some are also humidity-controlled. The weights change with time so that the masers are completely deweighted 90 days in the past in order to prevent any residual drift from affecting the timescale. The reference clocks are steered to the timescale by no more than 400 ps/day and with a time constant that varies from 10 to 60 days.

Reference system #2 is designated as lead reference, or MC2, to which all measurements can be corrected if necessary. However, most of the time the differences between these systems are about 1ns or less. Measurements between all clocks are made every hour; a second, independent high precision system measures the high performance clocks every 100s. The various clock vaults are located in several buildings that are separated by as much as 300m. The connecting cables are either low loss coaxial cables or fiber optic links. All are installed underground.

3. Time Dissemination

Various timed systems have been kept within narrow tolerances of the USNO Master Clock. The LORAN chains covering North America have been within about 100ns (rms), whereas the overseas chains had larger tolerances, the largest in the case of the Mediterranean chain (1micros). The GPS, with the correction given in the navigation message (constants A0 and A1), has been within 15ns rms during the period April through July 5, 1993. This refers to observations with the selective availability removed. Including selective availability, observed with a single frequency receiver, the rms error has been 69ns, with a maximum error of 291ns. These measurements include all available satellites with a 13-minute observation per pass.

By obtaining the small residual difference between UTC(USNO,MC) and UTC (BIPM) from the Automatic Data Service (ADS) of the USNO, a near real time access to UTC is, therefore, possible via the GPS at the level of accuracy given above. By averaging over all available satellite passes per day, a fixed station with a cesium frequency standard can increase this precision to below 10ns with appropriate filtering. The obtainable accuracy will usually be limited by the stability and calibration of the local antenna-receiver delays.

For highest accuracy, the USNO has extended the use of its two-way satellite time transfer instrumentation. Regular time transfers have been continued with the NIST in Boulder, Colorado, with the NRC in Ottawa, Canada, and with the USNO station in Richmond, Florida.

During 1992, experiments have also been conducted with the Technical University in Graz, Austria , and with OCA in Grasse, France. An additional high precision time reference station has been established on the island of Oahu, Hawaii, and initial two-way time transfers have been started with that station.

Some problems with the spread-spectrum modems have limited the obtained precision of these measurements to about 3ns. The mobile Earth station has been used to make relative delay calibration between USNO and several other sites. It is currently being planned to resume the time transfers to Europe, but approval from INTELSAT has not yet been received.

The instrumentation at the USNO consists currently of two 4.5m VERTEX antennas, the mobile Earth station, one VSAT, and a new "Fly-Away" small terminal that will be used for the quick calibration of remote stations because this terminal can be easily transported by air and assembled by one person in a few hours.

The ADS has been accessible to users via the INTERNET and via high speed mo dems. Recently, an automatic E-mail delivery system has been added. Users who cannot telnet into the ADS can now request files via E-mail. The requested file is then sent to the requester within 10 minutes. A list of available files, currently about 200, is also available in this way. As an example for such a request, a message with the file name or 'list' as subject, without any message body, would be sent to adsmail@tycho.usno.navy.mil (the numerical address of tycho is 192.5.41.239). A request for a file that does not exist will also produce the list.

4. Research

L. Breakiron analyzed the power spectra of cesium and maser clock data and found-short and long-term periodicities, some of which are related to temperature and humidity variations.

D. Matsakis investigated the use of stored-ion clocks for routine timekeeping purposes. Two phenomena, which may be related, limit the usefulness of the three prototype devices currently available at the USNO as long-term frequency standards. There is an apparent systematic drop in the frequency with time. One cause may be a change in the temperature of the mercury ions (determined by the temperature and pressure of the helium). Superimposed upon this is a set of frequencey jumps.

Matsakis also continued comparing the time transfer accuracy for the GPS system to that of VLBI for the Green Bank-Richmond baseline. All pre-hurricane data have now been analyzed, and it was found that the double difference had a peak-to-peak of 200ns since the new Mark III recorder was installed at the Green Bank 85' control building. Presumably, this error is entirely within the GPS comparisons, which were not conducted in common view. This accuracy, however, is sufficient for the pulsar investigations at Green Bank.

Matsakis began an investigation into the use of millisecond pulsars for timekeeping.

JET PROPULSION LABORATORY, PASADENA, USA

JPL has been actively involved in many areas that are key to IAU Commission 31; briefly, they can be outlined as follows:

(1) the determination of constants and the development of models and ephemerides for use by the community and IERS analysis centers;

(2) reference frame studies: 1) establishment of the JPL Radio Frame; 2) the establishment of the Dynamical Reference Frame of the Lunar/Planetary Ephemerides; 3) determination of ties between the various reference systems; 4) development of the fiducial-free approach using GPS data and the realization of an independent GPS terrestrial reference frame tied to the IERS.

(3) acquisition, reduction and analysis of VLBI data, including the TEMPO program, for Earth rotation, precession and nutation studies;

(4) reduction and analysis of lunar laser ranging (LLR) data for Earth rotation, press[Bion and nutation studies;

(5) acquisition, reduction and analysis of GPS data in Earth rotation and reference frame studies;

(6) intercomparisons of Earth rotation results from the various techniques;

(7) combination of Earth rotation measurements with a Kalman filter;

(8) analysis of the scientific implications of these measurements.

SHANGHAI OBSERVATORY, ACADEMIA SINICA, CHINA

Research work of high precise time synchronization have been carried out at Shanghai Observatory (SO), Shaanxi Observatory (CSAO), Beijing Observatory

(BAO), Beijing Institute of Radio Metrology and Measurement (BIRM), National Institute of Metrology (NIM), Institute of Geodesy and Geophysics (WTO) and Beijing University. National time service broadcasts standard time and frequency at PuCheng near Shaanxi Observatory. Some significant results were obtained in recent years.

Shanghai Observatory: Eight transportable hydrogen masers were developed in recent years. Some of them have been put into operation in Chinese VLBI network and other fields. The frequency stability is better than 3x10-15 for a several hours sample time. A new hydrogen maser has been used in Shanghai Observatory' s atomic time scale since last year. The comparison results between Hydrogen masers and the satellite clocks of GPS have been sent routinely to BIPM. The long term stability keeps within 30ns for two-months sample time and 60ns for ten-day sample. An extremely small atomic hydrogen maser (40x60x70cm), the third generation of II-maser. has been completed successfully last year at Shanghai Observatory.

The maintance of standard time and frequency at Shanghai Observatory is based on 3 HP commerical Cs clocks. UTC (SO) has been compared with GPS satellites clocks for international time comparison since 1990. The local atomic scale is compared routinely by Loran-C and TV techiques.

Shaanxi Observatory:Besides the astronomical research work, the time and frequency standards with more precise accuracy are kept here as a center of Chinese time service broadcasting time signals by means of short-wave and long-wave radio signals (BPM and BPL). Some research work related to precise time synchronization by means of using BPM. BPL. GPS and Screen satellite have been carried out routinely.

"Screen Satellite" time service system is based on the common view of the 714MHz signal of screen satellite. The precision of time service with BPM is about 1 ns and the precision of time comparisom is about 1 us. CSAO is also responsible for the reduction and publication of the Chinese Joint Atomic Time System which is composed of all commerical Cs clocks of five institutes (SO,CSAO,BAO,BIRM,WTO).

Beijing Observatory: A HP commercal Cs clock is used to control Chinese national radio broadcasting station which has a timing program for civil purpose. It also joined the Joint Atomic Time System.

Beijing Institute of Radio Metrology and Measurement: Commerical Cs clocks were Joined the Chinese Joint Atomic Time system. Some kinds of high performance crystal oscillators were developed and used in the field of scientific research.

National Institute of Metrology: It has established local time scale with HP Cs clocks. The Cs clocks were available to control the standard frequency in TV signals for the national TV broadcasting station. So it can provides time comparison by TV technique all over the China. The first laboratory' s Cs clock was developed at NIM and has the frequency accuracy of 3x10-13.

Institute of Geodesy and Geophysics: A HP Cs clock has been joined the Joint Atomic Time System. Time comparisons with BPM, BPL are carried out on a daily bais. The research work about the propagation characteristics of shortwave signal has obtained good results.

Beijing University: A laser pumping Cs frequency standard has developed over the past years and the stability of 1.2x10-11 t1/2 (t is sample time) by means of Allan Variance was obtained.

In China some research works are being discussed and developed as follos:

1. VLBI is one of applications for the use of atomic clocks. It also has the potential to provide more-accurate time comparison in the range of 10 to 100 ps.
2. Time comparison by means of GPS common view experiments.
3. "Broadcasting Satellite" time service.
4. Joint time keeping. by using Cs clocks and hydrogen masers in China.
5. The determination of more accurate time period by using pulsar Observations.
6. Further research in the field of time definition.

ISTITUTO ELETTROTECNICO NAZIONALE G. FERRARIS, TORINO, ITALY

1- Generation of UTC(IEN) and dissemination

The time scale UTC(IEN), which is the basis of legal time in Italy, has been maintained in agreement within 1.5 us with UTC in the period 1990-1993. At least three cesium clocks, of the five HP 5061 clocks avalaible, contributed to the generation of the international time scale TAI; in February 1993 an HP 5071A cesium clock with improved specification has been added to the IEN clock ensemble.

UTC(IEN) has been compared with the international time scale by means of two GPS receivers programmed for the BIPM common view Schedule.

In July 1992, a two-way time transfer experiment in the Ku band between the IEN and the Istituto Superiore delle Poste e delle Telecomunicazioni in Rome using Olympus satellite at 19 W, was started. Two mobile Alenia groundstations and Mitrex 2500A modems have been used.

In the fied of standard time signal dissemination, a date code, to be distribuited via modem on telephone lines, has been experimented since 1991 and will become operational in June 1993. The code format has been agreed among four European laboratories, IEN, SNT, TUG and VSL and is suitable to synchronize computer clocks. The syncronization precision and accuracy of this system, used both in a one-way and in a two-way mode and especially related to the CCITT V.22 modems behavior, has been investigated; timing accuracies of 100-120 ms (one-way) and 10 ms (two-way) have been found.

2- Algorithms for time reference generation

2.1 Individual instability estimation

The problem of estimating individual clock instability from comparison measurement has been deeply investigated proposing a general and consistent model for data analysis. For the case of 3 clocks a particular algorithm has been devised to individuate the minimum correlation solution; the analysis of the general case of N clocks with a suitable matricial formulation gives several information about clock instabilities and their correlations.

2.2 Ensemble time scale

In collaboration with the BIMP a study has been undertaken on the possibility of using pulsar data to improve the long-term stability of atomic time scale. The use of an "ensemble pulsar time" with an appropriate algorithm has been tested both on simulated data and with the real observation data at disposal. It turns out that with the real data now available an ensemble pulsar time with 10-14 instability for observation interval from one to almost three years is obtainable. Pulsar time can also be used as a flywheel to transfer atomic time accuracy from one epoch to another.

2.3 Application to telecommunication

In collaboration with the Italian telecommunication research group, a study has been undertaken and is still in progress on the best suited clock characterization tools to be used in case of the Synchronous Digital Hierarchy network.

3- Atomic frequency standard

3.1 Mg atomic beam

The Mg frequency standard prototype, realized at IEN, has been evaluated by comparison with UTC(IEN): the short-term frequency stability turns out to be : sigma y(t) = 1.10-11.t-12 (1<t<3000 s) without optical pumping.

The uncertainty budget is limited at 10-12, mainly due to the residual first order Doppler shift, whose evaluation is under way. All the other sources of uncertainty can be improved to the limit of 10-13 or better.

Frequency comparison between Mg and UTC(IEN) have been performed several time in 1992 showing a repeatability of 3.10-13.

3.2 Semiconductor lasers

In 1991 a research activity has started in the field of frequency stabilized semiconductor lasers, also in collaboration with the Metrological Institute G. Colonnetti (IMGC) and with the Polytechnic of Turin, at the following wavelengths: 780 nm (Rb), 852 nm (Cs) and 657 nm (Ca).

4- Infrared frequency standards

An OSO 4 molecular beam is under developement at IMGC: CO2 laser prestabilized on the P(46) of OsO4 via the Lamb-diptechnique has been set up as well as the vacuum system and the molecular source.

The frequency of the 81.5 um and of the 263.4 um NH3 lines have been measured demostrating also the feasibility of coherent frequency synthesis up to 3.7 THZ.

ROYAL OBSERVATORY OF BELGIUM (ROB).

A computed time scale deduced as a mean of three Cesium Beam Tubes, is operational since 1991; the mathematical clock is physically represented by a Rubidium; the physical and the mathematical clocks agree at the level of 2x10-13. In July 1992 one Hydrogen Maser CH1-75, produced by KVARZ, has been turned in continuous operation.

To get the performance of this unit KVARZ recommends to use,at the same site, two masers to perform an automatic auto-tuning of the cavity. As only one maser is in operation at the Royal Observatory of Belgium, in principle the auto-tuning is not operational. However, in the same laboratory, a BVA quartz oscillator is in operation since several years. Its short term stability is of the order 5 10-13 over time interval of 30 seconds; for this reason, KVARZ suggested to experiment to tune the CH-75 with an external frequency provided by the BVA. Experiments are in process.

HYDROGRAPHIC DEPARTMENT OF JAPAN, TOKYO

For the purpose of monitoring the relation between TAI and the dynamical time reduced from the orbital longitude of the Moon TDT (observed), the observation of occultations of stars by the Moon have been continued at the head office of Hydrographic Department of Japan (JHD) in Tokyo and three branch observatories, namely, Sirahama, Simosato and Bisei. About 1000 timing data including 600 photoelectric data were obtained every year.

TDT(Observed)-TAI obtained from the occultation observation for the epochs 1990.5, 1991.5 and 1992.5 (preliminary for this epoch) were 32.91s, 32.88s and 32.89s, respectively, with the mean error of +0.05s. Details are published in Data Report of Hydrographic Observations, Series of Astronomy and Geodesy as well as in the Japanese Ephemeris.

The services of the International Lunar Occultation Centre have been continued since 1981. The number of the data reported to the Centre in the years 1990 to 1992 amounts to 34,646.

Satellite Laser Ranging (SLR) observation has been carried out at Simosato Hydrographic Observatory since March, 1982. Total return signals obtained from Lageos I and II, Starlette and the Japanese geodetic satellite Ajisai in the years 1990 to 1992 amount to about 969,000.

JHD has also been carrying out SLR system (HTLRS) since 1987. HTLRS obtained in the years 1990 to 1992 about 53,000 returns for the four satellites at some sites within Japan including off-lying islands.

THE NATIONAL PHYSICAL LABORATORY, NEW DELHI.

NPL has been maintaining UTC with the help of three industrial Cesium clocks (all Hewlett-Packrd made) with one acting as the master clock. Time intercomparisons via GPS in common view mode are being regularly made and UTC(NP-I) inputs are being sent to BIPM.

NPL has been transmitting standard time and frequency signals at three carrier frequencies: 5, 10 and 15 MHZ. and disseminating standard time signals via Indian Domestic Satellite INSAT, with an accuracy of 10 micro second to the users. These broadcasts are on a down-link frequency of 2599.675 MHZ.

A development of hydrogen Maser was carried out by NPL in collaboration with NIST, Boulder, Colorado, USA. Work on development of indigenous Rubidium standards is also being carried out. In addition, NPL undertakes the calibration Work from Governament and private sectors for calibration of primary and secondary time and frequency standards.

NEUCHATEL OBSERVATORY, SWITZELLAND

The Observatory of Neuchatel has obtained the nomination of "Center of Excellence" of the European Space Agency (ESA) for the work in the atomic frequency standards. Two majors projects (a Space Rubidium Ultrastable Oscillator and a Space Hydrogen Maser) are presently under way. The pro- jects involve a cooperation between ESA, the Russian Space Agency and the Russian Academy of Sciences

A new basic research field has been introduced: the laser cooling technique applied to the development of the future primary atomic frequency standards.

CAGLIARI OBSERVATORY, ITALY

The Time and Frequency Laboratory of the Cagliari Observatory has been developed during the last years according to some field of interest related to atomic time scales, high precision synchronization and the relations between atomic and astronomical time.

Three commercial atomic standards are operated in good environmental conditions in order to generate a local time scale as uniform as possible. The performances of the local clocks with respect to UTC are continuously monitored by means of a GPS timing receiver, LORAN-C and TV links with other national Laboratories.

The synchronization data are regularly sent to BIPM in order to contribute to the computation of TAI.

In the late 1991 a decoder for geostationary television broadcasting satellites wase set up. It was operated several times during some european campaign, promoted by the Royal Observatory of Belgium to investigate some aspects of the precise orbit determination of geostationary satellites using TV synchronization links.

More recently the Laboratory was connected via computer link to the italian astronomical network ASTRONET, and through it to the main geographic networks such as Bitnet, Span etc. This is the first step for the creation of a national Time Scale, based on the atomic clocks kept in the principal laboratories in Italy (IEN, ISPT, CAO). A preliminary meeting about this topic was held in Turin in May 1993.

During 1993 a great effort was devoted by the staff of the Cagliari Observatory to put into operation the Telemetry Laser Station. At the end of June the first echoes were obtained from low-orbiting satellites, and it is foreseable that the complete operational conditions will be reached before the end of the year. This fact is of great importance in the field of time syncronization, because makes it possible for the Cagliari Observatory to participate in synchronization experiments by satellites, such as LASSO, and to start some investigation about relativistic effects on time scales.

REFERENCES

[1] Allan, W.D., 1993, Trends in International Timing, Proc. 7th European Frequency & Time Forum, Neuchatel.

[2] Bauch, A., Fischer, B., Heindorff, T., Schroder, R., 1993, The PTB's fourthCesium Atomic Clock, Proc. 7 th European Frequency & Time Forum, Neuchatel.

[3] Baumont, F., Grudler, P., Veillet, Ch., Wiant, J., Lewandowski, W., Petit, G.,1993, A Preliminary Report on the Comparison of Lasso and GPS Time Transfers, Proc.7 th European Frequency & Time Forun,Neuchatel.

[4] Collin, F.,Paquet, P., 1991, Realization of the Belgian Time Scale,Proc. 5thEuropean Frequency & Time Forum, Neuchatel.

[5] DeMarchi, A., 1993, The High C-Field Concept for an Accurate Cesium Bean Resonatur,Proc.7th European Frequency & Time Forum, Neuchatel.

[6] Csaba,Szekely,Drullinger, R.,1993, Diodo-Laser Pumped,Rubidium Cell Frequency Standars, Proc. 7th European Frequency & Time Forum Neuchatel.

[7] Drullinger, R.E., Lowe, J.P., Glaze, D.J., Shirley, J.,1993, NIST-7, The New US Primaty Frequency Standard, Proc. 7th European Frequency & Time Forum Neuchatel.

[8] Erjomin, E., Rjabov, S., Donchendo, S., Kupko, S., 1993, The Resultas of the Comparison of Frequency and Time Standard CH1-75 with Board Clock GPS"NAVSTAR" with the Help Receiver GEM SRS-01, Proc. 7th European Frequency & Time Forum Neuchatel.

[9] Godone, A.,Novero, C., Tavella, P., 1993, Repeteability of Magnesium Frequency Standar, Proc. 7th European Frequency & Time Forum Neuchatel.

[10] Hu, Z.C., 1992, Analysis of the Affect of the Receiving Parameters on the Precision of GPS Time Comparison, Publications of the Shaanxi Observatory,Vol.15 N.1.

[11] Imae, M., Aida, M., Goto T.,Miki, C., 1993, calibration experiments of GPS Timetransfer receivers and ionospheric delay measuring receivers between Japan and France, Proc. 7th European Frequency & Time Forum, Neuchatel.

[12] Lewandowski, W., Thomas, C., 1991, GPS Time Transfer, Proc.IEEE, Special ISSUE on Time,79,991-1000.

[13] Lewandowski, W., Petit, G., Thomas, C., 1993, Precision and accuracy of GPS Time transfer, IEEE,in press.

[14] Lewandowski, W., Moussay P., Cherenkov, G.T., Koshelyaevsky, N.B., Pushkin, S.B., 1993, GLONASS COMMON-VIEV TIME TRANSFER,Proc. 7th EFTF,1993, in press.

[15] Meyer, F., Granveaud, M., Laporte, B., Vernotte, F., Vincent, M., 1993, Improved Time Transfer Using a Geostationary Direct TV Satellite, Proc. 7th European Frequency and Time Forum, Neuchatel.

[16] Petit, G., Lewandowski, W., Thomas, C., 1993, Precise GPS Ephemerides for Time Transfer, Proc. 7th European Frequency & Time Forum, Neuchatel.

[17] Petit, G., Thomas, C., Tavella, P.,1992, An ensemble Pulsar Time, Proc. 24th PTTI, in press.

[18] Sennedot, D., Minguy, Y., Thomas, B., 1993, Use of LORAN-C Signals for Time/Frequency Applications, Proc. 7th European Frequency & Time Forum, Neuchatel.

[19] Tavella, P., Thomas, C., 1991, Comparative study of time scale algorithms,Metrologia 28, 57-63.

[20] Tavella, P., Thomas C., 1991, Report on correlations among the frequency changes of the clock contributing to TAI, BIPM report-91/4,49 pages.

[21] Thomas, C., 1992, Real-Time restitution of GPS time through a Kalman estimation, Metrologia,29, 397-414.

[22] Weyers, S., Schodrer, R., Heindorff, T., Bauch, A., 1993, Optical Punping in PTB's Frequency Standard CSX, Proc. 7th European Frequency & Time Forum, Neuchatel.

[23] Wu, G.C., 1992, Devolopment and expectation on PTTD, Publications of Shaanxi Observatory, Vol.15,N.1, 1-3.

[24] Yang, D.H., Wang, Y.Q.,1991, Preliminary results of an optically pumped Cs bean frequency standard,ACTA Metrologica Sinica, Vol.12,N.4, 297-300.

[25] Zhai, Z.C., 1990, Perfomance evaluation of the Shangai Observatory's new hydrogen maser,Annals of Shangai Observatory. VOL.11, 201-208.

[26] Zhai, Z.C., 1992, Devolopment and improvement for H-MASER,Publications of the Shaanxi Observatory,VOL.15,n.1, 4-9.

33. STRUCTURE AND DYNAMICS OF THE GALACTIC SYSTEM (STRUCTURE ET DYNAMIQUE DU SYSTÈME GALACTIQUE)

PRESIDENT: L. Blitz (USA)
VICE-PRESIDENT: J.J. Binney (U.K.)
ORGANISING COMMITTEE: J.B.G.M. Bloemen (Netherlands),
C. Cesarsky (France), A. Fridman (Russia),
G. Gilmore (U.K.), A. Kalnajs (Australia),
F. Matteucci (Italy), M. Mayor (Switzerland),
M. Morris (USA), Y. Yoshii (Japan)

1 Introduction

This report covers work published approximately from July 1, 1991 to June 30, 1993. The main reference has been the Astronomy and Astrophysics abstracts; some of the sections were written using the last 6 volumes available to each writer. Consequently, some of the references date back to 1990.

The intention has been to cover the field of galactic structure, and stellar dynamics. (Magneto)-hydrodynamics as applied to the Milky Way system has also been included. There is some inevitable overlap not only with the work of other commissions, but within some of the sections of this report. However, work which is clearly in the purview of other commissions, notably Interstellar Medium (Commission 34) and Galaxies (Commission 28) has not been included. It has not been possible to provide an exhaustive summary of the work done on the subject of galactic structure in the past triennium, and the report provides a subjective view by the individual authors of the most important work in each subfield. Nevertheless, the past triennium has seen a great deal of exciting new work done in the field including the apparent discovery of compact objects by means of gravitational lensing within the Milky Way (MACHOs), as this volume went to press.

Thanks go to all of the authors who provided input on what turned out to be a very tight schedule, especially to Art Upgren, Jim Stone and John Howe who provided sections or parts of sections even though they are not on the organizing committee and to Jim Morgan of the University of Maryland who helped with the formatting of the camera ready manuscript.

2 Basic Data

Arthur R. Upgren
Wesleyan University

2.1 Parallaxes

2.1.1 Trigonometric parallaxes from space

The ESA Astrometric Satellite Hipparcos finally succumbed to the ravages of twice-daily passages through radiation belts, and its data collection ended in March 1993, almost four years after launch. Preliminary reports of its performance were made at the Galactic and Solar System Optical Astrometry Observation and Application Workshop held at Cambridge, U.K. in June 1993, by L. Lindegren and J. Kovalevsky. They report a median standard error in proper motion of 4.6 milliarcseconds (mas) per year and in parallax of 2.2 mas from the first half of the observations.

J. Bergeron (ed.), Reports on Astronomy, Vol. XXIIA, 337–365.

They estimate errors from all observations will be about 2.2 mas/year and 1.8 mas for the proper motions and the parallaxes, the later representative of the true external errors. Thus the precision of its data equals or exceeds preliminary expectations. We are assured that the results will have a major impact on stellar and galactic astronomy for years to come when they are released in final form about 1996/97. In addition, about a million stars brighter than B = 11 have been observed photometrically by the Tycho mission. A link of this system to the extragalactic reference frame is planned.

2.1.2 Ground-based and other parallax observations

The ongoing programs of the U.S, Naval, Allegheny, Van Vleck, Sproul and McCormick Observatories are continuing, with improvements in precision at several places. A. R. Upgren reports that the emphasis of the Van Vleck parallax program concentrates more fully on nearby metal-poor subdwarfs and stars reported to be of intermediate metallicity, that may represent a thick disk population. C.C. Dahn reports that the Naval program has completed parallaxes of about one thousand stars. Its mission has been extended to include bright ($V < 10$) stars by using neutral-density magnitude-compensating filters, and very faint stars in the $16 < V < 22$ range by using a CCD detector. D. J. MacConnell of STI/CSC and W.H. Osborn of Central Michigan U. report first observations with the HST-FGS intended to determine parallaxes for high-velocity stars. R.B. Hanson (Lick Observatory) T. E. Lutz (Washington State U.) and W.F. van Altena (Yale) have nearly finished their joint project to improve the luminosity calibration of field subdwarfs by color and metal abundance, using data from the new edition of the Yale Parallax Catalogue.

2.1.3 Catalogues of parallaxes and nearby stars

The new fourth edition of the Yale Parallax Catalogue is being made ready for publication by van Altena, J.T. Lee and E.D. Hoffleit. It is scheduled to be sent to the L. Davis Press before the end of 1993. It contains 15,430 parallax observations for 7,888 stars. These numbers represent an increase over those given in the previous report in this series and elsewhere. Corrections between the various observatories, applied in earlier editions, have not been recommended by a number of investigators, especially since their physical cause cannot be identified, and hence will not be applied in this edition.

The third edition of the Catalogue of Nearby Stars (CNS3) is in final preparation by W. Gliese and H. Jahreiss. Its total of about 3800 stars represents a very large increase over the second edition of 1969 and its supplement of 1979. Part of the increase is due to an extension of the distance limit to 25 parsecs. As might be expected, most of the newly added stars are faint dwarf M stars.

Gliese, Jahreiss, Upgren and E.W. Weis are preparing a publication of the stars in the five lists of dwarf K and M stars found spectrophotometrically by A.N. Vyssotsky. Precise positions and proper motions as well as broad band photometry are now available for all of his 895 stars and stellar systems. At the Cambridge workshop cited above, Weis and Upgren reported a high-velocity bias present in the Vyssotsky stellar sample. This is disturbing because this sample is often taken as the best unbiased stellar group representative of the galactic disk. The cause of the bias is not yet clear but analysis is continuing.

The Hipparcos Input Catalogue was published and circulated in 1992. Its seven volumes list the 118,322 stars for which positions, parallaxes and proper motions are to be obtained by the satellite. Accurate B1950 and J2000 positions, cross references and much other data for each star make this a useful standard reference.

2.2 Proper Motions

The PPM Star Catalogue covering the northern celestial hemisphere was published in two volumes in 1991 by S. Roser and U. Bastian of ARI, Heidelberg. It contains the positions and proper motions of 181,731 stars north of -2.5 degrees for the equinox and epoch J2000. Similar data for the stars in the southern hemisphere has been obtained and will be published soon. This work contains data for 197,179 stars south of -2.5 degrees.

At Hamburg and the RGO the conventional plate adjustment solution and photographic photometry was completed for the CPC2; a catalogue containing 276,131 stellar positions and visual magnitudes for an average epoch of 1968 and an average standard error of 60 mas. The Pulkovo Observatory, in collaboration with the Sternberg Institute, ARI and Hamburg, the compilation of the proper motions of the FOCAT-S catalog was finished using AC and CPC2 data.

2.3 Photometry and Stellar Classification

Only one meeting covered the whole field of spectral classification during this triennium. This was "The MK Process at 50 Years: A Powerful Tool for Astrophysical Insight" sponsored by the Vatican Observatory and held at Tucson in September 1993 under the chairmanship of C. Corbally.

A very large number of studies were made of a very wide variety of stars and stellar types. They are reviewed in the report of Commission 45, among others. Those of the most general nature include a new spectral classification system by Keenan for the carbon stars bringing them into the revised MK system. Four subgroups of carbon stars are noted and designated R, N, CH and Hd (hydrogen deficient) C stars. New criteria have been identified that permit more realistic temperature sequences, as well as a luminosity classification for the first time.

Abt is classifying the 750 stars lacking MK classifications in the Bright Star Catalogue and a thousand members of visual multiple stars brighter than B = 8. Garrison is reviewing the MK standards over the entire HR Diagram. He is also classifying the nearest 800 G dwarfs and the 1500 brightest stars in the sky. With the promise from Hipparcos of good parallaxes for many of these stars for the first time, the classifications of Abt and Garrison will be especially valuable as luminosity determinations.

A number of objective-prism surveys have been followed up by classification at higher dispersion. This was due in part to the increased sensitivity of current spectrographs and detectors. Much of this work is summarized in Contribution of the Van Vleck Observatory no. 12: Objective-prism and Other Surveys.

New progress on catalogues includes the release by P. Renson, M. Gerbaldi and F.A. Catalano of their General Catalogue of Ap and Am stars, and a new catalogue by G. Cayrel of [Fe/H] determinations. The Institut d'Astronomie de l'Universite de Lausanne and the Observatoire de Geneve continue to maintain the General Catalogue of Photometric Data, with the Catalogue of Observations in the UBV System being kept up-to-date by Mermilliod. A new edition of the Catalogue of Am stars with known Spectral Types has been completed by Hauck.

3 The Stellar Component

Gerard Gilmore
University of Cambridge

The striking characteristics of studies of the stellar distribution in the Galaxy over the past triennium is the extension of statistical studies to include a very wide wavelength range, and to include kinematics. Fortunately, this burgeoning of activity is also matched by a burgeoning of excellent review articles and tutorial schools, as well as many sessions at regional IAU meetings. Thus extensive reviews cover the activity from many points of view, and will not be duplicated here.

One of the most significant publications has been the last planned paper in the Basle Halo Program "Model Comparison Program" (Fenkart & del Rio, 1991, *AASup*, **89**, 217). This major and long lasting survey has served as a model of modern statistical studies of Galactic structure, and their interpretation in terms of large scale Galactic structural properties. Important successor programs include those underway in Besancon, which incorporates both star counts and proper motion data (Bienayme *et al.*, 1992, *AA*, **253**, 389 and refs therein), a similar program based in Paris, at the MAMA group (Soubiran *AA*, **xxxx**), that in Japan, utilizing star counts in many passbands in several fields (Yamagata & Yoshii, 1992, *AJ*, **103**, 117), and the north pole survey in photometry and proper motions by Majewski in the USA (1992, *ApJS*, **78**, 87). Very considerable progress in this field is expected soon with the recent digitizations of the various Schmidt Sky Surveys, and the completion of the Lick and Yale proper motion surveys.

Extension of these optical surveys to the near infra-red has been a dramatic feature of late, with more promised with the recent release of the COBE data. Near infrared balloon, satellite and ground based surveys have been presented and analyzed by several groups. Studies of particular note include those by Blitz & Spergel (1991, *ApJ*, **379**, 631), Kent *et al.* (1992, *ApJS*, **78**, 403 and 1992, *ApJ*, **387**, 181), Ruelas-Mayorga & Teague (1992, *AASup*, **93**, 61 and 1992, *AA*, **259** 118), Nakada *et al.* (1991, *Nature*, **353**, 140), Terndrup *et al.* (1991, *ApJ*, **378**, 742), and Weinberg (1992, *ApJL*, **392**, L67). A feature of most of these discussions is their concentration of possible (in fact likely) non-axisymmetric structure in the stellar distribution in the central bulge. An ability to detect and focus on such deviations from the basic large scale structure is indicative of the growing sophistication of current Galactic studies. the basic features are now agreed, and a higher degree of sophistication in modeling is becoming justified. The current state of infrared surveys is well reviewed by Boulanger and Desert (*The Infrared and Submillimeter Sky after COBE*, eds. Signore & Dupraz, Kluwer, 1992).

Surveys of specific classes of identifiable objects have proered of late. Perhaps the most impressive such studies, both in the scale of effort required and the detailed analyses resulting, are the surveys for AGB stars in the Bulge. OH/IR surveys have been reported by Lindqvist *et al.* (1992, *AASup*, **92**, 43 and 1992, *AA*, **259**, 118) and are allowing extensive studies of the dynamics of the central stellar distribution (Weinberg *et al.*, 1992, *ApJ*, **384**, 81 provides one example).

Extension to other specific classes of tracer star, very often with specific dynamical aims in mind to complement the structural studies, has been achieved by many authors. Examples representing some of those studies include distant blue horizontal branch stars (Norris & Hawkins, 1991, *ApJ*, **380**, 104 and Arnold & Gilmore, 1992, *MNRAS*, **257**, 225), nearby AF stars (Knude, 1991, *AA*, **249** 88), X-ray stars (Kashyap *et al.*, 1992, *ApJ*, **391**, 667 and Favata *et al.*, 1992, *AA*, **256**, 86) and WR stars (Shara *et al.*, 1991, *AJ*, **102**, 716 and Alfaro *et al.*, 1992, *ApJL*, **386**, L47).

The picture of the Galaxy which is emerging from these many studies is detailed, justified, still uncertain, but testable against current and forthcoming reliable data. The many tutorial overview publications describing these results which have recently appeared include in particular two books in the ASP Conference Series, Galaxy Evolution: The Milky Way Perspective, ed. S. Majewski, and The Minnesota Lectures on the Structure and Dynamics of the Milky Way, ed. R. Humphreys. The proceedings of many more general conferences have appeared, with those of particular note including: The Globular Cluster - Galaxy Connection, eds. G. Smith and J. Brodie (ASP); Variable Stars and Galaxies, ed. B. Warner (ASP); Precision Photometry, eds. Davis Philip, Upgren & Janes (L Davis Press); Eleventh European IAU Regional Meeting, eds. Sanchez & Vazquez (CUP); The Stellar Populations of Galaxies eds. Barbuy & Renzini (Kluwer).

These conferences have been complemented by a major review in ARAA by Majewski (Galactic Structure Surveys and Evolution of the Milky Way, 1993, **31**, 575). It is a sign of a healthy subject that the number of reviews papers included in these conferences is sufficient to provide an adequate number of references for a normal report. The volume of science included is such that only by reading those articles can justice be done to the subject.

3.1 Stellar Initial Mass and Luminosity Functions

In the past few years studies of the stellar mass and luminosity functions have been dominated by heroic surveys and analyses. For stars with mass above about one-half that of the Sun the field remains dominated by the authority of Scalo's 1986 review (Fund Cosmic Physics **11**, 1). Very considerable progress has been made at lower masses however. New sky surveys have led to improved photometric studies (Tinney, Mould, & Reid, 1993, *ApJ*, **396**, 173) so that the apparent luminosity function for local disk stars is now apparently well defined to near the lowest luminous masses. All recent well calibrated surveys agree that the luminosity function declines at the lowest luminosities. It seems finally that long-standing suspicions that there is a substantial upturn in the space density of the lowest luminosity stars have been excluded from serious consideration. the apparent luminosity function is flat or declining at the lowest luminosities.

Conversion of the apparent luminosity function to a mass function requires detailed modeling of many parameters, especially the range in age, chemical abundance and binarism in the sample and in its calibrators. Kroupa *et al.* (1993, *MNRAS*, **262**, 545) quantify the effect of each variable. Kroupa *et al.*, in agreement with all other analyses, deduce that the mass function for low mass stars is adequately described as a power law in number of stars per unit mass interval, with power law slope about 1.1; that is, very low mass stars are the most common by number. However, their contribution to the mass or the luminosity of a mass function integrated over a realistically wide range is negligible. The possibility that brown dwarfs dominate the local mass density seems conclusively discounted.

The most robust test of these models and conclusions comes from study of open clusters and young star forming regions. Major statistical studies by Bryja *et al.* (1992, *ApJL*, **388**, L23), Simons & Becklin (1992, *ApJ*, **390**, 431), Hambly *et al.* (*MNRAS*, **253**, 1), Stauffer *et al.* (*AJ*, **101**, 980), among several other studies reach on average rather similar conclusions. An extension to the field stars of the Galactic halo (Richer & Fahlman, 1992, *Nature*, **358**, 383) while necessarily not extending to such low masses, further confirms the basic similarity of the stellar luminosity function in all environments.

While surprising and unexpected, as data and modeling improve, the similarity between determinations of the intrinsic stellar initial mass function in different environments tend to persist and surprise more than do the differences.

4 The Gas and Dust Components

Leo Blitz and John E. Howe
University of Maryland

4.1 Surveys and Reviews

Many surveys of the Galaxy have been reported during the past triennium, but the most notable change has been the large increase in infrared data from satellites (COBE [Wright, E.L., *et al.*, 1991, *ApJ*, **381**, 200], Spacelab, and the continuing analysis of IRAS data), and balloons (Okuda, H., *Infrared Phys.*, 1991, **32**, 365), as well as the traditional ground based radio surveys. There have also been a large number of meetings related to the distribution of gas and dust in the Milky Way. Publications from these meetings include *The Milky Way as a Galaxy*, R. Buser and I. King, eds.; *The Interstellar Disk-Halo Connection in Galaxies*, J.B.G.M. Bloemen, ed.; *The Evolution of the Interstellar Medium*, L. Blitz, ed.; and *Back to the Galaxy*, F. Verter and S. Holt, eds. Major new reviews of the HI and CO gas content of the Milky Way were written by Dickey and Lockman (1990, *ARAA*, **28**, 215), and Combes (1991, *ARAA*, **29**, 195).

Radio surveys include the Bell Labs HI survey (Stark, A.A., *et al.*, 1992, *ApJS*, **79**, 104) and the Pulkovo HI survey (Bystrova, N.V. & Yagudin, L.I., *Bull. Spec. Ap. Obs.* **24**, 107). Radio continuum surveys have been carried out at 6 cm at Green Bank (Langston, G.I., *et al.*, 1990, *ApJS*, **72**, 621), and Parkes (Griffith, M., *et al.*, 1991, *P.A.S.Aust.*, **9**, 243), 11 cm at Effelsberg, (Reich, *et al.*, 1990, *AASup*, **85**, 633; Fürst *et al.*, 1990, *AASup*, **85** 691, 805), 20 cm at the VLA (Becker, R.H., *et al.*, 1990, *ApJ*, **358**, 485; Helfand, *et al.*, 1992, *ApJS*, **80**, 211), 21 cm at Effelsberg (Reich, *et al.*, 1990, *AASup*, **83**, 539), and 800 cm (38 MHz) at Ooty (Rees, N., *MNRAS*, 1990, **244**, 233). Radio synthesis surveys at 21 cm and 74 cm (408 MHz) have continued at Penticton (Joncas, G. & Higgs, L.A., 1990, *AASup*, **82**, 113; Normandeau, M., *et al.*, 1992, *AASup*, **92**, 63). The first results of a new Hα survey of the galactic plane have also recently been published (Le Coarer, E., 1992, *AA*, **257**, 389).

4.2 Rotation Constants, Kinematics, and Global Structure

A fundamentally new method for deriving the rotation curve of the outer Milky Way was published by Merrifield (1992, *AJ*, **103**, 1552). Using the azimuthal variation of the angular variation of the atomic hydrogen scale height beyond the Sun, he showed that the rotation curve remains flat to 2.5 R_0. Dutra and Maciel, (1990, *Rev. Mex. Astron. Ap.*, **21**, 264), redetermined the rotation curve using type I, IIa, and IIb planetary nebulae, and showed that the curve determined from the first two types agrees with that determined form other population I tracers. Nincović derived a new formula for the planar velocity dispersions containing an asymmetric drift term (1992, *Ap. Sp. Sci.*, **187**, 159), and showed that it is possible to determine the slope of the circular velocity independent of the Oort constants.

Blitz & Spergel (1991, *ApJ*, **370**, 205) argued that the asymmetries in the atomic hydrogen distribution are best modeled as the result of gas on elliptical streamlines which is the dynamical result of an $m = 2$ distortion to the Galactic spheroid. Kuijken (1991, in *Warped Disks and Inclined Rings around Galaxies*, p. 159) preferred an $m = 1$ distortion to the potential as the origin of the asymmetry, and while there is general agreement that the disk is non-axisymmetric, there is continuing debate about its nature (Kuijken & Tremaine 1991, in *Dynamics of Disk Galaxies*, p.71 and 1994, *ApJ*, in press).

There seems to be more general agreement on the existence of a stellar bar in the central region of the Milky Way from an analysis of the dust distribution at 2.2 μm (Blitz & Spergel 1991, *ApJ*, **379**, 631; see also appropriate sections on stars and dynamics in this report). From Spacelab data, Kent *et al.* (1991, *ApJ*, **378**, 131; 1992, *ApJ*, **387**, 181) deduced the axisymmetric properties of the bulge, and disk, but argued that a non-axisymmetric bulge is necessary to explain the peak velocities of the CO and the HI distributions within a few hundred parsecs of the center. Similar conclusions were reached from modeling of Gerhard (1992, *Rev. Mod. Astron.*, **5**, 14), and CO data of Nakai (1992, *PASJ*, **44**, L27).

4.3 Local Galactic Structure

The past triennium has seen an expansion of the work done on local structures such as the cirrus and high-latitude molecular clouds. Although most of this work is more properly in the domain of interstellar medium studies, some is of relevance to studies of local galactic structure. For example, Verschuur (1991, *Ap. Sp. Sci.* has found evidence of large scale morphological waves in long HI filaments suggesting that the motion of the gas is controlled by magnetic fields within flux tubes. With other coworkers, Verschuur has also found evidence that the HI filaments, and the associated IR and CO emission can be offset from one another by as much as a degree (1992, *ApJ*, **390**, 514); it is unclear what this implies, however.

Blitz (1991, *Molecular Clouds*, p.49) has argued that based on kinematic evidence, there is a new class of molecular cloud characterized by a high negative velocity at high galactic latitude; three such clouds are known to exist. Lilienthal and Wennmacher (1990, *AA*, **235**, L13) have found a new upper limit of 240 pc to the distance of a high latitude HI filament toward SAO 30359. Jahoda *et al.* (1990, *ApJ*, **354**, 184) have identified and observed the Galactic line of sight with the smallest HI column density: $< 6.6 \times 10^{19}\ cm^{-2}$. Gamma-rays have been shown to correlate well with the Lynds dark cloud catalogue at intermediate latitudes (Feitzinger and Stüwe, 1990, *MNRAS*, **242** 395). A rarefied tunnel in the interstellar medium which appears to be an extension of the local bubble has been mapped in the direction of β C Ma and has been found to extend 300 pc in that direction (Welsh, 1991, *ApJ*, **373**, 556).

4.4 Vertical Structure of the Disk

A new word, 'worm', entered the literature, referring to vertical structures in the disk as a result of multiple supernovae. Worms are related to 'chimneys' which deposit their hot gas in the hot Galactic Corona. A catalogue of worms was published by Koo *et al.* (1992, *ApJ*, **390**, 108). Theoretical studies of chimneys and the associated 'blowout' into the halo were published by Li and Ikeuchi (1990, *ApJS*, **73**, 401), and by Igumentschchev *et al.* (1990, *AA*, **234**, 396). Dougherty and McKenzie (1990, *Proc IAU Symp. 140*, p. 185), argued that for determination of the gas density far from the plane, galactic wind models should replace hydrostatic models.

Lockman and Gehman examined the vertical structure of the Galactic HI layer and concluded that the vertical distribution of the gas is supported by turbulence and that there are several distributions of atomic hydrogen with different central densities and scale heights (1991, *ApJ*, **382**, 182). Blitz, on the other hand (1991, in *Proc. IAU Symp. 144*, p.41), found scant evidence for a molecular cloud population with a scale height greater than about 70 pc. In theoretical work by Franco *et al.* (1991, *ApJ*, **366**, 443), it was argued that radiation pressure can levitate the gas disk to high latitudes driving a soft Galactic fountain. The z distribution of HI in the longitude range $210 < l < 275$ was investigated by Yudeva (1991, *Bull. Spec. Ap. Obs.*, **27**, 109). Churchwell *et al.* found that the vertical distribution of ultracompact HII regions is similar to that of O stars in the Galaxy ((1990, *AASup*, **83**, 119).

Reynolds (1991, *ApJL*, **372**, L17), has shown that diffuse ionized gas is clumped into regions with an electron density of $0.08\ cm^{-3}$, and a volume filling fraction > 0.2 of a 2 kpc thick ionized layer. This layer has become widely known as the 'Reynolds layer.' UV absorption line measurements support a model where warm ionized gas is corotating with the disk to heights of 1.5 kpc, but where corotation breaks down at z heights between 1.5 and 3.5 kpc (Sembach, *et al.*, 1991, *ApJ*, **372**, 81). Savage *et al.* measured the scale height of Al III to be 1.0 kpc compared to 0.67 kpc for the HI scale height to the same set of stars (1990, *ApJ*, **361**, 107).

4.5 Spiral Structure

Several studies of global tracers of Galactic spiral structure were made during the past triennium. Zongwei *et al.* (1991, *ApJ*, **378**, 93) performed Monte Carlo simulations of the distribution of SNRs and HII regions to determine possible correlations with spiral arms or disk populations and conclude that SNRs are not necessarily concentrated in spiral arms in the numbers one would expect if they are preferentially born there. De-lin (1990, *Acta Astrophys. Sin.* **10**, 235) compared the radial and longitudinal distributions of HI and CO at subcentral points at $b = 0$ and $10° < l < 80°$ with current spiral arm models and found that strong CO emission exists in the interarm regions, there is no evidence for stronger HI and CO emission at the subcentral points, and the CO emission in general does not trace the spiral arms delineated by 21-cm HI emission or HII regions. Calculations of the flow of interstellar gas into the gravitational potential well of a spiral density wave at a 5 kpc galactocentric distance show that at gas densities of 0.05 – 0.2 cm^{-3}, a phase transition causes inhomogeneities along the wave which eventually forms individual clouds (Mishurov 1991, *Astronomicheskii Zhurnal* **68**, 65).

GMCs in the outer arm of the Galaxy with sizes and masses similar to the largest clouds in the inner Galaxy were found in a recent CO survey at $b = +1°$ and $65° < l < 71°$ (Digel, Bally, and Thaddeus 1990, *ApJL*, **357**, L29). The clouds lie at $R = 12$ kpc and imply a total molecular mass for the Galaxy at $R > 11$ kpc of 1–7 $\times 10^8 M_\odot$. An age gradient of HII regions and open clusters across and along the Sagitarrius–Carina spiral arm was found by Avedisova (1989, *Astrophysics* **30**, 83), as well as a decreasing pitch angle in the HII region/open cluster distribution correlated with the age of the objects. Murphy and May (1991, *AA*, **247**, 202) show from CO kinematic data and correlations with optical and radio tracers that recent massive star formation has occurred in massive GMCs 1–2 kpc away, located behind the Vela SNR and associated with the Vela spiral arm.

4.6 High Velocity Clouds (HVCs), Coronal Gas, Galactic Fountains and Winds

Several large-scale surveys of the high-velocity clouds were conducted during the past triennium, yielding important new results on the distribution and properties of these objects. Colgan, Salpeter, and Terzian (1990, *ApJ*, **351**, 503) searched for high-velocity emission and absorption features towards 63 continuum sources, detecting emission in 18 and absorption in none. They conclude that there can not be a significant mass of high-velocity HI which is undetected due to very low spin temperatures in low-density regions, and that there does not exist a large population of low column density HVCs. Wakker analyzed the results of a Dwingeloo whole-sky survey ($\delta > -17°$) for HVCs (1991, *AA*, **250**, 499; Wakker and van Woerden 1991, *AA*, **250**, 509) and identifies ten different population of HVCs distributed in 17 complexes, with a slope of -1.5 in the differential mass spectrum. He finds that the HVCs concentrate in the Galactic plane and are uniformly distributed in longitude, but are usually at large distances from the solar neighborhood. Higher resolution Westerbork observations of six fields show much substructure, but that 70% of the emission is extended over scales larger than a few arcminutes, suggesting an extended warm background component (Wakker and Schwartz 1991, *AA*, **250**, 484; Wakker, 1991, *AASup*, **90**, 495). Wakker summarizes much of the HVC work in *IAU Symposium 144* (p. 27).

Based on high resolution observations of Na I and Ca II lines toward nine stars in the direction of the Chain *A* high-velocity complex, Lilienthal and Meyerdierks (1990, *AA*, **240**, 487) set a lower limit of 200 pc above the plane for the height of the complex. Meyer and Roth (1991, *ApJL*, **383**, L41) detect weak Ca II absorption features in high resolution spectra towards SN 1991T in the galaxy NGC 4527 and attribute the features to low column density Galactic HVCs. HI emission spectra at 21 cm taken towards the Large Magellanic Cloud show high-velocity components well outside the LMC boundary and identify this gas as a local Galactic component (de Boer, Morras, and Bajaja 1990, *AA*, **233**, 523).

A number of observational and theoretical studies of Galactic outflow and inflow phenomena were reported. Mirabel (1991, IAU Symposium 144, 89) and Mirabel and Morras (1990, *ApJ*,

356, 130) found evidence for an infalling stream of HVCs towards the Galactic anticenter which interacts with the ambient gas, producing shocks and an energetic "anticenter supershell" of HI. Globally, the Galaxy accretes 0.2–0.5 $M_\odot$/yr of extragalactic HI in addition to a general disk-halo gas circulation (Mirabel 1989, *IAU Colloquium 120*, p. 396). Meyerdierks proposes that the North Celestial Pole Loop is the result of the expansion of a cavity produced by an infalling 350 $M_\odot$ halo cloud which has already collided with the Galactic disk (1991, *AA*, **251**, 269). Meyerdierks also concludes on the basis of HI observations that an intermediate-velocity cloud in the Chain *A* stream is the result of a high- and low-velocity cloud collision (1992, *AA*, **253**, 515). Kahn (1989, *IAU Colloquium 120*, p. 474) proposes that Type II supernovae remnants generate the gas and energy driving a Galactic fountain of intercloud gas. The infalling gas is seen as intermediate velocity clouds. Houck and Bregman (1990, *ApJ*, **352**, 506) show that clouds can be formed from the fountain flow but the details of the formation depend on whether the gas cooling time is longer or shorter than the sound crossing time in the halo gas. Shapiro and Benjamin (1989, *PASP*, **103**, 923) show that photoionization due to ionizing photons produced in the Galactic fountain flow itself is sufficient to reproduce the observed column densities of ionized metals in the flow. Spitzer and Fitzpatrick (1992, *ApJL*, **391**, L41) observed C IV and Si IV absorption features towards an O star 1500 pc above the plane and conclude that the velocities of the features show they are produced in infalling gas.

Li and Ikeuchi (1992, *ApJ*, **390**, 405) use a two-diminsional hydrodynamic code to model the Halo gas. Gas of temperature 0.5-1 $\times 10^6$ K and density 10^{-3}–10^{-2} cm^{-3}reproduces many of the observed ionized metal column densities. Lazaryan and Chibisov (1991, *Sov. Astron. Let.*, **17**, 208) propose a novel method for measuring the size of the Galactic radio halo based on a statistical analysis of synchrotron emission fluctuations. Cosmic rays may be important in driving a net outflow (1 $M_\odot$ yr^{-1} in the present epoch) of hot interstellar gas from the Galactic disk (Breitschwerdt, McKenzie, and Völk 1991, *AA*, **245**, 79).

4.7 The Distribution and Properties of the Gas in the Plane

A review of the structure of the Milky Way as seen in molecular emission is given by Thaddeus and Stacy (1990, *The EGRET Science Symposium*, p. 5). However, Adler and Roberts (1992, *ApJ*, **384**, 95) show cause for caution in identifying molecular cloud complexes from position–velocity diagrams and of assigning distances to them based on model rotation curves. Simulations of cloud complexes in model spiral galaxies show that many apparent correlations in position–velocity diagrams are in fact chance line-of-sight superpositions. Waller and Tacconi-Garman (1992, *ApJS*, **80**, 305) estimate distances to local CO clouds using the Massachusetts–Stony Brook CO survey data, and also find over 2100 lines of sight free of significant CO emission, useful as regions to study the diffuse intercloud medium and as windows for extragalactic observations. Characteristics of large-scale details in the Pulkovo HI sky survey are tabulated by Bystrova and Kozyrev (1990, *Bull. Spec. Astrophys. Obs.* **29**, 106). Delin (1992, *Acta Atrophys. Sin.* **12**, 13) finds in a statistical study of the large-scale correlations between CO and HI between $11 < l < 45$ that the CO and HI emissivities are linearly uncorrelated, even in regions corresponding to spiral arms in some models. Yasutomi and Tatematsu (1990, *PASJ*, **42**, 517) numerically modeled the dynamical evolution of GMCs in the disk and found that dynamical friction causes the GMCs to spiral in to the Galactic center. A minimum in the infall velocity occurs at $R = 5$ kpc, creating the 5 kpc molecular ring on a time scale of about 10^9 yr. Millimeter continuum observations from Antarctica of several southern clouds revealed an excess of emission at the longer wavelengths (Andreani *et al.*, 1991, *ApJ*, **375**, 148), attributed to emission from a cold dust component coexisting with the warmer dust seen in the IRAS survey. In the first of what will undoubtedly be one of many studies of the large-scale Galactic distribution of ionized carbon ([C II]) emission, Shibai *et al.* (1991, *ApJ*, **374**,

522) show that the [C II] and CO distributions are well correlated and that the total luminosity of the [C II] emission in the inner Galactic disk amounts to about 0.4% of the far-IR luminosity. They conclude that the [C II] is largely emitted from dense, partially-ionized photodissociated skins of molecular clouds traced by the CO, although extended low-density ionized gas may also contribute.

A wealth of information continues to be gleaned from the IRAS survey. Vedi and Walker (1991, *MNRAS*, **249**, 729) derive values of $A_V = 0.05$ towards the North Galactic Pole and $A_V = 0.08$ towards the South Galactic Pole. Bloemen, Duel, and Thaddeus (1990, *AA*, **233**, 437) decompose 60 μm and 100 μm maps, which trace the dust emission, into components associated with atomic (HI) and molecular (CO) gas. Stecker, Harding, and Skibo (1991, *AIP Conf. Proc. No. 220*, 20) find that the dominant feature in the radial distribution of IRAS far-IR emission coincides with the 5 kpc molecular ring. Spiral arm features are also apparent, and a strong correlation exists between the far-IR emission and CO and γ-ray emission.

Diplas and Savage (1991, *ApJ*, **377**, 126) studied the distribution of HI in the outer Galaxy from the Bell Labs HI survey data. They identify a confined component which exists only within $R \sim 18$ kpc and an extended component which exists out to 25–30 kpc in some directions. The average z distance of the HI in the direction of the Galactic warp is ~4 kpc at $R \sim 24$ kpc. Sodroski (1991, *ApJ*, **366**, 95) analyzed outer Galaxy GMC properties from the Goddard-Columbia Galactic plane CO survey and finds size and mass distributions similar to inner Galaxy GMCs, but the H_2 column density to CO intensity ratio (6.0×10^{20} molecules cm^{-2} K^{-1} km^{-1} s) is 2–3 times greater than for inner Galaxy GMCs. Fich and Silkey (1991, *ApJ*, **366**, 107) derived high nitrogen abundances in 18 HII regions at R = 12–18 kpc. Wouterloot *et al.* (1990, *AA*, **230**, 21) surveyed 1302 outer Galaxy IRAS point sources and found that 1077 have associated CO emission. Kinematic distances place these clouds out to $R = 20$ kpc and indicate that star formation is presently occurring in the outer Galaxy. The cloud distribution also shows the warped-disk shape seen in HI.

The radial distribution of diffuse ionized gas was investigated by Frail *et al.* (1991, *ApJ*, **382**, 168) using mean electron density measurements towards 18 pulsars. They find that n_e in the inner Galaxy is about 3 times that found locally, and propose that the warm ionized medium is primarily heated by O stars. The distribution of plasma turbulence was found by Hajivassiliou (1992, *Nature* **355**, 355) to agree morphologically with the soft X-ray background, possibly suggesting that the solar neighborhood is enveloped in an ionized bubble from a nearby supernova explosion. A warp in the outer Galaxy at $270° < l < 300°$ is also seen in ionized gas (Azcárate, Cersosimo, and Colomb 1990, *Rev. Mex. Astron. Atrofis.* **20**, 75). Cersosimo (1990, *IAU Colloq. No. 125*, 237) reports evidence for diffuse ionized gas in the 3 kpc arm, primarily but not exclusively associated with Pop I material, and some evidence of in interarm ionized gas connecting giant HII complexes in the Scutum-Crux and Norma arms. Lockman (1990, *IAU Colloq. No. 125*, 225) presents a Green Bank HII region survey, used with Southern Hemisphere data to investigate the radial distribution of HII regions. Oskanyan (1989, *Astrophysics* **30**, 75) shows that the distribution of Galactic HII regions shows a correlation with the distributions in Sc and SBc spirals, and estimates that the mass of neutral gas formed from "extinct" HII regions could be about 65% of the total mass of Galactic HI, if the lifetime of HII regions is only 6×10^5 yr.

4.8 Galactic magnetic fields

Heiles (1990, *IAU Symp. No. 140*, 35) reviewed the status of magnetic field strength measurements in HI regions and finds that the magnetic field pressure dominates or is comparable to other pressures in some dark clouds and in some HI shells. Trimble (1990, *IAU Symp. No. 140*, 29) also reviewed the observational parameters of the Galactic magnetic field at the same meeting. Field directions derived from Zeeman effects do not correlate well with those derived from Faraday rotation measures. Rand and Kulkarni (1990, *IAU Symp. No. 140*, 44) model the local magnetic field from available pulsar rotation measures and find a local field strength of $1.6 \pm 0.2\mu$G toward $l = 96°$ and a field reversal 600 pc toward the inner Galaxy. In the first Galactic quadrant, Clegg and Cordes (1990, *IAU Symp. No. 140*, 47) find evidence for field reversals along several lines of sight, and for a structure at $l > 60°$ and a distance more than 7 kpc causing excess rotation measures. This structure is also seen in Effelsberg 11 cm data (Junkes, Fürst, and Reich 1990, *IAU Symp. No. 140*, 63), and confirmed in later measurements by Clegg *et al.* (1992, *ApJ*, **386**, 143).

Andreasyan and Makarov (1989, *Astrophysics* **30**, 101) give a detailed description of field strengths and directions in spiral arms in the solar neighborhood and out to distances of 5–7 kpc. Independent studies of optical stellar polarization (Andreasyan and Makarov 1989, *Astrophysics* **31**, 560) and pulsar rotation measures (Costa, Hamilton, and McCulloch 1991, *Proc. Astron. Soc. Aust.* **9**, 131) both conclude that the Galactic field is concentrated in and directed along the spiral arms. Cordes, Clegg, and Simonetti (1990, *IAU Symp. No. 140*, 55) report that small-scale fluctuations in the Galactic magnetic field suggest that the energy in the turbulent magnetic component is comparable to the energy in the smoother large-scale field.

5 The Galactic Center

Mark Morris
University of California at Los Angeles

Improving techniques in radio, infrared, X- and gamma-ray astronomy are leading to an increasingly clear picture of the interplay of the various components of this region. Blitz *et al.* (1993 *Nature*, **361**, 417) provided a recent review of many of the topical areas in Galactic center research.

5.1 The Stellar Cluster

During the past three years, progress has been made in defining the properties of the central stellar cluster of the Galaxy, particularly in the inner parsec. However, many of the most important questions remain unanswered. On scales of hundreds of parsecs, it appears that the most luminous stars are concentrated toward the nucleus such that they occupy a thick disk extending about 100 pc to either side of the Galactic plane (Catchpole *et al.*, 1990, *MNRAS*, **247**, 479; Lindqvist *et al.*, 1992, *AA*, **259**, 118; Haller 1992, PhD Thesis, Univ. Arizona). Apparently, the Galactic center has repeated episodes of star formation in which massive (greater than a few $M_\odot$) stars are formed. It is not clear how these presumably massive stars could have formed such a thick distribution during their lifetimes. Lindqvist *et al.* (1992) point out that OH/IR stars follow the integrated 2 μm light, and are therefore useful tracers of the kinematics of the central cluster. Using a sample of 134 OH/IR stars lying close to the Galactic center (Lindqvist *et al.*, 1992, *AASup*, **92**, 43), they observe the Galactic rotation, and deduce a velocity dispersion of about 75 km s^{-1}. The mean values depend on the expansion velocity of the circumstellar envelopes of the OH/IR stars, which, in turn, reflects the stellar masses. More massive stars, having higher outflow velocities, are more concentrated to the galactic plane, and have a smaller velocity dispersion. Angular broadening measurements of 20 OH/IR stars near the Galactic center were reported by van Langevelde *et al.* (1992, *ApJ*, **396**, 686), and they used these measures to derive the distribution of interstellar scattering. A region of relatively large scattering was found to be nearly centered on Sgr A*, possibly as the result of density and turbulence enhancements at small galactocentric radii.

The question of of how star formation proceeds near the Galactic center was considered by Morris (1993, *ApJ*, **408**, 496). There are many factors inhibiting star formation there, or at least raising the Jeans mass, including strong tidal forces, strong magnetic fields, and large turbulent motions, so if stars do form within 100 pc of the Galactic center (and certainly within a few parsecs), they may tend to be quite massive. Massive stars forming within about 4 pc will leave remnants which can undergo gravitational settling and thereby sink to the inner 0.1-pc core in less than a Hubble time. As much as 10^6 $M_\odot$ of black hole remnants could have been gathered into the core by this process. Whether these remnants merged to form a single object, or whether they exist as a tight cluster occasionally interacting with stars by mergers remains an open question. Moneti *et al.* (1992, *MNRAS*, **258** 705) explored a number of IRAS sources near the Galactic center in the near-IR, and found a number of embedded sources which they regard as early-type ZAMS stars, so they conclude that star formation is ongoing on scales of 10 to 100 pc around the Galactic center.

5.2 Emission Line Stars and Winds

A number of the luminous stars in the Galactic center were found by Krabbe *et al.* (1991, *ApJL*, **382** L19; see also Burton and Allen 1992, *Proc. ASA* **10**, 55) to display near-infrared emission lines, most notably those of helium. It now appears that the collective winds from these emission line stars account for the "broad-line" region that had previously been observed to be more or less coincident with the central cluster of luminous blue objects comprising IRS16.

The winds from the IRS16 complex interact with the nearby red supergiant IRS7, creating a prominent cometary tail on this star directed away from the wind sources (Yusef-Zadeh and Morris 1991, *ApJL*, **371**, L59; Serabyn *et al.*, 1991, *ApJ*, **378**, 557; Yusef-Zadeh and Melia 1992, *ApJL*, **385**, L41). The tail shows a clear gradient in its velocity field, which provides useful constraints on the star's orbit and the interaction of the wind with the star (Serabyn *et al.*, 1991). The winds emerging from the IRS16 complex also appear to be interacting with nearby plasma, perhaps leading to a waviness in the "northern arm" and to a compression of the plasma there and in the "bar" which may cause enhanced emission at mid-IR wavelengths (Yusef-Zadeh and Wardle 1993, *ApJL*, **405**, 584). These authors also assess the earlier suggestion of Yusef-Zadeh *et al.* (1990, *Nature*, **348**, 45) that Sgr A*, if it is a massive compact object, might be focusing the nuclear winds into the plasma "bar", forming the apparently evacuated "mini-cavity" (Wardle and Yusef-Zadeh 1992, *Nature*, **357**, 308). The mini-cavity has also been found to be a strong source of iron-line emission by Eckart *et al.* (1992, *Nature*, **355**, 526).

A model for the combined winds from a cluster of stars in the central 0.8 pc was presented by Chevalier (1992, *ApJL*, **397** L39). He suggests that the wind passes through a shock at a radius of a few pc, and fills a hot bubble on a scale of 100 pc. The interaction of the collective winds with the ambient magnetic field might lead to some of the magnetized structures observed near the center.

5.3 The Radio Arc and Other Polarized Radio Structures

The Galactic center radio Arc remains an enigmatic structure, with less than complete consensus about its nature, other than that it is clearly a magnetic feature. Lesch and Reich (1992, *AA*, **264**, 493) noted the slightly inverted radio spectrum of the linear filaments of the Arc, and proposed that the emission arises from monoenergetic particles which have been accelerated by magnetic reconnection at the interface between neighboring filaments. Sofue *et al.* (1992, *PASJ*, **44**, 367) mapped the G0.18-0.04 region of the Arc interferometrically at 43 GHz, and concluded that the spectrum of the nonthermal radio filaments turns over between 4.7 and 43 GHz. The flat, thermal spectrum of G0.18-0.04 is confirmed.

Far-infrared polarization measurements of the thermal, arched filaments of the Arc were reported by Morris *et al.* (1992, *ApJL*, **399**, L63). The degree of polarization is large, and the small dispersion of the directions of the polarization E-vectors implies that the field is uniform on scales of 1 to 10 pc. The inferred direction of the magnetic field in the arched filaments is parallel to the filaments, perhaps as a result of sheer within the cloud.

The Radio Arc extends over a full 2° in Galactic latitude, according to Haynes *et al.* (1992, *AA*, **264**, 500). These authors also describe 3.5-cm observations of other polarized features extending perpendicular to the Galactic plane, including one which coincides with a previously reported radio "Thread" and another which coincides with the Western leg of the so-called "Galactic Center Lobe".

A new radio filament was reported by Gray *et al.* (1991, *Nature*, **353**, 237). This one, located further from the center than any other at l = 359.1, is unusual in having a few kinks, or cusps, along its length. As in the other "threads", the emission is polarized, and the structure appears to be magnetic in character.

5.4 Interstellar Gas and Dust: Sgr A West and the Circumnuclear Disk

The kinematics of gas in Sgr A West were investigated by several groups. Herbst *et al.* (1993, *AJ*, **105**, 956) imaged Sgr A West in the Br-γ line with relatively high spatial and spectral resolution. They identify a continuous, coherent stream of gas starting with the northern arm, and extending around the IRS16/SgrA* complex into the so-called "bar". They model the kinematics of this feature in terms of orbital motion in the field of a central point mass. Depoy (1992, *ApJ*, **398**, 512) reported that the distribution of forbidden FeII emission at 1.644 and 1.533 μm in Sgr A West is similar to that of NeII and Brα. The iron line emission, relative to Brα, is quite strong compared to normal HII regions.

Several studies have been carried out recently of a newly found, high-velocity (v_{LSR} = -170 to -190 km s^{-1}) neutral cloud superimposed upon, and foreground to, Sgr A*. Marr *et al.* (1992, *ApJL*, **400**, L29) observed HCO^+ absorption in this cloud with a millimeter interferometer, and concluded that some of this gas may be mixed in with the ionized gas of the "bar" in Sgr A West. Pauls *et al.* (1993, *ApJL*, **403**, L13) observed formaldehyde absorption in this high-velocity cloud, and they place it within the central 1-pc cavity of the circumnuclear disk.

The 1 to 7 pc circumnuclear disk (CND) was studied in increasing detail in the past triennium. Davidson *et al.* (1992, *ApJ*, **387**, 189) produced 50 and 100-μm maps of the CND made with the Kuiper Airborne Observatory with high spatial resolution. They offer models for the heating of the CND from its interior, and estimate the central luminosity to be 1 or 2 $\times$ 10^7 $L_\odot$. Also, they find that a substantial quantity of dust is emitting from within the central cavity of the CND. Jackson *et al.* (1993, *ApJ*, **402**, 173) carried out a variety of far-IR and radio spectroscopic observations of the inner 2 parsecs, and conclude that the CND is a nearly complete ring of clouds, with other kinematic components superimposed upon it along the line of sight. Inside the central cavity of the CND, Jackson *et al.* find evidence for several hundred $M_\odot$ of dense neutral gas, presumably the same gas seen in the cavity by Davidson *et al.*; they find that the northern arm of Sgr A West is an ionized rim upon this neutral gas, produced by the ultraviolet radiation from the IRS16/Sgr A* complex. The previously known molecular hydrogen emission at the inner edge of the CND was imaged by Burton and Allen (1992, *Proc. ASA* **10**, 55), who concluded that UV excitation is to be preferred over shock excitation as a means of accounting for the emission.

The heating and thermal equilibrium of the ionized gas in Sgr A West and the CND were considered by Maloney *et al.* (1992, *ApJ*, **401**, 559). They deduce that 7 $\times$ 10^6 $L_\odot$ is needed to account for the observed temperature and emission measure distribution in the "bar". This is comparable to the bolometric luminosity inferred from IR measurements, so the cooling emission from the hot gas near the center accounts for a major fraction of the bolometric luminosity emerging from this region.

5.5 Interstellar Gas and Dust: 10 to 1000 pc

It was pointed out by Spergel and Blitz (1992, *Nature*, **357**, 665) that the large, internal velocity dispersions in Galactic center clouds reflects a very high gas pressure, and that this gas pressure has a strong effect on the evolution of the medium. The hot, x-ray emitting gas found in the Galactic center is another indication of a high pressure. It should not be forgotten that magnetic pressures are at least comparable, and may in fact dominate the pressure balance in the Galactic center.

Some studies of individual molecular clouds in the Galactic center have appeared recently. A number of OH and H_2O masers were observed in Sgr B1 by Mehringer *et al.* (1992, *ApJL*, **401**, L87). They seem to indicate the ongoing formation of low- to intermediate-mass stars. An apparently expanding molecular cloud system was found to ring the nonthermal radio shell G359.1-0.5 (Uchida *et al.*, 1992, *ApJ*, **398**, 128). Uchida *et al.* present a model in which the expanding ring was initiated by the winds, radiation, and supernova ejecta from a massive central cluster of early-type stars. In a related paper, Uchida *et al.* (1992, *AJ*, **104**, 1533) describe a VLA study of HI absorption across G359.1-0.5 and some of the nearby radio sources. The HI absorption line velocities place G359.1-0.5 within the Galactic center region, along with the filamentary, radio "Snake", but a curious, quasi-axisymmetric nonthermal source referred to as the "Mouse" appears to be a foreground object.

The HII complexes Sgr D and Sgr E were studied in the radio continuum and recombination lines by Liszt (1992, *ApJS*, **82**, 495). He concluded that Sgr D is an HII region/SNR pair, while Sgr E is a cluster of compact HII regions centered on recently-formed, massive stars.

The Galactic center annihilation line source, 1E1740.7-2942, aroused a great deal of attention from X- and γ-ray astronomers during the triennium. It is located about 100 pc in projection away from the nucleus. Here, we note only a possible physical association between this potential stellar-mass black hole and a molecular cloud of the Galactic center population. Bally and Leventhal (1991, *Nature*, **353**, 234) and, with higher spatial resolution, Mirabel *et al.* (1991, *AA*, **251**, L43) both reported that the high-energy source lies near the peak of a molecular cloud of intermediate density, and both groups considered that the source may be fueled by accretion of cloud material onto a single, compact object. In another study using the VLA, Mirabel *et al.* (1992, *Nature*, **358**, 215) report the detection of radio jets emanating from a presumably compact and variable core in 1E1740.7-2942, allowing those authors to terms this object a "microquasar". Inasmuch as the jets may contain a pair plasma, the annihilation might be taking place in the radio lobes, where the positrons are deposited in the ambient medium.

A few large-scale surveys of molecular emission appeared in the recent triennium. Burton and Liszt (1992, *AASup*, **95**, 9) report a CO survey on a 2-arcminute grid over the Sgr A, B, and C source complexes. They revise their earlier model by concluding that the gas distribution in the inner few hundred parsecs is flat and untilted with respect to the Galactic plane, while the gas layer at larger radii flares and is warped, as in the earlier models. Zylka *et al.* (1992, *AASup*, **96**, 525) surveyed the 6-cm H_2CO absorption line out to a longitude of 4°, and along select strips, observed the 2-cm H_2CO line as well. Particular attention is paid in their discussion to the "Clump 2" cloud complex at $l = 3.2°$, which they decompose into "a number of kinematically well-defined sub-clouds."

5.6 A Radio Transient Source

An intense radio transient source popped up in December 1990 within Sgr A West near the Galactic center (Zhao *et al.*, 1992, *Science*, **255**, 1538). Following its detection, it increased in intensity for about a month, and them declined for 3 months, all the while maintaining a steep radio spectrum. Zhao *et al.* suggest that the varying emission arose from a synchrotron-radiating plasma associated with an X-ray binary system.

5.7 Sgr A*

Observational studies of the compact radio source at the Galactic nucleus, Sgr A*, and its near environment have led to some interesting suggestions, but solid breakthroughs in direct imaging will await the next triennium, when the full capacity of the VLBA is brought to bear on this source at wavelengths of 1 cm or less. The partially completed VLBA has already been used to reobserve Sgr A*: at 1.34 cm, Lo *et al.* (1993, *Nature*, **362**, 38) confirm the previously inferred elliptical Gaussian model, with a major axis size of 2.4 $\pm$0.2 mas; at 7 mm, Sgr A* is less than 0.4 mas (3.3 a.u.) in size (Backer *et al.*, 1993, *Science,* in press; Krichbaum *et al.*, 1993 *AA*, **274**, L37). Since the radio size of Sgr A* is strongly affected by scattering off of electron density fluctuations along the line of sight, even higher frequencies will probably be required to probe the intrinsic source size. A lower limit to the size of Sgr A* is about 10^{12} cm, derived from the observed limit to the refractive scintillation (Gwinn *et al.*, 1991, *ApJL*, **381**, L43). Sgr A* shows brightness variations, some of which may be attributable to intrinsic source variability (Zhao *et al.*, 1991, in *Relationships between AGN's and Starburst Galaxies*, ed: A. Filippenko, in press).

One of the most interesting findings of the past triennium is that Sgr A* has a strong excess of submillimeter emission (Zylka *et al.*, 1992, *AA*, **261**, 119; Dent *et al.*, 1993, *ApJ*, **410**, 650). The surviving explanation at present is that this is an ultracompact, self-absorbed synchrotron source. Serabyn *et al.* (1992, *ApJL*, **401**, L87) observed this excess at 222 GHz interferometrically, and confirmed the association of the submillimeter source with Sgr A*. A possible near-IR counterpart to Sgr A* was reported by Eckart *et al.* (1992, *Nature*, **355**, 526), although a non-physical positional coincidence cannot yet be ruled out for this source.

The mass of Sgr A* (or of any putative black hole in its vicinity) is still unsettled. Ozernoy (1992, in *Testing the AGN Paradigm*, p 40, p 44) continues to produce arguments that any black hole in our Galactic nucleus must have a relatively low mass, perhaps 100 – 1000 $M_{\odot}$. This notion is supported by Sanders (1992, *Nature*, **359**, 131), who points out that if the components of IRS16 are young stars which have formed recently within 0.1 pc of the center, then there cannot be a very massive black hole there, because its tidal forces would have prevented star formation. On the other hand, Melia (1992, *ApJL*, **387**, L25) finds that a million-$M_{\odot}$ model for Sgr A* involving a weak accretion flow out of the surrounding winds can reproduce the shape of the spectrum satisfactorily. In a related study, Melia *et al.* (1992, *ApJL*, **395**, L87) describe how source size measurements at sub-cm wavelengths will be important for determining the mass of Sgr A*. Current measurements already support a mass close to 10^6 $M_{\odot}$. If Sgr A* is a massive black hole, then it should exert a noticeable gravitational lensing effect on some of the stars in the central stellar cluster (Wardle and Yusef-Zadeh 1992, *ApJL*, **387**, L65).

6 The High-Energy Component

Hans Bloemen
SRON-Leiden / Leiden Observatory

This section summarizes studies of large-scale galactic phenomena in relation to gamma rays, X-rays, cosmic rays, and magnetic fields.

6.1 Gamma Rays

The past triennium has seen a major step in gamma-ray astronomy with the launch of NASA's Compton Gamma-Ray Observatory in April 1991. The EGRET and COMPTEL telescopes have both performed an all-sky survey during the first 15 months of the mission. The large-scale studies that are of interest here have just started, but first results are available.

Preliminary findings from EGRET seem to confirm that the galactic radiation at energies above $\sim$ 50 MeV, previously studied with the SAS-2 and COS-B satellites, is largely of diffuse origin. However, many small-scale discrepancies between the EGRET observations and the modeled emission can be seen. These may be discrete gamma-ray sources, but a diffuse origin cannot be excluded. They make up less than 10% of the total emission from the galactic disk (Hunter *et al.* 1993, *Proc. 23rd Int. Cosmic Ray Conf.*). Of particular interest is the excess seen at the galactic centre, not coinciding with the peak of the CO emission (Mayer-Hasselwander *et al.* 1993, *Proc. 23rd Int. Cosmic Ray Conf.*). It is only seen at energies above 300 MeV, but it needs to be verified whether this is possibly due to the decreasing angular resolution of EGRET with decreasing energy.

The COMPTEL telescope provides for the first time good possibilities to map gamma-ray emission in the 1-30 MeV range, which also contains nuclear gamma-ray lines. The first MeV image of the inner Galaxy obtained with COMPTEL shows a clear ridge of emission along the Galactic plane, but it is not obvious that this emission is largely of diffuse origin (Bloemen *et al.* 1993, *ApJS*,). The intensity distribution of this ridge distinctly differs from the expected distribution for a simple model of the diffuse radiation (*i.e.*, a model similar to that used in the COS-B and EGRET studies).

COMPTEL has for the first time mapped the entire galactic plane in the light of the 1.809 MeV line from ^{26}Al radioactive decay (Diehl *et al.* 1993, *Proc. Second Compton Symposium*). The origin of ^{26}Al is still an unsolved problem. Various populations have been proposed as sources of ^{26}Al, such as novae, supernovae of type II, and young massive stars (see *e.g.*review by Prantzos 1991, *Gamma-Ray Line Astrophysics*, p. 129). COMPTEL's imaging capability of the 1.809 MeV emission will be very important to distinguish among the potential candidates.

COMPTEL has recently reached a real breakthrough in gamma-ray line spectroscopy with the discovery of intense 3-7 MeV emission from the Orion complex, which can be attributed to the cosmic-ray induced lines from $^{12}C^{*}$ and $^{16}O^{*}$ at 4.44 and 6.13 MeV, and with the discovery of the 1.15 MeV ^{44}Ti line from the Cas-A supernova remnant.

The OSSE instrument aboard the Compton Observatory has performed numerous observations of the galactic plane and galactic-center region to measure the distribution of 511 keV positron annihilation radiation and to search for time variability of the emission (Purcell *et al.*1993, *ApJL*, **413**, L85). A strong line near 511 keV and a positronium-like continuum are present. No time variability of the line flux has been observed. The OSSE data can be described by a point source within a few degrees of the galactic center (or a galactic bulge component), superimposed on a disk component producing significant emission at longitudes up to $\pm 20°$.

Other observations of the 511 keV line (and the low-energy continuum) were presented by Harris *et al.* (1990, *ApJ*, **363**, 135) [combination of SMM observations between 1980 and 1987], Gehrels *et al.* (1991, *ApJL*, **375**, L13) [GRIS], and Chapuis *et al.* (1991, *Gamma-Ray Line Astrophysics*, p. 52) [HEXAGONE].

Skibo, Ramaty, and Leventhal (1992, *ApJ*, **397**, 135) and von Ballmoos (1991, *ApJ*, **380**, 98) have presented further modeling of the distribution in the Galaxy of the 511 keV emission and the implied positron annihilation.

6.2 X-Rays

The past triennium has seen the first fascinating results obtained with the ROSAT satellite. As part of the all sky survey, ROSAT has carried out the first complete scan of the galactic plane with unprecedented high sensitivity. First results for some sample areas presented by Motch *et al.* (1991, *AA*, **246**, L24) illustrated the fantastic richness of these data. Full analysis of the observations will definitely provide important advances in our understanding of the (extended) galactic X-ray sources and the diffuse X-ray background.

Burrows and Mendenhall (1991, *Nature*, **351**, 629) and Snowden *et al.* (1991, *Science*, **252**, 1529) have reported on the ROSAT detection of shadowing of the X-ray background in the 1/4-keV band by the interstellar cloud in Draco, located a few hundred parsecs from the mid-plane. This suggests that the million-degree gas responsible for the X-ray emission may form a thick disk or halo around the Galaxy.

Snowden *et al.* (1991, *ApJ*, **364**, 11) have shown that the variations in the Wisconsin B/C band intensity ratio is dipole-like and may be due to a temperature gradient in the X-ray emitting plasma of the local cavity. While the physical origin of such a temperature gradient is uncertain, the alignment of the dipole with the higher temperature (and absorbed) Loop I region may be significant.

Studies of the diffuse soft X-ray background (0.16-3.5 keV) as seen with the Einstein observatory were presented by Wu *et al.* (1991, *ApJ*, **379**, 564) and Micela *et al.* (1991, *ApJ*, **380**, 495). The intensity is found to be dependent on galactic longitude even when only high galactic latitude data are used. The Galactic ridge is not present at energies below 0.33 keV.

Garmire *et al.* (1992, *ApJ*, **399**, 694) have presented the definitive maps of the diffuse X-ray background (0.5-2.8 keV) as surveyed by the low-energy detectors of the A-2 experiment aboard the HEAO 1 satellite.

McCammon and Sanders (1990, *ARAA*, **28**, 657) have reviewed the soft X-ray background and its origins.

6.3 Cosmic Rays and Magnetic Fields

Galactic winds have played an important role in large-scale cosmic-ray propagation studies in this triennium. Breitschwerdt *et al.* (1991, *AA*, , **245**, 79) discussed the mutual interaction of hot gas and cosmic rays escaping from the Galaxy. They found that beyond about 1 kpc from the galactic plane, stationary outflow solutions do exist, provided that the intergalactic pressure is low. Pohl and Schlickeiser (1990, *AA*, , **239**, 424) have studied the propagation of cosmic rays under the influence of a galactic wind and investigated the potentials of radio-synchrotron observations to detect the presence of a galactic wind. Radio studies of cosmic rays in the Galaxy have been reviewed by Reich (1991, *The Interstellar Disk-Halo Connection in Galaxies*, p. 187). Webber, Lee, and Gupta (1992, *ApJ*, **390**, 96) and Bloemen *et al.* (1993, *AA*, **267**, 372) compared direct observations of cosmic rays near Earth with the predictions from galactic wind models. They found that the data are consistent with a convection velocity < 20 km $s^{-1}$and that the effective cosmic-ray scale height in the solar vicinity has to be $\lesssim 3$ kpc, independent of the presence of a galactic wind.

Ko, Dougherty, and McKenzie (1991, *AA*, **241**, 62) studied the structure and stability of a static galactic halo including cosmic-ray and Alfvén-wave effects. Their findings lend support to the idea that such a halo cannot be stationary and that a steady outflow is indeed to be expected. On a global scale, however, a hydrostatic equilibrium seems plausible (Stephens 1991, *The In-*

terstellar Disk-Halo Connection in Galaxies, p. 323). A galactic hydrostatic equilibrium with magnetic tension and cosmic-ray diffusion was studied by Boulares and Cox (1990, *ApJ*, **365**, 544). They explored the structure of the z-distribution of mass and pressure in the solar neighborhood for three gravitational potentials, differing in their content of dark matter in the Galactic plane. An essential feature is that cosmic rays, magnetic fields, and a significant component of the interstellar mass all have high z-distributions characterized by scale heights ≥ 1 kpc.

Hartquist (1990, *Evolution of the Interstellar Medium*, p. 99) has reviewed the role of cosmic rays in interstellar dynamics. Ptuskin and Soutoul (1990, *AA*, **237**, 445) have discussed the propagation of cosmic rays in a cloudy interstellar medium. Particle acceleration and non-thermal radiation from OB associations and superbubbles was studied by Bykov and Fleishman (1992, *Sov. Astron. Lett.* **8** and *MNRAS*, **255**, 269). The origin of cosmic rays was reviewed by Axford (1991, *Proc. 22nd Int. Cosmic Ray Conf.*, **6**, 406).

7 The Chemical Evolution of the Galaxy

Francesca Matteucci
University of Trieste

7.1 Light elements

The study of light elements (D, ^{3}He, ^{4}He, ^{7}Li, ^{9}Be and ^{11}B), which provides insights into both cosmological and Galactic history, has received a great deal of attention in the last three years. In particular, the measure of Be and B in metal poor stars of the Galactic Halo (Gilmore *et al.* 1991, *ApJ*, **378**, 17, Ryan *et al.* 1992, *ApJ*, **388**, 184 and Gilmore *et al.* 1992, Nature **357**, 379) has posed serious problems concerning the production of these elements by cosmic-ray spallation on interstellar CNO nuclei. This finding encouraged speculation either that ^9Be could have been synthesized during an inhomogeneous big bang, or that some Galactic process other than cosmic-ray spallation might be involved (Pagel 1991, Nature **354**, 267). However, the discovery of B in some of the same stars with B/Be ratio of about 10 (Duncan *et al.* 1992, *ApJ*, **401**, 484) has revived the argument in favor of cosmic-ray spallation reactions in the interstellar medium. Prantzos *et al.* (1993, *ApJ*, **403**, 630) tried to explain Be and B in metal poor stars by suggesting that during the halo phase galactic cosmic rays were more confined than today and had flatter spectra at low energy. Very recently, Pagel (1993, in Workshop on "Cosmical Magnetism, ed. D. Lynden-Bell, Kluwer, in press) suggested a chemically inhomogeneous model for the halo. However, none of the proposed models is able to explain the observed linear relation between oxygen and beryllium abundances.

Although Li represents a key element in cosmology as it is one of the few nuclei which are synthesized during the big bang, it is still unclear what its primordial abundance was and how it varied during the Galactic lifetime. D'Antona and Matteucci (1991, *AA*, **248**, 62) suggested novae and asymptotic-giant branch stars as possible stellar sites for ^{7}Li production. They also predicted an increase of a factor of 2 in the Li abundance during the last 4.5 Gyr. A recent paper (King 1993, *AJ*, **105**, 1087) seemed to confirm this prediction. A study of possible ^{7}Li production sites was presented also by L.E. Brown (1992, *ApJ*, **389**, 251), who examined numerical chemical models for the evolution of Li, Be, B and D. He concluded that the major producers of ^{7}Li are likely to be intermediate and high-mass stars ($M > 4M_\odot$). He favored a primordial Li abundance of ^{7}Li/H $\simeq 10^{-10}$. Reeves (1993, *AA*, **269**, 166) discussed the Li isotopic ratio in the present Galactic ga (Lemoine *et al.* 1992, *AA*, (MS1-642)) to derive information on the stellar source of ^{7}Li. He obtained a yield for the last 4.5 Gyr of $20M_\odot$ of ^{7}Li per Gyr per $10^{10}M_\odot$ of gas. This amount can explain the difference between Population I and II abundances. Hobbs and Thornburn (1991,

ApJ, **375**, 116) measured Li abundance in six extreme metal poor dwarfs. Their results agreed with previous ones showing an invariance of the Li abundance in metal poor stars, so favoring a primordial origin for most or all the Li observed in these oldest stars. The observational status of Li has been reviewed very recently by Spite and Spite (1993, in *Origin and Evolution of the Elements*, p. 201) where the reader is directed also for the most recent results on abundances and nucleosynthesis.

Steigman and Tosi (1992, *ApJ*, **401**, 150) studied the Galactic evolution of D and ^{3}He in order to infer their primordial values, which provide a valuable test for big bang nucleosynthesis. They inferred upper and lower bounds to the primordial abundances of D and ^{3}He and used these bounds to derive constraints to the nucleon/photon ratio and to the number of new, light elementary particles.

7.2 Heavy Elements

High dispersion spectroscopy to measure the abundances of heavy elements in stars of the solar neighborhood has continued to be one of the most studied fields in these last three years. The major work in this field appeared recently (Edvardsson *et al.* 1993, *AA*, **275**, 101) and it represents the result of 10 years work of coordinated observations and analyses by a team of six astronomers. They have studied the elemental composition of 189 F and G dwarfs in the solar neighborhood and derived the abundances of O, Na, Mg, Al, Si, Ca, Ti, Fe, Ni, Zr, Ba, and Nd. The relative iron abundance [Fe/H] and the abundance ratios relative to iron of most elements are estimated to be accurate with a standard deviation of 0.05 dex. They have studied the evolution of α-elements and odd-z elements, iron-peak elements and s-elements as a function of stellar age and orbital properties. They have confirmed that in general α-elements are overabundant in metal poor stars, but in addition they have shown that $[\alpha/\mathrm{Fe}]$ for these metal poor stars decreases with increasing galactocentric distance, implying that the star formation was faster and started first in the central parts of the Galaxy. The abundances of Na and Al have been found to increase at different rates with metallicity in the disk, showing a connection to their respective production in C and Ne burning in massive stars. Finally they have found a big scatter in the age-[Fe/H] relation and suggested that it could be due to inhomogeneous enrichment. On the other hand, Francois and Matteucci (1993, *AA*, in press) have suggested that part of this scatter can be due to the fact that some of the stars observed now in the solar region, might have been born at different galactocentric distances. The lack of a pronounced slope in the age-[Fe/H] relation and the scatter was evident already from a previous work of Strobel (1991, *AA*, **247**, 35).

Abundances in extremely metal poor stars have been measured by Ryan *et al.* (1991, *AJ*, **102**, 303). They have shown that the trend for α-elements, relative to Fe, continues the trend observed in halo stars with $[\mathrm{Fe/H}] \geq -2.5$. They have also shown that a large spread exists in the abundances of s-process elements in very metal poor stars. This has been confirmed by Molaro *et al.* (1993, *Origin and Production of the Elements*, p.476). Concerning the interpretation of the observed $[\alpha/\mathrm{Fe}]$ versus [Fe/H] relations, usually interpreted as a time-delay effect due to the different lifetimes of stars giving rise to supernovae II and Ia, an alternative idea was proposed by Edmunds *et al.* (1991, *MNRAS*, **251**, 33p). In particular, they have proposed that the observed trend could be explained by a "pseudo-secondary" behavior of Fe. On the other hand, Matteucci and Francois (1992, *AA*, **262**, L1) continued proposing the time-delay effect and tried to fit the data from Bessel *et al.* (1991, *ApJL*, **383**, L71) showing that the change in slope for the [O/Fe] ratio occurs at [Fe/H]=-1.7 instead of at -1.0. Their best model implies different time scales for the formation of the inner and the outer halo (see also Smecker-Hane and Wyse, 1992, *AJ*, **103**, 621, on the duration of the halo phase). This is again supporting the idea, also shared by Burkert *et al.* (1992, *ApJ*, **391**, 651), that the Galaxy formed from inside out. In particular, Burkert *et al.* presented a complex self-regulated chemo-dynamical model where heating and cooling processes in a multicomponent interstellar medium are taken into account. Ferrini *et al.* (1992, *ApJ*, **387**, 138) also presented a complex model for the evolution of the halo and disk in the solar neighborhood. Two phases for the gas are considered in the disk region (diffuse and molecular) and only one in the halo region (diffuse). Detailed chemical evolution is calculated confirming the time-delay interpretation for the $[\alpha/Fe]$ versus [Fe/H] trends. Josey and Tayler (1991, *MNRAS*, **251**, 474) discussed the oxygen production in the solar neighborhood and derived constraints on the infall history.

Concerning the r- and s- process elements, Mathews *et al.* (1992, *ApJ*, **391**, 719) studied the kinds of stellar environments responsible for r- and s- process nucleosynthesis by means of models of Galactic chemical evolution. The growth of r- process appears to be associated to the ejection rate of material from massive stars. Low mass type II supernovae are favored if the r-process is primary. The growth of s-process material is consistent with production in intermediate mass stars. The yields of both r- and s-process appear to require a neutron source independent of metallicity. Abundances of r-process elements in halo stars have been used as Galactic chronometers, in particular those of Th and Eu (see Francois, 1991 in *The Stellar Populations of Galaxies*, p.417, for the observed abundances and Pagel 1993, in *Origin and Evolution of the Elements*, p. 496, for the derivation of the age of the Galaxy). In particular, Pagel (1993) compared the data sets with different Galactic models and found that they agree with ages of the order of 15 Gyr. Sneden *et al.* (1992, *AA*, **246**, 354) derived abundances of Cu and Zn in a sample of field and globular cluster stars. They found that Zn abundances closely follow the overall metallicities whereas there are Cu deficiencies in all metal poor stars. Matteucci *et al.* (1993, *AA*, **272**, 421) studied the evolutionary histories of Cu and Zn by means of a detailed Galactic chemical evolution model. They concluded that a satisfactory agreement with observations is reached only if one assumes that a substantial fraction of Cu and Zn has been produced by supernovae Ia.

7.3 Abundance gradients in the Galactic disk

Abundance gradients of single elements and ratios of elements represent useful constraints to stellar nucleosynthesis and models of Galactic chemical evolution. In Wilson and Matteucci (1992, *ARAA*, **4**, 1) one can find an updated review on the interstellar abundances and Galactic abundance gradients. In this review new data are presented showing a gradient in $^{12}C/^{13}C$ and $^{16}O/^{18}O$ ratios with galactocentric distance. The presence of a gradient in the $^{14}N/^{15}N$ ratio is less clear and there is no measurable gradient of $^{32}S/^{34}S$ ratio. Theoretical predictions by Matteucci and D'Antona (1991, *AA*, **247**, L37), where the contribution of novae to ^{13}C and ^{15}N is taken into account, are compared to the data. The comparison seems to favor novae being the major progenitor responsible for the production of ^{15}N and part of ^{13}C. In the review it is also indicated that in the interstellar medium near the sun, the carbon isotope ratio is $\simeq$ 20% lower than the solar ratio. This shows that there has been only a moderate amount of enrichment of the nearby interstellar medium since the formation of the solar system. This result has been also confirmed by Centurion and Vladilo (1992, *AA*, **251**, 245). Friel and Janes (1993, *AA*, **267**, 75) measured the iron gradient in the Galactic disk from old open clusters. Their result agrees with the gradient derived from younger open clusters, suggesting little change of the abundance gradient with time.

Finally, in Rana (1991, *ARAA*, **29**, 129) one can find a review of the chemical evolution of the Galaxy. The author surveys the observational material concerning the G-dwarf metallicity distribution, the age-metallicity relation, gas and metallicity distributions in the Galaxy, the luminosity functions of stars and white dwarfs in the solar neighborhood, the distribution of pulsars etc..

7.4 Chemical Evolution of the Galactic Bulge

At the end of 1990, Matteucci and Brocato (*ApJ*, **365**, 539) predicted that the abundance ratios of α-elements/Fe in bulge stars should be larger than solar over almost the full Fe range. This prediction was based on the time-delay effect and on the fact that in order to fit the distribution of bulge stars with metallicity (Rich, 1992, in *The Center, Bulge and Disk of the Milky Way*, p.47 and Geisler and Friel 1992, *AJ*, **104**, 128) a fast time scale of star formation is required. In particular, the observed distribution of bulge stars shows that they form a distinct stellar population with age of the order of 10 Gyr and abundance range from -1.5 to nearly +1.0 dex. Very recently, Mc William and Rich (1993, preprint) presented the first detailed abundance study of galactic bulge K giants. From their analysis of the α-elements relative to iron they concluded that Mg/Fe and Ti/Fe are enhanced by +0.3 dex relative to the solar value over almost the full Fe range, in very good agreement with the predictions. However, they also found that Ca and Si closely follow the normal trend of disk giants. Therefore, more observational and theoretical work seems to be required before assessing this point. Several evolutionary scenarios for the Bulge have been proposed but we are still not able to tell if the Bulge was the first Galactic structure to be formed, as argued by Lee(1991, *IAU Symp. 149*, p. 446) on the basis of the ages of the Bulge RR-Lyrae, or if the Bulge is younger than the globular clusters, as argued by Rich (1992) on the basis of the Bulge asymptotic giant branch stars. Wyse and Gilmore (1992, *AJ*, **104**, 144) also discussed possible evolutionary scenarios for the Bulge. In particular, they argued that its chemical distribution is best explained if most of the original mass were lost, and discussed the fate of this gas. They stressed the point that a way to distinguish between different mechanisms of Bulge formation is to look at abundance ratios.

8 Stellar Dynamics

James Binney
Oxford University

8.1 The Dark Halo

The radial extent of the dark halo continued to be a topic of interest. Kulessa & Lynden-Bell (1992, *MNRAS*, **255**, 105) inferred that it has density $\rho \sim r^{-2.4}$ and mass $M \approx 1.3 \times 10^{12}$ at $r < 230$ kpc. Norris & Hawkins 1991, *ApJ*, **380**, 104 concluded from the radial velocities of horizontal-branch stars at galactocentric distances $r \approx 40 - 60$ kpc that the mass of the halo is probably proportional to r out to at least 40 kpc, with $M(40\text{kpc}) \gtrsim 5 \times 10^{11} M_\odot$. Allen, Schuster & Poveda (1993, *AA*, **244**, 280) computed orbits for similar high-velocity stars. Kuhn (1993, *ApJ*, **409**, L13) suggested that at least some of the dwarf spheroidal systems may not be gravitationally bound, but have unexpectedly long dispersal time on account of the nature of orbits in the dark halo. Some workers continued to explore the possibility that flat rotation curves arise not from extended matter distributions but from a failure of conventional dynamics at very small accelerations (Begeman, Broeils & Sanders 1991, *MNRAS*, **249**, 523).

Over the triennium a great deal of effort was devoted to the possibility of detecting the constituent objects of the dark halo by their ability to gravitationally lens background stars. Nemiroff (1991, *AA*, **247**, 73) presented a comprehensive analysis of lensing statistics for dark halos composed of sub-stellar objects; these should be detectable provided the objects have masses in the range $10^{-8} - 1\,M_\odot$. If the rate of lensing events that follows from four recently reported events (Alcock *et al.*, 1993, *Nature*, **365**, 621, Aubourg *et al.*, 1993, *Nature*, **365**, 623) is correct, the halo must consist predominantly of sub-stellar objects. Richstone & Gould (1992, *ApJ*, **388**, 354) showed that sub-solar objects in the halo can be accompanied by *extremely* few objects massive enough to burn hydrogen; the overall halo mass-to-light ratio has to be $\Upsilon \gtrsim 2000\, M_\odot / L_\odot$. Daly & McLaughlin (1992, *ApJ*, **390**, 423) concluded that infrared observations with SIRTF or ISO should be able to detect a dark halo made of substellar objects providing these have mass $m \gtrsim 0.005\, M_\odot$.

Hut & Rees (1993, *MNRAS*, **259**, 27P) argued that the relatively low mass ($m \lesssim 5 \times 10^6\, M_\odot$) of any black hole at the galactic centre might be incompatible with the proposal that the dark halo consists of massive $m > 10^5 M_\odot$) black holes; dynamical friction would drag such massive holes in to the centre at a significant rate, and, if they there merged into one another, the current limit of the mass of the central object would be exceeded. Moore (1993, *ApJ*, **413**, L93) discusses the limits that can be placed on the masses of halo black holes by virtue of their capacity to disrupt globular clusters.

Boccaletti, Pucacco & Ruffini (1991, *AA*, **244**, 48), Hjorth & Madsen (1991, *MNRAS*, **253**, 703), Spergel & Hernquist (1992, *ApJ*, **397**, L75) and Kandrup, Mahon & Smith (1993, *AA*, **271**, 440) discussed various aspects of violent relaxation. The origin of the $r^{1/4}$-law now seems fairly well understood, and the incompleteness of violent relaxation is well documented. The significance of the short Liapunov exponents characteristic of N-body systems remains controversial, although the experiments of Sakagami & Gouda (1991, *MNRAS*, **249**, 241) argue that it is not significant macroscopically.

Evidence accumulated that large-scale fluctuations may be important in stellar systems. Miller & Smith (1993, in *Back to the Galaxy*, eds. S.S. Holt & F. Verter, AIP) reported surprisingly large-amplitude dipole distortions at the centres of stellar systems. Weinberg (1993, *ApJ*, **410**, 543) showed that the large-scale modes of somewhat idealized models of stellar systems become highly excited as the systems' equilibria approach global instability, and that this excitation can greatly enhance relaxation.

Several workers extended the range of models of spherical stellar systems with analytic distribution functions (Gerhard 1991, *MNRAS*, **250**, 812, Louis, Cuddeford 1991, *MNRAS*, **253**, 414, Dejonghe & Merritt 1992, *ApJ*, **391**, 531, Louis 1993, *MNRAS*, **261**, 283, Dehnen 1993, *MNRAS*, **256**, *in press*). Saha (1991, *MNRAS*, **248**, 492) refined and extended the standard approach to the stability of spherical systems and showed that systems previously thought to be immune to the bar instability are actually very slightly unstable. Papaloizou, Palmer & Allen (1992, *MNRAS*, **253**, 129) showed that slowly rotating spherical systems are unstable to the formation of a tumbling bar.

The dark halo is generally expected to be non-spherical but it is still unclear whether it is axisymmetric or triaxial. There was continued progress in the construction of models of hot, non-spherical stellar systems, although are knowledge of such systems remains primitive. Gerhard & Saha (1991, *MNRAS*, **251**, 449) and Binney & Kumar (1993, *MNRAS*, **261**, 584) explored techniques for obtaining approximate integrals for general gravitational potentials and Dehnen & Gerhard (1993, *MNRAS*, **261**, 311) used one of these techniques to explore rather fully the possible internal dynamical structures of flattened, axisymmetric systems. Evans & Lynden-Bell (1991, *MNRAS*, **251**, 213) obtained similar models by solving the Jeans equations for systems with Stäckel potentials. Van der Marel (1991, *MNRAS*, **248**, 515) used the Jeans equations and assumptions regarding the orientation of the principal axes of the velocity dispersion tensors of the metal-poor and dark halos to constrain the axis ratios of these two components.

Lees & Schwarzschild (1992, *ApJ*, **384**, 491) and Schwarzschild (1993 *ApJ*, **409**, 563) showed that the halo could in principle be triaxial, although the dynamics of a triaxial halo would be fundamentally different from that of the classical Stäckel ellipsoids because the galactic potential is never quadratic in galactocentric distance. The phase space of a triaxial halo would be dominated by families of 'boxlet' orbits rather than the boxes and loops that fill the phase space of a Stäckel potential. It does not seem possible to construct strongly triaxial ($c/a \lesssim 0.6$) halos from such boxlets. Dejonghe & Laurent (1991, *MNRAS*, **252**, 606) adapted Eddington's classical solution of the integral equation for the distribution function, $\rho(x) = \int d^3\mathbf{v}\, f(E)$, to the case of triaxial bodies embedded in a triaxial Stäckel potential. Batsleer & Dejonghe (1993, *AA*, **271**, 104) presented a self-consistent two-integral model whose density distribution and distribution function can both be written in terms of elementary functions.

8.2 The Disk

Kuijken & Gilmore (1991, *ApJ*, **367**, L9) refined their earlier estimate of the surface density Σ_0 of the disk at the Sun, concluding $\Sigma_0 = 48 \pm 9\, M_\odot\,\mathrm{pc}^{-2}$. It appears that most, and perhaps all, of the material in the disk is accounted for by known stars, remnants and interstellar gas (Kuijken 1991, *ApJ*, **372**, 125, Bahcall, Flynn & Gould, *ApJ*, **389**, 234). Kuijken & Gilmore (1991) find that the surface density of dark halo material within 1.1 kpc of the plane is about $23\, M_\odot\,\mathrm{pc}^{-2}$.

The origin of the correlation between the velocity dispersions of stellar groups and their ages continued an active topic of debate. Burkert, Truran & Hensler (1992, *ApJ*, **391**, 651) revived the proposal that this correlation reflects the decay of turbulence and thus the scale-height of the interstellar gas layer. Sellwood & Kahn (1991, *MNRAS*, **250**, 278) discovered a self-propagating instability in N-body disks, which cuts grooves in the disks by heating stars that are in resonance with the current pattern speed. It may be possible to find evidence for this kind of episodic heating in the phase-space distribution of disk stars (Sellwood 1993, in *Galactic & Solar System Astrometry*, ed. L. Morrison, CUP). Toth & Ostriker (1992, *ApJ*, **389**, 5) studied the heating of the disk by infalling satellites. In the Cold Dark Matter cosmology, large numbers of low-mass halos are predicted, and these may have small or negligible core radii. On falling into the Galaxy, such halos would do appreciable damage to the disk, and Toth & Ostriker were able to place limits on their number density from the requirement that the disk remain pretty cool over several billion years.

The triennium saw major advances in the quality of the published observational data that can be compared with theoretical models of the evolution of the kinematics and chemical composition of the disk. (Fresneau, 1992, *AJ*, **104**,, 578; Yoss, Bell & Detweiler 1991, *AJ*, **102**, 975, Evardson *et al.*, 1993, *AA*, **275**, 101). Cameron & Torra (1991, *AA*, **241**, 57) determined Oort's constant A from the motions of young stars and a derived a low value, $A = 9.6$ km s^{-1}, after correcting for the effects of spiral structure. Jaschek & Valbousquet (1991, *AA*, **242**, 77; 1992, *AA*, **255**, 124; 1993, *AA*, **275**, 472) determined the solar motion by a variety of techniques, finding it to be at $V = 20 \pm 1$ km s^{-1}in direction $(l, b) = 52 \pm 3°, 22 \pm 3°)$.

Cuddeford & Binney (1993, *MNRAS*, **265**, 000) investigated the discrepancy between the slopes of the circular speed curve, $v_c(R)$, at the Sun that are determined (a) directly and (b) from the ratio σ_ϕ^2/σ_R^2 of the two velocity dispersions in the plane of the McCormick K and M stars. They showed that the classical relationship $\sigma_\phi^2/\sigma_R^2 = -B/(A-B)$ is seriously in error, but that correcting this error actually *increases* the discrepancy between the directly and indirectly determined values of dv_c/dR. Kuijken & Tremaine (1993, *ApJ*, **xxxx**, xxxx) suggest that the discrepancy may arise because the disk is distorted into an ellipse by a triaxial halo potential. If the disk is elliptical, the Sun would have to lie close to a principal axis, and if that axis were the disk's major axis (the minor axis of the halo potential), the local value of σ_ϕ^2/σ_R^2 would deviate from that predicted by the shape of the rotation curve in the observed sense. A more prosaic possibility is that the McCormick stars are for some reason unrepresentative: Meusinger, Reimann & Stecklum (1993, *AA*, **245**, 57) find $\sigma_\phi^2/\sigma_R^2 = 0.55 \pm 0.2$ in better agreement with the value predicted from the circular-speed curve. Antonuccio-Delogu (1991, *AA*, **247**, 45) fitted solutions of the Jeans equations to the radial run of velocity dispersion and found that the data were best fitted for the smallest allowed values of σ_ϕ^2/σ_R^2.

Robin, Crézé & Mohan (1992, *ApJ*, **400**, L25) show from deep CCD frames of the anticentre that the stellar disk stops abruptly 5.5-6 kpc beyond the Sun.

8.3 The Bulge

As detailed elsewhere in this report, the triennium saw an upsurge of work on the bulge. Kent (1992, *ApJ*, **387**, 181) presented an axisymmetric dynamical model of the bulge, and Minniti *et al.* (1992, *ApJ*, **393**, L47) determined the variation with radius of the velocity dispersion σ and rotation rate V of the bulge from the motions of K-giants. They find that σ decreases with radius as predicted by Kent (1992), while V is nearly proportional to r. Lindqvist, Habing & Winnberg (1992, *AA*, **259**, 118) determined the central angular velocity and mass profile of the bulge from the kinematics of OH/IR stars. Langevelde *et al.* (1992, *AA*, **261**, L17) found OH/IR stars at the galactic centre with velocities as large as -355 km s^{-1}, and concluded that they are probably on elongated but bound orbits.

An important development was the confirmation by Combes *et al.* (1990, *AA*, **233**, 82) and Raha, Sellwood and Kahn (1991, *Nature*, **352**, 411) that the bar instability in a rapidly rotating disk is liable to lead to the disk thickening dramatically to form a mildly elongated and moderately flattened bar. The bulge may have been wholly or partly formed through the central part of the Galaxy's disk thickening in this manner. This picture would explain why many bulge stars are less than 10 Gyr old, and why the bulge appears to be triaxial (Sellwood, 1993 in *Back to the Galaxy*, eds. S.S. Holt & F. Verter, AIP).

Little and Carlberg (1991, *MNRAS*, **250**, 161, *MNRAS*, **251**, 227) and Hernquist & Weinberg (1992, *ApJ*, **400**, 80) investigated the time-scale and effects of angular momentum transfer from a bar to the surrounding disk and halo. Analytical estimates indicate that a bar should feed energy and angular momentum into surrounding components, and this would be expected to lead to significant evolution of the bar. N-body experiments generally confirm the conclusions of analytic work, but suggest that the neglect in the latter of self-gravity causes the predicted angular momentum and energy transfers to be underestimated.

Two studies investigated the effect of triaxiality on the dynamical friction experienced by massive halo objects such as globular clusters. Pesce Capuzzo-Dolcetta & Vietri (1992, *MNRAS*, **254**, 466) studied the decay dynamical friction drives in the orbits of globular clusters belonging to a galaxy that resembles Schwarzschild's self-consistent triaxial system. Clusters on box orbits move through the dense central region of the galaxy and have decay times that can be as little as a tenth of the decay times of clusters on loop orbits of the same energy. Long Ostriker & Aguillar (1992, *ApJ*, **388**, 362) showed that a rapidly rotating bar, unlike Schwarzschild's non-rotating bar, would not drag globulars to the centre significantly more rapidly that a spherical bulge of similar radial density profile.

8.4 Star clusters

Capaccioli, Piotto & Stiavelli (1993, *MNRAS*, **261**, 819) find that the power-law indices, x, of globular cluster mass functions are better correlated with galactocentric distance than with metallicity, and interpret this finding as a consequence of the effect upon x of disk shocking of the clusters. Hut (1993, *ApJ*, **403**, 256) completed a survey of binary-single star interactions that covers the entire energy range of interest. Goodman & Hut (1993 *ApJ*, **271**, 277) calculated the distribution of binaries that is in statistical equilibrium with a given field-star distribution. Lee & Ostriker (1993, *ApJ*, **409**, 617) found that the density at which binary-driven re-expansion of the core of a model globular cluster occurs depends sensitively numerical details, such as the Coulomb logarithm and time-step length employed.

Theuns (1992, *AA*, **259**, 503) modeled encounters between clusters of 500 stars and molecular clouds. He found that catastrophic encounters with the most massive stars limit cluster lifetimes to < 100 My.

9 Hydrodynamics

James Stone
University of Maryland

Hydrodynamic modeling has become an increasingly powerful tool for studies of the formation, structure, and dynamics of the Galaxy during the past triennium. Some of the major results from such studies are summarized below.

9.1 Formation

Sommer-Larsen (1991, *MNRAS*, **250**, 356) described a self-similar analysis of the formation of the galactic disk by a protogalactic cooling inflow which reproduces a number of observations of the ages of stellar distributions. A simple chemical evolution model for a viscous galactic disk also reproduces the observed gas metallicity gradient (Hellsten & Sommer-Larsen 1992, *MNRAS*, **255**, 650). The self-regulated chemical and dynamical evolution of a hot protostellar disk has been examined by Berkert *et al.* (1992 *ApJ*, **391**, 651), who found that the observed vertical structure of the disk was reproduced in a natural way as a consequence of gravitational settling.

9.2 Halo

The formation, vertical structure, and hydrodynamic stability of the galactic halo continued to be an important problem during the triennium. Ko *et al.* (1991, *AA*, **241**, 62) and Ko (1991, *AA*, **242**, 85) studied the structure and stability of a static galactic halo including contributions to the total pressure from the thermal plasma, cosmic rays, and Alfvén waves. At large distances from the galactic plane, the model predicts a pressure much larger than that of the ambient intergalactic medium, implying the halo is not static but drives a steady wind. In the solar neighborhood, Boulares & Cox (1990, *ApJ*, **365**, 544) concluded magnetic fields, cosmic rays and a significant component of the interstellar mass all have high z-distributions, characterized by scale heights $\geq$ 1 kpc. By studying the propagation of cosmic ray nuclei in a convecting halo, Webber *et al.* (1992, *ApJ*, **390**, 96) set a limits on the thickness of the halo at less than 4 kpc and a galactic wind velocity of less than 20 km s^{-1} within this distance. Einaudi & Ferrara (1991, *ApJ*, **371**, 571) have given steady state solutions of the MHD equations applicable to galactic wind models, emphasizing the importance of conditions in the disk for determining the properties of the flow. Fichtner *et al.* (1991, *ApJ*, **371**, 98) have considered the effects of nonspherical flow geometries on galactic wind models. Shapiro (1991, *The Interstellar Disk-Halo Connection in Galaxies*, p.417) has summarized the results of new studies of the galactic fountain model, in which supernova heated gas in the galactic disk escapes into the halo, radiatively cools, and condenses into dense clouds which fall back to the disk. Shapiro & Benjamin (1991, *PASP*, **103**, 923) have reported that optically thin, steady fountain flows in which the effects of nonequilibrium ionization/recombination and photoionization are included can reproduce the observed properties of UV absorption and emission lines. Raymond (1992, *ApJ*, **384**, 502) has found reconnection in the random component of the galactic magnetic field in the halo may be a significant additional source of heating if models of impulsive energy release in solar microflares can be scaled to conditions in the halo. Danly (1992, *PASP*, **104**, 819) has reported significant variations in the infalling halo gas in the northern galactic hemisphere, indicating multidimensional effects may be important (Houck & Bregman 1990, *ApJ*, **352**, 506). In analyzing the evidence for the inflow of neutral gas toward the disk, Mirabel (1990, *Structure and dynamics of the interstellar medium*, p.396) concludes most infalling gas in the solar vicinity originated in the disk as part of the disk-halo circulation with only a moderate component originating from extragalactic sources. Wakker (1991, *The interstellar disk-halo connection in galaxies*, p.27) reviewed the observed properties of High Velocity Clouds (HVCs). Ferrara & Einaudi (1992, *ApJ*, **385**, 475) used a linear stability analysis of the MHD equations to study the formation of HVCs in various background equilibrium models for the halo. Comerón & Torra (1992, *AA*, **261**, 94) used two dimensional time-dependent simulations to study the oblique impact of an HVC on the galactic disk, suggesting the Gould Belt may have resulted from such a process. Meyerdierks (1991, *AA*, **251**, 269) provided extensive comparisons between observations of the North Celestial Pole Loop with a theoretical description of a cloud-Galaxy collision. Li & Ikeuchi (1992, *ApJ*, **390**, 405) have performed two dimensional simulations of galactic chimneys, calculating the distribution and ballistic motions of HVCs formed in the radiatively cooled gas.

9.3 Superbubbles

Following the remarkable observations of galactic "worms" and superbubbles catalogued by Koo *et al.* (1992 *ApJ*, **390**, 108), a number of authors have undertaken studies of the hydrodynamical growth and blowout of supershells from the disk. Igumentshchev *et al.* (1990, *AA*, **234**, 396) reported two dimensional hydrodynamical simulations of the blowout of supershells. Fragmentation of the shell by Rayleigh-Taylor instabilities in the blowout phase is evident. Tomisaka has reported two dimensional (1992, *PASJ*, **44**, 177) and fully three dimensional (1990, *ApJL*, **361**, L5) magnetohydrodynamic simulations of the blowout of superbubbles including the effect of an ordered magnetic field of 5 μG in the plane of the disk. Such fields are able to impede the growth of superbubbles across field lines, confining them to the disk ($z < 200$ pc) without restricting growth along the fields lines. These results have interesting implications for the filling factor of hot gas from superbubbles in the disk.

9.4 Dynamos and the Magnetic Field

An outstanding problem in galactic gas dynamics continues to be the operation of the dynamo, as evidenced by the large number of papers on this subject published during the triennium. The global properties of the galactic magnetic field were reviewed by Trimble (1990, *Galactic and Intergalactic Magnetic Fields*, p.29), while reviews of dynamo theory were given by Ruzmaikin (1990 *ibid*) and Krause & Wielebinski (1991, *Rev. in Modern Astro.*, **4**, 260). Vallée (1991, *ApJ*, **366**, 450) has matched observations of field reversals in the spiral arms of the Galaxy to azimuthal, radial, and vertical modes of turbulent dynamo theory, allowing a prediction of the mean field direction in the 'Perseus + I' arm at 12 kpc. Ferrière (1992, *ApJ*, **389**, 286; 1992, *ApJ*, **391**, 188) has investigated the action of supernova and superbubbles on generating radial field from azimuthal field (the α-effect). A detailed study of the $\alpha\omega$-dynamo by Parker (1992, *ApJ*, **401**, 137) indicates azimuthal fields are generated primarily in the gaseous disk, while the α-effect and field dissipation via rapid reconnection occurs in magnetic lobes in the halo inflated by cosmic ray pressure. By considering dynamo amplification of initially weak fields, Kulsrud & Anderson (1992, *ApJ*, **396**, 606) concluded it is difficult to generate the galactic mean field from an initially very weak seed field. new linear and nonlinear dynamo models for a galactic disk embedded in a halo were obtained by Brandenburg *et al.* (1992, *AA*, **259**, 453). A number of authors considered the stability of the magnetized galactic disk. Hanawa *et al.* (1992, *PASJ*, **44**, 509) investigated both

Parker and Jeans instabilities in the disk by means of a linear analysis. Matsumoto & Shibata (1992, *PASJ*, **44**, 167) used time-dependent MHD simulations to study the nonlinear growth of the Parker instability in three dimensions. Gomez de Castro & Pudritz (1992, *ApJ*, **395**, 501) used a number of observational data sets to argue that the motion of the Taurus molecular cloud and other structures in the Gould Belt has been initiated by an odd mode of the Parker instability.

9.5 Galactic Center

Stark *et al.* (1991, *MNRAS*, **248**, 14P) have argued that molecular clouds near the Galactic Center must be losing angular momentum rapidly either via dynamical friction or interaction with a barred potential, and must therefore be spiraling inward. Yasutami & Tatematsu (1990, *PASJ*, **42**, 517) reached a similar conclusion for clouds on a much larger scale. Gerhard (1991, *Dynamics of Disc Galaxies*, p.111) and Binney *et al.* (1991, *MNRAS*, **252**, 210) argue that a rapidly rotating bar in fact dominates the dynamics of cold gas near the Galactic Center. This conclusion is supported by comparison of the hydrodynamics of CO gas in the Galaxy with other galaxies known to be barred spirals (Nakai 1992, *PASJ*, **44**, L27). Chevalier (1992, *ApJL*, **397**, L39) has modeled the hydrodynamics of the wind from the galactic Center, and finds the associated shocked wind bubble can produce magnetic structures which may be related to those observed. The strong wind from Sgr A* appears to have ablated a remarkable "tail" of ionized gas from the cool supergiant IRS 7, as reported by high resolution VLA radio observations (Yusaf-Zadeh & Morris 1991, *ApJL*, **371**, L59; Yusaf-Zadeh & Melia 1992, *ApJL*, **385**, L41) and optical images (Serabyn *et al.* 1991, *ApJ*, **378**, 557). From the observed properties of molecular gas at the Galactic Center, Spergel & Blitz (1992, *Nature*, **357**, 665) conclude the ambient gas pressure is some two-and-a-half orders of magnitude higher than in the solar neighborhood. Wardle & Königl (1990, *ApJ*, **362**, 120) have modeled the generation of a wind at the Galactic Center by an accretion disk threaded by open axial magnetic field lines. Observed magnetic field strengths within 70 pc of the center are large enough to significantly affect the dynamics of the molecular clouds (Morris 1990, *Galactic and intergalactic magnetic fields*, p.361) and spiral density waves (Tagger 1991, it Dynamics of disc galaxies, p. 117). In a review of supersonic jets from many systems in the Galaxy, Padman *et al.* (1991, *Beams and jets in astrophysics*, p.484) have given a "tourist's guide" to possible hydrodynamic jets in the Galactic Center region.

10 Dynamics and Kinematics from Eastern Europe

Alexei Fridman
Institute of Astronomy RAS, Moscow

10.1 Kinematics and Linear Dynamics (General Questions)

A steady state galaxy model consisting of a halo and a disk was considered by S. Kutuzov (Abstracts of the conference: *Mathematical Methods in Structuring of the Structure and Dynamics of Gravitating Systems*, Petrozavodsk, June 15-18, 1993). The halo does not rotate and has an isotropic velocity distribution, the disk rotates and has an anisotropic asymmetrical velocity distribution. A phase density of the disk is assumed to be linear function of the angular momentum (area integral) with coefficients depending on the energy integral.

A linear analysis of the stability of a thin non-self-gravitating gaseous disk shows (A. Morozov, V. Mustsevoi, and A. Prosvirov, 1992, *Sov. Astron. Let.*, **18**, 20) that for a continuous radial dependence of the rotation rate, a second unstable mode may be excited in addition to the previously known centrifugal instability mode. It is suggested that one possible reason for the formation of branching arms in spiral galaxies may be the simultaneous excitation and superposition of the two modes.

E. Malkov, T. Nuzhnova, and B. Sagintaev (1991, *Sov. Astron. Let.*, **17**, 200) investigated the stability of gravitating elliptical disks embedded in a spherical subsystem with respect to large-scale bending and bar-like perturbations. It is shown that a sufficiently massive external subsystem can stabilize an elliptical disk of practically any eccentricity with internal flows corotating with the disk.

A theorem is proved on the number of instabilities of a uniform gravitating medium consisting of n components in relative motion (A. Fridman and Yu. Fridman, 1991, *Sov. Astron. Let.*, **17**, 74).

A two-parameter family of parabolic orbits in the meridional galactic plane was found by E. Timoshkova (1991, *Sov. Astron.*, **35**, 660) in the problem of perturbed stellar motion in a rotationally-symmetric field of a galaxy. The relationship between the family and stationary solutions is established. Analytic expressions for star co-ordinates and velocities are obtained in closed form for periodic and asymptotic motion along the trajectories.

The estimates of time of stare transitions between halo and core in spherical clusters as result of stellar encounters were given by V. Danilov (Kinematics Phys. Celest. Bodies, 1991, **7**, 75). Numerical investigation of grass-evolution equations for quasi-stationary and nonstationary (in regular field) stellar clusters has been performed. There are three possible types of "halo-core" structure evolution in the quasi-stationary stellar clusters depending upon the initial conditions. The results of comparison of the author's cluster model calculations with observations and numerical N-body experiments are discussed.

10.2 Nonlinear Dynamics (General Questions)

Solitary vortices in a stellar subsystem of a galactic disk are obtained on the framework of collisionless stellar hydrodynamics approximation (V. Korchagin and A. Ryabtsev, 1991, *AA*, **246**, 368).

V. Dolotin and A. Fridman (1991, *Sov. Phys. JETP*, **72**, 1) derived a nonlinear equation describing the dynamics of perturbations of small but finite amplitude of a solid-body rotating gaseous system of nonuniform density. In the case of short-wave perturbations (of wavelength λ much smaller than the Jeans wavelength λ_J, *i.e.*$\lambda \ll \lambda_J$) and in the opposite limiting case of long-wavelength perturbations $\lambda \gg \lambda_J$ the equation is reduced to the well-known nonlinear equation for Rossby waves with a vector nonlinearity — Charney-Obukhov (or Hasegawa-Mita) equation. The weak turbulence spectrum of this equation leads to known observable relation between the fundamental parameters of gas population of the Galaxy. The stationary solution of this equation describes dipolar vortices (modons). Possible observable manifestations of modons (double galaxies, double galactic nuclei and double stars) are discussed (1991)

V. Afanasiev, S. Dodonov, A. Sumin, A. Fridman (1991, *Asrton. Tsirk.*, **N1551**, 3) established by the numerical simulations of the dynamics of gaseous disk with a rotation curve corresponding the observed one for galaxy Mrk 1040, the generation of two anticyclonic vortices, the morphology and localization of which are close to that discovered recently by 6-meter telescope in SAO.

Effects of nonstationary action on the dynamical evolution of non-interacting particles, which are in a certain regular field, were studied by A. Baranov (1991, *Astron. Zh.*, **68**, 1160). Various forms of the disturbance of the spiral-wave type with slowly changing parameters have been considered. Qualitative features of the action of the non-stationary disturbance have been revealed. There is a category of captured stars which move together with the resonant zone and fill the phase space in the domains, not contiguous to each other. The noncaptured stars pass in the intervals between these domains and are not subject essentially to the action of the disturbance.

10.3 Structure, Kinematics, and Dynamics of the Milky Way

Seventeen giant (170-700 pc) star-gas complexes (SGCs) have been detected by T. Sitnik (*Dynamics of Galaxies and their Molecular Cloud Distributions*, 1991, p.91) within 3 kpc from the Sun. These SGCs include stellar groupings younger than (2-3) $\times 10^7$ years and molecular clouds with masses $10^5 - 10^6 M_\odot$, embedded into HI superclouds. An investigation of seven large-scale SGCs has shown, within them, an age gradient of stellar groups, equal to (0.3-1.2) $\times 10^7$ years for distances 270-500 pc. The sequential changing of the ages of stellar groupings across the Sgr-Car arm (with the youngest stellar groups, all the young H_2O masers and CO clouds located at the inner boundary of the arm) is evidence for the formation of this arm by a spiral density wave. This wave, propagating in individual HI superclouds with molecular clouds inside, stimulates star formation therein. Crude estimates of the spiral pattern angular velocity yield $\Omega_p \approx 17 - 25$ km $s^{-1}kpc^{-1}$ and the corotation radius $\approx 8.8 - 13$ kpc. Perhaps the Cygnus arm is lying near the corotation radius, since there is no age gradient across this arm. The direction of the changing ages is different in all the Cas-Per arm's SGCs. It cannot be excluded that in one of these SGCs a "reverse" age gradient is observed.

A new method of determination of the radio halo size is suggested by A.Lazaryan and G. Chibisov (1991, *Sov. Astron. Let.*, **17**, 208) based on a statistical study of synchrotron fluctuations where HII regions are used as distance indicators.

A. Sumin, A. Fridman, and V. Haud (1991, *Pis'ma Astron. Zh.*, **17**, 779) proposed on the basis of observational data the seven-component model of the Galaxy with sharp jump of the gas component's surface density (two orders of magnitude) near the outer edge of molecular ring. Within this model the partial contributions of each subsystem to the model rotational curve of the Galaxy were calculated. It turns out that the value of dimensionless growth rate of hydrodynamical instability for the gas component only, being much less than unity, is very close to that for the same gas in the gravitational field of the whole Galaxy. From this fact two conclusions follow: 1) The unstable gas component of the Galaxy lies near the boundary of the hydrodynamical instability in accordance with the Le Chatelier principle. 2) The stellar population of the Galaxy probably does not effect dynamically the generation of the spiral structure in the gaseous component. A. Sumin, A. Fridman, and V. Haud (*ibid.* 698) also proposed the three component model of the Galaxy and revealed that this model is dynamically close to the seven-component model, and therefore it makes possible to use this more simple model to study the different problems of the Galaxy dynamics.

The distribution of the peculiar velocity components in the galactic coordinate for 207 of the main sequence stars was considered by B. Fesenco (Abstracts of the conference: Mathematical Methods in Structuring of the Structure and Dynamics of Gravitating Systems, Petrozavodsk, June 15-18, 1993) with dividing of that sequence into 13 subintervals. The observed distribution is in good agreement with the normal low distribution. On the base of the Moscow high precision photometric catalogue the sample of 116 solar type stars is formed. The normal law distribution for the peculiar velocity components is confirmed. The velocity of the sun relative to the stars of that sample is equal to 34 ± 5 km s^{-1}. The value of the standard deviation of the peculiar velocity components averages 26 km s^{-1}. The distribution of the stars in fields of equal area is not in agreement with Poisson law.

34 INTERSTELLAR MATTER
MATIERE INTERSTELLAIRE

PRESIDENT: H.J. Habing; VICE-PRESIDENT: D. Flower; SCIENTIFIC ORGANIZING COMMITTEE: F.C.Bruhweiler, M.A. Dopita, E. Falgarone, R. Genzel, T.A. Lozinskayà, P.G. Martin, P.C. Myers, S.R. Pottasch, L.F. Rodriguez, M.R. Rosa.

1. INTRODUCTION. (H.J. Habing) This report covers developments in the field of commission 34 between summer 1990 en 1993. Once upon a time, twenty odd years ago, interstellar matter could be studied practically only inside our Galaxy; now emission lines of CO and H_2O have been detected in a galaxy, IRAS10214+4724, at z= 2.286. Other detections, of absorption lines seen against quasars, preceded these discoveries, but nevertheless, the still somewhat uncertain emission line measurements are a breakthrough.

Because interstellar matter is now studied deep in the universe, the subject has grown; but it has also grown because more astronomers are involved and many new observational possibilities are continuously realized; the HST is a recent example, millimeter interferometry another. It is the task of commisison 34 to monitor developments in its field of competence and to advise the General Assembly and the Executive Committee of the Union. Traditionally the commission also writes a full report of all the developments in its field of competence. It has become clear that the field is growing so rapidly that the traditional way of reporting is no longer adequate and that we have to rethink the format of this periodic report, why it is needed, and how we find a practicable way to produce a report of a sufficiently wide scope and of acceptable quality.

I thank those members of the Scientific Organizing Committee that contributed to this report at the expense of considerable amount of their time.

2. MOLECULAR CLOUDS. (E. Falgarone)

The evolution of the ideas in this field is largely driven by that of the observing capabilities (larger telescopes, higher frequencies, more sensitive receivers, airborne observations) and that of computers. Sensitive receivers in the millimeter and submillimeter domains allow much faster detections and large dynamical ranges are thus reached in maps of molecular lines. Observations at high altitude (balloons and plane) considerably increase the range of transparent windows through the atmosphere. High angular resolution reached by large telescopes, interferometers and the raise in frequency, provides information on the structure of clouds and star forming regions down to about 100AU. The rapid development of large array cameras in the near infrared provide unprecedented census of the content of young stellar objects (YSOs) over large areas of clouds. The observational improvements are powerfully complemented by the rapid development of parallel computers which allow some of the sophisticated numerical simulations highly desired to approach the complexity of the evolution of interstellar clouds.

2.1 - STRUCTURE

2.1.1 Scale invariant structure of molecular clouds

Large scale maps, mostly in CO and isotopic lines (up to the third rotational transition J=3-2) allow a better approach to the spatial distribution of molecular gas. The considerable increase of the dynamical range between resolution and map sizes and in integrated intensity reveals the self–similarity of the distribution of molecular gas in space and velocity–space. A large degree of connectivity is found at all scales (over four orders of magnitude in size) and the complex structure (filaments, knots....) of the molecular gas is quite reminiscent of what the IRAS sky survey has revealed in the far–infrared. The large scale maps also show the existence of widespread low intensity CO emission and that the decomposition in terms of clumps, a simplifying assumption for modelling, is not alway meaningful. A sensitive survey of a part of the galactic plane (Lee + 1990 ApJ 355 536) reveals the connectivity of the CO emission at very low levels. A new CO

J. Bergeron (ed.), Reports on Astronomy, Vol. XXIIA, 367–397.

survey of the outer arm of the Galaxy reveals giant molecular clouds comparable in size and mass to the inner Galaxy complexes but underluminous in CO (Digel + 1990 ApJL 357 L29). Yet unpublished very sensitive CO observations of the gas layer far out the galactic plane by Dame and Thaddeus also reveal that the CO gas layer extends at higher z than previously thought.

Large data sets now exist at high (or moderate but over larger areas) angular resolution in non–star forming regions. They allow for the first time a broad view of the link between the various components of the cold interstellar medium, and that with young stellar associations.

A survey of a fraction of the high latitude sky has been performed by Heithausen + (1993 AA 268 265). The detected CO clouds are not different than any other molecular clouds. It is the low column density (or low mass) tail of the cloud distribution.

Extensive work has been done on individual high latitude clouds, in MBM 12 (also called L1457 and L1458) in ^{13}CO(J=1–0) by Pound + (1990 ApJ 351 165) with the BTL antenna, in ^{13}CO(J=2–1) by Zimmermann and Ungerechts (1990 AA 238 337) with the KOSMA telescope, and IRAM–30m telescope (Zimmermann, PhD Köln University, 1993), in the Polaris flare by Heithausen + (1990 ApJL 353 L45), Grossman + (1992 AA 264 195), in the HI cloud associated with a molecular cloud in Ursa Majoris by Joncas + (1992 ApJ 397 165).

In the ρ Ophiuchus complex, large scale and very sensitive maps have been done in the ^{13}CO(J=1–0) line by Nozawa + (1991 ApJS 77 647) and in the ^{12}CO(J=1–0) line by de Geus + (1990 AA 231 137). HI, 100μm and molecular emissions are compared at large scale in de Geus and Burton (1991 AA 246 559) for the entire complex associated with ρ Ophiuchus.

In the Taurus–Auriga complex, smaller maps exist in L1495 (Kramer + 1991 AA 251 382) and NGC 1499 (or the California nebula) in Herbertz + (1992 AA 249 483)

In Orion, a ^{13}CO(J=1–0) and CS(J=2–1) survey has been performed with the NRO–45m antenna by Tatematsu + (1993 ApJ 404 643).

For star–forming regions, surveys of gas associated with clusters now exist in CO (Leisawitz 1990 ApJ 359 319) and at HI λ 21cm (de Geus and Leisawitz, 1991 ApJS 75 835). For the first time, the interaction of stars with the interstellar medium has been studied at large scale in the Scorpio–Centaurus complex by de Geus (1992 AA 262 258). It reveals that the influence of a young stellar cluster extends up to about 100 pc, even more.

New statistical approaches to the gas distribution over a wide range of scales have been developed along various lines of thoughts. Analysis in terms of "structure tree" by Houlahan and Scalo (1990 ApJS 72 133; 1992 ApJ 393 172), of fractals by Dickman + (1990 ApJ 365 586), Falgarone + (1991 ApJ 378 186), Zimmermann and Stutzki (1992 Physica A 191 79), Hetem and Lepine (1993 AA 270 451), of wavelet transforms by Henriksen (1990 ApJL 365 L27) and Langer + (1993 ApJL 408 L45) quantify the scale invariance and the correlations in the spatial distribution of molecular gas.

The self–invariant behavior of the cloud structure is possibly related to the turbulent nature of the flows in molecular clouds although other interpretations have been proposed as, for example, the structure generated by collisional fragmentation, Nozakura (1990 MN 243 543). See also section 1.4.

2.1.2 Density structure

Large maps of star–forming regions now exist in several millimetric and submillimetric transitions of many molecules, complemented by maps in the continuum which trace the dust emission up to submillimeter frequencies. Such maps allow in principle the determination of the density and temperature structures although the difficulties met at interpreting the data are often more serious than anticipated, due, for example, to radiative pumping of excited levels by infra–red photons originating in star–forming regions.

Important data sets in nearby molecular clouds are those of Loren + 1990 (ApJ 365 269) for cores in the ρ Ophiuchus complex, Castets + (1990 AA 234 469) and Dutrey + (1993 AA 270 468) in Orion A, Bachiller + (1990 AA 236 461) in Barnard 1, Wilson + (1990 AA 239 305) in DR21/W75, Montalban + (1990 AA 233 527) and Henning + (1992 AA 263 285) in Monoceros R2, and Boden and Heithausen (1993 AA 268 255) in MCLD 126.6+24.5.

The most impressive finding is that large densities and small scale structure down to the resolution of the observations are found everywhere. Clumping on scales smaller than the beam resolution is often inferred from modeling or from a comparison between average column densities and local densities responsible for collisional excitation of molecular transitions. Unexpected dense and small scale structures are found in clouds regions of low average column density by Falgarone + (1992 AA 257 715), in translucent clouds by van Dishoeck and Black (1991 ApJ 366 141), Gredel + (1992 AA 257 245), in cloud edges by Falgarone + (1991 ApJ 378 186). Dense cores are found and studied in many transitions and many molecules in high latitude clouds (Turner, 1992 ApJ 391 158; 1993 ApJ 405 229; in H_2CO 1993 ApJ 410 140, in NH_3 1993 ApJ 411 219; Vallée 1990 AA 233 553), in the reflection nebulae NGC 2023 (White + 1990 AA 227 200), in L1455 (Juan + 1993 AA 270 432) and in NGC2071 (Snell + 1991 ApJ 372 518).

Indirect evidence for very small scale structure in the least opaque interstellar clouds is provided by the sharp changes in the absorption line profiles of atoms and ions, in front of a visual binary for example (Meyer 1990 ApJL 364 L5).

Comparative surveys in various molecules like CCS, HC_3N, HC_5N, NH_3, (Suzuki 1992 ApJ 392 551), HCO^+ and ^{13}CO (Fukui + 1992 ApJ 398 544) reveal not only the complexity of the density structure (via collisional excitation) but also that of the chemistry (see for example Guélin and Cernicharo 1991 AA 244 L21).

Large molecular abundances are found in cirrus clouds while their low average column density (therefore low shielding from the UV field) would lead to opposite predictions (Meyerdierks 1990 AA 230 172; Herbstmeier + 1993 AA 272 514).

The unexpectedly large amounts of dense and warm gas derived from the earliest submillimeter observations of star-forming regions are confirmed by more recent data. Some of these observations have difficulty to be explained by the current models of photon-dominated regions (PDRs), even those which take clumpiness into account. Shock excitation is usually ruled out on kinematic arguments. Clumpiness is characterized by gas at densities larger than $10^5 cm^{-3}$ concentrated in few tenths of the volume with large density contrasts between the clumps and interclump medium. Densities as large as $10^8 cm^{-3}$ are derived in some sources under the assumption of pure collisional excitation of the observed transitions.

In M17, Stutzki and Güsten (1990 ApJ 356 513) find a clump-interclump density contrast of 23 derived from $C^{18}O$ and $C^{34}S$ observations. In W3, NGC 1977 and NGC2023, Howe + (1991 ApJ 373 158) find an even larger value of 10^2 for this contrast derived from 158 μm [CII] observations. Large column densities of ionized carbon in the Orion Nebula are confirmed by the first detection of the isotopic line ^{13}CII which allows a determination of the 158 μm [CII] line optical depth (Stacey + 1991 ApJL 382 L37).

Similar conclusions, namely the existence of large amounts of dense warm gas, are reached by Jaffe + (1990 ApJ 353 193) in NGC2023, on the basis of a comparison of CO emission in rotational lines up to $7 \rightarrow 6$ with the 158 μm [CII] emission, by Graf + (1993 ApJ 405 249) from ^{13}CO(J=6-5) observations of NGC 2024 , and in Orion from ^{12}CO(J=4-3) observations by Schulz + (1992 AA 264 629), from ^{12}CO(J=7-6) observations by Howe + (1993 ApJ 410 179) and from CO lines (up to $J=17 \rightarrow 16$) and 158 μm [CII] observations by Stacey + (1993 ApJ 404 219).

Very large densities are also inferred in massive star-forming cores from CS(J=10-9) observations by Hauschildt + (1993 AA 273 L23), from CS(J=7-6) observations by Plume + (1992 ApJS 78 505) and ^{13}CO and ^{12}CO(J=9-8) maps by Boreiko and Betz (1991 ApJ 369 382) who note that the lineshapes are similar for low and high J transitions. In IC1396, (a partially ionized globule) Serabyn + (1993 ApJ 404 247) also derive large densities from a CS multiline analysis.

2.1.3 Mass determinations

Observations of the shadows of high galactic latitude clouds against the soft X-rays background provide estimates of the total column density across these clouds (Snowden + 1991 Science 252 1529; Burrows and Mendenhall 1991 Nature 351 629). Conversion factor between the 100 μm surface brightness and total gas column density are obtained, independently of other tracers like CO.

Mass distribution of clumps are derived from many maps in the hope of finding a relation with the initial mass function for stars. Power laws are systematically found with spectral indexes $\sim$ -1.6 (Williams and Blitz 1993 ApJL 405 L75) and other works quoted above. The power laws for the clump mass spectra do not seem to depend on the linear resolution of the various observations.

2.1.4 Velocity structure: shape of lineprofiles

Gas flows within molecular clouds are highly turbulent. The observed linewidths are orders of magnitude larger than the expected thermal widths and many of the observational works quoted above allowed refinement of the determination of the scaling laws between the internal velocity dispersion of a given scale and its size.

Accordingly, earlier attempts to estimate the role of finite correlation length of the velocity on the radiative transfer have been developed Kegel + (1993 AA 270 407). The impressive result is that, as expected, correlations between the velocities at different positions in clouds, lower the photon escape probability, increase the excitation and mimic the effect of large densities. The impact on the density determination and measurements of isotopic ratios, for instance, may be large and the physical parameters derived from the line analysis should be taken with some caution. New attempts at finding a correlation length in the velocity field have been conducted with the NRO–45m antenna without convincing determinations (Kitamura + 1993 ApJ 413 221). Indeed the situation is yet more complicated since one of the characteristics of turbulence is the existence of velocity correlation at all scales.

Falgarone and Phillips (1990 ApJ 359 344) have proposed that the non–Gaussian wings observed with similar characteristics in non–starforming regions, over several orders of magnitude in size, are a signature of the intermittency of turbulence in molecular clouds. Another interpretation has been proposed for these wings by Elmegreen (1990 ApJL 361 L77), namely the steepening of Alfvén waves in clumpy media. Lambert and Crane (1990 ApJL 359 L19) found that the lineprofiles of various chemical species in absorption along the line of sight to ζ Oph (CN, CH, and CH^+) are quite different in terms of linewidths and linewings which suggests a link between the chemistry and the velocity field, not provided by a shock.

New radiative transfer computations in molecular clouds have been conducted in the frame of clumpy media by Burton and Hollenbach (1990 ApJ 365 620), Wolfire + (1993 ApJ 402 195). The remarkable smoothness of molecular line profiles when observed with an extremely large signal to noise ratio, poses the problem of the actual density structure of the emitting gas. It might consist of thousands of self–gravitating clumps of a few hundreds astronomical units (Tauber 1991 ApJ 375 635)

2.1.5 New gas components in clouds

Fluorescence allows the detection of widespread emission of H_2 in clouds where the ambient UV field is sufficient to excite the 2.2 μm line (Martin + 1990 ApJ 354 220; Burton 1990 ApJ 352 625).

Extended 3.3 μm emission has been detected in M17 Giard + (1992 AA 264 610) which traces the excitation of the C–H bound in molecular clouds.

Molecular clouds with no detected CO emission (therefore composed of H_2 mostly) have been found from comparison between dust 100μm emission and HI line emission (Blitz + 1990 ApJ 352 L13), although the number of such clouds may not be large according to the results of Heithausen's high latitude CO survey (quoted above). An HII region associated with a CO molecular cloud has been found beyond the optical disk of the Galaxy at an estimated distance of 28 kpc (de Geus + 1993 ApJL 413 L97 and Digel + 1994 ApJ).

A 158 μm [CII] ballon–borne line survey of the galactic plane by Shibai + (1991 ApJ 374 522) shows that at large scale the [CII] line distribution follows the CO distribution remarkably well. At small scale, they find diffuse extended emission whose origin is not yet clearly established.

2.1.6 A few persisting and new puzzles

The chemical complexity illustrated by studies quoted above and the line survey conducted between 330 and 355 GHz in Orion by Sutton + (1991 ApJS 77 255) is not yet satisfactorily explained by chemical models. The origin of the large observed CH^+ abundances in molecular clouds is not explained either (Gredel + 1993 AA 269 477).

Observations of ^{12}CO in absorption in front of BL Lac reveal the same velocity profile in absorption as in emission (Marscher + 1991 ApJL 371 L77). This result is confirmed by Lucas and Liszt (1993 AA 276 L3) in other absorption lines. The comparison of emission and absorption lines is a powerful tool (e.g. the work done in the HI λ =21cm line) and to prepare the extension of such a work in the CO line, Liszt and Wilson (1993 ApJ 403 663) have systematically searched for CO(1-0) emission in the direction of all extragalactic sources strong enough to allow aperture synthesis observations of CO.

2.2 - MAGNETIC FIELD

Direct measurements of the magnetic field intensity by Zeeman splitting have been extended to dark clouds. Using the centimeter lines of the OH radical Crutcher + (1993 ApJ 407 175) obtain two positive detections with B$\sim$ 20 and 10μG and a statistical analysis of the non–detections confirms that the average magnetic energy in molecular clouds is comparable to their average (non–thermal) kinetic energy.

The direction of the magnetic field is determined from polarization measurements in the far infrared in emission (Gonatas + 1990 ApJ 357 132) for Orion, in the visible range by absorption of starlight (Goodman + 1990 ApJ 359 363) in Taurus, Perseus and Ophiuchus. Studies of the polarization of the absorption of starlight in the near infrared unexpectedly show no deviation of the polarization direction compared to that derived in the optical (Goodman + 1992 ApJ 399 108). The spatial pattern of the direction of the polarization vector in clouds shows a well–defined mean direction except in clouds with embedded clusters (Myers and Goodman 1991 ApJ 373 509). However, the polarization vector is not always found at the same angle to the main axis of a cloud filament. The dispersion of the polarization direction in several complexes is compared to that of the direction of the gas filaments and is found to be slightly smaller (Goodman + 1990 ApJ 359 363).

2.3 - LINK BETWEEN UV FIELD, CHEMISTRY, DUST AND STRUCTURE.

The existence of large inhomogeneities in the density of molecular clouds is directly inferred from most of the observations quoted above, in all kinds of environments, and is requested to allow the penetration of UV photons deep within molecular clouds. The pervasivity of most of massive regions to UV photons is necessary to explain the observed extended emission of the 158 μm [CII] line (see references quoted above) further confirmed by the extended emission of the 492 GHz [CI] line, in S140 for example (Hernichel + 1992 AA 259 L77). This, and the existence of large amplitude motions induced at all scales in molecular clouds, bear profound impacts on chemical processes and the time evolution of dust grains.

Calculations of the penetration of UV photons in clumpy clouds have been performed by Boissé (1990 AA 228 483) for continuum photons.

To understand the fact that the observed abundances of organic molecules in molecular clouds are closer to computed "early time" abundances than the steady state ones, several groups have tried to couple the chemical evolution with either a time dependent UV irradiation or purely dynamical processes. Rapid turbulent mixing of material between diffuse (UV irradiated) and dense (shielded) regions deeply modifies the chemical abundances (Pineau des Forêts + 1992 MN 256 247; Chièze + 1991 ApJ 373 110). The motions of clouds relative to one another produce intermittent shadowing which also affects the chemical evolution (de Boisanger + 1992 ApJ 401 182; de Boisanger and Chièze 1991 AA 241 581).

Dense cores are made ephemeral structures by frozen-in magnetic field which may revert its evolution toward a protostellar condensation and modify the chemistry (Prasad + 1991 ApJ 373 123).

The chemical fluctuations which are observed at all scales (Goldsmith + 1992 ApJ 385 222 in NGC 2071) are possibly related to some of the above processes, but may be due also to fluctuations in the dust properties (explained as fluctuations in their size spectrum, see Boulanger + 1990 ApJ 364 136). These variations may be connected to the observed diversity of the extinction curves reported in the atlas of extinction curves by Fitzpatrick and Massa (1990 ApJS 72 163) and Papaj + (1991 MN 252 403). Comparisons of the far UV rise of absorption curves with CH and CH^+ absorption are made in Jenninskens + (1992 AA 265 L1).

Also possibly connected to the above, Pineau des Forêts + (1992 MN 258 45) and Le Bourlot + (1993 ApJL in press) have shown that there exists a possibility of reaching chaotic behaviour for chemical patterns in the parameter range of molecular clouds.

Many models have been built which try to reproduce the observed inhomogeneities of cloud properties at various scales, taking into account the small scale fluctuations of density and other properties. These fluctuations contrast with the large scale uniformity of some of the cloud properties in our Galaxy, and nearby comparable galaxies. The dust continuum emission of inhomogeneous clouds has been computed by Bernard + (1992 AA 263 258). Attempts to couple the chemistry and radiative transfer of line photons have been published (for homogeneous clouds) in the submillimeter and far infrared ranges by Le Bourlot + (1993 AA 267 233), in the millimeter range by Meixner and Tielens (1993 ApJ 405 216) and by Wolfire + (1990 ApJ 358 116) for clumpy PDRs. Radiative transfer has been computed for CO line photons in such clouds also by Tauber and Goldsmith (1990 ApJL 356 L63), Gierens + (1992 AA 259 271).

2.4 - DENSE CORES AND STAR FORMATION.

As for the less compact parts of molecular clouds, the main result brought in light by recent observations is the existence of substructure in the interiors of dense cores down to the resolution of the observations, with large density contrasts inferred from multitransition analysis at very small scales within the cores.

Observed dense cores in general are far from being thermally supported self- gravitating cores (Fuller and Myers 1992 ApJ 384 523), and models have been built which try to include non–thermal support in the balance equations (Myers + 1992 ApJ 396 631). In one case however, the classical picture of an isothermal self–gravitating core in which inside–out collapse has already started is consistent with observations (Zhou + 1990 ApJ 363 168 and 1993 ApJ 404 232 in the Bok globule B335).

Low–mass star formation has been shown to take place even in small isolated globules. Yun and Clemens (1992 ApJL 385 L21) have detected 14 outflows in a sample of 41 Bok globules, Duvert + (1990 AA 233 190) in IC396, and Reipurth + (1992 AA 256 225) have detected HH objects with extremely recent outbursts in the Bok globule B335.

Early stages of star formation have also been actively looked for in the continuum. Protostellar cold dust condensations are found in molecular clouds survey at λ =1.3 mm in ρ Ophiuchus (André + 1990 AA 240 321; Mezger + 1992 AA 265 743) complemented by a VLBI survey of the same cloud (André + 1992 ApJ 401 667), in NGC 2024 (Schulz + 1991 AA 246 570; Mezger + 1992 AA 256 631), in NGC 2264 (Tauber + 1993 ApJ 403 202). Multiwavelength photometry up to λ =450μm led to the discovery of a very cold dense fragment, probably gravitationally unstable (Chini + 1993 AA 272 L5). In several cases molecules are found to be depleted in these dust condensations which seems to confirm that some molecules disappear from the gas phase at low enough temperatures and build up molecular ice mantles which are observed in absorption in the near infrared (against intense IR sources, therefore in regions different from cold dense cores). These absorption measurements have allowed the first detections in interstellar space of highly symmetric molecules such as CH_4 (Lacy + 1991 ApJ 376 556), O_2 (Ehrenfreund + 1992 AA 260 431), C_2H_2 and other molecules (Evans + 1991 ApJ 383 674). CO ices have been observed in ρ Oph by Kerr + (1993 MN 262 1047) and Tielens + (1991 ApJ 381 181) in the direction of several protostars.

The issue of molecule depletion onto grains in cold dense cores is still controversial. In NGC 2024, molecular line observations (Mauesberger + 1992 AA 256 640) seem to confirm that the small structures detected in the submillimeter continuum of dust are depleted in molecules while in S106 Richer + (1993 AA 262 839) find that the column densities determined from the dust continuum emission are consistent with those derived from molecular lines.

A remarkable output of the powerful imaging capabilities of near IR cameras is the discovery of very rich young clusters of stars associated with dense cores like in NGC 2264 (Lada + 1993 ApJ 408 471). In L1630, Lada + (1991 ApJ 368 432) have conducted a systematic search for dense cores in the CS(2-1) line complemented by 2μm images of the same area which revealed new small clusters of young stars Lada + (1991 ApJ 371 171).

2.5 - DYNAMICAL EVOLUTION

2.5.1 Formation, lifetime.

Nothing really new is to report in that section. The dilemma is still whether clouds form via gravitational fragmentation (instabilities in large scale supershells of material swept up by supernovae remnants for example or in the galactic gas layer itself at still larger scale) or by collisional agglomeration of much smaller entities.

It is confirmed observationally that molecular clouds are not long-lived and the rate at which self-gravitating molecular material is assembled out of more dispersed gas is thought to be related to the star formation rate of the former generations of stars. The possible link between the observed fractal structure and the star formation rate and initial mass function is discussed in Larson (1992 MN 256 641). Large scale molecular observations of the Orion complex have shown the influence of stellar associations and self-gravity upon the structure of molecular gas. However, in a new paper (Miesch and Bally 1994 ApJ) show that all the statistical properties of the velocity field and brightness distributions are consistent with a turbulent description of the structure.

2.5.2 Virial balance

Many molecular clouds (or parts of molecular clouds) are now known to be far out of virial balance between self-gravity and internal kinetic energy. The internal pressure is found to be up to 100 times the thermal pressure, it is fluctuating between one cloud and another by 3 orders of magnitude and many molecular clouds are considered as transient structures temporarily confined at all scales by some pervading turbulent or magnetic (or both) pressure (most of the corresponding references have been quoted above).

The role of an external magnetic pressure is discussed in Bertoldi and McKee (1992 ApJ 395 140) and McKee and Zweibel (1992 ApJ 399 551). The fraction of molecular H_2 over HI is shown to be extremely sensitive to both the ambient UV radiation field and the external pressure if clouds are in virial balance (Elmegreen 1993 ApJ 411 170)

Thermal conduction and evaporation are discussed in McKee (1990 ApJ 358 375 and 392)

2.5.3 Ambipolar diffusion

Magnetic field is now observed, as said above, with an intensity which makes the magnetic pressure comparable to the kinetic pressure in molecular clouds. Many studies have been conducted to estimate the coupling of the neutrals to the ionized species and the role of charged dust grains is shown to be of fundamental importance (Nakano 1990 MN 242 535 ; Umebayashi and Nakano MN 243 103; Tomisaka 1990 ApJ 362 202).

A minimum wavelength for coupling of neutrals with the magnetic field in clumpy clouds has been found by Elmegreen and Fiebig (1993 AA 270 397) but such a length is highly dependent of the microscopic processes (collisions) at the origin of the coupling, and in particular very sensitive to the (basically unknown) grain size distribution of very small grains.

Mouschovias has addressed the issue of ambipolar diffusion in several contexts (1991 ApJ 371 296 and 373 169; 1992 ApJ 390 144; ApJ 390 166; ApJ 391 199) and Zweibel (1990 ApJ 362 545) the fundamental question of magnetic tangling in molecular clouds.

2.5.4 Turbulence and the growth of gravitational instability

Turbulence may prevent gravitational instabilities to grow. Not only is it an additional support against self-gravity but it is scale dependent and might make the largest scales more stable than the small ones. Bonazzola + (1992 JFM 245 1) have computed the turbulent pressure by using a method inspired by the renormalisation group technics. Léorat + (1990 MN 243 293) have developed numerical simulations of supersonic self-gravitating flows which show the disrupting power of turbulence for a bound structure.

The influence of magnetic field on the velocity fluctuations induced in the gas by its gravitational interaction with stars has been studied by Jacobi + (1990 AA 237 461). They find that the existence of magnetic field increases the energy contained in the velocity fluctuations of the gas to twice the energy of the magnetic field fluctuations.

2.5.5 Thermal effects

3-dimensional hydrodynamical simulations of collapse and fragmentation which take molecular line cooling into account, with therefore a detailed description of the equation of state, have been developed by Monaghan + (1991 ApJ 375 177). They show that, as expected, radiative cooling significantly increases the number of fragments.

Elphick + (1991 MN 250 617) have studied the consequences of the non-linearity of the equation for thermal stability. They derive the motions of the fronts separing the cold and hot stable phases and suggest that thermal instability may be at the origin of some of the complexity of the spatial organisation of the interstellar medium.

Conference proceedings, books

Low mass star formation and pre-main sequence objects 1990 ed. B Reipurth (Munich: ESO)

The evolution of the interstellar medium 1990 ed. L. Blitz (San Francisco: Astronomical Society of the Pacific)

The physics of star formation and early stellar evolution 1991 eds. C.J. Lada and N. Kylafis (Dordrecht: Kluwer)

IAU Symposium 147 *Fragmentation of molecular clouds and star formation* 1991 eds. E. Falgarone, F. Boulanger, G. Duvert, (Dordrecht: Kluwer)

Physical processes in fragmentation and Star formation 1991 eds. R. Capuzzo Dolcetta, C. Chiosi and A. di Fazio (Dordrecht: Kluwer)

Molecular Clouds 1991 eds. R.J. James and T.J. Millar (London: Cambridge University Press)

IAU Symposium 150 *Astrochemistry of Cosmic phenomena* 1992 ed. P.D. Singh (Dordrecht: Kluwer)

Infrared Astronomy with ISO 1992 eds. Th. Encrenaz and M.F. Kessler (New York: Nova Science Publishers)

Star formation in stellar systems 1992 eds. G. Tenorio-Tagle, M. Prieto and F. Sanchez (Cambridge: Cambridge University Press)

Back to the Galaxy 1993 eds. S.S. Holt and F. Verter (American Institute of Physics)

Protostars and Planets III 1993 eds. E.H Levy and J.I. Lunine (Tucson: The University of Arizona Press)

The structure and content of Molecular Clouds 1994 ed. T.L. Wilson (Berlin: Springer -Verlag)

3 CHEMISTRY (D. Flower)

There exists a vast literature on interstellar chemistry, and any attempt to cite individual research papers would be necessarily subjective and incomplete. Accordingly, reference will be made to books and conference proceedings which have appeared during the period under review and which guide the reader to the original papers.

The oft-cited volume honouring Alex Dalgarno, " Molecular Astrophysics" (Hartquist, T.W., ed.: 1990, Cambridge UP) comprises a selection of articles reviewing a variety of aspects of interstellar physics and chemistry. Two other review volumes appeared in the same year: Thronson, H.A. +, ed.: 1990, "The Interstellar Medium in Galaxies", Kluwer, Dordrecht and Watanabe, T. +, ed.: 1990, "Molecular Processes in Space", Plenum, New York. The former may be found particularly helpful to research students.

A welcome addition to the literature was the proceedings of no. 7 in the series of Manchester astronomical conferences (James, R.A. +, ed.: 1991, " Molecular Clouds", Cambridge UP. In the same year appeared the proceedings of a NATO Advanced Study Institute (Greenberg, J.M. +, ed.: 1991, " Chemistry in Space", Kluwer, Dordrecht) and of IAU Symp. No. 147 (Falgarone, E. +, ed.: 1991, " Fragmentation of the Molecular Clouds and Star Formation", Kluwer, Dordrecht). The latter covers topics whose elucidation is the objective of much of the research into the interstellar medium.

IAU Symp. No. 150 (Singh, P.D., ed.: 1992, "Astrochemistry of Cosmic Phenomena", Kluwer, Dordrecht) reviews much of the research in interstellar chemistry, but also has a broader remit. Spectroscopic aspects are additionally covered by Bohme, D.K., ed.: 1992, "Chemistry and Spectroscopy of Interstellar Molecules", University Tokyo Press, and star formation, once again, in Melrose, D., ed.: 1992, "Star Formation in Different Environments", Aust. J. Phys., 45, 387. Researchers into interstellar chemistry may also find useful the proceedings of the second biennial conference on molecular spectroscopic databases (Rothman, L.S. +, ed.: 1992, J. Quant. Spectrosc. Radiat. Transfer, 48, 469). The proceedings of the Faraday Symp. No. 28 on "Chemistry in the Interstellar Medium" (1993, Faraday Trans., 89, 2111) provide a timely and sometimes stimulating review of interstellar chemistry to the end of 1992. Finally, mention should be made of a volume on the closely related topic of "Astronomical Masers" (Elizur, M.: 1992, Kluwer, Dordrecht.)

4.1 INTERSTELLAR DUST (P. G. Martin)

In the time interval considered there have been review papers, compendia from meetings, and monographs which introduce research in this field. Among these are Dwek+ (1992,AnRevAA,30,11), and Mathis (1990,An Rev AA,28,37; 1993, Rpt Prog Phys,56,605); *Physics and Composition of Interstellar Matter* (Krelowski+ 1990, Nicolaus Copernicus University), *Dust and Chemistry in Astronomy* (Millar+ 1993, Institute of Physics), and *Infrared Cirrus and Diffuse Interstellar Clouds* (Cutri+ 1993, ASP); Hoyle+ (1991, *The Theory of Cosmic Grains*, Kluwer), Whittet (1992, *Dust in the Galactic Environment*, Hilger), and Wynn-Williams (1992, *The Fullness of Space: Nebulae, Stardust, and the Interstellar Medium*, Cambridge). The related field of interstellar chemistry was the subject of IAU Symposium 150, *Astrochemistry of Cosmic Phenomena* (Singh 1992, Kluwer); see also Cecchipestellini+ (1992,Nuovo Cimento-C,15,1047).

This report does not aim to be exhaustive of all literature in this field. It does emphasize, with some attempt at comprehensiveness in the current literature, *how* we learn about dust – as much as *what* we are learning – but, on the other hand, it does not pretend to distill all previous knowledge or catalogue all *applications* of dust, i.e., how dust influences or sheds light on various astrophysical phenomena.

4.1. SPECTRAL SIGNATURES. The surest way to verify a component material of interstellar grains is to identify a characteristic spectral feature. This might not tell exactly what form of grain is involved (size; free particle, or core or mantle, or part of composite grain) but it is a beginning. Much recent work is on carbon-bearing grains, which is reflected in the emphasis below.

4.1.1. Silicates. Virtually all cosmic Si is in the form of silicates, whose ubiquitous presence is known from the 10 and 20 μm features. The smooth broad profiles of these features indicate an amorphous structure, but the precise mineralogy remains unknown. The 8–23 μm spectrum of circumstellar silicates has been deduced from IRAS LRS spectra (Simpson 1991,ApJ,368,570); there are environmental changes, particularly at 10 μm (also Griffin 1993,MN,260,831). Band strengths and optical constants encompassing near infrared continuum absorption, accounting for inclusions of other materials in silicate grains, have been presented by Ossenkopf+ (1992,AA,261,567). SiH bands near 5 μm, being sensitive to the chemical environment, might indicate the nature of Si bonding (Moore+ 1991,ApJ,373,L31).

4.1.2. 2175 Å. Most explanations of the interstellar 2175 Å extinction "bump" require about 20–30% of the cosmic C. The bump has long been attributed to small particles of graphite, though the identification has never been fully secured with laboratory data. Lorusso+ (1993,SolidStateComm,85,729) have measured ultraviolet spectra of submicron sized graphitic particles; the spectral peak of the surface plasmon resonance shifts with particle size as anticipated. Aannestad vvv(1992,ApJ,386,627) has investigated the potential correlations of feature width, central position, and ultraviolet continuum components in a multicomponent extinction model. Voshchinnikov (1990,SovAL,16,215) examines the dependence of the profile on grain shape.

Hydrogenated amorphous carbon (HAC) is without ultraviolet feature when fully hydrogenated(Blanco+ 1991,ApJ,382,L97), but experiments on processing (dehydrogenation via ultraviolet irradiation or annealing) point to a viable explanation of the "bump" using amor–phous carbon grains (Blanco+ 1993,ApJ,406,739). High-ranking coals (e.g., anthracite) can also be sufficiently graphitized to produce the bump, while material like poorly graphitized coals could carry the underlying continuum (Papoular+ 1993,AA,270,5). The production and isolation of C_{60} (Kratschmer+ 1992,Carbon,30,1143) has opened a new avenue of research. Theoretical explorations have included the electronic transitions of C_{60} and other large molecules (Braga+ 1991,AA,245,232) and of small, hollow, onion-like car–bon particles (Henrard+ 1993,ApJ,406,92).

4.1.3. (Unidentified) Infrared Bands. The "unidentified" infrared emission bands (UIBs) – 3.3, 6.2, 7.7, 8.6, 11.3 μm and other features (e.g., Roche+ 1991,MN,252,282) – are thought to be due to stretch and bending CH modes. The particular bonding of C and H affects the details of the features, which become diagnostic of the carrier [e.g., 3.3 μm is the aromatic C-H stretch, whereas aliphatic groups like methyl (CH3) shift the feature to 3.4 μm]. Tokunaga+ (1991,ApJ,380,452) find two types of profile in high resolution observations of the 3.29 μm emission feature. A working model is a mixture of largely ionized polycyclic aromatic hydrocarbons (PAHs; see Szczepanski+ 1993,Nature,363,699).

In regions where the radiation field varies strongly in hardness and strength the influence of the environment on the size, structure, hydrogenation, and abundance of the various PAHs can be assessed (NGC7027: Sandford 1991,ApJ,376,599; Orion: Siebenmorgen+ 1992,AA,259,614; M17: Giard+ 1992, AA,264,610 and Chrysostomou+ 1992,MN,256,528; HD44179: Sloan+ 1993,ApJ,409,412; and NGC1333: Bregman+ 1993,ApJ,410,668). The 3.3 μm band seems absent inside ionized regions. The strengths of the 7.7 and 11.3 μm features are correlated; from the contributions of the features to the total infrared luminosity it is estimated that 1–5% of C is tied up in aromatic molecules like PAHs; and the precise nature of the carrier of the important underlying continuum is unclear (Zavagno+ 1992,AA,259,241).

Theoretical work suggests that the satellites of the 3.3 μm band are due to the an– harmonicity (Talbi+ 1993,AA,268,805). Laboratory work on infrared absorption spectra of some small isolated neutral and ionized PAHs (Szczepanski+) points to the importance of ionized species. Fluorescent emission spectra of aromatic and aliphatic C-H stretch modes have been recorded, supporting the PAH model (Shan+ 1991,ApJ,383,459). Experiments demonstrate the dependence of the peak C-H absorption wavelength on compactness of the PAH molecules and whether they are free-flying or condensed; astronomical data suggest a compact, condensed form (Flickinger+ 1991,ApJ,380,L43). The 3.40, 3.46, and 3.51 μm emission features in NGC 7027 are not due predominantly to aliphatic side– groups on PAHs (Sandford 1991,ApJ,376,599).

An alternative though related model is based on coal (Papoular+ 1993,FaradayTrans,89,2289; Papoular+ 1991,AA,247,215); the main, insoluble, organic, solid constituent (kerogen) exhibits all of the UIBs as well as an accompanying continuum. A structural distinction compared to PAHs is linking by oxygen bridges. Possible carriers of the 21 μm feature in preplanetary nebulae are discussed by Sourisseau+ (1992,AA,254,L1).

4.1.4. Extended Red Emission. In addition to scattered radiation, the light of reflection nebulae often shows a broad red emission band (extended red emission or ERE; see e.g., Sivan+ 1993 ApJ,404, 258). This has been interpreted as photoluminescence by HAC grains (Duley 1992,MN,258,773) or aromatic components of filmy quenched carbonaceous condensate (QCC; Sakata+ 1992,ApJ,393,L83). Accompanying sharp features in the Red Rectangle nebula (HD 44179) occur in a spatially distinct environment but it is suggested that the molecular carrier is produced from the HAC (Schmidt+ 1991,ApJ, 383,698). Similar ERE has been found in planetary nebulae, but only those that are carbon rich ($C/O > 1$); UIBs are also associated with carbon rich objects, but the carrier is not identical since some nebulae with UIBs do not have the ERE (Furton+ 1992,ApJ,386,587). A red luminescence band found in the Orion Nebula does not correlate with the 3.3 μm UIB feature (Perrin+ 1992,AA,255,271).

4.1.5. Ices. Extensive observations of ices have been made, mostly in molecular clouds. The threshold extinction at which water ice (3.08 and 6.0 μm bands) is deposited on dust shows wide variation from cloud to cloud (Williams+ 1992,MN,258,599). Within Taurus the band and the long-wavelength wing at 3.45 μm show good correlation with A_V and have the same (photodesorption-regulated) threshold (Smith+ 1993,MN,263,749); only about 9% of cosmic O is required in the H_2O ice. Studies have also been made of ices in the laboratory. Changes in the shape of the ice band on transition from crystalline to amorphous water ice (induced by keV ion-irradiation) are relevant to interpreting the thermal history of interstellar ice (Baratta+ 1991,AA,252,421). The 3 μm feature of a SiO condensate (not a silicate) containing trapped and adsorbed H_2O is found to match closely the band toward the Galactic center source IRS 7 (Wada+ 1991,ApJ,375,L17).

An absorption feature at 3.53 μm in W33A has been assigned to methanol; it is the second most abundant molecule (7% relative to H_2O) in the grain mant -les in this line of sight (Grim+ 1991,AA, 243,473). Consistent upper limits on the 9.8 μm feature have been set in other sources (Schutte+ 1991, ApJ, 382,523). Grain surface reactions or condensation directly out of the gas phase must contibute to formation, rather than simply ultraviolet irradiation of the ice mantle (as would be the case for formaldehyde which is not seen strongly). The 6.85 μm feature is not methanol, and so requires another compound.

CO frosts (4.67 μm), ubiquitous in the quiescent molecular cloud environment, have been reviewed by Whittet+ (1991,AARev,2,167); in some lines of sight the gaseous CO is significantly depleted. Details of the band profiles, studied in various mixed ices in the laboratory, clarify the production and nature of the solid CO (Palumbo+ 1993,AA,269,568; Tielens+ 1991,ApJ,381,181). Tielens+ (1991) and Kerr+ (1993,MN,262,1047) identify independent non-polar (e.g., CO) and polar (e.g., H_2O) matrices in the band profiles, which is confirmed with bands at 4.67 and 4.68 μm (Tegler+ 1993,ApJ,411,260). Solid CO_2 is detected at 15.2 μm toward the protostellar object AFGL 961, but not toward other sources with large column densities of water ice [$\tau(3.05\mu m) > 1$; Whittet+ 1991,MN,252,63].

Tegler+ (1993) have identified a broad absorption band at 4.619 μm as a CN-containing compound in a mixed ice. Sandford+ (1993,ApJ,409,65) propose that solid H_2 in H_2O-rich ices can be detected by an infrared absorption band at 2.417 μm. Detection of O_2 in a dirty ice matrix is discussed by Ehrenfreund+ (1992,AA,260,431).

4.1.6. 3.4 μm. The interstellar absorption band at 3.4 μm in diffuse interstellar dust indicates a C-H stretch in the CH_2 and CH_3 groups of a fairly complex carbonaceous material containing aliphatic functional groups (Sandford+ 1991,ApJ,371,607). The subpeak strengths correlate with visual extinction; the material could be an important repository of C in grains (3–40%). The material might be the residue from photoprocessing of more volatile ice mantles or by ion irradiation (§4.3.4; Sandford+; Strazzulla+ 1992,AA,266,434). A broad feature at 3.0 μm is attributed to O-H stretch, but from lack of correlation, apparently not in the same organic material. A 3.4 μm feature is seen in the carbon rich protoplanetary nebula CRL 616 (Lequeux 1990,AA,240,L19).

A distinctive subpeak in dense cloud lines of sight is indicative of diamond-like material, in sharp contrast with diffuse clouds (Allamandola+ 1993,Science,260,64); this dichotomy, two very different and independent solid hydrocarbons, is puzzling from the point of view of grain evolution.

4.1.7. Diffuse Interstellar Bands. A discussion of the diffuse interstellar bands (DIBs) under the rubric of dust might be inappropriate, since despite correlation in strength with E_{B-V}, the carrier does not lie in the solid particles producing the red- dening. Nor is there correlation with the small grains which dominate extinction beyond 1250 Å; in general the carrier behaves like a free neutral species in the gas and responds to the ionization level of the gas as if its ionization/dissociation threshold is somewhat higher than 5 eV (Herbig 1993,ApJ,407,142). Spectropolarimetry of the 5797 Å band shows the carrier is not aligned grains (Adamson+ 1992,ApJ,398,L69). Snow+ (1991,ApJ,382,189) suggest that the relatively narrow profiles of the 5780 and 5797 Å bands in HD 29647 might arise from unusual rotational excitation in molecules. Weakness in these bands in Orion has been attributed to sticking of the carriers to grains (Porceddi+ 1992,AA,260,391). Bands detected in circumstellar matter do not reflect the carbon/oxygen rich dichotomy; nor do they cor- relate with the presence of UIBs (Le Bertre+ 1993,AA,274,909; 1992,AA,255,288). McIntosh+ (1992,MN,255,P37; see also Adamson+1991,MN,252, 234) find that the carriers reside in the surface layers of dark clouds, not deep within them, and a different depth dependence for two different families of bands. Jenniskens+ (1993,AA,274,465) report complex structure in the strong 5780 and 6284 Å bands, possibly indicating a common carrier.

Salama+ (1992,Nature,358,42) use laboratory measurments to suggest that a species like the PAH pyrene cation $C_{16}H_{10}^+$ is responsible for the 4430 Å feature, using 0.2% of cosmic C. Laboratory spectra of the naphthalene cation $C_{10}H_8^+$ indicate discrete absorption bands close to the positions of several DIBs (Salama+ 1992,ApJ,395,301); however, another significant laboratory band is not observed astronomically (Snow+ 1992,ApJ,401,775).

Webster (1993,MN,262,831) develops a theory based on members of an unspecified class of molecular hydrides (perhaps fulleranes $C_{60}H_m$). Carriers with different m in different interstellar habitats might underlie the observational partitioning of the bands into families.

4.1.8. Lack of Signature. Absence of expected spectral features also offers constraints. For example, an 11 μm SiC feature is seen in carbon stars, but there is actually little SiC in the meteorites or the interstellar medium (Whittet+ 1990,MN,244,427) compared to what might have been expected.

Individual PAH molecules have distinctive ultraviolet absorption bands, but it is thought that in a mixture of PAHs no particular band will stand out (e.g., Lee 1993,ApJ,410,127); interstellar PAHs might contribute to the 2175 Å bump and the far ultra- violet rise (Joblin+ 1992,ApJ,393,L79; Verstraete+ 1992,AA,266,513). Ionized PAHs have continuum absorption into the visible which makes them easier to hide for a given energy budget of infrared UIB and continuum emission (Salama+ 1992,Nature,358,42). They contain 10–20% of the cosmic carbon.

The main evidence for amorphous carbon is the featureless infrared continuum emission in circumstellar envelopes of carbon stars; circumstantial evidence is that mentioned above for the relatives: PAH and HAC. Laboratory measurements (Colangeli+ 1992,ApJ,392,284) indicate a feature at 2400 Å that might explain that seen in a class of sources; however, by the same token, the lack of a distinctive feature there in interstellar extinction constrains the amorphous carbon; it is speculated that some modifications might reproduce the 2175 Å feature.

4.2. CONTINUUM. Continuum measurements, the traditional probes of dust, continue to benefit from ever wider spectral coverage.

4.2.1. Extinction. There are variations of the ultraviolet extinction for a given op- tical extinction, again apparent in the atlas of TD-1 extinction curves (Papaj+ 1991,MN,252,403). Simple components of the ultraviolet extinction are the bump, a linear rise, and a far ultraviolet non- linear term. Green+ (1992,ApJ,395,289) have measured the extinction from 1180 to 950 Å in the direction of ρ Oph; the rise below 1000 Å cannot be explained by a standard silicate-graphite model. Voschchinnikov+ (1993,AZ, 70,38) calculated the optical properties and extinction of silicates and graphite particles out to 100 Å.

Webster (1993,MN,262,59; 1992,AA,257,750) suggests that the variable component of extinction associated with the strong rise shortward of 1600 Å is attributable to fulleranes and their ions with various levels of hydrogenation in different environments. The relative strength of very broad structure (VBS) in the 5000–6000 Å region is correlated with that of far ultraviolet continuum extinction (Reimann+ 1991,AA,242,474). Jenniskens+ (1992,AA,265,L1) find the amount of non-linear rise to be proportional to the CH abundance.

4.2.2. Polarization. The wavelength dependence of optical and near-infrared polarization is given by the Serkowski relation with three parameters: strength p_{max}, peak position λ_{max}, and inverse width K. A linear dependence of K on λ_{max} seen overall also holds within individual regions with rather differing environments (Whittet+ 1992,ApJ,386,562). In the infrared there is excess polarization at 3–5 μm relative to the standard Serkowski formula; an infrared power law of index near 1.6 is a better representation; the infrared polarization does not respond to changes in the properties of the grains which give rise to dramatic variations in the wavelength dependence of polarization at shorter wavelengths (Martin+ 1992,ApJ,392,691). Chlewicki+ (1990,ApJ,365,230) calculate the wavelength dependence of interstellar circular polarization for organic refractories.

Ultraviolet spectropolarimetric observations obtained with WUPPE (Clayton+ 1992, ApJ, 385, L53; Wolff+ 1993,ApJ,403,722) along eight lines of sight show three different interstellar polarization wavelength dependences: Serkowski (larger λ_{max}), super-Serkowski, and in one star a bump near 2175 Å. A bare silicate grain model is successful at fitting the continuum wavelength dependence and the bump is attributed to small aligned graphite disks.

4.2.3. Scattering. Two diagnostics are albedo, ω, and asymmetry parameter of the phase function, g. Aspects of diffuse scattered Galactic light are reviewed by Henry+, Leinert, Lequeux, Onaka, and Witt in IAU Symposium 139 (Falgarone+ 1991, Kluwer).

In the ultraviolet there have been numerous measurements of both classical bright reflection nebulae and the diffuse scattered light in the Galaxy, often with discordant interpretations (summary in Hurwitz+ 1991,ApJ,372,167; Murthy+ 1991,ApJ,383,198; Onaka+ 1992,ApJ,379,532; Witt+ 1992,ApJ, 395,L5). One consistent result is a decrease in ω at the 2175 Å extinction bump. Interpretation of recent measurements of NGC 7023 (Witt+ 1993,ApJ,410,714; Murthy+ 1993,ApJ,408,97) seem to be converging on a decreasing ω and fairly constant g into the far ultraviolet, not unexpectedly. In the near infrared, ω might be larger than in some standard models (Sellgren+ 1992,ApJ,400,238).

4.2.4. Near Infrared Emission. The basic process underlying the infrared emission is conversion of ultraviolet photons. In a classical grain, production of near infrared emission requires a small size so that a single photon can "spike" the temperature transiently to a sufficiently high value; a closely related phenomenon occurs for large molecules, like PAHs (Sellgren+ 1992,ApJ,400,238). The quasi-absence of 3.3 μm PAH emission in the H II region M 17 and the strong near infrared emission from 2 to 20 μm indicate the presence of distinct very small grains (Giard+ 1992,AA,264,610). Electron-hole recombination radiation might provide additional emission excesses in all grains (Duley 1992,MN,258,773). Spiking has been modeled in detail by Siebenmorgen+ (1992, AA,266,501); they conclude from the absence of an interstellar 10 μm emission feature that less than 5% of the total silicate abundance is in very small grains ($a < 80$ Å). The 12 and 25 μm IRAS emission in globules can be explained by PAHs and tiny particles of gsaphite (see also Deluca+ 1993,MN,262,805; Lis+ 1991,ApJ,372,L107). Variations in the abundances of small particles are described by Bernard+ (1992,AA,263,258).

4.2.5. Far Infrared Emission. Visual reflection nebulae are analysed by Casey (1991,ApJ,371,183). In the 20-100 μm region the energy distribution of WX Ser suggests that the emissivity of silicates varies as ν^{α} with $\alpha \sim 1.5$ (Griffin 1993,MN,260,831). Such a flat dependence is seen in carbon stars and planetary nebulae too. In the interstellar medium, far infrared and submillimetre thermal emission is detected in dark clouds and star forming regions, decaying with index $\alpha \sim 1 - 1.5$ (e.g., Hoare+ 1991,MN,251,584). These are all less steep than predicted by the most straightforward models. Solutions being investigated include better treatment of the temperature distribution in quantum heating (Siebenmorgen+ 1992, AA,266,501), and fluffy and inhomogeneous grains (Ossenkopf 1991,AA,251,210; Rouleau+ 1991,ApJ, 377,526; Siebenmorgen 1993, ApJ,408,218). Polarized emission is detected at 1.3 mm (e.g., Leach+ 1991, ApJ,370,257), showing that grains can be aligned in the potentially less favorable conditions in dense molecular clouds.

4.2.6. X-ray Scattering. Forward scattering by dust particles causes a diffuse x-ray halo around background x-ray sources. The extent depends inversely on grain size (Klose 1991,AA,248,624), among other things (Mathis+ 1991,ApJ,376,490), and constrains the fluffiness of grains. Halos around point x-ray sources are discussed by Day+ (1991,MN,251,76), Garcia+ (1992, AJ,103,1325), and Predehl+ (1991, AA,246,L40; 1992,Science,257,935) and are broadly consistent with grain models based on extinction data.

4.3. GRAIN MODELS AND EVOLUTION. All of the building blocks must be assembled into a grain model in varying proportions to explain the observed characteristics of the interaction of radiation with dust and how they change with interstellar environment.

4.3.1. Depletion. That certain elements are in grains can be discerned by measuring depletion in the gas phase. Where dust is des– troyed, as in a shock (Bachiller+ 1991 AA,243,L21; Sophia+ 1993,ApJ, 413,251), depletion is undone; patterns of differential depletion with line width (e.g., two Fe for each Si) are suggestive of mineral composition (Spitzer+ 1993,ApJ,409,299). Depletion differences between line components might be indicative of selective accretion favoring refractory elements (Savage+ 1992, ApJ, 401, 706). The depletions of 26 elements in an average dense interstellar cloud correlate with condensation temperatures more closely than with the first ionization potentials (Hobbs+ 1992,ApJ,411,750).

The precise depletion depends on the cosmic standard adopted. Oxygen abundances from stars in young cluster are slightly lower than solar (Fitzsimmons+ 1992,MN,259,489), lowering the implied depletion. Still, the amount of O in grains is substantial (Cardelli+ 1993,ApJ,402,L17), much more than is bound in silicates. The case for N is less clear (see also Encrenaz+ 1991,SpSciRev,56,83).

4.3.2. Multicomponent Models. Most models have some graphite to explain the bump, silicates, and PAHs. Consider one simple but not unique illustration (Siebenmorgen+ 1992,AA,259,614): (i) large grains ($a > 100$ Å) of silicate and carbon, for the far-infrared emission and the linear rise in the extinction curve; (ii) small graphite particles ($a \sim 4 - 100$ Å), explaining the mid-infrared emission and the 2175 Å extinction bump; (iii) PAHs, producing the near and mid-infrared bands, part of the underlying continuum, and the non-linear rise in the far ultraviolet. Desert+ (1990,AA,237,215) and Rowan-Robinson (1992,MN,258,787) have other recipes. Whether different materials co-exist in composite particles is an open question. Other C can be in HACs or organic refractories (processed ices), probably as mantles. Grain components can be assigned a single size, or more realistically some size distribution. Underlying each of these models is an evolutionary scenario, often not fully elaborated, which gives rise to the grain components and links the chemistry of the interstellar gas and dust.

The mixture might change from place to place in the Galaxy (§4.3.4), including near the Galactic centre; there the ratio $A_V/\tau(9.7\mu m) \simeq 8$, about half the solar neighborhood ratio (Nagata+ 1993, ApJ,406,501); but over the range JHKL the spectral index of the interstellar extinction is ~2.3, only a bit steeper than locally. The mix can change in other galaxies too (e.g., Magellanic Clouds: Pei 1992, ApJ,395,130; Schwering+ 1991,AA,246,231); it is interesting that the increased far ultraviolet extinction in the SMC is accompanied by weaker 12 μm emission (Sauvage+ 1991,AA,237,296).

4.3.3. Circumstellar Origins. Grains are certainly supplied to the interstellar medium by stellar outflows, and some grains survive incorporation in the primitive solar system (§4.3.5). An overview of infrared and millimetre-wave observations in circumstellar envelopes is given by Omont (1993, JPhys-G, 19,39). Both carbon and oxygen-rich (silicate) dust are observed. New investigations have highlighted the possibility of PAHs (Buss+ 1991,ApJ,372,281; Latter 1991,ApJ,377,187; Ryter 1991,AnnPhy,16,507) and iron oxide grains (Rietmeijer 1992,ApJ,400,L39). Whether the rate of supply (e.g., Guglielmo+ 1993,AASupp,99,31) can keep up with destruction or explain the depletion patterns is debateable.

4.3.4. Evolution. The environmental dependence of the interstellar extinction curve offers clues to processes affecting grain evolution. Jenniskens+ (1993,AA,274,439) find that the linear rise is systematically less in dense regions (coagulation of small particles); that the bump is not weakened in dense media, but is sensitive to the presence of strong ultraviolet radiation fields; and that H II regions have abnormal behaviour of the bump width. Implications of deviations from the mean R_V-dependent extinction law are discussed by Cardelli+ (1991,AJ,101,1021), Clayton+ (1993,AJ,105,1880), Mathis+ (1992,ApJ,398,610), and Whittet+ (1993,ApJ,408,573); deviations are largest in the ultraviolet, involving small grains and the effects of grain coatings. Effects of changing the parameterized size distribution are given by Steenman+ (1991,ASpSci,184,9). The line of sight toward cluster NGC 1502 has an unusually small R_V, 2.4 (Tapia+ 1991,MN,253,649).

Vrba+ (1993,AJ,105,1010) study the ρ Oph cloud. The size distribution underlying the polarization changes systematically with R_V, grain size increases with optical depth (coagulation), and the polarization efficiency decreases with increased optical depth or grain size. The youngest dark clouds tend to have the largest grain sizes, suggesting that large grains are produced early on and that after $\sim 10^6$ yr grain evolution is mainly destruction of larger grains, at least for the outer several magnitudes of A_V. Near compact H II regions, both ionized and neutral regions, Hoare+ (1991,MN,251,584) suggest a decrease in dust to gas ratio compared to the diffuse ISM; see also Pismis+ (1991,MN,249,385). Where grains are shattered or destroyed, as in a supernova shock (Arendt+ 1991,ApJ,368,474), infrared emission is affected.

Grain growth in the dense interstellar medium occurs by coagulation (e.g., Rossi+ 1991,AA,251,587) and mantle deposition. Clearly the long term evolution in a galaxy (e.g., Wang 1991,ApJ,374,456) will be sensitive to what assumptions are made. Chokshi+ (1993,ApJ,407,806) studied the coagulation process in detail, concluding that sticking of small particles to larger ones could be significant on available timescales, but mutual coagulation of the larger particles would not occur. When smaller grains are removed efficiently by coagulation there is a dramatic effect on the visible and, particularly, the ultraviolet portion of the extinction curve, which seems to be in accord with observations of extinction and polarization (Martin+ 1990,ApJ,357,113; 1992,ApJ,392,691).

Icy mantles are seen to have grown in dense clouds, increasing the total mass of grains. Their presence is very important to chemical evolution (Hasegawa+ 1993,MN,263,589). The CO threshold might be controlled by heating by H_2 formation (Duley + 1993,MN,260,37). Subsequent processing of these ices by ultraviolet radiation and cosmic rays could lead, over many cycles in and out of dark clouds, to a buildup of organic refractory material, depending on the yield; explosive desorption is a proposed regulatory mechanism (Schutte+ 1991,AA,244,190). In the extreme, the material could become polymerized and carbonized (Jenniskens+ 1993,AA,273,583; see AA,274,653 for optical constants). Duley+ (1993,MN,260,415) explore direct deposition and subsequent evolution of HAC mantles with application to variations in infrared emission.

4.3.5. Interstellar Grains in Solar Nebulae. Interstellar material including grains is incorporated in stars and, more importantly for sleuthwork, the surrounding nebulae or disks. Some grains might be returned in altered form from these regions. There is ongoing examination of dust in circumstellar disks; for example, a small-particle 10 μm silicate feature much like that in comets or Galactic sources is detected in β Pic (Telesco+ 1991,ApJ,372,L29) and an- other Vega-excess system (Skinner+ 1992, MN,255,P31). Anomalous extinction near Herbig Ae/Be stars is analysed by Gorti+ (1993,AA,270,426).

But most work concerns relics of the early Solar nebula, aspects of which are treated in IAU Colloquium 126, *Origin and Evolution of Interplanetary Dust* (Levasseur-Regourd+ 1991, Kluwer), *Comet Halley: Investigations, Results, Interpretations* (Mason 1990, Horwood), *The Comet Halley Archive Summary Volume* (Sekanina+ 1991, NASA), Wyckoff (1991, Earth-science Reviews,30,125), and Buseck+ (1993 AnnRevEarthPlaSci,21,255).

Greenberg+ (1990,ApJ,361,260) and Tokunaga+ (1990,Icarus,86,208) consider the connection between comets and interstellar dust and ice. Analysis of dust impacts of very fine particles ($< 10^{-17}$ g) near the nucleus of Comet Halley is consistent with an aggregation picture (Sagdeev+ 1990,SovALett, 16,315). It is also interesting that in Comet Halley the "CHON" ma- terial (composed of these light elements, probably a less volatile residue) is intimately mixed with silicates on submicron scales (Lawler+ 1992,Nature,359,810).

A rapidly developing field is study of isotopically anomalous materials in meteorites, providing primary data on stellar and supernova nucleosynthesis (Harper 1993,JPhysG,19,81). This has extended to analysis of extracted fine-grained diamonds (e.g., Russell+ 1991,Science,254,1188; see also Nuth+ 1992,ASpSci,196,117) and microanalytical measurements of individual interstellar grains of graphite and SiC in primitive meteorites and collected interplanetary dust particles (Ott 1993,Nature,364,25). The isotopic compositions of these grains provide a record of stellar nucleosynthesis and of condensation processes near carbon stars (which will not be reviewed); the fact of their survival places constraints on conditions in the solar nebula and early Solar System (see also Bernatowicz+ 1991,ApJ,373,L73; Zinner 1991,SpSciRev,56,147; Zinner+ 1991,Nature,349,51; Amari+ 1992,ApJ,394,L43; Brown+ 1992,Science,258,970; Alexander 1993,GeoCosActa,57,2869; and Prombo+ 1993,ApJ,410,393).

The matrices of ordinary chondrites might contain a common carbon component of interstellar origin (Makjanic+ 1993,Meteoritics,28,63). The 3.4 μm absorption of Galactic center source IRS 7 is remarkably similar to that of the deuterium-rich organic polymer extracted from the Orgueil carbonaceous meteorite (Ehrenfreund+ 1991,AA,252,712). No C_{60} has been found (DeVries+ 1993,GeoCosActa,57,933).

Interstellar grains are even now penetrating into the solar system. The Ulysses spacecraft detected micrometre-size particles identified by their trajectories as being of interstellar origin (Grun+ 1993,Nature,362,428).

4.4. PHYSICAL PROCESSES. Once a grain model is specified, or altered, many observational characteristics can be examined and other potentially useful derivative properties, such as the Rosseland mean opacity (Ali+ 1992,ASpSci,188,109), can be computed.

The simplest grain models are based on electromagnetic scattering by Mie spheres, infinite cylinders, or homogeneous spheroids. Mishchenko (1991,ApJ,367,561; 1990,SovAstLett,16,946) shows how to average the extinction matrix over an ensemble of nonspherical dust grains, oriented axially in the interstellar magnetic field. Wright (1991,ApJ,375,608) has shown that two proposed modifications to the standard theory for the long-wavelength absorption by grains are incorrect (see also Duley+ 1992,MN, 255,243). It is possible to study less symmetric shapes and inhomogenous (e.g., composite or porous) particles using finite element techniques, the most popular of which is the discrete dipole array method. Example investigations are by Kozasa+ (1992,AA,263,423), and Perrin+ (1991,AA,247,497; 1992,CompRendusII,316,47).

The electric potential of grains of differing properties in various interstellar environments has been investigated by Taylor+ (1991,MN,248,148). Reynolds (1992,ApJ,400,L33) discusses photoelectric heating of the warm ionized medium by grains.

Grain surfaces provide sites for the formation of molecules. This process seems well established for H_2 in cold clouds but some aspects are still being studied (Duley+ 1993,MN,260,37). Production of H_2 from modified ice mantles by cosmic rays is explored by Averna+ (1991,AA,245,239).

5. H II REGIONS. (M.R. Rosa)

5.1. INTRODUCTION.

Research on H II regions (HIIRs) benefitted a lot from further gains in observational and computing capabilities. Large amounts of calibrated spectrophotometric data have been obtained, often combined with high spatial resolution and simultaneous coverage of sizeable spatial areas. Similarly, mapping of galactic and extragalactic objects in continuum and spectral lines at X-ray, UV, visible, IR, FIR, mm and radio frequencies no longer suffers from vast differences in spatial resolution and dynamic range. Usually, observational data are now readily transformed into astrophysical quantities and it has become common practice to support the interpretation of such data with dedicated models.

The working concept of classical, homogeneous and isolated H II regions (HIIR) is steadily being replaced by the perception of complex, inhomogeneous volumes of ionized gas mixed with coexisting hot-teneous and neutral-dense volumina and evolving into and interacting heavily with the ambient medium. HIIRs in external galaxies continue to be the focus of a large fraction of research activity in the field, although they have lost their exotic attributes - largely thanks to the very intensive study of the proto-typical 30 Doradus nebula.

In view of the evolution in the field the report has been restructured. Most of the ionized gas studied at the galactic center region proper apparently has little to do with HIIRs around O type stars, and the separate section has been dropped. HIIRs associated with the earliest and the very late phases of O star evolution are considered seperately. Extensive research on proto-typical HIIRs like Orion and 30 Dor has been summarized in a dedicated section. Physical conditions and their interpretation in terms of physical processes supported by models are an additional topic. The section on abundances encompasses work based on both, galactic and extragalactic HIIRs. The section on HIIRs in external galaxies then reflects largely our view of HIIRs and their environment as studied in other galactic systems.

Among relevant general books, proceedings, review articles and catalogues not cited at the beginning of the commissions report are:

Alloin+ 1993,The feedback of chemical evolution on the stellar content of galaxies,Observatoire de Paris; Barbuy+ 1992,The stellar populations of galaxies,IAU Symp 149,Kluwer; Baschek+ 1993,New Aspects of Magellanic Cloud Research, Lect.NotesPhys,416, Springer; Blitz 1991,The evolution of the interstellar medium,ASP Conf.Ser.,12; Bloemen 1991,The interstellar disk-halo connection in galaxies,IAU Symp 144, Kluwer; Burton 1993,Distribution and observed properties of the ISM, in Bartholdi+ 1993, The galactic interstellar medium,Lect.Notes 21st Adv.Course SSAA, Springer; Cassinelli+ 1993,Massive stars: Their lives in the interstellar medium, ASP ConfSer, 35. Edmunds+ 1992,Elements in the cosmos, 31st Herstmonceux Conf, CambridgeUP; Falgarone+ 1991,Fragmentation of molecular clouds and star formation,IAU Symp 147, Kluwer; Franco+ 1993,Star formation, galaxies and the ISM, CambridgeUP; Haynes+ 1991,The Magellanic Clouds,IAU Symp 148,Kluwer; Hollenbach+ 1990,The interstellar medium in galaxies,2nd Wyoming Conf, NASA-CP-30; Klare 1993,Stellar Evolution and interstellar matter,RevModernAstron,6,Springer; Lada+ 1991,The physics of star formation and early stellar evolution,NATO ASI Ser C, MathPhysSci,342,Kluwer; Leitherer+ 1991,Massive stars in star bursts,STScI SympSer, 5, CambridgeUP; Lozinskaya 1992,Supernovae and stellar wind in the interstellar medium, AIP, New York; Palous+ 1992,Evolution of interstellar matter and dynamics of galaxies, CambridgeUP; Thronson+ 1990,The interstellar medium in galaxies,2nd Wyoming Conf,ApSpLib,161, Kluwer; Van der Hucht+ 1991,Wolf-Rayet Stars and interrelations with other massive stars inING NEBULAE galaxies; IAU Symp 143, Kluwer; Proceedings of the 3rd TexMexConfAstrophys, 1992,PASP,103,pp 759-923.

5.2. STRUCTURE AND INTERRELATION WITH ENVIRONMENT

5.2.1 Surveys

Surveys conducted with improved sensitivity and spatial resolution continue to reveal new objects. Equally important, however, is the gain in insight into the structure of the ionized medium in the galactic disk and its interrelation with molecular, neutral, and stellar components.Searches of elusive HIIRs around early B stars have been conducted by Shestakova+ 1990,Sov.Astron.Lett.,16,462. Progress is reported on the deep MWG Ha survey with detection of an optical HIIR at 11.7 kpc distance in the Coalsack region (LeCoarer+ 1992,AA,257,389).

Optical HIIRs have been surveyed with the VLA (Fich 1993,ApJS,85,475). Other reports on radio surveys include: Gaylard 1991, 3rd Haystack Obs Conf on ISM; Azcarte 1991,ApSpSci,180,105; Whiteoak 1990,ProcAstrSocAust,8,274. Most of the 1800 sources listed in the latter survey are ultracompact (uc)HIIRs and OH stars (Whiteoak 1992,AA,262,251). Similarly the VLA 20cm continuum survey of northern galactic plane (1992 discrete sources) is dominated by ucHIIRs (Becker+ 1990,ApJ,358,485), as is the case in the galactic center region survey (Helfand+ 1992,ApJS,80,211). Galactic worms (118), probably the walls of superbubbles around evolved HII regions, have been catalogued by Koo+ 1992,ApJ,390,108. Radio surveys and pointed observations also include: Reich+ 1990,AAS,85,633; Fuerst+ 1990,AAS,85,691; Fuerst+ 1990,AAS,85,805; Azcarte 1992,ASpSc,194,225; Wu Yuefang+ 1992,ActaAstrSin,12,365; Abramenkov+ 1992,SovAstr,36,374.

5.2.2 Global structures. Global structures with high complexity are now regularly seen in radio and optical maps of HIIRs. Commonly this is interpreted as star formation activity at various stages between the early obscured IR luminous sources, the cloud evaporation and blister phase, and the late phases characterized by the formation of superbubbles by OB star winds and SNe (Pismis 1990,AA,234,443; Ward-Thompson+ 1991,MNRAS,248,670; Tateyama+ 1991,MNRAS,249,716; Pineault+ 1990,MN,246,169). The apparent absence of SNRs in the radio morphology of the Cygnus OB2 region might be a sign for the relative youth of this SF event (Wendker+ 1991,AA,241,551).

Pattern recognition techniques to distinguish optical images of HIIRs from SNRs have been employed by Inglis+ 1990,MN,246,358. A statistical study of HIIR/SNR correlations in the galaxy has been performed by LiZongwei+ 1991,ApJ,378,93. The disentangling of obscured complexes using radio and IR maps has been discussed by Van der Werf+ 1990,AA,235,407; the role of low frequency mapping in Kassim+ 1990, Low Freq.Astrophys from Space,Lect.Notes Phys, Springer.

Blister HIIRs at various aspect angles and obscuration have been observed eg. by Vallée+ 1991, AA, 250, 143 and Tapia+ 1991,AA,242,388. A 10 arcmin radio jet is seen in Orion B NGC 2024 (Subrahmanyan 1992,MNRAS,254,719), a region studied also at 1.3cm by Gaume+ 1992,ApJ,388,489. Multifrequency mapping of the S187 complex composed of HIIRs, HI clouds and molecular material has been reported by Joncas+ 1992,ApJ,387,591, and the Champaign phase of GM 24 was studied by Gomez+ 1993,ApJ,409,269.

5.2.3 Internal Structure of HIIRs.

Internal structure is best studied in Orion (see Section 4), but seems to be complex also in other HIIRs. Inhomogeneities and clumping are deduced from FIR, NIR and visual line density estimators as well as the radio continuum (Colgan+ 1991,ApJ,366,172). Related to density inhomogeneities is the continuing quest for direct proves of significant temperature inhomogeneities, which might be traced eg. by differing Te estimates from radio continuum and radio recombination lines in the case of W51 (Copetti+ 1991,MNRAS,250, 127). Bok globules and HH-like objects embedded in H II regions have been studied eg. by Duvert+ 1990,AA,233,190; Gyulbudagyan+ 1991,Astroph,33,528; Bohigas 1992,RevMexAA,24,765. Additional work on the problem of density and temperature inhomogeneities is reported in Section 5.

X-rays observed from the Carina and the Orion nebulae can be associated with the winds of Eta Car, resp. Thet1 Ori, and with a diffuse component, probably due to the multitude of unresolved T-Tauri stars (Koyama+ 1990,ApJ,362,215; Yamauchi+ 1993,ApJ,403,268). Kinematics of HIIRs have been studied by Clayton 1990,MNRAS,246,712; Pismis+ 1991,PASP,103,843. HIIRs with unusually narrow radio recombination lines were observed by Planesas+ 1991,RevMexAA,22,19. The propagation of waves in H II regions and the turbulence spectrum are considered theoretically by Krasnobaev+ 1991,SovAstrLett,17,396;

5.2.4 Dust associated with HIIRs. Dust is seen as continuum and band emission in the IR and FIR as well as through its extinction properties. Central issues remain the question of how much of the dust observed is actually located inside the ionized volumina, and whether or not the grain properties are different from the diffuse ISM and the dense cloud environments (O'Dell+ 1992,ApJ,399,L67).

Observations of scattered stellar light in M8 showing He+ lines in absorption have been used to derive Dust/Gas mass ratios (Sanchez+ 1991,RevMexAA,22,285). Peculiarities in UV extinction correlated with column densities of C IV and Si IV have been observed towards stars embedded in or seen through the M8 HIIR (Boggs+ 1990, ApJ,358,441). Deviations of dust extinction from mean R-dependance have been discussed by Mathis+ 1992,ApJ,398,610. De-alignment of grains in H II regions is substantial unless strong magnetic fields are invoked (Anderson+ 1993,AA,270,479)

Emission at mm and IR wavelengths of dust associated with HIIRs has been observed (Sievers+ 1991,AA,251,231; LuoShaoguang 1991,Chi AA,16,33), an account of the 3.29 mu dust emission feature in HIIRs has been given by Tokunaga+ 1991,ApJ, 380,452. In M17 dust seems to be depleted in the ionized gas phase (Giard+ 1992,AA,264,610). Dust grain processing inside the HIIRs has been investigated by Sorrell 1992,MN,255,594 and Siebenmorgen+ 1992,AA,259,614. Different grain size populations are also invoked by Pismis+ 1991,MNRAS,249,385. The iron oxide band ad 21 mu may contribute a good fraction of the total IRAS flux observed in HIIRs (Cox 1990,AA,236,L29). Fluorescence and scattering by PAH dust particles might account for far red emission bands observed across the Orion nebula (Perrin+ 1992,AA,255,271).

5.2.5 The Galactic Center.

Close to the galactic center HIIRs apparently are different from classical HIIRs. In the Arc region they might be ionized by impact with a strong magnetic field (Serabyn+ 1991,AA,242,376), a view not shared by (Maloney+ 1992,401,559). Radio recombination lines of Sgr A West indicate Te of 20 - 40 000 K (Roberts+ 1991, ApJ,366,L15). FIR lines and continuum of the arc filaments are consistent with stars of Teff 35 000 K - but the structure and the absence of such stars are difficult to explain (Erickson+ 1991, ApJ,370,L69). An extended region of ionized gas around the galactic center sources is seen in NIR He and H lines (Geballe+ 1991,ApJ,370,L73), a bow shock wind-wind collision indicated around IRS 7 (Yusef-Zadeh+ 1992,ApJ,385,L41).

5.3.0 EARLY AND LATE STAGES OF O STAR EVOLUTION

5.3.1 Compact (cHIIRs) and ultracompact HIIRs (ucHIIRs). The traditional phases, cHII and ucHII, from deeply embedded objects to fully developped classical HIIRs have received continuing interest. A broad description of the state of research on the population of about 2500 estimated ucHIIRs in the Galaxy has been given by Churchwell 1990,AARev,2,79. Bow shock models for ucHIIRs around O stars moving through the cloud medium can yield a unifying picture for their radio morphologies (Mac Low+ 1991,ApJ,369,395; Van Buren+ 1992,ApJ,394,534). Dust emission from ucHIIRs has been modelled by Hoare+ 1991,MNRAS,251,584 Observational work on the ionized component of individual cHIIRs has been reported by Zhou Zhen-pu+ 1990,Chin.AA,15,232; Estalella+ 1991,ApJ,371,626; Wood+ 1991,ApJ,372,199; Gaume+ 1991,ApJ,376,608;Fey+ 1992, AJ,103,234. Statistics of cHIIR in IRAS PSC data base have been evaluated by Zhou Zhen-pu+ 1990,Chin.AA,15,207. One of the only 4 high excitation optical cHIIR in the LMC has been described by Heydari-Malayeri+ 1990,AA,240,481, IR images of embedded young stars in M8 (Woodward+ 1990,ApJ,365,252), M17 (Lada+ 1991,ApJ,374,533), NGC 3603 and 30 Dor (Lapierre+ 1991 ASP (012.050) 155) indicate normal solar neighborhood IMFs.

5.3.2 Nebulae around stars. Ring nebuale around Wolf-Rayet stars and nebulosities ionized by other highly evolved massive stars have been the subject of studies aiming at the evolution of the nebulae and of the chemical abundances in the evolved stellar surfaces (Arnal+ 1991,AA,250,171; DeFreitasPacheco 1992,AA,266,36; Dopita+ 1990,ApJ,359,419; Esteban+ 1991,AA,244,205; Esteban+ 1992, ApJ, 390, 536; Heydari-Malayeri+ 1992,AA,236,L21; in Garmany 1990,ASP ConfSer 132: Smith+ p132; Rosa+ p135; see also in Van der Hucht+ 1991,IAU Symp 143: Dopita+ p371, Vilchez+ p379, Smith p385, Pakull P391, Montmerle p397, Cassinelli+ p421, Esteban+ p422; Niemela+ p425). The northern sky has been searched again for very low surface brightness ring nebulae by Miller+ 1993,ApJS,85,137. Objects with morphologies of wind-swept bubbles generally show ISM abundance ratios. Nebulae clearly composed of ejecta show O/H deficiencies and N/H, He/H overabundances with respect to the ISM. These abundance anomalies are typical for CNO cycled zones at the surfaces of evolved massive stars (Esteban+ 1992,AA,259,629). The dust content of WR ring nebulae is consistent with stellar material in ejection type nebulae and with ISM swept up material in wind blown bubbles (Marston 1991,ApJ, 366,181; Mathis+ 1992,ApJ,384,197).

5.4. PROTO-TYPICAL H II REGIONS

5.4.1 Orion Nebula.

The Orion nebula proper has received large attention again, stimulated by unprecedented capabilities in high sensitivity, high resolution imaging and spectroscopy at all wavelengths. On the smallest scales networks of ionized filaments, possibly shocked ionized gas, have been revealed in radio continuum maps (Yusef-Zadeh 1990,ApJ,361,L1). This is complemented by the extraordinarily high degree of structure at subarsecond resolution seen in particular in [SII] line HST WFPC images (Hester+ 1991,ApJ,369,L75). Further detail includes edges of ionization fronts (O'Dell+ 1991,PASP,103,824), shocks and proto-stellar disks (O'Dell+ 1993,ApJ,410,696). The radio morphology from scales of 10' to 0.1" is presented by Felli+ 1993,AAS,98,137).

Across the main body of the nebula ionization fronts at the near side are visible in stellar He I, Ca II absorption lines (O'Dell+ 1993,ApJ,403,687). Kinematical data from [OI] emission (O'Dell+ 1992,ApJ,387,229) and of radio recombination lines (Gosachinskij+ 1992,Sov.Atron.Lett,18,No 2) are used to study the ionization fronts at the rear side towards the molecular cloud. Motions have been mapped in [SIII] (Wen+ 1993,ApJ,409,262).

Physical conditions and extinction across the main body have been studied at 2" resolution by means of imaging spectrophotometry (Pogge+ 1992,ApJ,399,147). An area study of extinction and scattering in Orion based on 21cm and Hbeta maps indicates that most of the extinction occures in front of the nebula (O'Dell+ 1992,ApJ,399,L67), with possible ramifications on models assuming particle processing to explain extinction anomalies. Studies of neutral inclusions contain: high velocity resolution spectra of ionized knots and jets (Meaburn+ 1993,260,625); sub-parsec sized neutral cloudlets (Van der Werf+ 1990,ApJ,364,157); large scale interaction of the HIIR with the molecular cloud (Rodriguez-Franco+ 1992,AA,264,592)

A detailed "blister" model for Orion has been discussed by Rubin+ 1991,ApJ,374,564, Rubin+ 1991, PASP,103,834; new spectrophotometric observations and a corresponding model also by Baldwin+ 1991, ApJ, 374, 580. Extensive spectrophotometric data of faint emission lines (Osterbrock+ 1992,ApJ,389,305) indicate apparent depletion of Fe and Ni by a factor 5. HI and HeI lines agree rather well with recombination line theory.

At larger scales the assymetric blister type nature of the ionized region at the edge of a molecular cloud is revealed by mapping at low frequencies, where the large optical depths towards the core region allow for a direct determination of an electron temperature of 7865 K (Subrahmanyan 1992,MNRAS,254,291). On the other hand, intermediate band photometry of the visible part of the blister reveals symmetry about the Trapezium cluster (Greve+ 1993,AAS,99,577). Techniques to isolate line emission, nebular continuum and scattered stellar light in narrow and broad band imaging of H II regions are demonstrated on the example of the Orion nebula (Waller+ 1990,PASP,102,1217). At the largest scales Barnards Loop has been mapped between 12 and 25 MHz (Abramenkov+ 1992,Sov.Astron.J,36,246).

5.4.2 The 30 Doradus Nebula.

30 Dor, the proto-type of an evolved giant HIIR, has been mapped from radio to X-ray wavelengths. [OIII] and continuum images show that this 1 kpc diamtere HIIR is enclosed by a dust ring (Bruhweiler+ 1991,ApJ,370,551). Radio maps reveal embedded SNRs (eg. Sabalisck+ 1990,Rev.Mex.AA,21,507). X-ray images demonstrate that hot gas must be well mixed with the HIIR (Wang+ 1991,ApJ,370,541; Truemper+ 1991,Nat,349,579), likely due to both, stellar winds and SNe. Evidence of shocked gas is seen in the HIIR emission line spectum (Rosa 1993, in Klare 1993, cit, p 145). The related gas kinematics and stellar content have been revisited by Lortet+ 1991,AAS,89,185. Information on the stellar and dust contents can also be deduced from Astro-1 UV (Cheng+ 1992,ApJ,395,L29) and NIR images (Rubio+ 1992,AA,261,L29). The extinction towards 30 Dor has been studied using Paschen/Balmer line ratios by Greve+ 1991,AA,251,575.

A break-through in the quest for the nature of the 30 Dor cluster core R 136 was acieved by resolving it into a tight group of luminous early type stars of various evolutionary stages in HST FOC images (Weigelt+ 1991,ApJ,378,L1). The interaction of the evolving O star burst with the ISM has been set into context by (Walborn+ 1992,ApJ,399,L87). Work in progress on the stellar content of the entire 30 Dor OB cluster indicates that the IMF is not very different from the solar neighborhood. A CMD of 2400 stars, 150 of which have now spectroscopic classifications, has been presented by Parker 1992,PASP, 104,1107.

5.5. PHYSICAL PROCESSES AND MODELS

As quantitatively more high quality observational material became available, the past 3 years have seen a fresh impetus on activities to understand and model details of the physical processes in H II regions. Very detailed models of the Orion nebula are reported in Section 4.1.

Testing of the HI level populations on radio recombination line spectra showed that standard departure coefficients provide a sufficient description, while Lyα pumping seems to be of no importance (Wilson+ 1990,AA,238,331). At mm wavelengths (H36-H60) the same conclusion was reached by Gordon+ 1990,ApJ,365,606. Non-LTE effects were also studied by Cersosimo+ 1992,AA,239,287. Recombination line theory combined with radiative transfer in dusty nebulae and their impact on uncertainties in the observational determination of the primordial H/He abundance was investigated by Hummer+ 1992,MN,254,277. The recombination line spectrum of neutral Helium has been reassesed by Smits 1991,MN,251,316.

Photoionization models have been used extensively in order to provide ionization correction factors (Mathis+ 1991,AA,245,625), to determine the behaviour of ionic electron temperatures (Garnett 1992, AJ, 103, 1330), and to study effects of stellar atmosphere models on predicted emission line spectra (Evans 1991, ApJS, 76, 985). Peimbert+ 1991,PASP,103,815 investigated the effect of shock waves on the integrated spectra of HIIRs, and the use of NeIII/OII line ratios for He ICFs was looked at by Blum+ 1991,PASP,103,1182)

A model for PIG radio continuum spectra was produced by Pastor+ 1991,AA,246,551. Models for the IR emission from OB star environments have been presented by Leisawitz+ 1991,ApJS,77,451; escape of Ly a radiation from a multiphase ISM by Neufeld 1991,ApJ,370,L85. The radiative cooling of optically thin plasmas, eg. HIM surrounding HIIRs or located in wind/SN bubbles inside HIIRs, has been investigated by Schmutzler+ 1993,AA,273,318. Scattering of light in nebulae has been studied analytically by Kolesov+ 1991, Astroph.,33,No 2.

Density inhomogeneities have been included analytically into photoionization codes by Williams 1992, ApJ, 392, 99 and applied to Nova envelopes. Density sensitive line ratios in models of inhomogeneous nebulae have been studied by Safier 1992,ApJ,392,492 and a density gradient model was produced for IZw18 (Campbell+ 1990,ApJ,362,100).

The Bowen fluorescence mechanism has been studied further by Kastner+ 1990, ApJ,362,745; observationally by Lin+ 1993,MN,261,465. Excess [NIII] line emission in planetary nebulae seemingly does not arrise from Bowen type resonance fluorescence (Kastner+ 1991,ApJ, 381,L59) but rather from continuum fluorescence (Ferland 1992,ApJ,389,L63). Charge transfer reactions are reviewed by Shields 1990,in Molecular Astrophysics (Hartquist ed.),CUP,p461. The unidentified emission lines at 2.2 mu were studied further in PNs and cHIIR, and should correspond to ionization potentials of order 50 eV (Geballe+ 1991,MN,253,75).

5.6. ABUNDANCES AND GALACTIC GRADIENTS

Reviews on chemical composition include Aller 1990,PASP,102,1097; Dinerstein 1990, 2nd Wyoming Conf. on ISM,p 257; and Shields 1990,AnnRevAA,28,525.

A constant, but rather low, C/H was found across the Orion nebula by Walter 1991,PASP,103,830; invoking rather large Te fluctuations Walter+ 1992,ApJ,397,196 arrive at solar like CNO abundances on their set of new data from 22 positions in the nebula. A multi-position (17) analysis of the chemical composition of M17 was presented by Peimbert+ 1992,RevMexAA,24,155, and the ionization structure of a real nebula compared with photoionization models.

Anomalously high He+/H+ abundances have been reported from radio recombination line observations of DR-21 (Tsivilev 1990,Sov.Astron.Lett.,17,1) and W3 (Roelfsema+ 1992,ApJ,394,188). He+/H+ abundances from radio recombination lines have also been analyzed by Peimbert+ 1992,ApJ,395,484. The issue of primordial He/H abundance determinations using galactic and extragalactic HII regions has been looked at again (Pagel+ 1990,Proc.Astron.Soc.Aust.,8,243; Maciel+ 1991,RevMexAA,21,197; Olive+ 1991, ApJ,380,L1), and the role of local He enrichment from evolving massive stars was investigated (Pagel+ 1992,MN,255,325; Campbell 1992,ApJ,401,157).

In irregular galaxies N/O does not show systematical variations with O/H, the scatter in N/O is possibly due to the burst-like nature of the SF events (Garnett 1990,ApJ,363,142) and the self-enrichment of HIIRs (Pilyugin 1992,AA,260,58). N/H abundances in HIIRs in the outer galaxy are also relatively high (Fich+ 1991,ApJ,366,107). Ar/S in IZw18 is cosmic within 0.2 dex (McCall+ 1990,2nd Wyoming, 151).

The galactic abundance gradient has further been studied on a sample of small H II regions (Hunter 1992,ApJS,79,469). The chemical composition and abundance gradients in external galaxies were also the subject of many papers reported in Section 6. The abundance gradients now known for some 30 galaxies were discussed by Vila-Costa+ 1992,MN,259,121. Grids of photoionization models were again computed in order to study the role of possible IMF variations and stellar metallicity on the OII/OIII excitation sequence in extragalactic HIIRs (McGaugh 1991,ApJ,380,140).

Depletion of refractory elements affects the cooling and, at the high Z end, may lead to factor 2 overestimates of O/H (Henry 1993,MN,261,306). New transition probabilities for Cl and Ar (Raassen+ 1992,AAS,95,223) and new photoionization and recombination cross sections from the Opacity Project (Nahar+ ApJ,397,792) can be expected to have impacts on nebular models and chemical abundance determinations.

The comparison of LMC/SMC and galactic abundance ratios as derived from H II regions and F stars shows S/O, C/O to be in agreement for stars in SMC and Galaxy, C/O deficient in the SMC on the basis of HII data (Spite+ 1990, AA,234,67; Russel+ 1992,ApJ,384,508). O/H abundances in Virgo cluster spirals are substantially higher than those in comparable field galaxies, and the O/H gradients appear to be flatter (Shields 1991,PASP,103,916; Henry+ 1992,MN,258,321; Martin+ 1992,ApJ,397,463). The same trends are found in early type spirals in the field (Oey+ 1993,ApJ,411,137). S/O gradients in debate in previous years could not be found in M101 and M51 (Diaz+ 1990,Rev.Mex.AA.,21,223; Diaz+ 1991,MN,253,245), but might be related to ICF uncertainties.

5.7. H II REGIONS IN EXTERNAL GALAXIES.

Shields 1990,AnnRevAA,28,525 reviewed physical properties of HIIR in external galaxies and methods of abundance determinations, Kennicutt 1991, in Leitherer+ 1991,cit,p 157 the stellar contents.

The atlas of HIIRs in M101 by Hodge+ 1990,ApJS,73,661 triples the catalogued objects to 1264. For the same galaxy Scowen 1991,PASP,103,902 and Scowen+ 1992,AJ,104,92 have mapped and analyzed physical conditions of 600 HIIRs. Further catalogues yielding eg. HIIR luminosity functions, velocity dispersions, estimates of SF activity, of IMF parameters, of extinction properties and of chemical abundances in spiral have been produced by: Aparicio+ 1992,AA,260,77; Arsenault+ 1990,AA,234,23; Belley+ 1992,ApJS,78,61; Coradi+ 1991,AA,244,27; Courtes+ 1993,AA,268,419; Duval+ 1991,AA,241,375; Gonzalez-Serrano+ 1991,AA,242,334; Hodge+ 1990,PASP,102,657; Krienke+ 1991,PASP,103,661; Pastoriza+ 1993,MN,260,177; Rand 1992,AJ,103,815; Richer+ 1992,AJ,103,54; Roy+ 1993,ApJ,406,60; Rubin+ 1991,MN,252,550; Sakhibov+ 1990,Sov.Astron.,34,236; Sivan+ 1990,AA,237,23; Skillman 1991,PASP,103, 919; Strobel+ 1990,PASP,102,657; Strobel+ 1991,ApJ,383,148; Von Hippel+ 1990,AJ,100,403.

Spectrophotometric data for chemical abundance analysis were also obtained by Walsh+ 1993,MN,262, 27; Garnett+ 1992,AJ,104,1714; Burenkov+ 1990,Astroph.,32,135; Heydari-Malayeri+ 1990, AA, 234, 99; Price+ 1990,AJ,100,420; Roy+ 1991,AJ,101,825. On the basis of such data statistics of the HIIR population in spirals were the topics of contributions by Caldwell+ 1991,ApJ,370,526; Cepa+ 1990 ApSpSci 170,297; Garcia-Gomez+ 1991,AAS,89,159 and Ye 1992,MNRAS,255,32. Correlations of HIIRs with historical SN events have been looked at by Van Dyk 1992,AJ,103,1788.

As spatial resolution and sensitivity increase, studies on HIIRs in nearby galaxies begin to offer us views on the complex ISM analogous to what is seen locally. The complex N120 in the LMC for example turns out to be a sample of classical HIIRs, SNRs and wind blown bubbles (eg. Laval+ 1992,AA,253,213; McCall+ 1990,AJ,100,193). Molecular gas is found in Magelanic Cloud HIIRs (Israel+ 1991,AA,250,475) and in those in M33 (Scoville+ 1992,ApJ,385,512; Viallefond+ 1992,AA,265,437). Density inhomogeneities are revealed (Castaneda+ 1990,ApSpSci 171,203; Masegosa+ 1991,AA,249,505; Castaneda+ 1992,AA,260,370), diffuse X-ray emission not associated with distinct SNRs is observed (Chu+ 1990,ApJ,365,510), and large scale filaments and superbubbles are found outside the classical HII regions (Hunter+ 1990,ApJ,362,480; 1992,ApJ,391,L9).

The dust content of LMC HII regions was studied using photometry of embedded stars (Greve+ 1990,AAS,85,895). IRAS maps have been produced for HIIRs in M33 (Rice+ 1990,ApJ,358,418). Multi-frequency radio/IR/optical data were analyzed in order to evaluate the amount of dust heating by non-ionizing UV photons in HII galaxies (Dultzin-Hacyan+ 1990,AA,238,28), and to disentangle the thermal HIIRs from the non-thermal smooth disk in M101 (Graeve+ 1990,AA,238,39). High resolution mm continuum maps of the M 82 HIIRs were produced by Carlstrom+ 1991,ApJ,366,422)

Spectra often show high velocity features, sometimes summing up to 3000 km/s broad components of the nebular emission lines, likely due to the combined effect of stellar winds and SNe (Castaneda+ 1990,ApJ,365,164; Roy+ 1991,ApJ,367,141; Roy+ 1992,ApJ,386,498; YeTaisheng+ 1991,MNRAS,249,722. Velocity dispersions in HIIRs of NGC 1275 have been studied by Shields+ 1990,AJ,100,1805.

The stellar content of extragalactic HIIRs, most of which are 30 Doradus type evolved, giant objects around populous clusters of O stars and their late evolutionary WR phases, has been studied using properties of the integrated spectra or the UV flux (Bohlin+ 1990,ApJ,363,154; Ivanov 1991,MN,251,281; Terlevich+ 1991,Rev.Mex.AA,21,192; Hill+ 1992, ApJ,395,L37). Classification and number counting of individual stars or groups thereof have been reported by Debray 1990,IAU Symp 143,427; Drissen+ 1990,ApJ,364,496; Drissen+ 1993,AJ,105,1400 (based on HST FOC images); Rubio+ 1990,Rev.Mex.AA,21,249; and Deharveng+ 1992,AA,259,480; Dynamical constraints on star formation using HIIR data were deduced by Waller+ 1991, in IAU Symp 146, 187)

HeII emission, apparently of nebular origin, is found in serveral LMC/SMC and other extragalactic HIIRs (Garnett+ 1991,ApJ,373,458). Garnett+ 1991,PASP,103,850 discuss possible sources which include extremely hot massive stars, supported by new non-LTE stellar atmosphere models (Gabler+ 1992, AA, 265, 656), shocked gas, or strong X-ray radiators such as LMC X-1. Drissen+ 1991,AJ,101,1659 found 19 candidates for ring nebulae in M33, but no none of those HeIII regions.

Finally, arguments were brought forward that HIIRs in star forming galaxies at semi-cosmological red-shifts might be the metal line absorbers in QSO spectra (Gruenwald+ 1992,ApJS,78,153; Yanni+ 1992, ApJ, 391, 569).

6. SUPERNOVA REMNANTS (T.Lozinskaya and M.Dopita)

6.1. INTRODUCTION

Recent general reviews on Supernova and their remnants are to be found in IAU Coll. No 115; "High Resolution X-Ray Spectroscopy of Cosmic Plasmas", ed. Burton 1990, the 10th Santa Cruz Summer Workshop in Astronomy and Astrophysics: "Supernovae"; Ed. Woosley, S.R., the NATO Advanced Research Workshop on "Physical processes in Hot Cosmic Plasmas", 1990, the AIP Conf. Proc. No 211, 1990, and the ESO/EIPC Workshop on "SN1987A and other SNe", 1991. A major review of the interaction of supernovae and mass-loss stars with their environments is given in Lozinskaya T.A., 1992 "Supernovae and Stellar wind in the Interstellar Medium", AIP:New York. A review of the interaction of supernovae with circumstellar matter is presented by Chevalier R.A., 1990 in "Supernovae", p. 91-110. A review of supernova remnants in general is given in Strom R.G. 11th. European Regional Astronomy Meeting of IAU: "New windows to the universe", v.2, p.463-477.

The study of SNRs has been much enhanced in the period of review by the the combination of observations made in different on wavebands, such as the comparative morphological analysis of Pup A at radio IR, optical and X-ray frequencies by Arendt R.G. + 1990 (ApJ 310,266), the optical/X-ray/radio images of SNR 3C 400.2 by Winkler P.F + 1993 (ApJ 405, 608) or the comparison of the VLA radio and EINSTEIN

X-ray morphology of G33.6+0.1 by Velusamy T.+ 1991 (AJ 102, 676). These show clearly that the way to the future understanding of SNR evolution is through a mutli-disciplinary approach.

6.2 YOUNG SUPERNOVAE

6.2.1. Plerionic or Pulsar-Containing SNR

The Crab Nebula has been a continued focus of interest. CCD observations of the optical polarisation of the Crab Nebula have been used to study magnetic field structure in the synchrotron nebula (Hickson P., and van den Bergh S. 1990 ApJ 365, 224). Multifrequency high-resolution observations by Bietenholz M.F.+ (1990 Ap. J. 357, L13 ; 1991 ApJ 368, 231), show the magnetic field of the Crab Nebula is generally well ordered on a scale size of about 10", aligned radially along most of the periphery. Depolarization is caused by a network of filaments containing thermal gas, which are too small to be seen individually. The field in the Crab Nebula 'jet' is aligned along its length. The authors conclude that the jet is most likely the result of a high-velocity beam formed inside the nebula. On the other hand numerical simulation by Cox C.I.+ 1991 (MNRAS 250, 750) support the model in which jet is formed by the interaction of the SNR material with a cooled trail left by the progenitor star. Based on their model the authors conclude that Crab is situated within the warm component of the ISM and its the outer halo which has been proposed does not exist. Marcelon, M. +1990 (AA 228, 471) have shown from optical spectroscopy that the jet is expanding as a cylinder at 260km/s. CCD images have been obtained in the continuum at different polarization angles by Michel F.C.+ 1991 (ApJ 368, 463). These show that the dominant synchrotron emission comes from an hour-glass shaped structure symmetrically centered on the pulsar. This pattern can be provided by the pulsar magnetized wind organized azimuthally about the pulsar spin axis provided that the torus is slightly inclined to our line of sight. The overall polarization picture shows "scalloping" of the external regions of the synchrotron nebula and depolarization across its face. This can be understood if the bright filaments ejected by SN form a conductive Faraday "cage" which encloses the synchrotron nebula.

Bietenholz M.F + 1991(ApJ 373, L59) have measured the expansion of the synchrotron nebula. This rate of expansion is similar to that of line-emitting filaments. It is shown that the synchrotron component expands homologously but with acceleration so that it now seems to be larger than that of optical filaments. If confirmed, this implies that the relativistic gas is currently "bursting through" the net of filaments of thermal plasma.

Multifrequency VLA observations at different epochs allow one to obtain both the rate of expansion and changes in the radio spectra of the filaments (Velusamy T. + 1992 MNRAS 255, 210). The radio spectrum was shown to become steeper at the outer region of the shell suggesting shock acceleration of relativistic electrons. Images of the Crab nebula at 178 and 750 MHz have been reconstructed from lunar occultation data (Agafonov M.I.+ 1990, Astron Zh. tom 67, 549; Sov. Astron. 34, 275).

The thermal filaments have been studied by Hester J.J. + 1990 (ApJ 357, 539). They present 1.644 micron [FeII] and 1.57 micron continuum images. The ratio of IR to optical synchrotron emission show variations which appear to be related to position of the pulsar. MacAlpine G.M. and Uomoto A. 1991 (Ast. J. 102, 218) have measured the luminosity in the H- beta and HeI lines and in the synchrotron continuum. The mass of the line-emitting gas is estimated to be 1-2 M(sun) in agreement with previous measurements. Hennessy, G.S. + 1990 (ApJ 395, L13) have presented UV images made with UIT. Observations of the both the filaments and the continuum in the far UV have been made by Blair W + 1992 (ApJ 399, 611) using the Hopkins UV telescope (HUT). These suggest that the optical filaments are subject to a large variation in the ionisation parameter, although it is also possible that the C abundance is a factor of 10 above solar. An expansion velocity of 1100 km/s, was also found. Fesen, R.A. 1990 (ApJ 351, L45) has established the presence of dusty inclusions in filaments bright in [O I], [C I] and [S II]. Strom, R.G. + 1992 (Nature 358, 654) established from IRAS data that the dust temperature is 48K and the dust mass is $0.02M_{\odot}$.

Observations with IRAS have been combined with 21 cm data on the ISM around the Crab nebula (Romani, R.W. + 1990 Ap. J. 349, L51). These show an extensive bubble ($\approx$ 180 pc in diameter for an assumed distance of 2 kpc). This bubble is surrounded by a dense shell and is thought to have been blown by the progenitor stellar wind.

Finally, Lang, M.J. + 1990 (Nucl. Phys. B, Proc. Suppl. 14A, 165) and Vacanti G. + 1991 (ApJ 377, 467) have confirmed detection of the Crab nebula at TeV energies.

Despite the continuing interest in the Crab nebula, studies of other Crab-like SNR have continued to yield insights into the physics of this class of object. In particular, the X-ray spectra obtained with LAC of Ginga show that all, or most, Crab-like SNR have an X-ray spectral index near a value of 2. (Asaoka I. and Koyama K., 1990 P. A. S. Japan. 42, 625). The morphology of SNRs with pulsars has been considered by Bhattacharua D., 1990 (J. Ap. Ast. 11, 125)

The SNR 0540-69 in the LMC is one of the better examples of the complex phenomena presently thought to accompany a crab-like SN II expolosion. Caraveo P.A.+ 1992 (Ap. J. 396, L103) provided high-resolution optical imaging of this SNR. They discovered previously unseen structure smaller than the synchrotron nebula having a totally different shape, which could be interpreted as either a ring seen edge-on or else a jet. Chanan, G.A. + 1990 (ApJ 352, 167) measured the synchrotron nebula to be polarised by 5.6% at V.

Koo B.-C.+ 1990 (ApJ 364, 178) have detected an HI shell in the old, pulsar-powered SNR CTB 80. Its size and expansion velocity imply a dynamical age of about 8 10^4 yrs, close to the pulsar age 10^5 yrs. The shape of HI shell fits well that of previously known IR shell. Fesen, R.A.+ 1990 (AA 240, 376) presented a fully sampled data cube describing the kinematics of the core of CTB 80. The model suggested is that of an ellipsoidal shell expanding at a velocity of about 90 km/s , with a high systemic radial velocity 36+-10 km/s, excited by the central pulsar, PSR 1951+32. The timing properties of this pulsar were reported by Foster R.S.+ 1990 (ApJ 356, 243).

A search for fast-moving features in the optical remnant of 3C 58 by Van den Bergh S. 1990 Ap. J 357, p. 138) have not revealed any high velocities (v ¡ 500 km/s for bright crisp knots ¡1000 km/s for faint fuzzy features). The nebulosity may represent quasi-stationary circumstellar material acted on by passage of the fast-moving SN-shell.

The SNR G5.4-1.2 is a shell-type SNR having a pulsar and plerion just outside. It is assumed to be a more evolved version of CTB 80; the pulsar assumed to have already penetrated the shell. New polarisation observations of this SNR by Milne D.K. + 1992 (MNRAS 255, 707) reveal an extremely high high rotation measure. Kundt W., 1992 (Ast. Space Sci. 190, 159) considered the illumination of a similar object, the SNR G5.3-1 (=bird) by its pulsar PSR 1757-24.

Low-frequency 34.5 MHz observations of the Vela SNR by Dwarakanath K.S., 1991 J. Astrophys. Astron., 12, 199 gave a spectral index of -0.16+-0.02 for Vela X and -0.53+-0.03 for Vela YZ confirming that Vela X is a plerion, while Vela YZ is a shell-type SNR.

A number of recently discovered pulsar-SNR associations have been studied. PSR 1758-24 and G5.4-1.2, a remarkable pulsar-SNR association,has been investigated in detail (Manchester R.N. +1991 MNRAS 253, 7P; Kundt W., 1992, Ap. Space Sci. 190, 159; Milne D.K.+ 1992 MNRAS 255, 707). New observations of both the SNR G308.8-0.1 (previously known as an unusual SNR) and the pulsar PSR J1341-6220 (previously considered not to be related because of its location outside the SNR shell) show that the SNR to be shell-like and containing a short- period pulsar of age about 12 000 yrs (Caswell J.L. + 1992, ApJ 399, L151;Kaspi V.M. + 1992 ApJ, 399, L155). Kassim N.E.+ 1990 (Nature, 343, 146) have proposed a possible new association of the pulsar PSR 1800-21 with the SNR G 8.7-0.1 (W30). Finally, Wolszczan A.+ 1991(ApJ 372, L99) have discovered a young 267 ms pulsar within the radio shell SNR W44. Bignami G.F., Caraveo P.A., Mereghetti S., 1992 (ApJ 389, L67) report that the optical counterpart of 1E 1207.4-5209, the central X-ray source of a ring-shaped SNR PKS 1209-52 remains unseen down to V=24. Sanbonmatsu K.Y.+ 1992 (AJ 104, 2189) have given a distance determination for the SNR G27.4+0.0 and its central X-ray source.

6.2.2. Remnants of Type Ia Supernovae

Optical CCD interference filter imagery and long-slit spectroscopy of Kepler's SNR (Blair W.P+ 1991 ApJ 366, 484) revealed nonradiative (Balmer dominated) shock emission from knots. The shock velocity is estimated to lie in the range 1530-2000 km/s. An expansion time scale of 3200+-1200 yrs was found by Bandiera R. + 1991 (ApJ 374, 186) from the proper motion of 50 optical knots measured over a baseline of half century. The space velocity of 248+-12 km/s (for an assumed distance 4.5 kpc) confirms that the progenitor was a high-velocity object. The dynamics of the interaction of Kepler's SNR with the dense circumstellar medium, taken into account stellar wind of a moving progenitor and multiple shocks formation have been considered by Borkowski K.J. + 1992 (ApJ 400, 223).

Hatsukade I. + 1990 (PASJ 42, 279) discussed the X-ray spectrum of Kepler's SNR, and derived parameters of the hot non-equilibrium plasma. From this they were able to determine the abundance of iron, and to compared the results with those found for the Tycho SNR. Soft X-ray spectra of the SN 1006 (Leahy D.A. +1991 ApJ 374, 218) confirmed the theoretical "reverse-shocked ejecta" model of this SNR.

Smith R.C. + 1991 (ApJ 375, 652) analysed the optical spectra of six Balmer-dominated SNRs: Tycho, SN 1006 and four SNR of this class in the LMC. All but one in the LMC objects exhibit the broad (580 to 2300 km/s FWHM) Hα lines characteristic of collisionless shocks. The Hα line widths compared with theoretical models gave the shock velocity. These authors demonstrate that the intensity ratio of broad to narrow components is not a reliable diagnostic of shock velocity. Long K.S. and Blair W.P., 1990 (Ap. J. 358, L13) have identified Balmer- dominated filaments in RCW 86. The shock velocity derived from the wide Hα component (500-930 km/s) indicates that this SNR is rather young.

High-resolution radio observations of the Tycho's SNR at several frequencies (Dickel J.R + 1991AJ 101, 2151; Wood G. + 1992 AJ 103, 1338) have been used to determine the intrinsic direction of the magnetic field. In this remnant, the field shows cellular pattern with a net radial orientation and fairly low degree of polarization. A CCD image of Tycho SNR obtained in [Fe XIV] line failed to detect any emission (Teske, R.G. 1990 ApJ 362, 563).

6.2.2. Oxygen–Rich Type II Supernova Remnants

The general morphological structure of Cas A at 81 MHz is found to be similar to that at 1380 MHz; however the spectral index varies over the image (Woan G. and Duffett-Smith P.J., 1990 MNRAS 243, 87). These measurements show that Cas A does not contain a steep-spectrum compact component with a flux density higher than a few Jy, contrary to what had been suggested by earlier observations at metre wavelengths. The spectral index variations over a range of -0.64 to -0.92 have been used to study the relativistic electron population in Cassiopeia A (Anderson M.C. and Rudnick L. 1990, BAAS 22, 751; Anderson M. + 1991 ApJ 373, 146). Regions of steeper spectral index can be associated with compact features such as the "bow shocks"; whereas flatter spectra are associated with the bright radio ring. According to the model bow shocks are caused by dense clumps of ejecta newly encountering the shocked ISM, whereas the ring maps out the blast wave. Cas A shows a secular flux-density decrease. At the wavelength 7.9 m this has been measured to be (0.70+- 0.19)% /year over the period 1956-1991; half of the value derived by extrapolation of the frequency dependence of the rate of decline in flux (Vinyajkin, E.N. + 1992, Pisma Astron. Zh 18, 755). On the other hand, Hook I.M. +,1992 (AA. 255, 285) measured the rate of decrease at 81.5 MHz and now find no evidence for a dependence of the rate of flux decrease with frequency.

An upper limit of optical depth in the H165α and H166α radio recombination lines was obtained by R. L. Sorochenko and G. T. Smirnov 1993 (Pisma Astron. Zh. 19, 359). This result can be understood if the fossil HII region around Cas A has a higher temperature than that of a "normal" HII region or if the low-frequency cut-off in the Cas A spectrum is not entirely due to the interstellar medium.

A small region of the bright rim of Cas A was mapped in its 20 μmemission (Greidanus H. and Strom R.G., 1991 AA 249,521). This revealed no correlation between IR structures and fast moving optical knots, X-ray structures or radio condensations. In particuler, the detailed correlation between hot X-ray plasma and heated dust predicted by the theory is totally absent.

Fesen R.A., 1990 (AJ 99, 1904) discovered that Ne-rich filaments exist in Cas A. New clues to the nature of the Cas A progenitor were provided by the discovery that one knot in the NE ejecta 'jet' exhibits Hα and [NII] emission like those seen in quasi-stationary flocculi, as well as [OI], [OIII] and [SII] such as seen in fast-moving knots (Fesen R.A. and Becker R.H., 1991 ApJ 371, 621). These observations suggest that, at the time of the explosion the progenitor posessed a thin Hydrogen-rich outer layer. This places constraints on the evolutionary phase of the probable WR progenitor.

VLA observations of the SNR Puppis A at 327 and 1515 MHz by Dubner G.M., + 1991 (AJ 101, 1466) show signs of interaction of the shell with the inhomogeneous surrounding gas. They also found an excellent correlation between radio and x-ray morphologies and variations of spectral index over the image.

Hwang U. + 1992 (BAAS 24, 790) have presented high-resolution X- ray spectroscopy of the oxygen-rich SNR in the LMC, N132D, while Canizares, C.R. 1990 (IAU Coll. # 115, p136) has analysed EINSTEIN focal plane crystal spectrometer data of Pupp A, Cas A, and N132D in the LMC.

The SNR MSH 15-52 is an example of an oxygen-rich SNR containing a pulsar bright at X-Ray wavelengths. Trussoni E. + 1990 (AA 234, 403) provided X-ray EXOSAT observations of this system. Thorsett S.E. 1992 (Nature, 356, 690) has speculated on the possible identification of the PSR 1509-58 in the SNR MSH 15-52 with the "guest star" of A.D.185.

6.3 OLD SUPERNOVA REMNANTS.

6.3.1. Radio Observations.

The number of known SNRs known is growing rapidly, mainly thanks to radio observations. For the first time we have been able to observe in detail the birth of a radio SNR in SN 1987A (Staveley-Smith L.+ 1992 Nature 355,147).

The catalog of galactic SNR (Green D.A., 1991, PASP 103, 209) in its updated version (Green D.A., 1993, A Catalog of galactic SNR, MRAO, University of Cambridge, Department of Physics) now contains 182 objects. A WSRT radio survey at 327 MHz of the galactic plane region $44^{\circ} < l < 90^{\circ}$ has been provided by Taylor A.R.+ 1992 (AJ 103, 942) using WSRT at 327 MHz. A total 24 SNRs were identified, 11 of which are newly discovered objects. The excess of filled-center objects remarked upon, if confirmed, would be important for understanding of evolution of synchrotron emission of relativistic electrons either shock accelerated or ejected by a central pulsar. The southern galactic plane is being currently surveyed by the MOST operating at 843 MHz (Whiteoak J.B.Z., 1990, PAS. Australia. 8, 274), and Gorham P.W. 1990 (ApJ 364, 187) has provided a radio/infrared/optical study of candidate SNRs from the Clark Lake 30.9 MHz galactic plane survey. Of the 80 sources detected, more than half show one or more of the characteristics typical for SNRs. High-resolution 330 MHz VLA observations of 20 galactic SNRs have been reported by Kassim N.E., 1992 (AJ 103, 943). Trushkin S.A. + 1990 (Astrofiz. Issled. Izv. Spets. Astrofiz. Obs., 25, 84); Bull. Spec. Astrophys. Obs.-North Caucasus 25, 83) presented multifrequency radio observations with RATAN of 14 SNRs in the region $85^{\circ} < l < 135^{\circ}$.

A survey of HI 21 cm emission lines toward all northern SNRs by Koo B.-C+ 1991 (Ap. J.382 204) shows that among 103 objects observed four (G27.7-0.6, W51, CTB80, HB21)have definitely and a further 12 have probably related HI shells. Furthermore, 15 SNRs are associated with high-velocity (70 to 160 km/s) clouds, which are probably shock accelerated.

Several newly discovered objects and many well known galactic SNRs have been mapped at radio frequencies. These include the new composite SNR of low surface brightness G322.5-0.1. (Whiteoak J.B.Z. 1992 MNRAS 256, 121); high-resolution observations at 408 and 1420 MHz of G160.9+2.6 (HB9) by Leahy D.A.+ 1991 (AJ 101, 1033); new intensity maps at 4.75 GHz, linear polarisation and spectral index distribution on the CTA104A (Mantovani F. + 1991 AA, 247, 545); and observations of the SNR G73.9+0.9 embedded in a very complicated region inside the Cygnus supershell (Pineault S. + 1990 MNRAS, 246,169).

Radio observations of the "X-ray detected" SNR G156.2+5.7 (Reich W. + 1992, AA 256, 214) revealed a highly polarized non-thermal shell, having the lowest surface brightness at 1 GHz of all known SNRs. A weak HI shell detected may be associated with the SNR if this was created by a progenitor stellar wind.

New high-resolution VLA polarimetric observations of the filaments in the SNR IC 443 show the size scale of the largest region of locally organized magnetic field is about 85 arcsec = 0.6 pc (Wood G.A., + 1991 AJ 102, 224). NRAO observations of Cygnus loop at 408 MHz (Green D.A. 1990 AJ 100, 1927) reveal spectral index variations.

VLA observations of of the unusual SNR CTB 37A/B show that this does not have a jet but is instead superposed on a second, previously unidentified SNR G348.5-0.0 (Kassim N.E. +1991 ApJ 374, 212). These observations also reveal a remarkable "blow-out" region from the nearby SNR G348.5+0.1.

6.3.2. Molecular Observations.

Many objects display a SNR shock interactions with dense interstellar clouds, dense molecular material in interstellar clouds. A particularly fine example is provided by IC 443 which has been studied in HI by Moorhause A. + 1991 (MNRAS 253, 662). Dickman, R.L. + 1992 (ApJ 400, 203) present high resolution maps in CO and HCO+. An analysis of these gives a total molecular mass of 2000 $M_{\odot}$) and an expansion velocity of 25 km/s in the molecular gas. Bedogni R. + 1990 (AA, 231, 481); and Green D.A.+ 1992 (MNRAS 254, 686) find evidence of shocked molecular material associated with the SNR G33.6+0.1 in their CO and HCO observations.

Several SNRs have been found to possess molecular envelopes: for example, Uchida K.I.+ 1992 (ApJ 398, 128) identified a dense molecular ring surrounding the nonthermal radio shell G359.1-0.5 and Gomes Y. + 1991 (ApJ 377, 519) found a dense molecular envelope around the compact HII region G5.89-0.39 (W28 A2). Further studies of the molecular cloud associated with the SNR G109.1-1.0 have been provided by Tatematsu K. +1990 (ApJ 351, 157); Wang Z.+ 1992 (ApJ 388, 127); and Ni Chen-ping + 1990 (Acta Astron. Sin. 31, 121-127; English. Transl. Cin. Astron. Astrophys. 14, 422). The radio and X-ray semi-circular morphology of the shell of CTB 109 is shown to be a result of explosion at the edge of a dense interstellar cloud.

Since the majority of SN explosions occur in dense molecular complexes and/or OB associations the only way to properly understand their nature is to consider the evolutionary history of a whole area. A number of investigations have demonstrated the productivity of such approach:

From CO, HI and radio continuum observations, Tatematsu K.+ 1990, (AA. 237, 189) have reconstructed the kinematics of the SNR HB21 and its interaction with a "wall" of atomic and molecular gas and individual dense clouds in the complex region Cyg OB7. Nichols-Bohlin J. + 1993 (AJ 105, 672) considered the interstellar environment around the WR star HD 192163 which is supposed to have been influenced by a previous SN explosion in the Cyg OB1 supershell. HI and CO observations of the field containing the SNR HB3 (Routledge D. + 1991, AA, 247, 529) permits reconstruction of a complete model for the evolution of the complex region W3+W4+HB3, the associated atomic and molecular gas, and the Cas OB6 stellar population and star formation history. A CO emission shell has been found around SNR G54.4-0.3 (Junkes N., 1991 Proc. AS. Australia., 9, 315).The SNR appears to expand into the wind- blown bubble interior to this shell. Pineault S., + 1993 (AJ 105, 1060) have found an expanding HI shell around the SNR CTA1 and two cavities in the far IR as measured by IRAS. The SNR appears to have broken out into low-density regions.

Multiple SNe regulate the phase structure of the ISM both around OB associations and generally, in the gaseous disk of the Galaxy. This problem has been considered by Heiles C., 1990 (Ap. J., 354, 483); and Cioffi D.F. + 1991 (ApJ 367, 96). Zongwey Li, + 1991 (ApJ 378, 93) have discussed the statistical correlation of galactic SNRs and spiral arms.

6.3.3. Infrared Observations of Old SNRs

The IRAS mission opened a new tool for the study of both SNRs and their interstellar environments. Saken J.M., +1992 (ApJS 81, 715) have presented an IRAS survey of galactic SNRs. Arendt R.G., + 1992, ApJ 400, 562) have developed a new technique for analyzing for the IR emission of a SNR and applied it to the Cygnus Loop. They distinguish two components. The first is spatially correlated with the X-ray and arises from collisionally heated dust in the shocked gas. The second is spatially correlated with the optical filaments and arises both from dust which is radatively or collisionally heated within filaments and from IR lines emitted by the shocked gas. Graham J.R. + 1991 (ApJ 372, L21) and Graham J.R.+ 1991 (AJ 101, 175) discovered vibrationally excited molecular hydrogen coincident with and ahead of the optical filaments in the Cygnus Loop. Arendt R.G. 1991 (AJ 101, 2160) has discussed IRAS observations of a large area around the composite SNR G320.4-1.2. A pointlike source is prominent at 25 μmmap near the pulsar. Arendt R.G., + 1991 (ApJ 368, 474) have performed a quantitative analysis of the IRAS observations of Pup A. These enabled them to derive both the global properties of the SNR and details of dust heating, the grain size distribution, the dust mass etc.

Oliva E.+ 1990 (AA 240, 453) presented a detailed study of IR spectra of RCW 103 in the range 1.0-2.5 μm. IR maps and fluxes for the young SNR RCW 86 (Greidanus + 1990 AA 240, 385) were compared with other historical SNRs.The IR temperature is shown to decrease with age; densities derived from the IR and from X-ray data are comparable. However, X-ray masses are found to be an order of magnitude larger than the IR masses.

Using IRAS data, Gahm G.F.+ 1990 (AA 228, 477) have distinguished a void of $4.5^{\circ}x3^{\circ}$ in Lupus surrounded by a ring of enhanced emission which is coincident both in position and in shape with an extended X-ray source suggested by Riegler et al. to be a SNR 7to10 10^4 yr old. The SNR appears to be expanding within a larger bubble with a radius $\approx 7^{\circ}$.

6.3.4. X-Ray Observations of Old SNRs.

Reviews of X-ray observations, high resolution spectroscopy and plasma diagnostics of SNRs have been presented by Bleeker J.A.M., 1990 (Adv. Space Res. 10, 143), and Canizares C.R., 1990 (IAU Coll. No 115: High Resolution X-ray Spectroscopy of Cosmic Plasmas. p.136).

Seward, F.D. 1990 (ApJS., 73, 781) has gathered X-ray images of the 47 galactic SNRs observed by the EINSTEIN satellite into a definitive catalog. However, X-ray observations using ROSAT have now opened a new page in the study of SNRs (Smith A., 1990, ESA Bull., No.62, 59). For the first time a galactic SNR has been discovered at X-ray frequencies using ROSAT (Pfeffermann E.+1991AA 246, L28).

ROSAT PSPC and optical images of the SNR W44 by Rho J.-H. + 1992 (BAAS 24, 791) show the same centrally peaked X-ray morphology observed by Einstein, in contrast to the shell-like radio morphology. Similar results have been found for both W28 and 3C 400.2 (Long K.S.+ 1991 ApJ 373, 567). This is possibly the result of thermal evaporation of dense clouds, increasing the density of the hot gas (White R.L.+ 1991 ApJ 373, 543). The X-ray image of W44 also shows locally brighter emission and clumps along the optical filaments, suggesting both are produced by the interaction between the supernova shock and regions of enhanced ambient density. Observation of soft X-ray emission from the SNR G18.95- 1.1 (Ashenbach B.+ 1991 AA, 246, L32) revealed a shell-like morphology. This SNR was previously thought to be a composite one. It is located in a low density region created by the pre- supernova wind.

HEAO-1 A2 X-ray observations of the Cygnus Loop by Leahy D.A.+ 1990 (BAAS 22, 752 and ApJ 363, 547) gave the best spectra to date. A two component model gave an excellent fit to the data. Images of the Cygnus Loop at the X-ray energies above 1.5 keV (Hatsukade I., + 1990 ApJ 362, 566) show a center-filled morphology proving that a high-temperature low- density plasma fills the interior of the old remnant.

An image in the hard (2.5-25 kev) X-ray emission from the Vela SNR (Willmore A.P. + 1992 MNRAS 254, 139) displays a synchrotron nebula around the Vela pulsar, about 1 deg across, roughly aligned along the direction of the PSR's spin axis. This accounts for about 48total emission in 4-25 keV band. The power required to produce the relativistic electrons in the nebula is estimated to be 75rotational energy loss of the PSR.

Kaastra, J.S. + 1992 (AA 264, 654) have analysed the GINGA X-ray spewctrum of RCW 86. If this SNR can be identified with the supernova of AD 185, then its distance is about 1.0kpc, and the explosion energy was $2x10^{51}$ ergs.

Leahy D.A.+ 1992 (ApJ 374, 218) have finally discovered X-ray emission associated with the Gum nebula. Both the X-ray and optical properties of the Gum nebula are consistent with a SNR of an age of about $2x10^6$ yrs.

6.3.5. Optical/UV Observations of Old Supernova Remnants.

A number of detailed studies of the best investigated SNR Cygnus Loop have allowed us to better understand the shock-wave dynamics in an old SNR and its interaction with the ambient ISM. TAURUS observations in the Hα and [OIII] lines by Greidanus H.+ 1991 (AASupp 89, 15) allow study of the small-scale radial velocity structure of optical filaments. Detailed interpretation the data in terms of both "curved sheets" and "shocked clouds" was given by Greidanus H.+ 1992 (AA 257, 265). Shull P.+ 1991 (ApJ 383,

714) used Fabry-Perot scans of 61 fields and proper motions at 39 locations to show that the Cygnus Loop is expanding asymmetrically into the ambient gas, with rest-frame velocity as high as 380 km/s the near half and 150 km/s in the far half of the shell. This is explained as a result of the progenitor's location near a density discontinuity in the ISM and of the action of its wind on the ambient gas.

The Hα images of three regions in the Cygnus Loop obtained by Fesen R.A. + (AJ 104, 719) provide a nice insight into the outer filaments and the shock-cloud interactions. A faint Balmer-dominated filament is identified 30 arc min beyond the line of radiative filaments, resolving the nature of the weak X-ray, optical and nonthermal radio emission observed to the west of NGC 6960. Strongly curved Balmer- dominated filaments along W and SW edge apparently indicate shock diffraction caused by shock passage in between clouds. Shull P.+ 1991 (PASP 103, 811) obtained optical spectra of the near side of Cygnus Loop in the central part. One filament displays Balmer-dominated spectra.

Cornett, R.H. + 1992 (ApJ 355, L9) has presented UIT images of the Cygnus Loop. Blair W.P. + 1991 ApJ 379, L33) report the discovery of a fast radiative shock wave in the Cygnus Loop using the Hopkins UV telescope. Strong emission lines of CIII 977 Å, NIII 912 Å, and OVI 1038 Åwere seen. The gas appears to be excited by a 170 km/s shock wave with some contribution from lower shock velocities. Ultraviolet imaging observations of the Cygnus Loop have been reported by Blair W.P + 1991(ApJ 374, 202) and Cornett R.H. + 1992 (ApJ 395, L9). Blair W.P. +1992 (BAAS 24, 791) have obtained narrow band images of a number of SNRs with the Wide Field Planetary Camera on HST. Images of NGC 6995 obtained in the coronal lines [Fe X] and [Fe XIV] cannot be interpreted in terms of cloud evaporation and these tend to support the "curved sheets" model of Hester and Cox according to Teske R.G. 1990 (ApJ 365, 256).

Wang Z.R.+ 1992 (PASJ, 44, 303) detected hard X-rays from the SNR IC443 implying an electron temperature higher than 10 kev and a density smaller than 0.1 cm^{-3}. Teske R.G., 1991 (Ap J, 383, 233) detected both [FeXIV] and [Fe X] line emission on the northern rim of this SNR, implying a shock velocity =¡ 400 km/s.

The large angular size of the Vela SNR allows one to use interstellar lines in spectra of field stars to study the hot gas and the dynamics of the shell (Wallerstein G. + 1990, MNRAS 245,701; Wallerstein G. +1992, AJ 103, 1346). Using IUE spectra of the SNR and new spectra of HD 72088 Raymond J.C.+ 1991 (ApJ 383, 226) have found evidence for a thermally unstable 150 km/s shock wave and for a nonthermal contribution to the pressure in the Vela SNR.

6.4. EXTRAGALACTIC SUPERNOVA REMNANTS.

A multiwavelength study of the SNR N49 have been made by Vancura O. +1992 (ApJ 394, 158). This has been used to derive shock conditions and their variation within the SNR. Vancura O.+ 1992 (ApJ 401, 220) obtained far-UV observations (912-1860 A) of this remnant using the Hopkins UV telescope, detecting lines of OVI, OIV, CIV and HeII. The bulk of both CIV and optical emission originates in shocks with v¡=140 km/s but most of the OVI originates in optically faint, 190-270 km/s shocks traversing dense clouds. Chemical abundances for this and other Magellanic Cloud SNR were given by Russell, S.C. +1990 (ApJS 64, 93).

Chu Y.-H.+ 1992 (AJ 103, 1545) have performed echelle and imaging Fabry-Perot observations of the 30 Dor B SNR embedded in an HII region. The relation of progenitor star to HII and OB association sheds light on the evolution of a SNR in a star formation region. Scanning interferometer observations of the shell N186E in the LMC (Rosado M.+ 1990 AA 238, 315) were used to analyze the kinematics of the complex and a possible interaction between the old SNR with a younger one, N186D.

A new complete catalog of SNR candidates in M33 was given by Long, K.S. + 1990 (ApJS 72, 61). Smith R.C.+1993 (ApJ 407, 564) have been reported on optical emission-line properties and chemical abundances derived for the best candidates of this list.

6.5. DEVELOPMENTS IN THE THEORY OF SNR

A number of studies continue to provide insight into the evolution of SNR in a uniform or cloudy medium. White R.L.+ 1991(ApJ 373, 543) have found a new similarity solution that describes SNR evolution in an

ISM with evaporating clouds. Different model parameters produce remnants that are strikingly different from "ordinary" shells. In particular, the effects of clouds evaporation may explain SNRs which display shell-like radio emission and centrally peaked X-ray morphologies. Draine B.T. + 1991 (ApJ 383,621) demonstrated that SN events in dense clouds result in efficient conversion of the explosion energy to X-rays, which irradiate and extensively ionise the surrounding medium. Asvarov A.I.+ (1991 Pis'ma Astron. Zh. 17, 702; English transl. Sov. Astron. Lett. 17, 297) have considered the energy distribution of shock-accelerated electrons in a SNR. A model for the evolution of radioemission of the shell-type SNR in their adiabatic Sedov phase was given by Asvarov A.I. 1992 (Astron Zh 69, 753). It is assumed that the electrons are diffusively accelerated at the shock front but that the magnetic field is interstellar; adiabatic losses of the accelerated electrons are taken into account. Reynolds S.P+ 1992 (ApJ 399, L75) predict radio spectra resulting from the Fermi mechanism in the SNR of Tycho and Kepler and in SN1006. They find that the spectrum is slighly concave to high frequency, with a mean slope significantly steeper than the classical value of -0.5 , in excellent agreement with observation.

The evolution of a SNR in a strongly magnetized medium and with cosmic rays has been considered by Insertis F.M.+ 1991 (MNRAS 252, 82). Ferriere K.M.+ 1990 (BAAS 22, 750) and Ferriere K.M.+ 1991 (ApJ 383, 602) find that when the shock velocity falls below 110 km/s, ion-neutral collisions in the vicinity of the shock dissipate the waves which couple the cosmic rays to the thermal gas, impeding cosmic-ray acceleration. Kang H.+ (1992 ApJ 399,182) performed a simplified treatment of the effects of cosmic ray acceleration on the evolution of a SNR. Dorfi E.A., 1991, (AA 251, 597) made a numerical study of the evolution of SNRs including the non-linear effects of particle acceleration in shock waves to consider the production of Gamma rays and cosmic rays in SNRs with radiative cooling. Jones T.W.+ 1992 (ApJ 396,575) have also considered cosmic-ray- modified SNR shocks. In an attempt to explain asymmetrical SNR Fulbright M.S.+ 1990 (ApJ 357, 591) have discussed the efficiency of relativistic particle acceleration depending upon the obliquity between the shock and magnetic field; see also Mineshige S.+ 1990 (ApJ 355, L47).

Aslanov A.I. + (1990 AA 229, 196) computed the hard X-ray emission to be expected from young SNRs. Wei Cui+ 1992 (ApJ 401, 206) generated two-temperature models of old SNR with both ion and electron thermal conduction. Brinkmann W. 1992 (AA 254, 460) considered non- equilibrium, non-LTE ionisation in SNRs and estimated the relevant time scales in young SNRs.

The interaction between a pulsar and an associated SNR has been discussed from a theoretical viewpoint by Chevalier R.A+ (1992 ApJ, 396, 540) this has provided a complete picture of pulsar nebulae in supernovae and the effects of the pulsar's wind shocks in SNR; particularly relevant to observations of plerions.

The late evolution of SNRs and their fossil hot bubbles, taking into account magnetic field has been discussed by Slavin J.D. + 1992 (ApJ 392, 131)with a view to reexamining the filling fraction of the hot gas that can be sustained in an ISM regulated by SNe. Byckov K.V., 1991 (Astron.Zh. 68, 1181;English transl. in Sov.Astron. 35, No 6) has considered SNRs at the stage when the preshock pressure regulates the evolution. At this stage the intense radiative cooling stops, a new adiabatic phase begins, and the relative thickness of the swept up shell grows.

Igumenshchev I.V. + (1992 Astron, Zh. 69, 479; English translation in Sov.Astron. 36, No.3.) have made numerical studies of both a SN explosion in a disk-like envelope, assumed to be produced by nonspherical mass loss from the presupernova and of a SN remnant moving in homogeneous gas.

Investigation of the evolution of SNRs inside pre-existing wind- driven bubbles or cavities provide an appropriate starting point in the analysis of the observed properties of real SNRs. Tenorio-Tagle G.+ (1991 MNRAS 251,318) and Franco J.+ (1991 PASP, 103, 803) have performed 2D hydrodynamical calculations of a SN explosion within wind-driven bubbles assuming different density distribution in the ejecta.

The effect of axisymmetric or asymetric matter distributions on SNR evolution has been considered by a number of authors. Korycansky D.G. 1992 (ApJ 398, 184) has considered an off-center point explosion in a radially stratified medium in terms of the Kompaneets approximation. Bisnovatyj-Kogan G.S + 1991 (Astron.Zh. 68, 749; English transl. in Sov.Astron. 35, 370) made a numerical hydrodynamic investigation of adiabatic SNR evolution in three dimensions in media with asymmetric density distribution. The evolution of density and vorticity generated in the interaction of a SNR with an interstellar cloud has been discussed by Stone J.M.+ 1992 (ApJ 390, L17). They suggest vortex filaments may be the location of radio emission knots in young SNR. According to Franco J.+ 1993 (ApJ 407, 100) in OB associations, where sequential SN explosions take place the impact of supernova fragments may influence the evolution of collective supernova remnants.

Chevalier R.A.+ 1992 (ApJ 392, 118) have investigated the convective instabilities resulting from the interaction of ejecta with steep power-law density profile expanding into a relatively flat stationary power-law density profile. In the linear regime the solutions are all nonstable above a critical wavenumber and the growth rate is greatest at the position of the contact discontinuity. Two-dimensional numerical hydrodynamic computations make it possible to follow the instability into the nonlinear regime. Results obtained agree well with X-ray observations of the Tycho SNR, its radio morphology, and expansion.

35. STELLAR CONSTITUTION

President: P.Demarque (USA)
Vice-President: C.Chiosi (Italy)
Members: R. Canal, I. Iben, A. Maeder (past President), G. Michaud, K. Nomoto, A. Renzini, A.V. Tutukov, D. VandenBerg, G. Vauclair

1. INTRODUCTION

This report covers a few highlights of the research carried out during the last triennium. It focusses nearly exclusively on areas which have been touched by the recent developments in the equation of state and opacities of stellar matter. These developments, coupled with the remarkable and accelerating progress on internal dynamics and transport processes in the interiors of stars, have literally transformed the field in a very short time.

The new generation of stellar opacities and equation of state was a highlight of the IAU XXI General Assembly in Buenos Aires. Since then, these opacities, particularly the OPAL interior opacities due to Iglesias & Rogers (ApJ,371,408) which have been the most accessible to the community, have been used with remarkable success on a number of problems. One must also emphasize the equally important ongoing work on low temperature opacities which include the effects of molecules (Sharp A&AS,94,1 and Kurucz 1991). The task is still unfinished in the low temperature range, but it will have a profound impact on our understanding of stellar interiors.

The topics briefly covered in this report are: convection in stars, the physics of the interior of the Sun and sun-like stars, including pulsation, rotation and diffusion, and the evolution of low mass and massive stars.

Books published:

Solar Interior and Atmosphere, A.N. Cox , W.C. Livingston and M.S. Matthews, eds, University of Arizona Press,Tucson,1991

Challenges to theories of the structure of moderate-mass stars, Lecture Notes Phys. Vol.388, D.Gough & J.Toomre,eds.,Springer,Berlin,1991

The Sun, an introduction, by M.Stix,Springer,Berlin

J. Bergeron (ed.), Reports on Astronomy, Vol. XXIIA, 399–407.

The Physics of Astrophysics,Vol.I: Radiation and Vol.II: Gas Dynamics, F.H.Shu, Univ.Sci.Books,Mill Valley,CA

Proceedings of meetings:

Proc. IAU Symp. 155, Planetary nebulae, R.Weinberger & A.Acker.eds., Kluwer, Dordrecht, 1992

Proc. IAU Symp. 149, The stellar populations of galaxies, B.Barbuy & A.Renzini,eds., Kluwer, Dordrecht, 1992

Proc. IAU Symp. 151, Processes in interacting binaries, Y.Kondo, R.F.Sistero, R.S.Polidan, eds., Kluwer, Dordrecht, 1992

Proc. IAU Colloq. 137, Inside the stars, ASP conf.ser. Vol.40, W.W.Weiss & A. Baglin, eds., 1993

Proc. Cool stars, stellar systems and the Sun.7., ASP conf.ser. Vol.26, M.S.Giampapa & J.A.Bookbinder, eds., 1992

Vulcano workshop on Young star clusters and early stellar evolution, F.Palla, P.Persi, H.Zinnecker, eds., Mem.Soc.Astron.Ital.,62,705,1991

Workshop on astrophysical opacities, A.E.Lynes-Gray, C.Mendoza, C.J.Zeippen, eds.,Rev.Mex.Astron.Astrof.,Vol.23,special issue, 1992

Teramo Workshop on star clusters and stellar evolution, E.Brocato, F.R.Ferraro, G.Piotto, eds.,Mem.Soc.Astron.Ital.,63,1,1992

Proc. intern. symp. on origin and evolution of the elements, Tokyo, S.Kubono, T.Kajino, eds.,World Scientific, 1993

Articles in Annual Reviews of Astronomy & Astrophysics

Vol.29, 1991 Chemical evolution of the Galaxy, by N.C. Rana, p.129

Radioactive dating of the elements, by J.J.Cowan, F.-K. Thielemann, J.W. Truran, p.447

Seismic observations of the Solar Interior, by D.Gough and J.Toomre, p.627

Vol.30, 1992 Magnetic fields of degenerate stars, by G. Chanmugan, p.143

New developments in understanding the HR diagram, by C.Chiosi, G.Bertelli, A.Bressan, p.235

Evidence for black holes in stellar binary systems, by A.P.Cowley, p.287

Type Ia Supernovae as standard candles, by D.Branch, G.A.Tammann, p.359

Vol.31, 1993 Origin and evolution of X-ray binaries and binary radio pulsars, by F.Verbunt, p.93

2. CONVECTION IN STARS

2.1. Stellar convection zones

Important progress has been made in stellar convection, both in simulating numerically stellar compressible convection, and in introducing more realistic treatments of convection to stellar evolution calculations. New efforts are also aimed at including a description of the effects of magnetic fields in some of the models.

2.2. Numerical simulations of convection

One of the outstanding unsolved problems in calculations of stellar structure and evolution has been the treatment of convection in stars. It has recently become possible to perform increasingly realistic simulations of compressible convection in two and three dimensions (Chan & Sofia ApJ,336,1022; Cattaneo et al. ApJ,370,282; Chan & Serigawa 54.065.035; Stein & Nordlund,ApJ,342,L95; Hossain & Mullen Ap.J,380,631; Rast et al. ApJ,408,L53). Under some circumstances and provided they include an adequate treatment of radiative effects near the surface, these models can in principle be tested agaist observable stellar features such as granulation and effects on spectral lines (Kim & Fox,in Cool Stars,Stellar Systems & the Sun.7.,p.172).

2.3. New descriptions of convection in stellar models

One of the goals of these numerical simulations has been to replace the mixing length theory (MLT) formulation of convection, universally adopted in stellar structure calculations, by a physically more realistic description, and one that is free of the arbitrariness associated with the mixing-length parameter.

2.3.1. A parameterization of convection simulations

The first successful attempt has now been made by Lydon, Fox & Sofia (ApJ,397,701;ApJ,403,L79) to construct an evolutionary model of the solar interior which matches the solar radius without recourse to any adjustable mixing length parameter to describe the convection zone. The formulation of convection is based on the three-dimensional simulations of compressible, stratified convection of Chan & Sofia (ApJ,336,1022). Other physical assumptions in the models are fully consistent with the recent state-of-the-art standard models of the Sun described in the next section. The same formulation was subsequently applied by Lydon et al. (ApJ,1993) to the binary system Alpha Centauri whose two members are solar analogues with masses which bracket the mass of the Sun. They discussed their models within the context of the known uncertainties in the observed parameters of Alpha Centauri. Other recent evolutionary models for the two components of Alpha Centauri were constructed by Noels et al.(A&A,247,91) and by Edmonds et al.(ApJ,394,313). The latter also calculated the p-mode non-radial oscillation spectrum for both stars.

2.3.2. The Canuto-Mazzitelli model of convection

Another attempt to free the representation of stellar convection from the adjustable mixing length has been made by Canuto (ApJ,392,218) and Canuto & Mazzitelli (ApJ,370,295). These authors' approach is based on a theoretical model of turbulence, analogous to terrestrial atmospheric turbulence. In order to model the effects of compressibility, the scale of the convective region is used so that the mixing length is removed as an explicit parameter in the equations. The formulation has been applied to the Sun. The p-mode spectrum of the solar models has been investigated by Paterno et al.(ApJ,402,733). An attempt to reproduce theoretically the color-magnitude diagram of the old open star cluster M67 has been made by D'Antona et al.(A&A,257,539).

2.4 Magnetic convection

The problems of magnetic convection have recently been considered by Hanami & Tajima (Ap.J,377,694), Fox et al. (ApJ,383,860), Cattaneo & vainshtein (ApJ,376,L21), Fisher et al. (Apj,374,766) and Nordlund et al. (ApJ,392,747).

3. THE SOLAR INTERIOR, HELIOSEISMOLOGY AND SOLAR NEUTRINOS

The last few years have been particularly active in the area of solar modelling. Significant advances have been largely stimulated by helioseismology and neutrino experiments, and facilitated by spectacular advances in opacity calculations and in improvements in the calculation of the equation of state.

3.1. Helioseismology

Important advances:

3.1.1. The depth of the solar convection zone has been evaluated with precision by inversion of the p-mode data by Christensen-Dalsgaard, Gough and Thompson (ApJ,378,413).

3.1.2. Improved equation of state and OPAL opacities in the solar interior (Iglesias & Rogers ApJ,371,408;ApJL,371,L73) lead to agreement with observed p-mode spectrum, within the errors of the physics for the first time in standard models of the Sun. The sensitivity of the p-mode frequencies to the known errors in the solar age, its chemical composition, the equation of state, the atmosphere structure, the boundary conditions, the shell resolution of the model, and the nuclear reaction rates have been explored systematically (Cox, Guzik and Raby ApJ,353; Faulkner & Swenson ApJ,386,L55; Guenther et al. ApJ,387,372; Berthomieu et al. CESAM models A&A,268,775). A similar analysis was also performed for g-modes (Guenther et al.ApJ,392,328). Non-adiabatic effects on the p-mode spectrum have been explored by Balmforth (MNRAS,255,603;255,632;255,639), Cox and Guzik (1993,ApJ in press) and Guenther (1993,ApJ in press).

3.1.3. Concern with the effects of the diffusion of helium on the solar p-mode frequencies which were found to be affected at the same level as the current uncertainties in the other input physics. Detailed studies of helium diffusion in the Sun have been made by Guzik & Cox (ApJ,386,729); Goode et al. (ApJ,387,707); Dziembowski et al. (MNRAS,249,602); Kosovichev et al.(1993,MNRAS in press); Christensen-Dalsgaard et al.(ApJ,403,L75); Proffitt & Thompson (1993 ApJ in press); Guenther, Pinsonneault & Bahcall (1993,ApJ in press). For a comparison of diffusion coefficients used in stellar interior calculations, see also Chaboyer et al.(ApJ,388,372).

3.2. Solar neutrinos

The main improvements have been:

3.2.1. Upgraded energy generation rates by Bahcall & Pinsonneault (ApJ,); but see also the discussion of Turck-Chieze & Lopes (ApJ,408,347).

3.2.2. The inclusion of helium diffusion and non-adiabatic effects in the pulsation of the models (Guenther et al.1993, ApJ in press).

3.2.3. Comparisons with the preliminary gallium experimental results.

3.3. Seismology of solar analogues

The seismology of the evolving Sun has been considered by Guenther (ApJ,375,352). Edmonds et al. (ApJ,394,313) have explored the evolutionary status and seismology of solar analogues Alpha Centauri A & B. Guenther & Demarque (ApJ,405,298) have studied the evolutionary status and seismology of Procyon A. Other recent highlights of solar interior physics are discussed in the sections on convection and the solar-stellar connection.

3.4. Solar rotation curve by inversion

A great deal of effort continues to be concentrated on problems of the inversion of seismic data from the Sun, primarily for the purpose of deriving the state of rotation in the solar interior using the rotational splittings of p-modes (e.g.Schou et al. ApJ,385,L59).

A wealth of information on all aspects of solar interior physics and solar evolution can be found in the new compendium: Solar Interior and Atmosphere A.N. Cox , W.C. Livingston and M.S. Matthews, editors, University of Arizona Press, Tucson, 1991

4. THE SOLAR-STELLAR CONNECTION: ROTATION, DIFFUSION, AND LITHIUM DEPLETION

4.1. Lithium depletion by rotation

The depletion of lithium in the envelopes of the Sun and late-type stars of solar metallicity is a long standing problem which continues to attract much attention. Standard stellar models do not explain the observed depletion of lithium in the Sun and similar disk stars. The observations seem to be best explained by introducing a source of slow internal mixing, and many investigators believe that turbulent motions driven by rotationally induced instabilities are responsible for this mixing. Evolutionary models for rotating stars which include an explicit local description of rotational instabilities and relate the outward transfer of angular momentum from the star to internal mixing have been successful in explaining many of the trends in lithium abundances and surface rotational velocities observed in open star clusters (Pinsonneault et al. ApJ,367,239;ApJS,78,179; Thorburn et al. ApJ,415,150). A different approach to rotational depletion has been proposed by Charbonnel et al.(A&A,255,191). Other explanations as well, such as mass loss, have been advanced to explain the lithium observations (Boothroyd et al. ApJ,377,318; Dearborn et al. ApJ,394,L61; Swenson & Faulkner ApJ,395,654).

4.2. Angular momentum transfer and mixing

Theoretical discussions of the problems associated with angular momentum transfer in stars and its relation to element mixing are found in recent papers by Chaboyer & Zahn (A&A,353,173); Charbonneau (A&A 259,134); MacGregor & Brenner (ApJ,376,204); MacGregor & Charbonneau (ApJ,397,L63); Spiegel & Zahn (A&A,265,106) and Zahn (A&A,265,115). Other recent theoretical studies of stellar rotation include those by Clement (ApJ,406,651) and Durney (ApJ,407,367). Lithium dilution in F stars has been considered by Garcia & Spruit (54.065.047), Charbonneau & Michaud (ApJ,370,693) and Richer et al. (ApJS,82,329).

4.3. Lithium in halo dwarf stars

The problem of lithium depletion in halo dwarfs has special significance in the context of primordial abundances and cosmology. It is however even more difficult to study the rotational evolution of halo dwarfs because there are no young metal poor star clusters to guide theory in understanding the process of lithium depletion and its relation to rotation. Here, investigators must assume that the physical principles which model successfully disk stars can be equally applied to the stars in the halo which are very similar to disk stars, although they differ from them in two important respects, lower interior opacities and lower masses (Deliyannis et al.ApJ,414,740).

4.4. Helium diffusion

Diffusion can also affect the surface abundances of both lithium and helium in cool stars. Since halo dwarfs are too cool to exhibit helium lines, helium abundances cannot be estimated directly in these stars. But the abundance of lithium can be measured, and since both elements are expected to diffuse at nearly the same rate, lithium can be used as a proxy. Much effort has been extended in the last few years to understand the influence of helium diffusion on the structure and evolution of the Sun and sun-like stars. The effects of diffusion are expected to be most significant in metal poor halo stars. In these objects, which have the thinnest convection zones, an appreciable amount of helium could be drained from the convective envelope. This process is particularly effective when the convection zone becomes very shallow, as is the case for halo dwarf stars near the main sequence turnoff. The resulting hydrogen enrichment of the envelope could cause the star to turn toward the giant branch sooner in its evolution and thus lower age estimates of globular clusters. Some downward diffusion of helium near the center could also accelerate the exhaustion of the core in hydrogen. Calculations to estimate the effects of helium diffusion on the evolution of globular cluster stars have been carried out by Deliyannis & Demarque (ApJ,370,L89), Proffitt & Michaud (ApJ,371,584), Proffitt & VandenBerg (ApJS,77,473) and Chaboyer et al. (ApJ,388,372;ApJ,394,515). But rotational instabilities induce turbulence, which can inhibit diffusion in stars. This problem has recently been treated in some detail by Chaboyer in his doctoral dissertation (Yale University 1993), who considered evolutionary sequences of rotating stars which include molecular diffusion as well.

5. STELLAR EVOLUTION: LOW MASS STARS

5.1. Isochrones for halo and disk stars

New sets of evolutionary tracks for halo stars near the main sequence turnoff have been published, some including the effects of oxygen enrichment (VandenBerg ApJ,391,685;VandenBerg & Bergbusch ApJS,81,163), others oxygen and alpha capture element enrichment, as well as the effects of helium diffusion, both in the core and in the envelope (Chaboyer et al. ApJ,388,372;ApJ,394,515). They have been used to derive age differences and absolute ages for the globular clusters (see also Straniero & Chieffi ApJS,76,525). Metal-rich theoretical isochrones have also been published which include revisions of the chronology of disk star clusters in the Galaxy (Castellani et al. ApJS,78,517; Demarque,P. et al. AJ,103,151; Meynet et al. A&AS,98,477). Advanced phases of evolution are described by Lattanzio (ApJS,76,215) and Castellani & Castellani (ApJ,407,649).

5.2. Horizontal branch and post-horizontal branch evolution

5.2.1 Horizontal branch (HB) evolution Interest in HB evolution has increased in the last few years, stimulated by several factors: ultraviolet observations with IUE and now with HST; renewed discussions of the nature of the second parameter of HB morphology, and prospects of direct HB observations in more distant extragalactic objects such as M31; and finally the importance of understanding the evolutionary status of RR Lyrae variables and the dependence of their absolute magnitudes on metallicity. The long standing problem of the mass discrepancy between the masses derived for double mode RR Lyrae variables (pulsation masses) and evolutionary masses seems to have been resolved, basically as a result of the OPAL opacity increase, which has changed the predicted pulsation masses.

Recent studies of HB models address primarily the effects of oxygen or alpha element enhancement, and the "core breathing" phenomenon as the star approaches central helium exhaustion (Lee ApJ,373,L43; Castellani et al. ApJS,76,911; Dorman et al. ApJ,366,115;ApJS,80,701;ApJ,409,387; Caloi & Mazitelli A&A,271,139).

Concerning the RR Lyrae variables, Lee (ApJ,373,L43) considers the variables in Omega Centauri. Opacity effects on the pulsation calculations are estimated by Simon & Cox (ApJ,376,717), Cox (ApJ,381,L71), and Kovacz et al. (A&A,259,L46). Pulsation masses in M15 are discussed by Simon (ApJ,387,162). Evolutionary masses are calculated with OPAL opacities by Yi et al.(ApJ,411,L25).

5.2.2. Post-HB evolution

At the same time, post-HB evolution has received particular attention. The main focus is to gain a better understanding of the masses and composition parameters which determine how much time will be spent during post-HB evolution on the very red asymptotic giant branch and/or in luminous blue phases of evolution. Such a knowledge is essential for understanding the integrated spectral energy distribution of old stellar populations. It seems that a satisfactory explanation of the ultra-violet upturn in elliptical galaxies in terms of the advanced phases of evolution of low-mass metal rich stars is at hand (Horch et al. ApJ,388,L53; Castellani et al. ApJ,389,227).

6. STELLAR EVOLUTION: MASSIVE STARS

6.1. Convective core overshoot

Interest has continued to focus on the evolution of massive stars; in particular on the importance of convective core overshoot on their evolution near the main sequence and in the core helium burning and more advanced phases of evolution (Roxburgh,I.A&A,266,291; Brocato, Castellani A&A,258,397). Stothers & Chin are able to put some stringent limits on the extent of core overshoot in massive and intermediate mass stars (ApJ,381,L67;ApJ,383,820;ApJS,77,289;ApJ,390,136). They conclude that only a small amount of overshoot (less than 0.2 pressure scale height beyond the formal core boundary) is needed to match the observations when the OPAL opacities are used.

6.2. Evolution and pulsation

Other recent massive star studies include the work of Raiteri et al. (ApJ,367,228), Lamers et al. (ApJ,368,538), Brocato et al. (ApJ,410,99) and Spruit (A&A,253,131). The pulsational instability of OB star models constructed with the OPAL opacities has been investigated by Cox et al. (ApJ,393,272).

6.2. Apsidal motion test

Claret & Gimenez (A&A,244,319) have explored the effect of convective overshoot and choice of opacities on the apsidal motion constant of massive stars. They achieve an improvement with the available observations.

36. THEORY OF STELLAR ATMOSPHERES (THEORIE DES ATMOSPHERES STELLAIRES)

PRESIDENT: W. Kalkofen
VICE PRESIDENT: L. E. Cram
ORGANIZING COMMITTEE: Y. Cuny, D. Dravins, D. F. Gray, J. L. Linsky, R. Pallavicini, A. Peraiah, A. Sapar, M. J. Seaton, T. Takashi, R. Wehrse, L. A. Willson, C. Zwaan

Several members of the Organizing Committee as well as Dr. Lee Hartmann have written in their area of expertise for the Report of Commission 36 to the IAU. Their contributions are assembled here with minimal editing.
The topics treated in this report are:

Magnetic Activity in Stars (R. Pallavicini)
Stellar Activity Cycles (D. F. Gray)
Developments in Stellar Coronae (J. L. Linsky)
Dynamical Atmospheres and Winds (L. A. Willson)
Disk and Envelope Emission from Young Stellar Objects (L. Hartmann)
The Atmospheric Structure of Very Cool Stars (T. Tsuji)
Multidimensional Radiative Transfer (R. Wehrse)

Some of the individual reports make extensive reference to review papers and conference proceedings. To these should be added IAU Colloquium No 143 on *The Sun as a Variable Star*, the first IAU Colloquium specifically addressed to the so-called solar-stellar connection, the proceedings of which are to be published by Cambridge University Press and by Kluwer.
Below are the individual reports.

MAGNETIC ACTIVITY IN STARS (Roberto Pallavicini)

1. General Overview

Over the past decade it has become increasingly evident that magnetic fields play a fundamental role in the outer atmosphere of cool stars. It is not only that magnetic fields are directly responsible for solar-type activity phenomena like spots, plages and flares, but they also determine in a significant, perhaps dominant, way the structuring, energy balance and dynamics of the outer atmospheric layers. This new perspective in stellar atmospheric research has largely been a consequence of X-ray and UV observations from space (particularly with the *Einstein* and *IUE* satellites), as well as of new observational techniques, of vastly improved sensitivity and spectral resolution, developed for ground based optical and radio work. The new data have demonstrated the occurrence on stars of magnetic phenomena similar to those observable on the Sun, and have prompted the development of the so-called "solar-stellar connection", i.e. the close interaction of solar and stellar studies to the benefit of both disciplines.

In the last three years, research on stellar magnetic activity has continued to be actively pursued, as amply demonstrated by the many conferences, review articles and books which have been devoted to the subject. The number of research papers that have appeared in the major scientific journals is so large that it would be impossible to summarize them here. Rather, the emphasis will be placed on review articles and conference proceedings that present an up-to-date overview of the field. The reader is referred to these reviews for further details and complete bibliographical lists. It is also important to mention that the surveyed period has witnessed the start of operations of *ROSAT*, the Hubble Space Telescope and the *EUVE* satellite. These space missions, together with the continuing operations of *IUE*, and the availability of high quality photometric and spectroscopic optical data, are contributing enormously to our understanding of magnetic activity in stars.

2. Conferences and books

Several conferences have been devoted to stellar magnetic activity (or to specific aspects of it) over the past three years. It is interesting to note that the majority of them were dedicated to

J. Bergeron (ed.), Reports on Astronomy, Vol. XXIIA, 409–424.

the joint discussion of solar and stellar phenomena, in an effort to bring together researchers from two communities that were originally virtually separated. A good example is the conference on *Mechanisms of Chromospheric and Coronal Heating* (CH91) held in Heidelberg on 5-8 June 1990. Other examples are the IAU Joint Commission Meeting on *Solar and Stellar Coronae* (SC92) held in Buenos Aires on 27 July 1991, and the G.S. Vaiana Memorial Symposium *Advances in Stellar and Solar Coronal Physics* (SC93) held in Palermo on 22-26 June 1992. Along the same lines are also the *Seventh Cambridge Workshop on Cool Stars, Stellar Systems and the Sun* (CS92) held in Tucson on October 9-12, 1991, and the IAU Colloquium No. 130 *The Sun and Cool Stars: Activity, Magnetism, Dynamos* (MD91), held in Helsinki on 17-20 July 1990. The proceedings of these meetings offer a unique opportunity to compare observational results and theoretical models for both the Sun and other cool stars.

Other more specialized meetings were the Armagh Bicentennial Colloquium on *Surface Inhomogeneities on Late-Type Stars* (SI92) held in Armagh on 24-27 July 1990, and the NATO ASI on *Angular Momentum Evolution in Young Stars* (AM91) held in Noto on 17-21 September 1990. General reviews on magnetic activity in stars have been contributed by Zwaan (CH91), Rosner (HM91), Rodonò (AM91 and SI92), Baliunas (1991), Radick (1991), Feigelson, Giampapa and Vrba (1991), Montmerle (SC92), Pallavicini (1992a,b), Schrijver (CH91 and 1993) and Catalano (SC93). A catalog of chromospherically active binary stars, which updates an early version of 1988, has been published by Strassmeier et al. (1993).

3. Research highlights

a) Rotation, differential rotation, and dynamo action

In stars with outer convective zones (i.e. $\sim$ F0 or later), surface magnetic fields are believed to be produced by a dynamo mechanism that involves the interaction of rotation and turbulent convection. The present status of dynamo theory is discussed in MD91. As stressed by Brandenburg & Tuominen in those proceedings, the $\alpha\Omega$-dynamo can reproduce the general features of the solar cycle, but a completely satisfactory theory is still lacking. More important, the fact that many free parameters can be adjusted to fit one single object (the Sun !) does not provide a conclusive test of the theory. It is not even clear whether all types of stars should have the same dynamo pattern as the Sun. Theoretical arguments would rather suggest the opposite (Weiss SC93) and observational data exist that indicate a different behaviour including the existence of polar spots and of stars rotating more rapidly at the poles than at the equator (Eaton SI92; Vogt & Hatzes MD91). Recent advances in the study of stellar rotation, and its consequences on dynamo activity, have been reviewed by Bouvier (AM91) for pre-main sequence stars, by Stauffer (AM91) and Soderblom (AM91) for main-sequence stars, and by Gray (AM91) for giants. A "rotostat" mechanism has been proposed by Gray (AM91) to explain magnetic activity and rotational evolution in evolved stars. Among the cooler giants and supergiants, magnetic fields may play only a negligible role (Dupree HM91, Cuntz & Stencel HM91). Rosner (SC93) has discussed the problem of dividing lines in the HR diagram separating stars with magnetically confined coronae from stars with no coronae and large mass loss rates.

b) Measurements of stellar magnetic fields

It has become possible in recent years to measure photospheric magnetic fields in cool stars from the Zeeman broadening of magnetically sensitive lines. Since the observations are made by using unpolarized light, the intrinsic difficulty of opposite polarities cancelling each other in complex magnetic field topologies is eliminated. Recent advances in this field, including the discussion of possible systematic errors, can be found in Saar (MD91) and Solanki (CS92). The observations have provided magnetic field strengths and filling factors for a large number of cool stars, including some dMe flare stars and one pre-main sequence object. The inferred magnetic field refers to an average over the star surface. Instead, there is virtually no information on the magnetic field strength in localized active regions such as spots and plages. These magnetic field measurements are possible only for slowly rotating stars, since it is difficult to separate the small Zeeman broadening from rotational broadening in rapidly rotating stars. A technique has been developed to overcome this difficulty, by separating the contribution of magnetic regions in velocity ("Zeeman Doppler imaging"). Measurements of magnetic fields of hotter Ap stars have been reviewed by Landstreet (1992). Their magnetic field is much simpler than in cooler stars, with an overall structure that is roughly dipolar and extends over virtually the entire star.

c) Non-radiative heating

The long-standing problem of non-radiative heating in stellar chromospheres and coronae is far from solved (see CH91). The observational evidence points at the dominant role of magnetic fields, except perhaps for the quiet chromosphere that is more likely heated by acoustic waves (Ulmschneider CH91). MHD waves and/or electric currents are certainly required to heat the coronae of cool stars, but there is no general consensus on the detailed mechanism. The new X-ray data on stellar coronae made available by *ROSAT* (Schmitt CS92, SC93; Bromage CS92, SC93) confirm and extend previous results from *Einstein* (Vaiana et al. SC92, Sciortino SC93), but are unlikely to provide a definite answer to the heating problem. Clues could more easily be provided by spatially resolved observations of the Sun. Magnetic effects are also required to explain most of the observed chromospheric emission, particularly in active stars whose integrated flux is dominated by the contribution of magnetic regions. However, the so called "basal flux" in the Ca II and Mg II lines (Zwaan CH91, Schrijver CH91) may result from acoustic heating.

d) Photometric and spectroscopic observations of starspots

The presence of dark cool spots on the surface of many active stars (of the RS CVn and BY Dra classes) can be inferred from photometric variations, accompanied by color changes that indicate that the star is redder at the photometric minimum. In principle, the systematic monitoring of photometric changes permits to infer the spot size and location, and the variations of stellar rotation with latitude (differential rotation). In practice, things are much more complicated, and one has to worry about the uniqueness of the solution and systematic effects (Eaton SI92, Hall MD91). An alternative way to map surface inhomogeneities is to use spectroscopic observations, and the subtle changes induced by spots on a rotationally broadened line profile ("Doppler imaging"; Piskunov MD91, Vogt & Hatzes MD91, Collier Cameron SI92). This technique is starting to provide realistic maps of the otherwise unresolvable stellar disks. The existence of polar spots –that have been inferred from both Doppler imaging and analysis of photometric light curves– is however still controversial (Byrne SI92).

e) Chromospheric and coronal magnetic features

The presence of chromospheric plages and coronal active regions on stars can be inferred indirectly from rotational modulation of chromospheric and coronal emission and/or from the comparison of stars of similar spectral type, but very different activity levels (see reviews by Baliunas 1991, Pallavicini 1992a, Catalano SC93). Observations in the lines of Ca II and Mg II, as well as in UV lines and X-rays, show in fact a wide range of emission levels at each effective temperature. This is interpreted as being due to different fractions of the star surface covered by magnetic (plage) regions. Multi-component modelling of Mg II line profiles have been used to map the distribution of these regions on the stellar surface at different phases (Neff SI92). Similarly, rotational modulation and eclipses have been used in X-rays to infer the spatial structure of stellar coronae (Schmitt CS92, Foing SC93). The variability of coronal emission as produced by surface magnetic activity has been reviewed by Cheng & Pallavicini (SC92) and Pallavicini (SC93). The correlation between chromospheric/coronal emission and photometric variations (spots) is usually not very good, probably because of the high variability of the former on short time scales (Byrne SI92, Linsky SI92).

f) Stellar flares

Observations of stellar flares have been reviewed recently by Haisch, Strong and Rodonò (1991), Pallavicini (1992b) and Byrne (SC93). There is a strong similarity between solar flares and the typically more energetic flares that occur on dMe flare stars and RS CVn binaries (and occasionally also on other types of cool stars). This suggests that the process of energy release is basically the same in the two cases. Theoretical models of flares in stars and accretion disks have been discussed by Kuijpers (1992). A comprehensive summary of recent observations and empirical models of stellar flares is given in a book edited by Pettersen (1991) to which we refer the reader

(Univ. Arizona: Tucson), C.P. Sonett, M.S. Giampapa, M.S. Mathews, eds., p. 11.
Gray, D.F. 1988. Lectures on Spectral-Line Analysis: F, G, and K Stars, (The Publisher: Arva, Ontario).
Gray, D.F. 1994. Publ. Astron. Soc. Pacific, submitted.
Gray, D.F., S.L. Baliunas, G.W. Lockwood, & B.A. Skiff 1993. Ap.J. 400, 681.
Gray, D.F. & T. Nagel 1988. Ap.J., 341, 421-426.
Hossain M. & D.J. Mullan 1991. Ap.J. 380, 631.
Jankov, S. & B.H. Foing 1992. Astron. Ap. 256, 533.
Kim, Y.-C. & P.A. Fox 1992. Seventh Cambridge Worshop on Cool Stars, Stellar Systems, and the Sun (Ast. Soc. Pacific: San Francisco), M.S. Giampapa & J.A. Bookbinder eds., 172.
Kuhn, J.R. 1991. Adv. Space Res. 11, (4)171.
Larson, A.M., A.W. Irwin, S.L.S. Yang, C. Goodenough, G.A.H. Walker, A.R. Walker, & D.A. Bohlender 1993. Publ. Astron. Soc. Pacific, submitted.
Livingston, W. & M. Steffen 1988. Adv. Space Res. 8, 133.
Lou, Y.Q. 1992. Ap.J. 395, 682.
Pap, J.M. (ed.) 1994. The Sun as a Variable Star: Solar and Stellar Irradiance Variations, (Cambridge: Cambridge).
Radick, R.R., G.W. Lockwood, & S.L. Baliunas 1990. Science 247, 39.
Ribes, E., B. Beardsley, T.M. Brown, Ph. DeLanche, F. Laclare, J.R. Kuhn, & N.V. Leister 1991. The Sun in Time, (Univ. Arizona: Tucson), C.P. Sonett, M.S. Giampapa, M.S. Mathews, eds., 59.
Rosner, R., C.-H. An, Z.E. Musielak, R.L. Moore, & S.T. Suess 1991. Ap.J. Lett. 372, L91.
Smith, P.H., R.S. McMillan, & W.J. Merline 1987. Ap.J. 317, L79.
Sofia, S. 1994. in The Sun as a Variable Star: Solar and Stellar Irradiance Varia-tions, (Cambridge: Cambridge).
Spruit, H.C., . Nordlund, & A.M. Title 1990. Ann. Rev. Ast. Ap. 28, 263.
Stein, R.F., A. Brandenburg, & . Nordlund 1992. Seventh Cambridge Worshop on Cool Stars, Stellar Systems, and the Sun, (Ast. Soc. Pacific: San Francisco), M.S. Giampapa & J.A. Bookbinder, eds., 148.
Ulmschneider, P., K. Zähringer, & Z.E. Musielak 1991, Astron. Ap. 106, 9.

DEVELOPMENTS IN STELLAR CORONAE (Jeffrey L. Linsky)

This report summarizes the major developments (observational and theoretical) in the study of stellar coronae driven in large part by the analysis of the new data sets from ROSAT, EUVE, ASCA, Yohkoh, and HST, together with the continuing analysis of *Einstein* data and radio observations.

1. Introduction

During the past three years new and more powerful telescopes have allowed observers to study stellar coronae with higher energy resolution, observe fainter x-ray sources, observe large numbers of stars in the unexplored EUV region (80–700 Å), and survey the entire sky in x-rays and the EUV. These important capabilities result primarily from the new ROSAT, EUVE, and ASCA satellites. At the same time the *Einstein* archive has been mined further with the distribution of its archive on CDRoms, and the VLA, the new Australian Telescope, and VLBI networks are studying coronae at radio wavelengths. The spectacular Yohkoh x-ray solar images provide vivid evidence that the solar corona (and by implication stellar coronae) is highly dynamic with the heating and expansion of magnetic loops on many time scales (see papers in the 1992 October issue of *Publ. Astr. Soc. Japan* and [25]).

Launched on 1990 May 31, the Roentgensatellit (ROSAT) [46] includes two x-ray imaging detectors (the PSPC [37] and the HRI covering 0.1–2.4 keV) and the Wide Field Camera (WFC) for broadband imaging in the 60–200 Å region [38]. The all sky survey portion of the mission (August 1990 to February 1991) provided the first deep survey of the x-ray (see review by Schmitt in [25]) and EUV sky down to faint limits. Of the 384 sources listed in the WFC Bright Source Catalog [38] about 50% are known to be nondegenerate stars. Launched on 1992 June 7, EUVE [27] obtained broadband EUV fluxes in the 70–760 Å region during its all sky survey with about 350 stars listed in the initial catalog [28],and is obtaining EUV spectra and broadband fluxes during its pointed phase. A number of early papers based on EUVE data are in the 1993 September 10 issue of *Ap.J. Letters*. ASCA has been obtaining x-ray observations of stars with higher energy resolution and broader range (to 12 keV) than ROSAT since 1993 February 20. The G.S. Vaiana Memorial Symposium [25] contains a number of important review papers, including summaries of the ROSAT and Yohkoh work as of June 1992. For earlier comprehensive reviews of stellar coronae see [33] and papers in the proceedings of two symposia [12,17]. A recent comprehensive review of radio emission from stellar coronae is found in [19] and in the contribution by Drake in [25].

2. Late-Type Stars

Coronae are probably ubiquitous in late-type dwarfs and at least G-type giants as indicated by x-ray, radio, and EUV emission, and most recently by the detection with HST/GHRS of the Fe XXI 1354 Å line in the M dwarf AU Mic [29]. A ROSAT survey of very late M dwarfs [21] shows that the coolest M dwarfs, which have fully convective interiors, are just as x-ray bright as the early M dwarfs in which solar-type dynamos operate. Coordinated ROSAT and VLA observations of M dwarfs [21] show the good correlation of x-ray and radio luminosities [20]. An x-ray survey of late-type giants and supergiants from the *Einstein* archive shows many detections of F and K giants and confirms the existence of the x-ray dividing line located at middle-K spectral type [26]. The ROSAT all sky survey data [22] support this result. A WFC survey of nondegenerate stars within 10 pc detected 41/220 mostly F-M dwarfs [51], and Hodgkin and Pye [25] present a WFC survey of stars extending out to 25 pc. Hybrid-chromosphere stars are now detected as x-ray sources indicating that they have coronae [4,39].

Analysis of *Einstein* IPC archival data of the Pleiades [31] and Ursa Major [41] clusters provide new information on the age and rotational velocity dependence of coronal x-ray emission. The first survey of the Hyades by ROSAT has been published [45] and surveys of other clusters are under way [25].

M dwarf stars and especially dMe stars are bright, transient x-ray [35,23] and radio [3] emitters during flares, indicating large enhancements in the thermal and nonthermal electrons in their coronae. For a review of stellar flares see [34]. VLBI measurements have now resolved the radio emission from dMe stars [2], providing direct estimates of the brightness temperatures. Coordinated multiwavelength observations such as the RIASS campaign (ROSAT and IUE) [1] are needed to study quiescent and flaring properties of the corona and lower atmosphere together.

3. Binary Systems with Nondegenerate Components

Binary systems containing nondegenerate components that have hot coronae as indicated by their luminous x-ray and radio emission include the RS CVn, BY Dra, W UMa, and Algol systems. The ROSAT PSPC all sky survey contains 136 x-ray detected RS CVn systems [6] which can be modelled with a two-temperature coronal plasma at $\log T = 6.3$ and 7.2 [7]. There is an excellent correlation between the radio and *Einstein* IPC fluxes for a very large sample of RS CVn systems [8], despite the usual interpretation of the x-ray emission as thermal and radio emission as nonthermal [30]. Coordinated ROSAT, IUE, and VLA observations during the all-sky survey [16] confirm the earlier nonsimultaneous data results. The remarkable radio-x-ray correlation, which extends over five orders of magnitude and includes many classes of late-type stars [20], probably indicates that heating and particle acceleration occur in parallel. A systematic survey of the radio emission from classical Algol systems [48] shows that they have properties similar to the RS CVns, whereas the radio emission from young Algol-type systems (Serpentids) is probably from their winds [11]. A two-dimensional model of the magnetospheres of active late-type stars [32] explains the quiescent radio emission as gyrosynchrotron emission from trapped relativistic electrons, and time-dependent models of the nonthermal particle populations and the resultant radio emission are now available [5]. The first three-dimensional picture of a stellar corona has been constructed [43] from EXOSAT observations of AR Lac. Active binary systems are known to flare as seen in x-rays [23,10] and radio emission [3].

4. Early-Type Stars

The usual explanation for the luminous x-ray emission from O-type and early B-type stars is shock-heated gas in their winds (see review by Kudritzki and Hillier in [25]) rather than coronae, but an alternative corona-like model has been proposed [49] with magnetic plasma heated in current sheets. The *Einstein* IPC archive contains 74 early B-type stars that are x-ray sources with properties similar to the O-type stars and 34/1058 late-B stars, indicating that x-ray emission from these stars is either rare or from a late-type companion [18]. The ROSAT HRI and PSPC detected young late-B type stars, but the spread in L_x is large [36] and the origin of the x-ray emission remains a puzzle [42]. The *Einstein* archive data [41] and new ROSAT PSPC observations [44] contain very few x-ray detected single A-type stars and show a systematic decline in coronal x-ray brightness with increasing T_{eff} that may be due to strong coronal winds [44]. The magnetic chemically peculiar stars of both the helium-strong (early-B) and helium-weak (late-B) types are both radio [24] and x-ray sources [9]. The radio data may be explained by a magnetospheric model in which the stellar wind distends the magnetic field lines leading to gyrosynchrotron emission by relativistic electrons [24].

5. Theoretical and Spectroscopic Work

Analysis of the x-ray and EUV radiation requires accurate representations of the coronal plasma emissivities. Reviews of x-ray spectroscopy and the emissivity of stellar coronae can be found in [17]. For a recent summary of the important spectral lines in the 50–500 Å range and density-sensitive line ratios based on studies of the solar corona see [14]. One major complication in the analysis of stellar x-ray spectra is that solar (and by implication stellar) coronal abundances differ from those of the photosphere [13].

Analysis of low resolution x-ray stellar spectra obtained with the *Einstein* SSS and the EXOSAT TGS have been modelled with two-temperature coronae. *Einstein* IPC observations of 130 late-type stars can also be modelled with plasmas at two temperatures, including very hot coronae ($T > 10^{7.5}$ K) of RS CVn systems, but models with continuous emission measure distributions also fit the data and are physically more plausible [40]. A combined analysis of both the EUV and x-ray fluxes of many late-type dwarfs [50] leads to an upward revision of the temperatures in two-temperature coronal models by a factor of 2 to logT = 6.7 and 7.5. This may explain why the WFC has detected fewer M dwarfs than predicted on the basis of the earlier temperature estimates [38]. Higher resolution spectra from EUVE, ASCA, and future spacecraft are needed to test the validity of two-temperature coronal models. For a comprehensive summary of the energy budget requirements and theoretical mechanisms of chromospheric and coronal heating see [47]. Recent reviews of coronal heating mechanisms include those by Priest and Rosner in [25].

References

[1] Ayres, T.R., *et al.* 1994, submitted to ApJ.
[2] Benz, A.O. and Alef, W. 1991, *Astr. Ap.* **252**, L19.
[3] Bookbinder, J.A. 1991, Mem S.A.It **62**, 321.
[4] Brown, A., *et al.* 1991, *Ap. J.* **373**, 614.
[5] Chiuderi-Drago, F. and Franciosini, E. 1993, ApJ, in press.
[6] Dempsey, R.C., Linsky, J.L., Fleming, T.A., and Schmitt, J.H.M.M., 1993a, *Ap. J. S.***86**, 599.
[7] Dempsey, R.C., Linsky, J.L., Schmitt, J.H.M.M., and Fleming, T.A., 1993b, *Ap. J.* **413**, 333.
[8] Drake, S.A., Simon, T., and Linsky, J.L. 1992, *Ap. J. S.***82**, 311.
[9] Drake, S.A., *et al.* 1993, ApJ, in press.
[10] Doyle, J.G., *et al.* 1991, M.N.R.A.S. **248**, 503.
[11] Elias, N.M. and Güdel, M. 1993, *A. J.***106**, 337.
[12] Elvis, M. 1990, Imaging X-ray Astronomy: A Decade of *Einstein* Observatory Achievements (Cambridge: Cambridge U. Press).
[13] Feldman, U. 1992, Physica Scripta **46**, 202.
[14] Feldman, U., *et al.* 1992, *Ap. J. S.***81**, 387.
[15] Fleming, T.A., *et al.* 1993, *Ap. J.* **410**, 387.
[16] Fox, D., *et al.* 1994, A&A, in press.
[17] Gorenstein, P and Zombeck, M. 1990, High Resolution X-ray Spectroscopy of Cosmic Plasmas (Cambridge: Cambridge U. Press).

[18] Grillo, F., *et al.* 1992, *Ap. J. S.***81**, 795.
[19] Güdel, M. 1994, ApJS, in press.
[20] Güdel, M. and Benz, A.O. 1993, *Ap. J. (Letters)* **405**, L63.
[21] Güdel, M., *et al.* 1993, ApJ, in press.
[22] Haisch, B., *et al.* 1991, *Ap. J. (Letters)* **383**, L15.
[23] Linsky, J.L. 1991, Mem S.A.It **62**, 307.
[24] Linsky, J.L., Drake, S.A., and Bastian, T.S. 1992, *Ap. J.* **393**, 341.
[25] Linsky, J.L. and Serio, S. 1993, Physics of Solar and Stellar Coronae: G.S. Vaiana Memorial Symposium (Dordrecht: Kluwer).
[26] Maggio, A., *et al.* 1990, *Ap. J.* **348**, 253.
[27] Malina, R.F. and Bowyer, S. 1991, in Extreme Ultraviolet Astronomy, ed. R.F. Malina and S. Bowyer (New York: Pergamon), 397.
[28] Malina, R. *et al.* 1993, ApJ, in press.
[29] Maran, S.P., *et al.* 1994, ApJ, in press.
[30] Massi, M. and Chiuderi-Drago, F. 1991, *Astr. Ap.* **253**, 403.
[31] Micela, G., *et al.* 1990, *Ap. J.* **348**, 557.
[32] Morris, D.H., Mutel, R,L., and Su, B. 1990, *Ap. J.* **362**, 299.
[33] Pallavicini, R. 1989, A&A Reviews **1**, 177.
[34] Pallavicini, R. 1991, in The Sun: a Laboratory for Astrophysics, ed. J.C. Brown and J.T. Schmelz.
[35] Pallavicini, R., Tagliaferri, G., and Stella, L. 1990, *Astr. Ap.* **228**, 403.
[36] Patten, B.M. and Simon, T. 1993, to appear in ApJL.
[37] Pfeffermann, E., *et al.* 1987, in Soft X-ray Optics and Technology, Proc. SPIE **733**, 519.
[38] Pounds, *et al.* 1993, M.N.R.A.S. **260**, 77.
[39] Reimers, D. and Schmitt, J.H.M.M. 1992, *Ap. J. (Letters)* **392**, L55.
[40] Schmitt, J.H.M.M., *et al.* 1990a, *Ap. J.* **365**, 704.
[41] Schmitt, J.H.M.M., *et al.* 1990b, *Ap. J.* **351**, 492.
[42] Schmitt, J.H.M.M., *et al.* 1993, *Ap. J. (Letters)* **402**, L13.
[43] Siarkowski, M. 1992, MNRAS **259**, 453.
[44] Simon, T. and Drake, S.A. 1993, to appear in AJ.
[45] Stern, R.A., *et al.* 1992, *Ap. J. (Letters)* **399**, L159.
[46] Trümper, J., *et al.* 1983, Nature **349**, 579.
[47] Ulmschneider, P., Priest, E.R., and Rosner, R. 1990, Mechanisms of Chromospheric and Coronal Heating (Berlin: Springer-Verlag).
[48] Umana, G., Catalano, S., and Rodonò, M. 1991, *Astr. Ap.* **249**, 217.
[49] Usov, V.V. and Melrose, D.B. 1992, *Ap. J.* **395**, 575.
[50] Wood, B.E., Brown, A., and Linsky, J.L. 1993a, submitted to ApJ.
[51] Wood, B.E., *et al.* 1993b, submitted to ApJS.

DYNAMICAL ATMOSPHERES AND WINDS (Lee Anne Willson)

Advances have occurred in modeling convection in stellar atmospheres; also in modeling the atmospheres of pulsating stars, of stars with dusty winds, and of hot stars with non-spherical winds. These are reviewed below. Other advances reported during the current triennium but not reviewed here include studies of (a) heating of stellar chromospheres and coronae by various types of waves; (b) conditions for the propagation of acoustic and Alfvén waves in the solar and stellar atmospheres; (c) observational constraints on model atmospheres and winds for evolved stars; (d) relativistic outflows in supernovae and from around neutron stars; (e) winds generated in explosive events – novae and supernovae; (e) winds in interacting binary systems; and (f) the detailed hydromagnetic structure of the solar atmosphere and wind. The summaries below also do not include papers that are primarily interpretation of data or modeling for a single star.

Detailed simulations of convective motions in the solar and stellar atmospheres have been carried out and compared with observations by Dravins and Nordlund in a series of papers on "Stellar Granulation" (Dravins 1990, Nordlund and Dravins 1990). Methods for incorporating the dynamical effects of multiperiodic, aperiodic, or stochastic spectra of wave motions, such as those expected to be stimulated by convection or multi-mode pulsation, are discussed in papers by Gail,

Cuntz, and Ulmschneider (1990) and Koninx and Pijpers (1992).

For many cool giants and supergiants, winds are driven by a combination of the effects of pulsation and radiative forces on grains. It is important to include both pulsation (with a careful treatment of the thermal physics and post-shock nonequilibrium "chemistry) and the effects of dust if the systematics of observed mass loss for AGB stars (Miras) are to be simulated accurately, although for the shorter period, less luminous stars the mass loss process is less dependent on the dust (Bowen and Willson 1991). Sedlmayr and collaborators have been calculating details of the dust formation process in dynamical atmospheres with large excesses of C/O, and related calculations have been carried out by Dorfi and collaborators (Höfner & Dorfi 1992). In such atmospheres (which must be rare, as most observed carbon stars have C/O close to 1), outflows are unstable and may develop shocks even in the absence of pulsation (Fleischer, Gauger and Sedlmayr 1993) possibly leading to a natural explanation for the episodic mass loss observed in R CrB stars (Goeres and Sedlmayr 1992). Netzer and Elitzur (1993) find that for normal chemistry, for both M and C stars, dust driven winds are only possible if M > 10-7 Ms/yr, and dust- driven winds from carbon stars cannot exceed 10-4 Ms/yr; however, these results may depend critically on their assumption of a single grain size applicable to both types of stars. For the substantially warmer and less luminous RR Lyrae stars, Fokin (1992) has carried out numerical models including synthesis of Ha absorption line profiles; no mass loss occured, but models were not extended far enough out to allow for discovery of mass loss rates below about 10-9 Ms/yr. None of the cited dynamical models has yet incorporated all of the detailed physics that is known to affect the results: details of heating and cooling due to shocks and due to interaction of dust grains with ambient gas; effects of radiative forces on molecules; tracing the complete process of dust grain nucleation and growth; rigorous non-LTE radiative transfer and spectral synthesis; extension of the models far enough out to detect wind formation; and inclusion of both sub-photospheric driving and atmospheric response (including damping of the pulsation). However, models are increasingly including several or even most of these effects. Thus this remains an active area of investigation, with a wealth of observational data waiting to be interpreted more rigorously.

For hot stars, key papers concerning the 2D and 3D structure of the atmospheres and wind appeared during 1990-93. Bjorkman and Cassinelli (1993) find that equatorial disks form naturally as a result of wind driving in the presence of rotation if v/vcrit > .9 for O stars and > .5 to .6 for B2 stars, where the effect is maximal. Chen, Marlborough and Waters (1992) inverted observed v(r) to derive constraints on the form of any force(s) besides gravity, rotation, and gas pressure gradient that might be present. Mazzali (1990) explored effects of stellar rotation on the P Cygni line profiles. Kudritzki (1992) included both microturbulance and transsonic outflow in computing weak lines for O stars. Boyd and Marlborough (1991) computed polarization for rotationally distorted, radiation driven stellar winds for B[e] stars, while Taylor and Cassinelli (1992) pointed out that "polarization cancellation" could occur in certain cases for WR stars with axisymmetric, two-component winds. Poe, Owocki and Castor (1990) computed non-Sobolev, pure absorption models for the winds and found that these have no uniquely defined steady state. Owocki and Zank (1991) examined the effects of viscosity on such flows, and found that it could stabilize some flows. Hummer and Rybicki (1992) find that the criteria for using Sobolev theory in expanding winds are difficult to meet if transfer takes place mainly in the Voigt wings and partial or complete frequency redistribution is considered. Instabilities in hot star winds were also investigated by Owocki and Rybicki (1991) and Lobel, Achmad, de Jager and Nieuwenhuizen (1992), and Drew (1990) pointed out that apparent contradictions between mass loss rates and velocity laws derived from theoretical models and from Ha observations could be resolved if such instabilities give rise to density inhomogeneities near the star.

References

Bjorkman, J. E. and J. P. Cassinelli (1993). Formation Around Rotating Stars Due to Ram Pressure Confinement by the Stellar Wind. *Ap. J.* **409**, 429-449.

Bowen, G. H. and L. A. Willson (1991). From Wind to Superwind: The Evolution of Mass-Loss Rates for Mira Models. *Ap. J.* **375**, !53-L56.

Boyd, C. J. and J. M. Marlborough (1991). Continuum Polarization by Electron Scattering in Rotationally Distorted, Radiation-Driven Stellar Winds: B[e] Supergiants. Ap. J. 369: 191-199.

Chen, H., J. M. Marlborough, and L. B. F. M. Waters (1992). Dynamics of the Envelopes of Be

Stars in the Equatorial Plane. *Ap. J.* **384**, 605-612.
Dravins, D. (1990). Stellar granulation VI. Four-component models and non-solar-type stars. *Astr. Ap.* **228**, 218-230.
Drew, J. E. (1990). Relation Between Mass-Loss Rate and H-alpha Emission for O Star Winds. *Ap. J.* **357**, (No 2), 573-581.
Fleischer, A. J., A. Gauger, and Sedlmayr. (1992). Circumstellar dust shells around long-period variables. I. Dynamical Models of C-stars Including Dust Formation, Growth and Evaporation. *Astr. Ap.* **266**, 321-339.
Fokin, A. B. (1992). Shock waves and H-alpha profiles in the hydrodynamical model of RR Lyrae'. M. N. R. S. 256, 26-36.
Gail, H.-P., M. Cuntz, and Ulmschneider. (1990). Wave pressure in stellar atmospheres due to shock wave trains. *Astr. Ap.* **234**, 359-365.
Goeres, A. and E. Sedlmayr (1992). "The envelopes of R Coronae Borealis stars I. A physical model of the decline events due to dust formation", A & A 265, 216-236.
Höfner, S. and E. A. Dorfi (1992). Dust formation in winds of long-period vaariables II. dust energetics, optical properties, chemistry, time scales. *Astr. Ap.* **265 (No 1)**, 207-215.
Hummer, D. G. and G. B. Rybicki (1992). The Sobolev approximation for line formation with partial frequency redistribution. *Ap. J.* **387**, 248-257.
Koninx, J.-P. M. and F. P. Pijpers (1992). The applicability of the linearized theory of sound-wave driven winds. *Astr. Ap.* **265**, 183-195.
Kudritzki, R. P. (1992). The formation of weak metal lines in the atmospheres of O-stars. The role of microturbulence and transonic matter outflow. *Astr. Ap.* **266**, 395-401.
Lobel, A., L. Achmad, de Jager and Nieuwenhuizen. (1992). On the instability parameters of stellar atmosphers. *Astr. Ap.* **264**, 147-152.
Lucy, L. B. and D. C. Abbott (1993). Multiline transfer and the dynamics of Wolf-Rayet winds. *Ap. J.* **405**, 738-746
Mazzali, P. A. (1990). The effect of stellar rotation on the properties of a radiatively driven stellar wind. *Astr. Ap.* **238**, 191-206.
Netzer, N. and M. Elitzur (1992). The dynamics of stellar outflows dominated by interaction of dust and radiation. *Ap. J.* **410**, 701-713.
Nordlund, Å. and D. Dravins (1990). Stellar granulation III. Hydrodynamic model atmospheres. *Astr. Ap.* **228**, 155-183.
Owocki, S. P. and G. B. Rybicki (1990). Instabilities in line-driven stellar winds. V. Effect of an optically thick continuum. *Ap. J.* **368**, 261-263.
Owocki, S. P. and G. P. Zank (1991). The effect of viscosity on steady transonic flow with a nodal solution topology. *Ap. J.* **368**, 491-503.
Poe, C. H., S. P. Owocki, and Castor. (1990). The steady state solutions of radiatively driven stellar winds for a non-Sobolev, pure absorption model. *Ap. J.* **358**, 199-213.
Taylor, M. and J. P. Cassinelli (1992). Polarization cancellation in the two-component winds from Wolf-Rayet stars. *Ap. J.* **401**, 311-316.
Velli, M. (1993). On the propagation of ideal, linear Alfvén waves in radially stratified stellar atmospheres and winds. A & A 270: 304-314.

DISK AND ENVELOPE EMISSION FROM YOUNG STELLAR OBJECTS (Lee Hartmann)

The present paradigm for T Tauri stars is that of a low-mass, pre-main sequence star surrounded by a dusty disk [11,2]. The disk accounts for the infrared excesses of these stars, powered by a combination of accretion energy and absorption of light from the central star. Accretion probably also provides the energy source for the blue-ultraviolet continuum emission, either in a narrow boundary layer between the Keplerian disk and the slowly-rotating central star [4,9], or at the base of an accretion column [10]. The mass of a typical T Tauri disk is uncertain; the best estimates suggest masses at least the size of the minimum mass solar nebula $\sim 10^{-2} M_\odot$ [5].

Although the disk model for T Tauri stars seems fairly firmly grounded, there is at least one aspect of great concern, namely the "spectral index" problem. Defining the spectral index $s = dlogF_\lambda / dlog\lambda$, both steady accretion disks [11] and flat disks heated only by light from the central star [2,3] have

$s = -4/3$. In contrast, the median T Tauri spectral energy distribution has $s \sim -2/3$ [9], and a small but significant number of objects have $s \sim 0$ [3,5,9]. It has become common practice to *assume* that the observed infrared radiation arises in a disk, and assign an appropriate temperature distribution to it $T \propto R^{-m}$. Such a disk will then exhibit a spectral index $s = 2/m - 4$, so for example the "flat spectrum" sources $s \sim 0$ must then have $m = 1/2$, rather than the standard $m = 3/4$. It is then up to theory to try to explain why such a temperature distribution occurs.

So far theory has not really been up to the task, particularly for the "flat spectrum" sources. Gravitational instabilities of azimuthal quantum number = 1 in relatively massive disks have been suggested [15,1], but given the sensitivity of this instability to the outer boundary condition it is not clear how easily these modes are maintained, nor has it been demonstrated that the dissipation produced produces the required temperature distribution. Indeed, there are concerns that such a mechanism would require an unrealistically large accretion rate in the outer disk [9]. The main problem is that, for a "flat spectrum" source, the implied energy release per logarithmic radius interval is constant. This is difficult to achieve, especially in the outer radii of pre-main sequence disks, which are three to four orders of magnitude larger than the inner radii, and thus the gravitational potential energy per unit mass available at such large distances from the central star is small (unless the disk is very massive, which seems unlikely).

An alternative suggestion [9] is that gaseous disks should be "flared", i.e. concave toward the star, and so could be heated more by irradiation at large distances than a flat disk would be. Irradiation by the central star is an attractive mechanism for tapping the energy liberated deep in the gravitational potential well of the system, and redistributing it to much larger radial distances. However, the flared disk explanation suffers the difficulty that dust is expected to coagulate and settle to the midplane of the disk fairly rapidly. Since dust provides all of the absorptive opacity, the disk photosphere would be flat and no enhancement of heating would occur.

Recent suggestions to solve this problem involve irradiation of another dusty component by the central star. Safier [13] suggested that a dusty wind ejected from the disk could be the source of the infrared radiation. Natta [16] offered a variant of this idea, in which the envelope itself is relatively optically thin, but it can scatter a significant fraction of the light down into the opaque, dense disk; the enhanced temperature in the disk produces the required excess infrared radiation. Whitney & Hartmann [20] showed that the optical and near-infrared scattered light patterns of some young stellar objects imply extensive surrounding, possibly infalling, dusty envelopes. Some of these objects are "flat spectrum" sources, so that the scattered light observations [19] indicate that extended dusty envelopes as well as disks must be included in any consideration of the infrared emission. The final resolution of the spectral index problem is not yet at hand, but it is a good bet that the answer will involve some sort of dusty envelope.

The success of the disk paradigm for low-mass stars has naturally led to suggestions that disk formation might also be important for higher-mass stars. Several attempts have been made to model the spectral energy distributions of the young Herbig Ae/Be stars. Natta *et al.* [17,18] have focussed on "flat-spectrum" sources among the Ae/Be objects, modelling them by a combination of "flat-spectrum" disk and a tenuous outer envelope. Given the considerations mentioned above, these results are probably rather tentative, especially since all of these objects are heavily embedded in dust, with large, extensive scattered light envelopes at optical and near-infrared wavelengths.

The analyses of Hillenbrand *et al.* [8] and Lada & Adams [12] suggested that many Ae/Be stars do have massive disks, and that these disks accrete rapidly (as much as $\sim 10^{-6} - 10^{-5} M_\odot\,yr^{-1}$). These disks must have inner "holes" of $\sim 10 R_*$ to account for the general existence of a "bump" or peak in the infrared emission near $\lambda \sim 3\mu m$. These authors suggested that this "hole" may not be real, but may simply be where the dust gets destroyed by the high-temperature radiation fields of the star and disk (T$\gtrsim$1600K).

However, Hartmann *et al.* [7] pointed out that at the high accretion rates required, even if dust is destroyed the *gas* in the inner disk will be optically thick, and so no "hole" will appear in the spectral energy distribution. Furthermore, if the material is to accrete onto the star it is necessary to radiate much more energy in some wavelength band than is readily observable. For example, if the "inner edge" of the disk is at $10R_*$, but material is accreting all the way to $1R_*$, then 90% of the energy liberated is missing. It is difficult to conceive of putting this energy into high-frequency radiation; although such radiation would be difficult to observe directly because of the large extinction to these objects, one would expect to observe some line radiation of high-temperature species (He I or He II) in the optical which are not seen. Letting the material pile up at the "inner edge" of the disk does not seem to be acceptable either, given the high accretion rates required to match the observations.

Hartmann *et al.* [7] suggested that the infrared excess emission might be due to dusty envelopes, either around the optical object or an infrared companion. They suggested that the $3\mu m$ bump might be produced preferentially by transiently-heated small dust grains. Such grains have been invoked to explain the relatively high-temperature near-infrared thermal emission observed at large distances from B stars (e.g. [14]). A hard photon $\gtrsim 10eV$ can transiently heat a small grain to temperatures well in excess of radiative equilibrium, and some high-temperature emission can result before the grain cools off. This mechanism will only work for high-temperature central stars, and it is suggestive that the $3\mu m$ peaks are only seen in the Ae/Be stars, but not in the cooler T Tauri stars. In addition, some Ae/Be stars exhibit near- to mid-infrared emission bands attributed to PAHs (small clusters of polyaromatic hydrocarbon molecules) [6], supporting an model of a dusty envelope containing small grains for some (but by no means all) objects.

The dusty envelope model for Ae/Be stars is not without its own problems, such as how to account for the dynamical state of the dusty envelope on size scales $\sim 2000 - 200$ A.U., or whether the small grains can survive disruption by the intense radiation fields of the central stars. However, straightfoward application of the disk model raises comparable, if not worse, difficulties. Disk accretion is an attractive possibility for young, intermediate-mass stars, but this cannot be clearly demonstrated yet.

References

[1] Adams, F. C., Ruden, S. P., & Shu, F. H. 1989, ApJ, 347, 959
[2] Adams, F. C., Lada, C. J., & Shu, F. H. 1987, ApJ, 312, 788
[3] Adams, F. C., Lada, C. J., & Shu, F. H. 1988, ApJ, 326, 865
[4] Basri, G., & Bertout, C. 1989, ApJ, 341, 340
[5] Beckwith, S., Sargent, A., Chini, R., & Gusten, R. 1990, AJ, 99, 1024
[6] Brooke, T.Y., Tokunaga, A.T., & Strom, S.E. 1993, AJ, in press
[7] Hartmann, L., Kenyon, S.J., & Calvet, N. 1993, ApJ, 407, 219
[8] Hillenbrand, L.A., Strom, S.E., Vrba, F.J., & Keene, J. 1992, ApJ, 397, 613
[9] Kenyon, S. J., & Hartmann, L. 1987, ApJ, 323, 714
[10] Königl, A. 1991, ApJL 370, L39
[11] Lynden-Bell & Pringle 1974, MNRAS, 168, 603
[12] Lada, C.J., & Adams, F.C. 1992, ApJ, 393, 278
[13] Safier, P. 1993, ApJ, in press
[14] Sellgren, K. 1984, ApJ, 277, 623
[15] Shu, F. H., Tremaine, S., Adams, F. C., & Ruden, S. P. 1990, ApJ, 358, 495
[16] Natta, A. 1993, ApJ, in press
[17] Natta, A., Palla, F., Butner, H.M., Evans, N.J. II, Harvey, P.M. 1992, ApJ, 391, 805
[18] Natta, A., Palla, F., Butner, H.M., Evans, N.J. II, Harvey, P.M. 1993, ApJ, 406, 674
[19] Weintraub, D.A., Kastner, J.H., Zuckerman, B., & Gatley, I. 1992, ApJ, 391, 784
[20] Whitney, B. A., & Hartmann, L. 1993, ApJ, 402, 605

THE ATMOSPHERIC STRUCTURE OF VERY COOL STARS (Takashi Tsuji)

Most stars eventually evolve to very cool stars (red giants and supergiants, AGB stars), where products of stellar nucleosynthesis are mixed to the stellar surface and returned to interstellar space by large mass loss. On the other hand, very low mass stars appear to be very cool too, and they carry a large fraction of the mass of galaxies, possibly including the so-called dark matter. To clarify all these basic roles played by very cool stars, understanding of the atmospheric structure is crucial, since it provides the important basis for interpreting observations of various kinds.

Modeling the atmospheric structure for very cool stars, however, has been more difficult than for hotter stars, because of the complicated molecular opacities, among others. Recently, however, considerable progress has been achieved and it may be not long before classical model atmospheres of very cool stars will be quite sophisticated.

Molecular opacities

The opacity problem is of fundamental importance in the theory of stellar atmosphere, and especially molecular opacity is important in cool stars. Recent progress on molecular opacities and related topics is well reviewed by two meetings: one is "Workshop on Astrophysical Opacities", Caracas, July 1991 (Proceedings ed. Lynas-Gray et al. 1992) and the other is the 146-th IAU Colloquium on "Molecular Opacities in the Stellar Environment", Copenhagen, May 1993 (Proccedings ed. Jørgensen, 1993). Some of the impressive progress reported in these meetings is as follows. First, ab initio computations of molecular structures are now mature enough to provide accurate line positions and intensities for numerous transitions of astrophysically important molecules (e.g., H_2O, HCN, C_3). This will constitute a counter part of dedicated efforts done for atomic lines by Kurucz. Second, progress in experimental molecular physics also provides such basic data as thermodynamical, intensity, and spectroscopic data. Especially, the Fourier Transform Spectroscopy (FTS), pioneered by P. and J. Connes in the 1960's, is now revolutionizing laboratory molecular spectroscopy and providing important data of high precision for molecules of astrophysical interests. Third, helped by the recent computer technolgy, all these results made it possible to develop detailed line-by-line databases for molecular spectra. There is no doubt that the completion of such a database is of basic importance not only for our understanding of the atmospheric structure of cool stars but also for interpreting observational data of high precision to be obtained by new technology, both on the ground and in space.

Model Atmospheres of Evolved Cool Stars

Efforts of modeling the atmospheres of very cool stars continued. For example, the effects of sphericity have now been investigated with elaborate opacites both for oxygen-rich (Pletz et al., 1992) and carbon-rich (Jørgensen et al., 1992) giant stars. Now, classical model atmospheres of cool (super)giant stars may soon be regarded as quite sophisticated by the use of more accurate and realistic molecular opacities reflecting the recent progress mentioned above. The next problem is to test how well classical models represent real stellar atmospheres.

[Also, such a modeling of the stellar photosperes is an important prerequisite for modeling the outer atmospheres of cool evolved stars. For example, a classical issue if the radiative force on molecules could accelerate the mass-loss outflow in cool luminous stars is examined on the basis of such a modeling for the case of cool carbon stars (Jørgensen & Johnson, 1992). Further, diverse efforts for modeling the outer atmospheres are being done. Just to mention a few examples: formation of molecules and dust in the inner shell of oxygen-rich supergiants by considering the effect of chromospheric UV radiation (Beck et al. 1992), thermal structure of the outer envelope of supergiant stars by considering the cooling due to fine structure line emission (Rodgers and Glassgold, 1991), spectroscopic diagnostic of the inner shell of the dusty carbon star by the use of infrared molecular line profiles (Keady and Ridgway, 1993) etc.]

Red and Brown Dwarfs

Recent increase of the sensitivity in astrophysical observations made it possible to observe stars at the lower end of the main sequence and, further, possible substellar objects named brown dwarfs. Observers, however, claim that there is no model to explain their observations satisfactorily, and effective temperatures of the coolest M dwarfs estimated by different authors differ by as much as 500K! (e.g., Berriman et al., 1992; Kirkpatrick et al., 1993; Tinney et al., 1993). Also, despite extensive searches by various methods, no definite identification of brown dwarfs can be done yet (e.g., Stevenson, 1991; Burrow and Liebert, 1993). Is something wrong in our basic understanding of such very cool and very low luminosity objects? Clearly, more efforts of modeling are needed in this field, and understanding of the atmospheric structures is an important basis to predicting observable properties of these objects.

References

Beck H.K.B., Gail H.-P., Henkel R., Sedlmayr E., 1992, Astron. Astrophys., 265, 626

Berriman G.B., Reid I.N., Leggett S.K., 1992, Astrophys. J., 392, L31
Burrow A., Liebert J., 1993, Rev. Mod. Phys., 65, 301
Jørgensen U.G.,(ed.), 1993, Proccedings of the 146-th IAU Colloquium on "Molecular Opacities in the Stellar Environment", Lecture Notes in Physics, Springer-Verlag, in press
Jørgensen U.G., Johnson H.R., 1992, Astron. Astrophys. 265, 168
Jørgensen U.G., Johnson H.R., Nordlund Å., 1992, Astron. Astrophys. 261, 263
Keady J.J., Ridgway S.T., 1993, Astrophys. J., 406, 199
Kirkpatrick J.D., Kelly D.M., Rieke G.H., Liebert J., Allard F., Wehrse R., 1993, Astrophys. J., 402, 643
Lynas-Gray A.E., Mendoza C., Zeippen C.J.,(ed), 1992. Proceedings of the Workshop on Astrophysical Opacities, in Rev. Mexicana Astron. Astrof., vol.23, special issue
Plez B., Brett J.M. Nordlund Å., 1992, Astron. Astrophys. 256, 551
Rogers B., Glassgold A.E., 1991, Astrophys. J., 382, 606
Stevenson D.J., 1991, Ann. Rev. Astron. Astrophys., 29, 163
Tinney C.G., Mould J.R., Reid I.N., 1993, Astron. J., 105, 1045

MULTIDIMENSIONAL RADIATIVE TRANSFER (Rainer Wehrse)

Radiative transfer in three spatial dimensions (usually abbreviated "3D radiative transfer") is needed in the modeling and spectral analysis of objects that have both optically thick and optically thin parts and that do not exhibit point or axial symmetries. In addition, it turns out that it is often advantageous to use 3D radiative transfer methods also for cylindrical media (see below). The determination of radiation fields in such media is basically a six-dimensional problem: there are three spatial and two angle dimensions and there is the frequency. Usually, all points in the 6D space are coupled: the spatial coordinates by the transport operator, the angles and the frequency by the source function. For moving media there is additional coupling by the transport operator. Due to this high dimensionality, methods that are presently sucessfully used for plane-parallel or spherical media can hardly be generalized, and new methods –usually less accurate and much more time and memory consuming – have to be devised. An exception is the Monte-Carlo method (see below) that can be set up in a very similar way for all the various cases.

Note that multidimensional radiative transfer is required in astronomy e.g for the modeling of dust clouds, inhomogeneous gaseous nebulae, or accretion disks but also e.g. in the evaluation of remote sensing data or in the study of cloud formation in the earth's atmosphere.

For a static three dimensional non-refracting medium the radiative transfer equation is written

$$\frac{dI_\nu}{ds} = \chi_\nu(I_\nu - S_\nu) \tag{1}$$

($\mathbf{s}$ = vector in ray direction, I_ν = specific intensity at frequency ν, χ = extinction coefficient per unit pathlength, S_ν = source function). In a Cartesian coordinate system it takes the particularly convenient form

$$\mathbf{n}\cdot\nabla I_\nu = \chi_\nu(I_\nu - S_\nu) \tag{2}$$

($\mathbf{n} = \mathbf{s}/\mid \mathbf{s} \mid$= unit vector in ray direction), since here the angles between the ray directions and the coordinate axes are the same everywhere and therefore no derivatives with respect to angle occur. The transfer equation in various other coordinate systems has been derived by Jones and Bayazitoglu (1993). The existence and uniqueness of solutions has recently been shown by Johnson and Pitkäranta (1983) and in a more general way by Führer (1993).

For moving media –as in the 1D case– it is best in most cases to describe the radiation field in the comoving frame. Even for simple velocity fields the transfer equation becomes very complicated, in particular, derivatives of the specific intensity with respect to frequency and angles have to be added. Except for cases of pure absorption only solutions for accretion disks in which the advection and aberration terms are neglected have been obtained up to now (Baschek et al., 1993).

Since analytical solutions are not available (except for some trivial special cases) equation (1) has to be solved numerically. The following classes of methods have been used:

1. Monte-Carlo-methods

In this method (for details see e.g. Langer et al., 1978) the motion of photons through the medium is simulated. It is highly flexible and easy to program. Essentially all photon-matter interactions and geometries can be incorporated without problem. The disadvantage, as in the 1D case, is the low accuracy, in particular for very optically thin and very optically thick regions in spite of recent improvements in the convergence behavior (cf. Pozdnyakov et al., 1983).

2. Ray-tracing or characteristics methods

This method has been used e.g. by Efstathiou and Rowan-Robinson (1990), Steffen and Freitag (1991), Nordlund and others. It is very accurate in principle since it is based on the formal solution of the 1D problem. In practice, however, its performance is reduced by an extremely high number of necessary interpolations since the values of the absorption and scattering coefficients as well as of the source function are usually given on a prescribed grid whereas the rays pass between the grid points. In addition, the calculation of the source function in the case of high scattering fraction is a problem as it has been for grid methods.

3. Grid methods

In these methods either finite differences (Myneni et al., 1990; Stenholm et al., 1991) or finite elements (Turek, 1993, Turek and Wehrse, 1993) are used to convert the partial integro-differential equation (1) in a (usually very large) system of linear algebraic equations that is subsequently solved by iteration. The method recently developed by Turek seems presently to be the most sophisticated since (i) by upwinding it ensures stability, (ii) by using unstructured grids it can well adapt to the specific situation, (iii) by employing non-conforming finite elements the accuracy is further improved, and (iv) by utilizing preconditioning and a modern variant of the conjugate gradient method for the solution of the final linear system of equations even for very large scattering fractions a very good convergence rate can be achieved. In addition, a maximal absolute difference between the solution of the continuous and the discretized problem less than a prescribed number can be guaranteed.

In addition to these methods in simple cases (where e.g. only very few angle grid points are required and in one coordinate the discretisation can be coarse) also the line method (see Schiesser, 1990, for a detailed description of the general method and Ben Jaffel and Vidal-Madjar, 1989, for an implementation) can be applied.

References:

Baschek, B., Papkalla, R., Wehrse, R. 1993, in: Proc. 2nd. Haifa Conf. on Cataclysmic Variables, Regev, O., Shaviv, G., eds.
Ben Jaffel, L., Vidal-Madjar, A. 1989, Astron. Astrophys. 220, 306
Efstathiou, A., Rowan-Robinson, M. 1990, M.N.R.A.S. 245, 275
Führer, C. 1993, diploma thesis, Inst. f. Appl. Math., Heidelberg University
Johnson, C., Pitkäranta, J. 1983, SIAM J. Num. Anal. 20, 951
Jones, P.D., Bayazitoglu, Y. 1993, J.Q.R.S.T. 48, 427
Langer, S.H., Ross, R.R., McCray, R. 1978, Astrophys. J. 222, 959
Myneni, R.B., Asrar, G., Gerstl, S.A.W. 1990, Transp. Theory St. Phys. 19, 205
Nordlund, Å. 1982, Astron. Astrophys. 107, 1
Pozdnyakov, L.A., Sobol', I.M., Sunyaev, R.A. 1983, Sov. Sci. Rev. (Sect. E) 2, 189.
Schiesser, W.E. 1991, The Numerical Method of Lines, Academic Press
Steffen, M., Freytag, B. 1991, Rev. Modern Astron. 4, 43
Stenholm, L.G., Störzer, H., Wehrse, R. 1991, J.Q.S.R.T. 45, 47
Turek, S. 1993, Impact Comp. Sc. Eng., in press
Turek, S., Wehrse, R. 1993, Astron. Astrophys. submitted

37. STAR CLUSTERS and ASSOCIATIONS (AMAS STELLAIRES et ASSOCIATIONS)

PRESIDENT: J.-C. Mermilliod
VICE-PRESIDENT: A. Feinstein

ORGANISING COMMITTEE: R. Buonanno, J.J. Claria, G. Da Costa, G. Harris, K. Janes, G. Meylan, D. Vandenberg, J. Zhao

1 Introduction

The last three years have been very productive for cluster research. More than 450 references for open clusters and more than 800 for globular clusters have been collected from the literature and Astronomy and Astrophysics Abstracts (Vol. 52 to 56). The observations present very broad scope thanks to the new instrumental facilities like infrared arrays and the X-ray and IR satellites. Our knowledge of how and where open clusters form has been considerably advanced with the advent of infrared arrays images which has revealed the presence of compact, centrally condensed clusters of newly formed stars embedded in, or closely associated with dense cores of giant molecular clouds. The present report tries to survey the activity in these domains, summarize the results and trends, and collect the information on the observed clusters and associations. But, owing to the large number of references, it is not possible to quote every paper. A higher priority has been given to the new data obtained for star clusters, which should help everybody to be easily informed of the observations made during the past three years. Therefore the traditional form of the report has been largely maintained. Abstracts and papers in conference proceedings have usually not been quoted here to save place and no attempt has been made to include unpublished material. For complete bibliographic information, the Astronomy and Astrophysics Abstracts, chapters 121 (young stellar objects), 131 (star formation), 152 (stellar associations), 153 (open clusters), 154 (globular clusters), and 156 (Magellanic Clouds) remain the best source. To improve the efficiency of literature retrieval, it would be important that the first keyword refers to open, respectively globular clusters, whatever the observational data may be (proper motions, radial velocity, photometry and so on). The database for stars in open clusters (Mermilliod 1992) offers an up-to-date bibliographic search facility based on keywords. Murtagh and Adorf (1993) have given a short report of the astronomical literature publicly accessible on-line. The subject of low-mass star formation in southern molecular clouds has been reviewed in a comprehensive ESO report edited by Reipurth (1991), which provides extensive references to the literature.

Mermilliod J.-C. 1992, Bull Inform. CDS 40, 115
Murtagh F., Adorf H.-M. 1993, The Messenger, no 72, 45
Reipurth B. 1991, ESO Scientific Report no 11

2 Symposia, Colloquia and Workshops

No IAU symposia or colloquia directly related to star clusters have been held during this period, but several meetings have taken place.
- "Formation and evolution of star clusters", 102nd ASP Meeting, (Boston, July 14-15, 1990) edited by K. Janes (ASP Conf. Ser. vol 13, 1991);
- "Young star clusters and early stellar evolution" held in Vulcano (Sept. 16-20, 1991) edited by F. Palla, P. Persi and H. Zinnecker (Mem. Soc. Astr. It. 62 no 4, 1991);
- "Star clusters and stellar evolution", held in Teramo (Sept. 18-20, 1991), edited by E. Brocato, F.R. Ferraro and G. Piotto (Mem. Soc. Astr. It. 63 no 1, 1992);
- "The globular cluster-Galaxy connection" (1992 Santa Cruz summer Workshop) edited by J. Brodie and G. Smith, ASP Conf. Ser. vol. 48 (1993);
- "Dynamics of globular clusters" (held in July 1992), edited by S. Djorgovski and G. Meylan, ASP Conf. Ser. vol. 50 (1993).

Other meetings, at which the subject of star clusters figured more or less prominently, include:
- "The Magellanic Clouds", IAU Symp. no 148, (Sydney, July 9-13, 1990), edited by R. Haynes and D. Milne (Kluwer Academic Publishers);
- "Astrophysical ages and dating methods" (Paris, June 26-30, 1989), edited by E. Vangioni-Flam, M.

J. Bergeron (ed.), Reports on Astronomy, Vol. XXIIA, 425–444.

Cassé, J. Audouze and J. Tran Thanh Van (Editions Frontières, 1990);
- "Angular momentum evolution of young stars", Nato Advanced Research Workshop held in Noto (Sept. 17-21, 1990), edited by S. Catalano and J.R. Stauffer (Kluwer Academic Press).

3 Catalogues of cluster data

The database for stars in open clusters (Mermilliod 1992) has been developed continuously and a dozen copies have been distributed. Collaboration on this project is welcome. A solution is searched for to make it available more easily to the astronomical community. The 5th edition of Lynga's catalogue has not been updated, and it would be useful to elaborate a new version because better or new parameters for many clusters have been obtained since 1987. Garmany and Stencel (1992) have critically evaluated the reality of 25 OB associations ($55^o < l < 150^o$). They give lists of the member stars and their parameters, and of red supergiant candidates for association membership. Carpenter et al (1993) have produced a list of embedded clusters based on luminous IRAS point sources. Pasztor et al (1993) have searched for embedded clusters in the Cepheus-Cassiopeia region also from the IRAS point source catalogue. Fusi Pecci et al (1992) have produced a catalogue of 425 candidate blue stragglers in 21 globular clusters.
Due to the large amount of observational data available for galactic globular clusters and Magellanic Cloud clusters resulting, for a large part, from the extensive application of CCD photometry to the study of these objects, it would be important to develop specialized databases to keep this information publicly accessible in machine-readable form, through the now extended network facilities.

Carpenter J.M., Snell R.L., Schloerb F.P., Skrutskie M.F. 1993, ApJ 407, 657
Fusi Pecci F., Ferraro F.R., Corsi C., Buonanno R. 1992, AJ 104, 1831
Garmany C.D., Stencel R.E. 1992, A&AS 94, 211
Mermilliod J.-C. 1992, Bull Inform. CDS 40, 115
Pasztor L., Toth L.V., Balasz L.G. 1993, A&A 268, 108

4 Stellar associations and star-forming regions.

4.1 Observational trends

The list of observations in the various types of stellar associations and embedded clusters may seem small, although this field of astronomy has been subject to the most impressive development during the past years. It represents one of the most exciting research domains in the near future. This is partly due to the selection of thoses references dealing with the stellar component of associations and star-forming regions. Indeed a large part of the observation work is devoted to the mapping and investigation of the parent giant molecular clouds. While photometric (UBV, uvby, Walraven) or spectroscopic observations are still done to improve the member selection, find new and fainter members, and better define their properties (binarity, peculiarity, activity), the research tends to multi-disciplinarity. More attention is given to the galactic environement and the relation between the formed stars and the parent molecular clouds. Space observations comprising the X-ray, UV and IR wavelength domains have been obtained and will be published in the near future.
Interesting review papers have been written by Larson (52.131.277), Lada & Lada (54.131.223). Important contributions are presented in the NATO ASI series volume devoted to "The physics of star formation and early stellar evolution" edited by C.J. Lada and N.D. Kylfis (1991, Kluwer Academic Press) and in the III Canary Islands winter school of astrophysics "Star formation in stellar systems" edited by G. Tenorio-Tagle, M. Prieto and F. Sánchez (1992, Cambridge Univ. Press). The subject of star formation is reviewed by Shu (54.131.340) from a theoretician's point of view, and the observations are presented by Evans (54.131.341).
Cloud mapping with infrared arrays leaded to the discovery of several embedded clusters (53.131.063, 53.153.027, 54.153.065, 56.153.004) containing tens or even hundreds of young stellar objects. The four embedded clusters found in the giant molecular cloud Lynds 1630 and associated with the star-forming regions NGC 2023, 2024, 2068 and 2071 are centrally condensed. The stellar densities are high. Two clusters, NGC 2024 and 2071, show evidence for spatial magnitude segregation, with the brighter stars displaying a tendency to be more centrally condensed than the fainter sources. Verschueren (52.153.001) has investigated the cluster collapse that may happen before the gaz removal sets in.

These efforts result in a variety of designations both for the fields and the stars. It would be important to develop a coherent system of designation for each new cluster or star-forming region and maintain a catalogue of the stellar components with positions and cross-references for young stellar objects seen in visible light or discovered through their X-ray and/or infrared emissions. Although these new subjects give rise to much competition, systematic observations and exchange of information would be extremely profitable to the development of our knowledge to avoid repeating the errors made in the past study of open or globular clusters, which results now in a vast amount of data, but large gaps and disastrous incompleteness.
Observations of O and T associations are very close to the subject of star-forming regions and embedded clusters, so that the distinction between these domains may be revised.

4.2 Table 1. OB Associations

Ara OB1
56.152.009 uvbyβ phot.
CMa OB1
52.152.001 Walraven phot.
54.131.094 IRAS, IR em, O st.
Car OB2
Kaltcheva & Georgiev (1993) uvbyβ phot.
Cas OB9
53.152.004 Ap, Am st.
Cas OB14
53.152.004 Ap, Am st.
Cep OB2
52.113.041 UBVRI pg phot, Hα st.
53.131.136 IRAS, CO survey, IR src.
Cep OB3
53.113.032 uvby phot.
53.119.009 uvby light curve, eb.
56.113.003 uvbyβ phot.
Cep OB4
52.131.128 CO survey.
Cyg OB1
52.152.006 Glazar UV obs.
54.121.084 UV obs, spatial distr st.
54.131.245 IUE obs, sp, OB st, int/st gas.
56.152.008 IRAS, IR shell.
Cyg OB2 (VI Cygni)
52.152.008 uvby, IR phot.
52.117.057 radio, IR monit contact bin.
53.152.001 UBV CCD phot, sp cl, HRD, imf.
53.152.005 uvby, JHK phot, ext law.
56.121.035 IRAS, IR src.
Cyg OB3
54.131.245 IUE obs, sp, OB st, int/st gas.
Cyg OB4
53.152.004 Ap, Am st.
Lac OB1
52.115.010 sp, Hγ meas. temp, log g.
53.152.004 Ap, Am st.
52.153.011 HI distr.
Mon OB1
52.152.001 Walraven phot.
Per OB1
52.152.006 Glazar UV obs.
54.121.084 UV obs, spatial distr st.
Ori OB1
52.152.001 Walraven phot.
52.115.010 sp, Hγ meas. temp, log g.
53.152.007 rv, sb.
53.114.090 survey, Hα em st.
53.122.231 flare st.
55.121.056 Hα em st.
56.152.012 O abund.
Sco OB1
52.152.004 UBV, uvby phot, rv, sp cl.
52.152.006 Glazar UV obs.
54.121.084 UV obs, spatial distr st.
54.153.005 d, red, age, memb.
55.152.007 lf.
Sco OB2
52.152.001 Walraven phot.
Sco-Cen
56.002.102 UBVβ phot.
56.131.023 st, shell i/s mat.
56.133.001 IRAS src, ident.
Sct OB2
52.152.001 Walraven phot.
Vul OB4
53.152.004 Ap, Am st.

4.3 Table 2. R and T Associations

Cha I
52.131.171 IRAS map, IR phot.
53.121.016 IR phot.
55.121.019 lf.
55.131.116 review.
56.121.017 sp, X-ray src.
Hartigan (1993) new T Tauri st.
Cha II
55.131.116 review.
56.121.001 IRAS search, pms st.
56.131.036 d, HRD.
Hartigan (1993) new T Tauri st.
CMa R1
55.131.268 IRAS, IR em, st src.
Mon R2
52.131.065 mol outflow.
53.131.092 mol outflow.
56.131.040 sub mm/mm obs.

4.4 Table 3. Star forming regions / embedded clusters

NGC 712
52.131.165 CCD image.
55.131.079 HH obj.
IC 1396
53.131.136 IRAS search, yso.
Rho Oph
52.121.031 mm survey, yso.
52.131.201 DCO+ em, protostar.
53.121.020 rot mod, T Tauri st.
56.121.042 JHK survey, yso.
56.121.108 VLBI survey, yso.
56.131.088 UV i/s ext.
Vrba et al (1993) phot, sp cl, pol.
GM 24
53.153.027 opt, IR CCD images, IR ex.
55.153.029 radio, IR, opt obs, rev.
L1228
52.131.207 Hα em st.
L1251
Kun & Prusti (1993) d, memb.
L1551
54.131.369 mol outflow
55.131.258 Hα image, HH obj.
56.121.016 HH obj.
L1630
53.131.163 2.2μm survey, yso.
L1641
52.131.119 X-ray src.
53.121.009 HH obj.
54.131.073 mol outflow, yso.
L1642
56.153.066 UBVRI phot.
LkHα 101
55.153.028 IR survey, CMD, age.
Lupus
Hughes et al (1993) phot, sp, d.
Serpens
56.153.004 JHKnbL, "ice" image, 163 st obj.
R CrA
56.121.002 IRAS, en dist, pms st.
Rosetta
Block et al (1993) emb cl.
RSF 2 Cyg
53.131.319 UBVR phot, Hα em.
Taurus-Auriga
53.121.005 pm, sp.
55.121.012 IRAS IR survey.
55.121.041 IRAS, pms st.
56.121.018 PMS, age.
56.131.098 dyn, review.

4.5 References to 1993 papers on associations and star-forming regions

Block D.L., Geballe T.R., Dyson J.E. 1993, A&A 273, L41
Hartigan P. 1993, AJ 105, 1511
Hughes J., Hartigan P., Clampitt L. 1993, AJ 105, 571
Kaltcheva N.T., Georgiev L.N. 1993, MNRAS 261, 847
Kun M., Prusti T. 1993, A&A 272, 235
Vrba F.J., Coyne G.V., Tapia S. 1993, AJ 105, 1010

5 Open clusters

To make the text a little easier to read, I have dropped many references to the observations cited. It is however simple to find them below in section 5.10 presenting the cluster-by-cluster literature citations.

5.1 Photometry

CCD photometry of open clusters has been done for many objects. Many targets were known old open clusters (NGC 188, 2243, 2420, 2682, 6791, 7142) and the observations of these objects has produced improved colour magnitude diagrams that give more constraining tests of stellar evolutionary models. Colour magnitude diagrams for remote or faint, young (NGC 433, Tr 14, 16, Sher 1, Wes 2), intermediate-age (2266, 2355, 2627, 2658, 6603), old (IC 1311, Be 32, King 2, King 11), clusters not previously observed were obtained with CCD cameras. BVI CCD photometry has been used to identify very low-mass stars in the Pleiades and α Persei clusters. BV and RI filters are mainly used, while U is also sometimes observed. The main goal of these studies is to obtain colour-magnitude diagrams to estimate the distance and age of the clusters. However the lack of U-B colours limits the possibilities of reddening and membership determination. Both the discrimination of field stars and the determination of cluster parameters is more difficult and this makes the fitting of theoretical isochrones confusing, when the colour-magnitude diagrams have not been cleared of non-member stars. Comparison fields should be obtained more often to estimate the degree of contamination by field stars. The addition of red and near infrared filters produces better diagrams for red stars. CCD cameras have also been developed for the uvby (NGC 457, 1893, Be 94, Bo 1) and Washington (NGC 2266, 3680, Be 32) systems.
The photoelectric photometry in the UBV system is still active for clusters with large diameters. The tendency is now to measure a larger number of stars to make more complete studies than was the case in

the previous years (NGC 129, 436, 637, 1647, 2353, 5460, 5606, 5749, 5822, 6134, 6193, 6494, 6823, Ros 3, Tr 37, vdBH 99). Observations have also been published in the uvbyβ system for several clusters (Hyades, Coma Ber, NGC 1502, 2169, Tr 16, vdBH 99). Photographic photometry has often been combined with photoelectric measurements or with CCD observations (NGC 188, 225, 957, 1039, 5822, 6716). Infrared (JHK) observations of stars in very young open clusters have been carried out to study the variable extinction (1893, 2264, 6611).
Several photometric programmes aimed at studying the variablity of cluster stars have produced remarkable results. They are discussed below.

5.2 Spectroscopy

New estimates of the chemical composition and [Fe/H] of open clusters were obtained for dwarfs in nearby clusters, for turn-off stars in NGC 188, 752 and 2682, and for red giants in several old clusters (Friel & Janes 1993). The carbon isotope ratio in giant branch and clump stars of NGC 2682 (M67). Excess mixing is probably required to explain the anomalously low ratios observed in stars at the tip of the red giant branch and in the clump. Lithium abundances in the Hyades, Coma Ber, Orion, NGC 188, 2682, IC 4651 were determined. Much spectroscopic effort has been devoted to observe the CaII H-K and Hα strength to study the chromospheric activity of late-type dwarfs in nearby clusters (53.153.011, 53.153.050).

5.3 Proper motions

The determination of proper motions has been done for several clusters, although many studies were done for old favourite clusters (Hyades, Pleiades, Praesepe, α Persei, NGC 752, 2682, 7092). Very large numbers of stars were measured in the nearby open clusters (Pleiades, α Persei) to search for low-mass or very low-mass members and brown dwarfs. New clusters have also been studied (NGC 225, 1039, 2286, 6939, Tr 37). New or improved methods to compute membership probabilities have been proposed (52.153.002, 52.153.005, 52.153.008, 52.153.019, 54.153.013).

5.4 Radial velocity

This period was quite prolific in radial velocity determinations. Classical or CCD spectra were obtained for upper main sequence stars in several clusters (NGC 6231, Orion, Pleiades, IC 4665, Cr 228, Tr 14, Tr 16). Two groups showed that precision radial velocities for early type stars were now feasable by observing the B-type stars in the Pleiades (53.036.025, 54.111.004) and several northern young open clusters (54.153.056). Photoelectric radial velocities were published for red giants in open clusters by several groups (NGC 2360, 2423, 2682, 5822, 6134, 6811, 6939, IC 4756), and for 118 red giants in 18 open clusters were published (53.153.035). The large observational efforts devoted to M67 (52.153.020) produced orbits for 22 upper main sequence stars and red giants and resulted in the discovery of several additional spectroscopic binaries.

5.5 Variable stars

Several studies were devoted to the observations of light modulation to determine the period of rotation. Most targets were chosen in the nearby clusters (Pleiades, α Persei, Orion). Such studies are connected with chromospheric activity determination. The Be stars in NGC 869(h Per), 884 (chi Per) and 3766 were extensively observed and long-term variation patterns were brought to light. The programmes undertaken to search for short-period, contact binaries in old open clusters produced very promising results. Four contact binaries were discovered in To 2 (56.153.049) and eleven in Be 39, over a sample of 2000 stars (Kaluzny et al 1993).
Eclipsing and contact binaries in NGC 188, 752, 2682, 3293, 6871, 7209, IC 4665, and 4996 received also some attention. The major survey results published during these three years is the extensive coverage with 5200 CCD frames of NGC 2682 (M67) by Gilliland and collaborators (53.153.002). They reached an unprecedented precision and reported evidence suggestive of p-mode oscillations in a few stars. Serendipitous discovery of other kinds of variable stars was also made.

5.6 Blue stragglers

Most work on blue stragglers has been done in the old cluster M67. Spectroscopic analysis was conducted to determine the lithium abundance (53.153.045) and the abundance pattern (53.153.037) which is found to be similar to that of other stars in the cluster. A preliminary report of a long-term radial velocity monitoring of 9 blue stragglers (with Vsini < 100 km/s) shows that 5, and possibly 7, are binaries. An orbit has been obtained for F190 (55.153.021). Photometric monitoring of blue stragglers located within the instability region showed that two of them (F131 and F190) are variable (54.153.026), while F185 is not. The variability of F190 and the stability of F185 have been confirmed on the basis of a much more extensive CCD survey (55.153.017). Another variable blue straggler, F184, has been analysed in the latter reference. Five contact binaries among blue stragglers have been discovered by Kaluzny et al (1993) in Be 39.

5.7 Brown dwarfs

Infrared search for brown dwarfs in nearby open clusters has continued, from I POSS plates (55.115.007), R&I UK Schmidt plates (53.153.033) or IR array surveys (55.115.014). Selection based on a deep proper motion survey and photometric follow-up in RIJHK for a sub-sample of the stars identified has been done in the Pleiades (54.153.075). It resulted in the identification of a large list of low-mass stars with probable membership of the Pleiades. Spectroscopy has been obtained for one candidate in the α Persei cluster (55.115.012) to define the spectroscopic signature of brown dwarfs, as well as for one candidate in the Hyades moving group (55.115.011). Hamilton & Stauffer (1993) have obtained optical spectrophotometry for several very low-mass stars in the Pleiades. Most good candidates have masses close to 0.07 solar masses and are close to the border between very late M stars and brown dwarfs. Comparisons have been attempted with theoretical luminosity functions (52.153.004), and with isochrones (54.115.001). Many uncertainties are still attached to the conclusions due to the problems on the exact distance modulus, the temperature scale and bolometric corrections, and the difference between the models. McDonald and Clarke (1993) examined the dynamics of binary star formation and its implication for brown dwarfs in binaries.

5.8 Space observations

IUE spectra of solar-type stars in the Pleiades have been obtained (54.153.082). Observations from the IRAS satellite are used to search for very late-type stars in clusters, study star-forming regions associated with nebulosity (56.153.015) and detect embedded clusters (Carpenter et al 1993; Pazstor et al 1993). ROSAT observations in the Hyades have been analysed by Stern et al (56.153.039). Several X-ray sources were also discovered in the old cluster M67 (Belloni et al 1993).

5.9 References

Belloni T., Verbunt F., Schmitt J.H.M.M. 1993, A&A 269, 175
Carpenter J.M., Snell R.L., Schloerb F.P., Skrutskie M.F. 1993, ApJ 407, 657
Friel E.D., Janes K.A. 1993, A&A 267, 75
Hamilton D., Stauffer J.R. 1993, AJ 105, 1855
Kaluzny J., Mazur B., Krzeminski W. 1993, MNRAS 262, 49
McDonald J.M., Clarke C.J. 1993, MNRAS 262, 800
Pasztor L., Toth L.V., Balasz L.G. 1993, A&A 268, 108

5.10 Cluster-by-cluster literature citations

The present compilation has been prepared from the literature cited in Astronomy and Astrophysics Abstracts, vol. 52 to 56 inclusive, and the published literature until June 1993. This compilation contains only papers published in journals, and not contributions in conferences. Photometric observations imply generally determination of reddenings, distances, and ages. These items are therefore not specified.

Glossary of abbreviations

abs = *absolute*
act = *activity*
abund = *abundance*
anal = *analysis*
bin = *binary*
bd = *brown dwarf*

bs = *blue straggler*
ceph = *cepheid*
chem = *chemical*
chrom = *chromospheric*
cl = *classification*
col = *colour*
comp = *composition*
clus = *cluster*
CMD = *colour-magnitude diagram*
cont = *contact*
d = *distance*
diff = *differential*
distr = *distribution*
dyn = *dynamics*
eb = *eclipsing binary*
em = *emission*
en = *energy*
ex = *excess*
ext = *extinction*
form = *formation*
HH = *Herbig-Haro*
HRD = *HR diagram*
ident = *identification*
i/s = *interstellar*
imf = *initial mass function*
int = *integrated*
kinem = *kinematics*
lf = *luminosity function*
lm = *low mass*
mat = *matter*
meas = *measure*
mf = *mass function*
memb = *membership*
metal = *metallicity*
mm = *millimetric*
ms = *main sequence*
mod = *modulation*
monit = *monitoring*
obj = *object*
obs = *observation*
opt = *optical*
par = *parallaxe*
param = *parameter*
pe = *photoelectric*
per = *period*
pms = *pre-main-sequence*
pg = *photographic*
phot = *photometry*
plan neb = *planetary nebula*
pm = *proper motion*
pol = *polarization*
pos = *position*
red = *reddening*
rev = *review*
rg = *red giant*
rot = *rotation, rotational*
rv = *radial velocity*
sb = *spectroscopic binary*
sp = *spectral, -oscopic, -oscopy*
src = *source*
st = *star*
str = *strength*
struc = *structure*
temp = *temperature*
to = *turn off*
UV = *ultra-violet*
var = *variable*
vlm = *very low mass*
Wash = *Washington*
yso = *young stellar object*

C0000+671 (Be 59)
54.153.056 rv.
C0027+599 (NGC 129)
53.153.035 rv, rg.
56.153.006 UBV phot, sp cl, rv, rot.
C0039+850 (NGC 188)
52.153.011 sp abund.
52.153.062 BV phot, var st.
54.153.014 bin st form.
54.153.034 UBV pg phot.
55.153.005 age.
55.153.016 bs.
Meynet et al (1993) age.
C0040+615 (NGC 225)
54.153.036 UBV pg phot, sp cl, pm.
C0049+563 (NGC 281)
53.131.068 21cm survey.
C0048+579 (King 2)
52.153.021 UBVR CCD phot.
C0112+598 (NGC 433)
53.131.068 21cm survey.
55.153.014 UBV CCD phot.
C0112+585 (NGC 436)
53.153.009 UBV phot.
C0115+580 (NGC 457)
53.131.068 21cm survey.
54.153.056 rv.
54.153.077 pos.
Meynet et al (1993) age.
Fitzsimmons (1993) uvby CCD phot.
C0129+604 (NGC 581, M103)
54.153.056 rv.
54.153.077 pos.
Meynet et al (1993) age.
C0139+637 (NGC 637)
53.153.009 UBV phot.
54.153.056 rv.
C0140+616 (NGC 654)
54.153.056 rv.
Huestamendia et al (1993) UBV phot.
C0142+610 (NGC 663)
52.153.013 Hα em st.
54.153.079 uvby, JHK phot.
54.153.056 rv.
C0149+615 IC 166)
Friel & Janes (1993) rv, metal.
C0154+374 (NGC 752)
53.153.007 pm, memb.
55.002.129 pm, memb.
56.153.005 sp, iron abund.
Friel & Janes (1993) rv, metal.
Dzervitis & Paupers (1993) Vilnius phot.
Meynet et al (1993) age.
C0215+569 (NGC 869, h Per)
52.153.063 pos.
54.153.056 rv.
C2018+385 (Be 86)
55.153.012) UBV phot, CCD sp, WN st.
C0218+568 (NGC 884, chi Per)
52.153.063 pos.
54.153.056 rv.
Meynet et al (1993) age.
C2019+372 (Be 87)
52.153.028 sp anal, WR st.
54.153.054 opt sp, EXOSAT, X-ray.
C0230+573 (NGC 957)
54.153.056 rv.
54.153.073 UBV pg phot.
C0238+425 (NGC 1039, M34)
Ianna & Schlemmer (1993) pm, memb, UBV pg phot.
Meynet et al (1993) age.

C0247+602 (IC 1848)
54.153.056 rv.
C0302+441 (NGC 1193)
Friel & Janes (1993) rv, metal.
C0318+484 (Alpha Persei, Mel 20)
55.115.012 sp, bd.
55.153.010 pm, rv, CCD phot, memb.
56.116.017 G dwarf activity.
56.153.007 rv, sb.
Meynet et al (1993) age.
O'Dell & Cameron (1993) rot mod, per.
C0328+371 (NGC 1342)
53.153.035 rv rg.
C0344+239 (Pleiades, Mel 22)
52.111.008 par.
52.122.145 sp obs, flare st.
52.126.028 EXOSAT, WD.
52.153.057 struct.
53.153.005 pm, memb, low mass st.
53.153.011 BVRI CCD phot, sp, Hα em str.
53.153.021 struct.
53.153.025 rot mod.
53.153.033 vlm st, bd, lf.
53.153.047 flare st.
53.153.065 IR search bd.
54.111.004 rv.
54.153.039 pm.
54.153.075 pm, vlm st, bd.
54.153.082 IUE obs, F5-G5 st.
55.115.014 IK CCD search, bd.
55.122.149 flare st, sp.
55.153.002 rv, corona st.
55.153.013 super clus.
56.153.016 rv, sb.
56.153.018 pm.
Hamilton & Stauffer (1993) sp, lm st.
Meynet et al (1993) age.
Prosser & Stauffer (1993) var st.
Soderblom et al (1993) rot, chrom act.
White & Bally (1993) IRAS, i/s mat distr.
C0403+622 (NGC 1502)
54.153.079 uvby, JHK phot.
55.153.011 uvby phot.
C0411+511 (NGC 1528)
53.153.035 rv rg.
C0417+501 (NGC 1545)
53.153.035 rv rg.
C0424+157 (Hyades, Mel 25)
52.114.086 OI abund, rg.
52.153.004 low mass st, imf.
52.153.014 par, pm.
52.153.032 HeI sp anal.
53.153.011 BVRI CCD phot, sp.
53.153.031 pm, d.
53.153.050 sp, chrom act, Hα em.
54.153.029 pm, Ap st.
54.153.055 par, d.
54.153.082 IUE obs, F5-G5 st.
54.153.007 HST obs, bin st.
55.114.010 chem comp, rg.
55.115.007 JHK phot, bd.
55.153.022 uvby phot.
56.153.010 WD, mass, memb.
56.153.012 super cl.
56.153.039 ROSAT X-ray obs.
Mason et al (1993) speckle obs.
Petersen et al (1993) rv, sb.
White et al (1993) radio survey.
C0443+189 (NGC 1647)
53.153.035 rv rg.
56.153.009 UBV phot, sp cl, rv, rot.
C0445+108 (NGC 1662)
53.153.035 rv rg.
C0447+436 (NGC 1664)
52.153.019 memb.
C0501+237 (NGC 1758)
56.153.045 Vilnius phot.
C0504+369 (NGC 1778)
54.153.056 rv.
C0509+166 (NGC 1817)
Friel & Janes (1993) rv, metal.
C0519+333 (NGC 1893)
54.153.079 uvby, JHK phot.
Rolleston et al (1993) chem comp.
Fitzsimmons (1993) uvby CCD phot.
C0524+352 (NGC 1907)
53.153.035 rv rg.
C0525+358 (NGC 1912, M38)
53.153.035 rv rg.
C0532-054 (Orion, Trapezium)
53.153.015 rv, sb.
54.153.081 rot mod, period.
56.121.075 rot per.
Duncan (1993) rot.
King (1993) Li abund.
Meaburn et al (1993) sp, jet, knot.
Meynet et al (1993) age.
Shevchenko & Yakubov (1993) UBVRI phot.
C0546+336 (Ki 8)
Friel & Janes (1993) rv, metal.
C0548+217 (Be 21)
Friel & Janes (1993) rv, metal.
C0549+325 (NGC 2099, M37)
53.153.035 rv rg.
C0551+003 (NGC 2112)
Friel & Janes (1993) rv, metal.
C0558+233 (NGC 2129)
54.153.056 rv.
C0600+104 (NGC 2141)
Friel & Janes (1993) rv, metal.
C0605+139 (NGC 2169)
54.153.056 rv.
Maitzen & Lebzelter (1993) phot, sp, Ap, var st.
C0605+243 (NGC 2168)
56.153.030 UBV phot.
C0613-186 (NGC 2204)
55.153.006 JHK phot, rg.
C0622+198 (Bo 1)
Fitzsimmons (1993) uvby CCD phot.
C0622+198 (Bo 2)
Rolleston et al (1993) chem comp.
C0627+068 (NGC 2236)
54.153.089 UBV pe pg phot.
C0627-312 (NGC 2243)
52.153.018 UBV CCD phot, age, metal.
55.153.006 JHK phot, rg.
53.153.030 BV CCD phot, lf.
Friel & Janes (1993) rv, metal.
C0629+049 (NGC 2244)
52.153.037 i/s ext.

52.153.053 memb, kinem.
53.153.078 radius, age, mass.
54.153.056 rv.

C0638+099 (NGC 2264)
52.153.002 memb.
52.153.053 memb, kinem.
52.153.064 pos.
53.153.078 radius, age, mass.
54.153.012 Walraven phot.
54.153.056 rv.
54.153.063 sp abund, temp, log g, rot.
56.121.029 HH obj.
Lada et al (1993) IR image.
Tauber et al (1993) st form.

C0640+270 (NGC 2266)
54.153.031 UBV, Wash CCD phot.

C0642+003 (Do 25)
52.153.022) sp abund.

C0645-031 (NGC 2286)
52.153.067 pm, memb.
55.153.048 BV pg phot.

C0644-206 (NGC 2287, M41)
54.153.056 rv.
56.154.032 Fe, Ca, Na, Al, O abund.
56.154.067 Vilnius phot.
Meynet et al (1993) age.

C0646+004 (Bo 2)
Turbide & Moffat (1993) UBV CCD phot.

C0649+005 (NGC 2301)
53.153.028 CMD, temp, log g.
56.153.038 memb.

C0655+065 (Be 32, Biu 8)
54.153.030 UBV, Wash CCD phot.
55.153.011 uvby phot.

C0700-082 (NGC 2323)
53.153.035 rv rg.

C0701+011 (NGC 2324)
56.153.011 Wash phot, rg
Marie (1993) memb.

C0701-207 (To 2)
56.153.049 var st, cont bin.
Friel & Janes (1993) rv, metal.

C0712-102 (NGC 2353)
52.153.047 UBV phot, sp cl.
53.153.035 rv rg.

C0714+138 (NGC 2355)
54.153.033 UBV CCD phot.
Meynet et al (1993) age.

C0715-155 (NGC 2360)
52.153.009 rv, rg.
56.153.011 Wash phot, rg
Friel & Janes (1993) rv, metal.
Meynet et al (1993) age.

C0716-248 (NGC 2362)
54.153.050 I CCD phot, lf.
55.153.015 lf, mf, lm, st, corona.

C0718-218 (NGC 2367)
54.153.056) rv.

C0722-209 (NGC 2384)
54.153.056 rv.

C0724-476 (Mel 66)
55.153.006 JHK phot, rg.
Friel & Janes (1993) rv, metal.

C0731-153 (NGC 2414)
54.153.056 rv.

C0735+216 (NGC 2420)
52.153.005 memb.
55.153.006 JHK phot, rg.
Friel & Janes (1993) rv, metal.

C0735-119 (Mel 71)
56.153.011 Wash phot, rg

C0734-205 (NGC 2421)
56.153.002 UBV CCD phot.

C0734-137 (NGC 2423)
52.153.009 rv, rg.

C0738-315 (NGC 2439)
56.153.002 UBV CCD phot.
Meynet et al (1993) age.

C0742-254 (Rup 32)
Turbide & Moffat (1993) UBV CCD phot.

C0744-044 (Be 39)
56.153.011 Wash phot, rg.
Friel & Janes (1993) rv, metal.
Kaluzny et al (1993) CCD phot, var st, bs.

C0750-261 (Haf 19)
56.153.019 BV, GR CCD phot, cal seq.

C0750-262 (Haf 18ab)
56.153.019 BV, GR CCD phot, cal seq.

C0750-384 (NGC 2477)
55.153.006 JHK phot, rg.
56.153.011 Wash phot, rg.
Friel & Janes (1993) rv, metal.

C0754-299 (NGC 2489)
56.153.002 UBV CCD phot.

C0757-106 (NGC 2506)
55.153.006 JHK phot, rg.
56.153.011 Wash phot, rg.
Friel & Janes (1993) rv, metal.

C0757-607 (NGC 2516)
55.153.055 flare st.
Meynet et al (1993) age.

C0816-304 (NGC 2567)
56.153.002 UBV CCD phot.

C0835-297 (NGC 2627)
56.153.002 UBV CCD phot.

C0837+201 (NGC 2632, M44, Praesepe)
52.153.006 UBVRI phot, rv, corona st.
52.153.057 struct.
53.153.021 struct.
54.041.023 pos.
54.153.004 uvby phot.
54.153.017 rv, sb.
54.153.053 pm, memb, BV pg phot.
55.153.018 iron abund.
56.123.015 flare st.
Breger et al (1993) phot, var st.
Vidal et al (1993) CCD phot, var st.

C0838-528 (IC 2391)
54.153.071 memb, super clus.
55.153.055 flare st.

C0840-469 (NGC 2660)
55.153.006 JHK phot, rg.
56.153.011 Wash phot, rg.

C0841-324 (NGC 2658)
56.153.002 UBV CCD phot.

C0847+120 (NGC 2682, M67)
52.113.016 uvby CCD phot.
52.113.038 VRIc phot.
52.153.020 rv, sb, orb.
52.153.055 struct.

53.153.002 UBV CCD phot, var st, cont bin.
53.153.022 rv.
53.153.037 sp, abund, bs.
53.153.042 sp, rg, C isotope.
53.153.045 sp, bs, Li abund.
54.153.006 BVRIc CCD phot.
54.153.027 JHK phot, bs.
54.153.052 metal, temp, to st.
55.153.005 age.
55.153.006 JHK phot, rg.
55.153.016 bs.
55.153.017 CCD phot, var st, bs.
55.153.018 iron abund.
56.153.056 pm.
56.153.068 VRI CCD phot.
Belloni et al (1993) ROSAT X-ray src.
Friel & Janes (1993) rv, metal.
Meynet et al (1993) age.
Montgomery et al (1993) CCD phot.

C0914-364 (NGC 2818)
56.153.011 Wash phot, rg.

C0928-526 (NGC 2910)
56.153.002 UBV CCD phot.

C1001-598 (NGC 3114)
53.153.039 BV CCD phot, sp cl, rv.

C1022-575 (Wes 2)
54.153.049 UBV phot, sp cl, WN st.

C1028-595 (Cr 223)
54.153.028 UBV phot, DDO rg.

C1033-579 (NGC 3293)
56.122.035 uvby phot, var st.

C1036-589 (vdBH 99)
52.153.010 UBV, uvby phot, sp cl.

C1041-593 (Tr 14)
54.153.016 rv, sb.
Cudworth et al (1993) pm, memb, ubv pg phot.
Massey & Johnson (1993) UBV CCD phot, sp cl.
Penny et al (1993) rv.

C1041-641 (IC 2602)
55.153.055 flare st.

C1043-594 (Tr 16)
53.153.026 rv, sb.
Cudworth et al (1993) pm, memb, ubv pg phot.
Kaltcheva & Georgiev (1993) uvby phot.
Massey & Johnson (1993) UBV CCD phot, sp cl.

C1104-584 (NGC 3532)
Meynet et al (1993) age.

C1058-601 (Sher 1)
54.153.049 UBV phot, sp cl, WN st.

C1108-599 (NGC 3572)
53.134.028 plan neb.

C1109-604 (Tr 18)
52.153.022 UBVRI phot.

C1112-609 (NGC 3603)
Santos & Bica (1993) int sp, red, age

C1117-632 (Mel 105)
53.153.008 UBV CCD phot.
Santos & Bica (1993) int sp, red, age

C1123-429 (NGC 3680)
54.153.051 BV pg, CCD phot.
Friel & Janes (1993) rv, metal.

C1133-613 (NGC 3766)
54.112.096 Be st, light curve.

C1148-554 (NGC 3960)
56.153.011 Wash phot, rg.
Friel & Janes (1993) rv, metal.

C1222+263 (Coma Ber)
52.153.057 struct.
53.153.021 struct.
53.153.035 rv rg.
55.153.018 iron abund.
56.153.001 imf.
Bounatiro & Arimoto (1993) imf

C1250-600 (NGC 4755)
53.153.008 UBV CCD phot.
Santos & Bica (1993) int sp, red, age.
Meynet et al (1993) age.

C1254-646 (NGC 4815)
53.153.008 UBV CCD phot.

C1326-609 (Ho 16)
54.153.008 UBVRIc phot.

C1327-606 (NGC 5168)
53.153.008 UBV CCD phot.

C1404-480 (NGC 5460)
Meynet et al (1993) age.
Claria et al (1993) UBV phot.

C1424-594 (NGC 5606)
53.153.012 UBVRI phot.

C1426-605 (NGC 5617)
53.153.008 UBV CCD phot.
Santos & Bica (1993) int sp, red, age

C1431-563 (NGC 5662)
53.153.034 UBV phot.
Meynet et al (1993) age.

C1445-543 (NGC 5749)
56.153.067 UBV phot.

C1501-541 (NGC 5822)
52.153.009 rv, rg.
Friel & Janes (1993) rv, metal.
Twarog et al (1993) UBV pe, pg, DDO phot.

C1548-563 (NGC 5999)
Santos & Bica (1993) int sp, red, age

C1559-603 (NGC 6025)
Meynet et al (1993) age.

C1609-540 (NGC 6067)
Santos & Bica (1993) int sp, red, age
Meynet et al (1993) age.

C1614-577 (NGC 6087)
Meynet et al (1993) age.

C1624-490 (NGC 6134)
53.153.008 UBV CCD phot.
56.153.028 UBV, DDO, rv, rg.

C1636-432 (NGC 6192)
53.153.008 UBV CCD phot.

C1637-486 (NGC 6193)
55.153.024 UBVRI phot, col ex.

C1642-469 (NGC 6204)
53.153.008 UBV CCD phot.

C1650-417 (NGC 6231)
52.152.004 UBV phot, sp cl, rv.
55.152.007 lf.
54.153.005 param.
Meynet et al (1993) age.
Santos & Bica (1993) int sp, red, age

C1652-394 (NGC 6242)
Meynet et al (1993) age.

C1720-499 (IC 4651)
53.153.008 UBV CCD phot.
Meynet et al (1993) age.

C1743+057 (IC 4665)
54.153.037 rv, sb.
C1747-302 (NGC 6451)
53.153.008 UBV CCD phot.
C1750-348 (NGC 6475)
Meynet et al (1993) age.
C1753-190 (NGC 6494, M23)
52.153.003 UBV phot.
Meynet et al (1993) age.
C1800-279 (NGC 6520)
53.153.008 UBV CCD phot.
Santos & Bica (1993) int sp, red, age
C1815-184 (NGC 6603)
Bica et al (1993) UBV CCD phot.
Santos & Bica (1993) int sp, red, age
C1815-122 (NGC 6604)
54.153.056 rv.
C1816-138 (NGC 6611, M16)
56.153.015 IR src clus.
Santos & Bica (1993) int sp, red, age
C1816-138 (NGC 6618, M17)
53.132.051 JHK survey, emb src, yso.
C1825+065 (NGC 6633)
53.153.035 rv, rg.
54.153.027 JHK phot, bs.
C1830-104 (NGC 6649)
Berdnikov (1992) cep var, light curve.
C1836+054 (IC 4756)
52.153.009 rv, rg.
52.153.043 rv, rg.
C1848-063 (NGC 6705, M11)
53.153.008 UBV CCD phot.
56.153.046 Vilnius phot.
Brocato et al (1993) BV CCD phot, lf.
Meynet et al (1993) age.
Zakirov (1993) close bin, eb.
C1851-199 (NGC 6716)
52.153.048 UBV pe pg phot.
C1902+018 (Be 42)
53.153.040 UBVR CCD phot.
C1905+041(NGC 6755)
53.153.008 UBV CCD phot.
C1906+046 (NGC 6756)
Santos & Bica (1993) int sp, red, age
C1919+377 (NGC 6791)
55.153.004 BVI CCD phot, bs.
55.153.005 age.
Friel & Janes (1993) rv, metal.
Meynet et al (1993) age.
C1941+231 (NGC 6823)
52.153.037 i/s ext.
55.153.008 UBV pe, CCD phot.
56.153.003 memb, lf, cl mass.
C1936+464 (NGC 6811)
52.153.009 rv, rg.
C1939+400 (NGC 6819)
54.153.027 JHK phot, bs.
Friel & Janes (1993) rv, metal.
C1956+203 (Ros 3)
Turner (1993) UBV phot, sp cl, rv, Vsini.
C2004+356 (NGC 6871)
54.153.027 JHK phot, bs.
56.119.057 close bin, eb, phot.
C2008+410 (IC 1311)
55.153.009 BVR CCD phot.
C2009+263 (NGC 6882/5)
52.123.015 var st.
C2017+379 (Do 42)
52.153.044 UBVR phot, eb.
C2021+406 (NGC 6910)
53.131.319 UBVR phot, Hα em st.
54.153.027 JHK phot, bs.
56.153.044 Vilnius phot.
C2022+383 (NGC 6913, M29)
52.153.002 memb.
53.153.035 rv rg.
52.153.037 i/s ext.
54.153.056 rv.
52.153.053 memb, kinem.
C2030+604 (NGC 6939)
53.153.036 pm, rv, rg.
54.153.027 JHK phot, bs.
C2054+444 (NGC 6996)
52.153.050 Vilnius phot.
C2121+461 (NGC 7062)
52.153.065 uvby phot.
C2130+482 (NGC 7092, M39)
53.153.013 JHK phot.
56.154.030 UIT image.
Glushkova (1992) pm.
C2137+572 (Tr 37)
53.153.071 pm, memb.
C2144+655 (NGC 7142)
53.153.041 BV CCD phot.
54.123.011 var st.
Friel & Janes (1993) rv, metal.
C2151+470 (IC 5146)
54.153.056 rv.
C2152+623 (NGC 7160)
54.153.056 rv.
C2203+462 (NGC 7209)
53.153.014 pm, memb.
55.002.130 pm, memb.
C2220+556 (Be 94)
Fitzsimmons (1993) uvby CCD phot.
C2309+603 (NGC 7510)
53.153.038 BVI CCD phot.
C2322+613 (NGC 7654, M52)
52.153.049 uvby pg phot.
C2345+683 (King 11)
53.153.040 UBVR CCD phot.
C2354+564 (NGC 7789)
52.153.068 pol, memb.
54.153.027 JHK phot, bs.
Friel & Janes (1993) rv, metal.
C2355+609 (NGC 7790)
52.118.023 light curve ceph.
56.002.048 rv, ceph var.

5.11 References to 1993 papers on open clusters

Berdnikov L. 1992, IBVS no 3712
Belloni T., Verbunt F., Schmitt J.H.M.M. 1993, A&A 269, 175

Bica E., Ortolani S., Barbuy B. 1993, A&A 270, 117
Bounatiro L., Arimoto N. 1993, A&A 268, 829
Breger M., Stich J., Garrido R., Martin B., Jiang S.-h., Li Z.-p.,
Hube D.P., Ostermann W., Paparo M., Scheck M. 1993, A&A 271, 482
Brocato E., Castellani V., DiGiorgio A. 1993, AJ 105, 2192
Claria J.J., Lapasset E., Bosio M.A. 1993, A&AS 99, 1
Cudworth K.M., Martin S.C., DeGioia-Eastwood K. 1993, AJ 105, 1822
Duncan D.K. 1993, ApJ 406, 172
Dzervitis U., Paupers O. 1993, ASS 199, 77
Fitzsimmons A. 1993, A&AS 99, 15
Friel E.D., Janes K.A. 1993, A&A 267, 75
Glushkova E.V. 1992, Sov. Astr. 35, 466
Hamilton D., Stauffer J.R. 1993, AJ 105, 1855
Huestamendia G., del Rio G., Mermilliod J.-C. 1993, A&AS 100, 25
Ianna P.A., Schlemmer D.M. 1993, AJ 105, 209
Kaltcheva N.T., Georgiev L.N. 1993, MNRAS 261, 847
Kaluzny J., Mazur B., Krzeminski W. 1993, MNRAS 262, 49
King J.R. 1993, AJ 105, 1087
Lada C.J., Young E.T., Greene T.P. 1993, ApJ 408, 471
Maitzen H.M., Lebzelter Th. 1993, IBVS no 3856
Marie M. 1993, ASS 201, 71
Mason B.D., McAlister H.A., Hartkopf W.I., Bagnuolo Jr. W.G. 1993, AJ 105, 220
Massey P., Johnson J. 1993, AJ 105, 980
Meaburn J., Massey R.M., Raga A.C., Clayton C.A. 1993, MNRAS 260, 625
Meynet G., Mermilliod J.-C., Maeder A. 1993, A&AS 98, 477
Montgomery K.A., Marschall L.A., Janes K.A. 1993, AJ 106, 181
O'Dell M.A., Cameron A.C. 1993, MNRAS 262, 521
Penny L.R., Gies D.R., Hartkopf W.I., Mason B.D., Turner N.H. 1993, PASP 105. 588
Petersen D.M., Stefanik R.P., Latham D.W. 1993, AJ 105, 2260
Prosser C.F., Stauffer J.R. 1993, IBVS no 3826
Rolleston W.R.J., Brown P.J.F., Dufton P.L., Fitzsimmons A. 1993, A&A 270, 107
Santos J.F.C. Jr, Bica E. 1993, MNRAS 260, 915
Shevchenko V.S., Yakubov S.D. 1993, Sov. Astr. 36, 359
Soderblom D.R., Stauffer J.R., Hudon J.D., Jones B.F. 1993, ApJS 83, 315
Tauber J.A., Lis D.C., Goldsmith P.F. 1993, ApJ 403, 202
Turbide L., Moffat A.F.J. 1993, AJ 105, 1831
Turner D.G. 1993, A&AS 97, 755
Twarog B.A., Anthony-Twarog B.J., McClure R.D. 1993, PASP 105, 78
Vansevicius V. 1992, Baltic Astron. 1, 91
Vidal I., Belmonte J.A., Leung E. 1993, IBVS no 3822
White R.E., Bally J. 1993, ApJ 409, 234
White S.M., Jackson P.D., Kundu M.R. 1993, AJ 105, 563
Zakirov M.M. 1993, Kin. Fys. 9, 42

6 Globular clusters

6.1 Review papers

It is difficult to summarize the numerous results obtained in all the domains relating to galactic globular cluster research and published in more than 800 papers. Instead the present paragraph lists review papers covering several domains and the interested reader is asked to look at them for a presentation of the scientific achievement of the past years.

Many contributions on globular clusters can be found in the conference on "Formation and evolution of star clusters" representing most fields of present research on these objects. An interesting summary of some topics discussed at the 1992 Santa Cruz summer workshop and at the conference on "Dynamics of globular clusters" has been presented by Trimble (1993).

Recent observations have shown that globular clusters contain a significant binary population. a dramatic change from the conventional view of even a decade ago. A comprehensive review of the subject is

presented by Hut et al (56.154.056). Fusi Pecci et al (56.154.014) have reviewed the current status of the search and study of blue stragglers in galactic globular clusters and produced a catalogue of 425 candidate blue stragglers in 21 clusters.
Main sequence mass functions in globular clusters are of importance for the insight they provide into a number of astrophysical problems and are discussed by Richer & Fahlman (54.154.048). Techniques involved in deriving luminosity function from CCD observations are reviewed by Stetson (54.154.046).
The general subject of globular clusters as tests of stellar evolution has been comprehensively reviewed by Hesser (54.154.146). He addresses the spectroscopic tests of mixing within globular cluster stars and topics related to the colour-magnitude diagrams (binarity, blue stragglers, luminosity function), stellar evolution and cluster ages. Recent advances in the understanding of horizontal-branch stars are discussed by Lee (54.154.053), with special emphasis on the second parameter problem. Djorgovski (54.154.047) showed from the results of a multivariate statistical analysis that at least 4 (possibly 5 or 6) parameters are neded in order to fully describe the global properties of globular clusters.
The ages of globular clusters and sources of uncertainties are reviewed by Demarque et al (54.154.140). Advances in precision photometry and in the modelling of turnoffs indicate that the spread in age is at least 4 Gyr among galactic globular clusters. An approximate chronology of the Galaxy is presented on the basis of the current best understanding of cluster dating. The lowest age derived for NGC 6341 (M92) is 14 Gyr.
Some recent work on CNO abundances and $^{12}C/^{13}C$ ratios has been reviewed by Bell (54.154.111). It is necessary to know cluster oxygen abundance, because the location of the isochrones in the theoretical HR diagram also depends on it. Carbon isotopic abundances determined for 45 red giants in NGC 6121 and 6752 indicate that these low-mass, low-metallicity giants have undergone extensive mixing of their outer envelope with material processed through the CN cycle (54.154.133).
Absolute proper motions and space velocities for 14 clusters have been computed (Cudworth & Hanson 1993).

6.2 Selected highlights

A selection of observational facts is presented below. It can neither be exhaustive nor cover all aspects. I apologize for the missing references to important results not quoted here. Pleases refer to chapter 154 of the Astronomy and Astrophysics Abstracts for further bibliography.

Cudworth K.M., Hanson R.B. 1993, AJ 105, 168
Trimble V. 1993, Comments Astrophys. 16, 363 (no 6)

6.2.1 CCD photometry

As shown by the tabulation of globular cluster observations presented below, 74 globular clusters have received attention during this period. CCD photometry in various filter combinations (BV, BVI, UBVRI) was published for 39 clusters. These studies often present data for several thousand stars either to improved the morphology of the colour-magnitude diagrams or to determine the faint star luminosity function. These extensive studies have permitted to discover and observe many variable stars and blue stragglers. A radial distribution study of blue subdwarfs in ω Centauri based on CCD photometry showed that the blue subdwarfs are centrally concentrated with respect to subgiants and horizontal branch stars (55.154.022)

6.2.2 Binary stars

The analysis of the main-sequence width, based on high-quality CCD photometry, provided an estimate of the fraction of binaries ($\sim$ 8%) showing a sizeable effect of duplicity in the colour-magnitude diagram (53.154.036). A value of 10% photometric binaries has been observed in NGC 288 (56.154.072). Centrally condensed binaries have been invoked to explain the colour gradient found in clusters with central cusp (53.154.032). This result has been contested (55.154.012). Two possible causes of the observed colour gradient may be (1) a centrally concentrated population of faint blue stars, (2) a central deficiency of low-mass main-sequence stars. Two eclipsing binaries were found in NGC 6838 (M71) (54.154.076).

6.2.3 Blue stragglers

Among the interesting objects, one may notice the six SX Phe variables and three eclipsing binaries discovered among the blue stragglers in NGC 5466 (52.154.018). A significant population of blue stragglers has been found in NGC 6101 and shown to be more concentrated towards the cluster centre than subgiants of similar magnitude (54.154.055). Two variable blue stragglers were found in NGC 6838 (M71) (54.154.076).

6.2.4 Distances, ages and diameters

Very accurate relative distances have been obtained from the (log Period, infrared magnitude) relation of RR Lyrae variables (52.154.050).
A method based on direct superimposition of colour-magnitude diagrams has been proposed (52.154.017) to measure age difference among clusters with a similar metallicity. CCD photometry of Ruprecht 106 has shown that this cluster may be 4-5 billion years younger (52.154.030), a result confirmed by subsequent study (56.154.003).
The diameters of 98 cluster are found to correlate with the galactocentric distances, but not with luminosity (mass) or metallicity (54.154.003).

6.2.5 Luminosity function

The "bump" in the differential luminosity function of red giants, which arises as evolution up the red giant branch pauses, has been identified in 11 clusters and the variation with metallicity and age investigated (52.154.026). The dependence of the mass function slopes on physical parameters has been examined for a sample of 14 clusters with deep CCD luminosity functions. A consequence of the analysis is that a correction for the selective mass loss at the tidal radius must be applied to present-day mass functions (53.154.020). Richer et al (54.154.132) found that cluster evolution has been important in modifying the mass function in NGC 6397 and 6838.

6.2.6 X-ray and radio sources, pulsars

Radio counterparts of X-ray sources were identified in NGC 6712 (52.154.033), NGC 6712 and 7078 (52.154.044), NGC 6624 (56.154.002). A hard X-ray source has been discovered in the direction of the cluster Terzan 2 (54.154.101).
Ten additional millisecond pulsars were found in NGC 104 (47 Tuc) (54.154.135). Half of them are members of binary systems. Millisecond pulsars have now been found in 12 globular clusters during the four past years, and almost half of them reside in 47 Tuc.

6.2.7 Space observations

U-band PC HST images of the M15 core have been obtained and show a large core with $r_c = 2.2$ arcsec. This dimension may reflect postcollapse expansion (53.154.015). Far-ultraviolet images for NGC 1904 (56.154.029) and NGC 5139, 5272 and 6205 (56.154.030) were obtained with the Ultraviolet Imaging Telescope and allowed a complete identification of UV-bright objects in these clusters. Four clusters have been imaged at 2000 Å with a balloon-born telescope (55.154.048). High-resolution observations of the core of 47 Tuc with the FOC on the HST reveal a high density of "blue straggler" stars (54.154.138). Their presence in the dense core supports the hypothesis of their binary origin.

6.3 Cluster-by-cluster literature citations

Abbreviations specific to globular clusters

astr = *astrometry, -ic*
bhb = *blue horizontal branch*
B-W = *Baade-Wesselink*
col = *colour*
comp = *comparaison*
dens = *density*
detec = *detection*
FOC = *faint object camera*
fr = *frame*
gal mot = *galactic motion*
gal orb = *galactic orbit*
glob = *globular*
grad = *gradient*
high vel = *high velocity*
HST = *Hubble Space telescope*
id = *identification*
lin pol = *linear polarisation*
m/l = *mass-luminosity*
mod = *model*
morph = *morphology*
orig = *origin*

PC = *planetary camera*
prof = *profile*
prop = *property*
rad = *radial*
ref = *reference*
rgb = *red giant branch*
sd = *subdwarf*
segr = *segregation*
smr = *super metal-rich*
sp vel = *space velocity*
spat str = *spatial structure*
st cont = *stellar content*
surf phot = *surface photometry*
vel disp = *velocity dispersion*

C0021-723 (NGC 104, 47 Tuc)
52.154.006 VIc phot, rgb.
52.154.008 CO band obs, rg.
52.154.021 rad col grad.
52.154.025 C,N,O,Fe abund.
53.154.004 CaII str, abund.
53.154.039 C,N, abund, ms st.
54.154.027 VI CCD phot, rgb, metal.
54.154.125 Mg, Na abund, rg.
54.154.138 HST FOC obs, bs.
54.154.155 2 high vel st.
55.154.042 rel pm, memb.
56.154.012 HST PC obs, core.
56.154.013 CCD sp, abund.
56.154.016 memb, UV-bright st, "BS".
56.154.088 HST PC obs, X-ray src.
56.114.013 CN str, rg.
56.156.007 sp, rv, metal.
Carney et al (1993) BV phot, RR Lyr var.
Cudworth & Hanson (1993) abs pm, sp vel.
Rich et al (1993) IUE int sp.

C0050-268 (NGC 288)
52.154.053 age.
54.154.067 m/l ratio.
54.154.068 mod, abund, rgb.
56.154.072 BV CCD phot, bs, ms bin.
56.156.007 sp, rv, metal.
Guo et al (1993) abs pm, sp vel.

C0100-711 (NGC 362)
52.154.053 age.
54.154.068 mod, abund, rgb.
54.154.124 abs pm.
55.154.043 rel pm, memb.
Rich et al (1993) IUE int sp.

C0310-554 (NGC 1261)
56.154.015 BVR CCD phot.

C0512-400 (NGC 1851)
52.154.006 VIc phot, rgb.
52.154.043 VLA image.
53.154.004 CaII str, abund.
56.154.038 BV CCD phot.

C0522-245 (NGC 1904, M79)
55.154.054 BV CCD phot, metal, bs.
56.114.013 CN str, rg.
56.154.029 UIT image.
56.156.007 sp, rv, metal.

C0647-359 (NGC 2298)
53.154.006 search for RR Lyr var.
56.154.010 Wash phot, rg.
56.154.039 CCD sp, abund, Fe/H.

C0734+390 (NGC 2419)
54.154.021 BVI CCD surf phot.
56.114.013 CN str, rg.

C0911-646 (NGC 2808)
52.154.003 BV CCD phot.
53.154.040 21cm obs.

C1003+003 (Pal 3)
56.154.004 sp, CaII, abund, rg.

C1015-461 (NGC 3201)
Brewer et al (1993) UBVI CCD phot, bs.

C1126+292 (Pal 4)
56.154.004 sp, CaII, abund, rg.

C1207+188 (NGC 4147)
53.154.021 BV CCD phot.
55.154.007 abs pm, gal orb.
55.154.019 IR col prof.
56.114.013 CN str, rg.
56.154.004 sp, CaII, abund, rg.

C1223-724 (NGC 4372)
54.154.044 BVRI CCD phot.
55.154.016 lin pol, rg.
56.154.010 Wash phot, rg.

C1235-509 (Rup 106)
52.154.030 UBV CCD phot, bs.
54.153.032 BVI CCD phot.
55.154.032 young glob clus.
56.154.003 metal, age.
Buonanno et al (1993) comp mod, age.

C1236-264 (NGC 4590)
56.154.010 Wash phot, rg.

C1256-706 (NGC 4833)
55.154.016 lin pol, rg.
56.154.010 Wash phot, rg.

C1310+184 (NGC 5024)
53.126.036 new pulsar.
53.154.002 CCD phot.

C1313+179 (NGC 5053)
53.154.002 CCD phot.
54.154.065 orig bs.
54.154.128 CCD phot, struct.
56.114.013 CN str, rg.
56.154.004 sp, CaII, abund, rg.

C1323-472 (NGC 5139, Omega Cen)
52.154.031 RR Lyr var.
52.154.034 RR Lyr var.
52.154.047 CNO abund, rg.
53.154.022 CCD phot, ms st.
54.154.026 O, Fe abund, rg.
54.154.132 I CCD image, mf.
55.154.016 lin pol, rg.
55.154.022 rad distr, blue sd st.
56.154.007 CNO abund.
56.154.009 uvby CCD phot, ms st.
56.154.029 UIT image.

C1339+286 (NGC 5272, M3)
52.123.005 obs var st.
52.123.007 obs var st.
52.123.009 obs var st.
52.154.012 depp CCD phot.
52.154.046 VLA image.
52.154.062 astr ref st.
53.123.003 obs var st.
54.123.024 obs var st.
54.123.030 obs var st.
54.154.127 rv, vel disp.

55.154.005 IUE obs, UV st.
55.154.006 dens prof, lf.
55.154.048 UV image, bhb st distr.
55.154.057 abs pm.
56.154.030 UIT, IUE obs, UV-bright st.
56.155.014 O, Na, Va, Sc abund, rg.
Cudworth & Hanson (1993) abs pm, sp vel.

C1403+287 (NGC 5466)
52.154.018 CCD phot, bs, eb, var st.
54.154.067 m/l ratio.

C1452-820 (IC 4499)
Sarajedini (1993) BV CCD phot, bs.

C1514-208 (NGC 5897)
55.154.053 UBVI CCD phot, bs.
56.154.005 BV CCD phot, bs.
56.154.010 Wash phot, rg.

C1516+022 (NGC 5904, M5)
52.154.062 astr ref st.
54.154.126 BV CCD phot, RR Lyr var.
54.154.127 rv, vel disp.
54.154.132 I CCD image, mf.
55.122.014 K phot, RR Lyr var.
55.122.015 rv curve, RR Lyr var.
55.154.013 new var st.
55.154.028 sp, CN, CH band str.
55.154.048 UV image, bhb st distr.
56.114.013 CN str, rg.
56.122.126 B-W, RR Lyr var.
56.154.001 B pg phot, RR Lyr var, per.
56.154.004 sp, CaII, abund, rg.
56.154.021 O, Na, Fe abund, rg.
56.154.037 B pg phot, RR Lyr st, per.
56.154.040 Vi, JK phot, RR Lyr var.
Cudworth & Hanson (1993) abs pm, sp vel.
Smith & Norris (1993) sp, CN str, AGB.

C1524-505 (NGC 5927)
53.154.005 Wash phot.
56.154.057 BV pg phot, red.

C1531-504 (NGC 5946)
54.154.072 BV CCD phot.

C1608+150 (Pal 14)
55.154.008 deep BV CCD phot.
56.154.004 sp, CaII, abund, rg.

C1614-228 (NGC 6093, M80)
53.154.006 search for RR Lyr var.
56.154.020 IUE UV obs, col grad.

C1620-264 (NGC 6121, M4)
52.154.025 C,N,O,Fe abund.
52.154.046 VLA image, point src, pulsar.
54.154.127 rv, vel disp.
55.154.016 lin pol, rg.
56.154.004 sp, CaII, abund, rg.
56.154.013 CCD sp, abund.
56.154.032 Na, Al, O abund, rg.
Cudworth & Hanson (1993) abs pm, sp vel.
Smith & Norris (1993) sp, CN str, AGB.

C1620-720 (NGC 6101)
56.154.010 Wash phot, rg.

C1624-387 (NGC 6139)
53.154.001 phot, var st, RR Lyr.

C1629-129 (NGC 6171, M107)
52.154.048 spat str.
54.154.142 BV CCD phot.
55.154.018 BV pg phot, pm, memb.
55.154.038 deep BV CCD phot, lf.
56.154.004 sp, CaII, abund, rg.
Cudworth & Hanson (1993) abs pm, sp vel.

C1639+365 (NGC 6205, M13)
52.154.027 CCD phot, lf.
52.154.062 astr ref st.
53.154.009 C,N,O abund, rg.
53.154.014 Ca,Na,Fe abund, rg.
53.126.036 new pulsar.
54.154.127 rv, vel disp.
55.154.048 UV image, bhb st distr.
56.114.013 CN str, rg.
56.154.004 sp, CaII, abund, rg.
56.154.019 mass, st cont.
56.154.030 UIT, IUE obs, UV-bright st.
56.154.034 K phot, d.
56.155.014 O,Na,Va,Sc abund, rg.
Cudworth & Hanson (1993) abs pm, sp vel.

C1644-018 (NGC 6218)
54.154.039 pos, pm, UV st.
54.154.124 abs pm.
54.154.127 rv, vel disp.
55.154.007 abs pm, gal orb.

C1645+476 (NGC 6229)
53.154.010 BV CCD phot.

C1654-040 (NGC 6254, M10)
52.154.041 IR phot, i/s ext, d.
53.154.026 IR phot, rg.
54.154.127 rv, vel disp.

C1657-004 (Pal 15)
54.154.070 BV CCD phot.

C1658-300 (NGC 6266)
55.154.016 lin pol, rg.

C1707-265 (NGC 6293)
53.154.016 BV CCD phot.

C1711-294 (NGC 6304)
56.154.071 VK image, rg, HB st.

C1713-280 (NGC 6316)
56.154.071 VK image, rg, HB st.

C1716-184 (NGC 6333, M9)
53.154.006 search for RR Lyr var.
53.154.016 BV CCD phot.

C1715+432 (NGC 6341, M92)
52.154.007 lf, turn off st.
52.154.046 VLA image.
53.154.036 bin freq.
55.122.011 BV CCD phot, RR Lyr var.
55.122.014 K phot, RR Lyr var.
55.122.015 rv curve, RR Lyr var.
55.154.006 dens prof, lf.
55.154.023 BV pg phot, pm, sp vel.
55.154.048 UV image, bhb st distr.
56.114.013 CN str, rg.
56.122.126 B-W, RR Lyr var.
56.154.004 sp, CaII, abund, rg.
56.154.040 Vi, JK phot, RR Lyr var.
56.154.067 Vilnius phot.
Cudworth & Hanson (1993) abs pm, sp vel.

C1720-263 (NGC 6356)
54.154.027 VI CCD phot, rgb, metal.
Rich et al (1993) IUE int sp.

C1721-484 (NGC 6352)
Rich et al (1993) IUE int sp.

C1724-307 (Terzan 2)
54.154.101 X-ray src.
55.154.009 IR phot, rgb.

56.142.088 SIGMA obs, X-ray src.

C1732-304 (Terzan 1)
54.154.027 VI CCD phot, rgb, metal.
Ortolani et al (1993) BVRIz CCD phot.

C1732-447 (NGC 6388)
Rich et al (1993) IUE int sp.

C1736-536 (NGC 6397)
52.154.006 VIc phot, rgb.
52.154.022 G-band str.
53.154.004 CaII str, abund.
54.154.029 rv, vel disp, dyn.
55.154.020 uvby CCD phot, bin det.
56.114.022 near IR sp, rg.
56.154.008 rad distr bs, surf bright.
56.154.010 Wash phot, rg.
Cudworth & Hanson (1993) abs pm, sp vel.

C1735-032 (NGC 6402, M14)
53.154.017 HST FOC U,B image, Nova Oph 1938.

C1745-247 (Terzan 5)
52.126.043 eclipsing pulsar.
52.126.045 eclipsing pulsar.
53.117.135 bin MSP.

C1746-370 (NGC 6441)
Rich et al (1993) IUE int sp.

C1747-312 (Terzan 6)
53.154.023 ROSAT X-ray src.

C1755-442 (NGC 6496)
53.154.005 Wash phot.
Rich et al (1993) IUE int sp.

C1801-300 (NGC 6528)
54.154.027 VI CCD phot, rgb, metal.
55.154.041 deep BVRI CCD phot.
Rich et al (1993) IUE int sp.

C1806-259 (NGC 6553)
52.154.015 BVRI CCD phot.
54.154.027 VI CCD phot, rgb, metal.
55.154.049 sp 1 st, metal abund.
56.154.022 clump morph, age.

C1820-303 (NGC 6624)
52.154.043 VLA image.
55.154.040 CCD fr, pol, ext, dust.
55.154.052 mod, dyn, vel disp.
56.154.002 deep VLA image.
Rich et al (1993) IUE int sp.

C1821-249 (NGC 6626, M28)
52.122.082 post-AGB st.
52.154.019 BV pg phot, var st.
52.154.046 VLA image, point src, pulsar.
54.154.043 BV pg phot, pm.
Cudworth & Hanson (1993) abs pm, sp vel.
Rich et al (1993) IUE int sp.

C1828-323 (NGC 6637)
53.154.011 JHK phot.
Rich et al (1993) IUE int sp.

C1828-235 (NGC 6642)
Hazen (1993) phot, new var st, RR Lyr.

C1832-330 (NGC 6652)
53.154.023 ROSAT X-ray src.

C1833-239 (NGC 6656, M22)
52.134.035 neb plan.
52.154.016 i/s pol.
52.154.025 C,N,O,Fe abund.
52.154.035 i/s, st line prof.
53.154.014 Ca,Na,Fe abund, rg.
54.134.053 mod neb plan.
54.154.125 Mg,Na abund, rg.
55.154.016 lin pol, rg.
56.114.038 abund, CH st.
56.131.077 radio, IRAS obs, i/s mat.
56.154.013 CCD sp, abund.
56.154.067 Vilnius phot.
Cudworth & Hanson (1993) abs pm, sp vel.

C1850-087 (NGC 6712)
52.154.033 X-ray bin, opt id.
52.154.039 radio em, X-ray src.
52.154.043 VLA image, radio id.
52.154.045 gal mot.
56.154.011 UBV CCD phot, sp.
Cudworth & Hanson (1993) abs pm, sp vel.

C1906-600 (NGC 6752)
52.154.006 VIc phot, rgb.
52.154.038 rg, chrom act.
52.154.056 UV survey, IUE, hot st.
53.154.004 CaII str, abund.
54.154.125 Mg,Na abund, rg.
54.154.132 I CCD image, mf.
56.114.013 CN str, rg.
56.154.020 IUE UV obs, col grad.
Smith & Norris (1993) sp, CN str, AGB.

C1936-310 (NGC 6809, M55)
52.154.022 G-band str.
54.154.067 m/l ratio.
56.154.006 BVRI CCD phot.
56.154.010 Wash phot, rg.
56.154.021 abund.

C1942-081 (Pal 11)
56.154.004 sp, CaII, abund, rg.

C1951+186 (NGC 6838, M71)
52.154.027 CCD phot, lf.
54.154.127 rv, vel disp.
55.154.010 deep BV CCD phot.
55.154.014 test mod, dens prof, mf.
56.154.004 sp, CaII, abund, rg.
56.154.031 CN, CH str.
56.113.012 CCD Vilnius phot, i/s red.
56.154.067 Vilnius phot.
Cudworth & Hanson (1993) abs pm, sp vel.

C2059+160 (NGC 7006)
54.154.066 BV CCD phot.

C2127+119 (NGC 7078, M15)
52.154.006 VIc phot, rgb.
52.126.021 radio pulsar.
52.142.041 X-ray burst.
52.154.031 RR Lyr var.
52.154.043 VLA image, radio id.
52.154.046 VLA image, point src, pulsar.
53.126.051 pulsar obs.
53.154.004 CaII str, abund.
53.154.015 HST PC core image.
53.154.029 CaII em, bin st.
54.154.020 lf.
55.154.012 col grad.
55.154.025 mod, evol RR Lyr var.
55.154.052 mod, dyn, vel disp.
56.114.013 CN str, rg.
56.122.135 K1082, new type, var st.
56.154.010 Wash phot, rg.
56.154.026 pos, pm.
56.154.068 Vilnius phot, i/s red.
Cudworth & Hanson (1993) abs pm, sp vel.

Durrell & Harris (1993) BCV CCD phot, lf.

C2130-010 (NGC 7089, M2)
52.154.006 VIc phot, rgb.
53.154.004 CaII str, abund.
56.114.013 CN str, rg.
Cudworth & Hanson (1993) abs pm, sp vel.

C2137-234 (NGC 7099, M30)
53.154.036 bin freq.
56.114.013 CN str, rg.
56.154.010 Wash phot, rg.
56.154.020 IUE UV obs, col grad.
56.154.023 vel disp.
56.156.007 sp, rv, metal.

C2143-214 (Pal 12)
52.154.006 VIc phot, rgb.

C2305-159 (NGC 7492)
54.154.071 BV phot, bs, mass segr.

6.4 References to 1993 papers on globular clusters

Brewer J.P., Fahlman G.G., Richer H.B., et al 1993, AJ 105, 2158
Buonanno R., Corsi C.E., Fusi Pecci F.F., et al 1993, AJ 105, 184
Carney B.W., Storm J., Williams C. 1993, PASP 105, 294
Cudworth K.M., Hanson R.B. 1993, AJ 105, 168
Durrell P.R., Harris W.E. 1993, AJ 105, 1420
Guo X., Girard T.M., van Altena W.F., Lopez C.E. 1993, AJ 105, 2182
Hazen M.L. 1993, AJ 105, 557
Ortolani S., Bica E., Barbuy B. 1993, A&A 267, 66
Rich R.M., Minniti D., Liebert J. 1993, ApJ 406, 489
Sarajedini A. 1993, AJ 105, 2172
Smith G.H., Norris J.E. 1993, AJ 105, 173

7 Magellanic Cloud clusters

7.1 Large Magellanic Cloud clusters

A new catalogue of 1762 clusters in the Large Magellanic Cloud based on the ESO/SERC Southern Sky Atlas has been constructed by Kontizas et al (52.156.002). Bhatia et al (53.156.092) prepared a catalogue of 64 pairs of clusters. They give accurate coordinates and a photographic atlas. van den Bergh (53.156.133) summarized the properties of Magellanic Cloud clusters. They differ from those in the Galaxy in a number of important respects (radii, ellipticity, population size). Chiosi (53.156.158) discusses the main properties of the family of star clusters in the LMC. Walborn (53.156.033) presented a review of the stellar population and current observing programmes on the 30 Doradus region which is the region of the LMC which is now subjected to many studies, especially from space (X-ray, IR, HST). As depicted by the literature survey summarized in section 7.3, CCD photometry represents a large fraction of the observations conducted in the Magellanic Cloud clusters. Many studies aim at obtaining new colour-magnitude diagrams to infer the age, compare with theoretical isochrones and determine the luminosity function. CCD photometry is also used to discover and observe variable stars in Magellanic Cloud clusters: cepheids (52.156.057), RR Lyrae (52.154.024, 55.154.017, 56.156.008, Welch et al 1993) or other (eclipsing binaries, Be variables) (52.153.045, 55.154.055, Balona 1993). Integrated photometry has been performed by Bica et al (54.156.034). UBV photometry is now available for 624 LMC clusters. Elson (53.156.147) presented surface-brightness profile and colour-magnitude diagrams for 18 LMC star clusters. The profiles of the older clusters are well represented by models with a King-like core, while younger clusters show departure from such models. Olszewski et al (53.156.017) measured the radial velocity and equivalent width of the calcium triplet at $\lambda \sim 8500$ Å for more than 150 stars in 80 LMC clusters. They obtained an estimate of the metallicity and found little radial abundance gradient in the LMC. The structure of three LMC clusters has been investigated by measuring the core velocity dispersion from integrated light spectra (52.154.032). The inferred structure is similar to that of galactic globular clusters. The detailed structure and dynamics of NGC 1850 (Fischer et al 1993), NGC 1866 (55.154.015), NGC 1978 (56.151.010) have been studied by CCD images over a large field and radial velocities of supergiant members.
Theoretical isochrones have been compared to the colour-magnitude diagrams of LMC clusters to check for envelope semiconvection and core overshooting (54.154.032, 56.156.006) or canonical models (54.156.023).

Balona L.A. 1993, MNRAS 260, 795
Fischer P., Welch D.L., Mateo M. 1993, AJ 105, 938
Welch D.L., Mateo M., Olszewski E.W., Fischer P., Takamiya M. 1993, AJ 105, 146

7.2 Small Magellanic Cloud clusters

Kontizas (52.156.006) examined the morphology of 34 populous SMC clusters and measured the projected ellipticity. The inner-region clusters seem more elliptic than the outer ones. The SMC clusters are on the whole more elliptic than the LMC clusters.

7.3 Cluster-by-clusters literature citations

NGC 121
54.154.004 UBVRI phot seq.
NGC 152
Alvarado (1993) UBVRI pe phot, seq.
NGC 330
54.154.001 C,N abund.
54.154.069 C,N,O,Fe abund.
55.154.001 uvby CCD phot, metal.
55.154.003 Hα, uvby CCD phot, Be st.
55.154.055 short per var st.
55.156.047 IUE UV obs.
Caloi et al (1993) IUE obs, blue st.
NGC 456
53.156.056 sp cl.
NGC 460ab
53.156.056 sp cl.
NGC 465
53.156.056 sp cl.
NGC 1466
56.156.008 phot, RR Lyr var.
NGC 1711
52.153.060 CCD phot.
54.154.031 mf.
54.156.005 BV CCD phot.
Subramaniam et al (1993) st distr.
NGC 1712
52.153.045 CCD phot, var st.
52.153.061 CCD phot.
NGC 1722
52.153.061 CCD phot.
NGC 1727
52.153.061 CCD phot.
NGC 1831
56.156.006 BV CCD phot.
NGC 1835
52.154.032 vel dis, m/l ratio.
NGC 1841
52.154.024 BV CCD phot, RR Lyr st.
56.156.007 sp, vel, metal.
NGC 1850
Fischer et al (1993) dyn, dbl cl.
NGC 1866
53.122.001 CCD phot, sp, rv, var st.
53.122.012 B-W radius, cep var.
54.154.032 test mod.
55.154.015 V CCD image, rv, dyn.
Welch & Stetson (1993) search var st.
NGC 1978
56.151.010 BV CCD image, rv rg, dyn.
Alvarado (1993) UBVRI pe phot, seq.
NGC 2004
54.154.031 mf.
54.156.005 BV CCD phot.
54.156.023 BV CCD phot.
Balona (1993) CCD monit, short per var.
Balona & Jerzykiewicz (1993) UBV CCD phot.
Subramaniam et al (1993) st distr.
NGC 2031
Mould et al (1993) BR CCD phot, age.
NGC 2100
54.154.031 mf.
54.156.005 BV CCD phot.
Balona (1993) CCD monit, short per var.
Balona & Jerzykiewicz (1993) UBV CCD phot.
Subramaniam et al (1993) st distr.
NGC 2121
Alvarado (1993) UBVRI pe phot, seq.
NGC 2157
52.154.057 Cep var.
NGC 2164
53.154.008 BV CCD phot.
54.154.031 mf.
54.156.005 BV CCD phot.
Subramaniam et al (1993) st distr.
Welch et al (1993) BV CCD phot, var st.
NGC 2210
56.122.006 pe, pg, CCD phot, var st.
NGC 2214
54.154.025 CCD phot, rgb.
54.154.031 mf.
54.156.005 BV CCD phot.
56.154.065 UBVI CCD phot, age.
Subramaniam et al (1993) st distr.
Hen N59A
56.156.012 UBV phot, st cont.
Hen N90
53.156.031 opt, UV sp, O,B st.
Hen N159A
55.132.025 OB st, st cont.
56.156.002 UBVRI CCD phot
Lindsay 11
56.154.075 BR CCD phot, age.
30 Dor
53.156.033 review.
53.156.143 X-ray image.
54.156.002 st form.
55.156.031 UBV phot, sp cl.
56.153.008 HST PC image, 12 st.
56.153.020 HST UB phot.
56.156.013 ASTRO-1, UIT image.
56.156.015 IR survey, proto st.
56.156.018 UBV phot, sp cl, imf.
GLC0435-59
55.154.017 BV CCD phot, RR Lyr var.
53.156.017 rv, 150 st, 80 cl.
53.156.092 cat bin cl.
54.156.034 UBV int col, 64 cl.
56.156.007 sp, vel, metal.

7.4 References to 1993 papers on Magellanic Cloud clusters

Alvarado F., Wenderoth E., Alcaino G., Liller W. 1993, AJ 105, 2118
Balona L.A. 1993, MNRAS 260, 795
Balona L.A., Jerzykiewicz M. 1993, MNRAS 260, 782
Caloi V., Cassatella A., Castellani V., Walker A.R. 1993, A&A 271, 109
Fischer P., Welch D.L., Mateo M. 1993, AJ 105, 938
Mould J.R., Xystus D.A., Da Costa G.S. 1993, ApJ 408, 108
Subramaniam A., Sagar R., Bhatt H.C. 1993, A&A 273, 100
Welch D.L., Mateo M., Olszewski E.W., Fischer P., Takamiya M. 1993, AJ 105, 146
Welch D.L., Stetson P.B. 1993, AJ 105, 1813

8 Clusters in other galaxies

The review by Harris (1992) on globular systems in other galaxies emphasizes the results obtained during the past decade.

Harris W.E. 1992, Ann. Rev. Astr. Ap. 29, 543

8.1 Cluster-by-clusters literature citations

NGC 224, M31
52.154.002 IUE UV flux, 3 glob clus.
52.154.044 spect, metal.
53.157.023 CCD image.
53.154.028 metal 150 clus.
53.157.017 tidal radii glob clus.
56.157.005 age, metal.
56.157.006 spect smr cl.
56.157.044 clus lf.
56.157.185 JHK phot.
Battistini et al (1993) new clus.
NGC 598, M33
53.157.024 vel, kin, vel disp.
NGC 1275
52.157.102 young clus, Hα em.
NGC 1399
53.157.016 BV CCD phot, prop.
53.157.047 BV CCD phot, prop.
Ostrov et al (1993) metal grad, distr.
NGC 1404
56.157.046 glob syst.
NGC 3031, M81
54.154.028 BV pg phot.
NGC 4365
53.154.034 B CCD phot, lf.
NGC 4472
53.154.034 B CCD phot, lf.
NGC 4649
53.154.034 B CCD phot, lf.
NGC 5128
56.157.033 Wash phot, metal.
NGC 7814
56.157.106 detec glob clus.
Fornax
56.157.249 vel disp, metal.

8.2 References to 1993 papers

Battistini P.L., Bonoli F., Casavecchia M., Ciotti L., Federici L., Fusi Pecci F. 1993, A&A 272, 77
Ostrov P., Geisler D., Forte J.C. 1993, AJ 105, 1762

Commission 38: Exchange of Astronomers (*Echange des Astronomes*)
(Commission of the Executive Committee)

President: J.Sahade Vice-President: H.E. Jorgensen

Organizing Committee: D.M. Chitre, J.R. Ducati, G. Krishna, M. Morimoto, M.S. Roberts & G.F. Smith

Members: M.K. Aly, A.A. Al-Sabti, A.A. Boyarchuk, B. Caccin, A. Florsch, H.F. Haupt, Y. Kozai, D.A. Macrae, K.C. Leung, M. Marik, E.A. Müller, I.-S. Nha, S. Ninkovic, P.N. Okeke, P.M. Routly, G. Swarup, C.R. Tolbert, E.P.J. van den Heuvel, S.-G.Wang, F.B. Wood & S.-H. Ye.

Up to September 30, 1993, the grants given by Commission 38 in the triennium were the following:

Name	Origin	Destination
Benaglia P.	Instituto Argentino de Radioastronomia Villa Elisa (Bs. As.) Argentina	National Radio Astronomy Observatory Albuquerque, NM USA
Colombo E.	Instituto de Astronomia y Fisica del Espacio Buenos Aires, Argentina	Fred L. Whipple Observatory Amado AZ USA
Frolov P.	P.N. Lebedev Physical Institute Moscow Russia	Copenhagen University Observatory Copenhagen, Denmark
Godunova V.	Main Astronomical Observatory Kiev, Ukraine	Graz University of Technology Graz, Austria
Goldes G.	Cordoba Observatory Cordoba, Argentina	Marseille Observatory Marseille, France
Gu Z.-N.	Shanghai Observatory Shanghai, China PR	Royal Observatory Brussels, Belgium
Li Z.-S.	Shanghai Observatory Shanghai, China PR	Paris Observatory Paris, France
Nazarova L.	Sternberg Astronomical Institute Moscow, Russia	Royal Greenwich Observatory Cambridge, UK
Pritchard J.	University of Canterbury Christchurch New Zealand	Institut d'Astrophysique de Paris Paris, France
Qin Z.	Purple Mountain Observatory Nanjing, China PR	Max-Planck Institut für Radioastronomie Bonn, Germany
Sersen M.	Comenius University Bratislava, Slovak R	Inter-University Center for Astronomy & Astrophysics Pune, India

J. Bergeron (ed.), Reports on Astronomy, Vol. XXIIA, 445–446.

Singh H.P.	University of Delhi Delhi, India	University of London London, UK
Tong Y.	Dpt. Astronomy Beijing Normal University China PR	Instituto Nacional de Pesquisas Espaciais Sao Jose dos Campos Brazil
Urpin V.	Ioffe Institute for Physics & Technology St Petersburg, Russia	Tata Institute of Fundamental Research Bombay, India
Zhao J.	Shanghai Observatory Shanghai, China PR	Dominion Astrophysical Observatory Victoria BC, Canada

Ten additional requests had to be turned down because they did not meet the Commission guidelines, and another one did not provide all the required information.

Reports of the works performed under the Commission 38 grants were received from one grantee of the previous period and from five of the grantees of the present period.

40. Radio Astronomy (Radioastronomie)

PRESIDENT: M. Morimoto (Japan)
VICE-PRESIDENT: J.B. Whiteoak (Australia)
ORGANIZING COMMITTEE: D.C. Backer (USA), R. Fanti (Italy), R. Guesten (Germany), J. Moran (USA), J.M. van der Hulst (Netherlands), R. Davis (UK), J. Gomez-Gonzales (Spain), F.R. Colomb (Argentina), Ye Shuhua (China), P.G. Mezger (Germany), E. Gerard (France), T. Velusamy (India), L.B. Baath (Sweden), E.E. Baart (S. Africa), I. Gossachinskiy (CIS)

1 INTRODUCTION

Radio astronomy has been providing new frontiers to astronomy and this was still true during the last three years. In this report, we concentrate on discussing such new possibilities. Observational results and interpretations are discussed more in detail in the reports of other commissions and we will avoid mentioning these except when the results demonstrate that a new possibility is opening up.

2 RADIO TELESCOPES

2.1 Single Aperture Telescopes

2.1.1 Sub-millimeter wavelength telescopes

Sub-millimeter wave astronomy is growing very fast. Another major observing facility has been added to the three already existing.
The Caltech Submillimeter Observatory (CSO) telescope is operating routinely at 230, 345 and 460 GHz. A 5 x 5 element array receiver is planned for 490 GHz.
For the 15-m James Clerk Maxwell Telescope (JCMT) on Mauna Kea a 64-element array heterodyne receiver at 345 GHz is being developed at Cambridge University (MARS). For continuum observations, 91-pixel and 37-pixel bolometer arrays are being built for operation at 650 and 350 GHz respectively (SCUBA).
For the Swedish-ESO Submillimeter Telescope (SEST) a new SIS receiver covering the band 320-360 GHz was installed in 1992 (Nyman et al., 1990, Liege International Astrophysical Colloquium, p. 159). Its noise temperature is 200 K (SSB) and system temperatures as low as 420 K (SSB) have been realized
The first permanent sub-millimeter wavelength telescope for location at the South Pole, the 1.7-m Antarctic Submillimeter Telescope and Remote Observatory (AST/RO) instrument is fully assembled and being tested with a 490 GHz receiver in Boston before deployment in 1994 (Lo et al., 1991, IAU Symposium No. 148, p.112). Its first major task will be a CI line survey.
The Sub-millimeter Telescope (SMT), a joint project of the Max Planck Institut für Radioastronomie (MPIfRA) and the University of Arizona, was dedicated in September 1993 (Baars et al., 1990, Liege International Astrophysical Colloquium, p. 293). The spectroscopic systems have SIS receivers at 350, 450, and 690 GHz and acousto-optical spectrometers. He-cooled bolometers are available for continuum observations.
A 30-cm rocket borne telescope with a 12-element bolometer array is under construction in Japan for launch in 1994. It will observe dust emission in the Orion region at wavelengths of 0.5 and 0.25 mm .

J. Bergeron (ed.), Reports on Astronomy, Vol. XXIIA, 447–459.

2.1.2 Large millimeter to meter wavelength telescopes

A 100-m telescope (Green Bank Telescope; GBT) is under construction in Green Bank, USA by National Radio Astronomy Observatory (NRAO) as a replacement of the old 91-m transit telescope. It will use offset Gregorian optics to reduce diffraction caused by the subreflector and its support structure. This will greatly improve aperture efficiency and reduce side lobe level and noise temperature. It will be equipped with a laser survey system for actively controlling the surface, and will have a useful performance at mm-wavelengths. Its completion is planned for 1995 and it will be a very important addition to radio astronomy.
The 14-m telescope of the Five College Radio Astronomy Observatory (FCRAO) has been used extensively with a 15 element focal plane array (QUARRY), which operates in the 86-115 GHz band (Goldsmith et al., 1991, Bull. AAS, **23**, 1452). Efficient signal processing is provided by a new 15 x 1025 channel spectrometer.
The NRAO 12-m telescope has an SIS receiver at 100 GHz that has a receiver temperature of 19 K (DSB) and system temperatures as low as 130 K (SSB). The Schottky diodes in the 8 element (4 x 2) array receiver for the 215-245 GHz band have been replaced by SIS junctions with fixed tuning.
The surface of the 45-m telescope of the Nobeyama Radio Observatory (NRO) was reset in 1992, further improving its rms surface accuracy from 95 to 65 microns. It is now equipped with a 4-element SIS receiver at 115 GHz and a 7-element bolometer array at 150 GHz. A 25-element SIS array receiver is under development.
The surface accuracy of the 100-m telescope of MPIfRA and the 37-m telescope of Haystack Observatory were improved to have useful performance up to 100 GHz. The improvements will be of particular importance for mm-wave VLBI.
The size of the panels of RATAN-600 has been increased from 7.4 m to 11.4 m and the geometrical area in the zenith is about 15,000 m^2. Spillover was decreased to less than 3 K. FET and HEMT receivers give sensitivities better than 10 mK for 1 sec integration at cm wavelengths. Part of its surface accuracy was improved to 0.085 mm (rms) and a test for quasi-real time adaptive optics using simple one dimensional holography was successful.
The 54-m "Arecibo" style antenna of the Radio Measurements Institute in Armenia is situated in the Aragats mountains. The rms surface accuracy is 83 microns and operation up to 150 GHz is possible. The effective use of the facility has been hampered by the lack of receivers and by operational constraints.
The FCRAO in collaboration with several groups in Mexico is planning a 50-m, radome enclosed, telescope that will operate in the range of 80 to 350 GHz. The preliminary design consists of a parabolic primary reflector of 126 hexagonal panels that are actively controlled.
A new mirror system for correcting spherical aberration is being installed on the Arecibo 310-m telescope. Together with recent resurfacing and a new shielding fence, it will greatly enhance the performance of this powerful telescope. This will be important for planetary radar, pulsar timing and many other types of observations.
Constructions of medium sized (25-30 m) telescopes in Italy, China, Poland, and Russia are in progress. They will be used mainly for VLBI observations.

2.1.3 Survey telescopes in sub-millimeter and millimeter-wavelength astronomy

The Cologne 3-m telescope on Gornergrat, Switzerland, has a surface accuracy of 30 microns and receivers at 115 and 230 GHz. It is equipped with acousto-optical spectrometers (Tolls et al., 1990, Liege International Astrophysical Colloquium, p. 299).
The 0.9-m telescope of the Harvard-Smithsonian Center for Astrophysics in Cambridge (formerly at Columbia University) has been used during the past triennium to map molecular clouds in the outer Galaxy and M31 in the CO 1-0 line (Koper et al., 1991, *Ap. J.*, **383**, L11) and to search for gas far out of the plane in our Galaxy.
The 0.6-m telescope of the University of Tokyo and Nobeyama Radio Observatory, located at Nobeyama, has been dedicated to mapping the Galactic plane in the CO 2-1 line for comparison with the Columbia-CFA survey.
The pair of 4-m telescopes at the Nagoya University, Japan, is involved in a 600,000-point survey of the CO 1-0 line covering 1,500 square degrees of sky.
There is a plan in Japan to build a 2-m sub-mm telescope on the summit of the Mount Fuji.

2.1.4 Multi-beaming of radio telescopes

If more than one feed horn are arranged at the focus, the telescope will have more than one beam on the sky. This will ease the long-standing frustration of radio astronomers who would like to observe several positions on the sky simultaneously.
The main aims of multi-beaming are, a) to greatly reduce the time necessary to survey a certain area of sky, and b) to remove the effects of atmospheric fluctuations by comparison of adjacent beams. These two points are extremely important in mm-wave telescopes, where observations are severely limited by fluctuations in atmospheric emission and absorption, and where the telescope beams are usually very narrow.
For such arrangements to be effective, there will occur new technical problems. At present, the most important are the signal and data handling and feed designs for effective use of the focal plane. Developments of wide-angle optical system to get a focal plane like optical telescopes will become very important. After all such developments, eventually, all radio telescopes will have fields of view. Then new possibilities will open up.
Among some obvious examples of possibilities with multi-beaming are phase referenced VLBI observations of sources nearby in the sky, online correction of telescope pointing using a point source in the field, correcting and/or adjusting the telescope beam by coherently combining beam outputs, and so on.
The first attempt at multi-beaming was with the old 91-m transit telescope of NRAO in the early 60's. It is also extensively used with the 100-m telescope of MPIfRA.
At present, efforts are more concentrated on mm-wavelength operation. The first one was the 8-beam arrangement on the 12-m mm-wave telescope of NRAO. It used 8-feed horns with Schottky mixer receivers. Because of the limitation of receiver performance this system was not used extensively and is being replaced by SIS receivers.
At present, 15-beam arrangement on the 14-m telescope of FCRAO, an 8-beam system of the 12-m telescope of NRAO and 4-beam system on the 45-m telescope of Nobeyama Radio Observatory are in use. There are many more such systems being planned or about to be installed on telescopes. They are listed in Table 1.

Table 1. Multi-Beam Receivers

Institute	Frequency	Beams	Type	Comment
NRO	108-116 GHz	4	SIS	fixed tuning
	80-116 GHz	25	SIS	fixed tuning *
	40-50 GHz	6	SIS	planned
	1, 2 mm	7	bolometer	
MPIfRA	32 GHz	9	HEMT	*
	15 GHz	4	HEMT	*
	10.55 GHz	4	HEMT	
	4.75 GHz	2	HEMT	*
IRAM	230, 850 GHz	7	bolometer	
			SIS	planned
FCRAO	80-116 GHz	15	Schottky	
NRAO	215-245 GHz	8	SIS	fixed tuning *
JCMT	0.35, 0.6 mm	91	bolometer	*
	0.75, 1.1, 1.4, 2 mm	37	bolometer	*

* Under construction

The 64-element array receiver being developed at Cambridge for the JCMT will coherently combine the outputs so that beam correction etc. is possible.
With the RATAN-600, a new "array mode" is being designed to have a field of view up to the beam of single panel.

2.2 Antenna Arrays

In Table 2, parameters of Arrays/Interferometers are summarized.

2.2.1 Sub-millimeter wavelength arrays

The Smithsonian Submillimeter Array (SMA) is midway through construction and is expected to be completed by 1997. The initial instrument consists of six 6-m antennas distributed among 24 pads in 4 concentric layouts. The antennas will be placed along the sides of Reuleaux triangles, which have been found to provide the most uniform uv plane coverage within a circular boundary in the uv plane.
The array will be sited on Mauna Kea near the CSO and JCMT, and will be linked for combined operation.
The JCMT and CSO antennas have been linked to form the first sub millimeter wavelength interferometer. The baseline is 150 meters. Even with only a single baseline this instrument will be able to measure the angular sizes of compact objects, act as a spatial filter to study absorption lines against compact background objects and measure the relative positions of maser clusters. Initial results include the mapping of water vapor masers in the stellar envelop of VY CMa (Carlstrom, 1993, *Bull. AAS*, **25**, 859).

2.2.2 Millimeter wavelength arrays

The Berkeley Illinois Millimeter Array (BIMA) is in the process of being upgraded from 3 to 9 antennas. Two additional antennas have been funded and will be built within two years. Long baseline stations have been added, increasing the maximum baseline to 800 meters. Presently the array operates in the 85-115 GHz band but three more receivers will be added per antenna to cover the 70-90 GHz, 130-180 GHz, and 210-270 GHz bands. The new correlation spectrometer, based on the Bos chip, can handle a bandwidth of 800 MHz in a variety of modes with up to four separate spectra and up to 1024 channels/baseline.
The IRAM array on the Plateau de Bure was expanded to 4 antennas in 1993. The surface accuracy of the fourth 15-m antenna is 70 microns. Funds have been approved for 2 more antennas. The surfaces of these antennas will be made from machined aluminum rather than aluminum-coated carbon fiber.
The millimeter array of the Nobeyama Radio Observatory (NRO) will be expanded from five to six antennas. The new antenna is expected to have rms surface accuracy of 40 microns and 78 % aperture efficiency at 115 GHz (compared with 60 microns on the older antennas). A new correlator is under construction based on an LSI chip with 32 lags operating at 32 MHz clock rate. With time demultiplexed operation it will handle a 2 GHz bandwidth with 256 channels/baseline. The 45-m telescope will be added to the array, greatly enhancing its sensitivity.
NRO has begun to plan for a next generation millimeter wavelength array of approximately fifty 10-m antennas. Sites under consideration are in Hawaii, Tibet and northern Chile. Extensive site testing in northern Chile with phase monitoring instrumentation will begin soon.
A four year development plan for a Millimeter Array (MMA) has been submitted by NRAO to the NSF. The array would have forty 8-meter elements. Testing of the primary sites on South Baldy, New Mexico, and Mauna Kea, Hawaii, continue. The antenna design is in an advanced state and now consists of an uncounterweighted slant-axis parabola.

2.2.3 Centimeter to meter wavelength arrays

MERLIN was up-graded with the addition of the 32-m telescope at Cambridge. Now it has 7 telescopes and 21 baselines. Maximum baseline is now 218 km. Correlator was improved to have 2 x 16 MHz channels with 2 bit quantization. With a 12-hour integration, its sensitivity is 1,000K in brightness and angular resolution is 50 mas at 5 GHz.
The Giant Meter-wavelength Radio Telescope (GMRT) with thirty 45-m telescopes is under construction in India. It will work between 40 MHz and 1.4 GHz. Two of the antennas are now complete.

The Nobeyama Radio Heliograph was completed in 1992. It is a T-array of eighty four 80-cm antennas on the baselines of 500 m EW and 200 m NS. It produces 10" resolution maps of the Sun with a maximum time resolution of 50 ms. At present, the observing frequency is 17 GHz and an addition of a second frequency of 34 GHz is planned (also see Section 5.3).
For observing anisotropy of the Cosmic background radiation, an interferometer consisting of three 70-cm telescopes is being constructed at Cambridge.
The Compact Array of the ATNF now provides dual polarization, dual frequency operation in four frequency bands between 20 cm and 3.5 cm, on all of its 22-m antennas.

3 VLBI

3.1 Networks

The Very Long Baseline Array (VLBA) is a dedicated VLBI telescope, consisting of ten 25-m antennas distributed over the territory of USA, equipped with "state of the art" receiving, recording, correlation and image processing facilities. It will provide imaging observations with up to 0.1-mas resolution depending on observing frequencies (0.3 through 40 GHz).
The Joint Institute for VLBI in Europe (JIVE) is being established as the permanent basis for VLBI observations in Europe. The Institute is sited in the Netherlands and its first major project is to construct a new large correlator, capable of handling data from 16 stations. The Institute will also provide for processing space VLBI data. Planned completion is in 1997. This, together with the existing corrector in Bonn, will ease the bottleneck in processing.
VLBI networks on smaller scales are being established or planned.
A Chinese network consists of two 25-m telescopes, one in Shanghai and the second one in Urumqi (3,000 km north-west of Shanghai). Participation of existing telescopes in China and addition of new telescopes are being considered.
The Australia Telescope (AT) is being equipped with S2 type recording terminals and has telescopes at Narrabri, Coonabarabran and Parkes. These are used in conjunction with other antennas around Australia and overseas. This network will play a very important role especially for observing southern hemisphere sources, and in space VLBI.
KNIFE is a link between the Kashima 34-m and Nobeyama 45-m telescopes in Japan, mainly devoted to mm-wave observations. There is a plan to add more telescopes.
QUASAR is a project to distribute four 32-m telescopes in the former Soviet Union, mainly devoted to geodetic observations. Two antennas and associated equipment are being constructed and will be in operation soon.
The Italian Network now has two 32-m antennas and actively participates in EVN observations.

3.2 Millimeter-Wave VLBI

Mm-VLBI can provide angular resolution of up to 20 microarcsec using baselines on Earth. It can penetrate the dense plasma and see the cores of active galactic nuclei (AGN). The determination of precise positions of stellar maser emission will be also very important.
The first successful mapping with global baselines was reported in the late 80's (Bartel et al. 1988, *Nature*, **334**, 131). After intense efforts to overcome technical difficulties, increasing numbers of maps with 50-100-microarcsec resolution are being obtained at frequencies between 40 and 100 GHz. EVN and VLBA have included 40 GHz in their observing frequencies. Attempts to detect fringes on global baselines at 230 GHz are being made without overwhelming success.
Further improvements in mm-wave and digital technologies, increasing participation of large telescopes (especially up-graded 100-m telescope of MPIfRA and 37-m telescope of Haystack Observatory) and increasing enthusiasm will make mm-wave VLBI a very important topic in the coming years.

3.3 Space VLBI

The first successful experiment using a tracking and data relay satellite (TDRSS) as the space element demonstrated the technical feasibility and scientific potential of space VLBI. Two projects are in progress, both with a planned launch in 1996. One is RADIOASTRON using a Russian satellite and the other is VSOP (VLBI Space Observatory Program), using a Japanese satellite. Table 3 lists the parameters. The former has an elongated orbit which will be suitable for measuring source sizes and brightness temperatures, which are of fundamental importance to cosmology and the nature of cores of AGNs. The latter has a lower orbit and will give high quality images of AGNs

Both missions require very complicated and large scale international cooperation. Close cooperation is necessary in policy decisions, general and detailed planning, supply of dispensable items and so on. They will not only be exciting astronomical projects, but also interesting experiments in the sociology of science.

Table 3. Space VLBI Projects

Parameters	RADIOASTRON	VSOP
Orbit		
Period	28 h	6 h
Inclination	51.6 deg.	31 deg.
Apogee Altitude	41,000 km	20,000 km
Perigee Altitude	2,000-4,000 km	1,000 km
Telescope Diameter	10 m	8 m
Receivers	0.327, 1.6, 5, 22 GHz	1.6, 5, 22 GHz
Video Bandwidth	2, 4, 8 MHz	16, 32 MHz
Video Channels	1, 2, 4	1, 2
Total Data Rate	32, 64, 128 Mbps	128 Mbps

3.4 Recording Systems

There are 3 different recording systems based on different tapes and tape recorders. They are the Mk IIIa or VLBA system using lateral recording instrumentation recorders and developed in USA, the K4 system using helical scan data recorders and developed in Japan, and the S2 system using commercial video recorders developed in Canada.

They are not compatible with each other and this causes complications in the VLBI community. At present, a compromise is being sought in which tape copiers will enable transfers between the systems. Efforts to make clear definition and machine independent interface (Generic Interface) have been done.

The sensitivity of VLBI observations depends critically on bandwidth, which is limited by recording speed. This limitation has more serious effects on mm-wave observations, where long integration is not possible because of phase fluctuations in the terrestrial atmosphere.

The Mk IV system developed in Haystack can take with a data rate of 1 Gbps with a factor of 4 improvement in speed compared to the Mk IIIa.

A burst sampling method is being developed in Japan. In this method data are temporarily stored in a fast semiconductor memory and recorded on tape at a reduced speed. A pilot model having a 4 Gbps instantaneous data rate for 8 sec (roughly, the coherence time at 40 GHz) is being developed at Nobeyama Radio Observatory.

4 MM AND SUB-MM WAVE ASTRONOMY

Development of sub-mm wavelength astronomy is very fast. There are already 4 major facilities, of which one is in the southern hemisphere, and construction of an array telescope is in progress (see Section 2.1).

Observations of molecular and dust emission in star forming regions in and outside the Galaxy will become more active.

Molecular line observations form a major branch of mm and sub-mm wave astronomy. The discovery of high-redshift CO has stimulated a new area of extragalactic study (see Section 7).
Mm-wave VLBI is becoming a major field of research as it offers the highest angular resolution presently possible.
Improvements in SIS receivers still continue to lower noise temperatures and higher frequencies. (overview by Blundell and Tong, 1992, *Proc. IEEE*, **80**, 1702, Special Issue on Terahertz Technology).

5 RADIO ASTROMETRY

5.1 Reference System

Connected-element arrays and VLBI networks are now capable of determining source positions much more accurately than by optical means. The present accuracy is 10 mas for the reference system, but tens of microarcsec for relative positions between sources close to each other.
The lack of southern reference sources has degraded the accuracy of the reference system in the southern sky. The situation is being improved by VLBI observations from observatories in southern hemisphere.
It has not been easy to connect this new reference system to the old optical reference system, because the optical system is based on positions of stars, while the radio system is based on extragalactic sources.
The sensitivity of VLBI and MERLIN has been improved to make observations of radio stars possible. The accuracy of positions from such observations are sufficient to link radio and optical reference systems.

5.2 Dynamics of the Galaxy and galaxies

The improved MERLIN can measure positions of radio stars and pulsars to an accuracy of a few mas.
Using three 70-m telescopes of Deep Space Network (DSN), the positions of 11 stars relative to nearby extragalactic sources were determined using a technique of phase-referenced VLBI (Lestrade et al., 1992, *A. Ap.*, **258**, 112). The precision in relative position and parallax was 0.1-0.7 mas, and in proper motion was 0.04-0.4 mas/year. In the case of CrB, with the reference source only 0.5 degrees away and with 12 VLBI observations, parallax was determined to be 43.99 ± 0.07 mas (Lestrade et al., 1992, *A. Ap.*, **258**, 112). This must be the best known distance outside of the solar system.
The improvements of telescope sensitivity will make it possible to find a reference source closer in the sky than before. Multi-beaming of telescopes will make it possible to record simultaneously the reference source and the source of interest. There is a proposal to build two identical antenna clusters for phase referenced VLBI. Use of these new technique will improve the accuracy of position measurements to better than 10 microarcsec.
With this accuracy, proper motions and parallaxes of various objects in the Galaxy can be fairly accurately determined, and the dynamics inside the Galaxy studied more extensively.
Accurate timing measurements of pulsars also provide high accuracy in position and in some cases accelerations.
For 10-microarcsec accuracy in position measurements obtained ten years apart, it would be possible to detect a motion of 250 km/s for an object at a distance of 10 Mpc.
Thus radio astrometry in the near future will bring major impacts on dynamics of the Galaxy and galaxies.

5.3 Solar Radio Astronomy

For the Nobeyama Radio Heliograph, described in 2.2.3, simultaneous observations with Solar X-ray imaging satellite "Yohkoh" of active regions and flares are producing important results.

The 14-m telescope of Itapetinga, Brazil is fitted with a 5-horn receiving system at 22 GHz. With overlapping beam, this system can measure positions of solar bursts with a time resolution of milliseconds.
VLA and other arrays are also used to observe active regions and sometimes flares in microwave frequencies.
The Nançay Radio Heliograph is now producing interesting two-dimensional pictures of the sun at 150 and 450 MHz.

5.4 Radio Stars

The high sensitivity of VLA enables radio emission from stars to be readily detectable. The recent upgrade of MERLIN has provided the sensitivity to detect a star with radio brightness of 1,000 K and resolution of 50 a.u. at a distance of 1 kpc. It can detect and resolve the photospheric thermal emission of many late-type stars, and thermal and non-thermal emission associated with stellar winds of early-type stars etc. Improvements in sensitivity in VLBI observations will enable stars to be detected and resolved.
Stellar maser sources are becoming popular subjects for VLBI observations. Following the pioneering works of University of Massachusetts and Haystack (Rosen et al., 1978, *Ap. J.*, **222**, 132; Spencer et al., 1979, *Ap. J.*, **230**, 449), observations are being made with KNIFE, VLBA and EVN. Astrometric observations have revealed that the maser spots are near the star rather than in the exosphere.
The impact of radio star observations on the astrophysics of stars, especially on mass loss processes, will be very significant.

6 PULSARS

Pulsars, even after 25 years since their discovery, are still prime objects for studies of the nature and evolution of compact objects, testing various gravitational theories, interstellar scintillation, gravitational field and other properties of globular clusters, and so on.
Eight millisecond pulsars were discovered in the galactic disk, being roughly double the number known there. Outstanding among them is the very strong (1 Jy at 400 MHz) PSR J0437-4713 (Johnston et al., 1993, *Nature*, **361**, 613), which is binary in a 5.7 day orbit with a white dwalf companion (Bell et al., 1993, *Nature*, **364**, 603). It is the nearest and brightest millisecond pulsar known. Pulsed gamma-ray emission was detected by the Compton Gamma Ray Observatory for PSR B1706-44 (Thompson et al., 1993, *Nature*, **359**, 615) and PSR B1055-52 (Fierro et al., 1993, *Ap. J.*, **413**, L27).
The success of pulsar research is due to the continuous supply of new varieties of pulsars through discoveries in new surveys. The detection of millisecond pulsars has had a strong impact due to the high stability of the pulse rates (Taylor Jr., 1991, Proc. IEEE, **79**, 1054). The efficiency of surveys depends on the volume and resolution in a multi-dimensional parameter space, and requires handling a large amount of data and fast computations. For example, the Parkes pulsar processor samples every 0.3 milliseconds and has 256 frequency channels (Johnston et al., 1992, *Mon. Not. R. astr. Soc.*, **255**, 401). In some cases, searches are made not only for pulse rates and dispersion, but also acceleration of the rate.
Especially for millisecond pulsars, the effect of smearing of pulses due to dispersion is severe, and improvements in search techniques are required. There is continual progress in digital and computational technologies, and this will dictate the progress of pulsar research.
Of course, the performance of telescopes is as important as the data processing. The role of large telescopes is very important. The upgrade of the Arecibo telescope, completion of GMRT and GBT will add more power in the searching and studying of pulsars.
Extensive use of multi-beam arrangements in these large telescopes will also accelerate discoveries of new pulsars.
The improved performance of the MPIfRA 100-m has resulted in the detection of a pulsar at 36 GHz (Wielebinski et al., 1993, *A. Ap.*, **272**, L13).
With the detection of many more pulsars, especially with millisecond or sub-millisecond periods, the improved accuracies in timing observations, and the longer times over which observations are made, will provide opportunities for new areas of study such as a new time standard, a new distance scale, dynamics of the Galaxy, globular clusters, pulsar-blackhole pairs and so on. The season of discoveries will continue.

7 MOLECULES AND STAR FORMING REGIONS

Molecular line studies are still a major part of mm-wave astronomy.
New interstellar molecules continue to be detected. Molecules containing less-abundant atoms such as P, Na, Cl, Al, K, and Al are on the list, and Mg was added recently.
Studies of proto-stellar disks will highlight new frontiers such as sub-mm wave observations. The determination of isotopic abundances will be very important.
A surprise detection of strong CO line emission with $z = 2.3$ (see Section 9) will open up a new area of studies.
This rich field of discoveries now provides items for the reports of other commissions.

8 AGN

Improved angular resolution, image quality, sensitivity and larger numbers of samples result in a better understanding of core, jet and lobe configurations of AGNs.
However, even for mm-wave VLBI with global baselines, the "central engine" still cannot be proved, even for the nearest AGN's. We need space-space baselines or perhaps an entirely different approach.

9 HIGH-*z* GALAXIES

Brown and Vanden Bout have detected CO J = 3-2 emission with $z = 2.3$ in IRAS10214+4724 (1992, *Ap. J.*, **397**, L11). The CO luminosity is 20 times higher than the most gas-rich ultra-luminous infrared galaxies (Solomon et al., 1992, *Nature*, **356**, 318). The intensity of such line emission does not vary as D^{-2} but $(1+z)^3 D^{-2}$, where D is luminosity distance, and is easier to detect than previously thought. With further improvements in telescope sensitivity, many more detections will be made.
Then, there is no reason that other molecules do not to be detected. A new area of molecular and isotope chemistry is expected to open.

10 PHENOMENA AND OBJECTS OF COSMOLOGICAL IMPORTANCE

10.1 Gravitational Lenses

There are some 10 objects identified as gravitational lenses and more will be added. This has established that the gravitational lensing may be a rather widespread phenomenon.
There is one object, PKS1830-211, on which the detection of large intensity variations in two components is expected (Nair et al., 1993, *Ap. J.*, **407**, 46). This is a more direct evidence than just from the shape of the source, for the gravitational lens.
Determination of Hubble constant is possible from this time lag and geometry of the source. It gives a value between the present limits.
The motion of a lensed quasar is magnified by the gravitational lens. This will make measuring the motion of very distant object possible.

10.2 Sunyaev-Zel'dovich Effect

The Sunyaev-Zel'dovich (S-Z) effect is a decrease in cosmic microwave background radiation due to scattering by hot plasma in cluster of galaxies. Its successful detection was reported from single dish observations.
It can be used to probe intergalactic matter in the cluster of galaxies and also to estimate Hubble constant.
The first interferometric imaging observations using the Ryle Telescope (the upgraded 5-km telescope) have clearly shown the effect in a cluster, Abell 2218 (Jones et al., 1993, *Nature*, **365**, 320). It is much less susceptible to errors than previous observations due to confusing sources. The decrement of background brightness due to S-Z effect was about 1 mK. This result combined with X-ray observations supports a lower value of Hubble constant.

10.3 Cosmic Microwave Background Radiation

The Cosmic Background Explorer (COBE) clearly detected 10^{-5} fluctuations in background radiation on a scale size of 10 degrees in a frequency range of 30 and 90 GHz (Smoot et al., 1992, *Ap. J.*, **396**, L1) This is the first firm detection of anisotropies. There have been several ground-based experiments sensitive to different scale sizes, and a few reported detections.

10.4 Angular Size-*z* Relation

Angular size of a standard 'rod' falls inversely proportionally to the distance *D*, for small distance. For *z* approaching and exceeding unity, it falls off much more slowly than D^{-1}. It depends on the cosmological model, especially on deceleration parameter, q_0.
Many attempts have been unsuccessful because of uncertainties in the size of the rod etc. Using the size of the cores of AGNs, a clear indication of this "turn over" was seen (Kellermann, 1993, *Nature*, **361**, 134). The result corresponds to $q_0 = 0.5$.
More confirming evidence is necessary, but it appears that we are about to see the curvature of the Universe.

11 PROTECTION OF RADIO ASTRONOMY FREQUENCIES

Over the last three years there has been some improvement in the protection of radio astronomy frequencies. The most noteworthy success has been for the frequency band 1610.6-1613.8 MHz, containing the OH transition at 1612.231 MHz, of particular importance in studies of OH/IR objects. Over the last ten years, radio astronomy observations of this transition have been badly affected by satellite transmissions associated with the Russian GLONASS global navigation system. This system contains a number of satellites, each transmission using a 1-MHz carrier and additional 10-MHz band both entered at one of a series of frequencies separated by 0.5625 MHz between 1602.56 and 1615.50 MHz. A milestone occurred at the World Radio Conference held by the International Telecommunication Union (ITU) in March 1992, when the status of radio astronomy in the OH band was increased from secondary to shared primary. This meant that transmitting services sharing the same frequency band (such as navigation services which include GLONASS) would be required to operate without causing harmful interference to radio astronomy observations. In November 1992, as a result of negotiations between the Inter-Union Commission on the Allocations of Frequencies (IUCAF) and the Russian administration, a Joint experiment was undertaken in which a number of major observatories around the world made observations in the OH band while GLONASS transmitting frequencies were varied. In a summary of the observing results, it was recommended that GLONASS transmissions should be limited to center frequencies of 1608.75 MHz and lower as a first urgent step, then subsequently to frequencies of 1605.375 MHz and lower. It was further recommended that the GLONASS system should employ filtering to reduce out-of-band interference in both the OH bands at 1610.6-1613.8 MHz and 1660-1670 MHz. In September 1993, the Russian administration announced that it would reduce the effects of GLONASS transmissions in the OH band, with an aim to removing all interference to radio astronomy observations by 1998. Soon after this, observatories reported that GLONASS transmissions associated with six carriers in the OH band were no longer present.
There have been some activities in the area of international frequency protection, under control of the ITU. The International Radio Consultative Committee (CCIR), responsible for several Study Groups looking after the interests of various radiocommunication services (and including radio astronomy), was recently replaced by the Radiocommunication Bureau. The Study Group structure has remained unchanged. Study Group 7 has continued to seek improved protection for radio astronomy by producing appropriate documentation to be used as a basis for future changes to the radio regulations. A revised list of frequencies of the astrophysically most important spectral lines for radioastronomy, adopted at the 1991 IAU General Assembly as Resolution A2, has been included in a Study Group Recommendation. The Study Group is also producing documentation relative to interference to radio astronomy observations from satellite transmissions centered at frequencies outside radio astronomy

bands, and to the protection of radio astronomy observations carried out from locations on the Moon which are shielded from emissions produced on Earth. A handbook is being produced which discusses radio astronomy in the context of an ITU radiocommunication 'service'.

References

Reviews Relevant to this Report
Proceedings of IAU Symposia and Colloquia
Symposium. No. 156: Developments in Astrometry & Their Impacts on Astrophysics and Geodynamics.
Symposium No. 158: Very High Angular Resolution Imaging.
Colloquium No. 140: Astronomy with Millimeter & Submillimeter Wave Interferometers.

Advances in VLBI Techniques
Rogers, A. E. E., *Rev. Radio Science,* 1990-1992, p657, Pub. by URSI.

VLBI Technology
Proceedings of URSI/IAU Symposium on *VLBI TECHNOLOGY -Progress and Future Observational Possibilities-*, eds. T. Sasao et al., Terra Publishing Co., Japan.

Recent Pulsar Discoveries
Bailes, M. and Johnston, M., *Rev. of Radio Science*, 1990-1992, Pub. by URSI.

Table 2. Radio Arrays/Interferometers
(Order of operating frequency)

Telescope	Element	Frequency (GHz)	Baseline (km)	Institute
SMA**	12 x 6 m	230-860	3 x 0.5 (Y)	SAO (USA)
CSO-JCMT*	10 m & 15 m	300-600	0.16	Caltech (USA) & JAC (UK)
BIMA***	9 x 6 m	75-115, 210-270	0.3 EW x 0.2 NS	UC Berkeley & U. Illinois (USA)
OVRO***	6 x 10 m	75-115, 215-270	0.2 EW x 0.22 NS	Caltech (USA)
NMA***	6 x 10 m	80-120, 140-160, 220-240	0.57 EW x 0.54 NS	NRO (Japan)
IRAM***	4 x 15 m	80-115	0.288 EW x 0.16 NS	IRAM (France)
Radio Heliograph*	84 x 0.8 m	17	0.489 EW x 0.22 NS	NRO (Japan)
CAT*	3 x 0.7 m	13-17	1-5 m	MRAO (UK)
VLA	27 x 25 m	0.07, 0.3, 0.6, 1.4, 5, 8, 14.4, 23	21, 21, 19 (Y)	NRAO (USA)
MERLIN***	32m*, 76 m, 3 x 25 m, 2 x 15 m	0.4, 1.6, 5, 22	218	U. Manchester (UK)
RATAN-600	895 x 2 m x 11.4 m	2.3, 4.0, 4.8, 7.8, 11.1, 22	0.577	Space Research Inst (CIS)
Ryle Telescope	8 x 13 m	5, 15	1.2 EW	MRAO (UK)
AT Compact Array	6 x 22 m	1.4, 2.3, 5.5, 8.4	6 EW	CSIRO (Australia)
WSRT	14 x 25 m	0.327, 0.608, 1.4, 5	2.8 EW	NFRA (Netherlands)
Penticton	7 x 9 m	0.4, 1.4	0.6 EW	DRAO (Canada)
MOST	88 x 17.7 m x 11.6 m	0.843	1.6 EW	U. Sydney (Australia)
GMRT**	30 x 45 m	0.04-1.4	3 x 14 (Y)	TIFR (India)
Nançay Radioheliograph	2 x 10 m, 16 x 3 m, 24 x 5 m	0.15-0.45	3.2 EW x 1.2 NS	Meudon Obs. (France)

* New, ** Under construction, *** Upgraded

COMMISSION 41 : HISTORY OF ASTRONOMY
(HISTOIRE DE L'ASTRONOMIE) (a joint IAU-IUHPS Commission)

Report of Activities 1990 July 1 to 1993 June 30

PRESIDENT : S. Débarbat VICE-PRESIDENT : S..M. Ansari

During the XXIth General Assembly it was decided on a unanimous vote of those members present to make a formal resolution to be presented to the General Assembly in its closing session. This was done jointly by Commission 41 and Comission 5 and was

that the Union supports an initiative taken by Commission 41 and 5 (1) to establish a register of the whereabouts of all extant astronomical archives of historical interest ; (2) to impress on observatories and other institutions their responsibility for the preservation, conservation, and where possible cataloguing of such archives ; (3) to search for an institution that will allocate space and funds for maintening such a register and publishing it.

It was not felt that the proposal should take instruments into account at this stage, since the problems presented there are of a very different character ; but if the one project makes progress, the other should certainly follow in its train. The first objective was to stop the wholesale destruction of materials of historical value ; the second was to make the wherabouts of material better known to scholars. Many instances were reported at meeting of situations where a series of directors of institutes, librairians, and others invested great care in the preservation of materials over a long period, only to be followed by others entirely without a feeling for the past, prepared to sell or destroy their inheritance. The General Assembly endorsed the resolution. The following Working Group was established to look into the possibilities mentioned in the resolution : S. Débarbat (ex officio, President), S.J. Dick, E. Proverbio, B. Hauck (comm, 5), D. Dewhirst (comm. 5).

References

J.D. North : 1991 "Commission 41 at the Buenos Aires General Assembly of the IAU", Journal of the History of Astronomy 22, PP. 329-30, and Proceedings of the IAU/GA (Vol. XXIB, PP. 315-16).

The members of the *Working Group* had not possibility to meet but information was circulated and President of Com. 41 had several opportunities to discuss with people engaged in Archives or Libraries the last ones having archives under their responsability. Two main ideas have emerged from discussions. First the fact that, in view of having countries impressed about their responsabilities, a national register of archives be established and maintained in a given place. Secondly the fact that, in view of a better efficiency, a form to be given to the entries of register be studied. This would have, of course, to be made for computers and with appropriatly selected indexes. B. Corbin has informed, in Cicular number 2 (February 1993), the "JWG,Astronomical Libraries" about the IAU/GA resolution.

E. Proverbio (member of WG) published (December 1990) the Proceedings of a national meeting on the subject "Problemi del recupero e dell'ordinamento delle risorse strumentali e biblio-archivistiche negli Osservatori e Istituti/Dipartimenti di Astronomia e Astrofisica" (Pisa/Italy, November 1989). He organized (auspices of the Commissione per la storia dell'Astronomia) another meeting (Roma/Italy, April 1993) on "I problemi del recupero e valorizzazione del patrimonio strumentale, archivistico e bibliografico di interesse storico degli

J. Bergeron (ed.), Reports on Astronomy, Vol. XXIIA, 461–462.

Osservatori Astronomici e Astrofisici Italiani".

A Journée d'étude ("Préservation et utilisation des archives scientifiques") was organised (Th. Charmasson, Médiathèque de La Villette, Paris/France, March 1993). Among the three chosen examples "The astronomical archives" which were presented together with their present employement by N. Daliès and S. Débarbat.

Regarding the Register, several actions exist, have or will come to existence at national level such as the American Institute of Physics for USA, Deutches Museum for Germany, the Paris Observatory for France, History of Science Departement (University of Aarhus) for Denmark, Athens Observatory for Greece,...

Meetings of astronomical interest

Various meetings have been held in different places to which members of Com. 41 have greatly participated : - The 120th Anniversary of Konkoly Observatory (Budapest/Hungary, September 1991).- Current Problems and Future of Archeoastronomy (Budapest/Hungary, October 1991).- India and other asiatic astronomies (Hyderabad-Jaipur/India, December 1991).- Historical Astronomy Division of the American Astronomical Society (Atlanta/USA, January 1992).- Time and Astronomy at the Meeting of two Worlds (Warszawa-Frombork/Poland, April 1992).- Les savants et l'épistémologie vers la fin du XIXème siècle (Genève/Swizerland, May 1993).- Le travail scientifique dans les correspondances entre savants au tournant des 17ème et 18ème siècles (Paris/France, June 1992).- History of Women in Science, Technology and Medicine (Sopron/Hungary and Wien/Austria, August 1992).- Nature and History in the German enlightenment (UCLA Center for 17th and 18th Century Studies, Los Angeles/USA, May 1993).

The year 1993 marks the 200th anniversary of F.G.W./V.Y. Struve (1793-1864), the founder of Pulkovo Observatory, previously at Dorpat (now Tartu) Observatory. The following conferences are scheduled on that occasion : - Tartu (Estonia), August 26-29.- Pulkovo/Petersburg (Russia), September 7-9 and October 4-8. An international conference on Oriental Astronomy from Guo Shonjing to King Sejong is planned (Seoul/Korea, 6-11 October 1993).

Publications

Some *books* appeared having been mentionned to the Commission : - Claudius Ptolemäus : Der Sternkatalog des Almagest. Die arabisch-mittelalterliche Tradition, responsability of P. Kunitzch, 1990.- Xi'an Jiatong University West Han Dynasty Mural Tomb (1991) dicovered in 1987 and which houses a two thousand year celestial map in great danger of destruction ; the book is by Hu Lin-gui and Zhong Wan-mai.- Astronomie im Deutschen Museum (1993) by G. Hartl, K.Märker, J. Teichmann and G. Wolfschmidt following the re-opening of the re-installed Astronomie Galerie.- The Encyclopedia of Islam (New Edition, Leiden 1993) which contains a long entry on the stars ("fixed" and "planets").

Newsletters to be mentionned : - Center for History of Physics (American Institute of Physics, New York/USA).- Gruppo Nazionale per il coordinamento della storia della fisica (CNR, Italy).- Mitteilungen zur Astronomiegeschichte (Arbeitskreis Astronomiegeschichte in der Astronomischen Gesellschaft, Germany).- The Achievement Project (Oxford/UK).- H.A.D News, Historical Astronomy Division of the American Astronomical Society.

Articles have been published about Ancient Chinese Astronomy, by Purple Mountain Observatory ; a series of publications by numerous members of the Commission have concerned Australian and New Zealand Astronomy, Hebraic, Indian and Canadian subjects, Globes, Instruments, Observatories,... Other publications were relevant to astronomers from the past : Galileo, Denning, Pickering, Ross, Kaiser, Huygens, Schöner, Chandra,...

On the occasion of the XIXth International Congress of History of Science (Zaragoza/Spain) a Report (1989-1993) has been prepared and will be presented by J. North, in August 1993, Com. 41 being a joint commission of the IAU and the IUHPS.

42. CLOSE BINARY STARS (ETOILES BINAIRES SERREES)

PRESIDENT: Y. Kondo
VICE PRESIDENT: M. Rodonò
SECRETARY: R. S. Polidan
ORGANIZING COMMITTEE: P. P. Eggleton, A. Gimenez, E. F. Guinan, R. W. Hilditch, G. Hill, R. H. Koch, R. L.Mutel, P. Szkody, A. V. Tutukov, R. F. Webbink, A. Yamasaki, D.-S. Zhai

1. INTRODUCTION

At the time of this report in September 1993, about a year before the next General Assembly in 1994, the membership of Commission 42 stands at 357, an increase of forty-two members since the time of the last report prepared in September 1990. This is an increase of more than 10 percent, indicating the continuing vitality of our Commission in astronomical research. Many – possibly even the majority – of astrophysically intriguing objects involve interacting binary stars, which probably accounts for the ever increasing vigor in this field of research.

Among the significant events of the last triennial period is IAU Symposium 151 on "Evolutionary Processes in Interacting Binary Stars", which was held in August 1991 in Cordova, Argentina. The meeting was dedicated to Jorge Sahade, the Past President of IAU, to honor his outstanding contributions to our field for more than half a century. The proceedings of this symposium, edited by Y. Kondo, R. Sistero, and R. S. Polidan, was published from Kluwer Academic Publishers in 1992.

We are pleased to announce that Commission 42 has joined Commission 27 as a sponsor of the Information Bulletin on Variable Stars, which is edited by L. Szabados and K. Olah and published by Konkoly Observatory in Budapest, Hungary. For further information on this important newsletter, please see the report of Commission 27 in this volume.

A. Yamasaki has continued to serve as editor in chief of another newsletter of our Commission "Bibliography and Program Notes on Close Binaries". O. Demircan, T. J. Herczeg, V. G. Karetnikov, H. L. Malasan, H. Mauder, J. Papousek and C. D. Scarfe are serving as editors of this important publication for Commission 42.

I would like to thank all the contributors for their report – especially Ronald Polidan, whose help has been invaluable in putting this report in a final form.

2. STATISTICAL INFORMATION (A. Yamasaki)

The importance of the studies of stellar duplicity has been reported by a number of authors. The frequency of multiplicity (binary and multiple systems) is closely related to many astrophysical issues, such as star formation; stability and evolution of disks and planetary systems; evolution and interaction of binary and multiple systems; dynamical evolution of stellar systems; the search for very low mass companions; and birth rates of peculiar stars, cataclysmic variables, X-ray binaries, and millisecond pulsars.

J. Bergeron (ed.), Reports on Astronomy, Vol. XXIIA, 463–488.

The most reliable, unbiased determination of binary frequency has been given by studies of the solar neighborhood stars.

Solar-type stars Based upon CORAVEL radial velocity database, Duquennoy and Mayor (*A&A*,**248**, 485) studied the stellar multiplicity in an unbiased sample of 164 G-dwarf primaries in the solar neighborhood. Their results are: *i*) one third of the G-dwarf primaries may be real single stars (significantly more single stars than in the study of Abt and Levy, *ApJS*,**30**,273), *ii*) the period distribution is unimodal with a median period of 180 yr, *iii*) the mass-ratio distribution shows no maximum for q=1, but an increase toward small q (up to q=0.23) with a possible drop off at q=0.1 or less, and *iv*) though negligible contribution to the local mass density, the brown dwarfs may not be as rare as quoted by Campbell *et al.* (*ApJ*,**331**, 902). Mazeh *et al.* (*ApJ*,**401**,265) studied short-period binaries (P< 3000d) using the Duquennoy and Mayor survey, and found that the mass-ratio distribution is uniform and might even rise toward large mass ratio, confirming the different distributions between the short- and long-period binaries of Abt and Levy.

K and M dwarfs The observations and analysis by Tokovinin (*A&A*,**256**,121) of the radial velocities of a homogeneous sample of 200 G9-M3 dwarfs from the solar neighborhood indicate *i*) the absence of substellar mass companions (0.02-0.08$M_\odot$) with P<3000d, *ii*) the frequency of spectroscopic binaries is 10% while the systems with q>0.5 are more frequent, and *iii*) a non-monotonic distribution of the companion masses with a gap at 0.2-0.3$M_\odot$. Examining various surveys of M dwarfs within 20pc, Fischer and Marcy (*ApJ*,**396**,178) found that the incidence of companions to M dwarf primaries is 42% (this agrees with Henry and McCarthy, *ApJ*,**350**,334) and that the mass distribution of companions is flat, similar to the field mass function at low masses, supporting capture as the dominant mechanism.

B main-sequence stars Abt *et al.* (*ApJS*,**74**,551) obtained spectra of 74 B2-B5 IV or V stars, and analyzed the sample (including the earlier sample, Abt and Levy *ApJS*,**36**,241) obtaining the following results: *i*) systems with P>3.6d have companions that fit the Salpeter luminosity function and were formed by capture, *ii*) observed binary frequency of 74%, or, after allowance for incompleteness, 0.8 companions with mass> $2M_\odot$ per primary or 1.9 companions with mass> $1M_\odot$ per primary.

The identification of pre-main-sequence binaries is particularly important to investigate binary formation as well as evolution of circumstellar disks in single and multiple systems. Simon *et al.* (*ApJ*,**384**,212) identified 11 binaries and 2 triples among 28 young stars in Taurus star-forming region. They found binaries have formed by the time stars are 1Myr and that the observed multiplicity and distribution of separations are similar to those of the solar-type stars.

To study the binary evolution off the main sequence and the formation scenario of peculiar red giants like barium stars, CH and S stars, one has to understand normal field binary giants as fundamental knowledge. Statistical analysis of a sample of 213 spectroscopic binaries containing G or K giants by Boffin *et al.* (*A&A*,**271**,125) indicates that *i*) a circularization cut-off period around 70d, *ii*) some long-period binaries with small eccentricity are systems made of a giant and a white dwarf resulting a previous mass transfer episode, and *iii*) the observed mass function distribution fits a uniform mass-ratio distribution with giant mass of $1.5M_\odot$. Concerning S stars, the paradigm that binary S stars systematically lack Tc was furthermore confirmed by Jorissen *et al.* (*A&A*,**271**,463) with a CORAVEL monitoring.

The frequencies of binaries in clusters and associations have been studied by many workers. Morrell and Levato (*ApJS*,**75**,965) observed 96 brightest members of the Ori OB1 association and obtained the frequency of spectroscopic binaries with P<100d plus magnetic Ap-Bp stars to be 32%. Morrell and Abt (*ApJ*,**378**,157) measured the radial velocities of 15 bright B3-A2 members of IC 4665, giving a binary frequency of 27%, the smallest estimate to date. Kroupa and Tout (*MN*,**259**,223) found for Praesepe that the color-magnitude data are consistent with a large binary population and systems with B-V=0.8–1.0 prefer uncorrelated component masses. Radial velocities of red giants in five open clusters older than the Hyades allowed the overall percentage of binaries to be 27% (Mermilliod and Mayor *A&A*,**237**,61). Unusually low binary frequencies were confirmed for the Pleiades and the α Persei. Mermilliod *et al.* (*A&A*,**265**,513) obtained 13% for binaries with F5-K0, P<1000d in Pleiades (see also Rosvick *et al.*, *A&A*,**255**,130 and Steele *et al.* *MN*,**263**, 647). Morrell and Abt (*ApJ*,**393**,666) found the binary frequency of 14% for the α Persei cluster.

Since the discoveries of low-mass X-ray binaries and millisecond pulsars, binary stars in globular clusters have been studied extensively. Romani and Weinberg (*ApJ*,**372**, 487) derived the present main-sequence binary fraction in M92 and M30 to be less than 3%. Primordial binaries, low-mass X-ray binaries, and pulsars in globular clusters were discussed by Hut *et al.* (*A&A*,**241**,137), Taam and Lin (*ApJ*,**390**,440), Chen and Leonard (*ApJ*,**411**,L75), Wijers and van Paradijs (*A&A*,**241**,L37) and others.

Many important contribution papers on binary statistics appeared in "Complementary Approaches to Double and Multiple Star Research" (IAU Colloquium 135; eds. McAlister and Hartkopf) and in "Binaries as Tracers of Stellar Formation" (eds. Duquennoy and Mayor).

3. PERIOD CHANGES (T. J. Herczeg)

The number of period studies of various lengths and scopes, in the 3-year interval considered, can be estimated at 350 at least. This number does not include occasional short discussions in photometric and/or spectroscopic studies with the sole aim to fix the phases correctly but does include the published longer lists of times of minima (mostly photoelectric) with or without a short comment on the period. Perhaps the most active group in this work represents the decades-old Nurenberg- Izmir collaboration, now supplemented with a station at Eckental, Germany., Their latest list of minimum epochs has been published in *IBVS*, no. 3760. Photographic archival work of very great importance has been produced at the Sonneberg Observatory; other archives, such as at Harvard, Odessa, the Sternberg Observatory or at the Maria Mitchell Observatory have also contributed. Workers at the AAVSO increased their output, in particular concerning long-term behaviour of variables, among them several of cataclysmic type. The annual publication of the Cracow Observatory, Rocznik Astronomiczny, regularly publishes improved ephemerides for relatively less known eclipsing systems, based presumably on comprehensive period studies; the recent volumes

63 and 64 bring, respectively, 41 and 27 new photometric elements.

The following very restricted selection is oriented toward those types of objects which appear most important for current research.

Strictly periodic shifts in the observed orbital period are caused either by light-time effect or by apsidal rotation. Various cases of light-time effect, seldom definitely established owing to the small timing residuals, sometimes only 5 to 15 minutes, have been reviewed by Mayer (*BAC* **41**). Large amplitude light-time effects have been proposed for β Cep and σ Sco (Pigulski, Boratin, *A&A*,**253**; Pigulski, *A&A*,**261**). The corresponding orbital motions may explain most of the observed period changes of these β CMa-variables; the involved masses can be identified as close visual companions.

Among other methods of detecting additional members in a multiple system, the best approach is the spectroscopic one. Such a study has shown, for instance, that the symbiotic system CH Cyg is actually triple, with periods 14.5 and 2 years (Hinkle *et al.*, *AJ*,**105**). At the Toulouse Observatory, radial velocity work on stars with composite spectra also revealed several triple systems, see e.g. GINESTET *et al.* (*A&AS*,**91**).

A substantial number of systems with already known "classical" apsidal rotations have been revisited, some of them thoroughly rediscussed. Among them are CW Cep (Harvig, *Tartu Publ.* **53**, Stickland *et al.*, *Observ.* **112**, based on IUE spectra); Y Cyg (Stickland *et al.*, *Observ.* **111**, IUE spectra); V477 Cyg (Gimenez, Quintana, *A&A*,**260**, arriving at log k2 =-2.20±0.04); HS Her (Todoran, *AN* **313**, Khaliullina, *Astr. Circ.* **1552**); YY Sgr (Lacy, *AJ*,**105**); V523 Sgr (Lacy, *AJ*,**104**); V526 Sgr (Lacy, *AJ*,**105**). Further systems considered: CO Lac (Mosskovskaya, *Astr. Circ.* **1543**), ι Ori (Hilditch *et al.*, *Observ.* **111**), BD-1 1004 (Rao *et al.*, *JA&A*, **11**).

The puzzling case of the anomalously slow relativistic apsidal rotation in the system DI Her still defies explanation; for a recent view, see Khaliullin *et al.* (*ApJ*,**375**, possible perturbations by a third body). The superior accuracy of pulsar timing, on the other hand, makes detection and study of relativistic apsidal rotation as well as other general relativistic effects possible. Since the early discussion by Taylor of the first binary pulsar PSR 1913+16, half a dozen cases of apsidal motion have been found in binary pulsars in in eccentric orbits, through the efforts of Bayles, Lyne, Prince, Ryba, Taylor, Wolszczan, and others. The latest detection yet is PSR B1802-7, a neutron star + low-mass white dwarf system (D'Amico *et al.*, *MN*,**260**, also Thorsett *et al.*, *ApJ*,**405**). The second paper quoted here gives a diagram summarizing all 17 neutron star masses, known as of March, 1993; the error bars for X-ray systems are sometimes considerable.

Among relativistic effects probed by pulsar timing, of greatest importance is the proof of orbital decay (decreasing period) of PSR 1913+16 due to quadrupole gravitational radiation as required by Einstein's relativity theory; a recent review is by Taylor (*Phil.Tr. Roy.Soc.* **A341**). In another study of the orbital period changes of PSR 1913+16, Damour and Taylor (*ApJ*,**366**) considered the effect of the accelerations in the galactic orbits of the pulsar and the Sun; they found a 0.8% confirmation of the relativistic prediction.

Precise timing of the pulse period led in the case of the ms-pulsar PSR 1257+12 to a truly amazing result: the probable existence of two planetary masses in orbit around the pulsar (Wolszczan, Frail, *Nature* **355**).

Concerning the orbital periods of X-ray binaries, the variations proved measurable only in three cases as of yet: Cen X-3, Her X-1 and EXO 0748-676 (see Asai *et al.*, *PASJ* **44**). The respective values of $\dot{P}$/P are: -1.8$\times10^{-6}$/year, -1.3$\times10^{-8}$/year and -0.9$\times10^{-7}$/year. It may be assumed that these variations are caused by tidal torques.

Among close binaries without compact components, W UMa systems, with their relatively short periods, have been studied extensively. With only a few exceptions such as UZ Leo or V401 Cyg, these systems show irregular period changes; moreover, a particular mode, abrupt, discontinuous variations of the order of $10^{-5}\times$ P seems to be characteristic, as proposed early by F.B. Wood, Binnendisk, and others. Alternative possibilities, however, have been put forward: long-term light curve disturbances, starspots, cyclic, magnetically induced structural variations. Merits and difficulties of such attempts can be seen in studies of AH Vir (Demircan *et al.*, *AJ*,**101**), V502 Oph (Derman, Demircan, *AJ*,**103**) or the "spotted" system AG Vir (Bell *et al.*, *MN*,247), to mention a few. The important triple system 44i Boo (ADS 9494) approaches its periastron and, although abrupt period changes are certainly present, an apparent increase of the period may reflect, at least partly, the orbital motion; see Rovithis (*A&AS*,**86**) and observations by Burke, Jones, Oprescu, Grobel and others in recent issues of the *IBVS*.

RS CVn stars usually show highly irregular O-C diagrams; a case in point is a recent discussion by Kim (*AJ*,**102**) of the characteristic system AR Lac. In these binaries with intense surface and magnetic activity and established light curve asymmetries (migrating distortion) seemingly irregular but actually spurious period variations of a time scale of a few years may arise as emphasized years ago by D.S. Hall, Rodonò and others.

There is slow progress concerning the period changes of cataclysmic variables where the timing residuals show wide scatter probably due to strong disturbances of the light curve. Yet it seems that most CVs have relatively stable periods. The spectroscopic determinations of the periods and orbits are far more reliable; however, in many cases there is a real or spurious multiplicity of the periods. Much work was done about these systems as well as deriving periods for newly identified CVs; in both cases at least half a dozen successful studies are to be noted. For the SU UMa stars, the "superhump timing" was reconsidered by Molnar and Kobulitzky (*ApJ*,**392**). They found a significant correlation between the superhump period excess (over the orbital period) and the mass ratio of the components. Statistics concerning the orbital period distribution of dwarf novae, with possible inferences about the evolution of these systems, was given by Shafter (*ApJ*,**394**) and for AM Her stars, Ritter and Kolb (*A&A*,**259**).

4. LIGHT CURVE MODELLING (G. Hill)

We have been fortunate in that within this triennium there has been an IAU Conference devoted to light curve modelling (*Light Curve Modeling of Eclipsing Binary Stars*; ed. Milone; Springer-Verlag, 1993, termed LCMEBS hereafter) and a related conference in

Korea (*New Frontiers in Binary Star Research*; ASP Conference Series,**38**,1993, NFBSR). Contained within these volumes are summaries of all of the major modelling codes. In both volumes, Wilson (LCMEBS, p7; NFBSR, p91) points the way to future developments, not only within the WD code but for any close binary modeling code. He includes the treatment of narrow and broad band indices as well as polarization. The ever more realistic atmospheres of Kurucz (LCMEBS, p93) will contribute to these endeavors even though their use will strain the computational facilities of most institutions. In this regard the work of Milone, Stagg and Kallrath (*e.g.*, *ApJ*,**389**,590,1992; LCMEBS, p75;NFBSR, p172) who have applied the simplex algorithm to a vectorized WDLC is very important. With a vectorized code and excellent atmospheres it will soon be possible to compute realistic photometric indices. All of the major codes now model spotted stars (see LCMEBS) but as ever the limitations of uniqueness still bedevil the analysis. Generally the available light curves are undersampled so that the growth and cycle time and the evolution of a spot system is rarely covered. An excellent example of what can be done given good observations is found in the study of RZ Eri (Burki *et al.*; *A&A*,**256**,463, 1992) using the EBOP code (Etzel,LCMEBS p113).

In addition another model, termed the geodesic distribution binary system (GDDSYN) method, has been developed specifically for analyzing surface features on eclipsing binaries (Hendry and Mochnacki *ApJ*,**388**,602,1992). Future developments include line profile synthesis and Doppler imaging. A simple method to get starting values for these more sophisticated codes is given by Riazi (*AJ*,**104**,228,1992). It assumes spherical stars, no limb darkening or proximity effects. A problem which persists in the field is the theoretical component of what is really going on over the surfaces of the stars in the presence of strong incident radiation and the circulation currents that probably result. All of the models contain gross simplifications as to the surface temperatures and the effects of incident radiation on the actual structure of the atmosphere. The models would all benefit if a rigorous means of accounting for these effects was developed. Other related developments which aid in the analysis process involve Doppler imaging, eclipse mapping, cross-correlation, speckle and interferometric observations.

5. OBSERVATIONAL DATA (M. Rodonò)

This section maintains the traditional subsection division essentially guided by the wavelength range or method of observations. Such a subsection arrangement, though not ideal, is quite convenient because, especially for practical reasons, many studies are primarily guided by specific methods (photometry, spectroscopy, spectrophotometry, polarimetry) and specific wavelength bands (X-rays, EUV, UV, optical, IR, radio), which have been set up by, or made available to observers. An increasing fraction of observational studies, however, are being done within collaborative programs aimed at multiwavelength, multisite studies of close binaries. This tendency is obviously more pronounced when the principal study purpose concerns the physical characteristics of the individual stellar companions, especially the structure of their outer atmospheres, more than the orbit or global stellar parameters.

5.A Ground-Based Photometry (M. Rodonò)

It is virtually impossible for reasons of limited space, nor it would be a useful service to the general reader, to list here the huge amount of ground-based photometric papers published in the last triennium (July 1990 - June 1993), even by adopting the most sophisticated cryptic code. Actually, quite updated literature reference lists and databases are nowadays available or remotely accessible, such as the series of "Astr. & Astrophys. Abstracts" (Springer-Verlag) published every six months, the more specialized IAU Comm. 42 "Bibliography and Program Notes (BPN) on Close Binaries" (ed. A.Yamasaki), which is distributed to Comm. 42 members, the remotely accessible general literature database at Strasbourg Obs. (SIMBAD), the IUE archives at Goddard (NASA) and VILSPA (ESA), the high- energy database at Goddard (HEASARC, High Energy Astrophys. Science Archive Research Center), ESA's Space Science Department (SSD-ESTEC), and at the European Space Information System (ESIS-ESRIN). Any desired information on ground-based photometry and satellite data, sorted out by star name, subject, or author's name can be extracted from the above quoted literature sources. Hence, in the present report we will only try to illustrate general trends in ground-based photometry of close binaries that occurred in the past triennium.

Specific topics addressed by other sections of this report or by other IAU commissions (*e.g.*, novae and supernovae by Comm. 27) are disregarded, unless they have a vested close binary interest. Similarly, short notes or conference reports with preliminary results, that are or will appear in full papers, have been generally disregarded.

As usual, the most widely used photometric system has been the Johnson-Morgan UBVRI system (as well as the Cousins system), followed by the Stromgren uvby and the near-IR JHKLM systems, and by narrow-band Hα and Hβ photometry.

One of the highlights has been the increasing interest in the optical identification and study of CVs and low mass X-ray binaries discovered by the Einstein, Exosat, Rosat and Ginga X-ray observatories. A simple inspection of Comm. 42 BPN (Nos. 51 to 56) readily show the relatively large number of papers dealing with X-ray binaries and CVs with respect to classical close binaries. A fundamental reason for that has been the adoption of CCD detectors for photometry of faint sources down to about the 20th magnitude. Several close binaries and eclipsing binaries with periods of the order of 1-10 hours have been discovered (*IBVS*,**3489**; *MN*,248, 5p; *A&A*,242, 397; *PASP*,**103**, 300; see also papers presented at the 11th N.Am.Work. on Accretion-Powered Close Binaries), some of them in the central region of star clusters (M 15: *Nature* **351**,130; M 67: *IBVS*,**3586**; NGC 188: NATO-ASI on Active Close Binaries, p.891; NGC 288: *ApJS*,**82**,145; NGC 2004 and NGC 2100: *MN*,**260**,795; NGC 2346: *IBVS*,**3584**; NGC 5466: *AJ*,**100**,469 and *IAU Symposium 151*,483; NGC 6134: *A&AS*,**95**,429; Ton 2: *Acta A*,**42**,155; Pegasus: *AJ*,**100**,1151; Pleiades: *A&A*,**265**, 513; Praesepe: *ApJ*,**376**,514. The improved photometric accuracy, spatial resolution and limiting magnitude of CCD photometry have opened new research opportunities in the study of faint and cluster close binaries.

The binary nature of Nova Cyg 1978 = V 1668 Cyg (*MN*,**245**, 547) and of Nova Mon 1975 = A0620-00, the latter a black hole binary (*11th N. Am. Work. on Accretion-Powered Close Binaries*, p.7 and p.11), has also been discovered.

By using high-speed photometric techniques, a number of studies have addressed the question of short-period oscillations, flickering and flaring (*ApJ*,**359**,197; *ApJ*,**374**,340; *A&A*,**266**,237; *IBVS*,**3806**; *Acta A* **40**,129; *IBVS*,**3567**; *11th North Am. Work. on Accretion-Powered Close Binaries*, p.243 and p.279). Jets or jet- like ejection have also been found (*ApJ*,**354**,231; *Nature* **346**,637; *ApJ*,**357**,231).

Only a few interferometric studies have been done (*e.g.*, *ApJS*,**177**,161). This research topic certainly deserves more attention, but it is easy to anticipate that the completion of new large telescopes with interferometric capability, such as the ESO VLT (Very Large Telescope) and the Arizona U.-Italy LBT (Large Binocular Telescope), will provide new momentum to interferometric studies.

An increasing number of multiwavelength, multisite studies have been carried out (*e.g.*, YY Gem: *A&A*,**230**,419; V803 Cen: *MN*,**245**,140; V834 Cen: *ApJ*,**374**,744; HR 4049: *A&A*,**242**,433; Cyg X-2: *ApJ*,**355**,468; LMC X-3: *ApJ*,**364**,266; Sco X-1: *ApJ*,**396**,201; V 471 Tau: *ApJ*,**391**,773; τ Per: *AJ*,**101**,1821; V 711 Tau = HR1099: *A&A*, in press). These programs generally include concurrent photometry and spectroscopy, from ground-based and space-born telescopes, and are aimed at probing the vertical structure of stellar atmospheres or the system environment by monitoring specific emission diagnostics along a complete orbital or stellar rotation period, in order to disentangle permanent from transient features. The multisite approach to achieve round-the-clock coverage is essential in the case of unfavourable variability period or to follow transient or long-duration flare-like events. The multisite, multiwavelength programs have proven to be fundamental for solar-like activity and CV studies (cf. MUSICOS project, IAU Coll. 137, p.662).

The discovery and study of photospheric activity cycles by photometric methods in different types of close binaries (cf. *AJ*,**99**,1941; *A&A*,**237**,395; *A&AS*,**95**,55; and references therein) have triggered systematic monitoring programs of several dozen systems, which are performed also by unmanned Automatic Photoelectric Telescopes (APT). The majority of APTs are located in USA (*e.g.*, JPL, Mt. Hopkins, Phoenix, Villanova). Recently, however, two APTs were installed at Catania (Italy) and Seoul (S.Korea) Observatories. The importance of systematic photometry by automated telescopes is now widely recognized (cf., Surf. Inhom. in *Late-type Stars*, p.201; *New Frontiers in Binary Star Research*, p.45).

In addition to increasingly accurate light curves of known close binaries, the of which provides fundamental stellar parameters, the contribution of ground-based photometry can be appreciated by a simple inspection of even a likely incomplete list of newly discovered photometric close binaries and CVs. The new systems listed below, in addition to those already quoted, certainly deserve additional and extensive observations:

FO Aqr (*ApJ*,391,295); FS CMa (*IBVS*,**3602**); V1776 Cyg = Lanning 90 (*ApJ*,365,696); V365 Lac (*AZ*, **67**,1010); DO Leo (*PASP*,**102**,558); S Mus (*PASP*,**102**,551); BD+33 4070 (*IBVS*,**3814**); HD 2680 (*MN*,**261**,1); HD 46407 (*A&A*,**253**,407 and IAU Symp. 151,411); HD 89499 (*A&A*,**244**,310); HD 101191 (*MN*,**254**,404); HD 135775 (*MN*,**254**,404); HD 197010 = 1E2038.3-0046 (*IBVS*,**3536**); HD 223971 (IAPPP Comm. **44**,82); GX 339-4 (*MN*,**259**,395); H1907+690 (*ApJ*,**379**,715); LB 1800 = 4U0608-49 (*ApJ*,**355**,617); MS1603.6+2600 (*ApJ*,

365,686); PG0027+260 (*AJ*,**102**,272); SVS2831 Cyg (*AZ* **1547**,21); SVS2874 Cyg (*AZ* **1547**,17); 1E1615.0+3114 (*IBVS*,**3547**); in Magellanic Clouds: (*MN*,**250**,119); in M15: AZ 211 = X2127+119 (*A&A*,**270**,139); in NGC 2346 (*IBVS*,**3584**); in 47 Tuc (*Nature*,**360**,46); in Virgo (*AZ* **1548**,27); near BX Peg (*IBVS*,**3579**); in multiple systems (*PASP*,**104**,663).

5B. Ground-Based Spectroscopy (R. W. Hilditch)

The past triennium has seen two main developments in spectroscopy which are of major importance for the study of close binary stars. Firstly, new echelle spectrographs equipped with larger format CCD detectors have been installed on the world's major telescopes, thereby permitting studies of spectra at high resolution (< 0.1 Å) over a wide wavelength range (3000Å). For a given source brightness, observers can now achieve time resolutions comparable with that attainable only at very much lower resolution and spectral coverage a few years ago. Secondly, much attention has been devoted to the analysis of spectra involving techniques of deconvolution, doppler tomography, cross-correlation and spectral-line broadening functions. Hill (*ASP Conf.Ser*, **38**,127, 1993) provided an excellent summary of the techniques or procedures which he and co-workers have developed to determine radial velocities via cross-correlation from the spectra of close binary stars. By careful attention to such procedures it is now possible to obtain reliable radial velocity curves for both components in binaries even when the luminosity ratio is quite extreme. Light ratios of 10:1 are now relatively easy, and the procedures have been success- fully applied at 20:1 (Khalesseh & Hill *A&A*,**257**,199,1992). The consequences for all areas of close binary research are very significant since it allows much more accurate specifications of masses and absolute parameters for many types of evolved systems as well as for main-sequence systems with dissimilar components. No longer are we confined to studying double-lined systems with light ratios of order unity to determine such absolute parameters with precision.

Bagnuolo & Gies (*ApJ*,**376**,266,1991) have described new techniques, borrowed from medical physics, of separating the spectra of the two components in binaries from the observed spectra. This tomographic separation procedure has allowed these authors and others to disentangle such spectra very effectively in a wide range of binaries - O stars with similar components and with added contributions from colliding stellar winds, to Algol-type binaries with light ratios of 10:1. Thus a complete set of spectra, obtained to determine radial velocities of both components by cross-correlation for example, may then be subjected to much further analysis. Co-adding all the observed spectra, with appropriate doppler-shifts for both components, then results in separate and high S/N spectra of each stellar component. Not only does this work result in improved spectral classifications, but also it provides the opportunity to conduct quantitative spectroscopic investigations of line profiles for equivalent widths, rotational velocities, temperatures, surface gravities and chemical abundances.

Eclipse mapping and doppler tomography techniques have also continued to be applied in extenso by a number of observers/data analysts to establishing the structure of accretion discs in cataclysmic variables, and the distribution of star spots or zones in RS CVn systems and WUMa-type contact binaries. Good examples of such work are to be found in Dhillon *et al.* (*MN*,**252**,342,1991) for accretion disc mapping, and Strassmeier *et al.* (*A&A*,**247**,130,1991) for star spots. Much of this work concentrates on monitoring detailed changes in profiles of individual lines, requiring the use of large telescopes to attain good S/N in short time intervals.

Rucinski (*AJ*,**104**,1968,1992) has returned to the subject of the spectral-line broadening function applied to contact binaries. In his new spectroscopic study of AW UMa, he demonstrates the effectiveness of the singular value decomposition (SVD) technique to establish broadening functions (doppler profiles) which describe the intensity distribution across the surfaces of both stars as a function of radial velocity. One then requires a model of the binary system with appropriate intensity distributions and rotation laws to establish how well a model function compares with those derived from observed spectra. Clearly such techniques will be used in future in combination with photometric studies to determine more accurate descriptions of stellar surfaces.

If much effort has been put into new procedures for analysing spectra, then certainly no less effort has been spared in continuing with well-established practices of determining radial velocities, spectral-line profile variations as a function of orbital phase etc. from (typically) medium-resolution spectroscopy. It has been the practice, in the past two decades, for the contributor of this section of the Commission Report to provide a list of all the binary systems observed spectroscopically in the triennium. But the various abstracting and database services now available render the value of that task somewhat obsolete. A survey of the literature over the last three years demonstrates that three main themes dominate the publications of ground-based spectroscopic or spectrophotometric results - *(i)* time-resolved spectra of faint sources via 4-m class telescopes which are later co-added into orbital-phase bins by means of precisely-known photometric or X-ray ephemerides for further study, *(ii)* the "main-stream" determinations of radial velocities from medium-resolution spectra with the best techniques to obtain spectroscopic orbits of high precision, and associated line profile studies, *(iii)* searches for the existence of binaries amongst particular stellar types or clusters/regions.

This reviewer has noted some 460 publications which present significant spectroscopic investigations - a testament to the hard work of many astronomers world-wide. It is interesting to note that about 40% of these publications are concerned with detached normal or RS CVn-type binaries, such a large proportion being due to continuing surveys of solar-neighbourhood stars for binarity, late-type stars observable via radial-velocity spectrometers, or surveys of open clusters and recent star formation regions. Useful references for much of this work may be found in the proceedings of the conference "Binaries as Tracers of Stellar Formation" (eds. Duquennoy & Mayor, Cambridge UP, 1992) held at Geneva in 1992. A few new, and precise (2%), masses and absolute dimensions for detached systems have been determined since the extensive review of such data by Andersen (*A&A,Review* **3**,91,1991) (cf. for example, Holmgren *et al.* *A&A*,**248**,129,1991).

Cataclysmic variables and low-mass X-ray binaries constitute the second most popular spectroscopic targets at 25%. The majority of these papers discuss the properties of the accretion discs, streams and columns deduced from time-resolved spectroscopy via doppler tomography. But several also report radial velocities for the faint secondary components in these systems, thereby substantially improving the physical models for these interacting binaries, (*e.g.*, Friend *et al.* (*MN*,**246**,637,1990). Of particular importance is the LMXB system V404 Cyg (Casares & Charles *MN*,**255**,7,1992), which, with a mass function of 6.2 solar masses, may contain a black-hole secondary; and UU Sge, the central star of a planetary nebula, which is an eclipsing binary with spectroscopic orbits for both components determined for the first time (Pollacco & Bell *MN*,**262**,377,1993).

Semi-detached (Algol-type) and contact binary studies contribute 20% of the total, with significant contributions of precise determinations (5%) of masses, radii etc. being made (*e.g.* Tomkin *ApJ*,**387**,631,1992; Lu *AJ*,**102**,262, 1991; Samec & Hube *AJ*,**102**,1171,1991). These data provide real tests of mass-exchange/loss evolution models, not just in the Milky Way galaxy but also for such systems in other galaxies with different chemical histories (Bell *et al.*, *MN*,**250**,119,1991).

The flurry of activity on WR binaries in the early part of the triennium seems to have been superseded by attention to symbiotic stars in the latter part, with a total of 12% of papers published, many on MWC 560 and CH Cyg. A few papers discuss specifically the chemical abundances of the components of binary stars. Notable by their paucity, there are less than 10 papers reporting spectroscopy of high-mass X-ray binaries. Nevertheless, much improved determinations of the masses of the neutron stars and their OB-type companions in SMC X-1 and QV Nor are presented by Reynolds *et al.* (*MN*,**256**,631,1992; **261**,337,1993).

5.C Polarization Studies (R. H. Koch)

Between the previous reporting interval and the current one two major changes have occurred. The first of these is the slowly increasing number of polarimetric data in the spacecraft UV. While very few of these measures refer to close binaries themselves, the new results (*e.g.*, *ApJ*,**385**,L53, *ApJ*,**386**,562) are establishing a longer baseline for evaluation of the interstellar components of all measures. The levels of intrinsic polarization for well–behaved, single stars corrected for their interstellar components will provide a better reference against which to judge close binary polarizations. The second change has been increased use of modern polarimeters on large–aperture telescopes. This technological marriage has greatly enhanced S/N with essentially no sacrifices of other requirements. Several effects have followed from this cause. (1) Shortened duty cycles of observing have resulted in very large data sets. (2) Because of briefer duty cycles, it is now useful to search for phase–locked periodicities shorter than has been feasible up to now. (3) It is now possible to observe objects hitherto too faint. (4) Narrowed bandpasses permit spectropolarimetry for many objects so as to overcome the dilution characteristic of broad–band instruments.

It is not only close binary stars which benefit from this happy situation, but these systems do create some requirements for polarization data sets which are not characteristic of those for constant stars. For example, as noted in (1) and (2) above, data sets for periodic variables are now commonly voluminous and authors have recourse to publishing only a precis of the journal of results. A good example of this practice occurs in *ApJ*,**401**,628 wherein Stockman *et al.*. report on their concentrated program observing many CV's. No tabulated data are available to guide the reader in the reasoning of the authors. Even if data show no periodicities, archival availability should be ensured so that later workers may have access to them. This same practice underpins the extensive work of Donati, Semel, and Rees (*A&A*,**265**,669), whose paper powerfully illustrates item (4) above.

The past triennium has been characterized by extensive work on magnetic white dwarf CV's, Wolf–Rayet binaries, and symbiotic stars. By contrast, the familiar attention to conventional main sequence and giant pairs appears to have abated. It is not possible to refer here to all published studies; the observed systems are simply too numerous. Rather, citations will be confined to those available through June 30, 1993 and which have led to evaluations of binary magnetic, geometrical, or envelope parameters. For many of these objects, concurrent photometric and spectroscopic programs have been necessary. Below is a list of polarimetric analyses.

EZ CMa *ApJ*,**365**,L19, *ApJ*,**391**,L37, *ApJ*,**397**,277; V834 Cen *MN*,**256**,252; DH Cep *ApJ*, **366**,308; V444 Cyg *ApJ*,**359**,211, *ApJ*,**410**,342; V1500 Cyg *ApJ*,**371**,749; UZ For *MN*,**253**, 27; AM Her *MN*,**251**,28; θ Mus *ApJ*,**391**,L37; VV Pup *A&A*,**235**,245; Grus–V1 *MN*,**244**, 34P; H1907+690 *ApJ*,**379**,715; HD 226868/Cyg X–1 *ApJ*,**384**,249; Her X–1 *A&A*,**244**, L41; HR 1099 *A&A*,**232**,L1; LSS 3074 *IAU Symp. 151*,505; Sk–67D105 *IAU Symp. 151*,505.

Space limitation permits only a sampling summary of the large number of theoretical papers published over the last 3 years. The Glasgow methodology has advanced to the level that significant limitations in the original procedure have been removed in *ApJ*,**379**,663. This effort underpins those in *MN*,**260**,513 & 525 with specific application to Be stars. A generalization beyond the familiar evaluation of mass loss from expanding envelopes appears in *A&A*,**265**,663 and a more specific application to the energy supply in the jets of SS 433 is given in *A&A*,**259**,L43. Mention may also be made of the very ample treatment of polarized cyclotron emission from AM Her–type objects appearing in *ApJ*,**354**,625. Finally, the keV–band polarization signatures in emission lines from the disks of collapsed objects are treated in *ApJ*,**382**,125.

The only new review of current understanding of polarization in close binaries appears in *The Realm of Interacting Binary Stars*, p311. This is simply a statistical summary of the known incidence of polarized and unpolarized binaries in various evolutionary stages and the likely dominant polarizing mechanisms for each stage.

5.D X-Ray Observations (J. C. Raymond)

Extensive observations have been performed over the past few years with ROSAT, GINGA and SIGMA. ROSAT observations have been vital to X-ray source identifications, especially in crowded fields such as external galaxies or globular clusters. GINGA produces important X-ray timing results and studies of the iron line. SIGMA is important for observations of high energy X-ray tails. The ASCA satellite is just beginning to provide observations at higher spectral resolution.

Many exciting results have been obtained from observations of soft X-ray transients. J0422+32, discovered by GRO/BATSE, showed an X-ray spectrum similar to that of Cyg X-1, and it is likely to contain a black hole. Nova Muscae 1991 (GS 1124-683) also showed a luminous soft X-ray component with a high energy tail, and SIGMA observations showed the 511 KeV positron-electron annihilation line (Sunyaev *et al.* 1992 *ApJ*,**399**, L145). It also has a mass function which strongly suggests a black hole (Remillard *et al.* 1992 *ApJ*,**389**, L75).

Another exciting discovery in the past few years was the existence of exceptionally soft sources in the Large Magellanic Cloud (Greiner 1991 *A&A*,**246**, L17). These objects have blackbody temperatures below 0.1 KeV and high luminosities. While the nature of these systems is still open to question, Van den Heuvel *et al.* 1992 (*A&A*,**262**, 97) present a strong case that these are white dwarfs burning accreted hydrogen.

One general area of recent accomplishments has been a series of multiwavelength campaigns covering X-ray through radio wavelengths. Among other results, the campaigns on Cyg X-2 and Sco X-1 have helped to pin down the relationship between the position of a 'Z' source on the X-ray color-color diagram and its accretion rate (*e.g.* Vrtilek *et al.*, 1991 *ApJ*,**370**, 717).

5.E UV and EUV (G. E. McCluskey, Jr. and R. S. Polidan)

The IUE continued its role as the mainstay of ultraviolet astronomy and provided essentially all of the UV data on interacting binaries. To date, HST has observed very few interacting binaries, a situation which will hopefully improve in the near future. The ROSAT has provided limited but promising observations in the EUV spectral region for a number of systems and with the successful launch of EUVE in June 1992 this spectral region is now open to investigation in far greater detail than ever before. In June 1993 the first CD-ROM from the EUVE Science Archive was published.

The IUE observed many cataclysmic variables (CV). la Dous (*A&A*,**252**, 100) made a detailed study of IUE spectra of numerous CV, analysing the effects of orbital inclination on the lines and continuum. Harlaftis (*MN*,**259**, 593) compared the UV spectrum of Z Cha in normal and super outburst finding the accretion rate, which is higher for superoutbursts, to be a crucial factor. Selvelli (*ApJ*,**393**, 289) found that the recurrent novae T CrB probably does contain a white dwarf accretor rather than the previously suggested main sequence accretor.

Other CV studied were TX Col and TW Pic (Moucher *A&A*,250, 99), V795 Her (Prinja and Rosen *MN*,262, L37), QU Vul (Saizar *et al.* *ApJ*,398, 651), RW Sex, RW Tri and V Sge (Vitello and Shlosman *ApJ*,410, 815) and SU UMa, RX And and 0623+71 (Woods and Drew *MN*,245, 323).

Other interacting binaries studied with IUE include VV Cep (Bauer *et al.* *A&AS*,**90**, 175), υ Sgr (Dudley and Jeffrey *MN*,**247**, 400), AL Vel (Eaton *et al.* *AJ*,**100**, 799), ζ Aur (Eaton *ApJ*,**404**, 305), UU Cnc (Eaton *MN*,**247**, 62), 32 Cyg (Eaton *AJ*,**105**, 1525), ι Ori (Gies *et al.* *ApJ*,**403**, 752), Algol (Gonzalez-Riestra *et al.* EPIBS p.285), U Sge (McCluskey *et al.* *ApJ*,**378**, 281), τ Per, HR 2554 and HR 6902 (Schroder *A&A*,**236**, 165), Cyg X-2 (Vrtilek *et al.* *A&A*,**235**, 162), UW UMa (Wiggs and Gies *ApJ*,**407**, 252) and HD 47129 (Wiggs and Gies *ApJ*,**396**, 238). Stickland *et al.* continued their UV radial velocity curve program (*Obs*, **111**, 23; **111**, 167; **112**, 150; **112**, 277).

binary systems (*A&AS*,**96**, 255), including internal structure constants for the interpretation of observed apsidal motion rates (*A&A*,**244**, 319). Accurate binary star data around the TAMS seem to favour the existence of some amount of overshooting, at least within the mass range between around 1.5 and 2.5 solar masses (*ApJ*,**363**, L33). Below that value, convection in the core is too small and above it, the effect of overshooting may be negligible. A detailed study of Procyon indicated that no convective overshooting was necessary (*ApJ*,**405**, 298) and similar results were pointed out by Stothers & Chin using different tests (*ApJ*,**381**, L67; *ApJ*,**383**, 820). On the contrary, BW Aqr was shown by Clausen (*A&A*,**246**,397) to be better understood with overshooting models and Napiwotzki *et al.* (IAU Coll. 137, p.461) obtained similar indications from the width of the main sequence. The new opacities provide better agreement with observations, adopting smaller amounts of convective overshooting, but new observational data about actual metal abundances for well-detached binaries are still badly needed to perform a detailed comparison (IAU Coll. 137, p.177).

The importance of obtaining accurate dimensions for early-type stars and the involved difficulties have been emphasized (*Messenger*, **64**, 26; *Observ*, **111**, 14). Popper and Hill (*AJ*,**101**, 600) rediscussed several spectroscopic orbits of OB systems using cross-correlation techniques and warned against some systematic effects. A revision of the early-type systems BF Aur (*A&A*,**265**, 613) and NY Cep, with no secondary eclipse due to its high eccentricity (*ApSpSc*, **198**, 137) have been published. Accurate absolute dimensions have been obtained for the early-type binaries AH Cep (*A&A*,**236**, 409), V3903 Sgr (IAU Coll. 137, p.371), CW Cep (*A&A*,**241**, 98) and U Oph (*A&A*,**248**, 129), the two latter showing apsidal motion. Further, new radial velocity curves have been published using IUE measurements for the systems Y Cyg (*Observ*, **112**, 150; *BAAS*,**22**, 1292) and CW Cep (*Observ*, **112**, 277). Less massive detached binaries like EG Ser (*AJ*,**105**, 2291), HD 174853 (*AJ*,**102**, 1777) or RX Her (IAU Symp. 151, p.281; *IBVS*,**3403**) were also analysed.

The relatively unexplored region of the HR diagram around main sequence G-K stars is still being scrutinized spectroscopically by Popper (*ApJ*,**404**, L67). RU UMi is a late-type binary which may be a result of extensive mass transfer (*MN*,**260**, 478) and the orbit and physical parameters of V1073 Cyg (*A&A*,**265**, 597) and WX Cnc (*Ch.A&A*,**14**, 397; **15**, 214) have also been determined. A particularly interesting case is TZ For (*A&A*,**246**, 99), a fully-fledged red giant binary allowing accurate comparison with stellar models and a detailed tidal evolution analysis. Another interesting case, the system AI Phe, has been revisited (*ApJS*,**79**, 123). Many late-type eclipsing binaries are found to be active due to enhanced rotation induced by tidal synchronization. Three more papers by Popper of a series devoted to study spectroscopic orbits of close binaries with Ca II H and K emission have been published (*AJ*,**100**, 247; **101**, 220; **102**, 699), adding 15 binaries to the existing list. On the other hand, fundamental data for stars like V792 Her (*AJ*,**101**, 1489; see also *ApSpSc*, **182**, 1) and short-period binaries, like BH Vir (*A&A*,**237**, 148), RT And (*Ch.A&A*,**16**, 162) and ER Vul (*A&A*,**238**, 145) have been obtained. A second edition of the catalogue of chromospheric active binary stars by Strassmeier *et al.* (*A&AS*,**100**, 173) was published, including revised masses and radii. In some cases, these values are only estimations using methods like that by Hall (*AJ*,**100**, 554). Activity in the cool components of totally eclipsing Algol binaries was studied by Olson and Etzel (*AJ*,**106**, 342) and wind driven mass transfer, with special attention to the system Z Her, was discussed by Tout and

Hall (*MN*,**253**, 9).

Photometry and spectroscopy of very close early-type binaries, like SV Cen (*AJ*,**103**, 573) and LZ Cep (*Ir. Astron.J*, **20**, 32) or early-type Algols, like TV Cas (*A&A*,**257**, 199) has also been analyzed. To try to understand these and other interacting binaries, De Greve and De Loore have computed a series of evolutionary sequences (*A&AS*,**94**, 453; **96**, 653; **97**, 527). β Lyr has been revisited (*PisAZh*, **18**, 711; *A&A*,**266**, 307) as well as the evolutionary status of Algol (*MN*,**262**, 534) from a non-conservative point of view. The chemical evolution of mass exchanging binary systems was discussed (IAU Coll. 137, p.792; *MN*,**259**, 17) and absolute dimensions of Algol-type semidetached binaries were obtained for systems like TX UMa (*ApSpSc*, **201**, 177), RT Per (*Ch.A&A*,**14**, 437), RS Sgr (*BolAsArgAstr*, **36**, 28), U Cep (*ApJ*,**377**,278; *BAAS*,**22**, 1292) or V505 Sgr (*A&A*,244, 75; *ApJ*,387, 631). The system GO Cyg was found to be a reverse Algol (*ApSpSc*, **203**, 121) and CG Vir was suggested as a post-Algol phase system (*Ch.A&A*,**14**, 273). It was studied the effect of irradiation in the cool components of Algol-type binaries (*A&A*,**256**, 572) and other effects of proximity (NATO ASI, p.7). On the other hand, a classification survey has been started for classical Algol binaries (*BAAS*,**22**, 1334) while the various methods to estimate dimensions in this type of systems have been discussed (NATO ASI, p.121) and applied to compile a catalogue of the best available data (IAU Symp. 151, p.295). Cool Algols are introduced and discussed by Popper (IAU Symp. 151, p.395). Absolute dimensions of W UMa-type binaries have been obtained for the systems V781 Tau (*AJ*,**105**, 646), AA UMa (*Ch.A&A*,**14**, 389), SS Ari (*AJ*,**102**, 262), and RW Dor (*MN*,**255**, 285). With respect to cataclysmic binaries, Webbink (*11th NAmerWork CataclVar.*, p.117) has presented a spectroscopic and photometric survey with the objective of deriving absolute dimensions.

One of the areas of research showing important advances during the past triennium was the determination of stellar dimensions in binaries of the Magellanic Clouds (NATO ASI, p.61). Absolute dimensions were published for the semidetached system HV 2226 (*MN*,**250**, 119; IAU Symp. 151, p.501) and CCD light curves have been obtained for systems like HV 12484 (*MN*,**260**, 777), HV 2208 and HV 12634 (*MN*,**254**, 419). Specially interesting systems were found to be HV 2274 (*MN*,**258**, 527; IAU Symp. 151, p.509), a well-detached eccentric binary with apsidal motion, and the eccentric Wolf-Rayet eclipsing binary HD 5980 in the SMC (IAU Symp. 143, p.229). Some massive close binaries in the Magellanic Clouds were studied by Niemela and Bassino (IAU Symp. 151, p.497)

Another area of interest has been the study of binary members of open clusters. This is a promising research area due to the combination of isochrone and extinction determinations with stellar masses and radii (102 Summ. Sci. Meet. of the ASP, p.427 and p.433). Binary systems in clusters and associations which have been studied include H235 in NGC 752 (*BAAS*,**23**, 881; IAU Symp. 151, p.487), SS Lac in NGC 7209 (IAU Symp. 151, p.483; *BAAS*, **23**, 879) which is no longer eclipsing (!) or CW Cep, a member of Cep OB2 like NY Cep. Known eclipsing binaries in the old cluster NGC 188 have been re-analyzed (NATO ASI, p.891) and the population in M67 has been studied (*BAAS*,**22**, 1293) as well as some spectroscopic systems in the halo (IAU Coll. 137, p.158; *PASP*,**104**, 1108). A CCD study has discovered 12 short-period eclipsing binaries in the old cluster Be39 (*MN*,**262**, 49) supporting the theory of formation of W UMa binaries through magnetic braking. Moreover, several papers have pointed out that a larger amount of eclipsing binaries are found in globular clusters, and Pop II stars in general, than previously reported (*PASP*,**104**, 981; *ComAp*, **16**, 61; *A&A*,**244**, 310; IAU Symp. 151, p.483; *ApJ*,**389**, 590).

Of particular interest for tidal evolution theories is the study of late-type binaries, like the eccentric case of HD 212280 (*AJ*,**105**, 2265), the circular orbit of the above mentioned TZ For, the nearby Capella (*PASP*,**105**, 476) or the active RZ Eri (*A&A*,256, 463). From a theoretical point of view, the Tassouls continued with their effort to explain synchronization and circularization in close binaries (*ApJ*,**395**, 259). Observational and theoretical data are presented by Mathews and Mathieu (IAU Coll. 135, p, 244), Mazeh *et al.* (NATO ASI, p.145) and Claret and Gimenez (IAU Symp. 151, p.269) for well detached binaries, and by Hall (NATO ASI, p.287) for chromospherically active binaries. Tidal circularization among close binaries in the halo (IAU Symp. 151, p.491) and in M67 (p.471) has also been studied. Zahn (*A&A*,**252**, 179) has provided new insight into the still schematic treatment of overshooting as well as a general review (Bin. as Tracers of Stellar Form. p.253), while Maceroni and van't Veer (*A&A*,**246**, 91) contributed to the evolutionary study of angular-momentum-losing G-type binaries. The dynamical theory of viscous tides in binary systems (*A&A*,**257**, 783; IAU Coll. 137, p.374; *A&A*,**257**, 783) has been analyzed, while the effect of dynamic tides on variations of the orbit was analysed by Ruymaekers (*A&A*,**259**, 349).

During the past triennium, important advances have been produced in the study of apsidal motions of eclipsing binaries. A review of the status at the beginning of it was given by Gimenez (NATO ASI, p.137) and several systems have been added to the list of eccentric binaries with accurate dimensions and well determined apsidal motion rates, like CO Lac (*BAAS*,**23**, 835; ATs 1543), BW Aqr (*A&A*,**246**, 397), V477 Cyg (*A&A*,**260**, 227), U Oph (*A&A*,**248**, 129) or CW Cep (*A&A*,**241**, 98). Moreover some evolved, more difficult, systems have been revised like V1765 (*BAC*, **42**, 230) and V380 Cyg (*BAAS*,**22**, 1223). Photometric studies have been presented for YY Sgr (*AJ*,**104**, 796), HS Her (*ATs*, **1552**; *AN*, **313**, 183), BF Dra (*IBVS*,**3867**), GR Car (*IBVS*,**3835**), and α CrB (*IBVS*,**3876**). Many observers have contributed to the data base of eclipse timings for systems showing apsidal motion (*IBVS*,**3408**. 3494, 3495, 3552, 3900 or BBSAG Bulletins, in particular, No. 99). A new list has been produced (*BAC*, **42**, 98) of eccentric binary systems deserving further attention as potential apsidal motion candidates and Lacy has proposed a new method to solve for apsidal motion variations (*AJ*,**104**, 2213) which was applied to analyse YY Sgr (*AJ*,, **105**, 637), V523 Sgr (*AJ*,**105**, 630) and V526 Sgr (*AJ*,**105**, 1096). A preliminary comparison with models was presented (IAU Symp. 151, 277; IAU Coll. 137, p.469) showing a much better global picture than only a few years ago. The effects of dynamical tides on the rate of apsidal motion was studied by Smeyers *et al.* (*A&A*,**248**, 94) showing a possible way to explain some anomalous cases. From the point of view of the relativistic apsidal motion term, new studies have been made of DI Her (*ApJ*,**375**, 314) and AS Cam (*AJ*,**102**, 256; *BAAS*,**23**, 879) which still show important discrepancies with predicted values.

Reflection and limb-darkening has been studied by Hadrava (*A&A*,**257**, 218) and Rubashevskij (*SovA*, **35**, 626) who provides a method for determining nonlinear limb-darkening coefficients from model atmospheres and light curves while a new nonlinear limb-darkening law was proposed by Diaz-Cordoves and Gimenez (*A&A*,**259**, 227). For these studies, new model atmospheres produced by Kurucz (in Light Curve Modeling of Eclipsing Binary Stars, p.93) are now available. Discussions about brightness distribution on highly distorted stars were made by Barman (*BASIndia*, **19**, 59) and by Nakamura and Kitamura (*ApSpSc*, **191**, 267). A revision of the effective gravity-darkening in different types of binary systems was presented by Kitamura (NATO ASI, p.69).

7. EVOLUTIONARY PROCESSES

7.A Origin of Binary Stars (A. P. Boss)

The last three years have seen a strong growth in interest in the formation of binary stars, driven by events on four fronts: (i) improved understanding of the statistical properties of main-sequence (MS) binary stars, (ii) emerging statistics for the properties of binary pre-main-sequence (PMS) stars, (iii) probable detection of binary protostellar objects, and (iv) theoretical advances in understanding the classical as well as newly advanced hypotheses of binary star formation. Several volumes have appeared that focus directly on binary stars (*Binaries as Tracers of Stellar Formation* = *BTSF*; *Interacting Binary Stars* = *IBS*; *The Realm of Interacting Binary Stars* = *RIBS*; and *Complementary Approaches to Double and Multiple Star Research* = *CADMSR*); the latter in particular contains important papers too numerous to list in this short report. Reviews of the theories of binary star formation are given by Pringle (in *The Physics of Star Formation and Early Stellar Evolution*), Bodenheimer, Ruzmaikina, and Mathieu (in *Protostars & Planets III* = *PPIII*), Clarke (in *BTSF*), and Boss (in *RIBS*).

It has long been recognized that the dynamical properties of MS binary stars place crucial constraints on any theory of their origin (*e.g.*, Tout *MN*,**250**, 701). Duquennoy and Mayor (DM, *A&A*,**248**, 485) surveyed all G-type stars within 22 pc of the Sun and found a unimodal period distribution and a secondary mass distribution rising toward low ratios; a similar unimodal distribution was found by Fischer and Marcy (*ApJ*,**396**, 178) for M dwarfs. When analyzed according to the improved mass ratio algorithm of Mazeh and Goldberg (in *BTSF* and *ApJ*,**394**, 592), the short period binaries in the DM sample show a mass ratio distribution that rises toward high $q = M_2/M_1$ values (Mazeh *et al.* *ApJ*,**401**, 265), implying a different distribution for long and short period binaries. Abt, Gomez, and Levy (*ApJS*,**74**, 551) found that B stars have a secondary mass distribution that rises toward lower masses, excepting the short period systems.

Brosch (*MN*,**253**, 545) found a five star system in the shape of a diamond that appears to be an evaporating small cluster. McCarthy *et al.* (*AJ*,**101**, 214) used an infrared speckle camera to image a triple system (Gleise 22ABC) where both secondaries (0.12 and 0.18 $M_\odot$) appear to orbit stably about the primary (0.36 $M_\odot$) at distances of 5 and 40 AU, respectively.

PMS binaries also are yielding dynamical properties of great interest for origins. Binary PMS stars have been detected by Reipurth *et al.* (*A&A*,**235**, 197), Moneti and Zinnecker (*A&A*,**242**, 428), Chen *et al.* (*ApJ*,**357**, 224), Simon *et al.* (*ApJ*,**384**, 212), Koresko *et al.* (*AJ*,**102**, 2073), Haas *et al.* (*A&A*,**269**, 282), and Casey *et al.* (*AJ*, in press). Amazingly, the binary frequency appears to be *higher* for PMS stars than on the MS (Ghez, Neugebauer, and Matthews, in *CADMSR*; Leinert *et al.*, in *CADMSR*; Zinnecker, Brandner, and

Reipurth, in *CADMSR*; Reipurth and Zinnecker, *A&A*,, in press). This seems to require significant orbital evolution between the PMS and MS phases, and that the great majority of stars form in binary or multiple systems. The eccentricity distribution of PMS binaries covers the same wide range (e = 0.0 to 0.8) as on the MS (Mathieu, in *BTSF*).

Candidate binary protostars are beginning to proliferate, detected both through molecular line emission (Harjunpää, Liljeström, and Mattila *A&A*,**249**, 493) and infrared emission (Sandell *et al.* *ApJ*,, **376**, L17; Mundy *et al.* *ApJ*,, **385**, 306). About 10% of PMS binaries contain an infrared-bright companion with a separation of about 200 AU (Zinnecker and Wilking, in *BTSF*). The existence of binary protostars places the tightest constraint yet on the epoch of binary formation – evidently binary formation occurs during or immediately prior to star formation itself.

Theoretical work on binary formation has concentrated on the fragmentation mechanism, that is, on break-up during the protostellar collapse phase prior to the PMS phase. Finite-difference 3D hydro codes (Boss and Myhill *ApJS*,**83**, 311) have been used to calculate tidally-induced fragmentation into binaries (de Felice and Sigalotti *MN*,**249**, 248; Sigalotti and Klapp *MN*,**254**, 111) and fragmentation of initially prolate clouds into binaries (Boss *ApJ*,**410**, 157). Myhill and Kaula (*ApJ*,**386**, 578) found that clouds with initial power law density distributions will fragment if they are in rapid differential rotation. Initial conditions leading to hierarchical fragmentation and multiple system formation were found by Boss (*Nature*, **351**,298), Klapp, Sigalotti, and de Felice (*A&A*,**273**, 175), and Sigalotti (*A&A*, in press). Fragmentation during the standard isothermal test case continues to be a useful test for 3D hydro codes (Myhill and Boss, *ApJS*, in press).

Smoothed particle hydro (SPH) calculations by Bastien *et al.* (*ApJ*,**378**, 255), Arcoragi *et al.* (*ApJ*,**380**, 476), Bonnell *et al.* (*ApJ*,**377**, 553; *ApJ*,**400**, 579) and Bonnell and Bastien (*ApJ*,**401**, 654) have examined in great detail the processes through which initially cylindrical clouds can fragment into binary and multiple systems. Monaghan and Lattanzio (*ApJ*,**375**, 177) demonstrated multiple system formation in a strongly cooled SPH cloud, while Chapman *et al.* (*Nature*, **359** 207) showed that binaries could result from the collision of interstellar clouds.

Given that PMS binaries have a range of eccentricities, the question of generating high eccentricities becomes important. N-body simulations of the evolution of open star clusters show that cluster evolution is unlikely to excite initially small eccentricities to large values (Aarseth, in *BTSF*), so high eccentricities must be primordial. Binary formation is likely to occur inside protostellar disks, and the interactions between companion objects and protostellar disks have been studied by Lin and Papaloizou (in *PP III*), Artymowicz *et al.* (*ApJ*,**370**, L35), and Artymowicz (*PASP*,**104**, 769). Mathieu (in *IBS*) discusses observational evidence for both circumprimary and circumbinary disks. Lubow and Artymowicz (in *BTSF*) find that initially small eccentricities should grow to e = 0.5 or so for binaries embedded in disks. If the binary is still accreting mass from the disk, however, Bonnell and Bastien (*ApJ*, in press) find that the binary eccentricity decreases. The net effect on binary eccentricities remains to be determined.

Other formation mechanisms have been studied as well. Adams and Benz (*CADMSR*) present SPH calculations of the growth of $m = 1$ instabilities in disks with central stars, leading to companions containing about 1% of the disk mass. Savonije, Papaloizou, and Heemskerk (in *Astrophysical Disks, Ann. N.Y. Acad. Sci. 675*) also studied the growth of $m = 1$ modes (also Heemskerk, Papaloizou, and Savonije *A&A*,**260**, 161), but were unable to reproduce the rapidly growing modes found by Adams, Ruden and Shu (*ApJ*,**347**, 959) and hypothesized to lead to the formation of binary companions.

Clarke and Pringle (*MN*,**249**, 584 and 588) modeled the ability of stars with protostellar disks to interact and become gravitationally bound, and found that capture rates are likely to be too low for capture to be an important mechanism. McDonald and Clarke (*MN*,**262**, 800) showed that very low mass companions should be rare in systems formed by capture. Hills (*AJ*,**103**, 1955) studied the formation of transient triple systems resulting from encounters between close binaries and single stars, while Anosova and Orlov (*A&A*,**260**, 473) integrated trajectories for three-body systems and delineated systems leading to immediate escape.

7.B Mass Flow and Evolution (G. E. McCluskey, Jr. & E. F. Guinan)

The observational and theoretical investigation of evolution and mass flow in non-degenerate binaries continues at a rapid rate. Major publications are IAU Symposium 151, "Evolutionary Processes in Interacting Binary Stars" (EPBIS) ed. by Kondo *et al.*, "The Realm of Interacting Binary Stars" RIBS) ed. by Sahade *et al.* and "Structure and Evolution of Single and Binary Stars" (SESBS) by de Loore and Doom.

The evolution of massive interacting binaries continues to attract great interest, stimulated in part by the difficulties presented by the blue supergiant progenitor of SN 1987A. General reviews include Iben (*ApJS*,76, 55), de Loore and Doom (SESBS), and Vanbeveren (*SpScRev* **56**, 249). De Greve and de Loore (*A&AS*,**96**, 653) and de Loore and De Greve (*A&AS*,**94**, 453) calculated extensive evolutionary sequences for stars with initial masses from 9 to 40 solar masses. Case B evolution was adopted and it was assumed that 50% of the mass lost by the stars was lost to the system. Initial masses greater than 20 - 25 solar masses lead to Wolf - Rayet stars and then to massive X-ray binaries. Some lower mass systems become He-rich binaries. Vanbeveren (*A&A*,**252**, 159) calculated a series of models carefully accounting for stellar wind mass loss during hydrogen and helium core burning. For initial masses greater than about 50 solar masses, a powerful stellar wind during the luminous blue variable stage of H-shell burning prevents a dynamic phase of mass loss due to binary interaction. All primaries of more than 10 solar masses (initial) begin He-core burning as bare helium cores and evolve thereafter as single stars. Dynamic mass loss rates of 10^{-3} are reached. de Loore and vanbeveren (*A&A*,**260**, 273) investigated a binary model for SN 1987A. The B3 supergiant evolved from the original secondary star with an initial mass of 9 - 15 solar masses and mass ratio near unity. When the progenitor was in the H-shell burning phase, it accreted 8 - 10 solar masses from its evolving companion and eventually became SN 1987A. Extensive pre-supernovae scenarios were investigated by Podsiadlowski *et al.* (*ApJ*,**391**, 246). Progenitors of SN of types Ib, II and 1987A were produced. Hubeny and Plavec (*AJ*,**102**, 1156) studied a thick disk model for β Lyr and Mazzali (*A&A*,**254**, 241) proposed a two wind model. As shown by Kondo *et al.* (*ApJ*, in press) who survey

OAO-2, Copernicus, IUE, and Voyager observations of β Lyr, no existing model can explain the bizarre behavior of β Lyr across the electromagnetic spectrum.

The Algol-type and closely related interacting binaries have drawn considerable effort. A wide range of phenomena are encompassed by the term Algol-type interacting binary. McCluskey (RIBS, p.39) divides the Algols into three classes: low activity systems, *e.g.*, S Cnc or U Sge; active systems, *e.g.*, U Cep or TT Hya; and dynamic systems, *e.g.*, SX Cas or V367 Cyg, if the latter two are semi-detached systems.

Algol itself has been the subject of recent research. Using the VLBI to observe the radio emitting plasma in the system, Lestrade *et al.* (*ApJ*,**410**, 808) find that the K-star's corona is responsible for the non-thermal radio emission which, along with X-ray emission from Algol (*e.g.*, Stern *et al.* *ApJ*,**400**, 321) demonstrates the importance of magnetic activity. Richards (*ApJS*,**86**, 255) made a detailed study of Hα profiles in Algol and finds evidence for a low-density, transient, asymmetric, rotating accretion disk which contains a turbulent high-velocity substructure. Gas extends some distance out of the orbital plane between the close components near the hot star where the gas stream interacts with the disk and/or the stellar surface. Evolutionary calculations by Sarna (*MN*,**262**, 534) indicate that about 15% of the total initial mass and 30% of the total initial angular momentum have been lost to the system. Edalati *et al.* (EPBIS p.303) find that the low mass, short period Algol system R CMa underwent far more systemic mass and angular momentum loss than most Algols.

Polidan and Wade (EPBIS p.351) studied the carbon abundance of 5 massive Algols showing UV emission lines during primary eclipse. Four systems showed no detectable carbon while one did have carbon emission indicative of roughly normal carbon abundance.

Unama *et al.* (*A&A*,**249**, 217; *A&A*,**267**, 126) observed a number of Algols at microwave wavelengths. Algol, RZ Cas, TW Dra, RT Lac, V505 Sgr and HR 5110 have radio spectra similar to the RS CVn systems. This emission is probably gyrosynchrotron emission from mildly relativistic electrons. Olson and Etzel (*AJ*,**106**, 342) attribute the increase in observed photometric variations of Algol secondaries with increasing rotational velocity and decreasing orbital period to rotationally induced magnetic activity in part. Guinan (EPBIS p.245) notes the importance of MHD effects on cool components of all classes of interacting binaries. Elias and Mutel (RIBS p.335) review the radio observations of late-type components in binaries. A wide class of binaries have been detected with magnetic fields of roughly 100-1000 G or higher associated with Algol secondaries.

With regard to low-mass systems, Rucinski (RIBS p.111) comments that angular momentum loss via magnetic winds may rival or dominate nuclear evolution in the W UMa systems. Mass and energy flow between the two stars is still very poorly understood. Vilhu (EPIBS p.61) postulates that W UMa systems have detached systems with periods of 1-4 days as progenitors and that they evolve into FK Comae stars or blue stragglers (single stars). Eggleton (EPIBS p.167) discusses the increase in cool star activity with increasing rotation rate which is critical for mass and angular momentum loss.

Important problems remain for investigators of interacting binaries. The "accretion disk" concept is used to explain a wide variety of phenomena and is undoubtedly a simplistic model (Sahade RIBS p.17). The use of the Roche model for hot stars with powerful winds or for systems in the dynamical phase of mass loss is very questionable at best (McCluskey RIBS p.381; Sahade RIBS p.17; Sahade *et al.* RIBS p.399). Greatly improved models for systemic mass and angular momentum loss are required. Guinan and Gimenez (RIBS p.51) review the great importance oh MHD effects for cool binary components, effects almost invariably neglected in interpreting and modeling interacting binary observations. Herczeg (EPIBS p.307) finds that irregular, often abrupt, period variations play a substantial role in the overall observed period changes. An explanation still eludes us.

Progress in resolving these issues will be stimulated by multiwavelength studies and new technology and methods, *e.g.*, Doppler tomography (Kaitchuk EPIBS p.175), but primarily in the old-fashioned way - hard work!

7.C Eruptive Phenomena (P. Szkody and S. Starrfield)

The last three years have seen major advances in the understanding of the structure and eruptive phenomena associated with cataclysmic variables. Three detailed reviews of the entire field were recently produced (Livio and Shore each had an article in the Proceedings of the 22nd Saas Fee Advanced Course, 1993 in press; Cordova, in X-Ray Binaries, 1993 in press; Starrfield, in Reviews in Modern Astronomy, Klare, ed., 1992, p.73). Highlights of results on the accretion disk, the white dwarf, the secondary star, LMC novae, recurrent novae, novae abundances, and the outburst of Nova V1974 Cyg 1992 are presented here. A poster and oral session were devoted to the observations of V1974 Cyg at the January 1993 meeting of the AAS (*BAAS*,1992, **24**, 1189).

The most promising results on understanding the accretion disk have emerged from time-resolved studies of eclipsing systems, such as Z Cha (Harlaftis *et al.*, 1992, *MN*,**257**, 607; Wood, in CVs and Related Physics, 1993, in press), HT Cas (Wood, Horne and Vennes, 1992, *ApJ*,**385**, 294) and U Gem (Marsh *et al.*, 1990, *ApJ*,**364**, 637). Some of these studies involved using IUE or HST spacecraft to obtain ultraviolet data in conjunction with optical coverage in order to derive the spectrum of the white dwarf and the disk. Eclipse mapping procedures (comparison to computed models or maximum entropy mapping, reviewed in Horne, IAU Coll. 129, 1991, 3) were then used to derive the temperature and surface density structures of disks. The results all show that observed disks at quiescence have a flatter temperature distribution with radius than predicted by steady state models but the outburst disks are close to predicted models (T proportional to $R^{-3/4}$). At superoutburst, the Z Cha results indicate an extended vertical structure with a cool bulge on the disk edge at the position of the intersection of the mass transfer stream. Doppler tomography using the emission lines of U Gem at quiescence indicate an outer rim of disk emission, with the bulk of the S-wave emission from the stream impact area, but with velocities suggestive of a mixture of disk and stream material. In the non-eclipsing old nova system, GK Per, modelling of the IUE and optical light curves of its dwarf nova like outbursts with disk instability models determined a large inner disk, indicating that a magnetic field greater than 10 MG disrupts the inner portion (Kim, Wheeler and Mineshige, 1992, *ApJ*,**384**, 269).

Besides information on the disk, the above techniques have resulted in new information about the stellar components. The eclipse studies provide a temperature for the white dwarf and allow a study of its heating as a result of the outburst scenario. The Wood HST results on Z Cha at quiescence and outburst show that the white dwarf is only heated by 2000K during outburst and cools within 16 days. This result is surprising in that it implies no extensive boundary layer develops during outburst (standard accretion disk theory predicts the boundary layer luminosity should be roughly equal to the disk luminosity unless the white dwarf is rotating very rapidly). The lack of a prominent boundary layer luminosity was also concluded from the X-ray and UV observations of VW Hyi (Mauche *et al.*, 1991, *ApJ*,**372**, 659; Belloni *et al.*, 1991, *A&A*,**246**, L44). Further puzzles about the interaction of this boundary layer and the heating of the white dwarf during a dwarf nova eruption are evident in the two component temperature structure determined in the studies of the white dwarf in U Gem (Kiplinger, Sion and Szkody, 1991, *ApJ*,**366**, 569; Long *et al.*, 1993, *ApJ*,**405**, 327).

The heating effects on the secondary also invoke theoretical challenges. Doppler tomography on U Gem revealed a source of Balmer emission on the secondary star which was offset from the line of centers. Further study of the irradiation of the secondary in CH UMa, IP Peg and YY Dra (Davey and Smith, 1992, *MN*,**257**, 476) confirmed the same effect, which could not be explained by straightforward irradiation from a hot white dwarf plus a hot spot. Furthermore, detailed spectroscopy of the secondary in the long period dwarf nova DX And (Drew, Jones and Woods, 1993, *MN*,**260**, 803) revealed spectral abnormalities which may be related to prior evolution through the common envelope phase or large scale mass transfer from the companion.

Successful searches of the LMC for novae since 1990 have resulted in studies of LMC 1990 No. 1 (a fast ONeMg nova: Dopita *et al.* IAU Circular 5262), LMC 1990 No. 2 (a recurrent nova: Shore *et al.* 1991 *ApJ*,**370**, 193), LMC 1991 (a fast CO nova; it reached a V$\sim$ 9 making it one of the most luminous novae studied: Della Valle 1991, *A&A*,**252**, L9), and LMC 1992 (a moderate speed CO nova with an ultraviolet spectral development similar to OS And 1986: Shore *et al.* 1993, in prep.). Novae in the Clouds are all at the same distance with small reddening, so they are useful as fiducials for studies of Galactic novae. For example, a comparison of U Sco to LMC 1990 No. 2 shows that U Sco must be a Halo object (Shore *et al.* 1991, *ApJ*,370, 193) and it was possible to compare the outburst of Nova Her 1991 to that of LMC 1990 No. 1 and obtain the distance and reddening to Her 91 (Starrfield *et al.* 1992, *ApJ*,**391**, L71). A review of LMC Novae can be found in Starrfield *et al.* (1993; in New Aspects of Magellanic Cloud Research ed. Baschek, Klare, and Lequeux, p.181).

The studies of recurrent novae indicate that there are at least two classes of RN, those with giant secondaries such as V3890 Sgr (Gonzalez-Riestra 1992, *A&A*,**265**, 71; Harrison *et al.* 1993, *AJ*,, **105**, 320; Shore and Aufdenberg 1993, *ApJ*, in press), V745 Sco (Sekiguchi *et al.* 1990, *MN*,**246**, 78), RS Oph, and T CrB and those with evolved but compact secondaries such as U Sco, V394 CrA, and LMC 1990 No. 2. In both cases, however, theoretical calculations require that the white dwarf be very massive in order for the outbursts to occur with very short inter-outburst times. Schaefer (1990, *ApJ*,**355**, L39) has determined orbital periods for U Sco and V394 CrA and Johnston and Kulkarni (1992,

ApJ,**396**, 267) obtained a very low mass for the white dwarf in U Sco. Duerbeck *et al.* (1993, ESO Messenger, No. 71, 19) redid the study and report a high mass for the white dwarf. A similar problem with a possible low mass white dwarf exists for Nova V838 Her 1991 (Szkody and Ingram 1993, *ApJ*, in press). A study of T Pyx was inconclusive about the structure of the system (Schaefer *et al.* 1992, *ApJS*,**81**, 321).

Abundance studies have demonstrated that there are two compositional classes of novae: those that occur on carbon-oxygen white dwarfs and those that occur on oxygen-neon-magnesium white dwarfs. Theoretical studies of thermonuclear runaways on ONeMg white dwarfs show that such novae can produce important amounts of ^{22}Na and ^{26}Al (Weiss and Truran 1990, *A&A*,**238**, 178; Nofar *et al.* 1991, *ApJ*,**369**, 440; Starrfield *et al.* 1992, *ApJ*,, **391**, L71; Politano *et al.* 1993, in preparation). Observational studies of recent nova outbursts have shown that the following were ONeMg novae: Her 1991 (t_3 = 2.8days; Matheson *et al.* 1992, *BAAS*, **24**, 1260; Vanlandingham *et al.* 1992; *BAAS*, **24**, 1284; Harrison and Stringfellow 1992, *BAAS*, **24**, 1284), Pup 1991 (Shore *et al.* 1993; in preparation), and Nova Cyg 1992 (Hayward *et al.* 1992, *ApJ*,**401**, L101). Andreä *et al.* (1993, *A&A*,submitted) determined nebular abundances for 10 nova outbursts. Saizar *et al.* (1991, *ApJ*,**367**, 310) and Andreä *et al.* (*A&A*,**244**, 111, 1991) reported abundances for the very slow CO nova PW Vul and Saizar *et al.* (1992, *ApJ*,**398**, 651) reported abundances for QU Vul a slow ONeMg nova. Abundances for LMC 1990 No. 1 were reported by Dopita *et al.* (1993, in "New Aspects of Magellanic Cloud Research", ed Baschek, Klare, and Lequeux, Springer-Verlag, p.195). Analyses of GQ Mus (Pèquignot *et al.* 1993, *A&A*,**219**) and V443 Scuti 1989 (Anupama *et al.* 1992, *A&A*,**263**, 87) have appeared. GQ Mus was detected in the ROSAT Survey (Ögelman *et al.* 1993, *Nature*,**361**,331) which implies that it is still in outburst.

Hayward *et al.* (1992, *ApJ*,**401**, L101) reported the detection of Ne II 12.8μm in V1974 Cyg 1992 and showed that the line profile was produced by a very asymmetric ejection. Dinerstein *et al.* (1992, *BAAS* **24**, 1189) reported on the transition to the IR coronal phase, Elias *et al.* (1992, *BAAS*,**24**, 1189) used an optical interferometer to resolve the expanding shell 8 days into outburst ($\sim$ 5 mas). Garnavich (1992, *BAAS*,**24**, 1189), Bjorkman et al. (1992, *BAAS*,**24**, 1190), Gallagher *et al.* (1992, *BAAS*,**24**, 1189), Taylor *et al.* (1992, *BAAS*,**24**, 1189), and Austin *et al.* (1992, *BAAS*,**24**, 1284) all reported optical studies. Polidan *et al.* (1992, *BAAS*,**24**, 1191) reported on Voyager observations, Shore *et al.* (1992, *BAAS*,**24**, 1191) described the IUE and HST data, Shore *et al.* (1993a, *ApJ*, in press) showed that this nova emitted at a constant bolometric luminosity for at least the first 100 days of the outburst, and Shore *et al.* (1993b, *AJ*,submitted) described the kinematics of the ejecta using both IUE high dispersion data and two HST visits. Hauschildt *et al.* (1993, *ApJ*,submitted) described a non- LTE, spherical, expanding stellar atmosphere analysis of the early spectra and showed that there was a major structural change in the ejecta early in the outburst. Gehrz *et al.* (1993, *ApJ*, in press) described their KAO observations of the nova and provided an estimate of the neon and sodium abundances in the ejecta.

7.D Close Binaries with Collapsed Components (R. F. Webbink)

A startling new development of the past triennium has been the discovery of a possible planetary system surrounding a millisecond pulsar. Although the discovery first announced (PSR 1829-10: Bailes *et al.* 1991, *Nature*,**352**, 311) was subsequently retracted

(Lyne & Bailes 1992, *Nature*,**355**, 213), it set in motion a flurry of model-building which was subsequently vindicated by the discovery of a second, more compelling example (PSR 1257+12: Wolszczan & Frail 1992, *Nature*,**355**, 145). A rich variety of hypotheses have been put forward: simple survival of a prior planetary system (Bailes *et al.* 1991, *Nature*,**352**, 311); cannibalization of the central star in a solar-type system (Podsiadlowski *et al.* 1991, *Nature*,**352**, 783); spiral-in of a primordial planetary system in the merger remnant of a massive X-ray binary (Wijers *et al.* 1992, *Nature*,**355**, 393); formation from the supernova ejecta (Bailes *et al.* 1991; Lin *et al.* 1991, *Nature*,**353**, 827); condensation in the thick disk remnant of a close double white dwarf (or white dwarf/neutron star) merger (Podsiadlowski *et al.* 1991), or in a disk formed by accretion from the wind of a distant (now extinct) Be star companion (Fabian & Podsiadlowski 1991, *Nature*,**353**, 801), by deflation of the envelope of a Thorne-Zytkow object (Podsiadlowski *et al.* 1991), by tidal disruption of a low-mass stellar companion (Stevens *et al.* 1992, *MN*,**252**, 19P), or by the systemic outflow from an evaporating companion (Tavani & Brookshaw 1992, *Nature*,**356**, 320; Banit *et al.* 1993, Planets around Pulsars, p.167); evaporation of a stellar companion to planetary mass (Bailes *et al.* 1991; Krolik 1991, *Nature*,**353**, 829; Rasio *et al.* 1992, *A&A*,**256**, L35); and more. A wide range of issues raised by this discovery are encompassed in Planets around Pulsars (1993, ASP Conf. Ser. 36), in which the comprehensive review and critique of these scenarios by Podsiadlowski, and the discussion of the physics of planetary formation in a disk around a pulsar by Phinney and Hansen will be of interest to students of close binary evolution.

Uncertainties over the formation channel for low-mass X-ray binaries and their putative connection with binary millisecond pulsars continue to preoccupy theorists. These issues are surveyed in X-Ray Binaries and Recycled Pulsars (1992, Kluwer), and have been thoroughly reviewed by Bhattacharya and van den Heuvel (1991, *Phys. Rep.*, **203**, 1). Recent population synthesis studies by Romani (1992, *ApJ*,**399**, 621) lead to a very high space density of black hole X-ray transient sources with low-mass donors, in comparison with their neutron star counterparts.

The process of pulsar companion evaporation observed first in PSR 1957+20 continues to stimulate many investigations into the evolutionary implications of this phenomenon (Podsiadlowski 1991, *Nature*,**350**, 136; Fedorova & Ergma 1991, *Pis'ma AZh*, **17**, 999; Tavani 1991, *Nature*,**351**, 39; Ergma & Fedorova 1991, *A&A*,**242**, 125 and 1992, *A&A*,**265**, 65; Shaham & Tavani, 1991, *ApJ*,**377**, 588; Harpaz & Rappaport 1991, *ApJ*,**383**, 766; D'Antona & Ergma 1993, *A&A*,**269**, 219).

Questions of the frequency and *Nature*,of neutron star-neutron star and neutron star-black hole mergers have received new attention as sources of gravitational waves (Hils *et al.* 1990, *ApJ*,**360**, 75; Hils 1991, *ApJ*,**381**, 484) and possibly of gamma-ray bursts (Narayan *et al.* 1992, *ApJ*,**395**, L83; Mészáros and Rees 1992, *ApJ*,**397**, 570). Several significant efforts at modeling the merger process have been published (Kochanek 1992, *ApJ*,**398**, 234; Bildsten & Cutler 1992, *ApJ*,**400**, 175; Rasio & Shapiro 1992, *ApJ*,**401**, 226), and Tutukov and Yungelson (1993, *MN*,**260**, 675) have derived new estimates for the frequency of these events.

44. ASTRONOMY FROM SPACE

PRESIDENT: J.E. Trümper
VICE-PRESIDENT: G.G. Fazio
ORGANIZING COMMITTEE: B.F. Burke, J.B. Hutchings, E.B. Jenkins (Past President), S.D. Jordan, B.D. Savage, W. Wamsteker

1. Introduction

Joachim E. Trümper

The period covered by this report has seen very significant progress in the different fields of space astronomy. In the infrared and submillimeter region the cosmic Background Explorer has continued to dominate the space activities. The most significant result was the first detection of anisotropies in the cosmic background radiation. In the optical and UV region the Hubble Space Telescope, launched in spring 1990 has produced a wealth of significant scientific results despite the image aberrations which are expected to be corrected by the repair mission in December 1993. The astrometric Hipparcos mission reached its scientific goals by producing 1.5 milliarcsec positions and 1milliarcsec/year proper motions. In the EUV region the fist photometric all-sky survey 1990/91 performed by the Wide Field Camera on ROSAT (Röntgen satellite) has increased the number of EUV sources by almost two orders of magnitude (to ~ 400). A further step was made with the Extreme Ultraviolet Explorer (EUVE launched in 1992) providing increased survey sensitivity and high spectral resolution capabilities. Further spectroscopic data in the UV/EUV were obtained with the UV telescopes on the ASTRO shuttle mission and on the ASTROSPAS, a shuttle attended free flyer. In X-rays more than 50 000 new X-ray sources were discovered by means of the ROSAT all-sky survey in 1990/91 which also led to a much improved knowledge of the galactic and extragalactic soft X-ray background. In 1991-93 a few thousand ROSAT pointed observations, including very deep ones, produced a wealth of new results. The Broad Band X-ray Telescope (BBXRT) flown on the ASTRO space shuttle mission in 1990 concentrated on medium resolution X-ray spectroscopy. As the successor of Ginga which re-entered in fall 1991 the ASCA satellite (Advanced Satellite for Cosmology and Astrophysics), launched in spring 1993, provides new observational capabilities in imaging X-ray spectroscopy. In hard X-rays the Granat mission allowed to study the time variability of compact X-ray sources, in particular the X-ray transients.
Gamma-ray astronomy experienced a boost by the launch of the Compton Gamma-ray Observatory in 1991 delivering many new significant results on gamma ray bursts, pulsars, diffuse line emission and quasars.
Last but not least, the most significant event for solar physics from space was the launch and operation of the Yohkoh satellite which yielded a large database for the study of high-energy solar phenomena that accompany solar activity.
All in all the results of these new missions represent a strong stimulus for all fields of astronomy and astrophysics. Two general trends should be mentioned: first the role of coordinated multiwavelength observations involving both space and ground-based observatories is rapidly increasing. There have been a number of very productive campaigns of this kind, such as the RIASS programme (ROSAT-IUE) or the World Astronomy Department. Another important aspect is that space missions yield digitized and calibrated datebases which lend themselves to archiving. The databases of IRAS, IUE, HST, Einstein, EXOSAT, ROSAT etc., as well as the digital optical and radio surveys, have become very powerful tools for astrophysical research.

J. Bergeron (ed.), Reports on Astronomy, Vol. XXIIA, 489–524.

2. Infrared Space Astronomy

Giovanni G. Fazio

I. INTRODUCTION

During the period from 1990 to 1993 the data from the Cosmic Background Explorer (COBE) satellite has continued to dominate the results in space infrared and submillimeter astronomy. The most spectacular result was the first detection of an anisotropy in the cosmic background radiation (CBR). Continued improvements in the accuracy of the measurements of the CBR temperature have shown no deviation from blackbody radiation. Impressive maps of the entire galactic plane were produced at near-infrared wavelengths and in the spectral lines of [CII] and [NII]. Detection of the source F10214+4724 in the Infrared Astronomical Satellite (IRAS) Faint Source Survey data base set off a flurry of exciement and interest. F10214+4724 is a galaxy at a redshift of 2.286 and is the most luminous and massive single source known in the universe. The IRAS database, which has improved with continued analysis, when combined with ground-based and space observations at other wavelengths has produced many new results in the study of starburst galaxies and AGN, infrared luminous galaxies, the origin of elliptical and radio galaxies, protostars, galactic structure, and new views of interstellar dust. The Kuiper Airborne Observatory (KAO) continues to remain a premier facility for far-infrared and submillimeter astronomy. KAO obsrvations have produced important advances in our knowledge of star formation processes, the interstellar medium, the Galactic center, and stellar and planetary astronomy.

The COBE results are so extensive that they will be treated in a separate section of the report of this Commission. In this section special attention will be given to the IRAS and KAO results, as well as other results from balloon-borne telescopes and rocket observations.

II. THE INFRARED ASTRONOMICAL SATELLITE (IRAS)

Follow-up observations of galaxies with bolometric luminosities that are dominated by the far-infrared radiation detected by IRAS continue to constitute a major fraction of the published literature in extragalactic astronomy. A comprehensive review of this wide field of research is beyond the scope of this summary. This is merely an attempt to cover some highlights.

IRAS F10214+4724: Starburst in a Protogalaxy or Dominant AGN?

Observations of source F10214+4724 in the IRAS Faint Source Survey by Rowan-Robinson *et al.* (1991) set off a flurry of excitement and interest that has inspired numerous additional observations of this source, as well as searches for others like it. F10214+4724, at a redshift of 2.286, has an estimated luminosity of $\sim 3 \times 10^{14}$ $L_{\odot}$ and a mass of $\sim 3 \times 10^{11}$ $L_{\odot}$, making it the most massive and luminous single source known in the universe. The look-back time corresponds to approximately 80% of the age of the universe. The high excitation emission lines in the spectrum observed by Rowan-Robinson *et al.* (1991) indicate the presence of an AGN, but whether or not the source represents a protogalaxy undergoing its first burst of star formation, or a dust enshrouded AGN, is still up for debate. The mass of neutral molecular hydrogen was estimated soon after discovery of the galaxy, with the amazing detection of the CO 3–2 transition, first by Brown & vanden Bout (1991), and then confirmed by others (e.g., Solomon, Radford, & Downes 1992). These observations showed that F10214+4724 has an infrared to CO luminosity ratio twice that of most IR-luminous galaxies and about 30 times higher than that of normal spirals. Its IR colors and high IR/CO luminosity ratio indicate that F10214+4724 is similar to nearby ultraluminous galaxies that are known to be mergers. This galaxy is a primeval molecular galaxy with the mass of a large spiral, but with most of the mass in molecular gas rather than stars. It is the best candidate yet discovered for a forming disk galaxy or cluster of galaxies.

One reason for the great interest in F10214+4724, as for lower redshift ultraluminous infrared galaxies, concerns the dominant source of the extremely high far-infrared luminosity detected by IRAS. There is little doubt that **both** a luminous starburst and a non-stellar active nucleus are present in the source. The interesting question is— which component dominates the energy budget for the tremendous 10^{14} solar luminosities that this source generates? There is abundant evidence that F10214+4724 is a primeval galaxy undergoing rigorous star formation, including $\sim 6 \times 10^{11}$ $M_{\odot}$ of molecular gas (Solomon *et al.* 1992), and a UV-to-radio continuum energy distribution which is most simply interpreted as a powerful starburst (Rowan-Robinson *et al.* 1993). Likewise, there is substantial evidence for an embedded AGN: high excitation emission lines (Rowan-Robinson *et al.* 1991), strong polarization (Lawrence *et al.* 1993), and a very compact morphology in the light of the Hα emission line (Soifer *et al.* 1992). New results show [NII]/Hα and [OIII]/Hβ emission line ratios to be typical of those found in type 2 Seyfert galaxies (Elston *et al.* 1993). High spatial resolution near-infrared (rest-frame optical) imaging shows a double structure in the central 2 arcsec which suggests an interacting/merging system of two or more galaxies (Soifer *et al.* 1992). The contention is that this source may be a high redshift analog to the ultraluminous infrared galaxies observed in the local universe. Continuing studies of this exciting object will undoubtedly have a great impact on our understanding of galaxy formation and cosmology.

The Source of the Far-Infrared Emission in Starburst Galaxies and AGN

An important work which addresses the issue of the source of the far-infrared emission in active galaxies was published by Mouri & Taniguchi (1992). Using observations of [O I] λ6300, Hα, H_2 S(1), and Br γ emission lines in relation to the 25 μm to 60 μm flux ratios observed by IRAS, they constructed a useful model in which the evolutionary stage of starburst and AGN activity determines the relevant extent to which various sources of far-infrared radiation contribute to the total luminosity measured by IRAS. The far-infrared radiation emitted from AGNs is thus a mixture of a flat nonthermal component from the AGN and a steep thermal component from dust heated by young stars, with their fraction varying from one source to another, and likely changing in time as the various components evolve. Although more complex than a more simplistic model we might hope for, this result is likely a more realistic treatment of the complexities involved in interpreting the far-infrared emission from F10214+4724 and other sources which have evidence for AGNs and starbursts that coexist in the same galaxy.

Superwinds

A principal result to emerge from ground-based follow-up observations of infrared galaxies selected using IRAS color ratios is that many starburst galaxies have large-scale nebulae of ionized gas which often have the morphology and kinematics of a bi-polar outflow many kiloparsecs in size. In these "superwinds," the kinetic energy provided by supernovae and winds from massive stars in a central starburst drives a large-scale outflow that can shock heat and accelerate ambient interstellar and intergalactic gas. This work, pioneered by Heckman and his collaborators (e.g., Heckman, Armus, & Miley 1990; Armus, Heckman, & Miley 1990) demonstrates that the evolution of the interstellar medium, as well as the intergalactic medium, is profoundly affected by large-scale outflows on a galaxy-wide scale. The superwind process may also play a critical role in driving out much of the gas and dust associated with a starburst, diminishing the extinction at optical and ultraviolet wavelengths and thus allowing a previously embedded ultraviolet-excess AGN to dominate the bolometric luminosity of the source. This has important implications for current theories of the formation and evolution of classical quasars, radio galaxies and elliptical galaxies.

The Origin of Elliptical Galaxies

A long-standing argument against the hypothesis that most massive elliptical galaxies form primarily by merging spiral galaxies is the extremely high mass densities in the cores of the former compared to the later. A recent study of infrared ultraluminous galaxies (those with infrared luminosities comparable to classical quasars, $L_{ir} > 10^{12}$ $L_{\odot}$) suggests that these objects are elliptical galaxies in the process of formation (Kormendy, & Sanders 1992). The typical central densities of molecular gas, $\sim$ 100 $M_{\odot}pc^{-3}$, are as high as the stellar mass densities in elliptical galaxies, and the intense starbursts taking place in these sources are just what is expected during the dissipative collapse induced by a merger. The high gas density is transformed into a high stellar mass density, and the high metallicities of ellipticals are a natural outcome, since even the oldest ellipticals would form from starburst-enriched gas. These results have implications for searches for protogalaxies at high redshift because forming elliptical galaxies may be heavily enshrouded by dust and hence be best detected at infrared wavelengths.

The Origin of Radio Galaxies

One of the many early discoveries made by IRAS is that a significant fraction of radio-loud galaxies are luminous far-infrared sources (Golombek, Miley, & Neugebauer 1988). Radio galaxies have traditionally been thought to be giant ellipticals which are devoid of extensive quantities of gas and dust. However, a new CO($J = 1 \rightarrow 0$) line survey of powerful radio-selected galaxies with log $P_{408MHz} > 23.5$ W Hz^{-1} has doubled the number of previously known CO detections of powerful radio galaxies (Mazzarella, Graham, Sanders, & Djorgovski 1993). The computed H_2 masses range from $\sim 3 \times 10^9$ to $\sim 2 \times 10^{10}$ $M_\odot$, similar to the H_2 masses derived for other luminous infrared galaxies, but in stark contrast to the low H_2 gas masses found in radio-quiet IRAS-selected elliptical galaxies, 10^7–10^8 $M_\odot$ (Lees, Knapp, Rupen, & Phillips 1990). The results suggest that rich supplies of molecular gas may be ubiquitous in powerful radio-selected galaxies detected by IRAS, and the origin of powerful radio galaxies may be closely related to the genesis of dust enshrouded quasars and classical UV-excess quasars through the merging of gas-rich disk galaxies. The recent detection of weak, large-scale radio lobes associated with the eastern nucleus of the merging galaxy pair Markarian 463 bolsters the case for a unified model for the origin of quasars and powerful radio galaxies via an infrared-luminous phase (Mazzarella *et al.* 1991). The radio source in Mrk 463E is 20 kpc in extent, the largest known to be associated with a gas-rich spiral galaxy. Mrk 463 may exemplify a "missing link" or transition case between the confined (< few kpc) linear radio sources observed in Seyfert galaxies and the extended (hundreds of kpc) sources in classical quasars and radio galaxies.

Embedded AGNs in Infrared-Luminous Galaxies

Near-infrared spectroscopy provides a useful way to take advantage of the decreased extinction due to dust compared to optical wavelengths; the extinction, in magnitudes, at 2 μm is about one tenth that in the visual band. Using this tool, a number of workers have searched for evidence of embedded broad-line regions in infrared-luminous galaxies. Such sources have been found in the galaxies IRAS 20460+1925, IRAS 23060+0505 (Hines 1991), and IRAS 20460-1447 (Nakajima *et al.* 1991). This work is important because it demonstrates that there is a population of obscured QSO's, and that they can be selected by their warm infrared energy distributions and QSO-like luminosities. Nevertheless, a recent comprehensive study of 14 ultraluminous infrared galaxies by Goldader *et al.* (1993) found no new (embedded) broad-line regions that were not known from previous optical spectroscopy. Emission lines from present and past vigorous star formation dominate the infrared spectra of these galaxies. These results place important constraints on unified models which maintain that galaxies with classical type 2 Seyfert spectra may harbor type 1 Seyfert nuclei, but they are invisible because of obscuration due to dust.

A parallel effort to detect embedded AGNs in infrared-luminous galaxies took place using very long baseline interferometry at radio wavelengths (Lonsdale, Smith, & Lonsdale 1993). Over half of a sample of 31 infrared-luminous galaxies show high-brightness temperature emission, with $T_b > 10^5$ K and structure on scales of 5–150 milliarcseconds. The limits for non-detected sources are similar, consistent with the hypothesis that most of these galaxies have compact AGN cores at a level of a few percent of the total radio flux density. Structural information and energy budget considerations rule out a single supernova interpretation of the compact radio emission in these galaxies, although the possibility of several simultaneous, extraordinarily luminous radio supernovae within the central few hundred cubic parsec cannot be ruled out. Nevertheless, these results provide important new evidence for AGNs obscured by dust in a surrounding starburst. Whether or not the embedded AGNs dominate over radiation from hot stars in the total energy deposited into heating the dust observed in the far-infrared remains an active topic of research.

IRAS Protostars

IRAS showed that stars of roughly solar mass form out of clumps of molecular material ranging in mass and size from 1–100 $M_\odot$ and 0.02–0.2 pc. The temperature and density of these clumps have values around 15–25 K and 10^4 cm^{-3} (Beichman *et al.* 1986; Myers *et al.* 1987). The 1–100 μm spectral energy distributions (SED's) require a hot central star (T$\sim$3000 K), many magnitudes ($A_V \sim$ 10–50 mag) of intervening dust to absorb and re-radiate the stellar radiation, and a massive disk ($\sim 0.1 M_\odot$) of material located within a relatively empty, 1$\sim$ 100 AU, central void around the star (Myers *et al.* 1987). These early observational samples have been expanded to include hundreds of young stars with multi-wavelength studies of the Taurus (Kenyon *et al.* 1990; Beichman, Boulanger and Moshir 1992), Corona Australis (Wilking *et al.* 1992), Ophiuchus (Wilking, Lada and Young 1989) and the Chamaeleon clouds (Prusti, Whittet and Wesselius 1991; Gauvin and Strom 1992).

A combination of IRAS and ground-based observations of T Tauri stars strengthens the case for the existence of disks around many of these objects. Roughly 80% of T Tauri stars have passively heated disks contributing $\sim$ 20% or less of the total luminosity of the system (Rucinski 1985; Strom *et al.* 1988; Cohen, Emerson and Beichman 1989). The remaining 20% of the stars have substantial infrared excesses, suggesting the existence of active disks powered by accretion with infall rates as high as $10^{-5} M_\odot yr^{-1}$. Hildebrand *et al.* (1992) have shown that a star+accretion disk model can also satisfy the 1–1000 μm observations of Herbig Ae/Be stars, but with mass accretion rates, the disk masses and disk radii that are proportionately larger.

Estimates of disk masses for T Tauri stars range from 0.001–0.1 $M_\odot$ (Beckwith *et al.* 1990; Adams, Emerson and Fuller 1990), but are uncertain because of uncertainties in the mass absorption coefficient at long wavelengths. Additional support for the existence of massive disks comes from millimeter observations that show material in Keplerian rotation around the HL Tau (Sargent and Beckwith 1987, 1991), and a resolved disk around L1551-IRS-5 (Keene and Masson 1990) implying disk masses $\sim$ 0.1 $M_\odot$. The existence of a large amount of molecular gas and dust for a few million years has obvious implications for the formation of planetary systems (Strom *et al.* 1989; Keene and Masson 1990). Ground-based 10 μm observations of selected T Tauri stars set limits on the disk lifetimes in the range 1 to 3 million years (Strom *et al.* 1989; Skrutskie *et al.* 1990).

IRAS Luminosity Functions

The existence of samples of many tens of sources towards a variety of clouds permits a preliminary comparative analysis of the protostellar luminosity function under various physical conditions. The dark clouds such as Taurus, L1641 and Chamaeleon I share similar luminosity functions marked by a relatively small number of objects of intermediate luminosity. (Gauvin and Strom 1992; Strom and Strom 1991; Kenyon *et al.* 1989; Beichman *et al.* 1992. This observation presents a difficulty for simple accretion models which predict the existence of more high luminosity objects. A natural explanation for the observed excess of low luminosity objects is that infall occurs not onto the central star, but is held in the reservoir of a disk, to be released at a slower rate, and at a lower luminosity over the 1-3 million year lifetime of the disk (Kenyon *et al.* 1989; Beichman *et al.* 1992).

The luminosity function of the relatively diffuse dark clouds mentioned above differs from that in more concentrated clouds such as ρ Ophiuchus and CrA. These compact, dense clouds have a greater proportion of moderate and high luminosity ($> 5L_\odot$) Class I objects relative to Class II objects (Kenyon *et al.* 1990; see Zinnecker McCaughrean and Wilking, 1992, for a review). Possible explanations for the differences include a higher accretion rate or a formation of stars of different average mass, 2–3 $M_\odot$, in the denser clouds vs. 1–2 $M_\odot$ in the more diffuse clouds.

Galactic Structure Revealed by IRAS

The luminous carbon stars and mass-losing Mira variables detected by IRAS provide a nearly extinction free probe of Galactic structure that can be used out to distances beyond the Galactic Center. Analysis of IRAS source counts has shown that the Galaxy has a prominent warp similar to that seen in 21 cm maps

(Djorgovski and Sosin 1989). A key feature of the IRAS result is that the progenitors of these stars are intermediate and low mass stars many billions of years old, implying that the warp is not a transient feature of the interstellar medium, but a long-lived feature of the Galaxy. Gravitational perturbations induced by a massive dark halo can produce warps comparable to that seen by IRAS.

IRAS sources selected on the basis of their color or variability to be AGB stars clearly delineate a central concentration of stars toward the Galactic Center. Some authors attribute to a prominent Galactic Bulge (Habing *et al.* 1985; Rowan-Robinson and Chester 1987; Van der Veen and Habing 1990) while Weinberg (1992a) claims that the counts can be explained with only an axisymmetric exponential disk population. Detailed analysis of COBE results will play an important role in determining whether a spheroidal bulge really exists. Weinberg (1992b) also suggests that the Milky Way is a barred spiral galaxy with arms originating at the termination of a 5 kpc bar defined by variable, mass-losing giants. The contrast in the bar is high, about a factor of 7 in the counts between the minimum and maximum regions at a given radius. The bar seen in the IRAS counts may be related to the triaxial feature inferred from analysis of 2 μm ground-based observations (Blitz and Spergel 1991).

New Views of Interstellar Dust

Analysis of IRAS data have fundamentally changed our view of the interstellar medium. Instead of gas sprinkled with simple 0.1 μm silicate and graphite grains, IRAS has shown that there is a continuum between the atoms and molecules in the gas phase, macro-molecules, small dust grains and ultimately large classical grains. Variations in the relative abundances of these species trace directly the strength and hardness of the interstellar radiation field (ISRF) as well as grain formation and destruction processes.

The physical link between 12 μm emission and PAHs was greatly strengthened by the observation that the galactic distribution of the 3.3 μm PAH feature closely follows that of the 12 μm radiation found by IRAS (Girard *et al.* 1989). The new IRAS Sky Survey Atlas offers a three to five-fold improvement in sensitivity compared to earlier IRAS images at 12 and 25 μm and will facilitate study of this phenomena in a variety of environments.

Observations of the 12 μm radiation from other galaxies can also be used to learn about small grain physics. The large to small grain abundance ratio is relatively constant in normal galaxies of similar metallicity with a dispersion of less than 40% (Helou 1986; Helou, Ryter and Soifer 1991). Most of the variation in the $f_\nu(12\mu\text{m})/(f_\nu(60\mu\text{m}) + f_\nu(100\mu\text{m}))$ ratio is attributed to variations in the interstellar radiation fields in these galaxies which, at increasing levels, destroys small grains and gradually increases the contribution from classical large grains.

A number of groups have addressed the composition of interstellar dust using IRAS data (see Puget and Léger 1989 for a review). Clues to the nature of the small grains have come from variations in the strength of 12 μm emission as a function of interstellar radiation field (ISRF) strength within molecular clouds and HII regions (Boulanger *et al.* 1988). A comprehensive model of interstellar dust has been developed that fits the optical and UV extinction curves, the IRAS emission properties, and the longer wavelength submillimeter emission (Désert *et al.* 1990). The model invokes large grains, small ($\lesssim$0.01 μm) 3-dimensional grains, and smaller ($\sim$ 10 Å) planar grains (PAHs). Together the grains and PAHs may comprise more than 40% of all interstellar carbon.

III. RESULTS FROM NEAR-SPACE OBSERVATORIES

The Kuiper Airborne Observatory (KAO), a 91-cm telescope mounted in a converted C-141 aircraft capable of reaching altitudes of 12.8 km, continues to remain a premier facility for infrared and submillimeter astronomy. Important advances in our knowledge of star formation processes, the interstellar medium, the Galactic center, and stellar and planetary astronomy have resulted from KAO observations.

Molecular Clouds and Star Formation

Far-infrared polarimetry has led to the first mapping of magnetic fields in dense cloud cores (Hildebrand *et al.* 1993). Regions of uniform or smoothly varying polarization seen in many clouds indicate highly structured fields despite the presence of turbulence and fragmentation.

Starburst galaxies like NGC 253 and NGC 3256 have enormous rates of massive star formation occuring in their central, obscured regions. KAO observations of infrared fine structure lines (Carral *et al.* 1993) reveal a highly pressurized interstellar medium populated with numerous, small, dense, molecular clouds. The clouds and HII regions appear to be in pressure equilibrium with a supernova-shocked, hot, low density, all pervasive medium. A feedback mechanism may be indicated in which the pressure generated by the supernovae compresses the molecular clouds and triggers further massive star formation.

KAO maps of a number of pre-main-sequence stars suggest that much of the far-infrared radiation from young stellar objects does not arise from dust in a compact disk, but, in fact, from much more extened circumstellar clouds (Butner *et al.* 1990; Natta *et al.* 1992; Natta *et al.* 1993). In some cases the temperature distribution in these extended clouds suggests possible heating by a central disk rather than by a stellar photosphere.

Interstellar Medium

KAO maps of several galaxies in the 158 micron [CII] line using an imaging Fabry-Perot spectrometer have been made, leading to the first direct determination of the pressure in the cold atomic medium of external galaxies (Madden *et al.* 1993). Similar observations in a sample of gas rich galaxies have shown that warm dense gas may represent a substantial fraction of the total gas mass in the nuclear regions of galaxies (Stacey *et al.* 1991).

Galactic Center

Images of the [CII] line (Poglitsch, *et al.* 1991) and of the [OIII] fine structure lines, in conjuction with spectra of far-infrared lines from [OI], [CII], [SiII], [SIII], [NIII], and [OIII] and the 20 to 160 micron continuum at a number of positions both along and across the thermal arched filaments (Erickson *et al.* 1991 and Colgan *et al.* 1993) revealed that the dominating ionization mechanism for the thermal filaments in the Galactic Center Radio Arc is consistent with photoionization by local, young massive stars. Contrary to expectations, far-infrared polarization measurements indicate that the magnetic field in the thermal filaments just north of the dust ring is orthogonal to the field in the long non-thermal arcs that traverse and interact with the filaments. The field smoothly follows the curve of the filaments. (Morris *et al.* 1992).

Observations of the 63 micron [OI] fine structure line toward the SgrA West region (Jackson *et al.* 1993) showed a large amount of neutral gas in the cavity inside the circumnuclear molecular ring, possibly indicating an infall of gas toward the massive central object SgrA*. Measurements of the far-infrared lines of [SIII], [FeIII], [OIII], [NIII], and [NII] in Sgr A West are all consistent with photoionization by stars with an effective temperature of 35,000 K and do not require shocks or stellar winds for their production. The densities determined from the [OIII] and [SIII] lines suggest that the doubly ionized lines arise predominantly in the low density gas enclosing the high density 'mini-spiral' seen in radio continuum maps and lower excitation lines such as NeII (Erickson *et al.* 1993).

Far-infrared polarization measurements have shown that the magnetic field in the dust ring at the center of the Galaxy has the structure expected for a magnetic accretion disk in which angular momentum is removed centrifugally. This is the only direct observation of the magnetic field in any accretion disk. A streamer with a longitudinal field penetrates the northern portion of the ring.

Stellar Astronomy

Far-infrared spectroscopy of the [FeII] and [NiII] lines from SN 1987A at several epochs have probed the mantle of the supernova (Erickson *et al.* 1988; Haas *et al.* 1990, and Colgan *et al.* 1993). These observations provide evidence for mixing and clumping in SN 1987A, imply a spatial association between the dust clumps and the iron-group elements, and suggest segregation of the iron-group elements from the lighter elements such as Mg and O. A mysterious high-velocity feature at 3900 km/s was also found.

KAO observations of Nova Cygni 1992 (Nova V1974 Cygni) in May, 1992 resulted in identification of a number of atomic lines of hydrogen and neon (Gehrz *et al.* 1993). Observations in July 1993, showed that the original atomic lines had disappeared and been replaced by a Ne VI line. No infrared continuum was seen in any of the spectra, indicating that dust had not yet condensed from the materials ejected by the nova blast.

Far-infrared [OI] and [SiII] lines have been observed in the oxygen-rich, M2 Iab supergiant, alpha Orionis. Their fluxes are consistent with emission arising from dense, warm gas in the star's inner envelope and implies that nearly all of the available O and Si are in atomic form. This is the first detection of far-infrared, fine structure emission from the inner or transition region of a circumstellar envelope, where molecules and dust are expected to form.

Planetary Astronomy

A combination of laboratory measurements of H_2O trapped in SO_2 matrices, and KAO spectra of Jupiter's satellite Io, resulted in the discovery of small amounts of water on Io's surface (Salama *et al.* 1990 and 1993). This demonstrates that hydrogen is present there, and has important implications for our understanding of the chemistry on this satellite.

Balloon-Borne Astronomical Observations

The 1-meter balloon-borne telescope of the Tata Institute for Fundamental Research continues to produce high resolution far-infrared maps of H II regions including G351.6-1.3/G351.7-1.2, W31, and the galaxies NGC 4945, and the Circinus galaxy (Gosh *et al.* 1990a; Gosh *et al.* 1990b; Gosh *et al.* 1992).

Far-infrared observations of the Cygnus-X region in the galaxy were performed with a 60-cm diameter balloon-borne telescope developed at ETH Zurich in collaboration with the Geneva Observatory. Successful measurements were made on the dust emission and temperature of the molecular cloud complexes DR21/W75N and S106 (Holenstein *et al.* 1992).

The first large scale far-infrared [CII] line emission maps of the galactic plane were completed using balloon-borne telescopes equiped with Fabry-Perot spectrometers by a joint group from the Institute of Space and Astronautical Science (ISAS), Japan, the University of Arizona, and the Kyoto University, Japan (Shibai, H. *et al.* 1991). Extended, bright [CII] emission throughout the Galaxy. Possible candidates for this emission are extended photodissociation regions heated by the general interstellar radiation field, or extended low-density H II regions ionized also by the interstellar radiation field.

IV. FUTURE MISSIONS

The 1990's will see an enormous increase in space activities for infrared and submillimeter astronomy. Some of the major missions planned for the future are described below.

Infrared Space Observatory (ISO)

ISO is a 60-cm helium-cooled telescope with four focal plane instruments: a short wavelength camera, a multiwavelength photometer/polarimeter, and short and long wavelength spectrometers, covering the spectral region from 3 to 200 microns. ISO is an intermediate mission between IRAS and SIRTF in both schedule and capability. It will be launched by ESA in 1995 and will provide the first follow-up on the IRAS results and define further scientific questions for the later SIRTF mission.

Infrared Telescope in Space (IRTS)

IRTS is a helium-cooled telescope planned for launch in 1995 by Japan's Institute of Space and Astronautical Science. It will carry four experiments: a Fabry-Perot interferometer to map the C II and O I emission from the Galaxy, a submillimeter wavelength radiometer and a near-infrared photometer to measure the extragalactic background radiation, and a mid-infrared spectrometer. Because it shares a platform with other space experiments IRTS will obtain data for only 2 weeks.

Submillimeter-Wave Astronomy Satellite (SWAS)

SWAS is one of NASA's Small Explorer Class satellites, consisting of a 55-cm ambient temperature telescope with passively cooled detectors, to search for single lines of H_2O, O_2, C I, and ^{13}CO in molecular clouds and to qualify several critical submillimeter wave receiver and spectrometer elements to be used on later missions. The launch date is 1995.

The Near Infrared Camera and Multi-Object Spectrometer for the Hubble Telescope (NICMOS)

NICMOS is a second generation NASA instrument for the Hubble Space Telescope that is designed to provide imaging and spectroscopic observations at wavelengths between 0.8 and 3.0 microns. NICMOS contains cryogenically-cooled cameras and spectrometers to cover a wide range of scientific objectives, and has an expected lifetime of greater than five years.

Space Infrared Telescope Facility (SIRTF)

SIRTF, one of NASA's Great Observatories, is a 0.85-meter diameter cryogenically-cooled telescope which will be located in a solar orbit for a period of at least 3 years. SIRTF's three focal plane instruments will permit imaging and spectroscopy over the infrared spectrum from 3 to 200 μm with sensitivities of 100 to 10,000 times their predecessors. It will be operated as a facility for the entire scientific community.

EDISON

EDISON has been proposed to be a long-lived international infrared space observatory possessing a 1.7-m mirror cooled via radiation to approximately 20K. The observatory will be equipped with imaging and specroscopic instruments operating over the wavelength range 2–100 microns. High sensitivity, high angular resolution, and long life due to lack of cryogens will allow EDISON to undertake a comprehensive research programme impossible with any other existing or proposed facility.

Submillimeter Intermediate Mission (SMIM)

SMIM is a proposed submillimeter space telescope with a radiation cooled 2.5-meter diameter primary mirror. The focal-plane instruments include heterodyne receivers that will cover the band from 250–700 microns. Between 100 and 300 microns, a Fabry-Perot spectrometer would provide the spectral coverage. Cooling of the instruments would be provided by a 500-liter cryostat. The mission lifetime is projected to be about 1 year. The orbit attainable using the Delta launch vehicle would be a 24-hour elliptical (if an Atlas II were used, a heliocentric orbit would be preferred).

Pronaos

Pronaos is a 2-meter diameter balloon-borne telescope for submillimeter astronomy built in France by CNES (Toulouse). The telescope will accommodate two focal-plane instruments: a multiband photometer and a heterodyne spectrometer. The first flight will occur in 1994.

Far Infrared and Submillimeter Space Telescope (FIRST)

FIRST is a cornerstone mission of the ESA long term science program. It is a large (3 meter) diameter passively cooled telescope equipped with a combination of photometer/camera and very high resolution spectrometers for far-infrared and submillimeter wavelengths. FIRST was selected to be the ESA Cornerstone Mission 4, to be launched in 2006.

Stratospheric Observatory for Infrared Astronomy (SOFIA)

SOFIA will be a 2.5-meter ambient temperature telescope mounted in a specially modified Boeing 747 aircraft, which will replace the Kuiper Airborne Observatory around the year 2000. It will have capabilities throughout the spectrum from 0.3 microns to 1.3 mm wavelength. The planned flight program of 120 8-hour flights per year would support approximately 15 science instruments and 40 principal investigator teams annually. In addition, SOFIA will have a strong educational outreach program. SOFIA will be sponsored by NASA, with possible participation by the German space agency, DARA.

ACKNOWLEDGEMENTS

The author gratefully acknowledges the assistance of Charles Beichman, Edward H. Dunham, Edward Erickson, Carol Lonsdale, Joseph Mazzarella, and Gary Melnick in the preparation of this review.

REFERENCES

Adams, F.C., Emerson, J.P. and Fuller, G.A. 1990, *Ap.J.*, **357**, 606.

Armus, Heckman, and Miley 1990, *Ap.J.*, **364**, 471.

Beckwith, S., Sargent, A.I., Chini, R.S., Güsten, R. 1990, *Astron. J.*, **99**, 924.

Beichman, C.A., Boulanger, F., and Moshir, M.M. 1992, *Ap. J.*, **386**, 248.

Beichman, C.A, Myers, P.C., Emerson, J.P, Harris, S., Mathieu, R., Benson, P.J., and Jennings, R.E. 1986, *Ap.J.*, **307**, 337.

Blitz, L. and Spergel, D. N. 1991, *Ap. J.*, **379**, 631.

Boulanger, F., Beichman, C., Désert, F.X., Helou, G., Pérault, and Ryter, C. 1988, *Ap. J.*, **332**, 328.

Brown and vanden Bout 1991, *A.J.*, **102**, 1956.

Butner, H.M., Evans, N.J. II, Harvey, P.M., and Mundy, L.G. 1990, *Ap.J.*, **364**, 164.

Carral, P., Hollenbach, D. J., Lord, S. D., Colgan, S. W. J., Haas, M. R., Rubin, R. H., and Erickson, E. F. 1993, *Ap.J.*, in press.

Cohen, M., Emerson, J.P., and Beichman, C.A. 1989, *Ap. J.*, **339**, 455.

Colgan, S. W. J., Haas, M. R., Erickson, E. F., Lord, S. D., and Hollenbach, D. J. 1993, *Ap.J.*, submitted.

Désert, F.X., Boulanger, F., and Puget J-L. 1990, *Astron. & Astrop.*, **237**, 215.

Djorgovski, G. and Sosin, C. 1989, *Ap. J.*, **341**, L13.

Elston *et al.* 1993, in press.

Erickson, E. F., Haas, M. R., Colgan, S. W. J., Lord, S. D., Burton, M. G. 1988.

Erickson, E. F., Colgan, S. W. J., Simpson, J. P., Rubin, R. H., Morris, M., and Haas, M. R. 1991, *ApJ*, **370**, 69.

Erickson, E. F., Colgan, S. W. J., Simpson, J. P., Rubin, R. H., and Haas, M. R. 1993, in preparation.

Gauvin, L.S. and Strom, K.M., 1992, *Ap. J.*, **385**, 217.

Gehrz, R. D., Woodward, C. E., Greenhouse, M. A., Starrfield, S., Wooden, D. H., Witteborn, F. C., Sandford, S. A., Allamandola, L. J., Bregman, J. D., and Klapisch, M. 1993, *Astrophys. J.*, , in press.

Girard, M., Pajot, F., Caux, E., Lamarre, J.M, Serra, G. 1989, *Astr. Astrp.*, **215**, 92.

Goldader *et al.* 1993, in press.

Golombek, Miley and Neugebauer 1988, *A.J.*, **95**, 26.

Gosh, S. K., Iyengar, K. V. K., Rengarajan, T. N., Tandon, S. N., Verma, R. P., and Daniel, R. R. 1990a, *ApJ*, **353**, 564.

Gosh, S. K., Iyengar, K. V. K., Rengarajan, T. N., Tandon, S. N., Verma, R. P, Daniel, R. R., and Ho, P. T. P. 1990b, *ApJ*, **347**, 338.
Gosh, S. K., Bisht, R. S., Iyengar, K. V. K., Rengarajan, T. N., Tandon, S. N. and Verma, R. P. 1992, *ApJ*, **391**, 111.
Haas, M. R., Colgan, S. W. J., Erickson, E. F., Lord, S. D., Burton, M. G. 1990.
Jackson, J.M., N. Geis, R. Genzel, A.I. Harris, S.C. Madden, A. Poglitsch, G.J. Stacey, and C.H. Townes 1993, *ApJ.*, **402**, 173.
Habing ,H.J, Olnon, F.M., Chester, T., Gillett, F.C., Rowan-Robinson, M. and Neugebauer, G. 1985, *Astr. Astrp*, **152**, 11.
Heckman, Armus and Miley 1990, *Ap.J.S.*, **74**, 833.
Helou, G. 1986, *Ap. J.*, **311**, L33.
Helou, G., Ryter, C., and Soifer, B.T. 1991, *Ap. J.*, **376**, 505.
Hildebrand, R.H., Davidson, J.A., Dotson, J., Figer, D.F., Novak, G., and Platt, S.R. 10 Nov.1993, *Ap.J.*, in press.
Hildebrand, L.A., Strom, S.E., Vrba, F. J., Keene, J. 1992, *Ap. J.*, in press.
Hines 1991, *Ap.J.*, **374**, L9.
Holenstein, A. P., Schenker, G., Huguenin, D., and Kneubuhl, F. K. 1992, *Astron. Astrophys. Suppl. Ser.*, **96**, 115.
Keene, J. and Masson, C. 1990, *Ap.J*, **355**, 635.
Kenyon, S., Hartmann, L.W., Strom, K.M., Strom, S.E. 1990, *Astron. J.*, **99**, 869.
Kenyon *et al.* 1989.
Kormendy and Sanders 1992, *Ap.J.*, **390**, L53.
Lawrence *et al.* 1993, in press.
Lees, Knapp, Rupen, and Phillips 1990, *Ap.J.*, **379**, 177.
Lonsdale, Smith and Lonsdale 1993, *Ap.J.*, L9.
Madden, S.C., Geis, N., Genzel, R., Herrmann, F., Jackson, J.M., Poglitsch, A., Stacey, G.J., and Townes, C.H. 1993, *Ap.J.*, **407**, 579.
Mazzarella *et al.* 1991, *A.J.*, **102**, 1241.
Mazzarella, Graham, Sanders and Djorgovski 1993, *Ap.J.*, **409**, 170.
Morris, M., Davidson, J.A., Werner, W., Dotson, J., Figer, D.F., Hildebrand, R.H., Novak, G., Platt, S.R. 1992, *Ap.J.*, **399**, L63.
Mouri and Taniguchi 1992, *Ap.J.*, **386**, 68.
Myers, P.C., Fuller, G.A., Mathieu, R. D., Beichman, C.A., Benson, P.J., Schild, R. E., and Emerson, J.P. 1987, *Ap.J.*, **319**, 340.
Nakajima *et al.* 1991, *Ap.J.*, **373**, 452.
Natta, A., Palla, F., Butner, H.M., Evans, N.J. II, and Harvey, P.M. 1993, *Ap.J.*, **406**, 674.
Natta, A., Palla, F., Butner, H.M., Evans, N.J. II, and Harvey, P.M. 1992, *Ap.J.*, **391**, 805.
Poglitsch, A., G.J. Stacey, N. Geis, M. Haggerty, J. Jackson, M. Rumitz, R. Genzel, and C.H. Townes 1991, *ApJ.*, **374**, L33.
Prusti, T., Whittet, D.C.B. and Wesselius, P.R. 1991, *Mon. Not. Roy. Astr. Soc*, in press.
Puget, J-L and Léger, A. 1989, *Ann. Rev. Astron. & Astrop.*, **27**, 161.
Rowan-Robinson, M., and Chester, T. 1987, *Ap. J.*, **313**, 413.
Rowan-Robinson, M. *et al.* 1991, *Nature*, **351**, 719.
Rowan-Robinson, M. *et al.* 1993, *MNRAS*, in press.
Rucinski, S.M. 1985, *Astron. J.*, **90**, 2321.
Salama, F., Allamandola, L. J., Sandford, S. A., Bregman, J. D., Witteborn, F. C., and Cruikshank, D. P. 1993, *Icarus*, in press.
Salama, F., Allamandola, L. J., Witteborn, F. C., Cruikshank, D. P., Sandford, S. A., and Bregman, J. D. 1990, *Icarus*, **83**, 66.
Sargent, A. I. and Beckwith, S. 1987, *Ap. J.*, **323**, 294.
Sargent, A.I. and Beckwith, S. 1991, *Ap.J.*, **382**, L31.
Shibai, H., Okuda, H., Nakagawa, T., Matsuhara, H., Maihara, T., Mizutani, K., Kobayashi, Y.,Hiromoto, N., Nishimura, T., and Low, F. J. 1991, *ApJ*, **374**, 522.

3. Optical Astronomy from Space

J.B. Hutchings

This section summarises the publications arising from Hipparcos and the visible range results from the Hubble Space Telescope, between July 1990 and June 1993. Both telescopes have considerable history and also will continue to produce publications in this areas in the future, so the summary is arbitrarily restricted by the dates covered.

HIPPARCOS

Ths Hipparcos mission did not achieve its intended orbit at launch, but creative reprogramming of the science resulted in a highly successful mission. Naturally, much of the published literature deals with the orbit and revision of the methods necessary to analyse the data. Some 170 papers are found in the reporting interval. Of about 100 publications which are published in the refereed literature, 33 are found in a special issue of Astronomy and Astrophysics vol 258, May 1992. A summary of the mission was published in ESA SP-1154 (1992). There are also several papers in double stars in IAU Colloquium 134 (PASP conference series), in symposia 7,8 of the Cospar 1990 conference, and IAU symposium 141.

Overall, most of the papers are on data analysis methods, with another significant subset on the orbit and associated problems. There are also many discussions on the catalogues of stars associated with the mission. The main catalogue is the Hipparcos Input Catalogue, which was published in 7 volumes in March 1992 (Turon et al), and is in itself an excellent reference. Many early paper are on binary and multiple stars. While the final main product of the mission will be some time in the future, progress is very encouraging.

HUBBLE SPACE TELESCOPE

The Hubble Space Telescope has been in operation since mid-1990, with a slow start to science observations due to the spherical aberration of the primary mirror, and initial tracking jitter from solar panel oscillations. The aberration has most adversely affected the visible range science, since the main driver at optical wavelengths is spatial resolution. Here too much of the literature has been devoted to dealing with the image aberrations, and optical range science has been restricted to objects of relatively high brightness and low dynamic range.

The instruments delivering optical wavelength science are the FOS/RD, WF/PC, and FOC. While the FGS and HSP have been used for optical science, there are no scientific results from them in the literature yet. The innovativ `snapshot' program, which takes (mostly unguided) pictures from a large catalogue of related targets, to fill in unscheduled time, has yielded significant science.

In the reporting interval some 470 publications are found, which do not involve UV science. Of these, perhaps 1/2 are in refereed journals, of which some 10% refer to deconvolution methods for the images, and another 10% to the optical performance of the instruments. There are 22% extragalactic papers. The remaining topics, in decreasing order of publications, are galactic stars or objects, planetary science, gravitational lenses (snapshot program), globular clusters, and LMC objects.

Extragalactic targets have largely been the central regions of active galaxies and the central light cusps of nearby glaxies. Both focus on trying to understand the nature of galactic nuclei. Generally, in the case of AGN, there is evidence for a bi-conical region illuminated by the central light source, strongly supporting the existing ideas of an opaque torus surrounding the central source and broad-line region (e.g. NGC 1068, NGC 4151). Several galaxies have small light cusps that are close to being too small to arise from star clusters. Other unexpected results were crossed dust lanes in the centre of M51; a dust disk in the centre of NGC 4261 double nucleus in M31; young clusters near the cenre of NGC 1275.

Another category of source is jets, and the braided jets of 3C66B and the struture in the M87 and 3C273 jets are the more celebrated of these. The medium-deep parallel imaging survey has started producing published papers on faint objects in the distant universe. Other primary imagi targets have been distant radio galaxies, and Butcher-Oemler clusters.

Galactic objects have prominently included the Orion nebula, eta Carinae, R Aquarii. LMC objects include the R136 field in the centre of 30 Dor, the remnant and light echoes of SN1987a, planetary nebulae. The globular clusters include 47 Tuc and M15, and optical identifications of cluster X-ray sources have been made.

In all these targets, the HST resolution has been significantly better than ground-based, and major new morphological information has been obtained. Solar system work has mainly been devoted to the bright planets, with detailed imaging of Mars, Jupiter, and Saturn. The great white strom on Saturn in Oct/Nov 1990 was the subject of an extended special program.

There are many new unpublished results, and the HST publications are ramping up rapidly. It remains to be seen what difference the proposed optical fixes will produce in future work: it should alter the balance of HST science significantly.

Summaries of HST science, including UV as well as optical, are published in PASP (Kinney and Maran 1991, Maran and Kinney 1993) and also the conference proceedings from the Space Telescope Institute (Kinney and Blades 1991) and Chia Laguna, Cagliari (Benvenuti and Schreier 1992).

Selected reference list

Hipparcos references

Bec-Borsenberger, A.
Ephemerides of the 48 Hipparcos minor planets for the year 1993
Astronomy & Astrophysics Suppl 1 April 1993; 98(1); P. 77-125; 6 REFS

Canuto, E.; Donati, F.
The on-ground attitude reconstitution of the Hipparcos satellite
International Journal of Modelling and Simulation 1992; 12(4); P. 125-30

Seille, J.; Mazeau, P.; Garres, C.; Borde, J.
Environmental prediction and radiation effects of protons into solar generators: application to HIPPARCOS revised mission
RADECS 91: First European Conference on Radiation and its Effects on Devices and Systems (Cat. No.91TH0400-2)
Publ: IEEE; New York, NY, USA 1992; P. 307-11; 593; 0 REFS

Pannunzio, R.; Morbidelli, R.; Massone, G.; Lattanzi, M.G.; Sarasso, M.; Chiumiento, G.
Precise equatorial coordinates of double and multiple systems: an astronomical support to the Hipparcos mission
Astronomy & Astrophysics Suppl Nov. 1992; 96(1); P. 143-55; 12 REFS

Brosche, P.; Denis-Karafistan, A.; Denis, C.
Determination of the optical or physical nature of wide visual binaries by means of two-dimensional conditional probabilities
Astronomische Nachrichten 1992; 313(6); P. 341-7; 7 REFS

Fehrenbach, Ch.; Burnage, R.; Figuiere, J.
Radial velocities. VI. Ground based measurements for Hipparcos
Astronomy & Astrophysics Suppl Nov. 1992; 95(3); P. 541-79; 11 REFS

ESA's Report to the 29th COSPAR Meeting (ESA SP-1154)
Publ: ESA; Noordwijk, Netherlands 1992; 188

Turon, C. et al
The Hipparcos Input Catalogue
Bulletin d'Information du Centre de Donnees Astronomiques de Strasbourg
July 1992; no.41; P. 9-20; 31 REFS

Duflot, M.; Fehrenbach, C.; Mannone, C.; Burnage, R.; Genty, V.
Radial velocities. V. Ground based measurements for Hipparcos
Astronomy & Astrophysics Suppl Sept. 1992; 94(3); P. 479-517; 10 REFS

Makarov, V.V.
Periodical errors in the Hipparcos star abscissae, caused by oscillations of the basic angle
Pis'ma v Astronomicheskie Zhurnal 1992; 18(7); P. 630-7; 3 REFS

Vondrak, J.; Feissel, M.; Essaifi, N.
Expected accuracy of the 1900-90 Earth orientation parameters in the Hipparcos reference frame
Astronomy and Astrophysics Aug. 1992; 262(1); P. 329-40; 18 REFS

Turon, C. et al
Hipparcos Input Catalogue
March 1992; 7 vol. (2377+399+437)

Egret, D.; Didelon, P.; McLean, B.J.; Russell, J.L.; Turon, C.
The Tycho Input Catalogue. Cross-matching the Guide Star Catalog with the Hipparcos INCA Data Base
Astronomy and Astrophysics May 1992; 258(1); P. 217-22; 20 REFS

Scales, D.R.et al
Tycho photometry calibration and first results
Astronomy and Astrophysics May 1992; 258(1); P. 211-16; 10 REFS

Wicenec, A.; Bassgen, G.
Tycho background determination and monitoring
Astronomy and Astrophysics May 1992; 258(1); P. 206-10; 10 REFS

Hog, E. et al
Tycho astrometry calibration
Astronomy and Astrophysics May 1992; 258(1); P. 201-5; 11 REFS

Halbwachs, J.L.; Hog, E.; Bastian, U.; Hansen, P.C.; Schwekendiek, P.
Tycho star recognition
Astronomy and Astrophysics May 1992; 258(1); P. 193-200; 8 REFS

Bassgen, G. et al
Tycho transit detection
Astronomy and Astrophysics May 1992; 258(1); P. 186-92; 11 REFS

Hog, E. et al
Tycho data analysis. Overview of the adopted reduction software and first results
Astronomy and Astrophysics May 1992; 258(1); P. 177-85; 12 REFS

Pannunzio, R. et al
The treatment of Hipparcos observations of some peculiar double stars: anomalous cases
Astronomy and Astrophysics May 1992; 258(1); P. 173-6; 10 REFS

Mignard, F. et al
Hipparcos double star recognition and processing within the FAST consortium
Astronomy and Astrophysics May 1992; 258(1); P. 165-72; 10 REFS

Soderhjelm, S.; Evans, D.W.; van Leeuwen, F.; Lindegren, L.
Detection and measurement of double stars with the Hipparcos satellite: NDAC reductions
Astronomy and Astrophysics May 1992; 258(1); P. 157-64; 13 REFS

Evans, D.W.; van Leeuwen, F.; Penston, M.J.; Ramamani, N.; Hog, E.
Hipparcos photometry: NDAC reductions
Astronomy and Astrophysics May 1992; 258(1); P. 149-56; 10 REFS

Mignard, F.; Froeschle, M.; Falin, J.L.
Hipparcos photometry: FAST main mission reduction
Astronomy and Astrophysics May 1992; 258(1); P. 142-8; 8 REFS

Lindegren, L. et al
Positions and parallaxes from the Hipparcos satellite. A first attempt at a global astrometric solution
Astronomy and Astrophysics May 1992; 258(1); P. 134-41; 11 REFS

Turon, C.; Arenou, F.; Evans, D.W.; van Leeuwen, F.
Comparison of the first results from the Hipparcos star mappers with the Hipparcos Input Catalogue
Astronomy and Astrophysics May 1992; 258(1); P. 125-33; 23 REFS

van Leeuwen, F.; Evans, D.W.; Lindegren, L.; Penston, M.J.; Ramamani, N.
Early improvements to the Hipparcos Input Catalogue through the accumulation of data from the satellite including the NDAC attitude reconstruction description
Astronomy and Astrophysics May 1992; 258(1); P. 119-24; 17 REFS

Crifo, F.; Gomez, A.; Arenou, F.; Morin, D.; Schrijver, H.
Reliability of the Hipparcos Input Catalogue tested by the 'First look'
Astronomy and Astrophysics May 1992; 258(1); P. 116-18; 10 REFS

Lestrade, J.-F.; Phillips, R.B.; Preston, R.A.; Gabuzda, D.C.
High-precision VLBI astrometry of the radio-emitting star sigma CrB-a step in linking the Hipparcos and extragalactic reference frames
Astronomy and Astrophysics May 1992; 258(1); P. 112-15; 17 REFS

Arenou, F.; Grenon, M.; Gomez, A.
A tridimensional model of the Galactic interstellar extinction
Astronomy and Astrophysics May 1992; 258(1); P. 104-11; 17 REFS

Mennessier, M.O.; Barthes, D.; Boughaleb, H.; Figueras, F.; Mattei, J.A.
Computation of ephemerides for long-period variable stars for the Hipparcos mission
Astronomy and Astrophysics May 1992; 258(1); P. 99-103; 24 REFS

Bec-Borsenberger, A.
Solar system objects observed by Hipparcos
Astronomy and Astrophysics May 1992; 258(1); P. 94-8; 15 REFS

Grenon, M.; Mermilliod, M.; Mermilliod, J.C.
The Hipparcos Input Catalogue. III. Photometry
Astronomy and Astrophysics May 1992; 258(1); P. 88-93; 8 REFS

Jahreiss, H. et al
The Hipparcos Input Catalogue. II. Astrometric data
Astronomy and Astrophysics May 1992; 258(1); P. 82-7; 28 REFS

Turon, C. et al
The Hipparcos Input Catalogue. I. Star selection
Astronomy and Astrophysics May 1992; 258(1); P. 74-81; 82 REFS

Schrijver, H.
Comparison of Hipparcos results obtained on different dates on the same great circle
Astronomy and Astrophysics May 1992; 258(1); P. 70-3; 12 REFS

van der Marel, H.; Petersen, C.
Hipparcos great-circle reduction. Theory, results and intercomparisons
Astronomy and Astrophysics May 1992; 258(1); P. 60-9; 15 REFS

van Leeuwen, F.; Penston, M.J.; Perryman, M.A.C.; Evans, D.W.; Ramamani, N.
Modelling the torques affecting the Hipparcos satellite
Astronomy and Astrophysics May 1992; 258(1); P. 53-9; 6 REFS

Bernacca, P.L. et al
Early scientific results from Hipparcos and future expectations
ESA Bulletin Feb. 1992; no.69; P. 51-7; 0 REFS

Hog, E.; Kovalevsky, J.; Lindegren, L.
Hipparcos data reduction-construction of the Hipparcos Star Catalogue
ESA Bulletin Feb. 1992; no.69; P. 43-50; 0 REFS

Turon, C.
The Hipparcos observing programme: preparation of the Input Catalogue
ESA Bulletin Feb. 1992; no.69; P. 36-42; 0 REFS

Ratier, G.; van Katwijk, K.; Fade, G.; Perryman, M.A.C.
The Hipparcos payload's in-orbit performance
ESA Bulletin Feb. 1992; no.69; P. 27-32; 0 REFS

Heger, D.; McDonald, A.; Schutz, A.; Ojangurem, O.; Sollazzo, C.
ESOC's role in routine Hipparcos operations
ESA Bulletin Feb. 1992; no.69; P. 16-25; 0 REFS

Van der Ha, J.
Implementation of the revised Hipparcos mission at ESOC
ESA Bulletin Feb. 1992; no.69; P. 9-15; 0 REFS

Donati, F.; Sechi, G.
Method of comparison between determinations of the Hipparcos attitude
Astronomy and Astrophysics May 1992; 258(1); P. 46-52; 5 REFS

Donati, F.
Attitude determination in the Hipparcos revised mission
Astronomy and Astrophysics May 1992; 258(1); P. 41-5; 6 REFS

Lindegren, L.; Le Poole, R.S.; Perryman, M.A.C.; Petersen, C.
Geometrical stability and evolution of the Hipparcos telescope
Astronomy and Astrophysics May 1992; 258(1); P. 35-40; 8 REFS

Schrijver, H.; van der Marel, H.
Geometrical calibration and assessment of the stability of the Hipparcos payload
Astronomy and Astrophysics May 1992; 258(1); P. 31-4; 8 REFS

Lindegren, L. et al
The NDAC Hipparcos data analysis consortium. Overview of the reduction

methods
Astronomy and Astrophysics May 1992; 258(1); P. 18-30; 27 REFS

Kovalevsky, J. et al
The FAST Hipparcos data reduction consortium: overview of the reduction software
Astronomy and Astrophysics May 1992; 258(1); P. 7-17; 55 REFS

Perryman, M.A.C. et al
In-orbit performance of the Hipparcos astrometry satellite
Astronomy and Astrophysics May 1992; 258(1); P. 1-6; 45 REFS

Bec-Borsenberger, A.
Ephemerides of the 48 Hipparcos minor planets for the year 1992
Astronomy & Astrophysics Suppl April 1992; 93(1); P. 11-60; 3 REFS

Reviews in Modern Astronomy 4
Publ: Springer-Verlag; Berlin, Germany 1991; viii+286

Hunt, J.J.
Proceedings of the 3rd International Symposium on Spacecraft Flight Dynamics (ESA SP-326)
Publ: ESA; Noordwijk, Netherlands 1991; vii+539

Zhao Ming
On the celestial reference frame independent of the dynamical planes of the Earth
Acta Astronomica Sinica Sept. 1991; 32(3); P. 297-303; 11 REFS

Vondrak, J.
Calculation of the new series of the Earth orientation parameters in the Hipparcos reference frame
Bull Astron Inst of Czechoslovakia Oct. 1991; 42(5); P. 283-94; 34 REFS

Lenhardt, H.; Hering, R.; Walter, H.G.
Procedures for validating Hipparcos astrometric parameters
Manuscripta Geodaetica 1991; 16(4); P. 254-65; 24 REFS

Optical positions of radiostars. III
Costa, E.; Loyola, P.
Astronomy & Astrophysics Supp Aug. 1991; 89(2); P. 411-13; 6 REFS

Results and considerations on failure investigations of components used for space equipments
Tommaso, G.
Proceedings of the ESA Electronic Components Conference (ESA SP-313)
Publ: ESA; Paris, France 1991; P. 475-80; 662; 11 REFS

George Teleki Memorial Colloquium on Fundamentals of Astrometry. 100th Colloqium of the IAU
Astrophysics and Space Science March 1991; 177(1-2)

Simulators at ESOC: bringing experience to bear
Rowell, S.H.G.
Journal of the British Interplanetary Soc 1991; 44(7); P. 349-56; 4 REFS

Tautenburg proper motion fields
Scholz, R.-D.; Hirte, S.
Astronomische Nachrichten 1991; 312(1); P. 45-8; 11 REFS

Quasar link conditions for Hipparcos
Brosche, P.; Ducourant, C.; Galas, R.; Geffert, M.; Karafistan, A.
Astronomy and Astrophysics May 1991; 245(2); P. 669-75; 5 REFS

Infrared and Radio Astronomy and Astrometry. Symposia 7,8 and the Topical Meeting of COSPAR Interdisciplinary Scientific Commission E (Meeting 5) of the COSPAR 28th Plenary Meeting
Advances in Space Research 1991; 11(2)

Relative positions and proper motions of components of 32 triple stars from the Hipparcos Input Catalog
Anosova, Zh.P.; Orlov, V.V.; Todua, M.M.
Kinematika i Fizika Nebesnykh Tel 1990; 6(2); P. 8-12; 3 REFS

UBVRI photometry of G, K, M Hipparcos stars. II
Jordi, C.; Figueras, F.; Paredes, J.M.; Rossello, G.; Torra, J.
Astronomy & Astrophysics Suppl Feb. 1991; 87(2); P. 229-31; 8 REFS

A catalogue of QSOs and link stars for the HST link of the Hipparcos reference frame
White, G.L.et al
Proceedings of the Astr Soc Australia 1990; 8(4); P. 368-71; 24 REFS

Speckle interferometry of Hipparcos link stars. III
White, G.L. et al
Monthly Notices Royal Astrl Soc 1 Feb. 1991; 248(3); P. 411-13; 10 REFS

Results of the HIPPARCOS in-orbit payload calibration
Davies, P.E.; McDonald, A.J.C.
Journal of the British Interplanetary Soc 1991; 44(1); P. 37-48; 4 REFS

The Hipparcos mission-on the road to recovery
Hassan, H.; Clausen, K.; Perryman, M.A.C.; van der Ha, J.; Heger, D.
ESA Bulletin Nov. 1990; no.64; P. 59-66; 0 REFS

Ephemerides of the 48 Hipparcos minor planets for the year 1991
Bec-Borsenberger, A.
Astronomy & Astrophysics Suppl Dec. 1990; 86(2); P. 299-349; 3 REFS

On the problem of the HIPPARCOS reference frame
Lattanzi, M.G.; Bucciarelli, B.; Bernacca, P.L.
Astrophysical Journal Suppl July 1990; 73(3); P. 481-8; 33 REFS

HST References

O'Dell, C.R.; Wen, Z.; Hu, X.
Discovery of new objects in the Orion Nebula on HST images: shocks, compact sources, and protoplanetary disks
Astrophysical Journal 1993; 410(2, pt.1); P. 696-700+7 plates; 35 REFS

Schade, D.J.; Elson, R.A.W.
What do you gain from deconvolution? Observing faint galaxies with the Hubble Space Telescope wide field camera
Astronomical Journal April 1993; 105(4); P. 1581-6; 17 REFS

Hollis, J.M.; Michalitsianos, A.G.
Evidence for precession of the R Aquarii jet
Astrophysical Journal 1 July 1993; 411(1, pt.1); P. 235-8+1 plate; 23 REFS

Klaus, R.; Burkhardt, H.
Regularized multiframe restoration of Hubble Space and ground-based telescope images with parallel implementation
Proceedings of the SPIE - The International Society for Optical Engineering 1993; 1771; P. 124-34; 23 REFS

Roddier, C.; Roddier, F.
Combined approach to the Hubble Space Telescope wave-front distortion analysis
Applied Optics 1 June 1993; 32(16); P. 2992-3008; 17 REFS

Filippenko, A.; Ho, L.C.; Sargent, W.L.W.
HST observations of NGC 4395, the least luminous Seyfert 1 nucleus: evidence against the starburst hypothesis for broad-lined active galactic nuclei
Astrophysical Journal, Letters 1993; 410(2, pt.2); P. L75-8; 31 REFS

De Marchi, G.; Paresce, F.; Ferraro, F.R.
The structure of the core of 47 Tucanae from Hubble Space Telescope observations
Astrophysical Journal Suppl April 1993; 85(2); P. 293-314; 32 REFS

Barbieri, C. et al
Compact subarcsec structures of the double nucleus of NGC 6240 revealed with HST
Astronomy and Astrophysics June 1993; 273(1); P. 1-5; 19 REFS

Snyder, D.L.; Hammoud, A.M.
Image recovery from data acquired with a charge-coupled-device camera
Journal of the Optical Soc America A May 1993; 10(5); P. 1014-23; 31 REFS

Maoz, D. et al
The Hubble Space Telescope Snapshot Survey. IV. A summary of the search for gravitationally lensed quasars
Astrophysical Journal 20 May 1993; 409(1, pt.1); P. 28-41; 27 REFS

Bendinelli, O. et al
The first detection of a collapsed core globular cluster in M31
Astrophysical Journal, Letters 1993; 409(1, pt.2); P. L17-19+1 plate

Harris, H.C.; Hunter, D.A.; Baum, W.A.; Jones, J.H.
Photometric calibration of the HST Wide-Field/Planetary Camera. II. Ground-based observations of calibration fields
Astronomical Journal March 1993; 105(3); P. 1196-205, 1239-42; 7 REFS

Caldwell, J.; Xin-Min Hua; Turgeon, B.; Westphal, J.A.; Barnet, C.D.
The drift of Saturn's north polar spot observed by the Hubble Space Telescope
Science 16 April 1993; 260(5106); P. 326-9; 11 REFS

Jackson, N.; Sparks, W.B.; Miley, G.K.; Macchetto, F.
The radio and optical structure of 3C66B
Astronomy and Astrophysics March 1993; 269(1-2); P. 128-34; 17 REFS

Kroger, H.W.; Schmidt, G.K.; Pailer, N.
Faint Object Camera: european contribution to the Hubble Space Telescope
Acta Astronautica Nov. 1992; 26(11); P. 827-34; 0 REFS

Benedict, G.F. et al
Astrometric performance characteristics of the Hubble Space Telescope Fine Guidance Sensors
Publications of the Astron Soc Pac Oct. 1992; 104(680); P. 958-75; 12 REFS

Barnet, C.D.; Westphal, J.A.; Beebe, R.F.; Huber, L.F.
Hubble Space Telescope observations of the 1990 equatorial disturbance on Saturn: zonal winds and central Meridian albedos
Icarus Dec. 1992; 100(2); P. 499-511; 24 REFS

Westphal, J.A. et al
Hubble Space Telescope observations of the 1990 equatorial disturbance

on Saturn: images, albedos, and limb darkening
Icarus Dec. 1992; 100(2); P. 485-98; 30 REFS

Calzetti, D.; De Marchi, G.; Paresce, F.; Shara, M.
The center of gravity and density profile of 47 Tucanae
Astrophysical Journal, Letters 1993; 402(1, pt.2); P. L1-4+1 plate

Schmidt, G.D.; Elston, R.; Lupie, O.L.
The Hubble Space Telescope northern-hemisphere grid of stellar polarimetric standards
Astronomical Journal Oct. 1992; 104(4); P. 1563-7; 9 REFS

Guhathakurta, P.; Yanny, B.; Schneider, D.P.; Bahcall, J.N.
Globular cluster photometry with the Hubble Space Telescope. I. Description of the method and analysis of the core of 47 Tuc
Astronomical Journal Nov. 1992; 104(5); P. 1790-817; 57 REFS

Campbell, B. et al
Hubble Space Telescope Planetary Camera images of R136
Astronomical Journal Nov. 1992; 104(5); P. 1721-42+4 plates; 43 REFS

Maoz, D. et al
The Hubble Space Telescope Snapshot Survey. III. Further observations in search of gravitationally lensed quasars
Astrophysical Journal 1 Jan. 1993; 402(1, pt.1); P. 69-75; 29 REFS

Miley, G.K.; Chambers, K.C.; van Breugel, W.J.M.; Macchetto, F.
Hubble Space Telescope imaging of distant galaxies: 4C 41.47 at z=3.8
Astrophysical Journal, Letters 1992; 401(2, pt.2); P. L69-73+2 plates

Sandage, A.; Saha, A.; Tammann, G.A.; Panagia, N.; Macchetto, D.
The Cepheid distance to IC 4182; calibration of M/sub V/(max) for SN Ia 1937C and the value of H/sub phi /
Astrophysical Journal, Letters 1992; 401(1, pt.2); P. L7-10+1 plate

Lucks, M.
Detecting opportunities for parallel observations on the Hubble Space Telescope
Telematics and Informatics Summer-Fall 1992; 9(3-4); P. 331-47; 11 REFS

Korista, K.T. et al
Hubble Space Telescope Faint Object Spectrograph and ground-based observations of the broad absorption line quasar 0226-1024
Astrophysical Journal 20 Dec. 1992; 401(2, pt.1); P. 529-42; 43 REFS

ESA's Report to the 29th COSPAR Meeting (ESA SP-1154)
Publ: ESA; Noordwijk, Netherlands 1992; 188

Dols, V.; Gerard, J.C.; Paresce, F.; Prange, R.; Vidal-Madjar, A.
Ultraviolet imaging of the Jovian aurora with the Hubble Space Telescope
Geophysical Research Letters 23 Sept. 1992; 19(18); P. 1803-6; 17 REFS

Bahcall, J.N.; Hartig, G.F.; Jannuzi, B.T.; Maoz, D.; Schneider, D.P.
Faint Object Spectrograph observations of the large-redshift gravitational lens candidate 1208+1011
Astrophysical Journal, Letters 1992; 400(2, pt.2); P. L51-4; 23 REFS

Windhorst, R.; Mathis, D.F.; Keel, W.C.
Deep Hubble Space Telescope imaging of a compact radio galaxy at z=2.390
Astrophysical Journal, Letters 1992; 400(1, pt.2); P. L1-4+3 plates

Thomson, R.C.; Schade, D.J.; Elson, R.A.W.; Mackay, C.D.; Wilkins, T.N.
Deconvolution and photometry of Faint Object Camera images with point

Burgarella, D.; Paresce, F.
The structure of the inner arcsecond of R Aquarii observed with the Hubble Space Telescope
Astrophysical Journal, Letters 1992; 389(1, pt.2); P. L29-32+1 plate

Bahcall, J.N. et al
The Snapshot Survey: a search for gravitationally lensed quasars with the Hubble Space Telescope
Astrophysical Journal 1 March 1992; 387(1, pt.1); P. 56-68; 37 REFS

Lauer, T.R. et al
Planetary Camera observations of the M87 stellar cusp
Astronomical Journal March 1992; 103(3); P. 703-10, 1025-6; 33 REFS

Holtzman, J.A. et al
Planetary Camera observations of NGC 1275: discovery of a central population of compact massive blue star clusters
Astronomical Journal March 1992; 103(3); P. 691-702, 1020-3; 53 REFS

Kinney, A.L.; Maran, S.P.
The first year of observations with the Hubble Space Telescope
Publications Astron Soc Pac Dec. 1991; 103(670); P. 1237-49; 78 REFS

Franz, O.G. et al
Binary star observations with the Hubble Space Telescope fine guidance sensors. II. Bright Hyades
Astronomical Journal Jan. 1992; 103(1); P. 190-6; 10 REFS

Maoz, D. et al
A gravitational lens candidate discovered with the Hubble Space Telescope
Astrophysical Journal, Letters 1992; 386(1, pt.2); P. L1-3; 19 REFS

Hollis, J.M.; Dorband, J.E.; Yusef-Zadeh, F.
Comparing restored HST and VLA imagery of R Aquarii
Astrophysical Journal 1992; 386(1, pt.1); P. 293-8+4 plates; 22 REFS

Jakobsen, P.; Greenfield, P.; Jedrzejewski, R.
The Cramer-Rao lower bound and stellar photometry with aberrated HST images
Astronomy and Astrophysics Jan. 1992; 253(1); P. 329-32; 9 REFS

Magain, P.; Surdej, J.; Vanderriest, C.; Pirenne, B.; Hutsemekers, D.
Q 1208+1011: the most distant imaged quasar, or a binary?
Astronomy and Astrophysics Jan. 1992; 253(2); P. L13-16; 29 REFS

King, I.R. et al
The current ability of HST to reveal morphological structure in medium-redshift galaxies
Astronomical Journal Oct. 1991; 102(4); P. 1553-68; 31 REFS

Paltoglou, G.; Bell, R.A.
Theoretical colours and isochrones for some Hubble Space Telescope colour systems. II
Monthly Notices Royal Astron Soc 1 Dec. 1991; 253(3); P. 449-73; 50 REFS

Weigelt, G. et al
First results from the Faint Object Camera: high-resolution observations of the central object R 136 in the 30 Doradus nebula
Astrophysical Journal, Letters 1991; 378(1, pt.2); P. L21-3+3 plates

Hubble Space Telescope imaging of eta Carinae

sources
Monthly Notices Royal Astron Soc 1 Nov. 1992; 259(1); P. 104-14; 18 REFS

Rix, H.-W.; Schneider, D.P.; Bahcall, J.N.
Hubble Space Telescope Wide Field Camera imaging of the gravitational lens 2237+0305
Astronomical Journal Sept. 1992; 104(3); P. 959-67, 1268-9; 32 REFS

Bahcall, J.N.; Maoz, D.; Schneider, D.P.; Yanny, B.; Doxsey, R.
Hubble Space Telescope imaging of the large-redshift gravitational lens candidate 1208+1011
Astrophysical Journal, Letters 1992; 392(1, pt.2); P. L1-4+2 plates

Lauer, T.R. et al
Planetary Camera observations of the central parsec of M32
Astronomical Journal Aug. 1992; 104(2); P. 552-62; 32 REFS

Caldwell, J.; Turgeon, B.; Hua, X.-M.
Hubble Space Telescope imaging of the north polar aurora on Jupiter
Science 11 Sept. 1992; 257(5076); P. 1512-15; 11 REFS

Hershey, J.L.
A deconvolution technique for Hubble Space Telescope FGS fringe analysis
Publications of the Astron Soc Pac July 1992; 104(677); P. 592-6; 4 REFS

Blades, J.C. et al
Imaging of four planetary nebulae in the Magellanic Clouds using the Hubble Space Telescope Faint Object Camera
Astrophysical Journal, Letters 1992; 398(1, pt.2); P. L41-4; 23 REFS

Cheng, F.H. et al
The Hubble Space Telescope Observations of X-ray nova muscae 1991 and its spectral evolution
Astrophysical Journal 1 Oct. 1992; 397(2, pt.1); P. 664-73; 32 REFS

Elson, R.A.W.; Schade, D.J.; Thomson, R.C.; Mackay, C.D.
The massive stars in the R136 region of 30 Doradus
Monthly Notices Royal Astron Soc 1 Sept. 1992; 258(1); P. 103-6+2 plates

Arp, H.; Crane, P.
Testing the gravitational lens hypothesis in G2237+0305
Physics Letters A 10 Aug. 1992; 168(1); P. 6-12; 13 REFS

Workshop on the Martian Surface and Atmosphere Through Time (LTI/TR 92-02)
Publ: Lunar & Planetary Inst; Houston, TX, USA 1992; xvii+182

Weaver, H.A. et al
Inner coma imaging of Comet Levy (1990c) with the Hubble Space Telescope
Icarus May 1992; 97(1); P. 85-98; 26 REFS

Caldwell, J. et al
Titan: evidence for seasonal change-a comparison of Hubble Space Telescope and Voyager images
Icarus May 1992; 97(1); P. 1-9; 17 REFS

Maoz, D. et al
Gravitational lensing of quasars as seen by the Hubble Space Telescope Snapshot Survey
Astrophysical Journal 1992; 394(1, pt.1); P. 51-60 +1 plate; 32 REFS

Crane, P.; Stiavelli, M.
Resolution of the southern radio lobe of 3C 33 at 660 nm
Monthly Notices Royal Astron Soc 1 July 1992; 257(1); P. 17p-20; 15 REFS

The post collapse core of M15 imaged with the HST Planetary Camera
Lauer, T.R. et al
Astrophysical Journal, Letters 1991; 369 (2, pt); P. L45-9 + 2 plates
The core of the nearby SO galaxy NGC 7457 imaged with the HST Planetary Camera
Lauer, T.R. et al
Astrophysical Journal, Letters 1991; 369 ,2, pt.2; P. L41-4 + 1 plate

Stellar photometry with the Hubble Space Telescope Wide-Field/Planetary Camera: a progress report
Holtzman, J.A. et al
Astrophysical Journal, Letters 1991; 369(2, pt.2); P. L35-40+2 plates

NGC 1068: resolution of nuclear structure in the optical continuum
Lynds, R. et al
Astrophysical Journal, Letters 1991; 369(2, pt.2); P. L31-4+3 plates

HST imaging of the inner 3 arcseconds of NGC 1068 in the light of (O III) lambda 5007
Evans, I.N. et al
Astrophysical Journal, Letters 1991; 369(2, pt.2); P. L27-30+2 plates

Tests of the photometric accuracy of image restoration using the maximum entropy algorithm
Cohen, J.G.
Astronomical Journal Feb. 1991; 101(2); P. 734-7; 13 REFS

Photometric calibration of the HST Wide-Field/Planetary Camera. I. Ground-based observations of standard stars
Harris, H.C.; Baum, W.A.; Hunter, D.A.; Kreidl, T.J.
Astronomical Journal Feb. 1991; 101(2); P. 677-94; 17 REFS

Saturn's Great White Spot spectacular
O'Meara, S.J.
Sky and Telescope Feb. 1991; 81(2); P. 144-7; 0 REFS

A catalogue of QSOs and link stars for the HST link of the Hipparcos reference frame
White, G.L. et al
Proceedings Astron Soc Australia 1990; 8(4); P. 368-71; 24 REFS

Ground-based CCD observations of two OB associations in M31 obtained through replicas of two Wide Field/Planetary Camera filters
Hunter, D.A.; Winkelman, S.
Publications Astron Soc Pac Aug. 1990; 102(654); P. 854-64; 37 REFS

A uvby-like photometric system for the WF/PC of the HST
Kiselman, D.; Oja, T.; Gustafsson, B.
Astronomy and Astrophysics Nov. 1990; 238(1-2); P. 265-9; 11 REFS

Stellar photometry with the Wide Field/ Planetary Camera of the Hubble Space Telescope
Holtzman, J.A.
Publications Astron Soc Pac July 1990; 102(653); P. 806-34; 10 REFS

Maran, S.P.; Kinney, A.L.
The second year of observations with the Hubble Space Telescope
Publications Astron Soc Pac May 1993;

Hester J.J. et al
Astronomical Journal Aug. 1991; 102 (2); P. 654-7; 26 REFS

Binary star observations with the Hubble Space Telescope Fine Guidance Sensors.I. ADS 11300
Franz, O.G. et al
Astrophysical Journal, Letters 1991; 377 (1, pt.2); P. L17-20; 5 REFS

Faint Object Spectrograph observation of the low luminosity Seyfert galaxy NGC 1566
Kriss, G.A.; Harting, G.F.; Armus, L.; Blair, W.P.; Caganoff, S.; Dressel, L.
Astrophysical Journal, Letters 1991; 377(1, pt.2); P. L13-16; 36 REFS

FOS spectroscopy of resolved structure in the nucleus of NGC 1068
Caganoff, S. et al
Astrophysical Journal, Letters 1991; 377(1, pt.2); P. L9-12; 15 REFS

First results form the Faint Object Camera: high-resolution imaging of the Pluto-Charon system
Albrecht, R. et al
Astrophysical Journal, Letters 1991; 374(2, pt.2); P. L65-7+1 Plate

HST observations of 3C 66B: a double-stranded optical jet
Macchetto, F. et al
Astrophysical Journal, Letters 1991; 373(2, pt.2); P. L55-8+2 plates

Blue stragglers in the core of the globular cluster 47 Tucanae
Paresce, F. et al
Nature 25 July 1991; 352(6333); P. 297-301; 26 REFS

Ionization fronts and shocked flows: the structure of the Orion Nebula at 0.1"
Hester, J.J. et al
Astrcphysical Journal, Letters 1991; 369(2, pt.2); P. L75-8+4 plates

Faint Object Camera observations of a globular cluster nova field
Margon, B.; Anderson, S.F.; Downes, R.A.; Bohlin, R.C.; Jakobsen, P.
Astrophysical Journal, Letters 1991; 369(2, pt.2); P. L71-4+1 plate

First results from the Faint Object Camera: imaging the core of R Aquarii
Paresce, F. et al
Astrophysical Journal, Letters 1991; 369(2, pt.2); P. L67-70+1 plate

First results from the Faint Object Camera: SN 1987A
Jakobsen, P. et al
Astrophysical Journal, Letters 1991; 369(2, pt.2); P. L63-6+1 plate

First results from the Faint Object Camera: images of the gravitational lens system G2237+0305
Crane, P. et al
Astrophysical Journal, Letters 1991; 369(2, pt.2); P. L59-61+1 plate

First results from the Faint Object Camera: observations of PKS 0521-36
Macchetto, F. et al
Astrophysical Journal, Letters 1991; 369(2, pt.2); P. L55-7+2 plates

Hubble Space Telescope Wide-Field/Planetary Camera images of Saturn
Westphal, J.A. et al
Astrophysical Journal, Letters 1991; 369(2, pt.2); P. L51-3+3 plates

4. X-Ray Astronomy

Joachim E. Trümper

1. INTRODUCTION

The past triennium (July 1990 - June 1993) has seen enormous progress in X-ray astronomy. The following satellite/instruments were operating in orbit:
-Ginga (Japan/UK, 1987-1991):
X-ray spectra and time variability studies (2-30 keV).

-Mir-Kvant International Observatory (Russia, Germany, The Netherlands, UK, 1987 -):
broad band X-ray spectra and) time variability (2-1000 keV, X-ray imaging (2-20 keV).

-ROSAT (Germany, US,UK. 1990 -):
soft X-ray all-sky survey (0.1-2.5 keV) ~ 4000 pointed observations; X-ray spectra, imaging time variability. deep surveys.

-Granat (Russia. France. Denmark):
X-ray imaging . time variability(2-20 keV).

-BBR XT (US 1990):
X-ray spectroscopy.

-ASCA (Japan, US, 1993-):
broad band X-ray imaging (0.5-10 keV) spectrocopy.

2. X-RAY EMISSION FROM NORMAL STARS AND WHITE DWARFS

ROSAT: Discovery of X-rays from the bright and dark sides of the Moon; discovery of metal absorption in white dwarfs; discovery of self-absorption in hot star winds; detection of shocks in hot star winds; detection of X-ray emission from hybrid stars; detection of the latest X-ray emitting star (LHS3003); discovery of long-duration flares in dMe stars; detection of flares in hybrid stars; detection of flares in low-activity giants; detection of flares in F stars; detection of superflares on RS CVn systems; discovery of microflares in dMe stars; determination of coronal structure on a G-type dwarf; detection of eclipses in Algol and YY Gem; detection of an eclipse of the high-temperature component in AR Lac; detection of the strongest flare on a classical T Tauri star: ubiquitous distribution of weak line T Tauri stars over star forming regions (J. Schmitt, 1992. 1994).

Ginga: Detection of flare thermal emission up 18 keV in II Peg (Doyle et al.,1991).

3. X-RAY EMISSION FROM SINGLE NEUTRON STARS

ROSAT: Discovery of X-rays from the radio pulsars PSR 0950-08, PSR 0823+26, PSR 1823-13, PSR 1951+32 (associated with CTB80)
Discovery of pulsed X-rays from the radio pulsars PSR 1929+10, PSR 0833-45 (Vela pulsar), PSR 1706-44, PSR 1055-52, PSR 0656+15
Discovery of X-rays from the millisecond pulsar PSR 1957+20
Discovery of pulsed X-rays from the millisecond pulsar PSRJ 0437-4715 (Becker and Trümper, 1993)

Discovery of pulsed X-rays from the Gamma source Geminga - first detection of a single neutron star outside the radio band (Halpern and Holt, 1992)

4. CATACLYSMIC VARIABLES

Due to the ROSAT survey the number of X-ray detected CVs has doubled. The fraction of magnetic CVs is ≤ 80%, that of old novae and nova-like variables ≥ 20%. Two AMHer systems were found in the period gap (125.5 and 140 min). X-ray emission from an optically thick boundary layer could only be observed in an outburst of SS Cyg, but not in an outburst of 9 other sources (Bunk and Verbunt, 1994).

5. X-RAY BINARIES

Granat: discovery and localization of three galactic sources with emission features at approximately 511 keV (1E1740-2942, X-Ray Nova Muscae and Crab Nebula); three X-ray Novae were discovered (Nova Muscae, Nova Oph 1993, Nova Vela 1993); discovery of QPO from four black hole candidates: GX 339-4, X-Ray Nova Muscae, X-ray Nova Persei and Cyg X-1; mapping of the galactic center region in the spectral band 3-800 keV and the discovery of six X-ray sources in that region, including a new black hole candidate close to GX 5-1; variability was detected in a one arcmin region around SgrA in the 3-20 keV band. The source has a very low luminosity (~2×10^{36}erg/sec) which is hundreds of millions of times below the Eddington luminosity for a $10^6 M_0$ black hole.

Mir-Kvant: Broad band 2-400 keV spectra were obtained for five X-ray Novae (GS 2023+38, Nova Vul, Nova Per, Nova Oph 1993, Nova Vel 1993); prominent soft components and extremely hard X-ray tails were detected; during six years of imaging of the central part of our Galaxy nine new X-ray sources were detected.

ROSAT: Soft X-ray spectra of galactic binaries, 26 X-ray sources identified with the globular clusters in M31.

Ginga: Broad band spectra and time variability of galactic X-ray sources.

6. SUPERNOVA REMNANTS

ROSAT: Detection of X-rays from 56 previously known galactic radio SNRs during the all-sky survey; detection of X-rays from about 90 sources suggestive of new, radio-uncatalogued SNR candidates; three candidates have been confirmed as SNRs up to now, fist full size mapping with few arcmin resolution of the giant SNRs like the North Polar Spur and other giant radio loops; spatial and spectral mapping of the medium size remnants revealing strong brightness and temperature variations; detection of a shell of cold matter associated with the Vela SNR casting an absorption shadow across Puppis-A; detection of EUV emission from the X-ray bright portions of the Vela SNR, high resolution mapping of the Crab Nebula revealing enhanced X-ray emission from an inner toroidal region and from two opposite jet-like features apparently protruding from the pulsar, detection of X-rays from the region containing the highest velocity optical knots at the very outskirts of Cas-A (Aschenbach, 1993).

BBRXT: High resolution, broad band spectra (0.5-10keV) of 7 galactic SNRs and 2 SNRs in the Large Magellanic Cloud; spectra of Cas-A and Tycho indicate the need for multiple non-equilibrium emission models.

ASCA: High resolution, broad band CCD X-ray spectra (0.5-10keV) of galactic and extragalactic SNRs exhibiting an unprecedented wealth of emission lines.

7. SUPERNOVAE

ROSAT: First detection of soft X-rays from SN 1987A in 1991 (Beuermann et al, 1993); first detection of soft X-rays from SN 1993J in M81 as early as 6 days after the supernova explosion and measurement of the X-ray light-curve for the first 40 days (Zimmermann et al.,1993). Measurements were resumed on November 1, 1993, and soft X-rays have been detected again.

ASCA: Early measurement of the broad-band X-ray spectrum of SN 1993J; measurement of red and blue Doppler-shifted X-ray emission lines from the jets of SS433.

8. DIFFUSE GALACTIC X-RAY EMISSION

Granat: A map and the spectrum of the galactic center X-ray diffuse source was obtained. It was found that SgrA was weak in X-rays during the past 400 years.

ROSAT: Discovery of X-ray-shadowing by dense neutral clouds; exploration of the spatial distribution of hot and cold galactic gas (Snowdon et al.,1992); strong evidence for the existence of a 10^6 K hot galactic halo; discovery of soft X-ray emission from hot superbubbles, large evolved supernova remnants and HII regions (e.g. Loop I, Eridanus Enhancement, Orion-Monoceros Complex, Ophiuchus Region).

9. NEARBY GALAXIES

ROSAT Survey: 450 optically catalogued galaxies detected, high X-ray luminosity spirals found by correlations with the IRAS catalogue (Pietsch and Trümper 1992).

ROSAT: Mapping the diffuse emission of LMC, SMC; discovery of supersoft X-ray sources in LMC, SMC, M31 and other galaxies; discovery of ~ 400 X-ray sources in M31; discovery of soft X-ray halos in NGC 253, M101, NGC 891, NGC 4631, NGC 4565, NGC 4258.

10. ACTIVE GALACTIC NUCLEI

ROSAT: Large number (≈ 20 000) of X-ray detected AGN; cross correlation with radio surveys yield 2500 mostly optically unidentidied radio loud AGN; confirmation of soft X-ray excesses in Seyfert I and QSO spectra ; tight connection between soft X-ray excesses and the big blue bump in AGN spectra (Walter et al.,1993); detection of extreme X-ray luminosities in starburst galaxies; X-ray detection of high redshift quasars up to Z=4.32, detection of rapid X-ray variability (~ 800 s) in narrow line Sy1 galaxies (e.g.IR 310224-3809), nearly simultaneous optical and X-ray variability of the BL Lac object PKS 2155-304; absence of periodicity in NGC 6814 (the claimed 12800 s period is due to a nearby CV, Madejski et al.,1993).

Ginga: Confirmation of reflection/warm absorption models for AGN (Pounds et al.,1990).

ASCA: Line diagnostics confirms iron lines/warm absorber models for several AGN.

Granat: Detection of a thermal cut-off in the X-ray spectrum of NGC 4151.

11. CLUSTERS OF GALAXIES

Einstein and EXOSAT observations indicate a steepening of the cluster X-ray luminosity function, i.e. a deficit of luminous clusters at high redshift. This is consistent with hierarchical theories of cluster formation, in which merging processes play a central role. Early ROSAT results support the Einstein and EXOSAT findings.

Ginga obtained spectra of about 100 clusters. An analysis of the data indicates that the hotter, more massive and more luminous clusters contain relatively more gas and relatively less iron, so that the proportion of primordial gas is higher.

ROSAT observations show sub-structure in about 70% of clusters, suggesting that mergers are frequent, consistent with hierarchical theories of cluster formation. A comparison of radio and ROSAT images of clusters reveals irregular X-ray emission near the radio position in about 75% of the cases, consistent with the cause being a merger event (Henry and Briel 1993).

ROSAT observations of small compact galaxy (Hickson) groups show them to be rapidly evolving systems with massive haloes of dark matter.

ROSAT observations affirm the hypothesis that most clusters contain cooling flows, which result in a steadily increasing mass of cold dark clouds accumulating in the cluster core.

The radial distribution of dark and baryonic matter was measured using ROSAT cluster observations (A2256, A426 Coma, Perseus, Virgo clusters). The results of such observations should enable eventually constraints on the matter density in the universe to be established (Böhringer, 1993).

Results from the ROSAT survey of clusters yields the first three-dimensional spatial correlation function for an X-ray selected sample of clusters. The result is important for studies of the growth of matter fluctuations and galaxy formation in an evolving universe.

12. COSMOLOGICAL X-RAY BACKGROUND

COBE rejects the hot intergalactic medium hypothesis for the extragalactic X-ray background, (Mather et al., 1990).

Ginga shows that AGN spectra are not canonic power laws (Pounds et al., 1990).

Ginga and HEAO-1 fluctuation analyses find a factor 2-3 higher source counts than the Einstein Medium Sensitivity Survey, indicating the importance of intrinsic absorption.

ROSAT Deep Surveys reach a sensitivity more than a factor of 10 fainter than the deepest Einstein pointings and resolve more than 75% of the 1-2 keV background into discrete objects/fluctuations (Hasinger et al., 1993).

Optical Identifications in ROSAT deep fields are mainly AGN, a substantial fraction optically fainter than the limiting capabilities of current 4m-class telescopes (Shanks et al., 1991).

References:

Aschenbach, B., 1993, Advances in Space Research, in press
Becker, W., and Trümper, J., 1993, Nature 365, 528
Beuermann, K. et al., 1993, Astron Astroph. 281, L45
Böhringer, H., 1993, Advances in Space Research, in press
Bunk, W., and Verbunt, F., 1994 to be published
Doyle, J. et al., 1991, MNRAS 248, 503
Halpern, J., and Holt, S., 1992, Nature 357, 222
Hasinger, G. et al., 1993, Astron Astroph. 275, 1
Henry, J.P., and Briel, 1993, Advances in Space Research, in press
Madejski, G.M. et al., 1993, Nature 365, 626
Mather, J.C. et al., 1990, Ap.J. (Lett.) 354, L37
Pietsch, W., and Trümper, J., 1993, Advances in Space Research, in press
Pounds, K., 1990,Nature 344, 132
Schmitt, J.H.M.M., 1992 in "Solar and Stellar Cornoae" ed P. Pallavicini in Mem. Soc. Astr. It. 63, 563
Schmitt, J.H.M.M., 1994, Ap.J., Supp., in press
Shanks, T. et al 1991, Nature 353, 315
Snowden, S. et al., 1992, 179th American Astron.Soc. Meeting
Walter, R. et al., 1993, Astron.Astroph. 274, 105
Zimmermann, H.K. et al., 1993, Nature in press

5. Gamma-Ray Astronomy

Volker Schönfelder

In gamma-ray astronomy we are presently experiencing a period of exciting new results from the Compton Gamma-Ray Observatory of NASA. *Compton* was launched on April 5, 1991 by the Space Shuttle Atlantis and is since then extremely successful. It is the first satellite mission that covers the full space astronomy gamma-ray range from about 50 keV to 30 GeV - more than five orders of magnitude in photon energy. *Compton* contains four different instruments with complementary properties to cover the broad spectral range (see Schönfelder, 1990). These are:

OSSE (50 keV to 10 MeV), COMPTEL (0.75 MeV to 30 MeV),
EGRET (30 MeV to 30 GEV), and BATSE (50 keV to 10 MeV).

Compton has started its mission with an 18-month lasting full sky survey - the first one in gamma-ray astronomy at all. Since November 1992 extended observations of selected sky regions or targets are being performed. Most of the data from the survey phase have already been released for public access, and a vigorous Guest Investigator Program is now being performed.

A number of exciting discoveries has been made by *Compton* so far: the distribution of gamma-ray bursts on the sky - as measured by BATSE - has disproofed the previous widely accepted neutron star origin of bursts within the galactic disk (Meegan et al., 1992). Instead, solar system, galactic halo and cosmological origins are now being discussed. The puzzle about the origin of the burst has become more challenging than ever (Fishman, 1993).

The discovery of gamma-ray blazars by EGRET was a surprise. More than 30 of these objects have now been detected. Their high-energy gamma-ray emission seems to be produced in jets. In most cases, the luminosity at gamma-ray energies is higher than in any other part of the electromagnetic spectrum (Fichtel et al., 1993).

Also radio pulsars seem to have their peak-luminosity at gamma-ray energies. In addition to the previously known two gamma-ray pulsars Crab and Vela, pulsed gamma-ray emission has now been detected from PSR 1509-58, 1706-44, 1055-52 and Geminga, as well. Geminga may represent a new class of pulsars. It is only seen at x- and γ-ray energies, but not in the radio band. It is interesting to note that the efficiency for converting rotational energy into gamma-ray luminosity increases with the age of the pulsar (Harding, 1993).

Much progress has been achieved in the field of gamma-ray line spectroscopy: COMPTEL has generated the first complete map of the galactic plane in the light of the 1.8 MeV ^{26}Al-line. The distribution along the plane is not smooth, but contains bright spots, which might suggest a more local origin (Diehl et al, 1993). One of the spots coincides with the Vela supernova remnant, and the detected line flux is well within the expectations from a Type-II Supernova. Two other nucleosynthesis lines were detected for the first time: these are the 122 keV line from ^{57}Co by OSSE (from SN 1987A), and the 1.156 MeV from ^{44}Ti by COMPTEL (from Cas A). The detection of 4.4 MeV and 6.1 MeV line emission from the Orion complex by COMPTEL marks the first measurement of nuclear interaction lines from interstellar space. OSSE, at present, is sampling a composite image of the inner part of the galaxy in the light of the 511 keV annihilation line. This image looks significantly different from the 1.8 MeV ^{26}Al-image.

These highlights represent only a small sample of the *Compton* results. Investigations of a large number of x-ray binaries and transients, of the diffuse galactic gamma-ray emission and of several spectacular solar flares, are only mentioned here. The analysis of the *Compton* data is far from being completed and many more results are still to be expected.

References:

R. Diehl et al., Adv.Sp.Res. 13, No 12, 723-726 (1993)

C. Fichtel et al., AIP-Proc. 280, 461-472 (1993)

J. Fishman, AIP-Proc. 280, 669-680 (1993)

A. Harding, Proc. of 16th Texas-Symposium (Berkeley, Dec 1992)

C. Meegan et al., Nature 355, 143 (1992)

V. Schönfelder, Adv.Sp.Res. 10, No 2, 243-253 (1990)

6. Solar Physics from Space 1990-1993

Stuart D. Jordan

The period covered by this report was dominated by three developments: (1) a continuing study of the maximum phase of Solar Cycle 22, especially through the successful launch and operation of the **YOHKOH** mission by the Institute of Space and Astronautical Sciences, (ISAS) of Japan; (2) the continuing development of instruments and spacecraft for the Solar and Heliospheric Observatory (SoHO), a joint mission of the European Space Agency (ESA) and the American National Aeronautics and Space Administration (NASA); (3) the continuing, but so far unsuccessful, attempt of the international solar community to develop a 1-meter class optical telescope for operation in space. These and a few other developments are reviewed here.

The most significant event for solar physics from space during the period covered by this report was the successful launch and early operation of the **YOHKOH** satellite by ISAS. Launch occurred on 30 August 1991 from the Kagoshima Space Center, and operation to date has yielded a large database for the study of the high-energy solar phenomena that accompany solar activity. The satellite carries four major instruments: a hard X-ray telescope (HXT), a soft X-ray telescope (SXT), a set of wide band spectrometers (WBS), and a set of Bragg crystal spectrometers. The spacecraft and much of the instrument development were accomplished in Japan, where the mission operations control center is located. There was also substantial participation in instrument development in the U.K. and in the U.S.A. The ISAS ground station at Kagoshima and the NASA Deep Space Network provide the communications link with the satellite, which has an on-board bubble memory with an 80 Mb capacity. Excellent references for a more detailed description of all aspects of the mission, scientific as well as technical and programmatic, appear in Solar Physics, Volume 136 (all papers), and in Publications of the Astronomical Society of Japan, Volume 44, No. 5 (all papers).

The HXT on **YOHKOH** was designed to obtain observations of hard X-rays from solar flares with an unprecedented combination of angular, temporal, and energy resolution, in order to better study the process of electron bremsstrahlung in flares. From observations with an angular resolution of 5 arcseconds and a temporal resolution of 0.5 seconds, taken simultaneously in the four energy bands 14-23-33-53-93 keV, it has become possible to study in far more detail than was previously possible the generation and confinement of the high energy electrons known to be associated with flares. HXT data have already been obtained on hundreds of flares.

The SXT on **YOHKOH** has already obtained on the order of one million excellent soft X-ray pictures of the Sun from which future studies of solar active regions and flares will benefit. The soft X-ray regime is thought to be the primary energy range for radiative losses in many strong active regions and flares, so high resolution observations are essential for understanding the energy balance in coronal loops and the energy transfer process in flares.

The WBS instrument has already provided evidence for two types of gamma-ray flares, for significant time variations of the hard X-ray spectrum that will provided further insight into the acceleration of solar flare electrons, and for proton acceleration in some of the flares observed. This instrument provides light-curve observations in ten energy bands with a temporal resolution in the range 0.125-0.5 seconds, as well as energy spectra from three spectrometers with temporal resolution 1-4 seconds, over a broad energy range from 3 keV to 100 MeV.

Finally, the BCS is gathering data to address questions of plasma heating and dynamics during the impulsive phase of solar flares. The BCS observes the Sun in four wavelength bands between 1.7 A and 5.1 A. Altogether, the four instruments on **YOHKOH** have already yielded a database that will keep high-energy solar physicists busy for the entire decade, as they attempt to resolve some of the still outstanding problems of solar flare physics: mechanisms for the acceleration of electron beams, the presence and acceleration of proton beams, nonthermal versus thermal models for (perhaps different types of) flares, what are the major types of flares (more refined classification, as a function of dominant physical processes), and further constraining of energetic processes, perhaps leading to improved ideas on different classes of flare "trigger" mechanisms.

In parallel with this major mission, solar flare research continues on older data sets obtained largely from the now defunct Solar Maximum Mission (**SMM**) and **Hinatori**. When all of the space data on solar flares and solar activity have been combined with an equally impressive database obtained from the ground, especially in the radio regime (Dennis and Zarro, 1993), we can expect significant advances in our understanding of solar flares by the end of the decade.

The other major solar space mission that has been the focus of attention during the period 1991-1993 is the **SoHO** mission, a mission dedicated to the better understanding of the solar corona and the solar interior. **SoHO** remains scheduled for launch in the late summer of 1995. At the time of this writing the **SoHO** instruments are under final development, to be delivered during 1994 for integration into the spacecraft. If all goes as planned, the mission will operate for two or more years, collecting data from the first Lagrangian point between the Earth and the Sun where continuous solar viewing will be possible. A comprehensive summary of the mission, its instruments, and its scientific objectives appears in ESA SP-1104 (1989).

Major problems in coronal physics will be attacked with these data. The "Parker" mechanism of solar wind acceleration by the thermal pressure gradient can produce supersonic expansion of the wind, and come close to predicting some of the wind properties over the quiet Sun, but this mechanism, alone, is inadequate to explain solar-wind acceleration in the high-speed streams now known to be associated with coronal holes. Two coronagraphs on **SoHO** will attempt to observe the wind out to at least 10 solar radii, in order to better understand this acceleration process.

While it is understood that a hot corona will produce a negative thermal pressure gradient beyond some distance from the Sun, and thus contribute to solar wind acceleration, solar physicists still do not know what produces (i.e.) heats the corona in the first place, although theories abound. In order to provide a better observational base for evaluating theories of coronal heating, **SoHO** will carry two spectrographs which, together, will provide diagnostic spectra of high spatial and spectral resolution over a broad range of coronal and transition region temperatures and densities. This superior diagnostic capability will help to refine models of these solar regions, and further constrain theories on the underlying physics. A fifth instrument will provide EUV images of the Sun, to provide context and additional science data. Other instruments will measure particle concentrations in the solar wind.

The other major goal of **SoHO** is to provide a database that, in conjunction with data from a developing ground-based solar network (developed by the Global Oscillations Network Group, or **GONG**) will permit solar physicists to map the density, temperature, and velocity distribution of the solar plasma from the photosphere down to the center of the Sun. One instrument will observe full Sun oscillations with very high sensitivity, concentrating on the lowest frequencies that penetrate to the solar core. Another instrument, a Michelson Doppler interferometer, will measure line-of-sight velocities and magnetic fields to high angular resolution (1.4 and 4 arcseconds) as well as other parameters, to provide improved models of the more superficial layers of the Sun. A third instrument will measure variations in the solar irradiance on time scales of minutes up to the mission time to gain further insight into the very low frequency global oscillations.

If the news on **YOHKOH** is good, and on **SoHO** hopeful, it must be reported that the news on the efforts of the solar community to achieve a 1-meter class observing capability in space is bad. The hoped for Orbiting Solar Laboratory (**OSL**), planned during the 1980's by NASA, has been long delayed by budgetary constraints, and while it remains in the NASA long-range plan, most solar physicists are exploring other ways to achieve this capability sooner, if that is possible. Solar physicists in the United States have been invited by NASA to produce a plan for a Solar Research Base Enhancement program that could include a balloon bourn telescope of large primary diameter approaching 1-meter, but it is not yet clear whether this program will materialize. The ISAS in Japan is exploring the possibility of building a large aperture, possibly 80 cm, solar telescope (**Solar-B**) to be flown in space, though this remains in the planning stage as well at this writing.

Continual improvements in solar ground-based capability are partially closing the gap between the 0.1 arcsecond angular resolution goal possible from space and what can be achieved from the ground. Nevertheless, this writer is aware of no solar physicist who believes that 0.1 arcsecond angular resolution can be achieved from the ground with reasonable image stability over several minutes and over a field-of-view large enough to attack the fundamental problems of the role of the small-scale magnetic fields in the dynamics of the lower atmosphere. This very high angular resolution is necessary if we are to observe these dynamic atmospheric phenomena on the main scale of energy transfer in the photosphere and chromosphere, i.e., the atmospheric scale height and a continuum mean-free-path. The scientific need for this space-based facility remains.

In addition to the three areas of major activity mentioned, there are a number of other developments that could produce exciting new science over the rest of the decade. The first of the **CORONAS** missions is currently scheduled for launch in December of 1993, although there is a good chance that actual launch may occur in 1994, as Russia and Ukraine, the joint developers of the mission, restructure their joint space effort. This **CORONAS** mission will be directed to general studies of the physics of the outer solar atmosphere, will carry a 200 kg scientific payload, and is scheduled to perform EUV spectroscopy and white light coronagraphy from low-Earth orbit, as well as carrying a high-energy payload for the study of solar activity (Sagdeev, 1993).

The **Ulysses** mission of ESA and NASA, with an international payload, is also expected to provide exciting new insights into the solar atmosphere, by directly observing for the first time the Sun's high latitude regions. The pass nearly over the southern pole of the Sun is expected to occur during September of 1994, to be followed by a pass at about 80 degrees north solar latitude during August of 1995. Science investigations include in situ probing of the solar wind as the spacecraft moves in toward the Sun on its "boomerang" orbit from Jupiter, as well as detailed observations of the high latitude regions of the solar atmosphere by nine instruments during the high latitude passes. Phenomena ranging from "superspicules" to polar coronal holes will be observed directly, and insight should be gained into the high latitude components of the solar wind.

Another option for operating science payloads in space was implemented successfully with the deployment from a NASA Shuttle in April 1993 of a **SPARTAN** payload consisting of an ultraviolet coronagraph and a white light coronagraph. The **SPARTAN** is a completely autonomous subsatellite which, in this case, operated for 47 hours before recovery by the Shuttle. Preliminary results and coronal images were presented at the Solar Physics Divisional Meeting of the American Astronomical Society in July 1993 (not published) and should appear in the open literature around the end of 1993.

With more fine data expected from **Yohkoh**, a second, upcoming flight of **Spartan**, as well as **Coronas** and **Ulysses**, there is considerable excitement in the solar community over an organized observing campaign that could involve all four missions.

Technical development and application continues in at least three areas of importance to solar physics in space. The first is the switch to a new family of two-dimensional digital detectors, many built around microchannel plates and charge-coupled devices (CCDs). Continuing development of these detectors on sounding rockets, and further use on major space missions such as **SoHO** will continue. The second major technical development is the use of new thin-film coatings on normal incidence telescopes, coatings that significantly increase the reflectivity of a normal incidence surface at EUV wavelengths, at least over a limited wavelength range. This in turn makes possible the construction of normal incidence telescopes in this range, capable of achieving higher angular resolution than has proved possible with grazing incidence designs. Finally, the increased application of sophisticated computer networks and data storage techniques is making data access from major space missions increasingly available to users worldwide, a development that can only further the already highly international, global character of solar astronomy.

References

Dennis, B.R., & Zarro, D.M. 1993, **Solar Physics**, 146, 177.

Pub. Astron. Soc. Japan, 1992, 44, No. 5 (entire issue).

Sagdeev, R. 1993 (private communication).

Solar Physics, 1991, 136 (entire volume).

The Soho Mission, 1989, ESA SP-1104.

45. STELLAR CLASSIFICATION (CLASSIFICATION STELLAIRE)

PRESIDENT: D.J. MacConnell VICE-PRESIDENT: O.H. Levato
ORGANIZING COMMITTEE: R.A. Bartaya, C. Corbally, D. Egret, M. Golay (Past-President), M. Grenon, T. Lloyd Evans, H. Maehara, N.R. Walborn, R.F. Wing, K. Zdanavicius

1 INTRODUCTION (D.J. MacConnell)

This report covers research in the field of stellar taxonomy in the period July 1990 to June 1993. It is divided into several sections which were written by experts in each subfield. It is a pleasure to thank them for their efforts and cooperation.

Our field continues to thrive and flourish as evidenced by the activity reported here and by the number of active participants. Advances in instrumentation, data reduction tools, and data storage continue to make impressive strides and to have great impact on our specialty. With new opportunities come new challenges and responsibilities, however. High-quality data are gathering faster than ever before, and it behooves us to make certain that they are properly calibrated, used, and archived. We commend the establishment of the Spectroscopic Data Archives WG of Comm 29 chaired by Dr R.E. Griffin (see Inf Bull. Nos 67, 40 and 68, 14) in this respect.

In the last triennium, the spectrometers aboard the Hubble Space Telescope have started returning data of unprecedented quality which will have a large influence on UV spectroscopy for many years to come. On the other hand, during the next triennium, it is almost certain that IUE, the "workhorse" of UV spectroscopy for the past decade and a half, will cease to return data.

To conserve space, all references are given with only one name followed by a + sign if there are additional authors.

1.1 Meetings

There was only one meeting covering the whole range of stellar types in the period between IAU GAs XXI and XXII - "The MK Process at 50 Years: A Powerful Tool for Astrophysical Insight" - and for which IAU support was not sought. It was sponsored by the Vatican Observatory under the chairmanship of Dr C. Corbally and was held in Tucson (USA) in Sept 1993 just before the due date for this report. The proceedings are to appear in the ASP Conference Series. It will be referred to as "MK50" here.

IAU Symposia/Colloquia which Comm 45 sponsored or co-sponsored:

Coll 138: "Peculiar vs. Normal Phenomena in A-Type & Related Stars" Trieste July 1992
Symp 162: "Pulsation, Rotation, & Mass Loss in Early-Type Stars" Juan-les-Pins Oct. 1993
Coll 148: "Future Utilisation of Schmidt Telescopes" Bandung Mar. 1994
Symp 166: "Astronomical & Astrophysical Objectives of Sub-milliarcsecond Optical Astrometry" The Hague Aug. 1994
Symp 167: "New Developments in Array Technology & Applications" The Hague Aug. 1994

2 CLASSIFICATION USING SLIT SPECTRA (R.O. Gray)

2.1 Early-type Stars

2.1.1 Wolf-Rayet stars

A catalogue of the 157 known galactic Wolf-Rayet stars was published by Conti+ (52.155.027). New galactic WR stars were discovered by Moffat+ (54.153.049) and by Cohen+ (54.112.064). Crawford+ (A&A **251,** L39) suggested that the star We 21 is a massive Pop I WN8 star instead of a PN central star. Niemela+ AJ (**101,** 572) obtained spectra of many WR stars and reclassified them. Lewis+ (ApJ **405,** 312) carried out a spectroscopic study of an eclipsing WR binary. The spectral-type changes of the star AG Car were reviewed by Viotti+ (53.114.015). The luminous blue star, He 3-519, was shown by Hajian+ (ASP Conf Ser **35,** 240) to have WN features in the optical and UV. Underhill (ApJ **398,** 636) showed that a WC7 star (HD 192641) has ionized N. In a similar vein, Nugis+ (52.114.029) showed that the O VI feature, known to appear in WC stars, can also appear in WN stars. Moffat+ (53.112.003) discussed the spectral variablility of WR stars and the peculiar WN8 subclass. Méndez+ (54.114.069) discussed the features of the two types of WR stars - massive Pop I stars and the central stars of PN. Smith+ (52.115.002) suggested a "quantification" of the WC classification system and have reclassified 10 stars to fit this new definition. Vreux+ (52.114.033) and Eenens+ (54.114.096) discussed classification of WR stars in the IR and Lortet (53.002.003) the nomenclature of spectral types of WR stars.

J. Bergeron (ed.), Reports on Astronomy, Vol. XXIIA, 525–534.

2.1.2 O- and B-type stars

Garmany+ (53.114.071) presented new spectral types for 32 OB stars including two new O3 stars. Herrero+ (A&A **261**, 209) commented on the spectral types of 25 luminous galactic OB stars. HDE 269676 was revealed to be a massive star cluster (Heydari-Malayeri+ 53.156.134) with many of the brightest components being O-type stars. Bagnuolo+ (55.120.007) classified O-type binaries with the method of "tomographic separation." Spectra of the luminous stars in Tr 14 and Tr 16 were classified by Massey+ (AJ **105**, 980). Underhill (54.112.085) discussed the spectral type of the star λ Per. Garrison+ (54.114.045) have reclassified 200 of the brighter late B- and Bp-type stars on their refined MK system, while Abt+ are classifying about 1000 B-type stars in the Bright Star Catalogue. Conti+ (ApJ **406**, L21) have detected He II absorption lines in the K-band spectra of early O-type stars, and Morrell+ (53.114.070) discussed the unidentified stellar emission lines near 4500 Å in O-type main sequence spectra. Spectral types of a number of O9 - B9 supergiants were presented by Lennon+ (A&AS **94**, 569, A&AS **97**, 559) who discovered a new N-weak star, and Carpay+ (54.115.013) discussed the luminosity of a B2 hypergiant. Walborn+ (51.114.102) classified HD 93840 as a new extreme BN supergiant. Lopes+ (A&A **261**,482) have commented on the optical and IR spectra of 10 luminous peculiar B-type stars. The nature of a number of subluminous blue stars was discussed by Saffer+ (54.114.086), Schulz+ (53.114.083), Unglaub+ (52.117.002) and Theissen+ (A&A **273**, 524). Slit classifications of stellar candidates/blue stellar objects from a number of surveys were reported by Heber+ (54.114.087), Berg+ (55.114.016), Stepanyan+ (53.002.137), Wegner+ (AJ **105**, 660), Bixler+ (A&A **250**, 370), Usher+ (52.002.078), Beers+ (55.114.011), Moehler+ (52.114.030), Kilkenny+ (53.155.043), Wegner+ (52.114.028) and Jordan+ (A&A **273**, L27). A review by Shore (ASP Conf Ser **35**, 186) discussed the nature of the massive stars of the upper main sequence (the LBV, Of/WN, and the B[e] stars). Hack (52.114.157) reviewed the characteristics of WR, and dwarf, giant and supergiant O, B, and A stars. HD 160529 was declared a new LBV by Sterken+ (54.122.005) who tracked the spectral-type changes of this star during the last eight years. Similarly, the changes of HR Car, a LBV, were tracked by Hutsemékers and van Drom (54.112.024) and V439 Cyg by Polcaro+ (52.153.028). WRA 751 was discovered as a new LBV by de Winter+ (55.122.052). HR 446 was discovered to be a He-weak star (Hric, 52.114.043). Blue stars at high galactic latitudes were classified and analyzed by Dufton+ (A&A **269**, 201), Kilkenny+ (MNRAS **258**, 57), Arnold+ (MNRAS **257**, 225), Holmgren+ (MNRAS **258**, 521) and Conlon+ (A&A **269**, L1).

2.1.3 Be and Shell stars

The spectroscopic evolution of the shell star Pleione was reviewed by Goraya+ (52.112.089) and Guo Xiaozhen (Pub Beijing Astron Obs **16**, 7). HD 17520, an O9 V star, has undergone a transition to a Be-type star (Walter PASP, **104**, 508). The long-term variations of the Be star EW Lac were discussed by Guo Yulian (Pub Beijing Astron Obs **16**, 2), the spectral evolution of 27 CMa was reviewed by Ghosh (52.112.044) and that of MWC 560 by Pier (54.114.009) and Bopp+ (BAAS **24**, 801). HDE 245770, an X-ray/Be system, was classified as O9.7 IIIe by Giovannelli+ (Sp Sci Rev **59**, 1) and X Per, another such system, was studied by Reynolds+ (MNRAS **258**, 439). The spectra of a number of Be stars in the near-IR were described by Andrillat+ (52.112.003, 52.112.020), and the behavior of the OI line in Be stars was described by Jaschek+ (A&AS **97**, 781). Waters+ (A&A **272**, L9) suggested that τ Sco is a pole-on Be star, and HD 50138 was identified as a mild B[e] star (Jaschek+ 53.112.209). Two B[e] stars were studied in the IR by Jaschek+ (55.114.029). Jaschek+ (A&AS **95**, 535) defined 12 classes of Be stars according to the shape of the H-emission lines. Pollacco + (53.122.023) classified BG Cep as a Be star instead of a hot R CrB star.

2.2 A- and F-type Stars

Classifications were made for a wide variety of stars: Jaschek+ (54.112.035, 53.112.210) studied the Ae and A-type shell stars in the near IR, Corbally+ (IAU Circ 5452) noted the shell phase of HD 37258, Am stars were studied by Okyudo+ (51.114.086), Wegner+ (PASP **104**, 322) and Rao+ (52.116.053, Bull Astr Soc India **18**, 299, MNRAS **258**, 819). A few Am and Ap stars were classified in OB associations by Gray+ (IAU Col 138). Lagrange-Henri+ (52.112.025) carried out a systematic search for β Pic-like stars, while Holweger+ (54.112.062) and Stürenburg (A&A in press) in high-resolution studies of the λ Boo stars identified a number of them as showing β Pic characteristics. Jaschek+ (54.112.061) did a spectroscopic survey of 26 A-type stars with IR excesses detected by IRAS. Philip+ (53.114.067) identified features in high-resolution spectra which can help to distinguish FHB A-stars. Lodén (55.114.055) reclassified two stars previously classified as evolved Ap stars, while Renson+ (54.002.026) published a catalogue of Am and Ap stars. Noskova (Sov Astron Lett **18**, 283) classified a δ Scu variable, while Krzesiński+ (54.117.066) has classified the primary of CK Boo. López+ (52.153.032, A&A **273**, 482) identified the He I D_3 5876 Å line as an activity indicator in F-type stars. Bartkevičius (Baltic

Astron **1**, 194) published a catalogue of Pop II A-F supergiants, while Lu+ (ApJS **83**, 203) have published 2D-Frutti MK types for some of the stars in their catalogue of faint A, F and G stars at the SGP. Sowell (52.114.022) classified a sample of yellow supergiants from their Hα profiles. Abt+ are classifying all the northern A-type stars in the BSC and some southern A-stars as well.

The λ Boo stars received considerable attention during the past three years. Gray (54.036.126) discovered two new λ Boo stars in the field as well as two new shell stars, and Gray+ (IAU Col 138, AJ **106**, 632) discovered a λ Boo star in Ori OB1 and a number of mildly metal-weak A stars in Ori OB1 and Lac OB1. Stürenburg (A&A in press) has identified HR 7960 as a λ Boo star. Iliev (IAU Col 138) and Stürenburg (A&A in press) confirmed, using high-dispersion spectra and theoretical Balmer-line profiles, the reality of Gray's classification (45.114.012) of the λ Boo stars into normal and peculiar H-line types. Corbally+ (IAU Col 138) classified about 60 A-type stars in the halo; 10 of them show characteristics of λ Boo stars. Gerbaldi+ (IAU Col 138) have published a review on the λ Boo stars.

2.3 Late-type Stars

Keenan (54.071.033) investigated the constancy of the solar spectral type during the solar maximum and finds no change at 68 Å/mm. Corbally+ (55.155.058) have classified 299 early G dwarfs in the NGP and SGP regions. Garrison+ (BAAS **24**, 1207) classified a number of peculiar G stars. Bohlender+ (PASP **104**, 1152) reclassified two K0 giant stars on the basis of near-IR spectra. MacConnell+ (AJ **104**, 821) obtained confirming spectroscopy for distant, cool supergiants discovered in an objective-prism survey. This paper contains a description of luminosity criteria in the near IR, and a discussion of the classification of stars of unusual type. Garrison+ (53.122.073) presented spectra of Mira during recent maxima. Feng+ (53.114.075) classified 10 emission-line M stars, and Noguchi+ (PASJ **45**, 85) classified the near-IR spectra of 26 S-type stars. Optical spectra of two OH/IR sources were classified by Le Bertre (54.112.040), and Berdyugina+ (Sov Astron **36**, 425) used MgH as a luminosity criterion in the analysis of 12 red giants. Oliva+ (55.114.050) identify features in H-band spectra which can be used to classify K and M stars. Faraggiana+ (54.114.059) discussed the members of the class of super-Li-rich stars. Ruelas-Mayorga+ (55.155.076), Terndrup+ (52.155.013) and Sadler (55.155.039) discuss the classification of giants in the galactic bulge. Lu (53.114.060) classified 389 Ba stars some of which may be dwarfs. Briley+ (55.154.028), Smith+ (AJ **105**, 173) and Vanture+ (PASP **104**, 888) and Penny+ (MNRAS **257**, 89) have classified CN and CH stars in globular clusters. Lazaro+ (53.114.061) found HR 8795 to be exceptionally C deficient. Beers+ (55.114.013) found 50 stars showing anomalously strong G-bands.

Green+ (ApJ **400**, 659) discuss the use of the C_2 6191Å bandhead in C stars as a luminosity indicator. Green (PASP **104**, 977) classified 19 C-star candidates from the Case survey. Green+ (52.114.134) presented low-resolution spectra of 11 faint, halo C stars. ^{13}C-rich C stars (J-type) were classified by Chan (PASP **105**, 440), Evans (53.112.061) and Barnbaum+ (54.112.049). Abia+ (53.114.076) discovered a new super-Li-rich C star, IY Hya. Warren+ (MNRAS **261**, 185) have discovered some new dwarf C stars. Other studies of C stars involving classification include Le Bertre+ (52.112.011), Noguchi+ (51.114.091) and Le Van+ (ApJ **392**, 702). Heber+ (A&A **267**, L31) have found what appears to be a Pop I dwarf C star.

Kirkpatrick+ (ApJ **402**, 643) presented a sequence of spectra in the 0.6-1.5 µm region of well-studied M2 - M9 dwarfs. They discussed a number of temperature-sensitive features which can be used in classification. Red/near-IR spectra of 39 K and M dwarf standards and 38 secondary standards were presented by Kirkpatrick+ (54.114.094) and classified on the Boeshaar system. The spectra of 14 giant and higher luminosity stars are also presented, and spectral features which may be used to distinguish giants from dwarfs are listed. Davidge+ (ApJ **403**, L47) compared the properties in the 1.5 - 2.5 µm interval of a number of extreme M dwarfs. Several unidentified features, perhaps due to polyatomic molecules, are present. Other studies of red dwarfs include Mathioudakis+ (53.112.062), Fulbright+ (BAAS **24**, 799), Bessell (53.114.014), Schneider+ (54.114.073), Leggett+ (BAAS **23**, 1416), and Doyle+ (52.122.070).

Brown dwarfs were the subject of many studies even though they remain elusive. Boeshaar+ (53.114.099) have examined the spectra of a sample of very cool M dwarfs and giants in the 1.4 - 2.5 µm range. H_2O bands are insensitive to metallicity but saturate at M7. First overtone bands of CO, however, appear to continue to strengthen with decreasing temperature. Tinney+ (AJ **105**, 1045) also studied the H_2O bands of low-luminosity stars. Davidge+ (54.118.019) compared the near IR spectrum of the suspected brown dwarf binary Gl 473 with those of Gl 65AB and Gl 866AB, while Kirkpatrick+ (ApJ **406**, 701) compared the red/near-IR spectrum of the brown dwarf candidate GD 165B to the spectra of other low-luminosity objects. Magazzù+ (ApJ **404**, L17) suggested that the Li abundance should be preserved in brown dwarfs, but a search in low-mass stars led to a null detection. Irwin+ (54.115.031) discovered a star of exceedingly low luminosity. Kirkpatrick (BAAS **23**, 1396) has compiled a library of late dwarf spectra to provide a basis for the recognition of brown dwarfs. Other studies of brown dwarf candidates include: Stauffer+ (53.114.034), Ruíz+ (53.115.011), Rebolo+ (55.115.012), Zuckerman+ (BAAS **24**, 1158), and Graham+ (AJ **104**, 2016).

2.4 Pre-Main Sequence Stars

The following authors used spectroscopic observations to identify new T Tau stars and to estimate their spectral types: Gomez+ (AJ **104**, 762), Kogure+ (54.121.087), Miranda+ (A&A **271**, 564) and Gyul'budagyan (53.121.061). Spectral types of T Tau stars have been published by Franchini+ (55.121.021), Martín+ (55.118.017), Alcalá+ (A&A **272**, 225), and Martín (PASP **105**, 277). Spectral types of other emission-line PMS stars have been given by Herbig (52.121.030), Kolotilov (53.121.057) and Walter+ (52.121.057). Carter+ (54.121.083) established a classification sequence of emission PMS flare stars. A number of X-ray sources have been studied and turned out to be PMS stars of some sort: Nations+ (53.121.041), Walter (AJ **104**, 758) and Bouvier+ (55.142.023). The classification of a sample of PMS binaries has led to the discovery of a new T Tau binary, GG Tau (Martín+ IAU Col 135, 63). Herbig-Haro objects have been identified spectroscopically by Ogura (53.121.009, Proc Astron Soc Aus **9**, 295), and Ogura+ (53.121.003). Herbig Ae/Be stars have been classified by Yoshida (55.121.055), Persi+ (54.121.071), Welty+ (55.112.028), Hughes+ (53.112.024) and Andrillat+ (52.112.004). FU Ori stars were studied by Staude+ (53.121.014, ApJ **400**, 556), Kenyon+ (54.121.062, AJ **105**, 1505), Kolotilov (53.121.025) and Parsamyan+ (Sov Astron Lett **18**, 247).

2.5 Highly Evolved Stars

"Proto-planetary nebula" stars or "HR 4049" stars were a "hot" topic during the past few years. Parthasarathy+ (A&A **267**, L19) has identified an interesting star, SAO 244567, which seems to have turned into a PN within the last 40 years. The central star of the "Red Rectangle" has been studied by Waelkens+ (55.114.018), and Sloan+ (ApJ **409**, 412). Other papers dealing with the classification of PPN stars include Hrivnak+ (53.112.094), Geballe+ (52.112.012), Waelkens+ (54.122.059), Arkhipova+ (55.122.081), Conlon+ (ApJ **408**, 593) and Hu+ (A&A **273**, 185). Other post-AGB stars have been studied by Gonzalez+ (54.114.015, 028), Johnson+ (BAAS **23**, 1410), and Conlon+ (53.114.057). Sion+ (54.002.105) discuss the classifications of a number of highly evolved stars.

The central stars of PN were classified by Kaler+ (53.134.021), Tweedy+ (MNRAS **259**, 315), Bond+ (52.134.018), McCarthy+ (53.134.015), Werner+ (55.134.032), who report on the dramatic change in type of the central star of Longmore 4, Peña+ (A&A **265**, 757), Menzies+ (52.133.002), who discovered a probable new [WC11] star, Sun+ (Chin Astron Ap **16**, 322), and Napiwotzki+ (54.134.024, .056) who found a new class of hot central stars.

2.6 Binaries and Multiples

During the past three years, significant advances have been made in the ability to separate the spectra of binary stars. The Griffin subtraction technique has led to the identification of the secondary star of τ Per (Griffin+ A&A **254**, 289) as a rapidly rotating early A-type star which may be a λ Boo star. Griffin ("Binaries as Tracers of Stellar Formation", p 93) discussed the application of this technique to HR 6902 and ς Aur. Fekel (IAU Col 135, 89), Stockton+ (55.120.024), Cole+ (55.118.008) and Fekel+ (AJ **105**, 2265) used the method of "spectrum addition" to study a number of multiple stars and speckle binaries. Bagnuolo+ (BAAS **23**, 1378, 54.120.011, 55.120.007) have used "tomographic separation" to reconstruct the spectra of the components of SB2 systems. Perhaps even more exciting is the promise of being able to use dispersed speckle data to obtain the separated spectra of close binaries (see Kuwamura+ AJ **105**, 665). Other papers dealing with the classification of both members of a binary include Wegner+ (55.117.035), Bord+ (54.118.023) and Ginestet (54.120.005). Classification of binary systems containing a white dwarf can be found in: Fulbright+ (ApJ **406**, 240), Bergeron+ (ApJ **407**, 733), Jordan+ (54.126.046), Roberts+ (53.126.064), Beck+ (BAAS **23**, 1418), Oswalt+ (54.126.065) and Sion+ (54.126.066). Krzeminski+ (53.117.389) studied a low-mass X-ray binary. Classification of components in eclipsing systems is considered by Tomkin (55.119.011) and Grygar+ (54.119.032).

Numerous papers dealt with symbiotic systems and spectroscopic estimates of the spectral types of the component stars; see Munari (55.117.132), Barbá+ (55.124.045), Kaler+ (53.117.119), Bode+ (54.117.233), Schild+ (MNRAS **258**, 95), Kotnik-Karuza+ (A&AS **94**, 251), and Gochermann (54.117.076). IR spectroscopy of the symbiotic system V1016 Cyg was used (Rudy+ 52.117.133) to estimate the spectral type of the secondary star. Kaler+ (PASP **105**, 599) obtained spectra of a collection of PN nuclei and found that the nucleus of Ha 1-2 has the spectral characteristics of a symbiotic star. Gutiérrez-Moreno+ (PASP **104**, 1187) confirmed the symbiotic nature of 5 systems,

Calabró+(Ap Sp Sci **197,** 251) did the same for QW Sge, and Munari (54.157.147) found that the C star C-1 in the Dra dwarf galaxy was a symbiotic binary. Remillard+ (ApJ **396,** 668) found that the optical spectrum of S-154 in the LMC evolved from an LBV to a symbiotic star over the course of four years.

2.7 Variable Stars

Wade+ (53.119.043) have undertaken a spectral classification survey of about 100 classical Algol variables, and Walker+ (54.155.063) studied RR Lyr variables in Baade's Window with the ΔS method. Mantegazza (53.122.054) has found that consideration of the Ca II triplet and the Paschen lines allows the separation of RV Tau stars from Pop I supergiants. Much attention has been paid to R CrB stars: Kilkenny+ (51.122.182) classified UX Ant as an "F-type" R CrB star, Kilkenny+ (Obs **112,** 158) suggested V517 Oph is an R CrB star, and Evans+ (54.122.042) classified three stars as probable R CrB stars. On the contrary, Giridhar+ (52.114.014, J Ap Astr **13,** 307,) suggested that LR Sco was misclassified as an R CrB star. Pollacco+ (53.122.023) reviewed the criteria for R CrB stars and showed that DY Cen, possibly a unique transition object, displays marked differences from other members of the class. The spectroscopic evolution of R CrB itself was summarized in Cottrell+ (54.122.093), and the type of V482 Cygni was discussed by Rao+ (PASP **105,** 574). Bidelman (PASP **105,** 578) examined the 1893 spectrum of η Car and suggested a relationship to stars of the R CrB type.

The cataclysmic variable nature of a number of systems was confirmed spectroscopically by Szkody+ (55.117.038), Maza+ (PASP **104,** 1060), and Patterson+ (55.117.006). The nature of other CV's was examined by Abbot+ (ApJ **399,** 680), Dobrzycka+ (55.117.115), Mukai+ (52.117.091), and Kubiak (Acta Astr **42,** 177). Cutispoto+ (53.114.023) obtained spectral types of 8 southern stars found in EXOSAT X-ray images, while Mereghetti+ (A&A **263,** 172) determined the spectral types of 3 stars in the X-ray circle around an X-ray pulsar. Spectroscopic studies which led to the confirmation of a nova or dwarf nova nature and/or the determination of spectral types of components in a nova-like system include: Jablonski+ (55.117.186), Drew+ (MNRAS **260,** 803), Marsh (52.117.011), Szkody+ (55.117.046), Thorstensen+ (54.117.139), Shara+ (52.124.003), Anupama (55.124.002), Bruch+ (55.117.157), Bianchini (53.124.064), and Wenzel+ (Contr Astr Obs Skalnate Pleso **22,** 69). The nova-like object PU Vul was studied by Kanamitsu+ (53.124.110, .111), Tomov+ (54.117.223) and Belyakina+ (Bull Crim Ap Obs **81,** 25). Spectral classification of flare stars was carried out by Mirzoyan+ (55.122.149), Carter+ (54.121.083) and Honeycutt+ (PASP **104,** 1039). Optical spectroscopic chromospheric diagnostics in RS CVn stars and related objects were reviewed by Bopp (52.117.296).

2.8 Clusters and Associations

Spectral types of 105 members of Car OB2 were published by Garcia (ApJS, **87,** 197). Perry provided spectral types for stars in Sco OB1 and the nuclear cluster NGC 6231 (52.152.004), Massey and Thompson (53.152.001) classified the massive stars in Cyg OB2, red spectra have been used by Torres-Dodgen (52.114.136) to classify the brightest stars in Cyg OB2 and de Lara+ (53.131.082) classified stars in the Ser cloud. Levato and Malaroda have just completed the classification of 500 stars in the Ori OB1 association.

In his continuing study of clusters containing Cepheid variables, Turner obtained spectral types in NGC 1647 (AJ **104,** 1865), NGC 129 (Turner+ AJ **104,** 1132) and undertook a search for OB associations near certain long-period Cepheids (Turner+ ApJS **85,** 119). In addition, Turner (A&AS **97,** 755) has studied the open cluster Roslund 3. Faint stars in NGC 3114 were classified by Sagar+ (53.153.039), FitzGerald+ (52.153.047) obtained MK types in NGC 2353, Battinelli+ (55.153.014) in NGC 433, Landolt+ (52.153.010) in the van den Bergh and Hagen Cluster No. 99, and Lattanzi+ (54.153.036) in NGC 225. Lodén (Ap Sp Sci **199,** 165) has shown that HD 108353 is actually a multiple system or small cluster of stars of late-B spectral type. A "stellar diamond" in Virgo was studied by Brosch (54.152.011). Munari (55.153.053) discussed the involvement of the Torino and Asiago Observatories in a program of classification of stars in young open clusters. Types of stars in Berkeley 86 were obtained by Forbes+ (55.153.012) and by Polcaro+ (Mem Soc Astron Ital **62,** 933) in Berkeley 87.

Spectral types and emission-line strengths of young stars in the molecular cloud L1641 were obtained by Wouterloot+ (A&A **265,** 144). Block+ (A&A **273,** L41) obtained K-band spectra of two stars in a new cluster at the Rosette GMC CO peak. In a search for faint members of the Pleiades, Prosser+ (53.153.011) has classified 77 K and M candidate members, Hamilton+ (AJ **105,** 1855) classified four red stars in the Pleiades, and Mirzoyan+ (53.122.178) have classified 17 flare stars in the Pleiades. Kogure (55.121.056) classified 34 emission-line stars in Ori OB1b, and Prosser (AJ **105,** 1441) classified a number of faint stars in the fields of IC 4665 (55.153.010) and the α Per cluster. Hughes+ (AJ **104,** 680) classified stars in the region of the

Cha II dark cloud, Meyer+ (AJ **105**, 619) studied low-mass stars in the vicinity of σ Sco, and a study of PMS stars towards the Sco OB1 association was carried out by Piers+ (55.153.023).

2.9 Stars in Galaxies

The discovery and classification of WR stars in external galaxies has been particularly fruitful this triennium. WR stars were discovered and/or classified in the LMC by Testor+ (52.114.036) and Heydari-Malayeri+ (53.156.038) and in the SMC by Pakull+ (IAU Symp 143, 260). Heydari-Malayeri+ (A&A **258**, L13) discovered a peculiar Of?-like star in the LMC which they believe may be a transition object between Of supergiants and WR stars. The LMC star R127, originally classed as a WN star, is now an A-type star (Nota+ IAU Symp 143, 561). Morgan+ (54.156.035) have discovered a new WR star in the SMC. Spectroscopy of 4 WR candidates in IC 10 was obtained by Massey+ (55.157.076). A large number of WR stars in M31, M33, NGC 6822 and IC 1613 were classified by Armandroff+ (54.157.199). WR stars in M33 and M31 were classified by Schild+ (52.157.070, A&A **261**,419) and an Ofp/WN9 star in M33 detected. Types for a number of WR stars in NGC 300 were presented by Schild+ (A&A **266**,145).

Walborn and colleagues have classified a number early-type objects in the 30 Dor region including OB supergiants in compact groups associated with R127 and S Dor (53.112.019), Sk-69°203, a B0.7 Ia star, an O3 If/WN6-A object and two companions of SN 1987A (PASP in press). Walborn+ (ApJ **399**, L87) have classified a number of young O-stars in the LMC H II region N11. Parker (IAU Sym 149, 467, PASP **104**, 1107) has undertaken an extensive photometric and spectroscopic survey of the stars in the 30 Dor region and has classified most of the early-type stars. Parker (53.152.010) has also classified a number of stars in LH 58. Schild+ (A&AS **92**, 729) have also classified stars in 30 Dor. Parker+ (AJ **103**, 1205) classified stars in the LH 9 and 10 regions, and Conti+ (53.114.101) studied individual stars in the LMC. Wolf (53.156.068) has studied LBV stars in the MCs, and Parker+ (BAAS **24**, 1236) has found a new LBV in 30 Dor. Fitzpatrick (54.156.030) has set up recommended classification standards for B supergiants in the LMC and has classified about 100 stars on his system, and Humphreys+ (53.156.139) have studied a group of anomalous A-type supergiants in the MCs. Welch+ (AJ **105**, 146) studied the variable stars of NGC 1866, Sekiguchi+ (52.124.024) a recurrent nova in the LMC and Lawson+ (53.156.086) the R CrB stars in the LMC. Lortet+ (53.156.037) have obtained the spectra of 95 stars in the stellar association LH99; 40 are newly discovered O stars, including 3 O3-4 stars.

Li-rich red giants in the MCs have been classified by Lubowich+ (BAAS **24**, 1158). C stars between the MCs have been discovered by Demers+ (MNRAS **260**, 103), and four new proto-stars near 30 Dor were found by Hyland+ (MNRAS **257**, 391). Spectral types in the young SMC cluster N90 were obtained by Hutchings+ (53.156.031). SMC supergiants were studied by Nandy+ (52.156.014), SMC AGB stars by Reid and Mould (52.156.013), B[e] supergiants were discovered in the SMC by Heydari-Malayeri (52.112.007) and Zickgraf+ (A&A **260**, 205). Faint C stars in the SMC were studied by Westerlund+ (A&A **260**, L4).

2.10 General

Keenan (PASP **105**, 905) has developed a new spectral classification system for the C stars to bring them into the Revised MK system. The notation separates the four population groups: R, N, CH, and Hd (H-deficient) C stars. New criteria have been identified which allow more realistic temperature sequences and, for the first time, a rough luminosity classification. Keenan is also working on a second edition of the Perkins Catalog of Revised MK types for the Cooler Stars and gave a review on classification standards (54.114.099). Hoffleit (54.114.097) spoke about "Pre-MK classifications."

Walborn's continuing spectral classification studies of OB stars are based on both optical and space-ultraviolet spectra, and concentrate on the MCs. He is preparing an optical digital atlas of peculiar OB spectra to complement the earlier atlas of normal OB stars (51.002.144).

Drilling and Vijapurkar have nearly completed a project of MK classification of all non-emission OB+ stars in the Case-Hamburg-LSU OB-star surveys. Vijapurkar+ (ApJS in press) has classified 253 stars at 70 A/mm, and spectral types for 500 more stars are in process. This project has led to the discovery of more than 400 new O stars and early B supergiants and other stars of astrophysical interest.

Abt is classifying the 750 stars lacking MK classifications in the BSC, 1000 members of visual multiples brighter than B = 8. Abt+ are classifying at least one star in each of 264 known Trapezium systems. Levato+ are classifying a sample of Ap stars and are continuing their classification program of members of southern clusters and associations. Torres-Dodgen+ (PASP **105**, 693) investigated classification in the near-IR. Sato+ (52.114.018) classified 591 stars on the MK system. Lançon+ (A&AS **96**, 593) published a library of near-IR spectra sampling the HR diagram. A new *Atlas of Stellar Spectra* (Obs **113**, 100) has been prepared by a French team containing 470 spectra of specific types and MK standards.

Corbally+ have completed classifying slit spectra of NGP and SGP G-dwarf stars, and Garrison+ are publishing a classification study of the late B-type stars. Garrison+ will be publishing spectral types in the cluster NGC 3532. Garrison+ are reviewing the MK standards over the entire HR diagram; Garrison will present the results at MK50. He is also classifying the nearest 800 G dwarfs and the 1500 brightest stars in the sky. Corbally (Trans IAU **21B**, 328) discussed problems of accurate classification using digital spectra. Gray+ are continuing to investigate the astrophysical context of the λ Boo phenomenon. They also (BAAS **24**, 1279) are using MK classification of synthetic stellar spectra to calibrate MK types in terms of effective temperature, gravity, microturbulence and metallicity.

3 OBJECTIVE-PRISM AND SLITLESS CLASSIFICATION (C.B. Stephenson)

This report can be kept rather brief by making reference to the book "Objective-Prism and Other Surveys" (55.011.021; OPOS). While the purpose of that book was not precisely that of the present review, the fact is that it does summarize a very large amount of recent results in the field of objective-prism astronomy.

The review will be further shortened by my making no mention of several pilot programs carried out in recent years without any real surveys resulting; in any case, much of this is discussed in the following section on automated methods.

In my opinion, one of the most salutary things bearing upon objective-prism work that has happened in recent years is the rather large amount of follow-up work done on objective-prism surveys at higher spectroscopic resolution than that of the surveys. For example, see the report by Sion+ in OPOS; also Green+ (52.114.134). The quantity of the follow-up has been made possible entirely by the sensitivity of current spectrographs and detectors.

During the report period, the most significant astrophysical result to emerge is arguably the discovery of the second, third, and fourth dwarf C stars (54.114.090). Papers XII-XIV of the Case Low-Dispersion Northern Sky Survey (Pesch and others; ApJS) appeared listing blue stars and galaxies, A-F stars, and various emission objects.

At Abastumani Observatory, Kharadze+ have done MK classifications (166 Å/mm near Hγ) for several thousand stars in special regions (Bull Abast Obs No 72, 3-294). Kharadze+ published several hundred generally faint, early-type peculiar stars in the galactic anticenter (AJ **99**, 379). From Byurakan, several astronomers have published numerous lists of mainly blue objects plus follow-up spectroscopy, e.g. 52.002.045, .116; 53.002.149, .150; 54.002.088, .106, .107, .131, .133. This material has been appearing in Astrofysica.

At Warner and Swasey, Stephenson published two more lists resulting from his recently-completed photographic near-IR survey of the northern Milky Way: some 75 new, relatively faint S stars (AJ **100**, 569) and highly reddened stars likely to be distant and luminous (AJ **103**, 263).

V. Blanco and colleagues continued very productive in identifying new C stars, both in the Milky Way (52.155.043) and in the LMC (52.156.008).

Houk is nearing completion of classification for Vol. 5 of the *Michigan Spectral Catalogue* which will cover $-12° < \delta < +5°$ and will appear in 1994.

A considerable number of small sky-area Hα-emission surveys were published; any attempt for me to summarize them would neither be complete nor very useful. We sorely need a new edition of something like Wackerling's catalogue of emission-line stars (3.041.018).

4 AUTOMATIC SPECTRAL CLASSIFICATION (M.J. Kurtz)

A Special Session on automated spectral classification was held as part of the Comm 45 meetings during the General Assembly in Buenos Aires. Kurtz, Garrison, Kurucz, Golay, and Jaschek spoke; abstracts of the talks are in Trans. IAU XXIB.

The group from Tartu and Abastumani of Malyuto, Pelt, Shvelidze and collaborators continues to improve their methodology for the study of galactic structure. Recent work appears in Baltic Astron **1**, 526, as well as in the conference proceedings "Data Analysis in Astron IV", IAU Symp 161, and MK50.

Kurtz+ (in OPOS) have demonstrated their variance-weighted metric technique on a sample of B stars from the Michigan Survey plates. They show that the variance weighting substantially improves the classification over direct comparison methods. They introduce the technique of reflattening to eliminate residual errors in the continuum removal. LaSala has continued this work with a collection of slit spectra of G dwarfs and has automated classification accuracies similar to the B star sample of about one subtype. He presents his work in MK50.

Penprase (ApJS **83**, 273) has classified about 100 spectra using the unweighted distance technique, with photometry providing the first approximation. A group from Athens and Trieste (Hantzios+, in IAU Symp 161) have extended their criteria evaluation methods from low- to moderate-resolution (830 Å/mm) spectra to study some of the less crowded regions of the LMC.

Beauchemin, Borra, and collaborators at Laval University have improved their low-resolution classification based on synthetic photometry (AJ **103**, 405) and have applied a statistical cluster analysis to the data and have recovered the known types (AJ **105**, 1587).

Three groups are investigating the use of back-propagating, artificial neural networks for automated classification. Von Hippel will review this research in MK50 as well as talk about his own work in this area (also in a preprint to appear in MNRAS). He is using Michigan Survey plates, digitized at the APM, in collaboration with researchers at Cambridge; their mean error, using only the line information, is currently about 2 subtypes.

Weaver has applied the neural network technique to moderate-resolution, near-IR spectra of W-R stars (BAAS **22**, 848), O-M dwarfs, to get a mean classification accuracy of about 2 subtypes (BAAS **22**, 859), and to A stars, where he shows very good temperature and luminosity errors (in MK50). Smareglia+ will show results from their neural network experiments,with mean classification accuracies of about 2 subtypes, at Astronomical Data Analysis Software and Systems III.

5 CLASSIFICATION FROM EXTRA-ATMOSPHERIC SPECTRA (J. Rountree)

This reporting period has seen a considerable increase in efforts at classification in the far UV largely due to the continuing success of the IUE satellite. It is to be expected that this activity will grow in the future as more recently launched instruments, such as HST/GHRS, build up a substantial archive of standard stars. Classification in the far IR is currently limited to the IRAS archive but may also be expanded with the launch of ISO. Much attention is being paid to the methodology of classification in non-traditional wavelength regions as well as to the classification of individual objects. Recently published spectral atlases of UV spectra illustrate the morphology of stellar classes in this region.

5.1 Methodology

Prinja (52.114.050) measured the EWs of photospheric absorption lines in 300 O and B stars and verified the temperature dependence of Si III λ 1299 in O stars and of Si II λ 1265 in B stars; he found that C III/O IV varies with luminosity class in O stars. Fanelli+ (52.114.063) examined the spectral morphology of 218 O-K stars in the 1900-3200 Å region; they concluded that there is excellent temperature discrimination in the mid-UV but only moderate luminosity discrimination for classes III-V. Slettebak (priv. comm.) recently measured the EWs of 17 spectral lines in Be and B-type standard stars and concluded that, while there is no significant difference between the photospheric line spectra of Be and B normal stars of the same type, the wind lines are generally stronger in the Be stars. He found the ratios Si II/III, C II/III, Al II/III, and Fe II/III are sensitive to spectral type.

Faragianni+ (52.114.016, .101) demonstrated the importance of using IUE short-wavelength spectra in classifying λ Boo stars. Walborn (MK50) found that the O V λ 1371 wind profile is a signature of spectral type O3 in the UV.

Rountree+ (52.114.096, 53.114.079) defined a set of photospheric-line criteria for two-dimensional classification of B stars in the UV using high-resolution IUE spectra and MK standards. Imadache+ (52.036.258) devised an automated classification of low-dispersion IUE spectra based on multi-variate data analysis to detect abnormal objects.

5.2 Individual Objects

Hack+ (52.112.112) classified Herbig Be-Ae stars by comparing low-resolution IUE spectra with standards. Monterinos et al (52.115.019) used UV data to show that the spectral type of FG Sge is F6-7 I. Prinja+ (54.122.036) observed the LMC supergiant HDE 269216 and classified it as B8 Ia from the IUE line spectrum. However, the shallow continuum led them to conjecture that the star is an S Dor-type variable.

Tovmasyan+ presented observations with the Glazar telescope in the associations Per OB1, Sco OB1, and Cyg OB1 (52.152.006); in Ori (52.114.046, .147); in Cas, Gem, and Mon (54.114.042, .088) and in Pup (55.152.006). These observations at 1640 Å reveal the presence of a number of previously unclassified hot companions some of them subdwarfs (52.117.165, 53.117.020).

Bopp+ (53.117.107) classified three composite systems as F II + B interacting binaries by observing the UV spectrum of the hotter component. Parsons+ (ApJ **412**, 814) continued this work by classifying 22 late B and early A companions of late-type giants and supergiants. Evans pursued a similar program of determining the luminosities of Cepheids (53.122.070, 55.122.059) and yellow supergiants (AJ **105**, 1956) by classifying their hot companions. Akalin+ (55.117.095) used high-resolution IUE spectra to determine the spectral type of the hot component of δ Sge.

Walborn+ (ApJ **393**, L13) compared GHRS spectra of Mk 42 and R 136a with each other and with IUE standards to show that Mk 42 is of intermediate O3/WN type while R136a is a stronger WN. Landsman+ (ApJ **395**, L21) observed three UV-bright stars in ω Cen with UIT and found that the corresponding low-resolution IUE spectra were similar to those of normal B2-B6 stars but with weaker lines.

5.3 Atlases

Wu+ produced an enlarged and updated version of *The IUE Ultraviolet Spectral Atlas* (IUE *Newsletter* 43), a widely used compendium of low-relsoution standard spectra. Rountree+ prepared an atlas of B-type standard spectra (NASA Reference Pub 1312) for use in classifying high-resolution IUE spectra on their two-dimensional UV system. Walborn+ (MK50) worked on an atlas of high-resolution B-type spectra to illustrate the gradual disappearance of stellar wind features as a function of temperature and luminosity.

Doazan+ (52.002.091, 54.114.097) published an atlas showing the spectral characteristics of Be stars at far UV and at optical wavelengths. Ewald+ (52.114.104) described an atlas of high-dispersion Mg II profiles of standard stars of spectral types B0-G9.

5.4 Infrared Classification

Volk+ (54.133.006), using data from the slitless, low-resolution spectrometer on IRAS, classified 486 point sources into nine groups, based on their spectral morphology.

6 CLASSIFICATION USING MULTI-COLOUR PHOTOMETRY (M. Grenon)

In the following, references are restricted to papers presenting new calibrations or substantial sets of new data.

6.1 Wide-band systems

A catalogue of homogeneized UBV data covering 40 years of UBV photoelectric photometry has been produced by J.C.Mermilliod. It contains cross-identifications, spectral classifications and references for 102000 stars. Cousins BVRI photometry is used by Dawson+ (AJ **103**, 2063) to classify high-proper motion stars. Infrared IJHK photometry is used by Leggert to rank 322 red dwarf stars in metallicity (ApJS **82**, 351). The solar (R-I) index and the relation Teff/(R-I) for F-K dwarfs are revisited by Taylor (PASP **104**, 500).

6.2 Medium-band systems

Geneva system:

Hauck+ (A&A **269**, 403) produced Teff calibrations valid for the various types of chemically peculiar stars, *e.g.* magnetic Ap, He-weak, Hg-Mn, and Am using Geneva colours. The resolving power for stellar classification of the multidimensional space of Geneva colour-indices was investigated by Guarinos+ (1992, ESO Conf Proc **43**, 307). A catalogue of 400 photometric and spectroscopic standards, covering the whole HR-diagram, was produced by these authors. Intrinsic colours for O to early A-type stars are defined by Cramer (A&A **269**, 457).

Strömgren system:

Lodén is submitting a catalogue of uvby and Hβ data for 620 B-A3 stars observed in selected directions in the galactic plane. Gray+ (A&AS **87**, 541) published a catalogue of over 600 A-G supergiants. Gray extended the empirical calibrations of the Stömgren photometry to F and G supergiants (A&A **252**, 237); temperature, gravity and colour excess are precisely determined for Ia-II supergiants Cepheids of solar composition. The calibration was extended to population I A0-F5 supergiants by Gray (A&A **265**, 704); the accuracies are 150° in Teff and 0.02 in E(b-y). A combination of Strömgren and Cousins BVRI data will allow the determination of the [M/H] ratio for luminous A-G stars. The uvby system is being increasingly used for objects outside the original A-G5 range, namely for late-type giants and dwarf novae, *cf.* Echevarría+ (A&A **275**, 201).

Washington system:

Geisler+ (ApJ **102**, 1836) have significantly improved the metal abundance calibration. It is now valid for the whole range of [Fe/H] -4.0 to +0.5 and is applied to field and cluster giants of Population I and II.

DDO system:

The properties of the DDO system with respect to CNO abundance anomalies were tested by Tripiocco+ (AJ **102**, 744) with synthetic spectra. Metal-weak giants towards the galactic center were classified by Morrison (AJ **105**, 539).

Vilnius system:
The system was redefined for CCD detectors and used for faint object classification (A&AS **88,** 87, A&AS **95,** 51). Bartkevičius+ used it for the classification of Population II supergiants (Baltic Astr **1,** 216). An automatic method of absolute magnitude determination was proposed by Smriglio+ (A&A **266,** 634).
Near-infrared eight-color system:
Wing has used his narrow-band system, which measures continuum points and bands of TiO, CN, and VO to study the red supergiants of the LMC. More than 100 M stars in the central 25 square degrees have been observed, and Wing+ (BAAS **24,** 773) have used the rich LMC cluster NGC 2100 to calibrate the CN-based luminosities. An important result is the discovery of a large number of K supergiants (BAAS **24,** 785).

6.3 Other systems
The theoretical colours and isochrones for the Hubble Space Telescope FOC colour system are predicted by Paltoglou+ (MN **253,** 449). Empirical relations with other systems, namely the UBVRI, are investigated by Harris+ (AJ **101,** 677). The Hipparcos broad-band Hp and Tycho BT,VT system provide a new tool for the classification of the 96000 stars measured in the three bands. For O to K stars of known MK type, accurate E(B-V) are deduced. For M stars, an equivalent spectral type and a colour excess are precisely determined.

7 CATALOGUES AND ATLASES (D. Egret)

An increasing number of tools are now provided to astronomers world wide in order to gain access to catalogues and atlases in computer-readable form, either on magnetic media, or directly through the electronic networks (with the help of new data retrieval tools such as WAIS, gophers, WorldWideWeb, etc. see Heck+, "Adding Intelligence to Information Retrieval", Kluwer Acad Publ, 1993). The major data centers (CDS, Strasbourg, NSSDC, NASA-Greenbelt and a number of other regional data centers) are currently offering direct on-line access to the data as well as to complementary data analysis tools. This tends to replace the traditional means of data distribution (magnetic tapes, diskettes, CD--ROMs, or paper versions). Simultaneously, the editors of the major astronomical journals are moving to make tabular data more quickly available in digital form (CD-ROMs and "ftp" services being the adopted solutions). These efforts are complemented by ambitious projects of general information systems for astronomy such as ADS (Astrophysics Data System in the U.S.) and ESIS (European Space Information System) for which public releases have been made available in 1993. In the following we mention some of the most significant recent catalogues related to stellar classification.

7.1 Spectroscopic catalogues
Renson+ have now released their General Catalogue of Ap and Am stars (A&AS **89,** 429). And at the same time, Renson (Liège), Kobi and North (Lausanne) have proposed a database of Ap stars developed on a personal computer using the previous catalogue as a starting point (A&AS **89,** 61).
A new catalogue by Cayrel+ (A&AS **95,** 273) of [Fe/H] determinations appeared.

7.2 Photometric catalogues
The leading group in the field is located at Institut d'Astronomie de l'Université de Lausanne and Observatoire de Genève. They are maintaining an impressive collection of photometric data, known as the General Catalogue of Photometric Data (A&AS **85,** 989). Mermilliod is keeping the Catalogue of Observations in the UBV System (about 100000 stars) up-to-date. A new edition of the Catalogue of Am Stars with known Spectral Types has been issued by Hauck (Bull Inform CDS 40, 19).

7.3 Compilation catalogues and atlases
The publication of the HIPPARCOS Input Catalogue, with a printed version (ESA SP-1136, 7 volumes), and a computer version (with some revisions, available from the data centers) provides a wonderful compilation of the astrophysical data (including spectral classification and multicolor photometry) available for some 120,000 stars of the HIPPARCOS observing program. It is to be noted that a large fraction of the multicolor photometry represents new observations.
A CD-ROM version is announced for the beginning of 1994. Significant progress has been made toward filling in MK types for the 5th edition of the Bright Star Catalogue. In addition to a variety of MK classifications from various sources, the revised Perkins types for cooler stars (Keenan, 50.002.011) have been included. While the 4th edition of the BSC contained MK types for 87% of the catalogue stars, the latest version of the 5th edition (1993) includes such classifications for 93%.

COMMISSION 46: TEACHING OF ASTRONOMY (ENSEIGNEMENT DE L'ASTRONOMIE)

PRESIDENT: L. Gouguenheim
VICE-PRESIDENT: J. R. Percy
ORGANIZING COMMITTEE: J. Fierro, M. Gerbaldi, L. Houziaux, S. Isobe, C. Iwaniszewska, J.M. Pasachoff, R.R. Robbins, Aa. Sandqvist, D.G. Wentzel.

1. INTRODUCTION

It should be reminded that astronomy education is important for two main reasons. First, because it affects the quality of the training of the next generation of astronomers; and second because it affects the general public's understanding and appreciation of what astronomers do.

Commission 46 is a commission somewhat out of the ordinary and can in some ways be considered as an extended arm of the Executive Committee of the IAU. National Committees or Adhering Bodies in each member country nominate one person per country to be the national representative to the Commission. The National Representative maintains liaison between the Commission and the home country, writes a national triennial report on the development of astronomy education and votes on financial matters. Other IAU astronomers, with a special interest in education, can become regular individual members of the Commission. Non-IAU members can be invited by the Commission to serve for one triennial term, which is renewable.

Commission 46 considers that one of its major duties is to contribute to the rise of astronomy in the developing countries. The International School for Young Astronomers, the Visiting Lecturer Program, the Travelling Telescope are examples of such activities administered by the Commission. These items are discussed in more detail below.

2. INTERNATIONAL SCHOOL FOR YOUNG ASTRONOMERS (ISYA)

One of the most important activities of the Commission is the International School for Young Astronomers (ISYA) with Professor Donat G. Wentzel as the secretary assisted by Michèle Gerbaldi; it is held twice during each triennial period. The dominant goal of ISYA is to advance astronomy in astronomically developing countries. This is accomplished by increasing the visibility of astronomy in the host country, creating science and astronomy infrastructure by seeking lecturers and participants from physics to help the local astronomers both from within the country as well from abroad. The isolation of individual participants is broken down by ISYA, partly through its socializing activities and its stipulation that participants must come from several neighbouring countries.

The 19th IAU-sponsored ISYA was held in China July 19 - August 8, 1992, as reported in IAU Bulletin 69 and Commission 46 Newsletter #36. The ISYA met for two weeks in Beijing and one week at Beijing Observatory's Xinglong Observing Station. There were 30 participants, including 13 Chinese, 3 from Korea, 2 each from Hong Kong,

J. Bergeron (ed.), Reports on Astronomy, Vol. XXIIA, 535–538.

India, Indonesia, Vietnam, 1 each from Russia, Ukraine, Turkey, Israel, Holland, USA; 9 of them women.

Beijing Observatory supported the local costs, with additional support from the Chinese Astronomical Society, Chinese National Natural Science Foundation, the Third World Academy of Sciences and the Chinese Academy of Sciences. Travel support was provided by the IAU, ICSU, and UNESCO. Since nearly all faculty and participants succeeded in obtaining low fares, a substantial number of participants could be offered travel support.

Participants' background ranged from just-finished bachelor's degrees to almost finished doctoral thesis, from small universities with essentially no astronomy resources to established research centers. Accordingly, the lectures ranged from simple and didactic presentations up to a graduate-level survey providing mainly "research flavor": G. S. D. Babu (India) on galactic structure, M. Gerbaldi (France) on stellar spectroscopy, Zhai Disheng (Beijing Observatory) on evaluation of light curves and models for binary stars, D. Wentzel (USA) on concepts of MHD and solar physics, Wang Jing-Xiu (Beijing Observatory) on solar magnetic observations, and G. Bisnovatyi-Kogan (Russia) on stellar interiors and evolution. Highlights for participants from small institutions were occasional friendly scientific debates among faculty. The lectures were supplemented by practical exercises, including two evenings of observing at Xinglong, and many other scientific and social activities.

All foreign faculty stayed the entire three weeks, so that all could get to know the background of the participants and there were ample opportunities to meet with the participants. All participants except one also stayed the entire period. The ISYA was conducted in English. Students at first too shy to talk English soon lost all hesitancy. They made good friends among participants from other countries, and some scientific collaborations have resulted.

The 20th ISYA is being organized for January 3 to 21 1994 in Pune, India, at the Inter-University Center for Astronomy and Astrophysics (IUCAA). Preliminary plans envisage the 21st ISYA in Egypt in October 1994.

The rules governing the planning of and reporting on an ISYA have been revised.

3. VISITING LECTURERS PROGRAM (VLP)

The Visiting Lecturers Program (VLP) is intended to provide a series of visiting lecturers to astronomically developing countries in which such an effort can help establish a longer-term astronomical establishment. The IAU is to provide the lecturers' travel and the host institution supports the lecturer while giving a course for which students receive regular academic credit. The coodinator of the VLP is Professor Donat. G. Wentzel.

Three such VLP have been authorized. For the VLP in Paraguay and Peru, the goal was one Spanish-speaking visitor per VLP per year, sufficiently frequently so that each visitor can build on the material of the previous lectures. This has not been possible.

For Peru, the internationally publicized unrest during the last several years has made it impossible to find any visitors. The VLP has been cancelled. Nevertheless, astronomy at Universidad Nacional Mayor de San Marcos has become active during the VLP and remains so. The better physics students have been drawn toward astronomy, two with Licentiatura in Physics have a speciality in astrophysics, and three young astronomers are studying in Brazil. There is a Revista Peruana de Astronomia y Astrophysica, and ESA is offering a computer system.

The VLP in Paraguay at Universidad Nacional de Asuncion hosted Dr Armando Arellano Ferro from the National University of Mexico during September and October 1992. Additional support was provided by the Italian Dante Alighieri Society and Centro Latino Americano de Fisica, Brazil. Arellano Ferro gave a course on "optics and instrumentation" which included much fundamental astronomy and astrophysics. The photometer from the Travelling Telescope was used many evenings at a private observatory, and some students continued observing after the end of the course. Several former VLP students now teach astronomy in Paraguayan schools. One student is now continuing training in Mexico. Plans are under way for the next (and according to contract last) visiting lecturer during April - June 1994.

Since 1989 the China VLP has been dormant and no visiting lecturers have gone to China under IAU sponsorship and funding.

It is a pleasure to thank Don for the tremendous good work he is doing for astronomy education, in particular for both ISYA and VLP.

4. THE TRAVELLING TELESCOPE (TT)

The purpose of the Travelling Telescope (TT) is to provide astronomers, teachers and advanced students in the developing countries with practical training and experience in astronomy. It consists of a Celestron-8 telescope, instruments and accessories, in sturdy wooden shipping crates. John Percy is the project coordinator.

The instruments from the TT (photometer, spectrograph and camera) accompanied Armando Arellano Ferro to Paraguay for his VLP course on "Optics and Instrumentation in Astronomy" (see previous section). Eduardo Parini, a local amateur astronomer, generously made his Celestron-8 telescope available to the students and faculty of the VLP. Since autumn 1992, the students have been carrying out a program of photoelectric photometry of variable stars using the TT photometer. In the summer of 1993, a copy of the software package RPHOT, which takes, reduces, plots and analyzes photometric data was acquired. It was obtained for only a fraction of the normal price, thanks to the help and generosity of NOLTEN Software and OPTEC Inc. The package will be sent to Paraguay, and will become a permanent part of the TT instrumentation.

Sending only the instruments of the TT on an assignment is certainly simpler and cheaper than sending the telescope as well. Nevertheless, the complete TT package is available for ISYA's, VLP's etc. Applications for use of the TT should be sent to the President of Commission 46.

5. NEWSLETTER

The Commission publishes a newsletter approximately twice a year, which is mailed throughout the world. Between July 1990 and June 1993, the following issues have been published; #30 (November 1990), #31 (National Reports 1988-90; April 1991), #32 (Astronomy Educational Material Addendum 1988-90: Material in French and in Spanish; April 1991), #33 (September 1991), #34 (Astronomy Education and Instructional Aids; February 1992), #35 (August 1992) and #36 (February 1993).

Issue #34 and earlier were printed and distributed from Belgium, by Professor Leo Houziaux. It is a pleasure to thank Leo for this and his many other contributions to astronomy education. Issue #34 was an excellent 81 page compilation by Professor H. J. Augensen, made available to us by permission of Springer-Verlag, Heidelberg.

The newsletter is presently being edited, printed and distributed from Erindale Campus, University of Toronto, by Professor John R. Percy, who thanks his secretarial staff for doing most of the work.

The increasing cost of printing and mailing the newsletter requires the Commission to think very carefully about who should receive it; this will be a major topic for discussion by the Organizing Committee.

The newsletter welcomes any short items on astronomy which would be of broad interest to astronomy educators around the world.

6. IAU COLLOQUIUM #105 "THE TEACHING OF ASTRONOMY"

The proceedings of IAU Colloquium #105, sponsored by Commission 46, which were edited by Professors Jay M. Pasachoff and John R. Percy are now available in paperback, at a very reasonable cost, from Cambridge University Press.

This colloquium was extremely successful: the opportunity to gather 162 astronomers from 31 countries for four days of papers, panels and discussions was recognized as a valuable step toward worldwide interchange of ideas and experiences about teaching processes. Such an IAU-sponsored colloquium should be repeated at intervals: this will be another topic for discussion by the Organizing Committee.

7. MISCELLANEOUS

Thanks to Alan H. Batten, the chairman of the IAU Working Group for the Worldwide Development of Astronomy, and the financial support of IAU, a successful two-day meeting of schoolteachers from the Indian subcontinent, was organized in August 1993, in conjunction with the IAU Asian-Pacific Meeting in Pune, India. It is a pleasure to note that this meeting, similar to those organized by Commission 46 at each General Assembly, was held for the first time at a Regional Meeting, following the recommendations of our Commission.

COMMISSION 47: COSMOLOGY (COSMOLOGIE)

PRESIDENT: R. B. Partridge
VICE PRESIDENT: J.V. Narlikar
ORGANIZING COMMITTEE: J. R. Bond, A. Dressler, M. Geller, D. Jauncey, M. Rees, H. Reeves, K. Sato, S. Shandarin, P. Shaver, A. Szalay

INTRODUCTION
(R. B. Partridge)

As the following reports show, there has been a great deal of activity in cosmology in the past three years, particularly, I would say, on the observational side. For instance, measures of Hubble's constant H_o, both physical and geometrical, are pouring in (see Trimble's contribution below), and we should soon realize the promise of the Hubble Space Telescope in refining values of H_o. Interest in a non-zero cosmological constant Λ has resurfaced, driven in part by new observational results.

The gravitational lensing of distant sources has provided new means to probe both the properties of high redshift objects and the distribution of mass (both luminous and "dark" matter) in the lensing objects. As Fort suggests in his article below, observations of lensed sources may povide a value of H_o as well. A useful summary is provided by the volume *Gravitational Lensing* edited by Mellier *et al.* (Springer–Verlag, 1990).

A major recent trend in cosmology is the deepening and refining of surveys of extragalactic objects. I refer to both counts (see Colless below) and redshift surveys of galaxies, IRAS sources, QSO, etc. A great deal has been learned about the evolution and large-scale structure of the Universe from these systematic surveys (see Dekel's paper in our report of 3 years ago). This survey work is one mark of the maturing of our field.

Another, I would suggest, is the increasing interconnection between cosmology and microphysics. Limits on the number of neutrino families derived from nucleosynthesis calculations and the observations of primordial 4He have now been confirmed by accelerator experiments (see Denegri *et al.* , *Rev. Mod. Phys.* **62**, 1, 1990). The need for "dark" matter, probably non–baryonic, has been established by a number of astronomical observations (see *IAU Symposium 117*) and laboratory searches are now underway for possible candidate particles (see Sadoulet below).

Important information on the origin and evolution of large–scale structure in the Universe was provided by the U. S. COBE satellite (see G. Smoot *et al.* , *Ap. J. (Letters)*, **396**, L1, 1992). An instrument on that satellite detected, at a level of one part in 100,000, the fluctuations in the cosmic microwave background that astronomers had been searching for for more than two decades. The observations and their interpretation are summarized below. The Proceedings of the "Texas Symposium" held in Berkeley, California in Dec. 1992 will also be useful.

I will close by noting the role of advanced technology–large–aperture telescopes, CCD detectors, advanced cryogenic particle detectors and space platforms–in many of these advances in cosmology.

J. Bergeron (ed.), Reports on Astronomy, Vol. XXIIA, 539–564.

Cosmological Parameters
(V. Trimble)

The fundamental parameters are H (expansion rate), Ω (density in units of the critical density), and Λ (cosmological constant). Derivable from these are q_o (deceleration, k (dimensionless radius of curvature), and to (present age). The fraction of the critical density in baryons, Ω_b, is generally also regarded as important. The standard model thus defined is alive and well[1]. The triennium has seen a number of limits on fifth forces, distance dependence of G, alternatives to general relativity, and tired light effects[2], but no new positive results. Polarization of the community on large vs. small values of H has persisted; some progress has been made in constraining L from statistics of gravitationally lensed QSOs (see following paper); and it has become increasingly clear that none of the tricks yet thought of can reconcile Ω_b=1 with the apparent primordial abundances of H^2, He^3, He^4, and Li^7.

The traditional angular diameter test, interpreted as a constraint on q_o, favors $q \simeq 1/2$ from radio data[3] and nearly rules it out from optical data[4]. But disentangling evolutionary effects from cosmological effects remains sufficiently difficult as to cast doubts on all results[5]. Other geometrical tests present other problems, and must be properly corrected for effects of gravitational lensing, interstellar scintillation, and internal absorption in galaxies[6].

The most direct measurements of Ω come from analyzing the peculiar velocities of galaxies in clusters or on larger scales. Reported values range[7] from 0.03 to 1.0, with many equally carefully calculated ones in between[8]. Disagreement persists both on whether the necessary data actually exist and on whether the calculations have been done correctly[9]. We can say with some confidence that the universe is not closed by gravitational radiation (at some wavelengths), by black holes (of some masses) or by 17 keV neutrinos, though the tau neutrino remains intriguingly close to the necessary mass on some hypotheses[10] (see paper by Sadoulet here). Not surprisingly, less direct methods (incidence of cluster substructure, merger rates) also yield values from about 0.1 to 1.0.

Globular clusters remain a better constraint on to than either local stellar populations or the nucleocosmochronometers, because the halo came first and the disk apparently formed inside out[11]. There is a real spread in cluster ages[12], but only the oldest matter for our purpose. Standard physics and a solar mix of metals lead to cluster ages up to 16-19 Gyr[13]. Popular ways of decreasing this are non-solar mixes (high 0/Fe or alpha nuclei), inward diffusion of helium, and unexpectedly efficient convection[14]. Each allows a decrease of at most 1-3 Gyr (groups do not entirely agree). The modified ages remain impossible if both H and q_o are large. Moving the galactic center away is a less popular but possible solution[15]. A similar, though less sharp, conflict arises in interpreting the evolving colors of galaxies unless the universe is rather old[16].

The universe is not closed with baryons. The triennium began with some enthusiasm for $\Omega_b = 1$ and inhomogeneous nucleosynthesis, but this has gradually faded[17], and other ways of pushing on the standard model have not met with much success[18]. D/H and Li^7 both put limits on Ω_b,[19] which must fall between 0.01 and 0.09 for any reasonable choice[20] of H. The primordial abundance of He is currently the most worrisome quantity. No set of standard model parameters can accommodate Y_p below 0.237. This is within the error bars of most, but not all, recent determinations[21]. The standard model abundances are a viable starting point for galactic chemical evolution[22], and may permit everything in the universe to have the baryon/total mass ratio of typical rich clusters but not that of Hickson compact groups[23]. Among remaining alternative models, non-zero lepton number is perhaps most worthy of further exploration[24].

To reconcile globular cluster ages with $k = 0$ and $H \geq 60$ km/sec per Mpc requires a cosmological constant in the (dimensionless) range 0.74-1.0[25]. So large a value puts lots of volume at large redshifts, and so predicts large numbers of gravitational lenses and a rapid increase in numbers with z[26]. In fact lensing seems to be quite rare and the distribution with z rather flat, implying[27] $\lambda \leq 0.9$, though evolutionary effects still need to be taken into account[28]. Nevertheless, the cosmological constant is sufficiently back in favor to make recent calculations of look back time, luminosity distance, etc. in such universes a useful contribution[29].

The value of the Hubble constant is not (as Sandage wrote a generation ago) presently well known. At least 39 values were published in the triennium, based on nearly as many calibrators. The list, in order of publication (or at least arrival on the library shelves), with the references in the same order[30] is: 81-94 (PNe in Virgo), 46 ± 10 (SN Ia's), 78 or 64 (surface brightness fluctuations in Virgo galaxies), 70 ± 7 (Virial mass = stars + gas in Mich 160), 52 (Virgocentric infall = 168 km/s), 75-100 (SN Ia's calibrated in Virgo),

50 ± 11 (time delay in lensed QSO), 75-95 ± 20 (lensed arcs), 70 (luminosity function of globular clusters in Virgo E's vs. MW), 55 (decaying neutrinos as dark matter), 82 ± 7 (surface brightness fluctuations in Fornax and Eridanus), 42-69 (time delay in lensed QSO 0957, radio data), 85 (brightest stars in Virgo), 100 (PNe and surface brightness fluctuations), 92 ± 20 (Tully-Fisher in Coma), 63.5 (QSO 0957 angular diameter distance), 40 ± 9 (Sunyaev-Zel'dovich effect in A665), 80 (Cepheids), 90 ± 17 (Tully-Fisher), 60-80 (SN Ia model), 85 (PNe and Cepheids), 86 ± 14 (SN 1937C in IC 4182, calibrated on red supergiants), 61 ± 10 (SN Ia model), 65 ± 10 (SN 1972C in NGC 5253 calibrated on HII's), 80 ± 15 (Tully-Fisher), 85 (red and blue supergiants in Virgo), 60 ± 10 (SN II photosphere models), 80 ± 7 (many), 76 ± 9 (Virgo-Coma distance moduli), 44-63 (shape of galaxy luminosity function), 51-77 (brightest cluster galaxies), 45 ± 9 (SN 1937C in IC 4182 calibrated on Cepheids), 43 ± 11 (ScI angular diameters), 53 ± 10 (surface brightnesses of MW and Virgo), 45 ± 12 (angular diameters vs. M31), 10 (large scale streaming), 32, 68, or 87 (lensing of 0957), 61-66 (Tully-Fisher with halos removed), 54-64 (SN Ia model light curves). The unweighted median is 67 km/s per Mpc, and I am willing to be that this is within 35 km/s per Mpc of the truth. One thoughtful reviewer[31] has reached a similar conclusion.

References

1. P. J. E. Peebles *et al.* 1991, Nature 352, 768.
2. E. G. Adelberger *et al.* 1990, Nature 347, 261. C. S. Speake *et al.* 1990, PRL 65, 1957; T. P. Kirshner *et al.* 1991, Ap. J. 373, 665. A. R. Sandage J-M. Perlmutter 1990, Ap. J. 370, 455. E. Fischback C. Talmadge 1992, Nature 356, 207. C. M. Will 1992, Ap. J. 393, L59; S. Carusotto *et al.* 1991, PRL 69, 1722.
3. K. I. Kellermann 1993, Nature 361, 134.
4. E. Giraud 1990. A A 234, 20.
5. E. H. Gudmundsson . E. Rgnvaldsson 1990, M.N.R.A.S. 245, 463.
6. H. Yoshida M. Omote 1992. Ap. J. 388, L1. C. A. Hasivassilious 1991. Ap. J. 381, 1; B. Wang 1990, Ap. J. 383, L37.
7. P. S. Pellegrini L. N. Da Costa 1990; Ap., J. 357, 408. A. Nusser A. Dekel 1990, Ap. J. 405, 428.
8. A. J. S. Hamilton 1993, Ap. J. 406, L47; M. Plionis R. Valdarnini 1991, M.N. 249, 46.
9. T. Watanabe S. Inagaski 1991, P.A.S.J. 43, 413; J. A. Peacock 1992, M.N.R.A.S. 258, 581.
10. J. L. Mortara *et al.* 1993, PRL 70, 394; A. Dar 1991, Science 250, 1529.
11. S. A. Josey N. Arimoto 1992, A A 255, 105. G. J. Mathews D. N. Schramm 1993, Ap. J. 404, 468.
12. R. Buonano *et al.* 1991, A. J. 100, 1811. R. J. Dickens *et al.* 1991, Nature 351, 212.
13. E. Paez *et al.* 1990, A AS 84, 615. M. Catelan J. de Freitas Pachecho 1992, A A 258, L5.
14. C. R. Profitt G. Michaud 1991, Ap. J. 371, 584; Y-W. Lee 1992, A. J. 104, 1780. M. A. Wood 1992, Ap. J. 386, 539; P. R. Durrell W. E. Harris 1993, A. J. 105, 1420.
15. A. R. Walker 1992, Ap. J. 390, L81.
16. A. Buzzoni *et al.* 1993, Ap. J. 410, 499.
17. G. J. Mathews *et al.* 1990, Ap. J. 358, 36; T. Kajino *et al.* 1990, Ap. J. 364, 7. L. H. Kawano *et al.* 1991, Ap. J. 372, 1; B. Link 1992, PRL 68, 2425. G. J. Mathews *et al.* 1993, Ap. J. 404, 476.
18. E. D. Carlson 1990, PRL 65, 2225. N. Yu. Gnedin J. P. Ostriker 1992, Ap. J. 400, 1.
19. A. R. McCullour 1992, Ap. J. 390, 213; H. Reeves 1993, A A 269, 166. J. L. Linsky *et al.* 1993, Ap. J. 402, 694.
20. M. S. Smith *et al.* 1993, Ap. J. S 85, 219.
21. J. A. Baldwin *et al.* 1991, Ap. J. S. 74, 586; K. A. Olive *et al.* 1991, Ap. J. 380, L1. B. E. J. Pagel A. Kazlauskas 1992, M.N. 256, 49p; A. Campbell 1992, Ap. J. 401, 157. G. J. Mathews *et al.* 1993, Ap. J. 403, 65; J. Melnick *et al.* 1992, A A 253, 16.
22. B. V. Vainer *et al.* 1991, Sov. Astron. 35, 111.
23. A. Babul N. Katz 1993, Ap J 406, L47; N. Makino Y. Suto 1993, P.A.S.J. 45, L13. T. J. Ponman D. Bertram 1993, Nature 363, 51.
24. A. Bianconi *et al.* 1991, A A 341, 343.
25. M. Moles 1991, Ap. J. 382, 369.
26. M. Fukugita *et al.* 1990, M.N.R.A.S. 246, 24p.
27. E. L. Turner 1990, Ap. J. 365, L43; D. Maoz *et al.* 1993, Ap. J. 402, 69.
28. L. M. Krauss D. N. Schramm 1993, Ap. J. 405, L43.

29. B. Feige 1992, Astron. Nach. 313, 139.

30. G. H. Jacoby *et al.* 1990, Ap. J. 356, 373. G. Tammann B. Leibundgut 1990, A A 236, 9. J. L. Tonry *et al.* 1990, A. J. 100, 1416. L. Staveley-Smith *et al.* 1990, Ap. J. 364, 23. A. Sandage G. Tammann 1990, Ap. J. 365, 1. M. Fukugita C. Hogan 1991, Ap. J. 368, L11. G. Rhee 1991, Nature 350, 216; G. Soucail B. Fort 1991, A A 243, 23. W. E. Harris *et al.* 1991, Ap. J. S. 76, 115. P. Salucci D. W. Sciama 1991, A A 243, 341. J. L. Tonry 1991, Ap. J. 373, L1; D. H. Roberts *et al.* 1991, Nature 352, 43. J. L. Tonry *et al.* 1991, Science 253, 138; J. Peacock 1991, Nature 352, 378. M. Fukugita *et al.* 1991, Ap. J. 376, 8; R. Narayan 1991, Ap. J. 378, L5. M. Birkinshaw *et al.* 1991, Ap. J. 379, 466. B. F. Madore W. L. Freedman 1991, P.A.S.P. 103, 923. G. Gavazzi *et al.* 1991, Ap. J. 382, 19; T. Shigeyama *et al.* 1992, Ap. J. 386, L13. M. Dopita *et al.* 1992, Ap. J. 389, 27; M. J. Pierce *et al.* 1992, Ap. J. 390, L45. D. Branch 1992, Ap. J. 392, 35. M. Della Valle J. Melnick 1992, A A 257, L1. I. Ichikawa M. Fukugita 1992, Ap. J. 394, 61. M. J. Pierce *et al.* 1992, Ap. J. 393, 523; B. P. Schmidt 1992, Ap. J. 395, 366. G. H. Jacoby *et al.* 1992, P.A.S.P. 104, 599. S. van den Bergh 1992, P.A.S.P. 104, 861; R. Duemmler 1992, A A 264, 1. T. R. Lauer M. Postman 1992, Ap. J. 400, L47. A. Sandage *et al.* 1992, Ap. J. 401, L7; A. Sandage 1993, Ap. J. 402, 3. S. Kimeswenger *et al.* 1993. A A S 97, 501. A. Sandage 1993, Ap. J. 404, 419; E. R. Harrison 1993, Ap. J. 405, L1. G. M. Bernstein 1993. A. J. 105, 816; P. Salucci *et al.* 1993, M.N.R.A.S. 262, 392. I. Dominguez 1993, P.A.S.P. 105, 559.

31. J. Huchra 1992, Science 256, 321.

Gravitational Lenses
(Bernard Fort)

With the improvement of astronomical instrumentation, Gravitational Lensing (GL) has become a remarkable technique for observational cosmology. The number of publications in major journals is $\sim 10^2$ per year, so that here we can only report a small, representative sample to highlight some major issues.

The cosmological applications of GL were recently reviewed (Blandford and Narayan 1992), and the state of the theory and observations prior to 1991-2 is thoroughly presented in the first monograph on the subject (Schneider *et al.* 1992). In addition, there are the proceedings of biannual meetings on Gravitational Lenses (e.g. Surdej *et al.* 1994, Kayser *et al.* 1992).

To start from multiple quasars, there have been new subarcsecond seeing surveys with the HST and with ground based radio and optical telescopes. Several new candidates have been discovered : 1208+1011 (Bahcall *et al.* 1992a, Magain *et al.* 1992), HE 1104-1805 (Wisotzki *et al.* 1993), B1422+231 (Patnaik *et al.* 1993), and BRI0952-01, which is the most distant gravitationally lensed QSO with a redshift $z = 4.5$ (MacMahon *et al.* 1992).

GL statistics constrain the density parameter $\Omega_{10:12} < 0.02$ for intergalactic compact objects of mass $M_{ico} \sim 10^{10} - 10^{12} M_\odot$ (Surdej *et al.* 1993 and references therein), and $\Omega_{7:9} < 0.4$ for $M_{ico} \sim 10^7 - 10^9 M_\odot$ (Kassiola, Kovner and Blandford 1991); and QSO variability provides constraint $\Omega_{-3:-1.5} < 0.1$ for intergalactic compact objects of mass $M_{ico} \sim 10^{-3} - 10^{-1.5} M_\odot$ (Schneider 1993). The evidence against a large, $(\Lambda/(3H_o^2) \sim 0.9)$ cosmological constant (Fukuhgita *et al.* 1992 and references therein) has also been confirmed by HST observations (Maoz *et al.* 1993), and by statistical analysis of uncertainties (Kochanek 1993, ApJ, in press). Lensing of extended radio-sources can be a better cosmographical probe than the lensing of unresolved sources (Kochanek 1993). Clear statistical evidence that galactic cores are fairly small has been presented (Wallington and Narayan 1993, Kassiola and Kovner 1993).

There was further progress in investigations of individual interesting multiple QSO cases: — Radio observations of MG0414+0534 confirmed that one of the (formerly) three components was a close pair separated by 0.5", as it should be for a simple lens (Katz and Hewitt 1993); and optical deep imaging revealed the lensing galaxy (Schechter et Moore 1993). New VLA data (Patnaik *et al.* 1993) has been published on B0218+35.7 which is favorite candidate for the determination of H_o, due to its two compact and strongly variable components, its radio ring, and the smallest separation known, $0''.335$. Observations of UM673 provided a measure of the size of the intervening Lyman alpha clouds (Smette *et al.* 1992).

The K-band observations are particularly promising because the seeing is better, and high redshift galaxies can be more luminous. In particular, they made it possible to detect the deflecting galaxies in Q2016+112 (Heflin *et al.* 1991, Lawrence *et al.* 1993). There was also the important effort to extend the observations to near infrared in other cases, e.g Q2237+030 (Nadeau *et al.* 1991), PG1115 +080, H1413+117, and Q1429-008 (Hewitt *et al.* 1992, Annis and Luppino 1993), the radio ring MG1131+0456 (Annis 1992), and B1422+231 (Lawrence *et al.* 1993). It will also be necessary to perform IR spectroscopy of multiple QSOs which can have unusual R-IR spectrum like MG0414+0454 (Hewitt *et al.* 1992).

Two new radio rings , 0218+357 (O'Dea *et al.* 1992) and MG 1549+3047 (Lehar *et al.* 1993), raise their total number to 5. Rings provide a unique opportunity to model the mass distribution in the foreground galaxy and to reconstruct the extended source with an algorithm inverting the lens (Kochanek and Narayan 1992). It may be possible to observe $\sim 10^2\ deg^{-2}$ similar optical rings with the HST resolution (Miralda-Escude and Lehar 1992)

The first attempts (Kronberg *et al.* 1991) have been made to constrain mass distributions of galaxies using the facts that the change of polarisation of light by realistic GL is extremely small (Dyer and Shaver 1992), and that there is a tight coupling between the intrinsic polarization angle and the orientation in radio jets.

The procedure for determination of H_o requires direct measurement of the velocity dispersion of the lens galaxy. Such measurements have been undertaken for Q2237+0305 (Foltz *et al.* 1992) and Q0957+561 (Rhee 1991). The first discovered GL, Q0957+561, remains an attractive case for observing and modelling. The value of H_o obtained from modelling Q0957+561 is disturbingly low (Falco *et al.* 1991a), although an analyses of systematic errors (Kochanek 1991) relaxes this problem to $0 < H_o < 90 \pm 30$ (for other values of H_o see the paper by Trimble here). This lens appears to be quite complex, with the discovery of a large fold arc nearby (Berstein *et al.* 1993) and of a background group of galaxies at $z = 0.503$, in front of the QSO (Garret *et al.* 1992, Soucail and Angonin 1993). Incidentally, the magnification of the lens makes it possible to detect intrinsic X-ray emission of Q0957+561 (Jones *et al.* 1993). The effect of microlensing on estimates of the time delay has been investigated (Falco *et al.* 1991b). There is still a controversy on the value of the time delay, although the majority vote for $T = 1.45 \pm 0.04$ years (Press *et al.* 1992a,1992b, Oknyanskij and Beskin 1993, and references therein). If so, there is an interesting evidence that the compact radio source lags by 6.6 years behind the optical one, and that its size of the central radio source is not larger than that of the optical source, so that it can also be microlensed (Oknianskij and Beskin 1993).

The reported excess of foreground galaxies near highly luminous QSOs is still a matter of debate (Schneider 1992, Romani and Maoz 1992). However, there is a new interesting development: in 1990 Fugmann found correlation between galaxies from the Lick catalogue and $1Jy$ distant radio sources. Bartelman and Schneider (1992) studied this correlation further and found that it persists on an angular scale of 6 arcmin. They also found (Bartelmann and Schneider 1993a,b) that gravitational lensing by large scale structures can account for such correlations. An important factor in these correlations is the double magnification bias (Borgeest *et al.* 1991) which applies when the quasar sample has a double threshold, radio and optical, and the radio and optical fluxes are not correlated. Obviously the effect of smooth large-scale structure is too small, but superimposed granular distributions on the scale of clusters may significantly enhance the overall effect (Babul and Lee 1991, Jaroszinki 1991,1992).

Microlensing studies are becoming spectacular with large programmes of QSO monitoring (Surdej 1994) because it is a unique way to see whether there is some dark matter in the form of compact objects, ranging from massive black holes to brown dwarfs. The influences of source size (Refsdal and Stabel 1991), external shear (Mao 1992a), and transverse motion (Kayser 1990, Grieger *et al.* 1991, Wambsganss 1992, Witt 1993) have been theoretically investigated. It appears that for typical halos of galaxies, single microlens events are unlikely and result in a very complex pattern of moving caustics (Rauch *et al.* 1992). In particular, the effects of random motions of individual stars in a lensing galaxy are not negligible as compared to the effects of the the relative motion of the source, the lens galaxy and us (Kundic and Wambsganss 1993). Microlensing also provides information on the relative sizes of different regions in QSOs (red vs. blue continuum, or emission line regions), which in turn give critical constraints on accretion disk models (Wambsganss and Paczyński 1991a). However, puzzling observations do not allow to clearly disentangle intrinsic variability of QSOs from microlensing events in some cases: GC0248+430 (Borgest *et al.* 1991b), Q0957+561 (Schild and Smith 1991), 2237+03 (Racine 1992), Q2237+03 (Webster *et al.* 1991), AO0235+164 (Saust 1992), and 0414 +0534 (Angonin-Willaime *et al.* 1993).

The original idea of Paczyński (1986) to search for microlensing events from compact objects of the galactic halo in front of the large magellanic clouds (MACHO Searches) has been investigated theoretically (Nemiroff 1991, Gould 1993). Several monitoring programmes are under way in the southern hemisphere (Gould 1992, Griest *et al.* 1991, Vidal-Madjar *et al.* 1993), and similar surveys are suggested towards the galactic bulge (Paczyński 1991) and M31 (Crotts 1991). So far, the null detection within the accumulated period of observations do not permit us to constrain the density of compact objects for any range of masses.

There is also a proposal to search for a gravitational lensing effect of a massive black hole in the centre of our galaxy (Wardle and Yusef-Zadeh 1992). Another interesting proposal (Wambsganss and Paczyński 1992) is to search for possible milliarcsecond patterns in the radio image of multiple QSO, which can be produced by massive black holes in the primary lens galaxy.

In 1986 Paczyński pointed out that gamma-ray bursters at cosmological distances could be a signature of massive compact objects and Black Holes (Mao 1993). Ways to recognize various ranges of lens masses have been considered (Paczyński 1991b, Blaes and Webster 1992, Narayan and Wallington 1992, Gould 1993, Mao 1993, Wambsganss 1993)

Recent lists of large gravitational arcs surveys in rich clusters of galaxies confirm the frequent occurence of arcs in distant X-ray clusters (Surdej Ed. 1994). More than 20 cases are known but deep spectrophotometric data are published for only 12 of them : A370 (Kneib *et al.* 1993), A963 (Ellis *et al.* 1991), A1689 (Tyson *et al.* 1990), A2390 (Pello *et al.* 1991), A2218 (Pello *et al.* 1992), MS0302+1658 (Mathez *et al.* 1992), Cl0024+1654 (Kassiola *et al.* 1992), the cluster-lens MS2137-23 with the first radial arc (Fort *et al.* 1992), MS2053-04, MS1621+26, and MS0440+02 (Luppino *et al.* 1992, 1993), and the most distant cluster-lens Cl2236-04 (Melnick *et al.* 1993).

It is still difficult to learn much about the evolution of clusters and about the properties of the background sources from comparing theoretical predictions with the observed statistics (Miralda 1993a,b; Wu and Hammer 1993), since the sample is small and heterogeneous. Nevertheless, the subarcsecond widths of large arcs indicate a steep gradient of the potential which can be understood either in terms of a small core in a pseudo-isothermal distribution or (apparently preferably) of a deVaucouleurs radial profile (Hammer 1991).

This conclusion is in accord with the modelling of the radial arc in cluster MS2137-23 (Mellier *et al.* 1993) which constrained the core radius (30 *kpc*). The second fold arc system showed that the distribution of dark matter follows the ellipticity and orientation of the large faint halo of the bright elliptical galaxy. The bipolar potential of A370 has been similarly recovered by using newly discovered gravitational pairs (Kneib *et al.* 1993). The cluster Cl0024+1654 must also be dominated by a compact massive halo (Kassiola *et al.* 1993) despite the fact that it does not yet have a cD galaxy at its center. In these three clusters, a comparison of model predictions for extra images with deep photometry and high resolution observations appears to be quite a powerful method to recover the mass distributions. The analytic formulae for GL with elliptic mass distributions (Kassiola and Kovner 1993) arrived just in time for astrometric testing of multiple image predictions.

Besides the gross mass distribution, orientation patterns of arclets beyond the radii of large arcs reveal clumpiness of the potential near some bright elliptical cluster members (Pello 1992, Fort 1992). The triple arc in 0024+1654, broken by a perturbation of a cusp by a group of such E galaxies, permitted a direct estimate of their masses (Kassiola, Kovner and Fort 1992). The steepness of the cluster potentials can also be probed statistically by the frequency of rings around individual cluster galaxies (Kochanek and Blandford 1991).

Probable redshifts of individual arclets can be roughly constrained from their ellipticity and orientations (Kochanek 1990, Miralda 1991). The colour redshifts of arclets (Fort 1992) and redshifts of luminous arcs (Mellier *et al.* 1991, Pello *et al.* 1992) are in agreement with the deepest redshift surveys of field galaxies. Spectrophotometry of magnified arc(let)s would have a large impact on the studies of evolution and formation of distant objects (Mellier *et al.* 1991, Smail *et al.* 1993). The spectacular HST observation of a gravitational pair of compact blue objects in AC114 (Smail *et al.* 1993) and of A370 (Kneib *et al.* 1993) at CFHT may offer the opportunity to study objects with (unmagnified) $B \sim$ 29-30 (Miralda and Fort 1993).

Arclets can constrain the general geometry of the cluster potential, although the method is limited by poor statistics and by the effects of intrinsic ellipticity and orientation of the sources (Tyson *et al.* 1990, Kochanek 1990, Miralda-Escude 1991 a,b, Kaiser and Squires 1993). Deep weak-lensing measurements have been already started up to 15 arcmin from the center of Cl0024+1654 and in a blank field (Bonnet *et al.* 1993). Such observations should improve spectacularly with the availability of large CCD mosaics, since large scale mass inhomogeneities can produce coherent distortions of a few percent in a pattern on a scale of $\sim 1°$, and should constrain cosmological models for LSS formation (Blandford *et al.* 1991, Miralda-Escude 1991 and Kaiser 1992).

References

Angonin-Vuillaine, M.,C., Vanderriest, C., Hammer, F., Magain, P., 1993 A&A preprint

Annis, J.: 1992, ApJ 391, L17

Annis, J., and Luppino, G.,A.: 1993, ApJ 407, L72

Babul, A., Lee M.H.: 1991 MNRAS, 250, 407

Bahcall, J.N. *et al.* : 1992a, ApJ 392, L1; 1992b, ApJ 400, L51

Bartelman, M., Schneider, P.: 1993a, A&A 271, 421; 1993b, A&A, 268, 11

Berstein, G.,M., Tyson, T. A. , Kochanek, C.S.: 1993, ApJ 105, 816

Blandford, R., Narayan R.: 1992 Annu. Rev. Astron. Astrophys. , 30, 311

Blandford,R.D., Saust, A.B., Brainerd, T.G., Villumsen, J.V.: 1991, MNRAS 251,600

Blaes, O.M., Webster, R.L.: 1992, ApJ, 391, L63

Bonnet, H. *et al.* : 1993, A&A preprint

Borgeest, U. *et al.* , 1991a, A&A 243, 93; 1991b, A&A 251, L35

Crotts, A.P.S.: 1992 ApJ 399, L43

Dyer, C.C., Shaver, E.G.: 1992, ApJ 390, L5

Ellis, R., Allington-Smith, J., Smail, I.: 1991, MNRAS 249, 184

Falco, E.E. *et al.* : 1991a, ApJ 372,364 ; 1991b, MNRAS, 251, 698

Foltz, B., Hewett, P., Webster, R., Lewis, G.: 1992, ApJ 386, L43

Fort B.: 1992, Lecture note in Physics 406, 267, ref Kayser 1992
Fort, B., O. Le Fevre, O., Hammer, F.: 1992, ApJ 399, L125
Fukugita M., Futamase, T., Kasai, M., Turner, E.L.: 1992, ApJ 393,3
Garrettt, M.A., Walsh, D., Carswell, R.F.: 1992, MNRAS 254, 27
Gould, A. :1992, ApJ 392, 442 ; 1993, ApJ 404, 451
Grieger, B., Kayser, B., Schramm, T.: 1991,A&A 252,508
Griest, K., *et al.* : 1991, "The Macho collaboration" ApJ 372, L79
Hammer, F.: 1991 ApJ 383, 66
Hammer F., *et al.* : 1993 A&A, 250, L5
Heflin, M.B., Gorenstein, M.V., Lawrence, C.F., Burke, B.F.: 1991, ApJ 378, 519
Hewwitt, J.N., Turner, E.L., Lawrence, C.R., Schneider, D.P., Brody, J.P.:1992, AJ 104, 968
Jaroszinski M., 1991, MNRAS 249, 430 ; 1992, MNRAS 255, 655 ; 1993, ApJ 410, 21
Kaiser, N.: 1992, ApJ 388, 272
Kaiser, N., Squires, G.: 1993, ApJ 404, 441
Kassiola, A., Kovner, I.: 1993 ApJ preprint
Kassiola, A., Kovner, I., Blandford, R.: 1991, ApJ 381,6
Kassiola, A., Kovner, I., Fort, B.: 1992, ApJ 400, 41
Katz,C.A., and Hewitt, J. N.: 1993, ApJ, 409, L11
Kayser, T., Schramm, T., Nieser, L. Eds.: 1992, V 406, Springer-Verlag
Kayser, R.: 1990, ApJ 357, 309
Kneib, J.-P., Mellier, Y., Fort, B, Mathez, G.: 1993, A&A 273, 367
Kochanek K.S., 1990, MNRAS 247, 135; 1991a, MNRAS 371, 289; 1991, ApJ 382, 58, 1992b, ApJ 384,1; 1993 ApJ preprint
Kochanek, K.S., Blandford, R.: 1991, ApJ 375, 492
Kochanek K.S., Narayan , R.,1992, ApJ 401, 461
Kronberg, P., Dyer, C., Burbidge,E.N., Junkkarinenn, V.: 1991, ApJ 367, L1
Kundic, T., Wammbsganss, J.: 1993, ApJ 404, 455
Lawrence, C. *et al.* : 1992, MNRAS, 259, 5P
Lawrence, C., Neugebauer, G., Matthiews, K.: 1993, AJ 105, 17
Lehar J., Langston G.I., Silber, A., Lawrence, C.R., Burke, B.F., 1993 AJ 105, 847
Luppino, G.A., Gioia, I.M.: 1992, A&A 265, L9
Luppino, G.A., Gioia, Annis, J., Le Fevre, O.,Hammer, F.: 1993, ApJ preprint
MacMahon, R., Irwin, M;, Hazard, C., Corringan, R., 1992, Gemini 36,1
Magain, P.,Surdej, J.,Vanderriest, C.,Pirenne, B.,Hutsembekers, D.: 1992, A&A,L13
Mao, S.: a,1992, ApJ, 389, L41 ; b,1993 ApJ 402, 382,
Maoz, D. *et al.* : a, 1992, ApJ 394, 51; b, 1993, ApJ 402,69
Hjorth, P., Villemoes, L., Florentin-Nielsen, R.: 1992 A&A 255, L20
Mathez, G., Fort, B., Mellier, Y., picat, J.-P., Soucail,G.: 1992, A&A 256, 343
Mellier, Y., Fort, B., Soucail, G., Mathez, G., Cailloux, M.: 1991, ApJ 380,334
Mellier, Y., Fort B., Kneib, J.-P.: 1993, ApJ, 407, 33
Melnick, J., Altieri, B., Giraud, E.: 1993, A&A 271, L5
Miralda-Escude, J.: 1991a, ApJ 370, 1 ; 1991b, ApJ 380, 1
Miralda-Escude, J., Lehar, J.: 1992, MNRAS 259, 31P
Miralda-Escude, J.: a, 1993 , ApJ, 403, 497 ; b, 1993, ApJ 403, 509
Miralda-Escude, J., Fort, B.: 1993, ApJL preprint
Nadeau, D. *et al.* : 1991, ApJ 376, 430
Narayan, R.,Wallington,1992, S.: ApJ 399,368
Nemiroff, R.J.: 1991, A&A 247,73
O'Dea, C.P. *et al.* : 1992, AJ, 104,1320
Oknyanskij, V., Beskin, G., 1992, cf Surdej Ed. 1993
Paczyński, B.: 1986, ApJ 304,1; 1991a, ApJ 371, L63
Paczyński, B. 1991b, Acta Astron., 41, 257
Patnaik, A.R.*et al.* : 1992, MNRAS, 259, 1P; 1993, MNRAS 261,435
Pello, R. *et al.* : 1991, 366, 405; 1992, A&A 266,6

Press,W., Rybicki, G., Hewitt, J., 1992a , ApJ 385, 404; 1992b, ApJ 385, 416
Racine, R.: 1992, ApJ 395, L65
Rauch, K.P., Shude, M., Wambsganss, J., Paczinski, B.: 1992, ApJ 386, 30
Refsdal, S., Stabell, R.: 1991, A&A, 250, 62
Roberts, D., Lehar J., Hewitt, J.N., Burke, B.F.: 1991, Nature 352,43
Romani R.W., Maoz, D.,1992, ApJ 386, 36
Rhee, G.: 1991, Nature 350, 211
Saust, A.B.: 1992, A&A 266, 101
Schild, R.E., Smith, R.C.: 1991, AJ 101,813
Schechter P.L., Moore, C.B.: 1993 AJ, 105, 1
Schneider, P.: 1992, A&A 254, 14
Schneider, P., Ehlers, J., Falco, E.: 1992,"Gravitational Lenses", A&A Library, Springer-Verlag
Smail, *et al.* : I.: 1991, MNRAS 252, 1993a,b, MNRAS preprints,
Smette, A. *et al.* : 1992, ApJ 389, 39
Soucail, G., Angonin-Vuillaime, M., C., Vanderriest, C.: 1993 A&A preprint
Surdej J. Ed.: 1994, "GL in the Universe", Liege, June 21- 25, Sringer-Verlag, in press
Surdej, J. *et al.* :1993, AJ, 105, 2064
Tyson, A., Valdes, F., Wenk, R.A.: 1990, ApJ 349, L1
Van Drom, E. *et al.* : 1993, A&A preprint
Vidal-Madjar, A.: 1993,cf Surdej Ed. 1994
Wallington, S., Narayan, R.: 1993, ApJ preprint
Wambsganss, J.: 1992 ApJ, 392,424; 1993, ApJ 406, 29
Wambsganss, J., Paczynnski, B.: 1991, AJ, 102, 864 ; 1992, 397, L1
Wardle, M., Yusef-Zadeh, F.: 1992, ApJ 387, L65
Webster, R.L., Ferguson, A.M., Corrigan, R.T., Irwin, M.J.: 1991, AJ 102, 1939
Wisotzki,L., Kohler, T.,Kayser, R., Reimers D.: 1993, A&A preprint
Witt, H.J.: 1993, ApJ 403, 530
Wu, X.P, Hammer, F.: 1993 MNRAS 262, 187

High Redshift Galaxies and Galaxy Evolution

(Matthew Colless)

This report provides an overview of research into high redshift galaxies and galaxy evolution from an observational perspective over the period 1990 to 1993. Only references to the literature from this period are given in order to provide a snapshot of current research. Several related topics (e.g. high-redshift galaxies as probes of cosmological models, galaxy interactions and evolution at low redshift, and gravitational lensing) are dealt with elsewhere in this volume.

Counts and colours of faint galaxies

The general form of the number counts of faint galaxies in passbands ranging from U to K has been confirmed by several studies[58] to very faint limits (B≈26–29, K=≈20–23). Good agreement is obtained in the overlap regime between the wide-field photographic studies to B≈23 and the fainter small-field CCD studies[56,78]. The number counts are steeper in the bluer bands and flatter in the red, corresponding to a general blueing of the galaxy population at fainter apparent magnitudes. This is directly observed in the colour-magnitude distributions[78,99], although these studies suggest that some earlier work[51] may have overestimated the degree to which the galaxies become bluer at the faintest magnitudes. The very steep, almost Euclidean, slope for the bright (B<20) number counts[72] has been corroborated. In general, the counts in the blue are found to be steepest at bright magnitudes and then flatten out at fainter magnitudes to be only slightly steeper than the red counts[68,99]. Upper limits on the number of $z\sim 3$ galaxies have been derived from the scarcity of galaxies detected in B or R that are not detected in U due to the Lyman limit break[51,99]. A new development has been number counts in the K passband at 2.2μm. Complete counts[50] to K=17 and fainter counts[28,27] to K=23 have extended into the IR the trend to a flatter slope (and less evolution) fo- redder passbands.

At the faintest levels the surface density of galaxies becomes very high, reaching 10^5 deg^{-2} mag^{-1} at B≈26 and at K≈22. Some claims for a levelling off of the number counts at fainter magnitudes have been made[27], but these have yet to be confirmed. High angular resolution will be important in obtaining reliable results to yet fainter limits, as recent studies with the NTT[83] and HST[93] have shown.

Redshift surveys of faint field galaxies

Field redshift surveys[20,21] to B=22.5 showed that to this limit the bulk of the galaxies making up the factor of two excess in the number counts (compared to no-evolution models) are at rather low redshifts, typically less than z=0.5, and are not high-redshift primeval galaxies. The redshift distributions are similar to the no-evolution predictions, with no extra populations of local dwarfs or galaxies at high redshift, suggesting that the excess counts must result from a process resembling number density evolution. There is also a significant excess in the number of galaxies with enhanced star-formation, indicated by their [OII] equivalent widths[20,10]. Further magnitude-limited surveys[68,67,22,104] to B=24 and I=22 have extended these results, finding mean redshifts $\bar{z}\approx0.4$–0.6 and few galaxies with $z\geq1$.

The large excess of faint galaxies found in the counts together with the relatively low redshifts, particularly of the bluest objects, are difficult to reconcile with pure luminosity evolution of the galaxy population[68,21]. Possible problems with completeness and bias in the redshift surveys and an inadequate knowledge of the galaxy populations contributing to the luminosity function at low redshift have been cited as ways of reconciling the observations with mild luminosity evolution models[58,59]. Other authors invoke either a non-zero cosmological constant[113,67], strong merging of the galaxy population with associated star-formation[89,10], or a population of low-luminosity galaxies that go through a phase of rapid star-formation which is quenched by the exhaustion or expulsion of the gas supply, so that the objects fade beyond detection at low redshift[29,3]. High-resolution imaging of faint galaxies[49,67,23,104] suggests that interactions are a significant factor, but at present the nature of galaxy evolution at moderate redshifts remains an open question[58,52].

Galaxies in high-redshift clusters

There has been continuing interest in the properties of galaxies in high-redshift clusters. A second sizable catalogue of high-redshift clusters appeared[25], while several groups investigated the high fraction of blue galaxies in distant clusters (the Butcher-Oemler effect), both through photometric studies[87] and spectroscopically[36,37]. These studies confirmed the strong increase in the blue fraction with redshift and demonstrated that it is largely composed of emission-line and E+A galaxies that are much scarcer in clusters at low redshift. These galaxies appear to have been involved in recent or ongoing starbursts[37]. The Butcher-Oemler effect has also been studied in X-ray and infrared-selected cluster samples[47,2], confirm that the effect is not due to optical selection biases with redshift. Radio-selected groups[1] link the Butcher-Oemler effect in rich clusters to the field: all groups have a large blue fraction at high redshift, with the richer groups evolving most rapidly to redder colours at low redshift. There is now also evidence[36,2] for evolution of the *reddest* galaxies in clusters with z=0.5–1, with few cluster galaxies as red as present-day ellipticals by $z \approx 1$.

Some progress towards understanding the mechanisms of galaxy evolution in clusters has come from new high-resolution imaging, both from the ground[64] and from preliminary studies using HST[24,35]. These studies show a large fraction of the blue galaxies to be interacting, although whether interactions are the primary mechanism[64] or act jointly with ram pressure stripping and other effects[35] remains to be resolved. HST imaging has also turned up a serendipitous candidate for the highest redshift cluster[38]—a group of faint blue objects which may be associated with a QSO at $z \approx 2$.

Clustering of faint galaxies

Several recent observational studies[43,81,86,26,90] have determined the angular two-point correlation function $w(\theta)$ for faint galaxies with $B \lesssim 24$–26 and $R \lesssim 23$–24. The major finding of all these studies is a low $w(\theta)$ amplitude for such faint galaxies, with the blue galaxies more weakly clustered than the red. Although the observations are now well-attested, debate continues on their significance. Early claims that the low amplitude of $w(\theta)$ implied the faint blue galaxies were so weakly clustered that they could not be identified with normal galaxies at low redshift[43] have been disputed on the basis that the assumed evolution of gravitational clustering underestimated that actually observed in N-body simulations[77,114]. The other difficulty in interpreting these results lies in the unknown redshift distribution of galaxies at such faint magnitudes. Currently it is unclear whether $w(\theta)$ for faint galaxies can be entirely explained by strong evolution of clustering, or whether the faint blue galaxies are required to be at higher redshifts or more weakly clustered than expected. The deepest data on the *spatial* two-point correlation function $\xi(r)$ reaches $B \approx 22$ ($\bar{z} \approx 0.2$) and shows no evidence for weaker clustering[19].

High-redshift radio galaxies

The state of the controversy in 1990 concerning the age and star-formation history of high-redshift radio galaxies is captured by reviews in the Hubble Symposium[66,106,16]. The existence of an old (1 Gyr) stellar population was deduced from the SEDs and the small scatter in the K-z relation[66], putting the galaxies' redshifts of formation at $z \geq 5$. On the other hand, the 'alignment effect', whereby the optical and IR emission tends to be aligned along the radio axis, indicated a physical relation between the (young) radio source and the visible galaxy[8,16]. Other work showed that the spectra might also be consistent with younger ages[15,89].

Further imaging has supported the tendency for the IR continuum to align with the radio axis[44], but the elongation in the IR is generally less than in the optical, consistent with an aligned flat-spectrum component (contributing ~10% of the IR light) overlying a more symmetric red component[88,74]. This applies to 6C 1232+39, but for this object the blue continuum and Ly α are not contiguous as expected if they traced the star-forming regions[41]. Measurements[102,94] showing 5–20% polarization of the rest-frame UV continuum, aligned perpendicular to the radio axis, imply that a large fraction of this component is light scattered by dust or electrons from the AGN. Numerous models exist for the origin and scattering mechanism of this non-stellar light[30].

Recent work has shown that reddening and the contribution of emission lines have caused the IR flux (and hence the age of the stellar population) to be over-estimated[73,40]. In the case of B2 0902+34, at z=3.4, it is claimed that once corrected for such effects the SED is consistent with a galaxy observed during its initial burst of star-formation[42]. The lack of any 4000Å break in high-redshift radio galaxies also argues against the presence of an old stellar population[53].

Absorption line systems

The rarest QSO absorption line systems are the damped Ly α absorbers, which have high HI column densities ($N(HI)z \sim 2 \times 10^{20}$ cm^{-2}) and low velocity dispersions resembling those of present-day galactic HI disks[60,109,105,107]. The mass in these systems is comparable to the luminous mass in spiral galaxies today; however their incidence is 2–4 times greater than expected[60] if the size and number density of spirals were the same at $z \sim 2.5$ as at z=0. Abundances and dust-to-gas ratios in the absorbers are down by an order of magnitude compared to the local ISM[85,79]. The evidence for evolution in the properties of damped Ly α systems is weak[60], although an increase in the number density for $z \sim 3$ has been suggested[107]. Some galaxies associated with Ly α absorbers have been discovered through their Ly α, [OII], Hβ or CO emission[46,69,110,14,71], and appear consistent with large star-forming disks. Recent observations suggest that the clustering of Ly α-emitting galaxies around Ly α absorbers, and of the absorbers themselves, is strong over scales of several Mpc, in contrast to the weak clustering of faint galaxies[108,48].

Considerable observational effort has gone into the Lyman limit (LL) systems and metal line (MgII and CIV) absorbers which have column densities N(HI)$\sim 10^{15}$ cm^{-2}. Large samples have now been studied[61,100]. The MgII and LL systems show no evolution in comoving number density, but the CIV absorbers show strong evolution at z$\sim$2 due to a marked change in the strength of the line and the sample selection[100,98]. This may mark the era of halo formation[97], although contrary results indicate strong evolution of LL systems and CIV absorbers due to changes in the ionization level[61]. The complex velocity structure[63] of the line profiles on scales 50–300 km s^{-1} suggests either clouds within a halo or multiple smaller halos[62]. A decrease in the number of velocity components (clouds) at larger impact parameters is observed[62], supporting the halo model. The equivalent width of the MgII systems decreases with time[100], implying a decrease in the the number of clouds per halo[84]. The MgII and CIV systems are also clustered[84,100] on scales of several Mpc.

The galaxies associated with MgII absorbers have been identified[112,6,111] with a success rate of up to 80% and are generally bright, with L$\sim 0.2L^*$. The observed impact parameters to the QSO line of sight for the associated galaxies[6,100] are typically 30-50 h^{-1} kpc, although larger impact parameters up to 270 h^{-1} kpc have been found[112,80,39]. The former implies covering factors for the absorbing gas near unity; the latter much lower values. The impact parameter scales only weakly with luminosity[6,98]. By contrast, searches for MgII absorption in galaxies at known redshifts have failed in a significant number of cases, and find no deficit of fainter galaxies[5,39]. Recently the first z$\sim$1 galaxies associated with MgII absorbers have been reported[7]. These objects will undoubtedly yield important clues to the history of halo gas in normal galaxies[17,98], especially when low-redshift samples are further studied with HST[9].

Primeval galaxies

Despite extensive systematic searches[70,33,31], no population of 'classical' primeval galaxies (PGs, progenitors of normal luminous galaxies undergoing major bursts of star-formation at high redshift) has yet been identified. Several of these searches should have been able to identify 'classical' PGs out to z$\sim$5 if they were unobscured by dust[32].

Several examples of high-redshift galaxies not found through their radio emission or absorption line systems have emerged from searches for PGs[103] and gravitational lenses[75,4,76,82], and from studies of QSO companions[101,65] and hosts[54,55]. One particularly interesting object at z=2.3, F10214+4724, was first identified as an ultraluminous IRAS galaxy[91]. With an IR-dominated total luminosity of 10^3L*, and an H_2 mass estimated from CO observations[11,96] of 10^{12}M$_\odot$, F10214+4724 appears to be a good PG candidate. Further observations have detected submillimetre flux[18], shown the H_2 to be extended[57,92,13] and detected large amounts of neutral gaseous carbon[12]. Although the estimated H_2 and dust masses have been revised downwards[34,95], the data still imply an enormous star-formation rate (1000 $M_\odot$/yr)[45].

References

1. Allington-Smith J.R., Ellis R.S., Zirbel E.L. & Oemler A., 1993, ApJ, 404, 521
2. Aragón-Salamanca A., Ellis R.S., Couch W.J. & Carter D., 1993, MNRAS, 262, 764
3. Babul A. & Rees M.J., 1992, MNRAS, 255, 346
4. Bahcall J.N., Hartig G.F., Jannuzi B.T., Maoz D. & Schneider D.P., 1992, ApJ, 400, L51
5. Bechtold J. & Ellingson E., 1992, ApJ, 396, 20
6. Bergeron J. & Boisse P., 1991, A&A, 243, 344
7. Bergeron J., Cristiani S. & Shaver P.A., 1992, A&A, 257, 417
8. Bithell M. & Rees M.J., 1990, MNRAS, 242, 570
9. Bowen D.V. & Blades J.C., 1993, ApJ, 403, L55
10. Broadhurst T.J., Ellis R.S. & Glazebrook K., 1992, Nature, 355, 55
11. Brown R.L. & Vanden Bout P.A., 1991, AJ, 102, 1956
12. Brown R.L. & Vanden Bout P.A., 1992, ApJ, 397, L11
13. Brown R.L. & Vanden Bout P.A., 1992, ApJ, 397, L19
14. Brown R.L. & Vanden Bout P.A., 1993, ApJ, 412, L21
15. Chambers K.C. & Charlot S., 1990, ApJ, 348, L1

16. Chambers K.C. & Miley G.K., 1990, in Kron R., ed., Evolution of the Universe of Galaxies, Astron.Soc.Pac., San Francisco, p.373
17. Chokshi A., 1992, ApJ, 395, 21
18. Clements D.L., Rowan-Robinson M., Lawrence A., Broadhurst T.J. & McMahon R., 1992, MNRAS, 256, 35P
19. Cole S., Ellis R.S., Broadhurst T.J. & Colless M.M., 1993, preprint
20. Colless M.M., Ellis R.S., Taylor K. & Hook R.N., 1990, MNRAS, 244, 408
21. Colless M.M., Ellis R.S., Broadhurst T.J., Taylor K. & Peterson B.A., 1993, MNRAS, 261, 19
22. Colless M.M., Glazebrook K., Ellis R.S. & Broadhurst T.J., 1993, in Chincarini G. *et al.* , eds, Observational Cosmology, Astron.Soc.Pac., San Francisco, in press
23. Colless M.M., Schade D.J., Broadhurst T.J. & Ellis R.S., 1993, preprint
24. Couch W.J., 1993, in Chincarini G. *et al.* , eds, Observational Cosmology, Astron.Soc.Pac., San Francisco, in press
25. Couch W.J., Ellis R.S., Malin D.F. & MacLaren I., 1991, MNRAS, 249, 606
26. Couch W.J., Jurcevic J.S. & Boyle B.J., 1993, MNRAS, 260, 241
27. Cowie L.L., Gardner J.P., Hu E.M., Wainscoat R. & Hodapp K.W., 1993, preprint
28. Cowie L.L., Gardner J.P., Lilly S.J. & McLean I., 1990, ApJ, 360, L1
29. Cowie L.L., Songaila A. & Hu E.M., 1991, Nature, 354, 460
30. Daly R.A., 1992, ApJ, 399, 426
31. De Propris R., Pritchet C.J., Hartwick F.D.A. & Hickson P., 1993, AJ, 105, 1243
32. Djorgovski S., 1992, in de Carvalho R.R., ed., Cosmology and Large-Scale Structure in the Universe, Astron.Soc.Pac., San Francisco, p.73
33. Djorgovski S. & Thompson D.J., 1992, in Barbuy B. & Renzini A., eds, The Stellar Population of Galaxies, Kluwer, Dordrecht, p.337
34. Downes D., *et al.* , 1992, ApJ, 398, L25
35. Dressler A., 1993, in Chincarini G. *et al.*, eds, Observational Cosmology, Aston.Soc.Pac., San Francisco, in press
36. Dressler A. & Gunn J.E., 1990, in Kron R., ed., Evolution of the Universe of Galaxies, Astron.Soc.Pac., San Francisco, p.200
37. Dressler A. & Gunn J.E., 1992, ApJSuppl, 78, 1
38. Dressler A., Oemler A., Gunn J.E. & Butcher H., 1993, ApJ, 404, L45
39. Drinkwater M.J., Webster R.L. & Thomas P.A., 1993, preprint
40. Eales S.A. & Rawlings S., 1993, preprint
41. Eales S.A., Rawlings S., Dickinson M., Spinrad H., Hill G.J. & Lacy M., 1993, preprint
42. Eales S.A., Rawlings S., Puxley P., Rocca-Volmerange B. & Kuntz K., 1993, Nature, 363, 140
43. Efstathiou G., Bernstein G., Katz N., Tyson J.A. & Guhathakurta P., 1991, ApJ, 380, L47
44. Eisenhardt P. & Chokshi A., 1990, ApJ, 351, L9
45. Elbaz D., Arnaud M., Cassé M., Mirabel I.F., Prantzos N. & Vangioni-Flam E., 1992, A&A, 265, L29
46. Elston R., Bechtold J., Lowenthal J.D. & Rieke M., 1991, ApJ, 373, L39
47. Fabricant D.G., McClintock J.E. & Bautz M.W., 1991, ApJ, 381,33
48. Francis P.J. & Hewett P.C., 1993, AJ, 105, 1633
49. Giraud E., 1992, A&A, 257, 501
50. Glazebrook K., Peacock J.A., Collins C.A. & Miller L.A., 1993, preprint
51. Guhathakurta P., Tyson J.A.& Majewski S.R., 1990, ApJ, 357, L9
52. Guiderdoni B., 1993, in: *First Light in the Universe: Stars or QSOs?*, eds Rocca-Volmerange B., *et al.* , Editions Frontières, in press
53. Hammer F., Le Fèvre O. & Angonin M.C., 1993, Nature, 362, 324
54. Heckman T.M., Lehnert M.D., Miley G.K. & van Breugel W., 1991, ApJ, 381, 373
55. Heckman T.M., Miley G.K., Lehnert M.D. & van Breugel W., 1991, ApJ, 370, 78
56. Jones L.F., Fong R., Shanks T., Ellis R.S. & Peterson B.A., 1991, MNRAS, 249, 481
57. Kawabe R., Sakamoto K., Ishizuki S. & Ishiguro M., ApJ, 397, L23
58. Koo D.C. & Kron R.G., 1992, Ann Rev Astron & Astrophys, 30, 613
59. Koo D.C., Gronwall C. & Bruzual G., 1993, preprint
60. Lanzetta K.M., Wolfe A.M., Turnshek D.A., Limin Lu, McMahon R. & Hazard C., 1991, ApJS, 77, 1

61. Lanzetta K.M., 1991, ApJ, 375, 1
62. Lanzetta K.M., 1992, PASP, 104, 835
63. Lanzetta K.M. & Bowen D.V., 1992, ApJ, 391, 48
64. Lavery R.J., Pierce M.J. & McClure R.D., 1992, AJ, 104, 2067
65. Le Fèvre & Hammer F., 1992, A&A, 254, L29
66. Lilly S.J., 1990, in Kron R., ed., Evolution of the Universe of Galaxies, Astron.Soc.Pac., San Francisco, p.344
67. Lilly S.J., 1993, ApJ, 411, 501
68. Lilly S.J., Cowie L.L. & Gardner J.P., 1991, ApJ, 369, 79
69. Lowenthal J.D., Hogan C.J., Green R.F., Caulet A., Woodgate B.E., Brown L. & Foltz C.B., 1991, ApJ, 377, L73
70. Lowenthal J.D., Hogan C.J., Leach R.W., Schmidt G.D. & Foltz C.B., 1990, ApJ, 357, 3
71. Macchetto F., Lipari S., Giavalisco M., Turnshek D.A. & Sparks W.B., 1993, ApJ, 404, 511
72. Maddox S.J., Sutherland W.J., Efstathiou G.P., Loveday J. & Peterson B.A., MNRAS, 247, 1P
73. McCarthy P.J., Elston R. & Eisenhardt P., 1992, ApJ, 387, L29
74. McCarthy P.J., Persson S.E. & West S.C., 1992, ApJ, 386, 52
75. Mellier Y., Fort B., Soucail G., Mathez G. & Cailloux M., 1991, ApJ, 380, 334
76. Melnick J., Altieri B., Gopal-Krishna, Giraud E., 1993, A&A, 271, L5
77. Melott A.L., 1992, ApJ, 393, L45
78. Metcalfe N., Shanks T., Fong R. & Jones L.R., 1991, MNRAS, 249, 498
79. Meyer D.M. & Roth K.C., 1990, ApJ, 363, 57
80. Nelson B.O. & Malkan M.A., 1992, ApJS, 82, 447
81. Neuschaeffer L.W., Windhorst R.A. & Dressler A., ApJ, 382, 32
82. Pello R., Le Borgne J.F., Sanahuja B., Mathez G. & Fort B., 1992, A&A, 266, 6
83. Peterson B.A., D'Odorico S., Tarenghi M. & Wampler E.J., 1991, ESO
84. Petitjean P. & Bergeron J., 1990, A&A, 231, 309
85. Pettini M., Boksenberg A. & Hunstead R., 1990, ApJ, 348, 48 Messenger, 64, 1
86. Pritchett C.J. & Infante L., 1992, ApJ, 399, L35
87. Rakos K.D., Schombert J.M. & Kreidl T.J., 1991, ApJ, 377, 382
88. Rigler M.A., Lilly S.J., Stockton A., Hammer F. & Le Fèvre O., 1992, ApJ, 385, 61
89. Rocca-Volmerange B. & Guiderdoni B., 1990, MNRAS, 247, 166
90. Roche N., Shanks T., Metcalfe N. & Fong R., 1993, MNRAS, 263, 360
91. Rowan-Robinson M. *et al.* , 1991, Nature, 351, 719
92. Sakamoto K., Ishizuki S., Kawabe R. & Ishiguro M., ApJ, 397, L27
93. Schade D.J. & Elson R.A., 1993, AJ, 105, 1581
94. di Serego Alighieri S., Cimatti A. & Fosbury R.A.E., 1993, ApJ, 404, 584
95. Solomon P.M., Downes D. & Radford S.J.E., 1992, ApJ, 398, L28
96. Solomon P.M., Radford S.J.E. & Downes D., 1992, Nature, 351, 719
97. Steidel C.C., 1990, ApJS, 72, 1
98. Steidel C.C., 1993, in Galaxy Evolution: The Milky Way Perspective, Astron.Soc.Pac., San Francisco, in press
99. Steidel C.C. & Hamilton D., 1993, AJ, 105, 2017
100. Steidel C.C. & Sargent W.L.W., 1992, ApJS, 80, 1
101. Steidel C.C., Sargent W.L.W. & Dickinson M., 1991, AJ, 101, 1187
102. Tadhunter C.N., Scarrott S.M., Draper P. & Rolph C., 1992, MNRAS, 256, 53P
103. Thompson D.J. & Djorgovski S., 1991, ApJ, 371, L55
104. Tresse L., Hammer F. & Le Fèvre O., 1993, preprint
105. Turnshek D.A. & Bohlin R.C., 1993, ApJ, 407, 60
106. van Breugel W.J.M. & McCarthy P.J., 1990, in Kron R., ed., Evolution of the Universe of Galaxies, Astron.Soc.Pac., San Francisco, p.359
107. White R.L., Kinney A.L. & Becker R.H., 1993, ApJ, 407, 456
108. Wolfe A.M., 1993, ApJ, 402, 411
109. Wolfe A.M., Turnshek D.A., Lanzetta K.M. & Limin Lu, 1993, ApJ, 404, 480
110. Wolfe A.M., Turnshek D.A., Lanzetta K.M. & Oke J.B., 1992, ApJ, 385, 151

111. Yanny B. & York D.G., 1992, ApJ, 391, 569
112. Yanny B., York D.G. & Willliams T.B., 1990, ApJ, 351, 377
113. Yoshii Y., 1993, ApJ, 403, 552
114. Yoshii Y., Peterson B.A. & Takahara F., 1993, prerint

Results from the Cosmic Background Explorer (COBE) [1]

(G. F. Smoot and J. C. Mather)

Introduction

The structure of the Big Bang that started the expanding universe has been mapped for the first time by the Cosmic Background Explorer satellite (COBE). Designed, built, tested, launched, and analyzed by NASA's Goddard Space Flight Center and a nationwide science team[7], COBE captured the imagination with images of ripples in space-time, and tested the Big Bang theory with precise spectra of the heat radiation from the explosion. There are three main aims of the COBE: to find and measure the large scale fluctuations in the Cosmic Microwave Background Radiation (CMBR), to compare the spectrum of the CMBR with the theoretical prediction of a precise blackbody spectrum, and to measure or limit the cosmic infrared background radiation (CIB). These measurements are limited by foreground emissions. Interstellar free-free and synchrotron radiation, which perturb measurements of the CMBR anisotropy, are not well understood. Interstellar dust, which limits the precision of the spectrum distortion measurements, may include contributions from a cold dust component around 5 K with an order of magnitude more mass than the normal warm dust at 22 K. Interstellar ions and molecules, which emit in the far infrared, provide most of the cooling for the interstellar gas, and include the brightest emission line of the Galaxy (ionized carbon at 158 microns). Transiently heated dust grains and starlight show the structure of the Galaxy in new ways.

DMR and CMB Anisotropy

The origin of large scale structure in the Universe is one of the most important issues in cosmology. Currently the leading models for structure formation postulate gravitational instability operating upon a primordial power spectrum of density fluctuations. The inflationary model of the early Universe[2,21] produces primordial density fluctuations[22,24,39] with a nearly scale-invariant spectrum. Structure apparently forms as the result of gravitational amplification of initially small perturbations in the primordial mass-energy distribution, and non-baryonic dark matter seems necessary to provide sufficient growth of these perturbations. The determination of the nature of the initial density fluctuations then becomes an important constraint to cosmological models[8]. The discovery of the anisotropy in the cosmic microwave background radiation by the COBE DMR instrument[3,25,36,45] and the recent confirmation[18] mark the beginning investigation of these primordial fluctuations.

The Differential Microwave Radiometer (DMR) experiment is designed to map the microwave sky and find fluctuations of cosmological origin. For the 7° angular scales observed by the DMR, structure is superhorizon size so the spectral and statistical features of the primordial perturbations are preserved[32]. The DMR maps the sky at frequencies of 31.5, 53, and 90 GHz (wavelengths of 9, 5.7, and 3.3 mm). The frequency independence of the anisotropy is a strong argument that the anisotropy is in the cosmic microwave background radiation and not due to foreground Galactic emission or extragalactic sources. The confirming 'MIT' balloon-borne bolometer observations[18] have an effective frequency of about 170 GHz making the argument stronger. The typical fluctuation amplitude is roughly 30 μK or $\Delta T/T \sim 10^{-5}$ on a scale of 10° . The data appear consistent with a scale-invariant power spectrum with an uncertainty of ±0.6 in the exponent of the power law. The amplitude and spectrum are consistent with that expected for gravitational instability models involving nonbaryonic dark matter, and perhaps consistent with the measured large scale velocity flows. The angular power spectrum of the DMR maps was also computed, and is consistent with other methods of analysis of the maps[48,49].

The cosmic nature of the measured fluctuations was tested[4] by correlating the DMR maps with maps of Galactic emission, the X-ray background, Abell clusters, and other foregrounds. The "blamb" structure reported by the Relikt team[40] is not present in the DMR map.

Several medium and smaller angular scale experiments have announced new results. These include: the UCSB South Pole experiment[17,35], the Princeton Saskatoon experiment[44], the Tenerife collaboration[27,42], the CARA South Pole experiments: Python[13], and 'White Dish'[33], the Center for Particle Astrophysics MAX balloon-borne experiment[12,19,30], the MSAM balloon-borne experiment[9], the Owens Valley Radio-astronomy Observatory (OVRO)[31], the Roma balloon-borne experiment ULISSE[11], and the Australia Telescope[41]. In

[1]The National Aeronautics and Space Administration/Goddard Space Flight Center (NASA/GSFC) is responsible for the design, development, and operation of the Cosmic Background Explorer (COBE). Scientific guidance is provided by the COBE Science Working Group. GSFC is also responsible for the development of the analysis software and for the production of the mission data sets.

general, they report fluctuations at the 10^{-5} level. These measurements may distinguish models of structure formation or indicate the existence of cosmological gravity waves [10,37].

FIRAS and CMB Spectrum Observations

The Far Infrared Absolute Spectrophotometer (FIRAS) instrument compared the spectrum of the CMBR to that of a precise blackbody for the first time. It has a 7° diameter beamwidth, and covers two frequency ranges, a low frequency channel from 1 to 20 cm^{-1} and a high frequency channel from 20 to 100 cm^{-1}. Preliminary results[28] showed that the CMBR is consistent with a blackbody and that deviations are less than 1% of the peak brightness. The UBC rocket result[20] nearly immediately confirmed the FIRAS results. New FIRAS results show that limits on deviations from a blackbody are 30 times smaller: less than one part in 3000 of the peak intensity[14,15,29,47].

The absolute temperature of the cosmic background, T_o, was determined in two ways. The first uses the thermometers in the external calibrator and gives $T_o = 2.730$ K. The second calibrates the temperature scale from the wavelength scale, and gives 2.722 K for T_o. The adopted value is 2.726 ± 0.010 K (95% confidence[29]), which averages these two methods. Three additional determinations of T_o depend on the dipole anisotropy. The spectrum of the dipole anisotropy is sensitive to the assumed blackbody temperature. Since the velocity of the solar system with respect to the CMB is not known a priori, only the shape and not the amplitude of the dipole spectrum can be used. For FIRAS, this analysis[14] gives $T_o = 2.714 \pm 0.022$ K, while for DMR[26] it gives $T_o = 2.76 \pm 0.18$ K. The DMR data analysis keeps track of the changes in the dipole caused by the variation of the Earth's velocity around the Sun during the year. In this case the velocity is known, so T_o can be determined from the amplitude of the change in the dipole, giving $T_o = 2.75 \pm 0.05$ K.

The spectrum observations imply a tight limit on energy release in the early universe and strong support for the hot Big Bang model. From a redshift of about 10^6 to 10^3 no process can release electromagnetic energy at a level exceeding about 10^{-4} of that in the cosmic background radiation. Limits on the distortion parameters are $|y| < 2.5 \times 10^{-5}$ and $|\mu/kT| < 3.3 \times 10^{-4}$ with 95% confidence. The Comptonization parameter y restricts the possible thermal history of the intergalactic medium, which must not be very dense or very hot (less than $\approx$ 10 KeV). In addition, the FIRAS results limit energy release into the far infrared from Population III stars or evolving IRAS galaxies. In both cases, less than 1% of the hydrogen could have burned[47] after a redshift of 80, assuming $\Omega_{baryon}h^2 = 0.015$.

The FIRAS results can be combined and compared with other observations of the cosmic background spectrum. At this time the spectrum of the cosmic background is well described by a single temperature blackbody over four decades in frequency (or wavelength), without significant deviations. However, more precise measurements at long wavelengths could improve the COBE limits on μ by a factor of 10.

Non-cosmological results of the FIRAS include the determination of the mean far infrared spectrum of the Galaxy, and its decomposition into two components of dust emission and 9 spectrum lines[45]. The lines of [N II] and [C II] have been further interpreted[5,34].

DIRBE and the Cosmic Infrared Background

The primary objective of the Diffuse Infrared Background Experiment (DIRBE) is to conduct a definitive search for an isotropic cosmic infrared background (CIB), within the constraints imposed by the local astrophysical foregrounds, from 1 to 240 μm . Additional objectives include studies of the interplanetary dust cloud and the stellar and interstellar components of the Galaxy. Both the cosmic redshift and the reprocessing of short-wavelength radiation to longer wavelengths by dust act to shift the short- wavelength emissions of cosmic sources toward or into the infrared, and the CIB may contain much of the energy released since the formation of luminous objects. Measurement of the CIB would provide important new insights into issues such as the amount of matter undergoing luminous episodes in the pregalactic Universe, the nature and evolution of such luminosity sources, the nature and distribution of cosmic dust, and the density and luminosity evolution of infrared-bright galaxies.

The DIRBE approach is to obtain absolute brightness maps of the full sky in 10 photometric bands (1.2, 2.2, 3.5, 4.9, 12, 25, 60, 100, 140 and 240 μm). To facilitate discrimination and study of the bright foreground contribution from interplanetary dust, linear polarization is measured at 1.2, 2.2, and 3.5 μm , using a combination of orthogonal polarizers and spacecraft rotation. All celestial directions are observed hundreds of times at all accessible angles from the Sun in the range 64 - 124° . The instrument has a large field of view, 0.7° square, and the sky signal is continuously chopped against a zero-flux internal surface. A cold shutter allows measurement of instrumental offsets and internal stimulation of the detectors.

The photometric quality of the DIRBE data is excellent; when the full reduction of the cryogenic-era data

Consequences of the Discoveries of the COBE Satellite

(Nicolà Vittorio)

This report provides an overview of theoretical work on the cosmic microwave background and large scale structure formation over the period 1990 to 1993. Only the most recent references to the literature from this period are given in order to provide a snapshot of current research.

The Cosmic Microwave Background Spectrum

During the last 25 years, observations of the spectral distribution of the cosmic microwave background (CMB) have generally pushed towards demonstrating that the relic radiation spectrum is planckian, as foreseen by the hot Big Bang model. It is expected, however, that energy release by different phenomena (*e.g.* dissipation of shock waves and turbulence, annihilation of matter and antimatter, decay of unstable particles, evaporation of primordial black holes, etc) may induce distortion in the underlying planckian spectrum of the CMB. The amplitude of such distortions and the frequencies at which they may occur are related to the amount and the epoch of energy release. This is why the study of spectral distortions is so important in providing information on the universe at epochs otherwise inaccessible to observations.

The COBE-FIRAS experiment has measured the spectrum of the CMB between 30 and 600GHz. The spectrum is well fitted[1] by a black body with a temperature of $T = 2.735 \pm 0.060$. This result was soon confirmed by a rocket experiment[2] that in a similar spectral region found $T = 2.736 \pm 0.017$. The thermal nature of the CMB is a genuine prediction of the hot Big Bang model and the COBE-FIRAS results confirm this prediction. Moreover, the upper limit to the amplitude of possible spectral distortions between 30 and 600 GHz obtained by COBE-FIRAS can be used to place constraints on the thermal history of the universe at redshift $z \lesssim 10^4$. In particular, the lack of any distortions in the CMB spectrum allows us to reject the hypothesis that the soft X-ray background is produced by clumpy hot intergalactic gas[1]. Also, the lack of distortion may in principle constrain models for the evolution of clusters of galaxies (which contain hot gas) and bound the contribution that such objects may make to the observed X-ray background[3]. This is particularly relevant now that the ROSAT satellite has claimed to have resolved at least 75% of the X-ray background in individual sources[4]. COBE-FIRAS has recently improved[5]the upper limit on the spectеral distortion of the CMB and constrains the comptonization parameter y to be less than 2.5×10^{-5}.

In spite of its great success, COBE-FIRAS has left completely uncovered the low frequency zone ($\nu < 30GHz$), where various theoretical considerations indicate the possible existence of spectral distortions. This frequency region has so far been explored only with ground based observations[6,7]. The accuracy of these observations, especially at frequencies lower than $1GHz$ is limited by the detailed knowledge of the diffuse radiation from our own Galaxy. Phenomena occurring at large redshifts should appear as distortion at wavelengths longer than those covered by COBE-FIRAS. It has been recently discussed[8] that the largest distortion occurs at a frequency which depends upon the density of baryons, while the amplitude of the distortion at such a frequency depends both on Ω_b and the amount of energy released. Therefore future observations of possible spectral distortions at cm wavelength could provide information not only on the amount of energy release but also on the baryonic density parameter.

Cosmic Microwave Background Anisotropies

The COBE space mission has concluded a 20 years hunt for angular anisotropy of the cosmic microwave background (CMB). The cross correlation between the 53 GHz and 90 GHz COBE-DMR maps exhibits a signal which is consistent with the theoretical predictions of gravitational instability theory, where adiabatic density fluctuations generate CMB anisotropies through the Sachs-Wolfe[9] effect. In this context and assuming a flat universe, the COBE-DMR data provide a measure of the amplitude and shape of the primordial power spectrum of density fluctuations $[P(k) = Ak^n]$, on scales larger than $300\,h^{-1}$Mpc (as usual k is a comoving wavenumber measured in Mpc^{-1} and $\lambda = 2\pi/k$ is the corresponding comoving wavelength). Smoot *et al.* [10] found $n = 1.15^{0.45}_{0.65}$, consistent with scale invariant initial conditions. Under this assumption, $A = 6\pi^2 r_0^4 Q^2_{ps-rms}/5$, where $Q_{ps-rms} = (17 \pm 5)\ \mu K/2.7K$ is the ensemble averaged, rms quadrupole anisotropy[10] and $r_0 = 2c/H_0$ is the horizon radius.

It has been suggested[11] that the large scale anisotropy detected by COBE- DMR could be ascribed to hot gas in nearby superclusters. However, It has been argued[12] that this is unlikely, because of the lack of anticorrelation between the COBE-DMR and the X-ray background maps.

Assuming $\Omega_0 = 1$ in the data analysis is not crucial as long as we consider a flat universe. In fact the relative scaling of multipoles of order $\ell 2$ remains basically unchanged passing from a critical universe to vacuum dominated flat cosmologies [*i.e.* $\Omega_0 < 1$ and $\Lambda = 3H_0^2(1-\Omega_0)/c^2$][13,14,15]. Although suggested by the inflationary scenario, the assumption of adiabatic fluctuations is not mandatory. Primordial density fluctuations could be of the isocurvature variety, as in the pure baryonic, open models considered by Peebles[16] In this case however the shape of the autocorrelation function is expected to be much steeper than observed, for any reasonable choice of the primordial spectral index.

The analysis and interpretation of the COBE-DMR data require us to take into account the so called cosmic variance[17,18]. The pattern of the CMB anisotropy on the single, observable sky may be dominated by low order harmonics, which in turn are determined by density fluctuations on very large scales. This is true even for primordially scale invariant density fluctuations, as COBE observes the sky with low resolution ($\gtrsim 10^o$). There are two consequences of this effect. First, the statistics of the temperature fluctuation field on the single observable sky may be non-Gaussian[19,20,21] even if the primordial density fluctuations, out of which CMB anisotropies are generated, are indeed gaussian distributed as generally expected from the inflationary scenario[22,23]. This implies that observing a non-gaussian CMB pattern on the sky does not necessarily imply non-gaussian initial conditions for the density field. Second, as the pattern of the single observable sky is determined by the relative amplitude of low and high order multipoles, it may happen that a single realization of microwave sky can be consistent with the angular correlation function measured by COBE, even if the ensemble of the initial conditions is described by a spectral index n different from unity[24]. This is to say that while the best model that fits the data is a scale invariant model, we may not be able to reject values of the spectral index different from unity on the basis of a robust statistical argument.

For a flat, cold dark matter (CDM) dominated universe the COBE-DMR result implies a value of σ_ρ (the rms value of the density fluctuations on $8\,h^{-1}$Mpc scale) of the order of unity with an uncertainty of roughly 20% at the 1σ level[15]. Such a value of σ_ρ is too high for explaining the abundance of rich clusters[15], quasars[25] and the observed pairwise peculiar velocity on small scales[26].

On scales smaller than those probed by the COBE-DMR experiment it is now possible to directly estimate $P(k)$ from quite a number of large independent data sets. These are galaxies in the optical[27], IR[28,29] and radio[30] bands or clusters of galaxies.

Flat CDM models could in principle account for all these observations provided we consider primordial density fluctuations which are initially "quasi" scale-invariant, characterized by a spectral index $n \lesssim 1$ ("tilted models")[31]. These models can be justified in the framework of power-law inflation models[32,33], which also predict a substantial background of gravitational waves (GWB)[34,35]. Such a GWB contributes at least a fraction of the observed CMB anisotropy on large angular scales, including those sponsored by COBE-DMR. Only the remaining fraction can be ascribed to density fluctuations at last scattering. As a result, taking into account the effects of a GWB allows us to reduce the amplitude of density fluctuations, and nonetheless to reproduce the observed level of CMB anisotropy. In a CDM model, for example, σ_ρ can be smaller than unity and still be consistent with the COBE-DMR result[35].

An alternative way of being consistent with the COBE-DMR result and the small scale observations consists in assuming a flat model composed by a mixture of hot and cold dark matter[36,37]. Under the condition $\Omega_{HDM} + \Omega_{CDM} = 1$, increasing Ω_{HDM} reduces the amplitude of density fluctuation on small scale, once we normalize the amplitude of the power spectrum of these models to the COBE-DMR result.

A clear way of testing these models is to compare the theoretical predictions with the observations of CMB anisotropy at intermediate angular scales ($\approx 1^o$)[38]. The number of experiments probing the CMB anisotropy at these angular scales is rapidly increasing. To mention few of them we have the ULISSE[39] experiment at 6°, the Tenerife[40] experiment at 8°, the ACME-HEMT[41,42] experiment at 3°, the MAX[43,44] experiment at 1°. Other experiments at the intermediate and small angular scales are either ongoing or ready to go and in the near future should produce very interesting results.

From the theoretical side, the ACME-HEMT experiment has opened a discussion on the validity of the CDM model and in general on the need for a late reheating of the intergalactic medium. The argument goes as follows. Let us normalize the amplitude of density fluctuations to the COBE-DMR experiment and let us ask what the level of anisotropy is at $\simeq 3^o$. This is a quite large angular scale where we do not expect distortion of the primordial spectrum of density fluctuation. Because of this, the predictions at this angular scales should be independent of the "chemistry" of the universe (*i.e.* what kind of dark matter is dominant). Gorski *et al.* [45] concluded that the theoretical predictions, normalized to COBE- DMR, conflict with the upper limit ($\lesssim 35\mu K$) provided by the ACME-HEMT experiment. The only way to save the inflationary model would then be to invoke a secondary last scattering surface. Dodelson and Jubas[46] reach a different conclusion, taking into account the uncertainty in the COBE normalization. Gorski[47] and Muciaccia *et al.* [48] claim that the upper limits at intermediate angular scales may be inconsistent with the available data on large scale peculiar velocity field.

Taking a conservative point of view, it is fair to conclude that we are still far from being able to discriminate between theoretical models. For example, Bennet and Rhie[49] discussed theories for structure formation based on global monopoles and textures: they conclude that these theories are quite consistent with the COBE-DMR result.

The future is quite encouraging. Extensive mapping of the sky at the intermediate angular scales may help in disentangling the contribution to the CMB anisotropy from a GWB[50] to test the thermal history of the universe at redshift $z \lesssim 100$, and, possibly, to reconstruct the matter power spectrum from CMB anisotropy experiments. Since at the intermediate angular scales the effects of cosmic variance are strongly suppressed, there will also be a way of testing the gaussian nature of the initial conditions. All this is still to come. At the moment we can say that the COBE-DMR experiment has finally confirmed the theoretical prediction of a beautiful paper written 25 years ago by Sachs and Wolf[9].

References

1. Mather *et al.* 1990, Ap.J.Lett., 354, L37
2. Gush, H.P., Halpern, M., Wisshnow, H., 1990, Phys.Rev. Lett., 65, 537
3. Garmire, G.P., Nousek, J.A., Apparao, K.M.V., Burrows, D.N., Kraft, R.P. 1992, Ap.J. 399, 694
4. Hasinger, G., Burg, R., Giacconi, R., Hartner, G., Schimdt, M., Trumper, J., Zamorani, G., 1993, Astron.Astrophys., 275, 1
5. Mather *et al.* 1993, preprint
6. Bensadoun, M., Bersanelli, M., G. de Amici, Kogut, A., Levin, S.M., Limon, M., Smoot, G.F., Witebsky, C. 1993, Ap.J., 409, 1
7. Levin, S.M., Bensadoun, M., Bersanelli, M., G. de Amici, Kogut, A., Limon, M., Smoot, G. 1992, Ap.J., 396,3
8. Burigana ,C., De Zotti, G. and Danese, L. 1991, Ap.J., 379,31
9. Sachs, R.K., and Wolfe, M.A., 1967,Ap.J.,147,73
10. Smoot G.F., *et al.* 1992,Ap.J.Lett.,396,L1
11. Hogan, C. 1992, Ap.J. 398, 177
12. Boughn, S.P. and Yahoda, K., 1993 Ap.J. 412, L1, 93
13. Kofman, L., & Starobinskii, A.A., 1985, Sov.Astron.Lett., 11, 271
14. Górski, K.M., Silk, J., & Vittorio, N., 1992, Phys.Rev.Lett., 68, 733
15. Efstathiou, G., Bond, J.R., and White, S.D.M., 1992, MNRAS,258,1P
16. Peebles, P.J.E. 1987, Nature, 327, 210.
17. Abbott, L.F. and Wise, M. 1984, Ap.J.Lett., 282, 47
18. Bond, J.R. and Efstathiou, G., 1987, MNRAS, 226, 655.
19. Scaramella, R. and Vittorio, N. 1991, Ap.J., 375, 439
20. Falk,T., Rangarajan, R., M.Srednicki 1993, Ap.J.Lett., 403, L1
21. Luo,X., Schramm, D.N. 1993, Phys.Rev.Lett., 71,1124
22. Olive,K.A., 1990 Phys.Rep. 190, 307
23. Mukhanov, V.F., Seldman, H.A., Branderberger, R.H., 1992, Phys.Rep., 215, 203
24. Scaramella, R. and Vittorio, N. 1993, MNRAS, 263, L17
25. Efstathiou, G., and Rees, M.J., 1988, MNRAS, 230,5P
26. Davis, M., Efstathiou, G., Frenk, C., and White, S.D.M.,1985,Ap.J.,292,371
27. Vogeley, M.S., Changbom Park, Geller, M.J. and Huchra, J.P. 1992,ApJ, 391, L5
28. Fisher, K.L., Strauss, M.A., Davis, M., Yahil, A. and, Huchra, J.P. 1992, ApJ, 389, 188
29. Taylor, A.N. and Rowan-Robinson, M. 1992, Nature, 359, 396
30. Peacock, J.A. 1991, MNRAS, 253, 1p
31. Cen, R., Guedin, N.,Y., Kofman, L.A., Ostriker, J.P., 1992, 399, L11
32. Abbott, L.F. and Wise, M. 1984,Nucl.Phys.,B244,541
33. Lucchin, F., Matarrese, S., 1985,Phys.Rev.D,32,1316

34. Lucchin, F., Matarrese, S., and Mollerach, S. 1992, Ap.J.Lett., 401, L49
35. Davis, R.L., Hodges, H.M., Smoot, G.F., Steinhard P.J. & Turner, M.S. 1992, Phys.Rev.Lett, 69, 1856
36. Davis, M., Summers, F.J. and, Schlegel, D. 1992, Nature, 359, 383
37. van Dalen, A. & Schaefer, R.K., 1992, Ap.J., 398, 33
38. Gouda, N., and Sugiyama, N. 1992, Ap.J.Lett., 395, L59
39. P.de Bernardis, S. Masi, B. Melchiorri, F. Melchiorri, N. Vittorio, 1992, Ap. J. Lett., 396, L57
40. Watson, R.A., Gutierrez de la Cruz, C.M., Davies, R.D., Lasenby, A.N., Rebolo, R., Beckman, J.E., Hancock, S. 1992, Nature, 357,660
41. Gaier, T., Schister, J., Gundersen, J., Koch, T., Meinhold, P., Seiffert, M., Lubin, P., 1992, Ap. J. Lett., 398, L1
42. Schuster, J., Gaier, T., Gundersen, J., Meinhold, P., Koch, T., Seiffert, M., Wuencshe, C.A., Lubin, P., 1993, 412,L50
43. Meinhold,P., Clapp, A., Devlin, M., Fischer, M., Gunderson, J., Holmes, W., Lange, A., Lubin, P., Richards, P., Smoot, G., 1993, Ap.J.Lett., 409, L1
44. Gunderson,J.O., Clapp, A., Devlin, M., Holmes, W., Fischer, M., Meinhold,P., Lange, A., Lubin, P., Richards, P., Smoot, G., 1993, Ap.J.Lett., 413, L1
45. Górski, K.M., Stompor, R., Juszkiewicz, R. 1993, Ap.J.Lett., 410, L1
46. Dodelson, S., Jubas, J.M., 1993, Phys.Rev.Lett., 70, 2224
47. Gorski, K.M., 1992 Ap.J.Lett., 398, L5
48. Muciaccia, F., Mei, S. de Gasperis, G., and Vittorio, N., 1993, Ap. J. Lett., 410, L61
49. Bennet, D.P., and Rhie, S.H., 1993, Ap.J.Lett., 406, L7
50. Crittenden, R., Bond, J.R., Davis, R.L., Efstathiou, G., Steinhard, P.J., 1993, Phys.Rev.Lett., 71 324

RECENT PROGRESS ON THE DARK MATTER PROBLEM

(Bernard Sadoulet)

In last five years, considerable additional evidence[1] has been gathered on the fact that at least 90% of the mass in the universe is dark, by which we mean that it does not emit or absorb any form of electromagnetic radiation. Understanding this "dark matter" has become one of the more central problems in astronomy and cosmology. Once a subject of controversy among astronomers, its existence is now acknowledged by a large majority. The debate has shifted to the amount of dark matter in the universe and its nature. Its role in the formation of the large scale structure of galaxies is becoming a major study of cosmologists, and a number of observational tests have already limited significantly what it could be made of. Ambitious searches are starting, ranging from compact baryonic objects to elusive elementary particles which may have been produced in the Big Bang.

At the galaxy level, a large amount of dark matter is observed. In large spiral galaxies, it is often possible to measure[2] HI rotation curves out to many times the scale length characterizing the exponential decrease of the surface brightness away from the galactic center. The dark halo clearly dominates the dynamics and typical mass to light ratios $M(r)/L \sim$ 3-5 $M_\odot/L_\odot$ are obtained, increasing with radius even when no more stars are observed. Elliptical galaxies also contain large amounts of dark matter. While velocity dispersion measurement of stars probes a region where the dark matter is not dominant, the study of globular clusters[3] and of planetary nebulae typically show an increase of M/L from 3 in the inner part to 15 in outer parts. The extended X-ray emission[4] detected by X-ray satellites implies even larger values (>70).

The dynamic effect of dark matter is even more pronounced in clusters of galaxies. It has been known for some time that dispersion velocities of the many hundreds of galaxies which constitute rich clusters are often in excess of 1500 km/s. Such large values indicate very deep potential wells and M/L above 100. In many clusters[5], a large amount of gas is detected through its X-ray emission. There again, it is high temperature (~5 keV) X-ray gas which implies a similar depth of potential. In the last few years, a third piece of evidence has been gathered which also points to a very large amount of dark matter in clusters. Galaxy clusters gravitationally lens the light emitted by quasars and field galaxies in the background[6]. The mapping of the mass distribution through the many arclets seen in many clusters, indicates rather deep potential wells qualitatively similar to those observed with the two other methods.

The combination of all these observations makes it rather convincing that dark matter does indeed exist, unless the laws of gravity are violated on the large scale. If there is a wide consensus on this conclusion, there is still considerable debate on the amount of dark matter and the value of $\Omega = \rho/\rho c$, the ratio of the average density in the universe (which is dominated by the dark matter component) to the critical density. Summing the mass of individual galaxies and their halos leads to a value of Ω of the order of a few percent. The sum of clusters of galaxies gives $\Omega > 0.1$ to 0.2. It should be emphasized, however, that these methods are only able to give a lower limit on Ω, as they are insensitive to a smooth background density. Dynamic methods comparing the observed peculiar velocities to the density contrast are sensitive to the time available for acceleration and the growth of density fluctuations, and therefore to the value of the average density at the scale considered. Current measurements of velocity correlations and of velocity flows[7] hint at a large value of $\Omega(>0.3)$ at large scale. In principle, a correct measurement of Ω would have to rely on the measurement of the geometry. This is the aim of the classical cosmological tests. Although much more feasible than a decade ago[8] because of modern instrumentation, these tests continue to face the difficult problem of disentangling the geometry from the evolution of the standards used.

The average density is an important clue to the nature of dark matter. If $\Omega<0.1$, it may be made of ordinary baryonic matter, as that number of baryons in the universe is quite consistent with the measurement of primordial abundances of light elements and the standard nucleosynthesis scenario in a homogeneous universe. The attempts to circumvent this scenario by introducing inhomogeneities due, for instance, to a quark hadron transition have been so far unable to significantly change these limits[9]. If Ω is significantly higher, say greater than 0.2, we may be forced into the hypothesis that at least some dark matter is non-baryonic, that is, not made of ordinary matter.

Additional clues to the nature of dark matter may be obtained from the study of the large scale structure in the universe, formation of which is believed to occur through the gravitational collapse of primordial density fluctuations. It has been known for some time that dark matter is likely to play an important role and that in the simpler models it has basically to be non-relativistic (cold) at the time when density fluctuations begin to grow[10]. The opposite case, the so-called "hot dark matter" models[11] (based, for instance, on a light massive neutrino), lead to erasure of the density fluctuations on smaller scales by dark matter diffusion, and are not in qualitative argument with observations unless additional seeds (e.g., topological singularities) are introduced. The recent detection by COBE[12] of fluctuations in the microwave background temperature has brought even more interesting information. In the framework of primordial "adiabatic" fluctuations (i.e., initial fluctuations appearing in the same proportion in both the matter and photon components), the comparison of the COBE results with the density fluctuations observed in the large scale structure of the present universe[13] points to non-baryonic dark matter and roughly flat geometry (possibly through a non-zero cosmological constant). It is, however, possible to also account for the data, admittedly with more parameters to adjust, with baryonic dark matter in an "isocurvature" scenario[14] (where fluctuations in matter density are compensated by fluctuations in the radiation field).

As these considerations are likely to provide only indirect and somewhat model dependent indications, it is important to attempt to determine the nature of dark matter directly. A number of direct searches are already in progress.

In the case of baryonic dark matter, a number of possibilities are already excluded by observations.[15] The dark matter has to be in the form of condensed objects: brown dwarf or black holes. Both types can be combined under the name Massive Compact Halo Objects (MACHOs). Unless they are very massive, these objects can theoretically be detected by microlensing[16]: If one of these MACHOs happens to cross the line of sight to a star, say in the Large Magellanic Cloud, a temporary increase of the intensity will be observed. This increase would be symmetric in time, achromatic and non-repetitive, contrary to sporadic phenomena in stars. Three collaborations, Livermore-Center for Particle Astrophysics-Mount Stromlo, Saclay-Orsay-Observatoire de Paris-Observatoire de Marseille, and Princeton-Cracow, are now actively searching for such events, an effort which requires the regular observation of some ten million stars. Three microlensing candidates have been recently announced[17] by the first two collaborations.

As explained above, if $\Omega = \rho/\rho c$ is significantly greater than 0.1, we may be forced into the position that the dark matter is at least in part not made out of baryons. If we discard exotica such as a shadow universe or primordial black holes, the most attractive hypothesis then is that dark matter is made of particles that were created in the hot early universe and managed to stay around.

One of the well-motivated candidates is the axion. Such a particle has been proposed[18] in order to suppress the strong CP violation implied by the otherwise very successful Quantum Chromodynamics. It has not been observed, but it is interesting to note that the combination of laboratory and astrophysics experiments has constrained its mass in such a way that if it does exist, it must be cosmologically significant, accounting for a large fraction of the critical density. These "invisible" axions from the halo of our galaxy could in fact be detected through their conversion into monochromatic microwave photons inside a tunable microwave cavity in a large magnetic field. In the past ten years, two pilot efforts[19] explored the technology, but lacked about three orders of magnitude in sensitivity to reach a cosmologically interesting limit. A second generation experiment is currently in preparation at Livermore, which will begin to enter an interesting sensitivity range over one decade in mass (out of the three still allowed).

Without further information from a specific model, it is quite natural to assume that these dark matter particles were once in thermodynamic equilibrium with the quarks and leptons. In this case, their current density depends on whether they were relativistic or not at the time they decoupled from the rest of the universe. If they are light enough to be relativistic at that time, their density is just related to the decoupling temperature and is basically equal to that of the photons in the universe. This is, for instance, what is expected to have happened for light neutrinos, and a neutrino of 25 eV would give a Ω of the order of unity. Unfortunately such a neutrino is extremely difficult to detect in the astrophysical environment. It should be possible, however, to test this hypothesis in the laboratory once two neutrino oscillation experiments exploring this mass range are operational at CERN in 1994.

For particles which happened to have decoupled when they were non-relativistic, their density today is inversely proportional to their annihilation cross section[20]. A density close to the critical density leads to a cross section of the order of the Weak Interaction, indicating that the physics at the $W^{\pm}/Z^{\circ}$ intermediate vector boson scale (e.g., supersymmetry) may be responsible for the dark matter in the universe. This generic class of particles is usually called Weakly Interacting Massive Particles (WIMPs). A first generation of experiments[21] looking for elastic scattering of such WIMPs in the laboratory with solid state detectors proved that heavy Dirac neutrinos cannot be a major component of our galactic halo, and almost eliminated a class of WIMPs designed both to be dark matter and to account for the paucity of solar neutrinos. A second generation of laboratory experiments is being brought into operation; depending on the groups, they use larger masses of germanium detectors, large scintillating crystals of NaI, or novel "cryogenic detectors" working at millikelvin temperatures. While the first methods promise sensitivity gains of a factor 3 to 10, the cryogenic detectors allow an active rejection of the radioactive background (e.g., through the simultaneous measurement of the phonons and ionization produced in particle interactions[22]) and may give gains of one hundred or more. These experiments will begin to probe the rate region expected for the theoretically favored neutralinos, the lightest particles in supersymmetry.

References

1. See, e.g., the reviews by Trimble, V., Ann. Rev. Astron. Astrophys. 25, 425 (1987); Primack, J. R., Seckel, D., and Sadoulet, B., Ann. Rev. Nucl. Part. Sci. 38, 751 (1988); and Tremaine, S., Physics Today 45, 28 (1992).
2. e.g., Bosma, A., Ap. J. 86, 1825 (1981).
3. e.g., Mould, J. R., Oke, J. B., and Nemec, J. M., Ap. J. 92, 153 (1987).
4. Fabian *et al.* , MNRAS 221, 1049 (1986).
5. Fabricant, D. and Gorenstein, P., Ap. J. 267, 535 (1984); Jones, C. and Forman, W., Ap. J. 276, 38 (1984).
6. e.g. Tyson, J. A., Valdes, F., Wenk, R. A. *et al.* , Ap. J. 349, 2 (1990); Maoz et al., Ap. J. 394, 51 (1992); Fort, B. *et al.* , Ap. J. Lett. 399, L125 (1992); see also Fort's contribution above.
7. Bertschinger, E., Dekel, A., Faber, S. M., Dressler, A., Burstein, D., Ap. J. 364, 370 (1990); Dekel, A., Bertschinger, E., Yahil, A. *et al.* , Ap. J. 412, 1 (1993).
8. Sandage, A. and Tamman, G. A., in the proceedings of First ESO-CERN Symposium: Large Scale Structure of the Universe, Cosmology and Fundamental Physics, G. Setti and L. Van Hove, eds., CERN, Geneva, 1983, pp. 127, (1984); Loh, E. D. and Spillar, E. J., Ap. J. 303, 154 (1986); Loh, E. D. and Spillar, E. J., Ap. J. 307, L1 (1986).
9. Olive, K. A., Science 251, 1194 (1991).
10. e.g., Davis, M. *et al.* , Ap. J. 292, 371 (1985).
11. e.g., Shapiro, P. R. *et al.* , Ap. J. 275, 413 (1983); Schaeffer, R. and Silk, J., Ap. J. Lett. 281, L13 (1984).
12. Smoot, G., Bennett, C., Kogut, A., Wright, E. *et al.* , Ap. J. Lett. 396, L1 (1992) and contributions above.
13. Fisher, C., Davis, M., Strauss, M. A., Yahil, A. *et al.* , Ap. J. (1992).
14. Peebles, P. J. E., Nature 327, 210 (1987).
15. Hegyi, D. and Olive, K. A., Ap. J. 303, 56 (1986); Gunn, J. E. and Peterson, B. A., Ap. J. 142, 1633 (1965).
16. Paczynski, B., Ap. J. 301, 503 (1986) and Griest, K. *et al.* , Ap. J. 366, 412 (1991).
17. In Nature, October 14, 1993 issue.
18. Peccei, R. D. and Quinn, H., Phys. Rev. Lett. 38, 1440 (1977).
19. De Panfilis, S. *et al.* , Phys. Rev. Lett. 59, 839 (1987) and Phys. Rev. D40, 153 (1989); Hagmann, C. *et al.* , Phys. Rev, D42, 1297 (1990).
20. Lee, B. W. and Weinberg, S., Phys. Rev. Lett. 39, 165 (1977).
21. Ahlen, S. P. *et al.* , Phys. Lett. B 195, 603 (1987); Caldwell, D. O., Eisberg, R. M., Grumm, D. M. *et al.* , Phys. Rev. Lett. 61, 510 (1988); Caldwell, D. O. *et al.* , Phys. Rev. Lett. 65, 1305 (1990); Reusser, D. *et al.* , Phys. Lett. B235, 143 (1991).
22. Shutt, T., Ellman, B. *et al.* , Phys. Rev. Lett. 29, 3425 (1992); Shutt, T., Wang, N., Ellman, B.. Giraud-Heraud, Y. *et al.* , Phys. Rev. Lett. 29, 3531 (1992).

The Interplanetary Plasma and the Heliosphere Report of IAU Commission 49

B. Buti
Physical Research Laboratory
Ahmedabad 380 009, India

G.S. Lakhina
Indian Institute of Geomagnetism
Bombay 400 005, India

and

V. Krishan
Indian Institute of Astrophysics
Bangalore 560 034, India

Introduction

(B. Buti, President Commission 49)

The aim of the present report is to pinpoint the highlights of the scientific literature in the interplanetary and heliospheric plasma during the period 1990-1993. We have tried to cover these highlights in the following two reports: Nonlinear plasma processes in the Heliosphere and Plasma and MHD Phenomena in the Heliosphere. In order to manage within the allotted space, we have emphasized on a few hot and current topics in some details rather than just giving a bibliographical sketch of all the topics. The two reports are complimentary and two together hopefully cover most of the current topics.

J. Bergeron (ed.), Reports on Astronomy, Vol. XXIIA, 565–578.

Nonlinear Plasma Processes in the Heliosphere

B. Buti
Physical Research Laboratory
Ahmedabad 380 009, India

and

G.S. Lakhina
Indian Institute of Geomagnetism
Bombay 400 005, India

Introduction

This review deals with a few nonlinear plasma processes which are essential to properly interpret some of the observations in the interplanetary medium, planetary atmospheres, comets, etc. The topics covered here are: nonlinear waves and turbulence, magnetic reconnection, magnetic substorms and chaotic dynamics.

Magnetic Reconnection

Magnetic reconnection is a fundamental plasma process which converts magnetic energy into the plasma kinetic and heat energy. The process results in the change of magnetic topology, and therefore requires some kind of dissipation in the system to initiate its onset. This process is mainly responsible for inhibiting build-up of excessive amount of magnetic energy in the current sheets occuring in astrophysical plasmas. Magnetic reconnection is believed to play an important role in many astrophysical phenomena, like solar and stellar flares, X-ray emmisions from stars, radio jets, magnetospheric substorms, etc.

The classical Petschek model is generalized to the nonuniform and curved field. The reason for the absence of fast scaling in Biskamp's model is explained (Priest and Forbes, 1992). Futher, in 3D reconnection can takes place on any singular field line where the nearby field has X- type topology, and an electric field parallel to the field line exists. In the absence of 3D null points, reconnection can still occur by a process of magnetic flipping where plasma crosses the flipping layer but field lines rapidly flip along them by diffusion (Priest and Forbes, 1992). A nonlinear stochastic MHD model for the magnetic reconnection process where the magnetic field line stochasticity resulting from the overlapping resonance surfaces provide the anomalous resistivity has been worked out (Tetreault, 1992a,b). Heating due to collisionless magnetic reconnection has been studied by Moses et al. (1993). It is shown that parallel electric field can arise from plasma flows which violate the frozen-in field condition, these fields can play an important role in cosmic particle acceleration (Schindler et al., 1991).

Magnetic Reconnection at Sun

It is suggested that the phenomena of X-rays corona is due to reconnection of the field lines deformed continuously by the subphotospheric motions. Similar processes may be responsible for the naoflares from the stars (Parker, 1992). In an other study, it is found that the magnetic reconnection and energy balance of the coronal magnetic field in response to prescribed motion of the photosphere footpoints could lead to coronal heating (Vekstein et al., 1991). Magnetic reconnection process may even explain the formation of post flare loops (Forbes and Malherbe, 1991). It is suggested that the spicules are generated during reconnection of the magnetic field of the supergranule boundary cylinder layer (Pataraya et al., 1990). On the other hand, the opening of new magnetic flux on the Sun can lead to magnetic disconnection above the helmut streamers (Linker et al., 1992).

Magnetic Reconnection at Magnetopause

From the review of the reconnection versus viscous drag theory for solar wind interaction with the magnetosophere, it is concluded that steady state reconnections is highly unlikely, and that the relevant processes must be essentially time dependent, three dimensionsal and localized (Heikkila, 1990). Direct evidence for the reconnection of the open field lines of the tail lobe with Interplanetary Magnetic Field (IMF) in the vicinity of high latitude dusk magnetopause has been provided from the ISEE 2 data (Gosling et al., 1991). Further, the laboratory experiments on collisionless current sheet suggest that Flux Transfer Events (FTEs) may be formed because of 3D tearing at the magnetopause (Gekelman et al., 1991). Simulations using 2D MHD codes suggest that the ratio of magnetospheric to magnetosheath magnetic fields controls the excitation of different form of reconnection giving different signatures of FTEs (Ding et al., 1991). On the other hand, the particle simulations involving 2 1/2 D elctromagnetic codes predict that driven reconnection could occur in either of the two forms, a quasisteady single X line reconnection, and an intermittent multiple X line reconnection. The resulting heating leads to power law particle distributions (Ding et al., 1992a). The analysis of the tearing mode at the magnetopause has been extended to includes the effests of finite β, shear flow and the propagation at an arbitrary angle (Ding et al., 1992b; Wang and Ashour-Abdalla, 1992; Cao et al., 1991). It has been shown that the presence of background turbulence of lower hybrid or ion cyclotron type can act as external driver source which enhances the growth rate of the tearing instability (Das, 1992).

Magnetic Reconnection in the Magnetotail

The results of collisionless tearing studied by 2D particle codes, suggest that the ion tearing mode is stabilized when the cyclotron frequency based on the B_z field equals growth rate of B_z=0 tearing mode. External pertubations can, however, drive the mode unstable (Pritchett et al., 1991). On the other hand 3D MHD simulations indicate that crosstail current diversion is mainly due to localization of reconnection in a 3D configurations. The onset of reconnection eventually leads to plasmoid formation and ejection. A new feature is reduction of B_y on reconnected dipole like field lines. The presence of pressure anisotropy tend to stabilize the resistive tearing mode instability in the magnetotail (Birn and Hesse, 1991a,b; Hesse and Birn, 1991, 1992) Fast reconnection is initiated by a localized resistivity causing plasma jets and plasmoids ejection, and redistribution of the overall current system (Ugai, 1991; Hoshino, 1991; Scholar and Hautz, 1991) Data from ISEE 3 and Geotail mission indicate that plasmoids are "open" magnetic structures (Moldwin and Hughes, 1992).

Structure of the X line during steady state reconnection is studied by analytical and numerical approach. It is found that no steady state reconnection is possible when the influx of plasma into the reconnection region exceeds some critical value (Burkhart et el., 1991a). Magnetic reconnection in collisionless field reversals and the universality of iontearing is discussed by Kuznetsova and Zelenyi (1991) A kinetic theory for driven reconnection in the magnetotail has been developed. Driven reconnection occurs in two modes, namely, the exponential mode reconnection which is similar to the ion tearing instability, and the bursty mode reconnection which is rather short lived but occurs at a much faster rate (Lakhina, 1992a). The widely believed qualitative picture of the "sling shot", i.e., plasmoid acceleration by the tension of the open interplanetry magnetic flux, is shown to be misleading. The spontaneous magnetic reconnection resulting in plasmoid formation and acceleration represents sufficiently fast process. Simulations using 2D implicit electromagnetic code VENUS, show that tearing develops if B_n is small enough and ion gyroradius is of the order of sheet thickness. In 1D the instability does not saturate, instead plasmoids continue to be formed, accelerated and ejected (Otto et al., 1990; Zwinmann et al., 1990) On the analysis side, it has been argued that the principal axis analysis (PAA) of magnetometer data from a single satellite pass is insufficient to differenciate between magnetic closed loops and flux ropes model (Moldwin and Hughes, 1991).

Nonlinear Waves and Turbulence in the Interplanetary Medium

An over view of origin and evolution of the interplanetary MHD fluctuations observed in the solar wind including the turbulence due to cometary ion pick up processes is given by Roberts et al.(1991, 1992). A new model for SW turbulence is proposed by Lank and Matthaeus (1992). The importance of inwards propagating waves in transfering energy between different scale through nonlinear interaction with the outwards waves is discussed (Bavassano, 1992). Recently Buti (1991, 1992a) derived the evolution equation for the nonlinear Alfven waves in inhomogeneous plasmas and showed the effect of streaming on the solitary Alfven waves. In inhomogeneous but multispecies plasmas solitary Alfven waves have been investigated by Verheest and Buti (1992).

Using 2 and a 1/2 D fully relativistic PIC code, it is shown that electron acceleration by the lower hybrid waves excited by the ion rings can take place in the solar corona (McClements et al., 1990). During nonlinear evolution of parallel propagating waves, the wave steepening occurs due to beam ion bunching (i.e., nongyrotropic ions) caused by the trapping in the regions where waves have highest intensities (Akimoto et al., 1991). Quasi-linear evolution of ULF waves excited by cometary ions has been studied by Lee and Gary (1991). On the other hand, Huddleston et al. (1991) have used velocity diffusion due to observed level of turbulence to explain the development of the implanted ion distribution. The excitaion of lower hybrid instability by the pickup cometary ions (protons and water group) in the bow shock region and its quasilinear saturation has been discussed by Shapiro et al. (1993). Upper level of magnetic turbulence of nonresorant Alfven modes driven by multi ion beams is given by Verheest and Lakhina (1991). The results are found to be in good agreement with the MHD turbulence near comets. Large amplitude ion cyclotron wave could generate ion density holes which leads to the formation of intermittent double layers on the auroral field lines (Lakhina et al., 1992).

Analysis of ISEE 1 and 2 data indicate that small-scale ULF fluctuations at magnetopause are possibly due to nonlinear Alfven modes (Rezeau et al., 1993). Upstream of the earth's bow-shock, the nonlinear strong turbulence effects cause soliton like coherent wave packets to form and decouple from incoherent background beam excited weak Langmuir turbulence. They collapse to scales of 100 m and electric field $E \sim 2$ V/m (Robinson and Newman, 1991).

Chaotic Dynamics

Chaotic processes in plasmas are being investigated to understand the nature of plasma turbulence. The fractal structure of observed fluctuations could possibly throw some light on the source of observed plasma turbulence. From the existence and evolution of the multifractal structure of velocity fluctuations in the recurrent streams at 1 AU and at $\sim$ 6 AU, Burlaga (1991) concluded that the turbulence around 6 AU is intermittent. In the Voyager data of large-scale magnetic field fluctuations also, Burlaga (1993) found multifractals indicating intermittent turbulence. The transition from regular to chaotic motion of a particle in a sheared magnetic field due to loss of adiabaticity was shown by Büchner and Zelenyi (1991). Ion acceleration due to chaos has also been investigated by Büchner and Zelenyi (1990). Through two-time velocity correlations, Horton and Tajima (1991) showed that the chaotic motion of charged particles could be a potential source for the collisionless conductivity.

In solar wind, magnetosphere as well as cometary plasmas, there are more than 2 species e.g., solar wind besides electrons and protons has 5% of helium and comets Halley and Giacobini-Zinner were found to have water group ions. By using nonlinear dynamical techniques, Buti (1992b) showed that the presence of these heavy ions leads to the reduction of chaos in the system. There have been large number of attempts to study the chaotic behaviour of a variety of systems (cf. special section on Chaos and Stochasticity in Space Plasmas, Geophys. Res. Lett. 18, 1991; IAU Symposium 152 on Chaos, Resonance and Collective Dynamical Phenomena in the Solar System 1992).

Magentospheric Substorms and Nonlinear Dynamics

Several models of the substorm phenomena have been reviewed by Birn (1991), Fairfield (1992), and Lakhina(1992). It is generally concluded that near earth reconnection model provides the best frame work to fit in various observations of the substorms. That reconnection occurs is established, but the actual mechanism is yet to be worked out. Alternative view is to construct a coherent description of the substorm development by extracting some important components from various existing models. In this connection, the role of cross-field instability in substorm onset has been high lighted (Lui, 1991; Lui et al., 1991, 1992). Observations based on ISEE 1 and 2 data for energetic particles ($> 20keV$) and magnetic field, are found to be generally consistent with the predictions of the near earth reconnection model of substorm (Lin et al., 1991) Some recent studies also show that current disruption starts near-earth magnetotail $|X| < 20R_E$ and often within 15 R_E and the results are consistent with the near earth neutral model (Ohtani et al., 1992a,b). Furthermore observations from AMPTE IRM satellite (about 50,000 measurements) fit in the framework of near earth reconnection model (Baumjohann et al., 1992). However, there are some observations, like variations of plasma β values, which are thought to be consistent with ballooning instability model (Pu et al., 1992). However, it has been suggested that the ballooning modes can not affect the large scale 2d configuration (Lee and Wolf, 1992).

It is shown that development of ion Weibel instability can produce anomalous resistivity about 11 to 12 orders higher that the classical value (Lui et al., 1993).

Recent observations strongly indicate the formation of a thin current sheet at the center of plasmasheet during growth phase of the substorm (Mitchell et al., 1990; Pulkkinen et al., 1991; Sergeev et al., 1990). The stability of such forced thin current sheet agiainst ion tearing mode instability has been studied by Burkhart et al. (1992a,b) and Lakhina (1993). The trapped electrons have a strong stabilizing effect on the ion tearing instability. Formation of a thin current sheet and its stability against pressure anisotropy instabilty (kink type) is studied (Pritchet and Coroniti, 1992).

Many theoretical and numerical studies related to particle dynamics in the earth's magnetotail suggests that particle orbits can become chaotic. The effects of the chaotizations of the orbits on substorm process have been studied by several workers (Chen, 1992; Burkhart and Chen, 1992a,b; Pulkkinen et al., 1992). The collisionless conductivity using the chaotic single particle orbits has been calculated in the geomagnetic tail. The height integrated dissipative part of the collisionless conductivity governs the stochastic heating (Horton and Tajima, 1991). The effects of constant B_y on the collisionless conductivity produced by chaotic scattering and stochastic diffusion of particles in the current sheet for parabolic geometry is considered. The increase of B_y tend to strongly stabilize the tearing modes (Hermandez et al., 1993). Nonlinear particle dynamics in magnetotail gives rise to partitioning of phase space into different regions which are occupied by diffferent classes of orbits with separated time scales. This leads to differetial memory which affects the evolution of the particle distribution resulting in highly non-Maxwellian features (Burkhart et al., 1991).

The complexity of the solar-terrestrial phenomena such as solar wind magnetosphere coupling, solar activity, etc., can be viewed as the dynamical features of a nonlinear disssipative system. Such systems with complex behavior are known to have simple dynamical descriptions and are studied with a wide range of techniques. Among these techniques is the possibility of reconstructing the dynamics from experimental data and thus gaining insight into the complex nonlinear systems, independent of particular modeling assumptions. The analysis of the time series data of a single dynamical variable, using such techniques, can yield the characteristics quantities of the system, e.g., the fractal dimensions, Lyapunov exponents, etc. Therefore such studies have a lot of potential in exploring the processes connected with the substorms. Analysis of magnetospheic activity using time series data in the auroral electrojet indices (AE and AL) has shown low dimensionality (fractal dimension of 3.6) with a positive Lyapunov exponent. The dynamical equations may be constructed using the time series data of the variables obtained from the singular spectrum analysis. The key issues are the predictability of these dynamical models, and the relationship of the dynamical to the physical variables. (Sharma, 1992; Sharma et al., 1993; Vassiliadis et al., 1990, 1991).

References

Akimoto, K., D. Winske, T. G. Onsager, M. F. Thomsen, and S. P. Gary, *J. Geophys. Res., 96*, 17599 (1991).
Baumjohann, W. G. Paschmann, and T. Nagai, *J. Geophys. Res., 97*, 17173 (1992).
Bavassano, B., *J. Geophys. Res., 97* (1992).
Birn, J., *Rev. Geophys., Supplement, 29*, 1049 (1991).
Birn, J., and M. Hesse, *J. Geophys. Res., 96*, 1611 (1991a).
Birn, J., and M. Hesse, *J. Geophys. Res., 96*, 23 (1991b).
Büchner, J. and L. M. Zelenyi, *Geophys. Res. Lett., 17*, 127 (1990).
Büchner, J. and L. M. Zelenyi, *Adv. Space Res., 11*, 177 (1991).
Burkhart, G. R., J. F. Drake, J. Chen, *J. Geophys. Res., 96*, 11539 (1991).
Burkhart, G. R., and J. Chen, *J. Geophys. Res., 96*, 14033 (1991).
Burkhart, G. R., and J. Chen, *J. Geophys. Res., 98*, 89 (1992a).
Bukhart, G. R., and J. Chen, *J. Geophys. Res., 97*, 6479 (1992b).
Burkhart, G. R., J. F. Drake, P. B. Dusenbury, and T. W. Speiser, *J. Geophys. Res., 97*, 16749 (1992a).
Burkhart, G. R., J. F. Drake, P. B. Dusenbury, and T. W. Speiser, *J. Geophys. Res., 97*, 13799 (1992b).
Burlaga, L. F., *Geophys. Res. Lett., 18*, 1651 (1991).
Burlaga, L. F., *Astrophys. J., 407*, 347 (1993).
Buti, B., *Nonlinear and Chaotic Alfven Waves in Solar and Planetary Plasma Physics*, Ed. B. Buti, World Scientific, Singapore (1990).
Buti, B., *Geophys. Res. Lett., 18*, 1817 (1991).
Buti, B., *J. Plasma Phys., 47*, 39 (1992a).
Buti, B., *J. Geophys. Res., 97*, 4229 (1992).
Cao, F., and J. R. Kan, *J. Geophys. Res., 96*, 5859 (1991).
Chen, J., *J. Geophys. Res., 97*, 15011 (1992).
Das, A. C., *J. Geophys. Res., 97*, 12275 (1992).
Ding, D. Q., L. C. Lee, and Z. W. Ma, *J. Geophys. Res., 96*, 57 (1991).
Ding, D. Q., L. C. Lee, D. W. Swift, *J. Geophys. Res., 97*, 8453 (1992).
Ding, D. Q., L. C. Lee, C. F. Kennel, *J. Geophys. Res., 97*, 8257 (1992).
Fairfield, D. H., *J. Geophys. Res., 97*, 10865 (1992).
Gekelman, W., H. Pfister, and J. R. Kan, *J. Geophys. Res., 96*, 3829 (1991).
Gosling, J. T., M. F. Thomsen, S. J. Bame, and R. C. Elphic, *J. Geophys. Res., 96*, 14097 (1991).
Forbes, T. G., and J. M. Malherbe, *Solar Phys., 135*, 361 (1991).
Heikkila, W. J., *Space Sci. Rev., 53*, 1 (1990).
Hermandez J., W. Horton, and T. Tajima, *J. Geophys. Res., 98*, 5893 (1993).
Hesse, M., and J. Birn, *J. Geophys. Res., 96*, 5683 (1991).
Hesse, M., and J. Birn, *J. Geophys. Res., 97*, 10643 (1992).
Horton, W. and T. Tajima, *J. Geophys. Res., 96*, 15811 (1991).
Hoshino, M., *J. Geophys. Res., 96*, 11555 (1991).
Huddleston, D. E., A. D. Johnstone, A. J. Coates, and F. M. Neubauer, *J. Geophys. Res., 96*, 21329 (1991).
Kuznetsova, M. M., and L. M. Zelenyi, *Geophys. Res. Lett., 18*, 1825 (1991).
Lakhina, G. S., *J. Geophys. Res., 97*, 2961 (1992a).
Lakhina, G. S., *Memoirs of Geophysical Society of India, No 24*, 307 (1992b).
Lakhina, G. S., P. K. Shukla, and F. Verheest, *J. Geophys. Res., 97*, 8355 (1992).
Lakhina, G. S., *J. Geophys. Res., 98*, in press (1993).
Lank, G. P., and W. H. Matthaeus, *J. Geophys. Res., 97*, 17189 (1992).
Lee, D. Y., and R. Wolf, *J. Geophys. Res., 97*, 19251 (1992).
Lee, M. A., and Gary, S. P., *J. Geophys. Res., 96*, 21319 (1991).
Lin, N., R. L. McPherron, M. G. Kivelson, and R. J. Walker, *J. Geophys. Res., 96*, 19427 (1991).
Linker, J. A., G. Van Hoven, D. J. McComas, *J. Geophys. Res., 97*, 13733 (1992).

Lui, A. T. Y., C.-L. Chang, A. Mankofsky, H.-K. Wong, and D. Winske, *J. Geophys. Res., 96*, 11389 (1991).

Lui, A. T. Y., *J. Geophys. Res., 96*, 1849 (1991).

Lui, A. T. Y., R. E. Lopez, B. J. Anderson, K. Takahashi, L. J. Zaneti, R. W. McEntire, T. A. Potembra, D. M. Klumpar, E. M. Greene, and R. Strangeway, *J. Geophys. Res., 97*, 1461 (1992).

Lui, A. T. Y., P. H. Yoon, and C. L. Chang, *J. Geophys. Res., 98*, 153 (1993).

McClements, K. G., J. J. Su, R. Bingham, J. M. Dawson, and D. S. Spicer, *Solar Phys., 130*, 229 (1990).

Mitchell, D. G., D. J. Williams, C. Y. Huang, L. A. Frank, and C. T. Russell, *Geophys. Res. Lett., 17*, 583 (1990).

Moldwin, M. B., and W. J. Hughes, *J. Geophys. Res., 96*, 14051 (1991).

Moldwin, M. A., and W. H. Hughes, *J. Geophys. Res., 97*, 19259 (1992).

Moldwin, M. B., and W. J. Hughes, *J. Geophys. Res., 98*, 81 (1993).

Moses, R. W. J. M. Finn, and K. M. Ling, *J. Geophys. Res., 98*, 4013 (1993).

Ohtani, S., S. Kokubun, and C. T. Russell, *J. Geophys. Res., 97*, 3129 (1992a).

Ohtani, S., K. Takahashi, L. J. Zaneti, T. A. Potembra, R. W. McEntire, and T. Iijima, *J. Geophys. Res., 97*, 19311 (1992b).

Otto, A, K. Schindler, and J. Birn, *J. Geophys. Res., 95*, 15023 (1990).

Parker, E. N., *J. Geophys. Res., 97*, 4311 (1992).

Pataraya, A. D., A. L. Taktakishvili, and B. B. Chargeishvili, *Solar Phys. 128*, 33 (1990).

Priest, E. R., and T. G. Forbes, *J. Geophys. Res., 97*, 16757 (1992).

Priest, E. R., and T. G. Forbes, *J. Geophys. Res., 97*, 1521 (1992).

Pritchett, P. L., F. V. Coroniti, R. Pellat, and H. Karimabadi, *J. Geophys. Res., 96*, 11,523 (1991).

Pritchett, P. L. and F. V. Coroniti, *J. Geophys. Res., 97*, 16773 (1992).

Pu, Z. Y., A. Korth, and G. Kremser, *J. Geophys. Res., 97*, 19341 (1992).

Pulkkinen, T. I., D. N. Baker, D. H. Fairfield, R. J. Pellinen, J. S. Murphrec, R. D. Elphinstone, R. L. McPherron, J. F. Fennel, R. E. Lopez, and T. Nagai, *Geophys. Res. Lett., 18*, 1963 (1991).

Pulkkinen, T. I., D. N. Baker, R. J. Pellinen, J. Büchner, H. E. J. Koskinen, R. E. Lopez, R. L. Dyson, and L. A. Frank, *J. Geophys. Res., 97*, 19283 (1992).

Robinson, P. A., and D. L. Newman, *J. Geophys. Res., 96*, 17733 (1991).

Rezeau, L., A. Roux, and C. T. Russell, *J. Geophys. Res., 98*, 179 (1993).

Roberts, D. A., and M. L. Goldstein, *Rev. Geophys., Supplement, 29*, 932 (1991).

Roberts, D. A., M. L. Goldstein, W. H. Matthaeus, and S. Ghosh, *J. Geophys. Res., 97*, 17115 (1992).

Schindler, K., M. Hesse, and J. Birn, *Astrophys. J., 380*, 293 (1991).

Scholar, M., and R. Hautz, *J. Geophys. Res., 96*, 3581 (1991).

Sergeev, V. A., P. Tanskanen, K. Mursula, A. Korth, and R. C. Elphic, *J. Geophys. Res., 95*, 3819 (1990).

Shapiro, V. D., V. I. Shevchenko, A. S. Sharma, K. Papadopoulos, R. Z. Sagdeev, and V. B. Lebedev, *J. Geophys. Res., 98*, 1325 (1993).

Sharma, A. S., Nonlinear dynamics of space plasmas and predictive modeling, in *Reasearch Trends in Nonlinear Space Plasma Physics,* American Institute of Physics, New York, 1992.

Sharma, A. S., D. Vassiliadis, and K. Papadopoulos, *Geophys. Res. Lett., 20,* 335 (1993).

Tetreault, D., *J. Geophys. Res., 97*, 8531 (1992a).

Tetreault, D., *J. Geophys. Res., 97*, 8541 (1992b).

Ugai, M., *J. Geophys. Res., 96*, 21173 (1991).

Vassiliadis, D., A. S. Sharma, T. E. Eastman, and K. Papadopoulos, *Geophys. Res. Lett., 17,* 1841 (1990).

Vassiliadis, D., A. S. Sharma, and K. Papadopoulos, *Geophys. Res. Lett., 18*, 1643 (1991).

Vekstein, G. E., E. R. Priest, and C. D. C. Steele, *Solar Phys., 131*, 297 (1991).

Verheest, F., and B. Buti, *J. Plasma Phys., 47*, 15 (1992).

Verheest, F., and G. S. Lakhina, *J. Geophys. Res., 96*, 7905 (1991).

Wang, Z., and M. Ashour-Abdalla, *J. Geophys. Res., 97*, 8245 (1992).

Zwingmann, W., J. Wallace, K. Schindler, and J. Birn, *J. Geophys. Res., 95*, 20877 (1990).

Plasma and MHD Processes in the Heliosphere

V. Krishan
Indian Institute of Astrophysics
Bangalore 560 034, India

Introduction

The study of plasmas is of the utmost importance due to their universal presence and the way to grasp the universe is through a detailed understanding of the all-encompassing dynamical processes observed to-date in the sun and its environs. The small and the large scale flow patterns, the unearthing of the outer planetary magnetic fields, the linear and the nonlinear electrostatic and electromagnetic waves, acceleration, heating and radiation through a host of plasma-magnetic kinetic processes and the ubiquitous shocks have all pointed towards their kinship with extra-solar system phenomena. For example, the large inclinations of the rotation axes to the magnetic axes in Uranus and Neptune are reminiscent of the oblique rotator model of the stellar and pulsar magnetic fields. The mechanism of magnetic reconnection believed to be responsible for solar flares as well as for transfer of mass, momentum, energy and magnetic flux in planetary magnetospheres is found to be equally good at accelerating cosmic rays and enhancing gravitational collapse. The physics of degenerate plasmas is as essential for Jupiter as it is for white dwarfs and pulsars. Flow patterns on the solar surface, in planetary magnetospheres and in the large scale structure of the universe may have more in common than we are willing to believe in (Lanzerotti 1990; Krishan 1991). Cylindrical plasma structures whether in the form of solar coronal loops, cometary tails, galactic loops or extragalactic jets pose common questions of their equilibrium and stability. Therefore, any investment of effort in the study of solar system structures promises good returns. The collisionless shocks proposed in most of the astrophysical systems can be modelled in every detail, thanks to the in-situ observation of planetary bow shocks. Thus the citation index of solar planetary and interplanetary and heliospheric plasma phenomena reads like the story of the entire universe. In the following sections, a brief account of the work done in some of the solar system plasma physical problems is given. The account is complete neither in content nor in references, as limited space allotted by IAU forbids anything more ambitious.

Solar Wind

The solar wind blows in two modes: the quasi-stationary and the transient. The density and the velocity structure of the quasi-stationary high speed wind is consistent with its origin from the coronal holes, where, it is accelerated by wave damping (Grappin et al., 1991; Coles et al., 1991; Zhang et al., 1991), superthermal electron heat flux, macrospicules or by the energy released through network and internetwork magnetic elements (Parker, 1991). The quasi-stationary slow speed flows, on the other hand, are believed to be associated with the coronal streamers, edges of coronal holes and the heliospsheric current sheet or sector boundaries (Neugebauer, 1991). However, the mechanisms that prefer to accelerate α particles and other heavy ions (Bogdan et alo., 1991) to protons have not yet been delineated in a satisfactory manner. The determination of the relative contributions of the various possible mechanisms occurring on a variety of spatial and time scales as well as the ways in which they cooperate to produce a rather smooth flow awaits more theoretical and observational inputs. Further, the correlation between the solar wind speed near the earth and the magnetic geometry of the corona can be used to deduce the latitudinal distribution of wind speed at different phases of the sunspot cycle. The fastest wind appears to be centred around the warps in the heliospheric current sheet (Sheeley et al., 1991).

A quasi-stationary flow, when interrupted by a transient event like a coronal mass ejection (CME's), acquires a transient character. Since CME's result when prominence supporting magnetic field structure turns unstable, most of the properties of the transient wind reveal the closed magnetic fields and flows (Veselovski, 1990), configuration (Svestka, 1991) e.g., the bidirectional electron heat fluxc and proton streaming. The detection of singly ionized He ions with an over abundance of 3 orders of magnitude was interpreted to be a straight signature of the presence of prominence material, which reached IAU without thermalizing on the way (Mullan, 1991). The exceptionally low temperatures of $7 \times 10^4 K$ indicate a large expansion that a transient flow undergoes after CME's generated shock heating. The many characteristics of the transient flows like the low level of wave activity (Velli et al., 1991) strong magnetic fields (Goldstein Jr., 1991), heavy ion ionization, abundances and velocities and the flux tube expansion factor, interaction of fast and slow streams (Fainshtein, 1991) remain to be cracked (Neugebauer, 1991). The MHD simulations of solar wind turbulence reflect upon the nature of interplanetary fluctuations and their generation mechanisms (Roberts et al., 1991; Dryer et al., 1991).

The Solar Probe expected to cruise at an altitude of 3 solar radii may settle many of these issues. To clarify the relationship of enhanced fluctuations with coronal structures, collaborative observations of Solar-A satellite with Kashima 34m antenna have been proposed. A magnetohydrodynamic along with a kinetic treatment (Moses and Kennel, 1991) including a self-consistent interaction between the flows and fields as well as the waves and the particles is a prerequisite to a complete appreciation of the richness of the solar wind phenomena. The reader is referred to the review articles by March (1991) and Fahr (1991).

Magnetospheric Convection

The magnetospheric convection i.e. a large scale plasma flow in planetary magnetospheres is driven either by the solar wind (SW) as in Earth and Mercury or by the planetary rotation as in Jupiter and Saturn or by both the solar wind and the planetary rotation as in Uranus and Neptune. The interaction of the solar wind with the atmospheres of Venus and Mars is more akin to SW-comet intraction. The merging of the planetary and the interplanetary magnetic field (IMF) at the magnetospause plays the most important role in the transfer of mass, momentum, energy and magnetic flux on to a planet in SW-driven convection. Whether the merging is steady or sporadic is an issue of live controversy (Lookwood et al., 1993; Smith et al., 1992; Lockwood and Smith, 1992); the resolution of which resides in the latitudinal and the transient structure of the ionospheric flows (Newell, 1992; Escoubet et al., 1992).

The issues in the case of rotation driven convection are the transport of mass, momentum, energy and magnetic flux through microscopic turbulent diffusion processes across Io-Jupiter magnetosphere region. Inclusion of microdiffusion processes leads to a Lorentz like strange attractor solution, indicating the chaotic nature of convection (Summers and Mee, 1992). The quadrupolar convective processes dominated by centrifugal and magnetic field effects, resulting in highly asymmetric flows to and from the planet in the Triton torus - Neptune magnetosphere region have been discussed by anumber of authors in GRL (1990). From the plasma flow perturbations, it has been shown that the signatures of cylindrical flux transfer events (FTE) and wavy magnetopause (MP) (Silbeck and Smith, 1992) with its unearthlike rotation and magnetic field relationship and an earthlike wind. The large inclination of the rotation axis to the magnetic dipole axis in Uranus and Neptune (Akasofu et al., 1991), the trapped radiation belt system enclosing the major satellites of Uranus and the formation and stability of self-gravitating dusty epitons of Neptune are some of the major excitements in this field. the Phobos-2 measurements suggest that the SW-Mars and SW-Venus interaction is cometary like with mass loading of the solar wind. The acceleration of the heavy Martian ions could be similar to the terrestrial-auroral acceleration along with the local action of pick-up processes. A recent review of the earthly magnetospheric processes can be seen in Saunders (1991).

Dusty Plasmas

Dust particles acquire electric charge in the presence of a plasma, though it can also happen through photo-electron emission or absorption processes. The properties of dust particles then begin to depend strongly on their new status as charge carriers. For example, in the interstellar medium, charged dust grains when subjected to supernova associated shock waves may undergo significant changes in their destruction rate and overall dynamics. In the solar system, this new degree of freedom of the dust particles enables them to experience electromagnetic forces in addition to the gravitational forces, thus facilitating explanations of many an otherwise unexplained phenomena (Horanyi and Mendis, 1991). The electrostatically supported planetary dust rings are found to oscillate with 3 times the local Kepler frequency. The oscillation frequency structure of a dense ring is more complex. The lack of azimuthal symmetry in the co-rotating magnetospheric plasma can lead to a coupling with the vertical dust profile oscillations of orbiting dust rings (Melandso and Havens, 1991). The major properties of the spokes in Saturn's rings can be identified with those of magnetosonic waves which have group velocities larger than the Alfven speed. The magnetosonic waves can be driven in a fluid composed of small charged grains and ambient plasma. The impulse given to small grains by an electron beam or a meteor is enough to induce by collisions with large grains, the formation of a spray of dust. Dust rings around planets need to be continuously replenished since they lose particles due to several dynamical effects (Grun, 1991). Grain charge variations due to the modulation of photo electron current caused as the grain enters or exists the planetary shadow, are determined. The electromagnetic perturbation resonates with the orbital period and can modify the orbits size and eccentricity (Horanyi and Burns, 1991). The spatial and temporal variations of the plasma parameters and magnetic field, factors, which govern the charging of the dust and its subsequent motion, were included while developing numerical models to predict the dust distributions in P/Halley (Ellis and Neff, 1991).

Ulysses at Jupiter

During the seventeen day long encounter, Ulysses highlighted the intimate relationship between the solar wind and the Jovian magnetosphere by cruising through the previously unexplored dusk sector of the magnetosphere, where the magnetric field was found to be bent out of the meridian planes associated with the corotating planetary field. This was caused by the intense sweeping of the field into the magnetospheric tail by the solar wind as also indicated by the observed temporal variation of the trapped particle fluxes. Large fluxes of counter-streaming electrons and ions observed in the high latitude dusk side may be responsible for auroras in the earth-like polar caps of the planet. However, the density of the energetic particles suddenly dropped to solar wind values near 15 RJ, indicating the presence of open field regions. the Ulysses trajectory passed through the Io plasma torus, facilitating the electron density distribution measurements. The need for more detailed models of the magnetodisk current system including the effects associated with its inclination to the rotation axis and warping due to the solar wind stresses has been acknowledged while trying to model the Ulysses data. For preliminary results of the Ulysses encounter of Jupiter; papers describing volcanic activity on Io; observations of polar regions, magnetic fields, plasmas and waves, radio emission, energetic particles, their composition and energies, dust and highly configured plasma boundaries, see Science 257, 1992.

The Solar-Polar Expectations from Ulysses

After amply fulfilling the expectations of the planetary scientists, Ulysses now hurls towards the solar-polar regions to provide an out of the ecliptic view of the Sun. The curtain will rise in June, 1994. The complexities associateds with the looped and rotating equatorial magnetic fields are believed to be absent in the polar regions. Therefore, one expects to observe the low energy cosmic rays and be able to trace back clues to their origin which are otherwise lost in the labyrinth of the equatorial fields. For the same reason, it becomes possible to follow the polar-solar wind back into the Sun. Essentially Ulysses will observe from the poles what has been observed to-date from the equator. The list includes solar flares, radio waves, the extra solar system dust and gas, the polar heliosphere, the interstellar helium, the ubiquitous γ-ray bursts and even gravitational waves in addition to the direct measurements of the polar magnetic fields to confirm their simplicity (Wentzel and Smith, 1991; Appenzeller, 1992).

SOHO and Cluster Missions

SOHO - the Solar and Helioscopic Observatory and cluster consisting of a set of four spacecracfts are scheduled to be launched in the second half of 1995. SOHO has been planned to study the solar interior through helioseismology and solar irradiance variations as well as to delineate the heating and acceleration of the solar corona and the solar wind through spectroscopic measurements. Cluster, on the other hand, promises three dimensional time dependent measurements of the fields and flows and the associated current densities and vorticity in the near earth space plasma. A dream comes true for the plasma modeler! (Domingo Schmdt, 1991).

Shocks

Collisionless shocks have been observed by spacecrafts throughout the solar system. The physics of shocks depends on the inclination of their normal to the magnetic frield. This has given rise to the terminology of quasi-parallel and quasi-perpendicular shocks. In a collisionless plasma, particles communicate through electromagnetic fields. The collisionless coupling can take place through the electric field resulting due to the pressure gradient forces associated with intense electron heating and magnetic compression. This process is known as Laminar Coupling and is operative in quasi-perpendicular shocks. The Larmor Coupling is due to motion of ions in a magnetic field; the resulting inductive electric field couples the plasma particles. This is operative when a perpendicular shock is formed by the reflected ion beams or strong ion heating. The parallel shocks involve turbulent coupling which arises from the electric fields produced by the nonthermal particle distribution functions which form due to an impulse or a piston (Cargill, 1991). The quasi-parallel shocks involve all the three types of coupling. The decoupling of the piston and the shock has been questioned in the case of quasi-parallel and parallel shocks. Hybrid numerical simulations for hot plasmas indicate (i) parallel shock formation with hot electrons with no separation from the piston, (ii) no parallel shock formation with hot ion pistons, (iii) a quasi-parallel shock formation with severe coupling with the piston. These results emphasize a strong interaction between the driver and the consequent processes of particle acceleration and heating (Cargill, 1991). The collisions between pairs of quasi-parallel shocks is studied using hybrid numerical simulations. The two shocks are found to go through each other leaving behind a hot energetic plasma. The energization is more efficient for quasi-parallel shocks of comparable strengths (Cargill, 1991).

During an encounter of the AMPTE/CCE spacecraft, the earth's bow shock was seen to pass back and forth over the spacecraft with a typical period of oscillationof 20 seconds along with an upstream wave of the similar period propagating in the solar wind just before the multiple shock crossings. The magnetic field rotated through 360° for each pair of shock crossings. It is suggested that the plasma displacement associated with the upstream wave activity is the cause of the oscillating shocks (Strangeway and Zanetti, 1991). The quasi-parallel shocks are associated with a variety of magnetic field structues including the low amplitude, nearly sinusoidal, low frequency waves observed far upstream to the large amplitude, turbulent pulsations which are usually associated with the shock itself. Low frequency magnetric field turbulence is described in terms of a gas of Alfven solitons. An analysis of electric field turbulence does not play a significant role in the energetics of the quasi-parallel shocks. The plasma data from Giotto shows that both the inbound and outbound crossings of the cometary bow shock fulfilled the shock conditions and supports the observation that the normal component of the solar wind flow becomes subsonic at a point close to the shock transition. All these topics are covered in Adv. in Space Res. 11 (1991). For the inbound crossing of Comet Halley's bow shock, it is found that the transmission of the sunward-streaming hydromagnetic waves through the bow shock could lead to the enhancement of its wave amplitude by a factor of 2-3 and the generation of Alfven waves with the opposite helicity (Ip and McKenzie, 1991). Plasma and magnetic field observations from crossings of the Martian bow shock by the Phobos-2 spacecraft and the pioneer venus orbiter at Venus are compared with data from bow shocks of other planets and the AMPTE lithium release. It is found that the shock spectra of the inner planets are very similar in shape and their wave energy densities, when normalized to the upstream electron thermal energy density, are comparable. However, the Martian shock spectrum is more like the spectrum from AMPTE indicating the role of pick-up ions and their large gyro-radii (Dubinin et al., 1991; Moses et al., 1991). The Voyager data on Uranus's magnetic field in the vicinity of the shock reveal a series of whistler wave events, believed to be analogous to those whistler waves upstream of the Earth's bow shock that are driven by stream of electrons (Smith et al., 1991).

Shocks accelerate particles. The evolution study of ion distribution functions in the quasi-parallel bow shock, using the data from Prognoz-10 satellite shows that there exist many different plasma fluxes in the foreshock region. The foreshock consists of many shock-like structures which are caused by large amplitude MHD waves. Interaction of solar wind with these structures creates the beam ion distribution which gradually develops into the diffuse distribution. The 3-D ion distributions from the AMPTE-UKS ion instrument associated with magnetic field structures near the Earth's bow shock have been examined to search for origins of the observed ions from four source populations: reflected conserving magnetic moment, specularly reflected, magnetosheath leaked particles conserving magnetic moment and magnetosheath leaked particles accelerated parallel to the shock normal. The statistical analysis of the correlation of solar wind parameters with variations of the ion energy spectra shows that the differential particle fluxes in the energy range from 10-30 KeV correlate with the solar wind density thus supporting a solar wind source and that the spectral slope correlates with the solar wind velocity. For discussion of these topics, see Adv. in Space Res. 11 (1991). McKee and Draine (1991) have raised several issues crucial to the understanding of collisionless shocks e.g. the relative temperatures of the electrons and the ions; structgure of a quasi-parallel shock with a substantial amount of energy in accelerated particles, role and type of interstellar shocks responsible for accelerating the cosmic rays, the dependence of efficiency of acceleration on the inclination of the shock normal to the magnetic field and more. Computer simulations and observations of spectral lines by space telescopes at different electromagnetic bands should be able to answer some of these questions.

Planetary Radio Emission

Jupiter emits at radio waves in a variety of forms through thermal and nonthermal processes. A new mechanism whereby plasma waves at the local upper hybrid resonance frequency with anomlous dispersion are generated through Cherenkov instability of an electron beam and then converted into electromagnetic waves by scattering of inhomogeneities, has been proposed to account for most of the properties like conical beam pattern and frequency drift of the decameter S-radio emission of Jupiter (Boev et al., 1991). The radial profiles of Jupiters nonthermal emission at 6, 20 and 90 cm show the presence of a shoulder or flattening in the intensity at 2, 5 RJ due to absorption effects by the satellite Amalthea (Peter, 1991). In order to determine the sources of hectometer emission, a ray tracing study including refraction effects has been done with the conclusion that the hectometeric source locations at high magnetic altitudes in both the hemispheres fit best the emission pattern as observed by the Voyager space craft. The polarization properties of the hectometric emission can be explained by simultaneous radiations from two independent 100% oppositely polarized sources. A global model for the hectrometric (HOM) and decametric (DAM) Jovian emission in which the left and right HOM components are identified with the left and right non-Io DAM components, respectively, is outlined (Ortega-Molina et al., 1991). The origin of the nearly unique 100% elliptical polarization of Jupiter's decameter radio emission along with the polarization of cyclotron maser radiation, the dispersion relation of the rarefied plasma composed of energetic anisotropic electrons the growth rate of the maser and the brightness temperature are determined by Melrose and Dulk (1991). A qualitative model is proposed which invokes the observed HOM polarization characteristic and continuously radiating sources on field lines having footprints at all longitudes along a narrow band of high magnetic latitudes close to the northern and southern auroral zones. The study of Faraday effect on Jovian emission shows that the usual simple approximation where the rotation of the polarizaztion plane varies as the inverse of the frequency squared has to be modified to include a term which goes as the inverse of the fourth power of frequency in order to explain the recent observations. The additional frequency dependence arises from the path difference between the ordinary and the extra-ordinary waves and the high density gradient of the ionosphere (Boudjada et al., 1991). For millimeter wave emission and Bremsstrahlung X-rays from Jupiter, see J.G.R. 96, (1991). Lakhina et al. (1990) gave a model which is applicable for radio emissions from planetary magnetospheres. This model is based on calculations of radiation from accelerated Alfven solitons in homogeneous plasmas.

Auroral Hiss

A review of auroral hiss emissions observed at ground based stations and in the magnetosphere is presented. The most likely energy source of these emissions lies in the electrons and energies below 100 eV precipitating to the auroral ionosphere. The incoherent Cerenkov radiation amplified by the beam-plasma instability may provide a good interpretation of the auroral hiss emissions (Sazhim et al., 1993). The secular variation of the aurora is examined and compared, where possible, to sunspot data and magnetic activity data. The data provide confirmation of the anti-correlation of auroral occurrence in the polar regions with sunspot activity as a result of displacement of the auroral oval with changes in solar and magnetic activity (Silverman, 1992). The average response to an electron drifting through randomly fluctuating double layers aligned parallel to the ambient magnetic field is calculated and the thickness of the visual auroral arc is estimated to be $\sim$ 2.5 km with fine structure $\sim$ 250 m at electron energy of $\sim$ 350 eV (Prakash and Lysak, 1992). The effects of stochastic motion of the ions on the westward cross-tail current at the earthward edge of the plasma sheet is studied for the formation of stable inverted -V's and arcs in the night sector (Galperin et al., 1992).

In addition to the phenomena discussed above, the study of various plasma kinetic and MHD waves and instabilities in dusty and otherwise media is being pursued in an attempt to grasp the detailed workings of the radiation, heating and acceleration processes on the one hand and the equilibrium and stability of the hot and energetic plasmas on the other. The field of Interplanetary plasma and the Heliosphere with its insitu observations has been raising questions which challenge the versatality of a plasma physicist and stimulate the curiosity of an astrophysicist.

References

Akasofu, S.I. edt al., *Planet. Space Sci.*, *39*, 1259 (1991).
Appenzeller, T., *Science 257*, 1478 (1992).
Barrow, C.H., *Astron. Astrophys.*, *250*, 245 (1991).
Boev, A.G. et al., *Sov. Astron.*, *35*, 422 (1991).
Bogdan, T.J. et al., *J. Geophys. Res.*, *96*, 161 (1991).
Boudjada, M.Y. et al., *Astron. Astrophys.*, *247*, 255 (1991).
Cargill, P.J., *Adv. Space Res.*, *11*, 209, 241 (1991).
Coles, W.A. et al., *J. Geophys. Res.*, *96*, 13849 (1991).
Domingo, V. and R. Schmidt, *Europhys. News*, *22*, 213 (1991).
Dryer, M. et al., *Solar Phys.*, *132*, 353 (1991).
Dubinin, E. et al., *J. Geophys. Res.*, *96*, 11189 (1991).
Ellis, T.A. and J.S. Neff, *Icarus*, *91*, 280 (1991).
Escoubet, P. et al., *Geophys. Res. Lett.*, *19*, 1735 (1992).
Fahr, H.J., *Rev. Mod. Astron.*, *4*, 196 (1991).
Fainshtein, V.G., *Solar Phys.*, *136*, 169 (1991).
Galperin, Yu, I. et al., *Geophys. Res. Lett.*, *19*, 2163 (1992).
Goldstein, S.J. Jr., *Bull. Am. Astron. Soc.*, *23*, 960 (1991).
Gorkavyi, N.N. et al., *Sov. Astron. Lett.*, *17*, 1105, 1116 (1991).
Grappin, R. et al., *ann. Rev. Geophys.*, *9*, 416 (1991).
Grun, E., *Rev. Mod. Astron.*, *4*, 157 (1991).
Horanyi, M. and J.A. Burns, *J. Geophys. Res.*, *96*, 19283 (1991).
Horanyi, M. and D.A. Mendis, *IAU Coll.*, *116*, 1093 (1991).
Ip, W.H. and J.F. McKenzie, *Planet. Space Sci.*, *39*, 1045 (1991).
Krishan, V., *Mon. Not. Roy. Astron. Soc.*, *250*, 50, 157 (1991).
Lakhina, G.S., B. Buti and N.L. Tsintasadze, *Astrophys. J.*, *352*, 747 (1990).
Lanzerotti, L.J., Crawford Symposium on Magnetospheric Physics : Achievements and Prospects, 139 (1990).
Lockwood, M. and M.F. Smith, J., *Geophys. Res.*, *97*, 14841 (1992).

Lockwood, M. et al., *Nature, 361*, 424 (1993).
Luhmann, J.G. and L.H. Brace, *Rev. Geophys., 29*, 121 (1991).
March, E., *Rev. Mod. Astron., 4*, 145 (1991).
McKee, C.F. and B.T. Draine, *Science, 252*, 397 (1991).
Melandso, F. and O. Havens, *J. Geophys. Res., 96*, 5837 (1991).
Melrose, D.B. and G.A. Dulk, *Astron. Astrophys., 249*, 250 (1991).
Moses, S. and C.F. Kennel, *Adv. Space Res., 11*, 9 (1991).
Moses, S.L., et al., *J. Geophys. Res., 96*, 11221 (1991).
Mullan, D.J., *Astron. Astrophys., 248*, 256 (1991).
Neugebauer, M., *Science, 252*, 404 (1991).
Ortega-Molina, A. et al., *J. Geophys. Res., 96*, 11441 (1991).
Parker, E.N., *Astrophys. J., 372*, 719 (1991).
Pater, I. de, *Astron. J., 102*, 795 (1991).
Prakash, M. and R.L. Lysak, *Geophys. Res. Lett., 19*, 2159 (1992).
Roberts, D.A. et al., *Phys. Rev. Lett., 67*, 3741 (1991).
Saunders, M., *Adv. in Solar System Magnetohydrodynamics, 357* (1991).
Sazhum, S.S. et al., *Planet. Space Sci., 41*, 153 (1993).
Sheeley, N.R. Jr. et al., *J. Geophys. Res., 96*, 13861 (1991).
Siebeck, D.G. and M.F. Smith, *Geophys. Res. Lett., 19*, 1903 (1992).
Siebeck, D.G. and P.T. Newell, *EOS, 72*, 475 (1992).
Silverman, S.M., *Rev. Geophys., 30*, 333 (1992).
Smith, C.W. et al., *J. Geophys. Res., 96*, 15841 (1991).
Smith, M.F., *Planet. Space Sci., 40*, 1251 (1992).
Strangeway, R.J. and L.J. Zanetti, *Adv. Space Res., 11*, 219 (1991).
Summers, D. and Jian-Lin Mee, *Geophys. Res. Lett., 19*, 1899 (1992).
Svestka, Z., *Solar Phys., 135*, 491 (1991).
Velli, M. et al., *Geophys. Astrophys. Fluid Dyn., 62*, 101 (1991).
Veseleovskii, I.S., *Geomag. Accon., 30*, 248 (1990).
Wentzel, K.P. and E.J. Smith, *Europhys. News, 22*, 203 (1991).
Zhang, M. et al., *Geophys. Res. Lett., 18*, 1071 (1991).

COMMISSION 50: IDENTIFICATION AND PROTECTION OF EXISTING AND FUTURE OBSERVATORY SITES

PRESIDENT: Paul Murdin, Scotland; VICE PRESIDENT: Syuzo Isobe, Japan

1. GENERAL

Astronomers are already aware that measurements of the properties of observatory sites are critical, not only to rank sites in order of preference for the particular astronomical tasks to be undertaken, but also to design the most suitable instrumentation. Astronomers are increasingly aware of the deterioration of sites by man-made disturbances, and of the necessity to protect observatories.

The report period saw several publications and conferences devoted to these topics. IAU Colloquium 112 was organised by members of the Commission in 1988 in Washington DC (see *Trans IAU* XXI); the proceedings were published in 1991 (53.012.121; references in this form are to *Astronomy and Astrophysics Abstracts*). Study Group No 189 of the NATO Committee on the Challenges of Modern Society, for the Protection of Observatories, has published its reference report *The Protection of Astronomical and Geophysical Sites* (55.003.038). It contains chapters on light pollution, radio interference, pollution by satellites, space debris, and aircraft, and the legal avenues for the protection of observatories. *Radio astronomical seeing* was the subject of a URSI/IAU Symposium in Beijing in 1989; its proceedings also appeared in the report period (53.012.104, see 53.082.075-092). Problems of optical seeing are addressed by series of SPIE conferences: two published in the report period were on *Propagation Engineering* as part of a Technical Symposium on Optical Engineering and Photonics in Aerospace Sensing, 1990 (53.012.025) and on *Advanced Technology Optical Telescopes*, Tucson 1990 (see 52.082.019 to 025); SPIE conferences on *Adaptive Optics* also contain papers of interest.

2. IAU/UNESCO/ICSU MEETING

The IAU, including Commission 50, co-sponsored a meeting entitled *Adverse Environmental Impacts on Astronomy* at the UNESCO headquarters in Paris in 1992 July. The papers are in press. The Director General of UNESCO, Professor F Mayor, said in his address that UNESCO is sensitive to astronomers' concerns, both because the problems of environmental concern need to be addressed at the government level, both nationally and globally, and because UNESCO is involved in a major way in scientific programmes on the environment and the effects of human activities on the quality and harmony of the terrestrial environment. He said that UNESCO will do what it can to follow up the problems revealed. One possibility is that some observatories could be defined as World Heritage Sites and given protected status. As a result of the meeting, ICSU has set up a Working Group (Chairman Dr D McNally, former IAU Secretary) to take further the meeting's recommendations.

3. PROTECTION OF RADIO ASTRONOMY

Radio Regulations, setting the use to which bands of the radio frequency spectrum can be put, are formulated from decisions of the World Administrative Radio Conference (WARC) whose latest session was held in Spain in 1992 February. A big gain for radio astronomy was the upgrading of the OH allocation at 1610 MHz to primary status, giving added protection from sideband emission from the adjacent satellite transmission band, used for example by the GLONASS navigation satellites. Broadcasting satellites were made an allocation at 1452-1492 MHz, sufficiently far from the 1420

MHz 21 cm hydrogen line not to be a nuisance, and additional protection was footnoted for the 150.3-152 MHz band to protect against satellite-to-ground communications in an adjacent band. However High Definition TV was made an allocation at 21.4-22 GHz, near interstellar water, and this could cause problems in the future. WARC requested a study on sideband interference levels so that appropriate protection can be written into future allocations to protect passive (non-broadcasting) services like radio astronomy.

The European Science Foundation Committee on Radio-Astronomy Frequencies (CRAF) is made up of representatives of European radio astronomical observatories and is actively campaigning against the use of the CDMA spread-spectrum technique, which is used to protect privacy in radio transmission by spreading transmissions thinly over a wide band. CDMA is a new technique causing concern to radio astronomy, which is sensitive to interference at a lower level than other services.

4. LIGHT POLLUTION AND OTHER PROBLEMS FOR OPTICAL OBSERVATORIES

Light pollution is an increasing problem in all optical observatories located within, say, 100 km of a town or city (52.013.002). If the light pollution is local, like a lit roadway, then individual arrangements can be made for shielding, or in very favourable cases (as in Turku Observatory in Finland) possessing the right to switch off the lighting. Light pollution over large administrations is controllable only by wider legislation; the Canary Islands' Observatories legislation, which protects the sky to the standards recommended by the IAU, was implemented by Royal Decree in 1992, and the Protection Office of the Instituto dc Astrofisica de Canarias completed a control inventory of lighting on La Palma in 1993.

Light pollution has been quantified over the whole of Japan in the Star Watch programme (organised by the Commission Vice Chairman as a joint programme of the National Astronomical Observatory in Tokyo and the Japanese Environmental Agency). The technique uses simple photographic exposures of star trails obtained by thousands of school children and amateurs under standard conditions (54.082.060; see also *Teaching of Astronomy in the Asian-Pacific Region* Bulletin No 6, available from Dr Isobe). Other simple ways of determining night sky background with naked eye measurements (54.082.057) and a visual photometer (55.082.076) have been described by Upgren. The spectral contribution of light pollution has been quantified in Okayama (55.082.077), where the dominant light background below 700 nm is light pollution, but beyond 700 nm it is OH emission; so, optimistically ('looking on the bright side'), the site is competitive in this respect. Studies in Canada (54.013.095,100), in Israel (54.082.001), at Kitt Peak (52.082.053) and at Mauna Kea (52.082.054) show mixed results, with the more remote sites still remaining relatively unaffected. Light pollution can be propagated large distances (54.082.001), with dust being able both to increase and lessen the pollution (54.082.055), and with effects which are different for different types of astronomical observations (54.082.001).

The British Astronomical Association (the UK's amateur astronomers' society), affiliated to the US-originated Campaign for Dark Skies; it organised a European Colloquium on Light Pollution jointly with the UK Institution of Lighting Engineers at Reading in 1993 July. The ILE has published *Guidance for Lighting Engineers* on avoiding light pollution and is working on more detailed standards appropriate for different lighting projects in various environments. The Commission Internationale de l'Eclairage (CIE) has established, under the chairmanship of Dr D Schreuder, consultant to IAU Commission 50, a Technical Committee TC4.21 on the *Interference by Light of Astronomical Observations* and is also producing a design guide for lighting engineers.

French naturalists and astronomers have produced a Charter, part of the Cousteau Foundation's manifesto on the rights of future generations, aiming to preserve the night as a natural and inalienable element of man's environment on this planet. The Charter points to the threat to the natural balance of life as well as to the astronomical environment.

Italian astronomers have called for European Community action on light pollution. The

Societa Astronomia Italiana and the Unione Astrofili Italiani (the professional and amateur bodies in Italy respectively) have coordinated action to create in Italy zones of protection around astronomical observatories of radius up to 25 km. Local action has successfully improved lighting (Milano-Baroni), turned off lighting at a certain time (Oderzo-Perissinotto) and given observatories a say in the control of lighting (Treviso-Romano, Campo Catino-Di Sora, and Perigrine-Dalmeri).

Other intrusions on astronomical observing include aircraft condensation trails; their frequency and effect on solar observing has been studied at Carinthia, Austria (54.082.067). Refraction of the atmosphere affects accurate meridian circle observations; natural variability is large (52.082.048) but the effects of exhaust gases from a large incinerator in Tokyo are much worse (53.082.059).

5. SPACE ADVERTISING AND SOLAR REFLECTORS

FLAG was the name of a 300 square metre solar sail deployed from the Space Station MIR in 1993 February as a feasibility study for reflectors illuminating terrestrial locations for industrial processes and disaster control. The sail was unfurled and partly stabilised by rotation, but in the present experiment its pointing was uncontrolled. The success of the experiment was mixed, with sites in the south of France reporting the appearance of the sail as brief flashes of a bright 'star' in the sky.

Early in 1993, Space Marketing Incorporated, a company of Roswell, Georgia, USA, proposed to launch a Space Billboard 1 kilometer in dimension. The billboard would be attached as a sponsorship mechanism to a so-called ENVIRONMENTAL SPACE PLATFORM for space experiments. It would appear with a brightness and size rivalling the Moon, and would of itself interfere with optical astronomy at the start and the end of the night.

It was proposed that the billboard would be launched by NASA. ESA has undertaken not to launch any advertising satellite unacceptable to astronomers, but NASA does not appear to have taken a public stance on the environmental issue. The American Astronomical Society issued a statement strongly opposing the use of an earth orbiting spacecraft for advertising purposes, saying that the effects of such a spacecraft will interfere with the capabilities of most existing astronomical telescopes and the many ultra-sensitive telescopes under construction around the world.

For both these space activities, mislaunch might produce additional exposure for ground based telescopes. Fragmentation of the sails (10000 impacts daily was the estimate for the Space Billboard) would add to space debris and increase background throughout the night. The precedent for future billboards is a matter of concern. Both proposals were for devices in near-earth orbit. However the advantage of additional exposure would create pressure for high launches which would then be visible longer through the night. This would gravely interfere with astronomical research.

The Executive Committee of the IAU has asked the UN Committee on Peaceful uses of Outer Space, ICSU, the Space Agencies, and UNESCO to oppose satellite-borne solar reflectors which may damage scientific research.

6. STUDIES OF EXISTING AND POTENTIAL OBSERVATORY SITES

In the report period, astronomers published studies of the astronomical properties of a large number of existing and potential sites (Table 1). Particularly noteworthy for IAU members were the major survey of India in relation to possible sites for the National Large Telescope, a seeing survey of sites in the former USSR, and discussion of the potential use of the Antarctic area especially for continuous monitoring of solar and stellar oscillation, and infrared astronomy. Optimists will like the analysis of 55.082.062, which concludes that seeing performance in large telescopes and interferometers is likely to be better than expected from the standard Kolmogorov-Taylor theoretical models, especially at infrared wavelengths. Pragmatists should note how building design need not deteriorate local seeing (54.032.022, 52.082.025).

Table 1

USA
Flagstaff (Cloud cover and seeing measurements 55.82.013); Anderson Mesa (as a site for an optical array 52.082.072); Mt Hopkins (seeing 52.082.023); the MMT (seeing 52.082.019); New Mexico (VLA phase stability 53.082.077); Mauna Kea (mirror, dome and natural seeing at the CFHT 54.032.017, seeing properties 52.082.022); NRAO site survey in relation to a millimetre array 53.082.079

Europe
Israel (observational qualities and observing efficiency at Wise observatory 55.82.014); Spain (seeing and IR properties of the Canary Islands 52.082.045, 52.082.047, 54.013.078; a scidar seeing experiment from the Nordic Optical Telescope on La Palma 55.082.035; other NOT results 52.082.025; correlation of point source images at La Palma 52.082.006; interferometric measurements at La Palma 54.082.048; the refraction constant 53.082.037; geotechnical surveys of the LEST site on La Palma 55.082.017; effect of volcanic and Saharan dust 54.082.030; the wind spectrum on La Palma 53.082.074; atmospheric extinction at Calar Alto 53.082.072); Austria (Innsbruck 54.082.068); Italy (Cagliari observatory 52.082.013); France (high resolution solar imaging at the Pic Du Midi 53.013.065)

South America
Chile (studies for the ESO VLT at Cerro Pachon 52.082.038, 55.082.004, 55.09.012; differential image motion used to evaluate seeing 55.082.004; effect of the Pinatubo eruption on La Silla 55.082.06; speckle lifetime and isoplanicity at La Silla 53.082.020); Bolivia (extinction values and water vapour content at the Bolivia-Soviet Observatory 54.082.051, 55.82.072)

Asia
Japan (atmospheric extinction at Yonsei 55.082.081); India (infrared properties of Kavalur 54.082.072; general properties of Leh 52.082.04; surveys in relation to the India National Large Telescope Project for optical/infrared astronomy 54.082.015 to 23); Tibet (Yanbajing at 4300 m for air shower gamma ray detector 52.013.007)

Europe/Asia
Former USSR (seeing measurements at 10 sites, identifying Mt Sanglok in Tadzhikstan as of outstanding seeing 53.082.063, 54.082.014); Ukraine (in relation to SLR telescopes 54.082.052); Mongolia (52.082.034)

Antarctica
South Pole (ice used for detection of muons 53.013.097; prospects generally 53.013.104, 54.013.086; aerosol content (ice crystals) above the Amundsen-Scott Polar Station sites 55.082.037,038; atmospheric characteristics near the Belgrano II station in Argentine Antarctica 54.082.071)

7. SOCIOLOGICAL VULNERABILITY OF OBSERVATORIES

Observatories are affected not only by the natural and the artificial environments in which they lie but also by their environment within society. In Chile, the agreement establishing the European Southern Observatory has been renegotiated in political circumstances; this serious matter was resolved in 1993. The political changes in Europe have had their effect on astronomy as in other areas of life. As a result of the financial challenges associated with the re-unification of Germany, the Sonnenberg Observatory, known for the stellar variability patrol initiated by C Hoffmeister in 1925, has been threatened with closure. After the separation of the USSR into independent republics, Maidanak Observatory was nationalised by Uzbekhistan in 1993 May; at least temporarily the astronomers were sent away from the telescopes. Maidanak, identified as an excellent site by the seeing survey reported in 53.082.063, 54.082.014, contains telescopes built by republics outside Uzbekhistan, including Russia, Ukrainia, and Lithuania. Conditions for astronomers in troubled areas such as Armenia and former Yugoslavia are even more difficult. It is not for astronomers outside the countries concerned to comment on the politics surrounding these events. However all astronomers will join in hoping that troubled colleagues, their talented students (the astronomers of the future), the irreplaceable scientific archives, the exquisite equipment and the noble astronomical activities of the affected institutes, wherever located, will be protected as best as is humanly possible.

Paul Murdin, Royal Observatory, Blackford Hill, Edinburgh, Scotland.

Bioastronomy (Commission 51)

Introduction

Commission 51 covers a very broad sweep of science, from astronomical topics, some of which overlap interests of some other commissions, to earth sciences, chemistry, biochemistry and even cognitive sciences concerned with the study of intelligence and animal behaviour. It is not feasible to attempt a complete coverage of all topics, even after due allowance for areas of overlap with other commissions. I have therefore been selective, attempting to include topics in which studies of particular interest have appeared during the triennium.

R. D. Brown

1. *Planetary discs*: R. D. Brown

Commission 51 has interest in the study of formation of new stars and planetary systems from the viewpoint of formation of potentially inhabitable locations, study of such systems also being of interest to other Commissions (e.g. 16) from other viewpoints. Commission 51 is also interested in the interstellar medium, molecular clouds and circumstellar envelopes from the aspect of origin of life and of habitable locations. The possibility that some aspects of interstellar chemistry might also be an indicator of the incipient formation of habitable locations (protoplanets) encourages me to report on some observations of celestial masers that could provide evidence of some embryo planetary systems.

Following a series of observations at Pushchino (Russia) of the 1.35 cm H_2O maser in S140 (Lekht *et al.* 1993), the maser emission being localised in three narrow, symmetrically spaced intervals of radial velocity, it was proposed that this constitutes maser amplification in a Keplerian disk surrounding a young star. It was estimated that the dimension of the masing part of the disk is <30 AU, the mass of the central star being $<5\ M_\odot$ and that the temporal variations of the spectral peaks' radial velocity could be caused by the clumps which could be either giant planets or protosatellites of giant protoplanets.

An interferometric study (Planesas *et. al.* 1992) of a hydrogen atom maser discovered (Martin-Pintado *et al.* 1989) in MWC3 shows that it probably is a disk seen edge on, whose masing part has a radius ≈ 40 AU. Monitoring of this multiline maser (Gordon 1992; Thum *et al.* 1993) indicates that the disk should be in Keplerian rotation and the thorough modelling of the observed spectral profile (Ponomarev *et al.* 1993) gave estimates of the density, temperature, turbulence and optical depth of the masing part of the disk. In this case the estimated mass of the central star is high - 26 $M_\odot$ (Ponomarev *et al.* 1993).

In the last several years it has been possible to study methanol masers with the Australia Telescope, such interferometric observations giving better resolutions of groups of masers. The first publication (Norris *et al.* 1993a,b) reports ten alignments for which a plausible interpretation is that they are indications of such condensations and there are unpublished reports of further groups of aligned methanol masers. Typically the velocity gradients in these groupings are about 4 km s^{-1} over a linear extent of about 1000 AU (5 mpc). To sustain such a gradient in Keplerian orbit would require a central mass of about 5 $M_\odot$.

These alignments have been reported only for Class II methanol masers, which, as Cragg *et al.* (1992) have shown, need a pumping source of continuum radiation warmer than the gas kinetic temperature, and so are associated with a nearby infrared source which could be a protoplanetary condensation. If these interpretations stand up to further critical scrutiny then, when the maser excitation requirements are more quantitatively tied down by further modelling, we may have a particularly powerful probe of the physical conditions prevailing in such condensations.

J. Bergeron (ed.), Reports on Astronomy, Vol. XXIIA, 583–593.

References

Cragg, D. M., Johns, K. P. Godfrey, P. D., Brown, R. D.: 1992, *Ap. J.*, **259**, 203-208.
Gordon, M. A. 1992, *Ap. J.* **387**, 701.
Lekht, E. E., Likhachev, S. F., Sorochenko, R. l., Strelnitski, V. S., 1993 *Astron. Zhur.*, **70**, 731.
Martin-Pintado, J., Thum, C. & Bachiller, R.: 1989, *A & Ap.*, **222**, L9-L11
Norris, R. P. Whiteoak, J. B. Caswell, J. L. Wieringa, M. H.: 1993a, *Astrophysical Masers*, A.W. Clegg, G. E. Nedoluha (eds.) Springer-Verlag, Berlin Heidelberg, pp.203;
Norris, R. P. Whiteoak, J. B. Caswell, J. L. Wieringa, M. H. Gough, R. G. 1993b *Ap. J.*, **412:** 222.
Planesas, P., Martin-Pintado, J. & Serabyn, E.: 1992, *Ap. J.*, **386**, L23.
Ponomarev, V. O., Smith, H. A. & Strelnitski, V. S.: 1993, *Ap. J.*, in press.

2. *Habitable zones and Planetary Habitability* J. Heidmann

The eventual success of life detection in the universe is strongly dependent on where we search for it. In particular for advanced cosmic life, the first lead we follow now is where Earth-like planets are more likely to exist. Early deliberations on habitability of planets were assembled by Dole (1970). Here we summarise recent developments in the study of this problem under three headings:

Stellar habitable zones: a suitable star: The range of "solar type" stars has been extended to later K and M types (Doyle; Kasting *et al.*, 1993) and also to double stars. Planets must be not too close and not too far from the parent, hence the concept of circumstellar habitable zones (Huang, 1959, 1960).

Planetary climatic role: Much new work has been done on the fundamental role of planetary atmospheres. Recent studies have been made on the green house effects (Kasting *et al.*) and extensive studies, theoretical and observational, on three nearby planetary bodies: Mars (Kasting, 1991; McKay, 1991; Jakosky *et al.* 1993; Baker*et al.*, 1991: Paige,, 1992), Venus (Komatsu *et al.*, 1993; Robinson & Wood, 1993) and Titan (McKay *et al.*, 1993).

Environmental zones: Other habitability factors have been recognised. For instance, because of lethal effects from supernovae, only stars in the co-rotating region of spiral galaxies would be able to develop advanced forms of cosmic life. Their sky distribution is being evaluated as preferred search regions (Heidmann, 1993). Also, a fundamental role of our Moon in preventing large, chaotic variations of the inclination of the Earth axis, pernicious for life forms, has been recognised and subjected to detailed mathematical analysis (Laskar *et al.*, 1993; Laskar & Robutal, 1993). Eventual artificial activities such as terraforming Mars (McKay, *et al.*, 1991) have also been investigated.

Recent reports : *"Planetology & the origins of life"* and *"Habitable zones"* contain up-to-date information on habitable zones.

References

Baker, V. R., Strom, R. G., Gulick, V. C., Kargel, J. S., Komatsu, G., Kale, V. S., 1991, *Nature* , **353**, 589.
Dole, S. H., 1970, *"Habitable Planets for Man"*, 2nd ed. (Elsevier, New York).
Doyle, IAF 42nd Congress 92-615.
"Habitable zones", Ames, Res. Ctr., Jan 1994, ed. Doyle.
Heidmann, J., *Acta Astronautica* 1993 in press.
Jakosky, B. M., Henderson, B. G., Mellon, M. T., 1993, *Icarus*, **102**, 286.
Kasting, J. F., 1991, *Icarus* , **94**,1.
Kasting. J. F., Whitmire, D. P., Reynolds, R. T., 1993, *Icarus*, **101**, 108.
Komatsu, G., Baker, V. R., Gulick, V. C., 1993, *Icarus*, **102**,1.

Laskar, J., Joutel, F., Robutel, P., 1993, *Nature*, **361**, 615.
Laskar, J., Robutel, P., 1993, *Nature*, **361**, 608.
McKay, C. P., 1991, *Icarus* , **91**, 93.
McKay, C. P., Pollack, J. B., Lunine, J. I., Courtin, R., 1993, *Icarus*, **102**, 88.
McKay, C. P., Toon, O. B., Kasting, J. F., 1991, *Nature*, **352**, 489.
Paige, D. A., 1992, *Nature*, **356**, 43.
"Planetology & the origins of life", Cretell, Sept. 1993, ed. Raulin & Bruston .
Robinson, C. A., Wood, J. A., 1993, *Icarus*, **102**,26.

3. *Molecular clouds*: W.M. Irvine

The number of molecular species detected in the interstellar medium, not counting isotopic variants, has continued to grow over the past three years and now numbers nearly 100. In the expanding envelopes around evolved stars, which return material to the interstellar medium, astronomers have identified the CP radical, the second phosphorus-containing astronomical molecule (Guélin et al. 1990); SiN, the first astronomical molecule with an Si–N bond (Turner 1992); the first magnesium-containing interstellar molecule, MgNC (Kawaguchi et al. 1993); and the radical HCCN (Guélin and Cernicharo 1991). In interstellar molecular clouds there have been new identification or confirmations of previous tentative detections. These include the radical CCO (Ohishi et al. 1991, Brown et al. 1991), the H_3O^+ ion, which is a crucial intermediary in the production of potentially major oxygen-bearing species H_2O and O_2 (Phillips et al. 1992); the ion SO^+ (Turner 1992, Irvine 1993), the radical CH_2N (Ohishi et al. 1993). Other newly identified species include isomers of previously known interstellar molecules, including HCCNC and CCCNH, both isomers of cyanoacetylene (Kawaguchi et al. 1992a, Kawaguchi et al. 1992b); and the two cumulenes H_2C_3 and H_2C_4, which are isomers of the pervasive interstellar molecule cyclic-C_3H_2 and of diacetylene, respectively (diacetylene itself has not been detected in the interstellar medium because of its lack of a permanent dipole moment; however, it may be presumed to be present based on the detection of the corresponding nitrile, HC_5N).

Other previously know interstellar molecules have been found in new environments. Thus, both NO and NS have recently been detected for the first time in cold, dark clouds (McGonagle et al. 1990, McGonagle et al. 1993).

A potentially interesting probe of the pathways involved in molecular production is the ortho-para ratio for species with two or more identical nuclei. Such ratios have now been reported for H_2CO, H_2CS, C_3H_2, H_2CCO, H_2C_3, H_2C_4, CH_3C_2H, and CH_3CN. These results are summarised by Minh et al. (1993), who discusses the possible implications for exchange of material between icy grain mantles and the gas phase in interstellar clouds.

Models of interstellar chemistry, which previously have for the most part considered the time-dependent chemical evolution of an interstellar cloud under conditions of constant physical properties, have now begun to address more systematically the coupled physics and chemistry evolving clouds and star-forming regions. Recent reviews in this area, which contain many references to recent work, include Harquist et al. (1993) and van Dischoeck et al. (1993).

References

Brown, R.D., Cragg, D.M., Godfrey, P.D., Irvine, W.M., McGonagle, D., and Ohishi, M. (1991), *Origins of Life and Evolution of the Biosphere* , **21**, 399.
Cernicharo, J., Gottlieb, C.A, Guélin, M., Killian, T.C., Paubert, G., Thaddeus, P., and Vrtilek, J.M. (1991), *Astrophys. J. (Letters)* **368**, L39.
Cernicharo, J., Gottlieb, C.A., Guélin, M., Killian, T.C., Thaddeus, P., and Vrtilek, J.M. (1991), *Astrophys. J. (Letters)* **368**, L43.
Guélin, M., Cernicharo, J., Paubert, G., and Turner, B.E. (1990), *Astron. Astrophys.* **230**, L9.
Guélin, M., and Cernicharo, J. (1991), *Astron. Astrophys.* **244**, L21.
Hartquist, T.W. et al. (1993), *Q.J.R.A.S.* **34**, 213.
Irvine, W.M. (1993), in *Advances in Space Research,* in press.

Kawaguchi, K., Kaifu, N., Ohishi, M., Ishikawa, S-I., Hirahara, Y., Yamamoto, S., Saito, S., Takano, S., Murakami, A., Vrtilek, J.M. Gottlieb, C.A. Thaddeus, P., and Irvine, W.M. (1991), *Publ. Astron. Soc. Japan* **43**, 607.
Kawaguchi, K., Ohishi, M., Ishikawa, S-I., and Kaifu, N. (1992a), *Astrophys. J.* (Letters) **386**, L51.
Kawaguchi, K., Takano, S., Ohishi, M., Ishikawa, S-I., Miyazawa, K., Kaifu, N., Yamashita, K., Yamamoto, S., Saito, S., Oshima, Y., and Endo, Y. (1992b), *Astrophys. J. (Letters)* **396**, L49.
Kawaguchi, K., Kagi, E., Hirano, T., Takano, S., and Saito, S. (1993), *Astrophys. J. (Letters)* **406**, L39.
McGonagle, D., Ziurys, L.M., Irvine, W.M. and Minh, Y.C. (1990), *Astrophys. J.* **359**, 121.
McGonagle, D., Irvine, W.M., and Ohishi, M. (1993), in preparation.
Minh, Y.C., Irvine, W.M., Ohishi, M., Ishikawa, S., Saito, S., and Kaifu, N. (1993), *Astron. Astrophys.* **267**, 229.
Ohishi, M., Suzuki, H., Ishikawa, S-I., Yamada, C., Kanamori, H., Irvine, W.M. Brown, R.D., Godfrey, P.D., and Kaifu, N. (1991), *Astrophys. J. (Letters)* **380**, L39.
Ohishi, M., Irvine, W.M., McGonagle, D., Saito, S., and Yamamoto, S. (1993), in preparation.
Phillips, T.G., van Dishoeck, E., and Keene, J. (1992), *Astrophys. J.* **399**, 533.
Turner, B.E. (1992), *Astrophys. J. (Letters)* **388**, L35.
Turner, B.E. (1992), *Astrophys. J. (Letters)* **396**, L107.
van Dishoeck, E.F. et al. (1993), in *Protostars and Planets III*, ed. E.H. Levy, J.I. Lunine, and M.S. Matthews, University of Arizona Press, Tucson, pp. 165-191.

4. *Chemical evolution of Earth*: R. D. Brown

Three distinct scenarios have been proposed for the prebiotic chemical evolution of the Earth's biosphere: the Miller-Urey hypothesis of the generation of prebiotic molecules by lightning discharges in the reduced atmosphere (CH_4, NH_3, H_2O and H_2) of the young Earth; the cometary seeding hypothesis of Oro (Oro 1961, 1965; Brown 1977) ; the clay-based life and genetic takeover hypothesis of Cairns-Smith. The first of these has faced increasing difficulties as evidence has accumulated that the original atmosphere of the Earth was in an oxidised, rather than reduced, state.

Recent developments relating to the last of the three scenarios are summarised below. Here is outlined a recent proposal that increases the plausibility of the second of the scenarios and simultaneously turns a spotlight on giant planets as indicators of habitable zones in planetary systems.

When Oro made his suggestion it was generally felt that the amount of organic material arriving on the Earth's surface from cometary encounters would be relatively insignificant and swamped in the primeval seas. However Delsemme (1992) has now built a scenario, based on Safronov's (1969, 1972) demonstration that the giant planets, particularly Jupiter, were responsible for scattering asteroids. This prevented formation of another large planet and led to the accretion by the terrestrial planets of volatile planetesimal material that is more than sufficient to account for the mass of the present biosphere. Delsemme estimates that such exogenous origins of water would provide a liquid layer 15 km thick on the Earth's surface, amounting to 7.5×10^{21} kg, along with 1×10^{21} kg of carbon and a similar mass of gaseous atmosphere.

Delsemme summarises evidence, too extensive to reproduce here, that the Earth was formed by hot accretion and that any early atmosphere of original volatiles would have been evaporated and dispersed. The subsequent accretion of carbonaceous material by the Earth would have included prebiotic molecules such as aminoacids, purines and pyrimidines, probably (Anders 1971) generated by the Fischer Tropsch process in a gas of solar composition. The complexity of this chemistry, particularly as it relates to chondrites, has been discussed critically by Prin and Fegley (1989) and by Cronin *et. al.* (1988).

Anders, E.: 1971, *Ann. Rev. Astron. Astrophys.* **9**, 1-34.
Brown, R. D.: 1977, *Interdisciplin. Sci. Rev.*, **2**, 124-139.

Cronin, J. R., Pizarello, S., and Cruikshank, D. P.: 1988 *'Meteorites and the Early Solar System'*,Kerrige and Matthews (eds.) Univ. of Arizona Press, Tucson, pp.819-857.
Delsemme, A. H.: 1992, *Orig. Life and Evol. Bio.*, **21**, 279-298.
Oro, J. 1961 *Nature*, **190**, 389-390.
Oro, J. 1965 *"The Origins of Prebiological Systems and their Molecular Matrices'*, S. W. Fox 9ed.) Academic Press, New York, pp. 137-162.
Prin, R. G. and Fegley, B.: 1989 *'Origin of Planet and Satellite Atmospheres'*,136, S.K. Atreya *et. al.* (eds.) Univ. of Arizona Press, Tucson, pp.78-136.
Safronov, V. S. 1969 *'Evolution of Protoplanetary Cloud and Formation of the Earth and the Planets'*, translated in NASA TTF-667-1972.

5. *Developments in Chemical Origins of Life 1990-1993* A. G. Cairns-Smith

An enthusiasm of the late 80s for the idea that life on the Earth could have started with nothing but RNA molecules (acting as catalysts for their own replication) has been tempered by an appreciation of the difficulties of making the activated nucleotide necessary to set up such an "RNA world". Attention is now being paid to the idea that evolution might have got going with a somewhat different chemistry.

RNA-type molecules with modified backbone have been considered as primordial genetic material (reviewed by Schwartz 1992). Such research is of interest in its own right in helping to define the range of possible replicating molecules (reviewed extensively by Orgel, 1992). There have indeed been moves sideways, as it were, towards the consideration of models which might allow evolutionary processes to be set going in the laboratory based on molecules which may have little if any connection with our biochemistry. Both Rebek and von Kiedrowski and their colleagues have made different kinds of artificial replicating molecule (Rotello, Hong & Rebek, 1991; Terfort & von Kiedrowski, 1992) - a considerable landmark even if as yet their molecules are too simple to have any evolutionary potential.

Off the tack of primary genetic materials De Duve (1991) has made a case for sulphur, e.g. in thioesters, as well as iron, as having had more dominant biochemical roles to begin with. This fits with Wächtershäuser's (1990) model of a pre-evolution on the surfaces of iron sulphide minerals.

One place where such minerals are to be found are hydrothermal systems where hot solutions circulate within the Earth's crust, and the idea that life might have started in some such relatively protected environment has been re-considered in some detail (Holm, 1992).

The possible roles of comets in the origin of life (as providers and/or destroyers) have been similarly re-considered (Thomas, 1991/2). Perhaps the most interesting suggestion here is that comets may have been responsible not for providing biochemicals as such but, much more fundamentally, for giving the Earth its supplies of carbon, nitrogen and water, see Section 4.

References

de Duve, C. (1991). *Blueprint for a Cell.* Burlington, North Caroline: Neil Patterson.
Holm, N.G. (editor) 1992. Marine hydrothermal systems and the origin of life. Special issue of *Origins of Life and Evolution of the Biosphere*, **22**, 1, 242.
Orgel, I. E. (1992). Molecular replication. *Nature*, **358**, 203-209.
Rotello, V., Hong, J-I. & Rebek Jr, J. (1991). *J. Am. Chem. Soc.* **113**, 9422-9423.
Schwartz, A.W. (1992). Getting the RNA world started. In *Frontiers of Life* ed. J & K Tian Thanh Van, Cif-sur-Yvette: Editions Frontiers, pp 145-156.
Terfurt, A. & von Kiedrowski, C. 1992. Self-replication by condensation of 3-amino-benzamidines and 2-formylphenoxyacetic acids. *Angew. Chem. int. Ed. Engl.* **31**, 554-655.
Thomas, P.J. (editor) 1991/2. Comets and the origin and evolution of life. Special issue of *Origins of Life and Evolution of the Biosphere*, **21**, 265-436.

Wächtershäuser, G. (1990). The case for the chemotropic origin of life in an iron-sulphur world. *Origins of Life and Evolution of the Biosphere*, **20**, 173-176.

6. *Intelligence, Cognition and Communication in Other Species*: Diana Reiss

As we search for evidence of intelligent life elsewhere in the universe we face the question of whether we could recognise and evaluate intelligence in extraterrestrials. Reiss (198) and Marino (1993) have suggested that bioastronomers, biologists and psychologists share common problems in exploring non-human intelligence and communication in extraterrestrials or other non-human terrestrial species. Initially we face decoding issues which include 1) signal detection, 2) signal structural analysis, and 3) signal function. (Reiss, 1993a). Secondly, in investigating intelligence and evolutionary continuity of mental experience in other species we are challenged in finding appropriate measures and frameworks for assessing information processing strategies used by other species with body forms and environmental histories different from our own. Past views that humans are unique as a symbol and tool using species are currently being challenged as we have reports of highly divergent species such as chimpanzees (Gardner & Gardner, 1969; Fouts, 1973, Savage-Rumbaugh, 1986), African grey parrots (Pepperberg, 1981), California sea lions (Schusterman & Krieger, 1984), and bottlenose dolphins (Reiss & McCowan), 1993; Richards, Woltz & Herman, 1984) comprehending and/or producing symbolic codes. Several of these species have also demonstrated varying degrees of tool use. There is increasing evidence of tool use by non-human species (Seyfarth, Cheney, and Marier, 1980; Owings and Leger, 1980).

Evidence for vocal learning has been found in three highly divergent species, humans, birds (Marler, 1969) and dolphins (Reiss and McCowan, 1993). Research in this area strongly suggests that common strategies may be used in acquiring vocal repertoires across species.

Studies have demonstrated that chimpanzees, and orangutans as well as humans can recognise their reflections in mirrors (for review see Povinelli, 1987). Marino, Reiss and Gallup (1993) have found suggestive evidence of self-recognition in bottlenose dolphins. Diverse species such as elephants (Povinelli, 1989), and African grey parrots (Pepperberg, 1993) have demonstrated mirror guided behaviour in accomplishing tasks.

Evidence for conceptual quantitative abilities has been reported in the African grey parrot (Pepperberg, 1987) and the chimpanzee (Boysen and Berntson, 1989).

There is clearly much more complexity and plasticity found in the communication systems of other species that previously thought. More research is needed with more diverse species to more adequately access the extent to which referential communication, quantitative abilities and complex information processing has developed throughout the biological world.

References

Boysen, S.T. and Berntson, G,G. (1989). Numerical Competence in a chimpanzee (Pan troglodytes). *Journal of Comparative Psychology*, **103**, 23-30.

Fouts, R.S. (1973). Acquisition and testing of gestural signs in four young chimpanzees. *Science*, **180**: 978-980.

Gardner, R.A. and Gardner, B.T. (1969). Teaching sign language to a chimpanzee. *Science*, 1965: 664-672.

Marion, L.A., Reiss, D. and Gallup, Jr. G.G. (1993). A test of self recognition in dolphins: implications for comparative investigations of highly dissimilar species. In: *Self-awareness in Humans and Nonhumans*. S. Parker, R. Mitchell, and M. Boccia (eds.) Cambridge University Press.

Marino, L.A., (1993) SETI begins at home: Can we define and measure intelligence among species n this planet? *1993 Bioastronomy Symposium*, University of California, Santa Cruz.

Owings, D.H. and Lager, D.W. (1980). Chatter vocalizations of California ground squirrels: Predator and social role - specificity. *Zeitschrift fur Tierpsychologie*, **54**, 163-184.
Pepperberg, I.M. (1981) Functional vocalizations by an African abilities in the African parrot: labelling cardinal sets. *Ethology*, **75**, 37-61.
Povinelli, D.J. (1987) Monkeys, apes, mirrors and minds: The evolution of self-awareness on primates. *Human Evolution*, **2**, 493-509.
Povinelli, D.J. (1989). Failure to find self-recognition in Asian Elephants (Elephas maximus) in contrast to their use in mirror cues to discover hidden food. *Journal of Comparative Psychology*, , **103**, 122-131.
Reiss, D. (1988). Can we communicate with other species on this planet: pragmatics of communication between humanoid and nonhumanoid species. In *Biostronomy - The Next Steps* (G. Marx, ed.) Kluwer Academic Pubs. 253-264.
Reiss, D, and McCowan, B. (1993) Spontaneous vocal mimicry and production by Bottlenose Dolphins (Tursiops truncatus). evidence for vocal learning, *Journal of Comparative Psychology*. **107**, 301-312.
Reiss, D. (1993a) Studying the communication signals of bottlenose dolphins (Tursiops truncatus). *1993 Bioastronomy Symposium*, University of California, Santa Cruz.
Richards, D.G. Wolz, J.P. and Herman, L.M. Vocal mimicry of computer generated sounds and vocal labelling of objects by a bottlenose dolphin. *Journal of Comparative Psychology*, **98**, 10-28.
Savage-Rambaugh, E.S. (1986). *Ape language: From conditioned response to symbol.* New York: Columbia University Press
Schusterman, R.J. and Krieger, K. (1984) California sea lions are capable of semantic comprehension. *The Psychological Record*, **34**, 3-23.
Seyfarth, R.M., Cheney, D.L. and Marier, P. (1980). Vervet monket alarm calls; semantic communication in a free-ranging primate. *Animal Behavior*, **28**, 1070-1094.

7. *SETI*: J. Tarter

During the past triennium there have been nine SETI observational programs active around the world. Some of them are literally backyard operations, and others are major undertakings utilizing many large telescope facilities. Most, but not all, are conducted at radio frequencies and involve the US, Argentina, and Australia. The following tables summarize the characteristics of these nine programs, which are listed in order of the observational start date.

TABLE 1- OSURO; *Ohio State University Radio Observatory*

PI's: Prof. Robert Dixon and Prof. John Kraus

ANTENNA: Kraus design / 52 m parabolic equivalent

FREQUENCY COVERAGE: 1400 - 1700 MHz

POLARIZATION: single linear

FREQUENCY RESOLUTION: 100 kHz

SPECTROMETER: 3000 channel filter bank

SENSITIVITIY: 1.5×10^{-21} W/m^2

SEARCH STRATEGY: Sky Survey at sidereal rate. Data passing threshold stored for later analysis. Tracking mode is automatically triggered to follow an "interesting" signal until it sets.

TABLE 2 - *Small SETI Observatory*

PI's: Robert Grey

LOCATION: Chicago, Illinois

ANTENNA: 4 m satellite dish

FREQUENCY COVERAGE:

POLARIZATION: single linear

FREQUENCY RESOLUTION: variable from 100 Hz to 1 Hz

SENSITIVITIY: 10^{-21} W/m^2

SEARCH STRATEGY: Concentration on area of sky from whence the OSURO 1977 "wow" signal appeared to originate, also searches of M31 and M33.

TABLE 3 - META *Mega-Channel ExtraTerrestrial Assay*

PI's: Prof. Paul Horowitz, Harvard University

LOCATION: Oak Ridge Observatory in Harvard, Massachusetts

ANTENNA: 26 m f200 kHz

POLARIZATION: sequential, dual circular

FREQUENCY RESOLUTION: 0.05 Hz

SPECTROMETER: 8,388,608 channel FFT spectrometer with continuous self-test and mappable, redundant processors

SENSITIVITIY: 7 x 10^{-24} W/m^2

SEARCH STRATEGY:Sky Survey at sidereal rate over 68% of sky (-30 to +60 degrees declination), moving 1/2 degree per day. Reobservation within minutes for signals exceeding 20 sigma. Magic frequencies sequentially observed with respect toHeliocentric, Galactocentric and 3K Blackbody Background frames of rest.

TABLE 4 - META II

PI's: Prof. F. Raul Colomb

LOCATION: Instituto Argentino de Radioastronomia, Buenos Aires, Argentina

ANTENNA: 30 m fully steerable parabola

FREQUENCY COVERAGE: 1420 MHz +/- 200 kHz

POLARIZATION: sequential, dual circular

FREQUENCY RESOLUTION: 0.05 Hz

SPECTROMETER: META clone, 8,388,608 channel FFT spectrometer

SENSITIVITIY: 4 x 10^{-24} W/m^2

SEARCH STRATEGY: Half of observatory time spent on sky survey of southern sky for magic frequency signal in same manner as META in the North

---

TABLE 5 - TMTS *Ten Micron Targeted Search*

PI's: Prof. Albert Betz, University of Colorado, Boulder

LOCATION: Mount Wilson Observatory, Pasadena, CA

ANTENNA: 1.7 m (one of two)

FREQUENCY COVERAGE: 10 microns (30 THz +/- 60 km/s)

POLARIZATION: single linear

FREQUENCY RESOLUTION: 2.4 and 5 MHz

SPECTROMETER: Heterodyned, acousto-optical spectrometers with either 1024 or 2048 channels

SENSITIVITIY: 9×10^{-18} W/m^2

SEARCH STRATEGY: Targeted Search of 200 nearby, solar-type stars, looking for intentional CO2 laser signals

---

TABLE 6 - *Search for Interstellar Communication Channel*

PI's: Prof. David Blair, University of Western Australia

LOCATION: Parkes Observatory, NSW Australia

ANTENNA: 64 m

FREQUENCY COVERAGE: 4462.3 MHz (pi x HI)

POLARIZATION: two linear polirizations with three different frames of reference spread between them

FREQUENCY RESOLUTION: 100 Hz

SPECTROMETER: 2x512 channel correlator and HP3582A spectrum analyser

SENSITIVITIY: 2×10^{-24} W/m^2

SEARCH STRATEGY: Targeted search of -100 southern, nearby, solar-type stars

---

TABLE 7 - SERENDIP III: *Search for Extraterrestrial Radio Emission from Nearby Developed Intelligent Populations*

PI's: Prof. Stuart Bowyer, University of California Berkeley

LOCATION: Arecibo Observatory, Puerto Rico

ANTENNA: 305 m spherical dish

FREQUENCY COVERAGE: 423-435 MHz

POLARIZATION: single linear

FREQUENCY RESOLUTION: 0.6 HzSPECTROMETER: 4.2 million channel FFT spectometer

SENSITIVITIY: 6×10^{-25} W/m^2

SEARCH STRATEGY: Commensal search, non-directed sky survey at twice negative sidereal rate

---

TABLE 8A - HRMS SS: *High Resolution Microwave Survey Sky Survey*

PI's: Project Scientist for SS is Dr. Sam Gulkis, JPL

LOCATION: DSS13 at Goldstone DSN Complex in Mohave Desert

ANTENNA: 34 m beam-waveguide

FREQUENCY COVERAGE: X-band +/- 40 MHz

POLARIZATION: sequential, dual circular

FREQUENCY RESOLUTION: 20 Hz

SPECTROMETER: 2 million channel FFT spectrometer

SENSITIVITIY: 3 x 10^{-23} W/m^2

SEARCH STRATEGY: Sky Survey of northern hemisphere at 0.2 degree per second, using "racetrack" scans of invividual skyframes. Largest events above threshold along a scan line and co-located on adjacent scan lines are followed up with targeted observations.

TABLE 8B - HRMS TS: *High Resolution Microwave Survey Targeted Search*

PI's: The Project Scientist for HRMS is Dr. Jill Tarter, Ames Research Center

LOCATION: Arecibo Observatory, Puerto Rico

ANTENNA: 213 m spherical dish

FREQUENCY COVERAGE: 300 MHz in range 1300 to 2300 MHz

POLARIZATION: dual circular

FREQUENCY RESOLUTION: 1, 7 and 28 Hz simulataneously

SPECTROMETER: 14 million channel FFT spectrometer at 1 Hz, 2 million channel spectrometer at 7 Hz and .5 million channel spectrometer at 28 Hz

SENSITIVITIY: 3 x 10^{-27} W/m^2

SEARCH STRATEGY: Targeted Search of an initial list of 25 solar-type stars passing directly overhead at Arecibo searching for narrowband pulses and CW signals that may be drifting slowly in frequency.

TABLE 9 - *Columbus Optical SETI Observatory*

PI': Dr. Stuart Kingsley

LOCATION: Columbus, Ohio (home address)

ANTENNA: 10" Meade LX200 Schmidt-Cassegrain

FREQUENCY COVERAGE: Optical and near IR

POLARIZATION: -

FREQUENCY RESOLUTION: -

SPECTROMETER: -

SENSITIVITIY: -

SEARCH STRATEGY: This facility is so new that it has not yet begun a formal search program.

WORKING GROUP FOR PLANETARY SYSTEM NOMENCLATURE (WGPSN)

(GROUP DE TRAVAIL POUR LA NOMENCLATURE DU SYSTEM PLANETAIRE)
(Committee of the Executive Committee)

PRESIDENT: K. Aksnes
MEMBERS: R. Batson, A. Brahic, M. Fulchignoni, M.Ya. Marov
D. Morrison, T.C. Owen, V.V. Shevchenko, B.A. Smith
L.A. Soderblom
CONSULTANTS: J.M. Boyce, G. Burba, P. Masson

Since the IAU General Assembly in Buenos Aires in 1991, WGPSN has held its 21st and 22nd regular meetings in Bergen, Norway June 18-19 1992 and at Belgiraté, Italy June 19 1993. The Bergen meeting was attended by 8 members from the Working Group(WG) and 6 from The Task Groups(TG). At Belgirate, 6 WG and 4 TG members were present. In addition, an ad hoc nomenclature meeting was conducted in Liege, Belgium on 25 June 1992 to take advantage of the presence of 7 WG and TG members assembled there for another meeting. Much of the nomenclature discussions are now carried out by electronic mail which has taken over for correspondance by regular mail almost entirely. Telefax is used to reach the few members not reachable by e-mail.

During the triennium, most of the nomenclature work has concerned Venus on which planet thousands of new features have been discovered with the radar on the highly successful Magellan spacecraft.

For the first time in history the surface of an asteroid, 951 Gaspra, has been imaged, during a close passage by the Galileo spacecraft in October 1991.

RESOLUTIONS

1. The WGPSN

* Recognizes increasing requirements for names of small bodies in the solar system (including minor planets, comets, and planetary satellites)

* Notes the many common issues and procedures for establishing such nomenclature that have led to consultation and cooperation between the WGPSN and the Minor Planet Names Committee (MPNC) of IAU Commission 20

* Supports the initiative of Commision 20 in seeking closer cooperation between the WGPSN and the nomenclature activities of Commission 20, and therefore

* Recommends to the Executive Committee of the IAU that:

- A special joint committee or working group be established with representatives of Commissions 15, 16, and 20, and WGPSN

- This working group be charged to formulate a proposal for a unified procedure for establishing nomenclature for the planetary system including naming of minor bodies as well as nomenclature for surface features on solar system objects

- The working group present its proposal to the XXII IAU General Assembly

J. Bergeron (ed.), Reports on Astronomy, Vol. XXIIA, 595–603.

2. The WGPSN

* R e c o g n i z e s increasing requirements for names on planetary bodies and page limitations on planetary nomenclature in the IAU Transactions volumes

* N o t e s that the National Aeronautics and Space Administration or the U.S. Geological Survey may be willing to undertake regular publication of a Gazetteer of Planetary Nomenclature under the auspices of the WGPSN

* R e c o m m e n d s to the Executive Committee of the IAU that this Gazetteer be recognized by the IAU as an official documentation of planetary nomenclature approved by the IAU, in place of or in addition to that published in the IAU Transactions

NEW GUIDELINES FOR NOMENCLATURE

1. It is the sense of the WGPSN that namebanks should be established for transjovian planetary/asteroidal/"cometary" bodies. Names for these objects should be selected along lines that include the following:

(a) objects in 1:1 mean-motion resonance with Jupiter: heroes of the Trojan War.

(b) objects crossing (or approaching) the orbits of at least one of the giant planets and not in a stabilizing resonance: centaurs.

(c) objects crossing (or approaching) the orbit of Neptune and in a stabilizing resonance (other than 1:1) with it: mythological names associated with the underworld.

(d) objects with orbits sufficiently outside the orbit of Neptune that stability is presumed for a substantial fraction of the lifetime of the solar system: mythological names associated with creation.

2. Whereas the spacecraft engaged in planetary exploration have revealed worlds beyond our imagination, be it resolved that the WGPSN shall permit the names of these spacecraft to be commemorated in the planetary systems they have visited.

3. For the small bodies in the solar system, craters may be given names associated with the name and origin of the body itself, while distinct regions (Regio) may be named for the discoverer or associated people. Other features may receive names associated with the place of discovery.

NEW NOMENCLATURE

The WG is seeking IAU approval for the following 332 new names on Venus, 1 on Mars, 1 on Ganymede, 1 on Proteus, and 1 on the Moon.

VENUS

NAME	LAT	LON	DIAM (KM)	ATTRIBUTE
CRATER				
Abaka	-52.6	104.4	14	Mari first name
Abigail	-52.2	111.3	18.0	First name from Hebrew
Abington	-47.8	278.0	22	Frances; English actress 1737-1815
Adaiah	-47.3	253.3	19	Hebrew first name
Adamson	-14.8	29.7	20	Joy; Austrian author 1910-1980
Addams	-56.1	98.0	85	Jane; American social reformer 1860-1935
Aethelflaed	-18.2	196.4	20	English leader of the Mercians c.884-918
Aimee	16.2	127.3	18	French first name
Aita	8.9	270.7	14	Estonian first name
Ajana	-29.1	175.5	16	Altai first name
Akeley	8.0	244.3	25.0	Delia; American explorer 1875-1970
Akiko	30.7	187.1	25	Yosano; Japanese Tanka poetess 1878-1942
Aksentyeva	-42.0	271.9	40	Zinaida;Soviet geophys.,astron. 1900-1969
Alima	-45.9	229.3	13	Tatar first name
Alimat	-29.5	205.9	14	Osset first name
Alison	- 4.0	165.6	18.0	Irish first name
Alma	- 2.4	228.7	17.0	Kazakh first name
Anaxandra	44.2	162.2	23	Greek artist fl. c.228 B.C.
Andreianova	- 3.0	68.8	70	Elena; Russian ballerina c.1821-c.1855
Anush	15.0	86.5	13	Armenian first name
Asmik	3.9	166.4	17.0	Armenian first name
Audrey	23.8	348.1	15.0	English first name
Austen	-25.0	168.3	47	Jane; English novelist 1775-1817
Baranamtarra	18.0	267.8	25	Mesopotamian queen c. 2500 B.C.
Barrymore	-52.3	195.6	50	Ethel; American actress 1879-1959
Bascom	-10.3	302.2	36	Florence; American geologist 1862-1945
Bashkirtseff	14.7	194.1	36.0	Marie;Russian painter,diarist c.1859-1884
Bathsheba	-15.1	49.3	36	Hebrew queen c.1030 B.C.
Beausoleil	- 5.0	102.9	29	Martine; French earth scientist 17th Cen.
Beecher	13.0	253.4	45.0	Catherine;Amer. educator,author 1800-1878
Bernadette	-46.7	285.6	14	French first name
Blackburne	11.0	183.8	36	Anna; English biologist 1726-1793
Boivin	4.3	299.5	26	Marie;French medical researcher 1773-1847
Boleyn	24.5	220.0	70	Ann; English queen 1507-1536
Bonheur	9.8	288.6	55.0	Rosa; French painter 1822-1899
Bonnin	- 6.2	117.6	40	Gertrude;Am. reformer,writer 1875-1938
Boyd	-39.3	221.3	25	Louise; American explorer 1887-1972
Boye	-9.6	292.3	29	Karen; Swedish poet, novelist 1900-1941
Bradstreet	16.5	47.6	40	Anne; American poet c.1612-1672
Bryce	-62.6	197.0	25	Lucy; Austral. medical pioneer 1897-1968
Bugoslavskaya	-23.0	300.4	29	Yevgenia; Soviet astronomer 1899-1960
Caiwenji	-12.4	287.4	20.0	Chinese painter, calligrapher 907-960
Caldwell	23.6	112.0	48	Taylor; American author 1900-1985
Caroline	6.8	306.3	16	First name from French
Carr	-24.0	295.7	27	Emily; Canadian artist 1871-1945
Carter	5.3	67.2	20.0	Maybelle; Am. singer,songwriter 1909-1978
Centlivre	19.0	290.4	25	Susannah; English actress c.1667-1723
Cheryl	18.5	326.6	15	First name from French
Chloe	- 7.4	98.6	19.0	First name from Greek
Cohn	-33.2	208.0	23	Carola; Australian artist 1892-1964
Colleen	-60.8	162.2	15	Irish first name
Comnena	1.2	343.7	20.0	Anna;Byz. princess,phys.,writer 1083-1148
Conway	48.3	39.1	49	Anne Finch;Eng. natural scient. 1631-1679
Corpman	0.3	151.8	39	Elizabeth;Polish astr.,wife of Hevelius

VENUS (Cont.)

NAME	LAT	LON	DIAM (KM)	ATTRIBUTE
CRATER				
Cortese	-11.4	218.3	29.0	Isabella,It. physician,med. writer d.1561
Daphne	41.3	280.4	14	First name from Greek
Datsolalee	38.3	171.6	20	Washo Indian artist,basketmaker 1835-1925
de Ayala	12.3	31.9	20	Josefa; Spanish painter 1630-1684
De Witt	-6.6	275.5	20	Lydia; American pathologist 1859-1928
Delilah	-57.9	250.0	19	First name from Hebrew
Dinah	-62.8	37.0	19	Hebrew first name
Doris	2.3	89.9	16	First name from Greek
du Chatelet	21.5	165.0	20	Emilie;French math., physicist 1706-1749
Eileen	-22.8	232.5	15.0	Irish first name
Elena	-18.3	73.3	18.0	Italian first name
Elza	-34.4	275.9	19	Latvian first name
Erin	-47.0	184.8	15	Irish first name
Escoda	18.2	149.4	20	Philipino Girl Scouts organizer 1898-1945
Este	-34.2	238.7	20	Isabella;Italian archaeologist 1474-1539
Estelle	1.2	93.7	19	First name from Latin
Esther	19.4	21.8	18.0	First name from Persian
Eudocia	-59.1	201.9	29	Byzantine empress c. 401-460
Evangeline	69.7	223.0	15.0	First name from Greek
Evika	- 5.0	31.4	16.0	Tatar first name
Faiga	4.9	170.9	11	Anglo-Saxon first name
Farida	4.8	38.9	19	Azerbaijan first name
Felicia	-19.8	226.5	10	First name from Latin
Foquet	-15.1	203.5	40	Marie;French med. wr.,char. worker 17th Ce
Fossey	2.0	188.7	29	Dian; Am. zoologist,conserv. 1932-1985
Gabriela	-17.8	240.4	18	First name from Hebrew
Galindo	-23.3	258.7	23.0	Beatrix; It. physician, educ. 1473-1535
Gaze	17.9	240.2	30	Vera; Soviet astronomer 1899-1954
Gentileschi	45.2	260.6	20	Artemisia; Italian painter 1593-c.1652
Gillian	-15.2	49.9	15	First name from Latin
Gilmore	- 6.6	132.8	23	Dame Mary; Australian poet 1865-1962
Glaspell	-58.4	269.6	25.0	Susan; Am. playwright,novel. c.1876-1948
Godiva	-56.2	251.3	35.0	(Godgifu); Mercian noblewoman c.1040-1085
Goncharova	-62.9	97.8	33.0	Natalya; Russian artist 1881-1962
Grace	-13.8	268.5	19	First name from Greek
Gregory	7.2	95.8	20	Isabella; Irish playwright 1852-1932
Grimke	17.3	215.3	38	Sarah; American abolishionist 1792-1873
Guan Daosheng	-61.1	181.7	45.0	Chinese painter, calligrapher 1262-1319
Gudrun	10.6	326.3	15	First name from Norse
Gwynn	9.7	37.2	31	Nell; English actress,courtesan 1650-1687
Hannah	17.9	102.7	18	First name from Hebrew
Heidi	23.6	350.1	14	First name; form of Hester
Heloise	40.0	51.9	42	French physician, hospital founder c.1098-
Howe	-44.3	174.7	40.0	Julia; American biogr., poet 1819-1910
Hull	59.4	263.3	42	Peggy; American war correspond. 1889-1967
Ilga	-12.4	308.8	15	Latvian first name
Inira	-43.1	239.1	18	Eskimo first name
Irene	49.8	134.0	14.0	First name from Greek
Isabella	-29.8	204.2	165	Of Castile; Spanish queen 1451-1504
Jacqueline	-70.0	123.8	16	First name from French
Jennifer	- 4.6	99.8	19.0	First name from Greek

VENUS(Cont.)

NAME	LAT	LON	DIAM (KM)	ATTRIBUTE
CRATER				
Jerusha	-22.0	342.7	16.6	Hebrew first name
Jocelyn	-33.2	276.4	14	German first name
Johanna	19.5	247.2	18	Hebrew first name
Johnson	51.9	254.2	22.5	Amy; Australian aviator 1903-1941
Juanita	-62.9	89.5	17	Spanish first name
Judith	-29.2	104.3	17.5	Hebrew first name
Kaikilani	-32.8	163.1	20	First female ruler of Hawaii c.1555
Kanik	-32.5	249.9	16	Sakhalin first name
Kelea	8.9	25.6	25	Chieftess of Maui c.1450
Kenny	-44.3	271.1	50	Elizabeth;Australian nurse 1886-1952
Kingsley	-22.7	306.4	25	Mary; English explorer, writer 1862-1900
Kiris	20.9	98.8	14.5	Latvian first name
Kitna	-28.9	277.3	15	Kamchatka first name
Klafsky	-20.8	188.0	28.0	Katherina; Hungar. opera singer 1855-1896
Konopnicka	14.5	166.7	20	Marie; Polish author 1842-1910
Langtry	-17.0	155.0	50	Lillie; English actress 1853-1929
Laurencin	-15.4	46.4	30	Marie; French painter 1885-1956
Leah	-34.2	187.7	17.0	Hebrew first name
Lebedeva	45.2	49.6	42	Sarah; Russian sculptor 1881-1968
Leila	-44.3	86.8	19.0	First name from Arabic
Lenore	38.7	292.2	17	Greek first name (form of Helen)
Letitia	34.6	288.7	15	First name from Latin
Leyster	1.0	259.9	45	Judith; Dutch painter 1609-1660
Liede	-23.3	266.6	19	Estonian first name
Lilian	25.6	336.0	13.5	First name from Hebrew
Lineta	- 5.0	354.1	15.3	Latvian first name
Lockwood	-32.8	51.5	23	Belva; American lawyer,feminist 1830-1917
Lois	-17.9	214.7	18	First name from Greek
Lonsdale	55.6	222.4	45	Kathleen; Eng. physicist 1903-1971
Lydia	10.7	340.8	15.0	First name from Greek
Lyon	-66.5	271.0	20.0	Mary; American educator 1797-1849
Ma Shouzhen	-35.7	92.6	21	Chinese poet, painter 1592-1628
MacDonald	30.0	120.2	20	Flora; Scottish heroine 1722-1790
Madeleine	-4.8	293.2	19	French first name
Mahina	-2.0	182.2	15	Hawaiian first name
Manton	9.3	26.9	20	Sidnie; English zoologist 1902-1980
Maranda	4.9	169.8	15	Latvian first name
Margarita	12.8	9.2	13	Greek first name
Marie	-21.7	232.4	15.0	French first name
Martinez	-11.7	174.7	25	Maria; Pueblo artist, potter 1886-1980
Marzhan	-58.9	248.4	13	Karakal first name
Medhavi	-19.5	40.6	30	Ramabai;East Ind. aut.,humanit. 1858-1922
Melba	4.8	193.3	23	Dame Nellie; Austral. op. sing. 1861-1931
Merit Ptah	11.3	115.7	20	Egyptian queen, physician c.2700 B.C.
Michelle	-19.5	40.4	14	First name from French
Montagu	36.9	177.5	20.0	Mary; Eng. med. pioneer,writer 1689-1762
Montez	17.9	266.6	20	Lola; Irish dancer 1818-1861
Moore	-30.3	248.4	22	Marianne; American poet, editor 1887-1972
Morisot	-61.2	211.4	60.0	Berthe; French artist 1841-1895
Mowatt	-14.6	292.3	39	Anna; Am. actr., playwr.,author 1819-1870
Mumtaz-Mahal	30.3	228.3	35	Mogul empress, 1592-1631
Munter	-15.3	39.3	36	Gabriele; German painter 1877-1962

VENUS(Cont.)

NAME	LAT	LON	DIAM (KM)	ATTRIBUTE
CRATER				
Muriel	-41.7	12.3	18.0	First name from Greek
Nadine	7.8	359.1	18.0	First name from French
Nalkowska	28.2	290.0	22	Zofia;Polish novel., playwright 1884-1954
Naomi	6.0	70.1	18.0	First name from Hebrew
Nilsson	-75.9	278.0	32.0	Christine;Swed. op. sing.,viol. 1843-1921
Nin	-3.9	266.4	27	Anais; French novelist 1903-1977
Nina	-55.5	238.4	10.0	First name from Russian
O'Keeffe	24.5	228.7	72.0	Georgia; American artist 1887-1986
Oakley	-29.3	310.3	25	Annie; Am. sharpshooter,entert. 1860-1926
Olesnicka	18.3	210.8	33	Zofia; Polish poet fl. c.1550
Olga	26.2	283.8	17	Russian first name
Pamela	11.1	238.5	13	English first name
Peck	-28.9	294.2	30	Annie;Am. mountaineer, educator 1850-1935
Peña	-23.6	190.7	32	Tonita;(Quah Ah) Pueblo artist 1895-1949
Phaedra	35.9	252.7	15	First name from Greek
Phyllis	12.3	132.4	13	First name from Greek
Piscopia	1.5	190.8	27	Elena; Italian math., educator 1646-1684
Rachel	-48.7	13.5	11.5	First name from Hebrew
Richards	2.5	195.9	27	Ellen;founded sc. of ecology 1842-1911
Romanskaya	23.2	178.4	32.0	Sofia; Soviet astronomer 1886-1969
Rose	-35.2	248.2	15	German first name
Rowena	10.4	171.3	18	Celtic first name
Sabin	-38.5	274.8	33	Florence; Am. medical res. 1871-1953
Sabira	-5.8	239.9	15	Tatar first name
Salika	- 5.0	97.7	18.0	Mari first name
Samantha	45.5	281.4	15	First name from Aramaic
Sandel	-44.3	211.0	20	Cora; Norwegian author 1880-1974
Sanger	33.8	288.6	85	Margaret; American medical res. 1883-1966
Sarah	-42.4	1.7	19.0	Hebrew first name
Seymour	18.2	326.5	65	Jane; English quenn c.1509-1537
Shakira	3.0	213.5	19.0	Bashkir first name
Sirani	-31.5	230.1	30	Elisabetta;It. painter, etcher 1638-1665
Sitwell	16.6	190.2	30	Dame Edith; Eng. poet, critic 1887-1964
Sophia	-28.7	18.7	15	First name from Greek
de Staël	37.4	324.2	24	Anne; French histor., novelist 1766-1817
Stanton	-23.3	199.1	104	Elizabeth C.; Am. suffragist 1815-1902
Steinbach	-41.4	256.8	20	Sabina; German sculptor c.1250
Stephanie	-17.8	31.9	15	First name from Greek
Stowe	-43.3	233.2	78	Harriet B.; American novelist 1811-1896
Suliko	9.6	214.6	19	Georgian first name
Sullivan	- 1.3	111.0	22	Anne; Am. teacher of H. Keller 1866-1936
Surija	5.3	178.3	15.0	Azeri first name
Susanna	6.0	93.3	17.0	First name from Hebrew
Taira	-1.5	296.8	19	Osset first name
Tanya	-19.3	282.7	14	Russian first name
Tarbell	5.6	127.5	50	Ida; American editor, author 1857-1944
Taussig	- 9.2	228.8	25	Helen; Am. pediatr.,heart res. 1898-1986
Teasdale	0.8	76.3	35	Sarah; American poet 1884-1933
Teresa	-42.5	9.9	17	First name from Greek
Thomas	-13.0	272.6	26	Martha; Am. college president 1857-1935
Toklas	0.7	273.2	20	Alice; Am. writer, art patron 1877-1967

VENUS (Cont.)

NAME	LAT	LON	DIAM (KM)	ATTRIBUTE
CRATER				
Trollope	-54.8	246.2	27	Frances; English novelist 1780-1863
Truth	28.7	287.7	45.0	Sojourner; Am. abolishionist 1797-1883
Tsiala	2.9	100.0	17.0	Georgian first name
Tubman	23.6	204.5	50	Harriet; American abolishionist 1820-1913
Tussaud	21.8	220.9	25	Marie; Swiss wax artist 1760-1850
Udaltsova	-20.3	275.3	27	Nadezhda; Russian artist 1885-1961
Văcărescu	-62.9	199.8	32	Helene;Rumainian poet, novelist 1866-1947
Valadon	-49.1	167.0	30	Suzanne; French painter 1865-1840
Vallija	26.4	120.0	15.0	Latvian first name
Veronica	-38.1	124.6	14	First name from Latin
Virginia	-52.9	185.9	18.0	First name from Latin
Virve	- 5.1	346.8	18.5	Estonian first name
von Schuurman	- 5.0	191.0	28	Anna;Dutch ling., writer,artist 1607-1678
von Suttner	-10.6	234.9	20	Bertha;Austrian journ, pacifist 1843-1914
Wang Zhenyi	13.2	217.7	25	Chinese astron, geophysicist 18th Cen.
Warren	-11.8	176.5	50	Mercy; Am. poet,playwr.,histor. 1728-1814
Weil	19.4	283.0	22.5	Simone; French author 1909-1943
Wen Shu	-5.0	303.7	33	Chinese painter 1595-1634
West	26.1	303.0	29	Rebecca; Irish novel., actress 1892-1983
Wheatley	16.6	268.1	72.5	Phillis; American black writer 1753-1784
Whiting	-6.0	127.9	36	Sarah; Am.physicist,astronomer 1847-1927
Whitney	-30.1	151.3	45	Mary; American astronomer 1847-1921
Willard	-24.6	296.1	48	Emma; American educator 1787-1870
Winema	3.1	168.6	24	Modoc Indian heroine, c.1848-1932
Winnemucca	-15.4	121.1	30	Sarah;Piute interpr.,activist c.1844-1891
Wollstonecraft	-39.2	260.8	45.0	Mary; English author 1759-1797
Workman	-12.9	299.9	20	Fanny; Am. mountaineer, author 1859-1925
Yale	-13.4	271.3	20	Caroline; Am. educ. of the deaf 1848-1933
Yonge	-14.0	115.1	28	Charlotte; English writer 1823-1901
Zamudio	9.6	189.3	20	Adela; Bolivian poet 1854-1928
Zija	-3.5	265.0	18	Azeri first name
CORONA				
Abundia Corona	18.5	125.0	250	Norse goddess of giving
Allatu Corona	15.5	114.0	125	Akkadian earth goddess
Anala Corona	11.0	14.0	240	Hindu fertility goddess
Annapurna Corona	-35.5	152.0	300	Indian goddess of wealth
Aramaiti Corona	-26.3	82.0	350	Persian fertility goddess
Ashnan Corona	50.2	357.0	300	Sumerian harvest goddess
Atargatis Corona	-8.0	8.6	220	Hittite fertility goddess
Atse Estsan Corona	8.5	92.0	150	Navajo fertility goddess
Audhumla Corona	45.5	12.0	225	Norse primordial nourisher
Belet-Ili Corona	6.0	20.0	300	Mesopotamian nature/fertility goddess
Benten Corona	16.0	340.0	310	Japanese love/fertility goddess
Bhumidevi Corona	17.2	343.6	200	Hindu earth goddess
Bhumiya Corona	15.0	118.0	100	Hindu earth goddess
Blai Corona	-0.4	134.5	125	Celtic fertility goddess
Blathnat Corona	35.0	293.8	300	Celtic fertility goddess
Boann Corona	27.0	136.5	300	Irish fertility goddess
Bona Dea Corona	-24.0	157.5	275	Roman virgin/fertility goddess
Cailleach Corona	-48.0	88.3	125	Scottish Celtic fertility goddess
Carpo Corona	-37.5	3.0	215	Greek fertility goddess
Ceres Corona	-16.0	151.5	675	Roman harvest goddess

VENUS (Cont.)

NAME	LAT	LON	DIAM (KM)	ATTRIBUTE
CORONA				
Chiun Corona	18.3	340.5	150	Hebrew fertility goddess
Ciuacoatl Corona	53.0	150.9	100	Aztec earth goddess
Colijnsplaat Corona	-32.0	151.0	350	Teutonic fertility goddess
Copia Corona	-42.5	75.5	500	Roman goddess of plenty
Cybele Corona	-7.5	20.7	480	Phrygian fertility goddess
Derceto Corona	-46.8	20.2	200	Phillistine fertility goddess
Dhisana Corona	14.5	111.7	100	Vedic goddess of plenty
Eigin Corona	-5.0	175.0	200	Celtic fertility goddess
Epona Corona	-28.0	208.4	200	Celtic horse/fertility goddess
Ereshkigal Corona	21.0	84.5	320	Mesopotamian nature/fertility goddess
Eurynome Corona	26.5	94.5	200	Greek mother earth goddess
Fatua Corona	-16.3	17.7	375	Roman goddess of fertility
Fotla Corona	-58.5	163.5	150	Celtic fertility goddess
Gaia Corona	6.0	21.5	400	Greek earth/fertility goddess
Gefjun Corona	-33.5	98.5	300	Norse fertility goddess
H'uraru Corona	9.0	68.0	150	Pawnee earth mother
Habonde Corona	3.0	81.8	125	Danish goddess of abundance
Haumea Corona	54.0	21.8	375	Polynesian fertility goddess
Hepat Corona	-2.0	145.5	150	Hittite mother goddess
Holde Corona	53.5	155.8	200	German fertility goddess
Idem-Kuva Corona	25.0	358.0	230	Finno-Ugraic harvest spirit
Inanna Corona	-37.0	35.9	350	Semitic fertility goddess
Inari Corona	-18.0	120.3	300	Japanese rice goddess
Indrani Corona	-37.5	70.5	200	Hindu fertility goddess
Jord Corona	-58.5	349.5	130	Norse earth goddess
Kamadhenu Corona	21.0	136.5	400	Hindu goddess of plenty
Kayanu-Hime Corona	33.5	57.0	150	Shinto grain goddess
Khotun Corona	-46.5	81.5	200	Yakut goddess of plenty
Kuan-Yin Corona	-4.3	10.0	250	Chinese fertility goddess
Kubebe Corona	15.5	132.5	125	Hittite mother earth goddess
Kunapipi Corona	-33.9	86 0	220	Australian mother earth goddess
Kunhild Corona	19.3	80.1	200	German fertility maiden
Libera Corona	12.5	24.0	350	Roman fertility goddess
Lilinau Corona	34.0	22.0	200	Native American fertility maiden
Maan-Eno Corona	40.8	102.5	300	Estonian harvest goddess
Makh Corona	-48.7	85.0	200	Assyro-Babylonian goddess of fecundity
Mama-Allpa Corona	-27.0	31.0	300	Peruvian harvest goddess
Mari Corona	54.0	151.0	200	Cretan goddess of plenty
Maya Corona	23.0	98.0	225	Hindu mother earth goddess
Mayaeul Corona	-27.5	154.0	200	Mexican goddess of plenty
Mesca Corona	27.0	342.6	190	Irish fertility goddess
Miralaidji Corona	-14.0	163.8	300	Aborigine fertility goddess
Nehalennia Corona	14.0	10.0	345	Teutonic fertility goddess
Ninhursag Corona	-38.0	23.5	125	Babylonian earth goddess
Nintu Corona	19.2	123.5	75	Akkadian earth goddess
Nishtigri Corona	-24.5	72.0	275	Hindu earth mother
Nissaba Corona	25.5	355.5	300	Mesopotamian wisdom/fetility goddess
Ohogetsu Corona	-27.0	85.7	175	Japanese food goddess
Olwen Corona	37.5	67.5	175	Brythonic goddess of spring growth
Omeciuatl Corona	16.5	119.0	175	Aztec generative power
Purandhi Corona	26.1	343.5	170	Hindu goddess of plenty
Qetesh Corona	-20.5	343.5	150	Egyptian fertility goddess
Renenti Corona	32.7	326.2	200	Egyptian goddess of abundance
Rosmerta Corona	0.0	124.5	300	Celtic fertility/luck goddess

VENUS

NAME	LAT	LON	DIAM (KM)	ATTRIBUTE
CORONA				
Sarpanitum Corona	-52.3	14.6	170	Babylonian fertility goddess
Seia Corona	-3.0	153.0	225	Roman grain goddess
Selu Corona	-42.5	6.0	300	Cherokee corn goddess
Shiwanokia Corona	-42.0	279.8	500	Zuni fertility goddess
Sith Corona	-10.2	176.5	350	Norse harvest goddess
Somagalags Corona	9.3	348.5	105	Bella Coola earth mother
Sunrta Corona	8.3	11.7	170	Hindu fertility goddess
Tai Shan Corona	-32.5	95.0	175	Chinese fertility goddess
Tamfana Corona	-36.3	6.0	400	Norse fertility goddess
Tamiyo Corona	-36.0	297.5	400	Japanese goddess of abundance
Teteoinnan Corona	-38.5	149.5	125	Aztec fertility goddess
Thermuthis Corona	-8.0	33.0	330	Egyptian fertility/harvest goddess
Thouris Corona	-6.5	12.9	290	Egyptian fertility goddess
Toyo-uke Corona	-62.5	41.5	300	Shinto fertility goddess
Ukemochi Corona	-39.0	296.1	300	Japanese fertility goddess
Vasudhara Corona	43.2	2.7	160	Buddhist female Bodhisattva of abundance
Verdandi Corona	-5.5	65.2	180	Norse bestower of blessings
CHASMA				
Britomartis Chasma	-33.0	130.0		Greek/Cretan goddess of the hunt
THOLUS				
Mahuea Tholus	-37.5	164.7	110	Maori fire goddess

JUPITER
GANYMEDE

NAME	LAT	LON	DIAM (KM)	ATTRIBUTE
CRATER				
Teshub	-72.3	281.0	60	Elamite god of the tempest

NEPTUNE
PROTEUS

NAME	LAT	LON	DIAM (KM)	ATTRIBUTE
CRATER				
Pharos	----	-----		Island where Proteus reigned

MARS

NAME	LAT	LON	DIAM (KM)	ATTRIBUTE
VALLIS				
Marti Vallis	11.0	182.0	1700	Spanish word for Mars

MOON

NAME	LAT	LON	DIAM (KM)	ATTRIBUTE
CRATER				
Glushko	8.4	77.6W	43	V.P.; Russian space scientist 1908-1989

WORKING GROUP FOR THE WORLDWIDE DEVELOPMENT OF ASTRONOMY (GROUPE DE TRAVAIL POUR LE DEVELOPPEMENT MONDIAL DE L'ASTRONOMIE)

CHAIRMAN: A.H. Batten

The principal activity of the Group during the post triennium has been the organizing of a workshop for secondary-school teachers on the Indian sub-continent. This project arose out of discussions at the Group's half-day meeting in Buenos Aires, and was timed to coincide with the IAU Asian-Pacific Regional meeting held in Pune in August 1993. In concept, the workshop was similar to the one-day meetings organized at General Assemblies by Commission 46, and the Chairman is grateful to the President and several members of that Commission for help and advice. This meeting lasted two days and attracted between 40 and 50 teachers, mostly from India, but one each from Nepal and Bangladesh. Several participants had undertaken train journeys of four or five days duration in order to be present. Lecturers at the workshop (besides the Chairman) were J.V. Narlikar and N. Raghavan (both of India), S. Isobe (Japan), C. Iwaniszewska (Poland) and D.B. Hoff (U.S.A.). Unfortunately, M. Othman (Malaysia) was prevented from coming at the last minute. Regional meetings are, if anything, more suitable than General Assemblies for this kind of programme, and the Group hopes that teachers' workshops will become a common feature of regional meetings. The workshop was held at IUCAA in Pune, like the Regional Meeting itself, and the enthusiastic cooperation (including financial support) of the Director and staff of that institution is gratefully acknowledged. Financial support was also provided by the IAU.

The Chairman has sent out two newsletters in the past triennium and these have generated some correspondence. Sometimes help can be given by putting the correspondent in touch with existing IAU services. Several letters drew attention to the special problems being faced by astronomers on the African continent, and this led to the selection of "Problems of Astronomy in Africa" as the theme for a half-day meeting of the Group at the forthcoming General Assembly. At the time of writing, planning for this meeting is well advanced and we hope for contributions from several African countries. The Group is also developing relations with the UN Committee on Outer Space which is organizing a Space Science Workshop in Nigeria in late 1993, which the Chairman hopes to attend.

Another project of the group has been to develop relations with the small astronomical community in Viet Nam, following the initiative taken by Professor Kozai during his term of office as President of the Union. The Chairman has engaged in considerable correspondence and met two Vietnamese astronomers during the Regional Meeting in India. He is at present trying to arrange a visit to Viet Nam. A major topic of concern is opening up opportunities for Vietnamese to undertake graduate studies in developed countries.

Membership of the Group is at present ill-defined, although the Chairman has a list of those who have expressed interest in its activity. We are moving towards clarifying the criteria. In the meantime, he will be pleased to hear from any member of the Union who feels that the Group can be helpful, or who has ideas for future directions. One obvious function of the Group is to act as a clearing-house for information about the possibilities for research or study visits between countries. It can perform this function only if those in a position to offer opportunities will make them known to the Chairman. Response to a request for such information, made in the IAU *Information Bulletin*, has so far been disappointing.

J. Bergeron (ed.), Reports on Astronomy, Vol. XXIIA, 605.

MIX
Papier aus verantwortungsvollen Quellen
Paper from responsible sources
FSC® C105338

If you have any concerns about our products,
you can contact us on
ProductSafety@springernature.com

In case Publisher is established outside the EU,
the EU authorized representative is:
Springer Nature Customer Service Center GmbH
Europaplatz 3, 69115 Heidelberg, Germany

Printed by Libri Plureos GmbH
in Hamburg, Germany